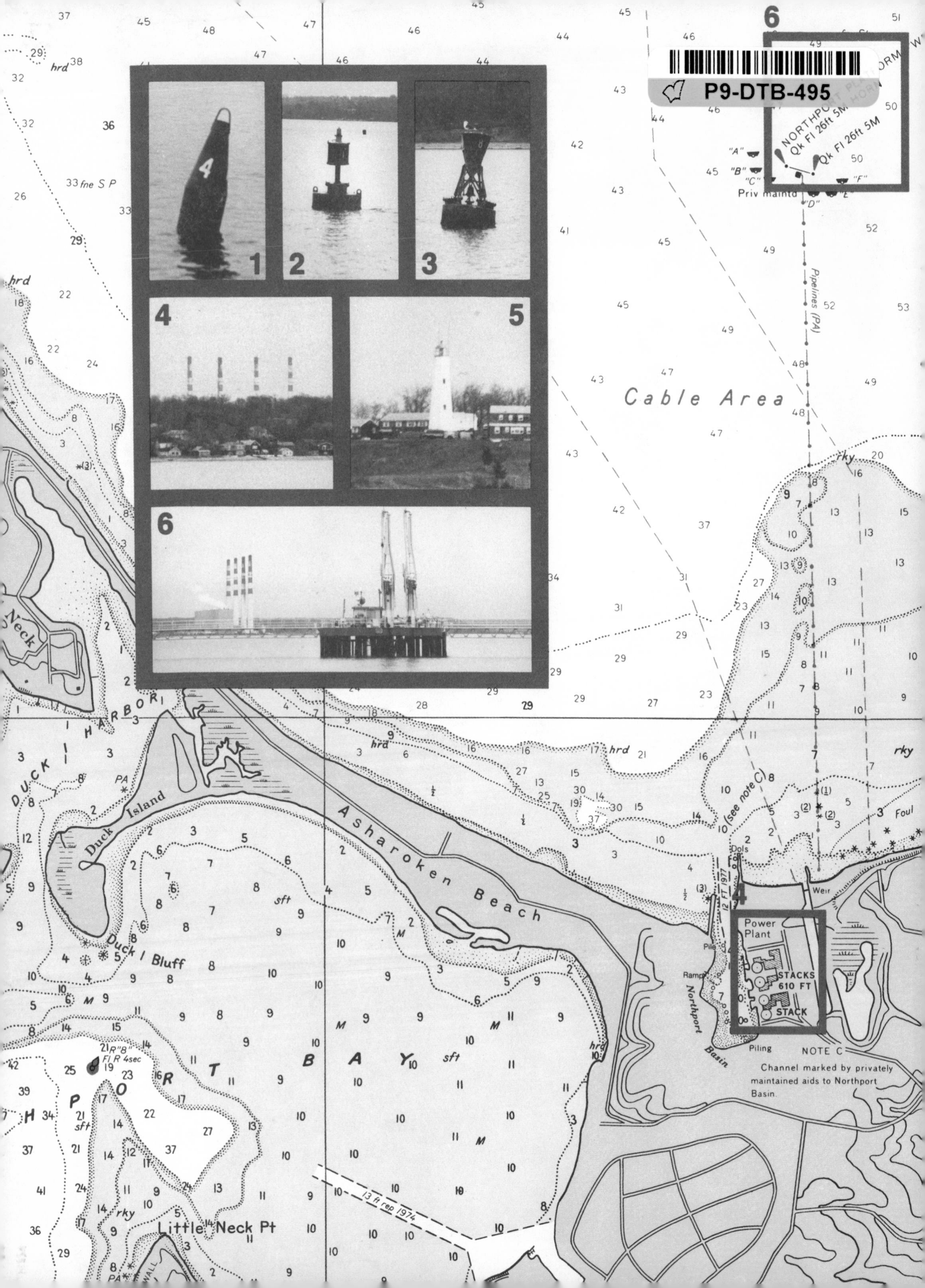

P9-DTB-495
1
2
3
4
5
6
NORTHPORT
Qk Fl 26ft 5M HORN
Qk Fl 26ft 5M
"A"
"B"
"C"
"D"
"E"
"F"
Priv maintd
Pipelines (PA)
Cable Area
hrd
rky
fne S P
sft
Foul
DUCK I HARBOR
Duck Island
Duck I Bluff
Neck
Asharoken Beach
(see note C)
12 FT 1977
Dols
Weir
Pile
Ramp
Piling
Power Plant
STACKS 610 FT
STACK
Northport Basin
NOTE C
Channel marked by privately maintained aids to Northport Basin.
R "8"
Fl R 4sec
NORTHPORT BAY
Little Neck Pt
13 ft rep 1974
PA

CHAPMAN

PILOTING

SEAMANSHIP & SMALL BOAT HANDLING

57TH EDITION

CHAPMAN PILOTING

SEAMANSHIP & SMALL BOAT HANDLING

BY ELBERT S. MALONEY

MOTOR BOATING & SAILING

HEARST MARINE BOOKS

NEW YORK

Frontispiece **Block Island's South Light—aid to navigation used by thousands of recreational boatmen, and famous landmark visited by hundreds of tourists each year.**

Endpapers **Aids to navigation and landmarks as they appear to the eye, and on a chart. Photographs are keyed to the appropriate chart symbols. Note the two views of the four smokestacks at Northport; #4 from inside Northport Harbor, and #7 as seen from the oil pumping station in Long Island Sound.**

57th Edition

Manufactured in the United States of America

Library of Congress Catalog Card Number 42-496-46
ISBN 0-688-05890-6 Standard
ISBN 0-688-06083-8 Presentation

57th Edition

1 2 3 4 5 6 7 8 9 10

Table of Contents

PART TWO: SEAMANSHIP

PART THREE: PILOTING

PART FOUR: GENERAL INFORMATION

FULL COLOR ILLUSTRATIONS

APPENDIXES

THE COMMANDANT OF THE UNITED STATES COAST GUARD
WASHINGTON, D.C. 20593

The U.S. Coast Guard has had a long interest in the safety of those who use the waters of the United States, dating from the days of the old Life Saving Service. Recent Federal legislation, efforts of Coast Guard personnel, and the cooperation of businesses and individuals in the boating community, have resulted in a signal decrease in fatalities associated with recreational boating. This was during a time which saw unprecedented growth, to more than 60 million, in the number of Americans involved in recreational boating.

At the same time, recreational boaters are still our biggest "customers" in terms of requests for assistance. Dangerous situations on the water are often the result of a skipper's inexperience, lack of preparation, and unfamiliarity with vessel or environment. The best preventive measures against such problems are common sense and education.

Common sense must be the natural ingredient in all activity on the water. Boating safety education is available from a variety of organizations to skippers and prospective skippers with the foresight to invest an ounce of prevention in their boating activity.

The volume you are holding, Chapman's Piloting, Seamanship, and Small Boat Handling, is widely recognized as one of the premier texts for small boat operators. The practical information contained here, which has been revised to reflect legislative, regulatory and other changes, will be useful to neophyte and seasoned skipper alike.

J. S. GRACEY
Admiral, U.S. Coast Guard

The Story of a Famous Book

With this, the 57th edition, *Piloting, Seamanship & Small Boat Handling* continues the tradition of excellence established almost 70 years ago with the 1st edition. As always, every word in the book has been reviewed to assure that the information is as current, comprehensive, and authoritative as each previous edition has been. It is these standards that have made this book the most renowned work in its field, and have earned it the reputation as the most famous reference on boating operations in the world.

The story of how this book became so famous and has endured so long has many parts. It begins with a dynamic personality, parallels the extraordinary growth of boating as one of America's favorite leisure activities, reflects the continuing need for boating safety, testifies to the talent and dedication of hundreds of people who have been directly involved with the contents of the 56 editions published through the years, and consistently benefits from the unqualified support of one of this country's great communications corporations.

Charles F. Chapman

No history of this book can be told without first telling the history of the man whose name has become synonymous with boating and who has influenced so many generations of boating enthusiasts. Charles Frederic Chapman was born in Norwich, Connecticut, in 1881. With easy access to the nearby Thames River, Chap, as he was affectionately known throughout his life, became interested in boats at an early age and quickly decided that his life's interests lay afloat. At Cornell University he studied naval architecture and mechanical engineering, but after graduation in 1905 his nautical career seemed beached when he took a job at the New York Telephone Company.

Although landbound, Chap refused to give up on his boyhood dreams. He soon joined the New York Motor Boat Club and bought his first motorboat. She was the *Megohm,* a trim 16-footer powered by a pint-sized, one-lung Detroit engine that produced all of two horsepower.

Chap was a true pioneer. He ventured forth on the waters of the East Coast at a time when only a relative handful of adventurers owned boats. He launched himself into his club's activities with his characteristic pace—full speed ahead.

As chairman of the Motor Boat Club's race committee he organized what soon became one of the most popular competitions in the sport—a 235-mile round trip on the Hudson River between New York City and Albany. With the temperamental engines and round-bottomed displacement hulls of the time, the race was more a matter of endurance than speed, as the early finishers took some 30 hours to complete the course. The success of this particular race opened the door for Chap to organize a number of other competitions among small and moderately priced boats—classes the racing rules had previously overlooked.

But Chap was not content with only the role of organizer. He began racing himself and soon was declared a rising star among motorboat pilots. In 1909 the boat Chap was skippering in an ocean race from New York to Marblehead, Massachusetts, caught fire and sank. While three men balanced in a tiny dinghy, three others clung to life by hanging on to its gunwales. Eventually, the entire crew was rescued by a passing schooner. No doubt the incident played a large part in Chap's lifelong dedication to the instruction of boating safety.

By 1912 he was brought to the attention of a man busy building his own reputation—William Randolph Hearst. Hearst owned the magazine *Motor Boating* and was looking for an editor. "It's yours, Chap. Take it and run it as you wish," was the assignment. And that is exactly what Chap did—for the next 56 years!

The year Chap took over the helm of the magazine it sold for 10¢ a copy and reached only a few thousand readers a month. Chap set his sights high, declaring "The boating business is a sleeping giant and I'm trying to wake it." *Motor Boating* soon began to grow in both stature and circulation. In 1982, now called *Motor Boating & Sailing,* the magazine celebrated its 75th Anniversary and a circulation of some 130,000 subscribers. Chap watched his "sleeping giant" of an industry grow at such a pace that today there are over 13 million registered boats on America's waterways.

As boating grew, so, unfortunately, did boating accidents and fatalities. In an effort to increase safe practices on the water and create a more informed group of participants, Chap used the pages of *Motor Boating* to start what he called "The Correspondence Course." Each month an article on a particular subject would appear and would end with a series of questions. Readers from all over the country sent in answers, and those who passed were given a certificate. This was to become the first formal boating

Charles F. Chapman

safety course in the nation, and it led in 1914 to a forerunner of the book you now hold in your hands.

But Chap also used his knowledge and energy elsewhere to further the cause of boating education. In 1913 and 1914 he was one of ten men who met first at the Boston Yacht Club and later at the New York Yacht Club to form the United States Power Squadrons. Over the years, many stories have evolved as to just why this group was founded, but probably the most interesting is Chap's own recollection. When interviewed in 1972 by *The Ensign* Magazine, the official publication of the USPS, Chap remembered:

> At the turn of the century, practically all boats, both pleasure and commercial, were powered by steam. . . . Navigation [laws] applied only to steam vessels and they were governed by a board of steamboat inspectors, who were very old seagoing men. These inspectors had no use whatsoever for small internal combustion-powered craft and it was their idea and fondest hope to gain control of these boats. A small group of us felt that the internal combustion powerboats should be protected from these steamboat men, and we formed this group to impress them with the fact that we would instruct the members on the rudiments of boat-handling and thus remove one of the objections which they had to small craft. That was really why USPS was formed.

1917: The First Piloting

And so the stage was set for Charles F. Chapman to begin the project that now serves as perhaps his most fitting monument—the publication of *Piloting, Seamanship, & Small Boat Handling.* As a prime mover in the establishment of motorboat racing as a national sport, as the editor of *Motor Boating,* as a founder of the U.S. Power Squadrons, and as a dedicated instructor of boating safety, Chap was associated with all aspects of the burgeoning recreational activity. But it was not recreation Chap had in mind during the early days of World War I when assistant Secretary of the Navy Franklin D. Roosevelt requested that he prepare an instructional manual for the Naval Reserve Forces. He did so in an incredible three days, and in 1917 the first edition of *Practical Motor Boat Handling, Seamanship & Piloting* was published. The book contained 144 pages in a 5-inch × 7-inch format and was a combination of articles Chap had run in his "Correspondence Course" and new material appropriate for the military boatman. The subtitle tells all:

"A handbook containing information which every motor boatman should know. Especially prepared for the man who takes pride in handling his own boat and getting the greatest enjoyment out of cruising. Adapted for the yachtsman interested in fitting himself to be of service to his Government in time of war."

In addition to preparing the manual, Chap offered the Navy the use of the Power Squadrons' "machinery ready to put into instant operation the training of great numbers of men required for the Naval Reserve Forces." Roosevelt accepted this offer with gratitude, and within a year more than 5,000 men who attended Squadron classes and used the instruction manual entered the armed services.

After the war Chap's energy showed no signs of slowing. He formed the National Outboard Racing Commission in 1927. He served for 25 years as chairman of the American Power Boat Association (APBA) Racing Commission, which sanctions such famous events as the President's Cup and the Gold Cup Regatta. In 1921 he teamed up with Gar Wood, the legendary designer/helmsman of fast powerboats, to race offshore from Miami to New York. Their record 47-hour and 15-minute run stood unbroken for three decades. In 1953, Chapman and Wood would be the first two men elected to the APBA's Hall of Fame.

By the 1920s *Motor Boating's* circulation and advertising lineage had grown dramatically. By 1922 Chap's book had undergone six revisions and in that year was retitled *Piloting, Seamanship & Small Boat Handling.*

By now, the emphasis of the book was clearly recreational boating, instruction, and safety. Still mostly a compilation of articles from the magazine, the book took on the appearance of a scrapbook of the sea, a compendium of nautical lore and a home instruction course in boat operation. It was used as the major reference by the USPS in the free courses it offered to the public (still a practice today), and was constantly revised to keep up advances in the boating industry and always-changing government regulations.

Through time, *Piloting, Seamanship & Small Boat Handling* has acquired numerous nicknames, including: "the Bible of Boating," "the Blue Book," "Volume Five" (because it was the fifth volume in a series of books published by *Motor Boating*), and just plain "Piloting" or "Chapman's." Today, its readers probably refer to it as "Chapman's" more than by any other title, even though the original author has not been directly involved with the book since his retirement in 1968, at the remarkable age of 86.

Help From Many Quarters

While Chap directed each new edition for almost 50 years, he counted on help from many assistants, and indeed, from the book's readers. With each new edition came suggestions from staff members, professional boat captains, airplane pilots, amateur sailors, cruising boatmen, and others who love the sea. He also relied on the cooperation of the U.S. Power Squadrons, the Coast Guard Auxiliary, the Army Engineers, the Coast & Geodetic Survey (now NOS, the National Ocean Service), the U.S. Navy, and most of all the U.S. Coast Guard. Scores of manufacturers of equipment, boats, instruments, and other nautical gear have always aided the staff in checking technical information and providing illustrations.

But through all the years, a handful of individuals have made contributions that may be almost as responsible as Chap himself for keeping this book the most popular, authoritative, and current book on boating published. Their names include William H. Koelbel, who worked closely with Chap for more than twenty years and wrote a number of chapters during that time; Morris Rosenfeld, the famed marine photographer whose art enhanced edition after edition; Morris's son Stanley, who continues the tradition; Gardner Emmons; Robert Ogg; and Dr. John Wilde.

Since 1966

Since 1966, the work of three men stands apart from that of all others. They are E. S. Maloney, Thomas Bottomley, and John Whiting. Each was hand-picked by Chap

they are men or women, can be referred to as YACHTSMEN or BOATMEN, the usage often depending on the size of the boat involved, but again without any clear distinction in boat length. The term BOATER is frequently used. He or she also may be called, informally, the SKIPPER or CAPTAIN.

Directions Aboard a Boat

The front end of a boat is its BOW; the other end is its STERN. PORT and STARBOARD (pronounced "starb'd") are lateral terms. Port designates the left side and starboard the right when on the vessel and facing the bow, or, to express it differently, when facing FORWARD. By turning around and facing the stern, one is looking AFT.

The bow is the forward part of the boat; the stern is the AFTER part. When one point is aft of another it is said to be ABAFT it; when nearer the bow, it is forward of the other. When an object lies on a line or in a plane parallel to the centerline of the vessel it is referred to as lying FORE-AND-AFT, as distinguished from ATHWARTSHIPS, which means at right angles to the centerline.

The term AMIDSHIPS has a double meaning. In one sense, it refers to an object or area midway between the boat's sides. In another sense, it relates to something midway between the bow and the stern. INBOARD and OUTBOARD, as directional terms, draw a distinction between objects near or toward amidships (inboard) and those away from the centerline or beyond the sides (outboard).

To express the idea of "overhead in the rigging," one says ALOFT. BELOW means the opposite direction. Note, though, that ABOVEDECK means on deck, and not actually above it as does "aloft." A person who is in or on a boat is ABOARD, or ON BOARD.

Terms Relating to the Boat

The basic part of a boat is its HULL. There is normally a major central structural member called the KEEL; beyond this; however, the components will vary with the construction materials and techniques. Boats may be of the OPEN type, or may be covered over with a DECK. As boats get above the smallest sizes, they may have a SUPER-STRUCTURE above the main deck level, variously referred to as a DECKHOUSE or CABIN (cabins also extend below the level of the deck). Small, relatively open boats may have only a SHELTER CABIN, or CUDDY, forward.

Terms Denoting Shape

SHEER is a term used to designate the curve or sweep of the deck of a vessel as viewed from the side. The side skin of a boat between the waterline and deck is called the TOPSIDES. If the sides are drawn in toward the center-

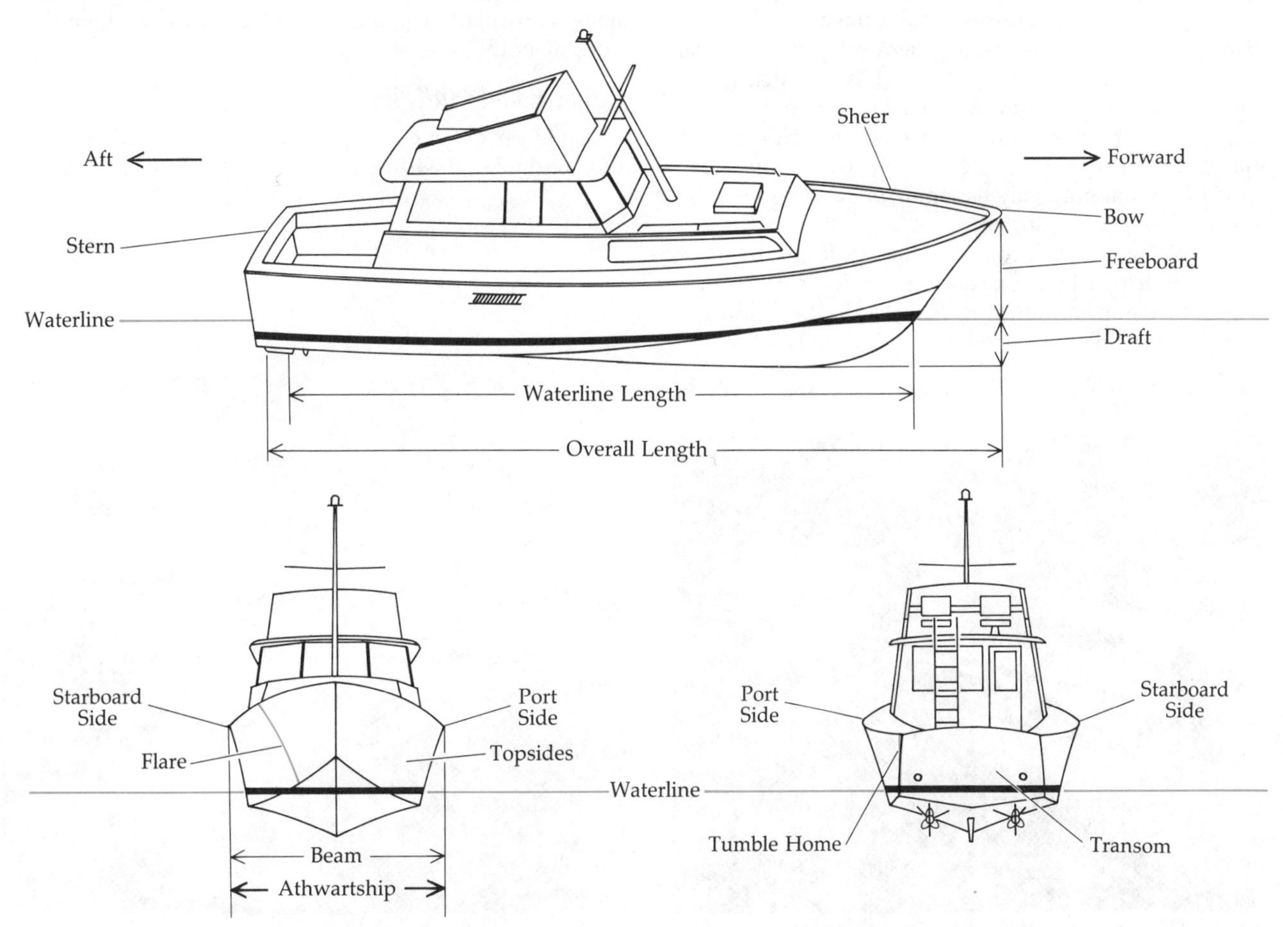

Figure 104 **These are terms that relate to basic boat dimensions and shapes, and directions aboard a boat. Note that port and starboard sides remain the same, no matter which way one is facing. Additional terms that apply specifically to sailboats are illustrated in Chapter 8.**

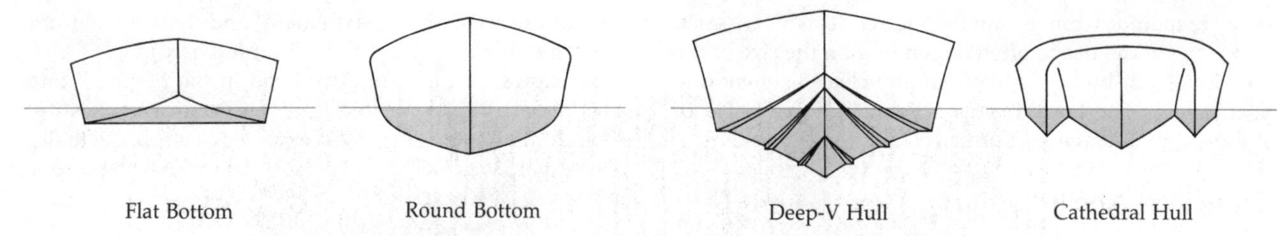

Figure 105 **A flat bottom boat is inexpensive to build; a round bottom provides a soft ride. The deep-V hull is used on high-speed offshore craft; cathedral hulls have good stability.**

line away from a perpendicular as they go upward, as they often do near the stern of a boat, they are said to have TUMBLEHOME. Forward, they are more likely to incline outward to make the bow more buoyant and to keep the deck drier by throwing spray aside; this is FLARE.

The bottom of a boat may be one of three basic shapes—FLAT, ROUND, or VEE—or it may be a combination of two shapes, one forward gradually changing to the other toward the stern. There are also more complex shapes such as CATHEDRAL-HULL, DEEP-VEE, MULTI-STEP, and others.

A DISPLACEMENT HULL is one that achieves it buoyancy (flotation capability) by displacing a volume of water equal in weight to the hull and its load, whether underway or at rest. A PLANING HULL, on the other hand, is one that achieves the major part of its underway load-carrying ability by the dynamic action of its underside with the surface of the water over which it is rapidly traveling; at rest, a planing hull reverts to displacement buoyancy. A SEMI-PLANING (or SEMI-DISPLACEMENT) HULL is one that gets a portion of its weight-carrying capability from dynamic action, but which does not travel at a fast enough speed for full planing action. It is often a hull that is round-bottomed forward, gradually flattening out toward the stern to provide a planing surface.

The two main types of MULTI-HULL craft are the CATAMARAN with two hulls of equal size held apart by rigid structural members, and the TRIMARAN with a principal central hull flanked by smaller outboard hulls.

Bows and Sterns

The STEM is the extreme leading edge of a hull; on wooden craft it is the major structural member at the bow. The stem of a boat is PLUMB if it is perpendicular to the waterline, or RAKED if inclined at an angle. The term OVERHANG describes the projection of the upper part of the bow, or stern, beyond a perpendicular up from the point where the stem or stern intersects the WATERLINE. EYE BOLTS or RING-BOLTS—to which all kinds of lines, ropes, and blocks may be attached—are frequently fitted through the stems of boats.

The flat area across a stern is called the TRANSOM. If, however, the stern is pointed, resembling a conventional bow, there is no transom and the boat is called a DOUBLE-ENDER. The QUARTER of a boat is the after portion of its sides, particularly the furthermost aft portion where the sides meet the transom.

Additional Hull Shape Terms

The lower outer part of the hull where the sides meet the bottom is called the TURN OF THE BILGE. If the boat is

Figure 106 **Hydrodynamic forces lift a planing hull partially out of the water to reduce drag and wave-making resistance. This makes high speeds possible without excessive power.**

Figure 107 **High speeds are often possible with multi-hull craft such as this catamaran, heeling well over in a good breeze; three-hull boats are trimarans.**

flat or vee-bottomed, the bottom and the topsides meet at a well-defined angle rather than a gradual curve—this is the CHINE of the boat. The more abrupt the angle of intersection of these planes, the HARDER the chine. SOFT CHINE craft have a lesser angle; this term is sometimes applied to round-bottom boats, but this is not correct. Some modern boats are designed with MULTIPLE CHINES (longitudinal steps) for a softer ride at high speeds in rough water; this is often referred to as DEEP-VEE design.

Larger round-bottom vessels may be built with BILGE KEELS—secondary external keels at the turn of the bilge that reduce the vessel's tendency to roll in beam seas.

The significance of the term DEADRISE can be appreciated by visualizing a cross-section of a hull. If the bottom were flat, extending horizontally from the keel, there would be no deadrise. In a vee-bottom boat, where the bottom rises at an angle to a horizontal line outward from the keel, the amount of such rise is the deadrise, expressed in inches per foot or as an angle.

With normal sheer, the deck of a boat, as viewed from the side, slopes up toward the bow, and the stern is at least level with amidships. REVERSE SHEER, however, has the deck sloping downward at the bow, or stern, or both.

FLAM is that part of the concave flare of the topsides just below the deck. If it curves outward sharply, it will both increase deck width and reduce the amount of bow spray that blows aboard. The CUTWATER is the forward edge of the stem, particularly near the waterline. STEM BANDS of metal are frequently fitted over the stem for protection from debris in the water, ice, pier edges, etc. The FOREFOOT of a boat is the point where the stem joins the keel.

If the boat has an overhanging stern, this part of the hull is the COUNTER. Her lines aft to the stern are her RUN; lines forward to the stem are her ENTRANCE (or ENTRY). The descriptions "fine" and "clean" are often applied to the entrance and run. The term BLUFF is applied to bows that are broader and blunter than normal.

Terms Relating to the Keel

The KEEL is the major longitudinal member of a hull. A metal fitting extending back from the underside of the keel for protection of the rudder and propeller is called a SKEG.

The RUDDER of a boat is the flat surface at or near the stern that is pivoted about a vertical or near-vertical axis so as to turn to either side and thus change the direction of movement of the vessel through the water. The upward extension of the rudder through which force is applied to turn the rudder is the RUDDERPOST (or RUDDERSTOCK). A STUFFING BOX keeps the hull watertight where the rudderpost enters. A stuffing box is also used where a PROPELLER SHAFT goes through the hull. The rudder is turned by the vessel's WHEEL, or by a TILLER—a lever attached directly to the rudderpost.

Deck Openings

HATCHES (sometimes HATCHWAYS) are openings in the deck of a vessel to provide access below. COMPANION LADDERS or steps lead downward from the deck; these also are termed COMPANIONWAYS.

COCKPITS are open wells in the deck of a boat outside of deckhouses and cabins. COAMINGS are vertical pieces around the edges of cockpit, hatches, etc., to prevent water on deck from running below.

Interior Terms

Vertical partitions, corresponding to walls in a house, are called BULKHEADS. WATERTIGHT BULKHEADS are solid or are equipped with doors that can be secured so tightly as to be leak-proof. The interior areas divided off by bulkheads are termed COMPARTMENTS—such as an engine compartment—or CABINS—such as the main cabin or after cabin. Some areas are named by their use; the kitchen aboard a boat is its GALLEY. The toilet area is the HEAD (the same name is also given to the toilet itself). OVERHEAD is the nautical term for what would be the ceiling of a room in a house. CEILING, on a boat, is light planking or plywood sheeting on the inside of the frames, along the sides of the boat. FLOORS, nautically speaking, are not laid to be walked upon, as in a house. They tie together the keel and the lower ends of the frames.

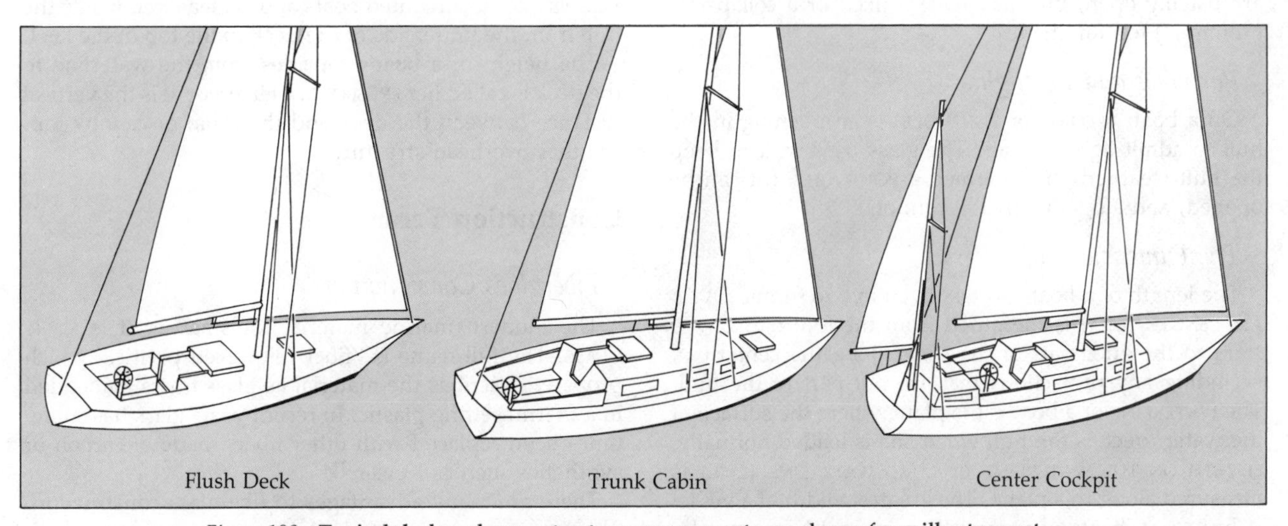

Figure 108 **Typical deck and superstructure arrangements are shown for sailboats; most powerboats are of the trunk cabin type, although some small cruisers may have a cuddy cabin with berths, head, and even a small galley under a flush deck forward of the cockpit.**

The FORECASTLE (pronounced fo'-c's'l), if any small craft can be said to have one, is the cabin furthest forward. The term is generally used today to mean the crew's quarters forward.

The term BILGE is also applied to the lower interior areas of the hull of a vessel. Here water that leaks in, or is blown aboard as spray, collects as BILGE WATER, to be later pumped overboard by a BILGE PUMP.

BERTHS and BUNKS are seagoing names for beds aboard a boat. Closets are termed LOCKERS, and a HANGING LOCKER is one tall enough for full-length garments. Chests and boxes may also be called lockers. A ROPE or CHAIN LOCKER is often found in the bow of a boat for stowing the anchor line or chain; this area is called the FOREPEAK. LAZARETTES are compartments in the stern of a vessel used for general storage.

When something is put away in its proper place on a vessel, it is STOWED. The opposite of stowing, to BREAK OUT, is to take a needed article from its locker or other secure place.

HELM is a term relating to the steering mechanism of a craft. An individual is AT THE HELM when he is at the controls of the boat; he is then the HELMSMAN.

Cabin Styles

A TRUNK CABIN is one that extends above the main deck level, but less than the full width of the boat so that walkways—side decks—are left on either side; a boat with such a cabin is a TRUNK-CABIN CRUISER. A RAISED-DECK CRUISER is one in which the cabin is formed by extending the topsides above the normal deck level—it extends for the full width of the hull, thus eliminating side decks. A SEDAN CRUISER has the main cabin on the same level as an after cockpit and opening out into it. A SPORT FISHERMAN is a motorboat, usually fast, with special equipment for offshore trolling for large game fish.

The BRIDGE of a vessel is the location from which it is steered and its speed controlled. On small craft, the term CONTROL STATION is perhaps more appropriate. Many motorboats have a FLYING BRIDGE, an added set of controls above the level of the normal control station for better visibility and more fresh air. These, also called FLY BRIDGES, are usually open, but may have a fixed or a collapsible ("Bimini") top for shade.

Portholes and Portlights

On a boat, a PORT (or PORTHOLE) is an opening in the hull to admit light and air. The glass used in it to keep the hull weathertight is termed a PORTLIGHT if it can be opened, and a DEADLIGHT if it cannot.

The Dimensions of a Vessel

The length of a boat is often given in two forms. OVERALL LENGTH (LOA) is measured from the fore part of the stem to the after part of the stern along the centerline, excluding any projections that are not part of the hull. The WATERLINE of a boat is the plane where the surface of the water touches the hull when she is loaded normally; LENGTH ON THE WATERLINE or LOAD WATERLINE (LWL) is measured along this plane. The greatest width of a vessel is her BEAM; boats of greater than normal beam are described as BEAMY. A vessel's DRAFT is the depth of water required to float her. Draft should not be confused with the term DEPTH, which is used in connection with larger vessels and documented boats and is measured inside the hull from the underside of the deck to the top of the keel.

The height of a boat's topsides from the waterline to the deck is called her FREEBOARD. HEADROOM is the vertical distance between the deck and the cabin or canopy top, or other overhead structure.

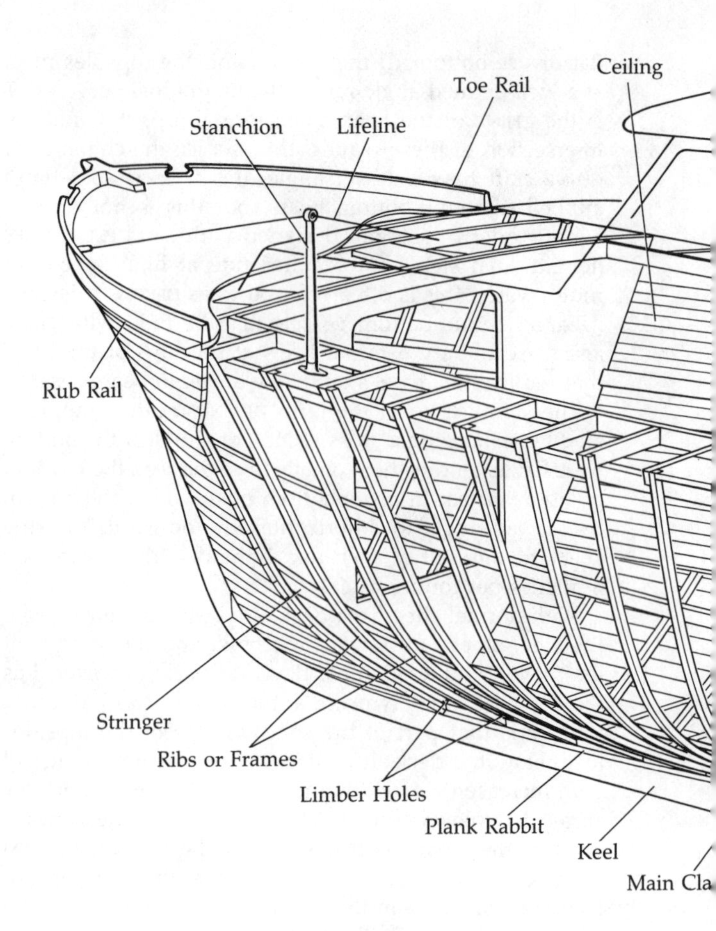

Figure 109 **Most construction terms are based on wood hulls such as this, but many of the features are found on fiberglass, steel, or aluminum boats. An ability to check construction details is useful when buying a boat; Chapter 12 lists details important to safety.**

Construction Terms

Fiberglass Construction

The modern marine material for small craft is FIBERGLASS. The full name is "fiber reinforced plastic," which properly describes the material as glass fibers embedded in a thermosetting plastic. In recent years glass has sometimes been replaced with other fibers made of carbon or synthetics such as Kevlar.™

There are many advantages to fiberglass construction. Surfaces can be of any desired compound curvature; multiple identical hulls, super-structures and lesser components can be made economically from reusable molds; and

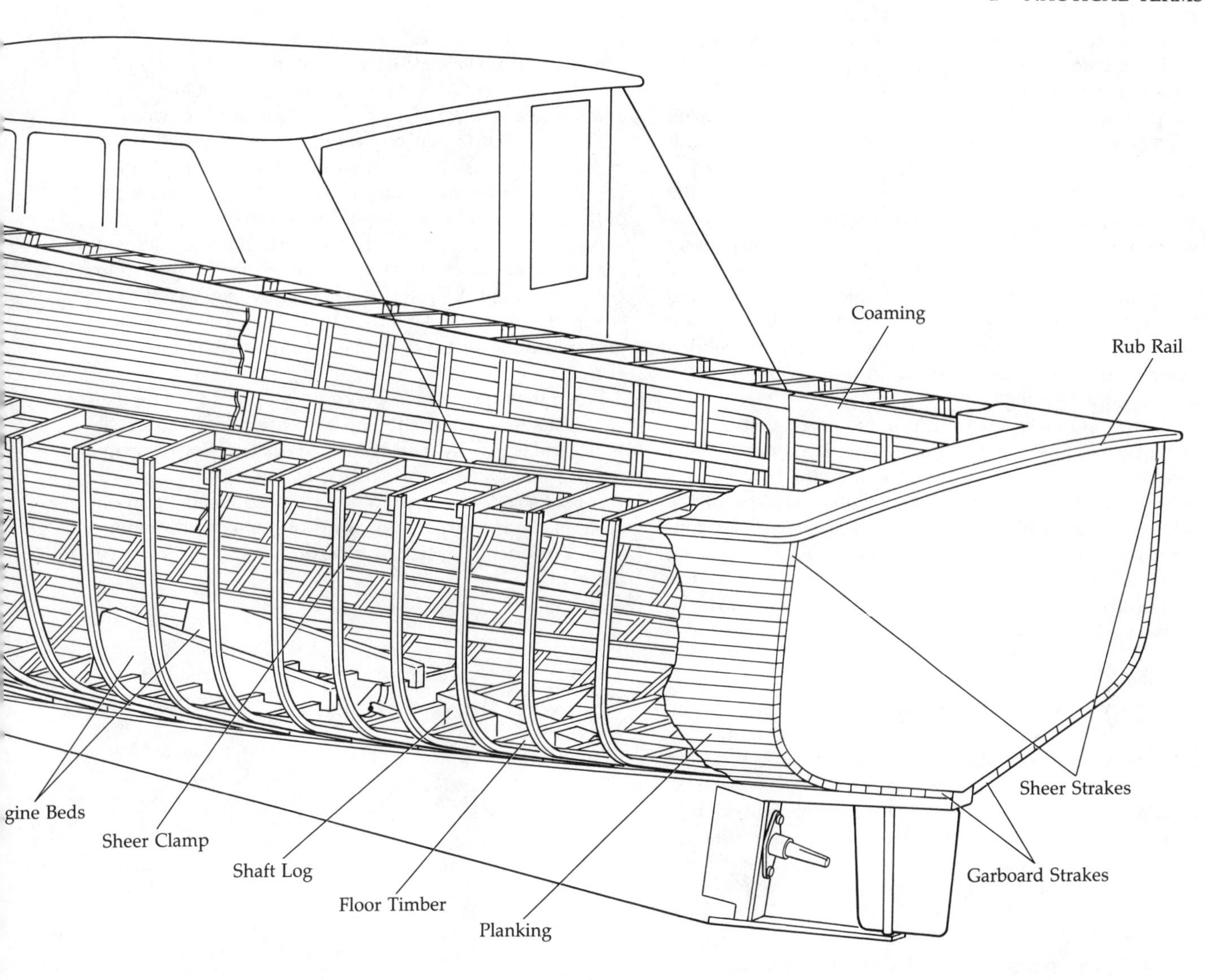

plastic hulls completely resist attack by marine organisms, although they do require anti-fouling paint.

The reinforcing material is usually in the form of MAT, ROVING, or CLOTH. This is LAID UP in alternate layers with POLYESTER RESIN which has been mixed with a CATALYST (and in some cases a fire-retarding chemical) that causes the resin to harden in a few minutes. The WORKING TIME of the mixture is controlled by the proportion of catalyst added and the temperature. Sometimes a CHOPPER GUN is used which sprays into the mold a mixture of resin and catalyst with the proper proportion of CHOPPED GLASS FIBER. The smooth and bright, shining color of the surface of a fiberglass boat comes from the GEL COAT, a thin finish layer of pigmented plastic which is placed in the mold as LAYUP commences.

Fiberglass is useful also in covering wood, either in new construction or in the repair of older boats, where its properties of ADHESION are important. EPOXY is another resin whose strength and adhesive properties are superior to POLYESTER, but it is more expensive.

Fiberglass construction has added a few new terms to the vocabulary of hull construction. Overhead surfaces are protected by HEADLINERS; HIGH-HAT SECTIONS are often used as stiffeners. BALSA-CORE and FOAM-CORE refer to methods of stiffening otherwise flexible panels of glass fiber and resin by separating two of them by perhaps an inch of the core material, producing a light, strong, and stiff structure.

Wood Construction

On wooden boats, FRAMES (also called RIBS) are set into the keel at a right angle, and covered with planking. Each continuous line of planking along the hull from bow to stern is called a STRAKE. The lowest strake, next to the keel, is the GARBOARD STRAKE. The GUNWALE (pronounced gun'l) is the upper part of the SHEER STRAKE, the top plank of the topsides. "Gunwhale" is *not* correct spelling.

When the topsides are carried substantially above the level of the deck, they are called BULWARKS, and at the top of the bulwarks is the PAIL. More common than bulwarks are TOE RAILS, narrow strips placed on top of the gunwale to finish it off and provide some safety for personnel on deck.

A heavier strake in the topsides extending beyond the

exposed face of the planking is termed a RUBBING STRAKE and is intended to protect the topsides from the roughness of piles, pier faces, etc. A strip of wood for the same purpose added externally to the planking is generally called a RUB RAIL or GUARD. Both of these may be faced with metal or plastic strips for their own better protection.

Many hulls are fitted with a SPRAY RAIL external to the planking just above the waterline. Such a rail usually extends about halfway aft from the stem and deflects downward any spray from the bow wave.

Various kinds of KNEES are used throughout the hull to connect members joined at an angle to each other. These may be of metal, but often a natural growth of wood is selected in which the grain runs in the desired direction for maximum strength.

A BREAST HOOK is a triangular reinforcing member, usually of wood but sometimes of metal, placed horizontally behind the stem of a boat to strengthen the bow.

When another timber is fastened along the top of the keel to strengthen it, or as a necessary part of the construction, this is the KEELSON (sometimes APRON). On some boats an extra piece is fastened externally to the bottom of the keel to protect it. This is termed a FALSE KEEL or WORM SHOE.

Fastenings

The term FASTENING is applied to screws or specially designed nails which hold the planks to the frames, as well as to screws, bolts, and similar items used to hold down equipment, cleats, chocks, etc. THROUGH FASTENINGS are bolts that go all the way through the hull or other base timbers, secured with a washer and nut on the inside; a BACKING BLOCK is generally advisable for spreading the stress to gain additional strength.

Deck Terms

If a deck is arched to aid in draining off water, it is said to be CAMBERED. Similarly, camber occurs on the tops of cabins, deckhouses, etc. The deck over the forward part of a vessel, or the forward part of the total deck area, is termed a FOREDECK. Similarly, the AFTERDECK is located in the after part of the vessel. The deck of a cockpit or interior cabin is called the SOLE.

Other Construction Terms

COMPOSITE CONSTRUCTION is the use of two or more materials in the hull of a vessel. Often this consists of wood planking over iron or steel frames.

LIMBER HOLES are passages cut into the lower edges of floors and frames next to the keel to allow bilge water to flow to the lowest point of the hull, where it can be pumped out. These limber holes must be kept clean so that drainage to the lowest point can occur; on some larger vessels a LIMBER CHAIN is run through these holes so that it can be pulled back and forth a short distance to clean them out.

Some boats are damaged by DRY ROT, a fungus attack on wooden areas. The name is incorrect, as moisture is essential for the growth and spread of the fungus. The wood is greatly weakened by the action of the fungus. Prevention of standing water in corners and pockets, and adequate ventilation, are the best defenses against dry rot; chemical treatments can be used to prevent rot, and to attack existing rot infestations and restore strength.

SEA COCKS are valves installed just inside THROUGH-HULL FITTINGS where water is taken in for engine cooling, operation of heads, etc.; they are important safety devices.

The BOOT TOP is that portion of the exterior hull at the waterline. It is usually finished with special paint—BOOT-TOPPING—of a color to contrast with the ANTI-FOULING PAINT of the bottom and the color of the topsides.

Areas that are varnished to a high gloss are termed BRIGHTWORK. (In naval usage, this term may be applied to polished brass.) A TACK RAG is a slightly sticky cloth that is wiped over surfaces that are to be varnished or painted just before the brush is applied, to remove all dust and grit.

When the rudder of a boat is exactly centered, one spoke of the steering wheel should be vertical; this is the KING SPOKE and is usually marked with special carving or wrapping so that it can be recognized by feel alone.

A boat's rudder is said to be BALANCED when a portion of its blade area extends forward of the axis of rotation. Such a rudder is easier to turn than one that is not so designed.

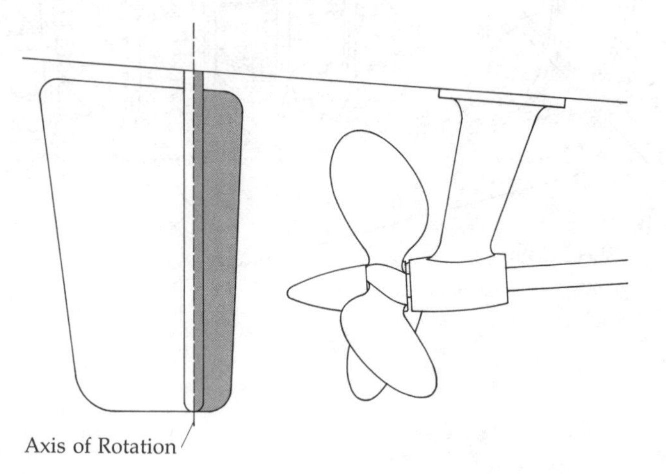

Figure 110 **Tinted area, forward of balanced rudder's axis of rotation, offsets much of the force required to turn the blade.**

Many motorboats will have a small SIGNAL MAST from which flags can be flown; see Chapter 25. These flags are hoisted on SIGNAL HALYARDS. When partially hoisted they are said to be AT THE DIP; when fully hoisted, they are termed TWOBLOCKED.

LIFELINES are used on larger craft at the edges of the side decks to prevent people from falling overboard. These lines usually consist of wire rope, often plastic-covered, supported above the deck on STANCHIONS. If made of solid material—wood or metal—they are called LIFERAILS. Boats may also have waist-high BOW RAILS of solid tubing for the same safety purpose. A PULPIT is an extension, usually a heavy plank with rails extending up from it, beyond the bow of a sport-fishing boat used for spearing or harpooning large game fish. Bow rail installations on sailboats are also often called pulpits, whether they extend forward of the stem or not. Many sailboats also have stem pulpits, and these are sometimes called PUSHPITS.

Vessels, including boats, are normally designed by a NAVAL ARCHITECT. In so doing, he prepares detailed PLANS and SPECIFICATIONS, including the LINES (shape) of her hull. CLEAN is a term applied not to a boat's condition, but rather to her lines. If the lines are FINE, so that she slips easily through the water, the lines are said to be clean.

A MARINE SURVEYOR is a person whose job it is to make detailed inspections (SURVEYS) of boats and ships to determine the condition of hull, equipment, machinery, etc.

Terms Used in Boating Activities

In addition to the numerous terms for the boat itself, there are many others to be learned in connection with boating in general. The list of these is virtually endless, and those that follow are only the more basic and often-used ones.

Docks, Piers, and Harbors

There is a difference between strict definition and popular usage for the term "dock." Properly speaking, a DOCK is the water area in which a boat lies when she is MADE FAST to shore installations—and "made fast" is the proper term rather than "tied up." A DRYDOCK is one that can be shut off from the surrounding water and pumped out in order to make repairs on a vessel's bottom. In boating, however, the term "dock" is usually applied to structures bordering the water area in which boats lie. A WHARF is a structure parallel to the shore, while a PIER projects outward from the shoreline. PILES (or PILINGS) are substantial elongated objects driven into the bottom to which craft may be made fast or which support a structure; a DOLPHIN is a group of piles driven close together and bolted or bound with wire cables into a single structure. (The name "dolphin" is also given to a species of marine mammal, and separately to a species of ocean game fish.) Piles are often used to form SLIPS out from a pier or wharf in which boats can be BERTHED; short CATWALKS (or FINGER PIERS) may extend out between the slips for easier access to the boats. A boat is made fast to a pile, *in* a slip or alongside a pier or wharf. To DOCK a vessel is to bring her to the shore installation and make her fast.

A HARBOR is an anchorage which affords reasonably good protection for a vessel, with shelter from wind and sea; strictly speaking, it applies to the water area only. PORT is a term that includes not only the harbor but all the facilities for freight, passengers, and services as well, such as wharves, piers or warehouses. A YACHT BASIN or MARINA is a protected facility primarily for recreational small craft.

JETTIES are embankments connected to the shore; when these are used to protect a harbor, and have no connection to the land, they are BREAKWATERS. GROINS are jetty-like dikes built out at roughly a 90° angle from the shore to prevent erosion of the beach.

Boats are HAULED OUT of the water on inclined planes

Figure 111 **A pier, above, is a structure that projects out from the shore; a wharf is one that is parallel to the shore.**

Figure 112 **A quick and simple means of launching a small boat is by a crane with slings.**

Figure 113 **Traveling lifts with slings can lift or launch boats of considerable size. The boats are transferred to or from cradles, which are placed about the boatyard as desired.**

at the water's edge called WAYS (also MARINE WAYS or RAILWAYS). The framework which supports a boat as she is hauled out is a CRADLE. Small and medium-sized boats are also lifted bodily out of the water and set on shore for storage or work with SLINGS, which are lowered, passed under the hull, and raised by CRANES or TRAVEL LIFTS.

Ropes and Lines

Generally speaking, the word ROPE is used but little aboard a boat; the correct term is LINE. "Rope" may be bought ashore at the store, but when it comes aboard a vessel and is put to use it becomes "line." MARLINESPIKE SEAMANSHIP is the term applied to the art of using line and the making of KNOTS, BENDS, HITCHES, and SPLICES. Chapter 13 is devoted to this important subject.

Lines used to make a boat fast to a shore structure are called MOORING LINES or DOCK LINES. BOW LINES and STERN LINES lead forward and aft from those respective parts of the boat. SPRING LINES lead from the bow aft or from the stern forward to prevent the boat from moving ahead or astern.

A PAINTER is a line at the bow of a small boat, such as a dinghy, for towing or making fast. The line by which a boat is made fast to a mooring buoy is called a PENNANT (sometimes PENDANT).

HEAVE is a nautical term for throw or pull. One HEAVES IN on a line when he pulls in the slack and TAKES A STRAIN on it. When a line is let out, one PAYS IT OUT; to lessen the strain on a line or let it out slightly or slowly, one EASES it. A line is SNUBBED when its outward run is checked; it is CAST OFF when it is let go. (A boat is cast off when all lines have been taken off the pier or other object to which it has been made fast.)

The BITTER END of a line is the extreme end, the end made fast when all line has been paid out. The middle part of a line not including either end is the BIGHT, particularly when formed into a loop. If a strain is put on a line heavy enough to break it, it PARTS.

HEAVING LINES are light lines, usually with a knot or weight at one end which makes it easier to throw them farther and more accurately. The knot that encloses a weight at the end of a heaving line is called a MONKEY'S FIST. BELAY has two meanings. A line is BELAYED when it is made fast without knotting to a CLEAT or BELAYING PIN; BELAY as a command signifies "stop" or "cease."

Ends of lines are WHIPPED or SEIZED when twine or thread is wrapped around them to prevent strands from untwisting or UNLAYING. Ragged ends of lines are said to be FAGGED. HAWSERS are very heavy lines in common use on tugboats and larger vessels.

TACKLE is a broad, general term applied to equipment and gear used aboard a boat. It has a specific use, however, to mean a combination of line and BLOCKS (pulleys) used to increase a pulling or hoisting force. The wheels or rollers of the blocks are the SHEAVES (pronounced shivs). When a line is passed through a block or hole it is REEVED; to RENDER is to ensure that the line will pass freely through a block or hole.

Lines have STANDING PARTS and HAULING PARTS. The standing part is the fixed part, the one which is made fast; the hauling part is the one that is taken in or let out as the tackle is used. Lines are FOUL when tangled, CLEAR when ready to run freely.

Anchors and Moorings

An ANCHOR is a specially shaped metal device designed to dig efficiently into the bottom under a body of water and hold a vessel in place despite winds and currents. Chapter 11 is devoted to anchors and their use. GROUND TACKLE is a general term embracing anchors, lines, and other gear used in anchoring. On boats, the anchor line may be referred to as a RODE. A MOORING is a semi-permanent anchorage installation, consisting of a heavy anchor (usually of the mushroom type), chain, a mooring buoy, and a pennant of nylon or manila.

The anchor carried on a boat for most normal uses is its WORKING ANCHOR. A heavier model carried for emergencies is termed a STORM ANCHOR, and a smaller, lighter anchor for brief daytime stops when the boat will not be left unattended is popularly called a LUNCH HOOK.

Anchors come in many styles—KEDGE, STOCKLESS, GRAPNEL, and modern LIGHT-WEIGHT types such as the DANFORTH, BRUCE, and PLOW. See Figures 1102 to 1107.

The term KEDGE is also applied to an anchor of any type that is used for getting a boat off when she has run aground. The kedge is carried out by a dinghy or other means and set so that a pull on the line will help get the boat off, or at least keep her from being driven harder aground; this process is called KEDGING.

When a boat is anchored, the ratio of the length of line in use to the distance to the bottom of the water as measured from the deck is termed SCOPE. An ANCHORAGE may be a carefully chosen, protected body of water suitable for anchoring, or an area specifically designated by governmental authorities in which vessels may anchor; special regulations may prevail in these areas.

Motions of a Boat

A vessel GROUNDS when she touches bottom, and if stuck there she is AGROUND. When a boat moves through the water she is said to be UNDERWAY. According to government regulations, a vessel is underway at any time that she is not aground, at anchor, or made fast to the shore. The direction in which she is moving is made more specific by stating that she makes HEADWAY (moving forward), STERNWAY (backwards), or LEEWAY (to one side or the other, as when pushed by a beam wind). A boat is said to be UNDERWAY WITH NO WAY ON when she is free of the bottom or shore but is making no motion through the water (adrift). A vessel has STEERAGEWAY if she is making enough speed through the water for her rudder to be effective. A vessel's anchor is said to be AWEIGH when it has broken out of the bottom and has been lifted clear.

The disturbed water that a boat leaves astern as a result of her motion is her WAKE. WASH is the flow of water that results from the action of her propeller or propellers. Both of these effects are commonly lumped together as "wake."

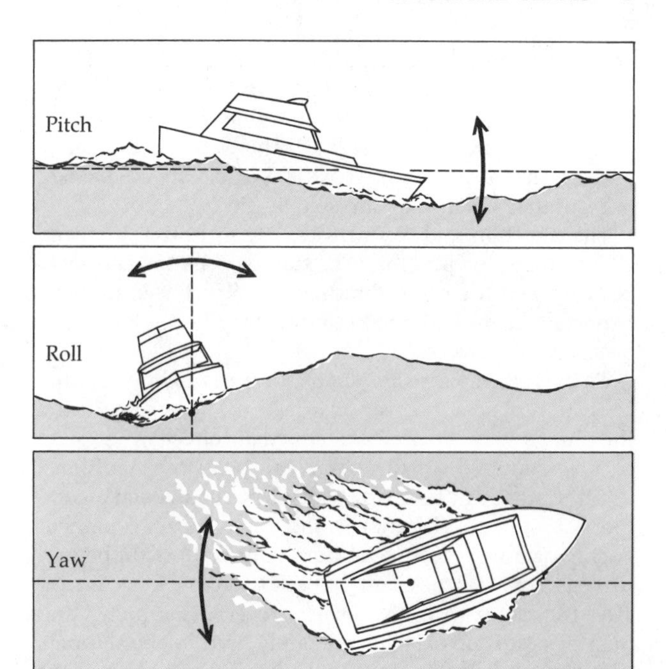

Figure 115 **Pitching, rolling, and yawing are normal motions in moderate seas, although they may combine to create uncomfortable results. In rough seas they can be dangerous.**

Figure 114 **Take care that your wake—the area of disturbed water behind you—does not disturb or damage other craft.**

Sidewise rotational motion of a boat in rough water is called ROLL; vertical motion as the bow rises and falls is termed PITCHING. A boat YAWS when she runs off her course to either side as she might if struck on the quarter by a following sea. If she yaws too widely and is thrown broadside into the TROUGH of the sea (between the CRESTS and parallel to them), she may BROACH (or BROACH TO), a dangerous situation that should be carefully avoided. In very rough seas a small boat can be thrown end-over-end; this is PITCHPOLING.

A boat may CAPSIZE without FOUNDERING. In the first instance she is knocked down so she lies on her side in the water or turns over; in the latter she fills with water from any cause and sinks. If a boat fills with water from over the side, she SWAMPS. Before his vessel is reduced to such straits, the wise skipper HEAVES TO, reducing headway and generally lying with the bow slightly off from meeting the waves head on. He may put out a SEA ANCHOR (or DROGUE) that will hold the bow at the most favorable angle; a sea anchor does not go to the bottom, it merely serves as a drag.

Trim

TRIM relates to the way a boat floats in the water. When she floats properly as designed, she is on an EVEN KEEL, but if she inclines to port or starboard, she LISTS. HEEL conveys the same idea as list, a sideward inclination from the vertical, usually temporary, particularly when under sail. If a vessel is too heavily loaded forward, she TRIMS BY THE HEAD (or is DOWN BY THE BOW). If her draft is excessive aft, she TRIMS BY THE STERN (is DOWN BY THE STERN).

Directions From the Boat

DEAD AHEAD refers to any point or object which the

vessel is approaching directly on a straight course. DEAD ASTERN is, of course, the opposite direction. An object that is at right angles (90°) to the centerline (keel) of a boat is ABEAM, or BROAD ON THE BEAM. If the boat passes near to it, it is said to be CLOSE ABOARD. Vessels are ABREAST of one another when they are side-by-side.

The direction midway between abeam and dead ahead is BROAD ON THE BOW, port or starboard as the case may be. Going in the other direction, an object seen midway between abeam and dead astern is broad on the (PORT or STARBOARD) QUARTER.

To express the direction of another vessel or object relative to our boat, we say that it BEARS so-and-so, using the phraseology above. For intermediate directions through a circle centered on the observer's boat, the traditional POINT system can be used, with 32 points making a complete circle. Each direction in the point system is named. A ship or other object seen from your craft might be said to bear TWO POINTS ABAFT THE PORT BEAM, ONE POINT ON THE STARBOARD BOW, etc. See Figure 116. Half-points and quarter-points do exist but are rarely used. Many skippers now use the "clock" system to indicate direction from the boat, with the bow at 12 o'clock, and the stern at 6 o'clock. The direction of a remote object is its BEARING.

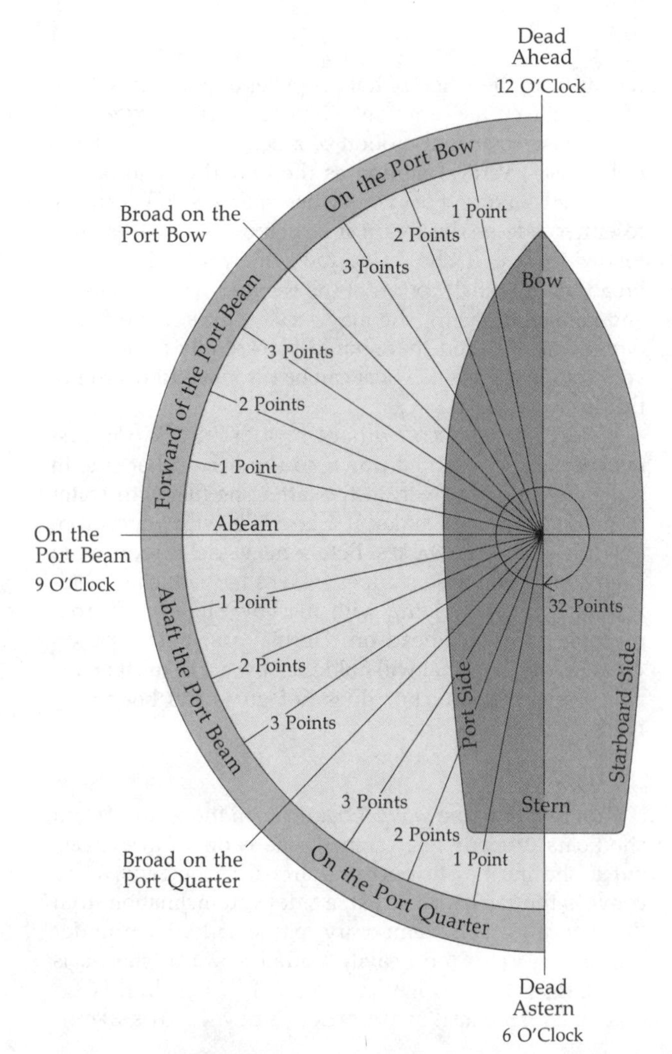

Figure 116 **Directions from on board are based on specific sections of the boat in the traditional point system. Skippers may also use the more modern clock system.**

WINDWARD means toward the direction from which the wind is blowing. A boat goes to windward, but in speaking of the side of a vessel and the parts on that side on which the wind is blowing, it is better to refer to the WEATHER side.

Opposite to windward is LEEWARD (pronounced loo'ard), the direction away from the wind, toward which it is blowing. The LEE side of a boat is the side away from the wind, and a boat makes LEEWAY when blown sideways off her course. However, the vessel finds shelter in the LEE of the land when she is under a WEATHER SHORE. See Chapter 8.

Terms Relating to Equipment

CHOCKS are deck fittings, usually of metal, with inward curving arms through which mooring, anchor, or other lines are passed to lead them in the proper direction both on board and off the boat; they should be very smooth to prevent excess wearing of the line. CLEATS are fittings of metal or wood with outward curving arms or horns on which lines can be made fast (BELAYED).

While cleats are generally satisfactory for most purposes on a boat, wooden or metal BITTS are recommended if heavy strains are to be taken. These are stout vertical posts, either single or double. They may take the form of a fitting securely bolted (never screwed) to the deck, but even better as in the case of a SAMSON POST, passing through the deck and STEPPED at the keel or otherwise strongly fastened. Sometimes a round metal pin is fitted horizontally through the head of a post or bitt to aid in belaying the line; this is a NORMAN PIN.

FENDERS are relatively soft objects of rubber or plastic used between boats and piles, pier sides, seawalls, etc., to protect the topsides from scarring and to cushion any shock of the boat striking the fixed object. Fenders, sometimes referred to in a landlubberly fashion as "bumpers," are also used between boats when they are tied, or RAFTED, together. FENDER BOARDS are short lengths of stout planking, often with cushion material or metal rubbing strips on one side. They are normally used with two fenders hung vertically to provide a wider bearing surface against a single pile.

LIFE PRESERVERS (also called PERSONAL FLOTATION DEVICES, PFDS) provide additional buoyancy to keep a person afloat when he is in the water. They take the form of cushions, belts, vests, jackets, and ring buoys. Life preservers must be of a Coast Guard–approved type to meet the legal requirements as set forth in Chapter 3.

A BOAT HOOK is a short shaft of wood or metal, with a hook fitting at one end shaped to aid in extending one's effective reach from the side of a boat, such as when putting a line over a pile or picking up an object dropped overboard. It can also be used for pushing or FENDING OFF.

A BOARDING LADDER is a set of steps temporarily or permanently fitted over the side or stern of a boat to assist persons coming aboard from a low pier or float. A SWIMMING LADDER is much the same except that it extends down into the water. A SWIMMING PLATFORM is a wooden shelf attached to a boat's transom just above the waterline.

GRAB RAILS are hand-hold fittings mounted on cabin

Figure 117 **Marine hardware includes all the chocks, cleats, deck pipes, stanchions, and other fittings that may be bolted to decks, as well as steering wheels and rigging gear.**

Figure 118 **A swim platform and ladder at the transom of a small boat make it easy to board from the water. Be sure the engine is shut off when people are swimming near the boat.**

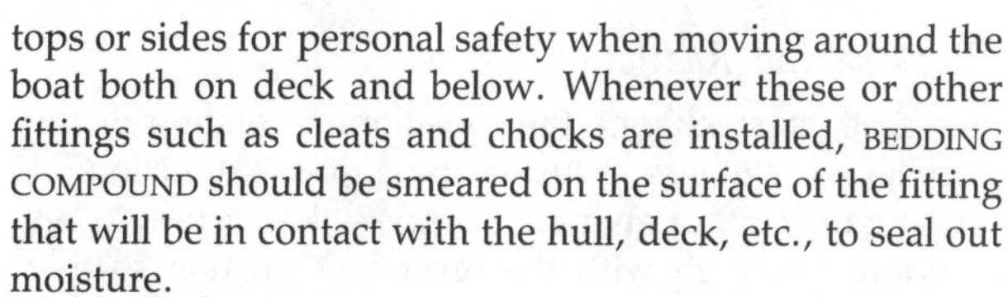

tops or sides for personal safety when moving around the boat both on deck and below. Whenever these or other fittings such as cleats and chocks are installed, BEDDING COMPOUND should be smeared on the surface of the fitting that will be in contact with the hull, deck, etc., to seal out moisture.

Magnetic COMPASSES are mounted near the control station in boxes or other protective casings known as BINNACLES. Compasses are swung in GIMBALS, pivoted rings that permit the compass BOWL and CARD to remain relatively level regardless of the boat's motion. To enable the helmsman to steer a COMPASS COURSE, a LUBBER'S LINE is marked on the inside of the compass bowl to indicate direction of the vessel's bow. Much more on compasses will be found in Chapter 15.

A LEAD (pronounced led) LINE is a length of light rope with a weight (the lead) at one end and markers at accurately measured intervals from that end. It is used for determining the depth of water alongside a vessel. A DEPTH SOUNDER is an electronic device for determining depths; often such devices are referred to as "fathometers"—actually that term is the trademark of one manufacturer of electronic depth sounders. Boats frequently used in shallow waters may be equipped with a long slender SOUNDING POLE with depth in feet and fractions marked off from its lower end.

A BAROMETER is often carried aboard a boat; it measures and indicates atmospheric pressure. A BAROGRAPH *records* changing atmospheric pressure. Knowledge of the amount and direction of change in atmospheric pressure is useful in predicting changes in the weather.

A BINOCULAR is a hand-held optical device for detecting and observing distant objects. Separate sets of lenses and prisms enable the user to see with both eyes at the same time. (The term is properly singular, as in "bicycle," but the plural form, "pair of binoculars," for a single unit is commonly used.)

A WINCH is a mechanical device, either hand or power operated, for exerting an increased pull on a line or chain, such as an anchor line.

Flags

Strictly speaking, a vessel's COLORS are the flag or flags that she flies to indicate her nationality, but the term is often expanded to include all flags flown. An ENSIGN is a flag that denotes the nationality of a vessel or her owner, or the membership of her owner in an organization (other than a yacht club).

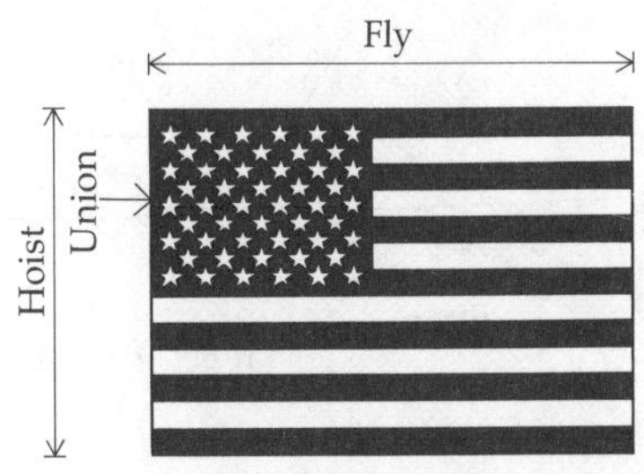

Figure 119 **The U.S. Ensign illustrates principal flag parts and dimensions. For most boat flags other than some pennants, the usual ratio between the fly and the hoist is 3:2.**

A BURGEE is a triangular, rectangular, or swallowtailed flag usually denoting yacht club or similar unit membership. A PENNANT is a flag, most often triangular in shape, used for general designating or decorative purposes.

The HOIST of a flag is its inner vertical side; also its vertical dimension. The FLY is its horizontal length from the hoist out to the free end. The UNION is the upper portion near the hoist—in the U.S. national flag, the blue area with the white stars. The UNION JACK is a flag consisting solely of the union of the national flag. See Chapter 25 for additional information on flags and how they should be flown.

Powerboat Terms

Certain basic boating terms apply specifically to boats equipped with some form of engine, whether INBOARD (mounted within the hull) or OUTBOARD (mounted on the transom and detachable).

Propulsion

THRUST for the movement of the boat through the water is achieved by the rotation of a PROPELLER (or SCREW) which draws in water from ahead and pushes it out astern. Boats with two engines and propellers are referred to as TWIN-SCREW craft. An often-used slang term for a propeller is "wheel." A few boats achieve their thrust from the reaction of a water jet.

A more recent development is the STERN-DRIVE (or I-O, or INBOARD-OUTBOARD, or OUTDRIVE). Here the motor is of the inboard type mounted within the hull at the stern; its external driving unit is similar to the LOWER UNIT of an outboard motor, and can also be tilted up. Power is applied to the propeller through two right-angle sets of gears.

Engine Mountings

An inboard engine is mounted on ENGINE BEDS, stout structural members running fore-and-aft. A SHAFT LOG is the device on a boat through which the PROPELLER SHAFT passes; the STUFFING BOX prevents water from entering at this point. On many boats, a propeller shaft is supported externally under the hull by one or more STRUTS.

In some boats, the engine is mounted so that the drive shaft extends forward to a gear box which reverses its direction to come out under the hull in a normal manner;

Figure 121 **On a stern-drive boat, the engine is inside the transom. The drive unit, which is outside the transom, houses reduction and reverse gears. It pivots from side to side, steering the boat by changing propeller thrust direction.**

This V-DRIVE permits a compact engine-in-the-stern arrangement.

Reverse and Reduction Gears

REVERSE GEARS change the direction of rotation of the propeller so as to give thrust in the opposite direction for stopping the craft or giving it sternway. REDUCTION GEARS are often combined with the reversing function so that the propeller, turning at a slower rate than the engine, will have increased efficiency. The gears of outdrive and V-drive units include the reversing capability and a reduction ratio if required.

Engine Accessories

A complete engine normally includes a FUEL PUMP to

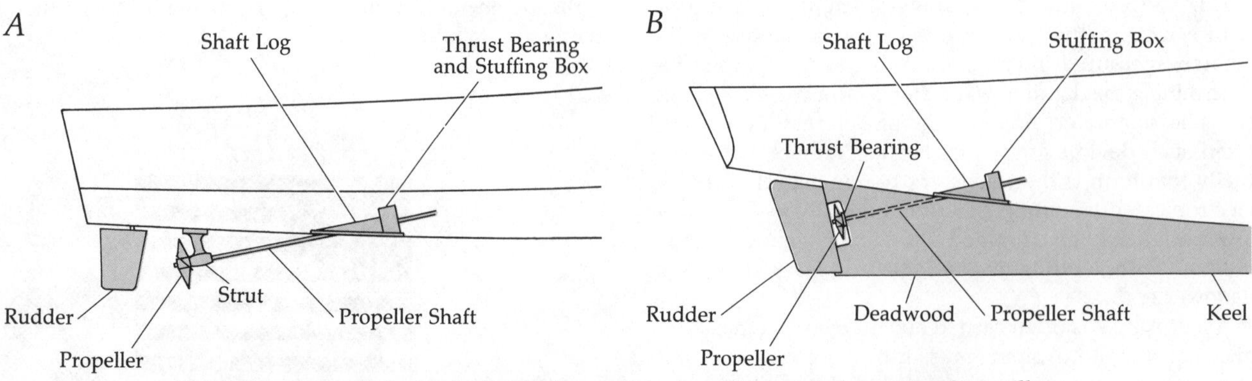

Figure 120 **On semi-planing inboard hulls, a strut supports the end of an exposed propeller shaft. On displacement hulls of heavy construction, the shaft runs through a "deadwood" above the keel. In each case a stuffing box is used to keep water from leaking up into the hull.**

feed fuel under pressure, an OIL PUMP to supply lubricating oil where needed within the engine, and one or more WATER PUMPS for circulating cooling water. An engine is RAW WATER COOLED if water for cooling is drawn through a hull fitting, circulated in the engine, and then discharged overboard. A CLOSED COOLING SYSTEM uses a separate quantity of fresh water or coolant to circulate solely within the engine; heat picked up by this water is transferred to raw cooling water in a HEAT EXCHANGER, or is dissipated in the water in which the boat is floating by means of an external KEEL COOLER.

Engine instruments usually include gauges for indicating oil pressure, cooling water temperature, and battery charging or discharging (AMMETER). A TACHOMETER indicates the speed of the engine in revolutions per minute (RPM), and may indicate the total revolutions or operating hours as a guide for performance of routine maintenance. Gasoline engines are fitted with FLAME ARRESTERS, screen-like metal fittings over the air intakes of carburetors so that any flame from a backfire will not come out dangerously into the ENGINE COMPARTMENT.

On many small outboard motors a SHEAR PIN is part of the propeller assembly. This small pin of relatively soft metal breaks apart—shears—when the propeller strikes a hard underwater object, thus saving the motor from more serious damage; the pin is easily replaced. Larger motors use a SLIP CLUTCH for the same protective purpose.

Sailboat Terms

Although some powerboat terms will also be applicable to many sailboats (especially those with an auxiliary engine), sailors have a language of their own.

All boatmen, even powerboat skippers, should have a working knowledge of the basic terminology of sailboats and sailing.

Basic Sailboat Components

The principal components of a sailboat are its HULL, SPARS, and SAILS. The hull is essentially as described before in this chapter, with certain modifications and additions resulting from the use of the wind as a means of propulsion. A portion of the wind force on the sails drives the boat forward, but another portion tends to make it move sideways. To counteract this tendency the exterior keel is much enlarged in area, or the boat is fitted with a CENTERBOARD, a relatively thin plate that can be put down vertically through the keel for greater lateral area, but which can be raised to lessen the boat's draft. To permit this raising and lowering action, the centerboard is normally pivoted on a pin near the forward corner. To maintain the watertight integrity of the hull, the centerboard is housed in a CENTERBOARD TRUNK which extends internally to above the waterline. On some very small sailboats, a DAGGERBOARD is forced down vertically in lieu of a pivoted centerboard.

Ballast

The sideward thrust of the wind, besides making leeway, also causes the sailboat to HEEL to leeward. To lessen such heeling and to increase stability, many larger sailing craft carry BALLAST—additional weight either EXTERNAL in the form of a heavy mass (usually lead) on the bottom of the keel, or INTERNAL within the hull. In smaller sailboats, the crew shifts its weight about within the craft, or just over its side, as a form of readily movable internal ballast.

Spars

The term "spars" is used broadly to cover masts, booms, gaffs, etc. MASTS are, of course, the principal vertical spars from which sails are SET. Masts are STEPPED when put in position, and are RAKED if inclined aft at an angle.

The horizontal spar along the lower edge of a fore-and-aft sail is a BOOM.

When a four-sided sail has its upper edge laced to a fore-and-aft spar, that spar is a GAFF.

Rigging

All the various lines about a vessel that secure and control the sails, masts, and other spars are its RIGGING. Lines or wires bracing the mast and certain other fixed spars comprise the STANDING RIGGING; those used for hoisting and adjusting the sails make up the RUNNING RIGGING.

Standing rigging—usually wire rope—includes SHROUDS, from the mast to the sides of the boat; FORESTAYS, which run from the mast to the bow; and BACKSTAYS, which, if used, generally run from the mast to the boat's stern.

The slack is usually taken out of a stay or shroud by the use of a TURNBUCKLE, a metal sleeve threaded right-handed in one end and left-handed in the other so that when turned in one direction eyebolts in both ends are drawn in and so shorten the fitting; when turned in the other direction the fitting increases its overall length and provides slack. The British term is RIGGING SCREW or BOTTLE SCREW.

Sails

CANVAS is a general term for a boat's sails, although modern sails are made of lighter synthetic materials. Sails DRAW when they fill with wind and provide power to drive the boat through the water. One MAKES SAIL when the sails are hoisted; one SHORTENS SAIL when the area of sails in use is reduced. A sail is REEFED by partly lowering it and securing it so that it continues to draw, but with reduced area and power. The sail is DOWSED (or DOUSED) when it is lowered quickly, FURLED when it is folded or rolled and then secured to a boom or yard. One LOOSES SAIL when UNFURLING it. A SUIT OF SAILS is a complete set for a particular boat.

Sailing Rigs

RIG is the general term applied to the arrangement of a vessel's masts and sails. Most recreational sailing craft are now fitted with triangular-shaped sails and are said to be MARCONI-RIGGED (or JIB-HEADED). If a four-sided sail with a gaff is used, the boat is described as GAFF-RIGGED.

The principal sail of a boat is its MAINSAIL. A sail forward of the mast, or ahead of the most forward mast if there is more than one, is a HEADSAIL; a single headsail is usually termed a JIB.

If a boat has only a mainsail it is a CATBOAT; she may be either Marconi-rigged or gaff-rigged. A boat having a single mast, with a mainsail and a jib, is a SLOOP. On some boats additional headsails are set from a BOWSPRIT, a spar projecting forward over the bow.

A modern CUTTER is a variation of the sloop rig in which

Figure 122 **Square-rigged vessels, such as the Coast Guard's *Eagle,* have rectangular sails set from athwartship yards. Most of the world's square-riggers are now training ships.**

the mast is stepped further aft, resulting in a larger area for headsails. A cutter normally sets two headsails—a FORESTAYSAIL with a jib ahead of it.

YAWLS have two masts, the after one of which is much smaller and is stepped abaft the rudder post. A KETCH likewise has two masts with the after one smaller, but here the difference in height is not so marked and the after mast is stepped forward of the rudder post. The taller mast is the MAINMAST; the shorter mast of either rig is the MIZZEN MAST, and the after sail is the MIZZEN.

SCHOONERS have two or more masts, but unlike yawls and ketches the after mast of a two-masted schooner is taller than the forward one (in some designs, the two masts may be the same height). Thus the after mast becomes the mainmast and the other the FOREMAST. Additional names are used if there are more than two masts.

Square-Rigged Vessels

Vessels on which the principal sails are set generally athwartships are referred to as SQUARE-RIGGED. These four-sided sails are hung from horizontal spars known as YARDS. Headsails are carried on as other rigs, forward of the foremost mast. Staysails may also be set between the masts, and the aftermost mast of a square-rigged vessel normally carries a fore-and-aft sail.

Sailing Terms

As sailboats cannot sail directly into the wind, the wind is almost always coming over one side or the other. If the wind is coming over the starboard side, the boat is said to be on a STARBOARD TACK; the PORT TACK is, of course, just the opposite. A boat TACKS when she comes up (heads) into the wind and changes the side over which it blows. In tacking, a boat COMES ABOUT. Changing tacks with the wind passing astern rather than ahead of the boat is called JIBING (pronounced with a long *i* as in ice); jibes (also spelled gibes or gybes) sometimes occur accidentally when sailing downwind (the wind nearly dead astern) and can be dangerous.

After the sails have been hoisted by their HALYARDS, they are trimmed with their SHEETS (JIBSHEETS, MAINSHEETS, etc.)—lines used to control their lateral position and movement. To TRIM is to haul in and tighten up on the sheet attached to a sail or its boom. As mentioned, sheets are "eased" when they are let out. Headsails normally have two sheets so that they can be trimmed on either tack; other sails have only one sheet.

Although a detailed discussion of sailing techniques is beyond the scope of this chapter, all boatmen should know that a sailboat is CLOSE HAULED (FULL AND BY) when she is sailing with the wind as far ahead as possible, about 45° off the bow in most boats (this is also termed sailing ON THE WIND, to WINDWARD, or BEATING). As the boat is steered further away from the wind, it BEARS OFF and is REACHING, with CLOSE reaching, BEAM reaching, and BROAD reaching occurring successively as the wind moves further aft. With the wind astern, the sailboat is RUNNING (or sailing DOWNWIND, OFF THE WIND, FREE, or SCUDDING).

See Chapter 8 for a discussion of sailboat seamanship, with additional specialized terms.

Rowboats, Oars, and Paddles

Although this book is primarily for the skippers of motorboats and sailing craft, it is good to know the essential facts about boats propelled by muscle-power. Dinghies are often rowed, and light motorboats are at times rowed or paddled to safety when the engine fails. The technique of paddling and rowing is an art in itself, best learned by practice under proper instruction, but various terms involved can be made familiar here.

Rowboats and Oars

ROWBOAT is the term generally applied to light craft propelled by one or more persons using OARS. These, of course, are long slender wooden shafts shaped into a round HANDLE (GRIP) at one end and a flat BLADE at the other; between these ends the shaft of the oar is the LOOM.

The transverse seats in rowing craft are called THWARTS. To support the thwart, a vertical piece is often fitted amidships, called a THWART STANCHION. STRETCHERS (also called FOOT BOARDS, although this creates confusion with an alternate term for bottom boards) are sometimes fitted athwartships against which the OARSMAN can brace his feet.

The terms ROWLOCK and OARLOCK are synonymous. These are the fittings in the gunwale that hold the oars when rowing. There are many styles of rowlocks; some are open, others closed; some mounted permanently, others removable; some attached to the oars. Wooden THOLE PINS (the ancient forerunner of rowlocks), driven into holes or sockets along the gunwale are still sometimes used.

Where the loom of the oar bears against the rowlock, it is often protected from excess wear by a LEATHER (which may be of leather or rubber). On the side of the leather toward the handle, there may be a BUTTON, a ring around the oar to keep it from slipping outward through the rowlock. Where the loom becomes the blade is the NECK (or THROAT) of the oar. The TIP of the blade, its outer end, may be metal-sheathed for protection from wear.

A person who is SCULLING stands in the stern and propels the boat by working a single oar back and forth, using either one hand or two hands.

Canoes and Paddles

CANOE is the term applied to very lightly built open craft of narrow beam and shallow draft that are paddled rather than rowed; they are normally double-ended.

A PADDLE is much like an oar except that it is shorter. The various parts of a paddle are named the same as for an oar. The grip is shaped differently as the paddle is used vertically rather than horizontally; the blade normally will be somewhat wider for its length in comparison with an oar.

Some paddles are double-ended (double-bladed) so that they may be more conveniently dipped alternately on either side of the canoe, an advantage when one person is propelling the boat rather than two.

Water Movements and Conditions

Tides and Currents

The word TIDE is certainly one of the most misused of nautical terms, so much so that its misapplication has come to be widely accepted. Properly speaking, it means only the rise and fall, the vertical movement, of bodies of water as a result of the gravitational pulls of the moon and sun. Commonly, but incorrectly, "tide" is also used to refer to the inflow and outflow of water as a result of changes in tidal level. CURRENT is the proper term for a horizontal flow of water, TIDAL CURRENT for flows resulting from tidal influences. It is proper to say a "two-knot current," but not a "two-knot tide." (A knot is defined on page 33.)

The incoming tidal current running toward the shore, or upstream in a river, is the FLOOD; the retreating, or downstream, current is the EBB. The RANGE of the tide (difference between the height of HIGH WATER and LOW WATER or HIGH TIDE and LOW TIDE) is not always the same as the days go by. SPRING TIDES occur when the moon is

Figure 123 **Current is the horizontal flow of water, and a buoy's leaning often is an indication of the speed at which the water is moving. In coastal areas, currents result from changes in tidal levels. In rivers, currents run from headwaters downstream to outlets in lakes or the sea.**

Figure 124 **Waves result from local wind action on the water surface. Large, deep-water swells may come from a great distance away; they crest over and become breakers as they move into shallower waters near the shore. Waves form over bars such as this one off a Pacific Coast inlet.**

new or full and have a greater range than those at other times. NEAP TIDES are those occurring at quarter moons; these have a less than average range. For more on tides and currents, see Chapter 20.

Waves, Swells, and Seas

Various terms are used to describe specific water movements or conditions of the surface. A broad term, WAVES, is frequently used for disturbed conditions on a body of water; these actually represent vertical movement of water particles regardless of their apparent forward motion. SEA is a general term often used to describe waves and water action on the surface. RIPS are short, steep waves caused by the meeting of currents. The confused water action found at places where tidal currents meet is also called a CHOP, a term also applied to small, closely spaced waves resulting from wind action on small bodies of water.

SWELLS (GROUND SWELLS) are long heavy undulations of the surface resulting from disturbances some distance away on oceans and sea. SURF is produced when waves coming toward the shore leave deep water, forming BREAKERS on the shore or beach as they CREST and curl over. A FOLLOWING SEA is one that comes up from astern, running in the same direction as the boat is going. A HEAD SEA is just the opposite. BEAM SEAS come from either side. CROSS or CONFUSED SEAS are irregular ones with components from two or more directions.

Navigation and Piloting

NAVIGATION is the art and science of safely and efficiently directing the movements of a vessel from one point to another. Most boatmen are primarily interested in one of its subdivisions, PILOTING, which is navigation using visible references, the depth of the water, etc. Other forms of navigation include CELESTIAL, taking SIGHTS by SEXTANT on the sun, moon, planets, and stars, used for long offshore passages; and ELECTRONIC, which may be used in varying degrees by inshore and offshore boatmen. DEAD RECKONING is the plotting of a vessel's position using courses and distances from the last known position. See Chapters 19 and 21.

Distances, Speeds, and Depths

On salt water, distances are measured in NAUTICAL MILES, a unit about 1/7th longer than the land of STATUTE mile. The international nautical mile is slightly more than 6,076 feet; the statute mile used in the U.S. on shore and fresh water bodies and along the Intracoastal Waterways is 5,280 feet.

Where nautical miles are used for distance, the unit of speed is the KNOT, one nautical mile per hour. Note that the "per hour" is included in the definition—to say "knots per hour" is incorrect. MILES PER HOUR meaning statute miles is the correct term for most inland fresh water bodies, as it is on shore.

In boating, depths are usually measured in feet, but offshore the unit FATHOM may be used—it is six feet.

Increasingly, metric units are coming into use—METERS and KILOMETERS (1000 meters) for distance; METERS and DECIMETERS (1/10 of a meter) for depths and heights.*

Courses

A COURSE is the direction in which a vessel is to be steered, or is being steered; the direction of travel through the water. A TRUE COURSE is one referred to TRUE (geographic) NORTH—it is the angle between the vessel's keel (centerline) and the geographic meridian when she is on course. MAGNETIC NORTH is the direction in which a compass would point if it were not subjected to any local disturbing effects. A MAGNETIC COURSE is one referred to magnetic north; it is almost always different from the true course. COMPASS NORTH is the direction of north indicated by a compass on a vessel; it may or may not be the same as magnetic north. A COMPASS COURSE is one referred to compass north. The HEADING is the direction in which a vessel's bow points at any given moment.

Variation and Deviation

VARIATION is the angular difference between true north and the direction of magnetic north at a given point on earth. Variation is "easterly" if the north mark of a compass card points to the east of true north, "westerly" if the opposite. DEVIATION is the error in a magnetic compass caused by local magnetic influences on the boat. It is easterly or westerly as the compass points east or west of magnetic north. COMPASS ERROR is the result of variation and deviation. See Chapter 15.

Bearings

A BEARING is the direction of an object *from* the observer and may be stated in terms of true, magnetic, or compass values. A RELATIVE BEARING is the direction of an object from the observer measured from the vessel's heading clockwise from 000° to 360°.

Plotting and Charting

In navigation, one PLOTS various data on a CHART (never

*1 nautical mile = 1.852 km; 1 statute mile = 1.609 km; 1 fathom = 1.83 m

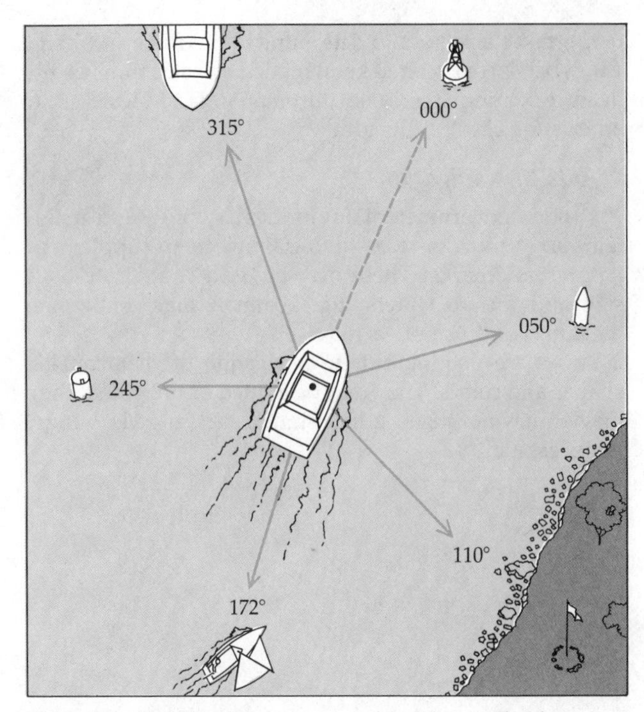

Figure 125 **Relative bearings are measured as angles from dead ahead clockwise around the boat. Directions are always shown as three digits, using zeroes if necessary (050°, not 50°).**

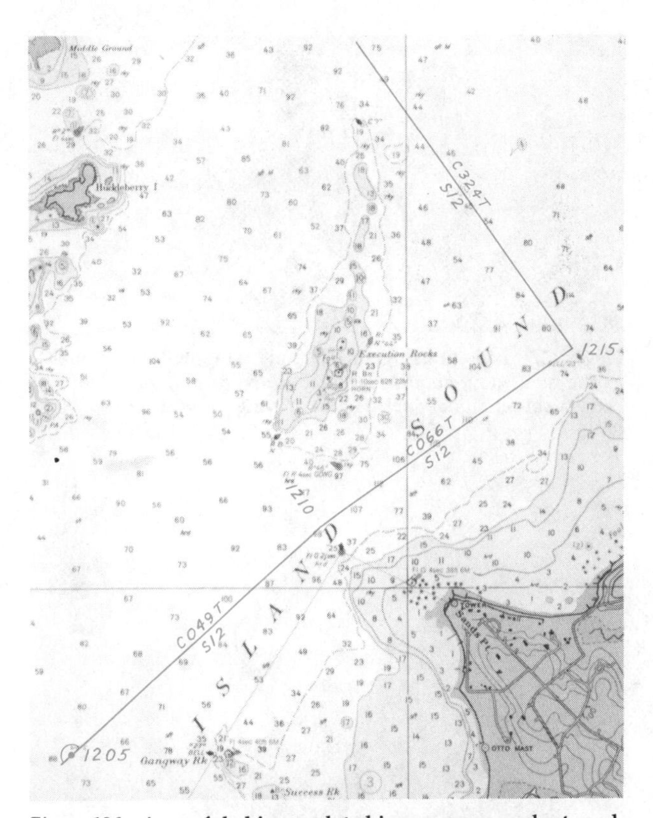

Figure 126 **A careful skipper plots his courses on a chart, and notes the time he passes aids to navigation. Other procedures for determining position are discussed in Chapter 21.**

map), takes BEARINGS to determine his position (FIX), and TAKES DEPARTURE from a known position to commence his dead reckoning. To CHART information is to record it on an existing chart of the area.

Aids to Navigation

Various governmental authorities establish and maintain AIDS TO NAVIGATION—artificial objects to supplement natural landmarks. These may be used to indicate both safe and unsafe waters; the boatman must know the meanings and use of each.

BUOYS are floating aids to navigation of characteristic shapes and colors. They may be lighted or unlighted; they may or may not have audible signals such as bells, gongs, whistles, etc.

Figure 127 **Aids to navigation include unlighted marks such as the nun buoy, *above*. A lighted buoy, *below*, may also carry a sound signal such as a bell, gong, or whistle.**

DAYBEACONS (not "markers") are fixed aids normally consisting of distinctive signs, the DAYMARKS, mounted on a pile or group of piles driven into the bottom. Daymarks will have a characteristic shape and color or colors.

LIGHTS (not "flashers") are aids consisting of a light source mounted on a fixed object (as distinguishable from lighted buoys). Each light will have a characteristic color and on-off pattern; most light structures will also have daymarks for daytime identification. The term "light" covers all such aids from the relativley weak minor light on a single pile to the largest seacoast lighthouse.

Detailed information on aids to navigation will be found in Chapter 16.

Miscellaneous Terms

There are many terms that do not fit neatly into one of the categories discussed thus far. When any part of a vessel's gear or equipment breaks or gives way, it CARRIES AWAY; an object GOES BY THE BOARD when it goes over the side (OVERBOARD). If a boat is STOVE (hull broken in from outside), she SPRINGS A LEAK OR MAKES WATER. When water is dipped out of a boat by hand the process is called BAILING.

The term CLEAR has many meanings. Upon leaving or entering a port, when traveling between countries, a ship must clear through customs authorities. She clears the land when she leaves it, clears a shoal when she passes it by safely. The bilges are cleared of water when they are pumped out. Fouled lines are cleared by straightening them out and getting them ready for use.

A boat STANDS BY when she remains with another craft to give her assistance if necessary. When used as an order, "stand by" means to be prepared to carry out an instruction. AYE, or AYE AYE, is a properly nautical (if somewhat military) way of acknowledging an order of instruction, indicating that it has been heard and understood, and will be obeyed.

Entries are made in a LOG (book) to record all events occurring aboard; more on this in Chapter 24. A PATENT LOG is a device to record distance traveled. One RAISES a light or landmark when it first becomes visible; he makes a LANDFALL when the shore is first sighted on coming in from sea. PASSAGE is generally construed to mean a run from one port to another; VOYAGE includes both the outward and homeward passages. WATCHES are periods of duty, usually four hours long, aboard a vessel; DOG WATCHES are shorter, two-hour periods between 4:00 and 8:00 pm (1600 and 2000 in the 24-hour system of time; see Chapter 19) used to ROTATE the watches so that a particular person does not have the watch for the same hours each day.

LUBBERLY is used to describe the doing of any task aboard a boat or ship in an ignorant or sloppy manner.

In nautical terminology, HARD means fully or completely as in turning a steering wheel hard aport (fully to the left) or HARD OVER (in either direction).

A BAR is a build-up, usually of sand, that reduces the depth of water; it may or may not be partially exposed above the surface of the water and it usually has an elongated shape. Bars frequently form across the openings (MOUTHS) of rivers, creeks, harbors, etc. Larger, broader areas of shallow water are often termed FLATS.

Figure 128 **Chafing gear such as this is used to protect lines wherever they come in contact with an abrasive surface.**

A REEF is a solid underwater obstruction, generally of coral formation, although artificial reefs have been constructed of junk cars, or other debris, to attract fish. A CHANNEL is a natural or dredged path through otherwise shoal waters.

The meaning of the word LAY depends on its usage. One LAYS AFT when he goes to the stern of the vessel. The lines of a boat are LAID DOWN before she is built. A boat is LAID UP when she is decommissioned for the winter in northern climates. A sailboat LAYS HER COURSE when she can make her objective without tacking. When an oarsman stops rowing, he LAYS ON HIS OARS. Referring to rope, LAY is the direction of the twist (right- or left-hand) and its tightness (hard or soft).

When a vessel is hauled out of the water she is SHORED UP with supports to hold her upright. If she is not supported properly, so that she is held amidships while the bow and stern settle, the boat will assume a shape that is described as HOGGED. In contrast, if the amidships portion of the keel is not adequately supported and droops, the boat is said to be SAGGED.

A vessel is said to HAIL from her home port; one HAILS another vessel at sea to get her attention, SPEAKS to her when communicating with her.

The direction in which a current flows is its SET and its velocity is its DRIFT. (The amount of leeway that a vessel makes is also its drift.) As discussed more fully in Chapter 20, SLACK is the period of time between flood and ebb currents when there is no flow in either direction. STAND is the period when there is no rise or fall in the tidal level; slack and stand usually do not occur simultaneously at any given place.

CHAFING occurs when a line or sail rubs against a rough surface and wears excessively. To prevent this, line is usually encased in CHAFING GEAR made of plastic or rubber tubing. To protect sails from chafing and wearing, BAGGYWRINKLE, consisting of short lengths of old line matted together, is placed on shrouds, spreaders, and other parts of the rigging.

A SHACKLE, or CLEVIS, is a roughly U-shaped metal fitting with a pin that can be inserted through a hole in one arm of the U and screwed or pinned in the other arm to close the link. Some special-purpose shackles close with a hinged SNAP or have a SWIVEL built in. A THIMBLE is a round or heart-shaped metal fitting with a deep outer groove around which line or wire can be eye-spliced. It protects the line from wearing on a shackle where it is joined to a chain, fastened to a deck fitting, etc.

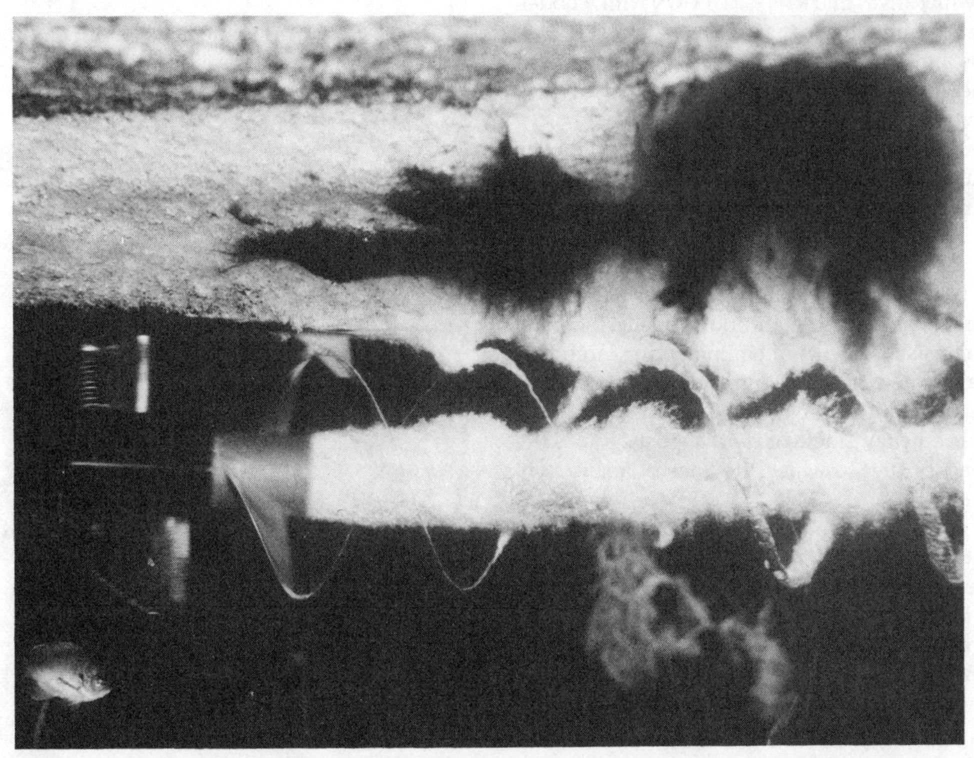

Figure 129 **Propeller cavitation, a partial vacuum formed at blade tips from excessive shaft speeds, can damage blades.**

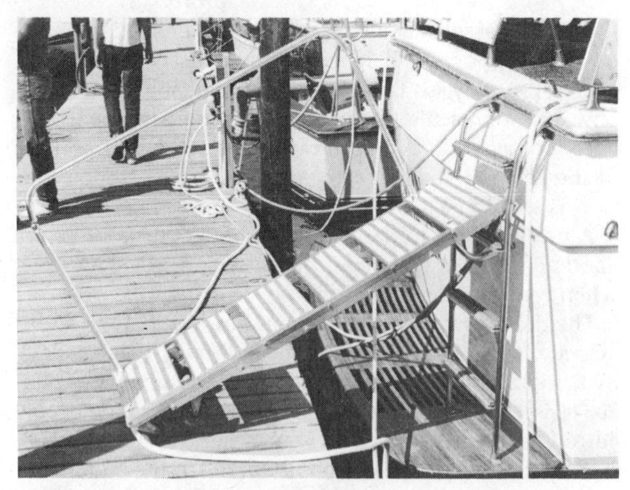

Figure 130 **A gangway, or gangplank, is a ramp from pier to boat. The unit shown here can convert to a series of steps.**

A FID is a smooth, tapered pin, usually of wood, used to open up the strands of a rope for splicing.

A WATERLIGHT is an electric light, often automatically operated, that is attached to a life ring with a short length of line. It is intended for use in man-overboard accidents at night.

The term CHARTER has usage both as a verb and a noun. To CHARTER a craft is to lease her from the owner for a temporary period of time. She may be chartered on a BARE BOAT basis, without crew, or as a TIME CHARTER with crew. The contract covering the use of the boat is termed the charter.

Ships and large vessels may have compartments below deck used solely for carrying cargo; these are HOLDS. The hatch giving access through the deck to such a hold is covered over with canvas and BATTENED DOWN with wooden or metal strips around the edges (BATTENS) to secure it against storms and water across the deck. The same term is used for the general securing of a boat against adverse weather conditions.

Another vessel is said to be seen HULL DOWN when she is at such a distance that her superstructure and/or masts are visible but the hull is not seen due to the curvature of the earth.

FLOTSAM is material floating on the surface of the water after a vessel has broken up and sunk. Larger pieces may be dangerous to the hull or shaft and propeller of small craft. JETSAM consists of items of equipment or cargo that have been deliberately JETTISONED (thrown overboard) to lighten a vessel endangered by heavy seas.

A BOLLARD is a heavy single or double post set into the edge of a wharf or pier to which the lines of a ship may be made fast.

CAVITATION occurs when a high-speed propeller loses its "bite" on the water, creating a partial vacuum, loss of thrust, and excessive shaft speed; continued cavitation can result in blade wear. On outboard motors where the propeller is relatively near the surface of the water and aft of the transom, a CAVITATION PLATE is mounted above the propeller to deflect downward the water discharge from the blades—a more correct term might be "anti-cavitation plate."

If the galley stove is of the type that requires its own vent above deck, the pipe for this is called a CHARLIE NOBLE.

A boat is SHIP-SHAPE when everything is in good order; WELL-FOUND when it is well equipped. BRISTOL-FASHION is a term used to describe an especially well-cared-for vessel. To UNSHIP an item of equipment is to unfasten or remove it from its normal working location or position. One SWABS (or SWABS DOWN) the deck when he washes it down with a mop, called a SWAB in nautical language.

The term GANGWAY is applied both to the area of a ship's side where people COME ABOARD (BOARD) and DISEMBARK, and to the temporary ramp or platform used between the vessel and the wharf or pier. In the latter usage, the terms GANGPLANK and BROW are also correct.

A WINDLASS is a particular form of winch, usually an anchor winch, with its DRUM on which the line is wrapped turning on a horizontal axis. If the drum's axis is vertical, the device may be called a CAPSTAN.

When an anchor line or chain is brought through the topsides rather than over the rail, it enters through a HAWSE HOLE and runs upward through a HAWSE PIPE. An ANCHOR CHOCK is a fitting for holding an anchor securely on a deck.

Small boats, such as dinghies and runabouts, are hoisted aboard larger craft by DAVITS, mechanical arms extending over the side or stern, or which can be swung over the side, plus the necessary lines and blocks. Large yachts sometimes have anchor davits.

When subjected to heavy strains in working her way through high seas, a vessel is said to LABOR. If she takes the large waves easily, she is said to be SEA KINDLY. A boat that takes head seas heavily and comes down hard on successive waves is said to POUND; this is more likely to occur with a hard-chined hull. If a boat takes little spray aboard when running into a choppy sea, she is termed DRY.

To FLEMISH DOWN a line is to secure it on deck in a tight flat coil roughly resembling a mat. When a line is laid

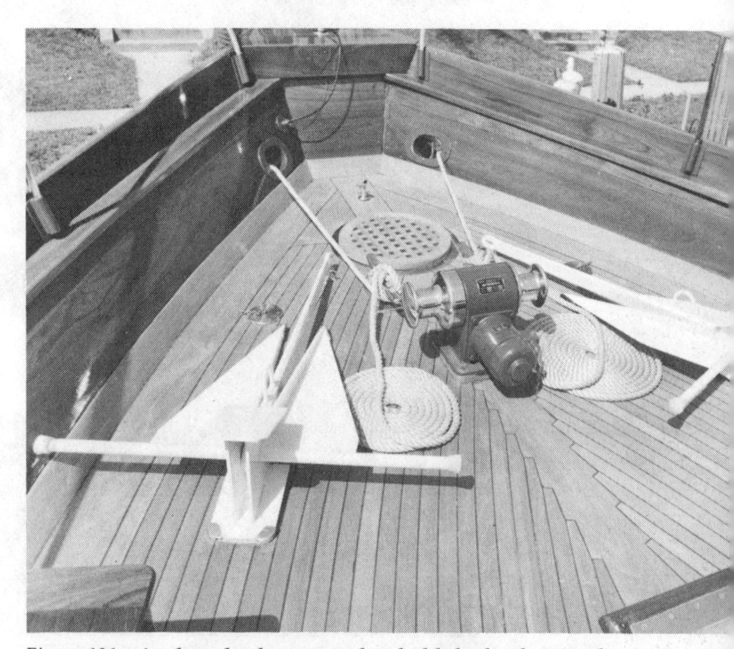

Figure 131 **Anchor chocks are used to hold the hook securely in place. The windlass is used to haul in the anchor rode.**

down in loose, looping figure-eights it is said to be FAKED (or FLAKED) down; each loop is a FAKE, see Chapter 13.

A boat is said to be OFF-SOUNDINGS when she is so far out from the shore (OFFSHORE) that depths (SOUNDINGS) cannot be conveniently measured for navigation; this is commonly taken as the 100-fathom line.* In contrast, a boat is ON-SOUNDINGS when she is within this line.

ADMIRALTY LAW is the body of law pertaining to ships and to navigation and commerce on the seas.

A CORINTHIAN is any non-professional in the field of boating.

Pollution Control Systems

In most inland and coastal waters, boats are required to have a sanitation system on board in order to control pollution. Standards have been set by the Environmental Protection Agency, and regulations have been issued by the Coast Guard, covering the certification and use of MARINE SANITATION DEVICES (MSDs), as described below.

The MACERATOR-CHLORINATOR takes the discharge of a marine toilet, grinds up the solids to fine particles, and treats the resultant waste with a disinfectant. HOLDING TANKS are tanks in which waste from the toilet is stored until it can be pumped out at a shore facility. Some models provide the option of overboard discharge for use in waters where this is permitted. RECIRCULATING toilets are usually portable units that are "charged" with a few gallons of water and special chemicals. This treated water, plus the wastes, are recirculated each time the unit is used. The toilet can be used for several days before pumpout is needed. INCINERATORS dry liquid wastes pumped into them, and reduce the resultant solids to ashes.

Documentation and Tonnage

Large ships and many commercial vessels are DOCUMENTED by the United States government (and foreign vessels by their governments). This procedure can also be used for medium- and large-size recreational craft. Such documentation is in lieu of REGISTRATION (NUMBERING) with state authorities. Before documentation can be considered, however, a discussion of "tonnage" is in order. The procedures and effects of documentation are covered in Chapter 2.

Admeasurement and Tonnage

The TONNAGE of ships and boats is determined in several different ways; it may be a measurement of either (1) weight, or (2) size without regard to weight. Persons unfamiliar with the term as it is used in connection with vessels are likely, in error, to think only of the common ton, which is a measure of weight—either 2,000 pounds in a short ton or 2,240 pounds in a long ton.* This is a part, but only a part, of the situation when ships are considered.

The determination of the "tonnage" of a vessel is called MEASUREMENT or ADMEASUREMENT, the latter term being found mainly in legal documents. (This measurement should not be confused with the measurement of racing sailboats for rating purposes, a quite different process.)

*On metric charts, the 200 meter curve is usually taken as the delineation.

Gross and Net Tonnage

The documentation law repeatedly refers to vessels of a certain tonnage, as for example when it states that a yacht of 5 net tons or over may be documented. NET TONS are derived from gross tonnage, which is a measurement of volume rather than weight.

GROSS TONNAGE is the total enclosed space or internal capacity of a vessel, calculated in terms of "tons" of 100 cubic feet each. This was agreed upon many years ago as the average space or volume required by a ton (by weight) of general merchandise. Gross tonnage includes all spaces below the upper deck as well as permanently closed-in spaces on that deck.

NET (or REGISTERED) TONNAGE is a measurement of the earning power of a vessel when carrying cargo. Thus to arrive at a net tonnage figure it is necessary to deduct from the gross the volume of such spaces as would have no earning capacity or room for cargo. For example, on a ship there would be deducted the volume of the fuel compartments, engine room, crew's quarters, bridge, etc. In the case of recreational boats and yachts, deductions would mainly be for engine compartments and control stations. Many charges against vessels, such as canal tolls, harbor dues, etc., are based on net tonnage.

Displacement and Deadweight Tonnage

Although not closely related to documentation or boats, a yachtsman should have some knowledge of several other forms of "tonnage" used in connection with ships.

DISPLACEMENT TONNAGE is the actual weight, in tons of 2,240 pounds, that a vessel displaces when floating at any given draft, such as "light" or "loaded." The displacement is calculated by figuring the volume of the vessel under the water in cubic feet and dividing by 35, as 35 cubic feet of sea water weighs 1 long ton.

DEADWEIGHT TONNAGE is the carrying capacity of a vessel figured by weight in terms of tons of 2,240 pounds. If her displacement was calculated when the vessel was "light" (with fuel and supplies but no cargo) and again when she was loaded (with the same fuel and supplies aboard), the difference would express the deadweight tonnage.

BUT DON'T OVERDO IT!

There is literally no end to the list of boating terms that could be included here, but practical limitations of space must prevail. What has been covered in this chapter will serve as a framework for the remainder of the book. As the boatman gains experience in this form of recreation, his vocabulary will broaden proportionately and naturally. It is hoped, however, that his enthusiasm for boating will not cause him to toss indiscriminate AVASTS, AHOYS, and BELAYS into every conceivable nook and corner of his conversation. The natural, proper use of correct terms is much to be desired; strained efforts to affect a salty lingo are conspicuously inappropriate.

41343
NY 6452 EY

2

Federal requirements for boat registration or documentation; the legal responsibilities involved in chartering, commercial operations, and international voyaging

BOATING LAWS AND REGULATIONS

Boating laws and regulations can be divided into three main categories. The first deals with boat ownership: registration, numbering, and documentation, all to be covered in this chapter. The second category is legislation relating to equipment requirements and standards, to be discussed in Chapter 3. The third category is the legal requirements for safe operation, the "Rules of the Road," as detailed in Chapters 4 and 5.

A quick comparison can be made between recreational boating and the operation of motor vehicles, an activity so much a part of our everyday lives. Where motor vehicle laws first concentrated on operation, with legal requirements for safety features a more recent development, much of the early boating legislation, other than Rules of the Road, was already aimed at equipment requirements.

The principal difference at the present time is that all car drivers must be licensed, but as yet there is no license or permit needed for the noncommercial skipper (except where some states require operator licenses on wholly inland waters within their boundaries, or certificates or other evidence of passing a boating safety course for youngsters under certain ages).

Federal Jurisdiction

This book will concentrate on federal boating laws and regulations rather than state and local ordinances, rules, etc., due to the wider application of the federal laws. State and municipal regulations will be touched upon later in this chapter, but only in broad terms because of the variations between different jurisdictions. You must familiarize yourself with the local requirements and restrictions for the waters that you use at home or out cruising.

An understanding of federal jurisdiction over navigable waters is essential for all boatmen. Federal jurisdiction covers the applicability of federal laws and regulations with establishment of aids to navigation by the Coast Guard, and charting by the National Ocean Service or Army Corps of Engineers. That jurisdiction does not deprive state and local authorities of their rights to regulate aspects of waterways use that *do not conflict* with federal law or regulation; local rules might relate to speed limits, restrictions on water skiing, etc. Many state boating laws also require safety equipment beyond that called for by Acts of Congress and Coast Guard regulations.

Limits of Federal Jurisdiction

"Navigable waters of the United States"—where federal law prevails—are defined in broad terms by law and are specifically delineated by Coast Guard regulations. These include coastal waters off the beaches as well as all bodies of water open to the sea or connected to the open sea by navigable rivers and channels. In addition to harbors and bays, this open-to-the-sea area includes major rivers as far as they can be navigated continuously from the sea—often a considerable distance for such great waterways as the Mississippi, Missouri, Ohio, and Tennessee rivers.

Also subject to federal jurisdiction are bodies of water that overlap or separate two or more states, even though they are not connected with the ocean. Inland lakes and rivers fit into this category when boats can use them to move persons or goods from one state to another; this is "interstate commerce" and subject to regulation by the U.S. Congress. An example is Lake Mead, which borders both Nevada and Arizona.

A third qualification places waters under federal jurisdiction if they can be used for travel to or from a foreign nation.

The construction of a dam does not change the navigable status of a river; building a dam and lock system may in fact extend the river's navigability rather than curtail it. The Coast Guard determines which bodies of water fall under federal jurisdiction, and such rulings are subject to challenge in federal courts. The determinations will be found in Part 2 of Title 33, Code of Federal Regulations; changes are published in the daily *Federal Register.*

Boundary Lines

The United States uses a number of boundary lines to indicate the seaward limits of its authority, each for one

Figure 201 **The construction of dams, and locks for passage of vessels around them, has extended far inland the navigable waters subject to federal jurisdiction. State and local laws, such as those governing speed, may apply in these waters together with the federal rules and regulations.**

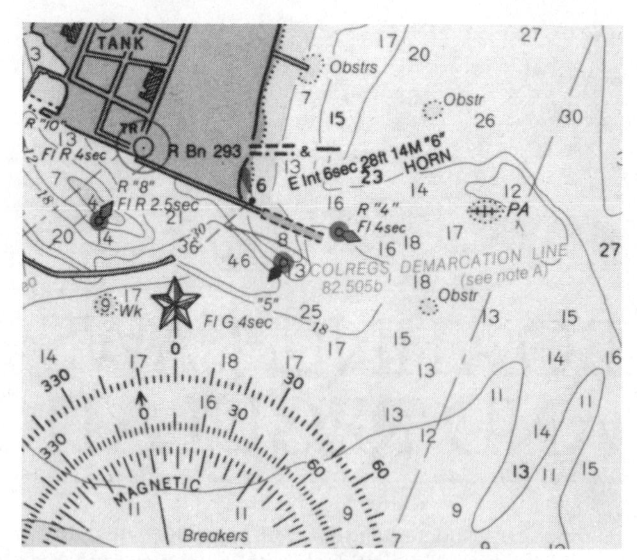

Figure 202 **At entrances to major rivers, bays, and similar water bodies, as well as along some stretches of shoreline, boundary lines have been established to mark the transition from International to Inland Rules of the Road. These are shown in magenta on charts as COLREGS dermarcation lines.**

or more regulatory purposes. There is a "base line" running generally along the shore and from headland to headland across the entrances to harbors, rivers, bays, etc. The waters for three nautical miles (5.6 km) seaward are termed the "territorial sea." Waters extending another nine miles out, to the "12-mile limit" (22.2 km), are designated the "contiguous zone." Waters out to 200 miles (370 km) from the base line comprise a "fishery Conservation Zone" for domestic laws and regulations, or "Exclusive Economic Zone" in international relationships.

Demarcation Lines for Rules of the Road Of greatest interest to boatmen are the lines that separate waters governed by the International Rules of the Road from those subject to the U.S. Inland Rules. (See Chapters 4 and 5.) When the 1972 International Rules—termed "72 COLREGS" by the Coast Guard as a contraction of their formal name, International Regulations for Preventing Collisions at Sea, 1972—came into èffect, "Demarcation Lines" separating International and Inland waters were established. In establishing these, it was found neither desirable nor practical to have an overall general line. Individual lines are defined instead, where practicable, by physical objects like fixed aids to navigation or prominent points on land, because they are readily discernable by eye rather than by instrument. The demarcation lines are set to be as short and direct as feasible, and as perpendicular as possible to vessel traffic flow. The general trend is from headland to headland, with highly visible objects used to define the line clearly. But sometimes the line is further inland than would be expected, and whole bays, harbors, and inlets are subject to the International Rules. *A boatman must at all times know which sets of Rules of the Road govern the waters he is in.* The demarcation lines are defined in detail in Coast Guard publication *Navigation Rules, International-Inland* (COMDTINST M16672.2A) and are shown on large-scale charts. See Figure 202.

Federal Laws and Regulations

All boating and shipping is subject to "laws" and "regulations." The former consist of Acts of Congress and provide basic policies and rules upon which more specific and detailed regulations may be based. Regulations have the advantage of flexibility because they can be created, modified, or revoked more easily than laws, whose changes are subject to the full legislative process.

Coast Guard Regulations

Most federal boating regulations are issued by the United States Coast Guard under authority given by Congress to the "Secretary of the Department in which the Coast Guard is operating." This rather cumbersome wording is necessary because federal laws permit transfer of this armed service from its normal peacetime Department (Transportation) to its wartime Department (Navy). The authority given by Congress to the Departmental Secretary is, in most instances, routinely delegated to the Commandant of the Coast Guard.

Coast Guard regulations cover a wide variety of matters including boat equipment and operation, lights for special-purpose vessels, aids to navigation, Inland Waters demarcation lines as mentioned above, and many other topics.

For technical questions regarding regulations, and information not usually available from local Coast Guard district offices, the Coast Guard has installed a toll-free telephone number in the Boating Technical Division of the Office of Boating, Public, and Consumer Affairs. The number is (800) 368-5647. Questions regarding specific cases or campaigns should be referred to the local offices.

Other Regulatory Agencies

Boatmen on some waters will also be affected by regulations issued by other federal agencies such as the U.S. Army Corps of Engineers and the National Park Service (see Chapter 9). The operation of radio and radar transmitters on boats on all waters is subject to the rules and regulations of the Federal Communications Commission as explained in Chapter 22.

Federal Boat Safety Act of 1971

The Federal Boat Safety Act of 1971 (FBSA/71) authorizes the "Secretary" to establish minimum safety standards for boats and "associated equipment," provides for numbering of all undocumented vessels equipped with propulsion machinery, directs that a "Boating Safety Advisory Council" be established, and authorizes financial assistance to the states for boating safety programs.

Definition of Terms

The FBSA/71 contains a number of definitions that must be understood, as they directly affect its applicability.

The word "vessel" includes every description of watercraft, other than a seaplane, that can be used as a means of transportation on the water. The term "boat" is limited to the following categories of vessels:

1. Those manufactured or used primarily for noncommercial use.
2. Those leased, rented, or chartered to another for the latter's noncommercial use.
3. Those engaged in the carrying of six or fewer passengers for hire.

The distinction between "vessels" and "boats" is necessary because certain sections of the FBSA/71 (notably those dealing with safety standards and equipment) pertain *only* to the category called "boats." Other sections (those dealing with numbering) pertain to a broader category—all undocumented vessels equipped with propulsion machinery. Still other provisions (the prohibition of negligent operation) pertain to *all* vessels. Note that there is *no upper size limit* in the definition of "boat" as used in the FBSA/71.

Despite its general aim toward regulating noncommercial boats and boating, craft carrying six or fewer passengers for hire (see the "passengers for hire" definition on page 50) are included, because these are not covered by the laws for vessels carrying more than six passengers for hire. Livery and charter boats are included in the FBSA/71 because their operation is closely akin to noncommercial recreational use even though they are a part of a commercial enterprise.

The term "associated equipment" means:

1. Any system, part, or component of a boat as originally manufactured, or as sold for replacement, repair, or improvement of a boat.
2. Any accessory or equipment for, or appurtenance to, a boat.
3. Any marine safety article, accessory, or equipment intended for use by a person on a boat.

The term "associated equipment" does *not* include radio equipment.

The word "State" means a State of the United States, the Commonwealth of Puerto Rico, the Virgin Islands, Guam, American Samoa, and the District of Columbia.

Applicability

The FBSA/71 applies to vessels and associated equipment used, or to be used, or carried on vessels on waters

Figure 203 **Coast Guard Publication CG-258 contains equipment requirements and other rules for pleasure craft based on Motorboat Act of 1940.**

Figure 204 **Federal requirements for recreational boats are covered in Coast Guard pamphlet CG-290. It is available from most local Coast Guard offices in English or Spanish editions.**

Figure 205 **Although it is not a federal requirement, sailboats without mechanical power must be registered in some states. All boats, including such sailboats, are subject to Coast Guard requirements for equipment such as personal flotation devices.**

subject to the jurisdiction of the United States. This Act *also* applies to every vessel owned in a "State" and *used on the high seas*—used on the waters beyond the territorial jurisdiction of the United States.

Certain limited categories of vessels are excluded from coverage by the FBSA/71; these include foreign vessels, military or public vessels of the United States (except for the aforementioned recreational-type craft, which are included), vessels owned by a state or political subdivision thereof and used primarily for governmental purposes, and ships' lifeboats.

Regulations

Although the Motorboat Act of 1940 is no longer generally applicable to the "boats" of the 1971 Act, the Coast Guard regulations which implemented the MBA/40 are retained and will remain in effect until replaced. (The MBA/40 remains in effect for commercial craft, such as fishing boats, less than 65 feet (19.8m) in length.)

Classes of Boats

For the purpose of applying graduated requirements for equipment as the size of the craft increases, the MBA/40 divides all motorboats into four "classes" based on length. This is "overall length" as measured in a straight line parallel to the keel from the foremost part of the vessel to the aftermost part, excluding sheer and excluding bowsprits, boomkins, rudders abaft the transom, outboard motor brackets, etc.

Class A less than 16 feet (4.9 m) in length

Class 1 16 feet and over, but less than 26 feet (7.9 m) in length

Class 2 26 feet and over, but less than 40 feet (12.2 m) in length

Class 3 40 feet and over, but not more than 65 feet (19.8m)

[Note: As mentioned, regulations derived from the authority of the FBSA/71 have no upper limit of length. A number of FBSA/71 requirements use a 20 foot (6.1m) length as a cut-off limit.]

Required Equipment

The regulations of the MBA/40, applicable until superseded by new rules, contain many provisions for required equipment, including fire extinguishers and other items (see Chapter 3). It is specifically stated in the FBSA/71 that regulations will be issued covering items not now required on board, such as ground tackle and navigation equipment, but this has not yet been done. Any such new requirements will not become effective without considerable study and wide publicity.

Numbering of Boats

The FBSA/71 provides for a system of boat numbering that is uniform throughout the U.S., although the Act permits the actual process of issuing certificates and number assignments to be done individually by the states. As of mid-1982 all but two had elected to perform this function within their jurisdiction; in New Hampshire and Alaska, boats are registered by the Coast Guard.

The FBSA/71 establishes broad standards for the numbering of vessels and provides for the issuance for more detailed regulations by the "Secretary." The individual states prepare their boat-numbering laws and regulations in accordance with federal standards and then submit them to the Secretary for approval. The Secretary can later withdraw approval if the state does not administer its system in accordance with the federal requirements.

Vessels Subject to Numbering The FBSA/71 requires the numbering of *all* vessels used on waters subject to federal jurisdiction, or on the high seas if owned in the U.S., that are equipped with propulsion machinery *regardless of horsepower.* Exempted are foreign boats temporarily in U.S. waters, documented vessels, ships' lifeboats, and governmental vessels other than recreational-type craft. (States may require the registration of documented boats, and the payment of fees, even though they are not numbered.)

A state numbering system may require the numbering of other craft (sailboats, rowboats, etc.) unless prohibited by federal legislation. States *may* exempt craft used solely for racing.

Special provisions are made for tenders (dinghies) carried aboard for other than lifesaving purposes.

"State of Principal Use" It is important to note that the FBSA/71 requires that a boat have a number obtained from "the State in which the vessel is principally used." This has been interpreted literally—where the boat is most often *used on the water;* i.e., neither the state where the owner lives nor the state in which the boat might customarily set on its trailer. The term "used" is taken as meaning the time that the vessel is "on" the navigable waters of the United States whether in motion, at anchor, at a mooring, or in a slip.

Certificate of Number The identification number issued for a vessel is shown on a CERTIFICATE OF NUMBER. This Certificate must be on board whenever the vessel is in use (with exceptions for small rental boats). The FBSA/71 specifies that a Certificate of Number must be of "pocket size." Each person using a boat must present its Certificate for inspection at the request of any law enforcement officer.

Federal law states that a Certificate may not be valid for more than three years; many states issue registrations for shorter periods. All Certificates may, of course, be renewed at or prior to expiration date.

Numbering Systems The specific details of boat-numbering are covered in Coast Guard regulations. The number pattern has two parts: a two-letter symbol identifying the state of principal use, and a combination of numerals and letters for individual identification. For some states, the two-letter abbreviation is the same as that used by the Postal Service; in other states, it is different.

The individual identification consists of not more than four arabic numerals and two capital letters, *or* not more than three such numerals and three such letters; fewer numbers or letters may be used. The letters "I," "O," and "Q" are not used because they can be mistaken for the numerals "1" and "0."

Number for Dinghies A state has the option of exempting dinghies from the numbering regulation. In such states, if used with a numbered vessel, a dinghy equipped with a motor of less than 10 horsepower need not be individually registered and numbered *if* "used as a tender for direct transportation between that vessel and the shore *and for no other purpose.*" The dinghy must display the number of the parent boat followed by the suffix "1" separated from the last letter by a space or hyphen as used in the basic number; for example, DC-4567-ED-1 or DC 4567 ED 1.

Figure 206 **A boat must be numbered in its "state of principal use" even though it is not the owner's state of residence, nor where the boat is stored on its trailer when out of the water.**

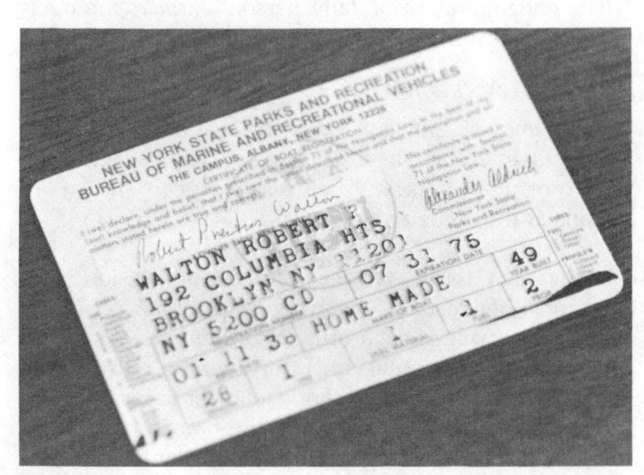

NEW YORK STATE PARKS AND RECREATION
BUREAU OF MARINE AND RECREATIONAL VEHICLES
THE CAMPUS, ALBANY, NEW YORK 12226

WALTON ROBERT P
192 COLUMBIA HTS
BROOKLYN NY 11201
NY 5200 CD
HOME MADE

Figure 207 **The state-issued Certificate of Number must be on board whenever a boat (except small rental craft) is in use.**

A dinghy used with a vessel *not* having a number, such as a documented boat, must be registered and have its own number as for any other boat if it is propelled by a motor of any horsepower.

Size and Style of Numbers The regulations require that boat numbers be clearly legible, and also contain specifications for style and size. The letters and numerals must be plain (no border or trim) vertical block characters not less than three inches in height. They must *not* be of slanting or script style; see Figure 208. The numbers and letters should all be of one color.

Contrast Numbers must provide good contrast with their background—black on white or white on black best meet this requirement. A backing or mounting plate may be used but must be of sufficient size to provide good contrast for the numbers, and thus good legibility, without regard to the hull color.

Display of Numbers The identification number must be painted on, or attached to, each side of the forward half of the vessel, and no other number may be displayed there. (This does not apply to serial numbers on small validation stickers that some states and the Coast Guard issue to note payment of current fees.) In most cases the numbers are quite near the bow, but since they need be displayed only on the boat's "forward half" it is sensible, on boats with considerable bow flare, to place the numbers far enough aft to be on a more vertical surface and thus more visible.

Numbers are placed on a boat so as to read from left to right on both sides of the bow. They may be on the hull or on a permanent superstructure. The digits between the two sets of letters in the total number *must be separated* from those letters by hyphens or spaces. The hyphen or blank space must be equal to the width of any letter other than "I" or any numeral other than "1"; see Figure 208.

Identification numbers must be maintained properly to ensure legibility at all times.

Names or insignia may be displayed on the bow of a boat, but they must not detract from the legibility of the registration numbers. Only one set of numbers may be displayed at any time. If a boat is transferred to a new state, the old numbers must be removed.

Applications and Renewals The addresses of state and other authorities to whom application for a Certificate of Number must be made are listed in Appendix C.

The same numbers are reassigned each registration period by a regular renewal Certificate. Upon transfer of ownership, if the boat continues in the same state of principal use, the old numbers stay on the craft and are assigned to the new owner.

Numbers may not be transferred from one boat to another, except for special dealers' and manufacturers' numbers issued for demonstrating, transporting, or testing boats.

Fees Under the FBSA/71, states may set their own fees for registering and numbering boats. Fees vary widely from state to state—in some jurisdictions they merely cover the cost of administering the numbering program; in oth-

Correct	Incorrect
ME 456 R ME–456–R	ME456R *ME 456 R* *ME 456 R*

Figure 208 **Correct style and spacing of registration numbers is shown at left. Letters and numbers without spacing, and use of italics or script characters, at right, is incorrect.**

Figure 209 **A common misconception is that numbers must be right at the bow of a boat. The legal requirement is that they must be on the forward half of a vessel, and actually they should be far enough aft from the stem to ensure ease of readability despite any pronounced flare of the bow.**

ers they are sources of revenue for the state.

The 1971 law specifically provides that a state may require proof of payment of state and local taxes before issuing a Certificate of Number. Withholding certification on any other grounds is *not* authorized, except that a title or proof of ownership may be required.

Reciprocity When a vessel is numbered in its state of principal use, it is considered as in compliance with the numbering system of any state in which it is *temporarily* used. There is no specific time limit stated in the federal law if such use is truly "temporary"; states often set their own time limits. This is of considerable advantage to boatmen who may cruise or trailer their craft to other state waters for vacation.

When a vessel is taken to a *new* "state of principal use," that state must recognize the validity of its number for at least 60 days before requiring new registration.

Notification of Changes If a boat is destroyed or abandoned, or sold, or used for more than 60 days in a state other than the one that issued its Certificate, or if it is documented, a report must be made to the issuing office within 15 days. The boat's Certificate is invalid and must be surrendered, and its numbers (and any validation stickers) must be removed. A report must also be made if the boat is stolen and when recovered after having been stolen.

An owner's change of address must be reported to the issuing office within 15 days. A change of motor need not be reported.

Operators' "Safety Certificates"

The FBSA/71 gives states the *optional* authority to require the operator of a vessel covered by the numbering provisions to "hold a valid safety certificate issued under terms and conditions set by the issuing authority."

No Coast Guard regulations have been issued for this purpose, and little use of the authority has been made by the states except in the case of young persons.

Termination of Unsafe Use

The Act also includes a provision that is unique to the general concept of boating laws and regulations: Termination of Unsafe Use. If a Coast Guard boarding officer observes a boat being used without sufficient lifesaving or fire-fighting devices, or in an overloaded or otherwise unsafe condition as defined in regulations, he may direct the operator to take immediate and reasonable steps necessary for the safety of those aboard the vessel. This order may include directing the operator to return to his mooring and remain there until the hazardous situation is corrected. Failure to comply with an order to terminate unsafe use of a vessel can result in arrest.

Federal regulations also designate certain portions of the Washington-Oregon coast as "regulated boating areas." When wave heights or surface currents exceed set limits, conditions are considered "unsafe" and under the Unsafe Use provision the Coast Guard may terminate boating in that area.

The Coast Guard may also forbid use of a boat that is "manifestly unsafe for a specific voyage on a specific body of water." This covers unsuitable design, improper construction or condition, or improper or inadequate operational or safety equipment.

Negligent Operation

The FBSA/71 states that "no person may use any vessel in a negligent manner so as to endanger the life, limb, or property of any person." Foreign vessels, governmental vessels, and other craft normally excluded from coverage under this Act are not exempted from the prohibition against negligent operations.

The endangerment of "any person" *includes* the operator of the vessel; no other person need be involved.

Boating Accidents

Sometimes of course, despite the best of care, boating accidents do occur. They may involve only a single boat, or two or more vessels. In all cases skippers incur obligations at the scene and afterward.

Duties in Case of Accident In case of collision, accident, or other casualty involving a vessel subject to the FBSA/71 (as well as such otherwise exempted vessels as foreign or governmental craft), the operator must, to the extent that he can without serious danger to his vessel or those on board, render necessary and practical assistance to other persons endangered by the incident. The operator must also give his name and address, and the identification of his vessel, to any person injured and to the owner of any property damaged.

The duties described here apply whether or not the incident resulted from apparent negligence. They also do not preclude any other duties required by law or regulation.

"Good Samaritan" Provision Any person who complies with the accident duties described above, or any other person who gratuitously and in good faith renders assistance at the scene of an accident or other boating casualty without the objection of any person being assisted, is protected by a provision of the FBSA/71. He cannot be held liable for any civil damages as a result of rendering assistance or for any act or omission in providing or arranging salvage, towage, medical treatment, or other assistance where he acts as an ordinary, reasonably prudent person would have under the same or similar circumstances.

Accident Reporting In case of collision, accident, or other casualty involving a vessel subject to the FBSA/71, the operator must make a report under certain circumstances. A report is required if the incident results in death, an injury requiring medical treatment beyond first aid, the disappearance of a person from a vessel under circumstances that indicate death or injury, or if there is property damage totaling more than $200.

If death occurs within 24 hours of the accident, or a person has a reportable injury or disappears from the boat, a formal report must be made within 48 hours; otherwise, within ten days.

Most state boating laws require that reports of boating accidents be made to a designated state office or official. If, however, there is no state provision for reporting such incidents, a report must be made to the Coast Guard Officer in Charge, Marine Inspection, nearest the site of the accident.

Coast Guard regulations on accident reporting list the information that must be furnished. The Boating Accident Report—CG-3865—may be used in reports to the Coast Guard. States normally use this form or one of their own patterned after it. See Figure 211.

Reporting requirements and forms are different for boats in commercial operation.

Accident Statistics The FBSA/71 also provides that as a part of an approved state registering and numbering system there shall be a casualty reporting plan in accordance with regulations issued by the Secretary. Detailed procedures for recording, compiling, and forwarding statistics are set forth in Coast Guard regulations.

National Statistics The Coast Guard is in turn charged with the responsibility of combining state reports with its own, to form annual national statistics on the number of registered boats by classes; accidents by type, cause, surrounding circumstances, etc.; and other related data.

Privacy of Information Information about the identity and ownership of vessels numbered under the Act is made available to federal and local officials as needed in any enforcement or assistance program.

In general, files relating to boat numbering are considered "public records," and information from them may be released to anyone, subject only to reasonable restrictions necessary to carry on the work of the records office.

On the other hand, individual "Boating Accident Reports" or extracts therefrom are *not* releasable. These are intended only to assist the Coast Guard in determining the cause of accidents and making recommendations for their prevention, and in compiling appropriate statistics. The privacy of accident reports permits the filing of full and accurate reports without the contents being used against an individual in civil suits.

Safety Standards The FBSA/71 authorizes minimum safety standards for "boats" and "associated equipment." Each standard must be reasonable, must meet the need for boating safety, and must be stated, insofar as practicable, in terms of performance. Under the Act's provisions, it is the responsibility of the manufacturer that all equipment on the boat when delivered meets any and all applicable safety standards of that date.

Penalties

The FBSA/71 provides for both civil and criminal penalties for violations of its provisions. It allows for variable penalties to meet situations of different degrees of seriousness, and provides for a flexible system of assessment and collection.

Figure 210 **Boating accidents must be reported if they involve more than specified limits of damage or personal property, even if only one boat is involved. In some areas reports are made to state authorities, rather than to the Coast Guard.**

DEPARTMENT OF TRANSPORTATION
U.S. COAST GUARD
CG-3865 (Rev. 5-81)

BOATING ACCIDENT REPORT

FORM APPROVED
OMB No. 2115-0010

The operator of a vessel used for recreational purposes is required to file a report in writing whenever an accident results in; loss of life or disappearance from a vessel; an injury which requires medical treatment beyond first aid; or property damage in excess of $200 or complete loss of vessel. Reports in death and injury cases must be submitted within 48 hours. Reports in other cases must be submitted within 10 days. Reports must be submitted to the reporting authority in the state where the accident occurred. This form is provided to assist the operator in filing the required written report.

COMPLETE ALL BLOCKS. *(Indicate those not applicable by "NA")*

NAME AND ADDRESS OF OPERATOR
AGE
OWNER TELE. NO.
OPERATOR TELEPHONE NO.
OPERATOR'S EXPERIENCE
THIS TYPE OF BOAT: UNDER 20 HOURS / 20 TO 100 HOURS / 100 TO 500 HOURS / OVER 500 HOURS
OTHER BOAT OPERATING EXP.: UNDER 20 HOURS / 20 TO 100 HOURS / 100 TO 500 HOURS / OVER 500 HOURS

NAME AND ADDRESS OF OWNER
RENTED BOAT: YES / NO
NO. OF PERSONS ON BOARD
FORMAL INSTRUCTIONS IN BOATING SAFETY: NONE / USCG AUXILIARY / U.S. POWER SQUADRON / AMERICAN RED CROSS / STATE / OTHER *(Indicate)*

VESSEL NO. 1

BOAT NUMBER | BOAT NAME | BOAT MAKE | BOAT MODEL | MFR HULL IDENTIFICATION NO

TYPE OF BOAT: OPEN MOTORBOAT / CABIN MOTORBOAT / AUXILIARY SAIL / SAIL *(ONLY)* / ROWBOAT / OTHER *(Specify)*
HULL MATERIAL: WOOD / ALUMINUM / STEEL / FIBERGLASS *(Plastic)* / OTHER *(Specify)*
ENGINE: OUTBOARD / INBOARD GASOLINE / INBOARD DIESEL / INBOARD-OUTDRIVE / OTHER *(Specify)*
BOAT DATA *(Propulsion)*: NO. OF ENGINES / MAKE OF ENGINE / HORSEPOWER *(Total)* / YEAR BUILT *(Engine)* / TYPE OF FUEL
BOAT DATA *(Construction)*: LENGTH / WIDTH *(Beam)* / DEPTH *(Inner Transom To Keel)* / YEAR BUILT *(Boat)*

ACCIDENT DATA

DATE OF ACCIDENT | TIME ___ AM ___ PM | NAME OF BODY OF WATER | LOCATION *(Give location precisely)*
STATE | NEAREST CITY OR TOWN | COUNTY

WEATHER: CLEAR / RAIN / CLOUDY / SNOW / FOG / HAZY
WATER CONDITIONS: CALM / CHOPPY / ROUGH / VERY ROUGH / STRONG CURRENT
TEMPERATURES *(Estimates)*: AIR ___ °F / WATER ___ °F
WIND: NONE / LIGHT *(0-6 MPH)* / MODERATE *(7-14 MPH)* / STRONG *(15-25 MPH)* / STORM *(Over 25 MPH)*
VISIBILITY: GOOD / FAIR / POOR
WEATHER ENCOUNTERED: WAS AS FORECAST / NOT AS FORECAST / NO FORECAST OBTAINED

OPERATION AT TIME OF ACCIDENT *(Check all applicable)*: COMMERCIAL ACTIVITY / CRUISING / APPROACHING DOCK / WATER SKIING / RACING / TOWING / BEING TOWED / DRIFTING / AT ANCHOR / TIED TO DOCK / FUELING / FISHING / HUNTING / SKIN DIVING OR SWIMMING / OTHER *(Specify)*
TYPE OF ACCIDENT: GROUNDING / CAPSIZING / FLOODING / SINKING / FIRE OR EXPLOSION *(Fuel)* / FIRE OR EXPLOSION *(Other than fuel)* / COLLISION WITH VESSEL / COLLISION WITH FIXED OBJECT / COLLISION WITH FLOATING OBJECT / FALLS OVERBOARD / FALLS IN BOAT / BURNS / HIT BY BOAT OR PROPELLER / OTHER *(Specify)*
WHAT, IN YOUR OPINION, CAUSED THE ACCIDENT: WEATHER CONDITIONS / EXCESSIVE SPEED / NO PROPER LOOKOUT / OVERLOADING / IMPROPER LOADING / HAZARDOUS WATERS / RESTRICTED VISION / FAULT OF HULL / FAULT OF MACHINERY / FAULT OF EQUIPMENT / OTHER *(Specify)*

PERSONAL FLOTATION DEVICES
WAS THE BOAT ADEQUATELY EQUIPPED WITH CG APPROVED LIFESAVING DEVICES? YES / NO
WERE THEY ACCESSIBLE: YES / NO
WERE THEY USED: YES / NO
WAS THE VESSEL CARRYING NON-APPROVED: LIFESAVING DEVICES: YES / NO
WERE THEY ACCESSIBLE: YES / NO
WERE THEY USED: YES / NO

FIRE EXTINGUISHERS
WERE THEY USED-*(If yes, list Type(s) and number used.)*: YES / NO / NOT APPLICABLE

PROPERTY DAMAGE *(Est.)*: THIS BOAT $ / OTHER BOAT $ / OTHER PROPERTY $
DESCRIBE PROPERTY DAMAGE
NAME AND ADDRESS OF OWNER *(Damaged Property)*

Figure 211 **The use of an official form will ensure that an accident report contains all the required information.**

Civil Penalties A basic civil penalty of not more than $500 may be assessed for any violation of the FBSA/71 or any regulation issued thereunder. If the violation involves a vessel, that vessel may be liable and may be proceeded against in the federal courts. A civil penalty of not more than $200 may be assessed by administrative action; no trial is required, but procedures are established for appeals.

Criminal Penalties Any person who willfully uses a vessel in violation of the FSBA/71 or any regulation issued thereunder may be tried in court and fined up to $1,000 or imprisoned for not more than one year, or both. A person who uses a boat in a "grossly" negligent manner is subject to the criminal penalties in addition to any civil penalties. ("Simple" negligence is subject to civil penalties only.)

Penalty for Failure to Give Aid The person in charge of a vessel in collision has duties to the other vessel and those aboard her. A skipper who fails to meet these responsibilities, without reasonable cause, may be found guilty of a misdemeanor and subjected to a fine of $1,000 or imprisonment for not more than two years, or both. The vessel may be proceeded against in federal court with the penalty assessed being divided equally between any informer and the United States.

Penalty for Operation While Intoxicated Federal law now prescribes both civil and criminal penalties for operating a vessel while intoxicated. The civil penalty can be as much as $1,000. If one is tried in court and convicted, the penalty can be a fine of up to $5,000 or one year in prison, or both. The Coast Guard is developing the necessary implementing regulations.

Many states have also adopted specific laws and penalties for operating a boat while "under the influence."

Written Warnings Coast Guard boarding officers may issue a *written warning* for certain minor violations of boating regulations; these are awarded in lieu of a citation that would lead to a monetary penalty. Citations, not written warnings, are issued for any violation involving danger to the occupants of a craft, or where three or more violations of any type are found in a single boarding.

Written warnings are issued only for the first offense of any year; subsequent offenses are subject to more severe penalties. Appeal procedures are provided if it is felt that the warning was not warranted.

Boating Safety Advisory Council

The FBSA/71 established the BOATING SAFETY ADVISORY COUNCIL of 21 members, each of whom is expected to have special knowledge and experience in boating safety. The Council consists equally of representatives from state and local governments, boat and associated equipment manufacturers, and boating organizations and the general public. All boating safety standards and other major safety matters are referred to the Boating Safety Advisory Council before being issued.

Regattas and Marine Parades

The Coast Guard has established a set of regulations for regattas and marine parades. In some areas the authority to regulate such events has been passed on to state authorities, but with the same general requirements and procedures.

The term "regattas and marine parades" includes all organized water events of limited duration conducted on a prearranged schedule; this covers races of all types.

An application must be submitted to the Commander of the Coast Guard District in which the event will be held at least 30 days in advance; late applications are normally rejected. The application must contain information as specified in Coast Guard regulations.

Approval is often followed, in turn, by "special local regulations" governing the conduct of the event, spectator craft restrictions, patrolling plans, etc. These are usually issued with Local Notices to Mariners, and carry specific penalties for violations.

Enforcement Authority

Federal boating regulations in general are enforced by the Coast Guard. USCG boarding vessels are identified by the Coast Guard ensign (see color illustrations following Chapter 27) and uniformed personnel. Upon being hailed by a Coast Guard vessel or patrol boat a vessel underway must stop immediately and lay to, or maneuver in such a way as to permit the boarding officer to come aboard; a search warrant is not required.

Many state and local governmental units have navigation regulations backed up with their own patrol boats. Skippers should be alert for posted signs or regulatory buoys indicating speed limits, "no wake," etc., and should stop immediately upon being hailed by an official craft.

Most law-enforcement vessels—Coast Guard and state or local—are equipped with a blue revolving light like those seen on police cars in many areas.

Park Regulations

The use of boats on many natural lakes and man-made reservoirs is subject to regulations issued by the National Park Service of the Department of the Interior. Other such bodies of water are controlled by state or local authorities. These rules and restrictions vary from one body of water to another, and each area must be individually checked to ensure that laws are not inadvertently violated.

State Boating Laws and Regulations

Federal laws and regulations now preempt state controls for boat and equipment safety standards, but the Federal Boat Safety Act of 1971 does allow states to impose requirements for safety equipment beyond federal rules, to meet uniquely hazardous local circumstances. The Act does *not* preempt state or local laws and regulations directed at safe boat operation. In addition, there are usually laws and regulations relating to boat trailers and their use.

The old saying that "Ignorance of the law is no excuse" applies to skippers of recreational boats. Each should know the requirements and restrictions of his state, and should take appropriate steps to expand his knowledge before cruising in other states. He should particularly be alert to varying state laws about boat trailering on highways.

Information on applicable state laws and regulations may be obtained from each state, usually from the same office that handles registration and numbering.

Documentation

Not all boats are numbered—many are documented. This is a process whereby official papers on the craft are issued by the Coast Guard in much the same manner as for large ships. The *numbering* requirements of the 1971 Federal Boat Safety Act do not apply to documented vessels; other provisions of this Act, however, do apply to all "boats" as defined therein.

Federal documentation is a form of national registration which serves to establish a vessel's nationality, eligibility to engage in a particular employment (commercial vessels only), and eligibility to become the object of a preferred ship's mortgage.

The Vessel Documentation Act of 1980, and the Coast Guard regulations derived therefrom, became effective 1 July 1982. These made major changes in the language, procedures, and forms used, some of which dated back almost two centuries. Although no substantive changes in documentation requirements were made by the Act, inflexible provisions of laws that had failed to keep up with modern practices and materials were replaced with simplified regulations that are more easily revised when changes are required. Many previous forms were eliminated outright; the remaining ones were simplified and now require much less time and effort to complete.

Basic Requirements

For any vessel, commercial or yacht, to be documented it must measure at least five net tons. Very roughly, this corresponds to a boat of 30 feet (9.1 m) or more in length, but this will vary with the type of boat.

The vessel must be owned by a U.S. citizen; a partnership, association, or joint venture, all of whose members are U.S. citizens; or a corporation meeting certain requirements of control by U.S. citizens. (Vessels owned by the federal or state governments may also be documented.) The captain and other officers, but not crewmen, of a documented vessel must be U.S. citizens.

Documentation of a craft used *solely* for recreation is *optional*, but vessels engaged in fisheries, Great Lakes trade, or foreign or coastwise trade *must* be documented.

Types of Documents

A CERTIFICATE OF DOCUMENTATION, Form CG-1270, is issued for all types of vessels. This common basic form is ENDORSED for the authorized use or uses of the vessel—registry (for foreign trade), fisheries license, Great Lakes license, coastwise license, or PLEASURE VESSEL LICENSE. (There may be more than one "commercial" endorsement, but these can not be combined with a pleasure vessel license.)

Advantages of Yacht Documentation

Privileges extended by documentation of vessels as yachts include (1) legal authority to fly the yacht ensign, which authority is not formally granted to other boats; and (2) the privilege of recording bills of sale, mortgages, and other instruments of title for the vessel with federal officials at her home port, giving constructive legal notice to all persons of the effect of such instruments and permitting the attainment of preferred status for mortgages so recorded. This gives additional security to the purchaser or mortgagee, and facilitates financing and transfer of title.

Documentation is advantageous for boats that cruise widely or spend major portions of the year in different states—it eliminates any concern over "state of principal use." Documentation may also be advantageous when entering foreign waters.

Yacht Admeasurement and Documentation

Are you the citizen owner of a yacht or boat of 5 net tons or more, used for recreation only? If so, you may elect to have it documented as a yacht by the U.S. COAST GUARD.

- Generally Speaking
- Optional Simplified Measurement Method for Pleasure Vessels
- Definitions
- Gross Tonnage
- Net Tonnage
- Notes
- Application for Simplified Measurement
- Formal Measurement
- How to Get Your Document
- Establishing Title
- Renewal of Your Document
- One Thing More
- Fees
- Further Information

Figure 212 **Much useful information on documentation is found in publications available from the Coast Guard.**

Obtaining Documentation

Before a vessel can be documented it must be MEASURED for its tonnage. ("Admeasured" is the more formal term, but it means the same and is gradually being dropped in favor of the simpler language.) Both gross and net tonnage must be determined.

Simplified Measurement A boat used *exclusively for pleasure* may be measured for tonnage under procedures that are far simpler than those required for commercial vessels. In brief, this method uses the numerical product of three dimensions: the overall length (L), overall breadth (B), and depth (D)—note that this is "depth," an internal dimension, and not the boat's "draft." The GROSS TONNAGE of a vessel designed for sailing is assumed to be $\frac{1}{2}$ (LBD/100); for vessels not designed for sailing it is calculated to be $\frac{2}{3}$ (LBD/100).

The gross tonnage of a catamaran or trimaran is determined by adding the gross tonnages for each hull as calculated above.

Where the volume of the deckhouse is disproportionate to the volume of the hull—as in some designs of houseboats—the volume of the deckhouse, calculated by appropriate geometric formulas and expressed in tons of 100 cubic feet each, is added to the gross tonnage of the hull as calculated with L, B, and D.

The NET TONNAGE of a sailing vessel is recorded as $\frac{9}{10}$ of the gross tonnage; for a non-sailing vessel, the multiplying factor is $\frac{8}{10}$. If there is no propelling machinery in the hull, the net tonnage will be the same as the gross tonnage.

Application for simplified measurement is normally by letter; there is no standard form unless one has been prepared by the local Regional Documentation Office located at Coast Guard District Offices, plus Philadelphia, Houston, and Portland, Oregon. The application must state the owner's name and address; the vessel's name and rig; her overall length, breadth, and depth, the name of the builder; and the vessel's model serial number and her official number, if previously documented. Dimensioned sketches, not necessarily to scale, should be submitted for a houseboat or a boat of unusual design. The owner can take his own measurements and complete the transaction by mail.

Formal Measurement The owner of a boat may elect to have formal measurement rather than use the simplified method; this is *required* if the vessel is to be used *commercially*.

If an owner is contemplating formal measurement, it is suggested that he first estimate the tonnage by using the simplified method. If the resulting net tonnage is less than 5 tons the vessel will probably be under 5 net tons when formally measured, and thus not eligible for documentation.

An application for formal measurement should be prepared in writing and submitted to the Coast Guard Officer in Charge, Marine Inspection at the Regional Documentation Office for the area in which the vessel is located. The information required is generally the same as for simplified measurement, plus the location of the vessel, because a Coast Guard official must come to the boat.

Under formal measurement procedures, a definite date and place should be agreed upon between the measuring officer and the owner, or his agent, for measuring and examining the vessel. There may be fees based on the time and distance of any necessary travel, particularly if the examiner is not located near the vessel.

Application for Document The vessel owner must submit an APPLICATION FOR DOCUMENTATION, Form CG-1258. This single page form—also used for changes in an existing document—has instructions on the reverse. (Some of the "instructions" are references to sections of the Code of Federal Regulation, but in general the form is simple to complete.)

When the Application Form is received by the documentation office, an "official number" will be assigned to the vessel and the owner notified. He should have the vessel properly marked (see below) and submit evidence of this action on Form CG-1322.

If the vessel is new, evidence of the facts of construction must be submitted on a Builder's Certification, Form CG-1261.

If a pleasure vessel license is sought for a used vessel for the first time, the owner has the option of presenting evidence of the complete chain of ownership, or merely a copy of the last prior registration—foreign, federal, or state—plus evidence that establishes title from that registration to the present owner. (If a used vessel has previously been documented, but such status has been dropped in the past, a complete chain of ownership evidence since the last prior documentation is required before a new document will be issued.)

Fees The fee for initial documentation of a pleasure vessel is $100. (There is no similar fee for documenting a commercial vessel.)

The fee for surrender or replacement, or simultaneous surrender and replacement, of a Certificate of Documentation is $50. This would occur, for example, because of a change in ownership, change of members in a partnership, change in tonnage, or change in home port. There is no charge for a change in owner's home address. The fee for a change of vessel name is $100, but this includes the above $50 fee for surrender and replacement of the document.

In addition there is a small charge of a few dollars for recording a bill of sale, a preferred mortgage, or similar papers.

Markings of a Documented Boat

Coast Guard regulations prescribe precisely how a documented vessel *must* be marked. This must be done

Figure 214 **A boat documented for pleasure use only must have her name and hailing port marked on some conspicuous part of the hull; this is normally done on the transom.**

and certified before the Certificate of Documentation will be issued.

Official Number The OFFICIAL NUMBER of the vessel, preceded by abbreviation "No.", must be marked by a permanent method in block type arabic numerals not less than three inches in height on some clearly visible *interior* part of the hull forward of amidships. (It is no longer required that the net tonnage also be marked next to the official number.)

Figure 213 **A documented boat must have its official number permanently marked in arabic numerals on a clearly visible interior member of the vessel, forward of amidships.**

Name and Hailing Port Vessels documented exclusively for pleasure must have the *name and hailing port* marked *together* in clearly legible letters not less than four inches in height on some clearly visible *exterior* part of the hull. (The requirement that these letters be Roman type has been deleted.) The "hailing port" can be the "home port" (port of documentation nearest the owner's home, or equivalent for partnerships or corporations) or the city (town, etc.) and state of the address used to determine the home port of the vessel. The state may be abbreviated but must be included. The name and port must be on the vessel itself by use of any means and materials which will result in durable markings. Having the name and port on the bottom of a dinghy carried athwartship on its side across a boat's transom, as is sometimes done, especially in the Pacific Northwest, is *not* a legally acceptable substitute for markings clearly visible on the documented craft itself.

Documented commercial vessels must have the name and home or hailing port marked in full on the stern, plus the name in full marked on both bows. All letters must be at least four inches in height; any clearly legible style is acceptable.

Renewal of Document

A Certificate of Documentation must be renewed every year. A Form CG-1280 must be filled out and submitted to the documentation offices at the vessel's home port before the last day of the month in which the Certificate expires. There is no fee for renewal, and the Certificate itself need *not* be mailed in.

Upon receipt of the Form 1280 showing that the Certificate of Documentation remains in existence and is accurate, the documentation officer will mail to the owner a RENEWAL DECAL, Form CG-1280-A, which is placed on the Certificate in the place provided.

Surrender of Document

A Certificate of Documentation becomes invalid and must be "surrendered" (returned to the documentation officer) under a number of circumstances, including, but not limited to: (a) a change in ownership of the vessel in whole or part; (b) a change in membership of an owning partnership; (c) a change in tonnage; (d) a change of name of the vessel; (e) the death of a "tenant by the entirety (one member of a group)" owning any part of the vessel; or (f) the discovery of a substantive or clerical error made by a documentation officer.

A Certificate does not become invalid but must be surrendered for reissue upon a change of address of the owner.

The surrender of a Certificate of Documentation requires the filing at the documentation office of a Form CG-1258 with the original Certificate; other forms may also be required as determined by the reason for surrender. A Certificate can be surrendered for replacement when all spaces for endorsement have been filled, or if the Certificate becomes mutilated and illegible.

Use of Documented Boats

If a vessel is given a document as a yacht, that paper will authorize its use for *pleasure only.* A yacht document does *not* permit the transporting of merchandise or the carrying of passengers for hire (see below), such as taking out fishing parties for a fee charged directly or indirectly. Any violation of this limitation may result in severe penalties against the craft and its owner, including forfeiture of the vessel. See also the section below on "Commercial Operation."

The documentation of a vessel as a yacht *does not exempt* it from any applicable state or federal taxes. Further, the fact that a boat is federally documented will not excuse the owner from complying with safety and equipment regulations of the state or states in which it is operated.

Vessels that are documented are neither required nor permitted to have a number issued under the 1971 Federal Boat Safety Act.

Commercial Operation

The federal government requires a licensed operator on board a *motorboat* or motor yacht under 300 gross tons engaged in trade or carrying passengers for hire. Also, licensed officers are required on *motor vessels documented as yachts* of 200 gross tons and over when navigating the high seas.

Carrying Passengers For Hire

Persons wishing to operate a motorboat carrying passengers for hire must have a license. There are two categories of licensing as determined by the number of persons carried—six passengers and fewer, or more than six. Neither of these should be confused with the licenses required for pilots and engineers on vessels of more than 15 gross tons or 65 feet length engaged in trade. Motorboats under 65 feet in length carrying *freight* for hire are not required to have licensed operators.

Definition of "Passenger" Of particular interest to boatmen who use their craft exclusively for recreation is the definition of a "passenger" in the Federal Boat Safety Act of 1971. A person on board a vessel is *not* considered by law to be a passenger if he is in any of the following categories: (1) the owner or his representative; (2) the operator; (3) bona fide members of the crew who have contributed no consideration for their carriage and who are paid for their services; or (4) any guest on board a vessel which is being used exclusively for pleasure purposes who has *not* contributed any consideration, directly or indirectly, for his carriage.

The last category is the most troublesome. It is emphasized that to avoid being in a "passenger" status—which would subject the boat and its skipper to special requirements—the guest must not contribute any consideration, directly or indirectly, for his passage.

When Is a Boat Carrying Passengers for Hire? It is of considerable importance that a skipper have some understanding of circumstances that might be considered by the authorities as "carrying passengers for hire." Some situations are obvious—such as taking persons on a fishing trip for a specified amount of money for each individual or for the group as a whole—but other cases are more borderline and must be examined carefully.

It has been held that when the owner is a businessman who takes associates or customers aboard for the purpose of creating good-will or negotiating business matters not related to the trip this *does not* put the vessel and operator in the category of carrying passengers for hire.

On the other hand, the Coast Guard has ruled that the following situations *do* constitute instances of carrying passengers for hire:

1. When the boat owner invites guests aboard and one

Figure 215 **A vessel documented for any type of commercial use must have its name on both sides of the bow, as well as the name and hailing port marked on the stern.**

or more of them brings a gift of food or drink to be used on the trip.

2. Where the cruise guest takes the skipper ashore for dinner and entertainment.
3. Where during a cruise one of the guests aboard buys fuel, or some similiar action.
4. If there is a prearranged plan or agreement in advance that the guests are to pay their proportionate share, or any specific amount, of the cruise expenses.

If the participants in an expense-sharing cruise are members of a bona-fide "joint venture," they are then all considered under the law to be "owners" and not "passengers" and thus no licensed operator would be required. It is essential, however, that this joint venture be legally sufficient and in writing. *Each person* must have *all* the incidents of ownership—the right to full possession, control, and navigation of the craft. A mere "share-the-expenses" agreement is not sufficient to establish a joint venture; rights of ownership and liability in cases of accident and suit are necessary.

If a boat is chartered with the crew furnished by the owner, it has been held that the craft is then carrying passengers for hire and must be in charge of a licensed operator.

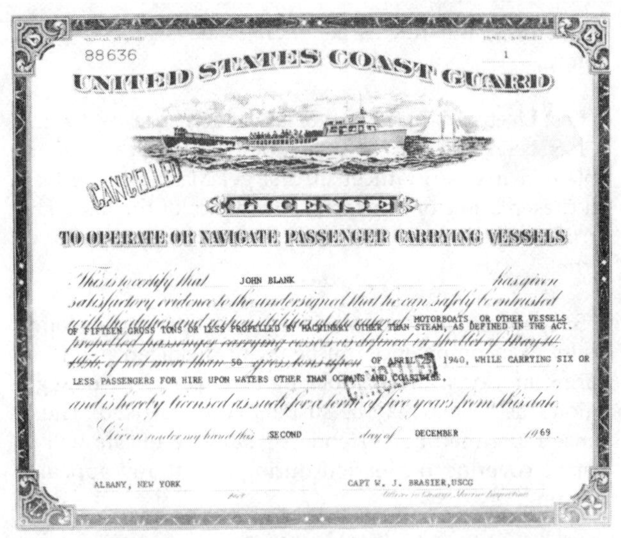

88636 1

UNITED STATES COAST GUARD

CANCELLED

LICENSE

TO OPERATE OR NAVIGATE PASSENGER CARRYING VESSELS

This is to certify that JOHN BLANK has given satisfactory evidence to the undersigned that he can safely be entrusted with the duties and responsibilities of operator of MOTORBOATS, OR OTHER VESSELS OF FIFTEEN GROSS TONS OR LESS PROPELLED BY MACHINERY OTHER THAN STEAM, AS DEFINED IN THE ACT OF APRIL 25 1940, WHILE CARRYING SIX OR LESS PASSENGERS FOR HIRE UPON WATERS OTHER THAN OCEANS AND COASTWISE.

and is hereby licensed as such for a term of five years from this date.

Given under my hand this SECOND day of DECEMBER 1969

ALBANY, NEW YORK CAPT W. J. BRASIER, USCG

Figure 216 **The Coast Guard issues licenses for the skippers of vessels that carry passengers for hire. It is a Motorboat Operator's License if six or fewer passengers are carried; if the vessel carries more than six passengers, it must be an Operator's License for inland waters, or an Ocean Operator's License if the vessel is used offshore. Licenses are issued by Officers in Charge, Marine Inspection, after an applicant has successfully passed the required examination.**

Licenses for Six or Fewer Passengers

Vessels equipped with propulsion machinery of any type, while carrying *not more than six* passengers for hire, must be "in the charge of" a person licensed for such service in accordance with regulations issued under the authority of the Federal Boat Safety Act of 1971.

Application for Licenses Licenses are issued by Officers in Charge, Marine Inspection, U.S. Coast Guard, but only at designated Regional Examination centers. Traveling Examination Teams *may* visit other ports or make specific trips to give examinations to *groups* of applicants. Much processing of applications for original and upgraded licenses, and renewals, is done by mail. Contact your local OIC, Marine Inspection or Coast Guard District headquarters, for further information.

An applicant must submit a sworn application to one of these officers, who will examine him concerning his character and fitness to hold such a license. The Officer in Charge, or his representative, investigates the proofs submitted concerning the applicant's character and ability, and determines whether his "capacity, knowledge, experience, character, and habits of life" qualify him for licensing as a Motorboat Operator.

The applicant must submit documentary evidence of at least one year's experience in operating motorboats.

Examination The applicant must pass an examination based on subjects any person operating a passenger-carrying vessel should know. The content of the exam, and whether it will be oral or written, varies with the issuing office and the waters concerned.

Questions deal generally with regulations governing motorboats, Rules of the Road for local waters, fire protection, lifesaving equipment, safe operation of gasoline engines, methods of operating and navigating boats carrying passengers for hire, and elementary first-aid.

As the Rules of the Road vary with the waters used, the holder of a license has the duty to familiarize himself with the prevalent rules if he should shift his operations to other waters.

Licenses to operate motorboats carrying passengers for hire are not issued to persons under 18 years of age. The applicant must be able to read and write.

Physical Fitness Required Applicants must be physically fit or a license will not be granted. Normally, they must be examined by a U.S. Public Health Service surgeon or a reputable physician. Bad hearing or eye-sight, color blindness, use of narcotics, insanity, or presence of certain diseases are grounds for rejection.

License Issuance and Renewal Licenses are issued and signed by Coast Guard Officers in Charge, Marine Inspection; see Figure 216. Every license (or certificate of lost license, see below) must be signed by the person to whom it is issued, who also places his thumbprint on the back. It is valid for five years, after which it can be renewed. There is no fee. Motorboat Operator Licenses will have inland and/or offshore geographical limits based on the area of the individual's boating experience.

To renew the license, the individual must present it to a Coast Guard Officer in Charge, Marine Inspection, together with a certificate of color vision, within one year after its expiration date. If more than a year has elapsed since expiration the person must take another examination. In most situations, a brief test on the Rules of the Road will be a part of the renewal process. Except under extraordinary circumstances, renewal cannot be made more than 90 days before the expiration date.

Re-examination An applicant who fails his examination may come before the same Officer in Charge, Marine Inspection, for another examination at a later date set by that officer, but not earlier than 10 days following the failure.

Lost Licenses In the case of loss of license, except where it has been revoked or suspended, a certificate can be obtained from an Officer in Charge, Marine Inspection, on presentation of satisfactory evidence of the loss. This certificate is then as valid as the license for the unexpired term.

Suspension or Revocation If any operator is found guilty of incompetence, misbehavior, negligence, endangering life, or willfully violating any safety law or regulation, his license may be suspended or revoked. The license must then be surrendered but a certificate will be issued covering the period during which any appeal is pending.

Licenses for More Than Six Passengers

Rules and regulations for small passenger vessels are prescribed in an Act of 10 May 1956, frequently called the "Ray Act." Under this Act, a "passenger-carrying vessel" means any vessel which carries *more than six passengers* and which falls into one of these categories:

1. Propelled in whole or in part by steam or by any form of mechanical or electrical power, and is of 15 gross tons or less.
2. Propelled in whole or in part by steam or any form of mechanical or electrical power and is of more than 15 but less than 100 gross tons and is not more than 65 feet (19.8 m) in length.
3. Propelled by sail and is of 700 gross tons or less.

A few exceptions are made, including foreign vessels, vessels operating on non-navigable waters of the U.S., vessels laid up and out of commission, public vessels, and lifeboats.

Operator Licenses A person holding only a license under the Federal Boat Safety Act of 1971, as described above, is not authorized to operate passenger-carrying vessels under the Ray Act. A separate license issued under the terms of this Act is required and its scope varies according to the waters navigated. For off-shore waters, an "Ocean Operator's License" is issued. Within the coastline and on the Great Lakes, it is an "Operator's License." In both cases, specific geographic limits are normally prescribed. Licenses are issued for a specified maximum gross tonnage, in multiples of 10, and are valid for five years.

Application is made to the nearest Coast Guard Officer in Charge, Marine Inspection. Physical fitness must be established and a written examination must be passed. Proof of appropriate experience is required.

Vessel Inspection

Vessels subject to the Ray Act must be inspected by an Officer in Charge, Marine Inspection, from whom applications for inspection may be obtained. These are among the vessels generally referred to in Coast Guard regulations and publications as "inspected vessels"; ordinary motorboats are categorized as "uninspected vessels."

Certificates of inspection are valid for three years. They describe the area in which the vessel may be operated, the maximum number of passengers that can be carried, the minimum crew required, minimum lifesaving and firefighting equipment to be carried, etc.

Chartering

The act of renting a boat is termed CHARTERING, except for runabouts and waterskiing boats hired on an hourly or daily basis. As a boatman, you may decide to charter another person's vessel, or you may want your boat to "earn her own keep" when you won't be using her. We will consider these two situations separately.

Chartering Another's Boat

If you have time for no boating activities at all except for a few vacation weeks each year, it may not make good economic sense to own and maintain your own boat for such limited use. Or you may have thoroughly explored the areas accessible to you with your own boat, and want to try areas further away. In either case, chartering is the way to go. Charters of both power and sailing vessels are available in almost all boating areas. Most charters begin and end at the same port, but some can be set up to cruise from "here" to "there" one-way.

Crewed or Bareboat Your charter can be either CREWED or BAREBOAT. Crewed means that the vessel's owner furnishes a captain—and often a mate/cook—who retains all responsibilities for the operation of the boat. You and your family or friends are "guests" aboard and have few responsibilities. You and your party may participate in the sailing and navigation activities if you like, but always under the supervision of the captain, who is often the vessel's owner.

Crewed charters normally include all provisions and supplies. Bareboat charters may or may not include food and beverages, or these may be available at an additional fee (this is recommended as it permits you to get underway without delay). The CHARTER AGREEMENT states exactly what is supplied by the owner, plus his responsibilities as well as yours as charterer.

On a bareboat charter you will be asked for references and for information on your boating experience. You will be checked out in the operation of the boat upon arrival, and you may be asked to take a captain along for the first day or so.

Charter fees vary widely with the size and type of boat, number of people, cruising area, season of the year, and other factors. You will probably be asked to put up a security deposit that will be returned after the charter, less any charges for damages or lost equipment.

Regulations and Insurance Regulatory requirements

vary with the chartering area. For a crewed boat the captain will normally have to be licensed, and in that case you should verify that he is. In U.S. and some foreign waters the vessel itself must have a certificate of inspection if the passengers exceed a specified number. If you take a bareboat charter for a family group you'll probably not be affected by any licensing requirements. If it is a share-the-expense party in U.S. waters, however, there may be a problem; see page 50.

Insurance on the vessel and liability coverage are handled by the boat's owner, but do verify this in the charter agreement—by all means, read the fine print.

Plan Ahead You'll usually need to make reservations well ahead of time. Normally a deposit is required—often 50% of the charter fee—and you should read the charter agreement's provisions about cancellations and refunds carefully.

Chartering Your Boat

If you want to charter your own boat, you can do this directly, or through an agent or broker. Any chartering must be under a written agreement; be sure that yours is complete and free of any pitfalls or loopholes—get professional advice. With an agent or broker you can get customers more easily and be free from much paperwork, and you'll also know that the charter is being handled legally and efficiently. You will probably also need guidance about charges for the basic charter and for the optional services. You can get help from a professional agent or broker, but you do not need to give him an exclusive hold on your boat.

Check Your Insurance Before making even the first move toward chartering your boat, *check your insurance policy*. It most likely prohibits chartering, but the restriction can be removed with an endorsement which requires the payment of an additional premium. You may also want to increase your liability coverage under these new circumstances.

Licenses and Regulations If you are going to captain your boat while chartering you will be "carrying passengers for hire." In U.S. waters this will require a Coast Guard license of the proper type for the waters used, the type and size of vessel, and the number of passengers carried. If you plan to carry more than six guests your boat must be inspected by the Coast Guard and meet much more stringent requirements than for your own use or when carrying six or fewer passengers; see page 50. In the waters of other nations, the requirements may vary, and there may be restrictions as to citizenship; check carefully.

Tax Aspects of Chartering The income and expenses of operating a boat for charter will materially affect your income tax status. Check appropriate government publications and seek professional advice.

International Voyages

When an American boat crosses the national boundaries of the United States to visit a foreign port, or a foreign yacht visits an American port, certain customs, immigration, and other regulations must be obeyed. As a result of various provisions and exemptions applying to yachts, not to vessels engaged in trade, the procedure has been so simplified that there is nothing in these legal requirements to deter a recreational craft from enjoying a cruise outside U.S. waters. There are severe penalties, however, for failure to observe the regulations that do apply.

Clearing and Entering

The terms "clearing" and "entering" are commonly used in connection with a vessel's voyage between ports of two nations. CLEARING involves obtaining permission to sail by presenting the ship's papers to a customs official. ENTERING relates to arrival, when the owner or master "enters" his vessel by having his ship's papers accepted by customs authorities. Thus a U.S. vessel might be required to *clear* from an American port and *enter* on arrival at a foreign port. Then for the return passage, the vessel would clear from the foreign port and enter upon reaching an American harbor.

If desired, BILLS OF HEALTH may be secured free of charge before leaving for a foreign port. This is not compulsory, but may make entry into the foreign port quicker and easier.

Exemptions for Noncommercial Craft Neither a licensed yacht nor an undocumented American pleasure vessel (not engaged in trade nor in any way violating the customs or navigation laws of the U.S.) is required to clear upon departure from the United States for a foreign port or place. Similarly a licensed yacht of any size or an undocumented American pleasure vessel (not engaged in trade nor in any way violating the customs or navigation laws of the U.S., and not having visited any other vessel "hovering" off the coast) is exempted from *formal* entry. These craft, however, *must* make a report to the proper authorities to cover such matters as the importation of items purchased while outside the United States.

Report on Arrival There are four separate legal aspects to entering this country from a foreign nation, each involving its own government agency and officials. These are (1) CUSTOMS, relating to bringing in dutiable merchandise; (2) IMMIGRATION, relating to persons' eligibility for entry; (3) PUBLIC HEALTH SERVICE, for preventing importation of human diseases; and (4) ANIMAL AND PLANT QUARANTINE, for preventing entry of contaminated plants, fruits, and vegetables, or infected animals.

On arrival at a foreign port, the owner or captain of a yacht (any size) should report to the authorities mentioned above, or to such of them as exist for this port. The "Q" flag (plain yellow) should be flown where it can be easily seen—starboard spreader, radio antenna, fishing outrigger, etc.—to indicate that the vessel desires to be boarded by customs and other government authorities. When reporting, the crew and guests must remain aboard until permission has been granted to land. Any additional or local regulations to be complied with, including details in connection with clearing from that port, will be supplied by the authorities.

Every vessel, whether documented or not, arriving in the United States from a foreign port or place must come into a port where the formality of entry can be accomplished. Only one person may get off the boat for the sole purpose of telephoning or otherwise notifying the authorities of the vessel's arrival; and no baggage or merchandise should be removed until the customs, immigration, and other officials have given their approval.

All boats, regardless of size, must report to the immigration authorities on return to a United States port. Any alien passengers aboard must be reported and a heavy penalty may be imposed for failure to detain passengers and crew if ordered to do so by the authorities. A report giving names, nationalities, and other information concerning any paid crew aboard must be made on a crew manifest.

In many U.S. ports a "one-stop" service has been established whereby for the simple situations of a noncommercial craft from nearby foreign waters a single government official from one of the agencies will represent all the authorities and bring all necessary forms to be filled in by the skipper. In other areas, all that is needed is a telephone call to a local or toll-free number, but whatever the requirements are where you return to the United States, *they must be complied with* to avoid possible penalties.

At the Canadian Border U.S. yachts going into Canada may secure CRUISING PERMITS with the right of free entry and clearance from 1 May to 1 October. These are issued without charge by the Canadian customs authorities at the Canadian port where the yacht first reports and must be surrendered when leaving the country. Provided the boat does not leave Canadian waters, she is then free to visit other Canadian ports until the permit is surrendered, though reports must be made at any port called where a customs officer is located.

CANADIAN LAWS AND REGULATIONS

See Appendix A for Canadian laws and regulations that cover boat licenses, construction standards, safe loading and engine power, ventilation, and reckless operation.

A Canadian boatman with craft of less than 5 tons can enter the border waters of the United States for a day's outing without applying for admission at a U.S. port of entry. He can obtain a Canadian Border Landing Card which doesn't require advance application and is good for repeated 24-hour visits throughout the navigation season. It is good only for border waters; if the Canadian boatman wants to go farther into the U.S., he must apply at a port of entry and submit to inspection.

Miscellaneous Provisions

Foreign yachts meeting certain requirements may receive from customs authorities in the United States a CRUISING LICENSE granting them special treatment while in United States waters.

Additional Information In general, customs duties are assessed on articles of foreign manufacture when they are brought into the United States. If, however, you purchased a foreign-made camera, radio, etc., in the United States and then took it with you when cruising outside U.S. waters, you would *not* be liable for any duty payment on return. The problem lies in establishing that the item has, in fact, been purchased in the United States. You should take such an article to a Customs Officer *prior to departure* and have it registered by serial number (or other distinctive, permanently affixed, unique markings) on Customs Form 4457; this will be signed by the Customs Officer. On return to the United States, you can present this form to the inspecting Customs Officer as proof of exemption from duty. A form will remain valid for re-use to identify the registered articles as long as it is legible.

Federal Water Pollution Law

The basic law covering water pollution is the Federal Water Pollution Control Act as amended and extended by the Water Quality Improvement Act of 1970, which was in turn amended in 1971. These laws cover far more than discharges from vessels, but they form the basis for all implementing standards and regulations.

Basic Provisions

The current laws relate to all watercraft, large and small, commercial and recreational. The law does, however, permit enforcing authorities to distinguish among classes, types, and sizes of vessels, as well as between new and existing vessels. There are provisions for waivers for groups of vessels, and for individual cases, where warranted, but it has been officially stated that very few waivers may be expected and only on the strongest justification for each situation.

Initially phased in over several years, the federal requirements are now applicable to all boats regardless of date of construction.

Performance Standards The general approach of the federal pollution legislation is the establishment of "standards of performance" for marine sanitation devices (MSD). These include all equipment on a vessel to "receive, retain, treat, or discharge sewage." The term "sewage" is defined to mean "human body wastes and the waste from toilets and other receptacles intended to receive or retain human body wastes."

The law states that performance standards must be consistent with maritime safety standards and marine laws and regulations. It specifically notes that in the development of standards consideration must be given to "the economic costs involved" and "the available limits of technology."

Present regulations divide waters into two categories: (1) freshwater lakes and reservoirs that have no navigable inlet or outlet, and (2) other waters, including coastal waters and tributary rivers, the Great Lakes, and freshwater bodies accessible through locks.

Freshwater bodies in the first category above are subject to "no discharge" rules. It should be noted, however, that craft equipped with a discharge-type MSD may be used on "no discharge" waters provided the device "has been secured so as to prevent discharge."

Type I marine sanitation devices are those that meet the initial standards—no more than 1000 fecal coliform per 100 ml of discharge and "no visible floating solids." A Type II device must conform to the revised higher standard—no more than 200 fecal coliform per 100 ml and no more than 150 mg per liter of "total suspended solids." All no-discharge MSDs are Type III; these include holding tanks and incinerator units.

On waters of the second category above, Type III no-discharge devices may, of course, be used. Originally it was planned that all discharge-type MSDs installed after 30 January 1980 would have to be models of Type II. Problems developed, however, in the design and manufacture of these higher-standard devices and it became necessary to issue a waiver. On 5 July 1978 the requirements were loosened to allow the installation of Type I MSDs, as well as units of Types II and III, on both new and existing boats of 65 feet or less length.

Enforcement The law specifically authorizes Coast Guard and other enforcement personnel to board and inspect any vessel upon the navigable waters of the United States to check compliance with the provisions of this section.

The Coast Guard has stated as a general policy that its units will not board vessels solely to check for the absence of a legal MSD. They will, however, make such an inspection whenever a boat is boarded for any other reason. Thus if your craft is boarded for a minor safety violation—such as improper navigation lights or suspected overloading—or as a routine procedure after receiving Coast Guard assistance, you can expect to be cited if you are not in compliance with the MSD requirements.

Penalties The penalty for operation of a boat without an operable marine sanitation device of "certified" type is a civil penalty of not more than $2,000 for each violation. Note that this is an administrative type of penalty and one that does not require trial and conviction in court. No penalty will be assessed until the alleged violator has been given an opportunity for a hearing.

The penalty rises to $5,000 per violation for manufacturers and dealers in such devices and vessels.

Preemption

A most important aspect of this federal anti-pollution law is "preemption." The Act provides that "no state or political subdivision thereof shall adopt or enforce any statute or regulation with respect to the design, manufacture, installation, or use of any marine sanitation device on any vessel subject to the provisions of this Section."

Allowable State Restrictions The 1972 amendments added a provision that a state may completely prohibit all sewage discharges, whether treated or not, but that such prohibition will not become effective until the Environmental Protection Administration has determined that *adequate pumpout and treatment* facilities exist; this determination must be made within 90 days after a state's request. The prohibition may apply to all or any part of a state's waters.

There is a further clause that allows EPA to prohibit the discharge of any sewage, treated or not, if a state can establish that such action is necessary to preserve water quality. This is expected to be applied only to special situations such as swimming areas, drinking water intakes, shellfish beds, etc. In some states this prohibition is extended to include "gray water," the discharge from galley sinks, from washbasins, in heads, and from showers.

Discharge of Oil

Federal regulations provide that "no person may operate a vessel, except a foreign vessel or a vessel less than 26 feet (7.9 m) in length, unless it has a placard at least 5″ × 8″, made of durable material, fixed in a conspicuous place in the machinery spaces, or at the bilge and ballast pump control station, stating the following":

DISCHARGE OF OIL PROHIBITED

The Federal Water Pollution Control Act prohibits the discharge of oil or oily waste into or upon the navigable waters and contiguous zone of the United States if such discharge causes a film or sheen upon, or discoloration of, the surface of the water, or causes a sludge or emulsion beneath the surface of the water. Violators are subject to a penalty of $5,000.

MASTER POWER
CONTROL PANEL

3

Legally required gear that your boat must carry; the additional items that add to your safety, as well as those that provide comfort and convenience

EQUIPMENT FOR BOATS

Although most manufacturers supply the basic equipment required by law, and often many other useful items, few new or used boats come with everything needed—or desired—for safe, enjoyable boating. You should select the additional gear that is best suited to your boating needs.

Categories of Equipment

Equipment for a boat can be divided into three categories for separate consideration. These groupings are:

Equipment Required by Federal, State, or Local Law This list is surprisingly limited. It has little flexibility; items are strictly specified and usually must be of an "approved" type.

Additional Equipment for Safety and Basic Operations This is gear not legally required but it includes all the items necessary for normal boat use.

Equipment for Comfort and Convenience This is the gear that is not necessary, but adds to the scope and enjoyment of your boating.

Factors Affecting Selection

In all three categories, the items and quantities will vary with the size of the craft and the use made of it, with the legally required items determined in most cases by the "class" of the boat (see pages 60-61), as set by the Motorboat Act of 1940 (MBA/40). Other factors, such as the amount of electrical power available on board, cost, etc., will also affect equipment selection.

Legally Required Equipment

The Federal Boat Safety Act of 1971 (FBSA/71—see Chapter 2, page 41) requires various items of equipment aboard boats. The specific details are spelled out in regulations stemming from the Motorboat Act of 1940, and are published in a Coast Guard pamphlet, "Federal Requirements for Recreational Boats"—CG-290. Table 3-1, pages 60-61, summarizes these requirements for boats in each class. The individual items are described in detail on the following pages.

Note that the FBSA/71 regulates the equipment of *all* boats *including sailboats without mechanical propulsion,* even though the regulations temporarily retained from the MBA/40 are in terms of "motorboats." The 1940 Act and its regulations remain effective for commercial vessels such as diesel tugboats which are not subject to the 1971 law.

State Equipment Regulations

Although the Federal Boat Safety Act of 1971 provides for federal preemption over state equipment requirements, a blanket exemption has been granted which allows all state and local regulations in effect when the FBSA/71 became law to retain in force for the time being. When federal requirements have been expanded beyond the regulations of the MBA/40, this exemption will be lifted and federal rules will prevail.

The FBSA/71 allows states and their political subdivisions to have requirements for additional equipment beyond federal requirements if needed to meet "uniquely hazardous conditions or circumstances" in a state or local area. The federal government retains a veto over such additional state or local requirements.

Required Equipment as a Minimum

Regard the legal requirements as a minimum. If a particular boat is required to have two fire extinguishers, for example, two may satisfy the authorities, but a third might be desirable, even necessary, to ensure that one is available at each location on board where it might be needed in a hurry. Two bilge vents may meet the regulations, but four would certainly give greater safety. Think "safety," not just the minimum in order to be "legal."

Non-approved Items as Excess

Some boatmen carry items of equipment no longer approved—older, superseded types, or items never ap-

DEPARTMENT OF TRANSPORTATION

COAST GUARD

EQUIPMENT LISTS

Items Approved, Certified or Accepted under Marine Inspection and Navigation Laws

COMDTINST M16714.3 (old CG-190)

Figure 301 **Coast Guard publication CG-190 lists approved items by category, acceptance number, manufacturer, and model number. More than 80 categories are listed.**

proved—as "excess" equipment in addition to the legal minimum, or more, of approved items. This is not prohibited by regulations, but it may give a false sense of protection. The danger of reaching for one of these substandard items in an emergency must be recognized and positively guarded against. Some obsolete items of safety equipment, such as carbon tetrachloride fire extinguishers, are actually hazardous on a boat, and should never be on board.

Figure 302 **An approved Type III personal flotation device worn by a water skier may be counted toward the required number to be carried on a boat. Ski belts are not approved.**

Lifesaving Equipment

USCG regulations require that *all* recreational boats subject to their jurisdiction, regardless of means of propulsion—motor, sails, or otherwise—must have on board a Coast Guard-approved PERSONAL FLOTATION DEVICE (PFD) for each person on board. (There are a few quite minor exceptions such as sailboards, racing shells, rowing sculls, and racing kayaks; foreign boats temporarily in U.S. waters are also exempted.) "Persons on board" include those in tow, such as water-skiers. The PFDs must be in serviceable condition.

More stringent requirements are placed on vessels carrying passengers for hire (any number) than are applied to recreational boats.

Types of Personal Flotation Devices

Coast Guard-approved personal flotation devices are marked with a PFD "Type" designation to indicate to the user the performance level that the device is intended to provide. There are five PFD Type designations.

Type I PFD is designed to turn an unconscious person from a face downward position in the water to a vertical or slightly backward position, and to have more than 20 pounds (9.07 kg) of buoyancy. A Type I PFD is, however, somewhat less "wearable" than the other types.

Type II PFD is designed to turn an unconscious person in the same manner as a Type I, and to have at least 15½ pounds (7.03 kg) of buoyancy. A Type II device is more comfortable for wearing than a Type I PFD.

Type III PFD is designed to keep a conscious person in a vertical or slightly backward position and to have at least 15½ pounds of buoyancy. Note that the Type III PFD has the same buoyancy as the Type II, but no turning requirement or protection for a person who becomes unconscious while in the water. These factors make it possible to design a comfortable and wearable device for activities where it is especially desirable to wear a PFD because the wearer is likely to enter the water. In-water trials have shown, however, that while a Type III jacket will provide adequate support in calm waters, *it may not keep the wearer's face clear in rough waters;* this limitation should be kept in mind in the selection and use of this device.

Type IV PFD is designed to be thrown to a person in the water and to be grasped rather than worn; it must have at least 16½ pounds (7.48 kg) of buoyancy.

Type V PFD is any device designed for a specific and restricted use.

Styles of Flotation Devices

Coast Guard regulations recognize a number of different styles of personal flotation devices and contain specifications that must be met or exceeded for each. Each device must be marked with its approval number and all other information required by the specifications.

Figure 303 **The most desirable type of PFD is the Type I life preserver. This is available both in the jacket model and bib type, *top left.* Old models without the slit down the front may no longer be counted toward the legal requirement. Type II buoyant vests, *top right,* are smaller and less costly than life preservers, and provide less buoyancy. They may be of any color. Inflatable vests are not acceptable as PFDs. Type III special purpose water safety buoyant devices, *below,* are produced in a wide variety of designs, including jackets, sport vests, water skiing vests, and hunting vests. Check for USCG approval before purchasing any such device.**

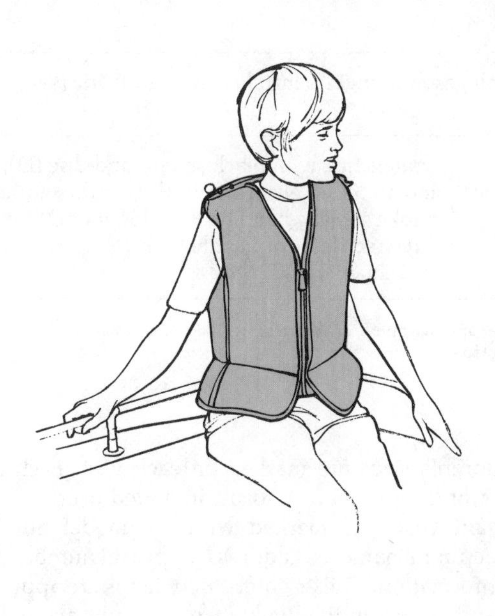

Life Preservers Type I PFDs consist of LIFE PRESERVERS of various styles, as shown in Figure 303. The flotation material of the jacket type is pads of kapok or fibrous glass material inserted in a cloth cover fitted with the necessary strps and ties. Kapok or fibrous glass *must* be encased in sealed plastic film covers; older life preservers without plastic inner containers are no longer acceptable.

Life preservers of the "bib" type may be of cloth containing unicellular plastic foam sections of specified shape, or they may be of uncovered plastic foam material with a vinyl dip coating. The bib type must have a slit all the way down the front and have adjustable body straps.

Life preservers come in two sizes—*adult,* for persons 90 pounds (40.8 kg) or more in weight, and *child,* for individuals weighing less than 90 pounds—and are so marked in large letters.

Life preservers will be marked with the type, manufacturer's name and address, and USCG approval number,

Minimum Required Safety Equipment for Boats to 26 Feet

Equipment	*Class A* *Less Than 16 Feet (4.9m)*	*Class 1* *16 Feet to Less Than 26 Feet (4.9–7.9m)*
Personal flotation devices	One Type I, II, III, or IV for each person.	One Type I, II, or III for each person on board or being towed on water skiis, etc., plus one Type IV available to be thrown.
Fire extinguishers		
When no fixed fire extinguishing system is installed in machinery space(s)	At least one B-I type approved hand portable fire extinguisher. Not require on outboard motorboats less than 26 feet (7.9 m) in length and not carrying passengers for hire if the construction of such motorboats will not permit the entrapment of flammable gases or vapors.*	
When fixed fire extinguishing system is installed in machinery space(s)	None	
Ventilation	At least two ventilator ducts fitted with cowls or their equivalent for the purpose of properly and efficiently ventilating the bilges of every engine and fuel tank compartment of boats constructed or decked over after 25 April 1940, using gasoline or other fuel having a flashpoint less than 110°F. (43°C). Boats built after 31 July 1981 must have operable power blowers.	
Whistle	Boats up to 39.4 feet (12 m)—any device capable of making an "efficient sound signal" audible 1/2 mile.	
Bell	Boats up to 39.4 feet (12 m)—any device capable of making an "efficient sound signal."	
Backfire flame arrester	One approved device on each carburetor of all gasoline engines installed after 25 April 1940, expect outboard motors.	
Visual distress signals	Required only when operating at night or carrying six or fewer passengers for hire. Same equipment as for larger boats.	Orange flag with black square-and-disc (D); and an S-O-S electric light (N); or three orange smoke signals, hand held or floating (D); or three red flares of handheld, meteor, or parachute type (D/N).

*Dry chemical and carbon dioxide (CO_2) or the most widely used types, in that order. Other approved types are acceptable. Toxic vaporizing-liquid type fire extinguishers, such as those containing tetrachloride or chlorobromomethane, are not acceptable.

Table 3-1

plus a date and place of individual inspection together with the inspector's initials. All are now required to be international orange in color.

Life Preservers are distinguishable from other lifesaving gear by the words "Approved for Use on All Vessels and Motorboats," which appear near other markings.

Buoyant Vests Type II PFDs include BUOYANT VESTS which come in many styles and colors. They use the same flotation materials as life preservers and have the same requirement for plastic inner cases, but vests are smaller and *provide less buoyancy*, hence somewhat less safety for a person in the water. Their design must be such as to float the wearer in the same safe position as required for life preservers.

Buoyant vests are made in three sizes—one for adults and two for children, medium and small; weight ranges for children's sizes are marked on each vest. Each PFD must be of the proper size for its intended user.

Buoyant vests are marked with the model number, manufacturer's name and address, approval number, and other information. These water safety items are approved by lot and are *not* individually inspected and marked.

Special Purpose Devices Personal flotation devices in this category may be either Type II or Type III if they are designed to be worn; they are Type IV if designed to be thrown.

The design of SPECIAL PURPOSE WATER SAFETY BUOYANT DEVICES is examined and approved by recognized laboratories, such as Underwriters Laboratories, see page 263; this approval is accepted by the Coast Guard. These devices are labeled with information about intended use, the size or weight category of the user, instructions as nec-

Minimum Required Safety Equipment for Boats 26 to 65 Feet

Equipment	*Class 2* *26 Feet to Less Than 40 Feet (7.9–12.2m)*	*Class 3* *40 Feet to Not More Than 65 Feet (12.2–19.8m)*
Personal flotation devices	One Type I, II, or III for each person on board devices or being towed on water skiis, etc., plus one Type IV available to be thrown.	
Fire extinguishers		
When no fixed fire extinguishing system is installed in machinery space(s)	At least two B-I type approved hand portable fire extinguishers, or at least one B-II type approved hand portable fire extinguisher.	At least three B-I type approved hand protable fire extinguishers, or at least one B-I type plus one B-II type approved hand portable fire extinguisher.
When fixed fire estinguishing system is installed in machinery space(s)	At least one B-I type approved hand portable fire extinguisher.	At least two B-I type approved hand portable fire extinguishers, or at least one B-II approved unit.
Ventilation	At least two ventilator ducts fitted with cowls or their equivalent for the purpose of properly and efficiently ventilating the bilges of every engine and fuel tank compartment of boats constructed or decked over after 25 April 1940, using gasoline or other fuel having a flashpoint less than 110°F. (43°C). Boats built after 31 July 1981 must have operable power blowers.	
Whistle	Boats up to 39.4 feet (12 m)—any device capable of making an "efficient sound signal" audible 1/2 mile.	Boats 39.4 to 65.7 feet (12–20 m)—device meeting technical specifications of Inland Rules Annex III, audible 1/2 mile.
Bell	Boats up to 39.4 feet (12 m)—any device capable of making an "efficient sound signal."	Boats 39.4 to 65.7 feet (12–20 m)—bell meeting technical specifications of Inalnd Rules Annex II; mouth diameter of at least 7.9 inches (200 m).
Backfire flame arrester	One approved device on each carburetor of all gasoline engines installed after 25 April 1940, expect outboard motors.	
Visual distress signals	Orange flag with black square-and-disc (D); and an S-O-S electric light (N); or three orange smoke signals, hand held or floating (D); or three red flares of handheld, meteor, or parachute type (D/N).	

essary for use and maintenance, and an approval number. Devices to be grasped rather than worn will also be marked "Warning: Do not Wear on Back."

Buoyant Cushions The most widely used style of Type IV PFD is the BUOYANT CUSHION. Although adequate buoyancy is provided, actually more than by a buoyant vest, the design of cushions does *not* provide safety for an exhausted or unconscious person.

This device consists of kapok, fibrous glass material, or unicellular plastic foam in a cover of fabric or coated cloth fitted with grab straps. Kapok or fibrous glass material pads must be encased in sealed plastic film covers.

Buoyant cushions are approved in various shapes and sizes, and may be of any color. Approval is by lot and number is indicated on a label on one side, along with the same general information as for buoyant vests. All cushions carry the warning against wearing on the back.

Life Rings Another style of Type IV personal flotation device is the LIFE RING BUOY, which may be of cork, balsa wood, or unicellular plastic foam. The ring is surrounded by a light grabline fastened at four points. See Figure 305.

Standard sizes are 20 inches (51 cm), 24 inches (61 cm) and 30 inches (76 cm) in diameter. Life rings of 18½ inch (47 cm) diameter have been approved under the "special purpose water safety buoyant device" category and are authorized for use on boats of any size that are not carrying passengers for hire.

Balsa wood or cork life ring buoys are marked as Type IV personal flotation devices and with other data generally similar to life preservers. They may be either orange or white in color.

Rings of plastic foam with special surface treatment will carry a small metal plate with approval information including the inspector's initials. Ring buoys of this style may be either international orange or white in color.

Some skippers paint the name of the boat on their ring buoys; this is not prohibited by the regulations.

Horseshoe Ring Buoys Another type of "marine buoyant device, Type IV" often found on cruising sail-

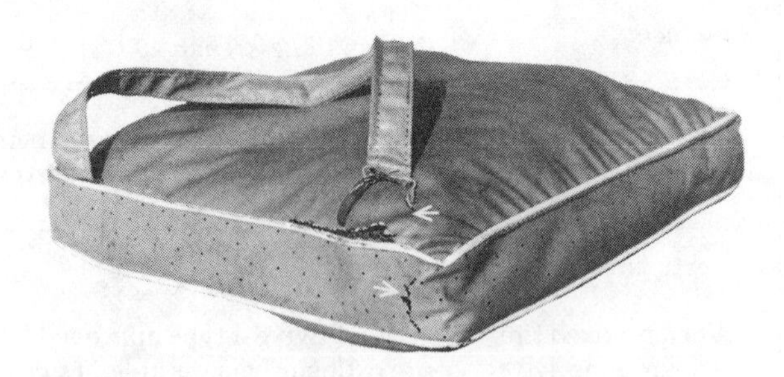

Figure 304 **Buoyant cushions can be used as Type IV "throwable" items. A worn one, *right*, should be discarded.**

boats and almost always on larger racing sailboats is the HORSESHOE RING BUOY for man-overboard emergencies. These are made of unicellular foam encased in a vinyl-coated nylon cover. The shape is aptly described by the device's name—it is much easier to get into in the water than a ring buoy. These are normally carried vertically in special holders near the stern of the boat; some holders are designed to release the horseshoe ring quickly by merely pulling a pin on a cord. To facilitate the recovery of the victim, these rings are often equipped with special accessories at the end of a short length of line—a small float with a slender pole topped with a flag (and, in some cases, with a radar reflector), an electric flasher light that automatically comes on when floating upright in the water, a small drogue to slow the rate of drift to more nearly that of a man in the water—all or a combination of these items.

Work Vests Buoyant WORK VESTS are classed as Type V personal flotation devices. These are items of safety equipment for crew members and workmen when employed over or near water under favorable conditions and properly supervised. Such PFDs are not normally approved for use on recreational boats.

PFD Requirements by Boat Size

Recreational boats *less than 16 feet in length* (and canoes and kayaks of any length) must have on board at least one approved personal flotation device of Type I, II, III, or IV for each person on board.

Recreational boats *16 feet or more in length* (except canoes and kayaks) must have on board at least one approved personal flotation device of Type I, II, or III for each person on board. Note that *buoyant cushions are not acceptable* as the basic type of PFD for vessels of this size. Recreational boats *16 feet or more in length must in addition* have on board at least one Type IV "throwable" PFD.

Size of PFD

Where the style of PFD that is carried comes in sizes, such as adult and children's life preservers, and the three sizes of buoyant vests, the PFD used to meet the requirements of the regulations must be of an "appropriate" size for the person for whom it is intended.

Stowage of PFDs

The regulations are very specific about the stowage of personal flotation devices. Any required Type I, II, or III PFD must be "readily accessible." Any Type IV PFD that is required to be on board must be "immediately available." These rules are strictly enforced.

The PFDs must also be "in serviceable condition."

Life Rafts

Boatmen who use colder northern waters, or cruise far offshore in any waters, should consider the purchase of a life raft to supplement other lifesaving gear and increase the chances of survival offshore. A raft with a canopy provides the best protection.

A life raft is not intended for use until an emergency occurs—but then it *must* work! It is of the utmost importance that life rafts be properly stored and protected, and that they be serviced periodically at an authorized facility.

Figure 305 **Type IV ring buoys are approved throwable devices available in standard sizes of 20-, 24-, and 30-inch diameters. Rings of 18½-inch diameter are acceptable on boats that are not carrying passengers for hire.**

Figure 306 **Horseshoe buoys also are approved throwable PFDs, and are popular on ocean cruising and racing sailboats.**

PFD Lights and Reflective Material

In order to better locate persons in the water at night, boats in *commercial* operation on certain waters must now carry personal flotation devices that are equipped with a light and reflective material on both front and back sides that will show up in the beam of a searchlight. Although such PFDs are not required for recreational boats, it's a good idea for increased safety! Kits are available for adding reflective material to existing PFDs.

Fire Extinguishers

All hand-portable fire extinguishers and semiportable and fixed fire extinguishing systems must be of a type that has been approved by the Coast Guard. Such extinguishers and extinguishing systems will have a metal nameplate attached with information as to the manufacturer, type, and capacity. (See below for information on "type" and "size.")

Types of Fire Extinguishers

Fire extinguishers for boats are described in terms of their contents, the actual extinguishing agent.

Dry Chemical This type of extinguisher is widely used because of its convenience and relative low cost. The cylinder contains a dry chemical in powdered form together with a propellant gas under pressure. Coast Guard regulations require that such extinguishers be equipped with a gauge or indicator to show that normal gas pressure exists within the extinguisher.

Carbon Dioxide Some boats carry portable carbon dioxide (CO_2) extinguishers; such units are advantageous as they leave no messy residue to clean up after use, and they cannot cause harm to the interior of engines as some other types may do. This type of extinguishing agent may be used where a fixed system is installed in an engine compartment, operated either manually or automatically by heat-sensitive detectors.

Carbon dioxide extinguishers consist of a cylinder containing this gas under high pressure, a valve, and a discharge nozzle at the end of a short hose. The state of charge of a CO_2 extinguisher can only be checked by weighing the cylinder and comparing this figure with the one stamped on or near the valve. See page 276 for further details on the maintenance of CO_2 extinguishers.

Vapor Systems Other chemical vapors with a fire extinguishing action are available. Halon 1301 is a colorless, odorless gas that stops fire instantly by chemical action. It is heavier than air and sinks to lower parts of the bilge. Humans can tolerate a 7% concentration, more than enough to fight fire, for several minutes. It is used in built-in systems actuated manually or automatically; additional tanks can be carried to restore protection after use.

Halon 1211, a closely related chemical, is used in hand-portable extinguishers.

Foam Fire extinguishers using chemical foam are legally acceptable on boats, but this type is rarely used as it leaves a residue that is difficult to clean up after use and may require a partial engine disassembly if dischared in the engine compartment.

Foam extinguishers are not pressurized before use and do not require tests for leakage. Such units contain water and must be protected from freezing. Foam extinguishers should be discharged and recharged annually.

Non-acceptable Types Vaporizing-liquid extinguishers, such as those containing carbontetrachloride and chlorobromomethane, are effective in fighting fires but produce highly toxic gases. They are *not* approved for use on motorboats and should NOT be carried as excess equipment because of their danger to the health or even life of a user in a confined space.

Figure 307 **Portable dry chemical fire extinguishers must have a pressure indicator to show their readiness for use.**

Classification of Extinguishers

Fires are classified in three categories: A (combustible materials), B (flammable liquids), C (electrical components), according to the general type of material being burned; see page 228. Fire extinguishers are similarly classified as to "type" and in addition are placed in size groups, I (the smallest) through V. (These USCG classifications B-I, B-II, etc., should not be confused with Underwriters Laboratories classification 2-B, 5-B, 10-B, etc., which may also appear on extinguishers. The UL classification provides a better guide to extinguisher capacity, but it has not been written into USCG regulations as yet.)

The primary fire hazard on boats results from flammable liquids and so type "B" extinguishers are specified. (Some "B" extinguishers have adequate or limited effectiveness on other classes of fires; some should *not* be used on other types of fires, such as foam extinguishers on class "C" electrical fires.) Only the two smaller sizes, I and II, are hand portable, size III is a semi-portable fire extinguishing system, and sizes IV and V are too large for consideration here.

The type and size of extinguishers for various classes of boats are shown in Table 3-1, pages 60-61.

Requirements on Boats

Class A and 1 boats must have at least one type B-I (5-BC or higher rating) hand-portable fire extinguisher, except that boats of these classes propelled by outboard motors, not carrying passengers for hire *and* of such open construction that there can be no entrapment of explosive or flammable gases or vapors, need not carry an extinguisher.

A fire extinguisher *is required* on outboard boats under 26 feet in length *if one or more* of the following conditions exist (see Figure 309):

1. Closed compartment under thwarts or seats in which portable fuel tanks may be stored.
2. Double bottoms *not* sealed to the hull, or which are not *completely* filled with flotation material.
3. Closed living spaces.
4. Closed storage compartments in which combustible or flammable materials are stowed.
5. Permanently installed fuel tanks. (To avoid being considered as "permanently installed," tanks must not be physically attached in such a manner that they cannot be lifted out with reasonable ease for filling off the boat; the size of the tank in gallons is *not* a specific criterion for determining whether it is a "portable" or a "permanently installed" tank.)

The following conditions do *not*, of themselves, require that fire extinguishers be carried (see Figure 308).

1. Bait wells.
2. Glove compartments.
3. Buoyant flotation material
4. Open slatted flooring.
5. Ice chests.

Class 2 boats must carry at least *two B-I* approved hand-portable fire extinguishers, or at least *one B-II* unit.

Class 3 boats must carry at least *three B-I* approved hand extinguishers, *or* at least *one B-II* unit *plus one B-I* unit.

Boats with fixed extinguisher systems in the engine compartment may have the above minimum requirement for their size reduced by one B-I unit. The fixed system must meet Coast Guard specifications.

Exemptions Motorboats propelled by outboard motors, while engaged in a previously arranged and announced race (and such boats designed solely for racing, while engaged in operations incidental to preparing for racing) are exempted by Coast Guard regulations from any requirements to carry fire extinguishers.

Keeping Safe

Don't merely purchase and mount fire extinguishers, and then forget all about them. All types require some

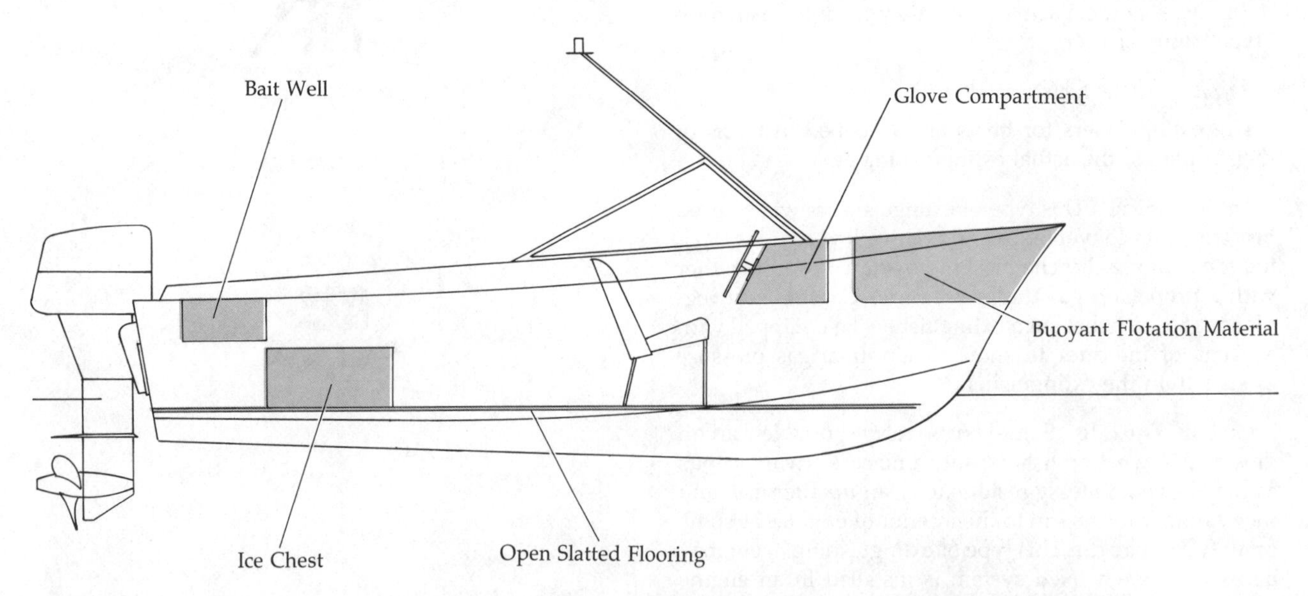

Figure 308 **A boat of open construction, with no enclosed spaces as defined by regulations, need not have a fire extinguisher on board, but one is still desirable for safety.**

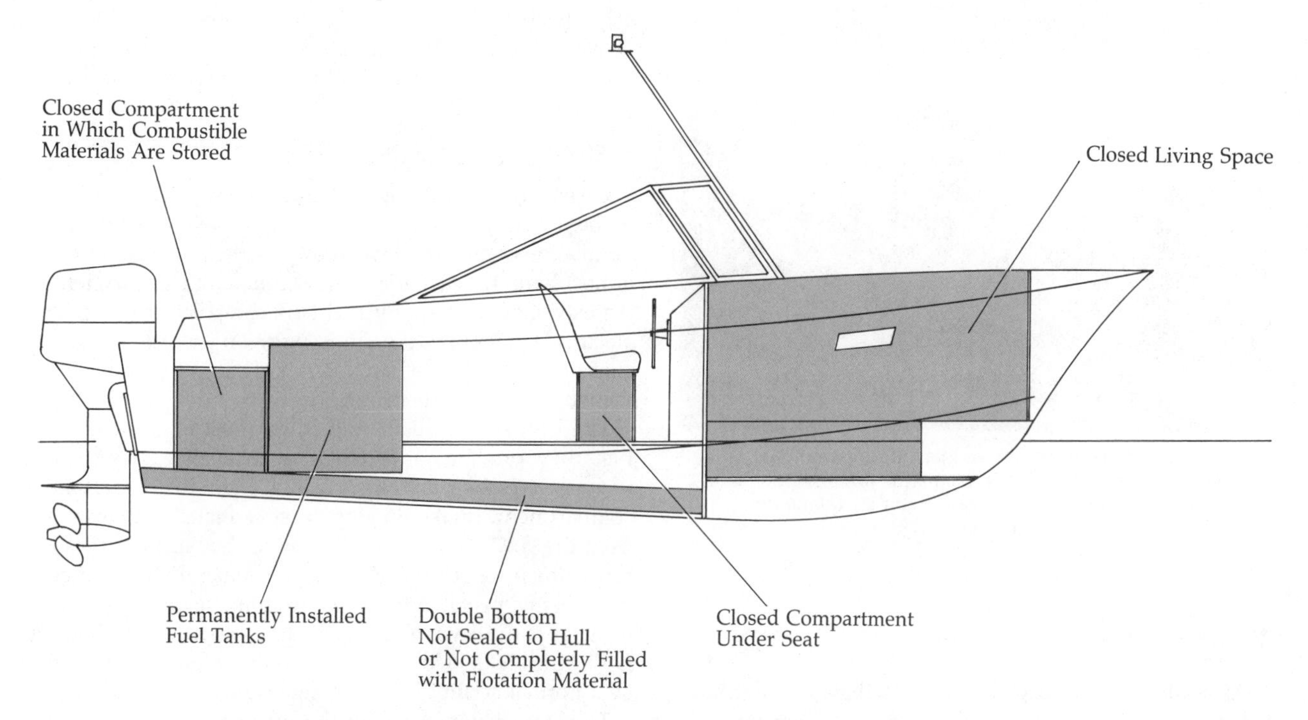

Figure 309 **Any one or more of the enclosed spaces identified *above* makes carrying a fire extinguisher a legal requirement. Boats powered by inboard engines, *below*, particularly those fueled by gasoline, often are fitted with a built-in CO_2 or Halon type fire extinguishing system. It may be set off manually, or automatically by temperature-rise sensors.**

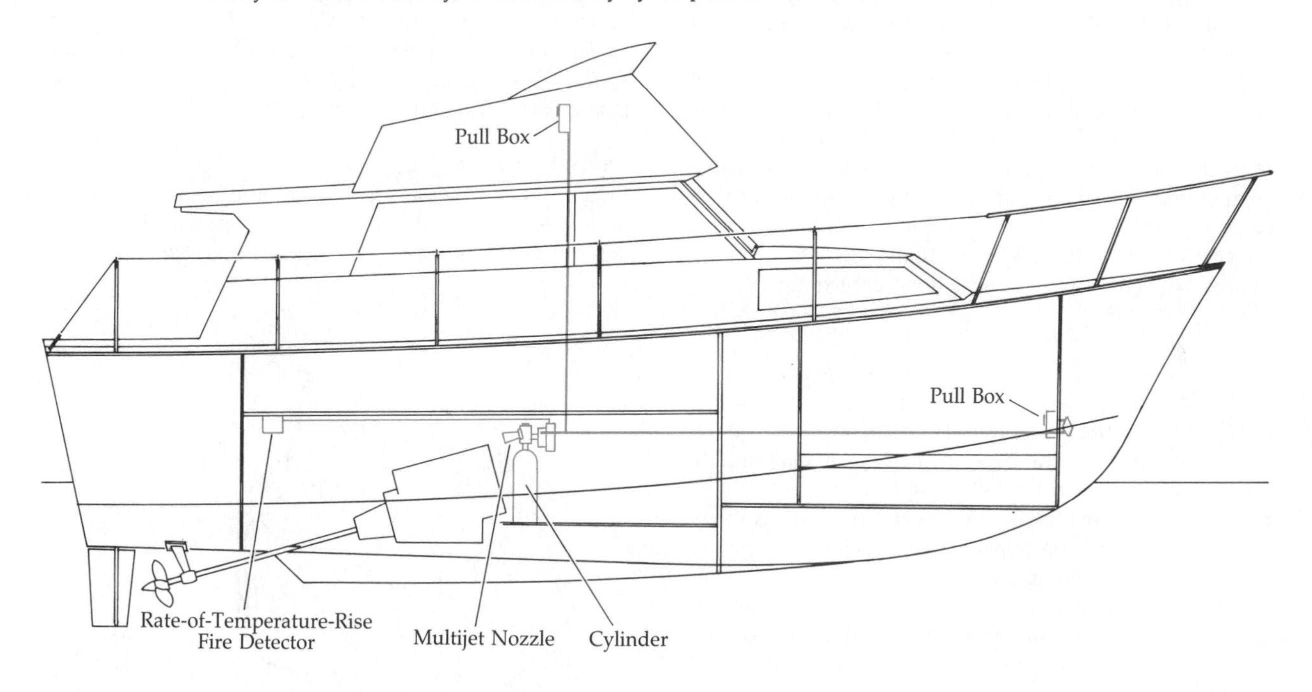

maintenance and checking; see page 276 for guidance in keeping up the level of fire protection you gain when first installing new extinguishers.

All members of your regular "crew," and also any guests that may be aboard, should know both the location and operation of all fire extinguishers.

Backfire Flame Control

Every inboard gasoline engine must be equipped with an acceptable means of backfire flame control. Such a device is commonly called a "flame arrestor" and must meet Coast Guard specifications to be approved. Accepted models will be listed by manufacturer's number in equipment lists or will have an approval number marked on the grid housing.

In use, flame arrestors must be secured to the air intake of the carburetor with an air-tight connection, have clean elements, and have no separation of the grids that would permit flames to pass through. A very few special engines or special installations are exempted from carrying flame arrestors.

Figure 310 **With some minor and technical exceptions, a gasoline powered marine engine must be fitted with a flame arrester—a metal grid over the carburetor air intake that prevents the exit of backfire flames.**

Ventilation Requirements

All motorboats, except open boats that use fuel having a flashpoint of 110°F. or less—gasoline but not diesel—must have ventilation for every engine and fuel tank compartment. Gasoline in its liquid and vapor states is potentially dangerous, but to enjoy safe boating, one need only follow Coast Guard regulations and good common sense.

The Coast Guard requires mechanical ventilation blowers on all boats built after 31 July 1980 under regulations derived from the Motorboat Act of 1940. Manufacturers must provide such equipment and no person may operate such a boat that has a gasoline engine for propulsion, electrical generation, or mechanical power unless it has an *operable* ventilation system that meets the requirements. Boats of "open construction" are exempted.

Open Construction Defined

The Coast Guard has prepared a set of specifications to guide the boat owner in determining whether his boat meets the definition of "open construction." To qualify for exemption from the bilge ventilation regulations, the boat must meet *all* of the following conditions:

1. As a minimum, engine and fuel tank compartments must have 15 square inches of open area directly exposed to the atmosphere for each cubic foot of *net* compartment volume. (Net volume is found by determining total volume and then subtracting the volume occupied by the engine, tanks, and other accessories, etc.)
2. There must be no long or narrow unventilated spaces accessible from the engine or fuel tank compartments into which fire could spread.
3. Long, narrow compartments, such as side panels, if joining engine or fuel tank compartments and not serving as ducts, must have at least 15 square inches of open area per cubic foot through frequent openings along the compartment's full length.

Be Safe—Be Sure

If your boat is home-built or does not meet one or more of the specifications above, or if there is *any* doubt, *play it safe and provide an adequate ventilation system.* To err on the safe side will not be costly; it may save a great deal, perhaps even a life.

Requirements for Older Boats

Except for those of open construction, a boat built on or before 31 July 1980 is required to have at least two ventilation ducts, fitted with cowls at their opening to the atmosphere, for each engine and fuel tank compartment. An exception is made for fuel tank compartments if each electrical component therein is "ignition protected" in accordance with Coast Guard standards, and fuel tanks are vented to the outside of the boat.

The ventilators, ducts, and cowls must be installed so that they provide for efficient removal of explosive or flammable gases from bilges of *each* engine and fuel tank compartment. Intake ducting must be installed to extend from the cowls to at least midway to the bilge or at least below the level of the carburetor air intake. Exhaust ducting must be installed to extend from the lower portion of the bilge to the cowls in the open atmosphere. Ducts should not be installed so low that they could become obstructed by a normal accumulation of bilge water.

To create a flow through the ducting system, at least when underway or when there is a wind, cowls (scoops) or other fittings of equal effectiveness are needed on all ducts. A wind-actuated rotary exhauster or mechanical blower is equivalent to a cowl on an exhaust duct.

Lack of adequate bilge ventilation can result in an order from a Coast Guard Boarding Officer for "termination of unsafe use" of the boat; see page 44.

Ducts Required

Ducts are a necessary part of a ventilation system. A mere hole in the hull won't do; that's a vent, not a ventilator. "Vents," the Coast Guard explains, "are openings

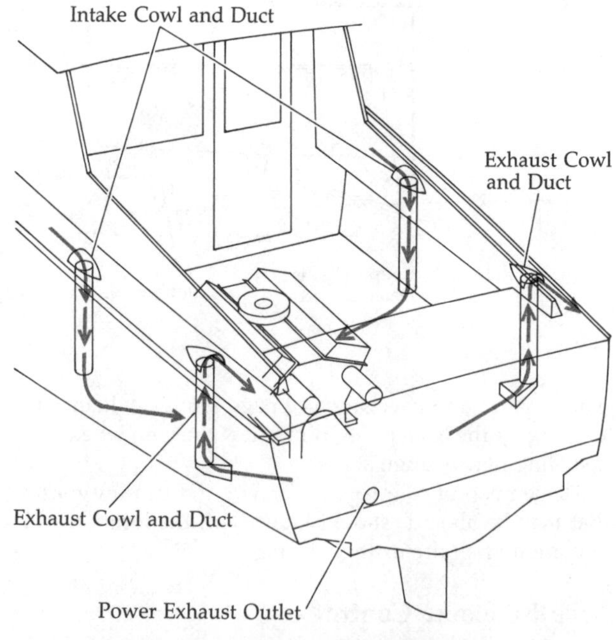

Figure 311 **Natural ventilation system used on some older boats has two intake and two exhaust ducts, with cowls. Exhaust ducts lead low, where fumes tend to sink.**

Ventilation Requirements for Small Powerboats

Net Volume (Cu. Ft.)	Total Cowl Area (Sq. In.)	Minimum Inside Diameter for Each Duct (Inches)	
		One-Intake and One-Exhaust System	Two-Intake and Two-Exhaust System
Up to 8	3	2	—
10	4	2 1/4	—
12	5	2 1/2	—
14	6	2 3/4	—
17	7	3	—
20	8	3 1/4	2 1/2
23	10	3 1/2	2 1/2
27	11	3 3/4	3
30	13	4	3
35	14	4 1/4	3
39	16	4 1/2	3
43	19	4 3/4	3
48	20	5	3

Note: 1 cu. ft. = 0.028 cu. m; 1 inch = 2.54 cm; 1 cu. in. = 6.45 cu. cm

Table 3-2

that permit venting, or escape of gases due to pressure differential. Ventilators are openings that are fitted with *cowls* to direct the flow of air and vapors in or out of *ducts* that channel movement of air for the actual displacement of fumes from the space being ventilated."

Size of Ducts

Ventilation must be adequate for the size and design of the boat. There should be no constriction in the ducting system that is smaller than the minimum cross-sectional area required for reasonable efficiency. Where a stated size of duct is not available, the next *larger* size should be used.

Small Motorboats To determine the minimum cross-sectional area of the cowls and ducts of motorboats having small engine and/or fuel tank compartments, see Table 3-2, which is based on *net* compartment volume (as previously defined).

Cruisers and Larger Boats For most cruisers and other large motorboats, Table 3-3, which is based on the craft's beam, is a practical guide for determination of the minimum size of ducts and cowls.

Ducting Materials

For safety and long life, ducts should be made of non-ferrous, galvanized ferrous, or sturdy high-temperature-resistant non-metallic materials. Ducts should be routed clear of, and protected from, contact with hot engine surfaces.

Positioning of Cowls

Normally, the intake cowl will face forward in an area of free airflow underway, and the exhaust cowl will face aft so that a suction effect can be expected.

The two cowls, or sets of cowls, should be located with respect to each other, horizontally and/or vertically, so as to prevent return of fumes removed from any space back into the same or any other space. Intake cowls should be positioned to avoid pick-up vapors from fueling operations.

Air for Carburetors

Openings into the engine compartments for entry of air to the carburetor are in addition to requirements of the ventilation system.

Requirements for Newer Boats

On boats built after 31 July 1980, each compartment having a permanently installed gasoline engine must be ventilated by a powered exhaust blower system unless it is "open construction" as defined above. Each exhaust blower, or combination of blowers, must be rated at an air flow capacity not less than a value computed by formulas based on the net volume of the engine compartment plus other compartments open thereto. The installed blower or blowers must be tested to prove that they actually do move air at a rate determined by other formulas also based on compartment volume.

The engine compartment, and other compartments open to it where the aggregate area of openings exceeds 2% of the area between the compartments, must *also* have a NATURAL ventilation system.

There must be a warning label as close as practicable to the ignition switch that advises of the danger of gasoline vapors and the need to run the exhaust blower for at least *four minutes* and then check the engine compartment for fuel vapors *before* starting the engine.

Ventilation Requirements for Large Powerboats

Two-Intake and Two-Exhaust System		
Vessel Beam (Feet)	*Minimum Inside Diameter For Each Duct (Inches)*	*Cowl Area (Sq. In.)*
7	3	7.0
8	3 1/4	8.0
9	3 1/2	9.6
10	3 1/2	9.6
11	3 3/4	11.0
12	4	12.5
13	4 1/4	14.2
14	4 1/4	14.2
15	4 1/2	15.9
16	4 1/2	15.9
17	4 1/2	17.7
18	5	19.6
19	5	19.6
Note: 1 foot = 0.305 m; 1 inch = 2.54 cm; 1 sq. in. = 6.45 sq. cm.		

Table 3-3

Natural ventilation systems are also required for any compartment containing both a permanently installed fuel tank and any electrical component that is not "ignition protected" in accordance with existing Coast Guard electrical standards for boats; or a compartment that contains a fuel tank that vents into that compartment (highly unlikely); or one having a non-metallic fuel tank that exceeds a specified permeability rate. The new regulations specify the required cross-sectional area of ducts for natural ventilation based on compartment net volume, and how ducts shall be installed.

The regulations above concern the manufacturer of the boat, but there is also a requirement placed on the operator that these ventilation systems be operable any time the boat is in use. Because such systems are so desirable, it is recommended that they be installed on *any* gasoline-powered motorboat or auxiliary; see also fuel vapor detectors, page 252.

Diesel-Powered Boats

Diesel fuel does not come within the Coast Guard's definition of a "volatile" fuel, and thus bilge ventilation legal requirements are not applicable. It is, however, a sensible step to provide essentially the same ventilation system for your boat even if it is diesel-powered, particularly because diesels require a high volume of air for efficient combustion. Diesel fuel does not explode, but it does burn; a broken fuel line can cause a fire that could result in the total loss of your boat.

Passenger-Carrying Vessels

Vessels that carry more than six passengers for *hire* are subject to special regulations. Consult the nearest Coast Guard Marine Inspection Office for details.

Visual Distress Signals

All boats of any type 16 feet or more in length, and all craft of any size carrying six or fewer passengers for hire, must be equipped with approved visual distress signals at all times when operated on coastal waters. Boats less than 16 feet in length must carry such signals when operating at night. There are a few limited exceptions, such as some boats engaged in racing.

Coastal waters are defined as the U.S. waters of the Great Lakes and the territorial seas (out to the "three-mile limit") plus connected waters where any entrance exceeds two nautical miles between opposite shorelines to the first point where the distance between shorelines narrows to less than two miles. Shorelines of islands or points of land within a waterway are considered in determining the distance between opposite shorelines.

Visual distress signals are classified for Day (D), or night (N) use, or for both (D/N). To be legal, a boat must have an orange flag with special square-and-disc markings in black (D); and an S-O-S electric light (N); or three orange smoke signals, hand-held, or floating (D); or three red flares of hand-held, meteor, or parachute type (D/N); Each signal must be legibly marked with a Coast Guard approval number or certification statement. It must be serviceable and not past any expiration date. (Pyrotechnic signals will have an expiration date 42 months after manufacture.) Combinations of various flare types and smoke signals are acceptable. (The flares must carry an approval; older launchers may be used if still available.) In some states, a pistol launcher for flares may be considered a firearm, and thus be subject to licensing and other restrictions; check your local authorities before purchasing such a device. Flares marked "Not Approved for Use on Recreational Boats" are for use only on commercial craft.

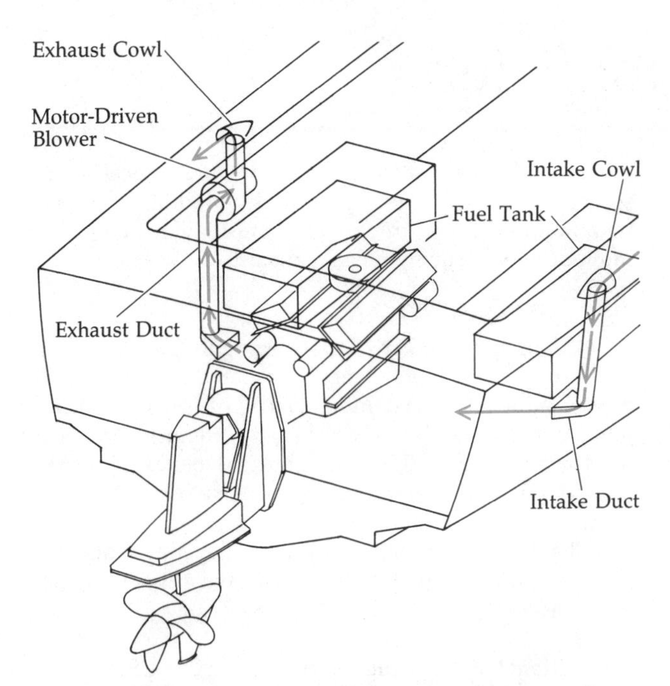

Figure 312 **Blowers are required equipment on new boats, and must be operated before the engine is started. If in an exhaust duct as shown, they should not interfere with the functioning of the duct as a natural ventilator.**

All visual distress signals must be stowed so as to be "readily available." The legal requirements are a *minimum;* prudent skippers will carry additional signals to meet their particular needs. Figure 518, page 105, illustrates approved distress signals.

Visual distress signals must not be displayed on the water except where assistance is needed because of *immediate or potential danger to the persons on board.* This allows tests and demonstrations to be made from the shore, but good judgment must be used to avoid false alarms; if there is any likelihood of reports being made to the local Coast Guard unit notify them by telephone or radio of your intended actions.

The use of pyrotechnic signals is not without hazard from the flame and hot dripping ash and slag. Hand-held flares must be held over the side and in such a way that hot slag will not drip on your hand. Pistol-launch flares should be fired downwind and at an angle of roughly 60° above the horizon, higher in stronger winds but never directly upward. In a distress situation do not waste signals by setting them off when no other vessel is within sight. The Coast Guard pamphlet "Visual Distress Signals for Recreational Boaters" contains the full legal requirements and much valuable information.

Other Required Equipment

Additional items of equipment required by law or Coast Guard regulations to be on board some or all boats include a whistle, a bell, navigation lights, and, in some instances, one or more day shapes. As with other equipment, these devices are graduated with the size of the vessel, and may vary with the waters on which they are used.

Since 24 December 1981, requirements for these items for boats on internal waters have been contained in the new Inland Rules of the Road. Requirements for all other waters are governed by the International Rules.

Details on required navigation lights and day shapes for both Inland and International Rules are covered in Chapter 4. Information on whistles and bells for vessels is in Chapter 5.

Equipment for Commercial Craft

Boats (including sailing and non–self-propelled vessels) used for carrying six or fewer passengers for hire (paying passengers), or for commercial fishing, are "commercial uninspected vessels" subject to the Motorboat Act of 1940 and Coast Guard regulations issued under the authority of that Act.

All non-recreational craft less than 40 feet in length, *not* carrying passengers for hire, such as commercial fishing boats, must have on board at least one life preserver, buoyant vest, or special purpose water safety buoyant device of a suitable size for each person on board. "Suitable size" means that sufficient children's lifesaving devices (medium or small, as appropriate) must be carried in addition to adult sizes.

All vessels carrying passengers for hire (six or fewer), and all other vessels subject to the MBA/40 which are over 40 feet in length, must have at least one life preserver of a suitable size for each person on board.

Each vessel 26 or more feet in length must also carry at least one life ring buoy in addition to the life preservers or other wearable lifesaving devices required.

All lifesaving equipment must be in serviceable condition; wearable devices must be readily accessible and devices to be thrown must be immediately available.

More than Six Passengers

Motorboats carrying more than six passengers for hire are subject to a separate Act and set of Coast Guard regulations. These vessels must be "inspected" and certified as being equipped with the specified minimum equipment such as lifesaving gear and fire extinguishers; see page 52.

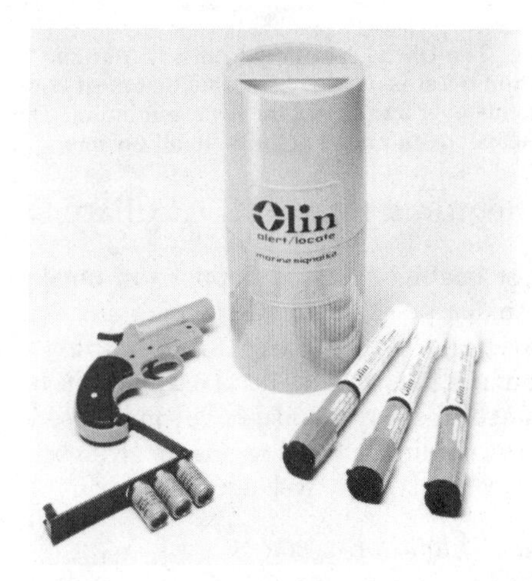

Figure 313 **Approved distress signals include hand-held, meteor, or parachute flares suitable for day or night use.**

Additional Equipment for Safety

Any newcomer to boating might think that he would be fully set to go if he ordered his new craft with "all equipment required by federal and state regulations." Actually, such is far from the real situation. Federal requirements stop too soon; state requirements—lacking uniformity—are not much help.

Even the smallest rowboat needs at least one length of line for making fast to a pier or mooring or anchor; and a 40-footer will need some six to eight dock lines—yet none is required by Coast Guard regulations. No one would venture far from shore without an anchor—but again none is required by federal law or regulation.

Possible Liability

Another factor is the possibility of liability in case of an accident. It is a mistake to think that the boating equipment specified by the government is the only equipment that you are *legally* bound to have aboard. Consider as well the far-reaching rules of negligence, as developed over the years in court cases. Technically, negligence is the unintentional breach of a legal duty to exercise a reasonable standard of care, thereby causing damage to someone. More simply stated, it is the failure to conduct oneself as a *reasonable* person would have under the circumstances. If, for example, a reasonable person would have carried charts, you may be liable to an injured party for an accident arising out of not having them available. If he would have had an anchor, compass, lines, tools, and spare parts, etc., you may be held liable for not having them when needed. It is no defense to argue that no regulations require you to carry them.

Figure 314 **The USCG Auxiliary Courtesy Marine Examination decal is awarded to boats that meet equipment requirements over and above the legal minimum. These requirements are an excellent guide to all owners.**

Items Required for USCG Auxiliary Decal

In most boating areas, a skipper can obtain a Coast Guard Auxiliary "Courtesy Marine Examination" (CME) of his boat, and receive a decal for the current year. The examination is free, and the decal indicates the boat meets equipment and safety standards beyond those legally required. The requirements for the decal, given below, make a good checklist for any well-equipped boat.

Personal Flotation Devices

An approved personal flotation device of the appropriate type is required for *each berth* on the boat; this may be more than the legal requirement of one PFD for each person on board at the time of a check. In addition, a boat with no bunks, or with only one, must still have a minimum of two lifesaving devices aboard.

Fire Extinguishers

The Auxiliary standards for fire extinguishers on smaller boats are likewise more demanding than the legal requirements. Although a boat under 26 feet, of "open construction," or one which has a built-in fire extinguisher system, need not carry an additional hand-portable extinguisher to meet the legal minimum, it must have one for a CME decal.

Sailboats of 16 feet or more in length, even without any auxiliary power or fuel tanks, must have at least one B-I extinguisher.

Additional Fire Extinguishers Even the USCG Aux requirements may not be enough to provide an extinguisher at all desirable locations:

1. The helm, where there is always someone when underway.
2. The engine compartment.
3. The galley.
4. Adjacent to the skipper's bunk, for quick reach at night.

Add estinguishers as necessary to meet the needs of your own boat.

Navigation Lights

The law does not require that a boat operated only in the daylight have navigation lights, but these must be fitted and in good working order to meet the Auxiliary's standards for all craft 16 feet or more in length. Proper lights for use both underway and at anchor must be shown. The decal will not be awarded if lights are grossly misplaced, even if they are operable.

A sailboat with an auxiliary engine must be capable of separately showing the lights of both a sailboat and a motorboat; the lights must be wired so that the display can be changed from one to the other.

Visual Distress Signals

To receive CME approval, *every* boat must have acceptable visual distress signals even if not legally required due to size or waters used. The Auxiliary requirements of types and numbers are the same as the legal requirements, except on inland rivers and lakes where any device suitable for attracting attention and getting assistance may be carried.

Anchor and Line

An anchor of suitable type and weight together with line of appropriate size and length is required for the CME decal. This is a valuable safety item should the engine fail and the boat be in danger of drifting or being blown into hazardous waters.

Bilge Pump or Bailer

For the CME decal, all boats must have a bilge pump

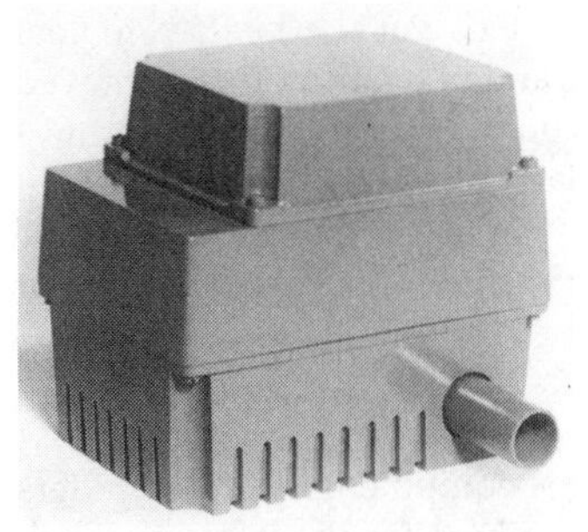
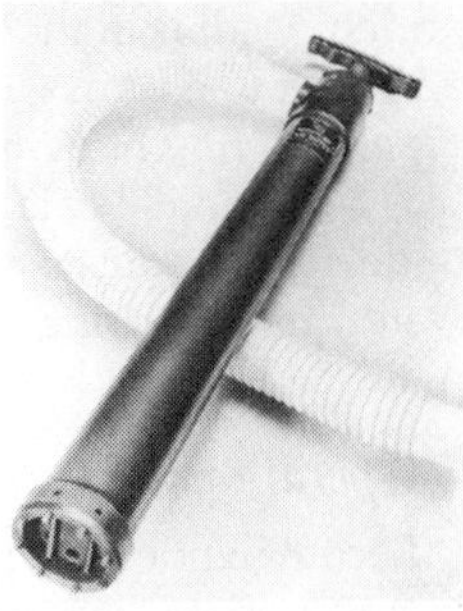

Figure 315 **Small electric bilge pumps are suitable for most boats, but these should be backed-up with a hand-operated pump for emergency use. Larger boats also may be fitted with a high-capacity pump belt-driven from a main engine.**

or bailer of a suitable size for the craft, and it must be in proper operating condition. This will normally be an installed electrical pump for boats more than roughly 18 feet in length, and a hand scoop-type bailer on smaller craft.

Smaller Craft Requirements

All Class A boats must carry a second means of propulsion. Normally a paddle or oars is used to meet this requirement, but an alternate mechanical means is acceptable *if* it has both a different starting source (battery) and a separate fuel source.

Requirements for Inflatables

An inflatable boat is eligible for a CME decal if it meets all requirements for a craft of its size, plus certain additional specifications such as a minimum of three separate air chambers that are *not* interconnected and an installed rigid transom (a strap-on outboard motor mount is not acceptable).

Sailboats

Sailboats without mechanical power, either installed or detachable, are required to have personal flotation devices for each person on board; the type of PFD varies with the size of the boat. Sailboats longer than 16 feet are eligible to become Coast Guard Auxiliary Facilities if they meet certain prescribed standards. These requirements can serve as a guide to all owners of sailing craft as to desirable safety equipment.

Sailing Facilities under 26 feet in length are required to have one B-1 hand fire extinguisher, and larger ones must have on board two such units. These sailboats must meet the CME standard of one approved lifesaving device on board for each berth, with a minimum of two such devices. Further, such craft must meet all standards for motorboats other than those relating to propulsion machinery, fuel systems, and ventilation of related compartments. This leaves in the requirements such items as an anchor with line, distress flares, and standards for galley stove installation and general electrical systems, as mentioned below—all matters appropriate to the safe operation of any sailboat.

Installation Standards

In addition to requiring the above items of equipment, the CG Aux program has established standards for the installation of fuel systems, electrical systems, and galley stoves. Information on gasoline fuel systems and electrical systems will be found in pages 266-273.

The Auxiliary requires that a galley stove be of a marine type and be installed so as to present no hazard to the craft and its occupants. Any common fuel is acceptable except gasoline and derivatives or distillates of naphtha and benzene. LP gas stoves or heaters which use an integral fuel container or one which fastens directly to the appliance are not acceptable for award of the decal.

Marine Sanitation Devices

The Courtesy Examination focuses on safety aspects of the boat. No check will be made for the presence or operating condition of an MSD, although the Examiner will be glad to provide information on legal requirements.

Equipment Recommended by the USCG Auxiliary

The CME checklist also contains a number of items recommended for the proper operation of a boat or for its safety—items beyond those required for award of the decal. The actual selections of items from this list will vary with the size and use of the boat involved.

This list of generally recommended items includes many that are required for "Facilities" and "Operational Facilities"—the Auxiliary members' own boats that have been brought to higher standards for use in their programs.

Anchors

The CME checklist recommends carrying a second anchor, in addition to the one required to pass the CME check. The additional anchor may be a lighter one for non-emergency daytime use. Chapter 11 of this book in general, and page 240 in particular, provide guidance in the selection and use of the proper size of anchor.The checklist also recommends that a length of chain be used between the anchor and the nylon line. Shackles used on either end of this chain should have their pins secured with safety wire.

Bilge Pumps

Boats more than 26 feet in length should have at least two means of pumping bilges; and their bilges must be clean and free of any oil or grease. Wood chips or any other debris that could clog pumps and limber holes must be removed. A manual bilge pump should be carried in every boat irrespective of any electrical pump.

Lines

Every boat should have mooring (dock) lines suitable in length and size to that particular craft. These should be of several different lengths for convenience in use. No generalization can be made as to the lengths, but the diameter should correspond roughly to that of storm or working anchor rodes.

A heaving line of light construction is desirable if the craft is large enough to need heavy mooring lines. The use of polypropylene line, or other material that is brightly colored and will float, is recommended for this purpose.

Life Rings

A ring life buoy with a length of light line attached is recommended for rendering assistance to swimmers or

Figure 316 **A searchlight can be used to spot buoys or other aids to navigation. Do not aim it toward the helm station of another vessel as it could temporarily blind the helmsman.**

Figure 317 **An old-fashioned hand-held lead line is a handy backup to an electronic depth sounder, and for an emergency such as checking depths all around a boat that is aground.**

acident victims in the water. This can also be used to float a heavier line across to a stranded boat. If of an approved type, this ring buoy may count as the required "throwable" PFD for all boats of Classes 1, 2, and 3. A water light (a device that automatically lights up in contact with the water) makes a ring or horseshoe buoy easier to spot at night.

Operational Equipment

Recommended operational equipment includes fenders in appropriate sizes and numbers for the boat involved. Not only will these be used in normal berthing, but they are also necessary if two boats must make fast to each other while underway or at anchor. A boathook will be found very useful for fending off, placing lines over piles, picking up pennants of mooring buoys, recovering articles dropped over the side, and many other uses.

A searchlight—installed on larger craft, hand-held on smaller boats—serves both as a routine aid in night piloting and as an emergency signalling device. A multicell flashlight or electric lantern can serve these functions, but not so well as a searchlight.

Navigation publications and charts should be carried aboard boats in accordance with their use; see Chapters 17 and 18 for further details. Compasses are discussed in detail in Chapter 15; here we need merely to note that one is desirable on almost any boat for emergency if not regular use. Piloting and plotting instruments are covered in Chapter 19.

A hand-held lead line is a useful, even as a back-up to the more complicated electronic depth sounder; one is particularly handy when one must probe around a stranded boat in search of deeper water. Also useful is a pole—a boathook will do very well—marked with rings at one-foot intervals; a mark of a different color or size should be added for the draft of the boat.

An emergency supply of drinking water—and perhaps food, too—should be carried on all boats. It may never be used, but when needed it can literally be a lifesaver. Supplies of this nature should be periodically freshened or replaced to ensure acceptable quality when needed. Distilled water should be carried for periodic replenishment of any storage batteries on board; this can, of course, serve as an emergency source of drinking water.

A first-aid kit is an essential item of safety equipment; see pages 278-280. The kit should be accompanied by a manual or separate book of instructions.

The list of tools and spare parts to be carried aboard must be developed by each skipper individually for his own boat. The items will be governed by the type of boat, its normal use, and the capabilities of the crew to use them. An item required on USCG Aux Facilities is one or more spare bulbs for the navigation lights. Items for all boats include simple tools, plugs, cloth, screws, nails, wire, tape, and other objects for the execution of emergency repairs at sea. Mechanical and electrical spares will be highly individualized by the particular boat and skipper.

Further Operational and Safety Items

A sea anchor or drogue, see page 219, and life rafts, preferably with a canopy, are much needed safety items for boats operating in unprotected waters.

Figure 318 **A good safety harness can be made with ½-inch nylon rope short-spliced to custom fit the wearer, whose permanent property it becomes. The safety line of ⅜-inch nylon has an eye splice at each end. One end is hitched through both loops, *above right*; the other end is takes a stainless steel carabiner of the type used by mountaineers. Do not trust your life to a snap of the type at the left in the picture, *below right*.**

Electric windshield wipers are excellent items of equipment when running into rainy or rough weather; safety is often much improved by their availability and use. The installation of washers that squirt fresh water on windshields will help greatly when spray, particularly of salt water, is received over the bow.

Many electronic items that will add to operational safety are described in Chapter 22 and 23. These include radiotelephones, electronic depth sounders, radio direction finders (RDFs), and fuel vapor detectors for engine rooms and bilges of gasoline-powered craft. Alarm systems are available to alert the skipper of dangerous conditions of engine overheating, low oil pressure, or high water in the bilge. All these items are in the "desirable," or even "necessary," category. Less vital equipment is discussed in the next major subdivision of this chapter.

A boat's cleaning gear may not be glamorous, but it is certainly operational. Specific lists will vary from boat to boat, but all may be expected to contain a swab (mop), bucket, sponges, chamois, metal polish and rags, and similar items.

Safety Harnesses

Although they are not legally required, safety harnesses should be provided for every member of the crew of any

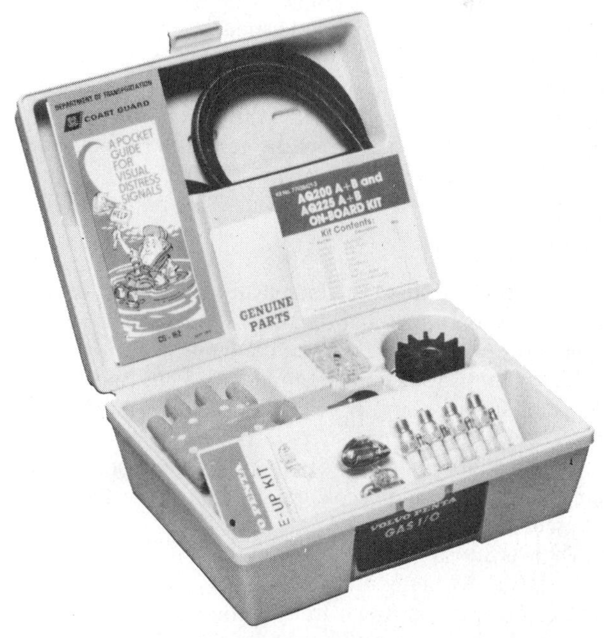

Figure 319 **Spare parts should include engine belts, filters, and cables, as well as items such as spark plugs. Also carry a spare for each type of navigation light carried on the boat.**

Figure 320 **A large capacity manual bilge pump such as this is capable of moving as much as half a gallon of water per stroke, without causing undue operator fatigue.**

Figure 321 **A good commercially available safety harness, *left,* should be used in rough weather, if the home-made type described on the preceding page is not constructed.**

sailboat that spends much time offshore. A few good commercial patterns are to be found, but many are clumsy, hard to adjust, and uncomfortable to wear and to lie down in. The photographs show a very simple design that consists of ½-inch nylon rope, formed by short splices to custom-fit the wearer, whose permanent property it becomes. The safety line of ⅜-inch nylon has an eye splice at each end. One end is hitched through both loops, and a stainless steel snap hook, of the sort used by mountaineers, is hitched to the other end. Do not use a snap of the type shown on the left in Figure 318, bottom right.

Bilge Pumps

Small boat skippers seldom encounter the sort of bilge pump that is best suited for them. Pumps used in daysailers throw too small a stream and are too tiring to operate, as they are usually mounted vertically and require repeated flexing of upper arm muscles. Many power boatmen use the engine or battery to power the pump, often relying, not so wisely, on emergency manual pumps that are no larger or more convenient than those found on daysailers.

There is an old saying that the most effective bilge pump is a frightened man with a bucket. Don't leave the bucket home, for there is much truth in the saying, but don't rely solely on the bucket, either. You should be prepared to keep yourself afloat indefinitely—without engine or batteries and without incurring excessive fatigue from physical effort—with the boat leaking several gallons per minute. With the right pump this is easily accomplished with the pump going no more than five minutes in every hour. To do this hour after hour you need a pump that throws at least half a gallon per stroke, and that can be operated using the larger body muscles.

One excellent arrangement is a large diaphragm pump mounted on a floor timber below the cabin sole and operated by a rocking lever that stows nearby when not in use and engages the pump through a slotted plate in the sole. (The plate should have a cover or things will fall through into the bilge.) The pumper can sit well braced on the sole and use all his upper body muscles to take several hundred strokes in a few minutes.

A diaphragm pump can pass fairly large objects, so the intake screen at the bilge pump can be coarse—say a piece of 1-inch copper pipe with the end closed over and a lot of ¼-inch holes bored in the sides—and still be easily kept clear, provided the intake hose is flexible and accessible and the bilge and limber holes are kept clear. Limber chains help.

Pump until the intake sucks air, counting the strokes, and record the count in the log. In normal circumstances pump once a day at a regular time. When sailing watch-and-watch the Watch Captain should *check* the bilge during each watch and record the results.

An automatic bilge pump should have an audible alarm or a bright indicator light where it will be seen. Tanks or hoses may spring leaks. Should this happen, the running pump can be noticed promptly and corrective action taken. An automatic bilge pump switch can also be rigged as an alarm for high water in that bilge. Mount it at a slightly higher level than the automatic switch that controls the pump, and connect it to a land bell, horn, or other type of audible signal.

Equipment for Convenience and Comfort

Many, but certainly not all, of the items of EQUIPMENT FOR CONVENIENCE AND COMFORT fall into the electronic or electrical categories. These items are covered more fully in Chapter 23 and include navigational aids such as automatic radio direction finders, radar, and sophisticated positioning equipment including Loran, Decca, and others. Speedometers and logs will provide speed and distance information that will assist in piloting. An automatic steering mechanism—often called an "autopilot"—adds greatly to the convenience of cruising. Loudhailers and intercom systems increase the convenience of operating or living aboard larger boats. Twin-engine vessels will benefit from engine synchronization indicators, and even more from equipment that will match the engine speeds automatically.

Electrical equipment items include pressure fresh-water systems, water heaters, electrical refrigeration, showers with drain pumps, plus heating and/or air-conditioning equipment.

All of these demand increased electrical power from the boat's system, either AC or DC. Auxiliary generating plants will permit the use of "shore-type" heavy-drain equipment underway or at anchor. Inverters are very useful items that permit the operation of moderate-drain AC equipment without the noise of running a gasoline or diesel engine-driven generator. Marine converters will supply the needed DC power without operation of the main engine if AC power is available; these will keep the boat's storage batteries at full charge when in port.

Outside the electrical-electronic field, equipment items for convenience and comfort are numerous and varied. Many skippers will want to have a boarding/swimming ladder for ease in going over the side and returning. A platform across the stern of larger boats is a great convenience for swimming or using a dinghy. The dinghy itself, perhaps with sails or a small outboard motor, may well be classed as an item of convenience equipment.

Fresh water making equipment—stills to convert salt water to potable water—will often add immeasurably to a boat's cruising endurance. Stabilizers are now available for medium and larger motor yachts to eliminate much of the rolling from offshore cruising.

An anchor windlass on the foredeck will remove a major part of the physical labor involved in "getting the hook up."

Weather instruments—barometer, thermometer, and anemometer, and perhaps a hygrometer for measuring the relative humidity—will add to the interest of boating. With study and practice, they can add to the safety, too, of a boat and its crew. A recording barograph will provide interesting permanent records of atmospheric, and thus weather, changes.

The spare parts inventory of a boat can be expanded beyond the limits of necessity into the area of convenience. Carrying spare propellors, and even shafts, can often greatly reduce delays at a strange port away from home if these major components should become damaged. These may be beyond the capability of the crew to install, but their ready availability can often reduce repair times from days to hours.

Diving gear, either of the simple mask-and-snorkle style or the more complex scuba type, can make boating in many areas more interesting and can at times be of real value in making underwater inspections and repairs. "Wet suits" may be required for colder water areas.

Boating Insurance

Marine insurance can be bought to cover loss or damage to the boat or its equipment, protect against liability for personal injury or property damage, provide medical payments in case of injury, and cover transportation of the boat while on land. Consult a good agent. Better yet, talk to two or three agents for quotations.

If you have an older wooden boat, don't be surprised if the agent shows no interest in insuring the hull. The cost of repairing wooden hulls can sometimes be higher than their value if they are more than a few years old.

4

Running lights and day shapes required by the new unified U.S. Inland Rules for vessels of all types; the International Rules that differ from those of the U.S.

RULES OF THE ROAD: LIGHTS AND DAY SHAPES

The RULES OF THE ROAD are sets of statutory requirements enacted by Congress to promote safety of navigation. The Rules consist of requirements for navigation lights and day shapes (discussed in this chapter); steering and sailing rules, and sound signals for both good and restricted visibility; and distress signals (Chapter 5).

There are different sets of Rules for international and inland waters, with some local variations in the latter. They all have their roots in regulations enacted by Great Britain in 1863, and by the U.S. Congress in 1864. The first International Rules of the Road were established in 1889; the most recent ones were adopted in 1972 (these are called "72 COLREGS" by the Coast Guard). These became effective 15 July 1977.

There are also different requirements for small and large vessels. References generally apply to boats of any size, unless specific sizes of vessels are stated to have differing requirements.

The 1980 U.S. Inland Rules

The 1972 International Rules were unique for the U.S., because for the first time they had the status of an international *treaty;* previous sets of International Rules had the status of an international *agreement.* A treaty is more binding on a country, offering less flexibility and requiring stricter compliance with its terms. A key provision of the 1972 International Rules was the requirement in Rule 1(b) that national rules for internal waters "shall conform as closely as possible" to the International Rules. The first result of this requirement was that the U.S. Inland Rules no longer applied out to the offshore "boundary lines" but were effective only behind new "demarcation lines" (See page 40, Chapter 2.) The longer-term effect was the passage by Congress of the INLAND NAVIGATIONAL RULES ACT OF 1980, which unified the previously separate Inland, Great Lakes, and Western Rivers Rules and their respective Pilot Rules. These became effective on all U.S. waters, except the Great Lakes, on 24 December 1981, and on the Great Lakes on 1 March 1983, following coordination with Canada.

Sequence of Consideration

Most recreational boatmen in the United States use their boats on waters of rivers, lakes, bays, and sounds that are governed by the U.S. Inland Navigational Rules, and for this reason primary consideration will be given to these Rules, with the International Rules being presented in terms of their differences with the Inland Rules. *Note carefully:* the term "Inland Rules" used in Chapters 4 and 5 refers to the 1980 U.S. unified rules, *not* to the Inland Rules that previously existed along with the now-repealed Great Lakes and Western Rivers Rules.

The International Rules of the Road apply, however, to all vessels once they clear the jetties or headlands at a harbor entrance. Indeed these Rules even intrude into harbors, bays, inlets, rivers, etc., in some areas along the New England coast and in the lower Florida Keys. Check your coastal chart for the location of the demarcation lines.

The consideration of the Rules of the Road in this chapter and Chapter 5 is not complete; it focuses on the points of greatest concern to boatmen.

Numbered Rules

Both the U.S. Inland Rules and the International Rules consist of major subdivisions; Part A - General, Part B - Steering and Sailing Rules, Part C - Lights and Shapes, Part D - Sound and Light Signals, and Part E - Exemptions. (Part E provides specific temporary and permanent exemptions for existing vessels, to permit a smooth transition from the previous requirements for lights and sound signals.)

Within both sets, the Rules are numbered from 1 to 38, with subparagraphs such as (a), (b), (c), and still lower levels indicated by (i), (ii), (iii). Rules that are identical or nearly identical in both sets carry the same numbers, so that parallel numbering can be continued in the Rules that follow. (Rule 28 is omitted from the U.S. Inland Rules.)

Indication of References

Numbers and letters shown in brackets are references to the applicable Rule. References to Annexes are indicated by a Roman numeral since the annexes to both sets are so designated; specific paragraphs and subparagraphs are noted by numbers and letters as appropriate. These references are shown to facilitate looking up the exact language of the Rules and Annexes in the Coast Guard publication *Navigation Rules, International - Inland.*

Why Navigation Lights and Day Shapes

On a vessel, NAVIGATION LIGHTS are lights shown that are of specified color (white, red, green, yellow, blue), arc, range of visibility, and location, as required by law and regulations. Their basic purpose is to prevent collision by alerting each vessel to the other's presence. Lights also indicate the relative heading of one vessel as seen from the other, and give clues to her size, special characteristics, and/or current operations. Most important is her orientation with respect to your boat—a fact you must know to determine who has the right-of-way.

Navigation lights are sometimes referred to as RUNNING LIGHTS (those shown underway) and RIDING LIGHTS (those shown while at anchor or moored), but this terminology is unofficial and does not appear in the Rules of the Road.

DAY SHAPES are objects of specified shape, size, and placement on a vessel, as required by rules and regulations; all are black in color. They serve some of the same purposes by day that navigation lights do by night. In normal operations and visibility no day shapes are required; the relative aspect and motions of two vessels can be determined by observation. However, day shapes are used to indicate special situations, such as being anchored or engaged in fishing, and some conditions not detectable by eye, such as a sailboat with her sails up but with her engine also in use (such a craft is regarded as a power-driven vessel and is not entitled to any of a sailboat's right-of-way privileges merely because her sails are up).

Figure 401 **During the day, when lights would be ineffective, day shapes are used to indicate special vessel status. Other information provided at night by lights—type, size, relative bearing—can be determined in daylight by direct observation.**

Importance of Knowledge

Knowledge of navigation lights is important to a small-boat skipper for two separate, but both important, reasons. First, he is legally responsible for his boat displaying lights of the proper color, intensity, location, and visibility. Unfortunately, he cannot assume that the manufacturer of his boat has installed proper lights; he must know the requirements of the applicable Rules of the Road and check that his boat fully complies. If it doesn't comply, it is he—not the boat builder—who will be cited by the Coast Guard, and possibly fined. He must also know which lights to turn on (and which not to turn on) and when.

Second, and perhaps even more important, he'll depend on his knowledge of navigation lights for the safety of his boat when operating at night. Quite often, all the information available about another vessel comes from interpreting her navigation lights. There are many lights and possible combinations; often a quick decision is needed, with no time to go "look it up in the book." *Know the lights* shown on both boats and ships. Even in the daytime and with good visibility, the day shapes carried by another vessel will give you information about her activities or limitations that you could not determine from simple observation. *Know day shapes* as well as lights.

The U.S. Inland Rules

The U.S. Inland Navigational Rules are applicable inside the demarcation lines separating inland and international waters (see Chapter 2), established at entrances to bays, sounds, rivers, inlets, etc. They are *not applicable* to waters of harbors that are specifically designated as not covered, as along coasts with many small harbors (New England Coast especially), or in the harbors of offshore land masses, such as Block Island and Catalina Island.

General

The Inland Rules apply to all vessels, regardless of nationality, on the inland waters of the United States, as described above. They also apply to U.S. vessels on the Canadian waters of the Great Lakes, to the extent that they do not conflict with Canadian laws or regulations. [1(a)]

Basic Definitions

Rule 3 contains general definitions that are used throughout the U.S. Inland Navigation Rules.

Vessel This term applies to every size and description of water craft, including nondisplacement craft (air-cushion vehicles) and seaplanes, used or capable of being used as a means of transportation on water. The term "vessel" includes all sizes without limit, from a dinghy to a supertanker.

Power-Driven Vessel Any vessel propelled by machinery.

Sailing Vessel Any vessel under sail, provided that propelling machinery, if fitted, is *not* then being used. A sailboat or motor sailer using both sail and engine simultaneously is a *power-driven* vessel for purposes of the Rules of the Road.

Underway A vessel not at anchor, made fast to the shore, or aground. This term applies whether or not the vessel is moving ("making way") through the water.

Vessel Engaged in Fishing Any vessel fishing with nets, lines, trawls, or other fishing apparatus that restricts maneuverability. It does *not* include vessels with trolling lines, or drift fishing with hand rods and lines.

Vessel Not Under Command A vessel which through some exceptional circumstances is unable to maneuver as required by the Rules, and therefore is unable to keep out of the way of another vessel. Typically, this category would apply to a boat drifting with an inoperative engine or steering system.

Vessel Restricted in Her Ability to Maneuver A vessel which from the nature of her work is restricted in her ability to maneuver as required by the Rules, and therefore is unable to keep out of the way of another vessel. Typically, this term applies to dredges, vessels engaged in laying or repairing submarine cables or pipelines, and vessels engaged in a towing operation that severely restricts the towing vessel and her tow in their ability to deviate from their course.

Visibility Vessels are deemed to be "in sight of one another" only when one can be observed from the other.

Restricted Visibility Any condition in which visibility is reduced by fog, mist, falling snow, heavy rain, sandstorms, or any other similar cause.

Rule 3 also contains other definitions of interest to boatmen. The term "Secretary" means the Secretary of the Department of the Federal Government in which the Coast Guard is operating. This is normally the Department of Transportation, but in time of war or national emergency, it could become the Department of the Navy by Presidential Executive Order.

Although separate sets of Rules of the Road for the Great Lakes and the Western Rivers no longer exist, these terms are still defined in the 1980 Inland Rules, because of certain exceptions and special provisions for specified waters. The "Great Lakes" are defined as the lakes and their connecting and tributary waters, including the Calumet River as far as the Thomas J. O'Brien Lock and Controlling Works (between mile 326 and 327), the Chicago River as far as the east side of the Ashland Avenue Bridge (between mile 321 and 322), and the St. Lawrence River as far east as the lower exit of the St. Lambert Lock. (These are essentially the same waters as defined in the former Great Lakes Rules, but the limits are stated differently and more explicitly.)

"Western Rivers" means the Mississippi River, its tributaries, South Pass, and Southwest Pass to the demarcation lines for COLREGS waters; the Port Allen-Morgan City Alternate Route; and that part of the Atchafalaya River above its junction with the Port Allen–Morgan City Alternate Route, including the Old River and the Red River. [13(1)] Note that the term "Western Rivers" does not apply to rivers on the West Coast of the United States—such rivers as the Sacramento or the Columbia; the Inland Rules *are* in effect for all rivers emptying into the Pacific Ocean.

Units of Measurement

Except for distances in nautical miles, linear units of measurement in the 1980 U.S. Inland Rules are given in the metric system. Specifications like the lengths of vessels, or the spacing and height of lights, will thus be in meters in this chapter and Chapter 5, with the corresponding length or distance in customary (English) units following in parentheses.

Navigation Lights

Navigation lights will be discussed for various categories and sizes of vessels in the sub-sections below; they are also summarized in Table A, page 593.

When Lights and Shapes Are Shown

Vessels are required to show the proper navigation lights from *sunset* to *sunrise* in all weather conditions, good and bad. During these times, no other lights that could be mistaken for lights specified in the Rules can be displayed, nor any lights that impair the visibility or distinctive character of navigation lights, or interfere with the keeping of a proper lookout. [20(a) and (b)]

The Inland Rules also state that navigation lights *must* be shown between sunrise to sunset in conditions of reduced visibility, and *may* be shown at any other time considered necessary. [20(c)]

Day shapes specified in the Rules must be displayed by day; although not specifically so stated, this has been taken to mean from sunrise to sunset. [20(d)]

Light Definitions

Masthead Light A white light placed over the fore-and-aft centerline of the vessel, showing an unbroken light over an arc of 225°, from dead ahead to 22.5° abaft the beam on both sides of the vessel. On boats less than 12 meters (39.4 ft) in length, the masthead light may be off the fore-and-aft centerline, but must be as close to it as possible. [21(a)]

As will be seen later, the term "masthead light" is something of a misnomer. More often than not, this light

COLOR ILLUSTRATIONS

Running lights for pleasure craft and commercial vessels, for both Inland and International Rules of the Road, are shown in the color section, pages 589-592.

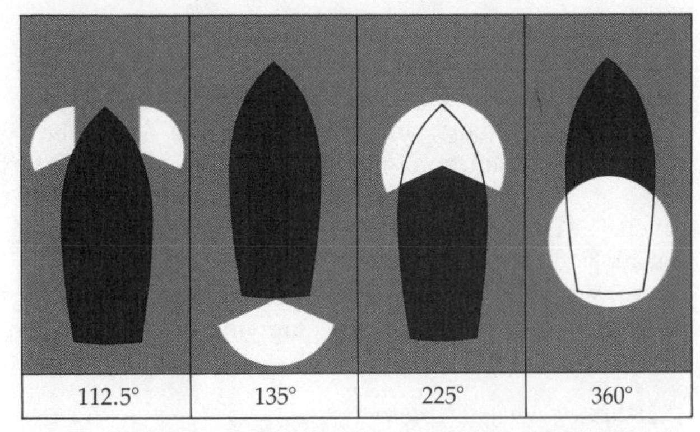

Figure 402 **Each type of navigation light must cover a standard arc of visibility so that some show only from ahead or along part of a vessel's side, others can be seen only from the stern area, and others show all around.**

is *not at the top* of a mast. On motorboats it is often on a short staff on top of the cabin. On sailboats it is usually part way up the mast and another light, the anchor light, is actually at the masthead.

Sidelights Colored lights—red on port and green on starboard—showing over an unbroken arc of the horizon of 112.5°, from dead ahead to 22.5° abaft the beam on each side.

Combination Light On a vessel of less than 20 meters (65.7 ft) length, the sidelights may be combined in a single fixture carried on the centerline of the vessel, except that on boats less than 12 meters (39.4 ft) in length, this combination light need be carried only as close to the centerline as possible. [21(b)] Note that the sum of the arcs of the two sidelights is exactly the same as that for the white masthead light.

Sternlight A white light showing over an unbroken arc of the horizon of 135°, centered on dead astern. [21(c)]

Towing Light A yellow light having the same arc as a sternlight, showing 67.5° to either side of dead astern. [21(d)]

All-Round Lights A light, the color determined by its use, showing over an unbroken arc of the horizon of 360°. [21(e)]

Flashing Light A light flashing at regular intervals, at a rate of 120 or more flashes per minute. [21(g)] This high flashing rate is used to lessen any possibility of confusion with "quick flashing lights" on aids to navigation.

Special Flashing Light A yellow light flashing at a rate of 50 to 70 flashes per minute, placed as far forward and as nearly as practicable on the centerline of a tow, and showing an unbroken light over a horizontal arc of not less than 180° nor more than 225° centered on dead ahead. [21(g)]

Lights for Varying Sizes of Power-Driven Vessels Underway

The basic navigation lights for a power-driven vessel of *less than 12 meters* (39.4 ft) in length, when underway, are a white all-round light, visible 2 miles, and sidelights, visible 1 mile. [23(c)] The single white light may be replaced by separate masthead and stern lights, and the sidelights may be separate, or a combination unit as noted above.

Masthead Lights

The masthead light of a vessel *12 meters (39.4 ft) or more but less than 20 meters (65.6 ft)* in length must be at least 2.5 meters (8.2 ft) above the gunwale. [I, 84.03(c)] On boats less than 12 meters (39.4 ft) long, the masthead light need only be at least 1 meter (3.3 ft) higher than the sidelights; it must be screened if necessary to prevent interference with the helmsman's vision. [I, 84.09(b)]

Vessels *more than 12 meters (39.4)* in length must carry a separate forward masthead light and a sternlight. The forward light need not actually be at the masthead, but must meet placement requirements given in Annex I.

On a vessel *20 meters (65.6 ft) or more* in length, the masthead light must be not less than 5 meters (16.4 ft) above the hull, *except* that if the beam of the vessel is greater than 5 meters, then the height must be not less than the beam, but need not in any case be greater than 8 meters (26.2 ft). [I, 84.03(a)] The term "height above the hull" is defined in Annex I.

The forward masthead light must be in the forward half of the vessel, except that in boats of less than 20 meters (65.6 ft) length, it need only be as far forward as possible. [23(a) (i)]

A second masthead light is required on vessels of *50 meters (164.0 ft) or more* length, and may be carried on vessels of lesser length [23(a) (ii)]. This light must be at least 2 meters (6.5 ft) higher than the forward light. [I, 84.03(a) (2)]

When two masthead lights are carried, the horizontal distance between them must be at least one-fourth of the length of the vessel, except that this distance need not in any case exceed 50 meters (164.0 ft). [I, 84.05(a)] *Exception:* Special provisions are made for Western Rivers vessels between 50 and 60 meters (164.0-196.9 ft) in length. [I, 84.05(b)]

Figure 403 **A masthead light, often not actually at the top of a mast, is white and covers an unbroken arc from dead ahead to 22.5 degrees abaft the beam on each side. A second masthead light is required on vessels more than 50 meters (164.1 ft) in length, and may be carried on vessels of lesser length.**

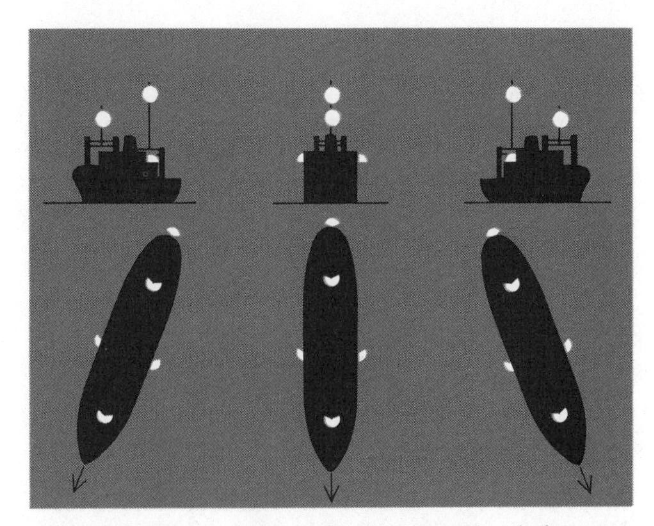

Figure 404 **When a vessel carries both forward and after masthead lights, these form a central range that is an excellent indicator of its heading, as seen from your boat. If the lower (forward) light is to the left of the after light, the vessel is heading to your port; if the lower light is to the right of the after masthead light, the vessel is heading to your starboard. When the two lights are directly in line with each other, the vessel is heading toward you. These lights will normally be seen well before the colored side lights.**

Value of Range Lights When two masthead lights are carried they form a RANGE, a valuable aid in determining the relative heading of another vessel when it is first sighted at night. Since these white lights are both brighter and higher than the sidelights, they normally will be seen well before the colored light or lights can be detected and read for their meaning.

The relative location of the two range lights will be the key. Should they be seen one directly over the other, the other vessel is heading *directly toward you* and danger of collison may exist. Should the lower (forward) range light be seen to the right or left of the higher (after) white light, the other vessel is on an oblique course and the angle can be roughly gauged by the horizontal separation between the lights.

Visibility Requirements Masthead lights must have a range of visibility as follows: Boats *less than 12 meters (39.4 ft)* long—2 miles; vessels of *12 to not more than 50 meters (39.4-164.0 ft)* length—5 miles; except that it may be 3 miles for boats less than 20 meters (65.6 ft) in length; vessels over 50 meters (164.0 ft) in length—6 miles. [22]

In all cases, masthead lights must be placed so as to be clear of all other lights or obstructions. [I, 84.03(f)]

Sidelights

Sidelights on a vessel *of any length* must be at least 1 meter (3.3 ft) lower than the (forward) masthead light. [I, 84.03(g)] On vessels of 20 meters (65.6 ft) or more in length, sidelights must be provided with black inboard screens to make the necessary cut-off of light at the "ahead" limit of each arc of visibility. [I, 84.09(a)] *Smaller vessels* also must have these screens if needed to meet the cut-off requirements, and to prevent excessive spillover from sidelights. A combination light using a single *vertical filament* bulb and a very narrow division between sectors need not have a screen.

Sidelights must have a range of visibility as follows: boats less than 12 meters (39.4 ft) in length—1 mile; vessels of 12 to 50 meters (39.4-164.0 ft) length—2 miles; and vessels more than 50 meters (164.0 ft) length—3 miles. [22]

Sternlights

The Rules and Annex I do not specify a vertical or horizontal placement for a sternlight, except that it be as nearly as possible "at the stern."

The required visibility for a sternlight is as follows: vessels *less than 50 meters (164.0 ft)* length—2 miles; vessels *50 meters (164.0 ft) or more* in length—3 miles. [22]

A power-driven vessel operating on the Great Lakes *may* show an all-round white light in place of the normal after-masthead light and sternlight. This must be carried in the same position and have the same visibility range as a second masthead light. [23(d)]

Figure 405 **Sidelights may be separate red (port) and green (starboard) lights showing over an unbroken arc from dead ahead to 22.5 degrees abaft the beam on each side. Note, *above*, that placement prevents excessive spillover of light when seen from dead ahead. Vessels less than 20 meters (65.7 ft) in length may carry a combination bow light, *below*, in place of the separate lights. Each half of the unit covers the same arc as separate sidelights, and in each case the total of 225 degrees is the same as that for the white masthead light.**

Figure 406 **A white sternlight with a 135-degree arc shines from dead astern to a point 22.5 degrees abaft each beam, matching the area not covered by masthead and sidelights. On boats less than 12 meters (39.4 ft) it may be combined with the masthead light as a single white all-round white light.**

Lights for Varying Sizes of Sailing Vessels Underway

A sailing vessel of any size underway—remember that if any propelling machinery is in use she is *not* a sailing vessel—will carry the same sidelights and sternlight as a power-driven vessel of the same length, but will *not* carry a forward white masthead light. [25(a)]

A sailing vessel *may* additionally carry at or near the masthead two all-round lights in a vertical line, with the same visibility range as a sternlight, the upper being red and the lower green. [25(c)]

Sailboats Under 20 Meters

If a sailboat is *less than 20 meters (65.6 ft)* in length, the sidelights and sternlight may be combined into a single fixture at or near the top of the mast where it can best be seen. (The Rules do not so state, but "The mast" can be presumed to be the mainmast if there are more than two masts.) [25(b)] The use of a combination masthead light, however, is *not* permitted when any mechanical power is being used, since there is no way to carry a white forward light higher than the colored sidelights. Carrying a three-color combination light also eliminates the use of the red-over-green optional light for sailboats, and a set of standard lights would be needed for use when the boat was under mechanical power.

A sailboat *less than 7 meters (23.0 ft)* long *should* carry normal sidelights and sternlight; but if this is not possible, it is sufficient to have an electric flashlight or lighted lantern ready at hand to be shown in time to prevent collision. [25(d) (i)] Again, if such a boat is operating under mechanical power, it must show the appropriate lights for a powerboat of its size.

Rowboats

Small boats propelled by oars may show the lights of a sailboat, or have handy an electric flashlight or lighted lantern to show to prevent a collision. [28(d)]

Lights for Vessels at Anchor

A boat *less than 7 meters (23.4 ft)* in length, when at anchor *not* in or near a narrow channel, fairway, or anchorage, or where other vessels normally navigate, need not show an anchor light. Also, there are locations on inland waters designated as "special anchorage areas," where anchor lights are not required for vessels *less than 20 meters (65.6 ft)* long. [30(g)]

These special anchorage areas are found frequently off yacht clubs and similar facilities. For the boatman, they offer the possibility of legally leaving his boat unattended at a mooring, without having an anchor light burning continuously or turning one on each sunset.

In anchoring situations other than those above, vessels *less than 50 meters (164.0 ft)* in length may show only a single all-round white light wherever it can best be seen, [30(b)], or the two anchor lights that are mandatory for larger vessels.

When two lights are used, the Inland Rules require a white all-round light in the fore part of a vessel at anchor, plus a second all-round white light at or near the stern, and lower than the forward light. [30(a)] The forward anchor light must be at least 4.5 meters (14.8 ft) above the other one. If the vessel is *50 meters (160.0 ft) or more* in length, the minimum height is increased to not less than 6 meters (19.7 ft.). [I, 84.03(k)]

Any vessel at anchor may, and a vessel of *100 meters (328.1 ft) and more* in length must, also use the available working or equivalent lights to illuminate her decks. [30(c)]

Standard Vertical Spacing

When the Rules require that two or three lights be carried in a vertical line, the lights must be spaced and positioned as follows: for vessels *less than 50 meters (65.6 ft)* in length, the vertical spacing must be not less than 1 meter (3.3 ft); for larger vessels, the spacing must be not less than 2 meters (6.6 ft). Where three lights are carried vertically, the spacings must be equal.

The lowest light on vessels *less than 20 meters (65.6 ft)* in length must be not less than 2 meters (6.6 ft) above the hull; on larger vessels, this height requirement is not less than 4 meters (13.1 ft). [I, 84.03(i)] This standard of spacing and positioning applies to all lights arranged vertically (except towing lights), including lights on fishing vessels, vessels aground, pilot vessels, vessels restricted in their ability to maneuver or not under command, and the forward lights of towing vessels. The lower of the two all-around lights required on a vessel engaged in fishing must be at a height above her sidelights not less than twice the distance between the two vertically arranged lights. [I, 84.03(j)]

Lights for Vessels Aground

A vessel that is aground must show the normal anchor light(s) plus two all-round red lights arranged vertically. [30(d)] A vessel *less than 12 meters (39.4 ft)* in length is exempted from this requirement.

Lights for Towing

Vessels of any size that are either towing another or

being towed obviously have, because of this condition, much less maneuverability than vessels proceeding singly. To indicate this situation at night, special navigation lights are prescribed for vessels towing or being towed.

Towing Alongside or Pushing Ahead

A power-driven vessel when towing another alongside must carry, in addition to her normal sidelights, two masthead white lights in a vertical line with standard spacing and positioning. These lights must be shown, but may be carried at *either* the position of the normal forward masthead light *or* that of the after masthead if used. [24(c)] However carried, the lowest after masthead light must be at least 2 meters (6.6 ft) above the highest forward light, to form an adequate range. (I, 84.03(e)] In lieu of the normal white sternlight she must carry two "towing lights" in a vertical line, positioned so as to be showing astern.

When a pushing vessel and a vessel being pushed ahead are "rigidly connected" as a *composite unit,* they are considered a single power-driven vessel and must position their masthead lights accordingly. This is *not* the same as a tug lashed to a barge ahead, no matter how tight the fastening cables are pulled in. [24(b)]

In these waters, vessels towed alongside or being pushed ahead are considered as a unit. On a vessel being pushed ahead, sidelights must be shown at the forward end plus a "special flashing light" as defined in Rule 21 (g). When towed alongside, a vessel shows sidelights at the forward end and a sternlight. [24(f)]

Exception On the Western Rivers (except below the Huey P. Long Bridge on the Mississippi River) and on waters specified by the Coast Guard, the prescribed lights for a vessel pushing ahead or towing alongside are only sidelights and two towing lights carried one above the other. [24(i)] It is said that masthead lights are not required on Western Rivers because of the number of low bridges which towboats must pass under.

Towing Astern

If the length of a tow astern, measured from the stern of the towing vessel to the after end of the tow, is 200 meters (656.2 ft) or less, the towing vessel must carry *two* masthead lights in the same manner as for pushing ahead or towing alongside, plus sidelights, a sternlight, and one towing light above the sternlight. If the length of the tow is greater than 200 meters (656. ft), *three* masthead lights are carried in a vertical line, rather than two. [24(a)]

A vessel towed astern carries sidelights and a sternlight, but no white forward lights. [24(e)] An inconspicuous, partially submerged vessel or object must carry special lights, rather than those just described. If it is less than 25 meters (82.0 ft) in beam, it must carry one all-round white light at or near each end; if it is 25 meters (82.0 ft) or more in beam, four all-round white lights at or near corners. If it exceeds 100 meters (328.1 ft) in length, it must have additional all-round white lights, placed so that the distance between such lights is not more than 100 meters (328.1 ft). These white lights must be visible for 3 miles. This rule exists to provide for the proper lighting of large liquid-filled bags (called dracones), log rafts, and other nearly awash objects. Vessels or objects towed alongside are lighted as a single vessel or object. [24(g)]

When it is impractical for a towed vessel to carry the prescribed lights, Rule 24(h) directs that all possible measures be taken to light the vessel or object being towed, or at least to indicate the presence of that vessel or object.

Small Boat Towing

Although the 1980 Inland Rules make no special provisions for small boats when towing, Rule 24(j) covers situations where towing is not the normal function of a vessel engaged in such work, and the towed vessel is in distress or otherwise in need of assistance. In this instance, all possible measures must be taken to indicate the relationship between the two boats, including the use of a searchlight to illuminate the towed vessel.

Lights for Special-Purpose Vessels

Specific lights or combinations of lights are prescribed for various categories of vessels, so they may be more easily identified at night.

Fishing Vessels

A vessel engaged in fishing, whether underway or at anchor, must show *only* the lights specified in Rule 26. Remember, however, that the definition of "fishing" does *not* include trolling lines.

A vessel engaged in *trawling*—dragging a dredge net or "other apparatus used as a fishing appliance"—must show two all-round lights vertically, green over white, plus sidelights and a sternlight when making way through the water. She must also carry a white masthead light abaft of and higher than the green light, if she is 50 meters (164.0 ft) or more in length; shorter vessels may, but do not have to, carry this light.

A vessel engaged in fishing *other than trawling* shows red-over-white all-round lights, again with sidelights and sternlight, if making way. If there is outlying gear more than 150 meters (492.1 ft) horizontally from the vessel, an all-round white light must be shown in that direction. This light must be not less than 2 meters (6.6 ft) nor more than 6 meters (19.7 ft) horizontally away from the all-round red and white lights.

In both cases just mentioned, the lower of the two vertical lights must be at a height above the sidelights not less than twice the spacing between the two vertical lights. The required visibility range is the same as for sternlights.

A vessel that is fishing close by other vessels also fishing *may* exhibit the additional signals described in Annex II. These signals would aid trawlers and purse seiners working in groups to coordinate their movements. [26(d)]

A fishing vessel *not* engaged in fishing will show the normal navigation lights for a vessel of her length [26(a)]

Pilot Vessels

A pilot vessel, when on pilotage duty, does not show normal lights for a vessel of her size, but rather two all-round lights in a vertical line, white over red, at or near the masthead. These lights have the same visibility distance requirement as a sternlight, which together with sidelights, must be shown when underway. At anchor the pilot vessels shows normal anchor lights as well as the white-over-red lights. [29(a)]

A pilot vessel *not* on duty carries only the normal lights for a vessel of her length. [29(b)]

Vessels With Limited Maneuverability

The U.S. Insland Navigation Rules prescribe special lights for various types of vessels that have limited maneuverability because of their condition or their operations.

Vessels Not Under Command

A vessel "not under command"—such as one with disabled engines, or one that for some reason cannot direct her movements—will show two red all-round lights vertically spaced where they can best be seen. Underway, such a vessel must also show sidelights and a sternlight, but not masthead lights. [27(a)]

Vessels Restricted in Maneuverability

A vessel of any size restricted in her ability to maneuver, except one engaged in minesweeping, will show three all-round lights with standard vertical separation, red-white-red. Underway or at anchor the vessel also will show the usual lights for that status. [27(b)]

A vessel engaged in a towing operation that severely restricts the tow's ability to deviate from its course will show normal towing lights, plus the red-white-red vertical lights. [27(c)] It is *not* intended, however, that vessels engaged in *routine* towing operations can declare that they are restricted in their ability to deviate from their course.

In addition to the red-white-red lights, a vessel engaged in dredging or underwater operations involving an obstruction to navigation will also show two vertically spaced red all-round lights on the side on which the obstruction exists and two green all-round lights, one above the other, on the side on which it is safe to pass; these lights shall be at the maximum practical horizontal distance, but never less than 2 meters (6.6 ft) from the red-white-red lights, and the upper light of these pairs must not be higher than the lower red light. A vessel of this category will not show anchor light(s) but will show underway lights, if applicable. [27(d)]

Diving Operations

Whenever the size of a vessel engaged in diving operations makes it impractical to exhibit all the lights prescribed in Rule 27(d) above, she must instead, show three all-round lights, placed vertically, red-white-red. [27(e)]

Minesweepers

A vessel engaged in minesweeping will show, in addition to all normal underway lights for a power-driven vessel, three green all-round lights. One of these will be at the foremast head and one at each end of the foreyard. These lights indicate that it is dangerous to approach closer than 1,000 meters (1,094 yds) astern or 500 meters (547 yds) on either side. [27(f)]

Smaller Boats

Boats *less than 12 meters* (39.4 ft) in length are not required to show the lights of Rule 27, but larger boats must comply. [27(g)]

Not Distress Signals

The various lights of Rule 27, especially those for vessels not under command, should be understood by other vessels as signals of limited maneuverability, not as signals of vessels in distress and requiring assistance. [27(h)]

Miscellaneous Provisions of the Rules

Rule 1(b) allows vessels equipped with lights meeting the requirements of the International Rules of the Road to use such lights on inland waters *in lieu* of the lights specified by the 1980 Inland Rules. Lights must be complete for one set of Rules or the other—no mixtures. See pages 88 and 89 for the applicable provisions of the International Rules.

This option is of value to the owners of boats that may be operated outside the Inland Rules demarcation lines at night, since it eliminates any need for making a change in lights when going from one body of water to the other. The navigation lights of the International Rules are permitted on inland waters, *but not vice versa;* a boat lighted for the Inland Rules may be in violation when proceeding seaward of the demarcation lines.

Exemptions for Government Vessels

Any requirements for lights and shapes—their number, placement, range or arc of visibility—do not apply to a Navy or Coast Guard vessel whose department Secretary certifies that its special construction—as for submarines and aircraft carriers, for example—makes this impossible. In such cases, however, their lights must comply as closely as feasible. [1(e)]

Rule 1(c) allows the use of special station or signal lights by naval ships and vessels proceeding in convoy. These additional lights, so far as possible, shall be such as cannot be mistaken for any lights of the Rules of the Road.

Special Submarine Lights

Because of their low height and close spacing, a submarine's normal navigation lights are sometimes mistaken for those of small boats, although submarines are large, deep-draft vessels with limited maneuverability on the surface. The Rules do not provide for any special lights for submarines at the surface, but under the authority of Rule 1(e), the United States has established a special distinctive light for its submarines; this light is in addition to all other required lights.

The distinctive light characteristic is an amber beacon with a sequence of one flash per second for three seconds, followed by three seconds of darkness. The light is located where it can be best seen, as nearly as possible all around

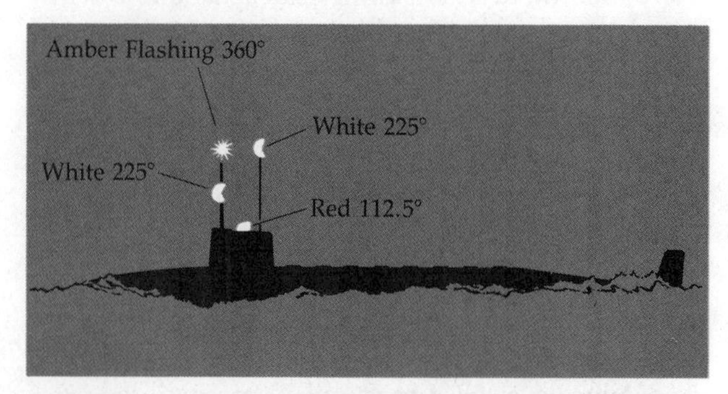

Figure 407 **A surfaced submarine shows an amber beacon that flashes once per second for three seconds, followed by three seconds of darkness, plus normal navigation lights.**

the horizon; it will be not less than two feet above or below the masthead light.

Signals to Attract Attention

The Inland Rules authorize a vessel to make light or sound signals to attract the attention of another vessel, provided that such actions cannot be mistaken for one of the normal lights or signals. A vessel may also direct the beam of her searchlight toward a danger, but not in such a way as to impede the navigation of any other vessel. [36]

Lights for Seaplanes

A seaplane is defined as including any aircraft designed to maneuver on the water. [3(a) and (e)] Seaplanes, while on the water, are thus in theory required to show the same lights as a ship of the same size. Because this will probably be impractical, the Rules offer an exception allowing them to show lights "as closely similar in characteristics and position as is possible." [31]

Air-Cushion Vehicle An air-cushion vehicle, when operating in the nondisplacement mode ("flying"), must show the normal-navigation lights for a vessel of her length, plus an all-round flashing yellow light where it can best be seen. [23(b)] There are no special lights for hydrofoil vessels.

Technical Details for Lights

Annex I to the U.S. Inland Rules also contains highly technical specifications for the "chromaticity" (color specifications) of navigation lights, and for the necessary luminous intensity to achieve the various ranges of visibility specified in the Rules. Specifications are also included for horizontal and vertical sectors, so that adequate light brightness is maintained fully over the specified arc, yet rapidly decreases outside the set limits. The requirements for horizontal sectors are the same for all vessels, but the vertical sector specifications for sailing vessels differ from the requirements for power-driven vessels. When buying lights, specify the type and size of vessel they are intended for.

Day Shapes

In the interest of collision prevention and greater overall safety on the water, the Rules of the Road prescribe a number of day shapes for certain situations and categories of vessels. Objects of specified shape and minimum size are used in the daytime, when lights are ineffective. The general requirements for the display of shapes are given in the numbered Rules, with details of configuration, size, and placement set forth in Annex I.

Ball Diameter of not less than 0.6 meter (2.0 ft).

Cone Base diameter not less than 0.6 meter (2.0 ft).

Cylinder Diameter not less than 0.6 meter (2.0 ft) and height twice the diameter.

Diamond Two cones, each as above, base to base.

If more than one shape is hoisted, they shall be spaced at least 1.5 meters (4.9 ft).

Figure 408 **A vessel under both sail and power, less than 12 meters (39.4 ft) in length, may show a conical day shape, point down. The day shape is required on longer vessels.**

Small Boats Vessels less than 20 meters (65.7 ft) in length may display smaller shapes scaled down to be commensurate with the vessel's size; the vertical separation may be correspondingly reduced.

Vessels Under Sail and Power

A vessel proceeding under sail in the daytime, when also being propelled by machinery, must carry forward, where it can best be seen, a black conical shape with its point downward. A vessel of *less than 12 meters (39.4 ft)* length is not required to exhibit this shape, but may do so. [25(e)] This day shape signals that such a vessel is *not* entitled to a sailboat's Rule 18 right-of-way privileges even though her sails are set.

Vessels at Anchor or Aground

Although large vessels must display certain shapes by day whenever they are at anchor or aground, this does not apply to *small boats* in most situations.

Vessels *under 7 meters (23.0 ft)* in length need show no shape if they are *not* at anchor in or near a narrow channel or fairway, anchorage, or where vessels normally navigate. Vessels *under 20 meters (65.6 ft)* in length need not display a shape when at anchor in a special anchorage area; and vessels *less than 12 meters (39.4 ft)* in length are not required to show shapes when aground or operating with limited maneuverability, except for diving operations. [27(g) and 30(f)]

In all other situations, vessels at anchor must hoist a ball-shaped daymark forward where it can best be seen. [30(a)] This is the signal displayed by the large vessels whenever at anchor. Similarly, a vessel aground must display three ball shapes in a vertical line. [30(d)]

Vessels Towing and Being Towed

Between sunrise and sunset, a power-driven vessel having a tow astern that is longer than 200 meters (656.2 ft) must carry a diamond shape where it can best be seen. [24(a)]

In such cases, the vessel being towed must also show a diamond shape where it can best be seen. [24(e)]

Vessels Engaged in Fishing

Vessels engaged in fishing by day must indicate their occupation by displaying, where it can best be seen, a shape consisting of two cones in a vertical line with their points together (in this case only, no vertical separation between the two shapes). If the vessel is *less than 20 meters (65.7 ft)* in length, a basket may be substituted for the two-cone shape. [26(b) and (c)] These shapes are used on vessels engaged in either trawling or in fishing other than trawling (but not on vessels trolling or drift fishing).

If its outlying gear extends more than 150 meters (492.1 ft) horizontally from the vessel, a vessel engaged in fishing, other than trawling, will show a day shape of a single cone, point up, in the direction of the outlying gear. [26(c)]

Vessels With Limited Maneuverability

A vessel not under command will show as a day signal two black balls one above the other. [27(a)]

A vessel "restricted in her ability to maneuver" will hoist three shapes in a vertical line, ball-diamond-ball. [27(b)]

A vessel engaged in a towing operation that severely restricts the tow's ability to deviate from its course, will show the ball-diamond-ball shapes, in addition to any other required day shapes. It is not intended, however, that vessels engaged in routine towing operations can declare that they are restricted in their ability to deviate from their course. [27(c)]

A vessel engaged in dredging or underwater operations when an obstruction to navigation exists will hoist two balls vertically on the side on which the obstruction exists, and two diamond shapes on the side on which another vessel may pass. [27(d)] If the size of a vessel engaged in diving operations makes it impracticable to exhibit these shapes, it shall show instead a rigid replica of the International Code flag "A" (white and blue, swallow-tailed) not less than 1 meter (3.3 ft) in height, arranged to ensure all-round visibility. [27(e)] This requirement for visibility in any direction will normally require multiple flags in a criss-cross or square arrangement.

While there is no official recognition in the International or U.S. Navigation Rules for the familiar "diver's flag"—rectangular red with one white diagonal slash—it is still advisable to fly it in the close vicinity of any persons engaged in diving as a warning to other craft to stay clear; this can be done by mounting the flag on a small float that moves along with the divers. Some states require that the red-and-white flag be flown when there are divers in the water; it should *not* be flown routinely when this condition does not exist. The "A" flag replica relates only to the status of a vessel and is not required, nor should it be displayed, if maneuverability is not limited, such as by having divers connected by hoses and lines. An "A" *flag*—made of cloth—is *not* an acceptable substitute for the "rigid replica" required by the Rules.

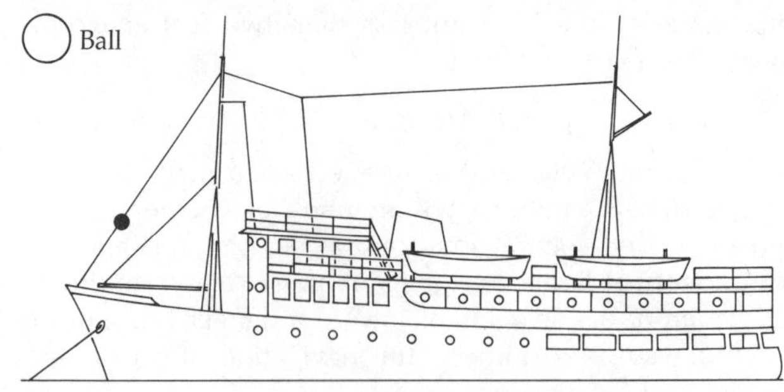

Figure 409 **When at anchor, vessels more than 20 meters (65.7 ft) in length must display a ball-shaped daymark.**

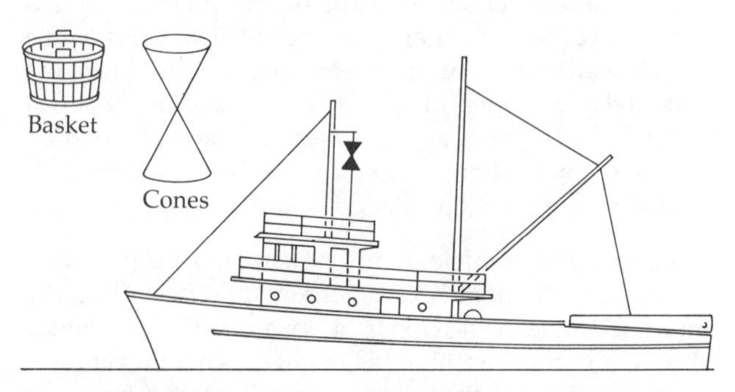

Figure 410 **Fishing vessels must display two cones; those less than 20 meters (65.7 ft) in length may show a basket instead.**

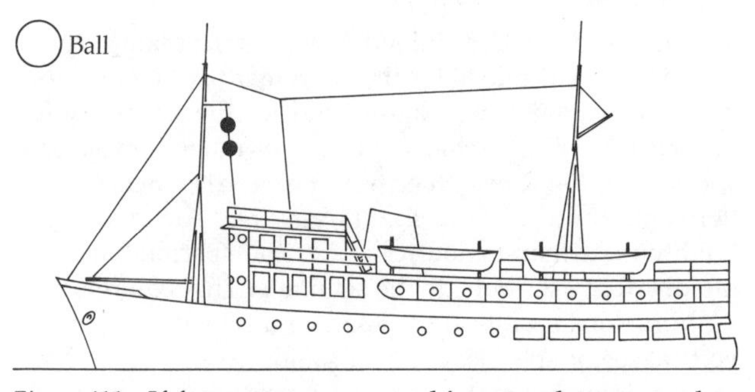

Figure 411 **If for any reason a vessel is not under command, it must display two black balls, one above the other.**

A vessel engaged in minesweeping will show three balls, one at the foremast head and one at each end of the foreyard; these have the same significance as the green lights shown at these positions at night. [27(f)]

Exemptions

To facilitate the change from the old Inland, Great Lakes, and Western Rivers rules to the 1980 Inland Navigational Rules, the new Rules provide temporary and permanent exemptions to required characteristics and positioning of lights. These exemptions apply to "any vessel or class of vessels, the keel of which was laid, or which was at a corresponding stage of construction" before 24 December 1980; newer vessels are not exempt and must comply with the new requirements. Where such exemptions are for a limited time, they are stated in terms of "years after the effective date of these rules" (24 December 1981). [38]

Vessels *less than 20 meters (65.6 ft)* in length that have lights as specified in the Motor Boat Act of 1940 are permanently exempt from meeting the new visibility ranges and color specifications. For *all vessels,* a permanent exemption says that lights need not be repositioned merely because of the changeover in specifications from customary units to metric units. The horizontal repositioning of masthead lights need not be done until 24 December 1990 for vessels over 150 meters (492.1 ft) in length, and shorter vessels are given a permanent exemption. An exemption until 24 December 1990 is also granted for the restructuring or repositioning of all lights to meet the specifications of Annex I. Power-driven vessels *12 meters (39.4 ft) or more, but less than 20 meters (65.6 ft),* in length are permanently exempt from the provisions of showing a masthead light forward, plus a sternlight aft, provided they show an all-round white light aft in place of the new requirements.

Pilot Rules for Inland Waters

The PILOT RULES for inland waters under the 1980 Inland Rules are found in Annex V. The Pilot Rules require that a copy of the Inland Navigation Rules be carried on board every "self-propelled vessel" 12 meters (39.4 ft) or more in length. [88.05] These are found in Coast Guard publication *Navigation Rules, International-Inland,* which is available for purchase from the Superintendent of Documents, Government Printing Office, Washington, D.C. 20402, or from local sales agents for charts and nautical publications. Here are Pilot Rules provisions for lights:

Law Enforcement Vessels

Law enforcement vessels of the United States, and of the States and their political subdivisions, may display a flashing *blue* light when engaged in direct law enforcement activities. This light must be located so as to not interfere with the visibility of the vessel's navigation lights. [88.11]

Lights on Dredge Pipelines

Dredge pipelines that are floating or supported on trestles must display lights at night and in periods of restricted visibility. There must be yellow lights along the pipeline, approximately equally spaced and no more than 10 meters (32.8 ft) apart, where the pipeline crosses a navigable channel; where the pipeline does not cross a navigable channel, the lights must be sufficient to clearly show the pipeline's length and course. Each light must be visible all around the horizon, flash 50 to 70 times per minute, and be located not less than 2 meters (6.6 ft) nor more than 3.5 meters (11.4 ft) above the water. There must also be two all-round red lights (not flashing) at each end of the pipeline, 1 meter (3.3 ft) apart vertically, with the lower red light at the same height above the water as the flashing yellow lights. [88.15]

Lights on Barges at Bank or Dock

Section 88.13 of Annex V describes the circumstances requiring lights on barges moored at banks or docks, and the characteristics of such lights. Basically, the Rules require white lights visible for 1 mile on a clear, dark night.

Modifications While Passing Under Bridges

Any vessel, when passing under a bridge, may lower any light or day shape if it needs to. When clear of the bridge, the lights or day shapes must be repositioned immediately. [88.09]

Penalties

The Inland Navigational Rules Act of 1980 provides that any person "who operates a vessel" in violation of that Act or any regulation issued thereunder—and this includes the various Annexes—is liable for a civil penalty of not more than $5,000 for each violation.

It should be noted that this penalty is not a "fine" requiring conviction in a federal court, but it is a civil forfeiture imposed by the Secretary or his delegate. The penalty may not be assessed until the person charged has received notice of the violation and an opportunity for a hearing. The Secretary or his delegate may remove or mitigate any penalty assessed. If a penalty, as finally determined, is not paid, the matter can be taken to a Federal District Court for collection and/or other appropriate action.

A portion of the Act also provides for withholding customs clearance for any vessel whose owner or operator has unpaid penalties. This provision is especially important for foreign vessels in U.S. waters; if penalty proceedings are not complete before the vessel's sailing date, clearance is normally granted after the posting of a bond.

Penalties Against Vessels

In maritime law, a vessel itself can be held to be in violation and be subject to penalties if operated contrary to any provision of the Inland Navigational Rules Act of 1980 or regulations issued thereunder, including the Annexes. The same maximum penalty, $5,000 for each violation, can be assessed, and in this case the vessel may be seized and proceeded against in any Federal District Court. The same procedures for due notice and hearings are available to the owner of a vessel charged with a violation as are available to individuals as described above. Withholding a required customs clearance is particularly appropriate in the case of a vessel charged with violations.

General Cautions

Although they are not strictly a part of the Inland Rules or their Annexes, several other navigation light matters should be considered by every skipper.

Misuse of Navigation Lights

Boats are often seen underway at night showing both running lights and the anchor light. This presents a confusing picture to another boat approaching from astern of the first vessel—two white lights such as would be seen from ahead of a larger vessel under the Inland Rules. This illegal situation is usually caused by carelessness on the part of the skipper—he turned on one too many switches! Check your lights each time you turn them on.

Another problem is the small Class A or 1 boat at anchor that shows her combination red-and-green light as well as her all-round white light aft. Here the illegal situation is caused by the manufacturer who installed only one switch for all navigation lights, making it impossible for the boat to show only the white light as an anchor light, without also showing the red and green sidelights. The solution is for the skipper to install a second switch, or a single switch with two separate "on" settings, so that the colored and white lights may be controlled independently.

Maintenance of Navigation Lights

Many boats are used only in the daytime, but the wise skipper checks his navigation lights once a month whether they have been used or not. He also carries one or more spare bulbs and fuses, each of the proper type and size.

International Rules of the Road

Careful study of the 1980 U.S. Inland Rules of the Road provides an excellent basis for understanding the "INTERNATIONAL REGULATIONS FOR PREVENTING COLLISIONS AT SEA, 1972"—the formal name for what are generally called the INTERNATIONAL RULES OF THE ROAD or " '72 COLREGS." The International Rules of the Road (IntRR) were made applicable to U.S. waters and U.S. vessels by Presidential Proclamation and subsequent Act of Congress. The 1972 IntRR consist of 38 numbered "Rules" (organized into "Sections" which in turn are grouped together into "Parts"), plus four "Annexes" (the IntRR do not have Pilot Rules, the fifth Annex of the U.S. Inland Rules). Some of these Rules and Annexes relate to lights and day shapes, and will be discussed in this chapter. Other Rules and Annexes are concerned with right-of-way, whistle signals, and distress signals; these are considered in Chapter 5.

Applicability

The International Rules, as established under U.S. authority, apply to vessels in two situations:

1. To all vessels in waters within United States sovereignty, outside the prescribed demarcation lines at entrances to bays, rivers, harbors, etc.; see page 40. In the absence of demarcation lines, they are also applicable *within* bays, harbors, and inlets, along specified stretches of coasts and up the connecting rivers to their limits of continuous navigation.
2. To all U.S. vessels on the high seas, not subject to another nation's geographic jurisdiction.

Although the IntRR state that they are applicable "upon the high seas and in all waters connected therewith navigable by seagoing vessels" [1(a)], there is a provision for a nation to prescribe its own rules for harbors, rivers, and inland waterways [1(b)]; this is the basis for the U.S. Inland Rules.

Manner of Presentation

The 1980 U.S. Inland Rules of the Road were derived from the 1972 International Rules, and match them rule for rule, and annex for annex in format. Considerable effort was taken to make the language identical wherever possible. Where there are differences, they exist because different operating conditions on the inland waters of the United States require rules for safety of navigation different from those applicable on the high seas. On inland waters, vessels are generally smaller in size and distances less. The International Rules, of course, do not contain the several special provisions for the Great Lakes or Western Rivers.

Thus, the consideration of the IntRR for lights and day shapes can be limited to the relatively few points of difference. *Where a difference is not noted, the International and 1980 U.S. Inland Rules are identical,* except for minor editorial modifications of no significance to vessel operation.

Navigation Lights

The requirement for showing navigation lights and day shapes is the same for waters governed by the International Rules as for those where the U.S. Inland Rules apply. See Table B, page 594.

Definitions

The IntRR definitions *do not include* a "special *flashing light*"—the yellow light used at the bow of barges pushed ahead on inland waters.

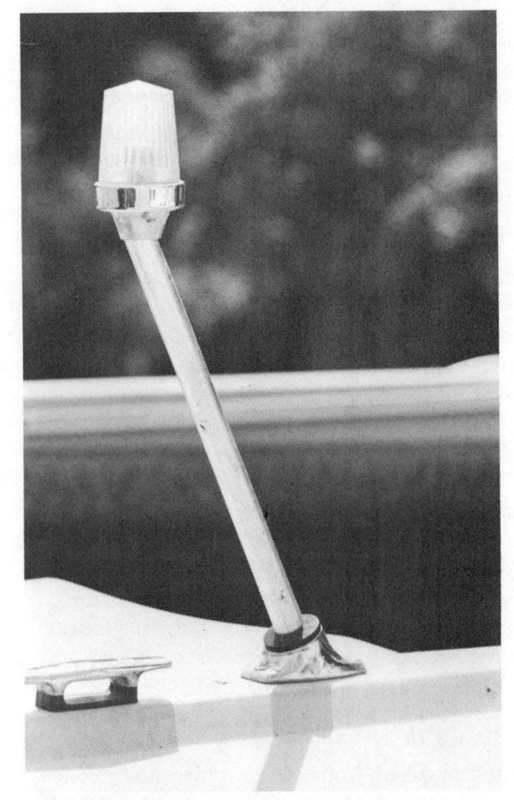

Figure 412 **In International Rules waters, the offset white all-round stern light permitted for small boats by the Inland Rules must be replaced by separate masthead and sternlights.**

The IntRR *do include* a definition not used in the Inland Rules. A "vessel constrained by her draft" is considered a power-driven vessel which, because of her draft in relation to the available depth of water, is severely restricted in her ability to maneuver.

Lights for Power-Driven Vessels Underway

International Rule (23(a) (i) *does not contain the exception* in the location of the forward masthead light for vessels *under 20 meters (65.6 ft)* in length; such a light thus *must* be on the vessel's forward half.

Lights for Vessels Anchored or Aground

The International Rules contain *no provision* for special anchorage areas as established in inland waters. The *only exception* to the requirement for an anchor light is for vessels *under 7 meters* (23.0 ft) in length anchored in areas free of water-borne traffic. Vessels under 12 meters (39.4 ft) in length are exempt from the requirement to show special lights and dayshapes when aground.

A vessel pushing a tow ahead or towing alongside must show a normal white stern light (rather than the two yellow towing lights in a vertical line, as required by Inland Rules.) [24(c)] A vessel pushed ahead will *not carry* the special flashing yellow light used at the bow of such barges on inland waters.

Lights for Vessels With Limited Maneuverability

International Rule 27(d) provides *additionally* that a vessel engaged in dredging or underwater operations and restricted in her ability to maneuver must, when underway, show masthead light(s), sidelights and a sternlight, together with the prescribed white, red, and green all-round lights required by other paragraphs of that Rule.

The exception for the special lights required by Rule 27 is *limited* to vessels *less than 12 meters* (39.4 ft) in length; *all* vessels over this length must comply. [27(g)]

International Rule 28 *permits* a vessel "constrained by her draft" to show three red all-round lights in a vertical line wherever they can best be seen. (This is the Rule which is omitted from the U.S. Inland Rules.)

Day Shapes of the International Rules

International Rule 25(e) *requires* the use of a cone, with point downward, as a day shape on *all* sailing vessels, when using machinery for propulsion while her sails are up. There are *no exceptions* in this rule for smaller sailboats.

The exception for showing an "anchor ball" is *limited* to vessels *under 7 meters* (23.0 ft) in length, when anchored in waters not subject to water-borne traffic. A similar exception applies to the requirement for showing three balls in a vertical line if aground in such a traffic-free area. [30(e)]

International Rule 28 *permits* a vessel "constrained by her draft" to show a day shape consisting of a cylinder.

Exemptions

International Rule 38 provides for temporary and permanent exemptions to the requirements for the characteristics and positioning of lights, to ease the transition from the 1960 Rules to the 1972 Rules. Where such exemptions are of limited duration they are stated in terms of years "after the date of entry into force of these Regulations"; this date was 15 July 1977.

Vessels in existence on 15 July 1977 had four years to install lights meeting the required range of visibility and color specifications. [38(a) and (b)] Any repositioning of lights because of the change from specifications in customary units to metric units need not be done at all, since there is a permanent exemption. [38(c)] Repositioning of masthead lights and sidelights to meet the requirements of Annex I to the 1972 International Rules need not be accomplished until 15 July 1986, except that a permanent exemption is granted vessels 150 meters (492.1 ft) or less in length with respect to the requirement for the location of the forward masthead light and the horizontal separation of the after masthead light, if carried. [38(d)]

Penalties

The 1972 International Rules of the Road do not in themselves contain any penalties for violation of their provisions. However, the Navigational Rules Act of 1977, which made these international regulations effective in U.S. waters and for U.S. ships, provides for civil penalties of not more than $5,000. Procedures are generally the same as for violations and penalties under the U.S. Inland Rules as described above.

Individuals and vessels can also be sued for damages resulting from violations of the International Rules as well as the U.S. Inland Rules of the Road.

1599
LIZELLE

5

Here are the actions to take, and signals to give, to avoid collision; sound signals required in "thick" weather when underway or at anchor

RULES OF THE ROAD: RIGHT-OF-WAY AND SOUND SIGNALS

Most of us have had the experience of walking down a street and meeting another pedestrian headed toward us. We turn to one side to avoid hitting him, only to have him dodge simultaneously to the same side. Perhaps there's a quick turn by both in the opposite direction, followed by a collision, or we manage to pass without hitting. To pedestrians on a sidewalk such actions are merely comical, and fairly simple to avoid. Between vessels on the water, they are serious indeed, and boatmen need to understand and follow the applicable Rules of the Road to avoid collisions.

Except for marked channels, there are no clearly defined paths for boats to follow. They have an open expanse of water on which to navigate, and their courses often cross the tracks of other vessels in the same waters. The caution needed on the water, even though traffic is much less than on land, is thus just as important as on sidewalks, streets, and highways.

Rules and Regulations

To prevent collision, carefully considered rules clearly state the duty of the skipper of any vessel encountering another vessel. These rules are of three general classes: the International Regulations for Preventing Collisions at Sea, 1972; the 1980 U.S. Inland Navigational Rules; and regulations issued by departments and agencies of the federal government. Most of these are promulgated by "the Secretary of the Department in which the Coast Guard is operating," or by the Commandant of the Coast Guard, under authority delegated to him by the Secretary.

The principal regulations are the five annexes to the Inland Rules, including Annex V, Pilot Rules. State and local authorities may also issue special regulations covering other matters related to this chapter, but these are too diverse to be considered in this book. In general, such special regulations conform to the Inland Rules and Coast Guard regulations, but elaborate on certain minor details. Skippers should know all regulations that apply to the waters they use.

The 1980 Inland Rules became effective on 24 December 1981 on all U.S. waters except the Great Lakes, and on those waters on 1 March 1983 following coordination with Canada.

Sequence of Consideration

The 1980 U.S. Inland Navigational Rules will be considered in detail here. They apply on waters *inside* demarcation lines at the entrances to most, but not all, harbors, bays, rivers, and inlets (see page 40). They probably thus apply to most recreational boatmen in the U.S.

The 1972 International Rules of the Road will be discussed in detail only where their provisions differ significantly from the 1980 U.S. Inland Rules. The Inland Rules are derived from the International Rules and closely parallel them; the various provisions are matched rule for rule and annex for annex. Many Rules are identical, but in a number of instances the content and language has been changed to reflect the generally smaller size of vessels and shorter distances involved in inland waters; in a few places an editorial change has been made for greater clarity. Annexes II and IV are identical; Annexes I and III have minor differences; and Annex V of the Inland Rules has no counterpart, since there are no Pilot Rules for the International Rules of the Road.

The consideration of the Rules of the Road in this chapter and Chapter 4 is not complete; it focuses primarily on points of greatest interest to boatmen.

Definition of Terms

The definitions of the Inland and International rules that were discussed in Chapter 4 on pages 78 and 79 are also equally applicable in this chapter. Additional definitions follow:

Whistle Any sound-signaling equipment capable of producing the prescribed blasts, and which complies with the specifications in Annex III. [32(a)]

Short Blast A blast of about one second's duration.

Prolonged Blast A blast of from four to six seconds duration. (The term "long blast" is no longer used in the Rules of the Road.)

Figure 501 **Many boating areas are relatively open expanses of water without specific channels. Boats may approach from one or more directions, often several at a time. Every skipper must know who has the right of way, and what signal to give, in each situation.**

"Privileged" and "Burdened" Vessels

For generations, skippers have used the terms BURDENED and PRIVILEGED to define the status of two vessels encountering each other. The privileged vessel is the one that has the right-of-way—the right to proceed unhindered by the other. The burdened vessel is the one that does *not* have the right-of-way, the one that must take any necessary action to keep out of the way of the privileged vessel.

These terms do not appear in the 1972 International Rules and the 1980 Inland Rules—but the concept remains. Officially, the privileged vessel is now the STAND-ON vessel; the burdened vessel is the GIVE-WAY vessel. Although these terms are used in the titles of Rules 16 and 17, they are not included in the definitions of either set of Rules. The new terms were adopted as being more descriptive of the required actions. The terms burdened and privileged will continue in use on an informal basis.

The U.S. Inland Rules

Steering and Sailing Rules

The basic purpose of any set of rules of the nautical road is to prevent collisions (indeed, this is the title of the International Rules). A major subdivision, Part B, termed "Steering and Sailing Rules," is particularly aimed at that goal with three subparts covering conduct of vessels in any visibility: conduct when in sight of each other [4–10], conduct in restricted visibility [11–18], and [19] Part D, Sound and Light Signals, which also largely focuses on collision prevention.

Technically, the right-of-way rules do not come into effect between two vessels until the possibility of a collision exists. Privilege and burden are not necessarily established when vessels first sight each other, but rather at the moment when a "risk of collision" develops. Although not an actual rule, a basic principle of collision prevention is that, where the depth of the water permits, two vessels should never get so close to each other that the risk of a collision can materialize. A collision between two vessels is almost impossible if each skipper fully and properly obeys the Rules of the Road.

Determining Risk of Collision

Every vessel shall use all available means appropriate *to* the prevailing conditions to *determine if risk of collision exists. If there is any doubt, such risk shall be deemed to exist.* [7(a)]

Every vessel must at all times maintain a proper watch by sight and *hearing,* as well as all available means appropriate in the prevailing conditions so as to make a full appraisal of the situation and of the risk of collision. [5]

An excellent method (and prescribed by the rules) of determining whether your boat is on a collision course with another is to watch the relative bearing of the other vessel. If this bearing does not change appreciably, either forward or astern, a risk of collision exists. [7(d)] The Rules of the Road then apply, and appropriate action should be taken.

Boatmen are cautioned, however, against presuming there is no risk, when there *is* an appreciable change in bearing. Each situation must be considered in light of its own conditions. [7(d)]

In reduced visibility, you must make proper use of any operational radar equipment on board, including long-range scanning, and of radar plotting or equivalent *systematic* observation of detected objects. [7(b)]

Do not make assumptions on the basis of scanty information, especially scanty radar information. [7(c)]

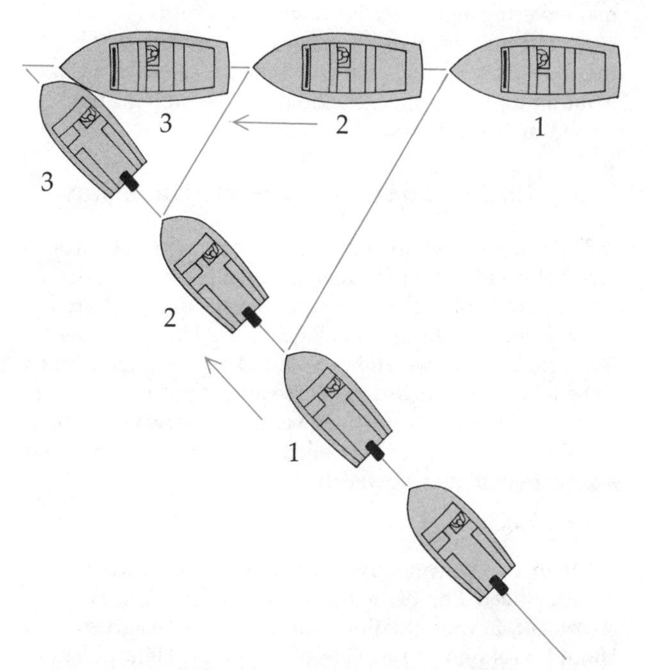

Figure 502 **The rules come into force whenever there is a risk of collision. In a crossing situation, keep a check on the relative bearing of the other craft. To be safe, there should be a distinct change in the angle, fore or aft; a constant or slow change in bearing indicates probable danger.**

Safe Speed

Every vessel must proceed at a safe speed *at all times.* [6]

Rule 6(a) lists specific factors to be considered in determining a "safe speed," including but not limited to:

The state of visibility;

Traffic density, including concentration of fishing or other vessels;

Your vessel's maneuverability, with special reference to stopping distance and turning ability;

At night, the presence of background lights such as those from shore, or from the back-scatter of your vessel's own lights;

The state of wind, sea, and current, and the proximity of navigational hazards;

The draft in relation to the available depth of water.

Rule 6(b) provides *additional* guidance for determining a safe speed for a ship or boat fitted with an operational radar.

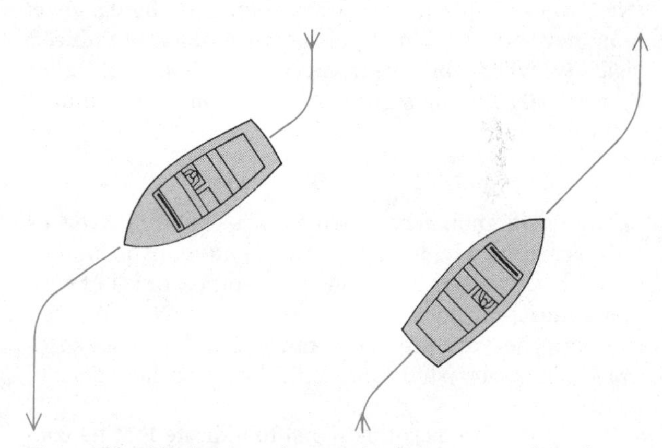

Figure 503 **All changes of course in the presence of another vessel should be so pronounced and definite that they will be noticed and properly evaluated by the other boat's skipper. Slight changes may fail to make your intentions clear.**

Actions to Avoid Collision

The rules specifically require that any action taken to avoid collision, if the circumstances allow, will be positive, made in ample time, and in keeping with good seamanship. Any change of course or speed should be large enough to be readily apparent to the other vessel visually or by radar; avoid a series of small changes. Your judgment—your "seaman's eye"—may tell you that only a slight change will be sufficient, but that change might well go undetected and leave the other skipper in confusion.

The action taken must be sufficient to result in passing at a safe distance; continue to check its effectiveness until the other vessel is clear. If you have to, you must slow down or *take all way off* by stopping or reversing. [8]

Whistle Signals

One- and two-blast whistle signals used between power-driven vessels encountering each other will be discussed in this section, along with the determination of right-of-way and the necessary actions to be taken by the vessels.

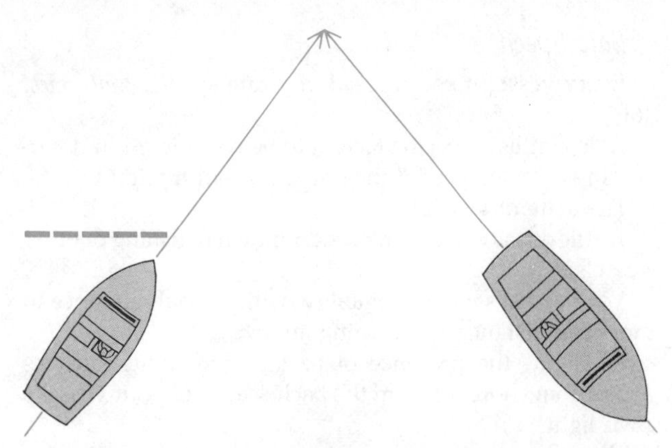

Figure 504 **In a confused situation, caused here by the boat backing into danger, a skipper uncertain of the other's intentions must sound the five-or-more-blast danger signal.**

These signals are to be used only when the vessels are in sight of each other and are meeting or crossing at a distance within half a mile of each other. [34] These signals must never be used in fog or other condition of reduced visibility, where the vessels are not visible to each other by eye; only the fog signals of Rule 35 may be sounded at such a time.

Danger Signal

In any situation where two vessels—power-driven or sail—are approaching each other, if one fails to understand the actions or intentions of the other, or is not sure that sufficient action is being taken by the other to avoid collision, the vessel in doubt must give the DANGER SIGNAL—*five or more* short and rapid blasts on her whistle. [34(d)]

A skipper also uses this signal to indicate that he considers the actions of the other vessel dangerous to either vessel—such as a negative reply to a proposal to pass in a certain manner. *Giving a danger signal does not relieve a vessel of her obligations or responsibilities under any Rule.*

Although not specifically covered in the rules, a careful skipper who either hears or sounds a danger signal will *at once* slow or stop his vessel until the situation is clarified for all concerned.

Equipment for Sound Signals

A vessel *12 meters (39.4 ft)* or more in length must have a WHISTLE and a BELL; a vessel of 100 meters (328.1 ft) or more in length must also have a GONG, the tone and sound of which cannot be confused with those of the bell. [33(a)]

A vessel *less than 12 meters (39.4 ft)* in length need not carry the sound-signaling equipment required on larger vessels, but if she does not, she must have "some other means of making an efficient sound signal." [33(b)] There are no requirements for small boats in terms of range (½ mile would be a reasonable minimum) or source of power—only in terms of efficiency—and sailboats have the same requirements as motorboats.

For vessels *over 12 meters (39.4 ft)* in length, however, Annex III contains detailed specifications for sound-signaling equipment, including different whistle tones, and ranges of audibility. Higher-pitched tones are used for smaller vessels; for those of *less than 75 meters (246.1 ft)* length, the band of frequencies is 250-525 Hz. Required ranges of audibility gradually increase, from ½ mile for vessels *12 but less than 20 meters (39.4-65.6 ft)* in length, to 2 miles for vessels *longer than 200 meters (656.2 ft)*; all distances are coordinated with specific sound pressure levels. [86.05] A vessel normally used for pushing ahead or towing alongside *may* carry a whistle with characteristics matched to the length of the longest towed-and-towing combination customarily operated, and use it even when operating singly. [86.15] The Annex specifies directional properties and positioning for whistles, and a limitation on the sound pressure level at the vessel's own "listening posts" (lookout or helm position). [86.07 and 86.09]

Specifications for bells and gongs include different sizes for smaller and larger vessels, and a minimum sound-pressure level. A bell on boats *under 20 meters (65.6 ft)* in length must be not less than 200 mm (7.9 inches) in diameter at the mouth. Bells on *larger vessels* must be at least 300 mm (11.8 inches) in diameter. The mass of the striker must not be less than 3% of the mass of the bell. [86.21 and 86.23]

The bell or gong, or both, on a vessel may be replaced by "other equipment" having the same sound characteristics, provided that the prescribed signals can always be sounded manually as well. [33(a)]

Maneuvering Lights

Whistle signals *may* be supplemented by a light signal synchronized with the whistle. This is a single all-round light, white or yellow, visible for at least two miles regardless of the size of the vessel. [34(b)] This MANEUVERING LIGHT must be placed in the same vertical plane as the "masthead" light(s), at least 1 meter (3.3 ft) *above* the forward masthead light, and at least 1 meter *above* or *below* the after masthead light. If a vessel carries only one masthead light, as will be the case for nearly all boats, the maneuvering light can be carried where it can best be seen, but not less than ½ meter (19.7 in) above or below the masthead light; remember that a maneuvering light is an all-round light and should not be significantly obscured in any direction. [I, 84.23]

Rules for Power-Driven Vessels Underway

The Rules of the Road recognize three types of encounters between two approaching vessels—meeting, crossing, and overtaking. These may be collectively referred to as "passing situations"; see Figure 505. The rules governing right-of-way, whistle signals to be given, and actions to be taken with regard to course and speed changes, are given below for power-driven vessels underway. Remember that a boat being propelled by both machinery and sails is treated as a power-driven vessel.

Meeting Situation

When two power-driven vessels are approaching one another head on, or nearly head on, this is a MEETING SITUATION. In this situation, unless otherwise agreed each should pass on the *port* side of the other. [14(a)] Neither vessel has the right of way over the other, and both must alter course to starboard, if necessary, to provide sufficient clearance for safe passage. This is exactly the same, it will be noted, as for two cars meeting on a narrow road.

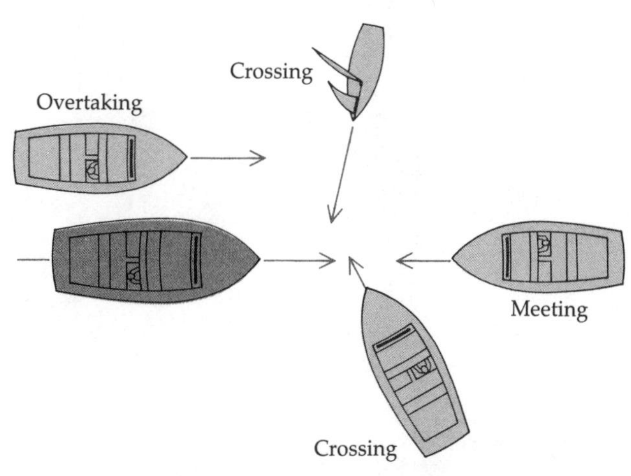

Figure 505 **The Navigation Rules recognize three types of "passing" situations—meeting, crossing, and overtaking. Specific procedures govern the actions and signals of both vessels in each case.**

For the rule of the meeting situation to apply, two requirements must be met: the vessels must be meeting in such a manner as to involve risk of collision, and they must be on reciprocal or nearly reciprocal courses. Such a situation is deemed to exist by day when a vessel sees the other ahead or nearly ahead, and by night when she can see the masthead lights of the other in a line or nearly in a line, or sees both sidelights. [14(b)] Although the rules do not set mathematical limits for the meeting situation, court interpretations and decisions over the years have established one point (11¼°) on either side of the bow as the practical boundaries within which vessels will be considered as meeting each other.

This rule does not apply to two vessels that will, if both keep on their respective courses without change of heading or speed, pass clear of each other.

When a vessel has any doubt that a meeting situation exists, she must assume the situation does exist and act accordingly; that is, change course to starboard. [14(c)]

In a winding channel, a vessel may first sight the other at an oblique angle rather than "head on, or nearly head on." Nevertheless, the situation is to be regarded as a meeting situation since they will be "head on or nearly head on" when the actual passing occurs. Each should keep to her own right side of the channel and neither has the right-of-way over the other.

Exception A power-driven vessel operating on the Great Lakes, Western Rivers, or other waters designated by the Secretary, and proceeding downbound with the current has the right-of-way over an upbound vessel. The downbound vessel proposes the manner of passing and initiates the whistle signals.

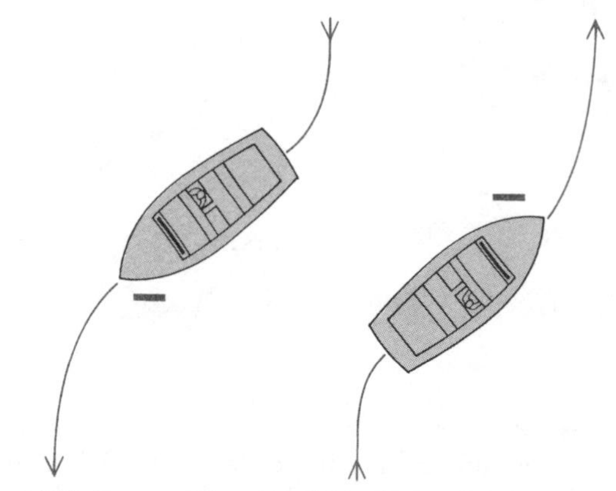

Figure 506 **In a meeting situation, *above* and *below* both vessels must sound one blast, give way to starboard, and pass port to port. Either vessel can signal first.**

Signals When power-driven vessels are within sight of each other and meeting at a distance of a half-mile or less, each vessel *must* sound whistle signals as follows:

One short blast, meaning "I intend to leave you on my port side."

Two short blasts, meaning "I intend to leave you on my starboard side."

Three short blasts, meaning "I am operating astern propulsion" (engines in reverse gear).

Upon hearing the one- or two-blast signal from the other, a vessel will, if in agreement, sound the same signal and take any steps necessary to effect a safe passage. If she doubts the safety of the proposed maneuver for any reason, she must sound the danger signal of five or more short, rapid blasts *and* take appropriate precautionary action until a safe passage agreement is made. [34(a)] The rule does not state which vessel should signal first—both boats are on an equal basis and either can make the first signal to the other.

Note that a vessel must *never* answer a one-blast signal with two blasts, or a two-blast signal with one blast. This is known as CROSS SIGNALS and is strictly prohibited. [V, 88.07]

When one vessel sounds the danger signal in a meeting situation, both vessels should immediately stop or reduce their forward speed to bare steerageway. [8(e)] Neither vessel should attempt to proceed or pass until agreement is reached through the exchange of the same whistle signal, either one blast or two.

Crossing Situation

When two vessels are not meeting head on, or nearly so, and each has the other forward of a direction 22½° abaft the beam, a CROSSING SITUATION exists.

Figure 507 **Vessels are crossing if they are not meeting head on or nearly so, and each has the other on a relative bearing forward of 22½ degrees abaft the beam, port or starboard. If there is a risk of collision, the rules give the right-of-way to boat A over boat B.**

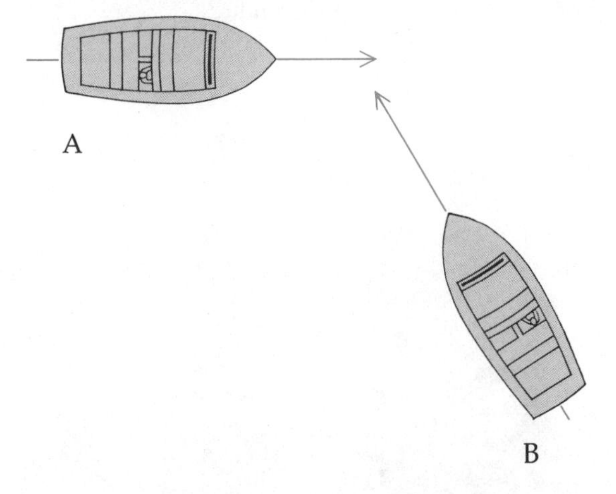

When two power-driven vessels are crossing, so as to involve risk of collision, the vessel that has the other on her own starboard side must keep out of the way of the other. [15] See Figure 507.

This leads to the unofficial definition of a boat's DANGER ZONE—a concept not explicitly stated in the Rules of the Road, but one very valuable for safety. This is the arc from dead ahead to 22½° abaft the *starboard* beam; see Figure 508. If you see another vessel within this danger zone, it most likely is the stand-on (privileged) vessel;

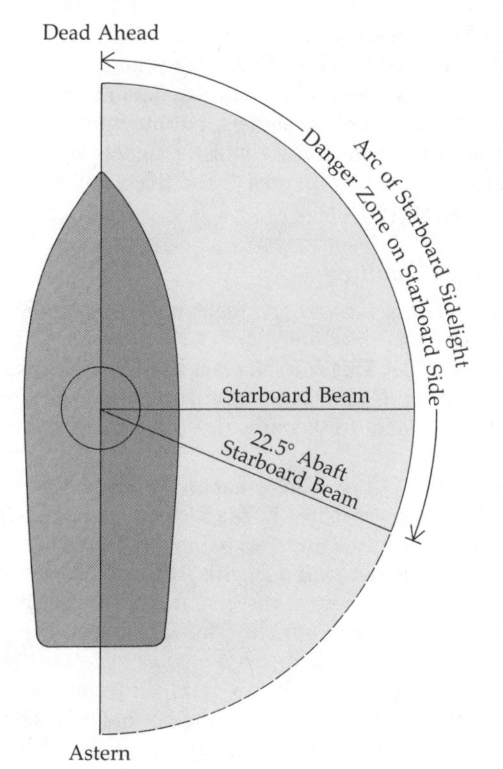

Figure 508 **A boat's danger zone is the area from dead ahead to a point 22½ degrees abaft the starboard beam. Any boat approaching yours in this area has the right of way.**

yours is the give-way (burdened) vessel and must change course or speed to avoid collision.

Exception On the Great Lakes, Western Rivers, and any other waters specified by the Secretary, a vessel crossing a river must keep clear of a power-driven vessel proceeding up or down that river. [15(b)] This situation is analogous to a road intersection on land, where the side street has a "Stop" sign to prevent interference with traffic on the main street.

Signals The 1980 Inland Rules *require* whistle signals in both meeting and crossing situations. If the vessels are in sight of each other and will pass within a distance of a half-mile [34(a)], one, two, and three short blasts are used with the same meanings as described above for meeting situations. Again, the Rules do not state which vessel should signal first, but by custom it is most often the vessel that is privileged, or believes herself to be privileged. A one- or two-blast signal is answered with the same number of blasts, if the other vessel is in agreement and will take the steps necessary to effect a safe passage. The danger signal of five or more short rapid blasts is used in the same way to indicate doubt in the safety of the actions taken or in the signaled intentions of the other vessel.

A maneuvering light *may* show one, two, three, or five or more flashes synchronized with the whistle blasts.

Note that this rule applies only if there is risk of collision. If two vessels will cross free and clear of each other, neither is privileged or burdened and whistle signals are thus inappropriate. In case of doubt, however, the safe procedure is to assume that the intentions of each are *not* known to the other.

Overtaking Situation

If one vessel is coming up on another, from astern of a direction 22½° abaft the beam of the other vessel, and making greater speed so as to close the distance between them, the vessel astern is said to be OVERTAKING the slower vessel; Figure 509. At night, this situation exists when the vessel astern can not see either of the sidelights of the vessel ahead. By day, the overtaking vessel cannot always know with certainty whether she is forward or abaft a direction 22½° abaft the other vessel's beam; if in doubt, she should assume that she is an overtaking vessel. [13(b) and (c)]

Notwithstanding any other provision in the Rules of Road, every vessel overtaking another must keep out of the way of the overtaken vessel. The overtaking vessel is burdened; the overtaken vessel is privileged. [13(a)]

The Rules state specifically that once a vessel is in an overtaking status, she remains so for the remainder of the encounter; no subsequent alteration of the bearing between the two vessels will serve to make the overtaking vessel a crossing vessel within the meaning of the rules, or relieve her of her duty to keep clear of the overtaken vessel until she is finally past and clear. In other words, should the overtaking vessel come up on the starboard side of the overtaken vessel, and move into her danger zone, the overtaking vessel does not by this movement become the privileged vessel with the right-of-way. [13(d)]

Signals If the overtaking vessel desires to pass on the port side of the vessel ahead—as would be normal if the slower vessel were keeping to the right of a channel (see Figure 510 left)—she must sound a two-short-blast signal. If the privileged vessel agrees with such a passing, she should immediately sound the same signal. The faster vessel then directs her course to port and passes as proposed and agreed.

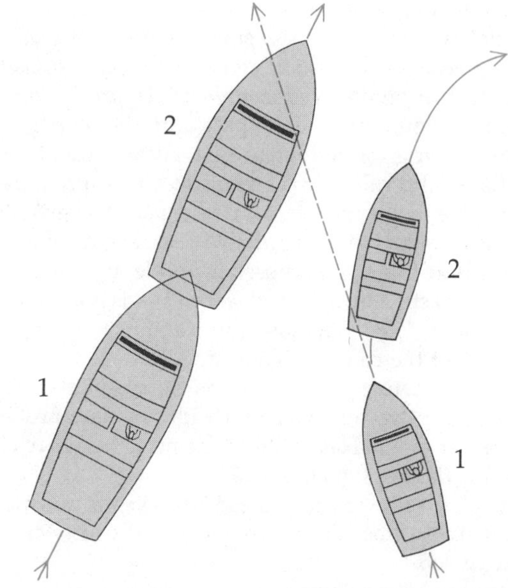

Figure 509 **An overtaking boat remains burdened, and does not become privileged, even though she moves up into the danger zone of the overtaken craft.**

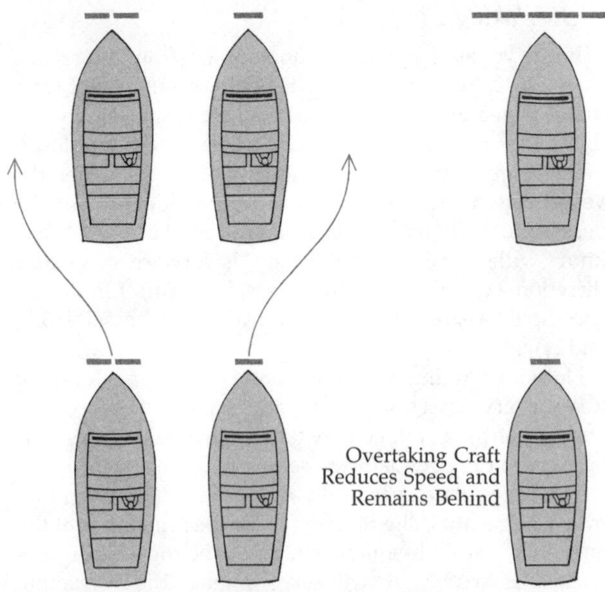

Figure 510 **An overtaking (burdened) boat signals first—with one blast if she wants to pass on the starboard side of the boat ahead, or two blasts if she wants to pass on the port side. The overtaken (privileged) boat replies with the same signal if the way is safe, or with the danger signal if it is not. The overtaking vessel must not attempt to pass until the proper signals are exchanged.**

Should the vessel astern desire to pass on the other's starboard side—not normal, but legally permissible (see Figure 510 center)—she must sound a one-short-blast signal, which is returned in kind by the privileged vessel if she consents. [34(c) and 9(e)]

If in either of the above situations the vessel ahead does not consider that the proposed passing can take place safely, she must immediately sound the danger signal of five or more short, rapid blasts. (The danger signal should also be sounded if the overtaking boat attempts an *unsafe* passing without first giving a whistle signal of her intentions.)

An overtaking vessel receiving a danger signal in reply to her stated intention to pass should immediately cease all actions related to passing, and reduce speed so as to close no further on the overtaken vessel; Figure 510 right. No attempt should be made to pass until the proper signals have been given and answered. Although not provided for in the rules, it is reasonable for a privileged vessel to give the danger signal, pause, and then give the *other* passing signal to the burdened vessel—saying in effect, "The side that you proposed is unsafe, but the other is OK." This action by the privileged vessel is not normal, but it is often logical because the leading vessel has a better view of the conditions ahead.

As in the meeting situation, cross signals—answering one blast with two, or two blasts with one—are prohibited. Answer a signal *only* with the same signal received, or with the danger signal. [I,88.07]

Although the slower vessel ahead is privileged, she must *never* attempt to cross the bow or crowd the course of the passing vessel.

Use of Radio Communication

The 1980 Inland Rules state that vessels which have reached agreement in a meeting, crossing, or overtaking situation by using VHF-FM radiotelephone (Channel 13), as prescribed by the Bridge-to-Bridge Radio-telephone Act, need not sound whistle signals, but may do so. If agreement is not reached, then whistle signals must be exchanged "in a timely manner," and these will govern the actions taken. [34(h)]

Narrow Channels

A vessel in a narrow channel must keep to the right and as close to the outer limit of the channel as is safe and practicable. This rule places a burden on a vessel on the *left* side to (1) be there only if she must, and (2) establish agreement for a starboard-to-starboard passage.

Exception On the Great Lakes, Western Rivers, and any other waters specified by the Secretary, a power-driven vessel operating in a narrow channel or fairway and proceeding *downbound with a following current* has the right-of-way over an upbound vessel and shall initiate maneuvering signals as appropriate. The vessel proceeding upbound against the current must hold as necessary to permit safe passing. [19(a)] The geographic coverage is limited so as to avoid tidal areas where the normal reversing current could cause confusion about which vessel has the right-of-way.

Don't Impede Other Vessels

A vessel less than 20 meters (65.6 ft) in length, or a sailing vessel, must not impede the passage of a vessel that can navigate safely only within a narrow channel or fairway. [9(c)]

A vessel engaged in fishing must not impede the passage of any other vessel navigating within a narrow channel or fairway. [9(c)]

A vessel must not cross a narrow channel or fairway in such a manner as to impede the progress of a vessel that must stay within that channel or fairway. The latter vessel must use the danger signal if there is doubt about the intentions of the crossing vessel. [9(d)]

If the circumstances allow, a vessel must not anchor in a narrow channel. [9(g)]

Rounding Bends in a Channel

A special signal is provided for a situation in which a vessel, power-driven or sail, approaches a bend or channel area where other vessels may be hidden by an obstruction such as the banks, vegetation, or structures. In this situation the vessel sounds one prolonged blast. This signal will be answered with one prolonged blast by any approaching vessel within hearing. [34(e)] If such an answer is received, normal whistle signals must be exchanged when the vessels come within sight of one another. If no reply to her signal is heard, the vessel may consider the channel ahead to be clear. All vessels in such situations must be navigated "with particular alertness and caution." [9(f)]

In each such situation, a vessel with limited maneuverability or one requiring most of the channel width, will often make a "security call" on VHF 13 or 16, or both, to announce its position and direction of travel. Such calls should be answered by any vessels approaching the area.

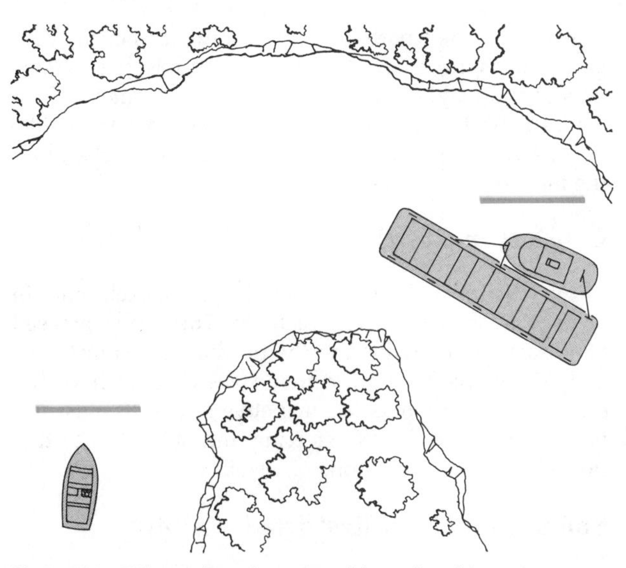

Figure 511 **When a boat is approaching a bend in a river, where a boat approaching from the other direction could not be seen at a distance of a half mile, it should sound a special signal of one prolonged (eight- to ten-second) blast. If no reply is heard the channel ahead should be safe.**

Leaving a Berth

A power-driven vessel leaving a dock or berth must sound a warning signal of one prolonged blast. [34(g)]

Note that, although it is not explicitly so stated in the Rules of the Road, it may be inferred that a boat just emerging from a slip or berth does not have the right-of-way over another vessel passing in the channel or nearby open water, even though she is in the other's danger zone. Her privileged status in a crossing situation is not established until she is "fully in sight."

Ferry Boats

Although there are no express provisions in the Inland Rules of the Road that give special privileges to ferry boats, the courts have repeatedly ruled that ferries are entitled to a reasonable degree of freedom of entrance to and exit from their slips. Boats should avoid passing unnecessarily close to piers, wharves, etc., where they may be caught unawares by the movements of other vessels.

Rules for Sailing Vessels Underway

The provisions of Inland Rule 12, which are stated below, apply only to situations involving two sailing vessels; neither may be under power or a combination of sail and power.

When two sailing vessels are approaching one another so as to involve risk of collision, one of them must keep out of the way of the other in accordance with the following conditions:

1. When each has the wind on a different side, the vessel with the wind on her port side must keep out of the way of the other.
2. When both have the wind on the same side, the vessel to windward must keep out of the way of the vessel that is to leeward.
3. If a vessel with the wind on her port side sees a vessel to windward and cannot determine with certainty whether the other vessel has the wind on her port or starboard side, she must keep out of the way of that vessel.

For the purpose of Rule 12, the windward side is the side opposite that on which the mainsail is carried (or in the case of a square-rigged vessel, the side opposite that on which the largest fore-and-aft sail is carried).

Signals Sound signals are not used in passing situations between two sailing vessels.

Encounters Between Sailing and Power Vessels

In general, a sailing vessel has the right-of-way over a vessel propelled by machinery or by both sail and machinery. The power-driven vessel must keep out of the way of the sailing vessel; passing signals are *not* given. [18(a)] There are, however, some exceptions and they should be thoroughly understood.

Should a sailing vessel overtake a power-driven vessel, the overtaking situation rule prevails and the sailing vessel is the burdened one, regardless of the means of propulsion.

Likewise, a sailing vessel is not the privileged vessel in an encounter with certain vessels engaged in fishing, a vessel not under command, or a vessel restricted in her ability to maneuver. [18(b)]

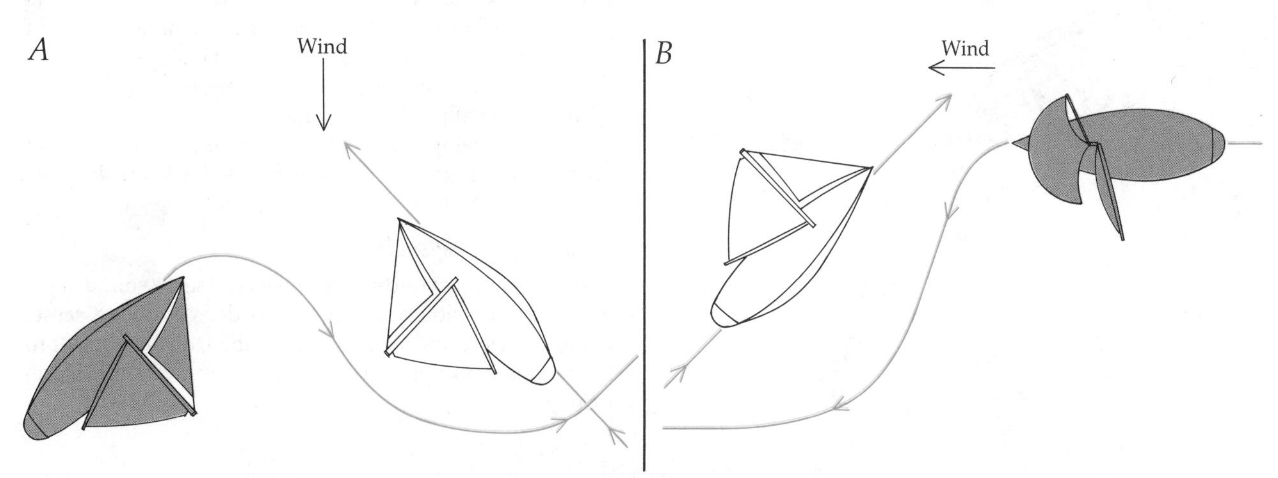

Figure 512 **Right of way between two sailboats is determined by the side of each vessel on which the wind is blowing, or the position of each in regard to the wind (windward or leeward) if both have the wind on the same side. The burdened vessel is shown in blue.**

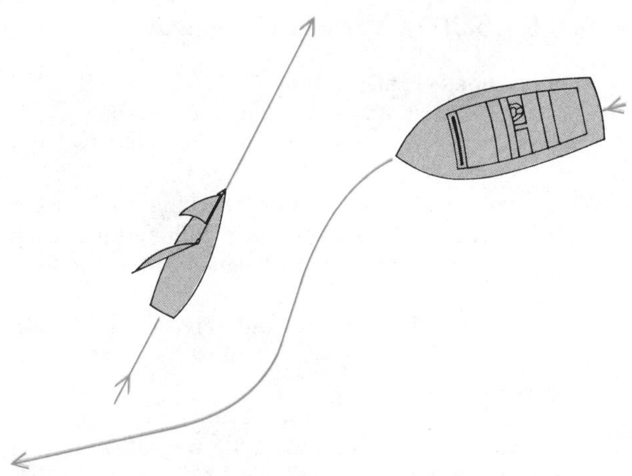

Figure 513 **In an encounter between a sailboat and one driven by power, the powerboat must give way and keep clear of the one under sail. However, if the sailboat is overtaking a powerboat, the sailboat is the burdened vessel.**

General Right-of-Way Provisions

Except for vessels specifically designated otherwise—such as vessels in narrow channels and those overtaking—vessels may be "ranked" for right-of-way as follows:

Vessels not under command
Vessels restricted in ability to maneuver
Vessels engaged in fishing
Sailing vessels
Power-driven vessels

Each type of vessel in this list must yield right-of-way to vessels listed higher, and will be privileged with respect to those lower on the list. [18]

Caution Required

Be aware that not all skippers know the Rules! Be prepared to take whatever steps are necessary to avoid collision, even if technically your boat has the right-of-way.

Vessel Traffic Services

Rule 10 directs, simply, that all vessels are required to participate with all applicable regulations. A VTS, established at selected parts to control traffic in heavily congested areas, normally provides traffic advisories to large vessels and may, at times, control their movements. Recreational boats are not normally subject to VTS regulations, but should be aware of the restrictions placed on the movement of larger vessels.

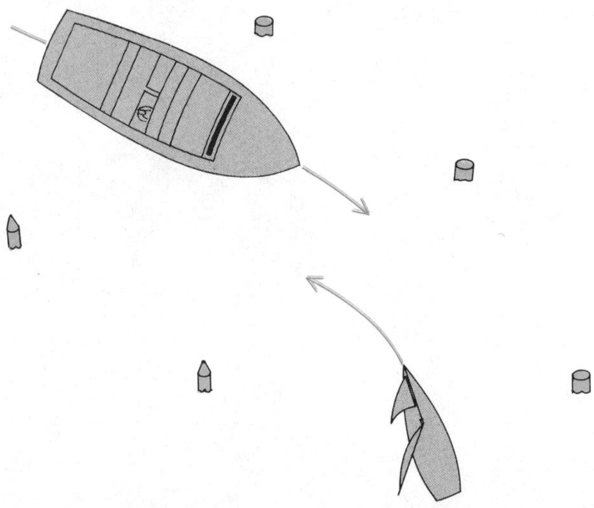

Figure 514 **A sailboat also must keep clear of a vessel that cannot navigate outside the limits of the channel it is using, certain fishing vessels, and those not under command.**

Conduct in Restricted Visibility

Rule 19 applies to vessels *not* in sight of each other in or near areas of restricted visibility. They must proceed at a "safe speed adapted to the prevailing circumstances and conditions"; a power-driven vessel must have her engine ready for immediate maneuvers. Vessels must also have due regard for the steering and sailing rules that prevail under *all* conditions of visibility.

Sound Signals in Restricted Visibility

Rule 35 prescribes sound signals to be given by different types of vessels in various conditions when in *or near* an area of restricted visibility. These are often termed "fog signals," but apply equally in any of the conditions included in the definition of restricted visibility. See page 79. They apply equally day or night. The same equipment is used for these signals as for maneuvering and warning signals, and the signals are the same for power-driven and sailing vessels.

Courts have held that fog signals should be sounded when visibility is reduced to the distance at which sidelights are required. These sound signals serve two purposes: they alert nearby vessels to the presence and rough position of the signaling vessel; and they may indicate her status (underway or not) or limitations of maneuverability (towing, being towed, sailing, fishing, etc.).

Power-Driven Vessels Underway

A power-driven vessel making way through the water must sound one prolonged blast at intervals of *not more than two minutes.* [35(a)] Such a vessel underway but stopped and making no way through the water must sound, at the same intervals, a signal of two prolonged blasts separated by an interval of about two seconds. [35(b)]

A vessel towing or pushing another ahead, a vessel not under command, a vessel restricted in her ability to maneuver (underway or at anchor), and a fishing vessel (underway or anchored) all sound the same signal—one prolonged blast followed by two short blasts. [35(c)]

Note that sailing vessels underway, vessels at anchor, and vessels being towed also may make some "other efficient sound signal" in lieu of the prescribed whistle blasts. [35(a) through (g)]

Sailing Vessels Underway

A sailing vessel underway makes the same sound signal in restricted visibility conditions as does a towing vessel, fishing vessel, etc., as just described above—one prolonged blast on her whistle followed by two short blasts. [35(c)]

Vessels at Anchor

A vessel at anchor *must,* at intervals of *not more than one minute,* ring the bell rapidly for about five seconds. If she is 100 meters (328.1 ft) or more in length, the bell must be sounded in the forepart of the vessel and immediately thereafter the gong must be sounded rapidly in the after part of the vessel. [35(f)]

A vessel at anchor *may* additionally sound a three-blast whistle signal—one short, one prolonged, one short—to give warning of her position, and the possibility of collision with an approaching vessel. [35(f)]

Special Anchorage Areas

Inland Rule 35(j) permits the omission of fog signals for vessels in a "special anchorage area" [see page 243], if they are less than 20 meters (65.6 ft) in length, or are a barge, canal boat, scow, or "other nondescript craft."

Vessels Aground

A vessel aground must sound the bell signal of a vessel at anchor (and the gong signal, if applicable) and additionally give three separate and distinct strokes on the bell immediately *before and after* the rapid ringing of the bell. [35(g)]

A vessel aground *may* also sound an appropriate whistle signal on detecting the approach of another vessel, if there is the possibility of collision. [35(g)] This rule is not specific, but it could be the short-long-short signal, or the letter "U" of the International Code (short-short-long) for "You are standing into danger."

Vessels Being Towed

If manned, a vessel being towed (or the last vessel if several are being towed in a string) must sound a fog signal of four blasts—one prolonged and three short blasts—at intervals of not more than two minutes. When possible, this signal should be sounded immediately after the signal of the towing vessel. [35(e)]

Exception for Small Craft

Boats *less than 12 meters (39.4 ft)* in length need not sound the above fog signals; but if they do not, they must make "some other efficient sound signal" at intervals not exceeding two minutes. [35(h)]

Pilot Vessels

A pilot vessel, when engaged on pilot duties in restricted visibility conditions, *may* sound an "identity signal" of four short blasts. This would be in addition to the normal signals of a vessel underway, underway with no way on, or at anchor. [35(i)]

Not Too Often!

The rules prescribe an interval of "not more than two minutes" (one minute for bell and gong signals) between the soundings of fog signals from a vessel. The timing, if not done by automatic means, should be measured to reasonable accuracy by the second hand on a watch or clock. Two minutes is a relatively brief interval and normally should not be further reduced by skippers concerned about collision. Use the interval between your blasts to listen for other vessel's blasts; it is as valuable for safety as blowing your own whistle or horn. If your signals are sounded automatically, interrupt them from time to time to make sure you are not signaling in synchronization with another vessel.

Basic Responsibilities

Two Rules of the Road are of such primary importance that every skipper should know them by heart. These are "the rule of good seamanship," and "the general prudential rule."

The Rule of Good Seamanship

Rule 2(a) is the broad, summing-up RULE OF GOOD SEAMANSHIP. It provides that nothing in the rules shall exonerate any vessel, or its owner, master, or crew, from the consequences of failing to comply with the rules, or of neglecting any precaution that may be required either by the ordinary practice of seamen or by the special circumstances of the case. Whatever you can do to avoid a collision, you must do!

Lookouts *Every* vessel—must at *all* times maintain a proper lookout by *sight and hearing* as well as by all other available and appropriate means, to keep fully appraised of the situation and any risk of collision. [5]

The need for a proper lookout is often taken rather too lightly on small boats. Under normal circumstances, the helmsman will satisfy this need, but he must be qualified, alert, and have no other responsibilities. The use of an automatic steering mechanism (an "autopilot"; see Chapter 23) is no justification for the absence of a human helmsman at the controls, observing all around the horizon and ready to take over immediately if needed. The use of radar at night or in fog does not justify the absence of an additional person as a lookout, stationed outside the bridge, usually forward, where he can hear as well as

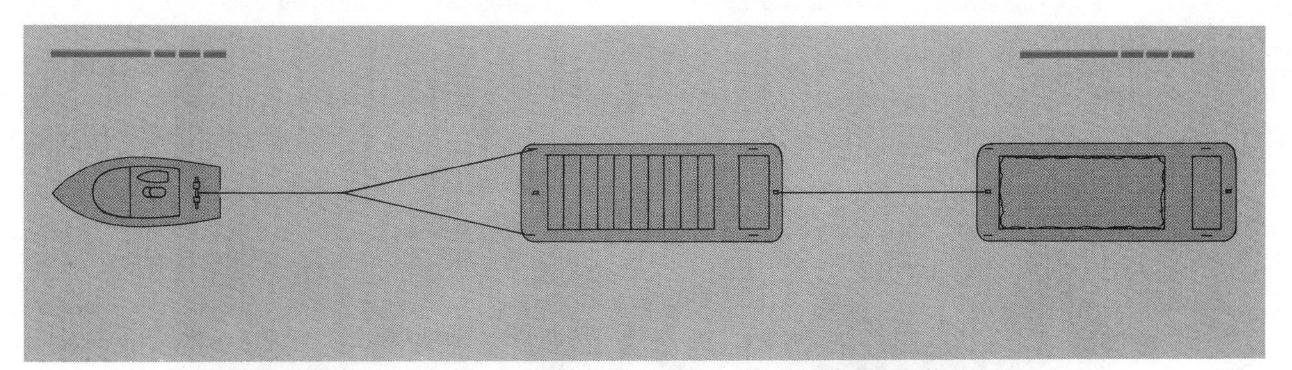

Figure 515 **A manned vessel toward astern (the last vessel if more than one is being towed in tandem) sounds a signal of one prolonged blast followed by two short blasts at intervals of not more than two minutes, immediately after the towing boat sounds its own signal.**

see. The noise level of most motorboats renders the helmsman totally ineffective as a listening watch.

It is not the intent of Rule 5 to require additional personnel forward if none is needed for safety. The burden of proof, however, will continue to be on each vessel involved in a collision to establish that a proper lookout could not have prevented the accident.

The General Prudential Rule

Rule 2(b) is often termed the GENERAL PRUDENTIAL RULE and is of great importance due to of its wide applicability.

In obeying and construing the Rules of the Road, due regard must be given to all dangers of navigation and collision, and to any special circumstances, including the limitations of the vessels, which may render a departure from the rules necessary to avoid immediate danger.

Interpretation of "Immediate Danger" Courts have placed considerable emphasis on the words "immediate danger" of Rule 2(b) in rendering their decisions. The basic principle is that the rules must not be abandoned whenever *perceptible* risk of collision exists, but only when *imperatively required* by special circumstances, as when the actions of one vessel alone will not avoid a collision.

Other Rules and Procedures

There are a number of other, somewhat unrelated, rules and procedures with which a skipper should be familiar.

Signals to Attract Attention

To attract the attention of another vessel, any vessel may make light and/or sound signals that cannot be mistaken for any signal authorized in the rules, or may direct the beam of his searchlight in the direction of the danger in such a way as to not embarrass any vessel. [36]

Use of high-intensity flashing ("strobe") lights for this purpose is illegal. They are an official distress signal.

Signals for Drawbridges

Coast Guard regulations—not a part of the Inland Rules or Pilot Rules—prescribe uniform and unmistakable whistle signals for drawbridge operation. A vessel wishing to open a drawbridge for her passage will sound one prolonged whistle blast followed within three seconds by one short blast; this signal has no other meaning in any set of Rules of the Road. The regulations also provide for making the opening signal by horn, bell, or shout, or by any similar device whose sound can be clearly heard. The vessel's signal should be repeated until acknowledged by the bridgetender before proceeding close to the bridge.

If the bridgetender will open the draw immediately, he responds with a similar one-prolonged, one-short blast signal. If the draw cannot be opened immediately the bridgetender will sound five short blasts. When this signal has been sounded from the bridge, a vessel is specifically prohibited from attempting to pass through the closed draw. When an open drawbridge is to be closed immediately, the bridgetender must first sound the five-short-blast signal. This signal must be acknowledged by the vessel by sounding the same five-short-blast signal; if the vessel does not acknowledge the bridgetender's signal, he must repeat it until that is done.

Some bridges have restricted hours of operation because of the volume of land traffic, typically all day or during morning and evening rush hours. During such periods, certain privileged vessels—government craft, tugs with twos, etc.—can request an opening by sounding five short blasts in lieu of the normal signal.

A vessel approaching a bridge that is already open still must give the opening signal. If no acknowledgment is received within 30 seconds, the vessel may proceed, with caution, through the draw.

Visual signals may also be used, if weather conditions may prevent sound signals from being heard, or if sound-producing devices are not functioning properly. (Sound signals may also be used *with* visual signals.) A vessel signals with a white flag of sufficient size as to be readily visible for a half-mile by day, or a white light bright enough to be seen at the same distance at night. This signal is raised and lowered vertically, in full sight of the bridgetender, until acknowledged. The tender signals that the bridge will open immediately by repeating the same signal. The signal from the bridge that it cannot be opened immediately, or if open must be closed immediately, is a red flag waved horizontally by day, or a red light swung back and forth horizontally at night, in full sight of the vessel. (Some bridges may have mechanical devices and/or flashing electrical lights to signify the same meanings.)

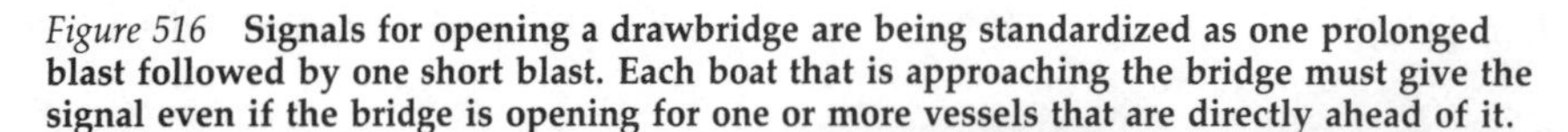

Figure 516 **Signals for opening a drawbridge are being standardized as one prolonged blast followed by one short blast. Each boat that is approaching the bridge must give the signal even if the bridge is opening for one or more vessels that are directly ahead of it.**

Radio communications may be used to request a bridge opening and to receive information about the bridge's status and the actions to be taken. (Bridges equipped with radios will have signs to that effect using either written statements or symbols.) If agreement is reached by radio, sound or visual signals are not required; both the vessel and the bridge must continue to monitor the selected channel—often 16, occasionally 13 or 12—until the vessel has cleared the draw. If radio contact cannot be maintained, sound or visual signals must be used.

A vessel wishing to pass through two or more bridges close together must give the opening signal for the first bridge, and after acknowledgement it must give the opening signal for the second bridge, and so on, until all bridges have acknowledged that they will be opening promptly.

If two or more vessels approach the same drawbridge at nearly the same time, from the same or opposite directions, *each vessel must signal independently* for the opening of the draw. The bridgetender need not reply to signals by vessels accumulated at the bridge for passage during a scheduled open period.

Signals for the operation of canal locks are often the same as those for bridges nearby, but not always. Consult the appropriate volume of the *Coast Pilot* or Navigation Regulations for the proper signals to use before approaching any lock.

Don't cause Unnecessary Openings Clearance gauges are maintained on many drawbridges; skippers should know the vertical clearance required for their craft. Coast Guard regulations provide both criminal and civil penalties for vessel owners and operators who cause unnecessary bridge openings because of "any nonstructural vessel appurtenance which is not essential to navigation or which is easily lowered." On the other hand, the same regulations provide penalties for any bridgetender who "unnecessarily delays the opening of a draw after the required signal has been given."

Pilot Rules for Inland Waters

As noted in Chapter 4, the Pilot Rules are now in Annex V to the 1980 Inland Rules. These require that a copy of the Inland Rules be kept on board every self-propelled vessel 12 meters (39.4 ft) or more in length.

Exemptions

The intent of the 1980 Rules is that any vessel, or class of vessels, the keel of which was laid, or which was at an equivalent stage of construction, on 24 December 1981, and which complied with older requirements, is exempted from the Annex III technical requirements for sound-signaling equipment until 24 December 1990. Vessels constructed since 24 December 1981 must comply with the requirements of the 1980 Inland Rules, including the technical details of Annex III.

Penalties

The penalties for violation of the Inland Rules are described in Chapter 4; see page 87.

International Rules of the Road

Certain of the International Rules relate to navigational lights and shapes; these were discussed in Chapter 4. This chapter will consider only those rules and annexes relating to right-of-way, conduct of vessels in restricted visibility, and sound signals.

Manner of Consideration

Study of the 1980 U.S. Inland Rules provides an excellent base for learning the International Rules of the Road. Hence, the International Rules will be considered in detail here only where they differ from the 1980 U.S. Inland Rules. Topics will be discussed in the same sequence.

Definitions

The International Rules contain one definition that is not used in the U.S. Inland Rules. A VESSEL CONSTRAINED BY HER DRAFT is a power-driven vessel which, because of her draft in relation to the available depths of water, is severely restricted in her ability to deviate from the course that she is following. [3(h)]

Steering and Sailing Rules

The rules for signals are very nearly identical in the two sets, but where they do differ, the difference can be considerable and significant.

Whistle Signals

The one- and two-short blast whistle signals in the Inland Rules under conditions of good visibility are *signals of intent and agreement,* whereas those of the International Rules are indications of *taking action to alter course.* The IntRR signals are often termed RUDDER SIGNALS; they *do not* require a reply, although in some instances the other vessel will take similar action and so sound the same signal. The three- and five-or-more short blast signals have the same meaning in both sets of Rules. The 1980 Inland

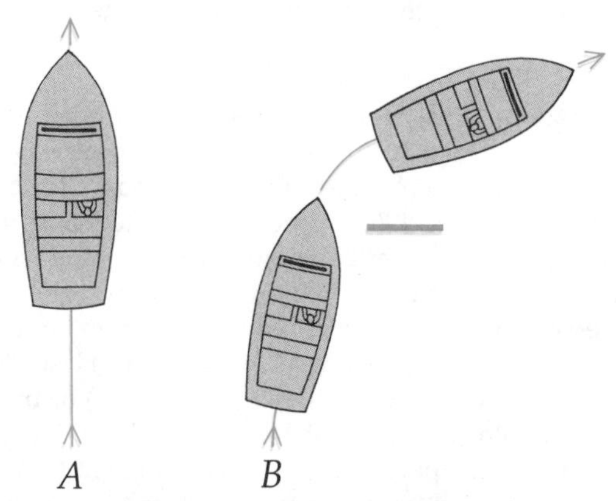

Figure 517 **Whistle signals of the International Rules indicate course changes rather than intent to pass. Such signals are not returned by the other vessel.**

Rules retained the former U.S. signals (rather than adopting those of the 1972 International Rules) because of their long established use, plus the belief that they were simpler, clearer, and more suitable for confined inland waters.

Equipment for Sound Signals

Annex III of the International Rules presently differs from that of the Inland Rules, but proposed International Rule changes may in several years eliminate these technical differences.

The International Rules allow somewhat higher-pitched whistles on vessels between 12 and 75 meters (39.4 and 246.1 ft) in length; the Inland Rules bar the use of some, but not all, small hand-held air- or freon-operated horns on craft 12 meters (39.4 ft) or more in length, because of their high-frequency sound; there are no frequency limits for shorter boats.

The International Rules *do not* contain the provision allowing a vessel normally used for towing alongside or pushing ahead to use a whistle matched to the length of the customary combination of towed and towing vessels (see page 101).

Sound-signaling equipment that is acceptable to the International Rules can also be used on inland waters, as provided for in Inland Rule 2(b) (ii).

Maneuvering Lights

A maneuvering light used in conjunction with whistle signals must be white; yellow is not acceptable as it is under the Inland Rules. The light must have a minimum range of 5 miles, and it need not be synchronized with the vessel's whistle although the same number of flashes are made as blasts are sounded. Each flash shall be of about one second duration, and the interval between successive signals must be at least 10 seconds. [34(b)]

Rules for Power-Driven Vessels

The rules for meeting and crossing situations are the same in both sets, except for whistle signals. The International Rules *do not have* the one-half-mile closest-point-of-approach requirement for signaling intent to pass, or the reply signals of agreement. They *do* require a whistle signal for each change of course, without reply signals from the other vessel. On winding rivers where the vessel's course follows the waterway, however, no whistle signals are used for the many necessary changes of course.

Overtaking Vessels

International Rule 9(e) for vessels in an overtaking situation applies *only* in a narrow channel or fairway and *only* if the overtaken vessel has to take some action to permit safe passing [9(e)(i)] (not in all waters and in all overtaking situations, as in the Inland Rules). The whistle signals are very different, and quite complex. [34(c)]

A vessel overtaking another will signal her intention by two prolonged blasts, followed by one short blast ("I intend to overtake you on your starboard side") or by two prolonged followed by two short blasts ("I intend to overtake you on your port side"). The vessel about to be overtaken signals her agreement not with the identical signal, as Inland Rules specify, but with one prolonged, one short, one prolonged, and one short blast—this is the letter "C" in Morse code and the international signal for "yes." In case of doubt, the vessel about to be overtaken will sound the five-or-more short and rapid blasts.

Use of Radio Communication

The International Rules *do not* contain provisions for the use of radiotelephone communications in place of whistle signals. This does not mean, however, that such additional procedures are not an excellent means of ensuring a safe passing.

Narrow Channels

International Rule 9(d) *differs* from its Inland Rules counterpart in that the sounding of a danger signal by a vessel that observes another crossing a narrow channel or fairway is optional, rather than mandatory.

Leaving a Berth

The International Rules *do not* include a signal for a power-driven vessel leaving a berth, as does Inland Rule 34(g). Such a signal, however, is not prohibited and would be a good idea, even on those U.S. coastal waters subject to the IntRR.

General Right-of-Way Provisions

The International Rules contain the *same* general right-of-way "ranking" of vessel types as the Inland Rules do (see page 100), *plus* a provision that any vessel except one not under command or restricted in her ability to maneuver must avoid impeding the safe passage of a vessel constrained by her draft, a category not included in the Inland Rules. [18(d)]

Traffic Separation Schemes

International Rule 10 contains *many details* about the actions of vessels operating in a Traffic Separation Scheme (TSS). This rule applies only to internationally recognized TSSs. All TSSs adjacent to the United States are plotted on charts. The use of a TSS is optional, but if a vessel uses it the provisions of Rule 10 must be adhered to. (Inland Rule 10 relates to Vessel Traffic Services, which are not the same as TSSs.)

Sound Signals in Restricted Visibility

The International Rules *do* prescribe a fog signal for a vessel constrained by her draft; this is the same "one prolonged blast followed by two short blasts," as prescribed for a sailing vessel, a towing vessel, a vessel engaged in fishing, and a vessel restricted in her ability to maneuver. [35(c)]

The IntRR *do not* provide for "special anchorage areas," where fog signals need not be sounded by small boats, and by barges, scows, and the like.

Exemptions

International Rule 38(g) exempts all vessels built before 15 July 1977 from the technical requirements in Annex III for sound-signaling equipment until 15 July 1986; newer ships and boats must meet these requirements.

Penalties

The International Rules of the Road contain no provisions for penalties for violation of the rules or the annexes. For all vessels in U.S. waters, and all U.S. vessels in any waters, the implementing action by Congress—the International Navigational Rules Act of 1977—does provide stiff penalties. These are the same as for violation of the 1980 U.S. Inland Rules; details on penalties are given on page 87.

Distress Signals

Because of their great importance, the accepted forms of DISTRESS SIGNALS have been written into both the International and U.S. Inland Rules of the Road. These are contained in Annex IV, which is identical in both sets of rules, except that the Inland Rules additionally list a white high-intensity "strobe" light flashing 50 to 70 times per minute.

Figure 518 lists the *officially* recognized signals, but in an emergency a skipper can, and should, use any means he can to summon help.

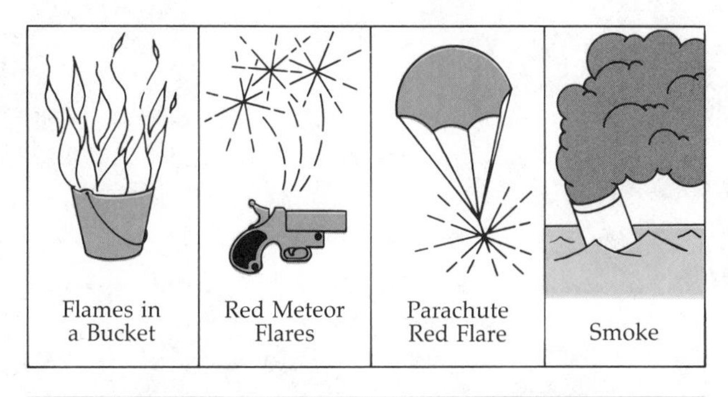

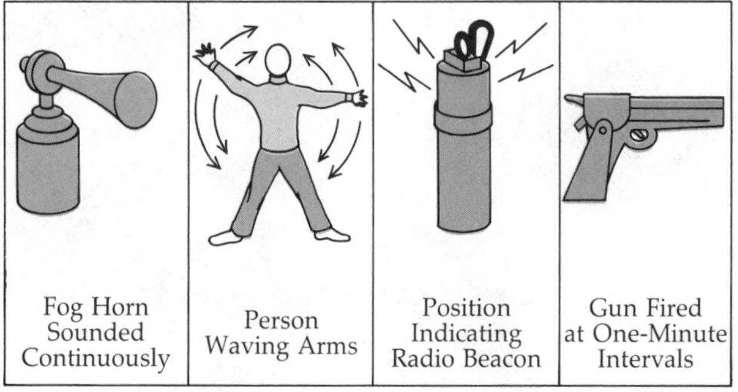

Figure 518 **Distress signals are contained in both the Inland and International Rules. These are the ones that are officially recognized, but in an emergency a skipper can use any means possible to summon help. In the U.S. this includes flying the national ensign upside down.**

Supplemental Signals

In the category of "supplemental signals," Annex IV lists a piece of orange-colored canvas, with either a black square and circle or other appropriate symbol, for identification from the air. (Actually, the circle should be solid, more properly termed a "disc"; and although "other appropriate symbol" is not defined, the word HELP in large black, block letters is often used.) Another supplemental signal is a dye marker, which can greatly enhance the chance of a person or other floating object being sighted in rough waters.

Unofficial Distress Symbols

An unofficial but well-recognized signal, especially in inland and coastal waters, is a red-orange flag of any size, waved from side-to-side.

Most American boatmen would recognize the flying of an inverted U.S. national or yacht ensign as a signal of distress. Remember that such a signal has no official sanction, however, because many national flags have no "top" or "bottom" and appear the same if turned upside down.

Use of radiotelephones for distress communications is covered in Chapter 22.

Know the Rules

It cannot be emphasized too strongly that: (1) the skipper of any boat must know the Rules of the Road relating to his boat and waters; and (2) a tight situation is too late for learning the rules! Study the rules thoroughly, and review them periodically. Making the right maneuver at the right time should be automatic.

6

Choosing the right boat and motor for your activities; how to handle outboard, stern-drive, and jet-drive craft; boat trailering

SMALL BOAT SEAMANSHIP

The pleasures of recreational boating are not related in any way to the size of one's boat, although size has a very real bearing on the type of boating activities that can be accomplished. While there are some elements of seamanship that apply to every type of boat, there are significant differences between the requirements for small, open outboard or stern-drive–powered vessels, and larger inboard-powered cruisers.

This chapter focuses on small craft to 26 feet (7.9 m) long, covering selection, equipment, and use of such boats. This is Coast Guard categories of Classes A and 1—about 96% of all registered boats in United States.

The Boat

Outboard boats, those vessels suitable for a detachable motor, come in wide ranges of size, design, construction material, and cost. At the lower end are the 8-foot (2.4-m) or smaller prams that can be car-topped or used as tender; at the upper end of the scale are the center console and cruiser hulls that can take a pair of motors of up to 200 or more horsepower each.

The Use of the Boat

The key to selecting the "right" boat for a particular skipper is the use or uses for which it is intended. Obviously the boat for water skiing is not the boat for trolling, and the boat for an afternoon outing may not be the one for a week's cruise. Probably no boating family will agree on just one or even two uses, and few families can afford a separate craft for each type of activity desired.

Figure 601 **An outboard motor clamps onto the transom; large, heavy ones are usually bolted in place; smaller motors, which are portable, may be removed when not in use.**

Compromise is inevitable, but if all possible factors are considered in advance the likelihood of disappointment is reduced. Be sure to take into consideration the type of waters on which the boat will be used—protected lakes and rivers, or coastal bays and offshore.

Hull Designs

There are two basic hull types, displacement and planing—and many variations of the latter. Displacement boats cruise *through* the water; planing hulls lift and skim *over* the surface.

Often it is difficult to make a sharp distinction between "displacement" and "planing" types of hulls. Planing hulls receive a large part of their support at normal speeds from the dynamic reaction of water against the bottom, and a lesser part from buoyancy which diminishes with increased speed but never quite disappears at any speed. Generally, planing begins when the water breaks cleanly away at the chines and transom. In these days of high horsepower, many cruisers have some planing action, and practically all runabouts are of the planing type.

Here are typical hull forms for small boats designed to take outboard or stern-drive power:

Flat Bottom (displacement type) Usually rowboats or skiffs 14 feet to 18 feet (4.3-5.5 m); used for fishing or utility purposes on shallow streams and small protected lakes. Generally heavy and roomy for their length, and slow.

Flat Bottom (planing type) Same size range as above, and used in the same waters, but their light weight allows them to step up onto plane with modest power and run at higher speeds than their displacement counterparts.

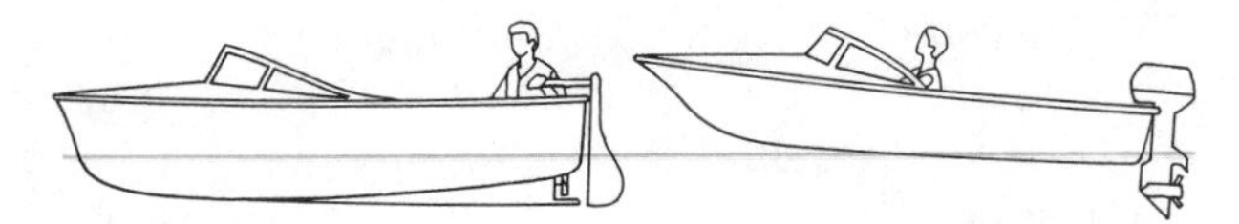

Figure 602 **Displacement hull, *left,* rides down in the water and has a very low top speed. Planing hull, *right,* derives lift from its forward motion, allowing it to ride more on top of the water, reducing resistance, and permitting higher speeds.**

Round Bottom (displacement) Dinghies, tenders, car-top boats, occasionally runabouts 12 feet to 18 feet (3.7-5.5 m). At slow speeds these hulls are often more easily driven and maneuvered than the flat-bottom craft. (Many light, round-bottom boats will also plane.)

Vee-Bottom Commonly used for runabouts, utilities, and cruisers when speed is a factor. Forward undersection is usually deep "V" in shape, but the "V" flattens out considerably by the time it reaches the transom. Longitudinal strakes help to provide lift and lateral stability,

Cathedral This term covers a number of wide-bodied hulls whose lateral stability is enhanced by having two or three hull shapes underwater, although they are not catamarans or trimarans.

Hydroplanes Generally used for racing. The bottom, which is flat, may be "stepped," i.e., divided into two levels, about amidships. The resultant notch reduces wetted surface, increasing speed.

Size and Loading

Because overloading a small boat can be exceedingly dangerous, *the safe limits for a particular boat must be known.* Coast Guard rules require that boats under 20 feet (6.1 m) (except sailboats and some special types) manufactured after 31 October 1972 must carry a "capacity plate" showing maximum allowable loads. Since August 1980, plates for outboard boats have shown a maximum horsepower for motor(s), maximum number of persons, and maximum weights both for persons only, and for motor, gear, *and persons;* see Figure 603. Plates for inboard, stern-drive and unpowered vessels omit the maximum power rating.

Boats manufactured before this regulation may carry no capacity plate, or one based on a formula that is no longer used. In either case, you can determine the maximum number of persons that can safely be carried by the following equation:

$$\frac{L \times B}{15} = \text{Number of Persons}$$

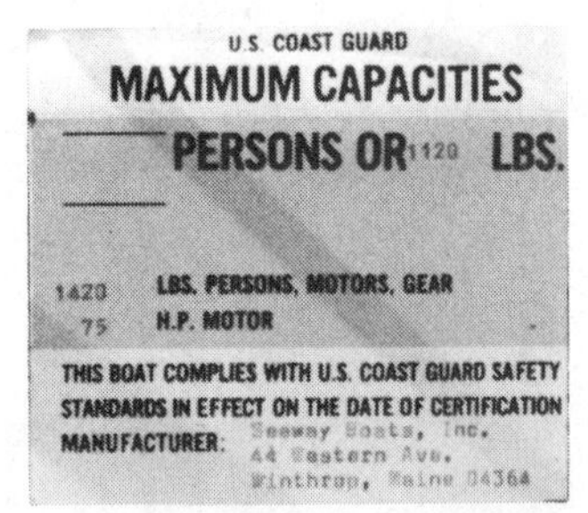

Figure 603 **Coast Guard capacity plate shows maximum allowable boat loads, and power for outboard motor(s).**

$$\left.\begin{array}{l} L = \text{Overall Length} \\ B = \text{Maximum Width} \end{array}\right\} \text{in Feet and Tenths of Feet}$$

Take the results to the nearest whole number. This is the number of persons that can be carried without crowding, *in good weather conditions.*

Also check the weight-carrying capacity of the boat to be certain it is adequate for the number of persons as determined above, taking into account their actual weight as well as the weight of engine, fuel, and equipment. The weight-carrying capacity of a small boat having a conventional hull can be determined by the equation:

$$\begin{aligned} 7.5 \times L \times B \times De &= \text{Allowable Weight, in Pounds} \\ L &= \text{Overall Length} \\ B &= \text{Maximum Width} \\ De &= \textit{Minimum} \text{ Effective Depth of the Boat} \end{aligned}$$

Measure De at the *lowest point* that water can enter. This takes into account low transom cut-outs or credits an acceptable engine well. Dimensions are in feet and tenths.

Add up the weight of your outboard motor, battery, fuel tank and fuel (gas weighs 6.6 pounds per gallon), and the equipment you carry, such as anchor, oars, radio, etc. Subtract this from the allowable weight found above. The remainder is the weight capacity available for persons.

In any particular loading situation, use the value from the number-of-persons formula, or the allowable weight formula, which sets the *lower* limit.

Special Design Features

Many boats will have special features for particular applications such as sport fishing. These invariably add to the cost of the boat, and are worth it only if the boat is to be used for this purpose.

A good safety feature is a self-bailing MOTOR WELL. Most craft have a transom that is "cut down" or lowered where the motor is to be attached; this is necessary so that the motor will be low enough that the propeller will be well below the bottom of the hull. Such a mounting provides efficiency for the motor, but may seriously jeopardize the safety of the craft by providing an easier point of entry into the hull for water. Safety can be maintained by an inner bulkhead forward of the motor that is *not* cut down, that is fully as high as the sides of the boat. The space aft of this bulkhead to the cut-down transom is the motor well, and self-bailing drains should be provided at each after corner.

Added Buoyancy

Boats less than 20 feet (6.1 m) in length manufactured since 31 July 1973 carry built-in flotation installed in accordance with Coast Guard regulations (except for sailboats and some special types). It is possible to add positive flotation to older boats to ensure the safety of your passengers. This can be in the form of sealed air chambers, or masses of plastic foam.

The buoyancy units should be located as *high* as possible in the hull so that the boat, if swamped, will remain in an upright position and not capsize. This provides much greater safety than a capsized hull; persons in the water will be able to hold on more easily and they may be able to recover some form of emergency signalling, bailing, or other needed equipment.

The Motor and Accessories

An outboard motor is a *detachable* power plant of one or more cylinders that is complete with drive shaft and propeller. The fuel tank and operating controls are usually separate, but on the smallest motors these may be mounted on the powerhead itself. It is a gasoline-fueled motor usually of two-cycle design (except for some low-power electric trolling motors), and it is normally watercooled. Starting may be either manual or electric.

Usually, the motor is mounted on a cut-out in the transom; a recent development (OMC Sea-Drive) is a motor installed on a bracket bolted behind a full transom. This eliminates the motor well found on most outboard hulls, and provides more usable space in the hull itself.

Stern-drives

Stern-drive power packages have a standard four-cycle gasoline inboard engine, or a diesel engine, mounted inside the hull, bolted through the transom to a drive unit that closely resembles the lower section of an outboard motor. Stern-drive engines are sometimes called "inboard-outboards," "outdrives," or "I-Os." They seek to combine the greater power and efficiency of an inboard engine with the directed-thrust steering, tilt-up capability, and other advantages of outboard propulsion.

Boats of stern-drive design can be included with outboards, except for matters directly relating to the engine itself. They are usually in the size category of medium and larger outboard boats, handle similarly, are used for the same general purposes, and can be trailered.

Figure 604 **A variation of the outboard is the OMC Sea-Drive, *above*, a motor designed for mounting on a bracket behind a full transom. This type of installation, available from the boat manufacturer or dealer, is used where an uncluttered cockpit is needed or desired. The engines are fairly large, and are rated by their piston displacement rather than by horsepower. Stern-drive boats, *below*, have the engine inside the hull; reverse gears are in the drive unit mounted outside the transom.**

Outboard Motor Selection

While horsepower always seems to be the first consideration in motor selection, factors such as weight, starting method, and even price may be even more important. With the wide range of motors available, however, it is not difficult to select a model suitable for almost any application.

Power Requirements

The horsepower required for a boat will depend upon size and weight of the boat (loaded), and the desired speed. A displacement hull of 14 feet (4.3 m) or so will serve adequately for lake and river fishing with a 10 horsepower motor; any greater horsepower would be wasted in an attempt to drive the boat faster than hull speed.

For water skiing behind a planing hull of 16 to 18 feet (4.9- 5.5 m), motors of 40 to 75 horsepower are suitable. A larger engine will make the boat go faster, provided the hull has the capacity for greater power, but speed does not increase in direct proportion to engine power. Very roughly, horsepower must be tripled or even quadrupled to double speed.

Outboard boats designed for cruising have planing-type hulls. These need enough power to get up on plane, a running attitude that is both more pleasant and more economical than pushing through the water in the displacement mode. While it is not necessary to provide a motor of the maximum permissible power, enough power is needed to get the boat onto plane when fully loaded, and to allow it to operate at a suitable cruising speed.

Figure 605 **Water skiing takes speed, requiring the use of a motor of 40 horsepower or more on a planing hull.**

No specific recommendation can be made—each situation is a case unto itself—but it's a good idea to make a trial run of any boat and engine combination, if possible, prior to closing a purchase deal. These trials should be made *with load,* either actual or simulated with weights, and figuring a bit of excess for later additions. If possible, try the rig in various conditions of wind and water that are typical for your boating area.

Electric Starting

All small outboard motors, and some medium-power models, are started by pulling a rope that is wound around the top of the flywheel. This is not practical for the larger motors, so electric starting systems are provided. The electrical system, with its battery and alternator, has the added bonus of making possible the use of electronic gear on the boat.

Electric starting does require equipment—and weight and cost—the starting motor, battery, control wiring, etc., but this is offset by the greater convenience that it provides. In general, motors less than 10 horsepower will be manually started; those more than 40 horsepower will have electric starting; motors between these sizes may have either method of starting.

Single or Twin Installations

Many outboard hulls are designed to allow the fitting of either one or two motors on the transom. There are definite advantages and disadvantages to each arrangement, given the same total horsepower.

The reason most often advanced for twin motors rather than a single one is the added safety the system provides. Properly maintained outboard motors are extremely reliable, but if one should fail, the other will bring the boat home.

Figure 606 **Twin outboard motors give a great degree of security from loss of power, but at a higher initial investment, more weight, and higher fuel consumption.**

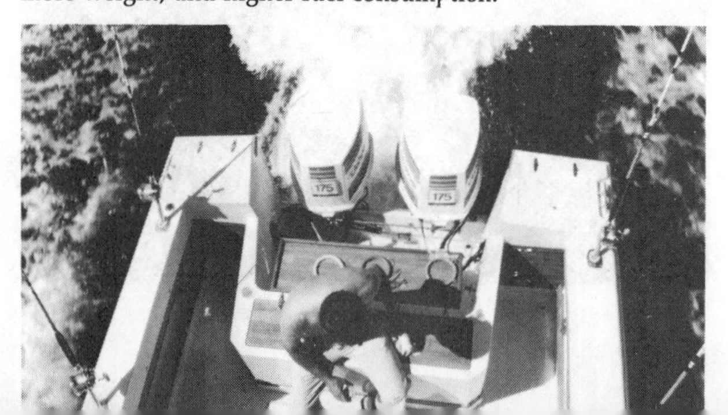

Disadvantages of the twin rig include the greater initial cost, which is about 1-⅓ times the price of a single large motor, an additional battery (or larger single one), more complex control systems, greater weight in the boat (about 50% more), greater underwater drag, and greater fuel consumption (not doubled, but greater by ⅓ to ½). Twin motors, and their batteries and fuel tanks, take up greater space within the boat than that needed for a single-motor installation.

A special case is the large outboard cruiser where use of two motors is required to meet horsepower needs, and the weight and space requirements are less important. In such an installation, twin motors will also allow the use of more efficient propellers with larger blade area.

If a boat is performing to certain standards with a single motor, what will result if a second motor of the same power is added (assuming the boat is rated to handle the double horsepower)? The added weight and drag, combined with hydrodynamic factors, will hold the speed increase to about 25%, although this will vary widely with specific installations. Fuel consumption, with both engines running at the same RPM, will be about 1½ times that of the single motor rig.

A combination to be considered is a single large motor adequate for all normal operation, plus a smaller motor of 4 to 10 horsepower. The smaller motor is used for trolling while fishing—large motors should not be run at slow speeds for extended periods—and for emergency backup. A 6 horsepower motor will move a medium-size outboard hull at 3-4 knots and get it home or to assistance.

Figure 607 **Because large motors are not designed to run at slow speeds for long periods, some boats carry a second, small motor that will push the boat at trolling speeds, and even bring it home if the large motor should fail.**

Propellers

Most outboard motors are sold complete with a propeller; this is a "stock" propeller, suitable for an average boat under average conditions. (Some motors, especially the larger ones, are offered with a choice of several propellers.) The stock propeller may not have the optimum diameter or pitch, or both, for a particular application.

Manufacturers publish tables of recommended propeller sizes for various applications, but these should be used only as initial guides. The answer lies in experimentation, unless you know another owner who has an identical boat and motor combination, whose style of boating matches yours, and is getting the best performance from his rig. In any case, the "right" propeller for any boat in a specific application is the one which allows the motor to turn up to its full rated RPM, but no more. It is necessary for the motor to turn to full rated rpm to develop full rated power; but if it will turn faster than that, the propeller is too small and full power is not being developed at the rated maximum RPM (which must never by exceeded, except for brief bursts).

If the boat is used for more than one type of activity, such as cruising, fishing, or water skiing, it may be that the same motor will require different propellers for the most efficient operation in each usage. As a spare propeller is an excellent safety item, the purchase of a more efficient one is not all "added expense"—the stock propeller becomes the spare.

Materials

Most stock propellers are aluminum, as are those normally purchased for optimum performance. Propellers of bronze are also available; these are stronger and more easily repaired, but are more expensive.

Propellers of stainless steel and other high-strength alloys are advantageous for special applications, such as racing, but are even more costly than bronze. Plastic is used for the propellers of some of the smaller motors, and strong, lightweight propellers of Teflon-coated steel are also available.

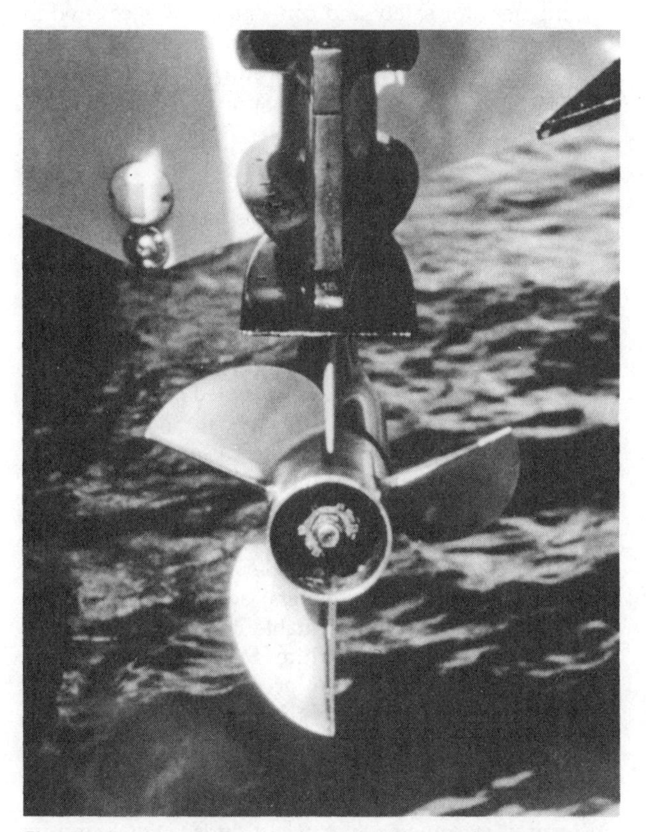

Figure 608 **One advantage of outboard boating is the ease with which a propeller can be changed in the case of damage, or when a change in boat use dictates a change in size.**

Number of Blades

The stock propeller for most outboards is three-bladed. For light loads and for higher speeds, a two-blade propeller may do a better job. For lower RPM use on heavy boats, it may be best to install a four-blade propeller. The number of blades is, of course, a factor closely related to diameter and pitch.

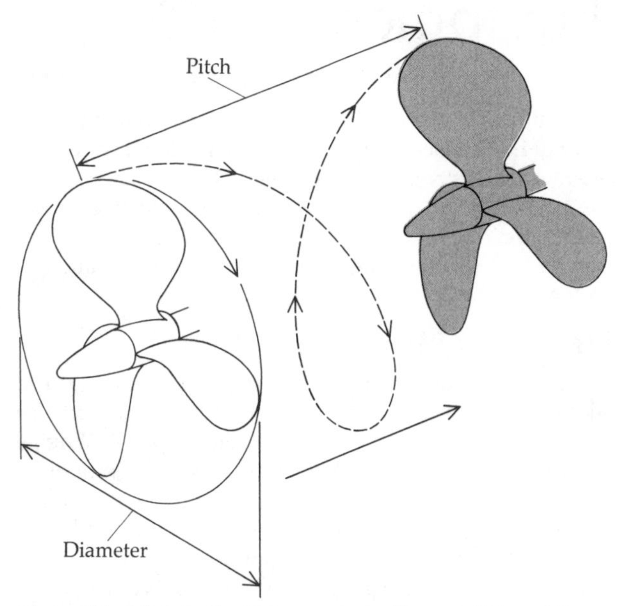

Figure 609 **Pitch is the distance a propeller would travel in one revolution if there were no slip. Diameter is twice the distance from the center of the hub to the blade tips.**

Pitch

The pitch of a propeller determines its "bite" on the water, and thus the RPM which the motor can turn up (this is also affected by propeller diameter). Motor RPM is of course related to boat speed—a lighter, faster boat will use a propeller of greater pitch than one that's heavy and designed for slower operation. Again, experimentation is really the only way of selecting the best match of propeller to boat.

If a single motor is "doubled up" with another similar to it, this will call for a change of propeller on the original motor. The faster speed obtained with the added horsepower will call for a greater pitch, probably one, but possibly two, inches more; again, only actual trials will tell for sure.

Shear Pins and Slip Clutches

Because outboard boats often operate in fairly shallow water, the drive shaft, gears, and other internal parts of the motor are subject to damage should the propeller hit an underwater object. To prevent such damage, the motors are equipped with either a SHEAR PIN (smaller motors) or a SLIP CLUTCH (larger motors) on the propeller shaft.

A shear pin is made of a relatively soft metal. It transmits the drive from the propeller shaft to the propeller, and is just strong enough for this. Upon impact with a rock or other hard object by a propeller blade, this pin is broken, sheared off near each end, and the impact is not transferred to the inner parts of the motor. To restore operation, it is necessary to remove the propeller and install a new pin, so always carry spares!

To overcome the nuisance of having to replace sheared pins, manufacturers developed slip clutches in which a rubber inner hub is used to transmit the drive power. Under normal loads there is no slippage, but upon impact with a hard object (or even with the sudden load of a too-quick shift of gears or advance of the throttle), the rubber hub slips somewhat to absorb the strain from the propeller blade. Obviously there is no internal component to be replaced after impact, and it provides adequate protection for the motors.

Equipment

The outboard boat must meet all legal requirements for a craft of its size (see Chapter 3), plus any additional equipment required by state or local regulations. Boatmen venturing into new waters subject to other regulations than "at home" should check with a marina operator, park ranger, or other local authority to avoid inadvertant violations.

Navigation Lights

Even if you do not intend to use your boat at night, it should be equipped with navigation lights in accordance with Inland or International Rules (see Chapter 4). Engine trouble, fuel problems, unexpectedly good fishing, or other situations may keep you out longer than you intended.

Nearly all outboard boats are delivered with a red-green combination light forward and an all-round white light aft. Such lights are legal, if they are properly used and *meet the requirements of the regulations.* All too often the after light is not high enough to meet the requirement for *all-round* visibility, especially when the boat is underway and the bow rises, or when the canvas top is raised. Some lights are on a telescoping stem so that they may be lowered when not in use during the daylight for convenience, and to protect the mounting from damage. Even when raised, these are usually too low to be seen from ahead. Be sure your boat can be seen from any direction at night; the white light is even more important than the colored lights as it will be seen at a greater distance. Have someone ashore, or on another boat, check as you run past at various angles, and then let your boat go dead in the water, with the stern light acting as an anchor light, to make sure it is not obstructed in any way.

Figure 610 **A stern light may be mounted on a telescoping staff so it can be kept low and out of the way during the day. When operating at night, be sure to extend the staff to full height so the light will be visible from all around the boat. See Chapter 4 for all details regarding running lights.**

Power-driven vessels not longer than 39.4 feet (12 m) have the *option* under both Inland and International Rules of the Road of carrying a single all-round white light, visible two miles, in lieu of separate forward and stern lights. This light *may* be placed off the fore-and-aft centerline if centerline mounting is not possible, *provided* that the sidelights are combined into a single unit which is on the boat's centerline or is located as nearly as possible under the same fore-and-aft line as the all-round white light.

Under International Rules only, a power-driven vessel that is less than 23 feet (7 m) in length and whose maximum speed does not exceed 7 knots—typically a lightly powered dinghy—may show only an all-round white light that is visible for two miles. Sidelights are not mandatory, but "shall, if practicable," be shown.

Equipment for Safety

Equipment for safety and convenience was covered in Chapter 3; some items will be covered again here, in their outboard boat applications, with other, more specialized items specifically related to outboards.

Lifesaving Devices

Outboards are more likely to capsize than are larger inboard boats; consequently there is a greater possibility that the occupants will find themselves in the water. Do not skimp on personal flotation devices; although buoyant cushions may meet the legal requirements, the more expensive life preservers and buoyant vests provide much greater protection. They are worn, rather than grasped, and so will keep an injured or exhausted person afloat. It should be an absolute "rule of the boat" that adult non-swimmers and all children wear life preservers or buoyant vests at *all times* when the boat is underway. At the higher speeds of most outboards, there may not be time to locate and put on a life jacket when a sudden emergency occurs.

"Special purpose life saving devices" *approved by the Coast Guard,* such as some hunters' jackets and some water skiing *vests,* are legally acceptable items for small boats not carrying passengers for hire. Ski *belts,* however, are not so approved and are not acceptable in meeting the legal requirements although their use is recommended while water skiing.

Paddle or Oars

Required in some state and local jurisdictions, and by the Coast Guard Auxiliary for their courtesy examination decal, a paddle or pair of oars should be on all outboard

Figure 611 **No matter how dependable a motor may seem, a paddle or pair of oars should be carried on small boats for emergencies.**

boats. They could be the only way of reaching safety in the event of motor failure. Most outboard craft row or paddle quite clumsily and with considerable effort, but it can be done, and the means to do so should be on board.

Anchor and Line

Every outboard should be equipped with a suitable anchor and line long enough to anchor in all areas where the boat is used. Obviously this requirement must be applied with reason when the boat is used off very deep coasts, but when purchasing the line it is better to get too much than too little. See Chapter 11 for detailed information on anchoring.

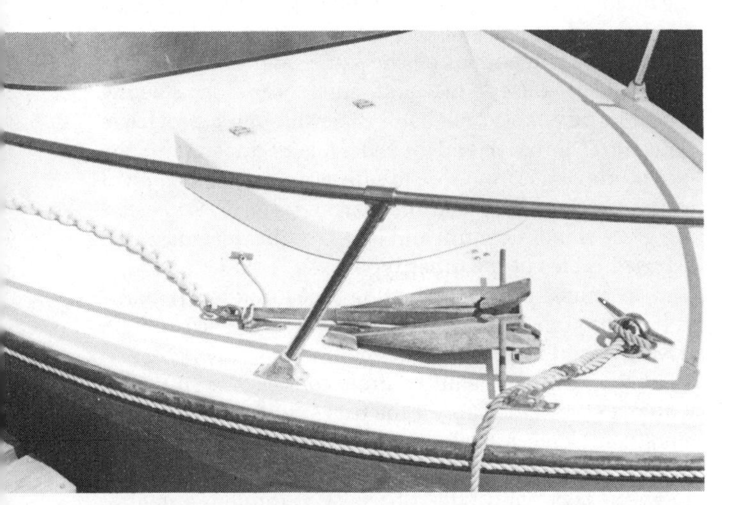

Figure 612 **Although not required by federal regulations, an anchor and its line should be carried and ready for use in an emergency. For greater security, a short length of chain can be connected between the anchor and the line.**

Bailer

All small boats should be equipped with a manual bailer. This can be a scoop purchased, or home-made from a household plastic jug. Transom drains are effective underway for many fast boats, but must have a means of positive closure when not in use. Manual or even electric pumps may be found on larger outboards. A large sponge is frequently convenient for getting that last little bit of water out of a small boat.

Visual Distress Signals

All boats 16 feet (4.8 m) or more in length operating in coastal waters or wider connecting bodies *must* carry visual distress signals; smaller craft need carry them only when operating at night. The signals must be Coast Guard approved, with units on board classified for day use and for night use, or a combination of both. See pages 68-69 for additional details.

Flashlight or Lantern

Every outboard craft should be equipped with a flashlight or electric lantern, whether or not it is intended to ever use the boat after dark. The light should be waterproof, and it should float if accidentally dropped overboard. Extra batteries, stored in a waterproof container, will often prove valuable in an emergency. Batteries in the flashlight or lantern, and the spares, should be renewed at the start of each boating season, regardless of their apparent condition.

Operational Equipment

No outboard boat should be without a magnetic compass. Select one of adequate size and quality, and install it properly. See Chapter 15 for additional information.

As most outboards are relatively open, it is particularly important that the compass be kept shielded from the sun's direct rays. When not in use, remove and store it out of the light or place a light-tight cover over its top.

Charts

Carry charts for the waters you use. Learn to use a chart with your compass, and do use it under favorable conditions even where it is not strictly necessary. This practice is excellent training, and will prepare you for using the chart with confidence and efficiency should an emergency arise. Make a point of keeping track of where you are at all times.

Tools

Every boat, no matter how small, should carry at least a few simple hand tools. The minimum should be a screwdriver, pair of pliers, and an adjustable open-end wrench. Consideration might be given to having two or more screwdrivers, and wrenches of different sizes. A spark plug wrench suitable for your motor is recommended.

Safety Chain

Most outboard motors are clamped to the craft's transom (some of the larger ones may be through-bolted). To protect against accidental loss overboard should the clamps loosen with vibration, a short safety chain (or cable) is often used between a connecting point on the motor and a ring securely fastened to the hull. Chains or lengths of steel cable, frequently plastic-coated to reduce rusting, may be inexpensively purchased for this purpose already equipped with snap fittings at each end. This is "cheap insurance" against the total loss of an expensive motor in deep water. With locking fittings, they also provide some security against theft.

Extra Fuel Tanks

The very smallest outboard motors have an integral fuel tank, but those of four or more horsepower now usually operate from remote tanks which are more convenient with their larger capacity. Boats of 18 feet (5.5 m) or so in length, with high horsepower motors, usually have built-in tanks.

It is often necessary to carry additional fuel to make long runs without stops. Select a spare tank or container carefully; it must be made for that specific use. The material of the tank and its design must be intended specifically for containing gasoline in a marine (often salt water) environment. Homemade or converted tanks are hazardous. Tanks from the manufacturer of your motor, or a reputable manufacturer specializing in marine tanks, are best.

Fuel tanks should be stored in a well-ventilated area, and secured against unnecessary movement. They should be protected from spray or rain. Inspect at least once each season. Dents and scratches can damage the plating that protects underlying steel from rust. A tank that shows any signs of rust should be replaced.

Steering and Other Controls

The simplest outboard motors have controls right on the motor itself. A steering handle, or tiller, is used to turn the motor from side to side for steering. A throttle may be built into the steering arm, or it may be a separate lever on the motor. A gearshift lever for forward, neutral, and reverse is located on the side of the powerhead. Controls for choking and carburetor jet adjustment are usually on the front of the motor.

Figure 613 **On small motors, all controls are on the powerhead. The shift controls are on the right side of this motor (to the left in the picture): the throttle is part of the steering arm. This installation is on a small sailboat.**

Remote steering and engine controls may be an option on motors of about 10 to 25 horsepower, and are standard on larger engines. This allows the operator to sit forward in the boat for better visibility, with a wheel for steering, and levers for throttle and shift control. A control also may be provided to change the angle of the engine against the transom to adjust the riding trim of the boat (see page 116).

Steering controls may consist of a continuous loop of metal cable (usually plastic-covered) running from one side of the engine forward, around a drum on the shaft of the steering wheel, and then aft to the other side of the motor. Pulleys are used to guide the cable, which may run forward on one side of the boat and aft on the other, or may run fore and aft on the same side.

Rod steering uses a rack and pinion gear at the steering wheel to push and pull a stiff cable through a protective housing. The other end of the cable is attached to the motor, moving it from side to side around its central pivot as the steering wheel is turned back and forth. Rod steering gives smooth control and is free of the hazards of the exposed cables of the other type.

In hydraulic systems, steering wheel motion is transmitted through fluid-filled lines to the stern, activating a push-pull rod attached to the motor(s).

Throttle and gear shift controls consist of a stiff cable within a sheath to transmit the back-and-forth motion required. In some models, the two controls are combined at the helm in a single lever which is pushed forward to go ahead faster and pulled backward, through a neutral center position, to go faster astern. This mechanism is somewhat more complex than separate throttle and gear shift controls.

Tachometers

A tachometer is important when running trials to select the optimum propeller, and it often is used as a reference for operating at cruising or trolling speeds. In the absence of a speedometer, a series of timed runs at regular increments of engine speed can be used to establish a speed curve for the boat. This curve then gives you boat speed through the water at any engine setting. See Chapter 19 for details on establishing a speed curve.

An electronic tachometer can be fitted to almost any outboard motor; it is a simple job requiring but little technical know-how and very little time. The instrument, when properly calibrated, indicates engine revolutions per minute.

Speedometers

As an alternative to developing a speed curve with a series of timed runs, a speedometer can be installed, and speed taken directly from it. This instrument also is quick and simple to install, and the more expensive models offer a reasonable degree of accuracy. Most all models can be calibrated by doing a series of timed runs, but even if a speedometer is not accurate in absolute terms, it can be used to determine the relative speeds obtained with various motor and propeller combinations. Calibration is necessary, however, when it is to be used in navigation.

Electronic Depth Sounders

Many outboard boats can carry an electronic depth sounder and put it to good use. Special transom mounts are available for smaller boats where through-hull mounting may not be practical. More on this very useful device is in Chapter 23.

Radiotelephones

Radios--either VHF/FM or Citizens Band--are often installed on outboards that have a battery for electric starting motors and an alternator to keep the battery charged. VHF equipment, with its smaller antennas, is especially suitable for smaller boats. Care should be taken to install equipment that is suitable for the boat and its intended use; consult an expert, not just a salesman, if you don't have the knowledge to make the choice yourself. In many applications, radios are safety equipment, as well as generally useful. See Chapter 22.

Boat Covers

Many outboard boats can be fitted with a cloth cover to keep out dirt when they are not in use, and to protect the interior during off-season storage. The covers, usually of Dacron or similar synthetic fabric, may be available in a style that covers your boat from stem to stern, plus the outboard motor; other styles may cover only the open parts of the boat. Covers are excellent for keeping out rain, and the dirt and leaves that might accumulate when the boat is on its trailer. Be sure that the cover provides adequate ventilation for the interior to prevent "dry rot" of interior woodwork, and other forms of fungus growth.

The cover must fit the boat, and must be capable of being adequately secured by means of snaps or a drawstring. (Snaps or other mechanical fasteners should be protected from corrosion by a non-staining lubricant; a silicone grease stick is excellent for this purpose.) Covers are of perhaps their greatest value when a boat is being trailered, but they must be fastened down adequately to prevent wind damage. In outdoor storage, a cover must also be adequately supported internally to prevent the formation of pools of rainwater that will stretch and eventually tear the cover, dumping the water into the boat.

For some outboards, particularly larger models, tops may be available to provide protection against rain, and shade in the hot sun, while the boat is in operation. Enclosure panels, with clear plastic inserts, also may be available, offering considerable protection against rain and spray, as well as allowing comfortable boating in cool weather. Such tops and panels must of course be removed before trailering the boat at highway speeds.

Registration and Numbering

Boats are subject to the registration requirements of their "state of principal use" or the Coast Guard; see Chapter 2. This means that all mechanically powered boats must be numbered, except for certain larger craft that may be documented. Most states have extended registration to cover "all watercraft," or "sailboats over 16 feet (4.88 m) in length," or other categories of boats.

Figure 614 **Addition of a top enclosure and panels can make a small boat suitable for weekend cruises or other short trips; these also keep the interior dry in inclement weather.**

Operation

An outboard boat can be used for many, widely varied purposes—fishing, water skiing, cruising, skin diving, and others. In *each* of these applications, greater safety and enjoyment will result from knowing and using the proper boating practices.

Boarding

Before a boat can be used, one must get on board it. There is a right way and a wrong way—several wrong ways—and the difference becomes increasingly important as the size of the boat decreases.

If you are boarding from a beach, climb in over the bow when possible. When boarding from a float or low pier, step aboard as close to the middle of the boat as you can. Crouch down so that your weight is kept low, thus helping to keep the boat stable. Grasp the gunwale for balance, if necessary, but do *not* step down onto the gunwale of a small outboard—you most likely will flip the craft, and wind up in the water yourself.

Do not carry bulky or heavy items while boarding. If you are alone, place such items on the pier where you can reach them once you are safely on board. If possible, have someone ashore hand you the equipment once you are in the boat.

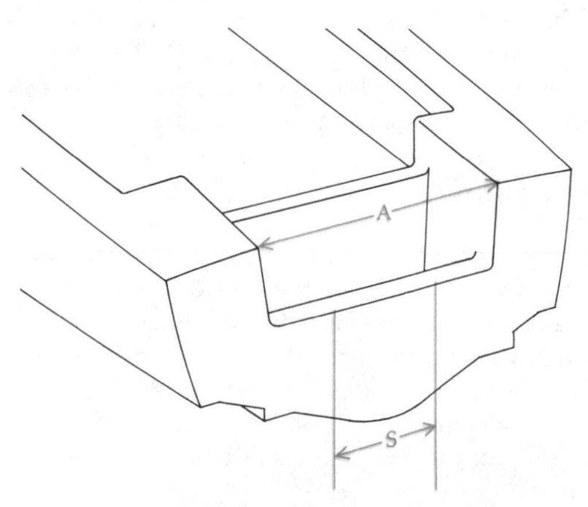

Figure 615 **Standards are issued by the American Boat & Yacht Council for the dimensions of transoms cutouts and motor wells, and for the center-to-center spacing of twin outboard motors.**

Motor Adjustment

Begin your outboarding by installing the motor properly—the manufacturer's diagrams will help in doing this. Seat the motor squarely on the center of the transom and securely tighten the bracket screws. Some motors come with a special mounting plate; with others you may want to use a rubber pad to reduce vibrations and to prevent the transom from being marred. Connect a safety chain or cable from the motor to the hull.

Twin motors are usually installed by a dealer, as are large single engines of the high horsepower range. Twin motors should have the proper spacing; 22 inches (55.9 cm) is recommended by the ABYC for motors up to 75 horsepower, 26 inches (66.0 cm) for engines 75.1 to 250 horsepower, and 32 inches (81.3 cm) for larger motors. Standardized motor well dimensions furnished to manufacturers provide this spacing.

If a combination of a large and small motor is used, the large one is mounted on the centerline, with the small "kicker" off to one side, usually on a bracket designed for such use. (Such a bracket allows the motor to be raised entirely clear of the water when not in use, reducing drag.) The low horsepower of the small motor won't present any significant steering problems from its off-center position.

Motor Height

Although there is a high degree of standardization in the design of transom cut-outs for motor mounting, it is wise to check that the lower unit of the motor is correctly located with respect to the bottom of the hull. Standard shaft lengths are 15 and 20 inches (38 and 51 cm), and 25 inches (64 cm) for the high-horsepower engines. There are some variations, however, and a transom can be modified if necessary to ensure proper motor positioning. The ANTI CAVITATION PLATE on the lower unit should line up with the transom chine, except for some installations on deep-V hulls.

If the motor is too high, a smooth flow of water will not reach the propeller and it will not be able to get a proper "grip" on the water. This may cause CAVITATION, in which the propeller spins in a partial vacuum, with possible damage to the engine from excessive RPM. If the motor is too low, drag will be increased due to both the greater area of the lower unit in the water and a distorted flow over and under the anti-cavitation plate. Attention to this detail of installation will provide both increased speed and decreased fuel consumption.

Thrust Line Adjustment

Most outboard motors have a tilt adjustment to vary the angle of thrust of the propeller. For best performance, the drive of the propeller should be in a line parallel to the flat surface of the water at the boat's most efficient operating angle, whether as a planing hull or in the displacement mode.

If the motor is in too close to the transom, the thrust line is upward from the horizontal, pushing the stern up and the bow down, making the boat "plow" through the water unnecessarily. On the other hand, if the motor is tilted too far out from the transom, the thrust line is below the horizontal, and the bow is forced up too high while the stern "squats." See Figure 616.

Boats differ in design, and loading conditions vary widely with any specific boat, so manufacturers make the adjustment of the tilt angle as easy as possible. On stern-drive engines and some larger outboard motors, an electro-hydraulic adjustment is provided; just push the button for "up" or "down."

Be aware of the inefficiency of an improper thrust angle, and do not neglect to change the adjustment to current operating conditions. The best angle may vary with

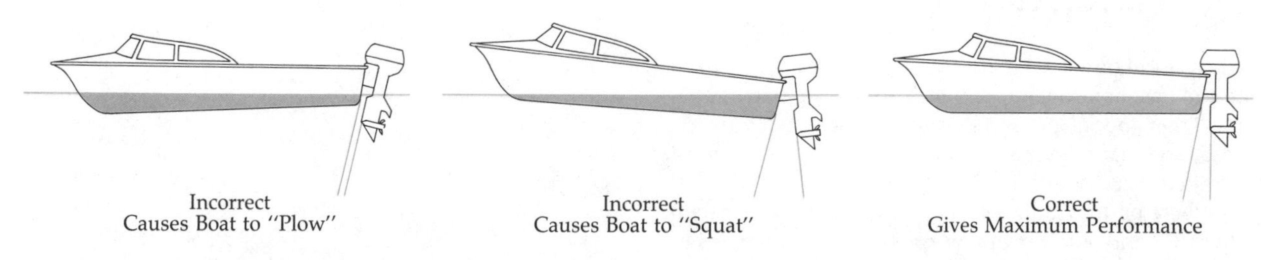

Figure 616 **Motor tilt must be adjusted so that the propeller thrust is horizontal when the boat is in running trim. On small motors adjustment is done manually by moving a pin to different holes in the mounting bracket. Most large motors have hydraulically operated tilt control.**

water conditions. On some boats, an adjustment with the motor tilted out (bow up) for smooth water may give more speed, and a better ride in rough water may be obtained with the motor trimmed in a bit. Only actual trials will tell for sure, however.

Special Applications of Tilt

While underway, the tilt feature of outboard motors and stern-drive lower units can be used to advantage for several special situations.

If the propeller becomes fouled with weeds, it is often easy to stop the motor, tilt it up, clear off the vegetation, lower the motor, restart it, and continue on your way. If a smaller motor shears its pin, it is often possible to replace it with the motor tilted up, making it unnecessary to unmount the motor completely to work on in the boat.

It is also possible to tilt the lower unit up far enough so the boat can clear shoal areas, but take care not to tilt it so much that the cooling water intake comes above the surface, and watch out that the propeller does not strike rocks or other hard objects that might damage it. Proceed slowly when using this technique.

Mixture Controls

Many outboard motors have controls for adjusting the fuel-air ratio to obtain optimum performance at low speeds, and, in the case of most smaller motors, for high speeds as well. Adjustment controls are outside the motor cover so they can be used while underway. Follow the instructions in your owner's manual for best results, and to minimize the possibility of engine damage due to improper setting.

Changes in fuel, air temperature, and altitude (such as going from mountain lakes to a lower elevation, or vice versa) may require changes in mixtures. Any adjustment should be made with the motor thoroughly warmed up, and the boat loaded to normal trim.

Most motors also have a manual choke for starting.

Fuels and Oils

Two-cycle outboard motors use a mixture of gasoline and lubricating oil. It is essential that both of these be of the correct type, and that the mixture ratio be that specified by the motor manufacturer.

Fuels

The drive for a cleaner environment has resulted in the availability of "no-lead" and "low-lead" gasolines, with other chemicals added to maintain octane ratings. In many cases these fuels have proven to be unsatisfactory, or even unsafe in marine engines; particularly the two-cycle outboards.

Outboard motors are generally of relatively low compression ratio, and can use fuels of quite modest octane rating. Marine white gasoline of the type that has been around for many years is fine. While lead-free, it does not contain the new additives that make modern automotive lead-free gasoline unsuitable. Some low-lead fuels may be used, with proper adjustment of ignition timing: check with a responsible representative of your motor manufacturer before attempting use of any low-lead or no-lead gasoline. Regular leaded automotive fuel is satisfactory for most outboard models. Keep in mind that marine engines normally operate under a constant load, unlike autos and trucks that accelerate and slow down, go uphill and downhill, with frequent changes of loads. The differences in the two sets of conditions are significant to engines and their fuels.

It is best to follow the fuel recommendations in the owner's manual for your motor.

Figure 617 **Outboard motors have one or more carburetor controls brought out to the front of the motor case to permit adjustment of fuel mixtures for various operating conditions. Be sure to follow the instructions in your motor manual.**

Oils

The typical outboard motor has its lubricating oil mixed into the gasoline fuel supply; there is no separate crankcase. Thus special properties such as "low ash" are required, and many normal features such as detergents must be limited or avoided.

The best oil to use in a two-cycle motor is one specifically made and sold for that purpose by oil companies or engine manufacturers. It will be certified as TC-W (two-cycle, water-cooled) and has no harmful compounds. Note the engine manufacturers do not *make* the oils which bear their names, but presumably they set the specifications for them. Using oil from the motor's manufacturer will ensure compatibility between oil and motor.

In an emergency, if two-cycle oil is not available, use the cheapest grade of automotive oil you can get; it is the least likely to have harmful additivies. Use oil of #30 weight, *not lighter*.

Oil-Fuel Mixture Ratios

Motor manufacturers specify the correct ratio of gasoline to oil for each of their models. For older motors the ratio was 24 to 1; for most modern motors it is 50 to 1. Motors made outside of the U.S. may have different requirements. The amount of oil to be used is sometimes given as ounces or fractions of a pint to be added for each gallon of gasoline.

Many marinas now have a special pump that dispenses pre-mixed gas and oil. In some instances the mixture is fixed, usually at 50 to 1; at other pumps, controls can be set to any one of several standard ratios.

Figure 618 **As a safety measure, portable tanks should be removed from the boat and filled on the gas pier or wharf. Wipe them clean and dry before putting them back aboard.**

Ratios of Oil to Gasoline

Fraction of Oil per Gallon of Gas	*Pints of oil added to gallons of gas*					
	Actual Ratio	*2 Gals.*	*3 Gals.*	*4 Gals.*	*5 Gals.*	*6 Gals.*
1/12	96:1	1/6	1/4	1/3	5/12	1/2
1/6	48:1	1/3	1/2	2/3	5/6	1
1/5	40:1	2/5	3/5	4/5	1	1 1/5
1/3	24:1	2/3	1	1 1/3	1 2/3	2
3/8	21:1	3/4	1 1/8	1 1/2	1 7/8	2 1/4
1/2	16:1	1	1 1/2	2	2 1/2	3
3/4	11:1	1 1/2	2 1/4	3	3 3/4	4 1/2

Table 6-1

Fueling Procedures

There are two aspects of fueling an outboard boat that must be given careful attention—safety and the proper mixing of the oil and gasoline.

Fueling Portable Tanks

When taking on fuel with any type of boat, there must be no smoking in the vicinity of the boat, and persons not participating in the fueling should be off the boat.

If possible, take portable tanks off the boat and fill them ashore. Make sure the hose nozzle is in contact with the rim of the tank fill opening to prevent generation of a static electricity spark which might cause an explosion.

Oil properly mixed with gasoline will not separate out later, but the initial mixing procedure requires careful attention; follow these steps for a three- or six-gallon tank:

1. Put in approximately one gallon of gasoline.
2. Pour in all the oil that is to be added.
3. Replace the cap on the tank and shake *vigorously*—you can't shake it too much!
4. Remove the filler cap, add the balance of the gasoline without delay, and replace the cap. If possible, shake the tank again; but a full tank is heavy and can't be shaken as vigorously as one with only one gallon in it.
5. Wipe off the outside of the tank, and return it to the boat (if it had been removed for filling) *after* any odor of fumes has disappeared.

For safety, portable tanks should be secured in the boat. A simple way is to provide wooden blocks on the hull or floorboards to prevent sideways or endways movement of the tank, with straps over the top of the tank to hold it down in rough going.

Filling Larger Tanks

If your fuel tanks are larger than six gallons, they are best left in the boat for filling. These, and permanently installed tanks, are filled after all doors, hatches, windows, etc., are closed to keep any gasoline vapors from getting below. After fueling and cleaning up, open all hatches and ports and allow time for ventilation to clear bilges and any enclosed spaces before starting the engine. See Chapter 12 for additional information on fueling safety.

The mixing of gasoline and oil for larger tanks that cannot be shaken can be done in several ways—all of which are time-consuming but very important, even more so than getting the exact fuel-oil mixture ratio.

One method is to use a metal funnel that has a screen filter (metal so that a grounding connection can be made from hose nozzle to tank). This funnel is inserted in the tank or fill pipe and as gas flows through it, oil is added at the same time. Try to control the flow of oil so the stream of gasoline blends thoroughly with it.

A second method is to mix the fuel and oil in batches in a separate five- or six-gallon container, using the procedure outlined for portable tanks, and then to pour the mix into the larger tanks. On tanks not larger than 15 gallons, it is permissible to mix the total amount of oil in one five- or six-gallon tank of gas, pour this into the large tank, and then add the remainder of the gasoline directly to the large tank. Do *not* pour oil directly into a tank; it will not mix properly.

Some fuel pumps are now metered to read in liters; 1 U.S. gallon equals 3.8 liters.

After filling tanks, check the fuel system for leaks, and carefully wipe up any spills—discard the rag ashore, don't stow it in the boat! Hose down with fresh water, if possible. Remember to ventilate the boat thoroughly before starting the motor; there is an explosion hazard with an outboard, even though the motor itself is outside the boat. Fumes from fueling are heavier than air, and can sink into the lower parts of a boat where any spark or flame can set them off. A power bilge blower is a good investment, particularly on an outboard cruiser.

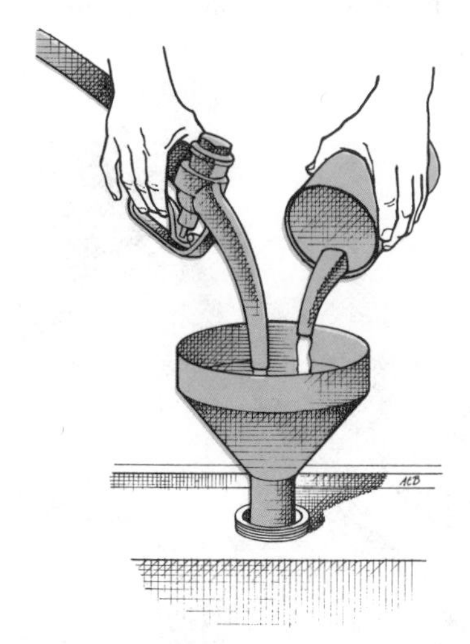

Figure 619 **When refuelling a large, fixed tank on an outboard, the best procedure is to add oil, along with the gasoline, through a screened metal funnel. Be sure the hose nozzle is in contact with the funnel, and the funnel is in contact with the fill plate. Do not pour oil alone into the tank.**

Boat Handling

Outboard and stern-drive boats steer with a directed thrust of the propeller. This makes for easy and positive steering, but if one has been accustomed to handling inboards, it has a distinctly different "feel," and one must become familiar with it.

The reaction of the boat to a turning of the motor thrust line will be different at slow speeds and at planing speeds; it will vary with hulls of different design and with motors of widely different horsepower ratings. Handle an unfamiliar craft with considerable caution until you understand its reactions to the wheel and throttle.

Note that when the motor is in neutral or stopped, there is a loss of steering efficiency, and this is particularly true if the boat is moving very slowly. It is not a major problem, but it must be considered in slow-speed maneuvering.

Steering

Steering an outboard is simple. With small motors, the operator sits in the stern of the boat and pushes the tiller handle on the motor *away* from the direction of the intended turn. Boats with larger motors are steered by a wheel in the same manner as a car; the boat will go in the direction the wheel is turned. Note that on any boat, the stern will swing *away* from the direction of the turn; watch out for this action when maneuvering in close quarters.

Two additional factors come into play with very large motors—outboards or stern-drives of about 100 horsepower or more: propeller torque, and the effect of the propeller pitch/diameter ratio.

Torque This is the tendency of the boat to heel to one side as a reaction to the powerful spin of the propeller; also the boat tends to head to one side, requiring a steering adjustment to maintain course. To counteract this, adjustable trim tabs are usually mounted under the after end of the cavitation plate. Effectiveness varies at different speeds, but the tabs can be set with an efficient range of about 600 RPM that covers normal operating speeds. For example, you could adjust the tab to provide "hands-off" steering from 2800 RPM to 3400 RPM, but above and below these speeds, steering corrections would be needed to maintain course.

Pitch/Diameter Ratio Effect A propeller tends to "walk" toward the direction in which it is rotating. For example, when going ahead, a right-hand turning propeller moves the transom to starboard; a left-hand propeller moves it to port. The greater the pitch for a given diameter, the greater the "walking" effect (and the more shallow the boat's draft the greater the walking action). With a little practice you can put this effect to good use in slow-speed, close-quarter maneuvering.

At high speeds, the pitch/diameter ratio can be felt in the steering, expressed as rim load on the steering wheel—the amount of pressure needed to turn the wheel. Again, the greater the pitch for a given diameter, the greater the rim load.

Thrust Line Adjustment When setting up large twin-engine outboards or stern-drives, the lower units should be canted about 5° toward the outside of the boat. This helps to break up the suction that tends to develop at the transom, particularly when coming out of a turn, hindering smooth response to steering.

Trim

The number of seats in a boat is *not* an indication of the number of persons it can carry safely. Overloading is a major cause of boating accidents—stay within the limits of your craft. The factors affecting trim become ever more critical as the load approaches capacity, and as the boats get smaller.

Distribute weight evenly so the boat will trim properly: level from side to side, and slightly down at the stern. It should *never* be down at the bow. Seat passengers as near

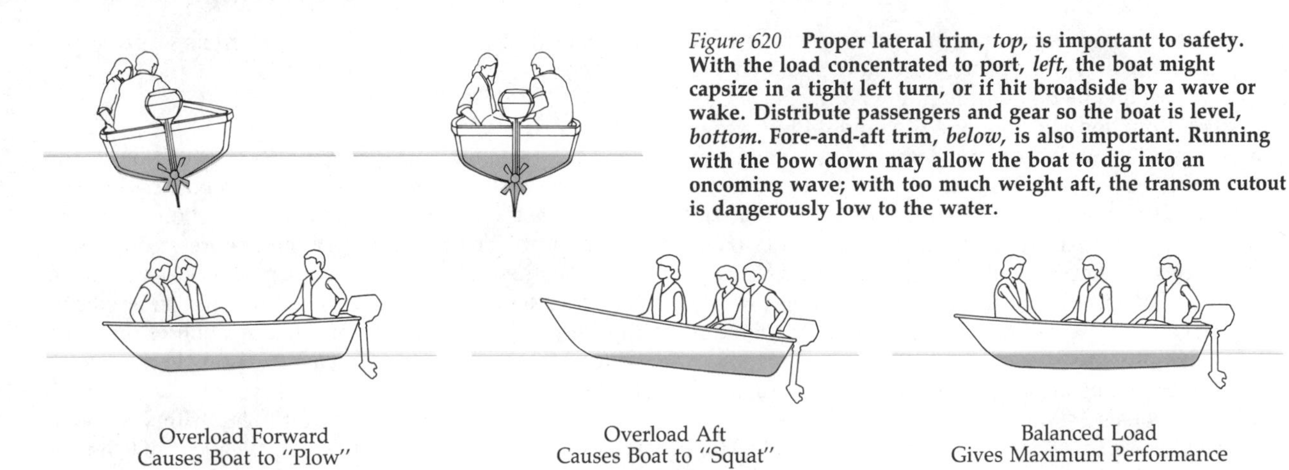

Figure 620 **Proper lateral trim, *top,* is important to safety. With the load concentrated to port, *left,* the boat might capsize in a tight left turn, or if hit broadside by a wave or wake. Distribute passengers and gear so the boat is level, *bottom.* Fore-and-aft trim, *below,* is also important. Running with the bow down may allow the boat to dig into an oncoming wave; with too much weight aft, the transom cutout is dangerously low to the water.**

the centerline as possible; they should not hang over the sides. Consider fore-and-aft trim as well: too many persons forward will bring the bow down, causing it to plow into the water; too much weight aft will bring the stern down too far, causing excess drag. This can be unsafe, and it increases fuel consumption.

Trim your boat as well as possible *before* getting underway. In smaller craft it is dangerous for people to try to change places or move about while the boat is scooting along briskly. If such movement is essential, slow or stop the boat first, remembering in rough weather to keep enough momentum to retain steerage control and to keep the boat headed into wind and waves. The person who must move should keep low and near the boat's centerline.

Stability

Outboards are often run at relatively high speeds, and their stability becomes an important safety issue. Some hulls will run straight ahead quite steadily, but have a tendency to heel excessively, or even flip over, when turned sharply.

The underwater shape of the hull is a definite factor in stability. Most outboard and stern-drive boats now available are of DEEP-V or MODIFIED-V form, or of the multiple-hull CATHEDRAL type. These all provide a high degree of stability in normal operation, but because they track so well in a straight line, they resist turning; attempts to make a sharp turn at high speed may cause BROACHING, flipping over sideways. This is also the case with some older, flat-bottom hull forms where there is a keel to provide directional stability.

Conversely, a flat-bottom boat without a keel has little directional stability, and may skid out sideways when a turn is attempted at excessive speed. Initially the boat will point off in the new direction, but actually continue to travel along what is essentially its old course.

In any case, the faster a boat goes, the more important it is to reduce speed to a safe level before starting a turn; never turn more sharply than necessary. Normal operation seldom requires a sudden, sharp, high-speed turn.

Reversing

Most outboard motors have a reverse gear that enables them to be backed down. Some small motors have no reverse, but pivot 180° for the same effect. Unless restrained, an outboard motor has the tendency to tilt itself up and out of the water when thrust is reversed. On many models, there is a manually operated REVERSE LOCK that must be latched into place to keep the motor down while engaged in backing maneuvers. For normal running, however, it is important that this latch be released so that the lower unit will be free to tilt up if it strikes an underwater obstruction.

On some large motors, especially those with electric shifting, anti-tilt latch operation is automatically interlocked with reverse gear operation.

Figure 621 **The reverse or anti-tilt latch prevents the lower unit from flipping up when the motor is put into reverse gear. It should be released when running forward normally so that the motor can tilt up if a submerged object is struck.**

Safe Operation

Exercise your knowledge of the Rules of the Road, but also exercise good judgment, whether you have the right-of-way or not. Small boats should stand clear as far as possible from large yachts and commercial vessels. Watch their wakes carefully. Don't count on ships to look out for you; they need more, much more, room to stop, are more difficult to turn, and sometimes have "blind spots"—

areas which are blocked to the helmsman's vision. Never assume that the pilot of a vessel sees you; keep well clear.

Leaving a Pier

In a small, light boat, it's usually possible to get away from the face of a pier by pushing off, waiting a moment until sufficient space opens up, then going ahead slowly with a very gently outward turn (a sharp turn could cause the stern to swing in and strike the pier).

The best practice, however, is to back out, and this should be done if wind or current is holding the boat against the pier. Turn the lower unit of the outboard motor or stern-drive away from the pier, and slowly apply power in reverse gear. Back out far enough so that the boat can be straightened out and moved away in forward gear without the stern swinging in against the pier.

Docking

Bringing an outboard safely in to a pier or wharf is relatively easy, but try to take advantage of any favorable influences of wind or current. Try to determine the *net effect* of these forces, as wind and current are often in different directions. In general, the combined effect will be roughly parallel to the pier, or roughly at right angles to it.

If this wind/current effect is parallel to the pier, head into it, as this allows the most positive control of the boat. Approach the pier in a gradual turn at the slowest speed possible, bring the boat up parallel with the face of the pier, and stop with a touch of reverse power. If the stern isn't quite in, turn the lower unit in toward the pier as you shift into reverse.

A bow line can be put ashore as the boat comes parallel with the pier, but under favorable conditions it is often possible to use no lines until the boat is snug against the pier.

If the wind or current is off the pier, approach at a relatively steep angle, heading as directly as possible into the wind or current. Then make a sharp, smooth turn just as the pier is reached to bring the boat parallel to the pier. Get your lines across as quickly as possible to prevent the boat from being blown away.

If the wind/current force is toward the pier, bring your boat parallel to the pier while still a few feet away from the face, and let this force bring you alongside. You may need to have fenders rigged if the force is fairly strong, and the water alongside the pier is choppy.

When there's no wind or current, approach the pier at a moderate angle—30°-45°—check forward motion with reverse power, and pull the stern in at the same time.

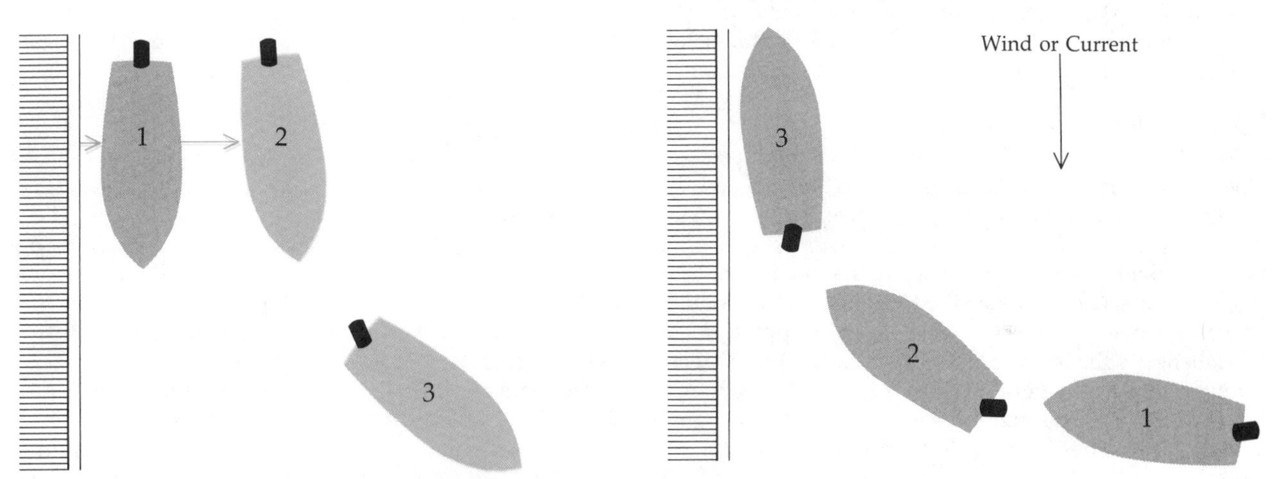

Figure 622 **A small boat may be pushed away broadside from a pier or wharf, *above left,* with the motor in neutral. When there is enough clearance, forward gear is engaged and the boat moves off. In most cases it is best to back away, *below left,* until the boat is well clear, then shift into forward and ease away from the pier. If possible, approach a pier against the wind or current (whichever is strongest) in a slow, gradual turn, *above right,* and bring the boat to a stop with a touch of reverse power. With wind or current off the pier, approach at a steep angle, *below right,* turning as the pier is reached, with lines ready to pass ashore.**

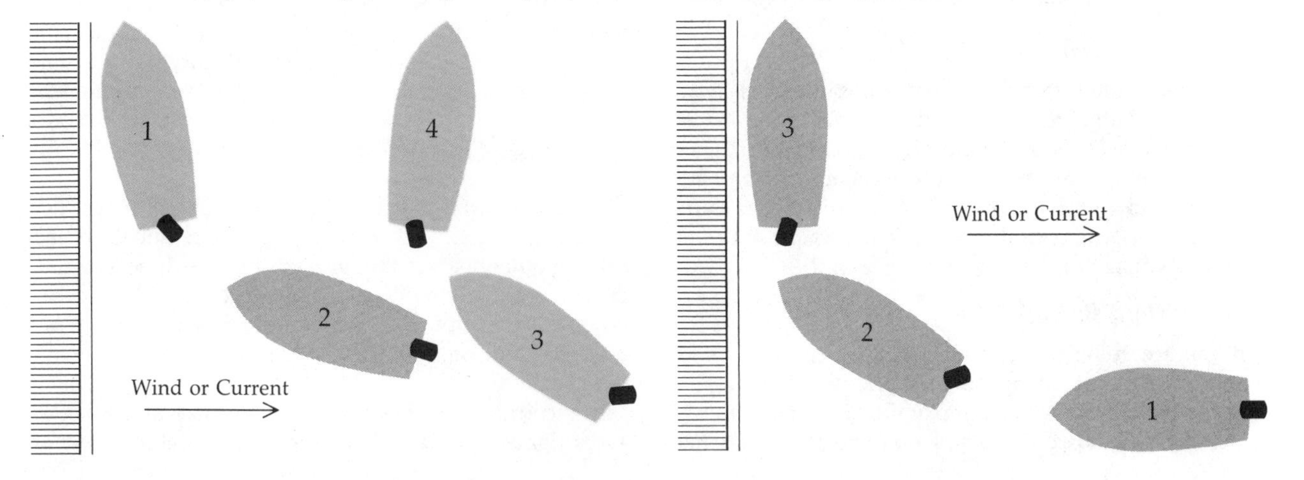

In all these maneuvers, keep speed slow and the boat under full control. Apply power in forward or reverse positively but smoothly to avoid stalling the motor. Reverse power must be applied gently if the anti-tilt latch has not been locked; otherwise the lower unit will pull itself up out of the water, and the motor will race without effect.

Here are some tips that will make dockside maneuvers safer and easier:

- Use fenders generously and pad your pier and pilings well.
- Keep your engine, controls, and steering gear in good shape.
- Keep your mooring lines in good condition, and use chafing gear whenever possible.
- Use your "amidship" or "springline" cleat for securing the first line when docking, and for the last line to let go when casting off. This can be a bight of line around a pile or cleat as it is an easy matter, when casting off, to let go of the free end and then haul the line aboard.
- Watch lines to see that they don't foul your propeller.
- Remember that a boat will pivot about a point about ⅓ of her length aft from the bow, and will not "follow the front wheels" as does a car. A boat cuts a path considerably wider than its beam when it goes around a turn.
- Practice maneuvers in calm, light-wind conditions first to get a good feel of the boat's responses. This will make it easier to keep control when in adverse conditions.

Stand Clear of Hazards

Should you come upon a tow in the bend of a river, or one maneuvering near shore, note whether the tow is swinging to starboard or port, and give it the room it needs.

You should not bring your boat near a moored vessel. If it is necessary, be exceedingly wary; strong river or tidal currents moving under a moored barge can pull a small boat beneath the surface. A cross current—one that is running at about a right angle to your course—requires great caution. If you get into such a situation, give any object, floating free or made fast, an extra wide berth.

No One on the Bow

It is extremely dangerous for anyone, especially children, to ride on the bow of a small boat with his legs dangling over the side. If someone slips off he can be run over by the boat before the skipper can take any action. It has happened often.

Watch the Weather

Weather is important to all boatmen, but especially to operators of outboard and stern-drive boats. Keep an eye on the weather, even if the morning forecast was for fair skies and gentle breezes. Learn to recognize threatening cloud formations. Train yourself to be sensitive to shifts in wind direction; a wind shift is often advance notice of a coming change in the weather.

Leave Word Behind

If you are going offshore or for a long run on inland waters, or even just fishing some distance away from home, tell someone ashore what your plans are in as much detail as possible. Should you become seriously overdue at home or at your destination, the Coast Guard, marine police, or similar organization can be sent to your assistance without delay.

The information left ashore should include: (a) where you plan to go, including any stops enroute; (b) list of persons aboard; (c) estimated time of return (or arrival at destination away from home); (d) alternate plans in case of bad weather; (e) communications equipment on the boat; (f) a good description of the boat.

Distress Signals

Know how to use distress signals you are required to have on board (see page 68). The signals must be in good condition, and ready for immediate use. In an emergency, and if you have used up your flares and smoke signals before help has arrived, you can signal to a boat that appears within sight by waving your arms. Stand (or kneel if rough water makes standing hazardous), and slowly and repeatedly raise and lower your arms outstretched to each side.

Never use "Mayday" on the radiotelephone unless there is a great danger and immediate assistance is required.

Local Knowledge

When in unfamiliar waters, take advantage of "local knowledge"; watch the operation of boats piloted by skippers who are at home in those waters, and don't hesitate to ask questions about possible hazards.

Slowing and Stopping

A common misconception is that a small boat can be maneuvered and stopped as easily as an automobile. This is not so, but much can be done with a boat if one takes time to learn it slowly. The new boat owner should practice leaving and returning to piers, and other maneuvers, until he has developed both skills and confidence; take it very cautiously and gently at first, and gradually build up to the procedures of experienced operators.

Always slow down gradually rather than pulling the throttle back quickly. All boats have a stern wave that will catch up with and pass the craft if it comes to an abrupt stop. This can bring water into the boat, especially if it has a low-cut transom and no motor well.

Fuel Consumption

There is no excuse for running out of fuel. Make tests and keep records to establish the rate of fuel consumption at various throttle settings (RPM as read on the tachometer). Follow this up with a check on the number of hours run since the last refuelling. Know the fuel consumption on a per-mile basis as an aid in planning cruises.

Courteous Operation

Keep your boat's speed under control at all times. Respect the rights, and comforts, of others afloat. When passing other boats in the same or opposite direction, give them a wide berth if possible, or drop down to slow speed. When passing an anchorage, throttle down to your slowest speed and keep an alert lookout for moorings, swimmers, debris, etc.

It is courteous to keep down your wake and wash to avoid damage to other boats. It's a wise action, too, for you are legally responsible for any damage that your wake causes to other boats or persons.

Emergencies

Various studies have shown the following to be the major causes of boating accidents:
1. Overloading, overpowering, and improper trim.
2. High speed turns, especially in rough water.
3. Failure to keep a sharp lookout for obstructions.
4. Going out in bad weather (or not starting for home soon enough when good weather turns bad).
5. Standing in a moving small boat.
6. Having too much weight too high in the boat, as when someone sits on the deck of a small outboard.
7. Leaks in the fuel system.
8. Going too far offshore.

A carefully matched boat, motor, and propeller, operated in accordance with the law and with courtesy, will go a long way toward eliminating accidents and worries. But some possibility of trouble always remains; be prepared to act in an emergency.

Squalls and Storms

If you are caught in a squall or heavy seas, slow down immediately, maintaining only enough power to head the boat into, or at a slight angle to, the wind and waves. Everyone should put on personal flotation devices, including the skipper. Don't try to smash your boat through the waves at high speed in an attempt to get home sooner—that's inviting disaster. Instead, take the seas as gently as possible on the bow. The hull may pound, and the waves may toss up a lot of spray, but you will have your best control over the boat.

Keep enough weight at the stern in stormy weather to hold the propeller in the water. Since an outboard has no rudder, control is lost if the propeller comes out. Should the motor quit, rig some sort of sea anchor (see pages 219-220) to hold the bow up to the seas.

Don't let water accumulate in the boat. If rain, spray, or waves bring water into the bilges, pump or bail it out immediately. Water inside a boat must not be allowed to build up to an amount that could affect the trim and stability of the craft.

Whenever safely possible, try to get to a protected harbor rather than attempt to ride out a storm. But do not let your desire for a harbor upset your judgement. Avoid a course that will put large waves on your stern; they can swamp the boat, although self-bailing motor wells reduce this hazard. There is also risk of "pitch-poling" (turning end over end) in a heavy following sea by driving too fast before it.

If you must turn in a heavy sea, watch for a lull in the wave formation and turn then, so you won't get caught broadside in a trough by an approaching wave. Always cross a large wave, or wake, at about a right angle; this lets you maintain the best control over your boat.

Capsizes

Stay with the boat if it capsizes! Almost invariably the temptation is to try to swim to shore if land is in sight, and almost always the shore is farther than it appears. Most outboard boats will remain afloat, even if filled with water; boats built since July 1973 must carry flotation that will keep them upright. A boat is a much larger and more visible object than a person in the water; stay with the boat and you will receive help much more quickly than if you swim off on your own.

Rescuing a Person in the Water

One of the leading causes of death in boating is drowning, and many of these fatalities result from people falling overboard. Know how to rescue a person overboard. Practice the maneuvers necessary to accomplish rescue; using a ring buoy or buoyant cushion to serve as the "victim."

As soon as someone falls overboard, maneuver the boat's stern away from him. Shift into neutral immediately (kill the boat's motor if you do not have a gearshift) and throw a buoyant cushion or life jacket near the victim—try to get it close, but don't try to hit him with it. Make sure that you are well clear of the person in the water before shifting into gear again.

Circle around quickly, selecting a course that will allow you to approach the person with the boat headed into the wind or waves. Approach him slowly, taking care to come alongside him, not over him. Stop the motor before attempting to get the victim aboard.

When alongside, extend a paddle or boat hook to him, or toss him one end of a line. *With the motor stopped,* lead him around to the stern, where the freeboard is lowest, if there is enough space at the transom for him to get aboard without hurting himself on the motor. If that is not feasible, help the victim aboard over the side as far aft as possible. In either case, the use of a boarding ladder will help. To avoid a capsize while he is coming aboard, other passengers should carefully shift their weight to the opposite side to maintain trim as much as possible. When helping a person aboard, hold him under the armpits and lift gently.

In Case of an Accident

If you are involved in a boating accident, you are required to stop and give whatever help you can without seriously endangering your boat or passengers. You must also identify yourself and your boat to any person injured or to the owner of any property damaged. See page 45 for additional details.

If you see an accident without being a part of it, you may render assistance without fear of liability. The Federal Boat Safety Act of 1971 contains a "good samaritan" section which provides that any person who renders assistance at the scene of a vessel accident will not be liable for civil damages from such action if he acts as a reasonably prudent person would have acted under the same circumstances.

Jet-Drive Boats

Because jet boats fall in the size range covered by outboard and stern-drive boats, their handling characteristics will be discussed here. The major feature of the jet is its instantaneous response in accelerating, stopping, or making any sort of turn.

Rather than having a propeller immersed below the hull, the jet system has a pump that draws water up into a

Figure 623 **The combination of high power and jet drive provides rapid acceleration. Because there is no torque effect, there is no tendency for the boat to be thrown off course. Steering is accomplished by deflecting the water jet from side to side.**

housing within the boat, and ejects it out the stern. This jet of water provides the propulsion without the torque effect of a large outboard or stern-drive unit. Steering is accomplished by turning the direction of the water jet, and the reverse mechanism is a "clam shell" that drops to divert the flow of water forward, toward the bow. If done suddenly, it stops the boat in its own length—a maneuver that should not be done from high speed without first warning any passengers aboard.

Because there is no gearbox as such, you can shift instantly from full ahead to full astern by using the "clam shell" with no strain on the engine. The direction of the water stream is the only thing that is altered in stopping or reversing.

In accelerating with a skier in tow, the throttle must be eased forward carefully, taking care not to strain the skier's arms or pitch him head-first into the water, with a jackrabbit start.

The pivot point of a jet boat is roughly 14 to 20 inches forward on the engine. Without rudder or lower unit drag, extremely sharp turns can be made; it is possible to reverse course in little more than the boat's length.

Getting into a tight space at a pier can be accomplished by "walking" a jet boat in sideways, using short bursts of power alternately in forward and reverse.

Steering is least effective at slow boat speeds, and on some models a small fin or rudder is used to supplement the action of the water jet.

Jet-drive boats with the best handling characteristics are those designed specifically for this means of propulsion. In general, a deep-V hull will perform better than one that is flat or nearly flat.

Boat Trailering

The addition of a trailer to a boat and motor "rig" provides considerable operational flexibility to the skipper. It allows him not only to travel to distant boating areas that would be otherwise inaccessible, but also to store his boat ashore at home—saving marina fees, and facilitating routine maintenance. Also, there's no chance for marine organisms to attach themselves to the hull; periodic and costly applications of anti-fouling paint are not needed.

Figure 624 **A trailer can greatly widen your scope of operations, as well as permitting off-season boat storage at home, but care must be taken in selecting the right one.**

Selecting and Equipping Trailers

Trailers, like boats and motors, come in many varieties. They may feature electric winches, self-leveling beds, finger-tip controls for loading and launching, etc. A light-duty trailer may suffice if all your boating is done locally; a heavy-duty one may be needed if you plan long hauls. The object is to get the right trailer for your boat and motor, and for your type of trailering.

Adequacy

The trailer for your boat must be adequate in both weight-carrying capacity and length. Under-buy in either of these aspects and your craft ultimately will suffer.

The trailer must be able to carry the weight of your boat, motor(s), *and the gear stowed on the boat.* Do not overlook your gear in making calculations of required trailer capacity! Your trailer must be rated to carry this load, and if the load is within 100 pounds (45 kg) of the rating stated on the trailer's capacity plate (usually found on the trailer frame), buy the *next larger model.* Don't underestimate weights; boat weights given by manufacturers may be for stripped models, so try to account for all equipment. Make a detailed list of each item you plan to carry in the boat, and its weight, and be sure to include a generous allowance for "miscellaneous." If possible, use a truck scale to weigh the trailer alone, then the trailer with boat and motor(s), and finally the rig complete with all gear. You may be surprised at the final gross weight!

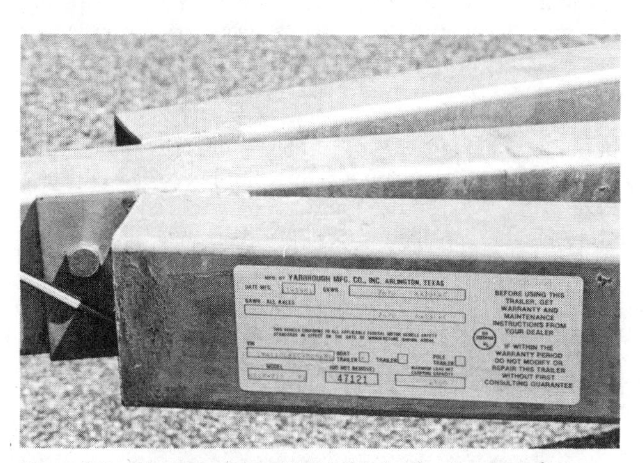

Figure 625 **The capacity plate on this trailer tongue, *above,* shows its load capacity, tire pressures to be maintained, and other data. As the full weight of an outboard motor or stern-drive is carried on the boat's transom, *below,* it is essential that there be adequate support under this portion of the hull.**

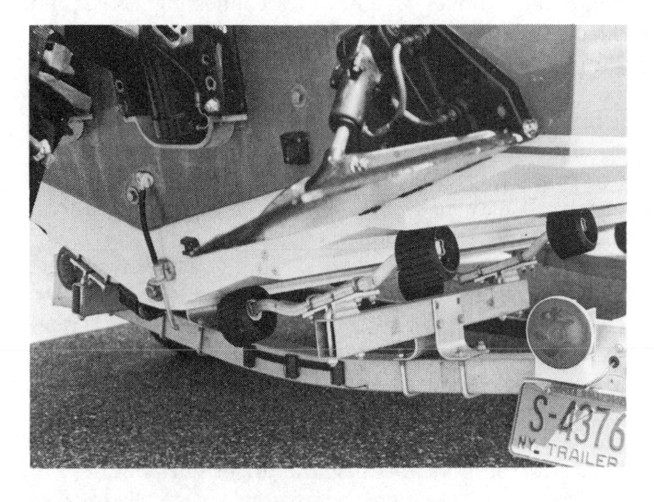

Length

Trailer length is critical because the boat's stern area, particularly the transom, must have adequate support. The boat must be able to fit on the trailer bed so that the transom is *directly* over the aftermost supports; see Figure 625, bottom. If there is *any* overhang, the hull will be distorted. This is also true for stern-drive craft, but it is particularly important for outboards, where the full weight of the motor(s) is on the transom. A "hook" (a downward bend) in the hull at the stern, caused by a too-short trailer, can affect both boat speed and general handling characteristics.

Supports

A boat's natural environment is the water. Here the hull is uniformly supported and there are no concentrations of pressure on the hull. A trailer can provide support only in limited areas, so the trailer bed must fit the contours of the hull as closely as possible. Plenty of padding and bracing may be needed.

Supports may take the form of rollers or of padded bars or stringers, or a combination of some of each. Some trailers, with multiple pairs of rollers on pivoted bars, are designed to conform to the hull shape automatically; these are the self-levelling models. Each type of support system has its advantages and disadvantages; the way the boat will normally be launched and loaded must be considered.

Rollers of hard rubber are widely used on trailers that tilt and allow the boat to move backward into the water by gravity. Their disadvantage is that each roller has very little area of contact with the hull with consequent high pressure in this area. The more rollers there are, the better, *provided* that the height of each is adjusted properly; the depth of the indentation into the roller by the keel is a rough guide as to the weight being carried at that point.

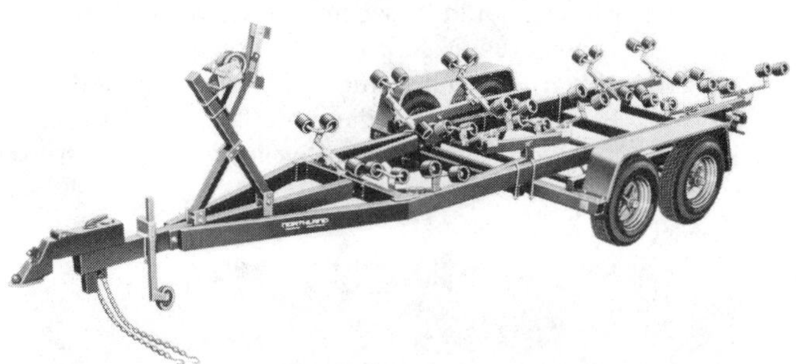

Figure 626 **A system of pivoted arms with pairs of rollers lets load-leveller type trailer conform to the shape of a boat's bottom automatically, after adjustments for length and beam.**

Padded bar rests provide the maximum area in contact with the hull, and thus the minimum of point contact pressure. These are excellent for boats that normally are lifted from the trailer, as by a crane's slings. They have considerable surface friction, however, and are generally unsuitable for sliding the boat off the trailer into the water—unless special features are provided.

Some trailers use a combination of rollers and padded bars, with a lever system that lowers the bars for launching and retrieving on the keel rollers, and raises the bars for support when the boat is fully on the trailer.

There also should be side supports to hold the boat firmly in position on the trailer bed, and a bow chock to keep the boat from moving further forward. All of these supports must be adjustable and positioned so that they can carry out their functions. Location of the bow chock, for example, should be adjusted so that the transom will be directly over its supports. The boat's position on the trailer also affects the weight at the trailer tongue and coupling. Any unbalanced condition, once the hull is properly mated to the trailer bed, is properly corrected by shifting the location of the axle and wheels, *not* the bow chock.

Safety Features

Safety must be a primary consideration in the selection of a trailer—safety for one's investment in boat and motor, safety for people who use them, and the safety of other vehicles on the road. A few more dollars spent initially on the trailer may prevent losses later.

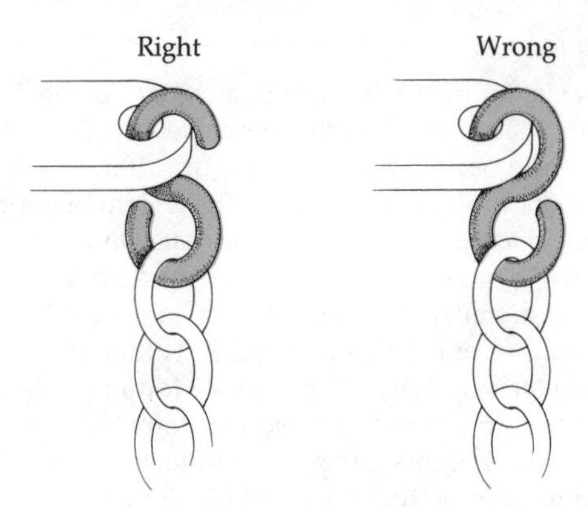

Figure 628 **There is a right way and a wrong way to hook up trailer safety chains. The open end of the hook should go up through the hole, not down through it, so the hook cannot bounce out. Cross the chains under the trailer tongue so they will support the tongue if the trailer coupling should fail.**

Hitches and Couplings

Typical specifications for COUPLINGS (on the trailer) and HITCHES (on the car) require larger hitch balls for larger and heavier trailers. Hitches for connecting boat trailers to cars may be Class I—up to 2,000 pounds (910 kg) gross trailer weight (GVW) and 200 (90 kg) pounds tongue weight, or the car maker's specifications, whichever is lower. (Specifications are often less for compact and foreign cars.) Class I hitches are *not* for tandem-axle trailers regardless of weight. Heavier rigs—up to 3,500 (1,590 kg) pounds GVW and 350 (150 kg) pounds tongue weight (or car manufacturer's specifications) require a Class II hitch. This should always be used instead of a Class I hitch whenever borderline conditions exist.

A trailer hitch should be secured to the vehicle's frame either with bolts (using lock washers) or by welding. A light-duty hitch attached to the rear bumper should *not* be used. For heavier loads, there are "weight distributing" or "load equalizer" hitches available. These lessen the load on the rear of the car, and permit it to operate at a more level attitude. They are necessary for trailers of more than 3,500 pounds (1,590 kg) gross weight, and are desirable for those in the 2,000- to 3,500-pound (910-1,590 kg) category. Equalizing hitches are available up to 10,000 pounds (4535 kg) GTW and 1,000 pounds (455 kg) tongue weight, but the car's rating must not be exceeded. Class III or IV hitches may be required. The use of added springs or adjustable air-filled shock absorbers will also help in carrying heavy tongue loads.

Figure 627 **The trailer hitch should be of a type that bolts or is welded to the car frame for safety. Do not use a hitch that is merely attached to the vehicle's bumper.**

Hitch balls on the car must match the coupler on the trailer tongue. Ball diameters start at 1-7/8-inch and generally increase in 1/8-inch increments. The diameter and GTW rating are normally stamped on each ball; larger balls have large-diameter threaded shanks. *Never use a mismatched coupler and ball.*

At least one SAFETY CHAIN between the trailer and the towing car is a must, and two is desirable. The chain's size must be adequate for the trailer with which it is used; minimum breaking strength should be 1½ times the maximum gross trailer weight. Any connecting hooks and attaching hardware should be of equivalent strength.

Brakes

Brakes must often meet requirements of state or local regulations for type and design. In general, brakes are necesssary for rigs of 1,500 pounds (680 kg) or more gross weight. It may be well to have even better brakes than the minimum legal requirement.

Many trailers have hydraulic "surge" brakes which act automatically as the towing vehicle slows. Although such brakes act *after* the car's brakes, the time lag is so brief there is no tendency for the rig to jackknife.

There are electric trailer brakes which act automatically as the car's brakes are applied, and also electric brakes that are manually controlled by the driver. Automatic brakes are desirable because the driver does not have to remember to apply them, but manually controlled brakes provide more flexible operation.

An excellent feature, required in some jurisdictions for larger trailers, is a "breakaway" device which will automatically apply the trailer's brakes if it becomes detached from the towing car, as in the case of a hitch failure.

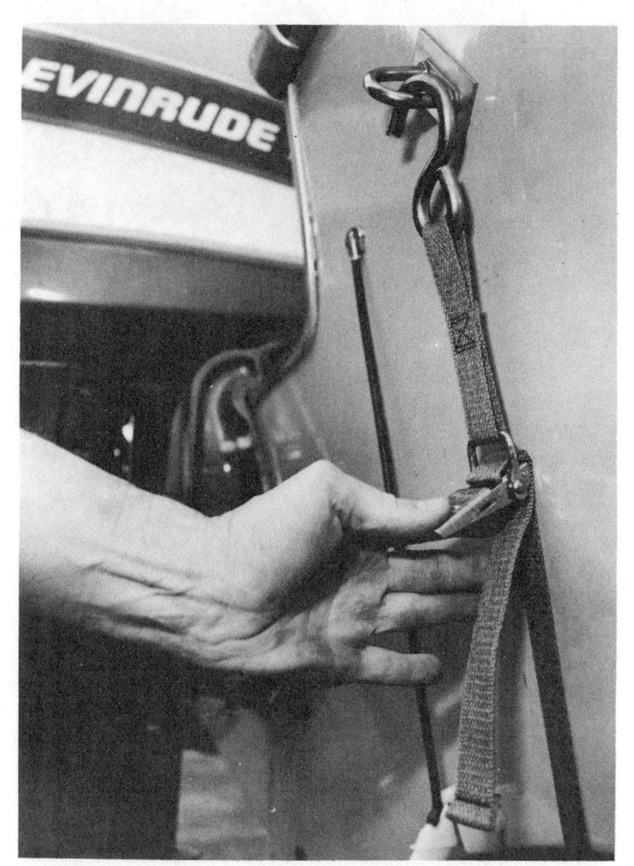

Figure 629 **When being transported, a boat must be held firmly on the trailer bed so that it cannot shift, or bounce on the bed. In storage at rest, tie-downs should be slacked off a bit to reduce any distorting strains on the hull.**

Tie-downs

Do not depend on the bow winch cable to hold the boat onto the trailer, and/or the bow down into its chock. There should be an additional TIE-DOWN of a lashing and turnbuckle. Moving aft, straps may be fitted across the hull to hold the craft firmly against the supports of the trailer, and to prevent any bouncing up and down on the trailer when on rough roads. There should be separate lashings with turnbuckles at each side of the transom. If a trailer as purchased does not have adequate hold-down straps and fittings, these should be added before any extensive trip is made.

Wheels and Tires

Tire failures on the road are one of the most prevalent and annoying troubles of trailering, and one of the least necessary. An understanding of the differences between car and trailer tires, and proper selection and maintenance, will virtually eliminate tire troubles.

Trailer wheels are usually smaller than those on the towing car (see page 131), and tire diameter is consequently smaller. This means they are more heavily loaded than the car tires, and must rotate faster. For high-speed, long-distance highway trailering, favor a larger wheel diameter over a smaller one, and select tires capable of meeting this service.

Electrical Systems

A trailer's electrical system powers its lights—tail, stop, license-plate, marker, and turn-signal—and in many cases a braking system. Good quality components and careful installation are essential for safety and long, trouble-free operation. Wires should be protected from mechanical abrasion, road splash, and other physical damage. It is important that there be a ground wire as a part of the electrical system between car and trailer; the coupling and hitch do not provide an adequate "return path" for electrical current. Connectors should be heavy-duty, and waterproof.

Check carefully all state and local regulations regarding the number and placement of lights on a boat trailer the size of yours. It is often recommended that tail, stop, license-plate, and turn-signal lights be mounted high—either on the boat or on a bar placed across the boat's transom—but in some jurisdictions these arrangements may not be acceptable.

Trailers are normally sold equipped with several light reflectors. Additional reflectors and/or reflector tape may be added as deemed necessary by the owner.

General Features

There are a number of optional features for trailers, the selection of which will depend upon the craft and its owner.

A winch is used with all but the very smallest of trailered boats. Designs and capacity vary considerably; both manual and electric models are available. Two-speed manual winches have a low-speed, high-mechanical-advantage mode of operation for heavy pulls, with a faster operating mode for use once the pull becomes easier. Electric winches are powered from the car's battery, and the motor makes the work of loading a heavy boat just a matter of pushing a button.

Figure 630 **Winches are used to gain added pull to get the boat back onto its trailer. Both manual units, shown, and electric winches are available with various gear ratios.**

All winches should have a hand brake and an anti-reverse latch or lock for safety. The winch cable may be of synthetic line, but steel cable is frequently used. If you use the latter buy a pair of heavy gloves also, because a steel winch cable must never be touched with bare hands. Any frayed or broken strands can cut hands badly.

The winch should be mounted on a stand that can be positioned fore and aft to match the positioning of the boat on the trailer bed. Vertically, a winch cable should come off the top of the drum on a level with (or slightly above, but never below) the boat's bow eye, to which the cable is attached. The winch should be capable of adjustment on its stand to meet this requirement.

Many trailer designs feature a bed or frame that is hinged so that it can tilt up for easier launching and reloading. With such a model you need not back the trailer wheels into the water, an action to be avoided if at all possible.

A convenient feature found on some models is a narrow walkway down one side of the frame inside the wheels. This permits you to get out over the water to attach the hook of the winch cable, rather than wade out and get wet feet. It is a simple feature, and often can be added by the owner if it isn't part of the basic trailer.

Figure 631 **A walkway can be added along one side of the trailer bed. This can make it much easier to guide the boat when it is being maneuvered on or off the trailer.**

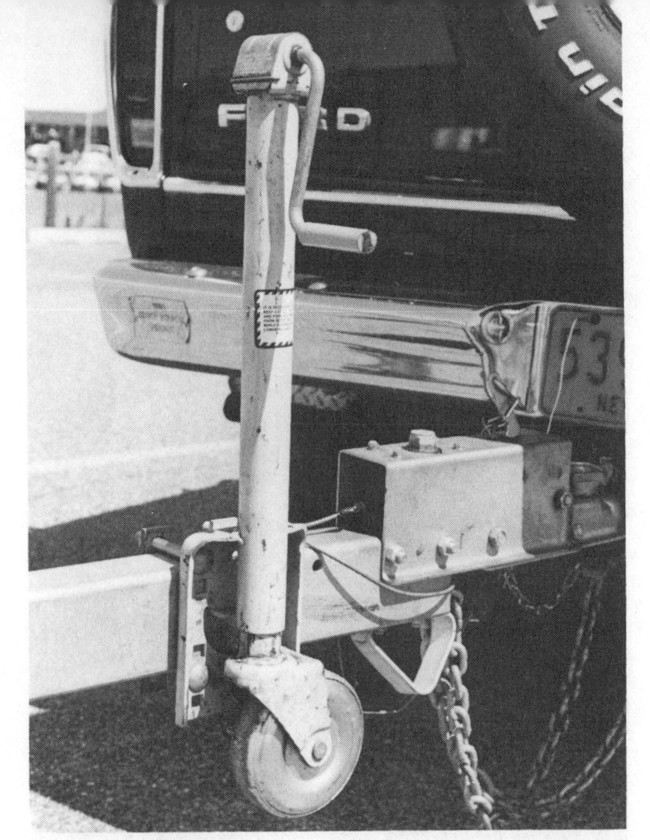

Figure 632 **A small stand at the front of the trailer, fitted with a wheel, allows a trailer to be parked in a level position and to be maneuvered free of the vehicle. The stand may or may not be adjustable, but it should pull or swing up to clear the road when the trailer is attached to the towing vehicle.**

Accessories

In addition to the basic features and options, there are accessories to make your trailering both safer and more enjoyable.

Some trailers include a front parking wheel, and one can be added in most cases when it is not standard equipment. A parking wheel should be on an adjustable shaft so that the height of the trailer tongue from ground level can be varied. The wheel will aid in maneuvering the trailer when it is detached from the car, and it will make hitching up an easier and safer job. The height can also be adjusted to give the proper angle to the boat when it is parked for storage.

Jack and Lug Wrench

Every over-the-road trailer rig should include a jack for the trailer; most automotive jacks are useless if the trailer should have a flat tire. The jack need not be elaborate–there are many simple types that are suitable.

As trailer wheels are generally smaller than those on cars, the size of the wheel lugs may be different. If the lug wrench for your car doesn't fit the trailer lugs, be sure to purchase and carry a wrench that fits.

Spare Wheel and Tire

One of the best investments that a trailer boatman can make is a spare wheel and tire. Service for the smaller tire sizes usually used on trailers, or a replacement if one is ruined, is seldom available on a highway. Having your own spare might save you many hours, or even days, should trouble occur on a trip.

A related item for possible emergency use is a set of replacement wheel bearings. Bearings can cause trouble unexpectedly, and they, too, can be difficult to locate in out-of-the-way places.

Other Gear

Every rig should have highway flares, flags, reflectors, and similar equipment to set out if a flat tire, overheated bearing, or other trouble should cause an emergency stop along the shoulder of a road. A pair of wheel chocks of suitable size and shape are handy for use with either the car's wheels (when launching or reloading) or the trailer wheels (when parking on sloping ground).

Although a cover for the boat is not strictly an accessory for a trailer, it is a very desirable item in connection with road travel. A properly fitted cover, secured for high speed travel, will keep out dirt, trash, and rain when on the road.

Insurance

Know *fully* the status of your insurance coverage in regard to pulling a trailer with your car. Some auto policies allow this, others do not, and some provide limitations on hauling trailers. Check your policy carefully and, if necessary, consult your insurance agent. If an endorsement is needed don't delay or neglect to get it, even at the cost of a small additional premium—be protected, especially in your liability coverage.

Car Requirements

A car used with a trailer may need some modifications in the way of "beefing up" to give fully satisfactory performance. Rear springs and/or shock absorbers may have to be replaced with heavier duty units because of the added weight transferred from the trailer tongue. The turn signal flasher unit may need replacement with one that can handle the additional load of the trailer signals.

Driving with a boat trailer behind the car is not as easy as ordinary driving, especially when the craft is high and wide. A second external rear-view mirror on the car's right side will make passing both safer and easier. If the boat is quite wide, it's best to install the external mirrors that project out far enough to give a clear view behind the boat; these are the mirrors often used on cars that pull house trailers.

If a heavy trailer is to be pulled, it may be desirable, or even necessary, to add a cooler for crankcase oil and transmission fluid. Vehicle manufacturers are becoming increasingly specific in the maximum loads that can be towed without modifications, and just what modifications are available—check your owner's manual and don't void your warranty.

A hitch on the front end of the car can be useful in some situations, if the bumper will accept such a hitch. Where a ramp is slippery—such as at low water states in tidal areas—or where the ramp is unsurfaced and soft, using a front bumper hitch permits the car to be turned around so the rear driving wheels will have better traction. Even when the vehicle has front-wheel drive, a front hitch permits better maneuverability of the rig in tight quarters, and the headlights provide illumination when launching or retrieving at night. Although a bumper hitch is suitable for this use, you should avoid putting heavy strains on it that could damage the car.

Multiple-Axle Trailers

As the length of a boat increases, so does its weight, and at an even greater rate. There is a definite upper limit to the weight-carrying capacity of trailer tires, and when the boat-plus-trailer gross weight exceeds the capacity of two tires, the only solution is the addition of another axle and pair of tires. Even with large tires such as 6.00-12 in a 6-ply rating, the gross weight limit for two is about 2,500 pounds (1,130 kg).

Because of their greater size and weight, trailers that have two or more axles are generally equipped with electric brakes and front parking wheel as standard equipment, and weight-distributing hitches should be used on the car.

Figure 633 **Many boats are of such size that they obstruct normal rearward vision. A large mirror on the right side of the vehicle, as well as one on the left, overcomes this. Cars may also require cooling system modifications.**

Trailer Operation

Safe, trouble-free trailering depends as much on proper preparations as on correct procedures when "on the road." Preparations are not complicated or difficult, but even a single item overlooked can make problems. It is an excellent idea to make up a check list, *and use it*. Such a list can be typed and then sealed in plastic; thus waterproofed it can be attached to the trailer tongue, where it won't be overlooked.

General Preparations

Start with the towing vehicle. It should be in good mechanical condition and fully serviced prior to a trip. The trailer, too, should be lubricated and given any other needed service well ahead of the time for departure.

All gear that is to be carried in the boat must be properly stowed. Soft and light items such as sleeping bags or cushions may be left loose, but heavy items or hard ones with sharp corners and projections, such as an anchor, must be secured firmly in place. If the motor has electric starting, the battery, although relatively small, is a particularly heavy item and must be held firmly in position, preferably in a fiberglass box.

Don't try to carry too much in a boat on a trailer, but don't overlook any essential items. Use a check list, and don't let the gross weight of the load creep up. Consider traveling with empty tanks and filling up at the other end—often 100 to 300 pounds (45-135 kg) or even more can be saved. Fuel weighs 6.6 pounds per gallon (0.8 kg/liter), and water is even heavier at 8.3 pounds per gallon (1 kg/liter). It may be necessary to carry some heavy items in the towing vehicle.

Weight Distribution

Give consideration to weight distribution in the trailer, switching around light and heavy items in the boat to bring the trailer into proper balance. In extreme cases, such as with craft of unusual design, you may need to relocate the axle and wheels with respect to the trailer frame, to achieve proper balance for the loaded trailer.

The optimum weight on the trailer coupling at the end of the tongue is approximately 7% of the gross weight, with any value between 5% and 10% being acceptable. For a larger boat-trailer combination of, say 3,000 pounds (1,360 kg), this means 150 to 300 pounds (70-135 kg) at the coupling—here a parking wheel with screw-action adjustment is an obvious necessity.

For smaller, lighter rigs, the weight on the tongue can be estimated when it is picked up from the ground. A more accurate measurement is desirable, however, and you can usually get one easily by using a household bathroom scale. This method is really a necessity as tongue weights approach and exceed 100 pounds (45 kg).

Effects of Improper Tongue Weight

The most readily apparent symptom of poor weight distribution is excessive sagging of the car's rear end. The load is too far forward on the trailer, the hitch is in danger of striking the road surface on bumps, and the trailer follows poorly because of the faulty balance.

Too light a tongue weight also can cause highway problems as there may even be an *upward* thrust on the hitch, as the car moves along. This, too, makes driving difficult as the trailer will tend to sway from side to side.

Check your tongue weight carefully before each trip, but once on the road observe how your trailer follows. It may be that your particular rig or hitch will ride better with a tongue load toward either the high or low side of the recommended range.

Preparing the Boat

Before you set out, tilt up the outboard motor and secure it firmly in position. If no travel latch is provided, a block of wood can be inserted between the motor and the transom, and the motor lashed down against it to keep it in place. The lower unit of a stern-drive may be in either the down or tilted position, as recommended by the boat manufacturer, but if tilted it must be secured in that position.

Tops and Covers

A folding canvas top should be in the down position; these are not made to take the wind speeds of highway travel—50 mph (80 km/hr) on the road will have the same effect as gale winds, or even more if moving into a head wind! Take down all flags, too; these are not made to stand the whipping of such strong "winds."

The boat cover should be on and securely fastened so that the wind, and dust and dirt, will not get in under its edge. If the trip is to be made on dirt roads, place a simple cover of canvas or plastic over the trailer winch to keep out grit that might interfere with its operation.

Securing the Boat

All tie-downs must be correctly tightened—not too much, not too little. They should not be so tight as to compress or otherwise distort the hull, but they must be tight enough so that there is no relative movement between the boat and the trailer supports; all vertical motion should be as a single unit, absorbed by the springs of the trailer. Additional tie-downs, not needed for storage, may be added, especially at the bow and transom; the anti-reverse latch of the winch should be checked.

Figure 634 **Boat tops are designed for cruising speeds, not the rigors of high-speed highway trailering; they should be lowered before the rig goes on the road. Well-secured boat covers, however, can be used to keep out dirt and rain.**

Tires and Wheels

Incorrect tire pressure is the biggest cause of trailering problems. Trailer tires use higher pressure than regular passenger car tires, and the pressure used must be right for both the size of the tire and the load carried by it (gross weight less tongue weight, divided by the number of tires).

Trailer tires must not be overloaded; the tire's "ply rating" is an important factor in determining the maximum load capacity. Table 6-2 gives the tire load capacity for various size tires and inflations at highway speeds. The highest figure on each line is the maximum inflation pres-

Load Capacity of Trailer Tires

Tire Size	Ply Rating	*Pounds of Tire Pressure (Measured Cold)*											
		30	*35*	*40*	*45*	*50*	*55*	*60*	*65*	*70*	*75*	*80*	*85*
4.80/4.00 × 8	2	**380**											
4.80/4.00 × 8	4	380	420	450	485	515	545	575	**600**				
5.70/5.00 × 8	4		575	625	665	**710**							
6.90/6.00 × 9*	6		785	850	915	970	1030	**1080**					
6.90/6.00 × 9*	8		785	850	915	970	1030	1080	1125	1175	1225	**1270**	
20 × 8.00-10	4	825	**900**										
20 × 8.00-10	6	825	900	965	1030	**1100**							
20 × 8.00-10	8	825	900	965	1030	1100	1155	1210	1270	**1325**			
20 × 8.00-10	10	825	900	965	1030	1100	1155	1210	1270	1325	1370	1420	**1475**
4.80/4.00 × 12	4	545	550	595	635	680	715	755	**790**				
5.30/4.50 × 12	4	640	700	760	810	865	**915**						
5.30/4.50 × 12	6	640	700	760	810	865	915	960	1005	1045	1090	**1135**	
6.00 × 12	4	855	935	**1010**									
6.00 × 12	6	855	935	1010	1090	1160	1230	**1290**					
6.50 × 13	6	895	980	1060	1130	1200	**1275**						

Table 6-2 **Figures in boldface are the maximum recommended inflation-load values for tires of each size at highway speeds.**

sure and load, and *must not be exceeded.* Remember that these are "cold" pressures, readings taken before the trip is started. Tire pressures will increase during the trip due to heating, but do *not* bleed out air to reduce pressure as this is a normal change anticipated in the setting of the cold pressure.

Tires that are under-inflated tend to bulge at the area in contact with the ground. As the tire turns, this area constantly flexes, putting an abnormal strain on the sidewalls and overheating the tire. Tire life is shortened, and actual failure can occur on the road.

Excess pressure is equally bad. In the worst case, the pressure will rise during driving to such a point that the sidewalls burst. At best there will be excessive wear in the center of the tread, and shortened tire life.

Know your loads; know your correct tire pressure—check tires *before each trip;* a special gauge may be needed.

Wheel Lugs

The lug nuts or bolts on each trailer wheel should be checked periodically for tightness; this is not necessary for every trip, but it should not be neglected. Looseness can cause uneven tread wear, and if a wheel should come off while driving, a serious accident is likely.

Wheel Bearings

Proper lubrication of wheel bearings is a must in trailer operation. "Bearing Buddies" permit you to add grease, forcing out any water that has gotten in during launching or retrieval. Although such adding of grease will increase the interval between complete removal and repacking, this should be done at least once a year for longer bearing life.

Tools and Spares

Make sure you have in your boat (or car) all necessary spares and tools—especially the spare wheel and tire (properly inflated) for the trailer, and the jack and lug wrench to use if making a change is required. Spare wheel bearings and the proper grease can prevent delays if trouble develops.

Be sure you carry a spare for each type of light on the trailer—these may burn out, and should be replaced immediately for safety

Laws and Regulations

You should be familiar with the laws and regulations

governing trailering in your state. If you are venturing farther than normal from home, however, don't assume that the requirements will be the same everywhere. Information on what is required usually can be obtained from police departments, auto clubs, or boat clubs, either locally or in the new area.

Hooking Up

Before mating the trailer's coupling to the car's hitch, the ball may be lightly coated with grease, if necessary, to reduce friction. After joining the two units make sure that the coupling is tightened properly and locked in accordance with its design. Connect the safety chains *immediately*, so they won't be forgotten. Most trailers use "S" hooks, Figure 628; a safer connection can be made with a "scissor" hook, or a shackle or clevis. If a single chain is used, it should be long enough to go around the hitch drawbar and back to the trailer to fasten onto the frame or to the chain itself near its beginning; double chains should be crossed. *The chains must have no more slack than is necessary for making sharp turns; they should not be long enough to permit the coupling to fall to the road if the trailer should suddenly become unhitched.* Next raise the parking wheel to its proper position for on-the-road travel, and make the electrical connection between car and trailer. (A neat trick is to have the electrical cord on the trailer be long enough that the connector can be placed *inside* the car trunk, to protect it from rain and road splash; the trunk lid will close easily over the wires without damage.

A Final Check

Before moving out make a final run-down of the check list. Check all lights with someone in the car depressing the brake pedal, and operating the switches for tail, license plate, stop, turn, and marker lights (if used). And don't forget to ask each member of the crew, "Have you forgotten anything?"

On the Road

Driving a car pulling a trailer can be quite different from driving the car only, especially the light, lower-power models that are now popular. There will be slower acceleration, and more time and distance will be required to stop.

If you are new to trailering, or if you have acquired a larger rig, take time to learn the driving characteristics of the combination. Practice in light traffic before taking to the crowded freeways and major roads on weekends.

Driving

The first rule in driving with a trailer behind is "No passengers in the trailer!" This practice is unsafe, and hence illegal in most jurisdictions. Do not even allow pets to ride in the trailer.

Drive as smoothly as possible—no sudden jerks in starting, and well-anticipated, easy stops. Jerks and sudden stops put added loads on the trailer connection, the boat itself, and the gear stowed in it. Avoid quick turns and sudden swerves; these can make the rig harder to handle.

Just as soon as you start out—before you leave your own property if possible—test your brakes and those of the trailer, if it is so equipped.

Watch the trailer in the rear-view mirror as much as possible, and keep an ear out for the development of unusual noises. If a car passes you with driver or passengers making hand signals to you, pull off the road and check to see what might be wrong.

Observe all speed limits, these are sometimes slower for cars with trailers, and for good reasons. Don't tailgate; your stopping distance is extended because of the trailer. Use added caution in passing; remember that your overall length is twice as much as normal, or more, and your car will have to be well ahead of the passed vehicle before there is room for you to get back into lane. Remember, too, that it will take longer to accelerate to passing speed. Keep in the proper lane; on multi-lane high-

Figure 635 **Begin a trip only after all preparations have been completed; use a check list to make sure nothing is left undone. If traveling a long distance, plan on stopping at intervals to check tie-downs, bearings, and the trailer stop lights and turn signals.**

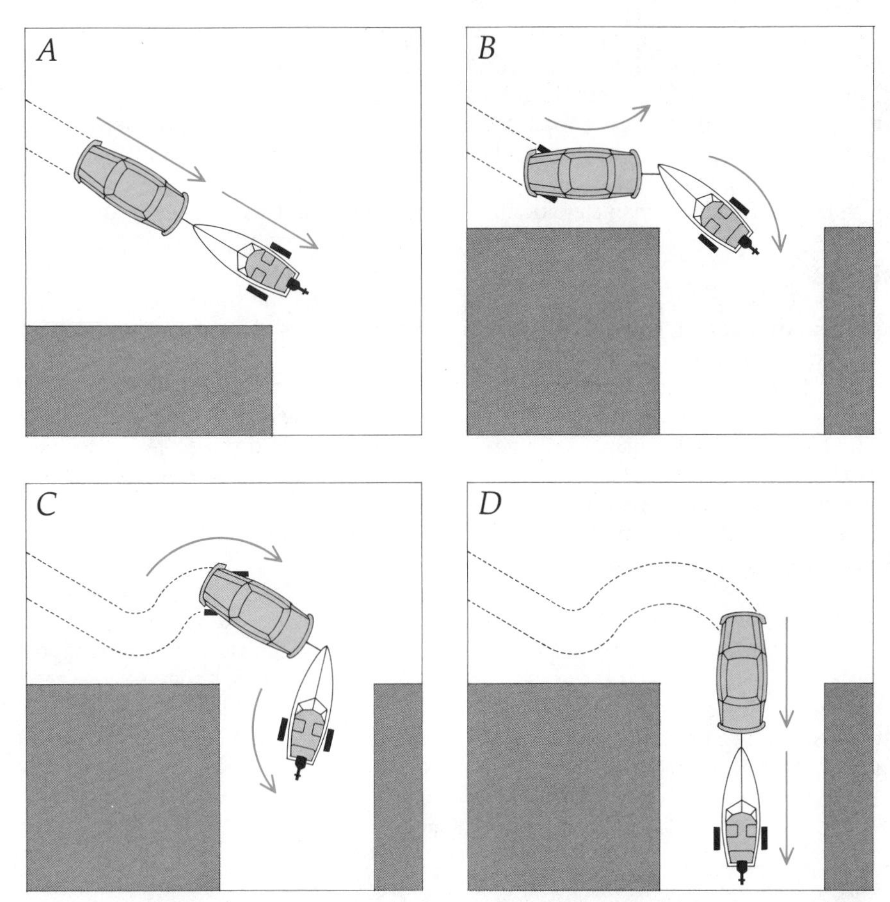

Figure 636 **The trailer always backs in the direction opposite to that of the car. At *A*, driver swings close to launching ramp, then at *B* cuts the car in toward the driveway. At *C* he cuts car wheels to the left and backs slowly into ramp as the trailer moves to the right. At *D* the driver straightens the car wheels to follow the trailer as it backs straight down ramp. It is best to practice in an empty parking lot.**

ways this will normally be the right-hand lane. If there is only one lane in each direction, and a pile-up of traffic begins to form behind you, it is courteous and a contribution to safety to pull to the side (where you can safely do so), and let the faster vehicles get past you.

On city streets, and in other restricted areas, you must swing a bit wider than usual in turns to allow for the trailer, which will tend to cut across the corner slightly rather than follow in the tracks of your car. It can be embarrassing to bump the trailer tires up over the curb when turning at a street intersection!

Stop and Check

On long highway hauls at relatively high speeds, it is important to stop occassionally and see how the rig is riding. For long trips you might stop after the first 5 to 10 miles (8-16 km), then again after about 50 miles (80 km), and then each 100 miles (160 km). The initial inspection is made *soon* after starting out, so any defect can be spotted before it becomes a serious problem. Make each inspection a thorough one. Stopping intervals, of course, must be adjusted to the availability of suitable locations, especially on heavily traveled roads.

If your trip begins in daylight but continues after dark, stop when it is time to turn on your vehicle's lights and recheck the trailer lights as well as those of your car.

Inspection Procedure Follow a regular procedure to make sure nothing is overlooked. Check the hitch for tightness, and the safety chains for the right amount of slack. Check tires visually; as mentioned, a pressure build-up is normal so don't bleed off air in the belief that it is excess. Feel all wheel bearings; warmth is okay, but if they are too hot to touch, you have a problem. Let them cool off, and then drive as slowly as possibly to a service station for disassembly and inspection. Be sure to check all tie-downs, and take a look inside the trailer at the stowed gear. Check all lights, especially stop and turn signals.

Check the car's tires and lights while you're at it, but do not check radiator water unless there have been signs of overheating. Then do so with great care if the cooling system is pressurized, as on most cars.

When Conditions Are Bad

Driving in rain, fog, or other conditions of reduced visibility is always a bit more hazardous, but it is especially so with a trailer in tow. Consider the added length of the rig, and its less favorable acceleration and stopping characteristics. High winds, particularly cross winds, present added hazards, and trailer rigs may be banned from some roads and bridges at these times. Be alert for caution or prohibition signs at toll booths and bridge approaches.

Pulling Tandem Axle Trailers

It has been found that the "following" characteristics of trailers with two axles can be improved in some cases by *not* inflating all four tires equally. If you are experiencing any weaving back and forth at highway speeds, try varying the tire pressures by 5 or 10 pounds with the front wheels softer and the rear wheels harder than normal pressure, or vice versa. No specific guidance can be given except to try various combinations and check for either improved or worsened handling characteristics.

At the Destination

When you arrive at your destination, make a final check just as you would at an intermediate stop. Even if the tie-downs are to be removed soon, for example, it is well to know whether or not they have tended to work loose. The same goes for the hitch and safety chain. Hot wheel bearings, or even warm ones, must be allowed to cool down before launching (see below). If bearings are too hot, it indicates a potential seizure problem. If at all possible, you should have the bearings serviced before the return trip.

Maneuvering the Trailer

There's a knack to maneuvering a car and trailer in close quarters, especially when backing up, that's best learned by practice. A large, empty parking lot is the best place for this; use empty boxes or similar objects to mark limits for parking, backing, and other maneuvers.

Learn how to back in a straight line, and how to turn while backing to put the trailer just where you want it. *A trailer always backs in a direction opposite to that of the car.* The trick is to turn you car's steering wheel in the direction you want the trailer to go. By turning the steering wheel to the right (clockwise), for example, the front end of the car will move to the left, but the trailer will back to the right. As the trailer moves into the desired path, reverse the direction of the steering wheel to let the car "follow" the trailer.

The key to success is *slowness;* make all moves slowly and watch the results carefully. If the trailer does not turn far enough, or turns so far as to tend to jackknife, just stop, pull ahead a few feet to straighten out, and try again.

Special Tips

When trailering a boat for any distance, it is a good idea to remove a fire extinguisher from the boat and carry it in the car (if the car is not so-equipped). All too often a passing boat on a trailer is a target for deliberately or carelessly discarded—and you may not be able to get an extinguisher in the boat. Don't forget to return the extinguisher to the boat before the boat is launched.

Launching and Retrieving

Both launching a boat from its trailer, and its retrieval (loading), must be carefully planned and executed to avoid damage to the boat or motor, or injuries to people.

Using a Ramp

Using a ramp is not the best way to get a boat into the water, but it is the method used most often. Ramps vary widely in their characteristics; many are surfaced with concrete or asphalt, while others are hard-packed dirt or sand—sometimes reinforced with wood or steel planking. Some are wide enough for only one launching operation at a time; others are wide enough for many rigs to use simultaneously.

The quality of a ramp depends on its slope, how far it extends into the water, and the condition of its surface. The angle of slope is not critical, but it should be steep enough so the trailer need not be backed down too far into the water and the wheel bearings submerged. If the slope is too steep, however, you may need an excessive pull to get a loaded trailer up and off the ramp. The ramp should extend far enough into the water that trailers can be backed down without running off the lower end, even in low water conditions. (Many surfaced ramps develop a sharp drop-off at their lower end; if the trailer has rolled past this point getting trailer wheels back up over this lip is a problem. A dirt or sand ramp must be firm enough to support the trailer and car wheels. A surfaced ramp must not have a coating of slime that could make footing dangerous, and provide inadequate traction for the car.

Preparations for Launching

Allow time for wheel bearings to cool down if there is any chance of their getting wet. If a hot bearing is put into cold water, the sudden temperature change creates a partial vacuum in the bearing housing, sucking in water that can corrode the bearings.

While the bearings are cooling, remove the boat cover, fold it, and store it in the car or boat. Tie-downs can be removed and stored in a safe place, but leave the winch line taut. If the motor or stern-drive lower unit has been in the down position during trailering, tilt it up. If the trailer's lights will be submerged, disconnect the plug to the vehicle's electrical system.

Remove or relocate equipment stowed in the boat so that the boat will trim properly when launched. If the fuel or water tanks are empty, or only partially filled, they can be topped up, unless there is a pier to move the boat to after it is in the water.

Use Two Lines

Attach both a bow and a stern line; a boat cannot adequately be controlled with only one line. Put over the side any fenders that might be needed if the boat will be moved to a pier or seawall once it is in the water.

"Preview" the Launching

Study the ramp and surrounding water area for any hazards, such as a slippery or too-short surface; estimate the wind and current effects. In case of doubt, don't hesitate to ask another skipper who has just launched or

Figure 637 **On many ramps, trailers must be backed well down so the boat can be floated free, but the best practice is to avoid immersing the trailer wheel bearings. Note that the bow and stern lines are being tended as the boat is launched. Note, too, that a crew member outside the car can tell the driver exactly when to stop.**

retrieved a boat. If you have time, watch the launching operations of others, noting any peculiarities of the ramp which may be new to you.

Now check the drain plug, if one is used. It is embarrassing to launch the boat, and have it start filling up with water! Look about to make sure that no item of preparation has been forgotten—then *check the drain plug again!*

The Actual Launching

Line up car and trailer so the backing process will be as straight and short as possible. On a wide ramp, give due regard to others and don't take up more than your share of space.

Back the rig down, preferably to the point where the trailer's tires, but not the axle and bearings, are in the water. (If ramp conditions are unfavorable, and you have a front bumper hitch, it may be advisable to use this for maneuvers on the ramp.) Set the parking brake on the car, and block a wheel on each side for safety. Have one person man the winch controls, and one or two take the bow and stern lines.

Release the trailer tilt latch, if it is this type. Tighten the winch brake and release the anti-reverse lock. *Do not* disconnect the winch cable from the boat. A craft then should slide easily off the trailer, its speed controlled by the winch brake; a push or two may be needed to get it started. Be sure the motor is tilted up so the propeller and skeg will not dig into the bottom as the boat slides down.

When the boat is floating free of the trailer, unhook the winch line. Move the boat aside and make it fast to a pier, beach it, or otherwise secure it temporarily. Return the tilted trailer bed to a horizontal position, and latch it in place. The winch line may be rewound on the drum if desired, or secured by catching the hook on a member of the trailer frame and taking up the slack.

Remove the car's wheel blocks, and drive it to an authorized parking area—give consideration to others by clearing the ramp promptly, and selecting a parking spot that will not take up unnecessary space or block others. If possible, hose down with fresh water any parts of the trailer that got wet during the launch. Use a padlock on the coupling or safety chain to prevent theft of the trailer. Place the chocks at the wheels of the trailer if it is detached from the car and the parking area has any slope.

At the boat, lower the motor or stern-drive unit to its operating position, and connect the fuel line if necessary. Complete any preparations needed for getting underway, such as transferring equipment to the boat from the car, load up your crew, and clear away from the launching area as rapidly as can be done with safety.

Figure 638 **In most cases a slight shove is all that is needed to start the boat sliding off the trailer rollers or pads.**

Figure 639 **Once free of the trailer, the boat can be pulled by bow and stern lines around to a pier for loading, while the trailer is driven to its parking space.**

Reloading on the Trailer

Beach the boat or make it fast to a pier, and get the car and trailer from the parking area. If the boat will not be used in the next day or so, it is a good idea to disconnect the fuel line of an outboard while the motor is running at a fast idle. Let it run until all the fuel in the motor is used up; this will help prevent the formation of gum and deposits in the carburetor and fuel lines.

Back the trailer to the water's edge, again stopping it, if possible, before the wheel bearings are submerged; make sure that the electrical plug at the vehicle is disconnected if the lights will go into the water. Set the car brakes and block the wheels. (A front bumper hitch is even more useful in retrieval where the extra pulling power of reverse gear can be used.) If the trailer bed tilts, release the latch and push the frame into the "up" position. Also tilt up the boat's motor or stern-drive unit, and work the boat into position to move onto the first rollers, with the keel of the boat in line with the trailer. Again, both a stern and bow line help in boat maneuvering. It may be possible for one of the crew to wade into the water to guide the boat into position. Run out enough winch cable to engage the hook in the boat's stem eye; watch out for kinks in the cable, and remember never to handle a steel line except with gloves.

Crank in on the winch line and the boat should come easily onto the trailer bed; a tilted frame will come down to the horizontal position by itself when the boat moves up it. Often a winch has a lower-geared speed which is useful for the initial pull to get the boat started, and a higher-speed mode to use once the pull gets easier.

Do not allow anyone to be in line with the winch cable. A

Figure 640 **In reloading, the winch line is simply attached to the bow eye, and the boat is cranked up onto the trailer bed; this bed tilts to aid in retrieval and launching. If a winch has two speeds, use low gear to get the boat started.**

cable under load snaps like a rubber band when it breaks, and can throw a hook or fitting great distances. *Keep clear!*

When the boat is fully on the trailer, latch down the tilt mechanism (if it is of this type), remove the wheel blocks, and move the rig clear of the ramp so that others can use it. Don't overlook the other preparations that must be made for road travel, but get away from the launching area to do them.

Launching by Crane

The use of a crane or travelling lift with padded slings is the launching method that is probably easiest on the boat. Done with care, it minimizes strains on the hull. Slings are usually provided by the crane operator, but check to make sure they are adequate before entrusting your boat to them. Know the proper placement of slings for a safe balance of your boat—your dealer can provide you with this information.

Use both bow and stern lines to control any swinging motion of the boat while it is in the slings, between the trailer and the water.

Figure 641 **Where a ramp is not available, a crane can be used to lift a boat from its trailer and lower it into the water.**

Land Storage

A trailered boat is normally stored on land between uses, spending more time on its trailer bed than in the water. For the sake of both boat and trailer, land storage must be done properly.

You can create at least partial protection against theft of boat and trailer by installing a special fitting that is secured to the trailer coupling with a keyed lock—if they can't "hitch up," they can't haul it away!

When the car is not pulling the trailer, the hitch ball can be covered with a small plastic bag secured by a short length of cord or elastic. This will prevent any grease on the ball from rubbing off on clothing, and will provide some protection against rust. Some owners unbolt the ball from the hitch and store it in the trunk or other protected spot.

Keep Water Out

Rainwater in the boat is undesirable for many reasons, but mainly because a collection of it can rapidly increase weight on the trailer, often beyond its capacity. A cover is desirable, but with or without it the tongue of the trailer should be raised enough so any water will run aft and out the drain hole. Crank down the parking wheel to raise the tongue, or, lacking this, block it up securely. This is one time you want to make sure the drain is *open.*

If a cover is used, give it adequate support so that pockets of rainwater cannot form, stretch the fabric, and possibly break through. Cross-supports under the cover and frequent fastening points around the edge will keep the cover sag-free; allow some "give" if the fabric is subject to shrinkage when wet.

Ease Strains on the Hull

Tie-downs kept taut for over-the-road travel should be slacked off, or even removed, in order to eliminate any strains and possible distortions to the hull that might occur during long storage periods.

Blocking up a Trailer

If a rig is not to be used for weeks for months, the trailer frame should be jacked up and placed on blocks to take the weight off the springs and tires. Use enough blocking to prevent distortion of the trailer frame—which in turn could distort the hull. The blocks need be just high enough to take the greater part of the load, but if the trailer wheels are clear of the ground, reduce air pressure in the tires by 10 to 15 pounds. Remember that you must have means at hand to restore pressure when the blocks are later taken out for the next trip; don't deflate tires unless you can pump them up again right there.

When the boat is to be stored for an extended period, it's best to remove the battery, electrical components, and even the motor—if possible—and store these items indoors for greater protection against theft and corrosion.

Frequent Checks

When the boat is stored and out of regular use, set up a schedule of weekly visits by yourself or someone else acting for you. Doing this at the same time each week will prevent forgetfulness or neglect. On each visit, check for any signs of deterioration, such as peeling paint. Don't let little problems grow into big ones before you take corrective action.

7

Principles of propeller and rudder action; their application in the handling of both single- and twin-screw inboard vessels

POWER CRUISER SEAMANSHIP

Some of the finest exhibitions of the art of boat handling are given by men who have never read a printed page on the subject of seamanship.

Witness, for example, the skill a commercial fisherman displays in manuevering his skiff into a tight berth in foul weather, bringing her against the wharf in a landing that wouldn't crack an egg. Such proficiency, develped over long years of meeting every conceivable situation, manifests itself almost as an instinct, prompting the boatman to react correctly whether he can think the problem out beforehand or not.

Sometimes the old-timer's methods might appear, to a novice, to verge on carelessness. That's probably because the experienced skipper understands in a practical way, however, just what the minimum requirements for safety are, so he wastes no time on non-essentials. If he uses three lines where a textbook says six, don't jump to the conclusion that he wouldn't know how to use more if they were needed.

So, in approaching a study of the principles of boat handling, keep in mind that the goal is to understand why your boat reacts the way it does when you're out there handling her. If she stubbornly refuses to make a turn under certain conditions, or persistently backs one way when you want her to go the other, you are less likely to make the same mistake twice if you understand the reasons behind her behavior.

Helmsmanship

One characteristic of a good seaman is the ability to steer well. This is called HELMSMANSHIP. It is a quality that cannot be learned entirely from book or in a classroom, but it has basic principles that can and should be studied.

Boat's Individuality

Boats are nearly as individualistic as people, and nowhere is this more noticeable than in their steering characteristics. Deep-draft and shallow-draft vessels handle differently. Boats that steer by changing their thrust direction—outboards and stern-drives—respond differently from boats steered by rudders, the response of heavy, displacement hulls to helm changes is quite unlike that of light, planing hulls.

The secret of good helmsmanship is to know your boat. If you skipper your own craft, this comes relatively quickly as you gain experience with her. If you take the wheel or tiller of a friend's boat, take it easy at first with helm changes, until you get the "feel" of the craft's response.

Steering by Compass

At sea and in many larger inland bodies of water, steering is generally done by compass. The helmsman must keep the compass lubber's line on that mark of the card that indicates the course to be steered. If the course to be steered is 100°, and the lubber's line is momentarily at 95°, the helmsman must SWING THE BOAT'S HEAD, with right rudder, 5 degrees to the right, to bring the lubber's line around to 100. *Remember, the card stands still while the lubber's line swings around it.* Any attempt to bring the desired course on the card up to the lubber's line will produce exactly the *wrong* result.

As a vessel swings with a change in course, the inexperienced helmsman tends to allow her to swing too far, from the momentum of the turn and the lag between his turning the wheel and the craft responding to it. That's why it is necessary to "meet her" so the vessel comes up to, and steadies on, the new course without over-swinging. In almost all cruisers, this requires that the helmsman return the rudder to the neutral, midships position *before* the craft reaches the intended new heading. He often will need to use a slight amount of opposite rudder to check the boat's swing.

A zig-zag course, yawing from side to side, brands the inexperienced helmsman. A straight course is the goal; this can be achieved, after the boat has steadied, by only slight movements of the wheel. Anticipate the vessel's swing and make corrections with a *little* rudder instead of letting her get well off the course before correcting.

A good helmsman will also turn the wheel slowly and deliberately. The actual manner of steering is again a characteristic of each boat, and is learned only by experience.

In piloting small boats it helps to pick out a distant landmark or other fixed object, if there is one, or a star at night, to steer by (but remember of course that stars change their positions over a period of hours). This helps keep a straight course, because the compass is smaller

Figure 701 **On open water, away from buoyed channels, the only way the person at the helm can maintain the proper course is by keeping an eye on the compass.**

and less steady than aboard a larger vessel. Drop your eye periodically to the compass to check your course. Avoid steering on cloud formations, because they move with the wind and change shape.

When steering toward an aid to navigation or other object, *don't forget to look back!* You may be steering quite precisely toward your objective, but the wind, or a cross-current, may be setting you to one side. In a narrow channel, you may soon be out of it and aground. Look back at the aids you have passed, or any prominent landmark, and be sure that you are on your proper track, not off to one side.

Basic Principles of Boat Handling

The proper approach to boat handling is a balance of study and practice. Learn all you can about the principles by which *average* boats respond under normal conditions; supplement this with experience in *your own boat* and others, and then learn to act so that the controlling elements aid you rather than oppose you.

In boat handling, there are three primary types of powered craft to consider—inboard single-screw, inboard twin-screw, and outboard (including stern-drive) whether single or multiple. The first two use a rudder or rudders in combination with a constant thrust from the propeller or propellers; the last uses directed thrust, usually without a rudder as discussed in Chapter 6. The handling characteristics of each type are different from the others, but as the single-screw, single-rudder inboard is the basic type on larger craft, we will consider it in considerable detail. Note that the principles apply equally for a sailboat being moved through the water by inboard auxiliary power.

No two boats will behave in identical manner in every situation. Just how a boat performs depends on many things—among them the form and shape of the hull's underbody; the construction; the shape, position, and area of the rudder; the trim; speed; weight; load; strength and direction of wind and current; and the nature of the sea.

Effects of Wind and Current

Wind and current are particularly important factors in analyzing a boat's behavior, because they may cause her to respond differently from what you would expect. As a case in point, when in reverse, many motorboats have a tendency to back into the wind despite anything that can be done with the helm.

Given two boats of roughly the same size, one of which (A) has considerable draft forward but little aft, and another (B), with relatively greater draft aft but more superstructure forward, you will find radical differences in their handling qualities. What governs is the relative area presented above water to the wind as compared with the areas in the water, both fore and aft.

Boat A, with wind abeam, might hold her course reasonably well when the bow of boat B would persistently fall off, requiring considerable rudder angle to hold her up. On the other hand, with wind and sea aft, B might go along about her business with little attention to the helm while A insisted on "rooting" at the bow and yawing off her course despite the best efforts of the helmsman.

As a general rule, the boat with low freeboard and superstructure but relatively deep draft tending toward the sailboat type (Figure 702*A*) will be less affected by wind and more by current than the light-draft motorboat (*B*) with high freeboard and deckhouses. The latter floats relatively high in the water and has little below the waterline to hold her against wind pressures.

In boats with the greatest draft and least freeboard aft, the greatest exposure to wind is presented by the relatively higher bow and cabin forward. Her bow will thus be affected by wind pressures more than the stern. With

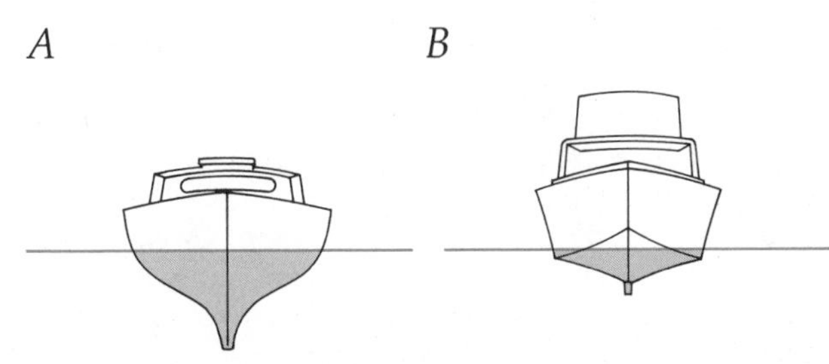

Figure 702 **A deep draft hull, *A*, will tend to be set off course by adverse water currents, but the low freeboard and superstructure offer little resistance to the wind. A typical semi-displacement hull, *B*, floats high in the water and a wind on the beam will result in considerable leeway.**

wind abeam her bow will tend to be driven to leeward more than the stern will, so she needs a certain amount of rudder angle to compensate.

For the same reason if she is drifting in a smooth sea with engine stopped, wind pressure on her bow will make the bow pay off so that the wind finally is brought abaft the beam. The action might also be compared to that of a sailboat with jib backed to windward.

But to show how boat behavior may vary, suppose it is raining and cockpit curtains are buttoned down on that same cruiser. Here a new factor is introduced: her windage aft is increased and may more than offset the effect of her windage forward. She might be very hard to handle in close quarters, because of the great amount of total windage compared to her draft.

Figure 703 **The flying bridge station provides an unobstructed view around the horizon on a vessel large enough to carry it without an adverse effect on the boat's lateral stability.**

Although flying bridges, Figure 703, on modern cruisers tend to add windage, they have an undeniable advantage in giving better visibility when maneuvering and when running at night or in strange waters.

Helmsman Must Develop Judgment

From these and other variations in behavior it is obvious that a boatman must develop judgement based on

Figure 704 **Approach a bridge with a narrow draw straight toward the center of the opening, so that if power should fail your course will tend to carry the boat through in the clear.**

understanding the boat he is handling and the forces acting upon her. Combinations of conditions are infinite, so he must be able to appraise the situation and act promptly.

Try to foresee possibilities, having solutions in mind before problems arise. If you're running down a narrow channel with a strong wind abeam, for instance, and you have a choice as to which side to take, the windward side is the better bet. If your engine stops, you'll have more of a chance to get an anchor down before going aground on the leeward side.

By the same token you wouldn't skirt the windward side of a shoal too closely. Or, suppose you are approaching a bridge (Figure 704), with a narrow draw opening, and a strong wind or current is setting you down rapidly on it. If you approach the opening at an angle and power or steering gear fails you'll be in a jam. But if you think to straighten out your course while still some distance off, so you will be shooting down the center of the opening in alignment with it, you'll be safer because your straight course will tend to carry the boat through in the clear.

In developing your "boat sense," draw from as many sources as possible. Watch how experienced boatmen and fishermen handle their boats, but allow for differences in your own boat when you try similar maneuvers.

The Terminology of Boat Handling

Before discussing actual problems of maneuvering, here are some terms you should know.

Remember that the PORT side of a boat is the *left side when you are facing forward;* the STARBOARD is the *right* side. This is easy to visualize while the boat has headway, but may be harder when the boat is reversing or if you are facing aft. Therefore, bear in mind that port is port *no matter which way the boat is going.* When a boat is "going to port" her bow is turning to port when she has *headway,* and her stern is turning to port when she has *sternway.* Figure 705 makes this clear. At *A* the boat has headway and her bow is turning to port. At *B* she is going astern and her stern is going to port.

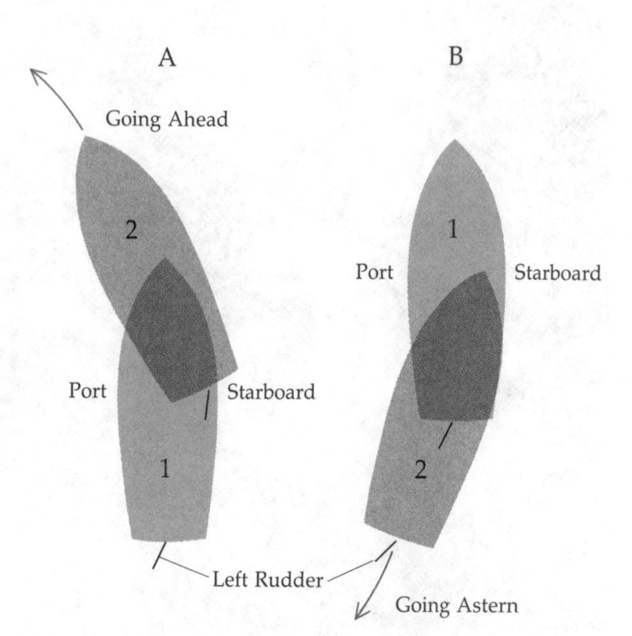

Figure 705 **With left rudder in a boat going ahead, as at A, stern is thrown to starboard, bow to port. With left rudder when going astern as at B, the stern is thrown to port.**

The terms "right rudder" and "left rudder" refer to how the bow would turn if the boat were making headway; the same term is used for the same position of the rudder even if the boat is moving astern. In Figure 705 the boat has LEFT RUDDER in both cases, and with headway, as at A, her bow goes to port. RIGHT RUDDER, which is not illustrated, is just the opposite.

A BALANCED RUDDER (Fig. 110) is one whose blade surface lies partly ahead of its rudder stock. An UNBALANCED RUDDER is one whose blade surface is fully aft of the stock. While the proportion of this balance area may be only 20 percent of the total rudder area, it exerts considerable effect in taking strain off the steering gear and making steering easier.

Right- and Left-Hand Propellers

Propellers are RIGHT-HANDED or LEFT-HANDED, depending on the direction of their rotation. The difference is important because rotatation has a great bearing on how a boat maneuvers, especially when reversing. To determine the hand, with the boat out of the water, stand astern of the propeller and look forward at the driving face of the blades, Figure 706. *If the propeller will turn clockwise when driving the boat ahead, it is* RIGHT-HANDED; *if counterclockwise,* LEFT-HANDED. The propeller at the left is left-handed; the one at the right is right-handed.

Many propellers on marine engines in single-screw installations are right-handed. In any maneuvering problems which follow, the propeller on a single-screw boat is right-handed unless noted otherwise.

In twin-screw installations the ideal arrangement is to have the *tops of the blades turn outward* when going ahead for better maneuvering qualities. As shown in figure 706, the port engine swings a left-hand "wheel," the starboard engine a right-hand wheel.

Don't be confused by the term right- and left-hand as applied to the engines that are driving the propellers. If you stand inside the boat, facing *aft* toward the engine, Figure 707, top, and the flywheel turns *counter-clockwise* (as in most marine engines) it is a LEFT-HANDED ENGINE and requires a RIGHT-HAND PROPELLER. A right-hand engine takes a left-hand wheel.

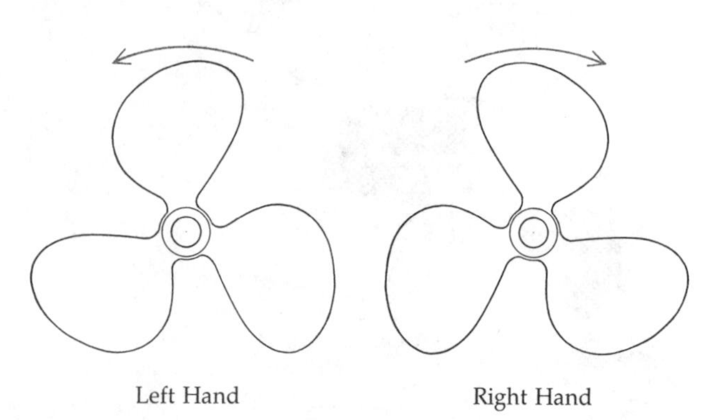

Figure 706 **Left-hand propellers turn counterclockwise and right-hand propellers turn clockwise, when viewed from astern, looking forward at the after side of the blades.**

Figure 707 **Direction of rotation is marked on flywheel housing of this marine engine, *above.* This engine has left-hand rotation, as shown by the counterclockwise direction of the arrow. The discharge screw current is given a spiral twist by the propeller blades, as this view of an outboard lower unit, *below,* shows. A left-handed inboard engine, driving a right-hand propeller, would create an opposite swirl.**

How the Propeller Acts

Mechanically powered boats are driven through the water by the action of their propellers, which act almost like a pump—drawing in a stream of water from forward (when going ahead) and throwing it out astern. This stream moving astern reacts on the water around it to provide the power for propulsion.

The water drawn into the propeller does not all actually flow from directly ahead, like a thin column of water, but for our purposes here it can be considered as coming in generally parallel to the propeller shaft. The propeller ejects it and as it does so imparts a twist or spiral motion to the water, Figure 707, bottom, its direction of rotation dependent on the way the propeller turns. This flow of water set up by the propeller is called SCREW CURRENT.

Suction and Discharge Screw Currents

Regardless of whether the propeller is going ahead or reversing, the part of the current which flows into the propeller is called the SUCTION SCREW CURRENT. The part ejected from the propeller is the DISCHARGE CURRENT, Figure 708. Discharge current is not only spiral in motion but is also a more compact stream and exerts a greater pressure than the suction current.

Placing the rudder behind the propeller, in the discharge current, increases the steering effect beyond what it would be if the rudder were elsewhere, and acted upon only by water moving naturally past the hull. A twin-screw cruiser thus has twin rudders, one behind each propeller, keeping their blades at all times more directly in the propeller's discharge current.

There is also a WAKE CURRENT—a body of water carried along by a vessel due to friction on her hull as she moves through the water. This has its maximum effect near the surface and little effect at keel depth.

Unequal Blade Thrust

Another factor also effects a boat's reaction to propeller rotation. While this factor is sometimes referred to as SIDEWISE BLADE PRESSURE, it is more properly an UNEQUAL BLADE THRUST, exerted by the ascending and descending blades of the propeller, Figure 709, top.

Here we are looking at the starboard side of a propeller shaft, inclined, as most shafts are, at a significant angle to the water's surface and to the flow of water past the blades. The actual pitch of the blades as manufactured, of course, is the same, but the water flows diagonally across the plane in which the blades revolve.

Figure 709, top, shows clearly how the effect of this is to increase the pitch of the descending starboard blade (right-hand propeller) as compared with the ascending port blade, when considered relative to the direction of water flow past the propeller.

The importance of this factor is reduced as the shaft angle is decreased, and naval architects sometimes take pains to have the engine installed as low as possible to keep the propeller shaft nearly parallel to the water's surface and to the flow of water past the blades. This contributes to propeller efficiency, and is a factor worth considering if it is consistent with other design requirements. Once a boat is built, shaft angle is difficult—usually impossible—to modify.

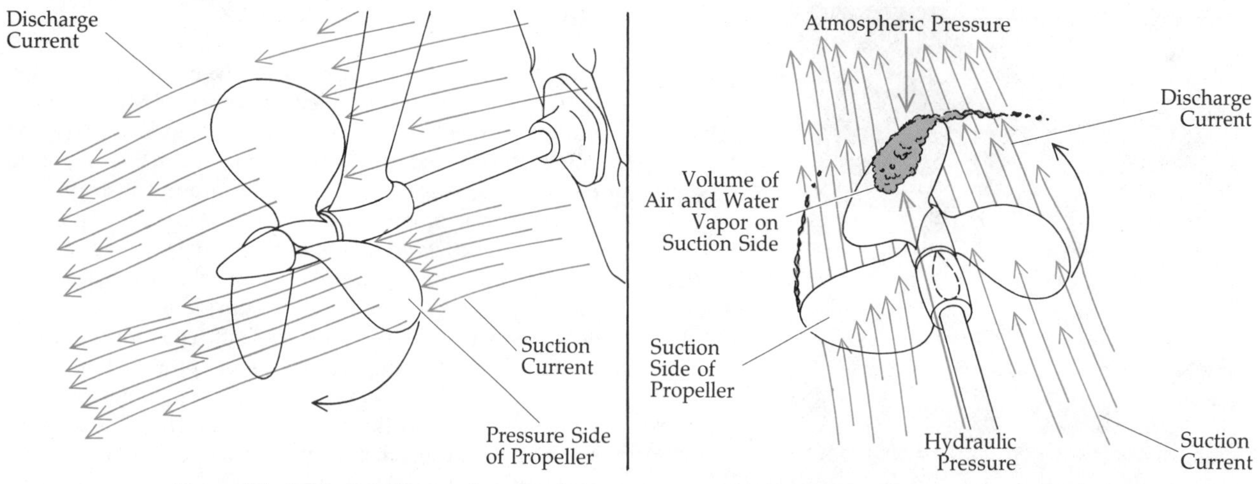

Figure 708 **With propeller turning ahead, suction screw current is drawn toward the propeller; the discharge current is driven out astern. Note the location of the rudder blade.**

Effect of Unequal Blade Thrust

The relatively greater blade pitch on the starboard side creates a stronger thrust on this side, causing the bow to turn to port. As far as this single factor is concerned, the stern of a single-screw boat with right-hand propeller thus naturally tends to go to starboard when the propeller is going ahead, and to port when it is reversing.

When such a boat has headway, therefore, its bow tends to turn to port and a certain amount of right rudder may be needed to maintain a straight course. To correct this tendency, a small trim tab may be attached to the after edge of the rudder and bent to an angle that provides the proper correction.

The effect of unequal blade thrust is so small in some cases as to be negligible, and quite pronounced in others. With left-hand propellers the effect is, of course, just the opposite of that described above.

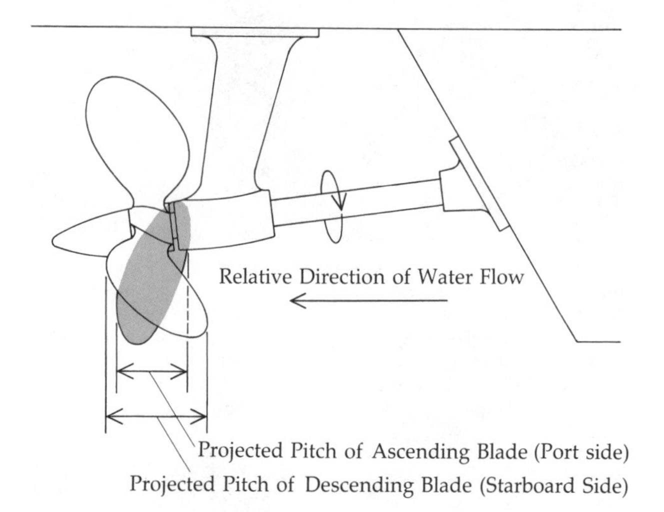

Figure 709 **Angularity of the propeller shaft, *above,* has the effect of increasing the pitch of the descending blade relative to that of the ascending blade, producing greater thrust to starboard. A left-hand wheel would produce greater thrust to port. Boats under sail, *below,* have no propeller discharge current to increase the rudder's effect in steering, and in some designs this places a limit on their maneuverability.**

How a Rudder Acts

The conventional arrangement for steering on most inboard boats is a vertical rudder blade at the stern pivoted on a stock, so that movement of the steering wheel or tiller throws it to port or starboard.

Steering wheels on motorboats are rigged so that they turn *with* the rudder, that is, turning the top of the wheel to port turns the rudder to port. For the boat with headway, this means that putting the wheel to port gives her left rudder, which kicks the stern to starboard, so the bow in effect moves to port. Conversely, turning the wheel to starboard gives her right rudder, throwing the stern to port so the boat turns to starboard.

Sailboat Rudder Action

Underway in a sailboat, or in an auxiliary when the engine is not on, water flows past the hull so that moving the rudder to one side of the keel creates both a resistance on that side and a current at an angle to the keel. The combined effect is that the sailboat's stern moves to port with right rudder, and to starboard with left rudder.

But note that when under sail *any control from the rudder is dependent on the boat's motion through the water.* If she is drifting, with motion relative to the bottom but none relative to the surrounding water, her rudder has no effect.

Only when the water flows past the rudder and strikes it at an angle does the boat respond. The faster she is moving, the stronger the rudder effect. It makes no difference how her headway has been produced—she may even be in tow—there is control as long as there is motion relative to the water.

Propeller Current's Action on Rudder

If a motorboat is "coasting" with her engine in neutral and propeller not turning, the situation is like that of the sailboat described above. If, however, the propeller is turning, the situation is quite different. Here the rudder blade is directly in the discharge current of the propeller, which is pumping a strong steam of water astern. Moving the rudder to one side of the keel deflects the stream to that side. The reaction which pushes the stern in the opposite direction is much stronger than it would be in the absence of that powerful jet.

At very slow propeller speeds the boat's headway may not be sufficient to give good control over the boat if other forces are acting upon her. For example, with a strong wind on the port beam, even with rudder hard over to port, it may not be possible to make a turn into the wind until the propeller is speeded up enough to exert a more powerful thrust against the rudder blade.

Here is a fundamental principle to remember. In close quarters a motorboat can often be turned in a couple of boat lengths by judicious use of power. If, for example, the rudder is set hard to starboard (right rudder) while the boat has no headway, and the throttle is suddenly opened and then closed, the stern can be kicked around to port before the boat has a chance to gather headway. The exact technique of turning in limited space will be described in detail later.

Figure 710 **When a boat's rudder is put over to make a turn, the stern is kicked away from the direction in which the rudder moves. Then, after sliding obliquely along the course from 1 to 2, it settles into a turn in which the bow follows a smaller circle (black line) than the stern (blue line).**

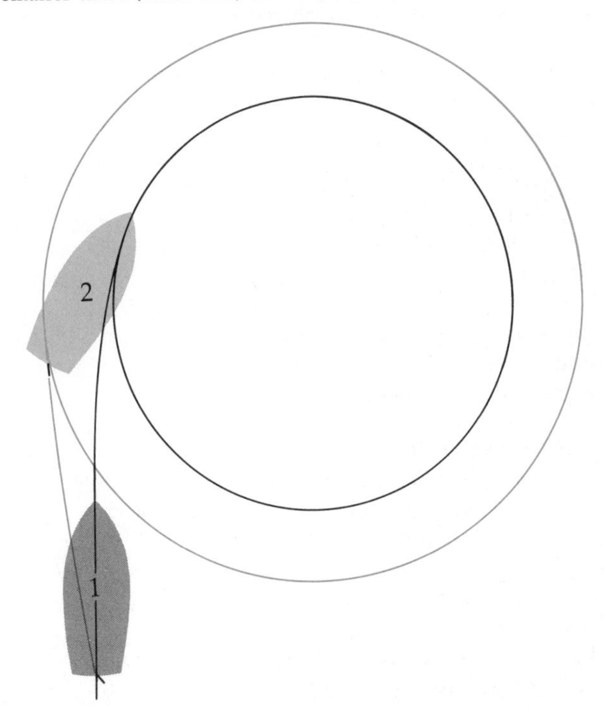

Turning Circles

When any boat has headway and the rudder is put over to make a turn (to starboard, let us say), the stern is first kicked to port and the boat then tends to slide off obliquely, "crab-wise." Its momentum will carry it some distance along the original course before settling into a turn, in which the bow describes a smaller circle than the stern, Figure 710. The pivoting point may be aft from the bow between one-fourth and one-third of the boat's length, varying with different boats and changing for any given boat with its trim.

While there is always a loss of speed in making a turn, the size of a boat's turning circle varies little with changes in speed, assuming a given rudder angle.

There is, however, a great difference in the size of turning circles for single-screw inboards as compared with outboards (or stern-drive boats) because, as has been noted, the shaft and propeller of the inboard are fixed on the centerline and cannot be rotated (Figure 711). The twin-screw inboard, on the other hand, provides excellent maneuverability, as will be seen later.

When the boat has sternway (reversing) the rudder normally would be turned to port (left rudder) to turn the stern of the boat to port, while right rudder should normally turn the stern to starboard in backing. Under certain circumstances, the effect of the reversing propeller may more than offset the steering effect of the rudder.

Water depth also has an effect on a boat's steering. Even though the keel may be several inches from the bottom, a boat's response to rudder action in shallow water is almost always sluggish.

Keep Stern Free to Maneuver

If a boatman understands the underlying difference between the steering of a boat and of a car, he will be aware of the need to keep the stern free to maneuver in close quarters. When he lies alongside a dock or another boat and wants to pull away, he should never turn the wheel hard over until the stern is clear.

To set the rudder to starboard while lying port side to a dock, for example, and then attempt to pull away by going ahead, Figure 712, tends to only throw the port quarter against the dock piles and pin it there, to slam into one pile after another, possibly damaging the boat.

The Tendency to Stray Off Course

Different boats require varying rudder angles to compensate for the tendency to fall off from a straight course. Depending on differences in construction, arrangement of rudder blade, and the hand of the propeller, the bow may fall off to port or to starboard.

For example a certain group of boats of almost identical design had a strong tendency to pull off course to port. In some of them, the condition was corrected simply by lowering the rudder blade, without changing its size or shape but merely by lengthening the stock an inch or two.

Before that correction the boats would make a quick and easy turn to port but would be obstinate in turning to starboard. As a matter of fact, unless steering controls were rigged with worm and gear to hold the wheels against pressure on the rudders, the boats would swing into a short circle to port the instant the wheels were left un-

Figure 711 **Pivoting drive unit of a stern-drive boat, *below*, permits a tight high-speed turn, while the inboard cruiser, *above*, with its fixed propeller, responds only to the rudder.**

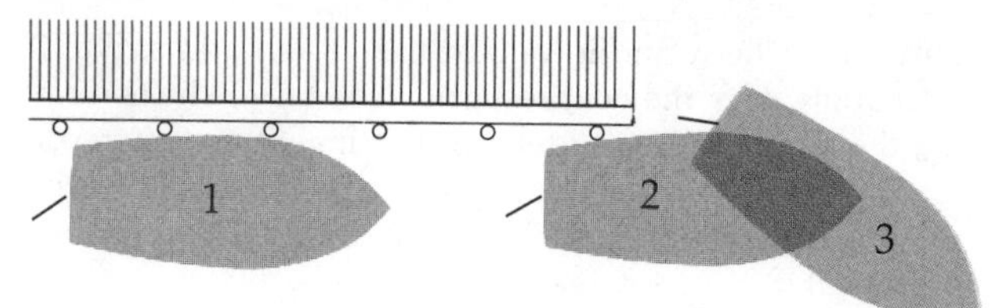

Figure 712 **With right rudder, as the boat moves from 1 to 2 to 3, the stern is driven against the piles, with risk of damage. Rudder should be amidships until boat is clear (or the craft should first be backed away from the face of the pier).**

attended. This exaggerated condition is far from normal, but it does illustrate the handling characteristics the helmsman must pay attention to, especially in an unfamiliar boat.

Response of Boat to Rudder and Propeller

Let us consider now a number of typical situations to see how the average inboard motorboat, or sailboat under power, will respond to propeller and rudder action. Remember that in each situation we have a "basic" boat—single right-hand propeller—unless it is stated otherwise.

No Way On, Propeller Not Turning

When a boat is "dead in the water"—making no way forward or sternward through the water, although she may be drifting with respect to the bottom—and her propeller is not in gear, turning the rudder has no more effect than turning a car's front wheels while it is parked. There must be a flow of water past the rudder for it to be effective, from either the boat's motion through the water or the propeller's discharge current, or both.

No Way On, Propeller Turning Ahead

Picture her first without any way on, engine idling, and rudder amidships. Because she is dead in the water there is no wake current to act on the propeller blades. Now the clutch is engaged and the propeller starts to turn ahead. Until she gathers headway, the unequal blade thrust (refer back to Figure 709) tends to throw her stern to starboard. As she gets headway, wake current enters the picture, increasing pressure against the upper blades, and this tends to offset the effect of unequal blade thrust.

What happens under identical conditions, except that the rudder is hard over at the time the propeller is engaged? In this case the propeller's discharge current strikes the rudder and exerts its normal effect of kicking the stern to port with right rudder, or stern to starboard with left rudder. With right rudder the kick to port would be much stronger than the effect of unequal blade thrust.

With Headway, Propeller Turning Ahead

After the boat has gathered normal headway, with rudder amidships, the average boat tends to hold her course in a straight line fairly well. From a purely theoretical standpoint, the unequal blade thrust, with a right-hand wheel, should tend to move the stern to starboard, and bow to port. On most boats, however, the effect is quite slight. Only in a comparatively few cases will unequal blade thrust have a pronounced effect on steering, and in these it can be corrected by a small rudder tab.

Now, the boat having headway, assume the rudder is put to starboard. The water flowing past the hull hits the rudder on its starboard side, forcing the stern to port. The propeller's discharge current intensifies this effect by acting on the same side, and the boat's bow turns to starboard, the same side on which the rudder is set. With rudder to port the action is just the opposite.

With Headway, Propeller Reversing

A boat has no brakes as does a car, so she depends on reversing the propeller to bring her to a stop. Assume that the boat has headway, rudder amidships, and the propeller is reversed. The rudder has little steering effect in this case, and unequal blade thrust of the reversed propeller tends to throw the stern to port. At the same time, the propeller blades on the starboard side are throwing their discharge current in a powerful column forward against the starboard side of the keel and bottom of the boat, with little on the port side to offset this pressure. This also adds to the forces moving the stern to port.

This principle explains why a boatman will bring a "basic" boat up to a pier, if wind and current allow a choice. The stern then is moved in toward the dock by the reversing propeller, instead of away from it.

With Rudder to Port Go back now to the case where the boat has headway and the propeller is reversed—this time with rudder over to port, let us say. Here the situation is more complicated. As before, we have the unequal blade thrust and propeller discharge current both driving the stern to port.

In addition there are two opposing factors. If the boat has much way on (ahead) her left rudder tends to throw the stern to starboard. However, her propeller suction current is being drawn in from astern in such a manner that it strikes the back of the rudder blade, tending to drive the stern to port.

Which combination of factors will be strongest depends on how much headway the boat has. If she has been running fast, the steering effect of left rudder will probably be the dominant factor and her stern will be thrown to starboard at first. As this steering effect weakens with reduced headway, with the propeller slowing her down, then the effect of the suction current is added to help the tendency of the stern to port. Eventually, with all headway killed, even the steering tendency to starboard is lost and all factors combine to move the stern to port.

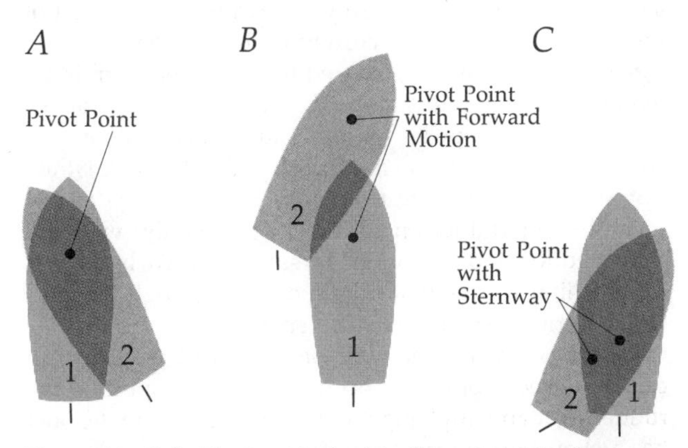

Figure 713 ***Left:*** **The boat is dead in the water at 1; when forward gear is engaged, the stern is kicked to starboard; the amount of kick depends on boat form and the amount of throttle given.** ***Center:*** **With forward motion when the helm is put over, the boat pivots around a point about ⅓ its length abaft the stem. Note that the stern swings through a wider arc than the bow.** ***Right:*** **When backing down with the rudder to port, the pivot point is about ⅓ the boat length forward of the stern, so the bow describes a wider arc than the stern.**

No Way On, Propeller Reversing

Now if the boat is lying dead in the water with no headway, rudder amidships, and the propeller is reversed, we again have that strong tendency of the stern to go to port as the discharge current strikes the starboard side of the hull. You see, in each case where the discharge current of the reversing propeller is a factor, the strong current on the starboard side is directed generally toward the boat's bow but upward and inward in a spiral movement. The descending blade on the port side, on the other hand, tends to throw its stream downward at such an angle that its lesser force is largely spent below the keel. Therefore, the two forces are never of equal effect.

Until the boat gathers sternway from her backing propeller it would not matter if the rudder were over to port or starboard. The discharge current against the starboard side is still the strong controlling factor and the stern is moved to port.

With Sternway, Propeller Reversing

Now visualize the boat gathering sternway as the propeller continues to reverse. Here arises one of the seemingly mystifying conditions that baffle many a helmsman during his first tricks at the wheel. The novice assumes that if he wants to back in a straight line his rudder must be amidships, just as it must be when he goes ahead on a straight course. But under certain conditions his boat may even respond to *right* rudder as he reverses by going to *port,* which is exactly what he doesn't expect. If he is learning by trial-and-error he comes to the conclusion that it depends on the boat's fancy, while rudder position has nothing to do with control.

Let's analyze the situation, however, to see if there is anything that can be done about it. Fortunately, there is.

Backing With Left Rudder Consider a boat in reverse with left rudder. Here there are four factors all working together to throw the stern to port. Unequal blade thrust is pushing the stern to port; the discharge current of the propeller is adding its powerful effect; and now we add the steering effect of the rudder acting on the after side of the rudder blade, against which the suction current of the propeller is also working.

Remember this condition well for it is the answer to why *practically every single-screw vessel with right-hand propeller naturally backs to port* easily although she may be obstinate about going to starboard when reversing.

Backing With Rudder Amidships Now, while backing to port, let's bring the rudder amidships and see what happens. Here we have eliminated the effects of suction current and steering from the rudder, leaving unequal blade thrust and the discharge current to continue forcing the stern to port.

Backing With Right Rudder Assuming further that we have not yet gathered much sternway, let's put the rudder to starboard and see if we can make the boat back to starboard as you might expect she should with right rudder. The forces of unequal blade thrust and discharge current still tend to drive her stern to port, but the suction current of the propeller wants to offset this. The effect of the discharge current is stronger than the suction so the tendency is still be port. Now, with sternway, the steering effect of right rudder is to starboard, but as yet we haven't way enough to make this offset the stronger factors.

Steering While Backing Just about the time we are about to give it up on the assumption that she can't be made to back to starboard, we try opening the throttle to gain more sternway. This finally has the desired effect and with full right rudder we find that the steering effect at considerable backing speed is enough (probably) to turn her stern to starboard against all the opposing forces. How well she will back to starboard—in fact, whether she will or not—depends on the design.

All of this means that if the boat will back to starboard with full right rudder, she may also be made to go in a straight line—but not with rudder amidships. There's no use trying. She will need a certain amount of right rudder depending both on her design and the speed. Some boats, while their backing to port is always much better than to starboard, can be controlled with a reasonable degree of precision by one who understands the particular boat he is handling.

In some cases, boats may even be steered backwards out of crooked slips or channels—not, however, without a lot of backing and filling if there is much wind to complicate the situation. Generally, the trick is to keep the boat under control, making the turns no greater than necessary to keep the boat from swinging too much.

In backing situations, set the rudder first and *then* add maneuvering power by speeding up the propeller.

With Sternway, Propeller Turning Ahead

There is one other situation to be considered, where we wish to kill sternway by engaging the propeller to turn ahead. Regardless of rudder position, unequal blade thrust with the propeller going ahead now tends to throw the stern to starboard, while the suction current is of no consequence. Unequal blade thrust may or may not be offset by the steering effect and the discharge current.

With rudder amidships, there is no steering effect and the discharge current does not enter our calculations. Therefore the stern will go to starboard. Now if you throw the rudder to port the discharge current of the propeller hits the rudder and drives the stern to starboard—even though the normal steering effect of left rudder sends the stern to port, with sternway. The powerful discharge current from the propeller going ahead is the determining factor.

If the rudder is put to starboard, the steering effect works with the unequal blade thrust, tending to move the stern to starboard, but the discharge current strikes the starboard side of the rudder and acts to kick the stern to port. Apply enough power that the force of the discharge current outweights the other factors, and the stern will go to port.

Propeller Action Governs

From the above analysis it is clear that in a single-screw inboard boat you must be constantly aware of what the propeller is doing, in order to know how best to use the rudder. What the propeller is doing is even more important than whether the boat has headway or sternway.

For example, suppose you are backing in a direction that brings your port quarter up toward a dock. The tendency if you were to try to steer away from it would be to use right rudder. But the efficacy of that would be doubtful, because of the boat's inclination to back to port.

Therefore, kick the stern away by going ahead with the propeller to set the rudder to port (left rudder, *toward* the dock), throw the reverse gear lever into the forward position, and go ahead strongly but briefly with the engine. The discharge current checks the sternway and, striking the rudder, throws the stern to starboard, clear of the dock.

Left-handed Propellers If your boat has a left handed propeller, you can generally reverse "port" and "starboard" in the foregoing discussions. But to be sure, make tests for all the possible situations.

Applying the Principles

Proficiency in motor boat handling is a matter of 10% knowledge of the principles, and 90% application of these principles. Therefore, the sooner you get afloat and start experimenting, the better. Absorb enough of the theory to have a broad idea of what to expect in practice; then get at the helm of a boat and actually learn by doing. It takes experience to learn all a boat's whims and traits—to know her strong points and use them wisely to overcome the weak ones.

Points to Remember

Here is a summary of basic ideas to keep in mind, again for a boat with a right-hand propeller.

- Remember that *the boat is always under better control with headway* than with sternway, because of the effect of the propeller's discharge current on the rudder.
- Until the boat has gathered headway the stern has a tendency to swing to starboard, even with rudder amidships, as the propeller starts to turn ahead. *With good headway, rudder angle is the principal factor affecting control.*
- *When backing, there is a strong tendency to go to port* regardless of rudder angle, except (ordinarily) with full right rudder and considerable sternway. To back in a straight line you need a certain amount of right rudder, which varies, usually, with the speed.
- *With no way on there is no rudder control,* yet the stern can be kicked rapidly to port or starboard by putting the rudder over and applying plenty of power before the boat has a chance to gather headway.
- *With a left-hand propeller, the boat's reaction will be contrary to that outlined for a right-hand propeller.*

The First Time Out

The first step in applying these principles is to take the boat out in open water—preferably with little wind and no current. These factors should be eliminated at first and studied separately later, after you understand the boat's normal reactions.

With unlimited room to maneuver and no traffic to worry about, try putting the boat through all the maneuvers discussed. Practice with every combination of conditions—with headway, sternway, and no way on; with right rudder, left rudder, and rudder amidships; with propeller turning in the direction of the boat's way, and also turning opposite to the direction of the boat's way.

In each of these maneuvers note whether the rudder or propeller has the greater effect; note, too, how changing the speed alters the boat's response. Put the rudder hard over each way and see what happens when considerable power is applied before the boat has any way on.

Practice turns to determine the boat's turning circle. Practice stops to determine the space required to bring the boat to a full stop with varying amounts of way, up to full speed.

Even if you are experienced in handling your own boat, follow these procedures when renting or chartering a strange craft, and to the extent possible when first taking the helm on a friend's boat.

Gaining Experience

Later, when these fundamentals are mastered, go out again and observe how wind and current and sea alter the situation. Note, when in reverse, how she tends to back into the wind until good steerageway is reached. How, if she is lying in the trough in a seaway, she tends to stay there with rudder amidships, because wave action is stronger than her natural tendency in smooth water to back to port.

You should, of course, pick your maneuvering grounds from the chart to avoid danger of grounding, but if you see the first wave of your wake stretching away on the quarter in a sharp inverted V, tending to break at the top, beware of shoal water on that side.

Watch too, how her stern starts to settle, and the bow comes up, if you do get into shoal water. If there's just enough water to permit her to run without grounding, a wave will pile up on her quarter, steepest on the shallower side, as the natural formation of the wake is disturbed.

Getting Underway

Whenever you are getting underway, assuming you are starting a cold motor, don't spend too much time at the dock or mooring "warming up." Cast off once your regular, brief getting-underway procedures are complete, and finish warming up under load, at about half speed. Long periods of idling are bad for the clutch, and the motor will warm up better and quicker with the propeller engaged.

Before casting off, check the oil pressure and see that the cooling water is circulating. Whether your engine is raw-water cooled or fresh-water cooled with a heat exchanger, there should be a visible flow of water overboard, usually through the exhaust line. Until this water flows, it is unsafe to get underway.

If the water doesn't circulate, investigate at once. Overheating can cause much damage, especially in a raw-water cooled engine if the cold water is suddenly picked up and pumped into the hot cylinder block. Cracking of the block could result.

Never race the motor as it idles, especially when cold. If the propeller is the right size and the throttle stop properly set, the motor should be able to take the propeller load when the clutch is engaged at idling speeds. Boats don't require transmissions with a change of speeds as in cars, because propeller slip automatically takes care of picking up the load gradually when the clutch is engaged.

After the clutch is in, open the throttle gradually. Roughly speaking, normal cruising should be done at about two-thirds or three-quarters throttle. The extra few hundred revolutions available at wide-open throttle seldom yield an increase in speed proportionate to the extra power required, and to the corresponding increase in fuel consumption.

One caution to observe when first learning to maneuver a boat is to avoid opeating at too high a speed. Later, when you are more familiar with the boat and its response to wheel and throttle, you can use more power in certain situations—to accomplish a quick turn, for example—without getting into a jam.

Checking Headway

Before attempting the maneuver of coming alongside a pier or picking up a mooring, experiment out in open water with the technique of checking the boat's headway. Having no brakes, you must reverse the propeller to bring the boat to a stop. Caution: *always approach docks or other craft at a very slow speed.* Failure of the clutch and reverse gears, or an unintentional killing of the engine, can result in embarrassment or damage.

Your experiments will show how effective the propeller is in killing the boat's headway. Generally speaking, the larger diameter propellers, acting on a large volume of water, will exert the greatest effect. Small propellers, poorly matched to heavy hulls, may churn up much water before they can overcome the boat's momentum.

Many fast boats can be stopped in an incredibly short distance. When their throttles are closed, the boat changes her trim suddenly and headway is quickly lost, even before the propellers are reversed.

Whenever you need to go from forward into reverse or vice versa, slow the engine down while going through neutral. If you make a practice of shifting from full ahead straight into full astern, look out for trouble. This is hard on the gear even if it doesn't fail—which it might do just when you're counting on it most.

Turning in Close Quarters

Turning a boat in a waterway not much wider than the boat's length often seems impossible to the novice, but it's really no more difficult than turning a car on a narrow road.

Take a look at Figure 714. Perhaps you've run to the head of a dead-end canal and must run around. Or the waterway is a narrow channel flanked by shoals. In the dead-end situation you must allow room so the swinging stern doesn't hit the canal bulkhead; in the narrow channel you must make a similar allowance to avoid throwing the stern into shoal water.

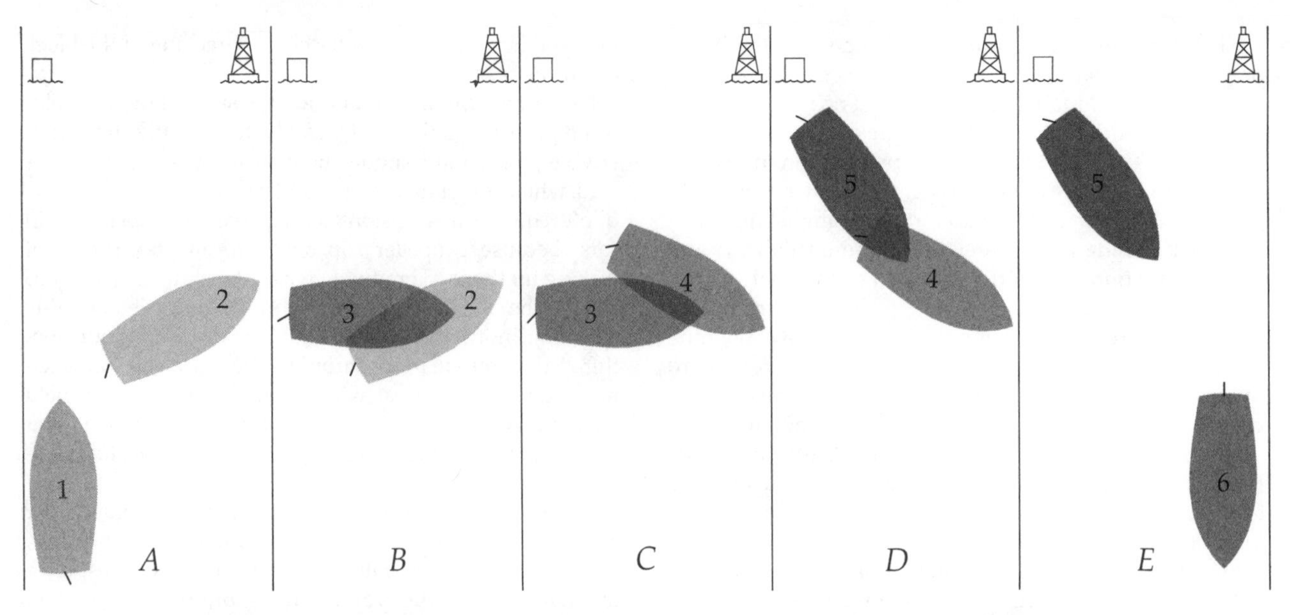

Figure 714 **Turning in tight quarters: At 1, boat starts turn at left side of channel, with allowance for room for the stern to swing to port. Headway is checked at 2 by reversing the propeller, with rudder kept to starboard throughout the maneuver. Unequal blade thrust and discharge current forces the stern to port as the boat backs from 2 to 3. Going ahead again, sternway is checked at 3 and propeller turning ahead kicks stern to port at 4 before the boat has a chance to gather headway. Reversing once more, headway is checked at 4 and boat backs to 5. If necessary, alternate going ahead and backing can be repeated, as governed by space available for the turn. At 5 the boat is in a position to go ahead with right rudder, bringing it amidships at 6, having reversed course 180 degrees.**

Referring to Figure 714, we start at *A*1, on the left side of the channel, running at slow speed. Putting the rudder hard over to starboard, the boat swings toward position 2, her headway being checked by reversing. *Always* execute this maneuver (if the propeller is right-handed) by going ahead to starboard, backing to port, to take advantage of the boat's natural tendencies.

As shown in *B*, *leave the wheel hard over to starboard* (right rudder). Normally you would expect to use left rudder in backing from 2 to 3, but this is unnecessary as the boat has no chance to gather appreciable sternway. As the reversing propeller stops the boat at 2, open the throttle for an instant and the stern will be kicked around to port to position 3.

Any attempt to shift from right rudder while going ahead to left rudder while going astern only results in extra gymnastics at the wheel, at a time when you want your hands free for the throttle and reverse gear lever. Leave the rudder as it is at this point.

Throttling up while engaging the clutch to go ahead, check the sternway at *C*3. Opening the throttle again, just for an instant, keeps the stern swinging to port toward position 4. Now the operation described in *B* is repeated, backing from position *D*4 to 5.

If you are maneuvering in a bulkheaded slip, allow plenty of room so your port quarter is not thrown against the head of the slip as you go ahead again at *E*5. The stern will continue to swing to port as the boat straightens away

Figure 715 **To back around to starboard, where space is limited, boat starts at 1 with right rudder and backs to 2 as it cannot turn short to starboard. At 2 the rudder is shifted to port, and the stern is kicked to starboard to position 3 by going ahead strong for a few revolutions. From 3 the boat is backed to 4 with full right rudder. Here the rudder is put to port, and the boat moves ahead to 5. Backing down from 5 it will need a certain amount of right rudder to maintain straight course.**

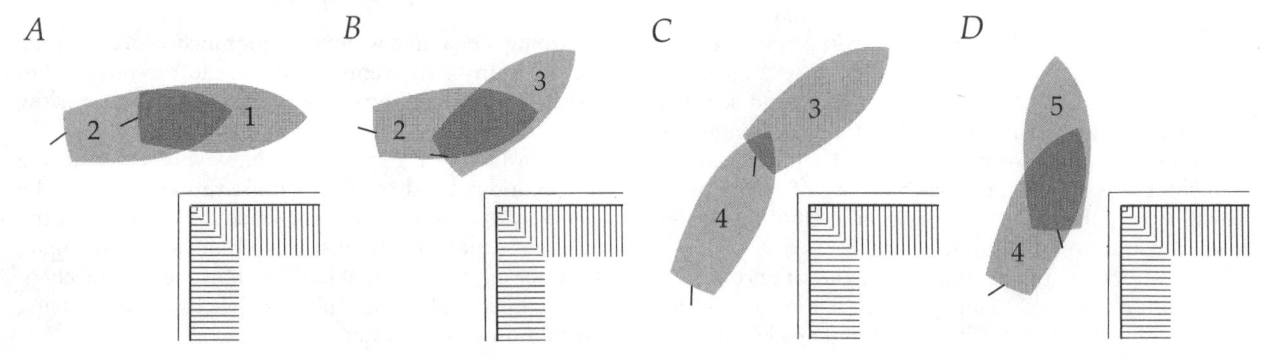

toward 6. At this point the rudder can be brought amidships, the boat having executed a 180-degree turn.

Backing Around a Turn

Figure 715 illustrates the successive steps required to work out another practical problem in boat handling. Suppose you are in a narrow slip or canal position (*A*1) and want to back around a sharp turn, to starboard.

Reversing with full right rudder, you will not be able to turn short enough to steer around the 90-degree angle. Most likely you will back to a position about as indicated at *A*2, gaining a little to starboard. Now (see *B*2) by going ahead with left rudder the stern is kicked over to starboard, placing the boat at 3.

Reversing once more, with full right rudder, the boat backs to *C*4, necessitating your going ahead once more with left rudder. This puts the boat at *D*5 in a position to back down on the new course.

Note that if the boat is expected to back down in a straight line from 5 she will need some right rudder.

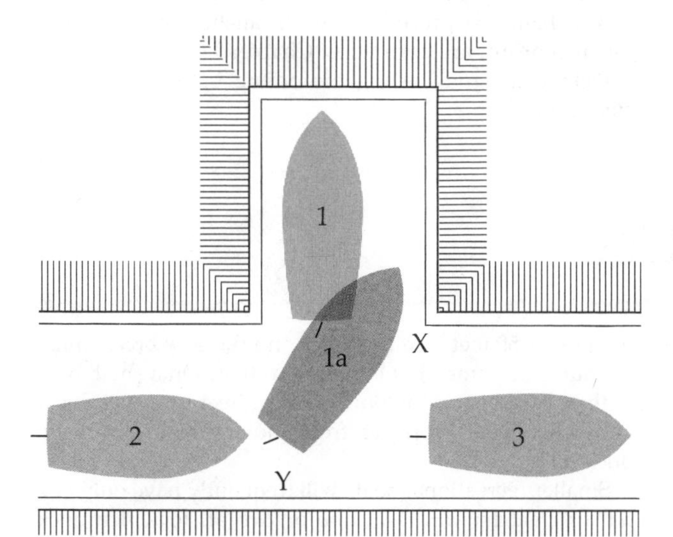

Figure 716 **To back out from a slip, the boat should be able to make a short turn to port from 1 to 3, when reversing with left rudder, but if bow swings wide it may hit at X and Y.**

Backing to Port From a Slip

Figure 716 illustrates a variation of the maneuver just described. Suppose the boat is lying in a slip, at 1, and you wish to back out into the channel in limited space. Figure 715 showed how you would need to work her around into a slip to starboard. In Figure 716 the problem would be exactly the same if she had to back around in the direction of 3.

However, if there is a choice and she can back out to port, with left rudder she will probably work around from 1 to 2 without any special maneuvering. To do it, her initial position in the slip must be about as shown with some clearance on her port side, but considerably more on the starboard side.

Clearance on the port side is needed because in reversing, her stern immediately starts to move to port. More clearance on the starboard side is necessary because the bow will swing over toward point X as she backs.

In fact, points X and Y are the ones to watch in executing this maneuver. If the boat doesn't turn as short as expected, and doesn't swing directly from 1 to 2, but takes a position as shown in 1a, the starboard bow is in danger of touching at X and the starboard quarter at Y. This could be corrected at 1a by going ahead with the propeller a few revolutions, using right rudder, to kick the stern over to port. This must be done carefully as there is little latitude here for headway.

Having straightened her up, you could then back with left rudder to 2 and finally square away with rudder amidships to proceed toward 3.

Leaving a Mooring

Yacht club and marina anchorages often have mushroom anchors or heavy concrete blocks and mooring buoys, to which boats are secured by a pennant or mooring line over the forward bitt. Boats at moorings are reached from shore by dinghy or the club launch.

Getting away from the mooring is among the simplest of maneuvers, yet there is a right and a wrong way to go about it. The principal hazards in doing it wrong are getting mooring line or dinghy painter fouled in the propeller, or colliding with a boat moored nearby.

If you use a dinghy to reach your boat, shorten up its painter so the propeller cannot pull any slack down into it, to be fouled around the shaft and blades. See that boarding ladders and fenders are brought on board and send someone forward to let the mooring line go. The pennant usually has a small buoy to float the end, making it easy to pick up again when you return.

In a river or stream with current, the boat will lie head to that current (unless the wind is stronger). But in a bay on a calm day there may be neither wind nor current to move the boat, even when the mooring line is let go. To go ahead under such conditions would almost certainly foul the mooring line in the propeller.

To avoid this, back away a few boat lengths, far enough so you can have the buoy in sight and give it ample room when you go ahead. Whether you back away straight or turn as you reverse depends on the position of neighboring boats.

Run slowly until you are clear of the anchorage, to avoid creating a nuisance with your wake. Then as you open up to cruising speed, drop the dinghy back just ahead of the crest of the second wave astern. If she is behind that crest she will tow too hard, running "uphill" with bow too high. Shorten the painter from this point until the pull on it lessens as she flattens out to a better trim, but don't have her run "downhill" either, if she tends to yaw off course. Exactly how you trim the dinghy painter depends largely on boat speed and sea conditions.

When There Is Wind or Current

When your mooring is in a stream or if a tidal current flows past it, the boat will be set back when the mooring line has been let go. This usually simplifies the problem of getting away; reversing may not be necessary.

In a wind (assuming the current is not stronger) the boat will be lying head to the wind. As she drops back from the mooring, whether or not the reverse gear is used, the bow will pay off to one side and shape the boat up to get away without additional maneuvering.

All boats with considerable freeboard have a strong tendency to "tack" back and forth as they lie at anchor in a wind, and the same is true to an extent at permanent moorings unless the mooring gear is so heavy as to retard this action.

If the bow is "tacking" in this manner and you want to shape the boat up to leave the buoy on one side or the other, wait till the boat reaches the limit of her swing in the desired direction and then let the pennant go. The bow will pay off rapidly as she catches the wind on that side.

Picking Up a Mooring Under Power

Returning to the anchorage, approach your mooring at slow speed. Note how other boats are lying at their buoys. They are heading into the wind or current (whichever is stronger) and your course in approaching your mooring should be roughly parallel to their heading. Stay clear of other moorings, or you may cut them or foul them. Shorten up on the dinghy painter; if possible station a crewmember at the stern to keep the painter from fouling the propeller.

Now slip the clutch into neutral when you estimate that you have just way enough to carry you up to the buoy. Station someone well forward on the bow with a boat hook, to pick up the pennant float. If you see that you are about to overshoot the mark, reverse enough to check the headway as the bow comes up to the buoy. If you fall short, a few slight kicks ahead with the propeller should suffice.

Don't expect the person forward to do the engine's work in holding the boat in position. Until the signal is given that the pennant eye has been secured on the bitt, keep the engine ready in case the boat tends to drop astern. Also, take care that the buoy does not chafe against the hull. If you do overshoot or fall off and your crewman forward cannot reach the pennant, just get clear and try again calmly. Caution—if yours is the only mooring in the anchorage, you will have to gauge the effect of wind and current on your boat as best you can to approach upwind or against the current or directly against any combination of these factors. Don't try to execute this maneuver with a strong wind abeam or astern. Stop your motor *only* after the pennant or mooring line is secure forward.

Boat handling procedures when anchoring, or getting the anchor up, are covered in Chapter 11.

Picking up and leaving a mooring under sail is discussed in Chapter 8.

Dock Lines and Their Use

Lines play an important part in the handling of vessels at a dock. Obviously, the larger the craft the more lines are likely to be called into play. It would be absurd for the skipper of a 30-footer to burden his boat with a spider web of springs, breasts, and other lines that would be appropriate only on a large ship.

Terminology

Familiarize yourself with correct terms and the functions of the lines.

Most boat owners speak quite loosely of bow and stern lines—and little else—depending on whether the line is made fast forward or aft and regardless of the direction in which it leads or the purpose it serves. To be correct, according to nautical terminology, there is only one BOW LINE. This is made fast to the forward cleat and run forward along the dock to prevent the boat from moving astern.

Conversely, a STERN LINE leads from an after cleat to a pile or bollard on the dock astern of the boat, to check her from going ahead. The special virtue of such lines as applied to small craft is the fact that, with but little slack, they can allow for considerable rise and fall of the tide.

BREAST LINES, on the other hand, lead athwartships nearly at right angles to the vessel and to the dock, to keep her from moving sidewise away from her berth. Large craft may use bow, waist, or quarter breasts, depending on where they are secured.

Naturally, breast lines on large vessels are more important than on small ones. If a vessel has a 100-foot beam, her bow is 50 feet from the dock and the bow breast must of course be somewhat longer than that. On a small boat with 10-foot beam, the bow is only 5 feet from the dock, and other lines keep her from moving away from the dock.

Smaller recreational boats will frequently have only one bitt or cleat forward and one or two aft to which dock lines may be made fast. Additional cleats along the sides, properly through-bolted, give good flexibility is using dock lines.

Dock or Mooring Lines

On recreational boats, DOCK LINES (also called MOORING LINES) are generally of nylon, either the usual twisted rope or the type with a braided core and cover. The line's size varies with the size of the boat. Typically, a 30- to 40-footer will use 5/8-inch nylon lines, with larger yachts going to 3/4-inch lines, and smaller boats to 1/2-inch or even 3/8-inch nylon for under-20-footers. Nylon has the desirable properties of stretch, long life, and softness on one's hands. (See more about lines in Marlinespike Seamanship, Chapter 13.)

Mooring a Boat

Eight dock lines that might be used to secure a boat are illustrated in Figure 717. They are not all used at the same time. What might be considered a typical satisfactory arrangement of mooring lines for a boat or yacht 20 to 65 feet long (6.1 m–19.8 m) is illustrated in Figure 718. The two spring lines are shown crossed and running to separate dock cleats or piles. This arrangement provides longer springs which can be drawn up rather snugly and yet allow for a rise and fall of tide. If only one pile or cleat is available, position your boat so that this point is opposite amidships and run both springs to it; see Figure 719.

The bow and stern lines should make roughly a 45 degree angle with the dock edge. The stern line is most easily run to the near-shore quarter cleat, but a better tie-up is made if it is run to the off-shore quarter cleat, as shown by the dotted line in Figure 718.

Breast lines here would be superfluous and should be omitted. An exception might be a slack line to the near-shore quarter cleat. This would be used solely for pulling the boat in against an offshore wind for easier boarding.

A small boat is often adequately secured with springs alone, using no breasts. Sometimes boatmen attempt to tie up to a dock with bow and stern breasts only, and get into trouble because they left too little slack in the line to allow for tide. If adjusted right for low water, breast lines may be so slack at high water that the boat gets a chance to catch in some stringer or projection of the dock when her bow or stern swings in.

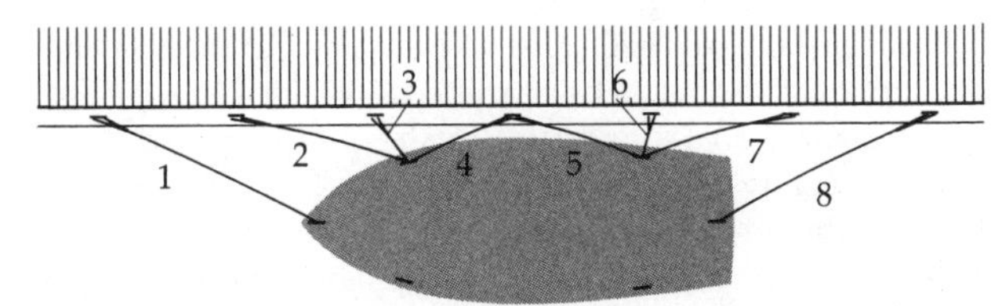

Figure 717 **Possible docking lines for a vessel include (1) bow line, (2) forward bow spring, (3) forward (bow) breast, (4) after bow spring, (5) forward quarter spring, (6) after (quarter) breast, (7) after quarter spring, and (8) stern line. A small boat will rarely need to use all of these.**

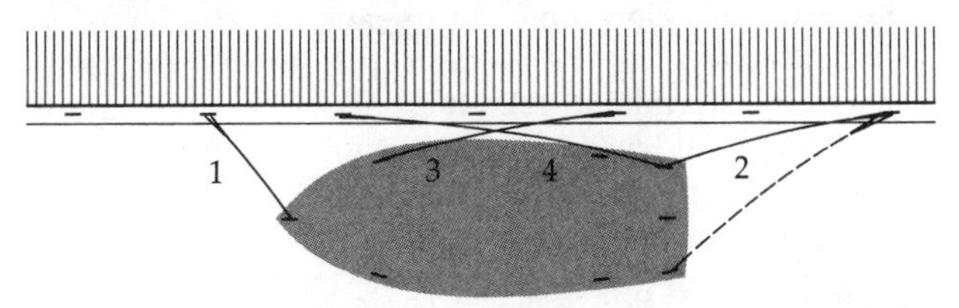

Figure 718 **Typical small boat mooring: Spring lines (3) and (4) are often crossed to gain a greater length for them, where tidal range is significant. The stern line (2) is often run to the offshore quarter cleat to gain greater length for it.**

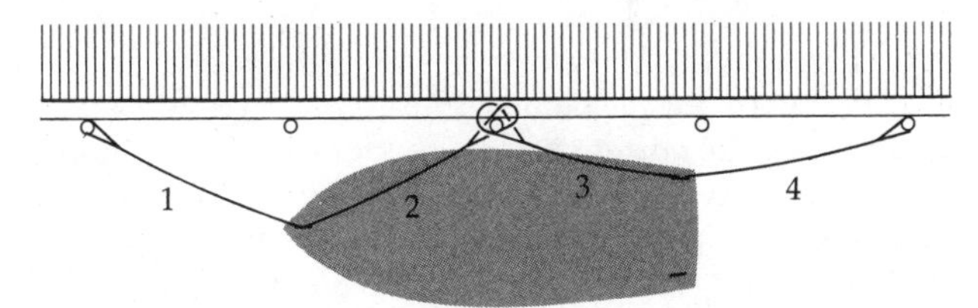

Figure 719 **Separate spring lines may be placed on the same pile, or a single line may be secured to bow, pile, and stern.**

If Lines Are Doubled

If two lines are used anywhere with the idea of getting double the strength of one, they must be of equal length when strain is put upon them. Otherwise one carries the load first and parts, and then the other follows suit.

Figure 720 **Where two lines are on the same pile, as shown *above,* the lower line can be removed without disturbing the upper line, by slipping it through the eye of the upper line. *Below:* If the lead is high from the pier or wharf to the deck, take an extra turn around the pile with the eye splice to prevent the line from slipping up off the top of the pile.**

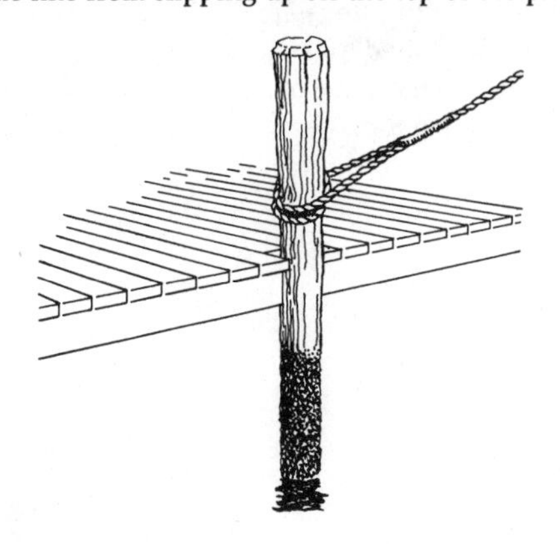

Two Lines on One Pile

Frequently, mooring lines have an eye splice of suitable size in the end that goes ashore, to be dropped over the pile or cleat. If you must drop a line over a pile where there is already a similar line from another boat, it is a simple trick to rig your line so that either can be removed without disturbing the other. Take the eye of your line and run it up through the first eye from below as shown in Figure 720, top.

If you find that another owner has dropped an eye over yours on a pile or cleat without "dipping" it through your eye, you can still get yours clear easily by reversing the process. Get a little slack in the other line and slip your eye up through its loop, and on over the top of the pile. Your line then can be dropped through the eye of the other, and you are clear.

If yours is to be the second line on the pile, it's both prudent and courteous to dip your line through the first loop. The other owner may not know the trick for getting his line clear without removing yours.

If a boat has much freeboard, the tide is high, and dock piles are relatively short, so that a mooring line leads down at a fairly sharp angle from deck to dock, there may be some risk of its slipping up over the top of the pile. An extra round turn of the eye splice taken over the pile will prevent this, Figure 720, bottom. If the eye in your mooring line is too small to go over the pile, pull a bight of line through the spliced eye to make a new "eye" of any necessary size.

Heaving Lines and Monkey's Fists

Good seamanship consists partly of knowing when a certain procedure is applicable to the size of the vessel you are handling, as when getting mooring lines from a vessel to a dock. The lines of a big ship are heavy hawsers, hard to handle, and impossible to heave. Therefore the deck crew makes use of heaving lines, which are light lines weighted at the end by a MONKEY'S FIST (an intricate woven knot which encloses a weight). This heaving line is bent to the hawser near the eye splice—*not* in the loop where it might be jammed when a strain is thrown on the hawser—and the line is sent from ship to dock, dock to ship, or ship to ship as soon as possible as a MESSENGER, The heavier line or hawser is then pulled over and made fast.

Small Craft Practices

On boats, heaving lines are more of a technique to be learned and then filed away in your memory for possible emergency use. Normally your crew can simply step ashore with the lines or hand them to someone there.

Figure 721 **A properly coiled mooring line can be heaved some distance, if you start with half the coil in each hand.**

Even if you must heave your regular mooring lines on a small boat, they will carry some distance if properly coiled, with half held loosely in the left hand of a right-handed person, the remaining half heaved by the right (Figure 721), all uncoiling naturally in the air without fouling into a knot and falling short. If necessary, a weight can be added to carry a line but this is seldom used. If you need to pass over a heavy tow line, however, then break out the heaving line principle and use it to good advantage.

Using Spring Lines

Spring lines are the most useful and important of dock lines. They can greatly facilitate docking or undocking in difficult situations, and their proper use is a mark of a knowledgeable skipper and crew.

Springs are useful in preventing undesired movement ahead or astern in a berth; they also keep a craft in position where it is secured to a fixed structure and there is a significant rise and fall of tide. There may be four spring lines—the FORWARD BOW SPRING, the AFTER BOW SPRING, the FORWARD QUARTER SPRING, and the AFTER QUARTER SPRING; see Figure 717.

Bow springs are made fast to the vessel at or near the bow, quarter springs at or near the stern. Forward springs lead forward from the vessel to the pier or wharf, and thus check any movement sternward. After springs lead aft from the vessel and check any movement ahead.

Remember that the terms FORWARD and AFTER relate to the *direction* in which the spring line runs from the vessel, and not to where it is made fast on board, this being indicated by the terms BOW and QUARTER.

Going Ahead on a Spring

We will see below how, by means of an after bow spring, the stern can be brought in to a dock by going ahead with the rudder set *away* from the dock; it is more easily done with line made fast farther aft along the side of the boat. This technique is especially useful when a boat must be gotten in to a dock against a stiff offshore wind, and hauling in on bow and stern breasts would require too much effort.

Ahead on a Quarter Spring If it is possible to secure an after spring at the boat's pivot point, she can be sprung in bodily to the dock and her parallel alignment (or any other position) can be held by means of the rudder as the propeller goes ahead, because the stern is free to swing as the discharge current acts on the rudder. This is more theoretical than practical on many smaller boats, because normally they have only the bow and stern (or quarter) cleats to secure spring lines to, while the pivot point will be well aft of the bow.

In rigging an after quarter spring, and going ahead with rudder amidships, the combination of forces is such that the boat will be sprung in nearly parallel to the dock. Her stern will come in till the quarter touches the dock, while the bow may stand off somewhat; care must be taken that current or wind does not take change and swing the bow out excessively. See Figure 722.

If the rudder is put to port at position 1 in the expectation that the bow will thus be thrown to port, the effect is negligible; the stern is prevented from being kicked to starboard by the taut spring. Putting the rudder to starboard, however, throws the stern in to port fast.

In general, it is highly undesirable to have a line from the stern to the pier or wharf, with nothing from the bow. A boat in such a situation is much less under the control of her helmsman.

Reversing on a Spring

The action is just the opposite if we reverse on a spring. Turning to Figure 724, note how the stern is swung sharply in toward the dock by the action of the forward quarter spring when reversing. Again, control is much less certain with stern movement restricted by a line and the bow free to swing to wind or current.

Now note what happens, as in Figure 725, when backing on a forward bow spring. The turning effect of the spring on the boat is not important here, and she'll go in nearly parallel to the pier.

Allow for Tidal Range

Boatmen on fresh water streams and lakes have no tidal problem when they tie up to a pier, but failure to consider tides in tidal waters can part lines, and even sink the boat.

Springs provide an effective method for leaving a boat free to rise and fall while keeping her from going ahead or astern, moving off fenders, or twisting in such a way as to get caught on dock projections.

The longer a mooring line can be, the more of a rise and fall of tide it can take care of with minimum of slack, Figure 726. Yet every line must be allowed slack enough so that all do not come taut together at either extreme stage of the tide. Generally the lines are MADE UP almost taut, as they don't change much with the weather and will stretch to take up changes in wind strength. This point must be carefully observed in narrow slips or close quarters. It is better to have long lines that are almost taut, rather than loose lines that would allow the boat to hit pier projections or other boats.

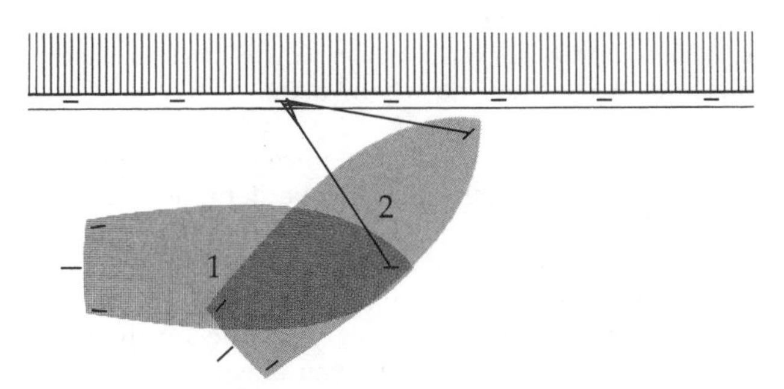

Figure 722 **By going ahead on an after bow spring, the boat's bow is pulled into the pier, but the stern springs away.**

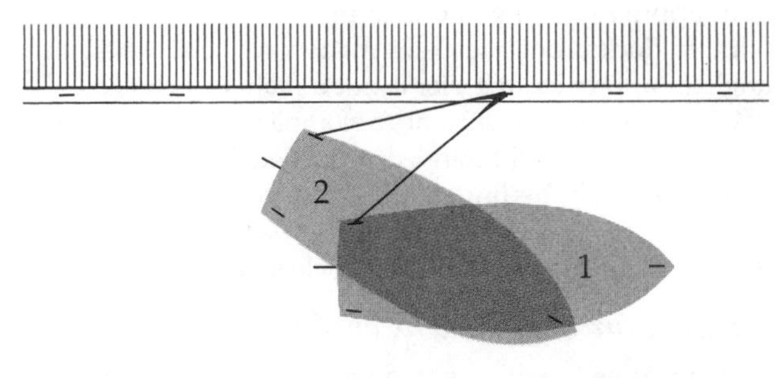

Figure 724 **When backing on a forward quarter spring, the stern swings in, but the bow swings out away from the pier.**

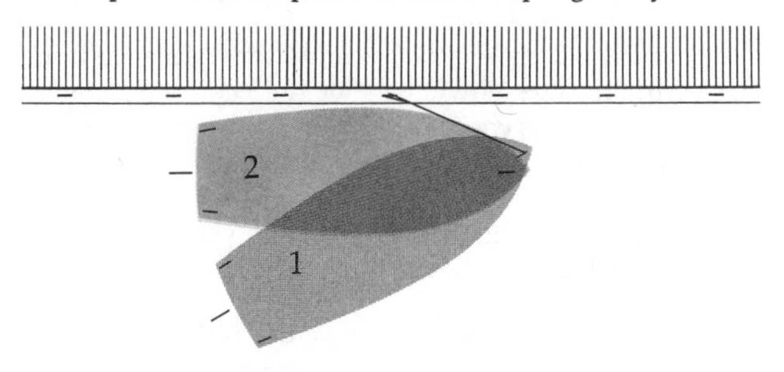

Figure 723 **If the rudder is turned away from the pier, the stern will swing in as power is applied.**

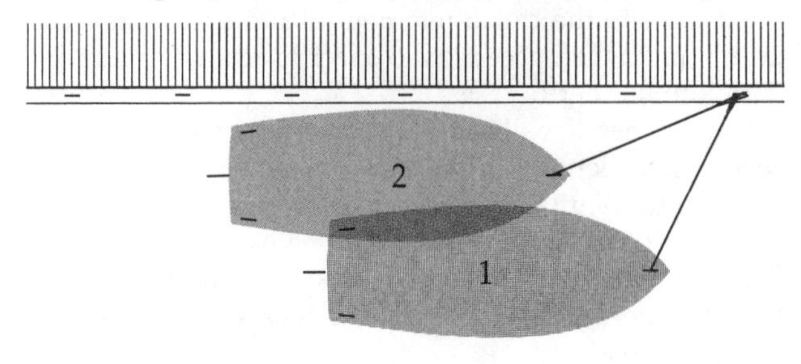

Figure 725 **By reversing on a forward bow spring, the boat can be sprung in nearly parallel to the pier or wharf face.**

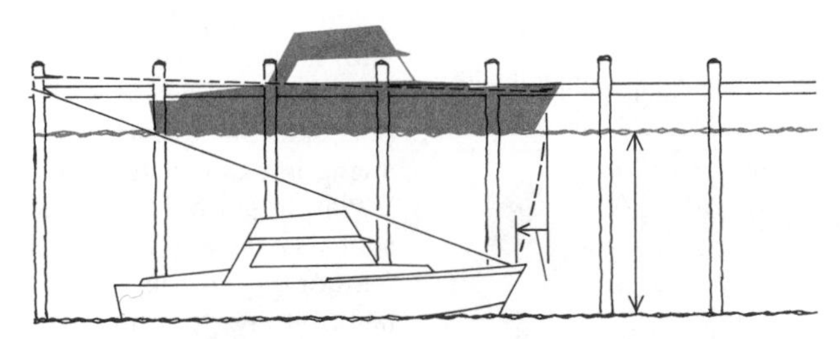

Figure 726 **Illustration shows how long spring lines act with great changes in tidal level. If water drops 11 feet from E to D, a 32-foot spring line (A to B) would cause the boat to move astern the distance from B to F. F represents the position of cleat C at high water. The principle is based on a right triangle, $32^2 = 30^2 + 11^2$, approximately.**

Landing at a Pier

Your knowledge of turning a boat in limited space will stand you in good stead when you tackle the problem of bringing her in neatly to a pier, wharf, or float.

As a matter of strict terminology, you should not refer to a "dock" when you actually mean "pier" or "wharf." A PIER projects out from the shoreline. A WHARF is a structure generally parallel to and not far from the shore. Technically, a DOCK is the adjacent water area, but in boating the term is often applied to the structure itself. The space occupied by a vessel, or assigned for her use, whether occupied or not, is her BERTH.

The action of coming alongside the structure, whatever it is, is referred to as DOCKING or MOORING. Departing is UNDOCKING, a term more often used with ships than boats.

The factors involved here are, in part, the same as in coming to a mooring, with certain modifications. For example, when you pick up a mooring you have wind and current to consider, generally determining the angle of your approach. Wind and current must also be reckoned with at docks, but there are many occasions when the angle of approach to a pier is not a matter of choice.

With no wind or current to complicate the situation, landing, with a right-handed propeller, is accomplished most effectively by bringing the boat in port side to the pier. Our previous analysis of propeller action when reversing makes the reason for this clear.

Keep Your Boat Under Control

Don't come in with a grand flourish at high speed, but throttle down gradually to keep the boat under control. When you consider that you have enough way on to reach the dock, slip the clutch into neutral and use the reverse gear as and when necessary to check headway as the boat goes into her berth.

If your speed has been properly estimated you will be several boat lengths from the dock when the clutch is moved into neutral. Coming in at too high a speed you will be forced to go into neutral a long way from the dock (losing maneuverability with the propeller disengaged) or face the alternative of trusting the reverse gear to check excessive headway.

This is a good time to exercise judgement—keeping the way needed for maneuverability, yet using no more than required. If you have been running at any speed, hold off briefly and let your own stern wave pass by your boat.

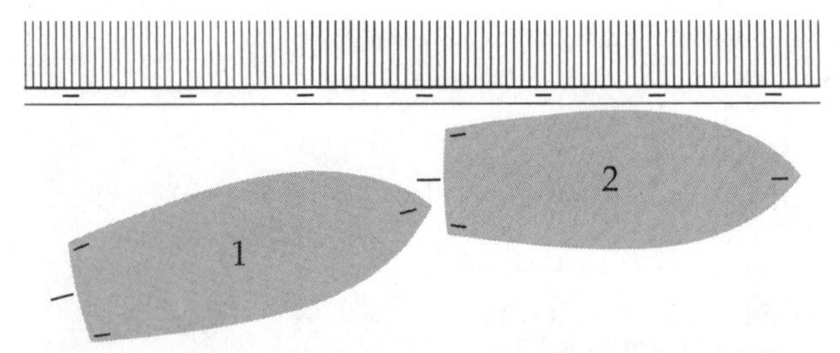

Figure 727 **With a right-handed propeller, approach a pier or wharf at an angle of 10 to 20 degrees for a landing, as shown. Discharge current of the reversing propeller will set the stern to port as at position 2, even though the rudder is amidships.**

As the reversing propeller throws its discharge current against the starboard side aft, the stern is carried to port. For this reason it is customary to approach, not exactly parallel with the dock, but at a slight angle—roughly 10 to 20 degrees. Then as she comes up to the pier her stern will be brought alongside, Figure 727, with the boat parallel to the dock, properly berthed.

In docking, have lines ready to run fore and aft, and have fenders in place to keep the boat from chafing against unprotected piles or dock edges.

Wind or Current Parallel to Berth

When a berth happens to be on the shore of a river or bank of a tidal stream—so that the current flows parallel to it—govern the direction of your approach by the current flow, even though this puts the starboard side toward the wharf. Heading into the current will enable you to keep the propeller turning over slowly—right up to the moment of reaching the wharf, if the current is strong enough.

The same is true if wind is the force you are opposing. If your course puts the wind over the stern, make a wide swing, starting far enough from the wharf to permit you to round up to leeward of it, coming in against the wind. To miscalculate here by not allowing room enough for the turn would be embarrassing, as you head for the bank with wind or current sweeping you downstream. Hence the importance of practice in open water, noting the size of turning circles under various conditions before attempting the actual docking.

In starting a turn like this, even at some distance from the berth, throttle down long before the wharf comes abeam. Otherwise your wash may carry along and leave you wallowing in your own wake just when you're trying to come alongside.

In coming up against wind or current, you can use that force to check your headway, instead of reverse.

Landings Downwind or With Current

If you can, avoid any landing in which wind or current are setting you down toward your berth. You would be too dependent on your reverse gear, and an error in judgement or a motor failure would put you on the spot.

Sometimes, however, space will not permit you to turn before docking. Suppose you are coming in to a canal lock with a strong wind astern. Hold your speed down to a minimum consistent with adequate control, and take the port side of the lock if you can, so that your stern will swing in against the lock face when you reverse.

Plan to get a line out from the stern or port quarter as soon as headway is checked. The boat can lie well enough temporarily to this line alone, whereas to get a bow line fast first and then miss making the stern line fast would be to risk being turned end-for-end by the wind.

Handling the Lines

If you have a couple of hands aboard, assign one to the bow and one to the stern to handle lines, with instructions not to make fast until headway is checked. The seriousness of checking the boat's way by means of a snubbed bow line instead of reversing the propeller is only too obvious. If single-handed you will have to work smartly, with a stern line fast to the after cleat coiled ready to carry ashore, and a bow line run in advance along the deck back to the cockpit. All docking lines of course would have to be led outboard of stanchions and shrouds, to be clear when taken ashore.

The problem is similar if you are making your landing with a following current. In either case, be ready with reverse gear on the approach, using it as strongly as necessary to hold the boat against its momentum and the push of wind or current. The propeller ordinarily should be turning over slowly in reverse for the last boat length or two of headway, the throttle being opened gradually and as needed to kill all headway at the right instant.

When Boats Lie Ahead and Astern

Let's vary the problem by assuming that boats are already lying at a wharf, leaving you little more than a boat length to squeeze into your berth. The technique is decidedly not the one you are accustomed to in parking a car at a curb.

Referring to Figure 728, top, boats A and B are already in their berths astern and ahead of the berth we (boat C) want to slip into. Boat A's position necessitates our going in at a greater angle than if the wharf were clear, and there is no room to go either ahead or astern once we have nosed up to position 2.

Consequently a person is stationed forward to take a line (an after bow spring) leading aft from the forward cleat to a pile or cleat on the dock. He takes a turn around the pile and holds fast while we go ahead with the propeller, setting the rudder to starboard. The spring prevents the boat from going ahead and the stern is thrown in toward the dock until the boat assumes her final position against the dock, as at 3.

As the boat swings in, the spring may be slacked off a little and often a fender or two will be necessary at the point of contact.

Using Wind or Current

In a case as in Figure 728, top, with conditions the same except for a wind from the south (assume boats A and B are headed east) we could have used the wind to advantage in bringing the boat in.

Under such conditions the boat could be brought up parallel to her final position at 3, allowing the wind to set her in to her berth at the dock. The bow, probably, would come in faster than the stern, but this would not matter.

During this maneuver, the engine would be idling and, if there were any tendency for the boat to go ahead or astern, it could be offset by a turn or two of the propeller as needed, to maintain her position midway between A and B.

Balancing Current Against Propeller Action

A variation of this problem is shown in Figure 728, bottom, where a current is flowing east. Here the boat is brought up to a position C1 parallel to her berth at C2. The propeller will have to be turning over very slowly in reverse (perhaps at the engine's idling speed) to hold her at C1 against the current.

Now if the rudder is turned slightly to port the boat will tend to move in toward the wharf, but the stern is likely to come in first. An after quarter spring should be run first, to hold the boat against the current as the propeller stops turning.

Only enough rudder and power should be used to work the boat slowly sidewise. Too much power will put her out of control and too much rudder may cause her stern to go to port too fast, permitting the current to act on the boat's starboard side.

Under identical conditions, except for a current setting west instead of east, the propeller would be allowed to turn slowly ahead, just enough to hold the boat against the current. Then, with rudder slightly to port, the boat would edge off sidewise to port, the bow coming in slightly ahead of the stern. The bow line or forward bow spring would be the first of the mooring lines to make fast in this case.

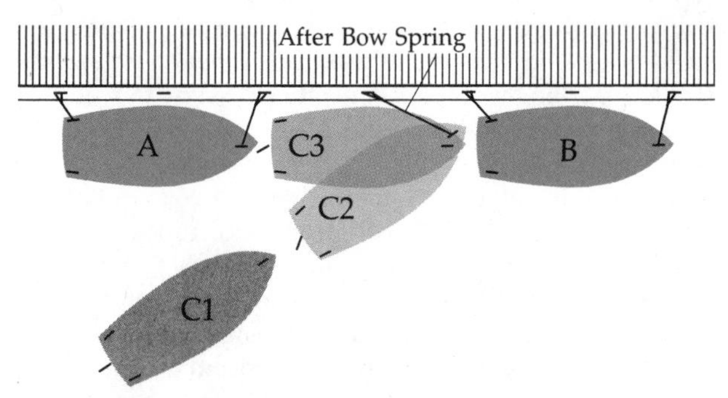

Figure 728 **Landing between boats A and B, *above,* boat C approaches at an angle. At position 2 a spring line is run aft to the pier or wharf from a forward cleat or bitt. Going ahead with propeller and right rudder, boat swings into berth at position 3. *Below:* The boat can be worked into her berth, using the current, by setting the rudder to port and using just enough power to offset the drift of the current.**

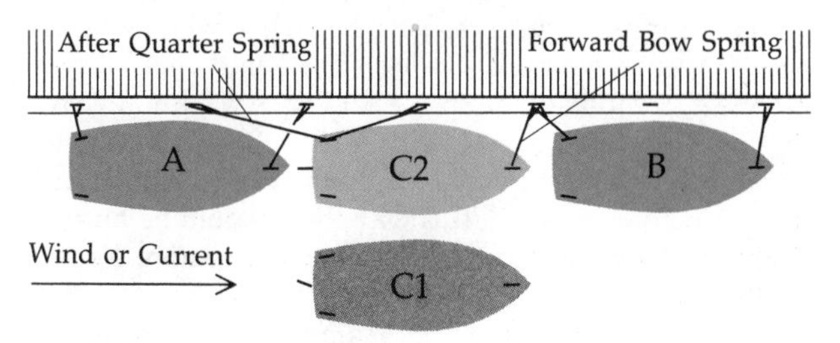

Landing on the Leeward Side

Sometimes you will run into situations where a pier juts out into the water, with wind blowing at right angles to it, giving you a choice of sides on which to land. If there is much wind and sea, the windward side can be uncomfortable as the boat will pound against the piles.

The rougher it is, the more important it becomes to take the leeward side. The wind will then hold her clear, at the length of her mooring lines. If there is much wind you will have to work smartly when bringing the boat up to such a berth, and it will help to have crew ready at both bow and stern to run the docking lines.

Bow Line First

The bow line of course must be run first and made fast. The stern line may present more of a problem since the boat has had to approach at an angle to allow for the wind's tendency to blow the bow off. The stern line can be heaved ashore by one person to another on the dock, after the latter has made his bow line fast. The bow line, if necessary, can be used as a spring to bring the stern in by going ahead with rudder to starboard, on the principle illustrated in Figure 728.

This maneuver, single-handed, is difficult. It can be accomplished by getting a bow line off to use as a spring, working the stern up to the dock with the power, and lashing the wheel if necessary to keep the rudder to starboard while you get a stern line fast.

Holding With One Spring

The use of a spring line as just discussed also works well for temporary holding on the lee side of a pier or wharf. Assume that you have run up along the leeward side of a wharf, to remain there for a short time while you pick up guests or perhaps put someone ashore. Instead of getting lines out to make fast fore and aft as you would for a longer stop, or expecting the crew to hold the boat against the wind's pressure, try using one line as an after bow spring.

After coming alongside, rig the line and go ahead easily till it takes a strain; then go ahead, with rudder turned away from shore, with just power enough to hold the stern up against the pier. This is a maneuver that you can accomplish single-handed without too much difficulty, by having your spring fast to the forward bitt as you come in, with the end ready to pass ashore for someone on the pier to drop over a pile or cleat.

Doubling the Spring

Single-handed, you might prefer to use the spring double, with the bight of line around the pile or cleat and both ends of the line fast to your bitt. Then when you are ready to leave, you can slip the engine into neutral, cast off one end and haul the spring back aboard without leaving the deck. This would be helpful if the wind were of some force, the cleat well back from the pier edge, and no one ashore to assist. This way there would be no risk of the boat's being blown off as you step ashore to cast off the line.

In any case, when the line has been cast off forward, the wind wil cause the boat to drift clear of the dock, the bow ordinarily paying off faster than the stern. Whether you go ahead or reverse to get under way will depend on other structures and boats nearby, and whether your course is to be up- or down-stream. The same maneuver is appropriate if current is setting in the direction that the wind is blowing.

Anchor to Leeward

Figure 729 illustrates a method by which boats may line up at a pier beam to beam, facing the pier. Each has a secure berth on the leeward side of the pier and many boats are accommodated by a small amount of dock space. The principal objection is the difficulty of stepping ashore from the bow.

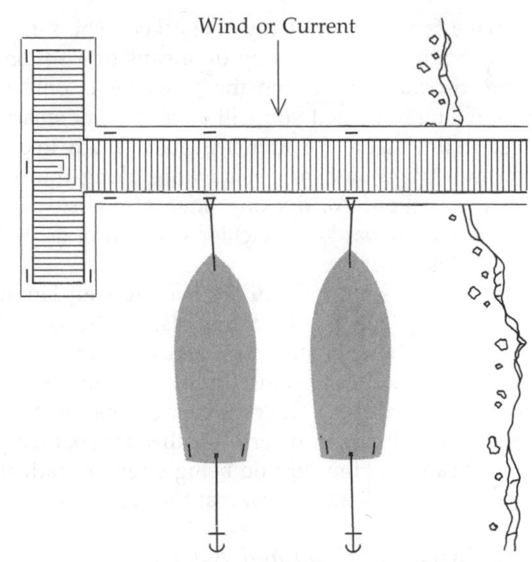

Figure 729 **Stern anchor is lowered as boat aproaches leeward side of a pier, where bow line is to be secured. The anchor rode should be about seven times the depth of the water.**

The arrangement shown makes a snug berth even in a hard blow, but a shift of wind might necessitate a shift of berth if the wind came abeam, throwing a heavy strain on the stern line and anchor.

The principle of getting into a berth like this is obvious from the diagram. Simply run up against the wind or current, propeller turning over just fast enough to give steerageway, and have a man let the anchor go over the stern, paying out line until the bow is close enough to get a line on the dock. The stern line is available as a check, though the helmsman should of course stop the boat without over-shooting his mark and hitting the dock.

After passengers have been discharged the bow line can be slacked and the stern anchor line shortened, though this means that one person must be aboard. If all are to go ashore, adjust the stern line so that it just checks the boat from touching the dock. Then if there is room, carry the bow line off at an angle to another pile. This increases the clearance at the bow, and the wind or current acting on one side will help to hold her clear.

The diagram is not to scale. Often the scope of the anchor line astern will need to be much greater than that indicated to provide holding power in a given depth. Five times the height of the deck above the bottom is a rough-and-ready rule to determine the scope of anchor lines, though this is qualified by many factors.

Using an Anchor to Windward

If you see that a berth on the windward side of a pier or wharf is inevitable and there is reason to believe that the hull will suffer even with fenders strategically placed, plan in advance to get anchors out to windward. This trick is not used nearly as often as it should be, especially if the boat will occupy the berth for a considerable time.

You will need a person on the bow and another on the stern to handle anchor lines, though in a pinch the helmsman can handle the stern line if he takes care to keep slack lines from fouling the propeller. Referring to Figure 730, top, the stern anchor is let go from the quarter when the boat is at 1, moving slowly ahead to be checked at 2, when the man forward lets his anchor go. If someone is tending the line aft, the helmsman can back a little as necessary to place the boat as the wind carries her down to her berth at 3.

Tend Lines Carefully

If the helmsman must leave his wheel and tend the stern line, however, then the boat can be jockeyed into position merely by adjusting the lengths of the various lines. This probably will necessitate hauling in some of the stern line while the bow line is slacked away. Throughout, tend the lines carefully. Dock lines can be run from the port bow and quarter.

Note that at 1 and 2 in Figure 730, top, the boat is not parallel to the wharf, but has been headed up somewhat into the wind. As she drifts in, after the engine is put in neutral at 2, the bow will come in faster than the stern. Be sure to consider the great strain the anchor lines are carrying with the wind hitting the boat abeam, especially if there is some sea running as well. The stronger the wind and the rougher the sea, the longer the scope required if the anchors are to hold.

If you plan to land on the windward side of a pier or wharf without using anchors to hold her off, allow for leeway on the approach and keep the bow somewhat up to windward if possible, checking headway while abreast of your berth. Then have fenders handy as she blows in.

When Current Sets Toward Dock

Let's suppose now a situation in which the boat must find a berth off a pier in such a position that the current will be flowing from the boat toward the pier, Figure 730, bottom. This is called a "Mediterranean moor" for the area in which it is widely used.

In this case it may be assumed that the outside of the dock, which ordinarily would be the natural choice for landing, must be kept clear for ferries, perhaps, or other boats, while the lower side of the dock is also restricted for some reason. The only remaining berth is above it, at A in the illustration.

Size up the situation to determine exactly where the anchor is to be let go, with certainty that you will have sufficient scope. If your anchor drags here, you will be in a bad spot as the current will set the boat down on the dock. Make sure the anchor really gets a bite.

The spot for letting the anchor go may be approached from any direction, as long as the boat is rounded up into the current. The procedure, if you wish to land bow to the dock with a stern anchor out, has already been outlined in Figure 729. This is not feasible above the pier, however, because of the difficulty of backing away into the current when leaving.

Tending Scope of Anchor Line

After the anchor has been let go, drop the boat back toward her berth at A, by paying out scope on the anchor line. The current alone will often do the work, though

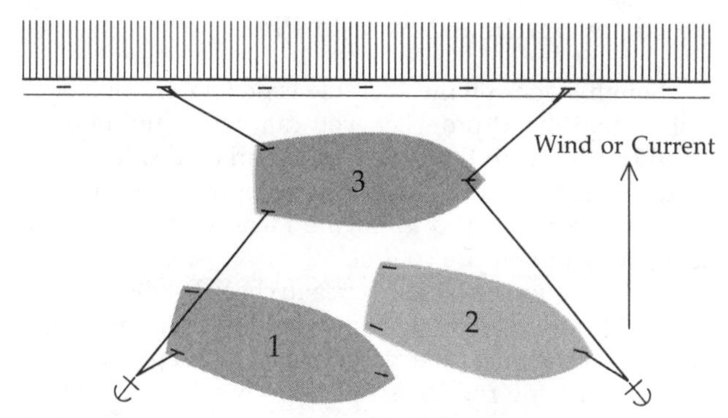

Figure 730 **Boat drops stern anchor at position 1, *above,* a bow anchor at position 2, then drops back on both rodes to position 3, far enough from the pier face to keep from pounding against it. Drawing is not to scale; actual rodes will be much longer. *Below:* Anchor is dropped from the bow, and the boat backs down to the pier. This is sometimes called "Med mooring," as the system is normal practice in the Mediterranean. The breast line here can be used to haul the boat to the pier at B, for transferring people to the shore.**

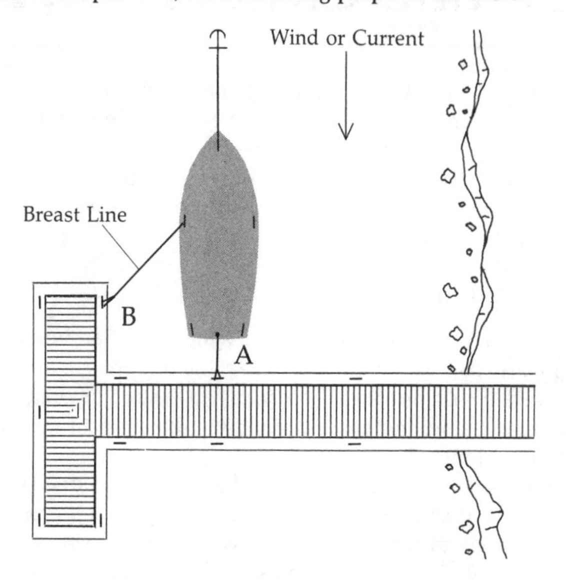

you can use the engine if wind tends to throw the boat out of line.

Just before the stern reaches the pier, secure a stern line to a pile, and adjust the boat's position by the lines as need be. If the anchor is holding securely (and you should leave the berth if it isn't) you can bring the stern close enough to the pier for passengers to step comfortably ashore.

The boat should not be left unattended in this position as she may safely be if berthed as in Figure 729. In Figure 730, bottom, where a convenient wing of the dock is available alongside the boat, run a breast line as shown. Then, before leaving the boat, shorten the anchor line to give clearance at the stern. Slacking the stern line will permit stepping ashore at B, after which the stern line can be adjusted from the pier.

The breast line is left slack but is available to haul the boat back in to B when boarding again after slacking the stern line. If the breast line is made fast too far aft you'll have difficulty hauling the stern in because of the "tacking" action of the boat in the current. Made fast forward of amidships, the boat will come in readily.

Landing Starboard Side to Pier

Though a dock on the port side is preferable, for a boat with a right-hand propeller, you can, with care, make a good landing with the dock on the starboard side.

Referring to Figure 731 the boat is approaching the pier at 1, nearly parallel to it, engine idling or turning over just enough to ensure control.

Just before you need to reverse to check headway, shift the rudder to full left (at 2) to swing the stern in toward the pier. If she does not respond, give the propeller a kick ahead while the rudder is full left. This definitely kicks the stern in and, as she swings, the reverse can be used to kill the headway.

Remember that here we have been considering a single-screw boat with a right-hand propeller. If our craft had a left-hand propeller, the starboard-side-to docking would be preferred, and a port-side-to landing would be the special situation.

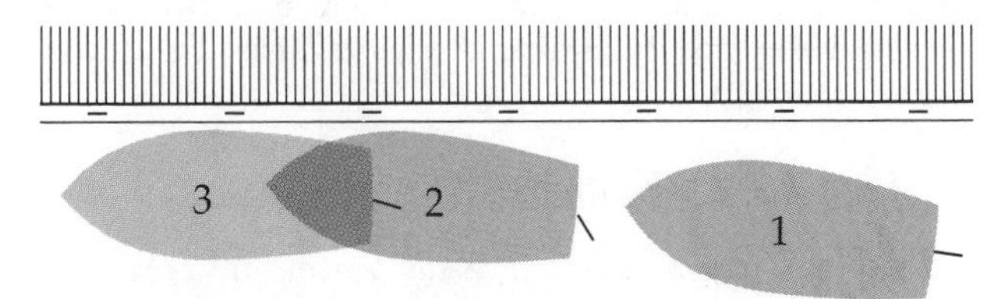

Figure 731 **When it's necessary to make a landing starboard side to the pier, with a right-hand propeller, approach slowly at 1, nearly parallel to the pier or wharf face. Rudder is shifted to full left at 2, and stern is swung to starboard with a short burst of power. Check forward motion at position 3.**

Getting Clear of a Berth

Getting safely away from alongside a pier or wharf can be either simple and easy or complex and difficult, depending on wind direction, the set of any current, and the nearness of other craft. Take advantage of any help you can get from the wind and current, and make good use of spring lines.

With Wind or Current Ahead

Consider first the situation in which wind and/or current are parallel to pier or wharf, and opposite to the docked boat's heading. The boat had landed against the current, using the easiest and most basic technique of coming alongside.

With smaller, lighter craft, all that may be needed is for someone at the bow to give a push off, preferably using a boat hook to extend his reach. A surer way to get clear, and a technique that larger boats will need, is to go ahead on an after bow spring with the rudder turned in toward the pier. The natural propeller action plus rudder action will swing the stern clear of the dock. The boat can then be backed down a short distance and will have sufficient clearance to go ahead and away. With either procedure, don't try to cut away too sharply lest your stern come back in and your port quarter strike the pier face.

For a different technique, see Figure 732, top.

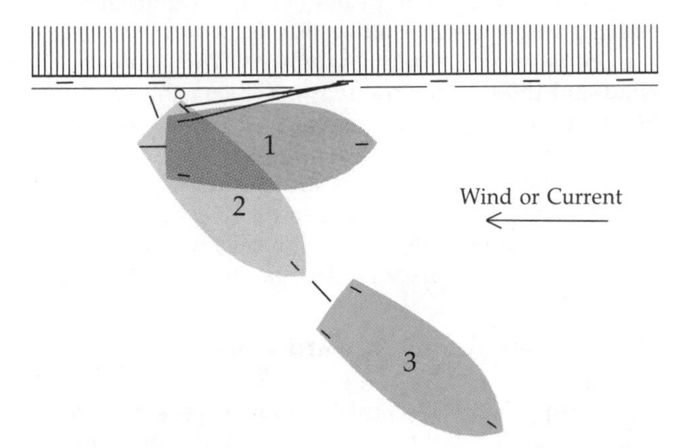

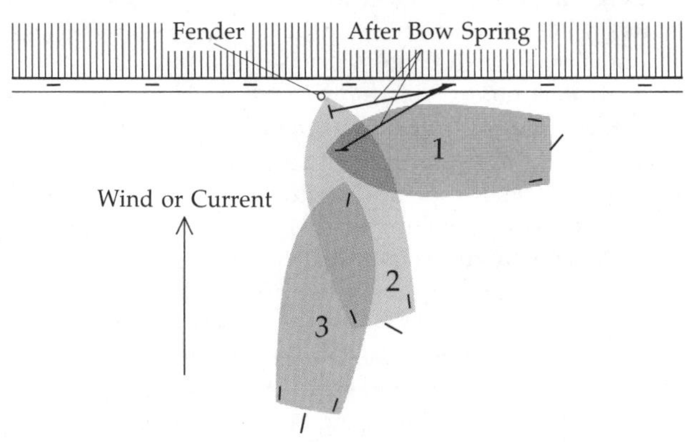

Figure 732 **When wind or current is ahead, *above,* back on a forward quarter spring to turn the stern in and the bow out, angling the boat to get clear by going ahead with rudder amidships. *Below:* Here is how an after bow spring is used to leave the windward side of a pier or wharf. Boat goes ahead on the spring from 1 to 2 with the rudder set toward the pier. At 3 the line is cast off, and the boat backed into the wind.**

From a Windward Berth

When wind or current tends to set the boat off the pier, you'll have little difficulty in getting clear when you are ready. In Figure 732, bottom, however, the wind (or current) tends to hold the boat in her berth against the windward side of the pier. This is a bad berth if there is wind enough to raise much of a sea.

If the spring line principle is not used in this case there may be a great deal of lubberly effort with boat hooks and manpower, and possibly damage to the boat. Correct use of an after bow spring is the seamanlike solution.

In Figure 732, bottom, all mooring lines have been cast off except the after bow spring. If the boat has been lying to bow and stern lines only, cast off the stern line and transfer the bow line to an amidships position on the pier to convert it into a spring.

Going ahead easily on the spring with right rudder tends to nose the bow in and throw the stern up into the wind away from the pier. If it does not respond with the rudder hard over, open the throttle to provide the kick necessary

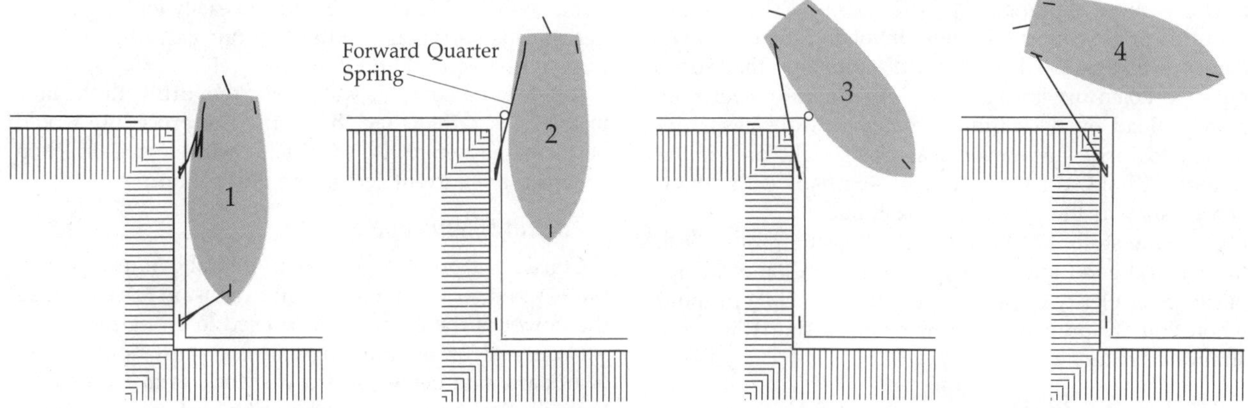

Figure 733 **A forward quarter spring can be used to help boat back out around the end of a pier. When the spring is taut, the boat reverses with full right rudder. Use fenders if necessary.**

to work the stern around. Depending on the nature of the pier and type of boat, you may need a fender or two at the critical spots between dock and boat.

This maneuver is continued until the boat has turned roughly to the position at 2. If there is a cleat about a quarter or a third of the way aft from the bow, use this instead of the bow fitting.

At 2, the boat has turned far enough so that the stem is not likely to slip, though you could allow the boat to turn further till her stern was squarely into the wind. Now you can get the spring line and fender aboard and back the boat away with rudder amidships.

When Single-Handed

This is another of those maneuvers that can be accomplished single-handed with little effort if done in a seamanlike manner. Use a doubled spring line, as previously described, and secure it to a cleat somewhat aft from the extreme bow, so you can reach it quickly from the helm.

With Wind or Current Astern

The procedure just outlined can also be used in getting away from a pier (or wharf) if the wind or current comes from astern.

Instead of merely casting off all lines and going ahead, get the maneuverable stern out away from the pier, shaping the boat up to go astern first, before going ahead on the course. Look again at Figure 732, bottom, but now assume that the wind is east or that the current sets westward (top of diagram being north). The after bow spring allows the stern to go out into the stream, to be kicked out if necessary by power, going ahead with right rudder. Often in such a case the stern swings out without aid from the engine.

When the boat has swung out anywhere from 45 to 60 degrees depending on the particular situation, the spring is cast off and the boat backs off far enough to clear the structure *before* you go ahead with left rudder.

Backing Around

Elsewhere we have discussed backing a boat out of her slip, turning either to port or starboard, before getting underway. A variation of this is illustrated in Figure 733, where the boat is pictured lying in a slip at right angles to the outer side of a pier.

The problem may be to back her around to have her lying along the outside of the pier, or the principle might be used as a method of backing clear of the slip before getting underway, especially if there is little room for maneuvering. In this case a forward quarter spring is used.

At 1 in the illustration, the boat is shown lying starboard side to the land. The first step is to make the spring ready from a point near the corner of the slip to the after bitt, either amidships if there is only one, or the bitt on the starboard quarter, if there are two.

With the spring ready, but left slack and tended, cast off the bow line, or slack it away to be tended by someone on shore, and back the boat easily with full right rudder. At 2, take a strain on the spring to prevent her from backing further, and cause her to pivot as the stern is pulled around to starboard, as in 3. As she continues to back toward 4, the boat assumes a position parallel to the outer face of the pier.

At 4, the shoreward end of the spring can be cast off and carried further up the pier to be made fast elsewhere as a stern line, if the boat is to remain in this new berth. In turn use the bow line to control the bow and prevent the boat's swinging too far; then secure it, perhaps to the cleat or pile formerly used to secure the spring.

This method is especially useful when there is a breeze off the structure that would tend to blow the boat away if maneuvering without lines. No human power is required and it can be accomplished in a leisurely and seamanlike manner. But as in any case where the stern is made fast to the shore with the bow free, special care must be used to not let the boat get out of control.

Getting Underway If the boat is to get underway after being backed around, no bow line is needed. When the boat has pivoted far enough, with starboard quarter near the corner of the pier, idle the engine while the shoreward end of the spring is cast off and brought aboard. Getting underway from such a position, the rudder must be amidships till the quarter clears the pier, then full left rudder turns her bow out into the stream.

Spring lines used this way should have a loop of convenient size in the end, formed either by an eye splice or a bowline, so they will slip quickly and easily off the pile or cleat yet stand the strain of holding the boat against power without slipping.

In this or any other maneuver involving the use of engine power against the spring, it is obvious that strain must be taken up slowly, easily. A surge of power puts a shock load on deck fittings that they were never designed to carry—may even tear cleats right out. If the fastenings hold, the line may part—perhaps with a dangerous snap as the parted ends lash out.

Once the strain has been taken up easily, proper deck fittings and good line of adequate size will stand the application of plenty of power. Bear this principle in mind when you are preparing to tow or in passing a line to a stranded boat.

Turning in a Berth

Turning a boat—called WINDING SHIP in the case of a large vessel—is easy if wind or current is used as an aid, or if the engine is used in the absence of these factors.

Figure 734 illustrates a boat lying, at position 1, with her stern toward the current. The problem is to turn her to head into the current. The first step is to let all lines go except the after bow spring.

Normally the effect of the current will then be sufficient to throw her bow in toward the dock and her stern out into the stream toward position 2. If any factor, such as a beam wind, tends to keep her stern pinned against the dock, kick the stern out by going ahead easily with right rudder. Keep a fender handy as a protection to the starboard bow.

Take steps to prevent the bow from catching on the dock as she swings, thus exerting a great leverage on the spring. A small boat may roll the fender a little and the boat can be eased off by hand, but in large craft it is customary to reverse the engine just enough to keep the bow clear, as shown in 3.

As she swings in with the current, the fender should be made ready near the port bow as shown at 4. At this point she may tend to lie in this position, depending on just how much strain there is on the bow spring. If she does not come alongside readily, even when helped by going ahead a little with right rudder, rig a forward quarter spring. Take a strain on the quarter spring and ease the bow spring, and she will set in to her new berth.

Larger vessels executing this maneuver would get a forward bow spring out on the port bow when the ship reaches position 3, rigging it not from the extreme bow

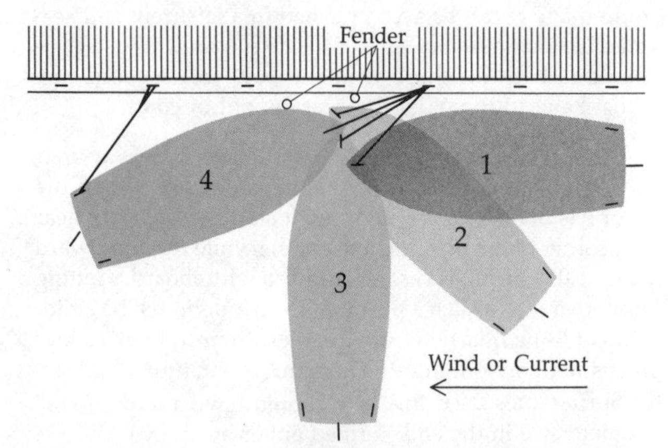

Figure 734 **A boat can be turned end for end, using just the current and spring lines. Procedure is described in the text.**

but a point further aft. Going ahead easily with right rudder on this spring alone (the first one cast off), the ship stays under control and eases in nicely.

In turning a boat this way, make the turn with the bow to the dock rather than the stern. The procedure would be the same where the wind is blowing in the direction in which the current sets in the illustration.

Turning With Power

Consider a problem similar to that above, except that there is neither wind nor current to assist in turning. Here the power of the engine can be used to swing her.

Go ahead on an after bow spring with right rudder (starboard side being toward the dock as in Figure 734) to throw the stern out away from the dock, using a fender to protect the bow. Allow the stem to nose up against the dock, using another fender if necessary to cushion it.

As the boat swings toward a position at right angles to the dock, ease the spring. With the bow against the dock, engine going ahead slowly, and rudder amidships, hold the boat in this position while the bow spring is cast off from on shore and re-rigged as an after bow spring on the port side. With right rudder again, the stern will then swing all the way around, fenders being shifted once more to protect the port bow.

(*Caution:* If nylon springs and lines are used, remember that they stretch and can store energy enough to pull the boat back in when power is removed. If nylon is stretched to the breaking point, it can snap back with lethal effect.)

This principle can be useful in dozens of situations when the theory is understood. In a berth only inches wider than the overall length of the boat, without room to go ahead and astern, you can turn the boat end for end with no line-hauling or manual effort of any kind.

If there happens to be wind or current holding her against the dock, the first half of this maneuver may be used to get the stern out into the stream, preliminary to backing away when ready to get clear of the dock.

Make Use of Fenders

We have noted from time to time in the foregoing illustrations the need to protect the hull with FENDERS in maneuvering around rough dock piles or wharf faces, or in lying alongside other craft, Figure 735, top left.

The standard types of yacht fenders are satisfactory if large and numerous enough. When a boat lies at a pier or wharf with no motion except the rise and fall of tide and the flow of current, even the lightest of guardrails may do.

Topsides that are heavily flared forward, tumblehome aft, high superstructure that reaches out practically to the deck edge, and comparatively light guard moldings all invite damage when the boat starts to roll and pound against piles.

Ordinary fenders (they are *not* called "bumpers") are generally made of a rubber-like plastic and filled with air under low pressure; some are sealed, others can be inflated to higher pressures with a hand pump. Most fenders are round, but some are square in cross section.

Fenders generally have an eye molded in at each end for attaching a short length of light line used to suspend them along the side of a boat. Some models have an axial hole through which a single piece of line is run, knotted at each side of the fender. A recent development is a patented design of fenders which, in addition to normal usage, can be fitted together end for end in tongue-and-groove fashion to make a longer fender.

Half a dozen substantial fenders are not too many to carry, and if the boat has few or no fender hooks, go over it with a critical eye to see that stout hooks or small cleats are placed where needed.

Fenderboards

If fenders hang vertically from the boat's side, they give protection against the face of a solid pier or wharf as the boat moves fore and aft. If the boat lies against vertical piles, vertical fenders will not stay in place during the boat's fore and aft movement. Hanging a fender horizontally by both ends may be satisfactory against piles; it is most effective when the fender is attached to the pile.

The solution to protecting the hull in many situations is with FENDERBOARDS. These are short lengths (approximately 4 to 6 feet) of heavy boards (2 inches by 6 inches is a common size) sometimes faced on one side with metal rub strips or rubber cushions. Holes are drilled and lines attached so that the board may be hung horizontally, backed by two fenders, as in Figure 735, top right.

Fenderboards give excellent cushioning between two or more craft rafted together as in Figure 735, bottom. One boat should put over the usual two fenders behind a fenderboard; the other puts over only its own two fenders. A second fender board should not be used as one board could tangle with the other.

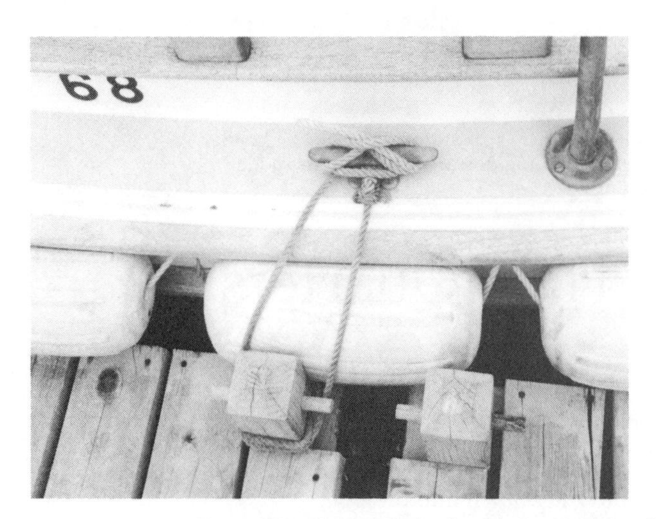

Figure 735 **Fenders, *left,* should always be used when maneuvering or mooring alongside a pier or wharf to prevent damage to the hull. A fenderboard, *right,* is a plank backed by a pair of fenders; its use is invaluable in cases where fenders alone cannot provide adequate protection to the hull. When rafting, *below,* one boat puts out fenders and fender board; the adjacent boat just adds fenders on its side.**

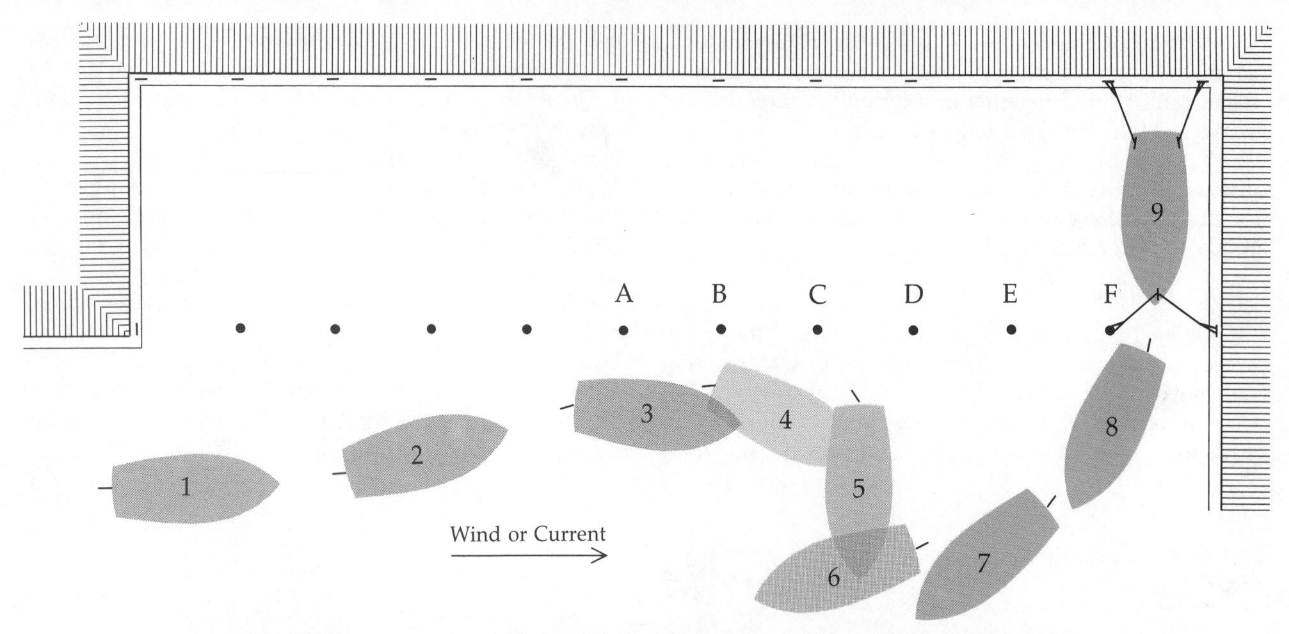

Figure 736 **Backing into a berth between piles in a basin, where space allows a turn ahead under power to maneuver the bow partly into the wind before backing down: Note the right rudder from 5 through 8 keeps the stern from swinging to port.**

Maneuvering at Slips in Tight Quarters

Many marinas and yacht clubs provide slips in which boats are berthed at right angles to a pier or wharf and made fast to piles, as shown in Figure 736. Normally, a short secondary pier, or CATWALK, extends out between each pair of slips.

A Berth Between Piles

Figure 736 illustrates a problem where a boat must be backed into such a berth in a basin. Complicating the situation is a bulkhead along the east (leeward) and south sides, which prevents approach from any direction except west, downwind.

If the space between piles and the south bulkhead is sufficient to allow the boat to make a turn, going ahead with right rudder, as in 2 to 3 to 4 to 5, she may round up this way, until at 5 she has headed partly into the wind.

If space, turning power, or wind prevent her from getting around any further than position 4, her situation is awkward because the wind will tend to blow her bow down to leeward faster than the stern, so her angle of approach will not conform with path 6 to 7 to 8. In that case, another technique would have to be worked out.

Assuming that she can round up under power to position 5 with right rudder, she has been placed so that she can back along the line 6 to 7 to 8 without shifting her rudder. Remembering that with full right rudder she might back practically in a straight line, note how her position at 5 allows the wind to blow the bow down to leeward.

At 7 her bow should still be somewhat up into the wind, and her starboard quarter close to pile F. Having her head up into the wind more than is shown at 7 is all right, because a kick ahead with left rudder will straighten her out to exactly the right angle. If that kick throws her stern too close to pile F, idling a moment allows the wind to carry her down into position.

As she passes pile F, pass a line to it—or pick one up if there's one attached to the pile—and secure it forward when a crewman at the stern has gotten another line on the dock cleat to windward. Then run out other dock lines.

Using Springs to the Piles

With a good breeze it might be impossible to get the bow around as shown in Figure 736(5). In that case use springs; Figure 737. Approach close to the line of piles A, B, C, and D, bringing her up alongside piles B and C, heading east, as at 1 in Figure 737. Now run an after bow spring, doubled, from the bitt around pile C and back to the bitt. Going ahead easily with left rudder will cause the stern to swing around to the south as at 2. Engine idling, she will swing completely around and if the spring is slacked a little she will lie in position 3, starboard side against piles E and F. You may need to back a little between positions 2 and 3.

Next rig a forward quarter spring on the starboard side to pile F and cast off the bow line to pile C. The reason this was doubled is now evident, as you can get the line back aboard by casting off the second hitch.

Reversing on the quarter spring throws her bow to the south and her stern in toward her berth as at 4. This spring can be carried forward by a crewman walking along the starboard side to the bow, where it becomes the forward bow spring to pile F when secured to the forward bitt.

Approaching Upwind

Figure 738 illustrates an approach against the wind. While the boat is still moving ahead at 2, left rudder throws the stern in toward the slip and pile F at position 3. Backing with right rudder brings her down to position 4, the bow sagging off to leeward somewhat, while the rudder is set to port and the propeller given a strong kick ahead. This throws the stern up to windward as at 5, shaping the boat up to back in with right rudder.

Between positions 4 and 6 you may need to go ahead and back several times, but in any case the combination of the engine kicking the stern to windward while the wind blows the bow to leeward can be used effectively to work her in.

Another alternative, at 3, would be to pass a line around pile F and hold on while the wind blew the bow to leeward just enough to shape the boat up to 5, to back in. A little reversing between positions 3 and 5 might be needed. At 5, of course, the line from the starboard quarter is cast off and carried forward to become the forward bow spring on the starboard side, secured to pile F for mooring.

Getting Clear of a Pile

In maneuvering around piles, you may be caught in a position where the wind and/or current hold the boat against the pile so she cannot maneuver. The solution is to rig a forward spring from the pile to a cleat aft, preferably on the side of the boat away from the pile, as shown in Figure 739. Then, by reversing with left rudder, pivot the boat around the pile and bring her bow into the wind or against the current to position 2, properly shaped up to draw clear by going ahead with power as the spring is cast off.

In getting clear from this point, you may need a little left rudder to keep the stern clear of the pile, but don't use enough to throw her stern so far over that the starboard quarter is in danger of hitting the other pile.

As a matter of fact, at position 2 the wind may catch the bow on the port side, easing the boat away from the pile, so you will be in a position to pull directly away with rudder amidships—always ready, however, to throw the stern one way or the other with the rudder, if necessary.

Making a Tight Turn

Sometimes a burst of engine power may help the boat to turn within the limits of a channel or restricted waterway where she couldn't turn at low engine speeds.

In Figure 740, for example, a boat is bound south in a narrow canal or stream with a strong northerly wind, with her berth on the west side. The south end of the illustration may represent the dead end of a canal or the stream may continue further. This is irrelevant except that if the stream widens further on it is obviously best to go on downstream to turn, then come back upstream against the wind to dock.

If there is room, wait till you are beyond your berth before you start your turn, as this allows more room for straightening out at the pier or wharf. As illustrated, you have slowed down, the boat is under control, and you gauge your distance from the east bank so that there is just sufficient room for the stern to swing to port with

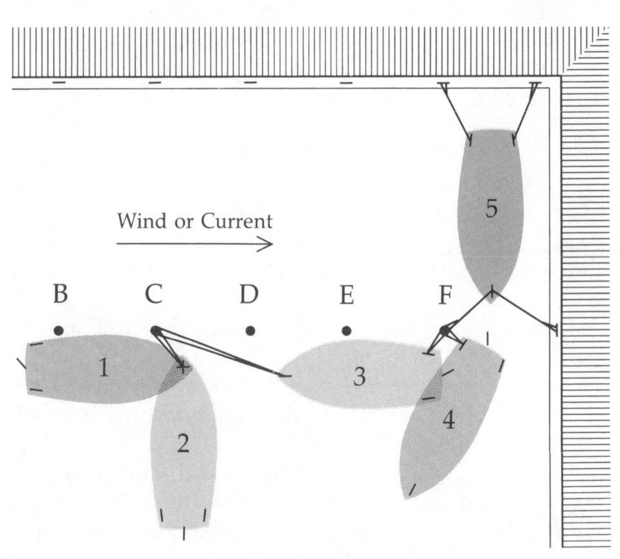

Figure 737 **Here springs are used on the piles to bring the bow into the wind when there is not enough room to turn into the wind, or where there is a strong breeze.**

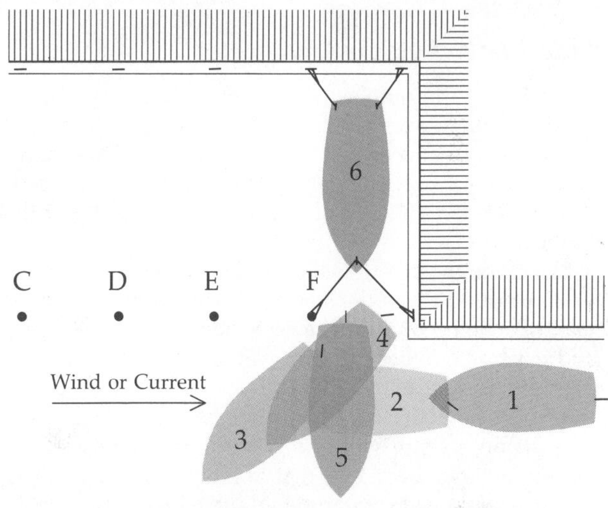

Figure 738 **In this case the arrangement of the bulkheads permits approach against the wind, making it easy to shape the boat up so it can be backed into its slip. Note at 4, left rudder and burst of power ahead swings stern to starboard.**

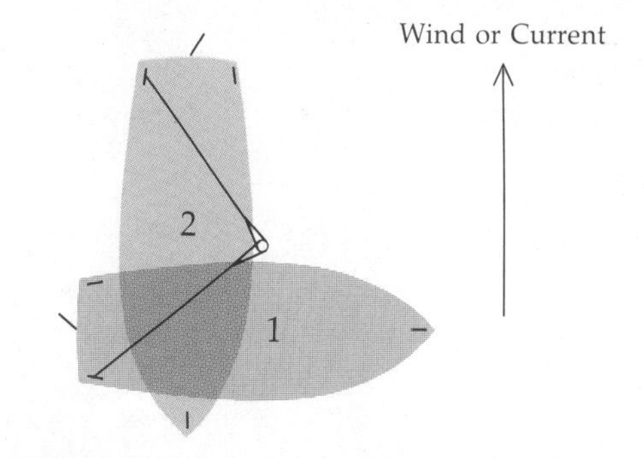

Figure 739 **If pinned against a piling, use a forward spring line to swing the boat around to the leeward side.**

right rudder at 2 without hitting the bank.

From experience in this maneuver, you know that in the absence of wind you can round up nicely through the track depicted in 1 through 6, with full right rudder at low speed. Now if you try the same thing at low engine speed with the northerly wind, instead of moving from 3 to 4, the wind catches you abeam and tends to hold the bow down to leeward as at 4a.

Using Power at Slow Speed

At 4a you would be in a bad way, forced to reverse to avoid hitting the west bank and fortunate if you could back out quickly enough to keep the bow from being driven down by the wind on the south bank, if this is a dead end.

Therefore at 2, if the boat is moving slowly, a sudden burst of power acting on the full right rudder will bring her around quickly. At 4 you must be ready to close the throttle quickly to keep the boat from gathering so much way that landing is difficult.

This is not a maneuver to be guessed at, but if you have occasion to make the same one repeatedly in bringing your boat in to her usual berth, you will soon know whether the strength of the wind will allow you to use these tactics or whether you must resort to some other procedure such as landing downwind and allowing her to turn at the dock on a spring.

Plan Maneuvers in Advance

Obviously the number of possible situations—considering the differences in boats and the strength, direction, and effect of wind and current—is almost infinite. Usually, however, applying one of the principles above modified as needed will permit a seamanlike handling of the problem.

Even though you know the principles, however, it pays to think ahead about the steps you will take. With a clear plan of action, you can take each step slowly and easily, and have time to keep the boat under perfect control. To avoid confusion, tell your crew the steps you plan to follow, the actions that each crew member will have to take, and the orders you will give when these actions are required. Even on occasions that call for swift and decisive action, you'll need calm, ordered judgement.

Use Common Sense

If your plan of action requires complete abandonment because of unforeseen conditions, don't hesitate to act accordingly. For example, if your plan for a clean approach to a pier or wharf has been upset by a freak current you couldn't calculate, back off and square away for another attempt. That in itself is good seamanship regardless of how others may judge your apparent "miss" on the first try. Common sense, if you act with deliberation, will enable you to work out a solution for any combination of conditions.

When you are at the helm and you have people on deck handling lines, give orders to each so that all action is under your control, instead of having two or three acting independently to cross purposes. This is especially imperative when your crew is not familiar with boats or with your method of boat handling.

Easy Does It

As explained, the maneuvering procedures are so simplified on a small boat that docking may involve only one person each at bow and stern, stepping ashore with their lines, with other lines passed as need be after these are secured.

One mistake that small-boat crewmen often make is that, in their eagerness to assist, they snub lines immediately when the skipper wants them merely to be tended and left slack, so that his boat is free to be maneuvered further with power. A case in point is where the boat is coming in to a pier and the bow man snubs his line and thus unintentionally swings out the stern, instead of allowing the boat to ease up and draw alongside as the skipper intended.

Caution In handling lines, whether they are mooring lines, anchor lines, or any other kind, *do not allow them to get over the side* in such a way that they will be sucked down into the propeller and wrapped around the shaft.

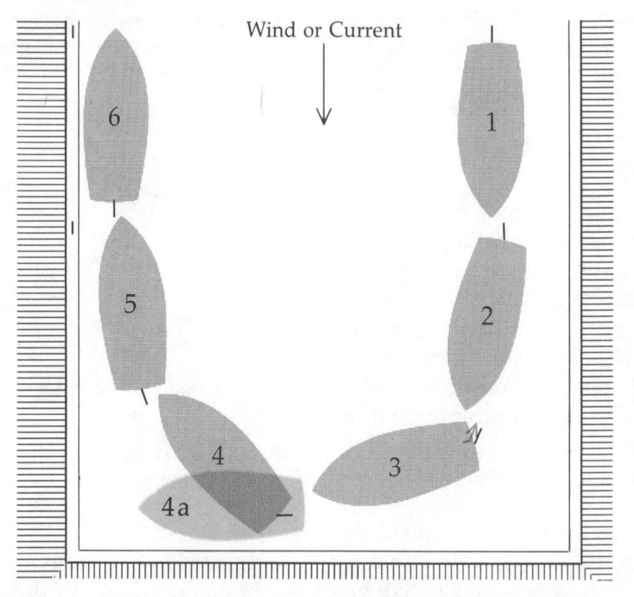

Figure 740 **Turning downwind in close quarters, with wind astern: Boat moves slowly to 3, where a brief burst of power helps it to turn to position 4, rather than 4a. At 4, close the throttle quickly to prevent overshooting berth at position 6.**

Handling Twin-Screw Boats

This chapter has thus far related almost entirely to a single-screw inboard boat. Although this permits a better study of the basic factors of boat handling, attention must now be given to twin-screw craft. In the twin-screw boat, the propellers are usually arranged so that the starboard wheel is right-hand, the port wheel is left-hand. The effect of this is to give a maximum of maneuverability.

Going ahead with the starboard wheel for a turn to port, the offset of the propeller from the center line adds greatly its effect of throwing the stern to starboard. Similarly, offset of the port wheel going ahead helps the steering effect when the port propeller is going ahead for a turn to starboard.

When reversing, the starboard wheel throws its discharge current against the starboard side of the hull to help the turn of the stern to port. Likewise, the port propeller reversing throws its stream against the port side of the hull to help the swing of the stern to starboard.

The important factors in turning and steering are thus combined by the outward-turning wheels. The steering effect is exerted in the same direction as the turning movement caused by the off-center location of the propellers.

Twin Rudders for Twin Screws

Twin-screw craft have two rudders, one directly behind each propeller in the discharge current, Figure 741. This provides much greater effectiveness than would a single center rudder.

Figure 741 **Maximum maneuverability of twin-screw boats is gained by placing twin rudders behind the propellers, where the discharge current can act against them.**

Basic Turning Maneuvers

Clearly, having two propellers gives a skipper the means of throwing one side or the other ahead or astern, independent of rudder control. In fact, much of a boat's maneuvering at low speed is done without touching the steering wheel as the two throttles are the key to the boat's control.

One Propeller Going Ahead or Backing

A boat's stern may be put to one side or the other by going ahead or backing on one propeller only without turning the other. Some headway or sternway in these cases accompanies the turn.

Referring to Figure 742*A*, a kick on the port propeller throws the stern to port, bow to starboard, as in a single-screw boat with right rudder. With twin rudders, right rudder helps this kick but is not necessary. A kick ahead with the starboard wheel throws the stern to starboard, bow to port, as shown at *B*. Reversing the port wheel only pulls the stern around to starboard and vice versa, (*C* and *D*).

When reversing with one propeller, the other being stopped, unequal blade thrust, discharge screw current and the offset of the working propeller from the center line all combine to throw the stern in a direction *away from* the reversing propeller.

Response to Rudder While Backing

A twin-screw vessel starting from a position dead in the water, with both propellers backing at the same speed, is at a great advantage over the single-screw vessel as she can be made to take any desired course by steering with her rudders. In the twin-screw vessel, opposite rotation of the propellers means that the forces which normally

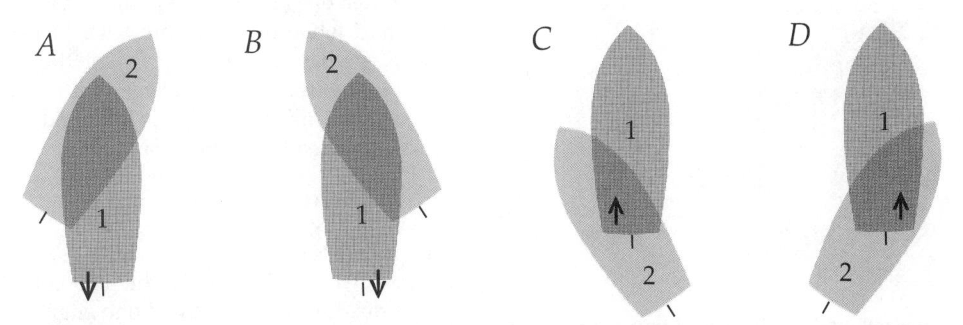

Figure 742 **Diagram shows actions of a twin-screw boat when power is applied to only one propeller, with rudder amidships. One propeller turning ahead swings the stern to its side; when one propeller is in reverse, boat backs to opposite side.**

throw the single-screw vessel off course are balanced out. Rudder action, however, will be less than when going ahead as the propellers' discharge currents are directed away from the rudder rather than onto them.

In addition to use of the rudders, the twin-screw vessel offers the possibility of using her throttles to speed up one motor or the other as an aid to steering while maintaining her sternway, or she can even stop one propeller or go ahead on it for maximum control in reverse.

While the twin-screw vessel backs as readily to starboard as to port, she is still subject to the effect of wind, waves, and current. The helmsman, however, is in a better position to exercise control over them.

Steering With the Throttles

In maintaining a straight course with a twin-screw vessel, the speed of the motors can be adjusted so that the leeward engine compensates for the effect of leeway. That is, the leeward engine can be turned a little faster to hold the bow up into the wind. There is a disadvantage, however, to using this technique as an undesirable "beat note" sound may be heard when the speeds of the two engines are not the same—when the two engines are not "in sync."

If the boat happens to sustain some damage to her steering gear, whether it be the rudder(s) or any part of the gear inboard, she can still make port by steering with the throttles (provided that the rudders are not jammed hard over to one side). One motor can be allowed to turn at a constant speed—the starboard one, let us say. Then opening the throttle of the port motor will speed up the port propeller and cause a turn to starboard. Closing the throttle of the port motor slows down the port propeller and allows the starboard wheel to push ahead, causing a slow turn to port. And we have also shown that she is at no great disadvantage when she finally maneuvers into her berth after getting into port as the throttle and reverse gears are adequate for complete control here too.

Turning in a Boat's Length

With practice, a twin-screw boat can easily be made to turn in a circle only a little larger than her length, as shown in Figure 743.

Turning to starboard, the rudder can be set amidships, while the port engine goes ahead and the starboard engine reverses. The engines will probably be turning at nearly, not exactly, the same speed. The boat drives forward more easily than it goes astern and the propeller at a given RPM has more propelling power ahead than astern. Therefore the RPM on the reversing starboard wheel may be somewhat higher than that on the port propeller, to prevent her from making some headway as she pivots.

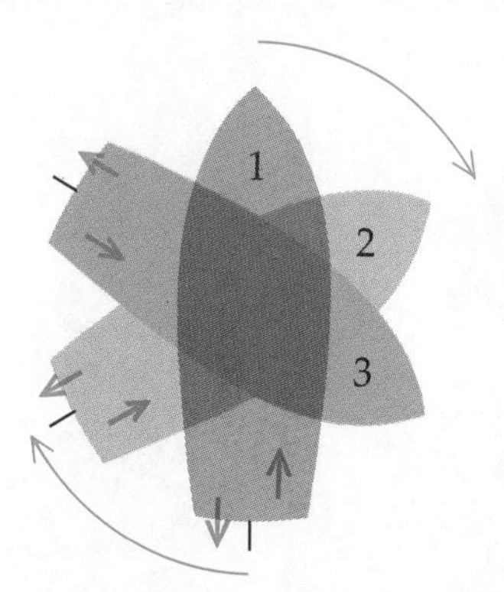

Figure 743 **A twin-screw boat can be turned in its own length, with rudders amidships, by going ahead on one engine while the other reverses. Changing engine speed on one propeller gives the boat headway or sternway as it turns.**

By setting the throttle for the reversing starboard engine this can be left alone and the port throttle is adjusted till the size of the turning circle is established. A rate of RPM can be found where she is actually turning in her own length. If the port engine is then speeded up a little, the circle is larger and she makes some headway. If the port engine is slowed down, the circle is also larger but she makes some sternway as the reversing starboard wheel pulls her around, stern to port.

Docking a Twin-Screw Boat

Landing at a pier (or wharf), a twin-screw vessel should approach at slow speed at an angle of 10 to 20 degrees, as is the case with a single-screw boat. When she has just way enough to carry her in nicely, the rudder is swung over to the side away from the dock to bring the stern in as the engines are thrown into neutral. To check headway as she comes up parallel to the structure, the outboard engine is reversed.

Landing either port or starboard side to the pier is accomplished with equal ease because of the starboard propeller turning counter-clockwise as it reverses in a port-side landing, or the port propeller turning clockwise in reverse on a landing starboard side to the pier.

Using Springs With Twin Screws

When a twin-screw vessel is lying in a berth and a spring is used to throw the bow or stern out as an aid to getting clear, one engine may be used. For example, with an after bow spring, going ahead on the outside engine only, throws the stern out away from the pier, in addition, the inside engine could be put in reverse.

Sometimes a twin-screw vessel is gotten clear of a wharf by rigging a forward bow spring and reversing the propeller on the wharf side to throw the propeller discharge current on that side forward between the boat and structure as a "cushion." Naturally this is most effective if the wharf under water is solidly bulkheaded rather than built on open piling. The discharge current from the inside wheel forces the boat away from the wharf and the line is then cast off. Further reversing on the propeller nearest the wharf while the other propeller turns ahead, as necessary, shapes the boat up to get the stern clear. The speed of the two motors will vary with conditions of wind and current and the rudder is left amidships till the boat is clear and ready to pull away.

Other Twin-Screw Maneuvers

If one propeller is stopped while the boat has headway, the bow necessarily turns in the direction of the propeller that is dead. Consequently, if one engine of a twin-screw power plant fails, and the boat must travel on the other, a certain amount of rudder angle on the side of the operating propeller is necessary in order to maintain a straight course, or some kind of drag must be towed on the side of the working propeller.

A basic principle in the maneuvering of a twin-screw vessel is to use the rudder primarily in relation to the direction of the vessel's movement through the water (that is, whether she had headway or sternway). Elsewhere, it will be recalled, the principle as given for the single-screw vessel was that the rudder should be considered in relation to the direction in which the propeller happens to be turning, regardless of whether the vessel has headway or sternway.

When a twin-screw vessel has headway, both propellers turning ahead, and a quick turn to starboard is desired, the starboard engine is reversed with right rudder. The fact that the vessel has headway means that the right rudder adds its steering effect to shorten the turn.

With sternway, both propellers reversing, if a quick turn to port is wanted, the port engine is thrown ahead, with left rudder. Again, due to the vessel's sternway, the rudder's effect is added to that of the propellers in causing a short turn to port. Return the port engine to reverse gear as the stern moves onto the required heading.

If a twin-screw vessel has considerable headway and her engines are reversed with rudder hard over, the stern will normally swing away from the rudder (to port with right rudder and vice versa) until the headway is overcome by the reversing engines. After she has gathered sternway, her stern tends to work toward the side on which her rudder is set. Then the vessel's stern in the situation just cited would eventually move to starboard with right rudder, after her headway had changed to sternway.

A

B

C

D

Figure 744 **In preparing to dock, this twin-screw cruiser has come to a stop just past its slip (at A), fenders are rigged, and crew members are standing by with stern lines. Helmsman backs down on starboard engine (B) while port engine is turning ahead slowly, to keep the transom aligned with the slip as the bow comes around. A stern line is passed ashore (C), both engines are reversed at dead slow speed as the boat backs straight into the slip (D), and a turn is taken around a cleat with the stern line. Both engines are shifted briefly into forward gear to stop the boat (D), and a bow line is taken ashore.**

3940

8

Hull and rig types, the theory of wind propulsion, boat handling under sail; the special problems of sailboat navigation and trailering

SAILBOAT SEAMANSHIP

Sailing is an old and a complex art, and sailors can spend a lifetime at it and still find that there is more to learn. It is also a simple and a fairly safe activity. A beginning sailor can have a fine time in a sailboat on his first day out, provided he takes a few precautions regarding the weather, and his safety on the water. Experienced powerboat people, as well as those new to watersports, are taking up sailing in greater numbers. While most of the seamanship and piloting chapters in this book apply equally to both sail and power boats, here are the terms used specifically on sailboats, and a description of basic equipment used to accomplish certain sailing seamanship maneuvers. Be sure to study the general nautical terms in Chapter 1 before adding the new ones below.

Sailboat Terms

The wind is the most important element in sailing and there are special words to describe the wind's changing character. The wind VEERS when it shifts in a clockwise direction (for example, south to southwest) and it BACKS counterclockwise. These variations in direction cause a sailboat to be LIFTED (in direction) if the wind change is favorable to her course and HEADED if it is unfavorable. The wind is described by the direction *from* which it blows: a wind out of the northeast is a northeast wind. The real direction and strength of the wind is the TRUE WIND, but this may not be the wind perceived aboard a boat while underway, because the true wind is combined with the boat's own speed and direction to form the APPARENT WIND. Apparent wind can be quite different from true wind, and is the more important wind to consider for sail adjustment, or TRIM.

Everything that the wind passes over has a WINDWARD and a LEEWARD side. The windward side is the part of the boat, island, or other object that the wind passes over first. If you look toward the wind you are looking to windward, or TO WEATHER. The LEEWARD side (pronounced lōō-r-d) is the protected side of a boat. A boat may be IN THE LEE of an island, or of another boat, and so receive less wind. To LEEWARD may refer to DOWNWIND (down the path of the wind); the opposite is UPWIND. A boat may change course TOWARD THE WIND (when the bow moves upwind), or AWAY FROM THE WIND (when the bow moves downwind).

Points of Sailing

A sailboat moves downwind with fairly simple aerodynamic forces, and is said to be RUNNING WITH THE WIND or SAILING FREE. Moving across the direction of the wind, a sailboat is REACHING; she may be on a BROAD REACH, a BEAM REACH, or a CLOSE REACH; these are the fastest points of sailing (see Figure 801).

A sailboat cannot sail directly into the wind, but sails upwind by means of a process called TACKING, which is a series of course changes, or TACKS, that serve to work the boat in a zig-zag pattern toward her destination upwind. If the wind is coming over the port side first, the sailboat is on the PORT TACK, no matter what her point of sail, and if the wind comes over her starboard side, she is on a STARBOARD TACK.

There are complex aerodynamic forces at work in WORKING TO WINDWARD. A well-designed and well-sailed boat is said to be WEATHERLY, or to POINT WELL, if she can sail to within 40° of the direction of the wind. A boat sailing as close to the wind as possible, and still maintaining her speed, is BEATING. She tends to HEEL noticeably on this point of sailing—she leans significantly leeward.

In going from a CLOSE HAULED course (a BEAT; close-hauled because the sails are hauled in "close" to the centerline of the boat) to a reach, a sailboat is said to FALL OFF the wind, and the helmsman tells the crew to let the sails out, or EASE them. This is an example of TRIMMING the sails; the trim must be checked constantly and adjusted for fast, efficient sailing.

A sail is trimmed with a SHEET—the line used to haul it in closer to the wind, or ease it, whichever is best for the boat's course. A boat's sails may be OVERTRIMMED for a particular wind or course if they are held in too tightly, and either the angle of the sails to the wind must be changed, or the course of the boat must be adjusted, to trim correctly.

If the sails are out too far or the boat is pointing too close to the wind, the forward edges of her sails will shiver, or even shake violently, and this is called LUFFING. A boat cannot sail efficiently when her sails are luffing, and she will need a change of sail trim or a change of course before she can pick up speed. If the sails are close-hauled and the boat's course is too close to the wind, she will not be able to sail well, and she will be PINCHING; the remedy is to fall off a little, steering away from the wind, so she will sail better.

Sometimes if the boat is running before the wind, sails well out, the helmsman may decide he'll get more speed if he runs WING AND WING, with, for instance, the main out to starboard and the jib out to port. Skillful steering will be needed to keep the sails full and to avoid an accidental jibe (see below).

Tacking and Jibing

The two most important maneuvers in sailing are TACKING and JIBING. Either one is a course change in which the sails move from one side of the boat to the other, from starboard to port tack or port to starboard tack. When tacking, the bow of the boat turns into and "across" the wind. In a jibe, the stern of the boat crosses the wind.

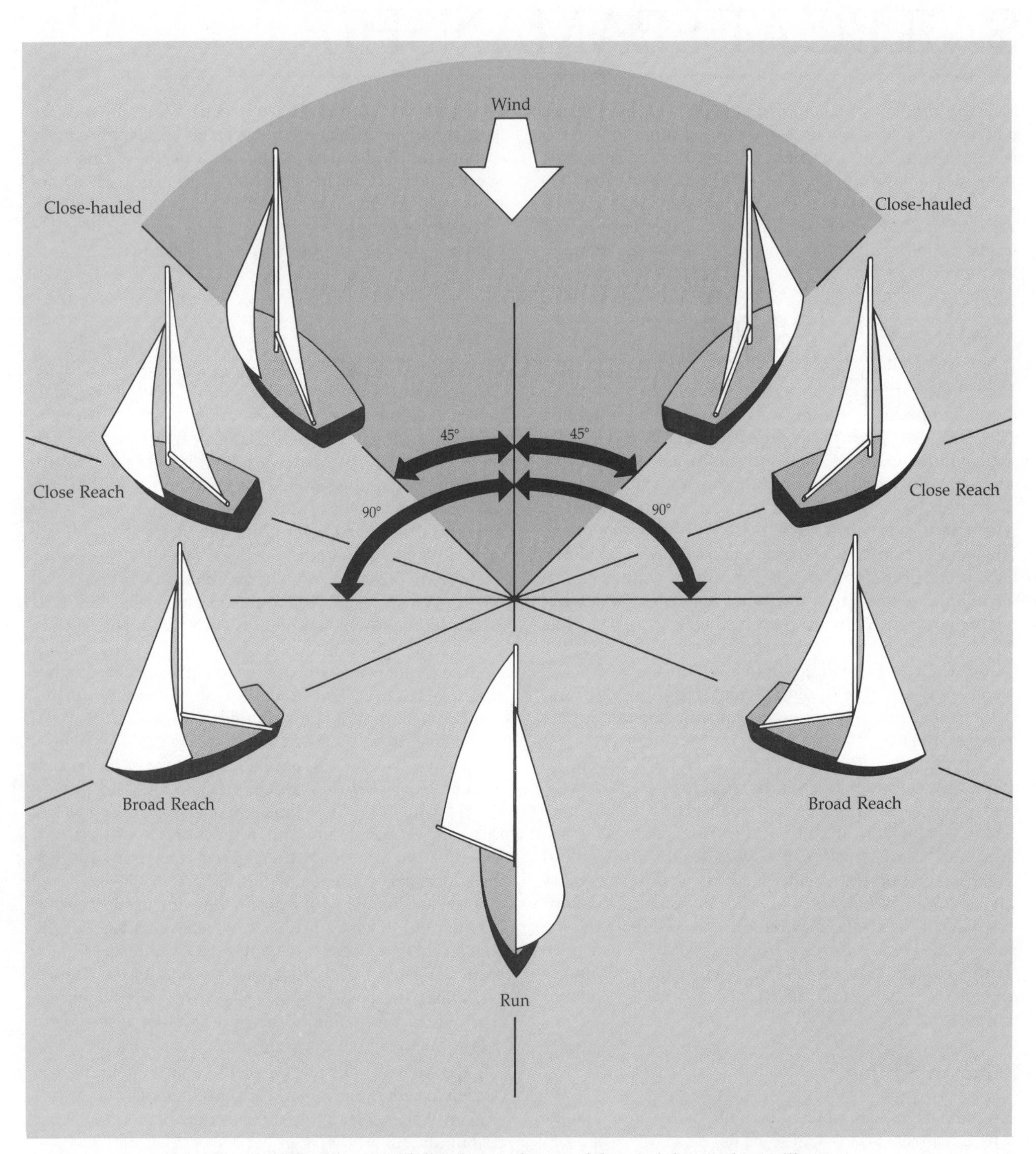

Figure 801 **A sailboat cannot sail directly into the eye of the wind, but modern sailboats usually can sail to within 45 degrees of the wind, or closer, when close-hauled. A reach is the fastest point of sail, with the sails eased about halfway out. A run is aerodynamically simpler, but can be the most dangerous point of sail. The sails are extended as far out over the sides of the boat as possible, and can swing across with tremendous force.**

Tacking

To tack, the helmsman says "Ready About!" to alert the crew to the activity about to begin. After the crew has responded "Ready!," the helmsman says "Hard-a-lee!," to indicate he is pushing the tiller hard to the lee side of the boat, the side the sails are on. He may actually use a wheel instead of a tiller to turn the boat, but the command "Hard-a-lee" remains the same.

The boat now turns sharply upwind and the wind is spilled from the sails completely. All the sails luff momentarily. The crew releases the jib. For an instant the boat is heading directly into the wind. Because of her forward momentum, she does not stop, but continues to turn in the same direction so that the wind now comes across the deck from the boat's other side. The sails begin to fill on the new leeward side. The jib is taken in and all sails must be checked to see if they need adjustment.

The boat is now sailing on a new tack. She has COME ABOUT and changed direction, usually no less than 80°.

If the maneuver is not executed smartly, or the craft does not have enough headway to carry the bow through the wind, she may stall with the wind dead ahead and then is said to be IN IRONS—no headway and therefore no steerageway. If she does not naturally fall off to either side, the remedy is normally to BACKWIND the mainsail or jib—push it out to one side to catch the wind until the boat turns away from it, and to angle the rudder so that the stern is directed away from the intended direction of travel.

When sternway is gained after being in irons, the tiller and rudder operate oppositely from their normal functioning. Getting out of irons smoothly may take a little practice; try intentionally getting into this situation and getting out quickly.

Jibing

To JIBE, the sailboat also changes course to bring the sails over to the other side, but does so while sailing

Figure 802 **Sailing wing-and-wing, this ketch is running with three sails trimmed way out to catch as much wind as possible. Care must be taken to avoid an accidental jibe.**

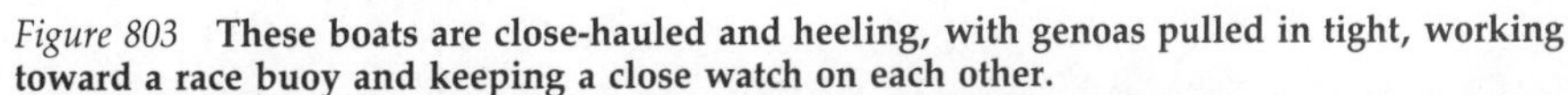

Figure 803 **These boats are close-hauled and heeling, with genoas pulled in tight, working toward a race buoy and keeping a close watch on each other.**

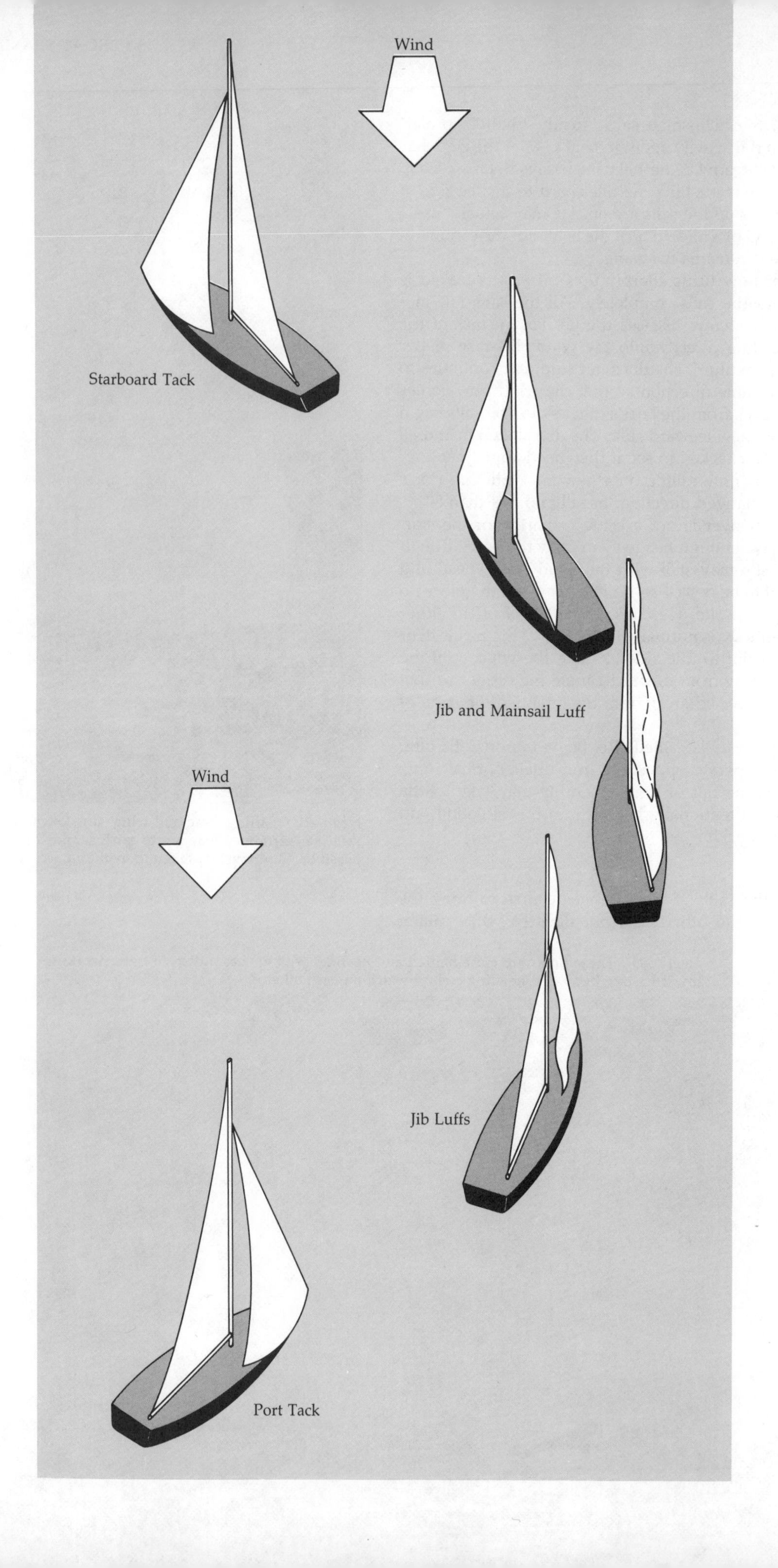
Wind
Starboard Tack
Jib and Mainsail Luff
Wind
Jib Luffs
Port Tack

Figure 804 **In the tacking diagram, *left,* boat begins to turn at the bottom figure. Sails begin to luff as the bow heads into the wind. Momentum carries the boat over onto the new tack (in this case the port tack) and sails are trimmed.**

Figure 805 **Jibing, *right:* In the lowest figure the boat begins to turn her stern into the wind, preparing to jibe. The wind catches the mainsail and whips it across the boat, under control of the main sheet. In this diagram the jib has remained on the starboard side, and has not been jibed.**

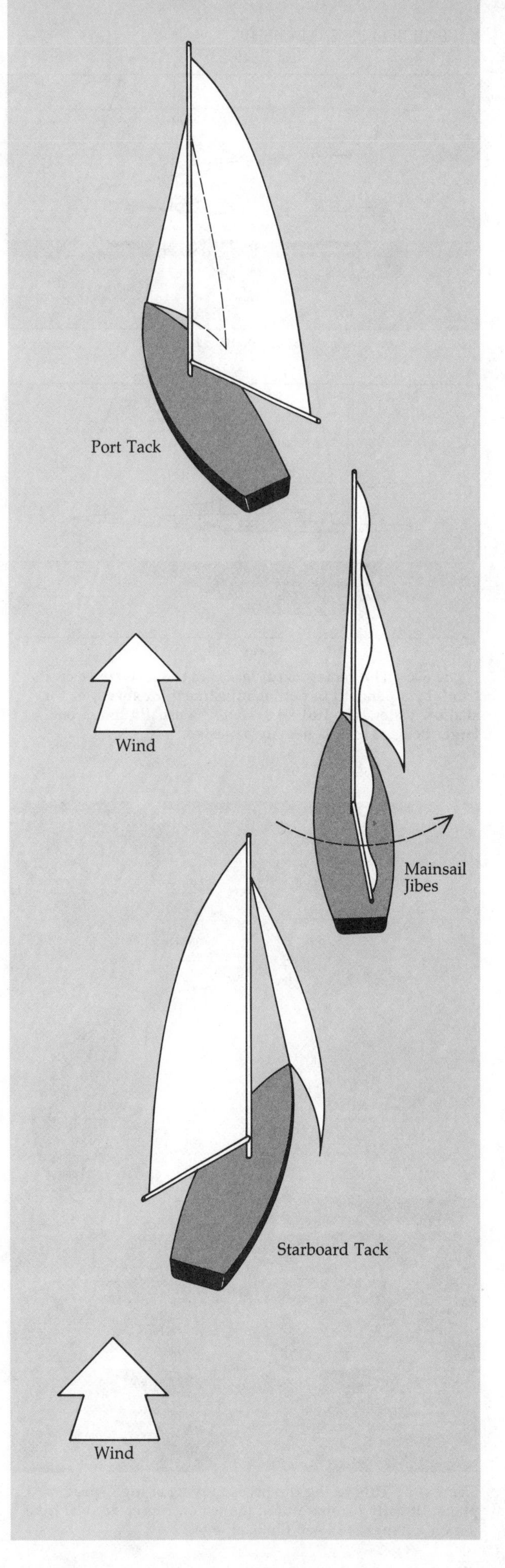

downwind; it is the stern, not the bow as in tacking, that swings across the wind in a jibe.

The helmsman says "Prepare to jibe!", and at that signal the crew begins to haul in the mainsail steadily until it is very nearly amidships. The helmsman waits for the sail to be brought in under control, says "Jibe Ho!", and promptly pushes the tiller away from the sails so that the wind catches the sails on the other side from aft. (A wheel would go the opposite way.)

As the wind catches the sails on the new side the crew quickly and smoothly eases the MAINSHEET (the sheet that trims the mainsail) and trims the JIBSHEET on the new leeward side.

When sailing downwind the sails have been eased way out, and must travel some distance across the boat to catch the wind on the other side. A jibe can be dangerous if every step of this maneuver is not carefully controlled.

In some jibes the change of course is almost imperceptible and only the sails will be seen to swing across the boat.

Accidental Jibe Whenever jibing, or close to a jibe, watch for an ACCIDENTAL JIBE—one for which the crew is not prepared. With inattentive steering the wind may catch the back side of the sail and throw the boom violently across the boat to the other side, risking serious damage to the rigging and to the heads of crewmembers.

Sailboat Hulls

A sailboat hull is different from any other hull in its need for a lateral resistance surface. For the boat to move ahead, the wind forces pushing the boat sideways must be thwarted, and converted to forces that move the boat forward. On a small boat this can be done by simply extending a board down into the water, and securing it there. There are several methods—LEEBOARD, CENTERBOARD, and DAGGERBOARD among them, as discussed in Chapter 1.

Ballast

A low center of gravity will help with the stability of all vessels, but as boats get larger they require an additional righting force in the form of BALLAST. (See page 29) This is usually a lead or iron keel, firmly attached to the very lowest part of the boat's hull. If it is lead or iron stowed in the bilge it is called INSIDE BALLAST.

The weight of a boat's ballast may be as much as half of the boat's overall displacement, and the keel may extend further below the water than the hull rises above it. A deep keel, however, prohibits a sailboat from entering shallow waters, and so a combination KEEL-CENTERBOARD hull has been developed where a centerboard operates through a shallow keel. A keel-centerboard boat might

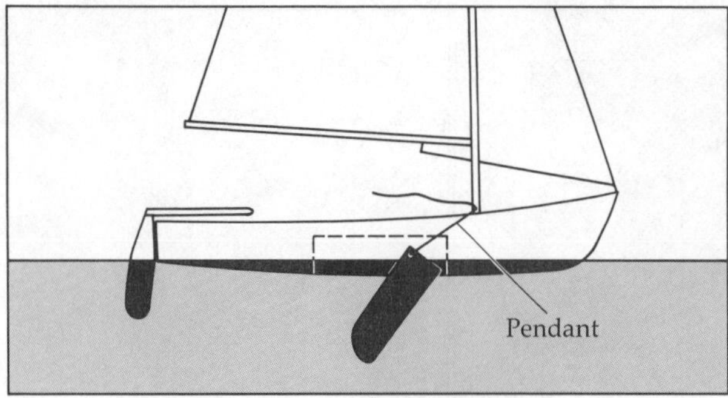

Centerboard Hull

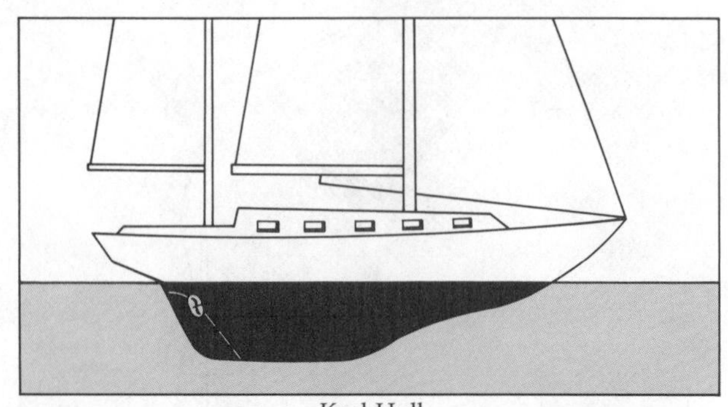

Keel Hull

Figure 806 **The centerboard, *top*, is raised or lowered in its trunk by a pendant to permit adjustment for speed, or for shallow water. The full keel, *above*, is usually found on larger boats, and it is heavily ballasted.**

Figure 807 **This racing dinghy is moving at high speed on a plane, literally leaping out of the water. Her crew must hike out on a trapeze to keep the boat level.**

DRAW (extend down below the surface of the water) nine feet with her board down, and only four with her board up.

A FIN-KEEL is one that looks like a fish's fin extending below the boat, and the boat usually has a rudder mounted some distance aft, often on an additional keel-like extension called a SKEG.

A long or FULL-LENGTH KEEL is usually found on cruising boats, because its directional stability is useful on long passages.

Wetted Surface

With a full-length keel, a sacrifice is made in maneuverability and often in speed through the water, however, compared to shorter keels with reduced WETTED SURFACE—less hull surface beneath the water.

A PLANING HULL is one that lifts out of the water and planes at high speeds. Although usually a feature of high-powered motorboats, planing is also possible for boats under sail, particularly light-weight, wide, and somewhat flat-bottomed boats, with a lot of sail area to provide the needed power. Higher speeds are possible on these planing hulls, because wetted surface is reduced, largely eliminating drag, and wave-making resistance of the hull is no longer a limiting factor.

Parts of the Hull That Support Rigging

The hull shape does not necessarily dictate what kind of rig will be above it, but it does determine most of the important handling characteristics of a sailboat. Parts of the hull must be made with extra strength to deal with the strains put on it by the rigging. This means that a certain hull will usually carry a certain rig best suited to it. Here are some parts of the hull that support the rig.

The CHAINPLATES are strong metal straps extending from the rail down toward the keel, on either side of the mast(s), to accept the rigging holding up the mast. The bottom of the mast usually rests on a part of the keel called the MAST-STEP. On smaller boats, the mast may be stepped directly on the deck or cabin top, where there is often a large hinge-like fitting called a TABERNACLE. This makes it possible to raise and lower the mast on boats that use canals with low bridges, or for trailerable boats, that must have a mast easily unstepped.

A BOWSPRIT may extend forward at the bow of the boat to increase the potential for sail area, and a BOOMKIN may extend off the stern to accept the backstay further aft (see below), thereby enlarging potential sail area again.

Self-Bailing Cockpits

A sailboat hull is designed to heel, and will sometimes take waves over the rail and into the cockpit. Any sailboat that will be taken offshore, large or small, should have a SELF-BAILING COCKPIT to get rid of that water promptly. In a small boat this may consist of holes at the stern that drain the water as the boat moves ahead. There may be suction valves in the boat's bottom, or there may be large drains in a cockpit whose bottom level is normally above the water level, providing natural drainage. The self-bailing feature is vital for a boat going offshore—and valuable on others as well—because the cockpit may fill with water in rough weather, and must be able to drain rapidly without assistance from the helmsman.

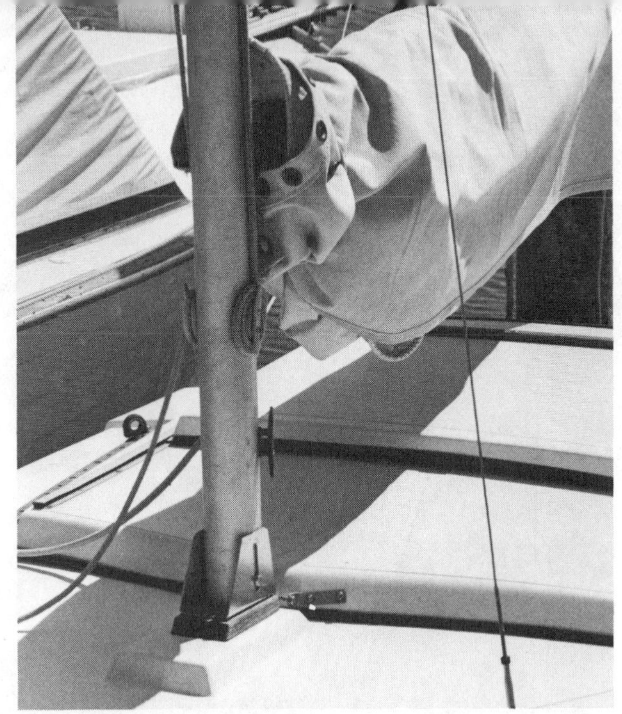

Figure 808 **This mast is mounted on a tabernacle on deck, to permit easy rigging and unrigging—a useful feature on boats that are trailered. Most masts are stepped through the deck or cabin, and rest on the floor timbers.**

Figure 809 **Many sailboats have a self-bailing cockpit with drains to handle water that may come aboard as a wave, or as rainwater. These drains must be kept free from debris such as leaves or rope ends. Small, inshore boats may have deep cockpits requiring a pump to keep dry.**

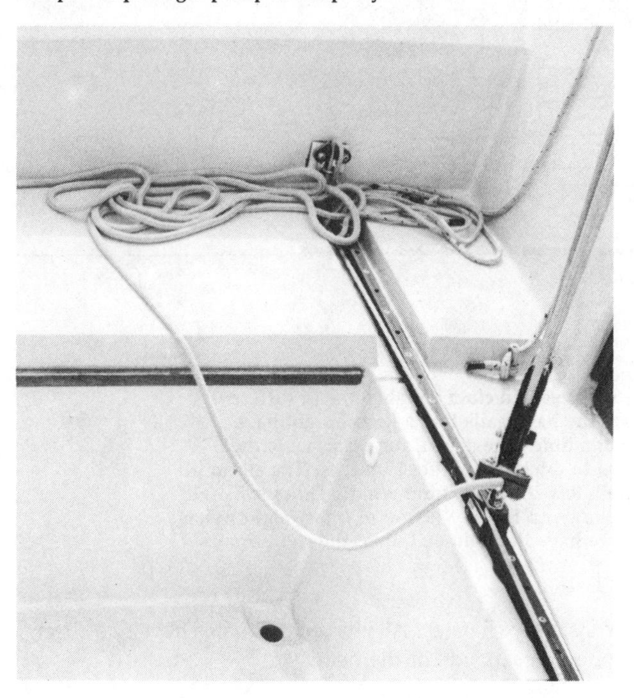

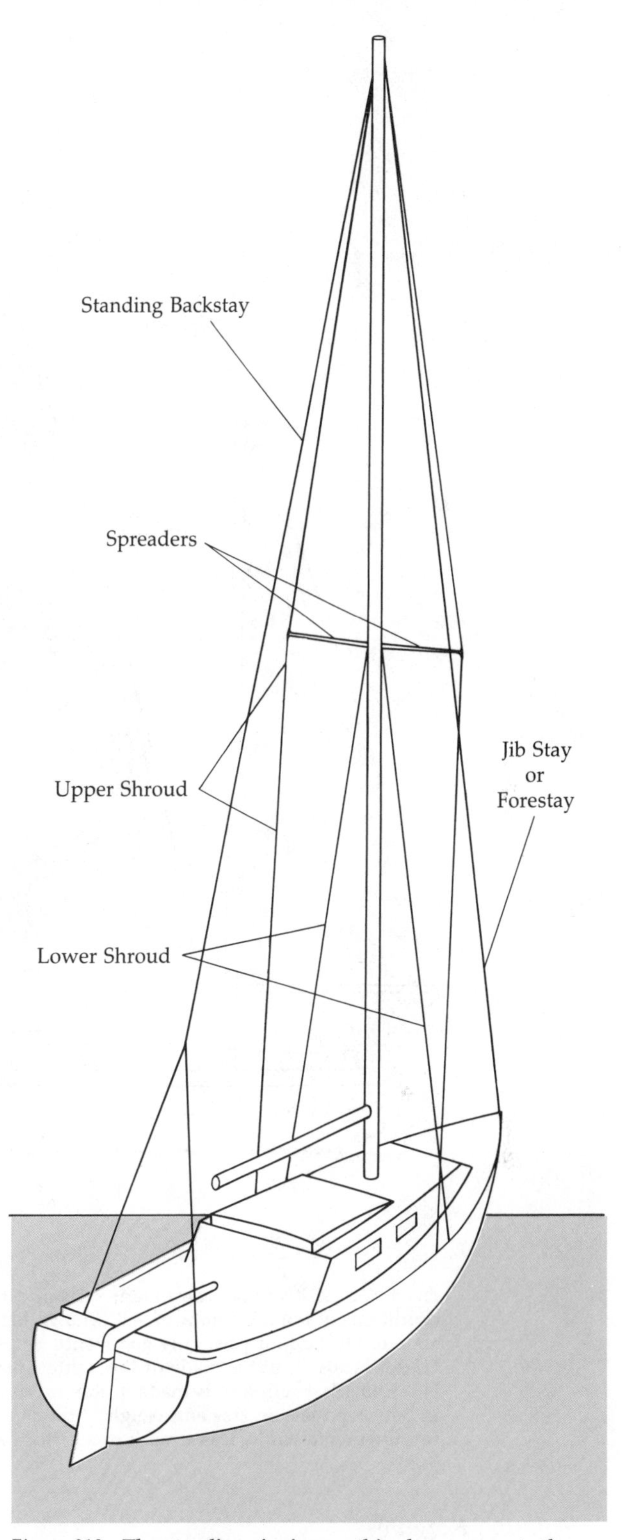

Figure 810 **The standing rigging on this sloop supports the mast under the tremendous strains placed on it by the sails.**

Rigging

The rigging looks complex on a sailboat, but the function of rigging is actually simple and easy to understand.

Standing Rigging

A mast is held up by wire SHROUDS and STAYS that go from the deck to fittings on the mast. Shrouds hold the mast on both sides, stays hold it fore and aft. (See Figure 810 showing UPPER and LOWER SHROUDS, FORESTAY, BACKSTAY, and SPREADER. Shrouds and stays are adjusted whenever necessary by TURNBUCKLES (or RIGGING SCREWS)—extremely strong screw-like fittings, threaded at both ends, that join stays and shrouds to the chainplates. Since only occasional adjustment is needed on this, it is called the STANDING RIGGING.

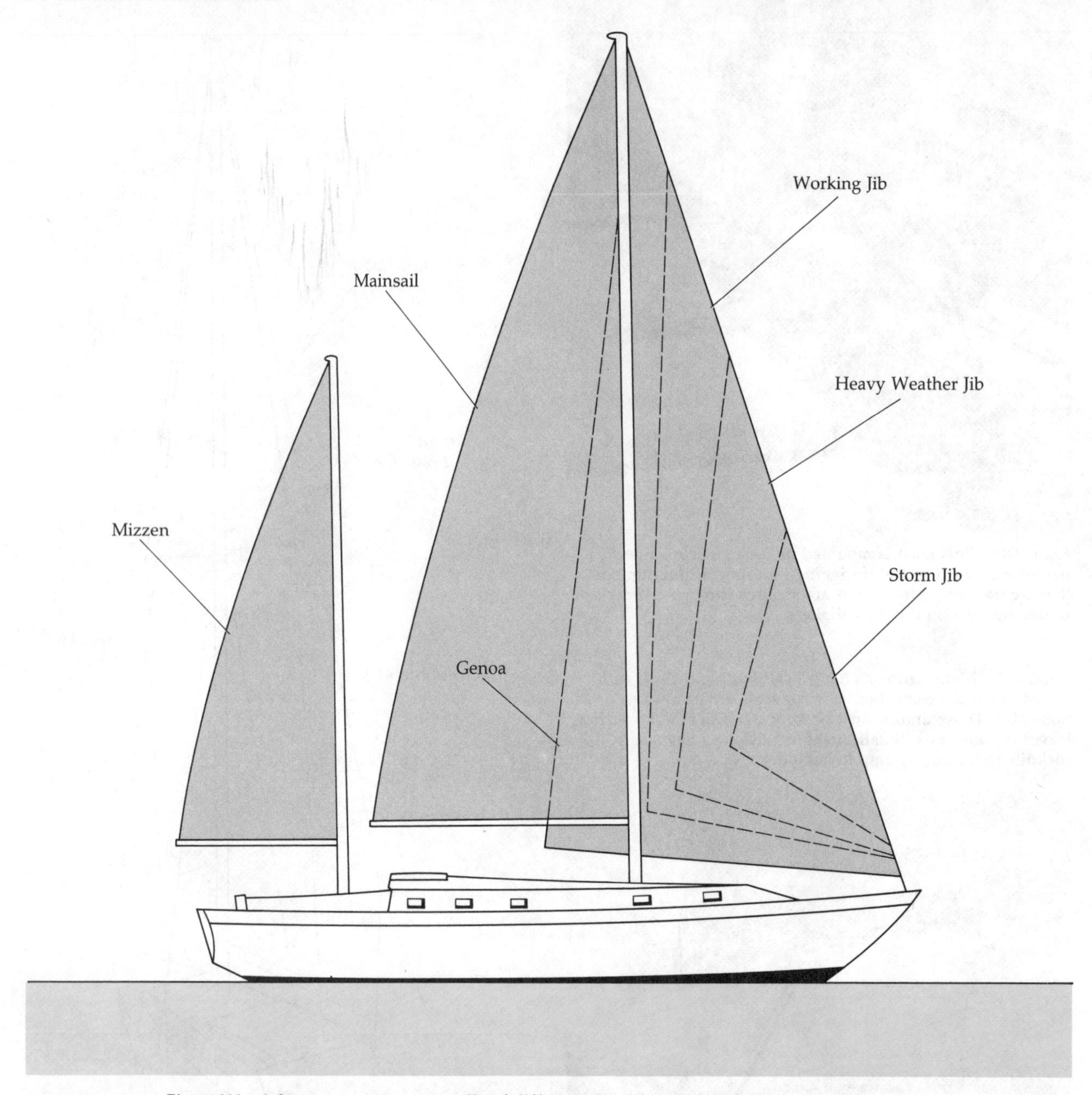

Figure 811 **A boat may carry many sails of differing sizes and cloth weights to suit different conditions of wind or point of sail. Shown here are the basic sails that a ketch might have aboard. Of the four jibs, only one would be used at a time. The genoa, for example, is the largest headsail, but is made of the lightest material to catch the slightest breezes. The storm jib is the smallest sail, and is made of the heaviest cloth to withstand storm winds. Other sails fall in between these in size and weight. Spinnakers, shown in Figure 815, are of lightweight nylon for downwind work. It is a mark of a sailor's skill to have the right sail up at the right time.**

Running Rigging

The ropes (and wires) that are used often in the handling of the sails are called collectively the RUNNING RIGGING. Most important of these are the halyards and sheets. A HALYARD, usually one for each sail, is a line used to raise the sail up the mast. It is from the old term to haul-yard. (A yard is a big spar at the top of a square sail, as seen on old square-rigged ships). Once the sail is raised, or HOISTED, it is controlled by a SHEET (which is not to be confused with the sail itself), usually made of Dacron rope. The sheet should be adjusted (trimmed) as necessary, to get the sail at the best possible angle to the wind, which constantly changes. All jibs not on booms have two sheets, one on each side of the boat.

Spars

A BOOM is a spar that (in most rigs) is attached to a mast at one end to hold the bottom edge and lower corners of a main or mizzen sail. Most boats have a main boom, and some have a mizzen boom also, each holding out the foot of the sail mentioned (see Figure 811 for description of each sail). A jib or staysail may also have a boom called a CLUB attached to an upright fitting on the deck. The point of attachment of all booms is an important hinge-fitting called a GOOSENECK, which must bear a tremendous amount of strain.

Some older rigs carry a spar similar to a boom on the upper edge of a four-sided sail. This spar is called a GAFF, and the boat carrying a gaff is a GAFF-RIGGED boat.

Deck Fittings

On the deck of a sailboat may be found a complex arrangement of tracks, fairleads, blocks, and winches. TRACKS hold the blocks and FAIRLEADS (simple turning points for running rigging) and permit adjustments to the lead of each sheet. The LEAD (pronounced lēēd) is the angle that the sheet takes from the sail; each sail has its correct leads for each wind condition and wind strength. During tacking, the lower block for the mainsheet is attached to the deck and may move back and forth on a track called a TRAVELER.

Winches A sheet must pass through blocks, or a series of blocks that may lend it mechanical advantage (see pages 297-298), which direct it to its specific cleat. There, to help handle the sometimes huge pressures exerted by the wind on the sails, is often found a WINCH. This is a mechanical revolving drum that turns only in one direction, always clockwise. By wrapping the sheet around the winch in this clockwise direction (called TAKING A TURN around the winch) a crewman turns a handle which through internal gearing provides a considerable mechanical advantage in trimming the sails. Winches are also used with halyards to hoist larger sails.

Ideally, two crew members will operate a winch, one GRINDING (turning the handle), and one TAILING (keeping a steady pressure on the slack end of the sheet coming off the winch). A SELF-TAILING WINCH makes it a job for one crew member, because the line is secured through a form of cam cleat at the top of the winch.

Figure 812 **A winch revolves in one direction and, with a few turns of the sheet on it, helps to trim the sails. A winch handle inserted in the hole at the top provides mechanical advantage. Two or three turns of line are usually all that are needed, perhaps four in a stiff breeze.**

Other Rigging

In addition to sheets and halyards, there are often other lines for fine-tuning the trim and shape of the sails. A BOOM-VANG is a removable block and tackle that serves to hold the boom down and flatten the mainsail. A GUY is a line, often a wire, set up to hold a spar (boom or spinnaker pole) in a certain position, and it may be called a FORE-GUY, or an AFTER-GUY, depending on the direction of its pull. A TOPPING LIFT is a line that holds up a boom when its sail is down or being set.

There are also small adjustment lines on the corners of the mainsail or main boom: a DOWNHAUL and an OUTHAUL, and each does to the sail what its name says. There are go-fast adjustments to be made: a CUNNINGHAM, just above the gooseneck, adjusts the tension of the leading edge of the sail, without affecting the dimensions of that sail, and thereby changes its shape; a LEECH-LINE, sewn into the leech of the sail, adjusts the tension in the after edge of a sail to prevent fluttering and to marginally alter sail shape; RUNNING BACKSTAYS may be set up on either side of the boom, from deck to mast, to help hold up the mast, and to help balance the strain placed on the backstay by the mast.

Other important rigging includes REEFING LINES, and, on some cruising sailboats, FURLING LINES. Reefing and furling are covered in the section on Miscellaneous Terms, later in this chapter.

Sail Plans

Over the years various rigs, or SAIL PLANS, have developed that determine certain characteristics of a sailboat, and how she handles. There is much discussion over the suitability of each sail plan for each particular use of a sailboat, whether for instruction, racing, cruising, or ease of handling. The debate should not obscure the fact that all rigs have their advantages and disadvantages, and it is only a matter of finding one that suits your kind of sailing.

Catboat, Sloop, and Cutter

The simplest rig is one-mast/one-sail, and this rig is characteristic of an old Cape Cod fishing sailboat called a CATBOAT, still seen in some small bays and harbors. Because of the proportionately large size of the one sail on a catboat, this rig can be surprisingly difficult to handle and is seldom seen on large boats.

By dividing the needed sail area into two sails, instead of one big one, much is gained in versatility and handling characteristics. One mast, two sails (main, and a sail before the mast called a HEADSAIL or JIB) makes a boat a SLOOP. A sloop may, in fact, have a number of sails on board, because although only two sails are set most of the time, different conditions or wind strengths may require different types of headsail (see Figure 811). A small WORKING JIB may be called for in a stiff breeze. A larger, overlapping GENOA is used for lesser winds, and the beautiful, colorful SPINNAKER is used in downwind sailing. The spinnaker is usually used in racing, requires certain special equipment, and handling one expertly is a skill to be

Figure 813 **A cutter often carries two headsails at once, and has a mast stepped further aft than on a sloop. Each sail is therefore somewhat smaller in area, and so easier to handle than the sails of a sloop of the same size.**

admired. Also useful on a large sloop are a SPINNAKER STAYSAIL, a BLOOPER (see Figure 815), and STORM JIB.

A CUTTER is similar to a sloop in that she has only one mast, but she has it stepped further aft, and can carry two headsails at once: a FORESTAYSAIL and a jib. She often carries a bowsprit to enlarge the FORETRIANGLE—the area in the sail plan in which the headsail fits. A cutter is considered slightly easier to handle than a sloop, and she is sometimes a little slower because she lacks the great upwind performer—the overlapping genoa.

Yawl and Ketch

Similar to a sloop in most handling characteristics is the YAWL. A yawl can become a sloop overnight with the removal of the small MIZZEN MAST and its rigging. The miz-

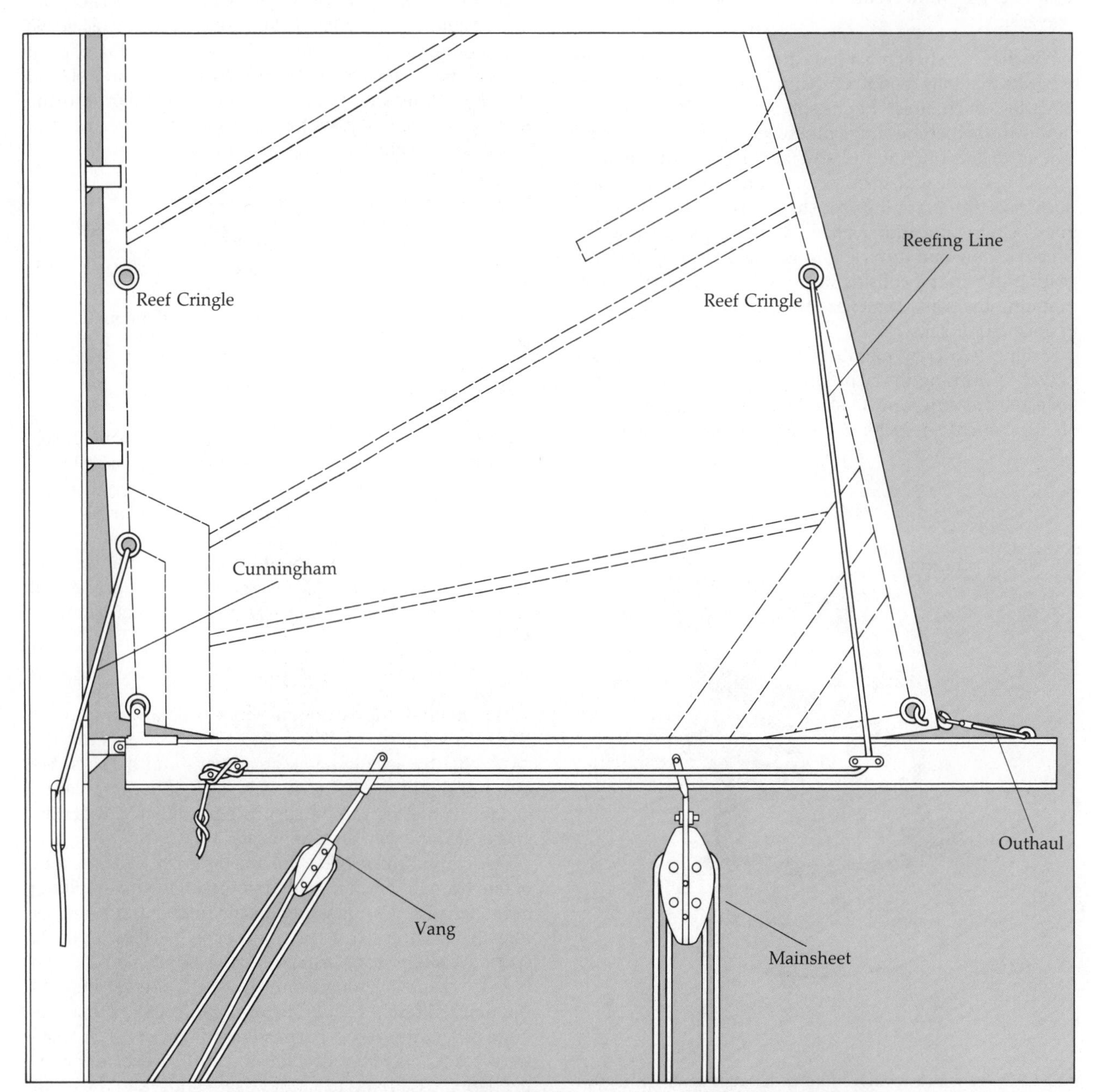

Figure 814 **The running rigging on the mainsail is used to make adjustments in the trim and shape of a sail. Mainsheet, outhaul, and reefing lines are found on most sailboats. Vang and Cunningham are seen more often on racing boats, to adjust the sail's shape. Reefing lines reduce sail area quickly in increasing winds. Sizes here are exaggerated for clarity.**

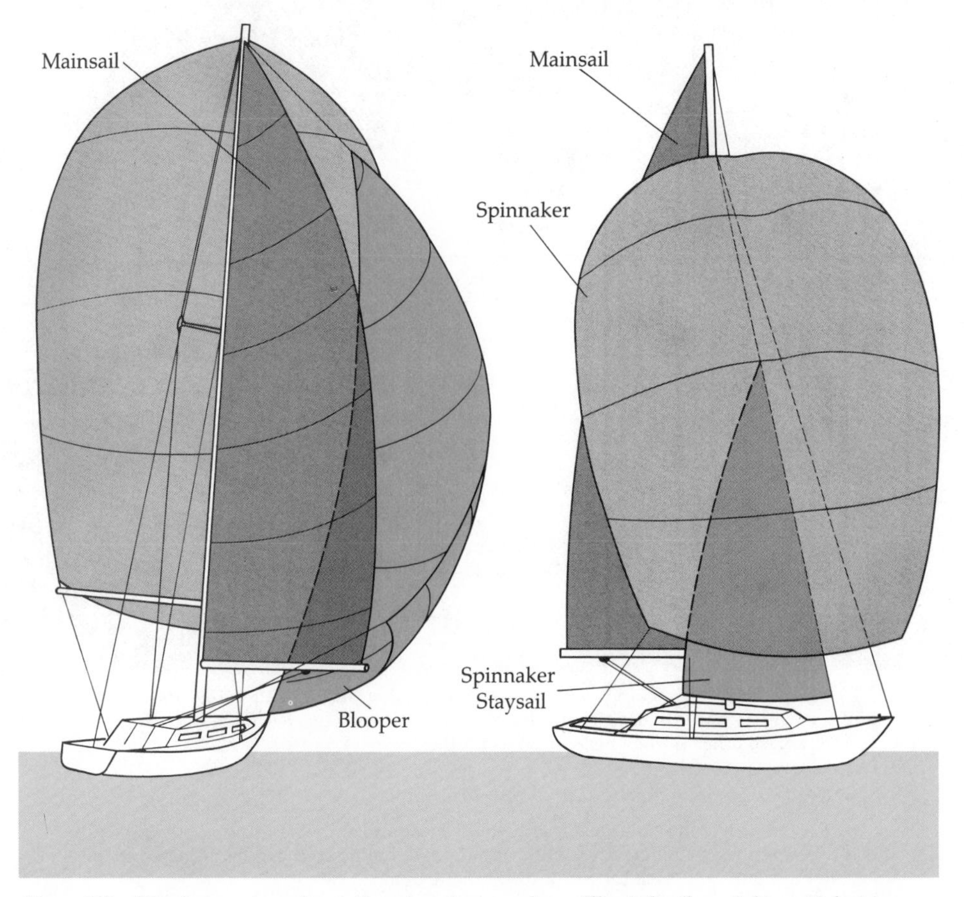

Figure 815 **The forces at work are the same for any size sailboat, be they eight or 80 feet in length. The catboat, *below left,* with her mast positioned near the bow, has about the simplest possible rig for a boat of this size. *Below right,* a sloop has only one mast, but may carry many sails, and have extremely complex rigging. The shadow of this racing sloop's staysail can be seen on the enormous spinnaker, an expanse of carefully shaped colorful nylon. Bloopers set alongside the spinnaker, *above,* help spread more sail in a race downwind.**

Figure 816 This ocean-going yawl has set a spinnaker and two staysails (forestaysail and mizzen staysail) in addition to the mainsail. The mizzen has been doused and furled, and another jib has been secured on the foredeck until it is needed.

Figure 817 A yawl, shown at left, has a small mast and sail, the mizzen, stepped quite far aft. Sometimes considered a useless appendage, the mizzen can be adjusted to balance the helm, and ease the work of the helmsman.

Figure 818 **Ketches are considered cruising boats because they are not as weatherly as sloops or yawls. The mizzen is large, and its mast is stepped further forward than a yawl's mizzen. It is forward of the vessel's rudder post.**

zen mast is usually stepped aft of the cockpit and the rudder post and it usually carries a relatively small sail. It has shrouds, spreaders, and boom, just like a small version of the mainmast ahead of it, and the boom usually extends out over the stern of the boat. The mizzen's function is chiefly to help balance the steering characteristics of the boat, and the mast is a vantage point for setting an additional small sail called a MIZZEN STAYSAIL, which almost always adds speed when reaching.

A KETCH is often confused with a yawl because she also has two masts, the mainmast forward and the smaller mizzen aft. There are debates over the factors of what distinguishes a yawl from a ketch, but the chief difference is that the mizzen mast on the ketch generally is forward of the rudder post, and is larger than a yawl's mizzen. It is not considered just an aid to steering, but is a motive force in the sail plan. Because the sail area is divided into more sails, each sail is smaller in area and more manageable then it would be on a sloop of the same size.

Schooner

In some ways considered outdated, the two-masted SCHOONER rig, where the tallest mast is the aftermost mast, or all masts are of equal height, is still seen quite often. There are also three- and four-masted schooners—and more—but they are rare. By dividing up the sail plan, the schooner offers the most variety of sail combinations, and is great for trade-wind sailing, where days go by with the wind blowing from nearly astern. Schooners do not go to windward as well as other rigs, however, and require complex rigging, and usually a large hull to justify the additional equipment.

Other Sail Plans

Beyond these basic rigs there are many other less common but often very useful rigs: LATEEN, WISHBONE, SPRIT, and GUNTER (see Figure 820). There are also a whole set of rigs and terms for SQUARE-RIGGED vessels, such as a BARK, BRIGANTINE, etc., but they are not considered practical for most boats with amateur crews.

The wishbone rig has achieved new popularity since it is used on BOARD SAILERS, a new type of sailing surfboard (see Figure 822). This rig is also now being used on some medium-size ketches and sloops.

Parts of a Sail

The parts of a sail—FOOT, LUFF, and LEECH, and its corners—HEAD, TACK, and CLEW, are shown in Figure 824. Of particular importance are the BATTENS in the leech, and the telltales. Battens are simple stiffeners that fit into pockets sewn into the leech (the after edge) of a sail—mainsail or jib. They serve to stiffen that loose edge and provide more usable sail area and a better sail shape. A

Figure 819 **This large gaff-rigged schooner, *Bill of Rights,* has six sails set. Starting from forward, the three jibs are the flying jib, the inner jib, and the staysail or jumbo jib. Then come foresail, mainsail, and main topsail.**

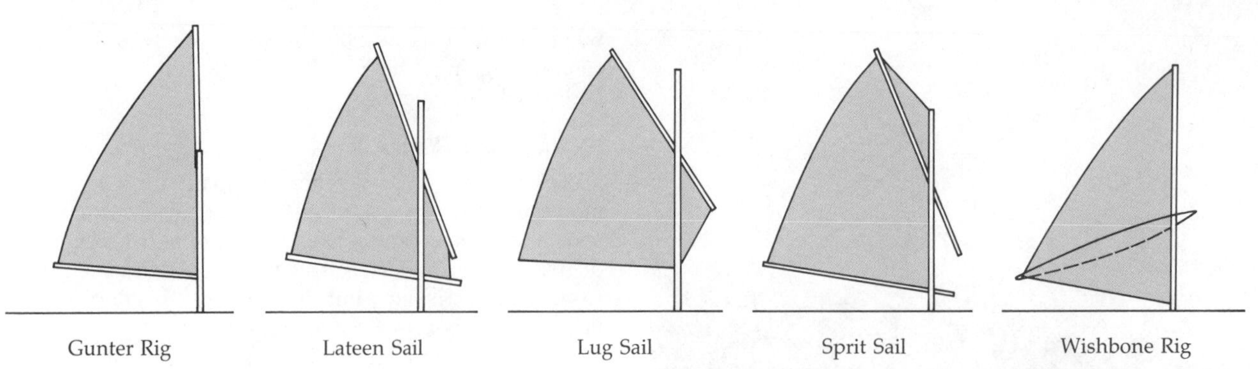

Figure 820 **Rigs seen on small boats are shown *above.* Most commonly seen is the lateen rig, found on board boats such as the Sunfish, and the wishbone rig, found on board sailers. These are in addition to the marconi (triangular mainsail—*Figure 817*) rig, and the gaff rig (*Figure 819*).**

Figure 821 **More people have learned to sail on a sailboard, *left,* than probably any other type of boat. They are easy to sail, fast, and portable, as well as fairly inexpensive.**

Figure 822 **New on the boating scene is the board sailer, or windsurfer, *below.* Neither surfboard nor sailboat, it is a little of both, and provides an inexpensive way to get out on the water and go fast. It has no sheets, no halyard, and no rudder, but steers by the fore-and-aft balance of the rig.**

Figure 823 **The *Joseph Conrad* at Mystic Seaport, Connecticut, is a beautiful example of a restored square-rigged ship of the last century. Requiring considerable expertise, and not terribly versatile, the square rig is seldom seen on pleasure boats.**

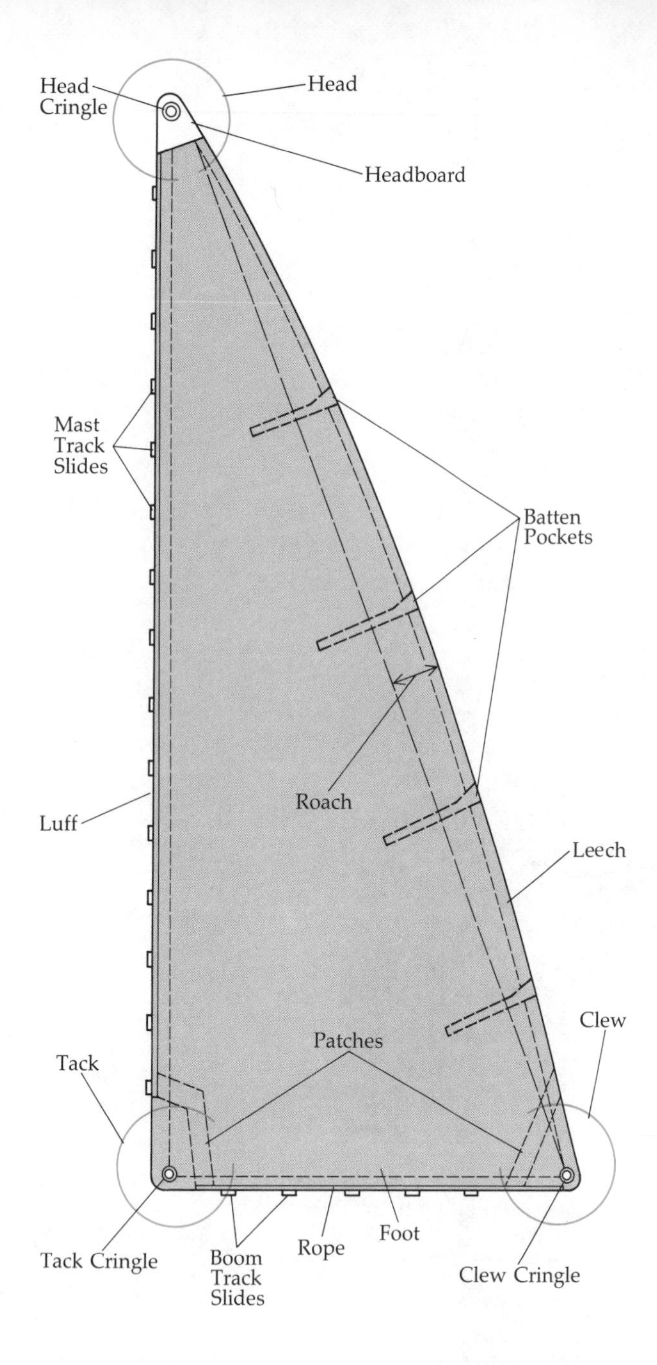

Figure 824 **Interest in sailing can start early. A gaff-rigged cutter and marconi sloop, *above,* race near shore under the close supervision of their skippers. A mainsail, *top left,* would seem to be a simple cloth triangle, but its construction is actually based on a complex equation of curves that the sailmaker uses to create proper belly, or shape. Made of dacron, it is reinforced at the edges and corners with patches of extra material, and given extra stiffness in the roach by battens. Reef points, *bottom left,* are sewn into the mainsail, and hang on both sides of the sail. To reef, these are taken in at tack and clew first, and then all are tied in under the foot of the sail (not around the boom), using reef knots, which are described in Chapter 13. The winding drum of a roller-furling jib mechanism, *below,* is shown with the jib rolled around it. This is a very good arrangment on cruising boats because it requires no hoisting, dousing, or traditional furling, and therefore no foredeck work. Care must be taken so that the harmful rays of the sun do not cause the dacron to deteriorate.**

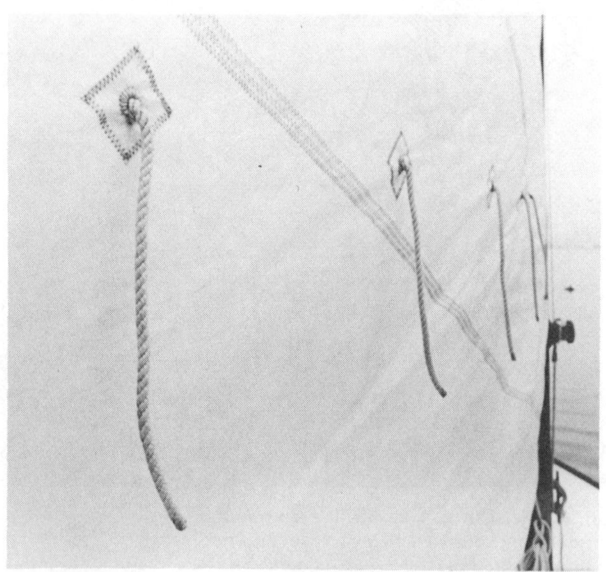

few sails are FULL-BATTENED, in which the battens run from leach to luff.

Miscellaneous Terms

Several important procedures on a sailboat have developed their own terminology. To learn in detail how to attempt these procedures, see Hearst Marine Books' *You Can Sail, Basic Sailing,* or *Practical Sailing.* Some of the terms you will use are given below.

A sail is HANKED ON, or BENT ON, the boom or stay, then loosed before it is hoisted when getting underway. When it is dropped, or DOUSED, it is then gathered and FURLED (tied down) to keep the wind from catching it and causing it to flog around. A sail is tied with SAIL-STOPS on the boom, or, for spinnakers and sometimes for jibs, bagged in its own bag. Another development is ROLLER-FURLING, where a sail is wrapped up around its luff (leading edge), much like a window blind only vertical. Whenever sails are not in use for more than a day or so, it is important that they be protected from the sun by sail covers or by being in their sailbag. Ultraviolet rays can deteriorate Dacron rapidly.

REEFING is an important method of reducing sail area in increasing wind. There are several reefing methods: traditional reefing, using REEF POINTS already attached to the sail, and JIFFY REEFING or SLAB REEFING, a modern simplification of older reefing. ROLLER REEFING (not the same as roller-furling) was once favored, using a special boom that could roll up a section of sail even under great pressure. It is used less and less because it is difficult to maintain a good shape on a sail that is roller-reefed.

A useful technique, usually reserved for heavy weather, when sailing upwind becomes impossible, is to HEAVE TO, as discussed in Chapter 10. Heaving-to is also useful whenever a short period of relative calm and steadiness is needed aboard, maybe for a peaceful lunch, or picking up something from the water (a dinghy or even a man overboard).

Why a Boat Sails

Sails have been used for thousands of years to drive boats through the water, even on the simplest rafts and dugout canoes of primitive tribes. The earliest rigs were square-rigged types, in which the sail set athwartships—across the boat. However, like the small boy's raft with a blanket sail, they had one serious limitation: they could sail only with the direction of the wind. To go to windward, they had to be rowed or find a favorable current to drift down. The art of using the wind's own power to propel a vessel against that wind had yet to be discovered.

Square, Gaff, and Fore-and-Aft Rigs

Centuries went by before a sailing rig was invented that enabled a boat to make any progress against the wind. The Arabs contributed an important development to the science of sailing when they worked out the LATEEN rig, a triangular sail supported at its head by a YARD that is attached to the mast, the near-midpoint of the yard to the near-top of the mast. A boom is used and extends forward of the mast to connect to the lower end of the yard. This rig is often used on SAILBOARDS, small sailing craft whose hulls resemble surfboards. These lateen sails were undoubtedly the forerunner of our modern fore-and-aft rigs.

For years, square sails drove all ships, including the famous 19th-century clipper-ships, which still hold many records for speed under sail. Displaced by the reliability of the steam engine, the big clippers have disappeared and with them the square rig, except for training ships, like the U.S. Coast Guard's *Eagle.* When the fore-and-aft sail came into vogue the GAFF RIG was most popular, the principal sails being spread from a gaff at the head of a four-sided sail and a boom at the foot, the luff being secured to the mast.

Theory of Wind Pressure

While the gaff rig was at the height of its popularity, the principles underlying the theory of sailing were not too well understood. But when the airplane made its appearance, with intensive studies of the action of wind on wing surfaces, the analogy between wing and sail was recognized: the sail can be considered a wing set vertically on the hull, driving the boat by giving a "lift" ahead, something like the lift upward in a plane. Sailmakers strove to shape their sails into a section resembling a bird's wing. No longer did sailmakers think solely in terms of pressure applied to the *after* side of the sail as the major factor in producing the drive needed to propel a hull. It became apparent that the reduction of pressure along the luff on the forward side of the sail was a big factor. As soon as that principle was generally understood, gaffs and long booms on racing sailboats began to disappear, giving way to the taller MARCONI rig with its triangular mainsail. New materials with greater strength also permitted the loftier rigs.

There are complex factors that bear on the aerodynamics of a sailboat working to windward. Many books are available on the subject but only a brief rundown of this science is possible here.

Effects of Wind Pressure

The first thing to realize in studying the principle by which a sail is made to drive a boat is that the wind blowing on a sailboat has several different effects. First, it will heel her over as the wind hits the sails. Then it tends to blow the boat sideways off her course, which is why she makes leeway. Third, there is the pressure differential of the wind passing along the curved surfaces of the sails, lower pressure on the forward (leeward) side of the sails, and a build-up of higher pressure on the windward side.

The naval architect studies all these tendencies and designs a boat to minimize the features that do not contribute to propulsion of the hull, and to develop the factors that do. To guarantee that a maximum of propelling effect will be gained from every square foot of sail, underwater lines of the hull are made as smooth as possible, of a shape that will allow it to slip through the water with a

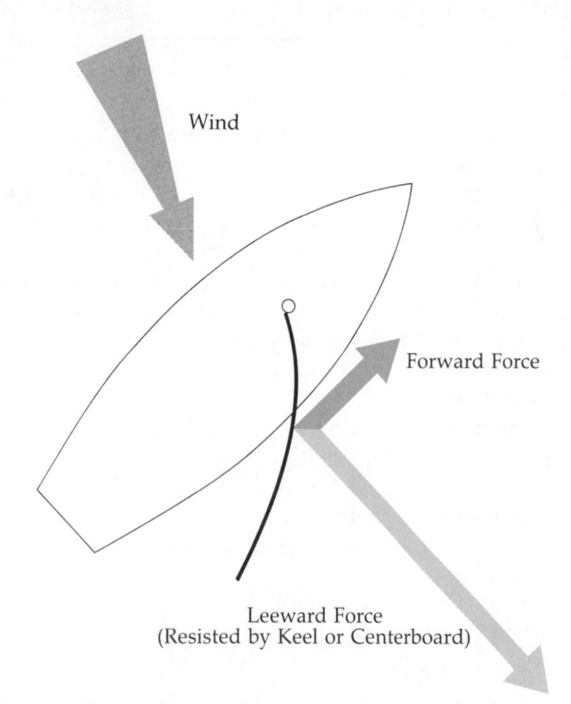

Figure 825 **The wind exerts the two forces of leeway and forward force on the boat. By design, the shape of the hull resists the leeward force, and the motion is forward.**

minimum of resistance, wave-making, and eddy-making. For example, lines that taper off to a point at the stern conform to one of the principles of streamlining, allowing the displaced water to close in smoothly at the stern without suction, drag, or needless disturbance. Other design factors enter in, such as carrying capacity, buoyancy in the bow and stern, and, finally, the number of berths desired in the accommodations in the hull.

Stability

Heeling, while it does not help in driving the hull forward, does have a considerable effect on the way the boat handles in winds of different velocities. The hull cannot help but heel to some extent, so the naval architect must design the underwater lines to offer a minimum of resistance to the water at different angles of heel.

Heeling must be kept under control; a boat needs stability to prevent her from capsizing under the pressure of wind or waves. A light, unballasted narrow hull, to picture an extreme example, sitting on top of the water, is easily knocked over. Witness a canoe under sail. Stability can be obtained in a light hull of little draft by increasing the beam. The broad-beamed catboats are good examples of this. But even broad beam does not always provide enough stability; once a broad, shallow-draft hull has been pushed down beyond a certain safe angle, she can still capsize, like a toy sailboat made of a flat board.

Stability can also be obtained in a relatively narrow deep hull. This is accomplished by putting weight low in the hull, lowering the center of gravity. If she has a deep keel she will carry a casting of iron or lead bolted to the lower extremity of the keel, sometimes in the form of a bulb. Such a boat, hit by a puff of wind, heels over, easily at first. But the further she heels, the greater the tendency that weight on the bottom of the keel is exerting to right the boat. Knocked down on her beam ends, she will right herself when pressure on the sails is eased. Landsmen often marvel when they see a yacht sailing "on her ear" (see Figure 826), masts inclined at a 45-degree angle and water boiling on deck around the rail. They feel an extra puff must certainly send her under. However, if she is properly designed, there are powerful forces acting to right her all the time.

A boat also needs stability to make maximum speed. A boat is normally designed so that heeling and sail forces are balanced to give a slight tendency to "round up" or turn into the wind. This is termed WEATHER HELM, and a little is desirable as it is safer than a tendency to fall off the wind and possibly into a jibe. Too much weather helm, however, requires carrying an excessive angle on the rudder, which then acts as a brake, slowing the boat's forward speed.

Figure 826 **Another force exerted by the wind on the sails creates the heel (leaning over) demonstrated here. Heeling is to be expected on all sailboats, but should the lee rail be submerged more than shown here, it is time to shorten sail.**

Leeway

The second tendency of wind on the hull and rig is to drive it off sideways, to leeward. Here's something that must definitely be counteracted; otherwise a sailboat would take on the characteristics of a raft driven before the wind. A shallow-draft hull, like a rowboat, presents very little vertical surface to resist being driven off by lateral pressure. So a pivoted centerboard can be lowered from its trunk to prevent the boat from slipping sideways or MAKING LEEWAY. A deep-keeled vessel has a similar expanse of surface below the water against which the water can act to minimize leeway. The amount of lateral resistance a keel or centerboard offers is an important factor in how good a sailer the boat will be, but it is not simply a case of "more is better." It is a complex decision for the designer.

Sail Shape

No sail is flat, even though it may be called "flat" for a particular wind condition; it is the sailmaker's art to cut and sew into each sail its own correct amount of curve or "fullness." This creates a windward side that is concave, and a leeward side that is convex. The wind reaches the

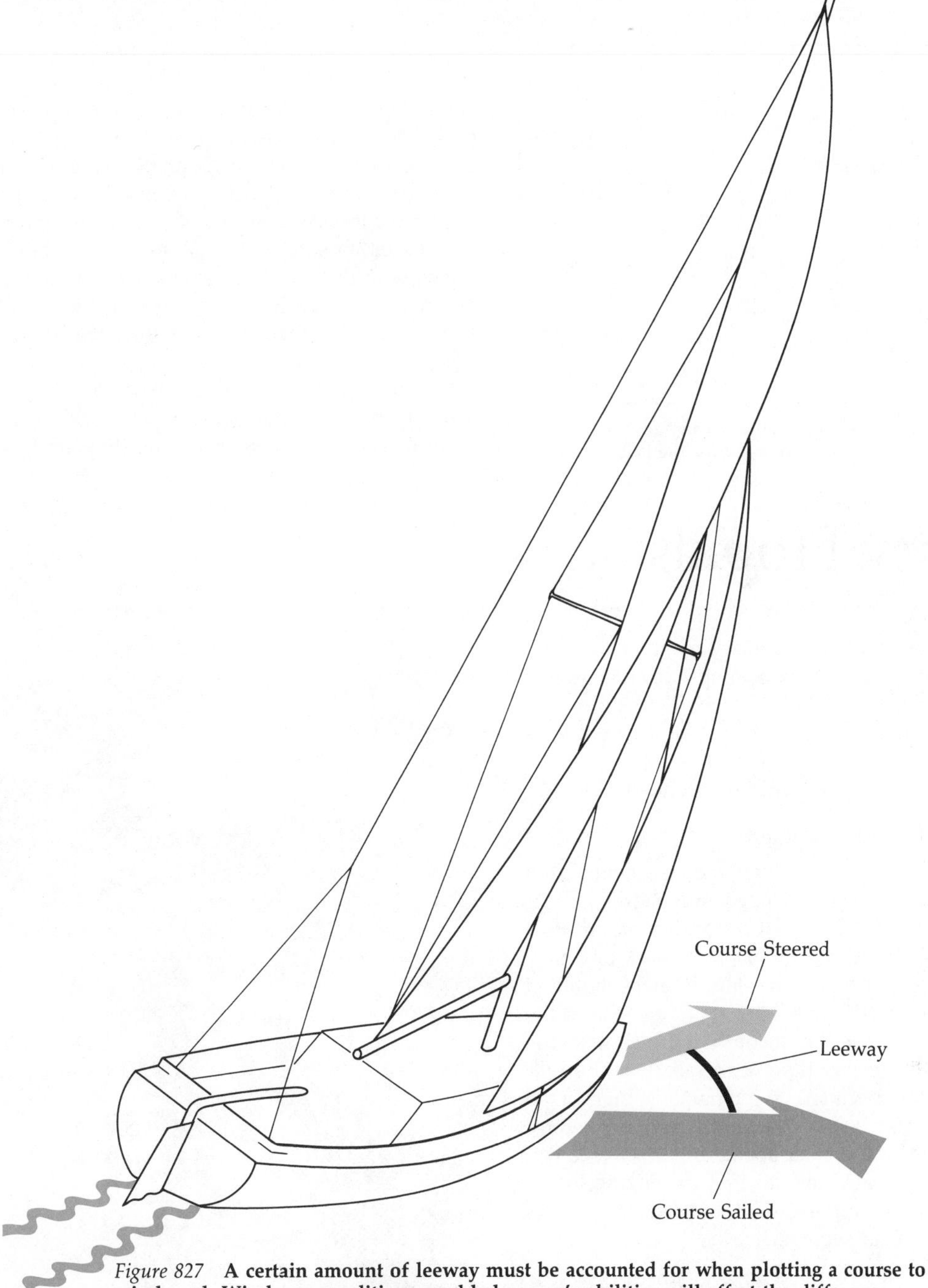

Figure 827 **A certain amount of leeway must be accounted for when plotting a course to windward. Wind, sea conditions, and helmsman's abilities will affect the difference between course steered, and course actually sailed (the course made good).**

luff of the sail and is split and altered slightly in its direction. This has the effect of slowing up the wind on the windward side, increasing pressure, and speeding up the wind on the leeward, convex side, dropping the pressure on that side of the sail.

This difference in pressures on the opposite sides of a sail creates complex forces, but one of the important ones is that the sail is pulled into the low pressure on the forward side of the sail, and acts to move the boat sideways and forward. (The sidewise component is resisted by the centerboard or keel and the resulting motion is mostly ahead.) On a large sailboat where the forces are great, this wind speed differential can be readily observed as a steady rush of redirected wind in the area between the mainsail and the jib, producing an important aerodynamic interaction called the SLOT EFFECT.

Sail Trim

The trim of each sail must be carefully made to get the optimum angle to the wind, and maintaining proper trim is the most important art in achieving fast, efficient sailing. Since the wind changes constantly, in direction and strength, the trim must be changed constantly as well, even if only by a few inches. A good crew is constantly keeping an eye on the wind, and perhaps a hand on the sheet, for any adjustment that could improve the angle of the sails to the wind.

Off the wind, other, simpler forces come into effect, and other sails may be used to take best advantage of these forces. Spinnakers—parachute-like, lightweight sails usually made of bright-colored nylon—are used on racing sailboats to achieve maximum sail area and speed downwind.

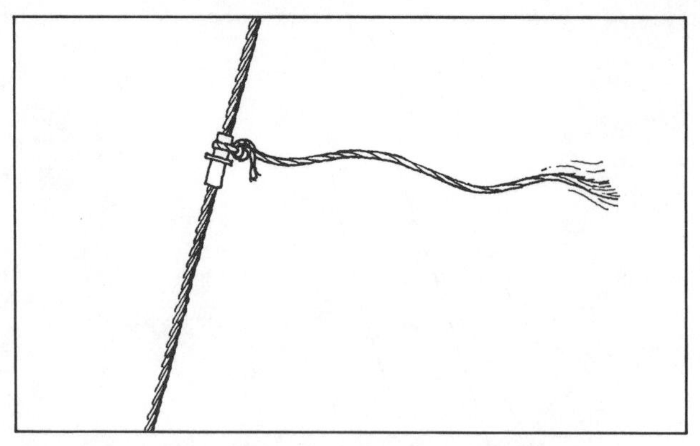

Figure 828 **A carefully positioned telltale can show not only wind direction, but direction of airflow over the sails.**

Telltales

A TELLTALE is any small device that indicates the apparent wind direction. There may be one at the top of the mast or on each shroud. Similarly, sail TICKLERS may be sewn into the sail itself. This last location provides an important indicator on a sailboat; with a short piece of yarn on either side of the sail, in any spot along the leading edge of the sail, a simple tickler can indicate the flow of wind in this critical area of greatest lift. With the sails trimmed so that the two yarns run parallel and point aft, the sail is dividing the wind properly, and providing a satisfactory angle for best aerodynamic lift. When you are learning to sail, these telltales and ticklers can be very helpful, and when you are racing they are invaluable.

Auxiliary Propulsion

All sailboats should have an auxiliary means of propulsion to use in calms. It could be a simple paddle stowed forward, an outboard engine on a special bracket or in a motor well, or an inboard engine installation. The means of auxiliary propulsion that your boat requires will depend on the size of the boat, and her intended use.

Types of Auxiliary Propulsion

Small open boats, and small or mid-size boats used strictly for racing, will usually carry just a paddle.

Boats that venture further from the harbor mouth may require a more powerful means of propulsion, and may carry an outboard motor for use in calms. It can be mounted on a tilting bracket on the transom, or on one side of the boat, or it can be installed in an outboard well near the stern of the boat. In either type of installation, the outboard motor is often mounted on a retractable bracket, so that the propeller may be lifted out of the water to reduce drag and allow the boat greater speed under sail (see Figure 829). Many sailboats have an inboard engine installation similar to that in a powerboat, but usually with an engine of much less horsepower.

Boat Handling with Auxiliary Propulsion

For a sailboat with an engine, docking and mooring are much the same as they would be for a motor-boat. A sailboat's weight-to-power ratio is greater, and her propeller is smaller, than that of a motorboat of like length, and so she will not stop as quickly when the engine is put into reverse. When her motor is running, even if her sails are raised, a sailboat must comply with rules of the road as a powerboat. (See Chapters 4 and 5.)

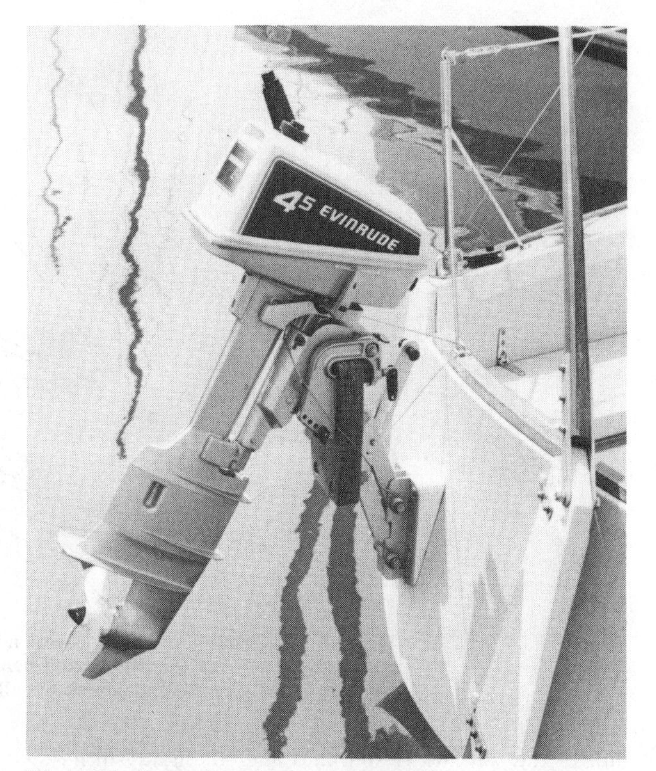

Figure 829 **An outboard motor is a sensible way to push a small boat along in calms, or into marinas. Since sailboats are easily driven through the water, only very low horsepower is needed. This motor bracket is in the up position, for stowage or when under sail; it drops down for motor use.**

Boat Handling Under Sail

For sailboats without auxiliary power, several special procedures must be followed when leaving the pier, leaving a mooring, pulling up an anchor, or returning to each of these at the end of a sail. See also Chapter 7 for basic boat handling skills.

Departing a Pier Under Sail

In most cases, the wind and current will permit a straightforward departure from a pier even under sail. When everyone is aboard and all gear is stowed, sails should be hanked on, with halyards attached, and made ready to hoist.

If the wind will help move the boat from the pier, it is a simple matter to take in the dock lines and raise the first sail. This is usually the mainsail. The boat will be maneuverable as soon as the sail is hoisted and sheeted properly, provided there is sufficient wind; steerageway may be gained by a shove off or along the pier. Other sails may be raised later when clear of the pier.

If the wind or current is pushing the boat onto the pier, departure may be more difficult. In extreme conditions, when it is rough or windy and therefore impossible to position the boat for a safe departure from the end of a pier, then either another boat, or an anchor set to windward, will be needed to pull the boat away from the pier.

Figure 830 **Arriving under sail at a pier requires practice to prevent crash landings. In the light breeze shown here, both sails can be left up and allowed to luff completely. Usually it is wiser to douse the sails just before reaching the dock, in case an unexpected puff of wind comes up to fill the sails.**

Coming into a Pier Under Sail

When approaching a pier for a landing under sail, the jib may be dropped some distance away, according to the skipper's judgement. This will slow the boat's speed but keep maneuverability. Lines and fenders should be prepared and ready to use, just as they would be for a landing under power.

The skipper must have an idea of how far the boat will coast after the last sail is lowered or allowed to luff, and before the boat loses maneuverability. This can be learned in open water by heading the boat up into the wind and allowing the sails to luff, and observing how far she carries her way. A lightweight, centerboard boat will stop in the water almost immediately, while a heavy keel boat may travel for several boat lengths before stopping, depending on wind and sea conditions.

When the boat is judged to be the right distance from the dock, she is brought into the wind with the sheets freed, sails luffing. The boat coasts in still with maneuverability. The right distance will vary with each boat, and in each set of wind and sea conditions. Ideally, the boat should lose way and come to a stop of her own accord within arm's reach of the pier, and lines can be neatly placed ashore and the boat secured.

Don't Be Ashamed to Try Again If the landing looks bad, the skipper should not hesitate to use the last of his maneuverability to turn away from the pier, and get away to make a fresh approach.

Departing a Mooring

Leaving a mooring under sail is just as simple as leaving under power, perhaps simpler because there is less worry of fouling the mooring in the propeller. The mainsail may be hoisted before casting off the mooring pennant, but the jib is often hoisted later, to keep the foredeck clear for the person throwing off the pennant.

Getting Off on the Right Tack

By waiting for a natural swing of the boat, by backwinding a sail as discussed below, or by holding the pennant far out to one side (and perhaps moving aft a bit while holding the pennant out) the person releasing the mooring can send the bow of the boat off in a chosen

Figure 831 **Leaving a pier under sail can be hectic. Here the crew is about to sheet in the jib, and the boat should gather way and begin to sail. Care should be taken that the boat does not drift back into the pier and damage the motor.**

direction, to port or to starboard, allowing the mainsail to fill on one side, and the boat to pick up speed.

Caution must be taken that the pennant is not released when the boat is exactly head to wind, because she is then "in irons" and may drift backwards "in irons" without steerageway. She may continue to drift without any wind in the mainsail until she fetches up on a boat moored astern, or in the shallows near shore. Procedures for getting out of irons are discussed on page 173.

Picking Up a Mooring Under Sail

When picking up a mooring under sail, the skipper's knowledge of the particular boat is called upon. Again,

Figure 832 **To pick up a mooring under sail, a slow approach, *above,* is made with the crew stationed forward. The boat must not approach too fast, but must have enough speed to maintain steerage. When the mooring is picked up and secured, *below,* sails should be dropped promptly. Practice is required and a good sailor makes it look much easier than it is.**

Figure 833 **This plow anchor is being raised by hand while the mainsail is already hoisted and allowed to luff. While the anchor is secured on its sprit or on deck, the boat will drift backwards in irons (see text) and must be skillfully handled to get off on the proper tack. Be cautious not to drift back on other boats when getting underway.**

wind, current, and sea conditions must be considered, and practice will pay off.

As the boat nears the mooring, station a crewman forward to pick up the pennant, or be prepared to go forward quickly yourself if you are alone. Turn the boat directly into the wind, and let the sails luff while the boat is still a few boat lengths away from the mooring buoy. She should have the momentum to keep going ahead and stop with her bow just at the pennant, but this takes practice. Just how many boat lengths to allow is part of the skill involved, and the distance will change from day to day depending on the wind and current and from boat to boat.

As with coming into a pier for a landing, if the maneuver isn't working right—just head the boat around, trim the sails, and try again.

Anchoring Under Sail

When anchoring a sailboat, the skipper must, of course, remember that most sailboats require more water than most powerboats, and he must be mindful of his draft when selecting an anchorage. Furthermore, it is important to remember that because of their keels, and because of their lofty rigging, sailboats are affected differently by wind and current than powerboats are. When anchored too close together, a sailboat and a powerboat will swing differently, and may collide. For this reason, and because of their greater draft, sailboats tend to anchor together in deeper water. See pages 243-245.

The procedures for setting and retrieving an anchor from a boat with only sail power are similar to those for setting and retrieving an anchor under power. However, it can be difficult to set the anchor properly without using the engine in reverse to back down on it and dig it into the bottom. (See Chapter 11.)

Setting an Anchor Under Sail

Approach the chosen anchorage under reduced sail, perhaps under mainsail alone. You will know what is best for your boat.

Follow the recommendations given in Chapter 11 on Anchoring. Instead of backing down in reverse gear while paying out scope, however, it is necessary to back down under sail.

Backing down is best done by backwinding the mainsail firmly and allowing the wind to catch it and back the boat down. If the boat spins and will not back in a straight line, straighten her up by pushing the boom out to the other side. This can be a tricky maneuver because the boat will start to sail away, and may not back down properly. It may be necessary to drop all sail, settle back to full scope, then raise sail again, briefly, to apply pressure on the anchor to set it.

Raising an Anchor Under Sail

To raise the anchor without the use of power in a small boat is as simple as pulling up the anchor line, then hauling up the anchor hand over hand. The sail should be already up and luffing, or ready to raise as soon as the anchor is up.

In larger boats, which might be too heavy to pull upwind while bringing in the anchor line, other measures will be necessary. The anchor line can be brought in on a windlass, or the boat can be sailed in short tacks up to the anchor, with her crew bringing the rode in by hand as the boat travels over it, until the anchor is at short scope. Then the crew must cleat the anchor line promptly, and the momentum of the boat will break out the anchor, and it can then be raised by hand.

Sailboat Navigation

Navigation under sail is essentially the same as navigation for motorboats, with a few complications thrown in. These complications result from the search for favorable winds, deeper water because of their deeper keels, and the sailboat's inability to travel directly into the wind.

When sailing into the wind, a sailboat will have a factor of leeway which must be considered for a good estimated position. This leeway will be increased or decreased somewhat according to the ability of the helmsman, and the weather conditions. Large seas will slow the boat and increase leeway.

Furthermore, since a sailboat cannot set a direct course to a destination to windward, she must tack upwind in short or long BOARDS (tacks). Each board will have a predeterminable compass heading, a factor of leeway, and estimated speed; its length must be calculated to achieve the most efficient progress in working to windward. These skills are covered in Chapters 19 and 21 on piloting. In racing, the length of these boards and the timing of these tacks become crucial in the windward leg.

Compass Problems

The compass on a sailboat operates much the same way as a compass on a powerboat, but with a difference.

Heeling Error

The magnetic field of a sailboat will change to one side as she heels steadily to one side, and the compass may be adversely affected. This is called HEELING ERROR, and is compensated by heeling magnets (see page 362). This is advanced compass compensation, difficult to assess and account for, and particularly important where pin-point accuracy is required on long passages; most sailors, however, need not be concerned with this problem.

Sailboat Right of Way

Because of their slower speed and reduced maneuverability, sailboats are given a special position in the Rules of the Road, and their own set of rules between boats under sail. While the Rules of the Road give a vessel under sail the right-of-way over a vessel under power, there are certain exceptions to this rule, and the Rules (Chapter 5) should be studied carefully. Between vessels under sail, the following rules apply, whether racing or not, to avoid risk of collision.

When two sailing vessels are approaching one another so as to involve risk of collision, the Rules of the Road state that one of them must keep out of the way of the

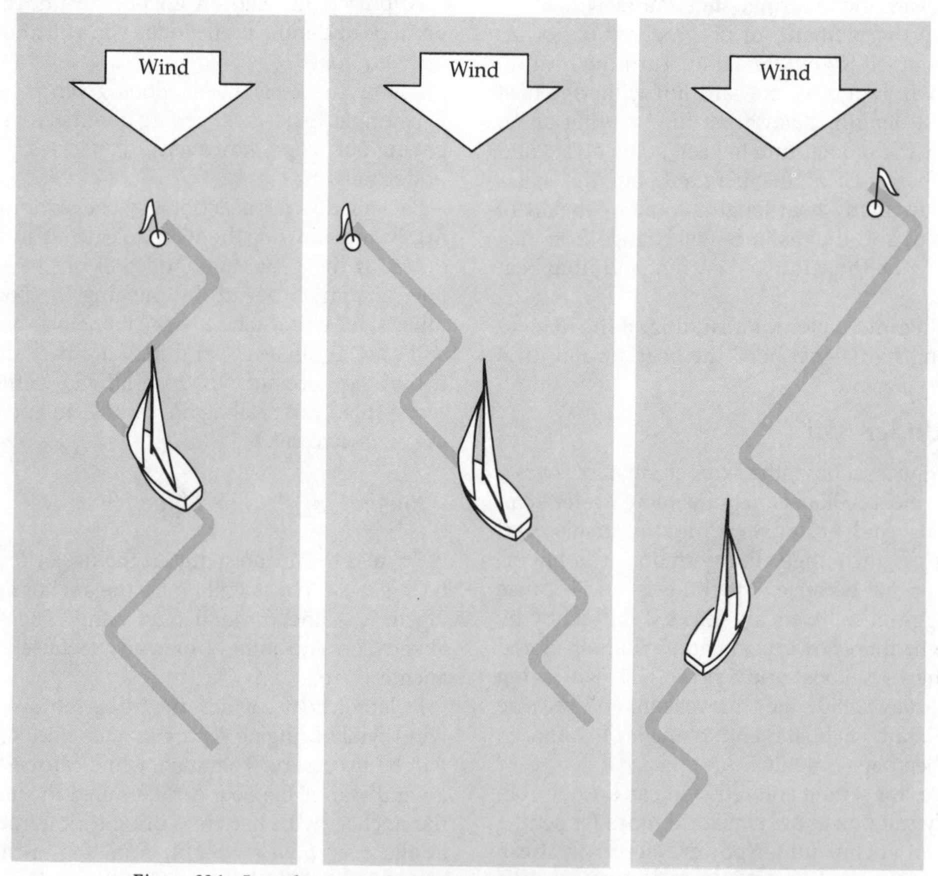

Figure 834 **In order to reach a destination upwind, a sailboat must tack continually and calculate her course and timing between tacks. On long voyages the navigator will be the most important person aboard.**

Figure 835 **Sailboats have special right-of-way rules amongst themselves, whether racing or not. In this photograph of a race, the boat in the foreground has the right of way because she is on the starboard tack, and the two boats ahead of her must stay clear or else be protested out of the race.**

other in accordance with the following conditions:

1. When each has the wind on a different side, the vessel which has the wind on the port side must keep out of the way of the other.
2. When both have the wind on the same side, the vessel which is to windward must keep out of the way of the vessel that is to leeward.
3. If a vessel with the wind on her port side sees a vessel to windward and cannot determine with certainty whether the other vessel has the wind to her port or starboard side, it must keep out of the way of that other vessel.

Note that these provisions are now identical under both the Inland and International Rules.

In addition, there are complex rules which apply to boats during a race having to do with overlap, proper course, room at a mark, and many others, not within the province of this book. The U.S. Yacht Racing Union in Newport, R.I., can supply the authoritative booklet on these rules.

Sailboat Trailering

Many sailboats can be carried on trailers and launched in the same manner as powerboats (see Chapter 6, pages 124-137). Indeed, some are specially designed for trailering, and are sold with a matching trailer.

While some sailboats can be rolled off the trailer into the water like a powerboat, it is more common for sailboats to be lifted off the trailer by a crane. The best supports for these boats are padded bars. For any sailboat, however, the pads or rollers must be adjusted to fit the contours of the hull, and to provide adequate support for the keel, or any area where there is a heavy concentration of weight—such as an auxiliary engine. This will prevent possible distortion of the hull, and is particularly important if the boat is stored for long periods on her trailer.

Sailboats are trailered with the masts unstepped and carried in a horizontal position. Special holders for the spars may be on the boat itself, or on the trailer. *Extra care must be taken when stepping or unstepping the mast to be sure there are no overhead wires in the vicinity that could come in contact with mast or rigging.*

Figure 836 **This daysailer is being pulled out of the water on her trailer. She requires no dockspace, no antifouling paint, and no chafing gear on mooring lines. However, she must be unrigged and carefully secured for highway driving.**

True Wind from Apparent Wind

Apparent Wind			*Boat Speed*							
			3 Knots		*4 Knots*		*5 Knots*		*6 Knots*	
	Degrees off Bow	*Velocity (Knots)*	*True Wind*		*True Wind*		*True Wind*		*True Wind*	
			Degrees off Bow	*Velocity (Knots)*	*Degrees off Bow*	*Velocity (Knots)*	*Degrees off Bow*	*Velocity (Knots)*	*Degrees off Bow*	*Velocity (Knots)*
Dead Ahead	0	Calm	180	3	180	4	180	5	180	6
	0	4	0	1	0	0	0	1	0	2
	0	8	0	5	0	4	0	3	0	2
	0	12	0	9	0	8	0	7	0	6
	0	16	0	13	0	12	0	11	0	10
	0	22	0	19	0	18	0	17	0	16
	0	30	0	27	0	26	0	25	0	24
	0	42	0	39	0	38	0	37	0	36
	0	60	0	57	0	56	0	55	0	54
Broad off the Bow	45	4	94	2.8	112	3	128	3.5	138	4.2
	45	8	65	6.2	74	5.4	84	5.7	94	5.6
	45	12	57	10.1	62	9.6	68	9.2	74	8.8
	45	16	54	14	57	13.5	61	12.9	65	12.5
	45	22	51	20	53	19.4	56	18.8	58	18.2
	45	30	49	27.9	51	27.3	53	26.7	54	26.1
	45	42	48	39.9	49	39.3	50	38.6	51	38.0
	45	60	47	57.9	48	57.2	49	56.8	49	55.9
Abeam	90	4	127	5	135	5.6	141	6.4	146	7.2
	90	8	111	8.5	117	8.9	122	9.4	127	10
	90	12	104	12.3	108	8.9	113	13	117	13.4
	90	16	101	16.2	104	16.5	102	16.7	111	17
	90	22	98	22.2	100	22.3	103	22.5	105	22.8
	90	30	96	30.1	98	30.3	100	30.4	101	30.6
	90	42	94	42.1	96	42.1	97	42.3	98	42.4
	90	60	93	60.1	94	60.1	95	60.2	96	60.3
Broad off the Quarter	135	4	154	6.5	158	7.4	160	8.3	162	9.3
	135	8	147	10.3	150	11.2	152	12	154	12.9
	135	12	144	14.2	146	15.1	148	15.9	150	16.8
	135	16	142	18.2	144	19	145	19.8	147	20.6
	135	22	140	24.2	142	24.9	143	25.7	144	26.6
	135	30	139	32.2	140	32.9	141	33.7	142	34.5
	135	42	138	44.1	139	44.9	140	45.6	140	46.4
	135	60	137	62.1	138	62.9	138	63.6	139	64.3

To Determine True Wind Direction:

On starboard tack, degrees of true wind off bow is the relative bearing of the true wind. Add this to the boat's true heading; the total is the true bearing of the wind (Subtract 360 degrees if the total is greater than 360).

On port tack, subtract degrees of true wind off the bow from 360 degrees to get its relative bearing. Add this to the boat's true heading; the total is the true bearing of the wind (Subtract 360 degrees if the total is greater than 360).

Table 8–1

True Wind from Apparent Wind

Apparent Wind			*Boat Speed*							
			7 Knots		*8 Knots*		*9 Knots*		*10 Knots*	
	Degrees off Bow	*Velocity (Knots)*	*True Wind*		*True Wind*		*True Wind*		*True Wind*	
			Degrees off Bow	*Velocity (Knots)*	*Degrees off Bow*	*Velocity (Knots)*	*Degrees off Bow*	*Velocity (Knots)*	*Degrees off Bow*	*Velocity (Knots)*
Dead Ahead	0	Calm	0	7	180	4	180	9	180	10
	0	4	180	3	180	4	180	5	180	6
	0	8	180	1	0	0	180	1	180	2
	0	12	180	5	0	4	0	3	0	2
	0	16	180	9	0	8	0	7	0	6
	0	22	180	15	0	14	0	13	0	12
	0	30	180	23	0	22	0	21	0	20
	0	42	180	35	0	34	0	33	0	32
	0	60	180	53	0	52	0	51	0	50
Broad off the Bow	45	4	146	5	151	5.9	155	6.8	158	7.7
	45	8	103	5.8	113	6.1	121	6.6	128	7.1
	45	12	80	8.6	87	8.5	94	8.5	100	8.6
	45	16	69	12.1	74	11.8	78	11.5	83	11.4
	45	22	61	17.7	64	17.3	67	16.9	70	16.5
	45	30	56	25.5	58	25	60	24.5	62	24
	45	42	53	37.4	54	36.8	55	36.2	56	53.4
	45	60	50	55.2	51	54.6	52	54	53	53.4
Abeam	90	4	150	8	154	8.9	156	9.8	158	10.7
	90	8	131	10.6	135	11.3	138	12	141	12.8
	90	12	120	13.9	124	14.4	127	15	130	18.8
	90	16	114	17.5	117	17.9	120	18.3	122	18.8
	90	22	108	23.1	110	23.4	112	23.7	114	24.2
	90	30	103	30.8	105	31	107	31.3	108	31.6
	90	42	100	42.5	101	42.7	102	42.9	103	43.2
	90	60	97	60.4	98	60.5	99	60.7	100	60.8
Broad off the Quarter	135	4	164	10.2	165	11.2	166	12.1	168	13.1
	135	8	156	13.8	158	14.8	159	15.7	160	16.6
	135	12	151	17.6	153	18.5	154	19.4	155	20.3
	135	16	148	21.5	150	22.4	151	23.2	152	24.1
	135	22	145	27.4	147	28.2	148	29	149	29.9
	135	30	143	35.3	144	36.1	145	36.9	146	37.7
	135	42	141	47.2	142	48	142	48.8	143	49.5
	135	60	139	65.1	140	65.9	141	66.6	141	67.4

9

Piloting and seamanship on rivers, canals, and lakes; the proper procedures for using locks

INLAND BOATING

Boating on inland lakes, rivers, and canals is different in some respects from that on coastal rivers, bays, and sounds, and along salt-water shores. The differences are not total; many aspects of safe recreational boating are the same on all waters. But where differences do exist they should be carefully considered.

The Pleasures of Inland Boating

Some of the finest cruising is found on the network of rivers and lakes giving access to areas far removed from tidal waters. Throughout the United States there are more than 30,000 miles (48,000 km) of waterways navigable by small boats. Using the Mississippi, the Great Lakes, the N.Y. State Barge Canal, and other waterways linking these with the Atlantic and Gulf Intracoastal Waterways, a boatman can circumnavigate the entire eastern portion of the United States, a cruise of more than 5,000 miles (8,000 km). Only a small portion of this would be in the open sea, although the Great Lakes are, of course, sizeable bodies of water, comparable to the ocean insofar as small craft cruising is concerned.

To these vast networks of interconnected waterways must be added the many isolated rivers and the thousands of lakes large enough for recreational boating. Many of these lakes have been formed behind dams in regions far removed from what is normally thought of as "boating areas." Today, there are few places where one would be surprised to see a sign reading "Marina" or "Boating Supplies."

Limitations

All cruising areas have their disadvantages and limitations. Some rivers and lakes may have shoals and rocks. Other rivers may have problems with overhead clearance; a fixed bridge might arbitrarily determine your "head of navigation."

On some of the principal inland waterways, the overhead clearance is severely limited—on some sections of the New York State Canal System it is only 15½ feet (4.7 m). Boatmen must unstep masts or fold down signal masts, radio antennas, and outriggers. Fixed overhead power cables are usually high enough to cause no problems to masted vessels, however, and their clearance is noted on charts. Unusually high water stages on rivers reduce overhead clearance by the amount of the rise over normal levels; use extra caution at such times. Charts for the Ohio River give vertical clearance for high water, based on 1936-67 levels.

River Boating

While rivers seldom offer vast open expanses where you will need the usual techniques of coastline piloting, they do require special piloting skills. Local lore often outweighs certain piloting principles that are the coastal skipper's law, because a river's ever-changing conditions put a premium on local knowledge. River navigation is thus often more an art than a science.

River Piloting

The fundamental difference is, of course, the closeness of the shore, usually with easily identifiable landmarks or aids to navigation. Knowing where you are, therefore, is *not* a problem. The skill of piloting here lies in directing your boat to avoid hazards, and on many rivers this is not a simple matter.

Water Level Changes

Inland waters are termed "non-tidal," but that doesn't mean they have no fluctuations in water level. The variations are apt to be of a seasonal nature as spring freshets, loaded with debris, flood down from the headwaters, overflow banks, and course rapidly on down to the sea. The annual changes in level can be astounding; at St. Louis the seasonal fluctuation in river level from winter-spring flood conditions to low levels in late summer and fall may be as much as 50 feet (12-15 m). In some smaller navigable streams, sudden hard rains may raise water level several feet (a meter or two) in a matter of hours.

Aids to Navigation

The larger inland rivers, those that are navigable from the sea, have aids to navigation maintained by the U.S.

Coast Guard. Many of their lights, buoys, and daybeacons are like those discussed in Chapter 16, but a few are of special design for their "inland" purpose. Many rivers under state jurisdiction have aids to navigation conforming to the Uniform State Waterway Marking System; see Chapter 16 and the color section.

The "Right" and "Left" Banks of a River Designation of a river's banks may at first be confusing. Left and right—port and starboard—banks are named relative to a vessel's course *downstream,* not to the vessel herself; the left (port) bank is the one on your left as you face in the direction that the river flows.

On the New York State Canal System, however, when regulations refer to the "starboard" side of the Canal, they mean the right side when entering from Waterford (near Albany). Thus the starboard side of the Champlain Canal is the east side; but on the Barge (Erie) Canal westward, the starboard side is the north side. Check your charts.

Mileage Markers Aids to navigation along many major rivers are conspicuously marked with mileages showing the distance from a designated reference point. It is always easy to get a "fix"; just relate the mileage on a daybeacon or light structure with the mileage figures on your chart. These mileage figures take the place of the arbitrary odd and even numbers used on coastal waters, and they are helpful in computing distance and speed. Statute, rather than nautical, miles are normally used.

Lights Most lights on major rivers, such as the Mis-

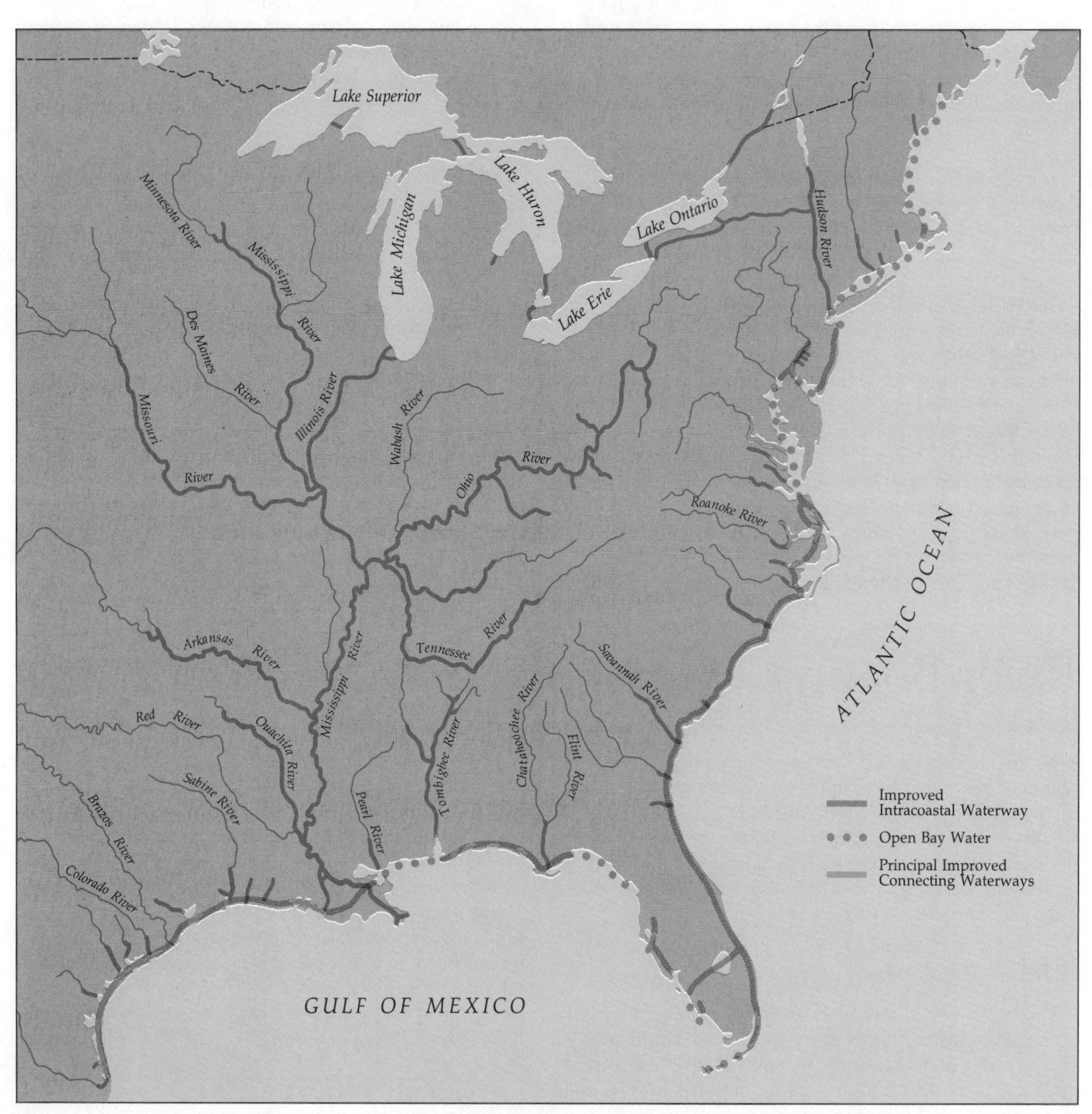

Figure 901 **The eastern part of the United States has an extensive network of inland waterways, including those that are natural, improved, or entirely man-made.**

sissippi, show through 360°—they are visible all around the horizon. Sometimes a beam projects in one direction only, however, or the intensity of an all-round light is increased in a certain direction by a special lens. The beam's width varies with the location; a narrower beam marks a more critical channel. Flashing lights on the right (proceeding downstream) side of the river show a single green or white flash; those on the left side show double red or white flashes. (These will change as the new IALA-B buoyage system is implemented. See Chapter 16.)

Generally speaking, lights on the "Western Rivers" (see Chapter 16) are placed strategically at upper and lower ends of "crossings" as marks to steer by, with additional lights between as needed. Where there are no crossings, lights along the banks serve as passing lights—lights that mark specific points along the banks. Your chart will show how a specific light should be used.

Daymarks The Mississippi River has two types of daymarks. PASSING daymarks on the right (proceeding downstream) are square green with green reflective borders; the left side has red triangles with red reflective borders. CROSSING daymarks are diamond-shaped, green or red as before, with small reflectors of the same color in each corner. See the color section.

Daymarks on some rivers consist of two white boards on posts or trees on shore, in the form of a large "X"; see Figure 902.

Buoys The skipper who is familiar with the basic U.S. system of buoyage will have no difficulty with buoyage on rivers. Buoys generally follow the same basic principles: black can-type buoys on the left and red nun-type buoys on the right when proceeding upstream from seaward. On the Mississippi and its tributaries, many buoys are unnumbered.

Buoys on the Mississippi River carry reflectors similar to those used on shore aids—red on the channel's left (when descending) and white or green on those on the right side. Lighted buoys show a white or green light if on the right side of a channel and a white or red light (double flashes) if on the left side. Buoys marking wrecks at the side of a channel show a quick flashing light of the appropriate color. Buoys marking channel junctions or obstructions that can be passed on either side are horizontally banded either black-over-red or red-over-black, and if lighted show an interrupted quick-flashing characteristic. (White lights will not be used after the new IALA-B buoyage system comes into effect; horizontally banded buoys, if lighted, will show a group composite (2 + 1) flashing light.)

Caution Required In all cases, check buoys by referring to your charts; some rivers deviate from conventional systems. On the New York State Canals, for example, there are red nuns on the starboard side, but the can buoys to port are white; in some places the buoys are floats with a red or white spindle on top.

Ranges On some rivers, like the Hudson and the Connecticut, where channels through flats are stabilized and maintained by dredging, ranges with conspicuous markers on shore will help you stay within narrow channel limits. Aligning the front and rear markers, you can hold your course safely in mid-channel despite any off-setting forces of current or wind.

Figure 902 **River piloting differs from that in coastal waters in that the skipper usually can tell at all times exactly where he is. The navigation problems consist mainly of avoiding underwater hazards such as shoals, sand bars, and snags. The daymark of these lighted aids to navigation, *above* and *below*, consists of two white boards in the form of an "X", plus a mileage sign with distance from a designated point.**

Color Illustrations Buoys and daymarks used on the Western Rivers are illustrated in color along with those used on other waters, see page 579.

Charts

River charts are commonly issued in the form of books, the pages covering successive short stretches of the river in strip form. A typical example is the bound volume published annually by the Mississippi River Commission, Vicksburg, Miss., to cover that river from Cairo, Ill., to the Gulf of Mexico. The scale is 1:62,500 (roughly 1 inch = 1 mile, or 1.6 cm = 1 km). All depth information (with a few exceptions) and detailed positions of rocks, reefs, shoals, and ledges that are shown on coastal charts are omitted. In their place, the course of the river is traced in blue tint between heavy black lines delineating the banks. A broken red line indicates the channel. Navigation lights are shown by a star and dot symbol with a number showing its distance upstream from a reference point—in this case, AHP, "Above the Head of Passes." The Head of Passes is the point not far above the mouth of the river where the river splits into separate channels leading through the delta into the Gulf. Mileages are given every five miles (statute), with small red circles at one-mile intervals. Mileages above and below the Head of Passes appear in red. A detailed alphabetical table lists all towns, cities, bridges, mouths of tributary rivers, and other important features adjacent to the main channel, with distances (AHP) in miles and number of the chart on which the feature appears.

River charts are usually on a "polyconic" projection (see Chapter 18), and elevations normally refer to a specified mean water level.

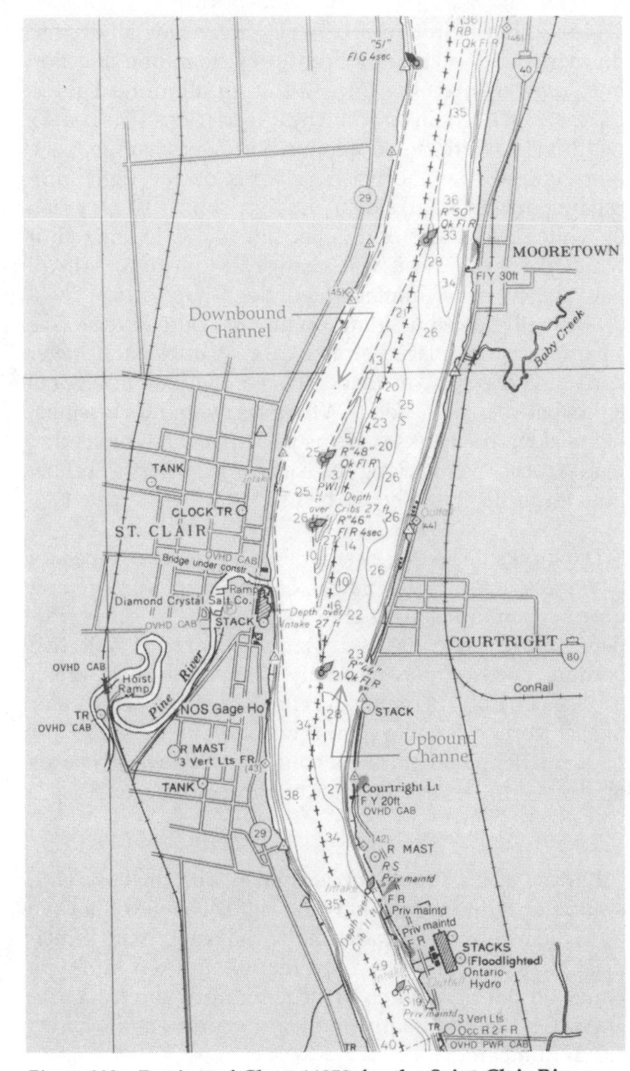

Figure 903 **Portion of Chart 14852 for the Saint Clair River; scale 1:40,000 on polyconic projection, in which distortion is small and relative sizes are correctly preserved. Depths are referred to the sloping surface of the river at specified lake level, elevation above the International Great Lakes Datum. Note the International Boundary, and separate upbound and downbound channels.**

Variations in River Charts River charts do not all use the same symbol and coloring scheme, so before using one, study its LEGEND. The legend gives aids to navigation symbols, abbreviations, topographic and hydrographic information, and often illustrations of characteristics peculiar to that chart. The symbols for buoys on charts of the Ohio River, for example, are unique to that waterway and may cause initial confusion to strangers unless carefully studied.

Upper and Lower Mississippi River charts show water in blue and man-made features in black; red is used for the sailing line, aids to navigation, and submerged features.

The book of Illinois Waterway charts shows the main channel in white with shoal areas tinted blue; land areas are colored yellow.

A charts booklet for Navigation Pools 25 and 26 of the Mississippi River shows hazard areas—depths under 6 feet (1.8 m)—in bright red shading, harbors or anchorage

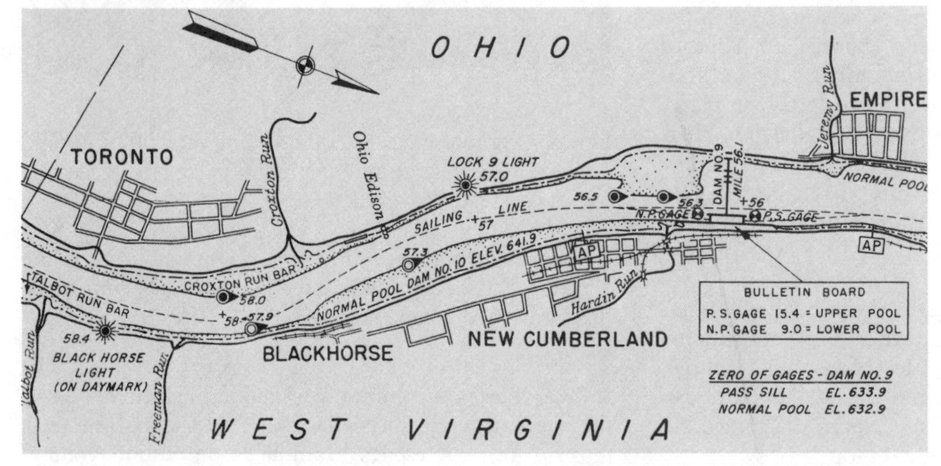

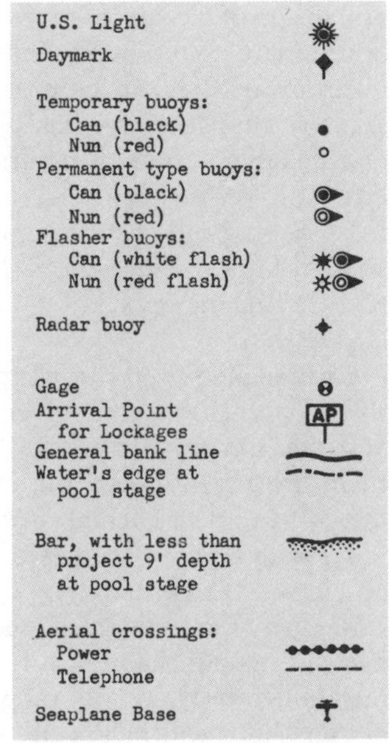

Figure 904 **On charts such as this one for the Ohio River, the arrow at the top indicates true north, which is seldom at the top of the chart. The arrow at the left, in the channel, indicates direction of river flow. Locations of bulletin boards that show river stages are also on the chart, along with the reference point for gauge readings. Symbols for nun and can buoys differ from those on National Ocean Service charts; these and other river chart symbols are shown in the legend at the *right*, excerpted from the chart. Note that lighted aids are called "Flasher" buoys.**

ILLINOIS WATERWAY - Illinois River					
Name of aid Character and period of light	Number of miles from Grafton	Bank or side of channel	Daymark Up	Daymark Down	Remarks
BEDFORD, ILL.	48.5	Right...	...	...	
Pilot Peak Bend Light Gp. Fl. W., 5 sec., 2 flashes.	47.8	Left....	TR	TR	Visible 360° with higher intensity beam oriented downstream.
BUCKHORN ISLAND DAYMARK	46.1	Right...	SG	SW	
Van Geson Island Light Gp. Fl. W., 5 sec., 2 flashes.	44.9	Left....	TR	TR	
Grand Pass Bridge Upper Light Fl. W., 4 sec.	44.1	Right...	SW	CW	Visible 360° with higher intensity beam oriented upstream.

Figure 905 **Charts for the Mississippi River System should be supplemented by USCG *Light List,* Volume V, which provides more details on aids than can be shown on the charts.**

areas with an anchor symbol, and the availability of fuel, water, marine railways, launching ramps, and repair and docking facilities.

Charts in the New York State Canals book, published by NOS, resemble the coastal charts described in Chapter 18. Land areas are screened gold tint, channels 12 feet (3.7 m) deep or more are white, with lesser depths light blue and depth contour lines for 6 and 12 feet (1.8 and 3.7 m), and black lines delineating the banks. Buoy symbols have magenta discs if they are lighted. Aids to navigation are numbered, and scattered depth figures give the skipper a good idea as to whether he dare venture out of the channel to seek an anchorage or for any other reason. Rocks and wrecks are indicated with standard symbols from Chart No. 1 (see Chapter 18). Arrows indicating current flow direction, used on many other inland charts, do not appear on those of New York inland waterways.

Missing from the typical river chart is the compass rose of coastal and off-shore charts which enables the navigator to lay a compass course. Instead river charts generally carry an arrow indicating true north; on strip charts, of course, north may not necessarily be at the top of the sheet.

Publications

Charts can convey a great amount of information, and you should always use the most detailed and latest editions you can get. Yet for space reasons, these charts cannot show all you need to know. Valuable additional data is published in the form of books and pamphlets.

Engineer Publications The Army Engineers at Vicksburg, Miss., publishes a pamphlet called *Mississippi River Navigation,* which contains not only interesting background information on Mississippi River navigation, but also a few pages of explicit instructions "For Part-Time Pilots."

Since channels, water levels, and other conditions on the Mississippi, Ohio, and other major rivers are constantly changing, the river boatman must keep posted on the latest information at all times. Army Engineers District and Division offices issue several regular publications about current conditions. These are variously termed *Divisional Bulletins, Navigational Bulletins, Navigation Notices, Notices to Navigational Interests* (weekly), and *Special Notices to Navigation Interests* (as required). These small publications show up-to-date river conditions such as channel depths and widths, estimated current velocities, controlling bridge clearances for specified stretches of rivers, construction projects and other hazards, and facilities like marine railways and lifts.

Coast Guard Notices The 2nd Coast Guard District, with Headquarters at St. Louis, issues *Local Notices to Mariners* with information on changes in aids to navigation, hazards, and other relevant Coast Guard matters. The USCG also issues *Channel Reports* noting the least depth found by Coast Guard cutters and buoy tenders on their river patrols.

Light List One of the most helpful documents to the river skipper is the Coast Guard's *Light List, Volume V,* covering all of the Mississippi River System. It tabulates lights, buoys, and other aids to navigation and gives mileages (in statute miles) from a specified point. Aids are described in greater detail than chart symbols and abbreviations can give.

Navigation "Tools"

A compass is of little use on most rivers; you will generally use binoculars to sight from one navigational aid to the next. In fog, recreational boat traffic thus comes pretty much to a standstill, although large commercial vessels normally carry on with radar. This does not mean you can dispense with compasses on all inland rivers; some are wide enough that you can continue in fog using a compass, speed curve, watch or clock, *and due caution.* Also, some rivers feed into large lakes.

Local Knowledge

Not even the best of charts and publications can "tell all" about a particular body of water. Rivers are particularly prone to seasonal or irregular changes, and if you are new to a specific stretch of river you should take every chance to ask experienced local people about hazards or recent changes.

River Currents

Broadly speaking, river currents, although they fluctuate in velocity, will always trend in one direction—from the headwaters to the mouth. Tidal rivers may nevertheless have strong tidal conditions at the mouth, which back the water up so that you can take advantage of a favorable current going *upstream*. On the Hudson, for example, an economy-minded skipper electing to run at 9 to 10 knots can time his trip to carry a favorable current all the way from New York City to Albany, 150 miles (241 km) inland. Tides at this latter city, even so far from the sea, may range in excess of 6 feet (1.8 m), although at other points far *downstream* the range will be only *half* that amount.

Current Strengths The strength of river currents varies widely from river to river, and from season to season for any particular river. Speeds on some sections of the Mississippi range from 1 to 6 mph (1.6 and 9.7 km/hr) under average conditions. At extreme high water stages, current strengths may be much greater—9 mph (14.5 km/h) or more in narrow and constricted areas.

River currents sometimes attain such speeds that navigation upsteam is not feasible, although capably handled boats can be taken down safely. In the St. Lawrence River's Galop Rapids, for instance, the current may run as fast as 13 mph (20.9 km/h). Some river boats have power enough to ascend certain rapids, but as a general rule rapids are best avoided in favor of the canals and locks that by-pass them, unless the skipper has local knowledge or engages the service of a local pilot.

"Selecting" Your Current River current characteristics are of the utmost importance to the masters of deep-draft commercial vessels. The surface current acting on a small boat may actually be contrary to that which grips a large ship's keel near the river bottom. Even surface currents vary from bank to mid-stream. Friction of the bank and bottom slows the water. The commercial skipper—to whom fuel costs and time of run are especially important—knows this difference. He uses the strength of the mid-stream current for his run downstream, and on the way back upstream he runs as close to the bank as he safely can, even turning into small coves, to take advantage of the countercurrents.

You may be less concerned with the economic factors that the professional pilot weighs so carefully, but you can profitably heed the same principles. You can cut running time and fuel costs by running courses that make the river's current work for you, or minimize its adverse effect. A 12-mph boat in a 4-mph current is making good either 8 or 16 mph, depending on direction of travel—a significant effect! Even for a 20-mph craft the difference between 16- and 24-mph speeds is 50%.

Channels at River Bends

Water at a bend in a river tends to flow across the stream on the outside of the bend, scouring out a natural channel there, while creating a bar out from the point around which it is turning. Unusual bottom contours or man-made structures may alter this tendency, but it generally means that river channels going around bends will shoal on one side and deepen on the other over time.

Study a river chart that gives depths (NOS charts do) and note this riverflow characteristic. Then when you are on a river whose charts do not give depths (the Mississippi and Ohio charts do not), you will have a better feeling for where to find deeper water. With this understanding of natural channels at bends you will be less likely to run straight courses from marker to marker, cutting corners and risking getting hung up on a bar.

The proper course at each bend is, of course, a curved line roughly following the trend of the river as a whole. If there are no aids to navigation to guide you, keep about a quarter of the river's width off the *outside* bank.

On some river charts, even the markings showing the topography ashore give clues to the river itself. Contour lines crowded close together near shore indicate a cliff rising steeply from the bank, with a good chance of deep water close under the bluff. The cliff may also be a landmark to steer by.

Channel "Crossing" Where a river bends in an "S" curve—a curve in one direction followed by a curve in the opposite direction—the straight section between the curves is termed a "crossing." On major rivers, these are usually marked by ranges or directional lights. In some instances, such as at low water stages, seasonal buoys may be added to mark the route of best water; these buoys may be found in pairs or singly.

"Eyeball" Piloting

Much of a river pilot's success depends on his acquired skill of interpreting what he sees. For example, no flat

statement can be made about what certain surface conditions reveal about relative depths of water. On one hand, where there is a chop in the channel there may be areas of smoother water over the bars, especially if there is any weed growth present. On the other hand there may be wind-current conditions where the channel will be comparatively smooth with ripples revealing the bars. The experienced pilot will know which condition prevails.

In some narrow river channels, with wind against current, a small sea builds up in which the larger waves disclose the deeper water and grow smaller until there are no seas at the channel's edge. Under any conditions, the experienced boatman will quickly learn to take advantage of nature's signs.

Watch Your Wake In unfamiliar waters, even your wake can give a clue to the safety of your course. As it rolls off into shallow water, its smooth undulations give way to a sharper formation, even cresting on the flats in miniature "breakers." When the waves reach a shoal or a flooded area where submerged stumps are close to the surface, the difference will show. If your wake closes up toward your stern, and appears short and peaked, sheer off, and fast, away from the side of the channel where this telltale signal appears.

Figure 906 **By staying in mid-channel, this small boat avoids the shoals that build up along the inside of river bends. A deep draft vessel may need to follow the outside edge of the bend in order to make sure there's sufficient depth for it.**

Figure 907 **The most often encountered hazards in river cruising are shoals, sandbars, and underwater rocks. In new man-made lakes, decaying trees and shrubs may be present.**

Shoals Shoaling is a serious problem in most rivers. Spring floods build up current velocities, stir up silt in river bottoms, wash away parts of river banks, and carry all this dirt downriver in suspension, to be deposited as mud flats and sand bars where the strength of the current lessens; these flats and bars become hazards to navigation. The Connecticut River is a good example of the shoaling problem; its channels must be dredged to authorized depths each year, at great expense. In the Mississippi River, a jellied mass of muck called "flocculation" is deposited as a sediment on the river bottom to a depth of 10 to 15 feet (3.0-4.6 m) each year. Deep-draft vessels will plow through it, and high water stages of the river flush it out into the Gulf, so the Army Engineers do not bother to dredge it at low water stages. Over the centuries, this is how the River's delta has been built up.

On a smaller scale, a bar or shoal forms at the mouth of almost any river where the current slows and its silt can no longer stay in suspension. Bad seas may build up here when strong outflowing currents are opposed by an on-shore wind. Disturbances like this can also occur inland at the confluence of two rivers.

Practice Makes Perfect As you develop a sense of river piloting, you'll become more conscious of the actual track you should make good over the bottom relative to shoals and the sides of channels. Where currents do occasionally flow diagonally across your course, you will allow for them instinctively and look astern frequently to help maintain that sense of position.

River Seamanship

Boatmen on inland rivers are subject to many of the general requirements for safety and good seamanship covered in other chapters. They must also be prepared for the special conditions that relate to specific bodies of waters.

Boat handling at piers and on entering or leaving slips may be complicated by swift river currents. Grounding hard on river shoals may be more troublesome than it is on tidewater, where the next rising tide will often free a boat without any other assistance.

Figure 908 **In some areas, piers must be floating to allow for large changes in water level at various times of the year. This is the Guntersville Yacht Club, Alabama.**

River Cruising

Cruising can be exceptionally enjoyable on rivers, but requires careful planning and execution.

Problems of where to make fast or anchor for the night are a part of cruising in any waters. On rivers you can often find shelter from blows close at hand, yet you will find areas big enough to work up a sizeable chop or even seas, especially when wind opposes current. You'll also need protection from wakes and wash of passing river traffic and you need to decide whether to seek the seclusion of quiet anchorages off the beaten track, or the activity associated with towns and cities.

New marinas on many of the inland waters—the Tennessee River network, for example—have been a boon to river cruising. They offer a place to make fast for the night, and easy access to fuel and supplies.

Safety Harbors and Landings The Tennessee is among the rivers that now provide safety harbors and landings for use in bad weather, mechanical difficulty or other emergencies; their locations are shown on charts. Safety harbors are usually coves off the navigable channel. Direction boards on shore indicate the entrance and cross boards mark the upper limits.

Safety landings are areas where the banks have been cleared of stumps, boulders, snags, or other underwater hazards so that boats can safely come to shore, with upper and lower limits marked by direction boards. These signs are color-coded on the Tennessee: white if a 9-foot (2.7 m) depth is available at all water level stages, orange if this depth is not available at lower levels.

The New York State Barge Canal provides terminals at intervals along its route; these are only occasionally in use by commercial vessels, but are otherwise available to recreational boats. The terminals are concrete with rather rough faces, so you will need fenders.

To allow for the seasonal range in water levels [or stages,] most Western River marinas and yacht clubs are afloat on strings of barges, so you may have a long climb up the river bank at low-water stages to get to town.

Anchoring

Your chart will often reveal a likely place to anchor for the night. A widening of the river may offer you the chance to get out of the reach of traffic, or a small tributary or slough may invite exploration (enter with caution and check depths as you go), or the natural configuration of river banks and bars may provide a natural "harbor" with complete protection. Islands occurring in mid-river often leave a secondary channel for small boats on the side opposite that used by deep-draft commercial vessels. When a river cuts a channel behind a section of bank, a "towhead" is formed. Sometimes these are filled in or dammed across at the upper end by river deposits, forming a natural protected harbor which can be entered from the lower end.

You may also be able to get your boat behind a pile dike—a structure designed to keep river banks from washing away. The dike juts into the river and, by tying up to one of the piles just inside the outer end, on the downstream side, you will be protected from passing traffic as well as from floating debris.

Use Caution Outside Channels When entering sloughs between islands or between an island and the bank, beware of submerged wing dams at the upstream end. To be safe, enter and leave from the downstream end.

When anchoring on the larger rivers near a sandbar or island—or when beaching a small boat there to go ashore—it's wise to pick the downstream rather than the upper end. If your anchor drags, or if you somehow get aground on the upstream end, the current will be pushing you harder ashore. Water at the downstream end is likely to be quieter and the eddies that normally exist there may help to free you.

Sandy bars that are exposed at low water may be quite unstable. Be careful when using them for camping or swimming.

Check Characteristics of the Bottom The character of the bottom varies widely on inland waterways. Particularly in their lower reaches, river bottoms are often soft mud, so you should carry at least one broad fluked anchor of a design that will dig down until it reaches good holding. An anchor with spidery arms and flukes would pull through mud and provide no holding power at all; but on a hard bottom it is likely to be more needed than the other type.

When anchoring over rocky bottoms or in areas full of snags and roots, it is best to rig a "trip line," a light line from the anchor's crown to a small buoy—an empty bleach bottle will do—at the surface. If the anchor snags and will not come free in a normal fashion, pick up the buoy and raise the anchor with the trip line, crown first. Some anchors have slip rings, shear pins, or other devices to make them snag proof; see Chapter 11 for additional information.

Leave an anchor light burning all night if there is any chance of other vessels being underway in your anchorage; if you are not sure whether you need an anchor light, be on the safe side and have one showing.

Making Fast to the Bank

In many areas, cruising boatmen make fast to the river's bank for a lunchtime break in the day's journey, or even overnight. Be cautious when doing this. First, check that the depth is adequate and the area is free of underwater obstacles; approach the shore slowly. Avoid vertical banks that may be in a stage of active caving; exposed tree roots in the bank may be evidence of recent erosion. Avoid rock riprap along the banks; it can damage a boat's bottom; see Figure 909.

Allowance for Water Level Changes On inland rivers your docking lines generally need not allow for the tidal changes common on the coasts, except, of course, on tidal rivers. On non-tidal waterways, however, there is always the chance of a change in level with hard thunderstorms or other heavy rainfalls. Observe the practices of local boats, and be guided accordingly.

Figure 909 **A typical Mississippi River bank riprapped with concrete to protect it against erosion. Use care near such banks to avoid damage to your boat's bottom.**

Tying up to a barge or float that will itself rise or fall with a change in levels is advantageous. Here, you need simply to leave enough slack in lines to accommodate the wake of a passing vessel.

River Cruising Problems

Boatmen who have cruised a given river may have conflicting reports of its hazards, or the absence of them. One may have made his cruise in the early spring, encountering high-water, flood conditions, racing currents, and floating debris, while another made his trip in September or October with low-stage water levels and slower currents so that he encountered few, if any, obstacles, except perhaps more numerous shoals. At extreme flood conditions, as a rule, river navigation is not recommended without the services of an experienced pilot.

Eddies and "Whirlpools" On the Mississippi, "sand boils" may be caused by sand piling up on the river bed. During flood stages these whirlpool-like disturbances can be so violent that they can throw a boat out of control. In more favorable months, they may be no worse than surface eddies—felt, but certainly of no danger to a boat.

Problems From Silt Some river boatmen have reported underwater bearings ruined, engine jackets filled with silt, and water pump impellers worn out at the end of a single river run. In all waters heavily laden with silt, you are wise to carry protection against it: raw water strainers, fresh water cooling systems, cutless-type underwater bearings, and pump impellers that will handle mud and sand better than bronze gears can.

Debris Floating and partially submerged debris such as tree trunks and branches are a hazard for small boats; keep a sharp lookout in waters where they have been reported. Floating debris is usually at its worst in the spring months, when flood water levels have swept away downed trees and other materials from above the normal water line.

Towed-under Buoys River currents sometimes flow so fast that buoys are towed under, leaving only a V-shaped eddy on the surface to reveal their location. Sometimes the buoy's top will be visible to a boat bound upstream, or the wake of a passing vessel will expose it momentarily.

The surface eddy of a towed-under buoy always points upstream as its "wake" divides downstream around it. Avoid *any* surface disturbance like this; a submerged obstacle is likely lurking beneath. Do not confuse the towed-under eddy with the condition of two currents converging at the downstream end of a middle bar. That condition may also show as a V-shaped eddy, but pointing *downstream.*

Misplaced Buoys Another hazard of river piloting at spring-time high water levels is that buoys can shift from their charted positions if they are dragged by floating trees or logs.

Special Hazards Some rivers present special hazards. The Army Engineers caution against regarding the Mississippi with insufficient respect. The lower river is very large, with low-water widths of 2,500 feet (0.8 km) and high-water (bank-full) widths up to 9,000 feet (2.7 km). The bank-full stages generally occur between December and July, most frequently in March or April. Low-water stages occur in the fall months.

The Mississippi's Lake Pepin typifies the kind of exposed area that you may encounter. This "lake," actually a broadening of the river proper, is 21 miles (33.8 km) long and up to 2½ miles (4 km) wide; sizeable seas can build up in such a body of water.

You can find many kinds of equipment at work along rivers, especially the larger ones. Hydraulic pipeline dredges may have lengths of floating pipeline you must avoid. Barges carrying bank-protection equipment may extend hundreds of feet out from shore; often in the swiftest part of the current. When passing any form of construction or maintenance equipment, keep *well clear* because of the hazard of being swept under it, and slow down to avoid damage from your wake. Regulations exist governing lights and day shapes for dredges and other "floating plant" working on river projects, and for the passing of such equipment by other vessels. Use your VHF radio to contact the working vessel and get permission to pass, and advice on how to do so safely.

Signals

Whistle signals for passing other vessels, or for use in conditions of restricted visibility, and other special situations are prescribed in the Inland Navigational Rules. Details are in Chapter 5.

With a few exceptions where there are two or more bridges very close together, the signal for requesting a drawbridge opening is one prolonged blast followed by one short blast; see page 102.

Passing Commercial Traffic

Most U.S. inland waterways handle considerable commercial traffic. Whether you are cruising or just out for the day, you must know how to handle your boat in a situation with a large commercial vessel. In a narrow channel, a big tug or tanker requires much of the available water. As she approaches, you'll see a sizeable bow wave built up ahead of her, and the water drawn away by suction to lower the level at her sides amidships. Give her as wide a berth as you possibly can, and be alert for violent motions of your boat as she passes.

Tugs with tows astern in narrow waterways present a real problem to approaching small boats. Fortunately, most of the river "towing" today is done by pushing scows and barges ahead of the tug. This keeps the whole tow under better control as a single unit. Passing at a bend is more dangerous than on a straightaway, and sometimes must be avoided entirely; the tug and its barges cannot help but make a wide swing; and where there is ample room to pass at the beginning of the turn, there may be none at all later. If small boats *must* pass at a bend, it is usually wiser for them to take the *inside* of the curve.

Jumbo Tows on the Mississippi Big rafts of Mississippi River barges bunched together in one vast tow may cover *acres* of water. You should never jeopardize their activities, regardless of right-of-way. Integrated tows may consist of a bowpiece, a group of square-ended barges, and a towboat (at the stern)—all lashed together in one streamlined unit 1,000 feet (305 m) or more in length. At night, the lights of the tow boat may be more conspicuous than the side lights on the tow far out ahead. To make barges more visible, the Inland Rules require that those pushed ahead must show a quick-flashing yellow light all the way forward on the centerline. Obviously, special caution is needed when running even well-lighted rivers

Figure 910 **Among the longest vessels in the world are the tows made of barges pushed by a towboat at the stern. Length often exceeds 1,000 feet (305 meters). Always pass tows with great caution, and avoid passing them in river bends.**

at night. The searchlights of commercial traffic may make it impossible for you to see anything at all. Shore lights and their reflections may add to the difficulty. Floating debris often can not be seen in time, if at all.

Give All Tows a Wide Berth On all rivers, you should give tows a wide berth. In particular, *stay away from in front of tows;* if you should lose power, the tow could probably never stop or steer clear in time. A commercial tow may at times need a half-mile to come to a full stop.

Small-boat whistle signals are usually inaudible at the noisy control station of large vessels some distance away. If you need to communicate use VHF radio, Channel 13 or 16 if you can. A passing is often verbally described over the radio as "one-whistle" or "two-whistle," just as if actual blasts were being sounded.

Watch Your Wake

Much inland boating is done on waterways that are quite narrow. Regulate your speed so that no *destructive* wake results; excessive wake can cause damage to other boats and to shore installations. Some wake is almost inevitable from motorboats, but it must not be "destructive"; keep your speed, and your wake, down when passing boats and shoreline facilities.

Canal Boating

Before construction of dams and locks, many of our rivers were unnavigable. Water coursed down valleys at the land's natural gradient, dropping hundreds of feet in not many miles, running too fast and encountering too many natural obstacles for safe navigation. To overcome such obstacles, engineers dam natural waterways at strategic spots to create a series of "pools" or levels that may be likened to a stairway. Good examples of how closely these can resemble an actual flight of stairs are at places like Waterford, N.Y., and the Rideau Waterway at Ottawa, where vessels descend or ascend a series of locks in immediate succession. This section of Chapter 9 covers boating on both completely man-made waterways, which might be termed "pure canals," and on "canalized" natural rivers.

Water Levels

To a river pilot, the term POOL STAGE indicates the height of water in a pool at any given time with reference to the datum for that pool. On many rivers, pool stages are posted on conspicuous bulletin boards along river banks so as to be easily read from passing vessels. Charts state the locations of these bulletin boards.

On the Ohio River, gauges at the locks of each dam show the depth of the pool impounded by the next dam downstream. For example, if the chart indicates a 12-foot gauge reading for a normal pool, a reading of 11.7 feet indicates that the next pool is 0.3 feet below normal elevation.

Locks and Dams

Without locks, dams on our inland waterways would restrict river cruising to the individual pools and would prevent through navigation except for boats light enough to be PORTAGED around the dams. Locks, in conjunction with the dams, permit boats to move from level to level. Locks vary in size, but since they almost invariably handle commercial traffic, their dimensions offer no restriction to the movement of recreational boats.

Principles of Locks and Locking

Locks are virtually watertight chambers with gates at each end, with valves admitting water to them as required. When a vessel is locked upstream to a higher level, first all the gates are closed and valves on the lock's downstream side are opened to let the water run out to the lower level. Then the downstream gates are opened so that the vessel can enter the lock. The downstream gates are then closed and water flows into the lock from above through another set of valves until the chamber is full to the upper pool level, with the boat rising along with the water. The upstream gates are now opened and the vessel resumes her course upstream. Locking a vessel down is the reverse of this process.

Water is not pumped at river locks; the natural flow is utilized, which is why some canals limit the number of lockings each day during droughts or annual dry seasons.

Variations in Lock Design

There are various kinds of locks, all of which accomplish the same result. Gates may swing to the side or roll back; sometimes they lift vertically and boats go under them.

Smaller locks with a minimal water level change may not have valves; water is let in or out of the chamber by opening the gates just a crack at first, then gradually widening the opening as the levels begin to equalize.

On one section of the Trent waterway in Canada, a hydraulic elevator lifts the boat in a water-filled chamber, and on another a marine railway actually hauls the boat out to get her up and over a hill. Through passage on a waterway like the Trent is obviously limited to boats within the capacity of the railway, both as to length and weight.

Whistle Signals

Signals are prescribed for vessels approaching a lock, to be answered by the lockmaster. The signals vary in different areas, so familiarize yourself with the signals that apply to the waterway you are using. At many locks, VHF/FM radios are used for direct communications between vessels and the lockmaster.

On the Ohio River, vessels sound a long and a short blast on the whistle from a distance of not more than one mile from the lock. Approaching boats must wait for the lockmaster's signal before entering.

Where locks are in pairs (designated as landward and riverward), the lockmaster on the Ohio may also use an air horn to give directions as follows: one long, enter landward lock; two longs, enter riverward lock; one short, leave landward lock; two shorts, leave riverward lock.

On the Mississippi, signs on the river face of the guidewall warn small boats not to pass a certain point until signalled by the lock tender. Boatmen use a signal cord near this sign, if their own horns are not loud enough, to attract the attention of the lock attendant. Similar arrangements are found on the Okeechobee Waterway crossing Florida, and elsewhere.

Signal Lights

Traffic signal lights at the Ohio locks resemble those you find on city streets—red, amber, or yellow, and green vertically arranged. Flashing red warns: do not enter, stand clear. Flashing amber cautions: approach, but under full control. Flashing green is the go-ahead: all clear to enter.

On New York State canals, a fixed green is the signal to enter; a fixed red requires the vessel to wait.

Precedence at Locks

The Secretary of the Army has established an order of priority for the users of locks controlled by the Engineer Corps, as follows:

1. U.S. Military vessels
2. mail boats
3. commercial passenger boats
4. commercial tows
5. commercial fishermen
6. noncommercial boats

In the descending order of precedence, the lockmaster also takes into account whether vessels of the same priority are arriving at landward or riverward locks (if locks are paired), and whether they are bound upstream or downstream.

Recreational boats *may*, at the discretion of the lockmaster, be locked through with commercial vessels if a safe distance can be maintained between them, and if the commercial vessels are *not* carrying petroleum products or other hazardous substances.

Locking Procedures

The concrete walls of locks are usually rough and dirty. Some older locks have metal-sheathed inside surfaces, but most are hard on small boats so keep your fenders ready. Ordinary cylindrical fenders pick up dirt and roll on the wall to smear your topsides. Instead, you should use fender boards consisting of a plank (generally 2 by 6 inches, several feet long) suspended horizontally outside the usual fenders hung vertically. They will normally work well amidships or where a boat's sides are reasonably straight.

Bags of hay have the same objection as cylindrical fenders, except on heavily flared bows and at the edge of the deck where they flatten down and work fairly well. Auto tires wrapped in burlap would be ideal except that their use is illegal in most canals (if they came adrift they would sink and probably foul the lock's valves or gates).

As you can't be sure which side of the next lock you will be using, it's wise to have duplicate fender systems for each side.

Entering and Leaving

It is imperative to enter locks, and leave them, at a slow speed with your craft under full control—especially when locking through with other boats. Sometimes, especially on fleet cruises, you will need to tie up two abreast at each lock wall. This is entirely practicable if all boats are intelligently handled.

Occasionally you will hear about boats being tossed about as water boils into the lock from open valves. Be alert for this possibility, but lock tenders on our inland waterways are almost always careful to control the rate of inflow to minimize any turbulence in the lock chamber, and you need not fear the locking process for this reason. With a light boat, however, use extra caution when locking through in company with large commercial vessels. A boat directly astern of a tug, for example, can be tossed around when the tug leaves and its big propeller starts to kick out its wash astern.

Approaching a Lock From Upstream

Be especially cautious when approaching a lock from the upstream side; follow the marked channel closely. Some years ago, a cruiser bound down the Hudson River missed the lock at Troy, which lies far toward the east bank, and went *over* the dam. Looking down a river from above a dam it is sometimes virtually impossible to see any break in the water's surface, yet there is no chance for such an accident if a skipper pays attention to the buoys as on his chart, or to the general configuration of the dam and locks if there are no buoys.

Actually there are some circumstances where it *is* proper to go *over* a dam. The Ohio River has a special type of dam that has a lock chamber on one side and "bear traps" on the other. Between them, there are movable wickets that can be held in an upright position during low water stages, and lowered at times of high water. When the wickets are up, vessels use the lock. With the wickets down, at high water, vessels run through the navigable

Figure 911 **Locks provide the means for boats to go "uphill" and "downhill" to waterways at different levels. Lock gates or valves are opened or closed to let water flow in or out, and action is controlled to minimize turbulence in the lock. On some waterways, locking through on a Sunday—in heavy traffic—requires the use of many fenders on each boat.**

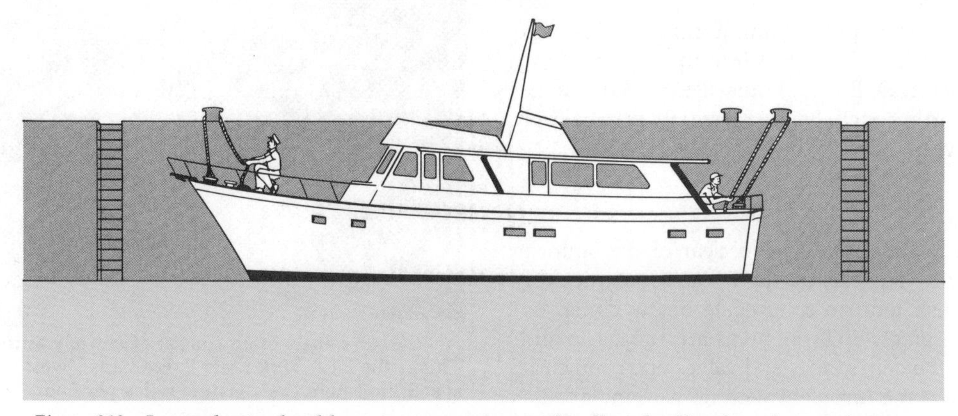

Figure 912 **Larger boats should use two separate mooring lines leading from bow and stern to separate posts on the lock wall; these are paid out or taken in as water level changes. On small boats, or larger ones with only one person aboard, a long single line may be fastened at one end of the boat with a bight around a mooring post at the top of the lock wall. The free end is taken in, or paid out, around a cleat at the other end of the boat as the water level changes.**

pass, over the dam without locking. By day, it is important to watch the bulletin boards at the locks to know whether the lock or pass is to be used. At night, control lights are shown at the guide walls. When the navigable pass is being used, a gauge reading on the powerhouse board shows the depth of water over the pass sill at the dam; the figures are in red on a white background, preceded by the word "Pass."

Have Enough Line

Another essential in locking is adequate line. Its diameter obviously depends on the size of your boat; and its length on the depth of the locks. Lines for locking through can be slightly smaller in diameter than your normal mooring lines; ½-inch manila is often used for boats in the 35- to 50-foot (10.7-15.2 m) range, ⅜-inch for smaller boats. (Manila is preferable to synthetic material because of its lower cost; locking lines will soon get too dirty for further use and must be discarded.)

In general, bow and stern lines should be at least twice the depth of the deepest lock that you plan to go through. This length permits running the line around a bollard on the top of the lock wall and back to your boat. Then at the lower level when you are ready to cast off you can turn loose one end and haul in on the other without assistance from above (which you are not likely to get); see Figure 912.

Use of Ladders and Other Methods

Lock walls often have ladders recessed into them, should someone be unlucky enough to fall in. On some canals, small boats may follow a ladder up or down, rung by rung, holding on with boathooks or short lengths of line. On other canals, rules do not permit this and the ladders must be kept clear for emergency use.

In addition to bollards at the top, some locks have posts recessed in the walls at intervals in a vertical line. Locking up, you can transfer your lines successively from lower posts to higher ones as you rise in this chamber.

In some newer locks, floating mooring posts built into the side walls move up and down with the water level. This makes for the easiest locking through of all, because you can use relatively short lines and they need no adjustment as your boat moves up or down.

Tend Your Lines Carefully

Whether water level is rising or falling, tend your lock lines carefully at all times. You will need one crewman forward and another aft; with only two people aboard, the helmsman can usually tend whichever line is nearest to him, generally the stern line. It's *extremely* dangerous to make fast to a bollard above and then secure the line to a bitt or cleat on the boat. If the water level drops unexpectedly, the boat can be "hung up" and seriously damaged, with possible injuries to those on board.

In some locks, coming up, special caution is required at the top of the lift. The boat may have been adequately fendered against the lock wall, but if lock chamber is nearly full the boat's gunwale may be above the wall and the topsides are then unprotected, particularly at a bow with flare.

Figure 913 **The easiest locking through is accomplished when the mooring posts are recessed into the chamber walls, and float up and down with water level changes. With this system, mooring lines are not adjusted during the locking process.**

The Army Engineers publish an informative illustrated leaflet entitled *Locking Through.* Write to the U.S. Army Engineer District, P.O. Box 59, Louisville, KY 40101 to get a copy. It sets forth, concisely, the essentials as they apply to the Ohio River.

Safety First

Recreational boatmen *must* keep clear of the spillway area below dams. The fishing there may be particularly good because fish tend to congregate below dams, but the hazards are great. Spillway areas are subject to sudden changes as the dam's gates and valves open and close; placid waters become turbulent with no warning at all as a heavy stream of water boils up unexpectedly from beneath the surface. Heed the warning signs and stay out of spillway areas.

Artificial Land Cuts

Not all waterways are merely improved versions of natural rivers and lakes. Artificial land cuts are often needed to interconnect navigable waterways and provide continuous passage between major bodies of water. Lake Champlain is accessible from the upper reaches of the Hudson River only because of a 24-mile (38.6 km) completely artificial waterway cut into the land. A dredged land cut is also found at the western end of the New York State Barge Canal, to connect Lake Erie with rivers in the middle of the state.

Narrow land cuts pose problems of their own. A typical cross section of a dredged canal might have a surface width of, for example, 125 feet (38.1 m), but a bottom width of only 75 feet (22.9 m) with a depth of 12 feet (3.7 m). If normal cruising speeds were maintained through such portions of a canal, the bank would be quickly washed down into the channel. In artificial waterways, therefore, speed limits are rigidly enforced. In New York State canals, the limit is 6 mph (9.7 km/h) in the land cuts and 10 (16.1 km/h) mph in the canalized river and lake sections. A telephone network between the various locks enables lock tenders to know when a boat is due; speeders are quite easily caught! Some sets of locks also operate on time schedules and it serves no purpose to go too fast, only to wait at the next lock.

Some land sections are cut through solid rock, perhaps only 100 feet (30.5 m) or less in width. This makes a virtual trough with ragged walls of blasted rock. In a confined channel such as this, the wakes of a group of boats may combine and reinforce each other in a synchronized pattern, building up out of all proportion to the wakes of each boat individually. Unless speed is held down as required by the circumstances, regardless of the legal limit, small boats can get out of control, with the possibility of being thrown into the rock walls and suffering serious damage.

Figure 914 **Canals often consist of entirely artificial land cuts. This is the New York State Barge Canal, west of Rochester; speed limit here is six statute miles per hour.**

Regulations

On the Western Rivers, no special permission or clearance is necessary for passage through the locks; there are, however, regulations to be observed; you can get copies of these from the Engineer District Offices at Chicago or St. Louis.

When boats are anchored at night along the side of the New York State Barge Canal, authorities require the use of a white anchor light if on the canal's port side, but a *red* anchor light if on the starboard side.

The New York Department of Transportation publishes a pamphlet describing the routes, vertical clearances, regulations, and general safety information for its inland waterways. For copies and a free map of the New York Barge Canal and Connecting Waterways, write to the Waterways Maintenance Division, State Department of Transportation, 1220 Washington Avenue, Albany, NY 12232.

Lake Boating

The term "lake boating" has extreme limits. The water area involved can range from the smallest of natural or man-made lakes and reservoirs to the Great Lakes, really inland seas. Some lakes are "protected waters" at all times; others can produce waves hazardous even to ships of considerable size.

An inland boatman can cruise hundreds of miles and never be far from shelter in case of bad weather, but he must still treat the larger lakes with respect. Lake Superior, for example, is the largest body of fresh water in the world. Deep, with rock-lined coasts and subject to storms and fog as well, it can present a challenge to even the saltiest skipper. On Lake Erie, smaller and comparatively shallow, seas in a gale cannot shape up in the normal pattern of open ocean waves. Instead they are short and steep, frequently breaking in heavy squalls. Even Lake Oneida in mid-state New York on the Barge Canal route can build up seas that challenge an offshore boat.

Water Levels

Lake levels vary from year to year and show a seasonal rise and fall as well, low in the winter and high in the summer. Monthly bulletins published during the navigation season show current and projected levels of the

Great Lakes, plus average and record levels. More current information is contained in weekly reports released to newspapers and radio and television stations.

In larger lakes, a steady, strong wind will lower levels to windward and pile up water on the leeward side. Barometric pressure can also influence lake levels, sometimes causing a sudden and temporary, but drastic, change in water level on any portion of a lake; this phenomenon is called a SEICHE.

Lake Piloting

Many smaller lakes do not require "navigation" by the skipper and thus do not have charts. On the larger lakes, however, shore lines may be many miles long and you should keep track of where you are at all times. On the largest lakes you need to plot your courses, so will use a compass, make time-speed-distance calculations, and even take bearings and fixes.

Great Lakes Charts

Charts of the Great Lakes, published by the National Ocean Service, are excellent, covering features of the navigable water in great detail. They show depths of water (the lesser depths in blue tint), safe channels, submerged reefs and shoals, aids to navigation, adjacent shorelines with topographic features and landmarks, types of bottom, and much related information. Scales vary from as large as 1:2,500 to 1:20,000 for harbor charts to 1:400,000 or 1:500,000 for general charts of individual lakes. The smallest scale used is for the general chart of all the Great Lakes at 1:1,500,000. The projection used for these charts is basically polyconic, but a few are additionally published with a mercator projection; see Chapter 18.

Great Lakes charts have a compass rose showing both true and magnetic directions with the variation stated. In parts of the Great Lakes the variation is zero, but in other locations great deposits of iron ore in the earth produce strong local effects; this is most pronounced along the north shore of Lake Superior, where variation has been observed to change from 27° to 7° within a distance of 650 feet (200 meters)!

Figure 915 **The many dams constructed for power and flood control have created lakes ideal for inland boating. This is a typical scene showing recreational boating on Lake Powell.**

Many lake charts show principal routes between major ports, giving the course in degrees (true) and distance in statute miles. Comparative elevations are referred to mean tide as calculated at New York City. Local heights and depths are measured from an established datum for each lake.

The Great Lakes are partly Canadian waters, and excellent charts for these areas are available from the Canadian Hydrographic Survey, Ottawa.

Coast Pilot 6 (Great Lakes Pilot)

To supplement the information on Great Lakes charts, get a copy of the current edition of the *Coast Pilot* 6, published by the National Ocean Service. This most useful volume provides full descriptions of the waters charted, laws and regulations governing navigation, bridge clearances, signals for locks and bridges, dimensions and capacities of marine railways, and weather information. Keep your copy corrected from *Local Notices to Mariners* issued by the 9th Coast Guard District, Cleveland, Ohio.

Great Lakes Light List

The U.S. Coast Guard publishes the *Light List, Volume IV,* covering the Great Lakes; see page 393.

This *Light List* identifies Canadian buoys and lights by the letter "C"; except for minor differences in design, the buoyage systems used for Canada and the United States are the same.

Caution Along Some Shores

In some lakes, such as Lake Champlain, depth becomes a factor. Certain shores may be particularly inviting because of their scenic attractions, but they may have depths of several hundred feet running right up to sheer cliff shores. If your engine fails, your anchor will be of little use in such deep water, and if the breeze is toward those cliffs you can be blown against them.

Rules and Regulations

Many smaller inland lakes do not fall under federal jurisdiction and are regulated by the appropriate states or agencies. Others are "navigable waters of the United States" where federal rules and regulations apply. Check on the status of unfamiliar waters, if you have any doubts.

Rules of the Road

The U.S. Inland Navigational Rules apply to all vessels on waters subject to federal regulations. The Canadian side of the Great Lakes and other border waterways are of course subject to Canada's rules.

Pollution Control Regulations

Inland lakes are often used as sources of municipal water supplies, or have their water quality maintained for other reasons. In new waters, check for any special regulations regarding overboard discharges.

TURBO CAT 31

10

Techniques for rough going at sea; how to handle emergencies such as stranding, man overboard, fires, and explosions

SPECIAL SEAMANSHIP TECHNIQUES

While boatmen would all prefer to be out on the water in pleasant conditions, they recognize that sometimes there will be bad weather; they may run aground, or need to assist another vessel in trouble. They may well encounter emergencies such as man overboard, fire, or the need to abandon a sinking boat. Every skipper should be thoroughly familiar with the special seamanship techniques required to safely cope with these situations.

Boat Handling Under Adverse Conditions

Perhaps the most frequent test of a skipper's abilities comes in how he handles his boat in adverse weather conditions. Boat size has little bearing on seaworthiness, which is fixed more by design and construction. The average power cruiser or sailboat is fully seaworthy enough for normal conditions in the use *for which it is intended.* Which means: don't venture into waters or weather conditions clearly beyond what your boat was designed for. Remember also that just what a boat will do is governed to a great extent by the skill of its skipper; a good seaman might bring a poor boat through a blow that a novice could not weather even in a far more seaworthy one.

Rough Weather

"Rough weather" is a relative term, and what seems a terrible storm to the fair-weather man may be nothing more than a good breeze to an experienced weather-wise boatman.

On large, shallow bodies of water like Long Island's Great South Bay, Delaware Bay and River, or Lake Erie, even a moderate wind will cause an uncomfortable steep sea with crumbling crests. Offshore or in deeper inland water the same wind force might cause moderate seas, but the slow, rolling swells would be no menace to small craft.

Know Your Boat

Boat handling under adverse conditions is an individual matter, since no two boats are exactly alike in the same sea conditions. When the going gets heavy, each hull design reacts differently—and even individual boats of the same class may behave differently because of factors like load and trim.

Each skipper must learn about his own boat, to determine how best to apply the general principles covered in the following sections. Reading this book and taking courses are important first steps. You can learn basic good seamanship by absorbing facts and principles, but you'll have to pick up the rest by using your book and classroom learning on the spot when the wind blows.

Preparations for Rough Weather

In anticipation of high winds and rough seas, a prudent skipper takes certain precautions. No single list fits all boats or all weather conditions, but among the appropriate actions are these:

1. Secure all hatches; close all ports and windows.
2. Pump bilges dry and repeat as required. ("Free" water in bilges adversely affects a boat's stability.)
3. Secure all loose gear; put away small items and lash down the larger ones.
4. Break out life preservers and have everyone on board wear them *before* the situation worsens; *don't wait too long.*
5. Break out emergency gear that you might need—hand pumps or bailers, sea anchor, or drogue, etc.
6. Get a good check of your position if possible and update the plot on your chart.
7. Make plans for altering course to sheltered waters, if necessary.
8. Reassure your crew and guests; instruct them in what to do and not to do, and give them something to do to take their minds off the situation.

Head Seas

The average well-designed power cruiser or larger sailboat should give little difficulty when running generally into head seas. If the seas get too steep-sided or if you start to pound, slow down by easing the throttle or shortening sail. This gives the bow a chance to rise in meeting each wave instead of being driven hard into it.

Figure 1001 **Reefed, with lee rail awash, the yawl,** ***above,*** **is being driven hard in moderate seas. An ocean racer, this sailboat can weather almost any condition of wind and sea if the skipper and crew are in good shape.** ***Below:*** **Youngsters and non-swimmers should wear PFDs at all times; everyone should don them when the sea gets rough.**

Match Speed to Sea Conditions

If conditions get really bad, slow down until you're making bare headway, holding your bow at an angle of about 45 degrees to the swells. The more you reduce headway in meeting heavy seas, the less the strain on the hull.

Avoid Propeller "Racing"

You must reduce speed to avoid damaging the hull or powerplant if the seas lift the screw clear of the water and it "races." It sounds dangerous—and it may be. First, there's a rapidly increasing crescendo of sound as the engine winds up, then excessive vibration as the screw bites the water again. Don't panic—slow down and change your course till these effects are minimized. Keep headway so you can maneuver your boat readily. Experiment to find the speed best suited to the conditions.

Adjust Trim

You can swamp your boat if you drive her ahead too fast or if she is poorly trimmed. In a head sea, a vessel with too much weight forward will plunge rather than rise. Under the same conditions too much weight aft will cause her to fall off. The ideal speed will vary with different boats—experiment with *your* craft and find out her best riding speed.

Change the weight aboard if necessary. On outboards, shift your tanks and other heavy gear. In any boat, direct your passengers and crew to remain where you place them.

Meet Each Wave as It Comes

You can make reasonable progress by nursing the wheel—spotting the steep-sided combers coming in and varying your course, slowing or even stopping momentarily for the really big ones. If the person at the wheel can see clearly and act before dangerous conditions develop, he should weather moderate gales with little discomfort. *Make sure the most experienced person aboard acts as helmsman!*

In the Trough

If your course requires you to run broadside to the swells, bouncing from crest to trough and back up again, your boat may roll heavily, perhaps dangerously. In these conditions in a power boat it is best to run a series of "tacks" much like a sailboat.

Tacking Across the Troughs

Change course and take the wind and waves at a 45 degree angle, first broad on your bow and then broad on your quarter. You will make a zig-zag course in the right direction, with your boat in the trough only briefly while turning. With the wind broad on the bow the boat's behavior should be satisfactory; on the quarter, the motion may be less comfortable but at least it will be better than running in the trough. Make each tack as long as possible to minimize how often you must pass through the trough.

To turn sharply, allow your powerboat to lose headway for a few seconds, throw the wheel hard over, then suddenly apply power. She will turn quickly as a powerful stream of water strikes the rudder, kicking you to port or starboard, *without making any considerable headway*. You won't be broadside for more than a minimum length of time. This is particularly effective with single-screw boats. With twin-screw power-boats the engine on the side in the direction of the turn may be throttled back or even briefly reversed.

Running Before the Sea

If the swells are coming from directly behind you, running directly before them is alright if your boat's stern can be kept reasonably up to the seas without being thrown around off course. But in heavy seas a boat tends to rush down a slope from crest to trough, and, stern high, the propeller comes out of the water and races. The rudder also loses its grip, and the sea may take charge of the stern as the bow "digs in." At this stage, the boat may yaw so badly as to BROACH—to be thrown broadside, out of control—into the trough. Avoid broaching through every possible action. Unfortunately, modern powerboat design emphasizes beam at the stern so as to provide a large, comfortable cockpit or afterdeck—this width at the transom increases the tendency to yaw and possibly broach.

Reducing Yawing

Slowing down to let the swells pass under your boat usually reduces the tendency to yaw, or at least reduces the extent of yawing. While it is seldom necessary, you can consider towing a heavy line or drogue (see page 219) astern to help check your boat's speed and keep her running straight. Obviously the line must be carefully handled and not allowed to foul the propeller. Do not tow

Figure 1002 **Sharp lines at the stern of this double-ender make for easier handling in a following sea. The average motorboat, with its wide transom, is much more difficult to keep from yawing as heavy seas pass under the transom.**

soft-laid nylon lines which may unlay and cause HOCKLES (strand kinks).

Cutting down engine speed reduces strain on the motor caused by alternate stern-down laboring and stern-high racing.

Pitchpoling

The ordinary offshore swell is seldom troublesome when you are running before the seas, but the steep wind sea of some lakes and shallow bays makes steering difficult and reduced speed imperative. Excessive speed down a steep slope may cause a boat to PITCHPOLE, that is, drive her head under in the trough, tripping the bow, while the succeeding crest catches the stern and throws her end over end. When the going is bad enough that there is risk of pitchpoling, keep the stern down and the bow light and buoyant, by shifting weight aft as necessary.

Shifting any considerable amount of weight aft will reduce a boat's tendency to yaw but too much might cause her to be POOPED by a following sea breaking into the cockpit. *Do everything in moderation, not in excess.* Adjust your boat's trim bit by bit rather than all at once, and see what makes her more stable.

Tacking Before the Seas

Use the tacking technique also when you want to avoid large swells directly astern.

Try a zig-zag track that puts the swells off your quarter, minimizing their effects—experiment with slightly different headings to find the most stable angle for your boat—but keep it under control to prevent a broach.

Running an Inlet

One of the worst places to be in violent weather is an inlet or narrow harbor entrance, where shoal water builds up treacherous surf that often cannot be seen from seaward. Inexperienced boatmen, nevertheless, often run for shelter rather than remain safe, if uncomfortable, at sea, because they lack confidence in themselves and their boats.

When offshore swells run into shallower water along the beach, they build up steep waves because of resistance from the bottom. Natural inlets on sandy beaches, unprotected by breakwaters, usually build up a bar across the mouth. When the swells reach the bar their form changes rapidly, they become short, steep-sided waves that tend to break where the water is shallowest.

Consider this when approaching from offshore. A few miles off, the sea may be relatively smooth while the inlet from seaward may not look as bad as it actually is. Breakers may run clear across the mouth, even in a buoyed channel.

Shoals shift so fast with moving sand that buoys do not always indicate the best water. Local boatmen often leave the buoyed channel and are guided by appearance of the sea, picking the best depth by the smoothest surface and absence of breakers. A stranger is handicapped here because he may not have knowledge of uncharted obstructions and so may not care to risk leaving the buoyed channel. He should thus have a local pilot if possible, or he might lay off or anchor until he can follow a local boat in.

If you must get through without local help, these suggestions may make things more comfortable: Don't run directly in but wait outside the bar until you have had a chance to watch the action of waves as they pile up at the most critical spot in the channel, which will be the shallowest. Usually they will come along in groups of three, sometimes more. The last sea will be bigger than the rest and by watching closely you can pick it out of the successive groups.

When you are ready to enter, stand off until a big one has broken or spent its force on the bar, and then run through behind it. Watch the water both ahead *and behind* your boat; control your speed and match it to that of the waves. An ebbing current builds up a worse sea on the bars than the flood does because the rush of water out works against and under the incoming swells. If the sea looks too bad on the ebb, it is better to keep off a few hours until the flood has had a chance to begin.

Departing through inlets is less hazardous than entering, as the boat is on the safe side of the dangerous area and usually has the option of staying there. If you do decide to go out you can spot dangerous areas more easily, and a boat heading into surf is more easily controlled than one running with the swells. On the other hand, the skipper of a boat outside an inlet may have to enter, and can only attempt it in the safest possible manner.

Heaving To

When conditions get so bad offshore that a boat cannot make headway and begins to take too much punishment, it is time to HEAVE TO, a maneuver whose execution varies for different vessels. Motor boats, both single and twin screw, will usually be most comfortable if brought around and kept head to the seas, or a few points off, using just

Figure 1003 **Boats that must cross a bar with breaking waves must avoid "pitchpoling"—being thrown end over end if caught driving down the face of a steep sea, burying the bow. This double-ended fisherman just misses being caught on the forward face of a breaker. This is no place for pleasure boats.**

enough power to make bare steerageway while conserving fuel.

Sailboats traditionally heave-to with the helm lashed downwind, to keep the boat headed up. A small, very strong STORM JIB is sheeted to windward to hold the bow just off the wind, while a STORM TRYSAIL is sheeted flat. This is a small, strong triangular sail with a low clew and a single sheet. A loose-footed sail, it is not bent to the boom, which is secured in its crutch. The jib-trysail combination balances the tendency of trysail and rudder to head the vessel into the wind against the effect of the jib to head her off, and the result is, ideally, that she lies 45 degrees from the wind while making very slow headway.

Lying Ahull This is the next step down as the wind increases. All sail is dropped and secured, the helm lashed to prevent damage to the rudder, and the vessel left to find her own way.

Sea Anchors

In extreme weather you may use a SEA ANCHOR, generally a heavy fabric cone or hemisphere with a hoop to keep it open at the mouth. Its leading edge has a bridle of a light lines leading to a fitting attached a heavier towing line. The apex of the cone (or the hemisphere's center) has a small, reinforced opening with a trip line, used to spill the bag and make it easier to haul the anchor back aboard. A sea anchor is not meant to go to the bottom and hold, but merely to present a drag that keeps the boat's head up within a few points of the wind as she drifts off to leeward. A sea anchor will usually float just beneath the water's surface.

Sea anchors may also be a series of plastic floats that fold compactly for storage and are said to be more effective than the traditional kind.

Whatever style of sea anchor you carry, test it in moderate weather a few times to make sure it is big enough to be effective. A small one may be easier to stow, but there's no point to having it if it won't do the job.

A close relative of the sea anchor is the DROGUE, which is towed astern to keep a craft from yawing extremely or broaching. A drogue has a larger vent opening to provide a lesser drag, because it is for use on a boat making way through the water.

Improvised Sea Anchors

In the absence of a regular sea anchor try any form of drag rigged from spars, planks, canvas, or other material at hand that will float just below the surface and keep your boat from lying in the trough.

Use of Oil With Sea Anchors

Sea anchors sometimes have an OIL CAN that permits oil to ooze out slowly and form a slick on the surface, thus preventing seas from breaking. The oil might also be distributed from a bag punctured with a few holes and stuffed with cotton waste saturated with oil. (See also the following section on the use of oil.)

Make sure this equipment—sea anchors, trip lines, oil bag lines, lead lines—is in good condition and not rotten from long disuse and stowage in the bottom of some locker.

Sea anchors work only if there is sufficient sea room because they permit a steady drift to leeward. When a vessel is driven down onto a lee shore (a shore to leeward of her), she must use her regular ground tackle to ride out a gale. It is imperative that you keep a constant watch to guard against dragging toward a lee shore. Use your engine to ease the strain during the worst of the blow, and give the anchor a long scope for its best chance to hold. Don't confuse "lee shore"—a shore onto which the wind is blowing, a dangerous shore—with being "in the lee" of an island or point of land, being on the sheltered, safer side.

Using Oil on Rough Water

Experienced seamen have long known the value of oil

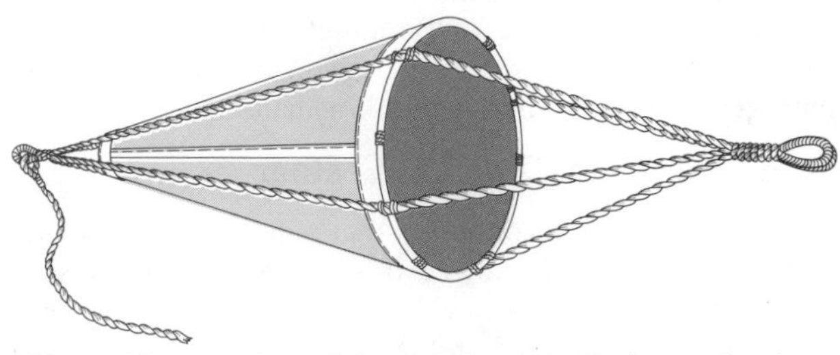

Figure 1004 **Tow line of a sea anchor is made fast to the ring at right; trip line is bent to the ring at the apex of the cone. This end has a small opening to let some of the water flow through.**

for modifying the effect of breaking seas. Oil is easily dispensed and quickly dispersed; the effect of even small amounts is significant. Here is a summary of information on its use:

1. On free waves, i.e. waves in deep water, its effect is greatest.

2. In a surf or in waves breaking on a bar, its effect is uncertain, as nothing can prevent the larger waves from breaking under such circumstances. Even here, however, it has some value.

3. The heaviest and thickest oils are most effective. Kerosene is of little use; but animal and vegetable oils and waste oil from the engines will all have a useful effect.

4. A small quantity of oil suffices, if applied so that it spreads to windward.

5. In cold water, the oil is thickened by the lower temperature and spreads more slowly, so will be less effective. The loss of effectiveness will vary with the type of oil used.

6. For a boat at sea, the best way to apply oil is to hang one or two small canvas bags over the side that can hold 1 to 2 gallons (3.8-7.6 liters) of oil; prick the bags with a sail needle to facilitate leakage of oil.

The position of these bags is determined by the circumstances. Running before the wind, hang them on either bow and allow them to tow in the water.

With the wind on the quarter, the effect seems to be less than in any other position, as the oil goes astern while the waves come up on the quarter.

Lying-to, hang them from the weather bow and another position farther aft, using sufficient line for them to draw to windward while your boat drifts.

7. To cross a bar with a flooding tidal current, pour oil overboard and allow it to float in ahead of your boat, which then follows with another bag towing astern. The oil is less dependable in this situation, however.

Entering across a bar with the tidal current ebbing, it is probably useless to try oil.

8. To approach a stranded boat, pour oil overboard to windward of her before going alongside. The effect in this case will depend upon the set of current and depth of water.

9. For a boat riding from a sea anchor in rough water fasten the oil bag to a line rove through a block on the sea anchor. The oil is thus diffused well ahead of the boat and the bag can be hauled on board for refilling.

10. Remember that the purpose of using oil is to spread a calming slick to windward and around your boat, so that she stays in the slick. If the oil goes off to leeward or astern, it is of no use.

Note: Discharging oil onto the water is a technical violation of Coast Guard anti-pollution regulations even though only a small quantity is used. Use oil only in an emergency situation involving immediate danger.

Seamanship in "Thick" Weather

Another "adverse conditon" that requires special skills is "thick" weather—conditions of reduced visibility, caused by fog, heavy rain or snow, or haze. Fog is probably the most often encountered and most severe.

Avoiding Collisions

Seamanship in fog is primarily a matter of avoiding collisions. Piloting and position determination, the legal requirements for sounding fog signals, and the meteorological aspects of fog are all covered elsewhere in this book. Here we will consider only the aspects of boat handling and safety.

Figure 1005 **It is best not to go out in foggy weather, but if you get caught in fog, you need to hear other boats, and be heard, in time to avert collision. Keep a sharp lookout for boats or any other hazards that might loom up suddenly.**

The primary needs of safety in conditions of reduced visibility are to *see and be seen*—to *hear and be heard.* The wise skipper takes every possible action to see or otherwise detect other boats and hazards, and simultaneously takes all steps to make his presence known to others.

Reduce Speed

You need of course to detect other vessels by sight, sound, or radar early enough to take proper action to avoid collision. Both the Inland and the International Rules of the Road require reduced speed for vessels in low visibility—a "safe speed adapted to the prevailing circumstances in conditions"; see page 93.

It is best to be able to stop short in time, rather than resort to violent evasive maneuvers to avoid a collision. The Rules of the Road require that, except where it has been determined that a risk of collision does not exist (by radar plot, perhaps), any vessel which hears another's fog signal apparently forward of her beam must "reduce her speed to the minimum at which she can be kept on course. She shall, if necessary, take all way off, and, in any event, navigate with extreme caution until danger of collision is over."

Lookouts

Equally important with a reduction in speed is posting LOOKOUTS. This is a requirement of the Rules of the Road, but it is also common sense. Most modern motor and sail boat designs place the helmsman aft of fairly far aft, where he is *not* an effective lookout, so you will probably need one or two additional people aboard as lookouts in thick weather.

Look . . . and Listen

Despite the "look" in "lookout," such a person is as much for *listening* as he is for seeing. A person assigned as a lookout should have this duty as his *sole responsibility* while on watch. A skipper should certainly post a lookout as far forward as possible when in fog, and, if the helmsman is at inside controls, another lookout for the aft sector is desirable. Lookouts should be relieved as often as necessary to ensure their alertness; if the crew is small, an exchange of bow and stern duties will provide some change in position and relief from monotony. If there are enough people on board, a double lookout forward is not wasted manpower; but the two should take care not to distract each other.

A bow lookout should keep alert for other vessels, listen for sound signals from aids to navigation, and watch for hazards like rocks and piles, breakers, and buoys. Note that in thick weather, aids to navigation without audible signals can indeed become hazards. A lookout aft should watch primarily for overtaking vessels, but he may also hear fog signals missed by his counterpart on the bow.

The transmission of sound in fog is uncertain and tricky. The sound may seem to come from directions other than the true source,and it may not be heard at all at otherwise normal ranges. See the discussion of fog signals in Chapter 16.

Stop Your Engine

When underway in fog in a boat under power, slow your engines to idle or shut them off entirely, at intervals, to listen for fog signals of other vessels and of aids to navigation. This is not a legal requirement of the Rules of the Road but it is an excellent, practical action.

In these intervals keep silence on the boat so you can hear even the faintest signal. The listening periods should be at least two minutes to conform with the legally required maximum intervals between the sounding of fog signals. Don't forget to keep sounding your own signal during the listening period—you may get an answer from close by!

When proceeding in fog at a moderate speed, slow or stop your engines immediately any time your lookout indicates he has heard something. The lookout can then have the most favorable conditions for verifying and identifying what he believes he has heard.

Radar and Radar Reflectors

Radar comes into its greatest value in thick weather conditions. If you have radar turn it on and use it, but *not* at the expense of posting a proper lookout.

Even if you do not have radar, you should carry a passive radar reflector (see page 547); this is the time to open it and hoist it as high as possible.

Cruising With Other Boats

If you are cruising with other boats and fog closes in, you may be able to take advantage of a procedure used by wartime convoys. Tie onto the end of a long, light line some object that will float and make a wake as it moves through the water. A life ring or a glass or plastic bottle with a built-in handle will do quite well. The object is towed astern with the boats traveling in single file, one object behind each craft except the last, and each bow lookout except the first can keep it in sight even though he cannot see the vessel ahead, or even hear her fog signal.

Anchoring and Laying-to

If the weather, depth of water, and other conditions are favorable, consider anchoring rather than proceeding through conditions of low visibility. Do not anchor in a heavily traveled channel or traffic lane, of course.

If you cannot anchor, then perhaps laying-to—being underway with little or no way on—may be safer than proceeding at even a much reduced speed.

Remember that different fog signals are required when you are underway, with or without way on, and when you are at anchor. By all means, sound the proper fog signal and keep your lookouts posted to *look* and *listen* for other craft and hazards.

Stranding, Assisting, and Towing

It is an unwritten law of the sea that a boatman should always try to render assistance to a vessel in need of aid. This is one of the primary functions of the Coast Guard, of course, but there are plenty of occasions when timely help by a fellow boatman can save hours of labor later after the tide has fallen or wind and sea have had a chance to pick up.

As often as not, giving assistance means getting a tow line to another skipper to get him out of a position of temporary embarrassment, or perhaps to get him to a Coast Guard station or back to port in case of gear failure. The situation can also be the other way around; *you* may go aground, or a balky motor may force you to ask a tow from a passing boat.

In either case, you should know what to do and why. Thus the problems of stranding, and their solutions, will be considered from both the viewpoint of being in need of assistance, and of being the one who must render aid to another vessel.

Stranding

Simple stranding, running aground, is more often an inconvenience than a danger and, with a little know how and some fast work, the period of stranding may be but a matter of minutes.

If grounding happens in a strange harbor, chances are you have been feeling your way along and so have just touched bottom lightly. You should be off again with little difficulty if your immediate actions do not put you aground more firmly.

Right and Wrong Actions

The first instinctive act on going aground is to gun the engine into reverse in an effort to pull off; this may be the one thing that you should *not* do.

In tidal waters the first thing to consider is, of course, the stage of the tide.

If the tide is rising, and the sea quiet enough so that

Figure 1006 **If you go aground on a falling tide, the boat may be left completely "high and dry." If so, try to brace the boat so that it will stay upright; it will be easier to refloat.**

the hull is not pounding, time is working for you, and whatever you do to assist yourself will be much more effective with time. If you grounded on a falling tide, you must work quickly and do exactly the right things or you will be fast for several hours or more.

About the only thing you know offhand about the grounding is the shape of your boat's hull, its point of greatest draft, and thus the part most apt to be touching. If the hull tends to swing to the action of wind or waves, the point about which it pivots is the part grounded.

Cautions in Getting Off

Consider the type of bottom immediately. If it is sandy and you reverse hard, you may wash a quantity of sand from astern and throw it directly under the keel, bedding the boat down more firmly. Take care also not to pump sand or mud into the engine through the engine's water intake.

If the bottom is rocky and you insist on trying to reverse off, you may drag the hull and do more damage than with the original grounding.

Also, if grounded forward, remember that reversing a single-screw boat with a right-hand wheel may swing the stern to port and thus the hull broadside onto exposed pinnacles or to a greater contact with a soft bottom.

How to Use a Kedge

The one *right* thing to do immediately after grounding is to take out an anchor and set it firmly; this is called a KEDGE; the act of using it is KEDGING.

Unless your boat has really been driven on, the service anchor should be heavy enough. Put the anchor and the line in the dinghy, make the bitter end fast to the boat's stern cleats and row out as far as possible, letting the line run from the stern of the dinghy as it uncoils. Taking the line out this way makes the oarsman's task much easier than if he drags the line with him through the water.

If you do not carry a dinghy you can often swim out with an anchor, providing sea and weather conditions do not make it hazardous to go overboard. Use life preservers or buoyant cushions—one or two of either—to support the anchor out to where you wish to set it. Be sure to wear a life jacket or buoyant vest yourself, to save your energy for the work.

If there is *no* other way to get a kedge anchor out, you may consider throwing it out as far as possible. Although this is contrary to basic good anchoring practice, getting a kedge set is so important that it warrants the technique. You may need to throw several times before you get it set firmly.

When setting out the kedge, consider the sideways turning effects of a reversing single screw and, unless the boat has twin screws, set the kedge at a compensating angle from the stern. If the propeller is right-hand, set the anchor slightly to starboard of the stern. This will give two desirable effects. When pulling the kedge line while in reverse, the boat will tend to back in a straight line. When used alternately, first pulling on the line and then giving a short surge with the reverse, the resulting wiggling action of stern and keel can be a definite help in starting the boat moving.

Getting Added Pulling Power

If you have a couple of double-sheave blocks and a length of suitable line on board, make up a HANDY-BILLY or FALL and fasten it to the kedge line. Then you can really pull! A handy-billy should be part of a boat's regular equipment.

During the entire period of grounding, keep the kedge-line taut. The boat may yield suddenly to that continued pull, especially if a passing boat throws a wake that helps to lift the keel off the bottom.

Two kedges set out at an acute angle from either side of the stern and pulled upon alternately may give the stern wiggle that will help you work clear.

If the bottom is sandy, that same pull with the propeller going ahead may wash some sand away from under the keel, with the desired result. If the kedge line is kept taut, try the maneuver with caution.

Move your crew and passengers quickly from side to side to roll the boat and make the keel work in the bottom. If you have spars, swing the booms outboard and put

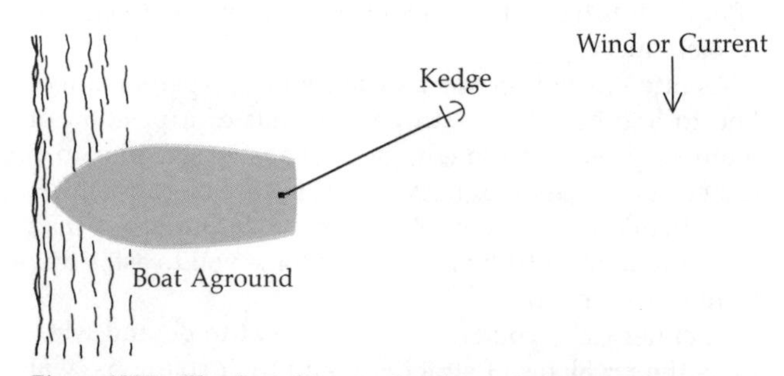

Figure 1007 **The first thing to do after going aground is to get out a kedge anchor to keep from being driven further aground. It also may provide a means of pulling free as waves or the wake from another craft lift your boat.**

people on them to heave the boat down, thus raising the keel line. Shift ballast or heavy objects from over the portion grounded to lighten that section and if you can, remove internal weight by loading it into the dinghy or by taking it ashore.

When to Stay Aground

All of the above is based on the assumption that the boat is not holed. If she *is* holed you may be far better off where she lies than you would be if she were in deep water again. If she is badly stove, you may want to take an anchor ashore to hold her on or pull her further up until temporary repairs can be made. As the tide falls the damaged hull may be exposed far enough to allow some ouside patching, if you have something aboard to patch with. A piece of canvas, cushions, and bedding can all be used for temporary hull patches.

What to Do While Waiting

While waiting for the tide to rise or for the Coast Guard to come, do not sit idle. Take soundings all around you. A swing of the stern to starboard or port may do more for you than any amount of straight backward pulling; soundings will tell you of any additional depth that can help you out.

If another boat is present, she may be able to help you even if she can not pull. Have the other skipper run his boat back and forth to make as much wake as is safely possible. His wash may lift your boat just enough to permit you to back clear.

If your boat is going to be left high and dry with a falling tide, keep an eye on her layover condition. If there is anything to get a line to, even another kedge, you can make her lie over on whichever side you choose as she loses buoyancy. If she is deep and narrow she may need some assistance in standing up again, particularly if she lies over in soft mud. Both the suction of the mud and her own deadweight will work against her in that case.

If the hull is unharmed, and you are left with a falling tide to sit out for a few hours, you might as well be philosophical about it—get over the side and make good use of the time. Undoubtedly you would prefer to do it under happier circumstances, but this may be an opportunity to make a good check of the bottom of your boat, or do any one of a number of little jobs that you could not do otherwise, short of a haul-out.

Assisting

If you are not stranded but able to help another boat that is, you must still know what to do and what not to do.

Take care that your boat does not join the other craft in its trouble! Consider the draft of the stranded vessel relative to yours. Consider the size and weight of the grounded boat relative to the power of your engines—don't tackle an impossible job; there are other ways of rendering assistance.

Getting a Line Over

It may seem easiest to bring a line in to a stranded boat by coming in under engine power, bow on, passing the line and then backing out again, but do not try it until you are sure that there is water enough for you. Make sure also that your boat backs well, without too much stern crabbing due to the action of the reversed screw. Wind and current direction will greatly affect the success of this maneuver. If conditions tend to swing your boat broadside to the shallows as she backs, pass the line in some other manner.

Try backing in, with wind or current compensating for the reversed screw, keeping your boat straight and leaving her bow headed out. In any case, after the line is passed and made fast, do the actual pulling with your engines going ahead, to get full power into the pull.

If a close approach seems unwise, you might drop the kedge anchor or anchors for the stranded boat, and then send the line over in a dinghy or else buoy it and float it over.

Making the Pull

If wind or current, or both, are broadside to the direction of the pull, keep your boat anchored even while pull-

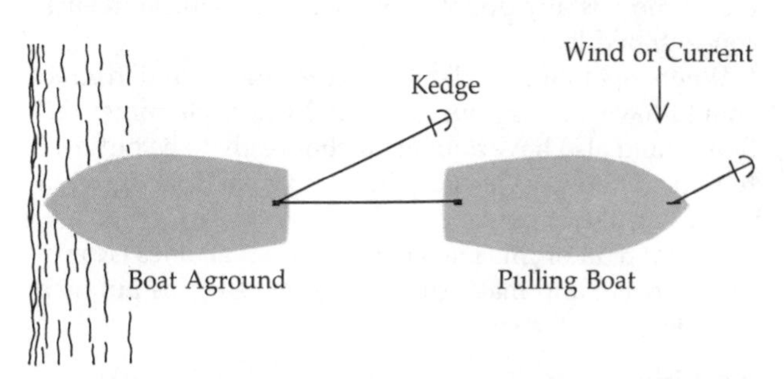

Figure 1008 **Most small boats pull with the tow line cleated astern, as shown *above*, which restricts maneuverability. For safe control, put out a bow anchor and take up on its line as the stranded boat comes free. This arrangement prevents the pulling boat from being carried into shoal water. It is generally desirable to pass a line from the assisting boat, *below*, to the one being aided. If the weather is rough, or for any reason the stranded boat cannot be approached closely, the rescuing vessel can float a line over using a life ring or a buoyant cushion to support it.**

ing and keep a strain on your anchor line. Otherwise, as soon as your boat takes the pulling strain (particularly if the line is fast to her stern) she will have lost her own maneuverability, and will gradually make leeway, which could eventually put her aground broadside.

If you are pulling with your boat underway, secure the tow line well forward of your stern so that while hauling, your boat can angle into the wind and current and still hold her position.

Tremendous strains can be set up, particularly on the stranded boat, by this sort of action, even to the point of carrying away whatever the tow-line is made fast to. Ordinary recreational craft are *not* designed as tugboats. The cleats or bitts available for making the line fast may well not be strong enough for such a strain and they are probably not located advantageously for such work. It is far better to run a bridle around the whole hull and pull against this bridle rather than to risk damaging some part of the stern by such straining. Use a bridle on your boat too, if there is any doubt of her ability to withstand such concentrated loads.

When operating in limited areas, the stranded craft should have a kedge out for control when she comes off. She should also have another anchor ready to be put over the side if necessary; to keep her from going back aground if she is without power.

Through all of this maneuvering, keep all lines clear of the propeller and make sure no sudden surge is put on a slack line.

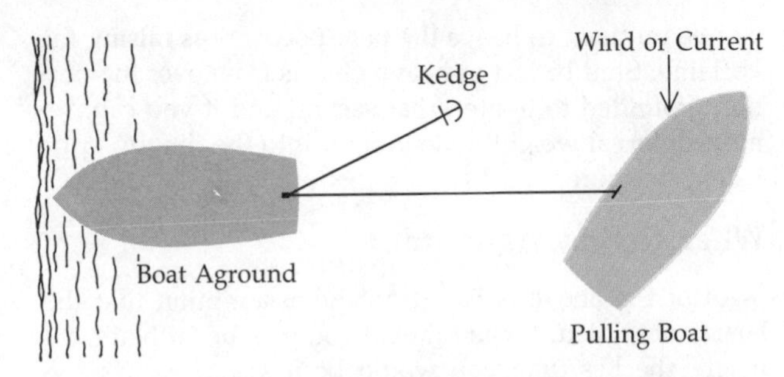

Figure 1009 **Above: to aid in maneuverability if a bow anchor cannot be set, make the tow line fast to a cleat forward of the stern on the upwind or up-current side of the pulling boat. Be sure the cleat is capable of taking the heavy strain.** ***Below:*** **The best procedure is to put a bridle around the hull or superstructure of both boats in order to distribute strain over as wide an area as possible. Be sure to pad pressure points to guard against chafing or scarring.**

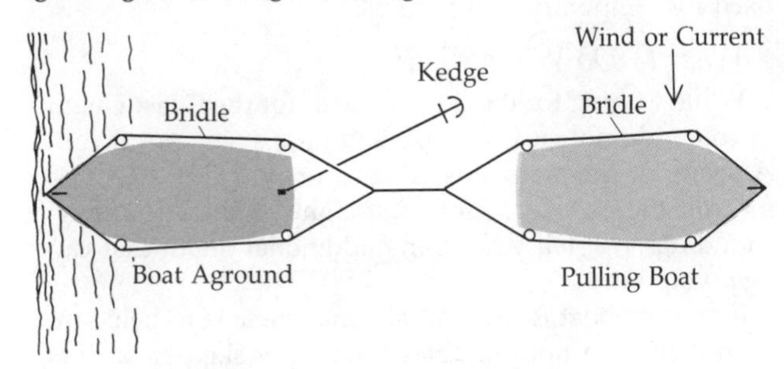

Towing

At some time you will probably need to take another vessel in tow. In good weather with no sea running the problem is fairly easy. It involves little more than maneuvering your boat into position forward of the other boat, and passing it a tow line.

Generally speaking, the towing boat should pass her tow line to the other craft. You may want to send over a light line first, and use that to haul over the actual towing line. A plastic water ski tow line that floats is excellent.

When approaching a boat that is dead in the water to pass it a line, do not be dramatic and try to run in too close if there is any kind of sea running. Just buoy a long line with several life preservers, tow it astern and take a turn about the stern of the disabled vessel, but don't foul its propeller in so doing. The crew of the vessel being assisted can pick the line up with a boathook from the cockpit with far less fuss than by any heave-and-catch method.

A boat's forward bitts are usually the more rugged, and so the line may be made fast at that point; then, with someone at the wheel to assist by steering, there is little more to do than to have an anchor ready to drop if the tow line parts or is cast loose for any reason.

Towing Lines

Three-strand twisted nylon has excessive stretch and dangerous snap-back action if broken under load; it should not be used if it can be avoided. Polypropylene line floats and is a highly visible bright yellow, but it has little elasticity and shock loads are heavily transferred to fittings on both the towed and towing craft. Further, it has less strength (requiring larger sizes), and is stiff, making it difficult to stow and handle; it is also particularly subject to abrasion damage. This line, in sizes up to ½-inch, is thus suitable only for light towing loads in protected waters.

The preferred line for towing is double-braided nylon. It has sufficient elasticity to cushion shock loads, but not so much as to create a snap-back hazard. Braid-on-braid line is stronger than three-strand twisted nylon of the same size and will not kink. Its disadvantage is that it does not float and must be watched to avoid entanglement in the towing vessel's propeller. It is the most expensive type of line, but for moderate- and heavy-duty towing it is well worth the difference.

Handling the Towing Boat

The worst possible place to make the tow line fast is to the stern of the towing boat, because the pull of the tow prevents the stern from swinging properly in response to rudder action, limiting the boat's maneuverability. The tow-line should be made fast as far forward as practicable, as in tug and towboat practice. If there is no suitable place forward, make a bridle from the forward bitts, running around the superstructure to a point in the forward part of the cockpit. Such a bridle will have to be wrapped with chafing gear wherever it bears on the superstructure or any corners, and even then it may cause some chafing of the finish.

Cautions in Towing

Secure the tow line so that it can be cast loose if necessary or, failing that, have a knife or hatchet ready to cut it. This line is a potential danger to anyone near if it should break and come whipping forward. Twisted nylon acts like a huge rubber band when it breaks, and there have been some bad accidents. Never stand near or in line with a highly strained tow line, and keep a wary eye out at all times.

If for any reason you must come near a burning vessel to tow it (for instance, to prevent it from endangering other boats or property) approach from windward so the flames are blowing away from you. Your light anchor with its length of chain thrown into the cockpit or through a window of the burning boat could act as a good grappling hook.

In any towing situation, never have people fend off the other vessel with hands or feet; even the smallest boats coming together under these conditions can cause broken bones or severed fingers. With large vessels the risk is the loss of a whole limb, or worse.

Never allow *anyone* to hold a towing line while towing another vessel, regardless of its size. They could receive badly torn tendons and muscles, they might lose the tow line over the side, or they could be dragged overboard.

Starting the Tow

Start off easy! Don't try to dig up the whole ocean, and merely end up with a lot of cavitation and vibration. A steady pull at a reasonable speed will get you to your destination with far less strain on boats, lines, and crews.

Towing Principles

When towing, keep the boats "in step" by adjusting the tow line length to keep both boats on the crest or in the trough of seas at the same time. Sometimes, as with a confused sea, this may not be possible, but the idea is to prevent a situation where the tow is shouldering up against the back of one sea, presenting maximum resistance, while the towing boat is trying to run down the forward slope of another sea. Then when this condition is reversed, the tow alternately runs ahead briefly and then surges back on the tow line with a heavy strain. If there is any degree of uniformity to the waves, the strain on the tow line will be minimized by adjusting it to the proper length.

As the tow gets into protected, quiet waters, shorten up on the line to allow better handling in close quarters. Swing as wide as possible around buoys and channel turns, so that the tow has room to follow.

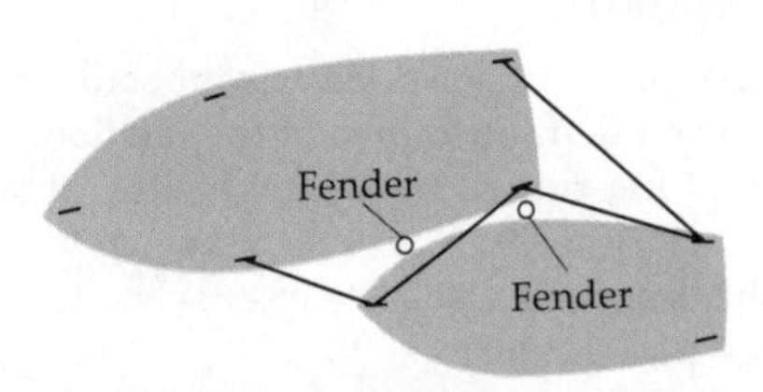

Figure 1010 **How springs are used when a boat takes a larger vessel in tow alongside: fenders are rigged at points of contact, and springs are made up with no slack in them. Both boats respond as a unit to the towing boat's rudder action.**

A small boat in tow should be trimmed a little by the stern; trimming by the head causes her to yaw. In a seaway this condition is aggravated and it is increasingly important to keep the bow relatively light.

It is easy for a larger boat to tow a smaller vessel too fast, causing it to yaw and capsize. Always tow at a moderate speed and make full allowance for adverse conditions of wind and waves.

In smooth water, motorboats may borrow an idea from tugs, which often take their tow alongside in harbor or sheltered waters, for better maneuverability. The towing boat should make fast on the other craft's quarter.

When Not to Tow

Towing can be a dangerous undertaking, as well as an expensive one, if not properly done. If you are not equipped for the job, stand by the disabled vessel. You may be able to put a line across and assist by keeping the other craft's bow at a proper angle to the sea until help comes.

Call the Coast Guard or other salvage agency and turn the job over to them when they arrive. Don't try to be a hero, as you are more than likely neither trained nor experienced in this type of work. No boat is worth a life!

Emergency Procedures

The average person tends to believe that emergencies and accidents will not happen to him. The seasoned skipper reasons otherwise. It is not likely to happen to him, but it *could,* and so he prepares himself just in case it does. He also prepares his crew and, to a lesser extent, any guests aboard his boat.

Value of Planning

An emergency is, by definition, not a planned event, but actions to cope with foreseeable ones can be rehearsed. Any emergency will be much more easily countered if there is an immediate routine for dealing with it.

Actual emergency circumstances of course will vary somewhat from the rehearsed ones, but a basic approach and assignment of duties are nevertheless a great help.

Need for Drills and Practice

The best plans are of little value if they have not been tried out and evaluated to the best extent possible.

When the plans have been tried out and modified as needed, the next step is to have periodic drills. Don't do it so often that it becomes boring or irksome, but do it at the beginning of each boating season, and once again on a staggered schedule later in the season. Make every effort to have the drills enjoyable without reducing their business-like nature.

Guests Aboard

Crew training naturally extends to preparation for emergencies. But do not overlook your guests; on most boats, they are at least "temporary crew." You may not be able to train them in drills, but they should certainly be shown where the life preservers are, and how to use them. Do this in a casual, relaxed manner so that they are informed, but not alarmed.

If the guests will be aboard for an extended cruise, expand their indoctrination to such items as the location and operation of fire extinguishers, and bilge pumps.

Man Overboard

Preliminary planning for a Man Overboard emergency may include some of the following preparations:

1. Equip all life preservers with whistles, attached by short pieces of light line, and encourage crew members to carry whistles in their pockets or attached to a lanyard along with a pocket knife.
2. Make up a flag float. These are required equipment on offshore racing sailboats, and they consist of a fiberglass or bamboo pole six to eight feet long with a flag at one end, a buoy near the other, and a lead ballast weight below the buoy to hold the flag aloft. Ideally the float should be permanently attached to a whistle-equipped life preserver, a water light, and a small drogue to prevent a strong wind from carrying the assembly downwind faster than a man can swim to catch it. In any case, have something prepared in advance to throw to a person in the water that can easily be seen by him and by you in good weather, allows him to make a loud noise to guide you in poor visibility, and can show a light—preferably the high-intensity strobe type—at night.
3. Carry an old magazine with many pages as a standard piece of rescue gear. Keep it in a spot known to the crew. A paper trail in the water can often lead you back to the scene of the accident when all else fails.
4. Think out the problems of getting a possibly unconscious person to the deck from the water, and carry the necessary gear aboard at all times.

Immediate Actions

At the cry *"Man overboard!"* fast—*immediate*—action is vital; every second counts, particularly at night, in heavy weather, or in cold water. The actions of the helmsman and other crew members should be automatic and without many shouted orders.

It helps if the Man Overboard cry is supplemented by an indication of which side he went over—port or starboard, or left or right. This focuses attention of those who did not see the accident and guides their actions.

Action by Helmsman

It is customary for the person at the helm to cut the throttle immediately, put the engines in neutral, and swing the stern to the side *away* from the person in the water to

Figure 1011 **Arrangements on offshore racing boats include horseshoe type throwable PFDs, with whistle, light, and flag float all attached so a person can be seen, and heard, in almost all conditions. Regular drills should be part of your routine, so everyone knows what to do if any person—even the skipper—goes over the side.**

avoid hitting him. Actually though, these instructions are of questionable value—a 30-foot boat doing 12 knots moves one boat-length in 1½ seconds. Thus even if the person fell off the bow, the stern would be past him before you could take effective action at the controls. If the victim falls overboard from the cockpit you will be well beyond him almost instantly. Don't waste any time with gear shifts; just slow down and go on immediately with the next steps of the rescue.

Post a Lookout

Keep the person in the water in sight at all times, or at least keep watching the spot where he was last seen. Designate a specific crewman to do nothing but watch the person or his last known location. If more than one crew member is available, assign two to this as their sole duty.

Throw some floating object overboard immediately, to mark the site and to give the victim something to grasp for added buoyancy. Ideally this would be a flag float, but any buoyant object is better than nothing, since the person in the water may have been injured in the fall or may panic in the water. Throw the object even though the person is known to be a strong swimmer; get it as close to him as possible without hitting him.

If another crewman is available, he should tear pages from a magazine or newspaper or from a roll of paper towels to create a trail to retrace in circumstances of poor visibility. In a pinch the helmsman can also do this.

At night train a light on the victim in the water and hold it there to guide the helmsman, unless you can see a water light thrown over at the time of the accident.

No "Heroes"

Don't let a "hero" jump over the side immediately to rescue the victim—then you'll have two persons rather than one to fish out of the water. The only exception to this would be in the case of a small child or an elderly or handicapped adult. If a rescuer must go overboard, he should *always* take a life ring or cushion or some other added buoyancy with him.

Maneuvering to Return

Whether you turn to port or starboard heading back to pick up the overboard victim is best determined by experiment in advance. Ordinarily a boat with a single right-handed propeller turns quickest to port, but some boats actually behave otherwise. The turning maneuver should be able to get you back to the scene, boat under perfect control, in less than a minute. You might find that reversing and backing will be more effective for your boat, but the odds are against it. Make a few tests. Tie together the ends of the lines on a fender so as to form a loop; experiment with recovering this until you know precisely what to do if the need arises.

Under Sail

If under sail at the time of the accident, *start the engine,* whether racing or not. Your maneuvers to return to the location quickly will vary with your point of sail and the wind's direction relative to the victim. Never tack. Come to a beam reach on the tack being sailed; sail off for about five boat lengths; jibe; sail downwind about three or four boat lengths; come to a beam reach and sail to a point where you can shoot up into the wind and just reach the person in the water. If your spinnaker is up, take it down before you return for the pickup, which is usually accomplished most quickly under power. Above all keep calm; you'll make a faster and safer pickup if you think out your plan so you can carry it out effectively.

Maneuvering for the Pick-up

Circumstances will tell you how best to approach the man in the water. Maneuvering into a position to windward of him will provide a lee as the boat drifts toward him. This danger in this technique, however, is that a large boat can drift down *onto and over* the person in the water, and you thus cannot use power for maneuvering because of danger from the propellers.

Although you give up having a lee made for the reboarding phase, a better procedure is probably to approach from downwind, to keep better control over the boat. After you have passed a line to the victim, put the engine in neutral to stop the propellers.

The exact maneuver you use to approach the victim depends upon your own judgment of conditions like seas, water temperature, physical condition and ability of the man in the water, boat maneuverability, and the availability of other assistance.

Generally, you should stop the boat a short distance from the victim, throw him a light line that floats or is buoyed by a cushion or life preserver, and pull him over to the boat. This is considerably safer than maneuvering the boat right up to him.

Getting the Victim on Board

While maneuvering your boat toward the victim, make all preparations for getting him back on board. Have a crew member who is physically capable ready to go into the water *if necessary.* He should remove any excess clothing and shoes and *always put on a life jacket;* he'll need the added buoyancy to save his strength to aid the victim. The rescuer should also have a light safety line to keep him in contact with the boat and to transfer to the accident victim if necessary.

How you get the victim back aboard depends on the design and equipment of your boat, on the person, and on the sea conditions. If you have a transom boarding platform use that; or rig your swimming ladder. Hang one or two short lines with a bowline in one end over the side for handholds or footholds in climbing back aboard.

Call for Help if Needed

If the victim is not immediately rescued, get on the radio with the urgent communications signal "Pan-Pan"—pronounced "Pahn-Pahn"—spoken three times before the call to get help from the Coast Guard and nearby boats. (Do *not* use "Mayday" as your craft is not in distress.) Continue the search until your vessel is released by competent authority.

Importance of Drills

Man Overboard drills are most important. You don't need anyone actually to go over the side; just use any floating object about the size of a human head as the "victim." For such drills, tell one crew member to stand aside and take no part in the "rescue" just as if it were

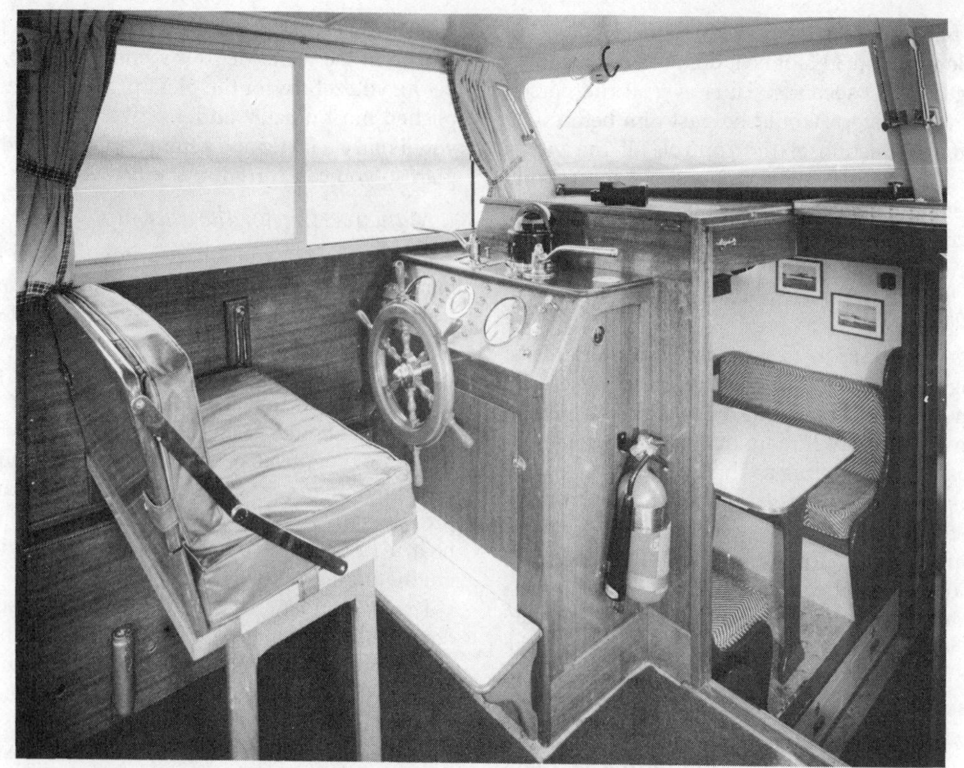

Figure 1012 **The risk of fire or explosion is greatest when the engines are being started, or have just begun to run. The helmsman should have an extinguisher within easy reach.**

he who was in the water. And don't forget that for some drills the skipper should be the non-participant—some day it just might be he who is the victim.

Always hold a critique after each drill. Discuss what was done right and what was done wrong, and how the procedures could be done better.

Fire Emergencies

Fire on a boat is a *serious* matter. Your surroundings are burning and you have nowhere to go except into the water.

Keep in mind that most fires are preventable. A boat kept in shipshape condition, with clean bilges and proper stowage of fuel and gear, is not likely to have a fire. Prevention requires constant attention; whenever you see a condition on your boat that might contribute to a fire, correct it at once.

Fires do occur despite a skipper's best efforts, so always keep firefighting gear accessible and in good working condition.

Explosions and Fires

Fires may start with either a bang—an explosion—or on a much smaller scale. If you have a gasoline explosion there usually is little you can do except grab a life preserver if you can and go over the side.

When clear of the danger, check about and account for all those who were aboard with you. Give whatever assistance you can to anyone in need or in the water without a buoyant device. Keep everyone together in a group—for morale and to aid rescue operations.

Fires and Extinguishment

Fires require four elements for their existence—fuel, oxygen (air), heat, and uninhibited chemical chain reactions; remove or interrupt any of these and the fire will go out. Many fires are fought by smothering (shutting off the flow of air) or by cooling them below the temperature that will support combustion. Others are extinguished by interrupting the chain reactions of the combustion process—this is the action of dry-chemical extinguishers so widely used on small boats.

Classes of Fires

Fires are classified into three categories:

Class A Fires in ordinary combustible materials such as wood, paper, and cloth, where the "quenching-cooling" effect of quantities of water or high-water-content solutions cools the burning material below the ignition temperature.

Class B Fires in flammable petroleum products or other flammable liquids, greases, etc., where the "blanketing-smothering" effect of oxygen-excluding media is most effective.

Class C Fires involving electrical equipment where the electrical conductivity of the extinguishing media is the first consideration.

Fire Extinguishers

Fire extinguishers are classified in the same "A", "B", "C" system as fires. Some types of extinguishers, however, have a suitability greater than their basic classification. Extinguishers required by law on boats are in the "B" category, but a carbon-dioxide or dry-chemical extinguisher also has value in fighting an electrical ("C") fire, and *some* dry-chemical units contain a different powder that is additionally effective on Class "A" fires. The newest type of B/C extinguisher contains Halon gas under pressure; it leaves no residue. Remember that the most effective extinguisher on typical "A" fires in paper, wood, and bedding is a bucket of water.

Fire extinguishers should be distributed around the boat in relation to potential hazards. One should also be near the boat's control station where it can be grabbed quickly by the helmsman. Another should be mounted near the skipper's bunk so that he can roll out at night with it in his hand. Other locations include the galley (but remember that water is best on an alcohol stove fire) and any other compartment at some distance from the location of other extinguishers. Fire extinguishers should be mounted where they are clearly visible to all on board.

Fighting Fires

Burning items such as wood, mattresses, and rags are best extinguished by water, as noted above. A bailer or bucket is thus a valuable piece of equipment. Throw burning materials over the side if you can.

If the fire is in a relatively confined space, close hatches, doors, vents, and ports to keep oxygen from feeding the flames. Do not reopen the hatch or door until your fire-fighting equipment is ready for use. If the fire is in a machinery space, shut off the fuel supply and activate the fixed fire-extinguishing system if you have one.

Remember that typical small marine fire extinguishers have a discharge time of only 8 to 20 seconds. Use them effectively from the very start of discharge. Aim at the *base* of the fire, not the smoke; don't just trigger the extinguisher and drop it.

Maneuvering the Boat

Now let us consider vessel maneuvers that will assist in extinguishing fires. Underway, wind from the boat's motion fans the flames so stop or reduce speed to reduce the wind effect. Also, it makes sense to keep the fire downwind—that is, if the fire is aft, head the bow of the boat into the wind; if the fire is forward, get the stern into the wind. You'll help keep the fire from spreading to other parts of the boat, and you'll keep smoke away from the boat.

Stow some life preservers well forward in case of fire aft, or in case you must abandon ship over the bow.

Summary of Actions

Consider taking these summarized steps, not necessarily in the order shown:

1. Apply the extinguishing agent by:
   a. Using a fire extinguisher,
   b. Discharging fixed smothering system, or
   c. Applying water to wood or materials of this type.
2. If practical, jettison burning materials (throw them over the side).
3. Reduce the air supply by:
   a. Maneuvering the vessel to reduce the effect of wind, and
   b. Closing hatches, ports, vents, doors, etc. if the fire is in an area where this action will be effective.
4. Make preparations for abandoning ship, which include:
   a. Putting on lifesaving devices, and
   b. Signaling for assistance by radio or any other means available.

Leaks and Damage Control

A boat floats so long as the water is pumped out faster than it leaks in. Modern fiberglass boats leak almost not at all (although a slow drip from the propeller shaft stuffing box ensures that the gland is not drawn up too tight), but if a through-hull fitting fails, or if you hit a submerged object while you are traveling fast, your fiberglass boat may have a serious leak.

Planning Ahead

Among preventive measures you can take against leaks are:

1. Install a manually operated diaphragm pump, as large as possible (you can get them with a capacity of a gallon per stroke), so mounted that you can operate it continuously for five or ten minutes without excessive fatigue.
2. Prepare an intake screen to fit in the end of the engine cooling water intake hose, so the engine's pumping capacity can be added to the bilge pump's should need arise. Keep this with your emergency gear.
3. Get a small, strong tarpaulin with corner grommets to which dock lines can be attached, so you can maneuver it to cover a damaged hull area from the outside and secure it there. Keep this with your emergency gear.
4. Fashion soft wooden plugs for each through-hull opening in your hull, tapered so they can be driven into place from the inside. Keep them with your emergency gear.

Standard Procedure

At first suspicion of damage that might cause serious leaking, switch on all electric bilge pumps *before* investigating. If inspection shows your fears to be groundless, switch them off again. They will not be damaged by a brief run while dry.

Crew Action

Depending on how many helpers you have aboard, assign someone to man your manual bilge pump and someone else to investigate the extent of the damage by checking the bilge. If conditions permit, make this investigation yourself. If leaking is rapid, form a bucket brigade if you have manpower enough. If there are three available bailers and only one bucket, break out a couple of cooking pots.

Emergency Pumping

If conditions warrant, close the engine water intake seacock, disconnect the water intake line, and connect your previously prepared intake screen, making sure there is enough water in the bilge to cover the intake well. Start the engine and *check* to be sure it is discharging water through the wet exhaust or other discharge line. Assign someone to frequently check the intake screen and be sure it is not obstructed. Vary the engine speed as required.

Two precautions must be observed. First, there must be enough water already in the bilge and flowing in to meet the engine's needs for cooling. Second, caution must be exercised to keep bilge dirt and trash from being sucked into the engine's intake. To lose power if the engine overheats might be disastrous. These two precautions will usually require that the intake hose be watched and tended at all times by a crew member assigned to that duty.

Operating USCG Droppable Pumps

Coast Guard aircraft have pumps that can be dropped in floating containers to boats that need them. Operating instructions are on and in the containers.

Emergency Repairs

Almost anything can be stuffed into a hole in the hull to help stop the inflow of water—cushions, pillows, bedding, spare sails—all of these will have some beneficial effect.

Material used to stop a leak will be more effective if it can be applied from the outside, where water pressure will aid in holding it in place. This will, of course, require that the boat be stopped while the temporary patch or plug is positioned and secured. If a hole must be plugged from the inside, the pillow, bedding, etc., can sometimes be held in place by fastening a batten or bed slat across it. If no better solution can be found, a member of the crew can be assigned to hold the plug in place. In any event, station a crewman or passenger to watch the patched hole and give immediate alarm if the patch or plug fails to hold.

Assistance

In any form of emergency, it is advisable to *alert* possible sources of assistance without delay, even before you are sure that you will need help.

Summoning Assistance

A boat's radio is usually the primary means for getting assistance, although it is far from being the only method. Use VHF Channel 16 (or 2182 kHz single side-band) and follow the procedures shown in Chapter 22. Make an urgent ("Pan-Pan") call to advise the Coast Guard and others of your problem. Don't put out a Mayday distress call unless you are in obvious danger of sinking, have an uncontrollable fire on board, etc. Don't panic and make a distress call under conditions that do not warrant such action, but also don't fail to alert others to your possible need for assistance soon.

Other distress signals are described on page 105. The Coast Guard requirements for carrying certain visual distress signals are discussed on page 68. Additionally, however, any device or procedure that attracts attention and brings help is a satisfactory distress signal.

Don't overlook the possible use of a small signal mirror equipped with sighting hole and cross-line target. The reflected mirror signals can be seen as flashes of light for many miles and may be just the device that could attract the attention of aircraft.

Distress may also be signaled by slowly and repeatedly raising and lowering arms outstretched to each side. This is a distinctive signal, not likely to be mistaken for a greeting. To be as effective as possible, this signal should be given from the highest vantage point on the boat with consideration given to color contrasts.

Helicopter Rescue

In more and more instances, Coast Guard assistance is being provided by helicopters rather than surface craft. This technique requires knowledge on the part of the boat skipper and is most effective with advance preparations.

Prior to Arrival of Helicopter

1. Listen continuously on VHF Channel 16, 2181 kHz, or other specified frequency if possible.
2. Select, and clear, the most suitable hoist area. For boats, sail and power, it probably will be necessary to tow a dinghy or raft due to lack of clear deck space with it stowed.
3. If hoist is at night, light pick-up areas as well as possible. Be sure you do not shine any lights on the helicopter and blind the pilot. If there are obstructions in the vicinity, put a light on them so the pilot will be aware of their positions.
4. Advise location of pick-up area before the helicopter arrives so that he may adjust for and make his approach as required.
5. Remember that there will be a high noise level under the helicopter, so conversation between the deck crew will be almost impossible. Arrange a set of hand signals between those who will be assisting.

Rescue by Hoist

1. Change course so as to permit the craft to ride as easily as possible with the wind on the bow, preferably on the port bow.
2. Reduce speed if necessary to ease the boat's motion, but maintain steerageway.
3. If you do not have radio contact with the helicopter, signal a "Come on" when you are in all respects ready for the hoist; use a flashlight at night.
4. Allow basket or stretcher to touch down on the deck prior to handling to avoid static shock.
5. If a trail line is dropped by the helicopter, guide the basket or stretcher to the deck with the line; line will not cause shock.
6. Place person in basket sitting with hands clear of sides, or strap person in the stretcher. Place a life jacket on the person if possible. Signal the helicopter hoist operator when ready for hoist. Person in basket or stretcher nods his head if he is able. Deck personnel give "thumbs up."
7. If necessary to take litter away from hoist point, unhook hoist cable and keep free for helicopter to haul in. *Do not secure cable to vessel or attempt to move stretcher without unhooking.*
8. When person is strapped in stretcher, signal helicopter to lower cable, hook up, and signal hoist operator when ready to hoist. Steady stretcher to keep it from swinging or turning.
9. If a trail line is attached to basket or stretcher, use to steady. Keep feet clear of line.

Figure 1013 **The Coast Guard is making increased use of helicopters as well as surface craft for rescue and assistance work. Note the hoist winch above the door. Text details the procedures to use if it is necessary to airlift a person off a boat.**

Abandoning Ship

Many boats involved in casualties have continued to float indefinitely. If it becomes necessary to abandon your boat due to fire, danger of sinking, or other emergency, don't leave the area. Generally a damaged boat can be sighted more readily than a person, and it may help to keep you afloat.

Keep in mind that distance over water is deceptive. Usually the actual distance is much greater than your estimate. Keep your head and restrain your initial impulse to swim ashore. Calmly weigh the facts of the situation such as: injuries to personnel, the proximity of shore, and your swimming abilities, before deciding upon your course of action.

Before abandoning your craft, put on your life preserver and give distress signals. Don't foolishly waste signaling devices where small likelihood of assistance exists. Wait until you sight someone or something. If your vessel is equipped with a radio telephone, a distress message should be sent. Every boat operating offshore should carry an EPIRB (see Chapter 23); this is the time to activate it.

Remember also that even though you have sent in a call for assistance, keep working to help yourself until outside help comes.

11

Selecting the appropriate anchors and ground tackle; anchoring procedures, how to prevent dragging, retrieval of snagged anchors, permanent mooring systems

ANCHORING

Of all the skills involved in seamanship, the art of anchoring is one the boatman must master if he is to cruise with an easy mind. Perhaps you have been just getting by with inadequate gear and bad practices acquired in home waters. Sooner or later, carelessness and lack of technical know-how will lead to difficulty—probably inconvenience, possibly danger.

The essence of successful anchoring is to "stay put," without dragging, whenever the anchor is set. And don't forget the need to respect the rights of nearby boats that could be fouled or damaged if your anchor drags.

Though anchoring skill may not be learned from the printed page alone this chapter should help the beginning skipper to get off to a good start and the seasoned boatman to verify or update his technique.

In quiet anchorages, in familiar surroundings, ground tackle (the gear used) and the methods employed are seldom put to test. Cruising into strange waters, finding inadequate shelter in an exposed anchorage during a hard blow, and unexpected variations in wind and current will surely take the measure of both tackle and technique.

The problem, then, breaks down into two principal parts—(1) the equipment we should carry, and (2) knowledge of how to use it.

Ground Tackle

Anchors have evolved over centuries from the simple stone used with a crude rope to the modern designs that have been carefully engineered to achieve the greatest holding power for the least weight. It did not take long for mariners to realize that what was needed was an anchor that would *dig in* and *hook* the bottom, rather than merely a weight that might drag across it. Simple wooden hooks were first added, later changing to iron. Still later a STOCK was installed perpendicular to the plane of the hooks so as to put them in a better position to get a bite of the bottom. Holding power in softer bottoms was improved by adding broad FLUKES to the hooks. In this century, anchors have highly specialized designs and derive their holding power by quickly and deeply burying themselves in the bottom.

Some Terms and Definitions

To prevent confusion in the use of terms, refer to the labeled illustration of the parts of an anchor, Figure 1101, and the definitions in the box on the following page. A popular version of the classic anchor was chosen to illustrate the parts. Subsequently we shall see how the proportioning and placement of parts have varied with the introduction of later designs.

The ANCHORING SYSTEM is all the gear used in conjunction with the anchor, including line, chain, shackles, and swivels. The anchor line, including any chain, is properly called the RODE.

Types of Anchors in Use

Scan a marine hardware catalog and, without experience, you may be confused by the diversity of designs. What you should be buying, essentially, is holding power; sheer weight is no index of that. On the contrary, scientific design is the key to efficiency, and a modern, patented anchor, if properly manufactured, stands at the top of the list today on a holding-power-to-weight basis.

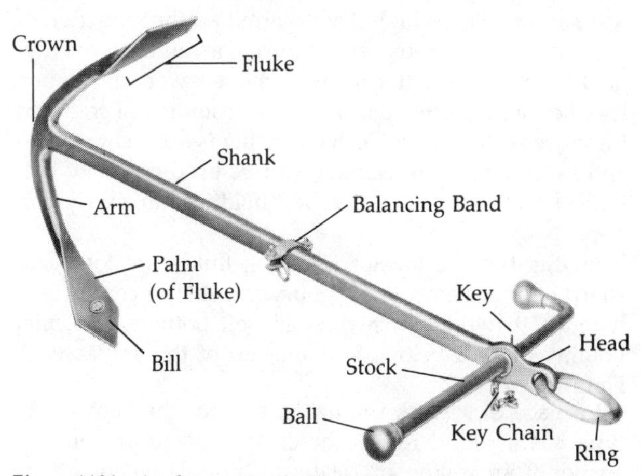

Figure 1101 **Anchor parts are identified on the traditional kedge anchor, but the terms are the same for most modern types.**

SOME BASIC DEFINITIONS

ANCHOR A device designed to engage the bottom of a waterway and through its resistance to drag maintain a vessel within a given radius.

ANCHOR CHOCKS Fittings on the deck of a vessel used to stow an anchor when it is not in use.

ANCHOR RODE The line connecting an anchor with a vessel.

BOW CHOCKS Fittings, usually on the rail of a vessel near its stem, having jaws that serve as fairleads for anchor rodes and other lines.

GROUND TACKLE A general term for the anchor, anchor rodes, fittings, etc., used for securing a vessel at anchor.

HAWSEPIPE A cylindrical or elliptical pipe or casting in a vessel's hull through which the anchor rode runs.

HORIZONTAL LOAD The horizontal force placed on an anchoring device by the vessel to which it is connected.

MOORING BITT A post or cleat through or on the deck of a vessel used to secure an anchor rode or other line to the vessel.

SCOPE The ratio of the length of the anchor rode to the vertical distance from the bow chocks to the bottom (depth plus height of bow chocks above water).

VERTICAL LOAD The lifting force placed on the bow of a vessel by its anchor rode.

Definitions (except scope) are from a code of standards and recommended practices adopted by the American Boat & Yacht Council.

Lightweight Type

The LIGHTWEIGHT TYPE introduced just prior to World War II was developed by R. S. Danforth. Its efficiency in war service was so high that it permitted the retraction of grounded amphibious vessels during assaults on enemy-held beachheads. After the war, models were developed specifically for recreational and small commercial craft and these are widely used. Such an anchor is excellent in mud and sand, and, with caution, can be used on rocky bottoms; however, it often does not hold well in grassy bottoms.

In this type of anchor, pivoting flukes are long and sharp so that heavy strains bury the anchor completely. It tends to work down through soft bottoms to firmer holding ground below, burying part of the line as well, Figure 1102.

In place of a stock through the head, the lightweight type has a round rod at the crown end to prevent the anchor from rolling or rotating. This placement of the stock does not interfere with the shank being drawn into the hawsepipes of larger craft for stowage. (Many skippers place protective rubber tips over the stock ends; others merely plug the ends of the rod, which is hollow, to prevent mud and sand from entering and so being brought on deck.)

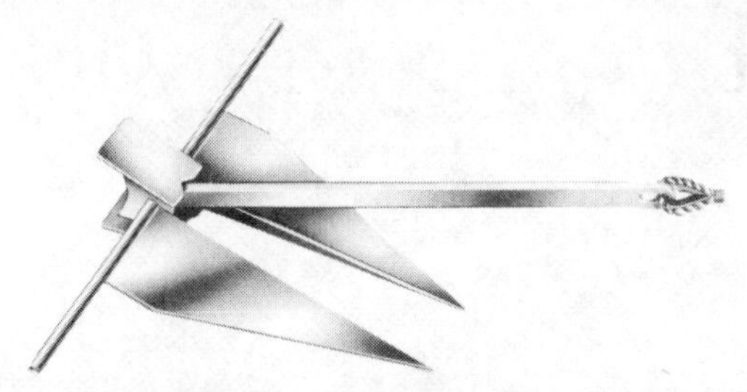

Figure 1102 **The Danforth lightweight anchor is available in both standard and "Hi-Tensile" models, both burying types with high holding power. Note the angled plates on crown.**

A key element in the high performance of Danforth anchors is the design of the crown. The two flat, inward-sloping surfaces force the thin, sharp flukes to dig into the bottom and penetrate deeply. The surfaces of the crown are placed away from the plane of the flukes in order to help reduce clogging with mud, grass, or bottom debris that might possibly interfere with the penetration of the anchor and its holding power.

For recreational and commercial small craft, standard Danforth anchors are made in approximate weights from 2-½ to 180 pounds (1.13-81.6 kg), the Hi-Tensile model in sizes from 5 to 90 pounds. For comparable weight, Hi-Tensile models are said to provide roughly 20% to 30% more holding power.

A number of other manufacturers produce anchors of the lightweight, burying type. In selecting an anchor, remember that all manufacturers have their own concepts of design, and "look-alikes" do not necessarily hold alike; compare data on holding power under similar conditions.

The Plow Anchor

The PLOW anchor is unique in design, resembling none of the other types. It was invented in England by Professor G. I. Taylor of Cambridge University; he called it the CQR (secure). It has found wide acceptance because of its demonstrated efficiency in a variety of bottoms. Opinions vary as to its effectiveness in heavy grass or weed, which is not surprising in that many weed growths resist penetration by any anchor. Similar models have since been developed and manufactured in the United States and other countries.

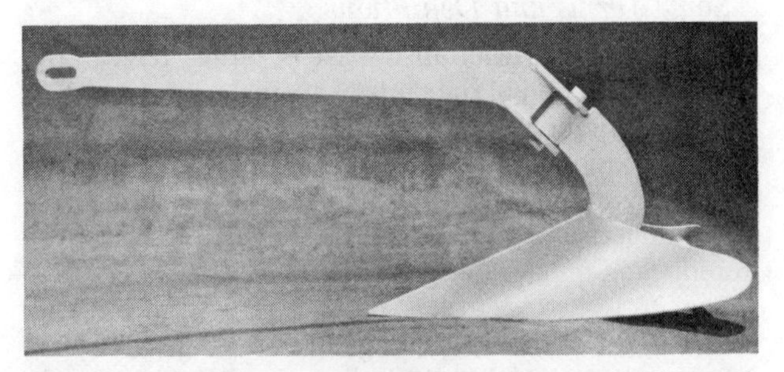

Figure 1103 **The plow anchor gets its name from the shape of its deep-burying flukes, pivoted at the end of the shank.**

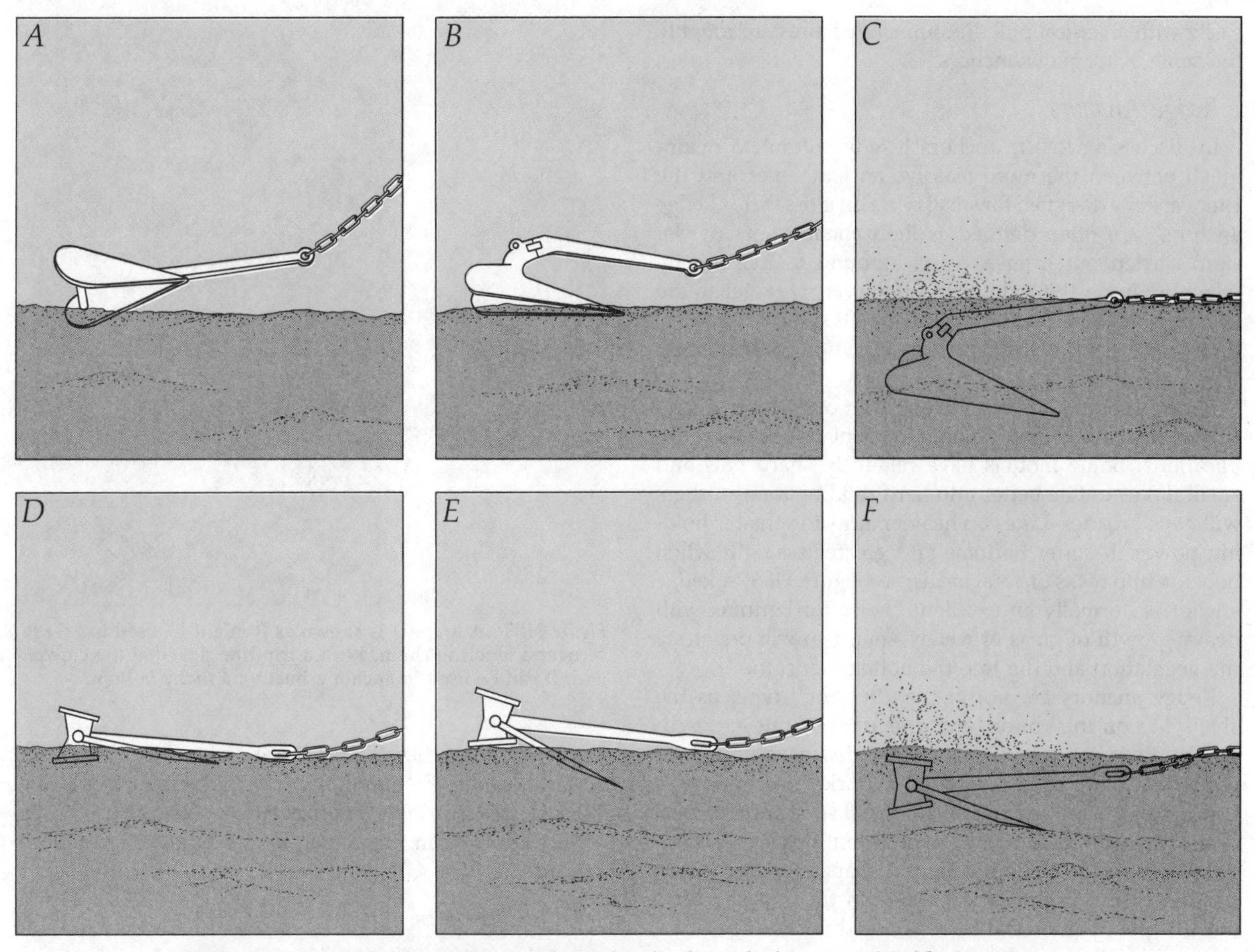

Figure 1104 **How burying anchors work: A plow lands on the bottom on its side *A*, gets a quick bite, *B*, and digs in deep *C* as it rights itself. The Danforth type lands with its flukes flat *D*, penetrating *E* as the strain comes on the rode, and burying the flukes, *F*.**

When a plow anchor is first lowered, the anchor lies on its side on the bottom; see Figure 1104. Then when a pull is put on the line, it rights itself after moving a short distance, driving the point of the plow into the bottom and finally burying the anchor completely if the bottom is soft. Suggested weights for CQR anchors vary from 5 pounds for a 16-footer to 60 pounds for 60-footers, with heavier sizes available for larger vessels.

Because of the pivoting feature of its shank, the plow anchor tends to remain buried over moderate changes in direction of pull on the line caused by wind or current shifts. There is no projecting fluke to foul the anchor line and the plow breaks out easily when the pull is vertical for raising the anchor. Plow anchors do not stow well on deck and are usually hoisted to a bow roller fitting where they are secured.

As with lightweight types, plow anchors are produced by a number of manufacturers and effectiveness may vary between different units.

The Bruce Anchor

Another anchor from the United Kingdom is the BRUCE. Originally developed for use with offshore oil and gas well drilling rigs, it has been scaled down for use with small craft in sizes from 2.2 to 110 pounds (1 to 50 kg).

A burying type, the Bruce anchor, Figure 1105, is designed to right itself no matter how it lands on the bottom while digging in within two shank lengths, it breaks out

Figure 1105 **The Bruce anchor is shaped so direction of pull can change through 360 degrees, once it is set, without breaking out, yet intentional breakout is easy on short scope. The anchor is said to work well in mud, sand, and rocks.**

easily with a vertical pull. Recommended sizes are roughly the same as for plow anchors.

Kedge Anchors

In discussing KEDGE anchors it is important to distinguish between the more massive ancient types and the later versions designed for small boats. In glossaries, "kedge anchors" are often defined as light anchors (of any design) carried out from a vessel aground to free her by winching in on the rode. Here, however, we refer to the kedge as an anchor with the more traditional type of arms, flukes, and stock as distinguished from newer lightweight types.

Kedge anchors are not widely used on modern recreational boats, but they do have their place in special applications. Some models have relatively sharp bills and small flukes to bite better into hard sand bottoms. Others will have broader flukes on heavier arms for greater holding power in softer bottoms and greater strength when hooking into rocks or coral heads; see Figure 1106. A kedge anchor is normally an excellent choice for bottoms with heavy growth of grass or weeds—one arm will penetrate the vegetation and dig into the bottom beneath.

Kedge anchors are not of the "burying" type, as the shank lies on the bottom and one arm remains exposed. On the other hand, a kedge's "hook" design recommends it, probably above all other types, on rocky bottoms where one fluke can find a crevice. Retrieval, with proper precautions, is not too difficult. (More about this later.) Modern kedge designs have a diamond-shaped fluke to lessen the risk of the anchor line fouling on the exposed arm, but this possibility must be considered if a change in direction of pull of a half-circle or more occurs.

Stock and Stockless Types

The plow and Bruce anchors are of the stockless type. Others are classed as being of the *stock type,* although the stock, as we have seen, may be at either the ring or crown end. Those with the stock at the crown end, such as the lightweight type, have a fixed stock. On nearly all designs where the stock is at the ring end, such as the various kedge anchors, the stock is loose and can be folded for better stowage. Frequently, a key is required to pin the stock of a kedge in its open position when set up ready for use, the key in turn being lashed in its slot to hold it in place.

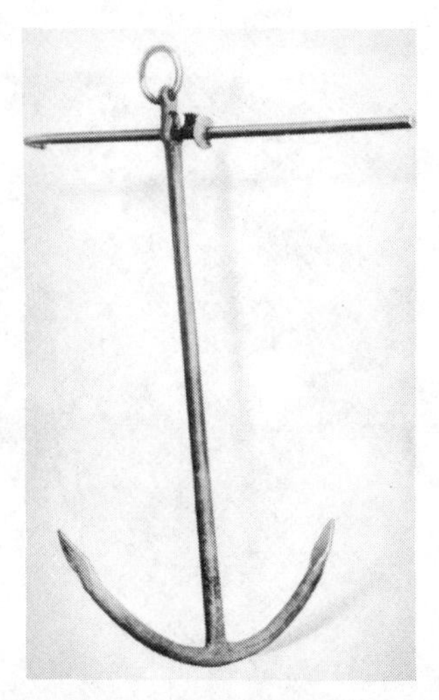

Figure 1106 **The kedge anchor must be heavier than a burying type anchor of equivalent holding power. Its thin arms and flukes make it the best type to use in weeds or grass. The model shown above has thin, sharp flukes for better digging into hard bottoms.**

Figure 1107 **A grapnel is shown as it might be used to recover a mooring chain. With a trip line rigged at the crown end, it can be used to anchor a boat on a rocky bottom.**

Other Anchors

Another design of anchor is frequently cataloged as NAVY TYPE. Inexperienced boatmen on seeing them on large ships sometimes conclude that they are best for all vessels, including small boats—this is not so. Ships use them because such stockless anchors can be hauled up into hawsepipes. The ratio of weight to holding power for these anchors is so great that if weight is held within reason for a small boat, holding power is far below safe limits.

GRAPNELS, though used by some commercial fisherman, are not recommended for general anchoring service aboard recreational boats. These are also stockless models with, as a rule, five curved sharp-billed claw-like prongs symmetrically arranged around the crown end of the shank. Eyes may be cast in both ends of the shank—at the head in lieu of a ring for attachment of a rode (if used as an anchor) and at the crown end for a buoyed trip line (see page 248). By dragging a small grapnel back and forth, a boatman may GRAPPLE for a piece of equipment lost on the bottom. See Figure 1107.

FOLDING anchors are those of a highly specialized design in which, at some sacrifice of holding power and strength, all parts fold against the shank into the smallest possible space for the most convenient stowage. In one stockless type, there are two pairs of flukes at right angles to each other, almost in the manner of a grapnel. In rocky bottoms they hook readily and may be rigged to pull out easily, crown first. Such anchors are often excellent in bottoms with heavy growths of grass or weeds as one or two arms penetrate the vegetation to get a bite into the bottom.

MUSHROOM anchors will be discussed later because their principal use is in conjunction with permanent moorings. Modified versions of the mushroom are manufactured for small craft such as canoes and rowboats, but their efficiency as anchors is at the lowest end of the scale.

Sea Anchors

All the anchors under discussion in this chapter are designed to keep a boat from drifting, by engagement with the bottom; SEA ANCHORS do not fall in this category. These are intended to float at or just below the surface, serving merely as a drag to hold a boat's bow (or stern) up toward the oncoming seas so as to prevent her from lying in the trough. Sea anchors are seldom used aboard recreational boats, and then only in the heaviest weather offshore, where there is room to drift to leeward. See Chapter 10 (page 219) for a discussion of sea anchors and their use in rough weather.

The Anchor Line

All of the gear, taken collectively, that lies between a boat and her anchor is called the RODE—whether it be synthetic fiber (like nylon), chain, wire, or a combination of fiber and chain.

Twisted Nylon

Nylon, in three-strand twist or double-braid form, is now by far the most widely used material for anchor lines. Manila is almost never seen today, and other synthetics, such as Dacron, polypropylene, and polyethylene, have less desirable characteristics. Chain makes a good anchor rode, but its weight, while desirable for anchoring, may necessitate your having a winch or other mechanical assistance on board to hoist it. On a small boat, the weight of an adequate length of chain, stowed in the bow, may be too great for proper trim.

For anchoring, nylon's greatest asset is its elasticity; it stretches a third or more under load. Its working elasticity of 15% to 25% is several times greater than that of manila. When a boat surges at anchor in steep seas, there is a heavy shock load on fittings and ground tackle unless provision is made to absorb it gradually. Nylon's elasticity does exactly that.

Some boatmen unwittingly lose part or most of the advantage inherent in nylon by buying too large a line. Within the limits of safe working loads, the smaller the diameter the better the elasticity for given conditions. A practical limit is reached when small diameters (though rated high enough for breaking strength) are not convenient to handle. Some experienced boatmen use nylon as light as 3/8-inch diameter on the working anchors of their 30- to 40-foot craft.

As mentioned in Chapter 13, nylon line is highly resistant to rot, decay, and mildew, but can be damaged by rust from iron fittings or a rusty chain. Nylon line should be stored out of direct sunlight to prevent a gradual deterioration from ultraviolet rays.

Braided Synthetic Line

Most nylon used for anchor lines is laid up by twisting three strands. Synthetics can, however, be laid up by BRAIDING. For anchoring (as well as mooring or towing) a

Figure 1108 **Braided nylon, with its elasticity that absorbs shock loads, is well suited for use as an anchor rode. It coils best when faked (flaked) down in a figure eight.**

braided outer cover of nylon surrounds a braided inner synthetic core—this is commonly called "double-braided" or "braid-on-braid" line. The result is a line of exceptional stability with no inherent tendency to twist because of the nature of its lay. Consequently, it can be fed down into rope lockers without fear of kinking.

When braided nylon is handled on deck, it is advisable to fake it down in figure-eight pattern, rather than the conventional clockwise coil used with twisted fibers. Because of the relatively smoother surface of braid, with more fibers to take the wear, chafe is less of a problem than it is with the twisted three-strand lay. Braided nylon retains an adequate degree of elasticity (14% at working loads, as against 25% for twisted nylon).

Manila Line

Although MANILA line holds knots and splices better than the more slippery synthetics, its lesser strength and greater susceptability to deterioration have been the causes of its replacement with more modern materials.

Chain

As the size of the vessel increases, so does the required diameter of a nylon anchor line. For yachts 65 feet or more, nylon would run to diameters of up to 3/4-inch or larger. This is getting to a size that is difficult to handle; the alternative is CHAIN.

From this it should not be inferred that chain is not also in use on smaller craft. Boats that cruise extensively and have occasion to anchor on sharp rock or coral often have chain; in some cases it is regarded as indispensable—as it stands chafing where fiber won't.

In larger diameters, the weight of chain makes a sag in the rode which cushions shock loads due to surging. Once the slack has been taken up, however, the shock on both boat and anchor is very much greater than with nylon. You must thus be sure to use adequate scope with chain, as will be discussed below.

The three kinds of chain most used as anchor rode are "BBB," "Proof Coil," and "High Test." Chain is designated by the diameter of material in the link, but the various types have links differing slightly in length. It is necessary to match the chain to the WILDCAT (a pulley designed for use with chain, also called a GYPSY) of the anchor windlass; the differences in link length are slight, but enough to cause trouble if there is a mismatch. (Most windlasses have a gypsy for either line or chain, but special models are available that can handle both.)

Any type of chain may be used for anchoring—BBB is slightly stronger than Proof Coil; High Test is significantly

Figure 1109 **On this large yacht, *left,* wildcats (lugs) on the winch engage the links of the all-chain rodes. Chain, leading down to the anchors through hawse holes, is shown *stoppered,* held by split-hook *devil's claws* that are attached to the winch. *Right:* A thimble, shackle, and eye splice are commonly used to secure rodes to the anchor ring, or to a shackle in a short length of chain between the line and the anchor.**

stronger than either. The selection of specific chain type and size for your boat involves several factors, the first of which is adequate strength. A safe standard to use is a WORKING LOAD—figured from the size of the boat and the conditions to be encountered—at 20% of the chain's breaking strength when new. But you must also consider weight. The chain must be heavy enough to provide a proper sag to cushion shock loads, but the length required for normal anchoring depths must not be so heavy, when stored on board, that it affects the boat's handling characteristics or even its safety. The weight factor may dictate a combination of chain and line.

Nylon-and-Chain

Today the consensus appears to be that for most average conditions, the ideal rode is a combination of nylon line and a short length of chain (6 to 8 or more feet; longer is desirable) between the line and the anchor.

One effect of chain in this combination rode is to lower the angle of pull, because chain tends to lie on the bottom. Of equal, perhaps greater, significance is the fact that modern lightweight anchors often bury completely, taking part of the rode with them. Chain stands the chafe, and sand has less chance to penetrate strands of the fiber line higher up. Sand doesn't stick to the chain, and mud is easily washed off. Without chain, nylon gets very dirty in mud.

Chain used in this manner may vary from 1/4-inch diameter for 20-footers up to 7/16-inch for 50-footers. It should be galvanized, of course, to protect against rust. Neoprene-coated chain is an added refinement, as it will not mar the boat, but such coating has a limited life in active use.

Securing the Rode

The complete anchor system consists of the anchor and the rode, usually made up of a length of line plus a length of chain. Each element of the system must be connected to its neighbor in a strong and dependable manner.

Eyesplice, Thimble, and Shackle

There are various methods for securing the rode to the anchor ring. With fiber line, the preferred practice is to work an EYE SPLICE around a THIMBLE and use a SHACKLE to join the thimble and ring. See Figure 1109, right. With nylon line you can use a galvanized metal thimble, or one of stainless steel; be sure to keep the thimble in the eye. A tight, snug splice will help, and seizings around the line and the legs of the thimble, near the V, will keep the thimble in the eye splice when the line comes under loads that stretch the eye. A better thimble, Figure 1110, is available in bronze alloy or plastic for use with synthetic rope. This is designed to hold and protect the line.

Where a shackle is used, it is a good idea to put a bit of silicone spray or waterproof grease on the threads of the shackle pin to keep it from seizing up over a period of time. *Be sure to safety wire the pin to prevent its working out accidentally;* see Figure 1111. Watch for corrosion if different metals are used in thimbles, shackles, and rings. Also beware of rust stains on nylon; cut out and resplice in new thimbles if the line becomes rust-stained.

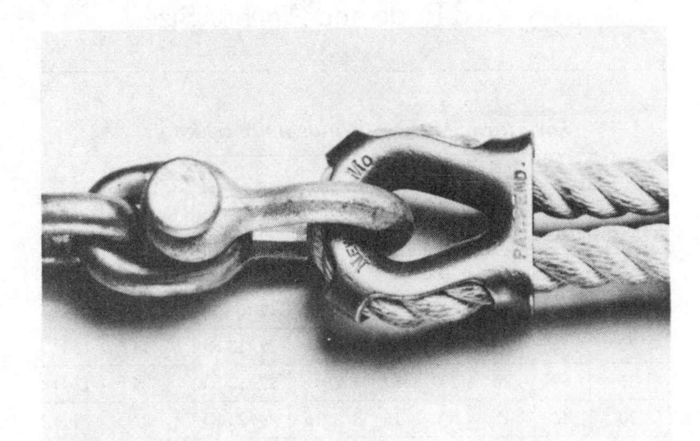

Figure 1110 **Newco thimble in bronze alloy (shown) or plastic is used with synthetic rode to prevent the line from jumping out of the thimble when the eye splice stretches under load.**

A thimble and shackle provide a ready means for backing up your line with a length of chain, if desired, shackling the chain in turn to the anchor ring. Shackles should be large enough so as not to bind against the ring.

Anchor Bends and Bowlines

Some skippers would rather fasten their line directly to the ring using an anchor bend, seizing the free end to the rode. Others use a bowline with an extra round turn around the ring, Figure 1112 top. In either case, these procedures make it easy to turn the line end-for-end occasionally, or to remove the line from the anchor for easy handling when stowing.

Turning a line end-for-end greatly extends its useful life, as the lower end which has chafed on the bottom becomes the inboard end seldom used with normal scope. Eye splices may be used at both ends of the rode, or added as necessary when the rode is turned.

Even where the regular working anchor is kept made-up with a combination of line and chain, you should know how to bend a line directly to an anchor. This is often the handiest way to drop a light anchor for a brief stop, or to make up a second anchor when a bridle or stern anchor is needed.

Use shackles to secure chain cables to the anchor; stout swivels, Figure 1112 bottom, are often an added refinement. As swivels are a weak point, they must be large. On an all-chain rode they are essential. Swivels, however, should not be used with twisted soft-laid synthetic lines; a hockle (kink) may be the result. Double-braided synthetic lines will not hockle, even though subjected to very heavy strains.

At the Bitter End

To guard against loss in case the anchor goes by the board accidentally, the bitter (inboard) end of an anchor rode should be made fast to some part of the boat. Sometimes you can do this by leading the line below, perhaps through a deck pipe, and securing it to a samson post or other strong timber. On sailboats you can secure it to a mast. On small boats where the entire length of rode is carried on deck, you can have an eye splice in the bitter end to fit a securely fastened ring or eye-bolt. In any case make sure the bitter end is secured.

Figure 1111 **Electrical wire is used here to show how the pin of an anchor shackle should be safety-wired. Bare, non-corroding wire is normally used for this purpose.**

Figure 1112 **An anchor bowline, *above,* with its extra turn around the ring, is a secure way to bend the rode to an anchor. Swivels, *below,* are required on all-chain rodes to prevent snarls. Be sure they are as strong as the chain itself.**

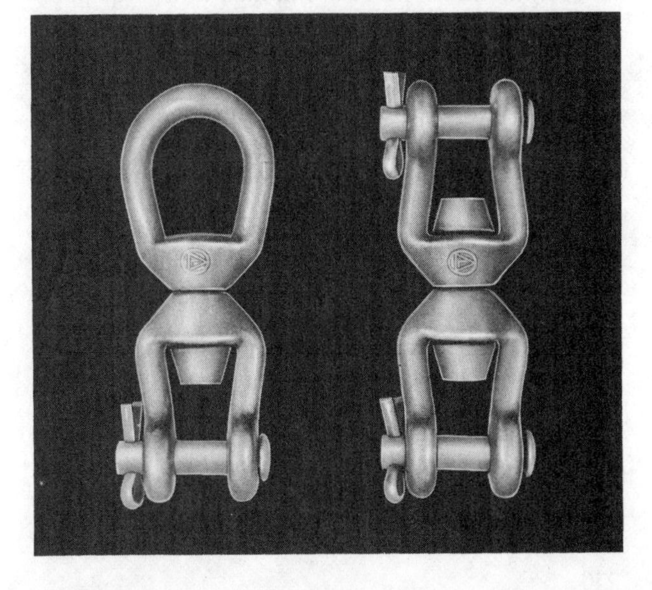

How Many—and How Heavy?

The number of anchors you should carry depends upon several things—the size of your boat, whether she is used only in sheltered waters or cruises extensively to many harbors, and, to some extent, the type of anchor.

Some small boats like runabouts and utilities have only a single anchor, but this cannot be considered adequate. Even discounting the possibility of fouling one anchor so badly that it cannot be retrieved, there are occasions when it is desirable to lie to two. Again, one anchor heavy enough for extreme conditions could be a nuisance in ordinary weather.

Many boats carry two anchors, proportioning the weight in the ratio of about 40% in one, 60% in the other. For cruising boats, three are undoubtedly better. This allows for two to be carried on deck—a light LUNCH HOOK for brief stops while someone is aboard, and a WORKING ANCHOR for ordinary service, including anchorages at night in harbor. The third might well be a big spare STORM ANCHOR, possibly stowed below, selected with an eye to its holding no matter what else lets go, even under extreme conditions of wind and weather.

Anchor Size and Holding Power

Down through the years there have been repeated attempts to reduce anchor weights to a simple formula or table based on boat length or tonnage. Recommendations have varied widely, gradually becoming lighter as more modern designs replaced old-fashioned kedge anchors. With the development of patented designs, however, came the problem of minor variations between manufacturers of anchors of the same general type. Thus any table of anchor size vs. boat size can only be a broad recommendation, to be modified for individual craft and local situations.

Stowage

A boatman's seamanship shows up in the attention he gives to stowing his ground tackle. Exactly how he goes about it depends to some extent on the kind of boating he does, the size of his boat, and the way it is equipped. In any case, unless his deck is uncluttered—with gear ready for immediate use, yet secured so that it cannot shift—he will never rate high as a seaman.

Ordinarily a cruising boat will carry one, sometimes two, anchors on deck, made up and ready for use. On some small boats, where it is not feasible to leave anchors on deck at all times, or in cases where lines are stowed below at the home berth, at least one anchor and line should be prepared and made ready before getting under way from your slip or mooring. Engines do fail and, when they do it's likely to be at an embarrassing moment, with wind or current setting you down on a shoal or reef. Then it's too late to think about breaking out gear that should have been ready at hand.

An anchor lying loose on deck is a potential hazard. If the boat happens to roll, it may slide across the deck, leaving scars in its wake and damage to equipment. Conceivably it could go over the side, taking line with it that

Suggested Rode and Anchor Sizes*

For Storm Anchor (Winds up to 60 knots)

	Beam		Rode		Anchor		
L.O.A. (feet)	Sail (feet)	Power (feet)	Nylon (feet)	Chain (feet)	Northill (pounds)	Standard	Hi-Tensile
10	5	5	100 1/4	3 3/16	12 (6-R)	8-S	5-H
15	7	7	125 1/4	3 3/16	12 (6-R)	8-S	5-H
20	8	9	150 3/8	4 1/4	27 (12-R)	13-S	12-H
25	9	10	200 3/8	4 1/4	27 (12-R)	22-S	12-H
30	10	11	250 7/16	5 5/16	46 (20-R)	22-S	20-H
35	12	13	300 1/2	6 3/8	46 (20-R)	40-S	35-H
40	13	14	400 5/8	8 7/16	80 (30-R)	65-S	60-H
50	14	16	500 5/8	8 7/16	105 (50-R)	130-S	60-H
60	16	19	500 3/4	8 1/2	105 (50-R)	180-S	90-H

For Working Anchor (Winds up to 30 knots)

	Beam		Rode		Anchor		
L.O.A. (feet)	Sail (feet)	Power (feet)	Nylon (feet)	Chain (feet)	Northill (pounds)	Standard	Hi-Tensile
10	5	5	80 1/4	3 3/16	6 (3-R)	4-S	5-H
15	7	7	100 1/4	3 3/16	6 (3-R)	8-S	5-H
20	8	9	120 1/4	3 3/16	12 (6-R)	8-S	5-H
25	9	10	150 3/8	3 3/16	12 (6-R)	8-S	5-H
30	10	11	180 3/8	4 1/4	27 (12-R)	13-S	12-H
35	12	13	200 3/8	4 1/4	27 (12-R)	22-S	12-H
40	13	14	250 7/16	5 5/16	46 (20-R)	22-S	20-H
50	14	16	300 1/2	6 3/8	46 (20-R)	40-S	35-H
60	16	19	300 1/2	6 3/8	80 (30-R)	65-S	35-H

For Lunch Hook

	Beam		Rode		Anchor		
L.O.A. (feet)	Sail (feet)	Power (feet)	Nylon (feet)	Chain (feet)	Northill (pounds)	Standard	Hi-Tensile
10	5	5	70 1/4	3 3/16	6 (3-R)	2 1/2-S	5-H
15	7	7	80 1/4	3 3/16	6 (3-R)	2 1/2-S	5-H
20	8	9	90 1/4	3 3/16	6 (3-R)	2 1/2-S	5-H
25	9	10	100 1/4	3 3/16	6 (3-R)	4-S	5-H
30	10	11	125 1/4	3 3/16	6 (3-R)	4-S	5-H
35	12	13	150 1/4	3 3/16	12 (6-R)	4-S	5-H
40	13	14	175 3/8	4 1/4	12 (6-R)	8-S	5-H
50	14	16	200 3/8	4 1/4	12 (6-R)	8-S	12-H
60	16	19	200 3/8	4 1/4	27 (12-R)	13-S	12-H

*Suggested sizes assume fair holding ground, scope of at least 7-to-1 and moderate shelter from heavy seas.

PLOW ANCHORS—Woolsey, manufacturer of the Plowright anchor, makes the following recommendations for winds up to 30 knots: for *working anchors*, 10′-21′, 6 lbs.— 22′-32′, 12 lbs.—32′-36′, 18 lbs.—36′-39′, 22 lbs.—39′-44′, 35 lbs. For *lunch hooks*, they advise stepping down one size. For *storm anchors*, up one size.

KEDGES—Holding powers vary widely with the type. Best to consult manufacturer for individual recommendations.

Table 11-1

Figure 1114 **Hawsepipes, commonly seen on large vessels, may be seen on some yachts, left. A circular pad protects the hull surface. Above: Because their shape makes them unsuitable for deck stowage, plow anchors are usually housed under bow pulpits.**

might foul a propeller. Every anchor on deck should be stowed in chocks which are available at marine supply stores to fit standard anchors. Lashings hold the anchors in the chocks. Hardwood blocks, properly notched, have often been used in lieu of metal chock fittings.

As an alternate to chocking on deck, anchors carried aboard sailboats may be lashed to bow rails or shrouds, off the deck, where there is no risk of their getting underfoot, and less risk of their fouling running rigging. Many boats with bowsprits or pulpits have a roller at the outboard end, to carry a plow or Bruce anchor in this outboard position as a regular working anchor. See Figure 1114 right. Some larger yachts have provision for hauling the anchor into a hawsepipe fitted into the topsides forward; see Figure 1114 left.

Stowing the Storm Anchor

As the big spare storm anchor is used only on rare occasions, you can carry it in some convenient location below, or in a lazarette or in stowage space below a cockpit deck, accessible through a hatch. Chocks here should be arranged to carry the weight on floors or frames, never on hull planking. If the big anchor gets adrift, it could easily loosen a bottom plank on a wooden boat, or break through a fiberglass hull.

The big risk in stowing a spare anchor away in some out-of-the-way corner is the possibility that other gear may be allowed to accumulate over and around it. Guard against that. The sole value of a storm anchor may some day depend upon your being able to get that big hook over quickly, bent to a long and strong spare rode that must be equally accessible.

Lunch hooks are small and seldom needed without warning, so there's justification for stowing them in some convenient locker. Keep them away from the compass, however, as they can be a potent cause of deviation (except, of course, for aluminum anchors).

Rope and Chain Lockers

Though small craft often carry their lines coiled on a forward deck or in an open cockpit, many cruising boats have a rope locker in the forepeak that can be used. Nylon dries quickly and can be fed down into lockers almost as soon as it comes aboard. Lockers must be well ventilated and arranged to assure good air circulation at all times. Dark, damp lockers are an invitation to dry rot and mildew. A vented hatch over the rope locker will permit exposure to a good flow of air.

The rode should always be ready to run without fouling. Line is often passed below through a deck pipe, slotted so that it can be capped even when the line is in use. Slots must face aft to prevent water on deck from finding its way below. Some cast mooring bitts are made with an opening on the after face, through which line can be passed below.

Chain won't soak up moisture and is easy to stow in lockers. Where weight of chain in the bow of a small offshore cruising boat is objectionable, it can be overcome by splitting a long rode into two or three shorter lengths, stowed where convenient and shackled together as necessary. The chain portion of a combination nylon-and-chain rode is ordinarily shackled in place for regular use, but nylon, if left on deck, should be shaded as much as possible from the sun to protect surface fibers from damage by ultraviolet rays.

Scope

Once you have chosen an anchor of suitable design and size to provide adequate holding power, you must consider SCOPE). It is a major factor that determines whether

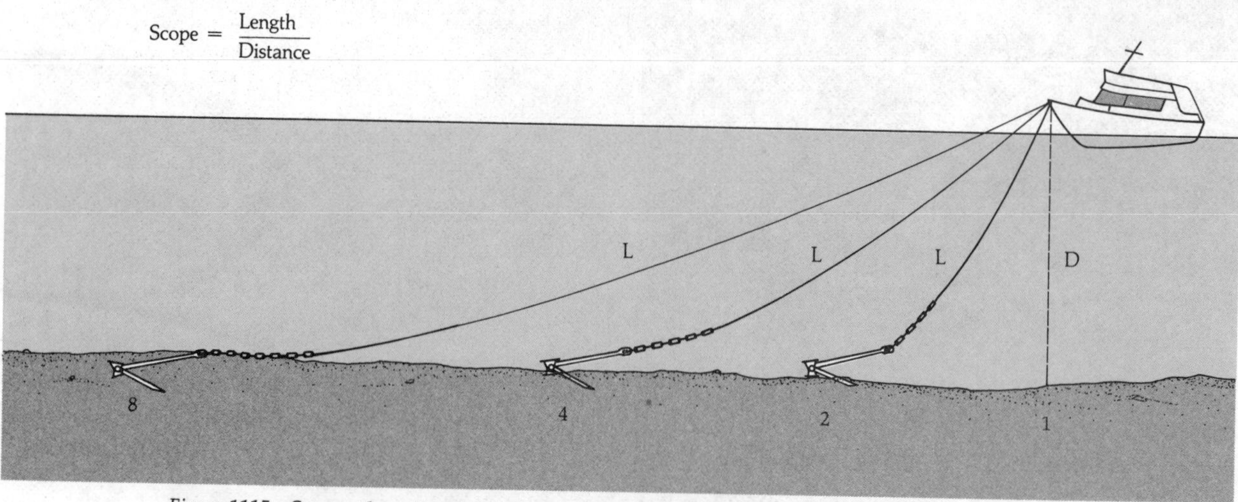

Figure 1115 **Scope, the ratio of rode length L to distance D from bow to the bottom 1 is critically important to safe anchoring. At 2 the rode length is twice distance D, but the angle of pull tends to pull the anchor free. At 4, with L four times the distance D, the anchor can dig in, but there is still too much upward pull on the rode. At 8 scope is 8:1, the short length of chain at the anchor lies flat on the bottom, and any pull helps to dig the anchor in deeper.**

you will, in fact, hold or drag. Too short a scope can destroy the efficiency of the best anchor.

Although some books use the term "scope" to refer to the length of anchor rode in use, most often it is recognized as the *ratio* of the length of the anchor rode to the height of the bow above the bottom of the body of water, as shown in Figure 1115. Note two important factors: the height of the bow chocks above the surface, and the range of the tide.

Let's assume you anchor in 10 feet of water with 60 feet of rode paid out. At first glance, this is a reasonable scope of 6:1. But if your bow chock is 5 feet above the surface, the ratio is immediately cut to 4:1 (60:15). Six hours later the tide has risen another 5 feet and now you have an actual scope of 3:1 (60:20), exactly half the original theoretical ratio, and much too slight for safety.

What is a proper scope? Under favorable conditions, 5:1 might be considered a *minimum;* under average conditions, 7 or 8:1 is regarded as satisfactory. Tests show that proper scope ratios range between 5:1 and 10:1, the latter for heavy weather. Even in a very hard blow, in an exposed anchorage, you will probably never need a scope of more than 15:1 with an anchor of suitable holding power. Effective scope for given conditions varies with the type of anchor.

In our hypothetical example above, the length of rode paid out should have been 140 (7:1) to 160 (8:1) feet; 100 feet (5:1) might be regarded as a minimum.

For maximum efficiency, all anchors require a low angle of pull—preferably less than 8° from the horizontal. With short scope, holding power is reduced because the angle of pull is too high, tending to break the anchor out. As the pull is brought down more nearly parallel with the bottom, flukes dig in deeper with heavier strains on the line. Surging, as a boat pitches in a sea, throws a great load on the anchor, particularly on short scope. With long scope, the angle of pull is not only more horizontal at the anchor, but the elasticity of a long nylon line cushions the shock loads materially.

Marking a Line for Scope

Granting that we know how much scope is required, how do we know when we have paid out enough? Estimates are risky. Plastic cable markers, Figure 1116, come in sets to mark various lengths (such as 25, 50, 75, 100, 125, 150, and 200 feet), and are attached by inserting them under a strand or two of the line. In daylight such markers are fine, but in the dark, the traditional markers of strips of leather, bits of cotton or flannel cloth, and pieces of marline with knots have the advantage of being able to be "read" by feel.

For all practical purposes, five or six marks at intervals of 20 feet (say from 60-140 feet) should be adequate. One practical method is to paint wide and narrow bands of a red vinyl liquid called Whip-End Dip at significant points, calling wide bands 50 feet, narrow ones 10. On chain rodes, as a measure of scope, some boatmen have painted links white at intervals.

If you anchor frequently in the same harbor areas, you may put a whipping to prevent chafe at two or three predetermined places.

Figure 1116 **Plastic numbered markers can be inserted in laid line to indicate the amount of line paid (veered) out.**

Anchoring Techniques

Thus far we have discussed only equipment, or ground tackle. Let's now consider the technique—the art of anchoring. Before you can think about *how* to anchor, however, you must decide *where* you'll anchor, and here, as in all phases of seamanship, a little foresight pays off handsomely.

Selecting an Anchorage

There will be times, of course, when you will stop briefly in open water, coming to anchor for lunch, a swim, to fish, or perhaps to watch a race—but, in the main, the real problem of finding an anchorage comes down to choice of some spot where there's good holding bottom, protection from the wind, and water of suitable depth. Such an anchorage is the kind you'd want for spending the night, free from anxiety about the weather.

Use the Chart

The chart is the best guide in selecting such a spot. Sometimes you will be able to find a harbor protected on all sides, regardless of wind shifts. If not, the next best choice would be a cove, offering protection at least from the existing direction of the wind, or the quarter from which it is expected. As a last resort, anchorage may be found under a windward bank or shore—that is, where the wind blows from the bank toward the boat. In these latter two cases, watch for wind shifts, which could leave you in a dangerous berth on a lee shore.

Anchorages are sometimes designated on charts with an anchor symbol. Areas delineated on the chart by solid magenta lines, perhaps with the water area marked with yellow buoys, may be designated as special anchorage areas, where lights are not required on vessels less than 20 meters (65.6 ft) in length; see Figure 1117. Never anchor in cable or pipeline areas or in channels, both indicated on charts by broken parallel lines in magenta.

Shallow depths are preferred for an anchorage, because a given amount of rode will then provide a greater scope for better holding. You must consider the range of tide, however, so that a falling level does not leave you aground or bottled up behind a shoal with not enough depth to get out at low water. You also must be alert to the special problems of reversing tidal currents, if such exist where you are anchoring.

Characteristics of the Bottom

The character of the bottom is of prime importance. While the type and design of anchor fluke have a direct bearing on its ability to penetrate, it may be stated broadly that mixtures of mud and clay, or sandy mud, make excellent holding bottom for most anchors; firm sand is good *if* your anchor will bite deeply into it; loose sand is bad. Soft mud should be avoided if possible; rocks prevent an anchor from getting a bite except when a fluke is lodged in a crevice; grassy bottoms, while they provide good holding for the anchor that can get through to firm bottom, often prevent a fluke from taking a good hold, except onto the grass, which then pulls out by its roots.

Sometimes bottoms which would otherwise provide reasonably good holding will be covered with a thick growth of vegetation that positively destroys the holding power of any anchor. Even if you happen to carry one of the fisherman's sand-anchor types, with its thin spidery arm and small flukes, expect it to pick up half a bushel of this growth. All you can do is clean it off and try elsewhere.

Characteristics of the bottom are normally shown on charts. By making a few casts with a hand lead armed with a bit of hard grease, you can bring up samples of the bottom as a further check. Chart abbreviations for some bottom characteristics are shown in Figure 1117 bottom.

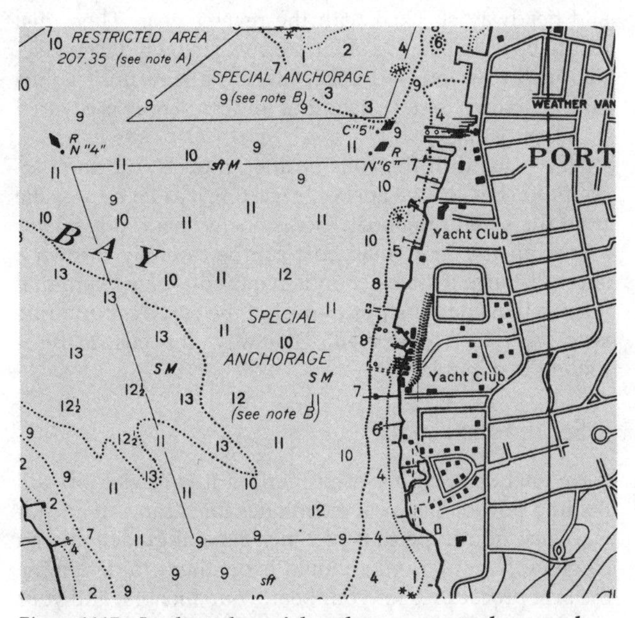

Figure 1117 **In charted special anchorage areas, *above,* anchor lights are not required on boats less than 20 meters (65.6 ft) in length. *Below:* Bottom characteristic abbreviations used on charts.**

Bottom characteristics:

Cl. clay	M. mud	Oys oyster	stk. sticky	gn. green
Co. coral	Rk. rock	hrd. hard	bk. black	gy. gray
G. gravel	S. sand	rky. rocky	br. brown	wh. white
Grs. grass	Sh. shells	sft. soft	bu. blue	yl. yellow

How to Anchor

Having selected a suitable place, and having the proper ground tackle on board, the next step taken is the actual process of anchoring: the approach, getting the anchor down, setting it, and making the anchor line fast. Each step must be done properly if a boat is to be secure.

Approaching the Anchorage

Having selected a suitable spot, run in *slowly,* prefer-

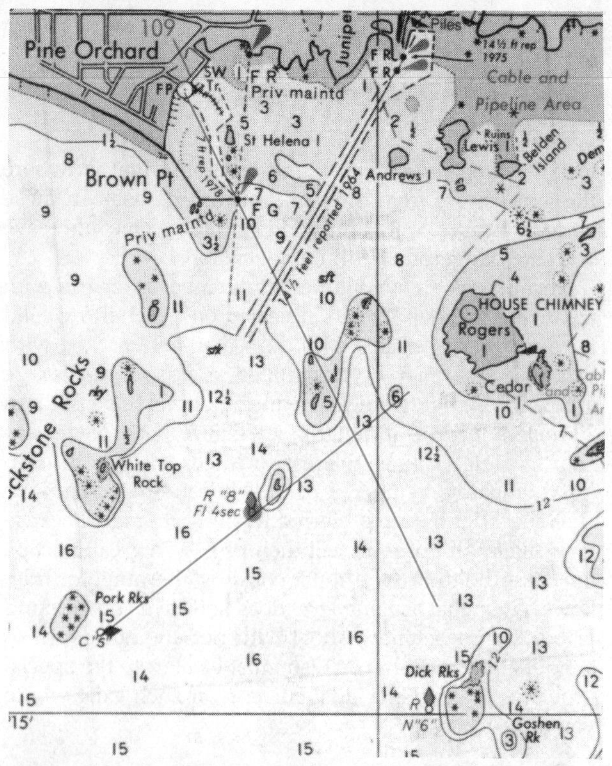

Figure 1118 **Charted range lights, or any charted objects that form a range, are useful in selecting an anchorage spot, and as a means of checking to make sure the anchor is not dragging. Ideally, anchor at intersection of two ranges.**

ably on some range ashore selected from marks identified on the chart, or referring your position to visible buoys and landmarks to aid you in locating the chosen spot. Use of *two* ranges will give you the most precise positioning; see Figure 1118. Later these aids will also be helpful in determining whether you are holding or dragging, especially if the marks are visible at night and it begins to blow after dark.

If there are rocks, shoals, reefs, or other boats to consider, give them all as wide a berth as possible, keeping in mind a possible swing of 360 degrees about the anchor with wind shifts or current changes.

Remember, too, that large yachts nearby may swing to a much longer scope than you allow—and, conversely, that you may swing much further than a smaller boat nearby lying on short scope. A vessel anchored by chain will normally have a shorter scope and thus a smaller swing circle. A boat on a permanent mooring will have the smallest movement of all. Observe how the boats that will be your neighbors are anchored or moored—and don't get into a situation of overlapping swinging circles.

The risk of fouling a neighboring boat is aggravated when, in a current, the deep-draft vessel holds her position while a light-draft boat swings to a shift of wind not strong enough to influence the other. Keel sailboats may lie one way in a light current, powerboats in another way.

The boat that has already established her location in an anchorage has a prior claim to the spot and can't be expected to move if you later find yourself in an embarrassing position. Consequently, allow room enough so that you can pay out more scope if necessary in case of a blow, without being forced to change your anchorage, perhaps at night.

The way other boats lie, together with the set of nearby buoys, will help to determine how you should round up to the chosen spot. Estimate the relative effects of wind and current on your own boat and come up *slowly,* against the stronger of these forces—in other words, heading as you expect to lie after dropping back on the anchor. Running through the anchorage, take care that your way is reduced to a point where your wake cannot disturb other boats.

Letting the Anchor Go

These preliminaries disposed of, you are ready to let the anchor go. Unless you must work single-handed, station one person on the forward deck. Enough line should be hauled out of the locker and coiled down so as to run freely without kinking or fouling. If previously detached, the line must be shackled to the ring, and the stock set up (if of the stock type) and keyed; see Figure 1119. Many an anchor has been lost for failure to attach the rode properly. Rodes, too, have gone with the anchor when not secured at the bitter end. Lightweight anchors are always ready for use and do not have to be set up, but always check to see that the shackle is properly fastened.

Despite what may be seen all too often, an anchor *should not* be lowered when your boat has *any headway.* In a motorboat, or a sailboat under power, the bow should be brought slowly up to the spot where the anchor is to lie, and headway checked with the reverse gear. Then, just as the boat begins to gather sternway slowly in reverse, the anchor is lowered easily over the side until it hits bottom, crown first. (Anchoring under sail only is covered on page 193.)

Never stand in the coils of line on deck and don't attempt to "heave" the anchor by casting it as far as possible from the side of the boat. Occasionally, with judgement, a light anchor in a small boat can be carefully thrown a short distance if such action is required—taking care that it lands in its holding position—but the best all-round rule is to *lower* it as described. That way, the possibility of fouling is minimized.

Setting the Anchor

An anchor must be set properly if it is to yield its full holding power. The best techniques for setting an anchor will vary from type to type; only general guidelines can be given here, and you should experiment to determine the best procedures for your boat, your anchors, and your cruising waters.

With the anchor on the bottom and the boat backing down slowly, pay out line (sometimes spoken of as VEERING) as the boat takes it, preferably with a turn of line around the bitt. When the predetermined scope has been paid out, snub the line quickly and the anchor will probably get a quick bite into the bottom. A lightweight, burial type like a Danforth is frequently set with a scope as short as 2 or 3, especially in a soft bottom. Anchors such as a kedge or grapnel seem to set better with a scope of 5 to 8.

Figure 1119 **In some cases it is possible to set up a kedge anchor with its stock in place so the anchor can be let go quickly. Its chain rode leads outboard from the windlass, through a hawsepipe, up to the ring. The davit tackle can be hooked into the balancing band of the shank. A gypsy is provided for a fiber rode on the port side of the windlass.**

Sometimes the anchor may become shod with a clump of mud or bottom grass adhering to the flukes; in these cases, it is best to lift it, wash it off by dunking at the surface, and try again.

After the anchor is set, you can pay out or take in rode to the proper length for the anchorage, and for the prevailing and expected weather conditions. Scope must be adequate for holding, but in a crowded anchorage you must also consider the other boats.

When you must work single-handed, get your ground tackle ready to let go long before you arrive at the anchorage. Bring the boat up to the chosen spot and then lower the anchor as the boat settles back with wind and current, paying out line as she takes it.

Regardless of the type of anchor, after you have paid out full scope, reverse your engine to apply a back-down load in excess of any anticipated strains. This is particularly important if your boat is to be left unattended.

You must make a positive check that the anchor is holding, and not dragging. There are several ways to do this. If the water is clear enough that you can see bottom, you can detect any movement easily. If you cannot see bottom (which is generally the case) select two objects on the beam that form a natural range and watch for any change in their relationship—if none occurs, your anchor is holding. An even simpler method is possible if you are using a buoyed trip line from the crown of your anchor (see page 248). When you are applying reverse power to test the anchor's holding, the float on this line should continue to bob up and down in one spot unaffected by the pull on the anchor rode. If you see the float making a path through the water, however, you can be sure your anchor is dragging. In warm, clear tropical waters, it is an excellent practice to put on a mask and fins and "swim the anchor," checking visually how well it is buried.

Making Fast

After the anchor has gotten a good bite, with proper scope paid out, make the line fast and shut off the motor.

On boats with a forward bitt (sampson post), an excellent way to secure the anchor line is to make two full turns around the bitt, and then finish off with a half-hitch around each end of the pin through the bitt. The bitt takes the load and the pin secures the line; this way, the line is more easily taken off the bitt than with a clove hitch or any other hitch.

Where a stout cleat is used to make fast, take a full turn around the base, one turn over each horn crossing diagonally over the center of the cleat, and finish with a half-hitch around one horn. See Figure 1310.

The fundamental idea in making fast is to secure in such a manner that the line can neither slip nor jam. If the strain comes on top of a series of turns on a cleat, then you will find it nearly impossible to free if you want to change the scope, without first taking the strain off it.

If you must shorten scope, clear the bitt or cleat first of old turns or hitches. Don't throw new ones over the old.

A trick worth using when the sea is so rough that it is difficult to go forward on deck—especially if you are single-handed—is to set up the anchor in the aft cockpit, lead the line forward on deck through a closed chock and then back aft to the cockpit; if there are stanchions for lifelines, the lead of the rode from chock to anchor must obviously be outside them. When you're ready to let go the anchor can be dropped on the weather side from the cockpit, with the line running through the bow chock and secured on a bitt or cleat aft.

Hand Signals

Anchoring, like docking, is one of the situations where it's a great help to have another hand aboard. A problem is communication between the person on the foredeck and the skipper at the helm. With engine and exhaust noise, it's usually difficult for the skipper to hear, even though the man on deck can. Wind often aggravates the problem. If the skipper is handling the boat from a flying bridge, he can usually hear better and, from his higher position, can see the trend of the anchor line.

In any case, it helps to have a pre-arranged set of hand signals. There is no need for standardization on this, as long as the helmsman clearly understands the crew's instructions. Keep the signals as simple as possible. Motion of the hands, calling for the helmsman to come ahead a little, or back down, can take the most obvious form. Pointing ahead or aft will do it. Simply holding up a hand palm out—a "policeman's signal"—may be used to signal a "stop" to whatever action is then taking place.

Increasing Holding Power

With any burying type of anchor the holding power increases with both an increase in depth of burial and in fluke area. With one patented anchor it is possible to increase both factors. The Bruce Cable Depressor is intended to be shackled between a Bruce anchor and its rode. In soft bottoms, use of one Cable Depressor is claimed to provide a holding power approximately double that of the anchor alone; two units in line ahead of the anchor is said to triple the holding power. There is no increase in the break-out pull required as the Cable Depressor and anchor pull clear sequentially. A 22-pound Cable Depressor is available for use with larger boats.

When the Anchor Drags

Let's assume now that you have anchored with a scope of 7:1, have inspected the rode, and taken bearings, if possible, as a check on your position. Though the wind has picked up, you turn in, only to be awakened near midnight by the boat's roll. Before you reach the deck you know what has happened: the anchor is dragging and the bow no longer heads up into the wind.

This calls for instant action, not panic. A quick check on bearings confirms what the roll indicated. You're dragging, with wind abeam. Sizing the situation up swiftly, you note that danger is not imminent; there is still plenty of room to leeward and no boats downwind to be fouled. Otherwise you would have to get underway, immediately, or be prepared to fend off.

The first step in trying to get the anchor to hold is to let out more scope. Don't just throw over several more fathoms of line; pay it out smoothly, with an occasional sharp pull to try to give it a new bite. If you're dragging badly and can't handle the rode with your hands, take a turn around the bitt and snub the line from time to time. If this doesn't work, start the engine and hold the bow up into the wind with just enough power to take the strain off the rode. This gives the anchor a chance to lie on the bottom and perhaps get a new bite as you ease the throttle and let the boat drift back slowly. If you haven't held when the scope is 10:1, get the anchor back aboard and try again with your larger storm anchor, or in another spot.

Sentinel—or Buoy?

Suppose, now, that you have no spare storm anchor to fall back on. Can anything be done to increase your holding power? Here we enter an area of controversy with something to be said for two quite different techniques. One has the objective of lowering the angle of pull on the anchor line, and the other the lessening of the shock loads on the anchor or on the boat itself. We'll consider both procedures in turn.

For generations, cruising boatmen have known the device known as a SENTINEL, or KELLET. See Figure 1120. In principle, the sentinel is nothing more than a weight sent more than half-way down the rode to lower the angle of pull on the anchor and put a greater sag in the line that must be straightened out before a load is thrown on the anchor. Working only with what came readily to hand in such a case, boatmen have shackled or snapped their light anchor to the main anchor line, and sent it down the main rode with a line attached to its ring, to be stopped at a suitable distance. A pig of ballast or other weight would do as well, provided it could be readily attached.

The other school of thought would use a buoy, Figure 1121, rather than a weight, claiming that, properly used, the buoy can carry most of the vertical load in an anchoring or mooring system, limiting the basic load on the boat to the horizontal force required to maintain the boat's position. The argument is advanced that the buoy permits that boat's bow to ride up easily over wave crests, rather than being pulled down into them, with excessive loads on both rode and anchor.

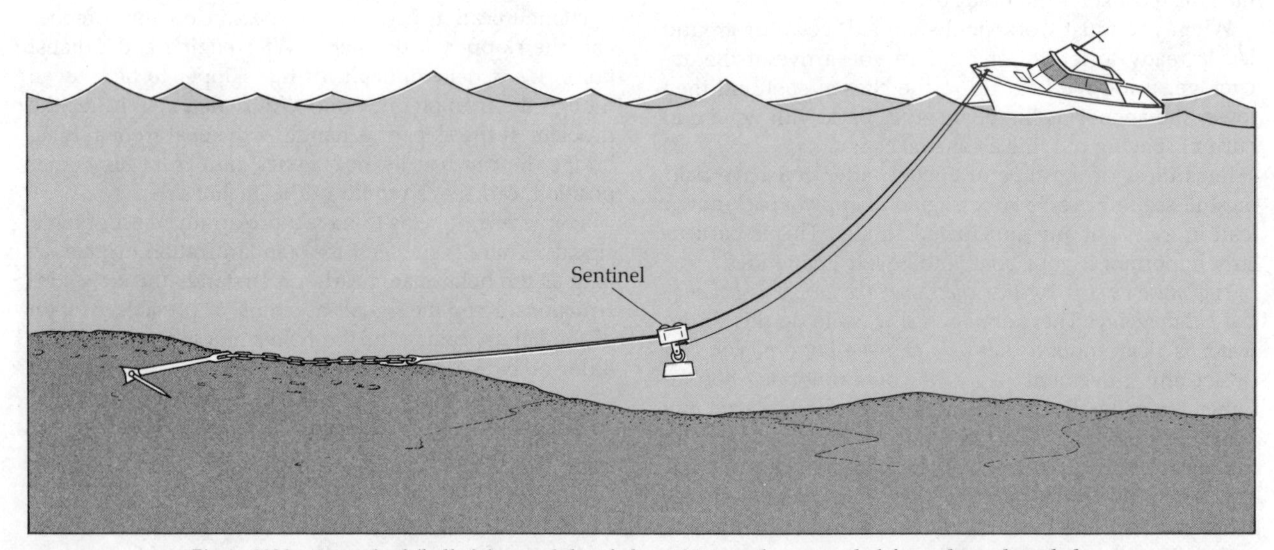

Figure 1120 **A sentinel (kellet) is a weight of about 25 pounds suspended from the rode to help keep the pull on the anchor as horizontal as possible to prevent dragging in rough weather.**

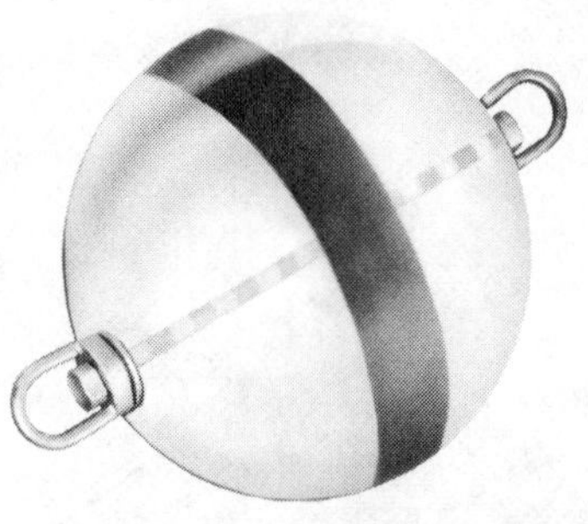

Figure 1122 **A plastic foam buoy: note how the strain is transmitted through a solid rod. Some models have the two connections at the same end of the buoy.**

What to Watch

From these opposing views, certain conclusions may be drawn. To be most effective, a buoy, if used, should preferably be of the type found in a permanent mooring system, Figure 1122, where its efficiency is undoubted. Its connection into the system should be as positive as it would be in a mooring buoy, all strain being carried directly by the rode—in short, no "weak link" here! Provision should be made for carrying a proper buoy as part of the emergency equipment, rather than trusting to a makeshift device improvised under stress of weather.

On the other side of the picture, if the sentinel is to be used, it should be done *with ample scope,* and every precaution taken to avoid chafing the main rode.

An Alternate System

The chain and sentinel techniques can be combined. Carry a boat-length of substantial chain and a 25-pound pig of lead with a ring bolt cast in it. Stow them away somewhere in lieu of ballast. When the chips are down, with breakers to leeward, shackle the chain to your biggest and best anchor, and the chain in turn to your best and longest nylon rode, with the ring of the pig lead shackled in where chain and nylon join. It cannot be anything but an improvement over the same long scope of nylon without benefit of the extra length of chain and added weight. This would seem to eliminate the twin problems of chafe (at the sentinel) and any tendency to hold the boat's bow down in the surge of pitching seas.

Getting Underway

When you are ready to WEIGH ANCHOR and get underway, run up to the anchor slowly under power, so that the line can be taken in easily without hauling the boat up to it. Ordinarily the anchor will break out readily when the line stands vertically.

As the line comes in, you can whip it up and down to free it of any grass or weeds, before it comes on deck. If the anchor is not too heavy, wash off mud by swinging it back and forth near the surface as it leaves the water. With care, the line can be snubbed around a bitt and the anchor allowed to wash off as the boat gathers way, preferably sternway. Two things must be watched: don't allow the flukes to hit the topsides, and be careful that water flowing past the anchor doesn't get too good a hold and take it out of your hands.

Although nylon anchor line will not be harmed by stowing without drying, it is undesirable to carry this additional moisture below decks. Coil the line loosely on deck and allow it to dry, but expose it to sunlight no longer than necessary.

In all anchor handling, try to avoid letting the anchor hit the hull at any time. Whether your boat is made of fiberglass, metal, or wood, some gouges, dents, or nicks may result. Guests are often eager to "help" by getting

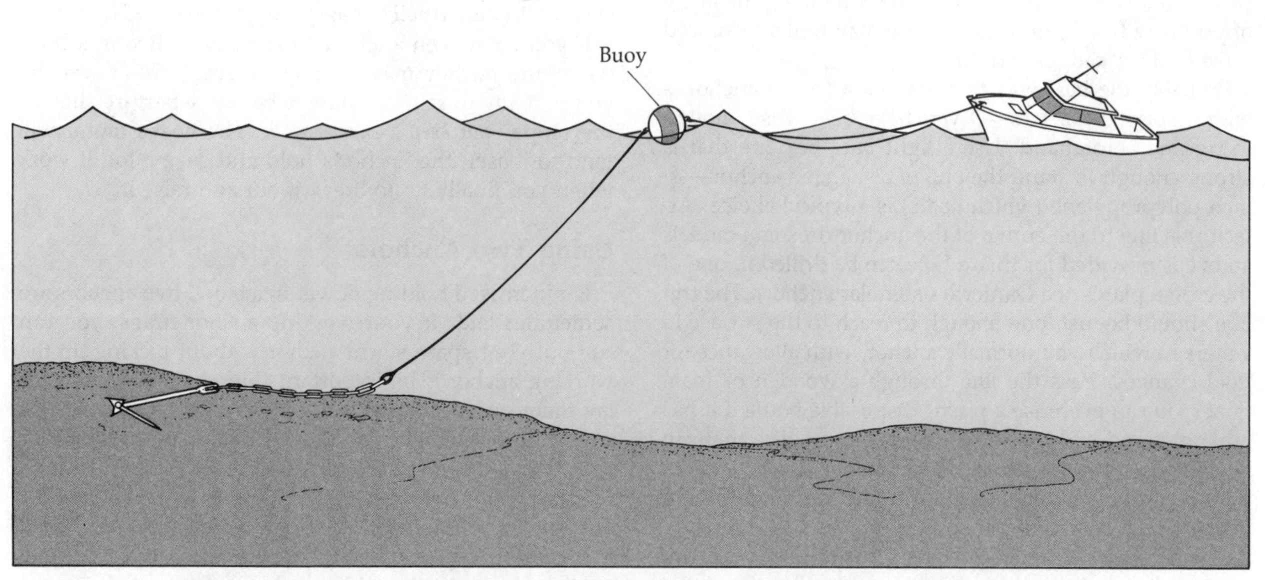

Figure 1121 **A buoy on a rode helps to act as a shock absorber as it allows boat's bow to easily ride up wave crests without excessive strain being transmitted to the anchor itself.**

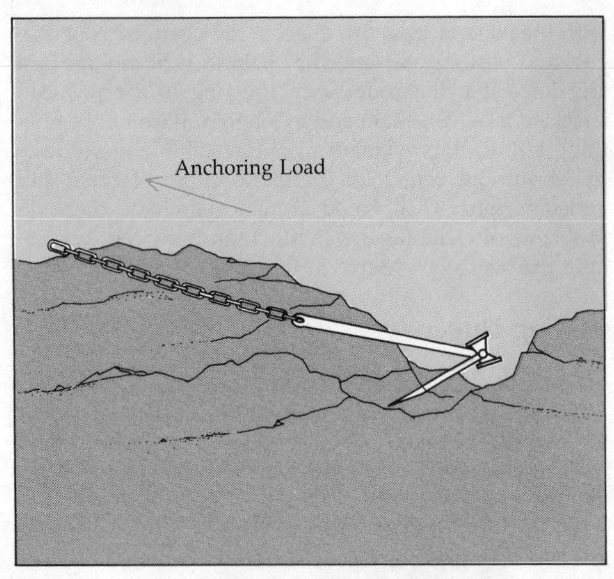

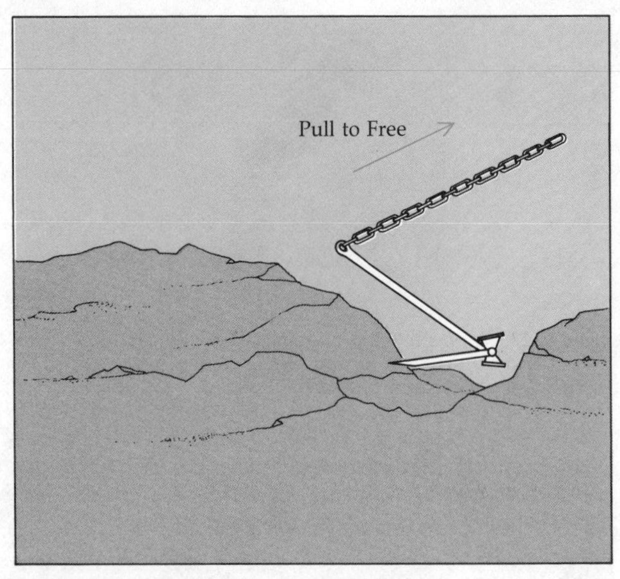

Figure 1123 **If an anchor fouls on a rocky bottom, the first attempt to clear it should be by reversing the original angle of pull, *left,* with moderate scope, to draw it out, *right.***

the anchor up, but unless they have had some experience it's better to handle this part of the job yourself. Handle and stow lines carefully. If a bight or end of line slips over the side it is almost certain to run back under the bottom and get fouled in the propeller.

In a boat under sail alone, have your mainsail up before you break the anchor loose. The same procedure is used as stated above but there is no motor to help. However, it is possible to use your sails to assist. See page 193 for details.

Clearing a Fouled Anchor

If an anchor refuses to break out when you haul vertically on the line, snub it around the bitt and go ahead with the engine a few feet. If the anchor doesn't respond to this treatment, it may have fouled under some obstruction. To clear it, try making fast to the bitt and running slowly in a wide circle on a taut line. Changing the angle of pull may free it, or a turn of line may foul an exposed fluke (if it's a kedge) and draw it out.

Probably the best way to break out a fouled anchor is with a BUOYED TRIP LINE—if you have been wise enough to rig one beforehand. Use a light line, but one that is strong enough to stand the pull of a snagged anchor—3/8-inch polypropylene (which floats) is a typical choice. Attach this line to the crown of the anchor (in some models an eye is provided for this; a hole can be drilled in one of the crown plates of a Danforth or similar anchor). The trip line should be just long enough to reach to the surface in waters in which you normally anchor, with allowance for tidal changes. Pass the line through a wooden or foam float (you can even use a plastic disposable bottle if it has a handle) and end the line in a small eye-splice that can be caught with a boathook. If the anchor doesn't TRIP in a normal manner, pick up the trip line and haul the anchor up crown first.

If you haven't rigged a trip line, sometimes you can run a length of chain down the anchor line, rigged so that another boat can use her power to haul in a direction opposite to that in which the anchor line tends, thus changing the angle of pull 180°. With kedges, if one fluke is exposed, a chain handled between dinghies can usually be worked down the rode to catch the upper arm and draw the anchor out, crown first.

If the anchor is not fouled in something immovable but merely set deeply in heavy clay, you can generally break it out by securing the line at low water and allowing a rising tide to exert a steady strain. Or, if there is a considerable ground swell, snub the line when the bow pitches low in a trough. There's some risk of parting the line this way, in case the fluke is fouled worse than you think.

In rocky bottoms, the first thing to try is reversing the direction of pull, opposite to that in which the anchor was originally set, using a moderate amount of scope; see Figure 1123.

There is a type of anchor in which the ring is free to slide the full length of the shank. Properly rigged, it is claimed to be virtually snag-proof. See page 251.

If you have been anchored for a day or two in a brisk wind, the anchor may be dug in deep. Don't wait till you're ready to sail; 20 minutes before departure shorten the scope—*but keep a sharp watch.* The boat's motor will tend to loosen the anchor's hold and save a lot of work when you finally go to break it out and raise it.

Using Two Anchors

For increased holding power in a blow, two anchors are sometimes laid. If your working anchor drags, you can run out your spare storm anchor without picking up the working anchor. The important thing to remember is to lay them out at an angle, not in line, to reduce the risk of having one that drags cut a trough in the bottom for the other to follow; see Figure 1124.

Special care is necessary for a boat with two anchors out if she is subjected to extreme wind shifts, as might occur with the passage of a squall line. A change of pull on the anchor lines of 180°, more or less, can bring the two rodes into contact with each other in such a way

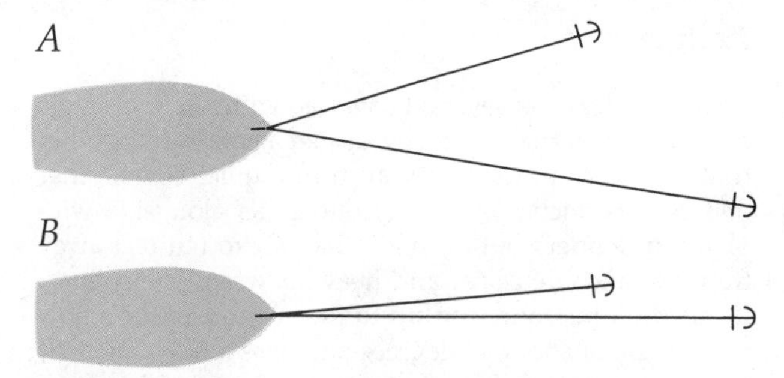

Figure 1124 **If two anchors are set out ahead of the boat, it is best to have the rodes at an angle as at *A* rather than in a straight line as at *B* to reduce the possibility of their fouling as the boat swings to wind or current.**

neither anchor will hold or reset if pulled out. In some situations, one anchor is actually safer than two!

To Reduce Yawing

Deep-draft sailboats usually lie well head to the wind but motorboats often "tack" back and forth at anchor. Skiffs, with high freeboard and little draft forward, are among the worst offenders in this respect.

You can stop yawing by laying two anchors, lines leading out from either bow, making an angle of about 45° between them. To do this, get one anchor down first and have a helper tend the line carefully as you maneuver the bow off to one side before letting the other go. Then you can settle back on both lines, adjusting scope as necessary.

With good handling, you can get two anchors down single-handed. The easiest way is to settle back on one anchor, making fast when the proper scope has been paid out. Then go ahead easily with the propeller, rudder over enough to hold the line out taut so you can keep an eye on it at all times. When the line stands out abeam, stop your headway, go to the foredeck and let the other anchor go, then drop back, snub the line to set the anchor, and then adjust the lines to equal scope.

If a dinghy is available, the second anchor can be carried out in it, lines being adjusted as required after both anchors are set.

Guard Against Wind or Current Shifts

Sometimes you will need to anchor where the tidal current reverses, or wide wind shifts are likely. Here it is wise to set two anchors as security against an anchor breaking out and failing to set itself again.

The anchors are set 180° apart with the bow of the boat at the midpoint between them; see Figure 1125. With both lines drawn up tight, the bow remains over essentially the same spot and swinging is limited to the boat's length. This "Bahamian mooring" works best for a reversing tidal current with the wind blowing across the current so as to keep the boat always on one side of the line between the two anchors.

When setting a second anchor for use as described above, set the up-current anchor in the conventional way and then back down till double the normal scope is out. After the down-current anchor is set, adjust the scope at the bow chocks until both are equal. When going ahead with a rode tending aft, take care not to foul the propeller.

If the two-anchor technique is used in a crowded anchorage to limit swinging radius, remember that other nearby boats may lie to one anchor only, so the risk of having the swinging circles overlap is increased.

Stern Anchors

In some anchorages, boats lie to anchors bow and stern. The easiest way to get these down is to let the bow anchor go first, and then drop back with wind or current on an extra long scope (15-18 times the depth), drop the stern anchor, and then adjust the scope on both as necessary,

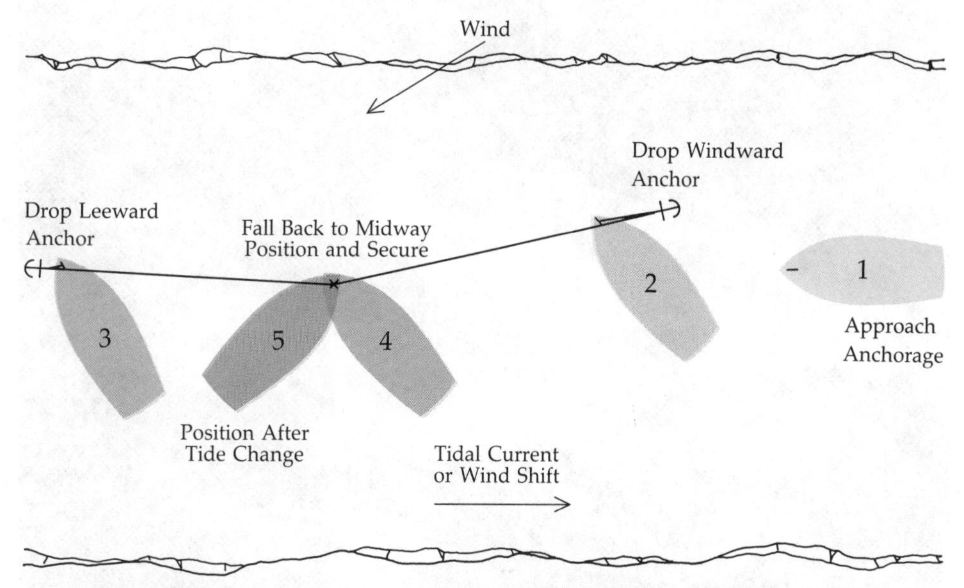

Figure 1125 **When anchoring in a narrow waterway with reversing tidal current, two anchors should be set from the bow as shown. Adequate scope should be used on each anchor, with the rodes adjusted so there is no slack in one when the other is taut.**

taking in line forward. In tidal waters make allowance for increasing depth as the tide rises. The value of this arrangement is generally restricted to areas where permanent moorings are set explicitly for this purpose, as in narrow streams, or on occasions where there is no risk of getting a strong wind or current abeam. Under such conditions, the strain on a vessels' ground tackle could be tremendous.

Sometimes a stern anchor will be useful if you seek shelter under a windward bank. Let the stern anchor go from aft, carefully estimating scope as it is dropped, and pay out more scope as you run up toward the bank or beach. Bed a second anchor securely in the bank, or take a line to a structure or tree ashore.

The stern anchor will keep the stern off and prevent the boat from ranging ahead. But, again, *watch that stern line while the propeller is turning!*

At Piers and Wharves

A berth on the weather side of a pier or wharf is a bad one, as considerable damage can be done to a boat pounding heavily against piles, even with fenders out. Anchors can help to ease the situation in a case where such a berth is unavoidable. Keeping well up to windward, angling into the wind as much as is practicable, have a man let one anchor go on a long scope off the quarter (the port quarter if you'll lie starboard side to the pier). As he pays out scope, run ahead and get another off the port bow, judging positions of both so you can drop down to leeward toward the pier on equal scope, with lines tending off at a 45 degree angle. Properly executed, this maneuver will prevent your vessel from hitting the pier, and the lines you then carry ashore will be needed only to prevent the boat from moving ahead or astern. See also Chapter 7, pages 158-159.

Rafting

At a rendezvous several boats frequently lie to a single anchor. Sometimes as many as ten boats raft together (rather too many for safety even in a quiet cove). After one boat is anchored, the second pulls alongside with plenty of fenders out on both. Stay six to ten feet away from the anchored boat and heave bow and stern lines. If this can't be done, run up to the anchored boat's bow at an angle of about 45 degrees and pass a bow line first, then your stern line. Make sure you have no headway when lines are passed. As soon as the bow line is aboard the anchored boat, stop your engine so that there will be no chance of going ahead, breaking the anchor out.

Allow your boat to drift astern until transoms align, except sailboats. Because of the dangers of spreaders tangling in a rocking situation, it's best to line up rafting sailboats so that the rigging is clear at all points aloft. Then let the bow swing off and pull the sterns in close so it will be easier to step from one boat to another. To keep the boats in line and fenders in position, run a spring line from the stern of the arriving boat to a point well forward on the anchored boat.

If a third boat makes fast, the anchored boat should be in the middle; if more tie up always alternate them, port and starboard of the anchored boat. Each succeeding boat should use the same technique, always with a spring from the stern of the outboard boat forward to the one next inboard. Keels of all boats in the group should be nearly parallel. Naturally, boats should raft only when there is little wind and a relatively smooth surface. When four or more are tied together, it is a good precaution for the outboard boats to carry additional anchors out at a 45 degree angle. When it's time to turn in for the night, every boat should have its own separate anchorage.

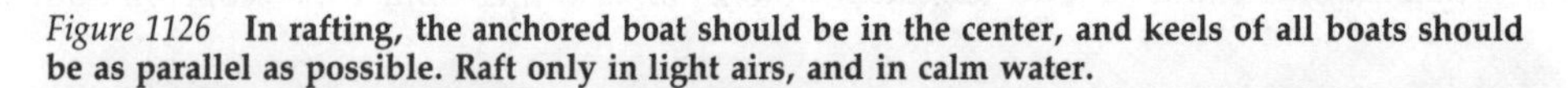

Figure 1126 **In rafting, the anchored boat should be in the center, and keels of all boats should be as parallel as possible. Raft only in light airs, and in calm water.**

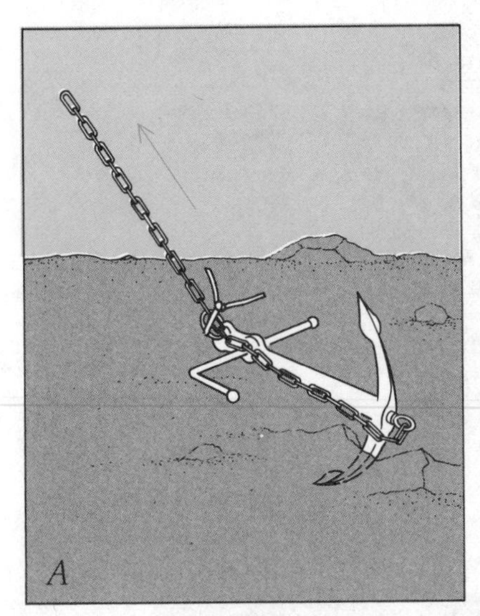

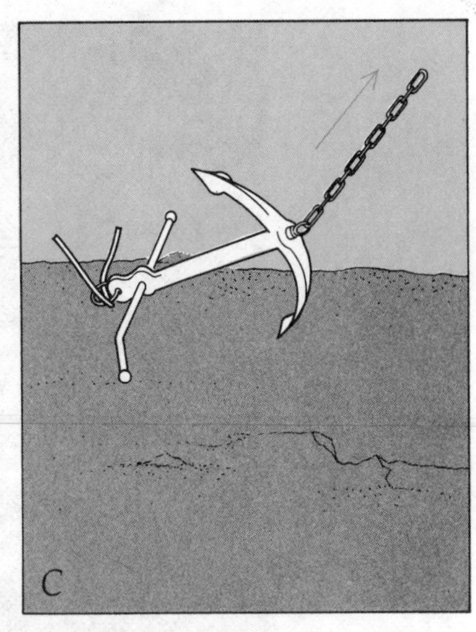

Figure 1127 **Scowing an anchor: The rode is attached to the anchor crown, led back along the shank, and lashed at the ring. If the anchor is snagged (*A*), an upward pull (*B*) parts the lashing and the anchor can be drawn out crown first, (*C*).**

Special Situations

Although the foregoing sections have covered nearly all the problems in anchoring, there are still a few that might be termed "special situations."

Anchoring at Night

When anchoring overnight, if you have no ranges to check your position (or if those you have are unlighted) you can rig a drift lead (the lead line will do). Lower it to the bottom, leave some slack for swinging, and make fast. If it comes taut, you've dragged. Don't forget to pick it up before getting underway.

In general, vessels anchored at night must show an anchor light, two lights if over 50 meters (164.0 ft) in length. Anchor lights are not required for vessels under 20 meters (65.6 ft) in a "special anchorage area," however, or for craft less than 7 meters (23.4 ft) when *not* in a channel, fairway or where other vessels normally navigate. See details in Chapter 4, page 82.

Requirement for "Anchor Ball"

With the same exceptions as noted above for anchor lights, a vessel anchored during daylight hours must hoist a black ball shape where it can best be seen. This is not less than 0.6 meter (23.6 in) in diameter for ships, but may be of lesser size for small craft. See Chapter 4, page 85.

On Rocky Bottoms

Earlier we spoke of certain steps that could be taken to clear a fouled anchor. Avoid rocky bottoms or those with coral heads; they are hazardous at best, regardless of the type of anchor used. Before leaving a boat unattended, apply a test load to the anchor well in excess of any expected load.

If you normally anchor in rocky bottoms or suspect that the bottom is foul in the area where you must anchor, it is better to forestall trouble. One time-tested device is the buoyed trip line described on page 248.

An alternate scheme is to SCOW the anchor, Figure 1127, by bending the rode to the crown, leading it back along the shank, and stopping it to the ring with a light lashing of marline. With sufficient scope, the strain is on the ring and not on the lashing. When hove up short, the strain is on the lashing. When this parts, the anchor comes up crown first.

There are anchors that have a slotted shank in which the ring can travel freely from end to end. If the anchor should snag, the theory is that when the boat is brought back over the anchor, the sliding ring can slip down the shank so the anchor will be drawn backwards. The Sure-Ring, Figure 1128, has an added refinement—a unique ring design that permits the rode to be bent to an eye in the shank for overnight or unattended anchoring. A generally similar design is the Benson "Snag Proof" anchor; here, too, the ring can slide up to the crown end if the anchor must be backed out.

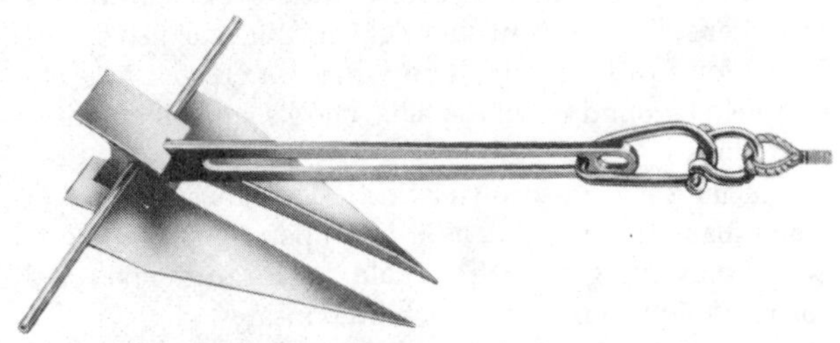

Figure 1128 **In the Danforth Sure-Ring model, a ring slides in a slot down the shank so the anchor can be drawn out backwards if it snags. For use when snagging is unlikely, the rode can be shackled directly to the eye at the end of the shank.**

Kedging off a Shoal

The term KEDGING is applied to the use of a light anchor (not necessarily kedge-type) carried out to deep water in a dinghy to haul a stranded boat off a shoal. If you ever have to resort to this, coil the line down carefully in the

Figure 1129 **Split plastic tubing can be used to cover line where it passes through chocks to protect against abrasion.**

dinghy so that it pays out freely from the stern as you row. The dink might well become unmanageable in a wind or strong current if an attempt is made to pay the line out from the deck of the stranded boat. See also Chapter 10, pages 222-224.

Some Cautions

Take extreme care to avoid chafe on fiber lines. Wherever the line comes in contact with chocks or rails and rubs back and forth under continuous strain, outer fibers may be worn to such an extent as to seriously weaken the line. Mooring pennants are particularly susceptible to this, as are dock lines.

When lying at anchor, you can "freshen the nip" by paying out a little more scope from time to time, or protect the line at the point of chafe with strips of canvas wrapped around it. Modern chafing gear is available in the form of plastic or rubber sleeves, Figure 1129, which can be snapped over the line, centered in a chock, and seized with thongs on either side to prevent shifting. Chafing gear also comes in the form of lacing line or a white waterproof tape, to be applied around the line at bitts, chocks, and other points of contact.

The increasing use of relatively light anchor rodes of small diameter (as small as ⅜-inch) points up the necessity of preventing chafe. The chafe that a ¾-inch line could tolerate might render a ⅜-inch line unsafe.

Chocks, Bitts, Cleats, and Other Fittings

Chafe is aggravated wherever a fitting has a rough surface to accelerate abrasion of the fiber. Even small nicks and scratches in a chock can damage a line by cutting fibers progressively, one at a time. Serious weakening of a line develops when it is forced to pass around any fitting with relatively sharp corners, such as a square bitt with only a minimum rounding of the edges, especially when the bitts are too small for their job. Theoretically the ideal bitt is round and of generous diameter. The best chocks are those of special design with the largest possible radius at the arc where the line passes over it.

Mooring bitts especially must be fastened securely. The best bitt is the old-fashioned wooden bitt, Figure 1130, long enough to have its heel fastened solidly to the boat's keel. If a cast fitting is used on deck, it must be through-bolted and the deck below reinforced with a husky backing plate.

Damage From Ultraviolet Rays

The outer layers of all types of rope are subject to some degree of damage from ultraviolet rays of the sun. With nylon line of relatively large diameter (upwards of ¾-inch)

Figure 1130 **A substantial, well-rounded samson post is shown behind the drum of an electric anchor windlass. Note the horizontal norman pin that keeps line from slipping off.**

such damage is probably negligible. As the diameter decreases the problem becomes proportionately more serious; in 3/8-inch nylon it is an important factor because so much of the line is in the outer fibers. Taken special care to shield such lines from unnecessary exposure to direct sunlight. Often the rode can be fed down into a locker. If it must be carried on deck, shade it or assign it to less critical uses and get new anchor line.

Maintenance

Adequate ground tackle is *essential* to the safety of a vessel of *any* size. Unfortunately, time and use take their tolls. Proper inspection and care are required to get the greatest possible life out of any component of an anchoring system.

Fiber Lines

Keep fiber lines free from sand and grit; dry manila before stowing. Use a low-pressure hose to wash off grit, or slosh the line overboard, tied in a loose coil. Don't use a high-pressure nozzle—it may force grit deeper into the line.

Lines must be straight before any load is applied. Placing a strain on a kinked line can damage or break the fibers. To a lesser degree, sharp bends are harmful; blocks should always have sheaves of adequate diameter; see page 297.

Periodically, lines in regular use should be turned end-for-end; most chafe and wear comes on the anchor end. If you use one size for all anchor and mooring lines, you can put a new spare anchor line aboard each spring, put the former spare into regular use, and make the oldest anchor line into dock lines after cutting out any chafed sections.

Special techniques are required in working with nylon line, as when making up eye splices, or even unreeling it from a coil. These techniques are covered in detail in Chapter 13.

Anchors and Chain

Galvanized anchors are usually coated by the hot-dip process which leaves a tough protective finish, normally requiring no care except washing off mud. Occasionally they are freshened up in appearance by a coat of aluminum paint. You can also buy a spray coating that is heavy in zinc, a "cold galvanize." This coating will temporarily improve the appearance of anchor and chain that are starting to rust, but it has poor wearing qualities when compared to hot-dip galvanizing.

Figure 1131 **Bulb cast into the shackle end of this mushroom mooring anchor adds to its weight and holding power.**

If the chain comes up fouled with mud, give it a thorough cleaning. Some larger yachts have a faucet and hose connected to a fresh or salt water pressure system for this purpose.

Periodic Inspection

Inspect all lines periodically, particularly anchor lines. Check the effect of abrasion, cuts, rust on nylon, broken or frayed yarns, variations in strand size or shape, burns, rot and/or acid stains, and fiber "life." Nylon line may fuzz on the surface although the yarns are not broken. This seems to act as a cushion, reducing further outside abrasion.

Compare the line to new rope. Untwist and examine the inside of strands; they should be clean and bright as in new rope. If powder or broken fibers appear, the line has been overloaded or subject to excessive bending. Nylon may be fused or melted, either inside or out, from overloads.

Most old-timers in yachting know cordage. If you are in doubt, replace your line or get an expert's opinion. For a more detailed discussion of marlinespike seamanship, see Chapter 13.

Permanent Moorings

PERMANENT MOORINGS, as distinguished from ordinary ground tackle in daily use, consist of the gear used when boats are to be left unattended for long periods, as at yacht club anchorages. See Figure 1132. The traditional system often consists of a mushroom anchor, chain from the anchor to a buoy, and a pennant of stainless steel or nylon from the buoy to a light pick-up float at the pennant's end.

Mushroom anchors, especially the type with a heavy bulb cast in the shank, Figure 1131, can, through suction, develop great holding power under ideal conditions if they have enough time to bury deeply into bottoms that permit such burying. Unfortunately, ideal bottom conditions are not always present. Often large cast concrete blocks, similar to those used by the Coast Guard to moor buoys, are put down in lieu of mushroom anchors.

Complicating the problem is the fact that anchorages are becoming increasingly crowded so that boats cannot have adequate scope because of overlapping swinging circles. Add to this the threat of abnormally high hurricane

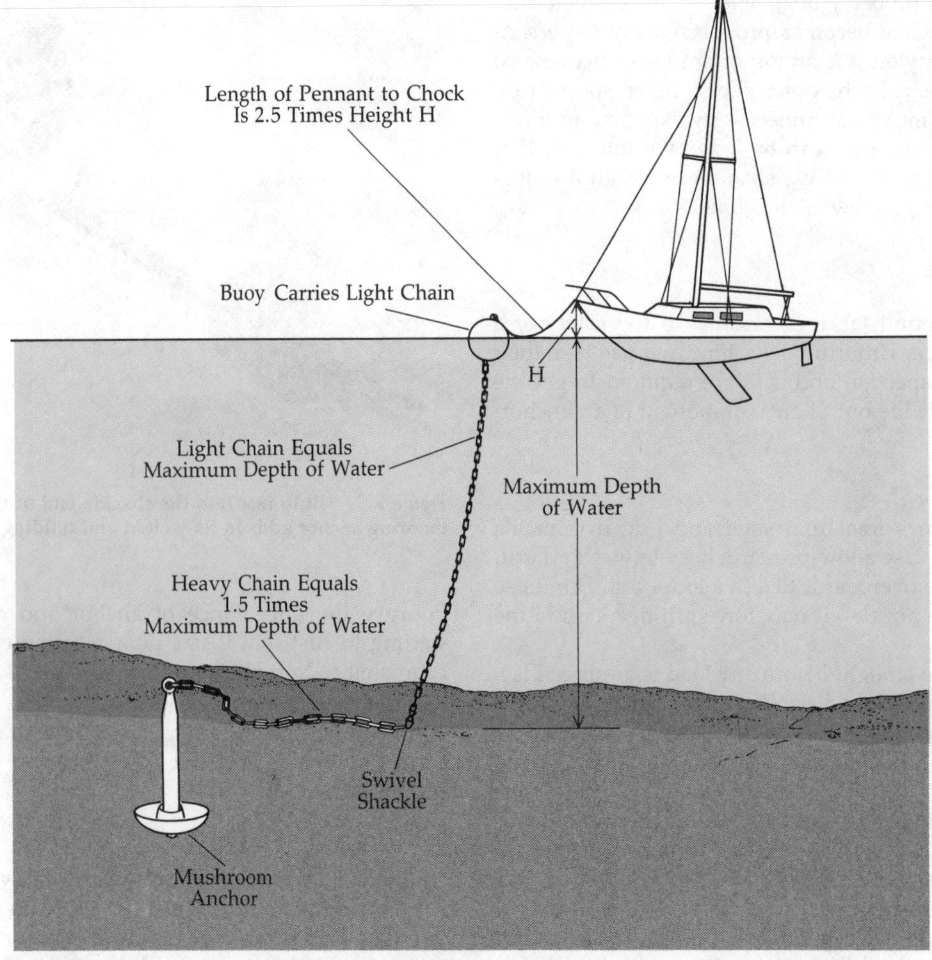

Figure 1132 **Diagram of mooring practice recommended by the Lake Michigan Yachting Association and approved by the U.S. Coast Guard: Total scope is the length of heavy chain, plus length of light chain, plus length of pennant. Minimum space between moorings should be 1.25 times total scope plus boat length.**

Manhasset Bay Yacht Club Mooring System

	Overall Boat Length (Feet)	*Mushroom Anchor Minimum Weight (Pounds)*	*Heavy Chain*		*Light Chain*		*Pennant*			*Total Minimum Length of System, Chocks to Mushroom Anchor (Feet)*
			Length (Feet)	*Diameter (Inches)*	*Length (Feet)*	*Diameter (Inches)*	*Minimum Length (Feet)*	*Diameter Nylon*	*Stainless Steel (Inches)*	
Motorboats	25	225	50	7/8	20	3/8	20	7/8	9/32	70
	35	300	35	1	20	7/16	20	1	11/32	75
	45	400	40	1	20	1/2	20	1 1/4	3/8	80
	55	500	50	1	20	9/16	20	1 1/2	7/16	90
Racing Sailboats	25	125	30	5/8	20	5/16	20	7/8	9/32	70
	35	200	30	3/4	20	3/8	20	1	11/32	70
	45	325	35	1	20	7/16	20	1 1/4	3/8	75
	55	450	45	1	20	9/16	20	1 1/2	7/16	85
Cruising Sailboats	25	175	30	3/4	20	5/16	20	7/8	9/32	70
	35	250	30	1	20	3/8	20	1 1/2	11/32	70
	45	400	40	1	20	7/16	20	1 1/2	3/8	80
	55	550	55	1	20	9/16	20	2	1 1/2	95

System is based on maximum water depth of 20 feet; for greater depths, length of light chain should at least equal the expected maximum.

Table 11-2

tides which reduce scope to a ratio allowing no safety factor, and you have the explanation for the devastation wrought by several hurricanes along the Atlantic Coast.

Systems Used by Typical Yacht Clubs

The problem faced by the Manhasset Bay Yacht Club at Port Washington, N.Y., is typical. Here about 200 boats are moored in a limited space. If each boat could use a length of chain equal to 5 to 7 times the depth of water (maximum 30 feet), safety would be assured, but this would require a swinging radius of several hundred feet for each boat, which is not possible. After exhaustive study the Manhasset boatmen prepared a set of recommended standards, given in Table 11-2. A generally similar system was adopted by the Lake Michigan Yachting Association; see Figure 1132.

Guest moorings are often available at yacht clubs and at some marinas. The launch men will know which of those not in use for the night are heavy enough to hold your boat. As a rule, it's easier and safer to pick up such a mooring. In some places, a charge may be made.

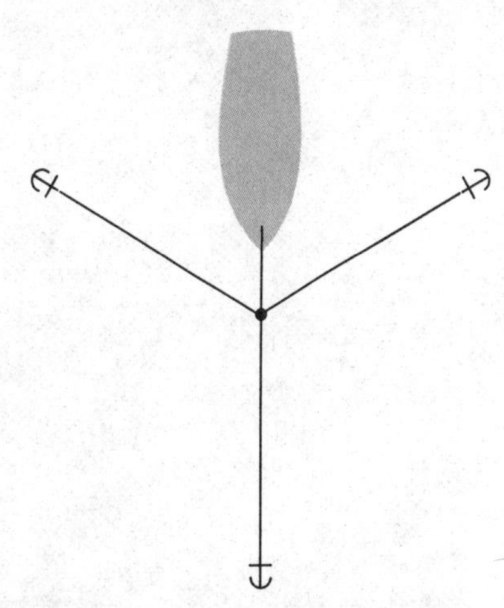

Figure 1134 **A modern permanent mooring system designed to be an improvement over the standard single mushroom anchor uses three lightweight anchors bridled 120 degrees apart. A relatively short rode limits swing.**

A Multiple-Anchor System

One hurricane that ravaged the North Atlantic Coast swept through an anchorage in the New York area and tore almost every boat from her moorings. Only two survived. What these two had in common was an "unconventional" mooring system—multiple (3) anchors bridled to a common center, with chain and pennant leading from that point to the boat.

Figure 1134 shows the obvious advantages of the system. Regardless of how the wind shifts, the boat swings through a small circle, despite the advantage of a relatively long total scope from boat to anchor. The short rode up from the three-way bridle minimizes any "tacking" tendency. Always there will be one or two anchors to windward, the strain tending in the same direction. Using modern lightweight anchors instead of mushrooms, the greater the load the deeper they bury. Mushrooms often need a relatively long period to bed in securely, but lightweight anchors will dig in almost immediately. Using a safety factor of 1.5, each anchor in the three-anchor system should have a holding power equal to the design holding power of a permanent mooring.

Figure 1133 **Salisbury plastic Chafe-Gard is used here to protect mooring lines from abrasion at the bobstay of this auxiliary. It is lashed to the lines with thongs.**

Mooring Buoys

To comply with the uniform state waterway marking regulations (applicable on waters under state control), mooring buoys should be white with a horizontal blue band. Buoys used in any mooring system should be of a type that transmits strain *directly through* the buoy, using chain or rod; see Figure 1122. Buoys perform a useful function in removing much of the vertical load; the pennant is under a more nearly horizontal load, and the boat's bow is freer to lift to heavy seas.

Stainless steel pennants are often preferred because of failures traceable to the chafing of fiber pennants. When fiber is used for the pennant, protect it with chafing gear, especially if there is a bobstay against which it can saw. See Figure 1133.

Annual Inspection Necessary

Because ordinary moorings often need a period of time for silting in before achieving their full holding power, annual inspection of chain, links, shackles, and pins should be made early in the season—never disturbed just about the time storm warnings are issued. On the other hand, it is also wise to give the pennant from buoy to boat a double-check in mid-season, just before the August-September months when hurricanes most often strike.

12

The procedures, organizations, installations, and equipment that contribute to your boating safety

SAFETY AFLOAT

Good seamanship begins with knowledge and use of safe practices in all aspects of boating. For the skipper, this includes knowing his duties and responsibilities, as well as having an understanding of the construction of boats, their equipment, operation, and maintenance.

While not allowing it to detract from his enjoyment of boating, the wise operator practices safety at all time while afloat, and studies it frequently ashore. He recognizes the importance of safety, and it is always in the back of his mind. He views safety not as an arbitrary set of rules, but as the practical application of special knowledge and common sense.

Most boating accidents and difficulties arise from ignorance, and could have been avoided. A person does not knowingly put his life, the lives of others, and the safety of valuable property in jeopardy, but he may do so through lack of knowledge.

Duties and Responsibilities of the Skipper

The primary responsibility of a skipper is the safety of his vessel and the people aboard. This applies to all sizes of boats, on all waters, and at all times.

Leadership

Leadership and discipline are subjects rarely considered by recreational boaters, but they are valuable assets, particularly should an emergency occur.

Discipline means prompt and cheerful obedience to laws and regulations designed primarily for safety. It also means a square deal to shipmates; the skipper who expects discipline of his crew must likewise discipline himself.

Discipline does not mean a long string of commands with a crew constantly scurrying about the deck. There can be discipline on board the smallest yacht without there being any apparent show of it. Real discipline is a function of leadership and leadership can be exercised in dungarees on board even the smallest boat.

Leadership is based on three things: (1) each skipper must know himself, his abilities, and his limitations; (2) he must know his job, know it so well that he doesn't have to think about the details of doing it; (3) he must know his crew and his boat, and what he can reasonably expect of them in an emergency.

Foresight

Next to leadership comes forehandedness or foresightedness. A first-class skipper doesn't wait for an emergency to arise; he has already formulated solutions to any emergency he may face. "Plan ahead" is one of the best pieces of advice you can be given in boating.

Even the most experienced skipper will admit that he has at times been fooled by unexpected effects of wind or current. Dangerous situations can develop with great suddenness—so even when all looks well, watch out! Have an answer to every threat, and a plan to take you out of every danger.

Vigilance

Next in importance is vigilance. The skipper must see intelligently all that comes within his vision, outside and inside his boat. And his vigilance must extend beyond this to *foreseeing* situations as well. A pilot's rule that he must be able to get into alternate fields as well as the airport of his destination holds meaning for the boat skipper, too.

Common Sense

One more checkpoint: common sense. The successful skipper has a sense of proportion and of the fitness of things. We can all recall cases where through lack of common sense, we did things that later looked rather ridiculous. Use your head!

Check of Equipment

Before you get any boat underway, check to see she is really ready to sail. Not only should all equipment required by law be on board and in proper condition for use, but all navigational and other equipment should be at hand. Check water and fuel tanks, inspect ground tackle, check stores, and complete all those other small jobs that can be done easily at the mooring or safe in a slip, but not at sea in an emergency.

Keep a "pre-departure check list"—and other check lists, too; see Chapter 24. Don't trust to memory or routine to ensure that *everything* necessary is done—have a list and use it.

Physical Condition of the Skipper

Also relevant to a boat's safe operation is the physical condition of its skipper. The constant vigil which is necessary requires his complete possession of his faculties and a sense of physical well-being. No one should expose his boat or his people to danger, except in extreme emergency, unless he is in good physical and mental condition.

Although the partaking of alcoholic beverages is commonly a part of many boating activities, the prudent skipper abstains or severely limits his intake while underway or when getting underway is anticipated in the next few hours. Alcohol adversely affects memory, balance, night vision, and muscular coordination, and tests have shown that the effects of sun and wind encountered in boating tend to aggravate these effects. Think before you drink—especially when boating.

Avoiding Risk

Do not permit any of your crew or guests to take needless risks. If someone must do a dangerous job that could result in his being swept overboard insist that he wear a life jacket.

Carefully observe the current weather and check the forecast before getting underway. If the weather doesn't look good, stay in port until conditions improve. A day on the water is not worth the risk of your boat or a life.

Maintaining a Lookout

A small craft cannot readily use the lookout routine of the large vessel, but the vigilance and caution of that routine should be acknowledged. Make it a matter of pride that you as skipper will observe any danger before anyone else does; this is definitely your responsibility.

Specific Responsibilities

The responsibilities of the skipper or of the individual on watch are as follows:

- Safe navigation of the boat.
- Safe and efficient handling of the boat in the presence of other boats.
- Safety of personnel and materiel on board.
- Rendering assistance to all in danger or distress.
- Smart handling and smart appearance of the boat.
- Comfort and contentment on board.
- A good log.

The Crew

Recreational boating is usually a family affair—indeed, one of its most pleasing aspects is that essentially the whole family can share in the activities.

One aspect of boating that should not be overlooked is the *training* of the crew. No boat of any appreciable size should depart its slip or mooring for an afternoon's run or a cruise measured in days without an adequately trained crew, *especially an alternate for the skipper.* One other person should be fully competent to take the helm at least under all normal conditions of wind and waves, and really should be able to do so under adverse weather conditions. The mate should be capable of bringing the boat alongside a pier or of anchoring it. A sudden incapacitating illness or accident may thrust heavy responsibilities upon the mate without warning. If that person has been trained ahead of time, the emergency will be significantly less drastic and much more easily handled.

Training the Crew

A training program need be neither onerous nor unpleasant. It should, however, be planned and carried out on a formal basis so that the skipper can be *sure* that everyone knows what he has to know. The secret to a successful family crew training program is to make it *fun,* and not let it take the pleasure out of pleasure boating.

With just a little imagination a man-overboard drill can be made into a game without losing one bit of its effectiveness. With patience and adequate opportunities for practice, the mate and older children can be developed into skillful helmsmen. Every member of the crew should know where the fire extinguishers are, *and should have had actual experience in using them.* It is well worth the small price of recharging one or two extinguishers to gain the experience of putting out a fire with one. (By the way, skipper, have *you* ever actually put out a fire with an extinguisher of the type now on your boat?) This practice should not, of course, be carried out on the boat, but it is easily done ashore.

The mate, at least, and preferably several others of the crew, should know how to place the communications radio in operation, how to change channels, which channel is to be used for emergencies, and what to say. All persons aboard, crew or guests, should know where the life preservers are stowed *and the proper way to wear them.*

Instruction Plus Practice

A proper training program consists of instruction plus practice. First, learn for yourself what should be done and how, and then pass this information on to your crew. For routine matters such as boat handling, give each crew member the chance to become proficient—it may be quicker and easier on your nerves always to bring the boat alongside a pier yourself, but you should have the patience to

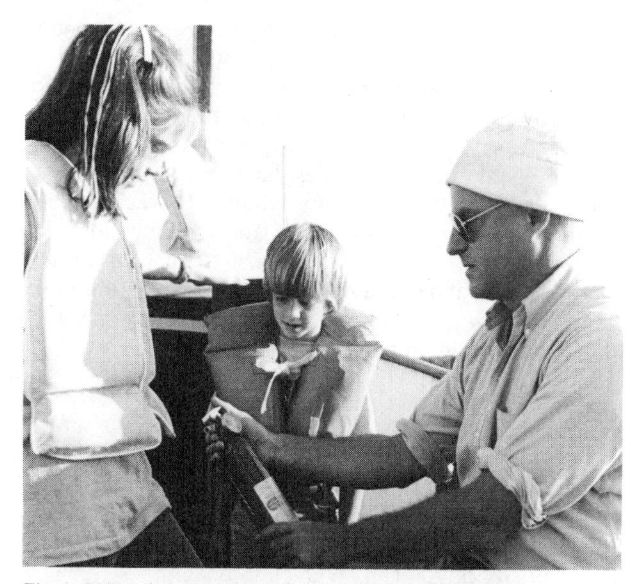

Figure 1201 **It is not enough that approved type fire extinguishers be on board in appropriate locations; every responsible member of the crew, and every guest, should know where they are, and how they are to be used.**

let others learn how by doing. For emergency procedures, have both planned and unannounced drills.

All aboard, *including you,* should be trained to do the right thing instinctively and quickly in a real emergency.

Important Things to Do Underway

Here is a short checklist of things to do:

- Check and plot the boat's position frequently when in sight of land or aids to navigation; be certain of the identification of the objects used to fix position.
- Take soundings and heed their warning.
- Note the effect of wind and current on the boat, especially in close waters or when maneuvering near other boats.
- Do not follow other boats blindly; steer a safe course and do not assume that the other craft is on a safe course.
- When in doubt about your position, slow down; do not wait until the last minute.
- Remember that the other skipper may not see you, and always be alert to take immediate steps to prevent a collision.

Basic Operating Procedures

Fueling

Fueling a boat properly is an essential element of good seamanship. Make sure you have enough fuel on board, and if any is needed, put it in safely. Certain precautions must be *carefully* and *completely* observed *every time* that a boat is fueled with gasoline. (Diesel fuel is non-explosive, but it will burn; the step-by-step procedures given below should be followed with both fuels.)

Before Fueling

1. Make sure that your boat is secured to the fueling pier. Fuel before dark, if possible.
2. Stop engines, motors, fans, and other devices that can produce a spark. Open the master switch if the electrical system has one. Put out all galley fires and open flames.
3. Close all ports, windows, doors, and hatches so that fumes cannot blow aboard and below.
4. Disembark all passengers and any crew members not needed for the fueling operation.
5. Prohibit all smoking on board and nearby.
6. Have a checked fire extinguisher close at hand.
7. Measure the fuel in the tanks and do not order more than the tank will hold; allow for expansion.

While Fueling

8. Keep nozzle or can spout in contact with the fill opening to guard against static sparks.
9. Do not spill gasoline.
10. Do not overfill. Filling a tank until fuel flows from the vents is *dangerous.*
11. For outboards, remove portable tanks from boat and fill on shore.

After Fueling

12. Close fill openings.
13. Wipe up any spilled gasoline; dispose of wipe-up rags on shore.
14. Open all ports, windows, doors, and hatches; turn on bilge blower. Ventilate boat at *least* four minutes.
15. Sniff low down in tank and engine compartments. *If any odor of gasoline is present, do not start engine;* continue ventilation actions until odor can no longer be detected. Check for any drips and liquid fuel.
16. Be prepared to cast off lines as soon as engine starts; get clear of pier quickly.

Figure 1202 **Fires and explosions are most likely to occur while refuelling, or in the first few minutes after refuelling. Take every precaution detailed in the text to minimize this danger. Ground nozzle firmly against deck fill plate.**

Loading

Overloading is a major cause—perhaps the greatest single cause—of accidents in smaller boats and dinghies, and also has significance for medium-size boats. Overloading is particularly hazardous because people do not fear it as they do fires and explosions. Many a skipper, cautious in his handling of gasoline, will unknowingly load his craft far beyond safe limits. The number of seats in a boat is *not* an indication of the number of persons it can carry safely.

Determining Capacity

"Loading" and "capacity" are terms primarily related to the weight of persons, fuel, and gear that can be safely carried. The safe load of a boat in persons depends on many of its characteristics, among them hull volume and dimensions; what it is made of; for outboards whether there is an effective engine well inboard of the transom notch where the engine is mounted; and how heavy the engine is.

USCG Capacity Plates Boats under 20 feet in length (6.1 m)—inboard, outboard, and stern-drive other than sailboats and certain special types—manufactured after 31 October 1972 must carry a capacity plate specified by Coast Guard regulations issued under the Federal Boat Safety Act of 1971. (Boats built before this date *may* carry a capacity plate designed by the Boating Industries Association.) These capacities are computed from rather complex formulas. Plates on outboard craft show both maximum weight for persons and, since 1980, the maximum number of persons, and the total maximum weight for motor, fuel, gear, *and persons*. Plates for inboard and stern-drive craft show only the maximum weight for fuel, gear, and persons. Your boat's capacity plate must be mounted where you can see it when you are getting underway.

Limits on capacity plates apply for boating in good to moderate weather conditions. In rough waters keep the weight well below the limit. The presence of a capacity plate does *not* relieve you of the responsibility for sound judgment. You should know probable future weather conditions as well as those prevailing when starting out.

Safe Loading

Remember that people represent a "live" load; they move about and affect a boat quite differently than static loads like the engine or fuel tank. If your boat's capacity is fully utilized, or the weather gets rough, distribute the load evenly, keep it low, and don't make abrupt changes in its distribution. Make any shift in human or other weights only after stopping or slowing.

Horsepower Capacity

A second aspect of outboard boat capacity is the maximum horsepower motor that it can *safely* carry; this capacity is exceeded perhaps as often as the weight-carrying one. The safe maximum horsepower is included on both the older BIA and the newer USCG regulation capacity plates.

You need not use the maximum safe horsepower; most boats give satisfactory and more economical service with motors of less horsepower. A larger engine does not always mean more speed; it does mean that the increased weight of the engine, and of its fuel and accessories, will significantly reduce the number of people that can be safely carried.

Figure 1203 **Hand gear to a person seated in a small boat; do not attempt to step aboard carrying heavy or bulky items.**

Boarding a Boat

There is a safe way to step aboard a small boat—outboard or dinghy—and an unsafe way. In boarding from a pier, step into the boat as near to the center as possible, keeping body weight low. If you're boarding from a beach, come in over the bow. Keep lines tight or have someone steady the boat.

Never jump into a boat or step on the gunwale (edge of the hull). If you must take a motor or other gear aboard, place it on the edge of the pier where you can easily reach it from the center of the boat. Better yet, after you are in the boat, have someone on the pier hand it to you.

Pre-Departure Actions

Another measure of good seamanship is the procedure you follow before actual departure. If you will be gone for overnight or longer, you should repeat many of these steps each day before getting underway.

Final Weather Check

Few boats are so large that their skipper can ignore weather conditions. You should make a final weather check close to depature time. Know the time of best weather broadcasts, radio and TV, in your home area (these broadcasts vary widely in their scope and suitability for boating); know the telephone numbers of agencies to call for last-minute information. When away from your home waters, find on arrival at each overnight stop how you can get early morning forecasts before departure.

The Marine Weather Services Charts published each year by the National Weather Service (see pages 322 and 323) give information on major radio and TV weather broadcasts as well as the location of day and night storm warning displays. NWS is also steadily expanding the number of VHF-FM radio transmitters that broadcast continuous weather information on 162.55, 162.40, or 162.475 MHz.

Pre-Departure Check List

Prepare your own pre-departure check list *and use it.* The following items apply generally; you may not need them all and you may add some of your own. Check that:

1. All safety equipment is aboard, accessible, and in good working condition, including one Coast-Guard-approved lifesaving device for each person.
2. The bilge has no fuel fumes or water. Ventilate or pump out as necessary.
3. Horn and all navigation lights operate satisfactorily.
4. All loose gear is stowed securely. Dock lines and fenders should be stowed immediately after getting under way.
5. All guests have been properly instructed in safety and operational matters—both dos and don'ts.
6. Engine oil levels are adequate, both crankcase and reverse-reduction gears; water level is sufficient in closed cooling systems. After starting engines, check overboard flow of cooling water.
7. Fuel tanks are as full as you need. Know your tank capacity and fuel consumption at various RPMs, and the cruising radius this gives. Make sure there is enough fuel aboard for your anticipated cruising, plus an adequate reserve if you must change your plans for weather or other reasons.
8. There is a second person on board capable of taking over for you if you are disabled.

Float Plan

Before departing, tell a responsible relative or friend where you intend to cruise and when you expect to make port again; make sure he has a good description of the boat. (Do not attempt to file this "Float Plan" with the Coast Guard; they do not have the manpower to keep track of boats as the FAA monitors aircraft flights.) Tell him of any changes in your cruise plans, so he can tell the Coast Guard where to search and what type of boat to look for if you are overdue. Be sure to tell him when you return, to prevent false alarms about your safety.

Forms are not required for float plans, but they do make it easier to record the necessary information and ensure against omissions. A format for a Float Plan is printed in the Coast Guard pamphlet "Federal Requirements for Recreational Boats" (formerly designated as CG-290); similar formats are in many books and periodicals. Forms are available also from many marine insurance offices.

Safety Organizations

There are many public and private organizations devoted to the promotion of boating safety. Several teach boating safety; others are concerned with the material and operational aspects of boats and boating.

Educational Organizations

United States Power Squadrons The USPS is a volunteer organization of more than 53,000 members organized into over 453 local Squadrons. These units, located throughout the United States and in some overseas areas, offer educational programs of basic safety and piloting to the public, and more advanced courses to members.

It is a nonprofit, self-sustaining private membership organization dedicated to teaching better and safer boating. Despite the word "Power" in its name, the USPS today includes many sailboat skippers in its ranks, and has added the phrase "Sail and Power Boating" to its literature. The boats of members can usually be identified by the USPS Ensign with its blue and white vertical stripes; see color section.

More than 34,000 boatmen and families receive free instruction in the USPS Boating Course every year. For information on local classes, write USPS Headquarters, P.O. Box 30423, Raleigh, NC 27612 or call 800-336-BOAT.

The Canadian Power Squadrons are organized in a manner similar to the USPS, with modifications to fit Can-

Figure 1204 **The United States Power Squadrons is an organization of volunteers dedicated to making boating safer through education. Boats skippered by USPS members are usually identified by the distinctive flag with blue-and-white vertical stripes, shown here at the starboard spreader. The flag at the masthead identifies the owner as a USPS officer; in this case the blue trident on a white field means Rear Commander.**

ada's laws and customs. Additional information on the USPS appears in Chapter 27.

United States Coast Guard Auxiliary The USCG Aux is a voluntary civilian organization of owners of boats, private airplanes, and shore radio stations; it is a non-military group, although administered by the U.S. Coast Guard. It promotes safety in the operation of small boats through education, boat examinations, and operational activities.

Although organized along military lines and closely affiliated with the regular Coast Guard, the Auxiliary is strictly civilian in nature; it has no law enforcement powers. See page 588 for illustrations of USCG Aux flags and uniform insignia.

The Auxiliary carries on a program of courses free of charge except for the cost of materials. Members take specialty courses to increase their knowledge and abilities.

Courtesy Marine Examinations are a well-known program of the Coast Guard Auxiliary. Specially trained members conduct annual checks on boats, but only with the consent of the owner. Boats that meet a strict set of requirements are awarded a distinguishing decal. Figure 314. If a boat fails to qualify, the owner is urged to remedy any defects and request a reexamination. No report of failure is made to any authority.

The Auxiliary promotes safety afloat by assisting the Coast Guard in patrolling regattas and racing events. In many areas, the Auxiliary also participates in search and rescue for vessels that are disabled, in distress, or have been reported overdue.

Information on public classes of the Auxiliary or membership in the organization may be obtained from a local Flotilla or by writing to Coast Guard District Headquarters, or you can call 800-336-BOAT; see Figure 2603, page 595.

More information on the USCG Aux appears in Chapter 26.

American National Red Cross The Red Cross offers programs of water safety education through its more than 3,000 local chapters in all areas of the United States. The swimming and water safety skills taught by the Red Cross range from beginning through advanced swimmer, survival swimming, and lifesaving. The small craft safety courses include canoeing, rowing, outboard boating, and sailing.

Red Cross first-aid training is also available to the public through local chapters. A wise skipper would do well to learn the proper action to be taken in cases of bleeding, fractures, stoppage of breathing, shock, burns, and other common emergencies.

Textbooks used in Red Cross training programs may be purchased from larger bookstores or through local chapters; all are valuable reference publications. Available texts from the Red Cross include *Swimming and Water Safety, Lifesaving and Water Safety, First Aid, Basic Canoeing, Canoeing, Basic Rowing, Basic Outboarding,* and *Basic Sailing;* all are reasonably priced.

The Red Cross publishes a variety of safety pamphlets and posters, and produces informational and instructional films. These are available free to interested groups. Information on classes, publications, films, and other safety activities can be obtained from local chapters.

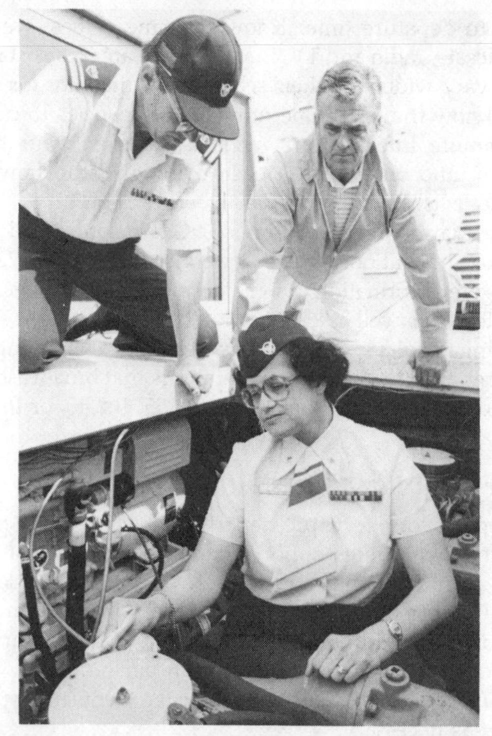

Figure 1205 **Owners can request the free Courtesy Marine Examination given by the Coast Guard Auxiliary. Boats that meet all legal requirements, plus certain additional standards, are awarded the "Seal of Safety" sticker for the current year.**

Boat and Equipment Organizations

In addition to the national volunteer groups working toward greater safety afloat through education, there are organizations related to the boating industry that promote safety through standards for boats and equipment, including its installation and use.

American Boat & Yacht Council, Inc. The ABYC is a nonprofit public-service organization founded to "improve and promote the design, construction, equipage, and maintenance of small craft with reference to safety." Membership is open to companies and individuals.

The American Boat & Yacht Council develops and publishes "Safety Standards" which are recommended specifications and practices for making small boats as free from dangerous defect or deficiency as possible. Standards are stated in terms of desired performance. They are prepared by Project Technical Committees formed as broadly based groups of recognized authorities. All technical reports and safety standards are advisory; the Council has no powers of enforcement.

The ABYC does not "approve" boats, equipment, materials, or services. Some standards refer to other standards or to testing laboratories. Standards are reviewed periodically. New and revised standards are sent out as supplements to the complete compilation published in loose-leaf form. Included are standards for such diverse matters as "good visibility from the helm position," "lifesaving equipment," "engine exhaust systems," "electrical grounding of DC systems," "lightning protection," "sewage treatment systems," and "navigation lights." The full set of recommended practices is published as *Safety*

Figure 1206 **Local chapters of the American National Red Cross teach swimming, lifesaving, and other water safety subjects to persons of all ages. Every person who goes boating should try to become a competent swimmer.**

Figure 1207 **The American Boat & Yacht Council, Inc., is a non-profit, public service organization founded to increase safety in the design, construction, equipment, and maintenance of small boats. The compilation of its "Safety Standards" is worthwhile reading for skippers of all boats.**

Standards for Small Craft, available from the Council at P.O. Box 806, 190 Ketcham Ave., Amityville, NY 11701. This is a good book to have on hand for reference.

Design, safety, and construction standards for boats and motors are also prepared by Boating Industries Association–National Marine Manufacturers Association (BIA-NMMA); the American Bureau of Shipping (ABS); Society of Automotive Engineers (SAE), Lloyds of London, North German Lloyds, and Veritas. Boats manufactured in any country may be built to conform to standards of any of these organizations.

Marine Department of Underwriters Laboratories, Inc. Underwriters Laboratories is a not-for-profit corporation that dates from 1894 and now has a Marine Department with testing facilities at Tampa, Florida. It serves industry and the boating public by conducting safety investigations and tests of marine products, by developing Marine Safety Standards, and by preparing special Marine Supplements to UL Electrical Safety Standards.

The principal activity of the Marine Department is testing boating equipment for safety, a process that begins with manufacturers voluntarily submitting product samples. These are then tested for compliance with appro-

Figure 1208 **Boating products that have passed a safety evaluation by the Marine Department of Underwriters Laboratories are "Listed." The UL label, *above,* can be used in advertising the items, and may be found on them in the form of stickers or tags. Listed products are re-tested periodically to ensure that they continue to meet the UL safety requirements. Fire Protection Standards for Motor Craft, No. 302, *below,* is a safety bulletin issued by the National Fire Protection Association. A copy should be studied by every boat owner, and carried aboard as a useful reference document.**

NFPA No.
302

An American National Standard
ANSI Z120.1-1974
May 14, 1974

MOTOR
CRAFT
(PLEASURE & COMMERCIAL)
1972

NFPA

Copyright © 1974
All Rights Reserved
NATIONAL FIRE PROTECTION ASSOCIATION
470 Atlantic Avenue, Boston, MA 02210

1.5M-4-76-FP(11.5M)
Printed in U.S.A.

priate safety requirements, and evaluated for overall design and construction in relation to their use. After a product has successfully completed the evaluation and complied with all of the UL requirements, the Marine Department conducts a follow-up investigation at the factory to confirm that the manufacturer's production controls comply with UL requirements.

A device that passes all its tests is "listed" by Underwriters Labs and may carry both on the product and in its advertising the UL "Listing Mark," consisting of the Laboratory's name or symbol (see Figure 1208), the product name, a control number, and the word "Listed." The name of the device is included, with the name of the manufacturer, in the annual UL Marine Product Directory. UL does not approve or disapprove anything. Listing is an expression of UL's good-faith opinion, based on tests, that the item meets minimum applicable safety standards. It is not a warranty of quality or performance, nor are all listed products of the same class necessarily equivalent in quality, performance, or merit.

Only products commercially available are eligible for listing. New products may be submitted in their model stage for evaluation, but no final listing action is taken until production units are submitted.

The presence of a UL or UL-Marine label on any device means that a production sample has been successfully evaluated relative to safety requirements, and nothing more. The label may, however, be the basis on which "authorities having jurisdiction" grant approval for use. Such authorities include individuals making judgments for their own purposes, industry people making judgments for components of original equipment installations, marine surveyors for insurance purposes, and administrators of regulations making judgments required by law.

National Fire Protection Association This organization's activities extend far beyond boating and marine interests to include all aspects of the science and methods of fire protection. NFPA issues codes, standards, and recommended practices for minimizing losses of life and property by fire.

NFPA does not approve, inspect, or certify any installations, procedures, equipment, or materials, and it does not approve or evaluate testing laboratories. It does prepare, by coordinated action of committees of experts, codes and standards for the guidance of all persons in the matter of fire protection. Frequently, NFPA codes and standards are written into law or regulations by various governmental units.

Of interest to boatmen is NFPA's booklet Fire Protection Standard No. 302 for Motor Craft (Pleasure and Commercial), available from the National Fire Protection Association, 470 Atlantic Ave., Boston, MA 02110.

Construction for Safety

Safety afloat starts with the design and construction of the boat itself. The typical skipper may not build his own boat, but he should be able to recognize proper design features and sound construction characteristics. He should be reasonably familiar with safe and unsafe aspects of boat construction—the points that should be checked in determining whether or not a particular boat, new or used, should be bought—and thus avoid those that fail to measure up to desirable standards.

A boatman thinking of buying a used boat should seriously consider the services of a qualified MARINE SURVEYOR, who is an expert in determining the condition of the hull, engine, and equipment. The surveyor may well discover defects that the would-be buyer overlooked and his impartial survey is sound protection for a purchaser; the modest cost is well justified. Surveys are often required before a boat-buyer can get insurance, particularly on older boats.

Hull Construction

Modern boats, built by reputable manufacturers, are basically seaworthy if they are honestly constructed to proven designs. This points out the proper course in the selection of a good, sound boat. If you have no experience in selecting a boat, get the guidance of someone who is better qualified; be sure to talk to several people to avoid personal biases.

Where possible, two means of exit should be provided from compartments where people may congregate or sleep. Thus in a small boat with only one cabin and cockpit, a forward hatch is a desirable feature.

Construction Materials

For seaworthiness, materials suitable for boatbuilding and high-quality construction are both essential. Use fire-retardant materials wherever possible. Boats may be built equally well of wood, fiberglass, steel, or aluminum. Combinations of materials may also be used, such as a wooden superstructure on a fiberglass hull. Note too that boats can also be built poorly in any of these materials.

Safety on Deck

A medium-size or large craft may have side decks on which individuals move forward to the bow. It is essential that four safety features be present.

First, these decks must be wide enough for secure footing, even though the space may be measured in inches.

Second, if the deck is made of a material that is slippery when wet (and most are), anti-slip protection must be provided at critical points. This may consist of a built-in roughness in fiberglass decks, grit added to deck paint, or special non-skid strips attached to the deck with their own adhesive backing. Any place where someone has slipped—or could slip—needs non-skid protection.

Third, there must be enough hand-holds—places to grab and hold on—that no one is ever beyond a secure grasp when moving forward and aft, or when coming aboard or leaving the boat. The old saying "one hand for the boat and one for yourself" is an excellent one—even in calm waters; at all times, hold on to something strong enough to bear your weight.

Fourth, no boat large enough for them should be without lifelines, or liferails. They are often carried forward to join low metal rails (pulpits) at the bow and stern to enclose the deck completely.

Owners of offshore cruising sailboats would be wise to meet the requirements set by the United States Yacht Racing Union: fixed bow and stern pulpits (unless lifelines are arranged to substitute adequately for a stern pulpit); pulpits and stanchions through-bolted or welded to the deck; taut, double lifelines with an upper wire secured to pulpits and stanchions; pulpits and upper lifeline at least 24" (61 cm) above deck except in the way of shrouds when the lifelines are permanently attached to the shrouds. The lifelines need not be affixed to the bow pulpit if they terminate at, or pass through, 24" high stanchions set inside of and overlapping the bow pulpit. These requirements are compatible with those of the American Boat & Yacht Council for lifelines and rails on both power and sail boats.

Just because a boat has lifelines doesn't mean you can rely on them. The stanchions should be well-made and through-bolted. On lightly framed wooden boats, make the stanchion bolts long enough to reach below the clamp (inner longitudinal structural member) into backing plates located there. On fiberglass boats you must rely on the quality of the deck construction, but check to be sure that each stanchion base bolts through a generous backing block under the deck.

Even if your lines and stanchions pass muster, don't conclude they are the best point of attachment for the snap of a safety harness in bad weather. A dock line stretched between the mooring cleat on the foredeck and the cleat on the stern has a number of advantages as a point of attachment, as it allows you to move about the deck without having to stop, unsnap and snap on again as you must do at each stanchion if you rely on the lifelines.

Superstructure and Stability

The belief that a boat can carry weight anywhere as long as she appears to trim right can be dangerous. Addition of more superstructure, such as a flying bridge, is often particularly bad, as is shifting heavy weights such as motors, ballast, tanks, and machinery. Where stability is concerned, no shifts of major weights or changes in design should be considered without the advice of a qualified naval architect. Amateur boatbuilders sometimes ruin boats and reduce their safety factors by deviating from architect's specifications for weight distribution.

Through-Hull Fittings

Through-hull fittings which pass through near or below the waterline should have SEA-COCKS installed to permit positive closing. It is desirable to have solid, non-corrod-

Figure 1209 **Anti-slip protection on deck walkways and at boarding areas is a must. Teak strips, *top left,* have natural non-skid qualities. *Top right:* On a boat where small children are carried, netting rigged along lifelines is an excellent safeguard against their falling overboard. Lower and upper lifelines, *bottom left,* are carried right to the bow pulpit of this sailboat. Turnbuckle connections used to tension the lifelines are safety wired, with the wire taped to prevent injury to personnel. Every through-hull fitting, *bottom right,* should have a seacock to cut off entry of water should a hose fail, or to work on equipment served by the fitting.**

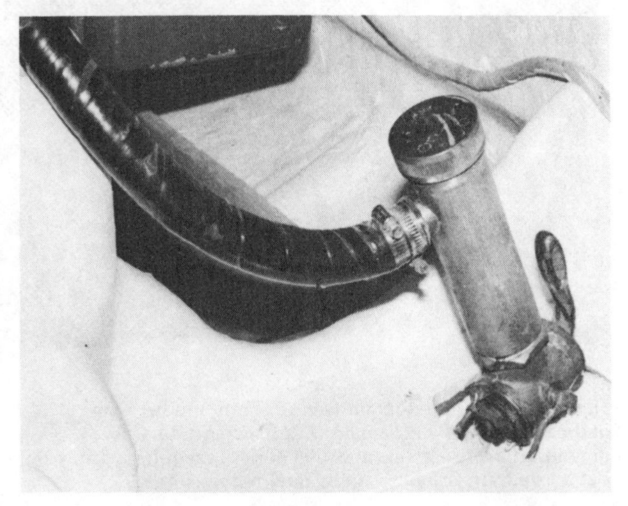

ing metal pipe extend from the sea-cock to a level above the waterline, from which point flexible hoses may be run to the pump or other accessory. Where hose is placed over pipe nipples for connections, use double hose clamps.

There are now available through-hull fittings and sea-cocks that have been safety evaluated and "listed" by the Underwriters Laboratories. ABYC *Safety Standard H-27* furnishes additional information on these.

Engines and Fuel Systems

Fires and explosions on boats are usually traceable to the engine room or galley, where improper equipment, faulty installation, or careless operation and maintenance is the direct cause. All these causes are under the owner's control.

Engine Installation

Engines should be suited to the hulls they power. Extremes of either underpowering and overpowering can be dangerous. Original installations and subsequent changes require the advice of a naval architect or one thoroughly familiar with that phase of boat design.

Amateur conversions of automobile engines for marine use may be a source of trouble because of basic differences between the two types. Lubrication and cooling are the chief problems. Select an accepted type of marine engine or a commercially produced conversion instead.

Pans should be installed under the engine to catch any oil and grease drippings and prevent these from getting into the bilge. Remove accumulations promptly for fire safety.

Fuel Systems

The primary aspect of fuel system safety is the prevention of fires and explosions. Not to be overlooked, however, are the reliability aspects of the system that will ensure a continuous flow of clean fuel to the engine. The discussion that follows will be principally concerned with boats using gasoline engines; exceptions and requirements relating to diesel fuel systems will be covered separately.

The Coast Guard has a set of complex technical requirements for fuel systems of all boats under 20 feet (6.1 m) in length having gasoline engines for mechanical or electrical power or for propulsion. These are requirements placed on manufacturers of new boats, but they can serve as an excellent guide for all boat owners. These standards can be found in the *Code of Federal Regulation,* available in

Figure 1210 **Carburetors on marine engines (other than outboards) must be equipped with flame arrestors. A downdraft carburetor such as this does not require a drip pan, but an updraft carburetor (now rare) requires one.**

the Reference Department of larger local libraries, as Chapter 33, Part 183, Subpart J.

Fire Prevention

A boat's entire fuel system must be liquid- and vapor-tight with respect to the hull interior. Gasoline fumes mixed with air are an explosive combination. These fumes, several times heavier than air, settle to the bottom of the bilge. A concentration of gasoline in air as low as 1¼%—a half-teacup, a few ounces of gasoline—can create enough explosive vapor to totally destroy a large boat.

The obvious answer is *prevention*—make it impossible for gasoline, either in liquid or gaseous form, to get into the bilge in the first place. Then keep the bilge clean and ventilate the engine compartment thoroughly, and there will be nothing that can be ignited.

Leakage of liquid gasoline into the bilge can be prevented by a properly installed fuel system that uses strongly built gasoline tanks, approved tubing for fuel lines, leak-proof connections, tight fittings, and lengths of flexible metallic fuel hose to take care of vibration. As discussed earlier, keep gasoline vapors created when the tanks are filled from finding their way down through open hatches and companionways. Finally, do not permit sparks and flames in the engine room.

Carburetors

Carburetors must be equipped with flame arresters for protection against backfire; see Figure 1210. If not of the downdraft type, they should also have a pan covered with a fine mesh screen attached under the carburetor to collect any drip. Leakage will be sucked back into the engine if a copper tube is run from the bottom of the pan to the intake manifold. Keep flame arresters clean both for safety and best engine operation.

Fuel Tanks

Permanent fuel tanks must be installed securely, to prevent any motion in a seaway. Using portable tanks belowdecks is not good practice.

Fuel tanks must be constructed of a metal that is compatible with both the fuel and a normal marine environment. Copper, copper alloys of certain specific compositions, hot-dipped galvanized sheet steel, and terneplate steel are all used for gasoline tanks. In each case, construction, installation, and inspections should be in accordance with standards of the American Boat & Yacht Council.

Gasoline tanks must not be integral with the hull. The shape of the tank should be such that there are no exterior pockets that would trap moisture after the tank is in its installed position. Tank bottoms must not have sumps or pockets in which water could accumulate.

Internal stiffeners and baffle plates are used to provide necessary rigidity and resistance to surging of liquid in the tanks. Baffles must meet certain specified design criteria to prevent formation of liquid pockets in the bottom of tanks and vapor pockets in the tops.

There should be no drains or outlets for drawing off fuel from the tank, nor outlets or fittings of any type in the bottom, sides, or ends of tanks. Fuel lines to the engine should enter the tank at its top and internally extend nearly to the bottom.

Fuel Filler Pipes and Vents

One of the most essential requirements in the proper installation of a fuel system is that there be a completely tight connection between the gasoline tank and the filling plate on deck. The fuel fill must be located so any spillage will go overboard, *not inside the hull,* and tight pipe connections between deck plate and tank will prevent leakage or spillage below. Fill pipes, at least 1½-inch (3.8 cm) inside diameter, should run down inside the tank nearly to the bottom to lessen the production of vapors.

A suitable vent pipe for each tank should lead *outside the hull;* vents should *never* terminate in closed spaces such as the engine compartment or under the deck. Minimum inside diameter of vent pipes should be 7/16 inch (11 mm). Where a vent ends on the hull, the outlet should be fitted with a removable flame screen to protect against flashbacks from the outside sources of ignition. In boats that normally heel, such as auxiliary sailboats, it may be necessary to have dual vents with the port tank vent led to the starboard side and vice versa.

Non-metallic hose may be used as a coupling between sections of metallic fill piping. The hose must be reinforced or of sufficient thickness to prevent collapse. A grounding jumper wire must be installed across the non-conducting section to provide a complete electrical path from the deck fitting to the fuel tank, which in turn is grounded.

A fuel level indicator, if used, should be of an approved type. If the fuel supply for an auxiliary electric plant is not drawn from the main tank, the separate tank should be installed with its own filler and vent.

Fuel Lines

Fuel lines should be of seamless copper tubing, run in sight as much as possible for ease of inspection, protected from possible damage, and secured against vibration by soft non-ferrous metal clips with rounded edges. A short length of flexible tubing with suitable fittings should be used between that part of the fuel line that is secured to the hull and that part which is secured to the engine itself; this will prevent leakage or breakage as a result of vibration.

Reinforced non-metallic hose may be used for the full distance from the tank shutoff valve to the engine only if the line is fully visible and accessible for its entire length. Wherever used, non-metallic hose must be dated by the manufacturer and not used for a period longer than that recommended by the maker.

Tube fittings should be of non-ferrous drawn or forged metal of the flared type, and tubing should be properly flared by tools designed for the purpose.

A shut-off valve should be installed in the fuel line directly at the tank connection. Arrangements should be provided for ready access and operation of this valve from outside the compartment in which tanks are located, preferably from above deck. Where engine and fuel tank are separated by a distance exceeding 12 feet (3.7 m), an approved-type manual stop-valve should be installed at the engine end of the fuel line to stop fuel flow when servicing accessories.

Valves should be of non-ferrous metal with ground seats installed to close against the flow. Types which depend on packing to prevent leakage at the stem should not be

Figure 1211 **Portable fuel tanks used with outboard motors are equipped with a quick disconnect fitting on the fuel line. This is designed to minimize fuel leakage when detached.**

used. For fuel and oil lines a type of diaphragm packless valve is available which is pressure-tight in any position. A UL-listed electric fuel valve shuts off the flow of gasoline whenever the ignition switch is turned off; it can also be independently turned off in an emergency.

Fuel Pumps and Filters

Electric fuel pumps, where used, should be located at the engine end of fuel lines. They must be connected so that they operate only when the ignition switch is on and the engine is turning over.

A filter should be installed in the fuel line inside the engine compartment, properly supported so that its weight is not carried by the tubing. Closure of the filter must be designed so that opening it for cleaning will not cause fuel to spill. It should also be designed so that the unit can be disassembled and reassembled in dim light without undue opportunity for crossing threads, displacing gaskets or seals, or assembly of parts in an improper order, resulting in fuel seepage after assembly. Fuel filter bowls should be highly resistant to shattering from impact and resistant to failure from thermal shock.

Fuel Systems for Outboards

Fuel tanks and systems permanently installed in the hull of outboard-powered boats should be designed, constructed, and installed in accordance with the above principles. No pressurized tanks should be built into or permanently attached to hulls. A quick-disconnect coupling may be used between motor and fuel line but, when disconnected, it must automatically shut off fuel flow from the tank. Arrangements should be provided so that making and breaking the connection can be accomplished with a minimum of spillage.

Be extremely cautious in using plastic containers for storing gasoline; they can accumulate a static electricity charge that is not drained off by grounding connections. A static electricity-generated spark could ignite vapors in the container.

Plastic containers should meet fire retardance and fuel-compatibility requirements of ABYC. The containers should carry the approval of a national listing agency or a major city fire department. *Do not use any other plastic container for gasoline or other flammable liquid.* Stow any portable fuel container in a well-ventilated place, protected from both physical damage and exposure to direct sunlight. Any leakage or vapors must drain overboard.

Be extremely careful that not even a single drop of gasoline finds it way into the bilge. Proper bilge ventilation is a strict legal requirement and it is a vital safety measure; see Chapter 3.

Fuel Systems of Diesel-Powered Boats

Fuel systems of diesel-powered boats generally conform to safety standards for gasoline-fueled vessels, with a few exceptions as required or permitted by the nature of the fuel and characteristics of diesel engines.

Tanks for diesel fuel may be iron or steel as well as nickel-copper. They may be painted if not galvanized externally; iron tanks must not be galvanized internally. Diesel tanks may have a sump or pocket in the bottom for collection of water but, if so, must have a positive means for removal of any accumulations from above deck. Tanks for diesel fuel may be integral with the hull.

Fuel lines for diesel oil may be of iron or steel pipe or tubing in addition to the metals approved for gasoline. A return line to carry excess fuel from the engine back to the tank is needed.

Additional Information

Many additional technical details on gasoline and diesel fuel systems for boats will be found in ABYC *Safety Standards H-24* and *H-25*.

Underwriters Laboratories Marine Division has evaluated and accepted for listing a number of tanks, filters, valves, and related equipment by various manufacturers.

Exhaust Systems

Exhaust lines and pipes should be installed so that they cannot scorch or ignite woodwork. Where necessary, gratings can be used to prevent gear from touching the line. Exhaust systems must be gas-tight, and constructed and installed so that they can be inspected and repaired readily along their entire length. Any leaks must be rectified *at once* to prevent escape of exhaust gases into various compartments. *Carbon monoxide is a deadly gas.*

A "wet" exhaust system requires a continuous flow of cooling water (from the heat exchanger or the engine block) discharged through the exhaust line, entering as close to the manifold as possible. Exhaust systems of the "dry stack" type (no cooling water) operate at considerably higher temperatures and their use is relatively rare except on larger vessels.

An exhaust system should be run with a minimum number of bends. Where turns are necessary, long-sweep elbows or 45-degree ells are recommended. The exhaust system must not cause back pressure at the exhaust manifold greater than that specified by the engine manufacturer. A small pipe tap, located not more than 6 inches (15.2 cm) from the exhaust manifold outlet, should be provided for measuring back pressure.

The exhaust system should be designed to prevent undue stress on the exhaust manifold, particularly where an engine is shock-mounted. All supports, hangers, brackets, and other fittings in contact with the exhaust should be noncombustible and so constructed that high temperatures will not be transmitted to wood or other combustible materials to which fittings are secured.

The exhaust piping or tubing should have a continuous downward pitch of at least ½ inch per foot (1 cm per 24 cm) when measured with the vessel at rest. It must be designed and installed to eliminate any possibility of cooling water or sea water returning to the engine manifold through the exhaust system.

Exhaust Systems for Sailboat Engines

Exhaust systems for auxiliary sailboats should be designed to prevent sea or cooling water from running back into the engine when these are installed close to or below the water line. A riser must reach above the waterline to allow a steep drop-off of at least 12 inches (30 cm); the remaining piping should have a downward pitch to the outlet of at least ½ inch per foot (1 cm in 24). The high point should be a goose neck, with cooling water injected aft of this point; see Figure 1212. The vertical dry section must be either adequately insulated and provided with a metallic bellows flexible section, or water-jacketed. As an alternate arrangement, installation of a water-trap silencer is recommended.

Flexible Exhaust Lines

Steam hose or other non-metallic material may be used for exhaust lines where greater flexibility is desired. Every flexible line of this type must be secured with adequate clamps of corrosion-resistant metal at each end. Hose used for this purpose must have a wall thickness and rigidity sufficient to prevent internal separation of plies, or collapse. Full-length non-metallic exhaust tubing may be used in wet exhaust systems providing it is water-cooled throughout its length and is not subjected to temperature above 280°F (138°C). Tubing used for exhaust service should be specifically constructed for that purpose and so labeled by the manufacturer. Tubing should be installed in a manner that will not stress or crimp the inner or outer plies.

Additional Information

ABYC Safety Standard P-1 provides much additional information on "wet" and "dry" exhaust installations. It should be studied by anyone building or rebuilding a boat equipped with an inboard engine.

Electrical Systems

A boat's entire electrical installation should comply with the best safety practices. Requirements of marine installations are more exacting than other applications where salt- and moisture-laden atmospheres are not prevalent. Wiring and other electrical equipment should be installed correctly in the beginning and kept safe by frequent inspection.

Information on electrical systems will also be found in Chapter 22, on pages 515 and 516.

The sources of electrical power for engine starting, lights, pumps, electronic gear, and other electrical accessories are usually lead-acid storage batteries kept charged by an alternator integral with the main engine. Larger vessels equipped with electrical cooking and air-conditioning will have a separate engine-driven generator producing AC electrical power similar to shore current. The discussion

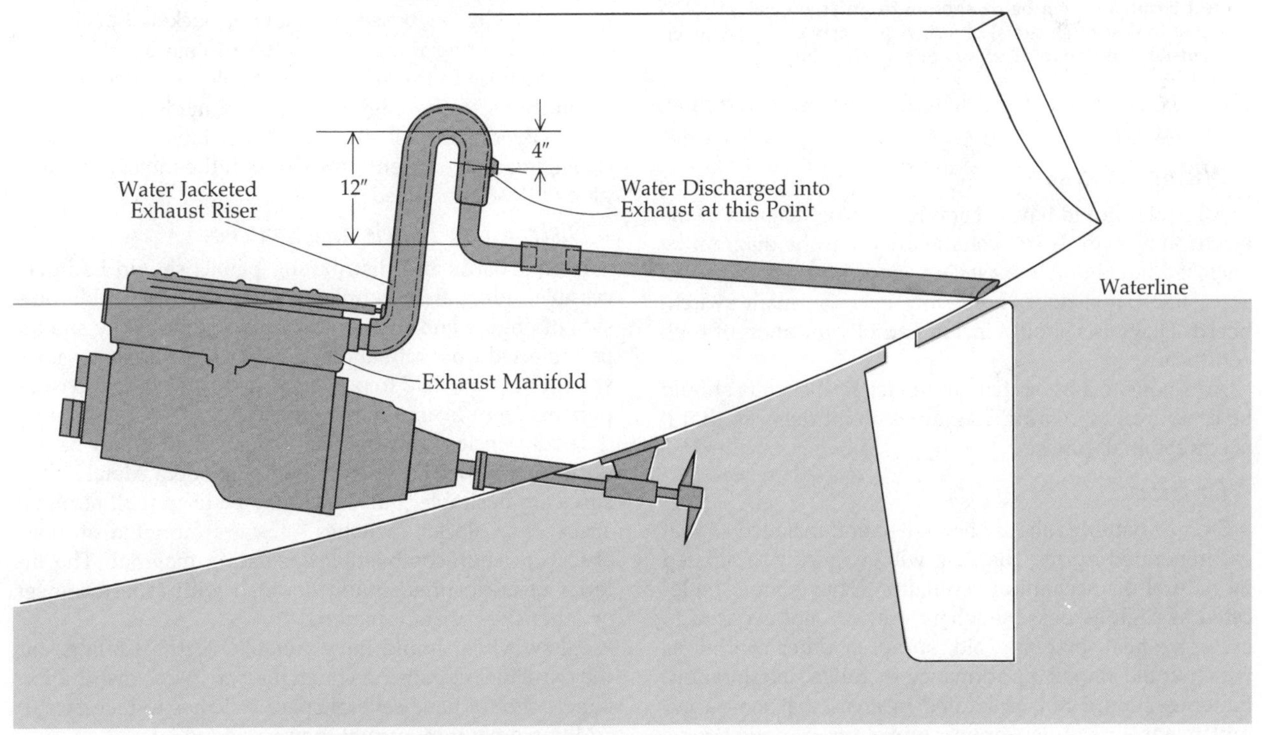

Figure 1212 **The ABYC standard for small craft exhaust systems has special provisions for auxiliary sailboats. Where an engine is located below the waterline, a high-rising loop must be installed in the exhaust line to bring it above the waterline.**

in this section will be directed to the more-common 12- and 32-volt DC electrical systems found on all boats, though they may also be equipped with AC power.

All direct-current electrical systems should be of the two-wire type with feed and return wires parallel throughout the system. Conductors should be twisted together where their magnetic field would affect compasses or automatic pilots. Where one side of a DC system is grounded, connections between that side and ground should be used only to maintain that side at ground potential and should not normally carry current.

The Coast Guard has placed electrical systems requirements on manufacturers of boats under 20 feet (6.1 m) in length that use gasoline for propulsion or for on-board mechanical or electrical power. These are in Chapter 33 of the *Code of Federal Regulations,* Part 183, Subpart I; they

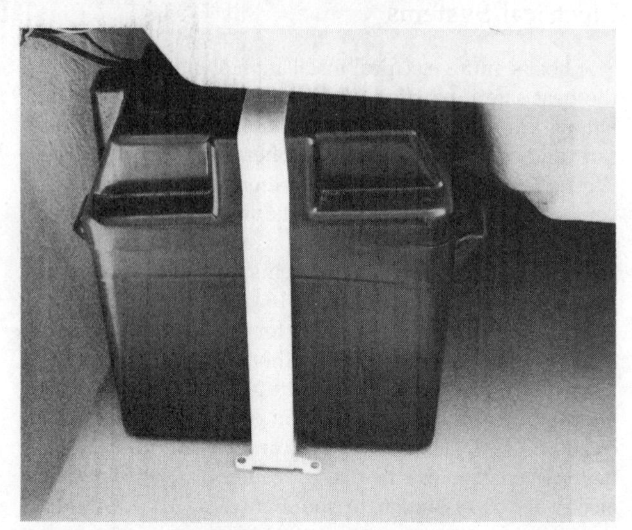

Figure 1213 **A storage battery should be boxed or covered to protect terminals from being shorted by an accidentally dropped tool or other metal object. A properly designed cover prevents accumulation of vapors during charging.**

can serve as a valuable guide to the owners of craft of all sizes and ages.

Wiring Diagrams

All boats should have a complete wiring diagram of the electrical system. It is recommended that the diagram be enclosed in a suitable plastic envelope and secured to an accessible panel door, preferably near the main switchboard. Diagrams should include an identification of each conductor.

Additions to and/or changes in electrical circuits should be drawn on the wiring diagram without delay so that it is correct at all times.

Batteries

Storage batteries should be located and installed so that gas generated during charging will be quickly dissipated by natural or mechanical ventilation. They should be located as high as possible where they are not exposed to excessive heat, extreme cold, spray, or other conditions which could impair performance or cause deterioration. Batteries should not be located in areas that may accumulate gasoline fumes, or near any electrical or electronic devices. Engine-starting batteries should, however, be located as close to the engine as practicable.

A battery must be secure against shifting as well as against vertical motion that would allow it to pound against the surface it rests on. Batteries should be chocked on all sides and supported at the bottom by non-absorbent insulating supports of a material that will not be affected by contact with electrolyte. These supports should allow air circulation all around. Where the hull or compartment material in the immediate vicinity of the battery is aluminum, steel, or other material readily attacked by the acid, the battery should rest in a tray of lead, fiberglass, or other suitable material.

Batteries should be arranged to permit ready access for inspection, testing, watering, and cleaning. A minimum of 10 inches (25 cm) clear space above the filling openings should be available.

A non-conductive, ventilated cover, or other suitable means, should be provided to prevent accidental shorting of the battery's terminals; Figure 1213.

An emergency explosion-proof switch capable of carrying the maximum current of the system (including starter circuits) should be provided in each ungrounded conductor as close as possible to the battery terminal connection. This switch should be readily accessible; a switch with a remote operating handle that can be opened or closed from the bridge or other location outside the compartment containing the batteries is desirable.

Alternators and Motors

Generators, alternators, and electric motors intended for use in machinery spaces or other areas which might contain flammable vapors must be designed and constructed to prevent such devices from becoming a source of ignition. These units must be of a type approved for marine service.

Generators, alternators, and electric motors must be located in accessible, adequately ventilated areas as high above the bilge as possible, in low or pocketed positions.

Ground connections from battery to starter should be made as close to the starter as possible. All other ground connections to the engine should be made at this same point. By making all ground connections at one point, damaging stray-current flow through the metal of the engine will be minimized.

Distribution Panels and Switches

Switchboards and distribution panels should be in accessible, adequately ventilated locations, preferably outside of engine and fuel-tank compartments. They should be protected from rain and spray; where necessary, panels should have a shield to protect terminals or electrical components from overhead dripping.

Totally enclosed switchboards and distribution panels of the "dead-front" type are recommended. Metal enclosures are desirable, but wood may be used if all terminal strips, fuse blocks, switches, etc., are mounted on non-absorbent, non-combustible insulating material. The interior of enclosures should be lined with asbestos sheet or other fire-resistant material.

All switches should have suitable electrical ratings for the particular circuits on which they are used; distribution panels should have several spare switches to take care of additional equipment that may be added later.

Switches used in fuel-tank and machinery spaces, or other hazardous locations, must be designed to prevent

Electrical Wire Gauge Selection

Current (Amps.)	Length of Wire (Feet)					
	10	15	20	30	40	50
5	16	16	16	14	14	12
10	16	14	14	12	10	10
15	14	14	12	10	8	8
20	14	12	10	8	6	6
25	10	10	10	8	6	6

Table 12-1 **ABYC standards for wire gauges for non-critical loads; see text. All wires should be stranded as this type suffers less from mechanical fatigue.**

ignition of flammable or explosive vapors. Such switches must be of a type approved for marine use.

Wiring

Conductors used for general wiring throughout the boat should be approved for marine use to ensure adequate mechanical strength, moisture resistance, insulation, and current-carrying capacity for the intended service.

The table of wire gauges, Table 22-1, page 517, is for an allowable voltage drop of 3% on a 12-volt system. For lighting and other non-critical purposes, the wire size shown there may be reduced to the gauges shown in Table 12-1 (wire gauge number varies *inversely* with conductor size). The "length" for these tables is the distance from the source to the load, one way, measured along the route followed by the wires; this is much greater than the direct distance—don't estimate; measure.

Route all wiring as high above the bilge as possible with consideration given to protection of wire and connections from physical damage. Wiring should be secured throughout its length, at intervals not exceeding 18 inches (46 cm); in a manner that will not crush or cut insulation around the conductor. Non-metallic clamps are excellent; if metallic clamps are used, they must be lined with a suitable insulating material. All non-metallic materials must be resistant to the effects of oil, gasoline, and water.

Unless absolutely necessary, wire splices should not be made in any circuit vital to the operation of the boat nor in the navigation light circuit. Splices, where made, should be taped to relieve all strain from the joint. If solder is used, it should be a low-temperature resin-core radio solder. Acid-core solder must *not* be used.

Terminal Connections

Terminal connections should be designed and installed to insure a good mechanical and electrical joint without damage to conductors. Metals used for terminal studs, nuts, and washers should be corrosion-resistant and galvanically compatible with the wire and terminal lug.

Terminals may be of solderless type, preferably with ring-type ends. Solderless terminals should be attached with crimping tools specified by the manufacturer of the terminal. The holes in ring-type terminals should be of a size that will properly fit the terminal stud. Formed and soldered wire terminal connections should not be used. It is recommended that terminal lugs include a means of clamping the wire insulation for mechanical support. A short length of insulated sleeving over the wire at each terminal connection is desirable.

No more than four conductors should be secured to any one terminal stud. Where additional connections are necessary, two or more terminal studs should be used and connected together with copper jumpers or straps.

Wires terminating at switchboards, junction boxes, or fixtures should be arranged to provide a surplus length of wire sufficient to relieve all tension, allow for repairs, and permit multiple wires to be fanned at terminal studs.

Additional Information

Further guidance and recommendations for DC electrical systems for boats are in *Safety Standards E-1, E-3, E-9, E-10, A-20,* and *A-21* of the American Boat & Yacht Council. Excellent information will also be found in *Your Boat's Electrical System, 1981-82 Edition,* published by Hearst Marine Books.

AC Electrical Systems

More and more boats are being wired for 117-volt AC electrical systems so that larger, home-type appliances and equipment may be used afloat. Power may be obtained from dockside connections only, or also from an engine-driven generator while underway. From a safety viewpoint, it is important to recognize the increased hazards from this higher-voltage wiring and devices.

All component parts of a 117-volt AC system must be designed, constructed, and installed to perform with safety under environmental conditions of continuous exposure to vibration, mechanical shock, corrosive salt atmosphere, and high humidity. The system must provide maximum protection against electrical shock for people on the boat, in the water in contact with the boat, or in contact with the boat and a grounded object on shore.

Appliances and fixed AC electrical equipment used on boats must be designed so that current-carrying parts of the device are effectively insulated from all exposed metal parts by a non-conductive material suitable for use in damp locations.

Further technical specifications for AC electrical systems on boats will be found in *ABYC Safety Standard E-8.* Portions of standards *E-1* and *E-3* are also applicable.

Information on AC electrical systems will also be found in *Your Boat's Electrical System, 1981-82 Edition.*

Electrical Shock Hazards

Voltage is only one factor in determining whether a shock will be fatal, injurious, or only startling. Many people have received shocks from spark-plugs, where the voltage is 10,000 or more, with no more harm than a few bruised knuckles. The actual factors that determine the extent of harm are: (1) the amount of current that flows through the body; (2) the path of the current through the body; and (3) the length of time that the current flows.

The effects of different levels of current are not exact and will vary with the individual and the path of current through his body. Unfortunately, there is no way of predicting how much current will flow in a given situation—play safe; work only on de-energized circuits. Make certain no one can make contact with a 117-volt shore-power circuit and the water.

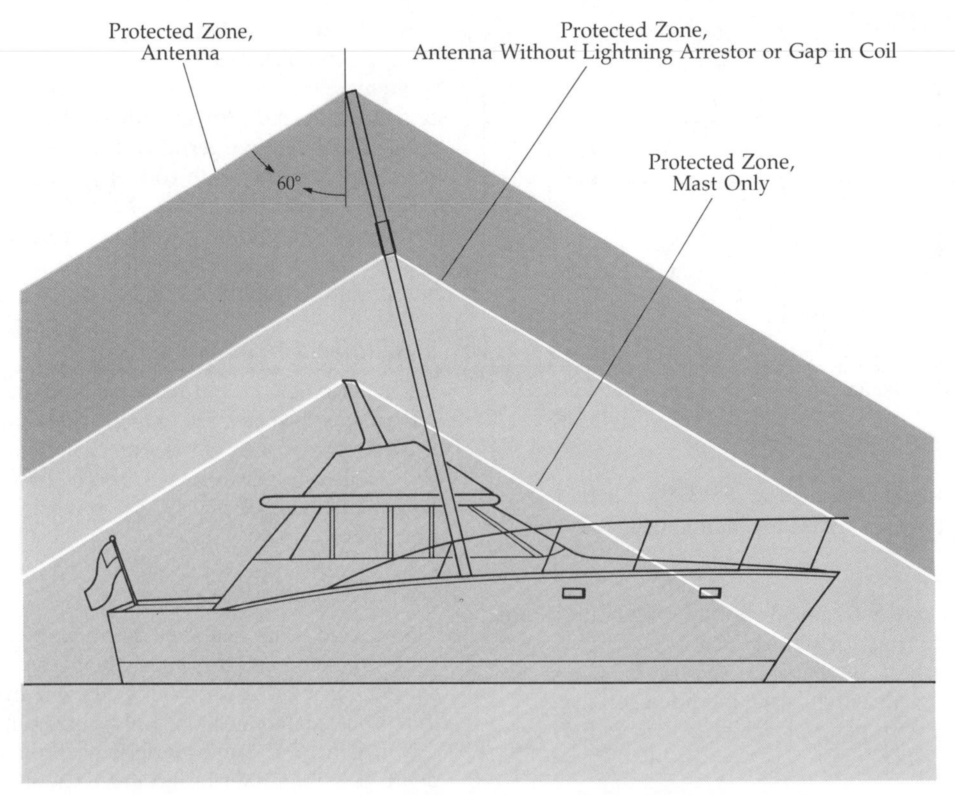

Figure 1214 **Under certain conditions (see text) a grounded radio antenna can be used to provide a "cone of protection" from lightning.**

Lightning Protection

You only infrequently hear of a boat, power or sail, being struck by lightning, but this can, and does, happen. You can add to both your physical safety and your peace of mind by getting some basic information and taking certain precautions.

Protection Principles

A grounded conductor, or lightning protective mast, will generally divert *to itself* direct lightning strokes which might otherwise fall within a cone-shaped space, the apex of which is the top of the conductor or mast and the base a circle at the water's surface having a radius approximately twice the conductor's height. Probability of protection is considered to be 99.0% within this 60-degree angle as shown in Figure 1214. Probability of protection can be increased to 99.9% if mast height is raised so that the cone apex angle is reduced to 45 degrees.

To provide an adequately grounded conductor or protective mast, the *entire* circuit from the masthead to the ground (water) connection should have a conductivity equivalent to a #8 gauge wire. (If this wire is stranded, no strand should be smaller than #17 gauge. Any copper strip used should not be thinner than #20 gauge, 0.032 inch, 0.8 mm.) The path followed by the grounding conductor should be essentially straight with no sharp bends.

If there are metal objects of considerable size within a few feet of the grounding conductor, there will be a strong tendency for sparks or side flashes to jump to them from the grounding conductor. To prevent such possibly damaging flashes, an interconnecting conductor should be provided at all likely places.

Large metal objects within the hull or superstructure of a boat should be connected with the lightning protective system to prevent a dangerous rise of voltage due to a lightning flash.

Protective Measures

For powerboats, a radio antenna may serve as a lightning rod if it has a transmitting-type lightning arrester or means for grounding during electrical storms, and if the antenna height is sufficient to provide a cone of protection adequate for the length of the boat. Antennas with loading coils are considered to end at a point immediately *below the coil* unless the coil has a suitable gap for bypassing lightning current. This protection can be obtained *only* with an antenna of metal rod or tubing. Fiberglass whip antennas with an internal wire conductor—almost universally used for VHF, CB, or SSB—will *not* provide lightning protection.

Sailboats with metal standing rigging are adequately protected if all rigging is bonded together and connected to ground. A wooden mast must have a heavy metal conductor running down the mast from a short "lightning rod" at the masthead; this conductor is tied into the grounding system.

Metal objects situated wholly on a boat's exterior should be connected to the grounding conductor at their upper or nearest end. Metal objects within the boat may be connected to the lightning protective system directly or through the bonding system for underwater metal parts.

Metal objects that project through cabin tops, decks, etc., should be bonded to the nearest lightning conductor at the point where the object emerges from the boat and again at its extreme lowest end within the boat. Spotlights and other objects projecting through cabin tops should be solidly grounded regardless of the cone of protection.

A ground connection for lightning protection may consist of any metal surface, normally submerged, which has an area of at least one square foot. Propellers and metal rudder surfaces may be used for this purpose; a radio ground plate is more than adequate. A steel hull, of course, constitutes a good ground connection to the water.

Protection for Personnel

The basic purpose of lightning protection is safety for people, so everyone on board should take the following precautions.

Stay inside a closed boat as much as possible during an electrical storm.

Avoid making contact with any item connected to a lightning protective conductor, and especially in a way that bridges two parts of the grounding system. For example, do not touch both the reverse lever and spotlight control at the same time.

Stay out of the water.

Additional Information

ABYC Safety Standard E-4 covers lightning protection for all types of small craft.

Equipment for Safety

Some safety equipment is required by laws and Coast Guard regulations. These requirements are generally written in such broad language, however, that boatmen need additional knowledge and guidance.

Boats need to have additional items of equipment beyond the legal minimum, to receive the approval decal of the Coast Guard Auxiliary Courtesy Marine Examination. See Chapter 3.

Still further pieces of equipment are desirable for safety, and convenience. Many of these are discussed in Chapter 3 and other chapters. This chapter will supplement equipment considerations covered elsewhere in the book.

Lifesaving Equipment

As noted in Chapter 3, each boat must, by law, have a personal flotation device (PFD) for each person on board, and water skiers towed behind are counted as being "on board." A wise skipper does not skimp on the quality or quantity of his PFDs, making sure that he has enough even if a few unexpected guests show up.

Where the type of PFD used comes in different sizes, such as adult and children's life preservers, or the three sizes of buoyant vests, the PFD must be of an appropriate size for the person who is to wear it.

Although the regulations allow a choice from three or four categories of PFDs, the safety-conscious skipper will choose Type I. This type provides the maximum buoyancy and will hold a person's face clear of the water so he can breathe even though unconscious.

Pumps and Bailers

There are no federal requirements that a boat (if used exclusively for recreation) be equipped with a bailing device, but all boats should be equipped with some form of pump or bailer. Too many skippers place their *full* dependence on electrical bilge pumps, and perhaps on only one of those. Once water rises in the bilge high enough to short out the battery, the pump is out of action. (Incidentally, on most boats the batteries are placed low in the bilge—excellent for stability, but not for safety if the boat starts taking on water.)

Only the smallest boats should depend on a hand bailer or bucket. Other boats should be equipped with a manually operated pump of generous capacity. Such an item can be either a fixed installation or a portable pump stowed where it can be quickly and easily reached when needed.

Distress Signaling Equipment

A boat's radiotelephone is probably the most often used means of summoning assistance in an emergency. Use of the radio for this purpose is covered on pages 528-530. The legal requirements for carrying visual distress signals are covered on page 70.

Without special equipment, a distress signal can be made to other vessels in sight by standing where one is clearly visible and slowly raising and lowering one's outstretched arms.

Signal flares may be projected several hundred feet into the air by a special type of pistol, or to a lesser height from a pocket "tear gas" gun that resembles an old-fashioned fountain pen. The laws of several states regulate possession of these and similar explosive projectile devices; check carefully before putting one aboard, but if permissible, these are effective and desirable items of safety equipment.

There are also hand-held flares of the single- and double-ended types, the latter producing a bright flame at one end for night use and a dense smoke at the other end for day distress situations. Before using any flare, read all instructions carefully so as to avoid personal injury or aggravation of the emergency. Read carefully *before* any emergency arises.

Special lanterns are available that will blink the three-short, three-long, three-short flashes of S-O-S.

Distress signaling equipment of any type should be stored in protective, waterproof containers and stowed where it is readily available for use in an emergency. Containers should be opened and contents inspected at least annually; flares have an expiration date and must be replaced periodically.

Miscellaneous Items of Safety Equipment

There are many small items of equipment that increase safety and convenience.

All but the smallest boats should be equipped with several *flashlights* and at least one battery-powered ELECTRIC LANTERN. Flashlights should be distributed where they will be quickly available in an emergency. One should be within reach of the skipper's bunk. There should also be a flashlight in the guest cabin, if any, and another on the bridge. Batteries in all units should be checked monthly to be sure of sufficient power. All should be completely waterproof, and lanterns should be the kind that will float if dropped overboard.

An installed searchlight of adequate intensity is desirable on medium-sized and large boats. It is useful both in routine navigation and docking, and in emergencies such as man-overboard situations or assisting another boat in distress. On smaller boats, the electric lantern can serve as a hand-held searchlight.

Boats of medium or large size should have on board some means of cutting lines in an emergency.

TOOLS and REPAIR PARTS for the engine and major accessories can be considered parts of a boat's essential safety equipment. Personal experience and conversaton with seasoned skippers will offer the best guidance to just what tools and parts should be carried aboard any particular boat. Insofar as possible, tools should be of rust-proof metal; tools susceptible to rust or corrosion, and all spare parts, should be given a protective coating, and stowed to protect them from adverse effects of a marine environment.

All but the smallest boats should carry EMERGENCY DRINKING WATER and FOOD supplies. The amount will vary with the number of persons normally on board. The type of food is not important—anything will do when you get hungry enough! Non-perishability over a long period is important, but it might be wise once each year to consume your emergency supplies (a shipwreck party?) *after* they have been replaced with a fresh stock. Such items can usually be purchased from marine supply stores or from Army-Navy surplus stores; food should be simple and compact, but sustaining and energy-producing.

FIRST-AID KITS are essential safety items; they will be considered in more detail later in this chapter.

NAVIGATION EQUIPMENT, CHARTS, and GROUND TACKLE are all items of safety equipment, but each is considered in detail elsewhere in this book; see index.

Safety Aspects of Other Equipment

Several items generally found on boats are not safety equipment per se, but have definite aspects of safety about

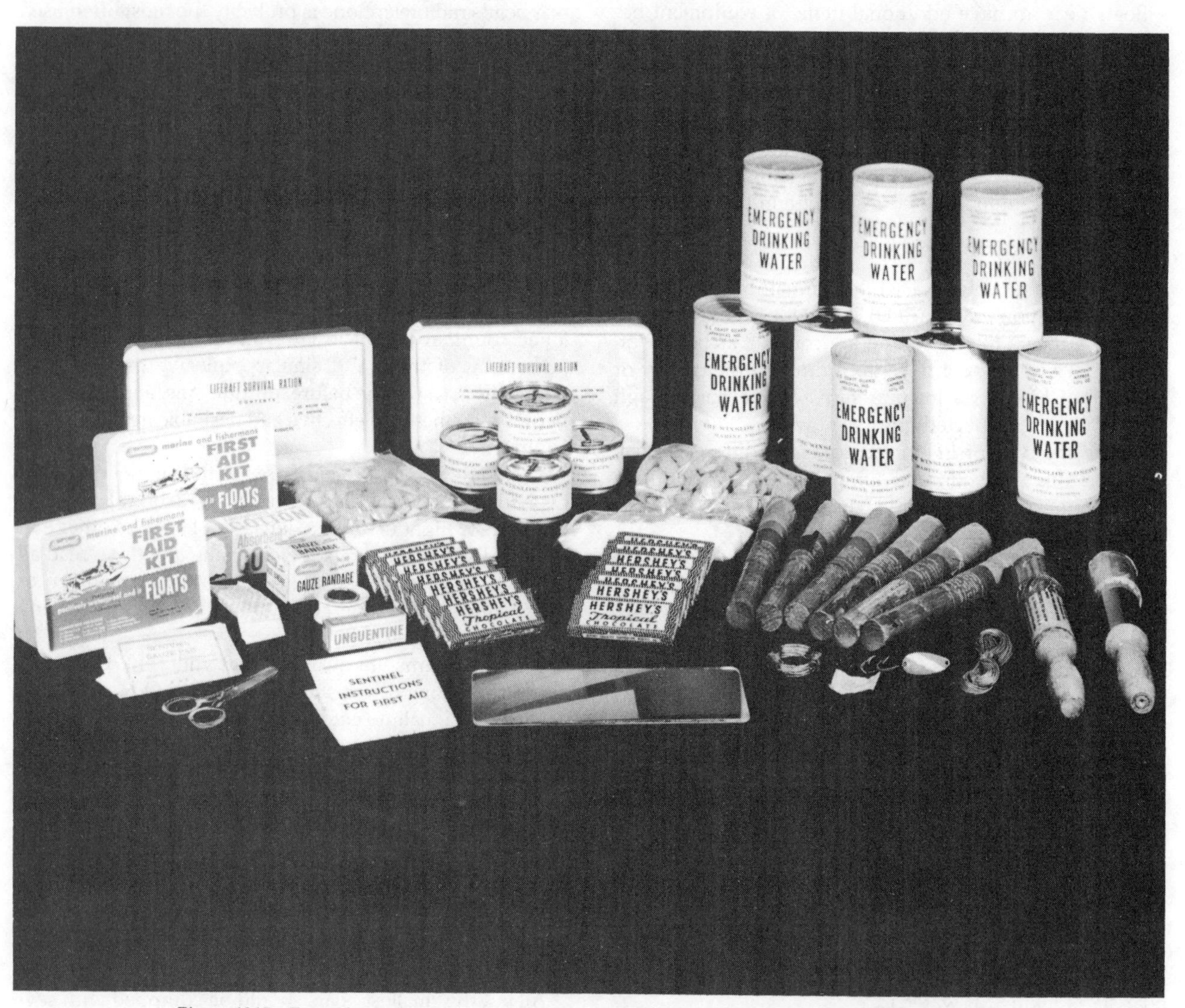

Figure 1215 **Every boat should carry a supply of emergency food and water, with type and amounts based on size of boat and crew, as well as the normal cruising area. Prepared kits of long shelf life products, as illustrated, can be purchased, or the owner can assemble his own package of emergency supplies.**

Figure 1216 **Sailboat galley stove is hung so that cooking surface remains level when boat heels. Note guard rails that keep pots or pans from sliding off the front, rear, or sides of the stove, and the hefty grab rail on the counter next to it.**

their design, installation, or operation that must be considered by boat owners.

Galley Stoves

Galley stoves should be designed, manufactured, and approved for marine use. Types of fuel that are ordinarily used include alcohol, kerosene, electricity, and liquified petroleum gas (LPG); a few boats are now using compressed natural gas (CNG) as stove fuel. Stoves fueled with coal, wood, diesel oil, or canned heat (solidified alcohol) are only rarely seen. Gasoline is *not* a safe stove fuel and should never be used on a boat.

Electricity is probably the safest source of heat for cooking, but an auxiliary generating plant is required to produce the large amounts of AC power required. Because of the inexpensive and simple nature of the equipment, alcohol stoves are widely used on boats despite their rather high fuel costs; with moderate and reasonable precautions, such installations can be quite safe. Plain water will extinguish alcohol fires.

LPG stoves are excellent for cooking, but can present a serious safety hazard unless installed and operated in accordance with strict rules. An excellent guide to a safe LPG installation will be found in the requirements of the Coast Guard Auxiliary Courtesy Marine Examination.

Stoves should be permanently and securely fastened in place when in operation. Portable stoves are not recommended; if one must be used, it should be secured while in use. Adequate ventilation should be provided to prevent too much heat when a stove is operated for extended periods of time. All woodwork or other combustible material around a stove, including smoke stacks, must be effectively protected with non-combustible sheathing; see Figure 1216; 1/8-inch (3.1 mm) asbestos board covered by sheet metal is recommended.

Fuel for alcohol and kerosene stoves may be supplied to the burners either by gravity or pressure systems, provided fuel tanks cannot be filled while the burners are in operation except where the supply tank is remote from burners and the filling operation will not introduce a fire hazard. A removable or accessible liquid-tight metal drip pan at least 3/4-inch (1.9 cm) deep should be provided under all burners. Pressure tanks should have suitable gauges and/or relief valves.

Refrigeration

Ice is used on some boats as a means of keeping food fresh and for cooling beverages. It offers no safety hazards, but water from melting ice should be piped overboard rather than into the bilge. Fresh-water drip tends to promote dry rot and/or odor. A collection sump, with pump, may be used if desired.

Mechanical refrigeration is normally of the electric-motor-driven compressor type similar to units found in the home. On some boats a compressor is belt-driven from a main engine.

Safety aspects of mechanical refrigeration include use of a non-toxic and non-flammable refrigerant, non-sparking motors, safety valves on high-pressure portions of the system, and general construction adequate to survive the rigors of marine service.

Kerosene and bottled-gas refrigerators have a constant small open flame. This can produce an explosion hazard, and the use of such absorption-type units is presently excluded from the ABYC Safety Standard A-6.

Heaters

Cabin heaters are sometimes used on boats in northern waters; gasoline should never be used as a fuel in these, and portable kerosene or alcohol heaters are not recommended. Built-in electrical heaters are safe; portable electrical heaters should be used only if secured in place while in operation.

Any heater discharging combustion products must be vented through a stovepipe and a "charley noble" set in a water-cooled deck plate or other fire-preventive fitting. Many sailors favor coal- or charcoal-burning heaters, despite the mess they create and the problems of storing fuel, because of the even, dry heat they deliver.

LPG (liquid petroleum gas) heaters should have an automatic device to shut off the fuel supply if the flame is extinguished; pilot lights should not be used.

Many models of air-conditioning equipment used on boats are of the "reverse cycle" or "heat pump" type and can supply warmth rather than cooling when needed, provided that the surrounding water is not too cold. This is a thoroughly safe heating method.

Maintenance for Safety

Continual attention is needed for some of the material aspects of safety; for others, periodic checks at weekly, monthly, or annual intervals is sufficient. The important thing is that these checks of safety equipment be made *regularly* when needed; use of a check list is strongly recommended.

Keep Your Boat Clean

Cleanliness is an important aspect of safety. Accumulations of dirt, sawdust, wood chips, and trash in the bilge will soak up oil and fuel drippings and become a fire hazard. Such accumulations may also stop up limber holes and clog bilge pumps. Keep your bilge *absolutely* free of dirt and trash; check frequently and clean out as often as needed.

Lifesaving Equipment

Lifesaving equipment usually requires no maintenance, but it should be carefully inspected at the beginning of each boating season and again near its mid-point. Check for cut or torn fabric, broken stitches, and other signs of deterioration. *Do not delay in replacing below-par lifesaving devices;* attempt repairs *only* where *full* effectiveness can be restored. In case of doubt, ask the Coast Guard.

Fire Extinguishers

Installed fire-extinguishing systems should be checked at least annually, more often if recommended by the manufacturer. An excellent procedure is to inspect at the beginning of the year's boating activities, and again at mid-season. Portable fire extinguishers should be checked at least every three months; monthly checks are not too frequent.

Dry Chemical Extinguishers

Pressurized dry-chemical extinguishers have a gauge that should be checked for an indication of adequate pressure—needle in center or green area of scale. Do not, however, merely read the gauge; tap it lightly to make sure that it is not stuck at a safe indication. If it drops to a lower reading take the extinguisher to a service shop for recharging. Even if the gauge reads O.K., take the unit out of its bracket and shake it a bit to loosen the dry chemical inside to keep it from settling and hardening. Any areas on the exterior of the cylinder showing rust should be cleaned and repainted. *Do not* test by triggering a short burst of powder. The valve probably will not reseat fully, and the pressure will slowly leak off.

CO_2 Extinguishers

Carbon dioxide extinguishers must be checked annually by weight. Any seals on the valve trigger must be unbroken. Portable units and built-in systems have a weight stamped at the valve. If the *total* weight is down by 10% or more of the *net* weight of the contents, the cylinder must be recharged. Weighing is normally the only check for CO_2 fire extinguishers, but it must be done accurately. Portable units can be checked by the skipper if he has suitably accurate scales. Built-in systems are best checked by a professional serviceman with his special knowledge and equipment.

CO_2 cylinders also should have the date on which the cylinder was last hydrostatically pressure-tested. This should be done every 12 years if the cylinder is not discharged. If the extinguisher is used, it must be pressure-tested if this has not been done within the preceding five years. If a used CO_2 extinguisher is purchased, it should be discharged, hydrostatically tested, and recharged before it is installed aboard. The pressure in these cylinders is tremendous, and if the cylinder has been damaged through rust or corrosion, it is just like a time bomb, and you do not know when it may go off!

Carbon dioxide extinguishers should not be installed where bilge or rain water will collect and cause rust. Any exposed metal should be painted at each annual inspection. If a cylinder becomes pitted from rust, have it hydrostatically tested for safety. Replace any damaged hoses or horns on this type of extinguisher.

Halon-Type Extinguishers

Automatic systems using Halon-1301 must be checked by weighing the cylinder and comparing actual weight with that stamped on it. Check every six months or more often if recommended by the manufacturer; a loss of only several ounces is significant. Use accurate scales; bathroom scales are not precise enough.

Periodic Discharges

It is a good policy to discharge a fire extinguisher periodically even though it is not needed for fighting a fire. An effective way of doing this would be to discharge one of the portable units each year on a regular rotation basis. This should be done in the presence of the whole crew and, as discussed earlier, preferably in the form of a drill, putting out an actual small fire—off the boat, of course—in a metal pan or tub. Probably 99% or more of all persons who go boating have never discharged a fire extinguisher, let alone used one to put out an actual fire. There is not time to read the instructions carefully and practice after a fire has started on your boat. A word of warning with respect to CO_2 extinguishers: *never* unscrew the hose from the cylinder and then discharge it openly.

Make sure that when fire extinguishers are removed for testing or practice discharge they are serviced by a competent shop and reinstalled *as soon as possible*. It is also wise not to denude your boat of fire protection by removing all extinguishers at the same time for servicing; do a half or a third at a time.

Log Entries

Make an entry in the boat's log of all inspections, tests, and servicing of fire extinguishers. This will help keep these essential checks from being overlooked and may prove valuable in the event of insurance surveys or claims.

Engine and Fuel System

Check the engine and fuel system frequently for clean-

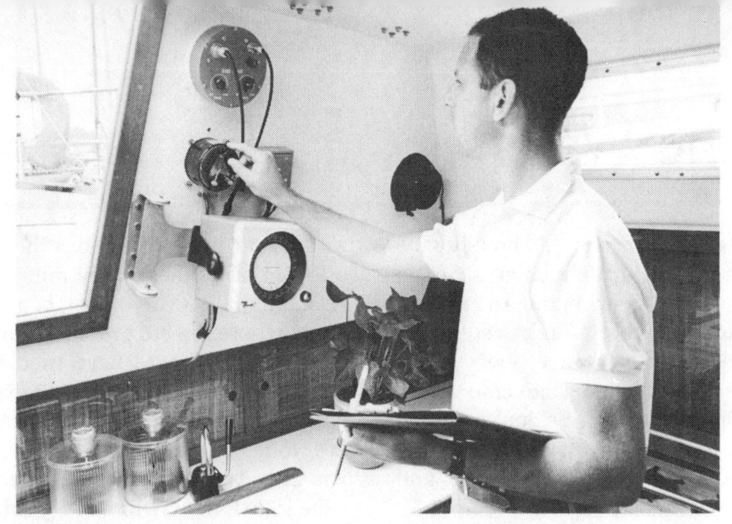

Figure 1217 **When inspecting any system such as a boat's wiring, use a check list to make sure nothing is overlooked.**

liness and leaks. Wipe up any oil or grease drippings and stop leaks as soon as possible. *Take immediate action in the case of any gasoline leaks.* Do not use the boat, and disconnect the leads from the battery (with all loads turned off so that no spark will jump) so that the engine cannot be started.

Check the entire fuel system annually inch by inch, including fuel lines in areas not normally visible. Look for any evidence of weepage of fuel, or external corrosion of the lines. If any suspicious joints or lengths of tubing are found, call in a qualified mechanic without delay.

Bilge Ventilation System

Although it is the manufacturer's responsibility to install a powered ventilation system on certain types of boats, it is the operator's responsibility to ensure that the system continues to perform properly. Even if yours is an older boat, built before these requirements came into effect, maintain your bilge ventilation system in top operating condition.

Electrical Systems

A boat's electrical system should be inspected thoroughly every year, including all wiring in areas not normally visible. This should be done by a person qualified to evaluate what he finds—by you if you have the necessary knowledge and experience, or by an outside expert if needed. Search for any cut or chafed insulation, corrosion at connections, excessive sag or strain on conductors, and other visible signs of deterioration. A leakage test should be made by opening each circuit at the main distribution panel and measuring current flow when all loads are turned off. Ideally there should be no current flow; current of more than a few milliamperes indicates electrical "leakage" that should be tracked down and corrected without delay.

Connections at the storage batteries need special attention. Disconnect them and remove all corrosion with a wire brush; replace and tighten them, then apply a light coat of grease or other protective substance.

Bonding Systems

If all through-hull fittings, struts, shafts, etc., are electrically connected by an internal bonding system, this wiring should be checked annually. Especially careful checks should be made where connections are made to the protected fitting or other metal part; connections in the bilge are subject to corrosion and development of poor contacts with high electrical resistance.

The skipper can make a visual check of the bonding system, but an electrical expert with specialized equipment is needed for a thorough evaluation. Should any signs of corrosion at points of connection between bonding wires and through-hull fittings be noted, a complete electrical test is recommended.

With respect to possible electrolysis (stray current corrosion) damage, a bonding system with one or more poor connections may be worse than no system at all. Electrolysis can and does cause weakening of through-hull fittings, bolts on struts and rudder posts, etc., that could result in serious safety hazards.

Hull Safety Maintenance

Boats normally kept in the water should be hauled out periodically for bottom cleaning and repainting. This occasion should be an opportunity for a *safety* inspection of hull and fittings below the waterline. On wood boats, check hull planking for physical damage and for any general deterioration from age. Check fiberglass hulls for any cracks, especially at points of high stress. Call in an expert if you find any suspicious areas.

Through-Hull Fittings

Each time your boat is hauled, check to see that through-hull fittings and their sea-cocks are in good condition and operate freely. Include fastenings susceptible to damage from electrolytic action.

Underwater Components

Check underwater fittings annually. This includes shafts, propellers, rudders struts, stuffing boxes, and metal skegs. Stuffing boxes should be repacked as often as necessary to keep them from leaking excessively, shafting checked for alignment and excessive wear at strut bearings, and propellers examined to see if they need truing up.

Follow-Up of Inspections

Nothing is gained if prompt and thorough follow-up actions are not taken on findings of periodic safety inspections. *Maintenance related to safety must not be delayed. Do not operate a boat that has a known safety defect.*

Safety in the Water

Anyone who goes boating regularly should know how to care for himself in the water. Fortunately, most boatmen and their families do know how to swim, but this may be limited to taking a few strokes in a calm, relatively warm pool, with safety and rest only a few feet away. Give special attention to staying afloat under adverse conditions of water temperature and waves. Seldom is it necessary, or even desirable, to swim any distance; the problem is staying afloat until help arrives. The skipper and crew should all have instructions in "drown-proofing."

Safety *in* the water can well start with a swimming and/or lifesaving class at the local "Y" or community recreation center. Local American Red Cross chapters also give instruction in these subjects.

Lifesaving Devices

A person in the water following a boating accident should, of course, have a life jacket or buoyant cushion. It is important that these devices be worn or used properly. Buoyant cushions are far from the best flotation device, but are widely used because of their convenience and low cost. Buoyant cushions are *not* intended to be worn. The straps on a buoyant cushion are put there for holding-on purposes and also to aid in throwing the device. Because they must be grasped, cushions are not suitable for small children or injured persons, and are not desirable for non-swimmers. Cushions should never be worn on a person's back since this tends to force his face down in the water.

The lifesaving device providing the greatest safety is the canvas jacket with flotation material in sealed plastic pouches. Stow these life preservers in several locations about your boat, so fire or other disaster cannot cut you off from all of them; they must also be easily accessible. As mentioned, everyone on board, including guests, *must* know where the life preservers are stowed. You and your regular crew should have tried them on and should be able to get into them quickly even in the dark (practice while wearing a blindfold). Most importantly, order everyone into life preservers *in advance* of their need, if you can. At night, anyone on deck should routinely wear a life jacket; non-swimmers, young children, and persons physically handicapped should wear one at all times when not below decks—a person with a cast on an arm or leg goes down like a rock without added buoyancy.

Never carry non-approved, damaged, or condemned lifesaving devices as "extras"—someone might grab one of them in an emergency.

Swimming Tips

Even with the best of lifesaving equipment aboard, you may find yourself in the water without the aid of a buoyant device. If your boat stays afloat or awash, *stay with it*. Search vessels and aircraft can spot a boat or its wreckage far more easily than an individual whose head only is above water.

While swimming you may find temporary relief from fatigue by floating or by varying your style of swimming. Cold or tired muscles are susceptible to cramps. A leg cramp can often be overcome by moving your knees up toward your chest so that you can massage the affected area. Save your breath as much as possible; call for help only when there is definitely someone close enough to hear you.

First-Aid Afloat

At the beginning of this chapter we considered certain responsibilities of the skipper. Another of these is caring for minor injuries and illnesses of his crew and guests. Your boat should have at least simple first-aid supplies and equipment, plus a manual of instructions for their use; you should have had basic first-aid instruction and CPR training, such as given by the Red Cross.

But no one who is not educated and properly qualified to practice medicine should attempt to act as a doctor. There are, however, many instances where availability of a first-aid kit, some knowledge on the part of the boatman, and a ready reference book have materially eased pain or even saved the life of a sick or injured person on a boat. All skippers should be prepared to render emergency first aid, doing no more than is absolutely necessary while getting the victim to a doctor or hospital as rapidly as possible.

First-Aid Kits

There are many published guides to the proper contents of a marine first-aid kit. Probably each of these has

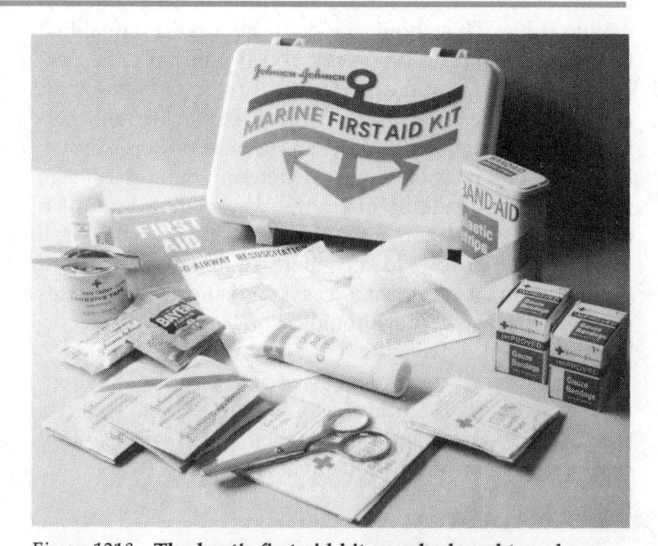

Figure 1218 **The boat's first-aid kit may be bought ready-made, or home assembled using items recommended in the text. It is often advisable to supplement a store-bought kit with those recommended items that are not supplied with it. A skipper also should consider taking a first-aid course.**

merit; the list that follows can, and should, be modified for the size of your craft, the area to be cruised, the hazards likely to be encountered, and any special personal needs.

The first-aid kit must be accessible and each person aboard should be made aware of its location.

Basic First-Aid Materials

There are certain basic first-aid materials that should be carried aboard all classes of boats. Bandages should be ready for instant use—each dressing individually wrapped, a complete dressing in itself. Each should be sterilized and designed to provide protection against infection and contamination.

Antiseptic liquids are best packaged as individually packaged applications.

Basic first-aid materials should also contain burn treatment compounds to take care of sunburn or burns caused by other hazards. Scissors have multiple uses.

Pre-packed first-aid kits, available at most drug stores and marine supply stores, are often not adequate; you should check their contents against recommendations that follow. Fill in significant shortages separately, and store them in or near the kit.

The Container

The container for a boat's first-aid kit should not be made of a metal that will rust, nor of cardboard that could soak up moisture. A plastic box of the type used for fishing tackle or tools is a good choice. It should close securely and be moisture-tight. It should *never* be locked, although it may be sealed with tape. Keep it on a high shelf out of the reach of small children. The kit may be lightly secured in place for safety in rough weather, but it must be portable and capable of being quickly unfastened and taken to the scene of an emergency.

Contents (Instruments)

First-aid kits should contain both simple instruments and consumable supplies. In the first category, the following are recommended:

Scissors—small and sharp; if there is room for two pairs, one should be of the blunt-end surgical type. Also a packet of single-edge razor blades.

Tweezers—small, pointed; tips *must* meet exactly.

Safety Pins—assorted sizes.

Thermometer—inexpensive oral-rectal type, in a case.

Tourniquet—use only for *major* bleeding that cannot be controlled by pressure; follow instructions *exactly*.

Eye-Washing Cup—small, metal.

Cross Venti-Breather—resuscitation device for drowned or asphyxiated person; aids in mouth-to-mouth resuscitation; follow directions on box.

Hot-Water Bottle and Ice Bag—these items, or a single unit that will serve both purposes, can do much to ease pain. Use with caution, following instructions in the first aid manual or medical advice book; improper use can aggravate conditions, sometimes seriously.

Contents (Supplies)

The first-aid kit should contain approximately the following types, sizes, and quantities of consumable supplies. Within reasonable limits, substitutions can be made to reflect local availabilities and personal preferences.

Bandages—1", 2", and 4" (2.54, 5.08, and 101.6 cm) sterile gauze squares, individually wrapped. Bandage rolls, 1" and 2" (2.54 and 5.08 cm). Band-Aids or equivalents in assorted sizes, plus butterfly closures or Steri-Strips. A minimum of one unit of each item, more on larger boats with more people aboard.

Triangular Bandage—40" (1016 cm) for use as sling or major compress.

Elastic Bandage—3" (7.62 cm) width for sprains or splints.

Adhesive Tape—Waterproof, 1" by 5 or 10 yards (2.54 cm by 4.57 or 9.14 m).

Absorbent Cotton—standard size roll, for cleaning wounds, padding, etc.

Applicators—cotton-tipped individual swabs (Q-tips or equivalent) for applying antiseptics, removing foreign objects from cuts, eyes, ears.

Antiseptic Liquid—Tincture of iodine or merthiolate; or zephiran chloride, aqueous solution, 1:1,000. Antiseptic may be in bottle, but preferably in individually packaged applications. Also a 4 oz. (118.2 ml) or larger bottle of antiseptic (70% alcohol) solution, or Phisohex solution, or Betadine solution (Providone-iodine 10%).

White Petroleum Jelly—small jar, or preferably tube; plain, not carbolated, for small burns, dressings. An antiseptic burn cream may be substituted. This may be supplemented by a tube or bottle of sunburn remedy and a package of sterile petroleum-jelly-saturated gauze squares.

Antiseptic Ointment—1 oz. (29.5 ml) tube of Bacitracin or Polysporin, or BPN triple antibiotic ointment, as recommended by your physician (prescription required); a topical antibiotic ointment to help prevent infection in minor cuts and abrasions. Spread over affected area twice daily after thoroughly cleansing with Betadine solution.

Antiseptic-Anesthetic First-Aid Spray—(Medi-quik or Bactine) for superficial skin wounds, minor cuts, abrasions, minor burns, insect bites. Cleanse affected areas thoroughly prior to spraying.

Nupercainal Anesthetic Ointment—1 oz. (29.5 ml) tube; recommended for sunburn, non-poisonous insect bites, minor burns and cuts, painful hemorrhoids.

Pain Killer—aspirin, aspirin substitute, or Motrin. For more severe pain—aspirin and Codeine, or Demerol (prescription required).

Sleeping Pills—Seconal or equivalent as prescribed by your physician; use with caution as directed. If sleeplessness is related to pain, give aspirin or Darvon Compound.

Antibiotic—use *only* if there will be a delay in reaching a doctor and infection appears serious (e.g. associated with fever and chills). Use broad spectrum antibiotic, e.g. Ampicillin or a Cephalosporin, or similar drug, as prescribed by your physician.

Ophthalmic (Eye) Care—if necessary to flush out eye, use Eye Stream (Alcon Lab, Inc.) 4 fl oz., or sterile solution of 1 tsp. salt in 1 pint (0.5 l) of water; do *not* use sea water.

Antihistamine—Benadryl (Diphenhydramine) or Pyribenzamine, or chlorpheniramine. Use for allergies—allergic rhinitis, hives, asthma, and allergic reactions to insect bites and stings. If asthma, wheezing, or bronchospasm develops, use adrenalin 1:1,000% solution (prescription required) 0.3 cc, injected subcutaneously as directed. May be repeated in 20 minutes. Adrenalin, syringe, and needle may be packaged as an insect sting emergency kit—prescription required.

Ammonia Inhalants—crushable ampules; for relief of symptoms of faintness.

Seasickness Remedy—Dramamine, Marezine, Bonamine, Bonine. Should be taken one hour prior to embarcation. The dose as recommended by the manufacturer may be repeated every 24 hours for duration of journey. Dramamine may be repeated every 6 hours.

Anti-Acid Preparation—Maalox liquid or tablet, for heartburn and indigestion.

Laxative—Use the one of personal preference. Milk of Magnesia is excellent, and also acts as an antacid. Fleets disposable enemas, desirable to have on board.

Anti-Diarrhea Drug—3 oz. (88.7 ml) bottle of Paregoric (prescription required) or 8 oz. (236.5 ml) bottle of Kaopectate.

Figure 1219 **Burning sloop shown on page 256 was towed alongside bridge, *above*, where shore-based fire-fighting gear was used to extinguish flames.**

First-Aid Manual

A compact first-aid manual should be packed in the top of the first-aid kit. A larger medical-advice and first-aid book may be carried in a book shelf, but there should always be a small booklet of instructions in or immediately adjacent to the container of instruments and supplies.

First-aid manuals can be obtained from the American Red Cross and other safety agencies. Books relating specifically to medical problems afloat and their management can be obtained through book stores and at many marine supply stores.

Replenishment

A first-aid kit missing some items does not provide a full measure of safety afloat. As supplies are used, replenish them promptly. Further, some drugs do not last indefinitely. They may lose their strength, become unstable and toxic, or evaporate to increased concentrations that could be harmful. Ask your physician about the safe shelf-life of prescription drugs and other supplies carried in your first-aid kit. Replace questionably old supplies whether used or not.

If instruments or the first-aid manual are lost or damaged, replace them quickly.

Medicine Cabinet

Most cruisers have medicine cabinets in the head where additional first-aid supplies may be stored. This cabinet should supplement rather than replace the first-aid kit as outlined above, because its contents are not portable to the scene of the emergency. It should contain the items that normally are stocked in a medicine cabinet at home, such as aspirin and upset stomach remedies.

Dental First Aid

There is a first-aid kit available that can help eliminate the pain and discomfort caused by toothaches, broken teeth, lost fillings, loose caps and bridges, sore and bleeding gums, etc. This is the ORx Dent-Aide Dental Emergency Kit containing pain-killers, powder for temporary fillings and the base to go with it, a dental mirror, tweezers, cotton pellets, and other needed items. Also included is a 56-page step-by-step instruction book.

Administering First Aid

It is the skipper's responsibility to acquaint himself with the proper first-aid procedures. This can be done through classes given by local chapters of the American Red Cross in a standard first-aid course. In any case, you should follow these two rules when administering first aid:

Rule One Always take first aid to the victim. You may aggravate an injury by attempting to move an accident victim. You will add to a sick person's discomfort by requiring him to move about. Limit movement of the affected individual to that absolutely necessary to get him clear of a hazardous area.

Rule Two When giving first aid, take your time. Remember, there are only three instances when speed in giving first aid is required:

1. When the victim has stopped breathing
2. When there is arterial bleeding
3. When the victim has been bitten by a poisonous snake.

The measures required in these instances are taught in the standard Red Cross first-aid course available locally; get a group of your fellow boaters together and attend one.

It should be noted that we have eliminated splints from the first-aid kit. The average first-aider is not qualified to set a broken limb; it is much safer to place the limb in pillows or blankets, tying them securely around the fractured member so that no further damage will be done. And always remember that first-aid should stop at first aid . . . which is administering aid and comfort to avert further complications until professional medical attention can be obtained.

Radio Advice

In many emergency situations, a call to the Coast Guard can bring medical advice for treatment of serious injuries or illness. If within VHF range, make initial contact on Channel 16, followed by a shift to Channel 22. If farther offshore, make initial calls on 2182 kHz, then shift to 2670 kHz; or on a 4, 6, or 8 MHz Coast Guard frequency using single sideband.

Basic Guides to Boating Safety

1. Carry proper equipment—know how to use it.
2. Maintain boat and equipment in top condition.
3. Know and obey the Rules of the Road.
4. Operate with care, courtesy, and common sense.
5. Always keep your boat under complete control.
6. Watch posted speeds; slow down in anchorages.
7. Do not *ever* overload your boat.
8. See that lifesaving equipment is accessible.
9. Check local weather reports before departure.
10. Inspect hull, engine, and all gear frequently.
11. Keep bilges clean, electrical contacts tight.
12. Guard rigidly against any fuel system leakage.
13. Have fire extinguishers instantly available.
14. Take maximum precautions when taking on fuel.
15. Be sure to allow adequate scope when anchoring.
16. Request a USCG Auxiliary Courtesy Marine Examination.
17. Enroll in a U.S. Power Squadrons or Coast Guard Auxiliary boating class.

13

Selection and use of rope and lines; how to make useful knots and splices; blocks and tackle

MARLINESPIKE SEAMANSHIP

From anchor rodes to sheets, vangs, and whippings, anyone helping with the work of a boat becomes involved with lines, knots, and splices—and their proper use. The knowledge and the hand skills are as important to a powerboat skipper as to the owner of a sailboat. They comprise a subject recognized for its fascinating lore, for the beautiful artifacts that are a part of it, and for its sheer practicality. Working with CORDAGE—the collective name for ropes, lines, and "small stuff"—is an ancient skill that now takes in new technologies.

The information that follows is intended to give you a picture of

- the construction, materials, and characteristics of various kinds of modern rope;
- The tools, techniques, and special words used in working with rope;
- how knots, bends, hitches, and splices are made and used;
- lines in use: coiling, anchoring, docking, and fastening things, as well as care and inspection.

The nautical name for this special information and skill is MARLINESPIKE SEAMANSHIP. MARLINE is a two-strand twine once used extensively with rigging; a MARLINESPIKE is a useful tool, often one blade of a sailor's rigging knife.

"Rope" vs. "Line"

Rope is bought as rope. But once it is in use aboard a boat it is called *line,* or by the name of the rigging part it has become. Sailors will tell you that there aren't many ropes aboard a ship or boat. There is bolt rope, at the foot or luff of a sail, or a tiller rope, or a foot rope, or a few other rare ones. Everything else is a line.

Rope Materials

Like a sailor's knife, rope goes back thousands of years. For centuries rope was made from natural fibers, especially flax, hemp, and manila (from the wild banana plant). Most rope was *laid,* usually three STRANDS twisted together. Each strand is composed of three YARNS, each of which is composed of multiple FIBERS; see Figure 1301. Sailors worked with knots and splices that have been in use for centuries, based on the standard way rope was made by twisting fibers.

Since the mid-20th century, there have been more technological changes in rope than in the preceding thousand years. New synthetic fibers have altered the form of rope and even the way particular splices or knots are made and used.

Although you still see some cordage made from the old materials of hemp or manila, it is increasingly hard to buy. The new kinds of cordage are so superior that a sailor has little reason to pay any attention to most of the old materials, although he may have difficulty in making the right choice among the new forms.

The most obvious change is in the form of rope itself: probably more than half the rope used in recreational boating in the U.S. is plaited, braided, or double-braided (a core inside a cover), compared to the traditional, three-strand laid rope. Small stuff, used for whippings and seizings is also greatly changed. Shock cord and even monofilament line have special uses, all made possible by the materials that have come on the market in recent decades.

Synthetic Materials

The first important synthetic fiber to appear was polyamide (nylon). Nylon rope is strong (more than twice as strong as the best yacht manila for the same size), has useful qualities of elasticity or controlled stretch, and has gone through successive improvements. The rope can be given various degrees of softness or hardness, and some variations in surface textures, to fit its intended uses. With its shock-absorbing elasticity it is particularly well suited for dock lines and anchor lines.

High-tenacity polyester fiber (Dacron, Terylene, Duron, Fortrel, A.C.E., and Kodel are trademarks) is made into rope that is virtually as strong as nylon but has one important difference: the rope can be made to have very little stretch. This property makes polyester fiber superior for special purposes like the running rigging on sailboats, where elasticity is undesirable. In manufacturing, polyester rope can be given varied finishes: woolly, smooth, or textured to make it easy to grip, as required.

Aramid fiber (Kevlar), the newest material used for marine rope, combines strength and strong dimensional stability (near zero stretch). It is comparatively expensive and, not surprisingly, is used chiefly on competitive sailboats.

Polypropylene rope is least expensive among the synthetics; it is about as strong as manila, but tends to deteriorate rapidly from exposure to the ultra-violet component of sunlight. Its main advantage is that it floats, so it is suited to some commercial fishing applications as well as to water-ski tow ropes and dinghy painters. Any other use aboard a boat is an economy measure and may be unwise. For appropriate special purposes it should be large size (compared to nylon) and renewed frequently.

Rope Construction

The construction of three-strand laid rope has changed little in thousands of years except for varying surface textures. The new materials and new types of rope-making machines have caused one particularly significant development in rope construction, however: plaited and braided ropes.

The various kinds of rope "geometry" are worth looking at in some detail, starting with the oldest.

Three-Strand Rope

Some knowledge of the anatomy of three-strand rope makes it easier to work, especially in splicing and in finishing off the ends (a subject discussed at the end of this chapter).

If you hold the rope so that the end is away from your body, the strands will be seen to have a twist that is clockwise. This is called "right-hand lay." Most rope is right-hand lay, and has been since the pyramids were built.

If you look at the twist in just one of the three strands, it is a left-hand twist. Finally, with all but the smallest ropes the fibers have an opposite twist within the strand.

Because some tension is put on the rope in manufacturing, these opposite twists, of strand and rope, tend to keep the rope from UNLAYING, or untwisting.

The reason for noting the lay of the rope is this: when you splice you will be working with these twists, sometimes retwisting a short length of strand with your fingers, and SPLICING (interweaving) the strands back into the normal lay of the rope. Since virtually all laid rope is right-handed, your fingers should learn to work with the twist, whether you are tying knots, splicing, or coiling.

Laid rope comes in several degrees of hardness or stiffness. The technical terms are SOFT LAY, MEDIUM LAY, and HARD LAY. For boating purposes you would use only medium lay. SAILMAKER'S LAY is a variation, but unless you're making sails you won't need it. Very soft lay is sometimes on sale, and is superficially attractive—silky soft like milkweed seed pods. It's of little use on a boat because it kinks easily. Hard-laid rope, used commercially, may be too difficult to work with to be of any value on a boat.

You may read about, but seldom see, FOUR-STRAND ROPE and LEFT-HAND LAY ROPE; both are for special purposes. And there is one more twist: to make very large cables, such as those that tugs or large ships use, three right-hand lay ropes can be twisted together, left-hand lay; this is called CABLE LAY.

Most wire rope is made the same way: by twisting strands. However, the more intricate strandings, with 7 × 7 and 1 × 19 strands, are seen most often. In 7 × 7, seven strands, each in turn consisting of seven wires, are twisted together; in 1 × 19, 19 single strands make up the rope. This 1 × 19 wire rope is stronger but less flexible than the 7 × 7 type, and is used primarily for standing rigging.

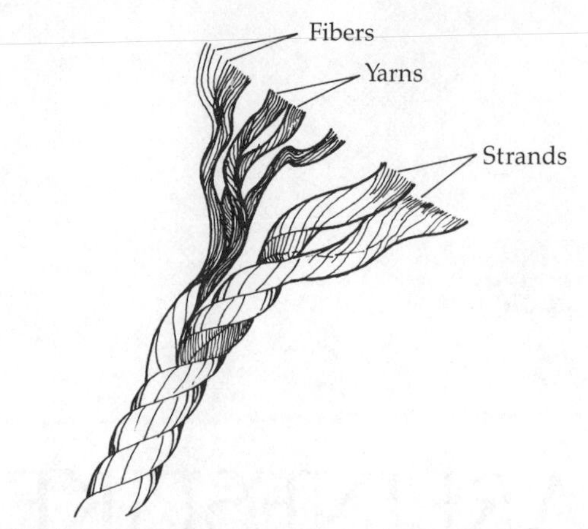

Figure 1301 **Laid rope is made up of three strands twisted clockwise in a right-hand lay; the fibers that make up each strand are twisted to the left.**

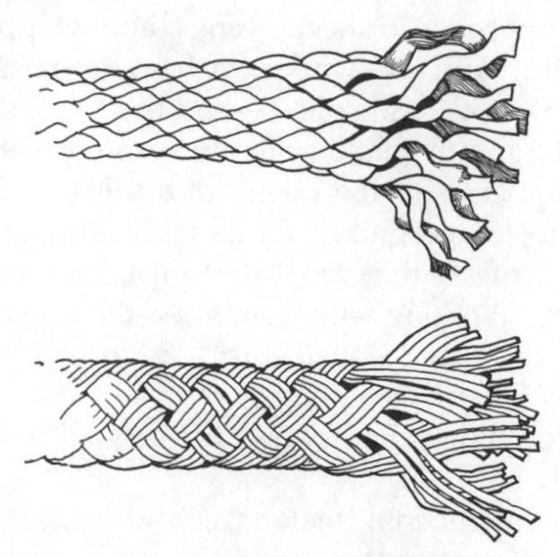

Figure 1302 **Braided line may be single-braid, *above,* or eight-part plaited type. Single braid in smaller sizes is used for flag halyards, sail bag ties, and similar purposes.**

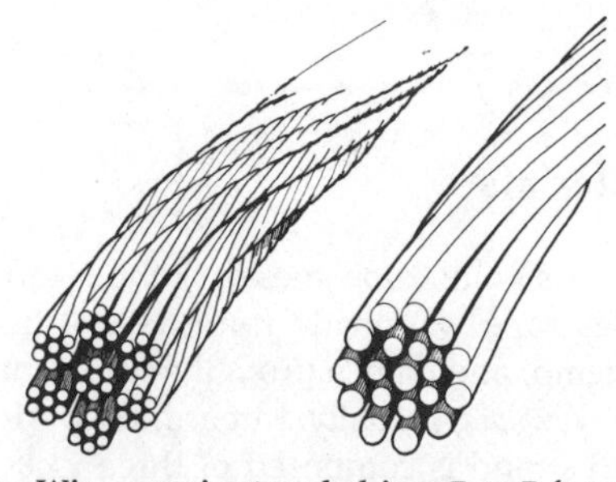

Figure 1303 **Wire rope is stranded in a 7 × 7 format, left, which can be used in running rigging, or 1 × 19 format, *right,* the type used for sailboat standing rigging.**

Braided Line

Small diameter single-braided or plaited line (¼-inch and smaller) is used for flag halyards, to tie the ends of sailbags, and for special purposes such as awnings. Most single-braid line can't be spliced unless it's a hollow braid. Plaited line, with eight parts, can be spliced.

Double-Braided Double-braided line is a widely used type. It has a braided core inside a braided cover. For low-stretch applications the cover is polyester; the core is often a mixture of fibers with greater stretch. This construction has advantages for many purposes, but it requires special techniques for splicing. (See Figure 1320.)

Double-braid is very flexible because both the cover and the core are composed of small strands. Its stretch depends on the materials used. Its resistance to abrasion is excellent, because the wear is distributed over many strands. Braided line tends to coil evenly and unlike laid line is virtually kink free.

On sail and power boats you will often see double-braid nylon used for dock lines because of its smooth running and easy-on-the-hands characteristics; occasionally it will be used for anchor lines. You'll see double-braid polyester used for sheets and halyards on racing sailboats.

Other Types of Rope

In addition to stranded and braided rope, several other forms of flexible materials have special uses aboard boats. These materials are not traditional cordage, but they do some of the same things rope or small lines will do.

Shock cord, which is multi-strand rubber with a synthetic cover, can stretch twice its length. Its uses are endless; to hold topping lifts out of the way, to make quick lash-ups for furling sails, as gill-guys to hold halyards away from the mast, and to hold a pair of oars on a cabin top or books in a book rack. Shock cord is not as strong as rope, its extreme elasticity makes it unsuitable for many purposes, and it cannot be spliced. Its ends are usually made up with plastic or metal clips to prevent fraying; eyes (actually looped ends in the shape of eye splices) are made the same way. A loop at one end of a short piece of shock cord and a toggle inset at the other end makes a good tie to keep a coil of line in place.

Webbing, woven of nylon or polyester, is used for sail stops (replacing the traditional sewn strips of doubled sailcloth) and for dinghy tie-downs as well as hold-downs for boats on trailers. Webbing is very strong, and when used for sail ties it holds well with a square knot or a slippery reef knot. Webbing is either single or hollow flat (double).

Monofilament is another kind of line found on boats. Its main use is for fishing lines and a heavier piece makes a good lanyard for a pocket stopwatch or small calculator. Monofilament fish line must be knotted only in certain specific ways or it will invariably come untied.

Finally, there is what the old salt, sitting on a hatch cover making up whippings, servings and seizings, calls "small stuff." Waxed nylon line and flat braid (a plaited form also called parachute cord) is used for whippings and to prevent chafe. Marline, a tan-colored two-strand string used for servings, is now scarce. Cod-line, a similar three-strand material, is useful if you can get it.

Purchasing Rope

In the course of boat ownership, you will need to make decisions about what size and type of rope to buy for new dock lines, or how to improve the rigging when replacing a sheet, vang, or halyard. In addition to the factors already reviewed concerning elasticity and the form of rope, you need to consider other points, including size.

Figure 1304 **Double-braided line has an inner core within a cover. By using synthetic fibers of different types, the amount of stretch can be controlled.**

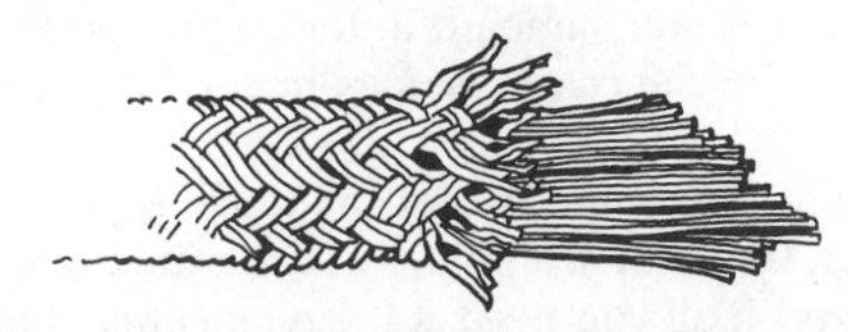

Figure 1305 **Shock cord is made up of rubber strands within a braided cover. It is generally used in short sections with loops and toggles at the ends.**

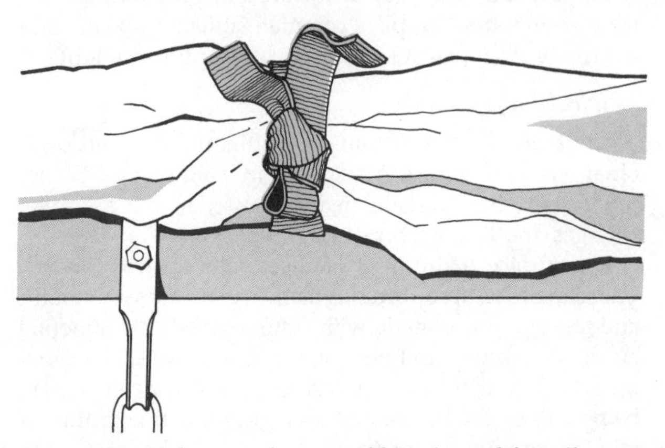

Figure 1306 **Nylon or polyester webbing is used for sail stops, for lashing a dinghy on deck, or as tie-downs to hold a boat snuggly on its trailer.**

Strength

Experts, concerned with safety and dependability, say that the working load for laid line should not exceed 11% of its TENSILE STRENGTH—the load, in pounds of "pull," at which the rope would break. Braided rope is normally specified for use at, or under, 20% of its tensile strength. As rope ages it inevitably suffers from abrasion and sunlight. Braided rope has the advantage that the core, providing a major part of the strength, is protected. All parts of laid rope suffer from wear.

In practice, a common-sense approach is often the determinant. A small size of rope might be strong enough for a jib sheet, but a larger size will be easier to grasp with the hands.

Size

The size of rope, in the United States, is customarily given as its diameter in inches, particularly in the recreational-boat sizes. But since most of the rest of the world measures rope by its circumference and in millimeters, you will see the "nominal" sizes and metric comparison tables more and more. In addition, manufacturers and sellers of rope know very well that a piece of braided rope, held loosely, is thicker than one stretched out. Because of this, they market only such rope by its weight, which is a more accurate measure of the amount of material. At some marine stores, however, you will find all types of rope sold by length, often in pre-packaged kits.

Availability

The retail availability of special kinds of rope can be a problem for boatmen. An inland hardware store may stock several kinds and sizes of rope—suitable mainly for farmers. Preferences and needs vary among commercial fishermen, powerboat owners, and sailors. Laid rope for sailboat rigging is still standard in the central portion of the U.S., while on the coasts various braided ropes are more popular.

Sometimes there are special solutions. If you can't find waxed flat braid for whippings at a marine dealer, talk to a sailmaker. If all you need at the moment is a small size, waxed nylon dental tape is roughly similar. If you plan on making up a polypropelene dinghy painter that will float, and thus stay out of the way of propeller blades, try a source that supplies commercial fishing boats; and get the black rope—it's the most resistant to sunlight.

Selection

There are dozens of brands, configurations, and combinations of materials, particularly in braided rope. Sometimes one characteristic, reduced stretch, for example, changes another, such as ease of splicing.

The surface textures of braided polyester ropes vary. Those made with continuous filaments are shiny, smooth, and strong. Those made with spun yard (short filaments) are softer, fuzzier, and not quite as strong. Select the shiny ones for halyards, and the fuzzier ones for sheets, where being easier on the hands and gripping the drums of winches are important qualities.

Some skippers who do a lot of night sailing use lines of different surface texture, with different feel for different purposes such as sheets and guys.

Color Coding

The use of color coding for racing and cruising sailboats is common. The lines may have a solid color, or be white with a colored "tracer." A recommended standard is:

Mainsail, sheet and halyard—white,
Jib/Genoa—blue,
Spinnaker—red,
Topping lifts—green,
Vangs and travelers—orange.

Size and Strength

For most uses, the appropriate size of rope is one that is large enough to be comfortable in the hands under normal working situations. A flag halyard, which has very little strain on it, can be thin to reduce wind resistance.

Sail halyards must be strong enough to take the tremendous strain of the sail filled with wind, with minimum stretch, be resistant to abrasion at the sheaves, and still offer the least possible wind resistance.

The proper size of a line may be determined by still another factor: chafing. A dock line or anchor line, for instance, is used many times, sometimes chafes seriously and sometimes not at all, and yet should always be strong enough to hold under extreme storm conditions. A mooring pennant needs to be many times stronger than is required for the worst possible storm conditions, partly because the big fall storm may come at the end of the season, when the line has already been weakened by various small chafings. (See Chapter 11 for more details on anchor and mooring lines.)

Matching Rope Size and Blocks

If a line is a bit too large for a block, a fairlead, or a chock, you run two risks: binding or jamming, and chafe. Sometimes the solution is not a smaller line but a larger fitting.

If a line is too small for a particular block it can get caught between the sheave and the shell—another kind of jam. (See Figure 1321 for the parts of a block.) If a wire sheave is adapted for rope, the sheave must be dressed down for a proper groove. (Be aware that for comparable sizes, rope will have only a fraction of the strength of wire.)

The diameter of the sheave in a block can also be important. If a line makes a 180-degree change of direction over a very small sheave, the stress on the line is obvious. Sometimes naval architects use two small blocks at the top of a mast to solve this problem—each block gives the halyard a 90-degree turn. Similarly, a fairlead that makes a change in direction in a line results in a certain amount of friction; if the line goes through the fairlead at an extreme angle, the friction is greater. In such a situation a block might well be better than the fairlead, to make line handling easier as well as to minimize abrasion.

Use Quality Rope

The problems of cheap rope may be hidden, but they are real. Some polypropelene rope, at the bottom of the cost scale, is made from large-diameter filaments. Rope with finer filaments costs more, but lasts twice as long and is much stronger. Hard-lay nylon, which may be old, is so difficult to work with that it may not be worth its low price.

As you work with the variety of lines aboard a boat, you will form the habit of constantly checking for chafe and other problems: a splice that is beginning to fray, a chock with a rough corner that will weaken a dock line by chafing, a block whose sheave tends to be stiff and needs cleaning and lubricating. Frequent inspection of gear and rigging, and replacement, lubrication, or other suitable repair is an important part of good seamanship.

Information on inspection details and procedures is given near the end of this chapter under "Use and Care of Rope."

Tools

For most everyday work with rope you need only your fingers and a good knife. But many marlinespike jobs need special tools, a couple of professional techniques, and some standard materials.

Here is what you would find in an experienced sailor's "ditty bag," and some of the extras that might be in the tool locker or perhaps on a home workbench—for most kinds of marlinespike seamanship:

A good knife, preferably the type called a rigging knife, with a built-in marlinespike and probably a hasp that is used for opening and closing shackles. The blade should be sharp although often a keen but slightly jagged edge works well on synthetic rope. If you don't have a sharpening stone a file from the engine tool locker will keep the blade sharp.

A large fid or marlinespike, for separating the strands of laid rope when splicing. You can usually splice small sizes of laid line, if it's fairly soft, using only your fingers. But a wooden fid or metal spike, used to separate the strands and to get the strand you are working with through quickly and easily, makes any splicing easier, and is usually a necessity for larger sizes of rope. Special fids will be needed for splicing double-braided rope, different sizes matched to the diameter of the rope. Several types of hollow fids and other special tools are on the market.

Sailmaker's needles (keep them in a small glass or plastic bottle) for making whippings and repairing sail slide attachments and jib hank fastenings. Keep several sizes of needles on hand.

A sailmaker's palm to push the needles through the rope. Even if you don't expect to repair sails, a palm is a good-hand-tool. Keep the leather soft by applying Neatsfoot oil twice a year.

A pair of scissors.

A pair of pliers —simple sharp-nosed ones are best.

Waxed sail twine (tape and/or cord) and perhaps some old-fashioned brown marline for a traditional-looking project. As you'll see at the end of this chapter, waxed nylon twine or lacing tape is used to make whippings on the ends of lines. Lacing tape, which is flat and plaited, is also used for sail fittings; even if you never sew a sail you'll need twine or tape for lashings. Many people also keep a piece of beeswax in their ditty bags, for waxing twine.

A candle and a book of matches. Heat is one way to melt and seal the ends of synthetic rope. The flame in a galley stove will do it, and for shore work (or if your boat has 117-volt AC power) a hot-knife is best; see below.

An electric rope cutter. Not everyone needs the professional's tool, which plugs into an electrical outlet and uses a hot wire to *melt* the rope (quickly, smoothly, and sealing the ends). However, many amateur sailors who like to work with rope own one. Perhaps the handiest tool for the boat owner is an electric soldering gun, with a special cutting head obtainable from hardware stores. This knife-like head is also useful for melting and smoothing the fine strands of rope that are left after a splice is made. Electric rope cutters or hot knives require 117-volt AC power.

Plastic tape for temporary whippings. White tape with some stretch to it, black electrical tape, waterproof sail repair tape, or first-aid tape will do. These are all adhesive tapes, unlike the nylon lacing tapes referred to earlier.

Liquid, quick-drying plastic for dipping rope ends into as a substitute for a twine whipping or other end finish for a line. They are called Whip-It, Whip End Dip, or by similar names.

Keep all the tools and materials well enclosed against moisture, which can make scissors and knives useless. A canvas ditty bag with plastic wraps for the metal tools will do, as will a tight-closing plastic box.

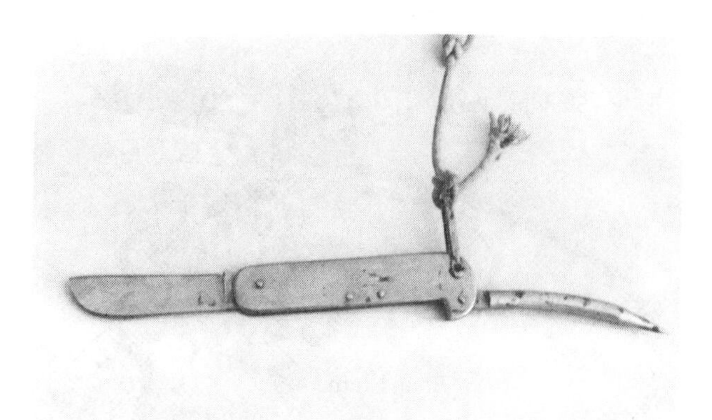

Figure 1307 **The combination knife and marlinspike, *above*, is the tool most often used in marlinespike work; this one has seen a lot of service. Many sailors carry separate knives and fids. Traditional sailmaker's palm, *below*, is needed to push needles through heavy sailcloth when making repairs. Strong waxed thread is used in this operation.**

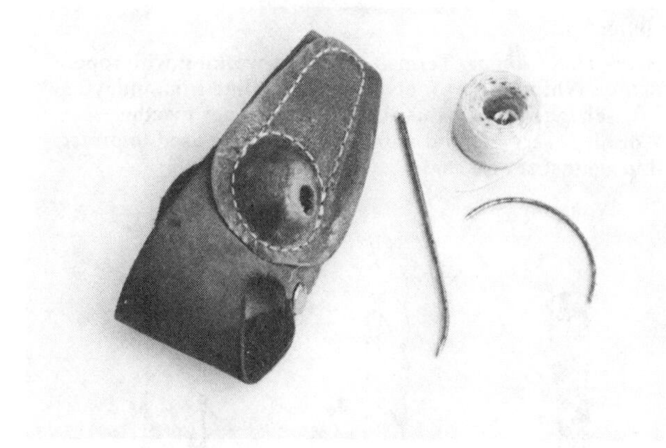

Use of Ropes and Lines

So far we have been looking at rope and line primarily as the raw materials of marlinespike seamanship. Now we turn our attention to their use, and you will notice additional special language that is used aboard boats.

Definition of Terms

Many of the special terms are best understood by handling line or by looking at illustrations. The STANDING PART is the long end of a piece of line. If you loop the working part back on itself, you form a BIGHT. If the bight is around an object or the rope itself, it's a TURN; if it's a complete circle, it's a ROUND TURN. The extreme other end of the line is the BITTER END. BENDS, HITCHES, KNOTS, and even SPLICES are all technically KNOTS, but to the purist there are differences. A splice, of course, was only a form of interweaving of the strands—until braided rope came along with a new form.

A WHIPPING is small twine or tape wrapped tightly around and through the end of a rope to keep it from unravelling. A SEIZING is a similar wrap-around, but for another purpose, such as binding two parts of a line together in a piece of rigging, or binding a sail hank to the sail. PARCELLING is a more complex wrap-around, combining twine and tape, to take wear or prevent chafe. The illustrations allow you to compare these somewhat similar methods. See Figure 1308. A new word is "LOCKING," which means sewing through the throat of a braided line eye splice to hold it in place.

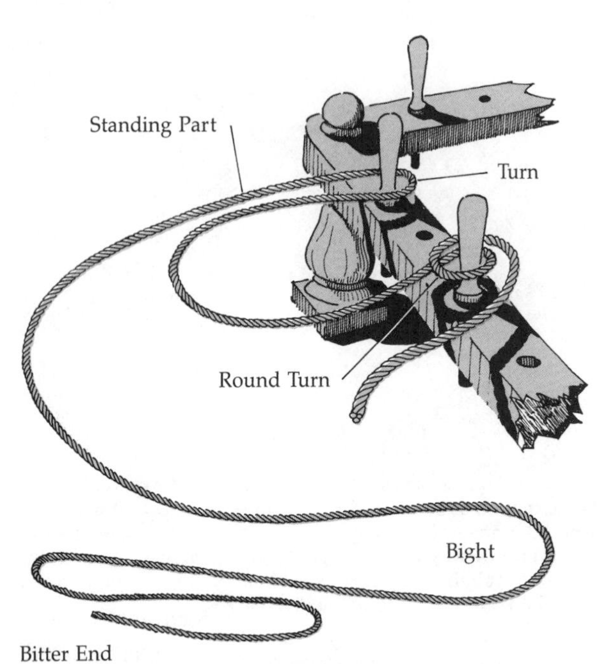

Figure 1308 ***Above:*** **Terms used when working with rope.** ***Below:*** **Whipping prevents the end of a line from unlaying,** ***left;*** **seizing,** ***center,*** **is used to bind two lines together. Worming, serving, and parcelling,** ***right,*** **are used to protect line against severe chafe.**

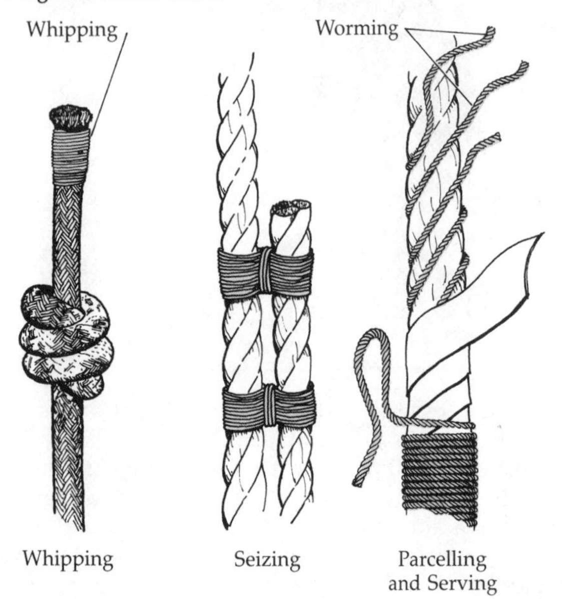

Handling Lines

When you handle lines aboard a boat, you soon learn some basic techniques; one of these is how to coil a line.

Coiling a Line

Laid line, having its natural twist built in, should always be coiled clockwise so it won't kink, buckle, and tangle. If it's a free dock line, either end can be used to start a coil. Otherwise, always start at the secured end—where the halyard or sheet is cleated, for example, and work toward the free end. Never start at the free end—you'll end up with a twisted, awkward coil that's anything but docile. If the line to be coiled is loose and untangled, it's easy. If it looks the least bit tangled, "overhaul" it by running it through your hands from one end to the other, so it's ready to be coiled.

Now, start by holding the line in your left hand. With an easy sweeping motion, using your right hand, bring each coil to your left hand and take it in with your fingers. The coils are always clockwise, remember. An even sweep of your arm, the same distance each time, will result in coils of the same size. Sometimes your thumb is used to add or control a twist. It takes practice.

Don't try to wind up the line over your elbow, unless you're a landlubber in the backyard doing clothesline.

With braided line, which has no lay or twist to it, don't coil it! Simply hand a length from your right hand to your left hand and you will see a figure-eight develop. From this eight, the line will always pay out without kinking or fouling. But since coiling line is a frequent task aboard boats, sometimes with one kind of line and sometimes with another, get in the habit of making clockwise coils no matter what type of line you are handling.

Stowing a Coiled Line

What you do next with the coil or figure-eight depends on circumstances. If the line is the end of a jib sheet that will be used again in a few minutes, or the unused part of a dock line that you need to keep ready, just turn the coil so the free end is down and lay it on the deck, in the corner of the cockpit, or wherever is convenient.

If the coil is to be hung up, which it will be in most cases, there are several ways to do it. A completely free coil of line, such as a fairly heavy spare anchor line that is to be stowed for awhile, can be tied up—using shock cord gaskets, or rope beckets made up for the purpose, or small pieces of spare line. A newer, convenient method is to use cinch straps of Velcro self-adhering tape. If it is a very heavy and long line you would do best to lay it onto the deck in figure-eights—you couldn't hold the coil in your hand anyway. On a ship this is called FAKING DOWN.

A free coil, such as an unused dock line that you may use again soon, is often handled as follows: take a short arm's length of the end of the line and wrap the coil with three, four, or five turns and then, using the last length of line, pass a loop through the entire end of the coil. Hang the coil up, on a cleat or hook, using this loop; see Figure 1309. When you want to use the line, the loop comes out quickly and the whole coil is readily available.

A halyard is handled somewhat similarly, except that:

1. You don't wrap the coil. Leave it free, for quickest availability when the sail is to be dropped.

Figure 1309 **Lines not in use, *left*, are coiled neatly and hung out to dry. They must uncoil freely, without tangling. *Right, above:* When sails are hoisted, halyards are made up in coils and hung on a cleat at the base of the mast. A flemish coil, *below*, is an attractive way to deal with the free end of a line, but don't leave it too long.**

2. You use the STANDING PART of the line, not the free end, to make the pass-through loop, and to keep it from slipping loose give it a turn or two as you hang it on the cleat.

There is another way to deal with a free end of line somewhat decoratively: a flemish coil; see Figure 1309. This is easy to do on deck or dock and looks yachtlike, but if you leave it too long the coil will pick up dirt and moisture and leave a soiled mark on deck when it's taken up.

Of course an eye splice is most useful in one end, if not both, of a dock line. A good practice is to make the eye at least three times larger (diameter) than the size of any piling or bollard it is likely to be used on. Five times is better. The reason: a small eye splice, under strain, tends to pull apart at its throat.

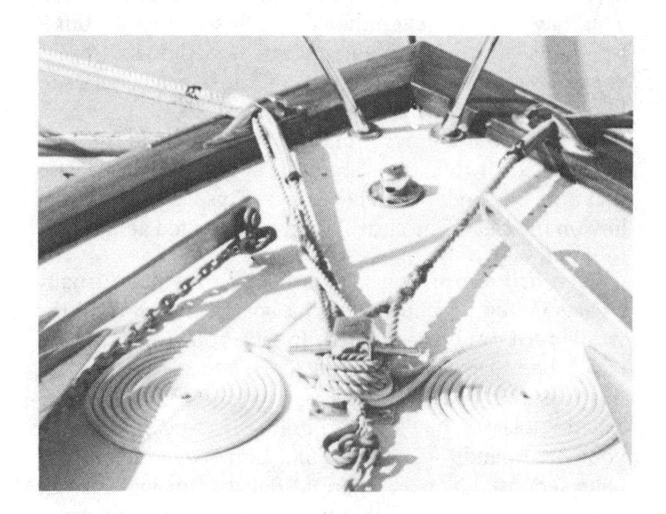

Knots

The first principle in dealing with laid rope is that it has a twist, and you must work with it, not against it. This applies to splicing and knotting as well as to coiling.

You will also feel how a little friction of one part of the rope against another holds it fast. That's the second principle for any kind of rope: in knots, splices, bends, or hitches, the pressure of the rope against itself, as it tightens, is what does the holding.

The third principle is a rule, a definition: a good knot is one that can be made almost automatically (your fingers learn to do it, just as they learn to write your name), that holds securely in the usage it is meant for, and that can

Figure 1310 **In cleating a line, start with a turn around the cleat. The half hitch that completes the fastening is done with the free part of the line at left. The line can be freed for adjustments without taking up on the standing part at right.**

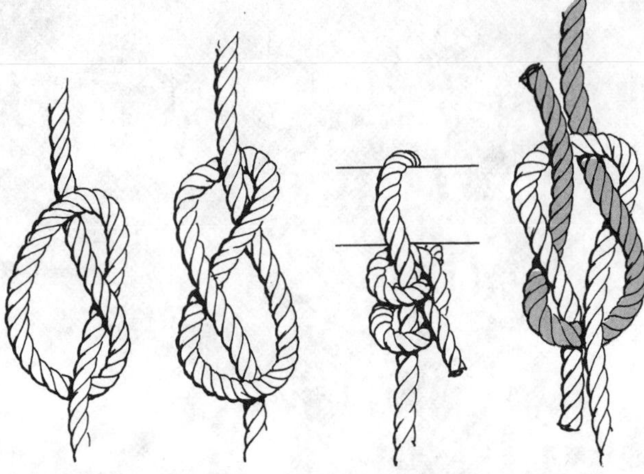

Figure 1311 **Knots, *left to right,* are the overhand, the figure-eight, two half-hitches, and a granny. The granny is shown as an example of what not to do; it can jam, or it may come untied when it should remain secure.**

be unfastened or untied readily. With practice you will be able to tie knots in the dark or in more than one position.

Some knots are adjustable; others are not. At times, the choice of which knot to use depends on the size or character of the line involved—there are variations in the way line works, as will be mentioned in this chapter.

Most of the following information about specific knots concerns their efficient use. The illustrations are for identification; the paragraphs below give additional information about tying or using each one.

Cleat Hitches

One of the simplest knots, certainly the most used aboard a boat, involves nothing more than turns around a cleat. You may not have even thought of this as a knot—but it is. When fastening a line to a cleat, as with many things at sea, a small error at the start can cause problems later. Look closely at the illustration, Figure 1310, and see how the dock line at the right comes in at an angle to the base of the cleat. It then goes around the base of the cleat so that it passes under each horn once. This keeps the strain low on the cleat. The cleat should be mounted at an angle to the direction of strain.

Even half a turn, plus a firm hold on the line, usually creates enough friction to hold a boat at a dock until the whole turn can be completed. In fact, you should always take a half turn around a cleat when a load is coming on the line; this is an essential part of good line handling called SNUBBING. Never try to hold a load-bearing line in your bare hands—it can slip and burn your fingers and palm seriously. Note in Figure 1310 that snubbing is nearly a whole turn; and also that the cross-over that will make the figure-eight should be made after the line goes around *both* horns of the cleat.

One and a half or two figure-eights are enough; more would add no security and just take time to undo later.

There are two ways to complete the hitches on a cleat: you can leave the last turn free, perhaps keeping it under your eye if it is a jib sheet, for example. In this way the line can be instantly thrown off if necessary.

For more security you can turn the last hitch over, so it is tightened to bind against itself. This might be best for a dock line or halyard, but is not safe if instant release may be necessary. In a small sailboat, in puffy winds, you probably wouldn't cleat the main sheet all the way—you'd take one turn and hold it tight in your hand.

Similar figure-eight turns are sometimes used on mooring bitts.

Useful Knots

The following are the most-used knots aboard recreational boats, listed approximately in order of complexity and ease of learning, not in order of usefulness.

Overhand Knot Shown here chiefly because it's simple, and helps explain the next knot. Use it sparingly; it's almost impossible to untie after it is tightened. (It is useful to hold the end of a winch line on a trailer hitch.)

Figure-eight Knot With one more turn into the overhand knot you have a **stopper knot,** to keep the end of a line from running through a block or a small grommet. Easy to untie.

Two Half-Hitches For fastening a line to something else, such as a grommet in the corner of an awning. Quick and easy, but for many special purposes there are better knots you will prefer. A single half-hitch also has some special uses.

Reef Knot or Square Knot If you are tying a bundle, this knot works. If the line is under constant pressure, and if both ends are the same size, it can still be untied. if it is made with a strip of canvas or webbing, as in using gaskets to furl a sail, it is a useful knot and easily untied even when wet. If it is used to tie two lines together, to make a longer line, it is a mistake. Use the reef knot sparingly.

Granny This is a knot that too many people tie automatically when they are trying to tie a square knot. They should teach their fingers to go the other way (see above), because the granny is a useless knot. Sometimes it slips, sometimes it jams. It has no value at all on a boat.

Slippery Reef Knot Half a square bow-knot. Good for furling a sail. This is always easier to untie than a square knot.

Clove Hitch Commonly used to tie a line to a piling. This is often a mistake. Although the clove hitch has the advantage that it is easily adjustable, when first made or later, the same characteristic makes it likely to slip—espe-

Figure 1312 ***Left to right*** **are the reef (square) knot, slippery reef knot, clove hitch, and buntline hitch. Note how a pull at the free end of the bight in the slippery knot is all that is needed to loosen the knot.**

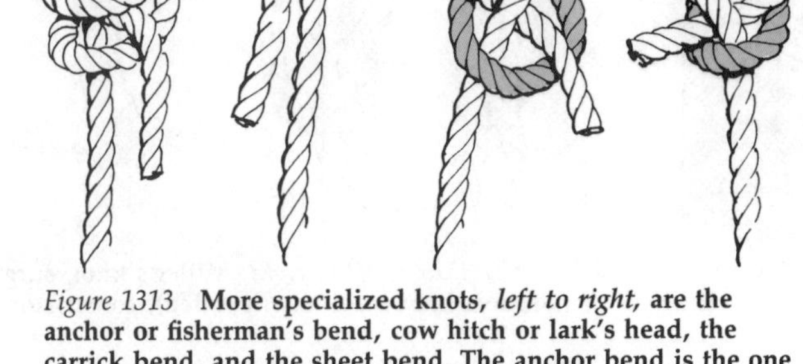

Figure 1313 **More specialized knots,** ***left to right,*** **are the anchor or fisherman's bend, cow hitch or lark's head, the carrick bend, and the sheet bend. The anchor bend is the one most often used, as described in Chapter 11.**

cially with slippery nylon line. See Figure 1316 for a way to take an extra half-hitch to make the clove hitch more secure. You'll use the clove hitch less often once you learn better ways to do the same thing. One good use: if you use the same line from the bow to a piling and then to the stern, the clove hitch is ideal. Both ends are taut and you have the adjustability feature.

Buntline Hitch This is excellent for fastening a halyard to a shackle. It is its own stopper knot and won't jam in a block as an eye splice might. Sometimes called the "inside clove hitch," or the "studding sail tack bend," this is easy to tie and untie. Another good use is on a trailer winch snap shackle—a modern use for a hitch from the days of square-riggers.

Anchor Bend Also called fisherman's bend. This is a standard way to fasten an anchor line to the ring of an anchor. It is excellent for making up a spare anchor, and it can be used in many other situations. The double loop reduces possible chafe and makes the half hitches more secure. Of course there are other ways to make an anchor line fast, including a bowline or an eye splice over a thimble, plus a shackle (see Chapter 11 on Anchoring, Figure 1111).

Cow Hitch or Lark's Head A small existing loop (usually an eye splice) turned inside itself when you want to fasten a line to a large piling. It is also useful to fasten such a loop to a ring, provided the other end of the line is free; see Figure 1313. Many people use this hitch to fasten a jib sheet to the clew of a jib on a small sailboat.

Carrick Bend This is one of the traditional ways to fasten two lines of the same size together. It looks beautiful in a drawing but under strain it changes appearance. The carrick bend is probably best for fairly heavy, stiff lines of the same size.

Sheet Bend This is an excellent way to tie two lines together especially if they are of different sizes or textures. As you will realize when you practice tying knots, some are just variations on a theme. The sheet bend upon a close look turns out to be actually a bowline using two lines instead of one continuous piece. You can use part of a sheet bend to fasten onto a loop (see below).

Becket Hitch This is really the second half of a sheet bend, applied to a loop, such as an eye splice or a bowline. If an extra round turn is taken, it's called a double

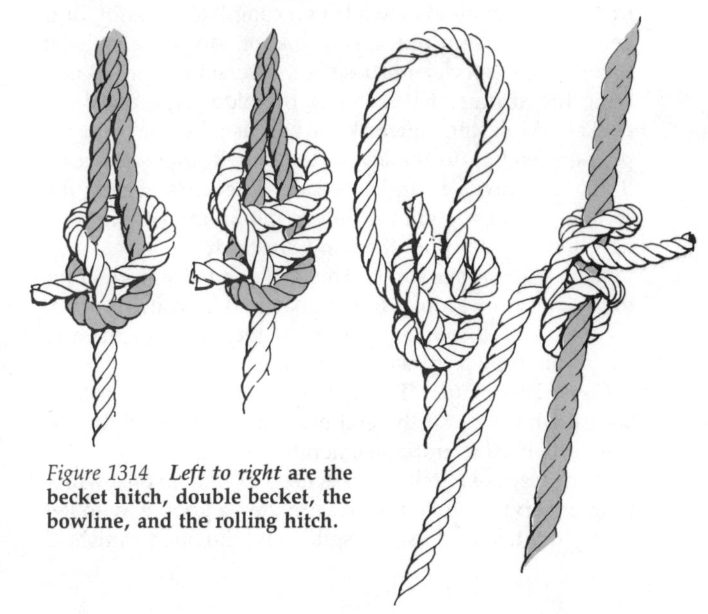

Figure 1314 ***Left to right*** **are the becket hitch, double becket, the bowline, and the rolling hitch.**

becket hitch. This version is especially useful when the added-on line is smaller than the one with the loop.

Bowline This is the most useful of all knots aboard a boat. Once learned (and practice is necessary) it is easy to make, never slips or jams, and can always be untied. Two bowlines, one on each line, are an excellent combination when you need to tie two lines together.

Rolling Hitch To tie a small line to the standing part of a larger one, so it won't slip, use the rolling hitch. You can use this hitch to hold a jib sheet while riding turns are removed from a winch, to haul on any line, or to make an adjustable loop for an awning tie-down. The rolling hitch is also used to attach a line to a round wooden or metal object with the least possibility of slipping sideways.

Miller's Knot This is the sailmaker's knot to tie up the end of a sailbag. It can be readily undone, even when tied with the small lines used for that purpose, while a square knot or two half-hitches can prove all too balky after a long period of being tightly tied.

Surgeon's Knot Tying the ordinary square or reef knot, even around a parcel, often requires a helper—someone to hold a finger on the half-formed knot until it is com-

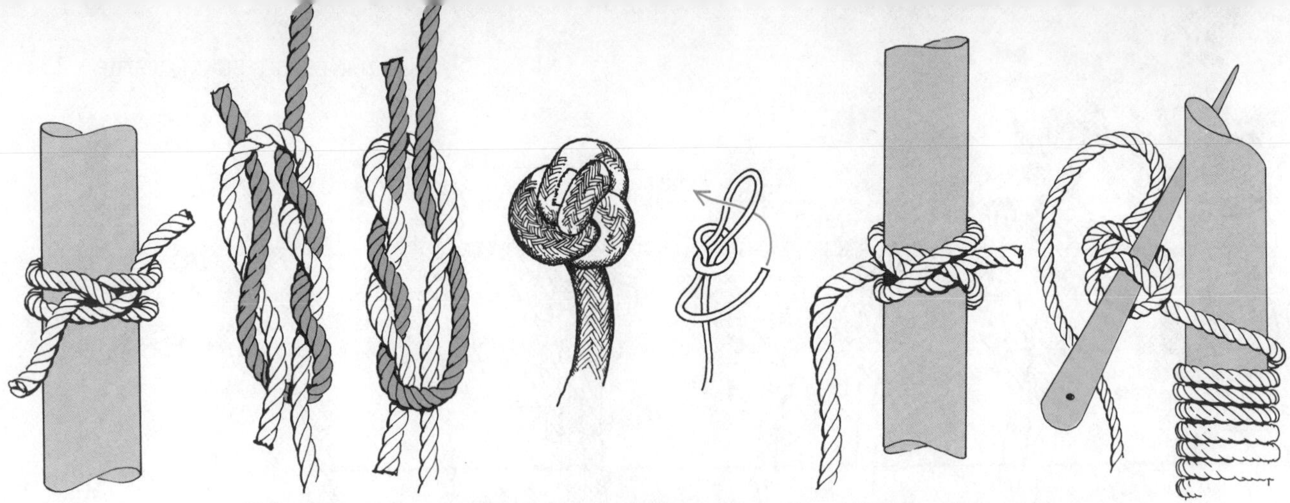

Figure 1315 ***Left to right:*** **Miller's knot, surgeon's knot, reverse surgeon's knot, Ashley's stopper knot (note how it is tied), constrictor knot, and the marlinespike hitch.**

pleted. If you take the extra turn of a surgeon's knot, the friction/tension holds it while you complete the knot. Surgeons, of course, call it a suture knot. Aboard a sailboat there is one special purpose for a version of this knot: tying the strings at the end of the older type of batten pocket. While the square knot will usually shake loose, perhaps leading to the loss of the batten, there is a cure. If you tie a **reverse surgeon's knot**—the extra twist in the second half of the knot is what's important—it will hold tightly no matter how the wind shakes it.

Ashley's Stopper Knot This makes a hefty stopper for the end of a line, looks seamanlike, and is easily undone. It takes a little longer to tie than the figure-eight, and should be drawn up with some care.

Constrictor Knot This is very useful to make a temporary whipping on the end of a line. It draws tight; indeed it is hard to untie—sometimes a virtue.

Marlinespike Hitch This is used to take up on a whipping or serving while it's being made—a good way to get it very tight. Withdraw the spike, and the hitch vanishes.

Using the Right Knot

The knots described, and shown in detail on the illustration pages, can be divided into two groups: basic knots used generally on boats, and a few specialized knots that come in handy at certain times.

You have seen that some knots are echoes of others, and that more than one knot is sometimes used for the same purpose. Here are some reasons: a knot may work well with rough-textured rope but tend to slip when tied with new slippery synthetic line. One knot may work well with small line, while another works out better with heavier and stiffer line. The sheet bend can be tied in a hurry; it takes more time to make two bowlines when tying two long lines together, as for a tow. One bowline, with half the sheet bend tied into it (the becket bend) has the advantage that it can be unfastened quickly.

Some of the potential problems are most easily solved by alternate methods. An anchor bend, for example, is customarily given some extra security with a short seizing or a constrictor knot made with sail twine; see Figure 1316, left. On other occasions, where the knot won't be in use as long as an anchor line might be, the free end can be passed through a strand of laid line (see Figure 1316, center), or an extra half hitch can be used; see Figure 1316, right—a clove hitch with extra half hitch.

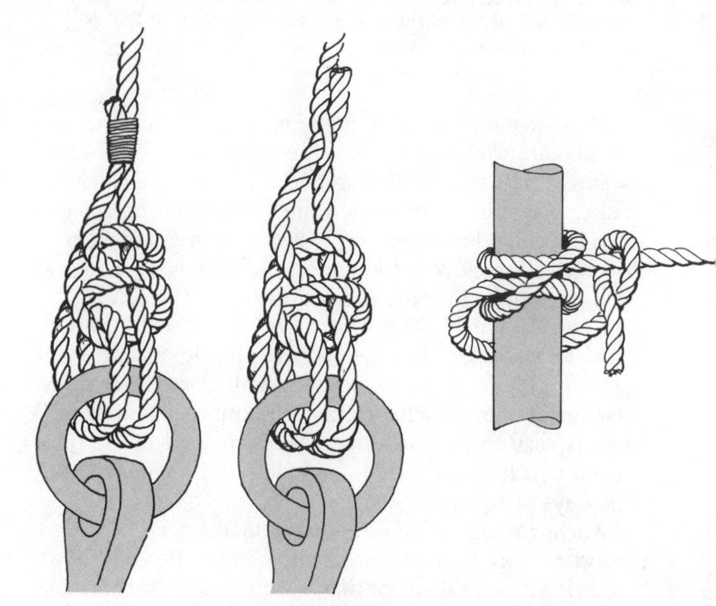

Figure 1316 **Twine seizing, *left,* is used to give extra security to an anchor bend. A variation, *center,* is to pass the free end through the standing part of the line. A clove hitch, *right,* is finished with a half hitch around the standing part of the line for added security.**

How Knots and Splices Reduce Rope Strength

	Knot or Splice	*Efficiency (Percentage)*
Knots	Normal Rope	100
	Anchor or Fisherman's Bend	76
	Timber Hitch	70–65
	Round Turn	70–65
	Two Half-hitches	70–65
	Bowline	60
	Clove Hitch	60
	Sheet Bend	55
	Square or Reef Knot	45
Splices	Eye Splice on Thimble	95–90
	Long Splice	87
	Short Splice	85

Table 13-1

Splices and Splicing

Most stock sailboats come with all basic rigging supplied; it is not until something new is added or something old is replaced that a splice is even thought about. Power and sailing craft may have a made-up anchor line (usually not as long as it should be), but probably will not have enough dock lines. Marine dealers often sell dock lines of various lengths already made up, or you may want to make your own. A sailmaker will often be a good source for new jib sheets, other lines, or small stuff. But in spite of the fact that many boat-owners do not need to do much splicing, it is a good idea to know the basics. You can be a wise customer, and you can make the occasional emergency repair if you know the rudiments of splicing.

Splicing, like knotting, is a finger art. In small to moderate sizes of laid ropes you can make splices without tools, although a knife and fid are extremely helpful. If you are splicing heavier line or braided line, simple tools are necessary—a fid or marlinespike for laid line, and a special fid for braided line.

Splicing Principles

As you can easily see in the illustrations, or by looking at a splice in laid line that is already made, the principle is simple: three strands are tucked over and under so that they interweave with three other strands. If you are splicing the ends of two lines together in a SHORT SPLICE, the result is obviously thicker. A short splice, therefore, won't go through a block of the correct size for the diameter of the line. A LONG SPLICE is the solution—part of each strand is taken out and the tapered result makes a thin splice (with less strength). Most boat owners, however, would replace a broken line rather than splice it.

Short Splice

To start a short splice, unlay the strands of both rope ends for a short distance, about ten turns of the lay. Tape or fuse the six strand ends, or whip them, to prevent unlaying. A seizing is often made around each of the ropes, or each is wrapped with a piece of tape, to prevent strands from unlaying too far. These seizings or tape will be cut as the splice is completed.

Next "marry" the ends so that the strands of each rope lie alternately between strands of the other as shown in Figure 1317A. Now tie all three strands of one rope temporarily to the other; see Figure 1317B. (Some omit this step; it is not absolutely essential.)

Working with the three free strands, remove temporary seizing from around the other rope and splice the strands into it by tucking the strands, working over and under successive strands from right to left against the lay of the rope. When first tucks have been made, snug down all three strands. Then tuck two or three more times on that side.

Next cut the temporary seizing of the other strands and the rope and repeat, splicing these three remaining strands into the opposite rope.

Figure 1317E shows how the short splice would appear if not tapered, after finally trimming off the ends of strands. Never cut strand ends off too close. Otherwise when a heavy strain is put on the rope, the last tuck tends to work out, especially with synthetics.

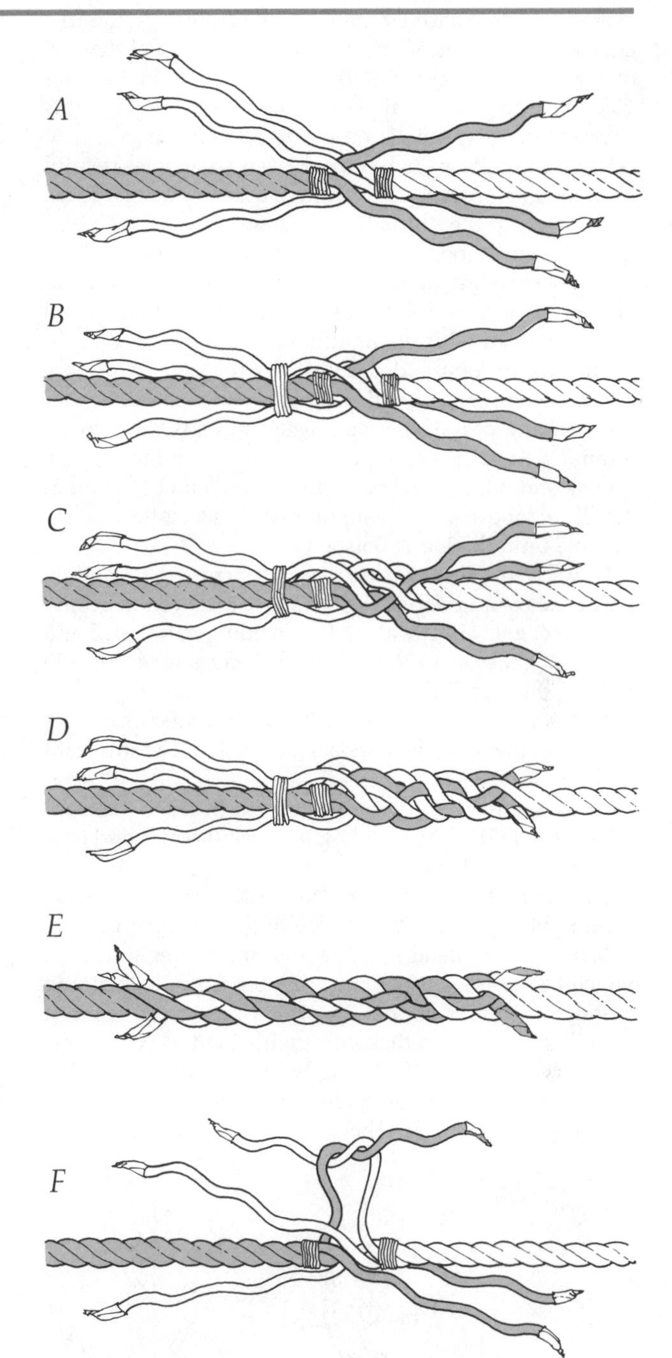

Figure 1317 **A short splice is the strongest commonly-used method of joining two lengths of line without a knot.**

Another method which some find easier, is to start as in Figure 1317A and tie pairs of strands from opposite ends in an overhand knot; see Figure 1317F. This, in effect, makes the first tuck.

Eye Splice

Although the short splice is the easiest to learn, an EYE SPLICE is much more often needed. The same principle—

interweaving—applies, but there is one point where an error is easily made.

Start the eye splice by unlaying the strands of the free end six to ten turns of lay. Now tape or heat-seal the end of each strand to prevent its unlaying while being handled; whipping can be applied to the strand ends, but this is rarely done as this is only a temporary intermediate action if the ends are to be tapered. It is sometimes helpful to place tape around the unlaid strands every four to six inches to help the "turn" in the strand.

Next form a loop in the rope by laying the end back along the standing part. Hold the standing part away from you in the left hand, loop toward you. The stranded end can be worked with the right hand.

The size of loop is determined by the point X (Figure 1318B) where the opened strands are first tucked under the standing part of the rope. If the splice is being made around a thimble, the rope is laid snugly in the thimble groove and point X will be at the tapered end of the thimble. The rope may be temporarily taped or tied to the thimble until the job is finished.

Now lay the three opened strands across the standing part as shown in Figure 1318A so that the center strand b lies over and directly along the standing part. Left-hand strand a leads off to the left, right-hand strand c to the right of the standing part.

Tucking of strand ends, a, b, and c under the three strands of the standing part is the next step. Get this right and the rest is easy.

Always start with the center strand B. Select the topmost strand (2) of the standing part near point X and tuck B under it. Haul it up snug but not so tight as to distort the natural lay of all strands. Note that the tuck is made from right to left, against the lay of the standing part.

Now take left-hand strand a and tuck under strand (1) which lies to the left of strand (2). Similarly, take strand c and tuck under strand (3), which lies to the right of strand (2). Be sure to tuck from right to left in every case.

The greatest risk of starting wrong is in the first tuck of strand c. It should go under (3), from right to left and look like the drawing. The way you do it is to flop the whole thing over in your hands before making the tuck of strand c. You'll notice only one free strand (that's c) untucked, and only one of the original strands in the standing part that doesn't have a strand under it. Be sure you make the third tuck in the right direction.

If the first tuck of each of strands a, b, and c is correctly made, the splice at this point will look as shown in Figure 1318B.

The splice is completed by making at least two additional tucks in manila rope or four full tucks in synthetic rope with each of strands a, b, and c. As each added tuck is made be sure it passes over one strand of the standing part, then under the strand next above it, and so on, the tucked strand running against the lay of the strands of the standing part. This is clearly shown in Figure 1318D, the completed splice. Note c, c^1, and c^2, the same strand as it appears after successive tucks.

Suggestions The splice can be made neater by tapering. This is done by cutting out part of the yarns from the tucking strands before the finishing tucks. In any case, the first two or three tucks in manila or three or four tucks in synthetics are made with the full strands. (Synthetics are often slippery and thus require at least one extra tuck.) After that, some prefer to cut out a third of the yarns, and make the last tuck. This produces an even taper. After the splice is finished, roll it on deck under foot to smooth it up. Then put a strain on it and finally cut off the projecting ends of the strands. Do not cut off the "tails" of synthetic rope too short.

The loose fibers may be fused with a flame or a rope-cutting tool such as a soldering gun, but be careful not to melt or set fire to the rope.

The eye splice is often made on a metal or plastic thimble. When used this way it is necessary to work the splice very tightly and it is almost always desirable to add a whipping, using a needle and waxed nylon twine or tape (see instructions for whipping at the end of this chapter).

Another way to make a synthetic rope splice tight is to place it in boiling water so that the rope shrinks.

In all splicing, careful re-laying of the rope—so that

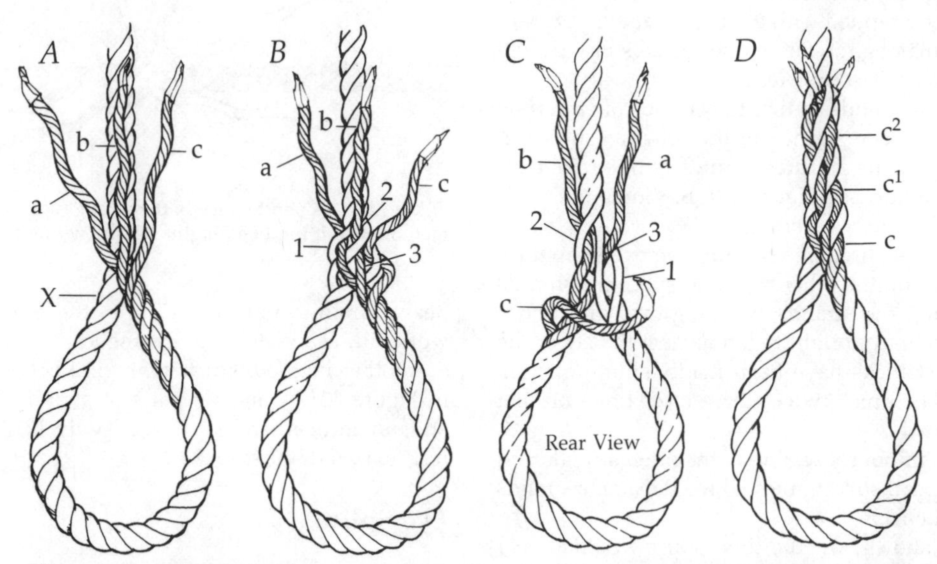

Figure 1318 **An eye splice forms a permanent loop in a line; it may be made around a thimble to guard against chafe.**

every strand is under the same even tension—is important.

One other splice is seen frequently on boats: the BACK SPLICE at the end of a line. This makes a good, neat finish to a line, but it has a major disadvantage: if you want to unreeve the line at the end of the season, or to replace it, the back splice won't go through the blocks. See Figure 1319. A well-made whipping makes an equally neat finish (see Figures 1324 and 1325). You can tie a figure-eight or Ashley stopper knot to keep the end from going through the block when you don't want it to.

Double-Braided and Wire Rope

An eye splice in double-braid looks difficult—but is relatively simple to learn. The technique has almost nothing to do with other kinds of splicing—it is just a logical way to use the cover and core of the braided line, since both are hollow. A thimble can be inserted during the splicing process. The eight steps of the procedure are described, with explanatory illustrations that show how special tools are used; see Figure 1320.

Splices in wire rope, best left to professionals, and wire-to-rope, surely left to professionals, use techniques somewhat similar to laid rope splices, but differing in details.

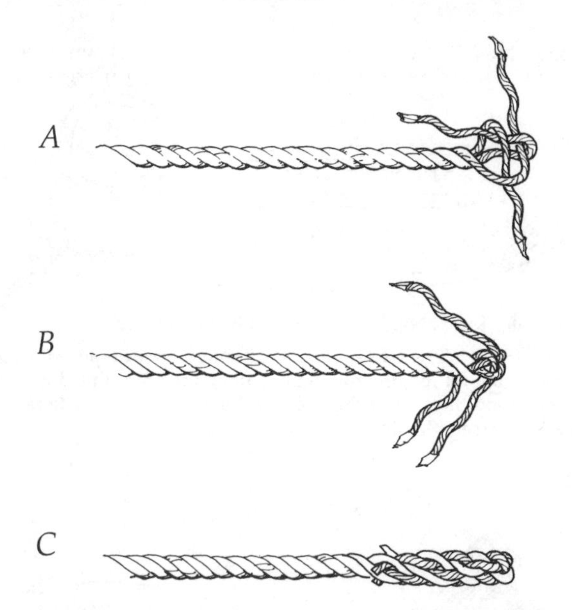

Figure 1319 **Back splice is started by bringing each strand over the one to its left, and under the next one, as in *A*. The strands are then tucked as shown in *B*, for at least three strands, *C*. Longer strands can be used, and trimmed in thirds to provide a tapered finish.**

Use and Care of Rope

Earlier you took a look at proper ways to coil a line, to flake it down in figure-eights, and to fasten to a cleat. In addition, there are some other pointers on using and caring for rope.

Keep Rope Clean Dirt, sand, oil, and acids are destructive to both natural and synthetic rope. To wash rope, put it in a mesh bag or pillow case—that way it won't foul up your washing machine. Mild washing powder won't hurt either kind of rope.

Don't Let It Kink Kinking is not only annoying, but it can also give problems that lead to breakage. Sometimes coiling new three-strand rope first against the lay, and then with the lay (clockwise), helps to get it so it will coil and run smoothly. Rope, including braided rope, should be taken off its original spool in a direct unwinding pull—not looped off over the top of the spool. As mentioned earlier, larger, stiffer ropes may need different handling from small, flexible lines. Double-braid can develop a twist when used off and on winches. This can lead to kinking. Overhaul the line and take out the twist, as needed.

Make Small Repairs Promptly If an eye splice is coming undone or a whipping breaks, fix it with care before relying on the line again. An incomplete eye splice can be dangerous, and an unwhipped line end will keep right on coming apart until it is fixed.

Guard Against Chafe and Abrasion You should "change the nip" on an anchor line every few hours, instead of letting the wear come on the same place over an extended period. You can use a leather chafe guard, which is best, or, if your line is small enough, one of the kind made from split garden hose. You can whip or serve a place in a dock line that goes over a gunwale or through a chock. Using small braided tape or cord is good for such a whipping because if it wears through it breaks and you can see it. Special chafing tape is also available. Some boat owners pre-whip dock lines and anchor lines at a number of convenient places and then adjust the lines so the whippings will be in the chocks.

Avoid Friction Damage Slipping, on a power winch, can result in friction heat that will damage a line.

Use the Right Size Line in a Proper Manner A line that jams in a block can tear itself apart. Heavy strains that do not break a rope can nevertheless weaken it. Both continual stretching and sudden shocks are damaging.

Inspect your Lines and Rigging Even with the improvements in synthetic laid line, it is worthwhile to open the lay and inspect the interior strands, to see if the fibers have started to break or powder. Checking eye splices, whipped ends, and places where lines go through sheaves, blocks, or fairleads is the best way to detect fraying, chafe, and other deterioration.

Rope Inspection Procedures

Inspection of manila rope once meant unlaying a number of strands, looking for broken fibers. The same procedure works for synthetic line, but the details are different. Frayed strands, powdered fibers inside the rope, and stiffness are the signs of serious deterioration. In addition, one should go over the entire length of line, looking for cuts and nicks, exterior signs of abrasion, and burns.

A slight abrasion fuzz on synthetic line, laid or braided, acts as a protective cushion. Pulled or cut strands in braided line are more serious; they can affect 2% of the line strength for each strand involved.

Eye splices should be inspected, inside and out, for distortion of the thimbles as well as a tendency for the splice to come undone.

A

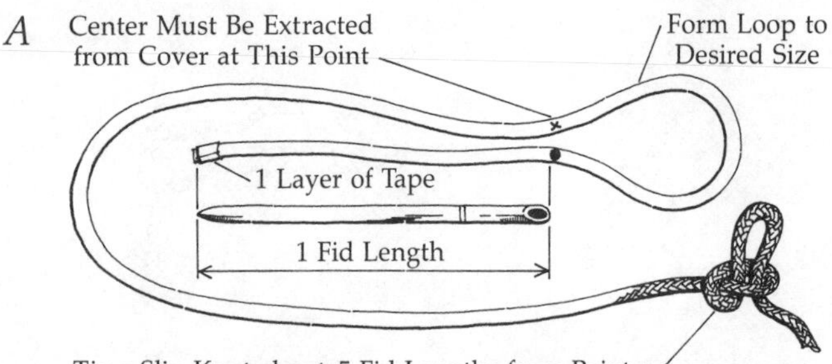

Figure 1320A **Tightly tape end with one layer of tape. Mark a big dot one fid length from end of line, and from this dot form a loop the size of the eye you want. Mark an X where the loop meets the dot.**

B

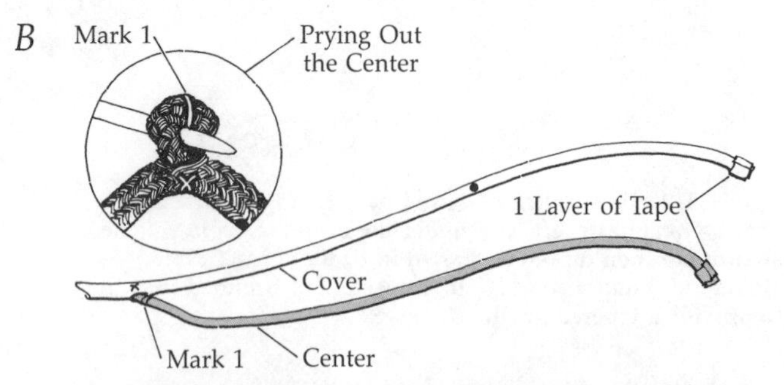

Figure 1320B **Bend the line sharply at X, and spread the strands apart firmly to make an opening so the center can be pried out. Mark one big line on the center where it comes out (this is Mark #1), and use your fingers to pull all the center out of the cover from X to the end. Pull on the paper tape inside the cover until it breaks back at the slip knot, and pull it out. Put a layer of tape on center end.**

C

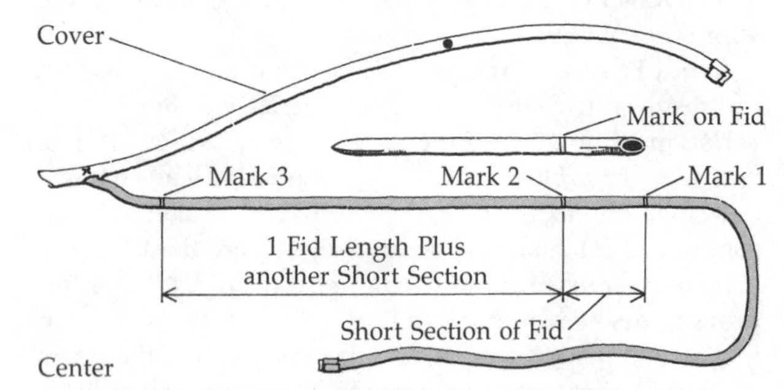

Figure 1320C **Pull out more of the center. From Mark #1 measure a distance equal to the short section of the fid, and mark two heavy lines (this is Mark #2). Mark #3 is three heavy lines at a distance of one fid length plus one short section of the fid from Mark #2.**

D

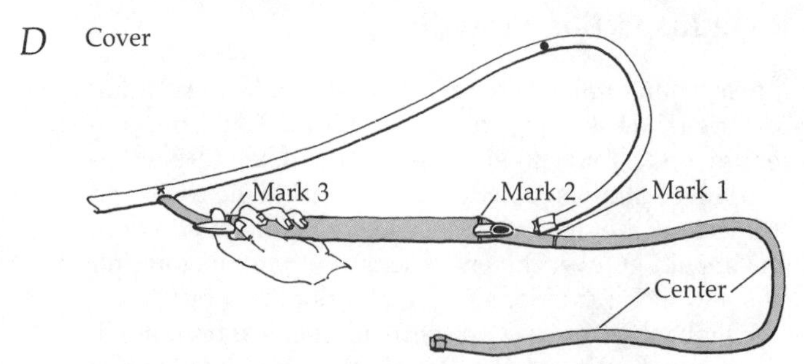

Figure 1320D **Insert fid into center at Mark #2, and slide it lengthwise through "tunnel" until point comes out at Mark #3.**

E

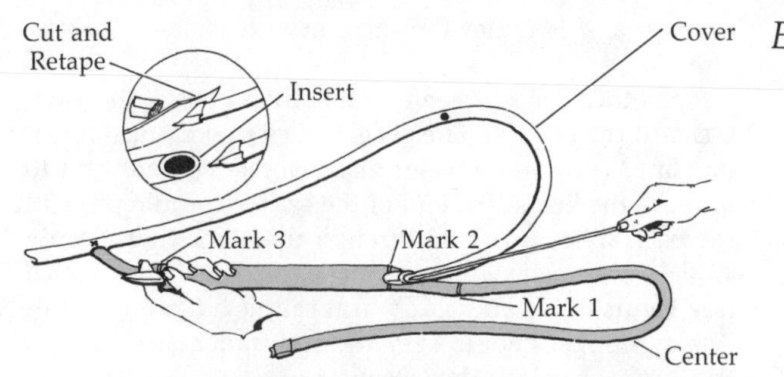

Figure 1320E **Cut across taped end of cover to form a point, and retape tightly with one layer of tape. Jam this point into open end of the fid; jam pusher into fid behind the tape. Hold center gently at Mark #3 and push both fid and cover through center until dot almost disappears at Mark #2.**

F

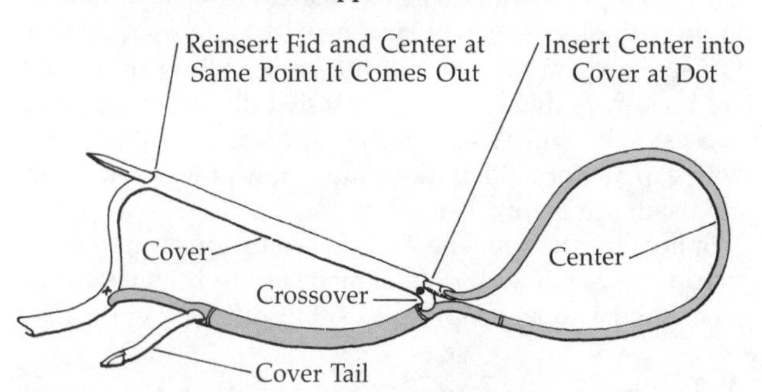

Figure 1320F **Note how center tail must travel through the cover. It must go in close to dot, and come out through the opening at X. On large eyes several passes may be needed for fid to reach X. When this occurs, simply reinsert fid at the point it came out and continue to X. To start, insert fid in cover at dot and slide it through tunnel to X. Form a tapered point on center tail, jam it into the open end of fid, and push fid and center through the cover. After fid comes out at X, pull all center tail through cover until tight, then pull cover tail tight.**

G

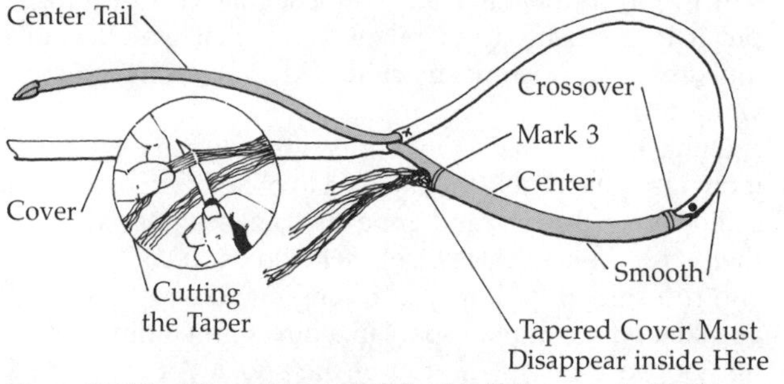

Figure 1320G **Unravel cover tail braid all the way to Mark #3, and cut off groups of strands at staggered intervals to form a tapered end. Hold loop at crossover in one hand, and firmly smooth both sides of loop away from crossover. Do this until the tapered tail disappears inside Mark #3.**

H

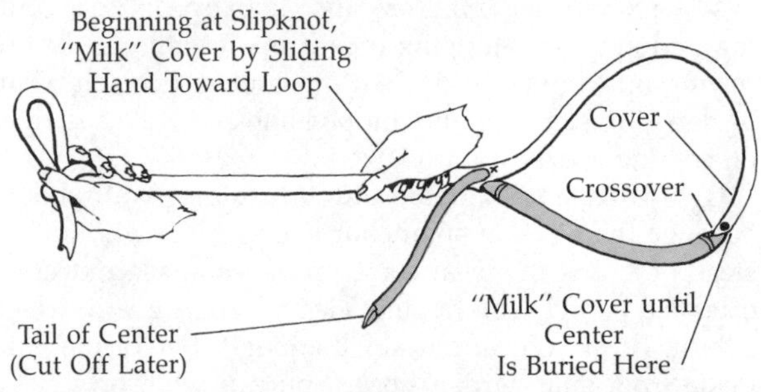

Figure 1320H **Hold rope at slipknot, and gently begin to "milk," or slide, the cover slack toward the loop. You'll see the center begin to disappear into the cover. Go back to the knot, and continue sliding more cover firmly until all center and the crossover are buried.**

Whippings and servings, of course, should be inspected regularly for excessive wear, so they can be renewed. Several times a season halyards should get a special inspection where they run through the block at the masthead when in use, because here the small motion always in the same place around the block, plus sharp bends in the line, lead to breakdowns. Anyone who has had a sail drop suddenly because of this in a strong wind knows the wisdom of checking halyards.

Any three-strand lines that have had severe kinks or hockles can be assumed to have lost 30% of their strength. Never use force to straighten a kink. Instead, turn the line in your hands or trail it astern.

An old-fashioned cure for wear on a line is still useful with synthetic rope: reversing the line end for end. In other cases a small repair, a whipping to take the chafing, or a slight shortening of the line to make a new eye splice may be the cure.

Blocks and Tackle

The use of a BLOCK AND TACKLE (pronounced tay-kle) or, to use a higher sounding name, mechanical appliances, on board a small boat is generally limited to sailboats. Any competent seaman should have a basic knowledge of it, however, because its use enables one man to do the work of many.

A block and tackle is used where hoisting sails, as well as setting them, requires some means for one or two people to provide the strength of many. No matter how small the sailboat, the sheets usually run through one or more blocks, which means we have a mechanical appliance.

To see how this aids, go sailing in a 20-foot boat, in a moderate breeze, and bend a line to the boom. While underway, attempt to trim in the sail with your improvised sheet. It will come in, but it will be a struggle, so try it with the regular system of blocks and tackle and you will see with what ease the sail comes in.

About the most common use on a motorboat is in hoisting your dinghy.

Terminology

A BLOCK consists of a frame of wood or metal inside of which is fitted one or more SHEAVES (pulleys—the word is pronounced shiv), and is designated according to the number of sheaves it contains, such as single, double, or triple. The size of the block to be used is, of course, determined by the size of the rope that will run through it. If a fiber rope is being used, the size of the block should be *about* three times the *circumference* of the rope and the sheave diameter about twice the circumference. Therefore, for ⅝-inch rope (about 2 inches in circumference), the block could be 6 inches (three times the circumference) and the sheave diameter 4 inches (twice the circumference). This is an approximation. See Table 13-2, block sizes and rope diameter, for recommended sizes.

Wire rope is also used, but usually only as halyard leads on sailboats. This should be stainless steel and sheaves should be as large as possible for long rope life. Make sure the rope cannot squeeze between the sheave and the cheeks of the block or your line may jam.

The term TACKLE is used for an assemblage of FALLS (ropes) and BLOCKS. When you pass ropes through the blocks, you REEVE them and the part of the fall made fast to one of the blocks, or the load, as the case may be, is known as the STANDING PART, while the end upon which the force is to be applied is called the HAULING PART. To OVERHAUL the falls is to separate the blocks; to ROUND IN is to bring them together; and CHOCK-A-BLOCK or TWO-BLOCKED means they are tight together.

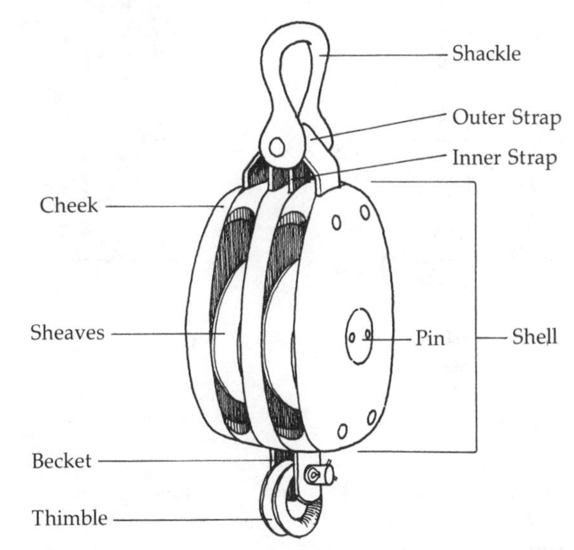

Figure 1321 **Parts of a block.**

Kinds of Tackle

Tackles (Figure 1322) are named according to the number of sheaves in the blocks that are used (single, two-fold, three-fold purchases), according to the purpose for which the tackle is used, or from names handed down

Rope and Block Sizes

Size of Block (Diameter of Shell) (Inches)	*Diameter of Rope (Inches)*
3	3/8
4	1/2
5	9/16 - 5/8
6	3/4
7	13/16
8	7/8 - 1
10	1 1/8
12	1 1/4
14	1 3/8 - 1 1/2
16	1 5/8

Table 13-2

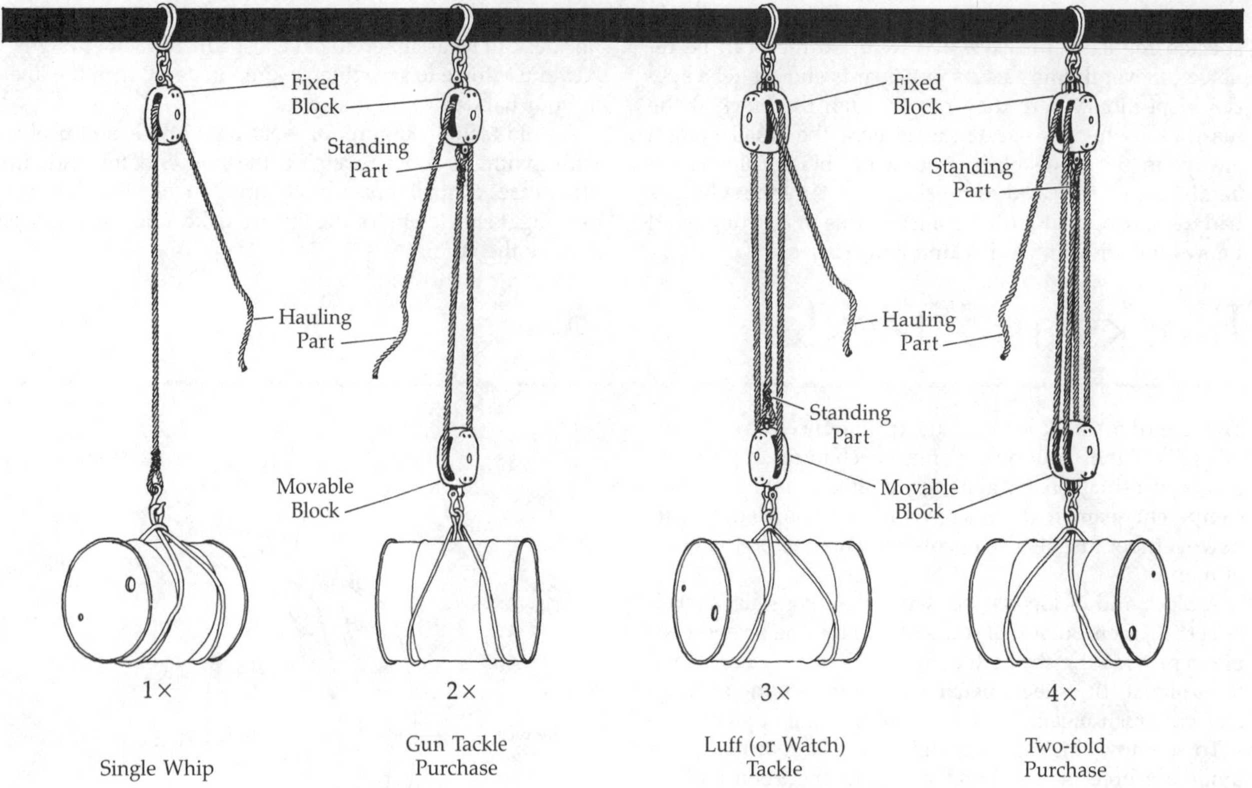

Figure 1322 **Some of the common tackles used on boats.**

from the past (luff-tackles, watch-tackles, gun-tackles, Spanish-burtons, etc.). The tackles that may be found aboard cruising boats are:

Single Whip A single fixed block and fall—no increase in power. The gain is only in height of lift or change in direction of pull.

Gun Tackle Two single blocks. If lower block is movable, double force is gained. If upper block is movable, triple force is gained.

Luff Tackle A double hook-block and single hook-block. Force gained three if single block is movable, four if double block is movable.

Two-Fold or Double Tackle Two double sheave hook-blocks. Force gained four or five, depending upon application.

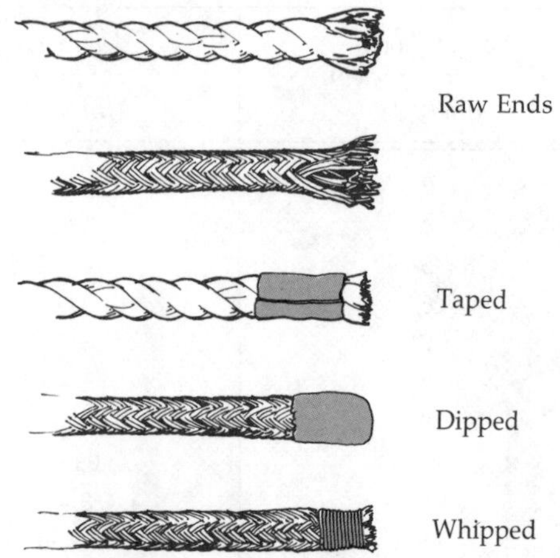

Figure 1323 **Ends of laid or braided line can be taped, dipped in a sealer, or whipped to prevent unlaying.**

Calculating Power of a Tackle

The force gained in all these tackle combinations is theoretical only, as the friction of the blocks is ignored. The usual method of compensating for this friction, and calculating the actual force required, is to add 5% to the weight of the object (W) for each sheave in the tackle before dividing by the number of falls. For example: in the two-fold purchase of Figure 1322, if the weight of the drum is 500 pounds, 5% is 25 pounds; 4 sheaves × 25 pounds = 100 pounds. Thus the total W to be considered is 500 + 100 = 600 pounds. With four falls, the force needed to lift the drum would be 600 ÷ 4 or 150 pounds, rather than the theoretical 125 pounds if friction was disregarded—a significant difference. The actual friction will, of course, vary with the type of bearings and pins, state of lubrication, and sheave diameter.

There are, of course a number of other purchases, the heaviest commonly used aboard ship being a three-fold purchase, which consists of two triple blocks.

It must be remembered that the hauling part on a three-part block should be reeved around the *center* sheave. If this is not done, the block will cant, causing it to bind, and in extreme cases to break.

The Bitter End

If a piece of rope is cut, it will start to fray and unravel at the end—sometimes in seconds. If it is taped before cutting, or if the cutting is done with an electric cutter that melts the fibers, the end won't fray, at least not immediately.

But a rope end still needs protection for almost any use. There are a variety of ways a line can be protected; some

Figure 1324 **Plain whipping steps.**

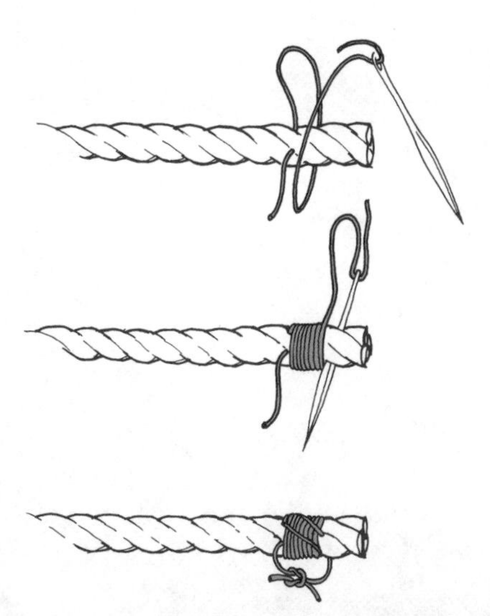

Figure 1325 **Sailmaker's whipping is started by stitching the cord through the line, top. Wind six or more turns around the line, center, and stitch back through it. Then bring the cord back over the turns along a groove between strands. Stitch through a strand to the next groove, and bring the cord back along this. Stitch through the next strand to the final groove, bottom, and finish with a square knot in the cord.**

are good if you're in a hurry, some fit one situation to do with line handling, others are right for other purposes.

A careful taping of the end of a line is useful but hardly looks seamanlike; one might say it's a good way to handle a coil of spare rope that you're storing in a locker before putting it to use; see Figure 1323. A rope end with adhesive tape is at least ready for a more sophisticated finish.

The air-drying liquid plastics mentioned earlier can also be used for rope ends; some even have color coding.

Braided line is more difficult to finish with a traditional yet functional end. An eye splice on a braided dock line is an obviously correct solution. The end is protected and the eye splice is ready to use. A whipping, preferably one made very tight with a sailmaker's needle, results in a usable finish.

With laid line, however, the marlinespike arts of the sailor provide more opportunity. A variety of whippings, knots, or a back splice are among the choices.

To make a plain whipping, you need the right size of waxed nylon twine or lacing and your hands; see Figure 1324.

A far better whipping, called a sailmaker's whipping, requires a palm and sailmaker's needle, as well as the cord or lacing twine; see Figure 1325 for instructions. This whipping will last longer as well as look better.

A standard way to make a good-looking end for a line—provided it doesn't have to go through a block, is a back splice. Properly tapered, perhaps polished off with a little extra whipping, it lasts a long time; see Figure 1319.

Two fancier rope ends, from many that are available, are illustrated here without instructions. You can learn to make them—as you can learn other ornamental and useful knots and rope work—in any of several advanced books on marlinespike seamanship.

First, the wall-and-crown (see Figure 1326), a good-looking finish that combines two simple knots.

Second, the manrope knot—which is really a doubled wall-and-crown. Finished with a back splice it looks good on the end of a bucket rope—and helps you hold the bucket.

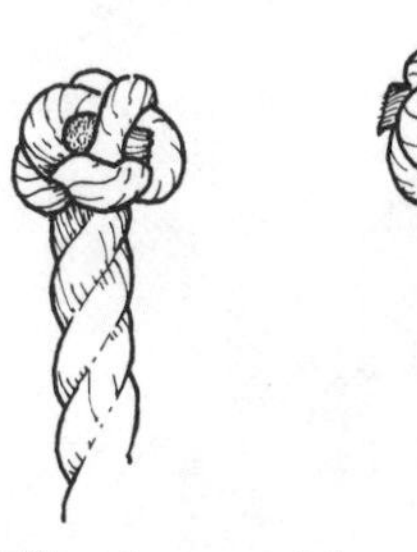

Figure 1326 **Wall-and-crown, *left,* and manrope knot, *right,* are decorative ways of finishing the bitter end.**

14

The basic forces that make weather; storm systems and their propagation; how to forecast weather from your own local observations

WEATHER

Weather is an important element of your boating; it can add to or detract from your enjoyment, and it can have a vital bearing on your safety. You need not be qualified as a weather forecaster, but you must be able to understand and properly use forecasts prepared by experts. And you must be able to correctly interpret local weather signs—wind shifts, changes in cloud patterns or ocean swells.

The Elements of Weather

Temperature, precipitation, and wind are the weather elements of most concern to boatmen. Waves, of course, vitally affect small boats, and these are a consequence of wind. Temperature is more related to comfort than safety (except where it relates to the formation of fog) and will be considered in this chapter as an accompanying effect. Precipitation is of major concern only when it restricts visibility. Winds are thus our concern—where they will blow from, and how hard.

To focus our attention on weather that will affect most recreational boatmen, we will limit our considerations—except where noted otherwise—to the mid-latitudes and the subtropic and tropic zones of the *northern* hemisphere.

Sources of Information

As the skipper of a small craft you really need two different kinds of weather information: what to expect in the next hour or so, and what weather the next two or three days will bring.

Forecasts

Weather forecasts are prepared in the U.S. by field offices of the National Weather Service (NWS), an element of the National Oceanic and Atmospheric Administration (NOAA), Department of Commerce. In Canada, forecasts are prepared by the Atmospheric Environment Service. Other countries have government agencies with similar responsibilities; there is a full and rapid flow of information across national boundaries.

Modern Weather Maps

The DAILY WEATHER MAP, with its many complex STATION MODELS of conditions at various cities at a given time, prepared in Washington and distributed by mail is still available, but in this age of instantaneous electronic communications, it is received most quickly in facsimile equipment. Although newspapers in some larger cities still print a weather map, it is usually a FORECAST MAP, predicting conditions around the country at about the time the paper will normally be read.

The more common weather map of today is seen on your TV screen as a part of a news program; this may be of existing conditions or a forecast, or both. Many cable TV systems now carry a channel of continuous weather information. TV weather maps may omit many details but are useful in evaluating probable weather changes for your area.

Weather Broadcasts

Almost all broadcast radio stations, AM and FM, transmit weather conditions and forecasts at fairly regular intervals. These are based on official information, but vary widely in level of detail and timeliness. You should be able to get some weather information with even the smallest pocket receivers in most boating areas.

The best sources of weather information by radio are the *continuous broadcasts* by NWS stations on frequencies that can be picked up by any marine VHF-FM transceiver (and on small, inexpensive "weather radios" for just such use). Frequencies, in probable order of use, are WX-1, 162.550 MHz; WX-2, 162.400 MHz; and WX-3, 162.475 MHz. In Canada, other frequencies are used; these vary with the geographic area, check locally for the specific channel.

The recorded messages are basically for marine interests and thus provide excellent information for boatmen, including current conditions at numerous local points, the general weather pattern, marine forecasts for the local and adjacent areas, and tide data (if applicable). Tapes are updated periodically and special warnings are spliced in whenever appropriate. Broadcasts are made 24 hours a day; antenna heights and transmitter power are normally adequate to give a range of 40 miles, often considerably more. Frequencies are assigned to minimize interference between stations, but this does occur at times.

Weather warnings, although usually not routine forecasts, are broadcast by Coast Guard Group radio stations when issued and at regular times; information on broadcast content and schedules are published in *Local Notices to Mariners.*

High-Seas Weather Information

Weather reports and forecasts are also available for oceanic regions from several sources. Major Coast Guard radio stations have scheduled weather broadcasts, as do the high-seas Marine Operators (ship-to-shore telephone service). Brief weather messages are included in the transmissions of the time and standard frequency stations, WWV (Ft. Collins, Colorado) and WWVH (Hawaii), on 5, 10, 15, and 20 MHz. Information on various ocean areas is broadcast at specific numbers of minutes after each hour; detailed information can be found in publications of NWS, or the National Bureau of Standards, which operates these stations.

Radiofacsimile Weather Maps

A vessel at sea that is equipped with a FACSIMILE RECEIVER (or a printer attached to a single-sideband receiver) can obtain up-to-date weather maps of a wide variety of types—current and forecast conditions, surface and upper level information, and other special types. These are increasingly popular on ocean-going yachts, but you'll need experience to obtain useful information from them.

Weather Signs

Weather forecasting is a science that has improved in recent years, but every skipper knows that it is not infallible. Get your forecasts by any and all means possible, but keep an eye on present local conditions, *and especially on changes in them.* A forecast of good, safe conditions for your general area does *not* preclude temporary local differences that could be hazardous. (Keep alert also for special warnings broadcast by the Coast Guard on VHF Channel 22 following a preliminary announcement on Channel 16.)

Weather signs and their interpretation are discussed on pages 324-330.

What Makes the Weather

Weather doesn't just happen; there are basic causes and effects. An appreciation of these influences will lead to a better understanding of weather systems and their movement—the basis for weather forecasting.

The Earth's Atmosphere

The earth is enveloped in its ATMOSPHERE—a mixture of gasses, mostly carbon dioxide and oxygen, that extends upward with decreasing density for many miles. Half of all this air is in the lower 3½ miles, and we are not concerned with that portion which lies above about 20 miles.

The atmosphere consists of five layers, of which only the lower two, the TROPOSPHERE and the STRATOSPHERE—and the boundary area between them, the TROPOPAUSE, will be of interest to us. The height of the troposphere varies with latitude, averaging 11 miles at the equator but only five miles above the poles. This is the area which contains our weather systems, and most of our clouds.

Heat From the Sun

Although the sun is some 93 million miles away, and only an infinitesimal fraction of its energy falls on the earth, this is enough to make the earth habitable and to establish our climate and weather patterns. The atmosphere absorbs more than half the sun's energy that reaches the earth, but 43% does penetrate to the surface. The energy that is not reflected from clouds is partially absorbed and partially reflected from the land or water. Much of the solar energy that reaches the earth's surface is re-radiated as heat rays of longer wavelength. These are trapped in the lower atmosphere much the way the air in a greenhouse is warmed to grow plants. Various types of surfaces absorb and reflect different percentages of the incoming solar rays; these differences result in both climatic differences between geographic regions and local differences between land and sea areas.

The atmosphere moderates the sun's effect, filtering out excess (and harmful) rays by day, and holding in heat at night. Without our atmosphere, we would have to cope with extremes like those on the moon, +200° Fahrenheit or more in sunlight, −200° or less in darkness. Cloud cover directly affects the cooling off of the air at night as clouds reflect back a portion of the heat rising from the earth's surface; on clear nights, more heat is lost.

Circulation Patterns

Under a fundamental law of physics, heated air expands; it thus becomes less dense (lighter) and tends to rise. Air is primarily heated by contact with the earth's surface. This air rises and is replaced by colder, heavier air, which in turn warms and rises. The process is essentially continuous, resulting in vertical currents over wide areas.

The nearer the equator, the more surface heat there is to warm the air. Here it rises and flows towards the poles above a flow of replacement colder air from the poles toward the tropic regions. This simple pattern is, however, modified by the earth's rotation. The northward flowing air at high altitudes is bent eastward and by the time that it reaches about 30° latitude it has started to build up an area of higher pressure forcing some of the flow downward and back towards the equator. This steady flow, bent westward by the earth's rotation, becomes the reliable "northeast trade winds" of the northern sub-tropical zone. The portion that continues northward toward the pole eventually descends as the "prevailing westerlies" of the higher mid-latitudes. Between these regions there are the "horse latitudes," near 30°, where winds are weaker and less constant. Near the equator, where the warmed air rises and turns north or south, the region of weaker winds is called the "doldrums."

Local Wind Patterns

Temperature differences—heating and cooling air—cause localized breezes as well as global winds. Land areas heat up more quickly than water areas during hours of sunlight; air rising over the land is replaced by air coming in from seaward—these are the refreshing "sea breezes." Consequently, at night the land loses its heat more quickly, the water's surface becomes the relatively warmer area, and "land breezes" flow towards the sea. In both cases, these breezes are felt quite close to the surface; there is a counter-flow of air at higher elevations to complete the local circulation pattern.

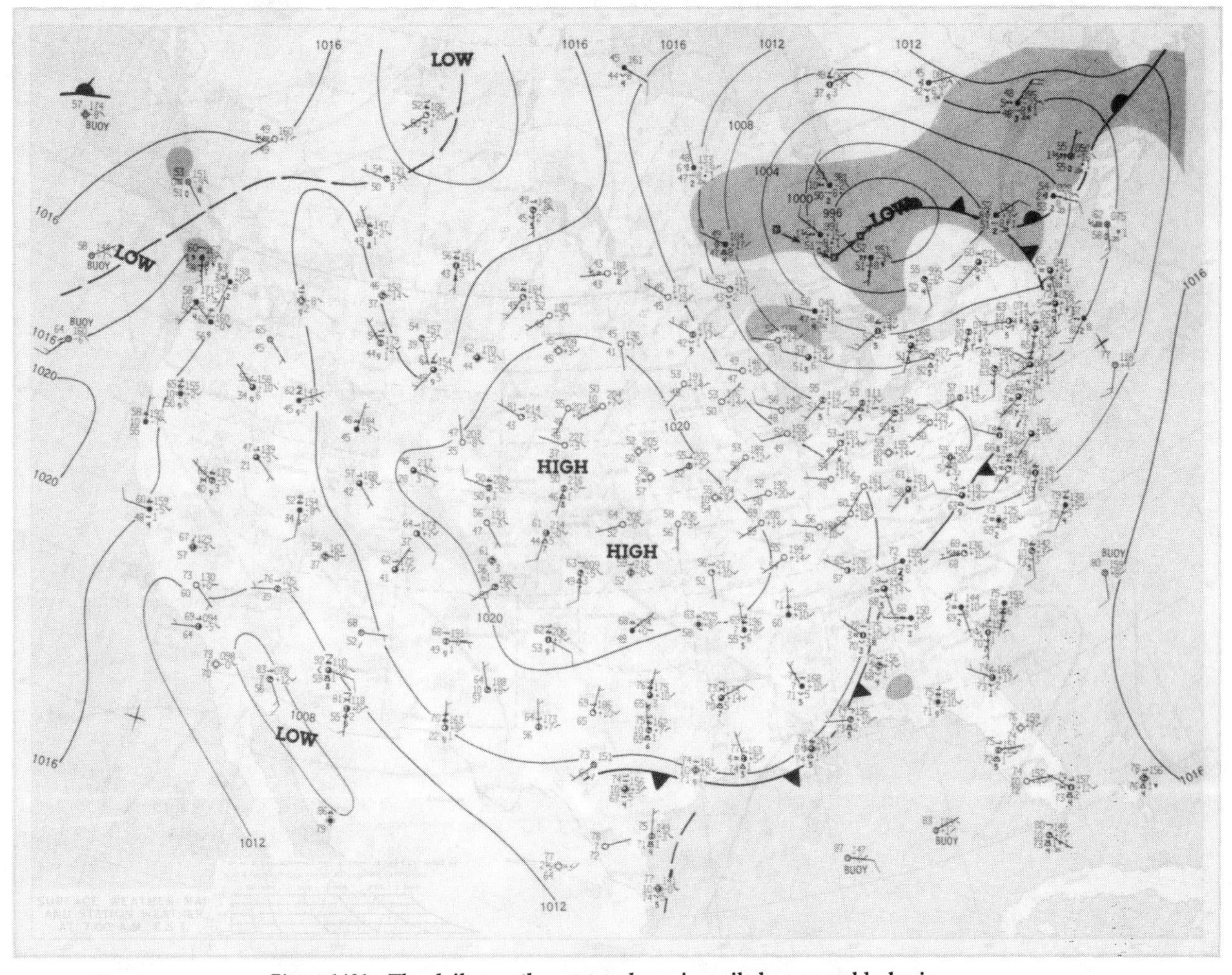

Figure 1401 **The daily weather map, *above,* is mailed on a weekly basis from the National Weather Service to subscribers. The back of each Sunday map carries a full illustration and explanation of all the symbols used. Facsimile weather maps can be received at sea on special radio equipment, *below,* or a printer that attaches to a single-sideband radiotelephone. A variety of types of weather maps are broadcast, and it takes both study and experience to learn how each is used in forecasting.**

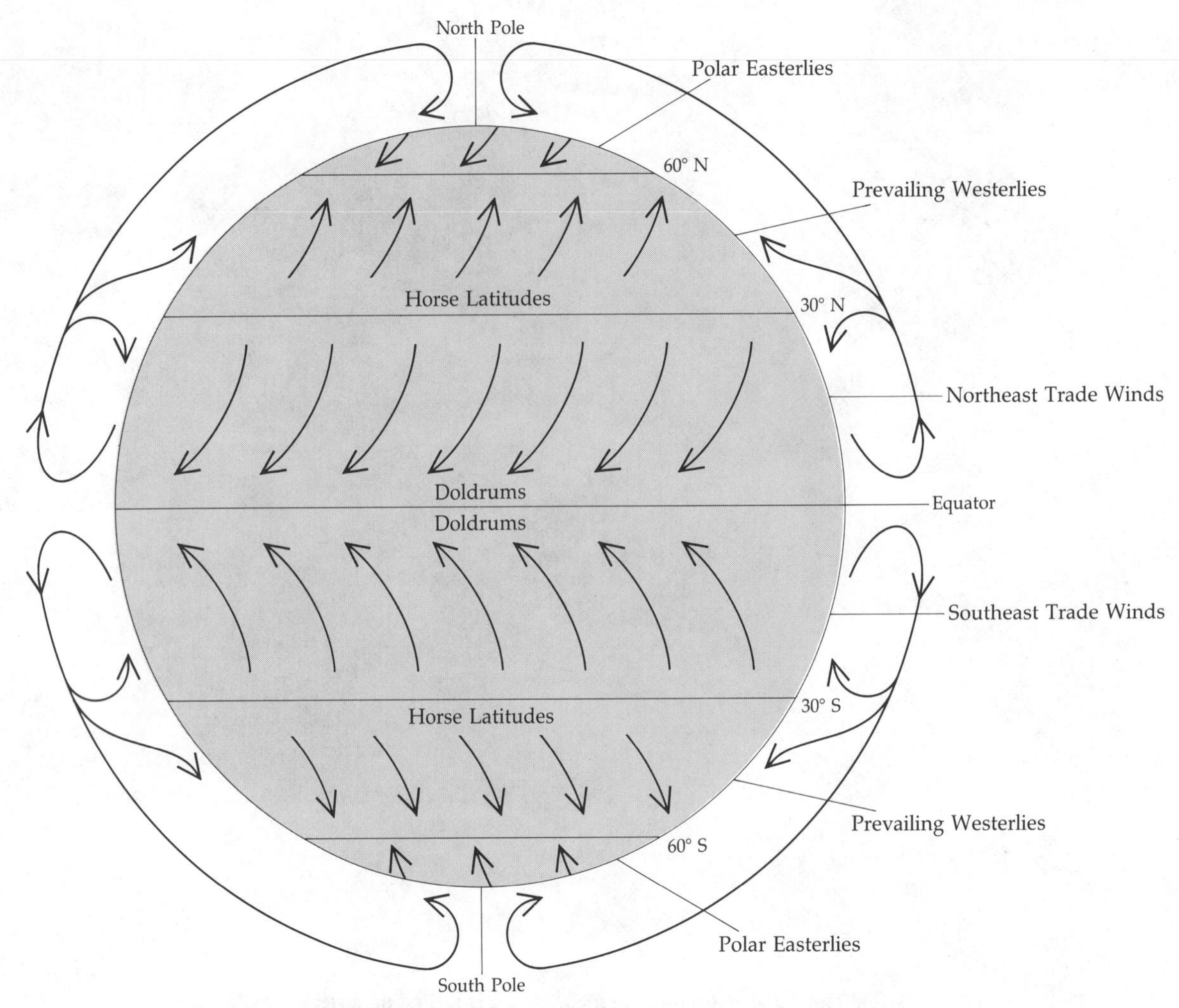

Figure 1402 **These idealized global air circulation patterns are based to a large extent on the heated air that rises along the equator and descends as cold air at the poles, and at latitudes approximately 30 degrees from the equator.**

Clouds

The most visible manifestation of weather is clouds; winds and temperature can be felt, but not seen. Many cloud patterns are not only beautiful and interesting to watch, but they can also be meaningful in interpreting weather conditions and trends.

Moisture in the Air

There is more water in the air than you might think; it evaporates from the oceans, lakes, and rivers and can exist in the air in any of three physical states—vapor, liquid, and solid.

A term you will often hear is RELATIVE HUMIDITY—the amount of water vapor present in the air as a *percentage* of the maximum possible amount for air at that temperature. This maximum amount decreases with a decrease in temperature; the relative humidity of a mass of air increases as its temperature falls, even though the actual amount of moisture is unchanged. When the relative humidity reaches 100% and the air is further cooled, some moisture condenses into a visible form. The temperature at which this occurs is the DEW POINT.

Formation of Clouds

When warm moist air rises from the earth's surface, it expands and cools. The relative humidity increases and when it reaches 100% clouds are formed; there may also be precipitation in the form of RAIN (liquid) or SNOW, SLEET, or HAIL (solid). When moisture in the air condenses into a visible form at or very near the surface, this is FOG.

Moist air is also sometimes cooled by horizontal movement, as from warmer water surfaces to cooler land surfaces; or movement up the slope of a hill or mountain. A warm air mass may be pushed up over a cold air mass, cooling to the point where clouds form and rain falls.

Types of Clouds

There is an international system of cloud classification. The different cloud types have descriptive names that depend mainly upon appearance, but also sometimes upon the processes of formation as seen by an observer. Despite an almost infinite variety of shapes and forms, it is still possible to define ten basic types.

High Clouds (Above 18,000-20,000 Feet)

Cirrus (Ci) Detached clouds in the form of white, delicate filaments, or white or mostly white patches or narrow bands. These clouds have a fibrous (hair-like) appearance, or a silky sheen, or both.

Cirrocumulus (Cc) Thin, white patch, sheet, or layer of cloud without shading, composed of very small elements in the form of grains, ripples, etc., merged or separate, and more or less regularly arranged; most of the elements have an apparent width of less than one degree.

Middle Clouds (From About 7,000 Feet up to 18-20,000 Feet)

Cirrostratus (Cs) Transparent, whitish cloud veil of fibrous (hair-like) or smooth appearance, totally or partly covering the sky, and generally producing halos around the sun or moon.

Altocumulus (Ac) White or gray, or both white and gray, patch, sheet, or layer of cloud, generally with shading, composed of layers, rounded masses, rolls, etc., which are sometimes partly fibrous or diffuse, and which may or may not be merged; most of the regularly arranged small elements usually have an apparent width of between one and five degrees.

Altostratus (As) Grayish or bluish cloud sheet or layer of striated, fibrous, or uniform appearance, totally or partly covering the sky, and having parts thin enough to reveal the sun at least vaguely, as through ground glass. Altostratus does not show halo phenomena.

Low Clouds (From Near Ground up to About 7,000 Feet)

Nimbostratus (Ns) Gray cloud layer, often dark, the appearance of which is rendered diffuse by more or less continuously falling rain or snow, which in most cases reaches the ground. It is thick enough throughout to blot out the sun. Low, ragged clouds frequently occur below the layer, with which they may or may not merge.

Stratocumulus (Sc) Gray or whitish, or both gray and whitish, patch, sheet, or layer of cloud which almost always has dark parts, composed of tessellations, rounded masses, rolls, etc., which are non-fibrous (except for VIRGA—precipitation trails) and which may or may not be merged; most of the regularly arranged small elements have an apparent width of more than five degrees.

Stratus (St) Generally gray cloud layer with a fairly uniform base, which may give drizzle, ice prisms, or snow grains. When the sun is visible through the cloud, its outline is clearly discernible. Stratus does not produce halo phenomena except, possibly, at very low temperatures. Sometimes stratus appears in the form of ragged patches.

Cumulus (Cu) Detached clouds, generally dense and with sharp outlines, developing vertically in the form of rising mounds, domes, or towers, of which the bulging upper part often resembles a cauliflower. The sunlit parts are mostly brilliant white; their bases are relatively dark and horizontal. Sometimes cumulus is ragged.

Cumulonimbus (Cb) Heavy and dense cloud, with a considerable vertical extent in the form of a mountain or huge towers. At least part of its upper portion is usually smooth, or fibrous or striated, and nearly always flattened; this part often spreads out in the shape of an anvil or vast plume. Under the base of this cloud, which is often very dark, there are frequently low ragged clouds either merged with it or not, and precipitation, sometimes in the form of virga.

Variations

Note that cloud groupings are not absolute or exclusive. Portions of cumulus and cumulonimbus become middle clouds; at times altostratus and the tops of cumulonimbus may be at the heights of high clouds.

COLOR ILLUSTRATIONS

Many of the cloud types described here are shown in the color section, pages 584-585.

Fog

Fog is merely a cloud whose base rests upon the earth, either land or water. It consists of water droplets, suspended in the air, each droplet so small that it cannot be distinguished individually, yet present in such tremendous numbers that objects close at hand are obscured.

If we are to have innumerable water droplets suspended in the air, there must be plenty of water vapor already in that air. If droplets are to form from this vapor, the air must somehow be cooled so the vapor will condense. If the droplets are to condense in the air next to the earth, the cooling must take place at the surface of the earth. If the fog is to have any depth, successively higher layers of air must be cooled sufficiently to cause condensation in them also. Fog forms from the surface up. Thus, the land or water must be colder than the air next to it; and this must be colder than air above.

Air is said to be SATURATED with water vapor when its water-vapor content would remain unchanged if it were placed above a level surface of pure water at its own temperature. It could absorb no more moisture. The amount of water vapor required to saturate a given volume of air depends on the temperature of the air, and increases as the temperature increases. The higher the temperature the more water vapor the air can hold before it becomes saturated.

If a mass of air is originally in an unsaturated state, it can be saturated by cooling it down to a temperature at which its content of water vapor is the maximum containable amount, that is to say, to the dew-point temperature. Or we can saturate it by causing more water to evaporate into it, thereby raising the dew-point temperature to a value equal to the air temperature. In regard to

Figure 1403 **Fog forms when warm, moist air is cooled to the point where the water vapor condenses out of it.**

the latter process, unsaturated air, as it passes over rivers and lakes, over the oceans or over wet ground, picks up water vapor and has its dew point raised. Also, rain falling from higher clouds will increase the amount of water vapor in unsaturated air near the earth.

Four Types of Fog

We can distinguish four types of fog: RADIATION FOG, formed in near-calm conditions by the cooling of the ground on a clear night as a result of radiation of heat from the ground to the clear sky; ADVECTION FOG, formed by the flow of warm air over cold sea or lake; STEAM FOG, or SEA SMOKE, formed when cold air blows over much warmer water; and PRECIPITATION FOG, formed when rain coming out of warm air aloft falls through a shallow layer of cold air at the earth's surface.

Note that both steam fog and rain fog are basically the result of evaporation from relatively warm water, a process which increases the dew point. These fogs can be placed in the single category of WARM-SURFACE FOGS; radiation fog and advection fog can come under the single heading of COLD-SURFACE FOGS.

Radiation Fog

There are four requirements for the formation of radiation fog (sometimes called GROUND FOG). First, the air must be stable, that is, the air next to the earth must be colder than the air a short distance aloft. Second, the air must be relatively moist. Third, the sky must be clear so that the earth readily can lose heat by radiation to outer space. This enables the ground to become colder than the overlying air, which subsequently is cooled below its dew point both by contact with the ground (conduction of heat) and by radiative loss of heat to the ground.

The fourth requirement is that the wind must be light to calm. If there is a dead calm, the lowest strata of air will not mix with the ones above, and fog will form only to a height of two to four feet. If there is a slight motion of the air (a wind of three to five knots) and, hence, some turbulent mixing, the cooling is spread through a layer which may extend to a height of several hundred feet above the ground. With stronger winds, the cooling effect is distributed through so deep a layer that temperature does not fall to the dew point and fog does not form.

The likelihood of radiation fog can be predicted with considerable accuracy; see pages 329-330.

Radiation fog is most prevalent in the middle and high latitudes. It is local in character and occurs most frequently in valleys and lowlands, especially near lakes and rivers where you may be cruising. The cooled air drains into these terrain depressions; the lake or river aids the process of fog formation by contributing water vapor, which raises the dew point.

This type of fog, which may be patchy or uniformly dense, forms only at night. Shortly after sunrise it will start to evaporate ("burn off") over the land, the lower layers being the first to go. It is slow to clear over water, since the temperature of the water does not vary nearly as much from night to day as does the temperature of the land.

Advection Fog

Radiation fog bothers us chiefly in the late summer and early autumn; ADVECTION FOG can be an annoyance at any season. (Advection means "transport by horizontal motion.") Winds carrying warm, moist air over a colder surface produce advection fog. The dew point of the air must be higher than the temperature of the surface over which the air is moving. Thus, the air can be cooled below its dew point by conduction and by radiation of heat to the colder surface. Other requirements for the formation of advection fog are: (a) the air at a height of 100 feet or so must be warmer than the air just above the surface; (b) the temperature of the surface—be it land or water—must become progressively colder in the direction toward which the air is moving. Advection fog may form day or night, winter or summer, over the sea.

Coastal Fog For boatmen, the most bothersome variety is COASTAL FOG. When steady winds blow landward and carry warm oceanic air across cold coastal water, the resulting fog may blanket a great length of coastline and, especially at night, may extend many miles inland up bays and rivers. It can be seen on land as it blows past street lights. For example, consider the Pacific Coast, where the water close to the land is often colder than the water well offshore. The prevailing winds in summer are onshore and the air, since it frequently has come from mid-Pacific, is usually nearly saturated with water vapor. The same thing happens when southerly winds carry air across the Gulf Stream and thence northward across the colder Atlantic coastal waters. Advection fog also may form over the larger inland lakes whenever relatively warm and moist air is carried over their colder surfaces.

Advection fog is generally much harder to dissipate than

is radiation fog. A wind shift or a marked increase in wind velocity usually is required. Unlike radiation fog, sunshine has no effect on advection fog over the water.

Steam Fog

On the Mississippi and Ohio rivers, steam fog is a particular hazard to late evening or early morning boating in autumn. When cold air passes over much warmer water, the lowest layer of air is rapidly supplied with heat and water vapor. Mixing of this lowest layer with unmodified cold air above, under certain conditions, can produce a SUPERSATURATED (foggy) mixture. Because the water is much warmer than the air, vertical air currents are created, and we observe the phenomenon of steaming. Hence, this type of fog is called "steam fog"; it forms in just the same way as steam over a hot bath.

Sea Smoke In winter when cold air below about 10°F. (−12°C) blows off the land and across the adjacent coastal waters, steam fog may be widespread and very dense. It is then SEA SMOKE. Along North American coastal waters, steam fog occurs most frequently off the coasts of Maine and Nova Scotia, and in the Gulf of St. Lawrence, where it can be a serious navigational hazard. However, its occurrence is not restricted to the higher latitudes. Off the east coast of the United States it has been observed as far south as Florida, and it also occurs occasionally over the coastal waters of the Gulf of Mexico.

Precipitation Fog

When rain, after descending through a layer of warm air aloft, falls into a shallow layer of colder air at the earth's surface, there will be evaporation from the warm raindrops into the colder air. Under certain conditions this will raise water vapor content of the cold air above the saturation point and PRECIPITATION FOG, or rain fog, will form.

Distribution of Fog

In the United States, the coastal sections most frequently beset by fog are the area from the Strait of Juan

Figure 1404 **Cold air passing over warm water picks up enough heat and moisture to form steam fog. It is most prevalent, morning and evening, during autumn months. It is found most often on inland rivers as well as small lakes and ponds.**

de Fuca to Point Arguello, California, on the Pacific Coast, and the area from the Bay of Fundy to Montauk Point New York on the Atlantic Coast. In these waters the average annual number of hours of fog occurrence exceeds 900—more than 10% of the entire year. In the foggiest parts of these areas, off the coast of Northern California and the Coast of Maine, fog is present about 20% of the year.

Going southward along both the Atlantic and Pacific coasts, the frequency of fog decreases, more rapidly on the Atlantic Coast than on the Pacific. The average annual fog frequency over the waters near Los Angeles and San Diego, for example, is about three times that in the same latitude along the Atlantic Coast.

Seasonal Frequencies

The time of maximum occurrence of fog off the Pacific Coast varies somewhat with the various localities and, of course, with the individual year. In general, however, over the stretch from Cape Flattery, Washington, to Point Arguello the season of most frequent fog runs from July through October, with more than 50% of the annual number of foggy days occurring during this period. However, along the lower coast of California from Los Angeles southward, the foggiest months are those from September through February, and the least foggy are from May through July.

On the Atlantic side, off the coast of New England the foggiest months are usually June, July, and August, with a maximum of fog generally occurring during July, in which some fog is normally encountered about 50% of the time. Off the Middle Atlantic Coast, however, fog occurs mostly in the winter and spring months, with a tendency toward minimum frequency in summer and autumn. Along the South Atlantic Coast (from Cape Hatteras to the tip of Florida) and in the Gulf of Mexico fog rarely creates a problem for a boatman. It is virtually non-existent during the summer, and even in the winter and early spring season (December through March), when it has maximum frequency, the number of days with fog rarely exceeds 20 during this four-month period.

The Great Lakes as a whole tend to have fog in the warmer season. The explanation for this is to be found in the comparison of the lake temperatures with the air temperatures over the surrounding land. From March or April to the beginning of September the lakes tend to be colder than the air. Hence, whenever the dew-point temperature is sufficiently high, conditions favor the formation of advection fog over the water.

The greatest fogginess occurs when and where the lakes are coldest in relation to the air blowing off the surrounding land. On Lake Superior, north-central Lake Michigan, and northwestern Lake Huron the time of maximum frequency is late May and June; elsewhere it is late April and May. Since the lake temperatures become colder from south to north and from the shores outward, the occurrence of fog increases northward and towards the lake center.

Figure 1405 **Sea smoke in Great Harbor, Woods Hole, Massachusetts. Water temperature at the time of this photograph was + 31.6° F. (−0.2° C), and air temperature, 30 feet above sea level, was + 5° F. (−15° C). The wind was from the northwest at about 20 knots.**

Highs and Lows

The global circulation of the atmosphere resulting from unequal heating results in the building up of areas of above average BAROMETRIC PRESSURE, (see below), with corresponding areas of below-average pressures. These are the HIGHS and LOWS that are seen on weather maps, usually designated by large letters "H" and "L."

General Characteristics

Highs, also called "anti-cyclones," generally bring good weather. Lows, also called "cyclones" (but not to be confused with the misuse of this term for some severe local storms), bring bad or unsettled weather. Highs and lows are rather large areas of "weather" measured in many hundreds of miles across. Circulation around an area of high pressure is clockwise; around lows, it is counterclockwise. (Circulation is the opposite in the southern hemisphere.) Winds are generally weaker in highs than in low pressure systems.

The Formation of Highs

Areas of high pressure are formed by the descent of cold, dense air toward the surface in polar regions and in the horse latitudes (see Figure 1402). As air flows outward toward areas of lower pressure, the rotation of the earth changes its direction so that a clockwise circulation is established. Highs form sequentially in the north polar region and move southward; as they reach the latitudes of the prevailing westerlies they are carried first southeastward, then eastward, and often finally northeastward.

The existence of continents and oceans distorts the theoretical picture of the formation of highs and lows; the actual process is quite complex. High-pressure "breeding zones" form in relatively specific areas rather than in broad zones; these areas change between summer and winter.

The Formation of Lows

The formation of low pressure cells (see page 313) is quite different from that of highs. On the boundary between warmer air and cooler air, a horizontal wavelike situation develops; see Figure 1409. This grows and becomes more and more distinct and may even "break" as an ocean wave does on the beach. The boundary between the two areas of air of different temperature is termed a FRONT, a major weather phenomenon that will be considered in detail in the following section.

Small, local low-pressure cells may also develop over deserts or other intensely heated locations. Here the air heated at the surface expands, rises rapidly, and creates an effect of low pressure; air rushes in from outside this area with a counterclockwise swirling motion. Lows can also form on the lee side of mountain ranges as strong winds blow across the peaks.

Air Masses and Fronts

A general knowledge of air masses and fronts will make weather forecasts more understandable, and more useful.

Air Masses

AIR MASS is another term for a HIGH-PRESSURE CELL, a build-up of air descending from high-altitude global circulation. (There is no corresponding feature or name for the low-pressure equivalent.) Here, however, we are dealing with a volume of air, a physical mass, rather than the pressure conditions in it. It is a large body of air covering as much as several hundred thousand square miles in which the conditions of temperature and moisture are essentially the same in all directions horizontally.

Basic Characteristics

Air masses derive their basic characteristics from the surface beneath them. They are characterized as continental or maritime, polar or tropical, and as warmer or colder than the surface over which they are then moving. Figure 1406 shows the principal air masses which affect North America and their general normal movement.

Maritime and continental air masses differ significantly in their characteristics of temperature and humidity; thus they bring different kinds of weather to us. Oceans suffer less extreme variations of heat and cold than do continents, and maritime air masses change less with seasons. As a result a maritime air mass moving over land tends to moderate any conditions of excess heat or cold.

Cold air masses are characterized by unstable internal conditions; the air next to the earth's surface warms and tries to rise through the overlying colder air. On the other hand, a warm air mass remains relatively stable; air cooled by contact with the colder ground tends to sink and warm air above tends to stay there or rise. These conditions result in stronger, gusty winds within a cold air mass, but weaker, steadier winds in a warm air mass. Visibility is better in a cold air mass; rains are often in the form of thunder showers from cumulus clouds, rather than the drizzle from the stratus cloud forms of a warm air mass.

Weather Fronts

A general knowledge of air masses is important as it leads directly to our consideration of weather fronts. A FRONT is the boundary between two different air masses, one cold and one warm; the bodies of air do *not* tend to mix, but rather each moves with respect to the other. The passage of a front results in a change of weather conditions at that location, frequently a change for the worse.

Cold Fronts

With a COLD FRONT, the oncoming cold air mass pushes under the warm air mass and forces it upward; see Figure

1407. In the northern hemisphere, cold fronts generally lie along a NE-SW line and move eastward or southeastward. The rate of movement is roughly 400-500 statute miles (500-800 km) per day, more in winter and less in summer.

A strong, rapidly moving wintertime cold front will bring weather changes that are relatively more intense but briefer in duration. If the front is slow-moving, precipitation may not fall until after the front has passed; if the cold front is weak, there may be no precipitation at all.

A SQUALL LINE may form just ahead of a strong cold front. Squall lines are very turbulent with strong winds that may endanger small craft. A squall line appears as a wall of very dark, threatening clouds. Winds shift and increase in strength abruptly; rain may fall heavily, blown nearly horizontal by the winds.

The approach of a cold front is indicated by a shift of the wind towards the south, then southwest. Barometric pressure readings fall. As the front approaches, clouds lower and build up; rain starts slowly but increases rapidly. As the front passes the wind continues to veer, westward, northwestward, and then northerly. After passage the sky clears quickly, temperatures drop, pressure builds up quickly, and the wind may continue to veer to the northeast. For a few days at least the weather will have the characteristics of a high.

Warm Fronts

A WARM FRONT occurs when an advancing warm air mass reaches colder air and rides up over it; see Figure 1408. Warm fronts are generally oriented in directions NS, NW-SE or WE and change their direction more often than cold fronts do. The rate of movement is slower, 150-200 miles (240-320 km) per day, and thus warm fronts are eventually overtaken by the next following cold front.

Warm-front weather is milder than that of a cold front and may extend several hundreds of miles in advance of the actual front. Clouds form at low levels and rainfall is

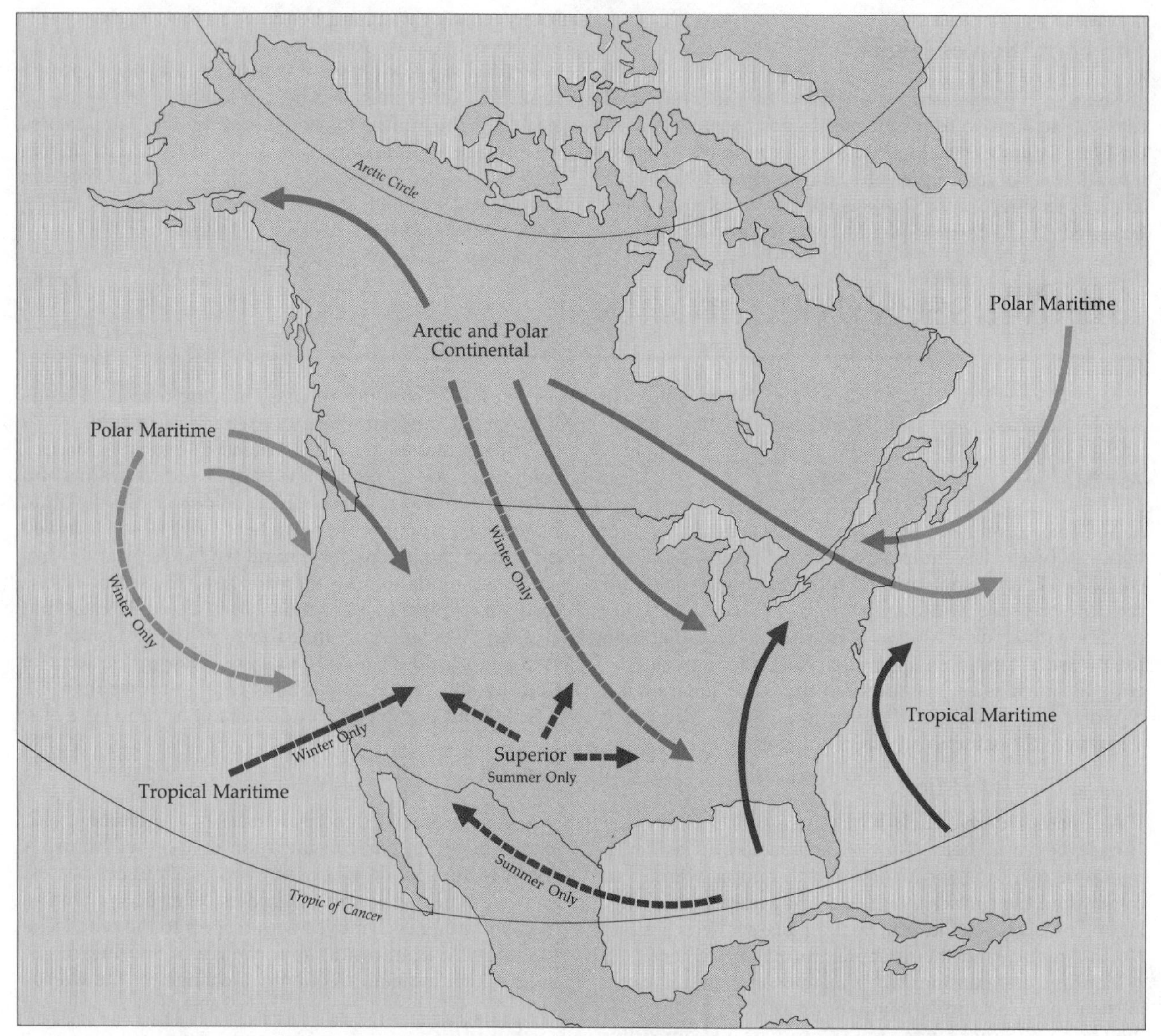

Figure 1406 **These are the sources and direction of movement of the air masses (highs) that influence weather in North America. The properties of an air mass, and its conflicts with adjacent masses, are the causes of weather changes. Note that some are seasonal.**

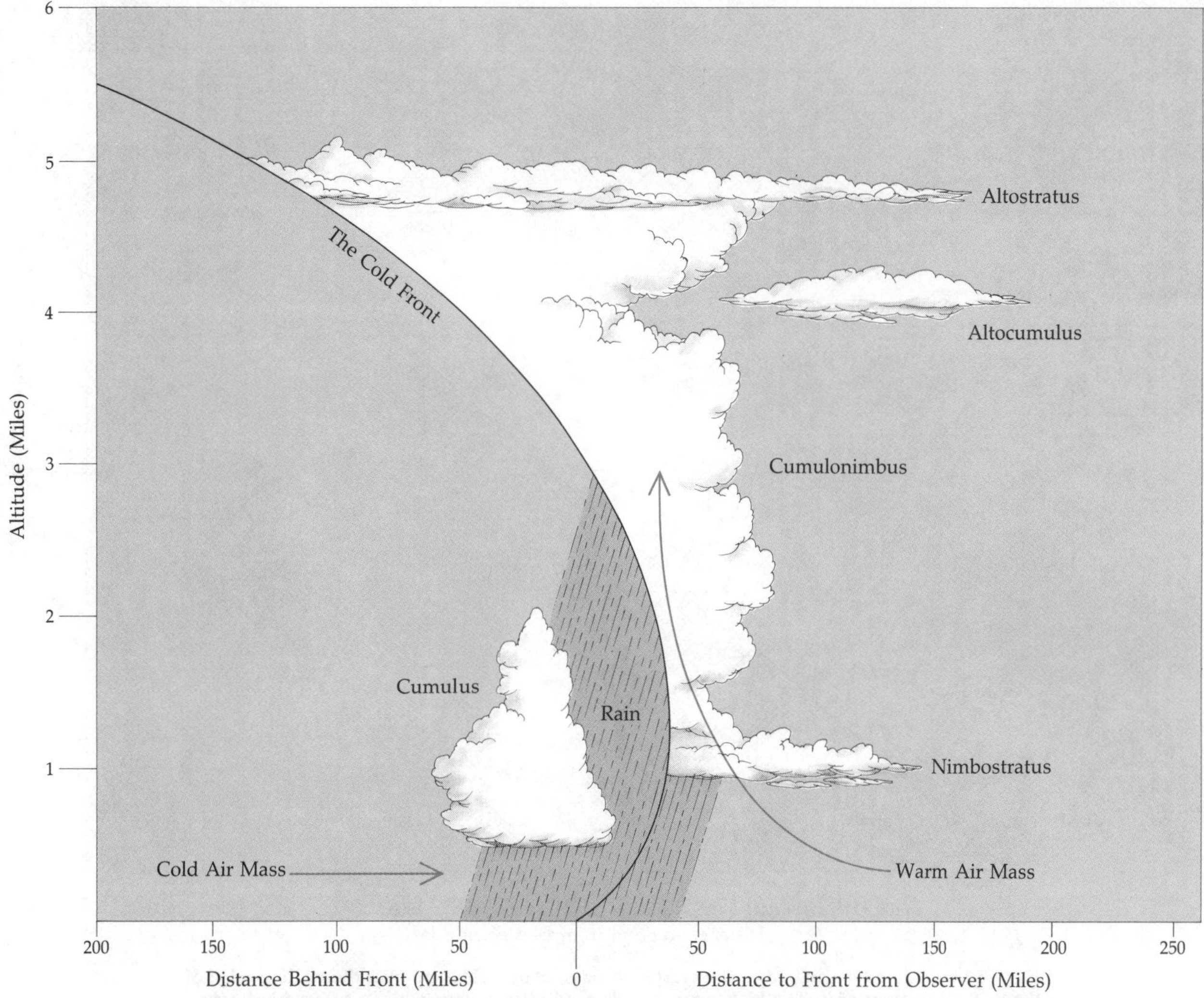

Figure 1407 **The bottom of an advancing cold front tends to wedge under the warm air ahead of it. When temperature differences are significant, rapidly rising moist warm air ahead of the cold front forms towering cumulus clouds above a nimbostratus base.**

generally more moderate but extended in time. The approach of a warm front is signaled by a falling barometer (but falling more slowly than for an approaching cold front), a build-up of clouds, and the onset of rain or drizzle. After the front passes there will be cumulus clouds and temperatures will rise; the barometer will also slowly rise.

Stationary Fronts

Occasionally fronts will slow down to the point of little or no forward movement. A STATIONARY FRONT hanging over you will bring conditions of clouds and rain much like those of a warm front.

Occluded Fronts

An OCCLUDED FRONT is a more complex situation where we have warm air, cold air, and colder air; it occurs after a cold front overtakes a warm front because of its faster movement, and lifts the warm air mass off the ground. Either the warm front or the cold front is pushed upward from the earth's surface; see Figures 1408*A* and *B*. The appearance on a weather map is that of a curled "tail" extending outward from the junction of the cold and warm fronts; this is a low-pressure area with counterclockwise winds.

Storms

Although a boater is interested in whatever the weather is and what is forecast, his real attention is focused on the possibility of STORMS. While sunny skies and fair winds are eagerly awaited, it is the approach of a storm that causes concern for the safety of his craft and crew.

Extra-Tropical Cyclones

The principal source of rain, winds, and generally foul weather in the United States is the EXTRA-TROPICAL CYCLONE. There are other storms—the hurricane, the thun-

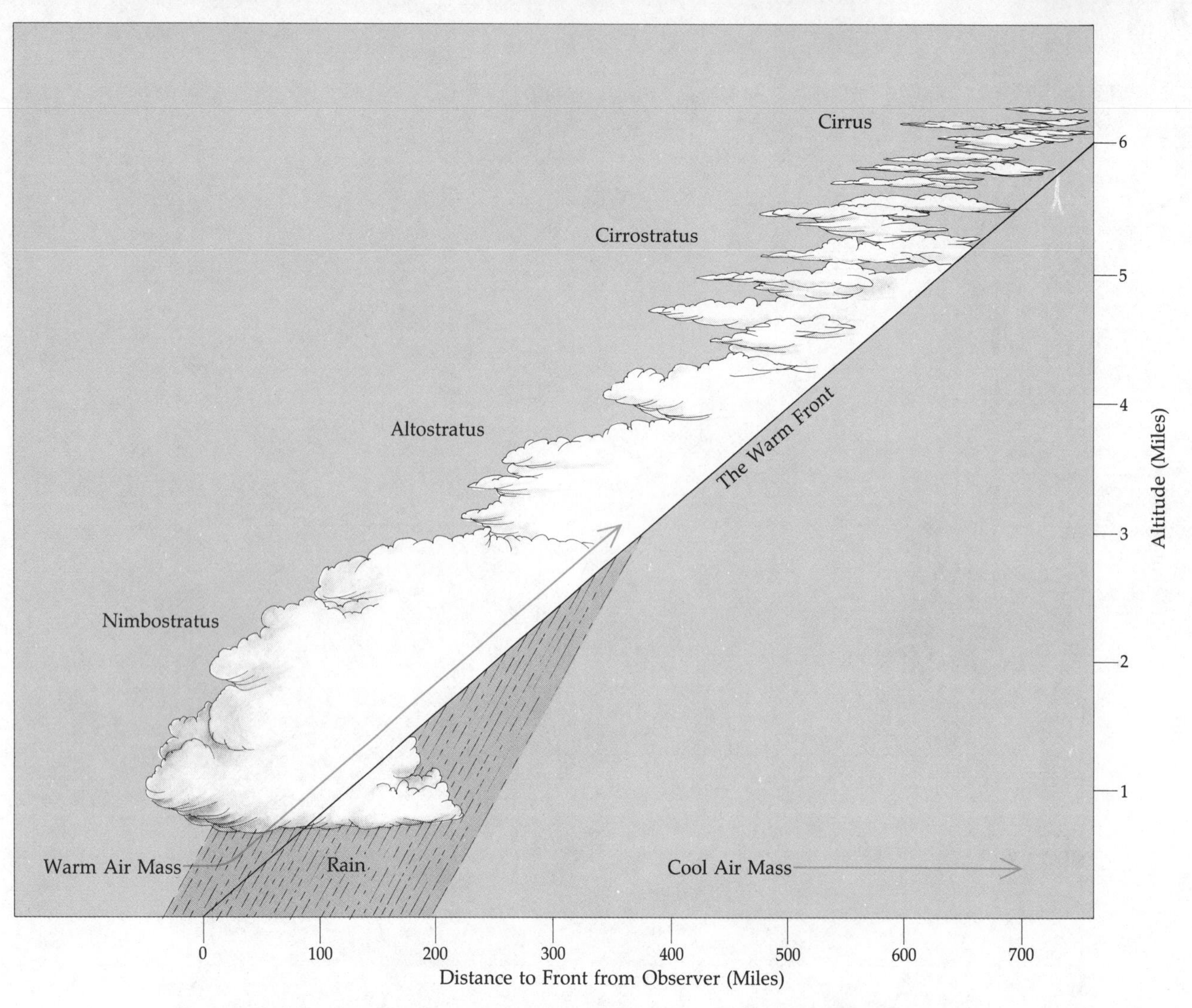

Figure 1408 **The advancing edge of a warm air mass, *above,* rides up over the trailing edge of the cold air in front of it. The high cirrus clouds can indicate approach of a warm front a day or more before it reaches a given area. Occluded fronts, *below,* develop when a cold front overtakes a warm front. If the air behind the cold front is not as cold as the air ahead of the warm front, it will tend to ride up over the colder air as at A. If this air is colder as at B, it will wedge under the cool air that was ahead of the warm front.**

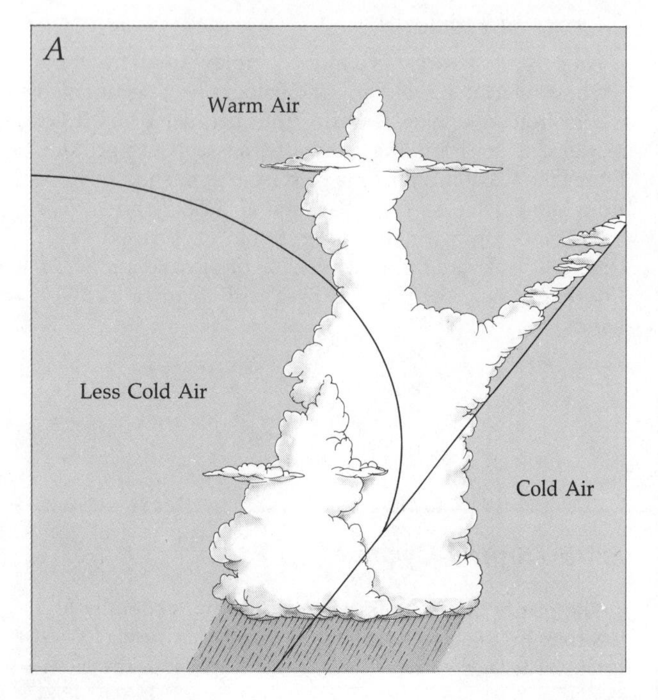

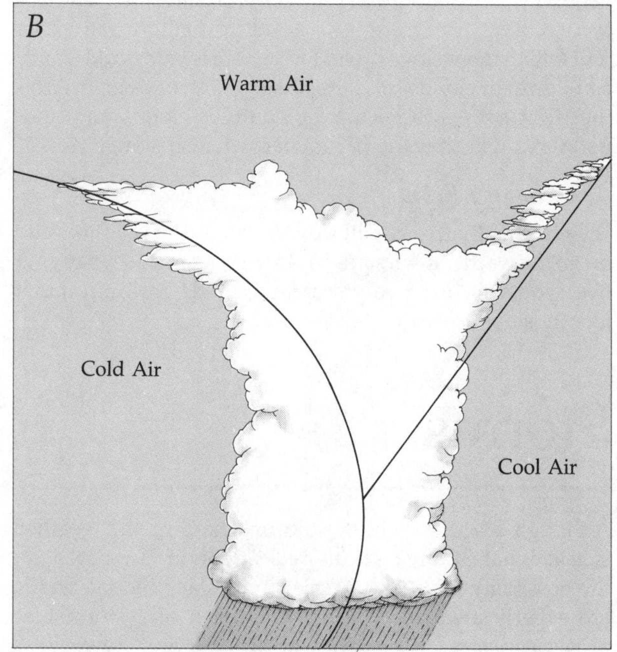

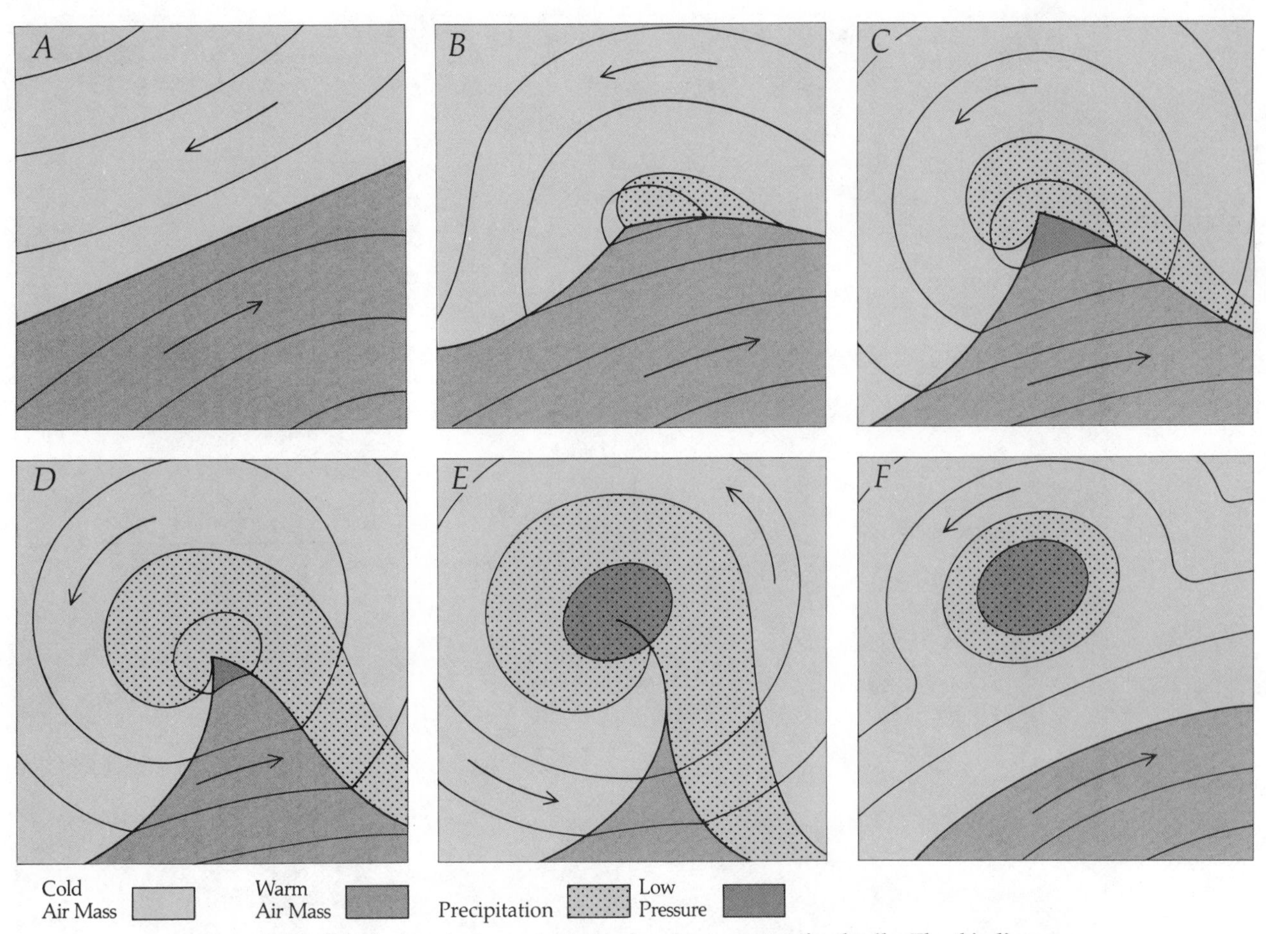

Figure 1409 **Development of an extra-tropical cyclone: see text for details. The thin lines are isobars that pass through points of equal barometric pressure. Arrows indicate wind.**

derstorm, and the tornado—that are usually more destructive, but the extra-tropical cyclone is the ultimate cause of most of our weather troubles. Such a storm (in the northern hemisphere) is defined as a traveling system of winds rotating counterclockwise around a center of low barometric pressure and containing a warm front and a cold front.

Development of an Extra-Tropical Cyclone

The development of an extra-tropical cyclone is shown in Figure 1409. At *A* there is a warm air mass, typically moist tropical air, flowing northeastward and a cold air mass, typically polar continental, flowing southwestward. They are separated by a heavy line representing the front between them. At *B* the cold air from behind pushes under the warm air; the warm air rushes up over the cold air ahead of it. A cold front is born on the left, a warm front on the right. Where they are connected, the barometric pressure is lowered and the air starts circulating counterclockwise around this Low. At the rear of the cold front, a high-pressure area develops. At the same time, the whole system keeps moving in a general easterly direction; the dotted area represents rain. When warm, moist air is lifted, as it is when a cold air mass pushes under it or when it rushes up over cold air ahead, it cools by expansion. After its temperature has fallen to the dew point, excess water vapor condenses to form first clouds and then rain.

At *C* the storm is steadily developing, with the low pressure area intensifying more and more, the clouds and rain increasing, and the winds becoming stronger. At *D* the storm is approaching maturity; note how the gap has narrowed between the fronts.

By the time *E* is reached, the cold front has begun to catch up with the warm front and an occluded front is formed; the storm is now at its height. When the situation at *F* is reached, the storm has begun to weaken and soon the weather will clear as a high behind the cold front reaches your area.

Extra-tropical cyclones often occur in families of two, three, or four storms. It takes about one day (24 hours) for this disturbance to reach maturity with three or possibly four days more required for complete dissipation. In winter, these storms occur on the average of twice a week in the U.S.; in summer they occur somewhat less frequently and are less severe. Their movement is eastward to north of east at a speed in winter of about 700 miles (1,125 km) per day and in summer perhaps 500 miles (800 km). Such storms usually cover a large area; they can affect a given locality for two days or more.

Thunderstorms

Extra-tropical cyclones are the principal source of wind, rain, and generally foul weather for large areas. There are also local, small-scale but intense THUNDERSTORMS which

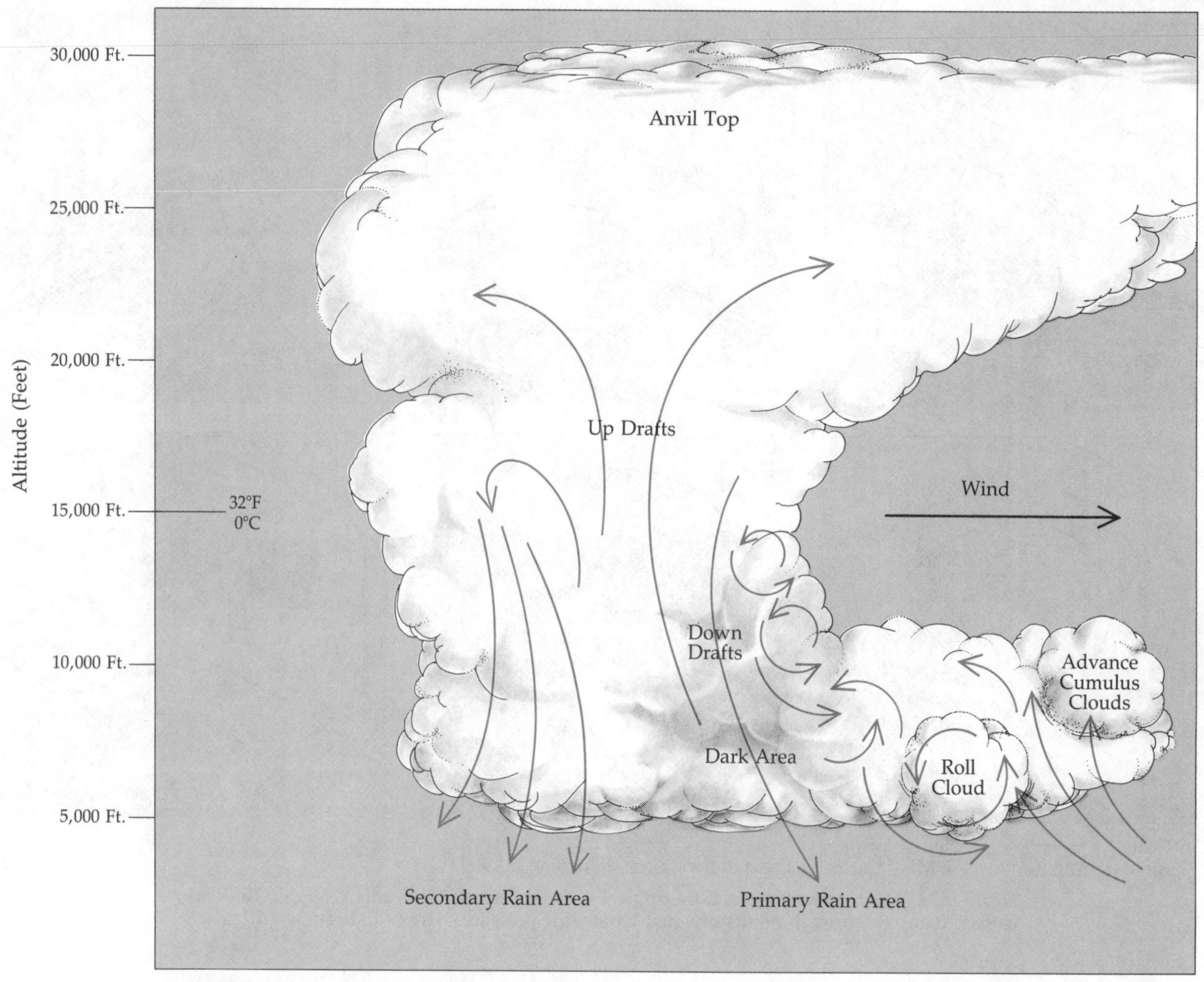

Figure 1410 **Typical thunderstorm is based on a cumulonimbus cloud that may tower from 25,000 to 50,000 feet high; the anvil top that develops at the top of the cloud is caused by high-altitude "jet stream" winds.**

you must be prepared to handle. These are often part of a larger storm, linked to the cold front of the cyclone. If the air comprising the warm sector is sufficiently unstable, however, which is often the case in summer, thunderstorms together with general rain will occur during the approach of a warm front as the warm-sector air ascends over the cold air ahead. Thunderstorms and rain also develop on their own in the middle of an air mass, unaided by frontal activity, again usually in the summer.

A thunderstorm is a storm of short duration, arising only in a cumulonimbus cloud, attended by thunder and lightning, and marked by abrupt fluctuations of temperatures, pressure, and wind. A LINE SQUALL, now usually referred to in meteorological terms as a SQUALL LINE, is a lengthy row of thunderstorms that may stretch for 100 miles (160 km) or more.

Incidentally, a SHOWER (as opposed to gentle, steady rain) is a smaller brother to the thunderstorm, though the rainfall and wind in it may be of considerable intensity. It is the product of relatively large cumulus, or small cumulonimbus, clouds separated from one another by blue sky; a shower is over quickly, and it is not accompanied by thunder and lightning.

Characteristics of a Thunderstorm

In all cases, the prime danger signal for a thunderstorm is a cumulus cloud growing larger. Every thunderstorm cloud has four distinctive features, although you may not always be able to see all four as other clouds may intervene in your line of sight. Figure 1410 is a drawing of a cumulonimbus cloud; it shows these four features diagrammatically. Starting at the top, notice the layer of cirrus clouds, shaped like an anvil and consequently called an ANVIL TOP, leaning in the direction toward which the upper wind is blowing. This tells us the direction in which the storm is moving. (The anvil top development is illustrated in Figure 1411; these three photographs were taken from the same position over a time interval of only one-half hour!)

The next feature is the main body of the cloud—a large cumulus of great height with cauliflower sides. It must be of great height, as it must extend far above the freezing level if the cirrus anvil top is to form. (Cirrus clouds are composed of ice crystals, not water droplets.) The third feature is the ROLL CLOUD formed by violent air currents along the leading edge of the base of the cumulus cloud. The fourth and final feature is the dark area within the storm and extending from the base of the cloud to the earth; at the center this is rain, at the edges hail and rain.

Requirements for Thunderstorm Formation

There are three requirements for the formation of thunderstorms. One, there must be strong upward air currents such as those caused by a cold front burrowing under and lifting warm air or by the heating of air in contact with the surface of the earth on a summer day. Two, the air parcels forming the storm must be buoyant relative to their neighbors outside the storm and able to keep on ascending higher and higher until they pass the freezing level. Third, the air must have a large concentration of water vapor. The most promising thunderstorm air is of tropical maritime origin; whenever it appears in your cruising area, especially when a cold front is approaching, you need to be watchful.

Frequency of Thunderstorms

Thunderstorms occur most frequently, and with the greatest intensity, in the summer over all parts of the U.S. While they may strike at any hour, they are most common in the late afternoon and early evening over inland and coastal waters. The surrounding land has been a good "stove" for many hours, heating the air to produce strong upward currents. Over the ocean, well away from shore, thunderstorms more commonly occur between midnight and sunrise. Finally, thunderstorms are most frequent and most violent in subtropical latitudes. The southeastern part of the U.S. may have four thunderstorms per week in summer.

Ahead of a thunderstorm the wind may be steady or variable, but as the roll cloud (Figure 1410) draws near, the wind weakens and becomes unsteady. As the roll cloud passes overhead, violent shifting winds, accompanied by strong downdrafts, may be expected. The wind velocity may reach 60 knots or more. Heavy rain, and sometimes hail, begins to fall just behind the roll cloud. The weather quickly clears after the passage of the storm, however, which brings cooler temperatures and lower humidity.

If the cumulonimbus cloud is fully developed and towers to normal thunderstorm altitudes, 35,000 feet (10.7 km) or more in summer, the storm will be violent. If the anvil top is low, only 20,000 feet (6.1 km) or so, as is usual in spring and autumn, the storm will be less severe. If the cumulonimbus cloud is not fully developed, particularly if it lacks the anvil top, and the roll cloud is missing, only a shower may be expected.

Lightning

A build-up of dissimilar electrical charges occurs within a vertically developing cumulonimbus cloud, and between the cloud and the earth below. The earth normally has a negative charge. Portions of the cloud become charged electrically, both positive and negative. When the build-up of opposite charges becomes great enough, a LIGHTNING FLASH occurs within a cloud or between the cloud and ground. Actually what we think of as a flash is really a series of strikes *back and forth* over a period of roughly two tenths of second; a flash to the surface usually starts with a faint "leader" from the cloud followed instantly by a massive strike upward from the surface; about two-thirds

Figure 1411 **Growth of an anvil top: The first photo was taken at noon, the second at 1220, and the third at 1230.**

of all lightning flashes are within clouds, however, and never reach the surface. A lightning flash is almost unbelievably powerful—up to 30,000,000 volts at 100,000 amperes; it happens so quickly that it is essentially explosive in nature.

The sudden, vast amount of heat energy released by a lightning flash causes the sound waves termed THUNDER.

Lightning protection for boats is discussed on page 272.

Anticipating Thunderstorms

The possibility of a thunderstorm may be first noticed from static crashes on an AM radio receiver. You can plot the approach, once the cumulonimbus cloud is visible, with a series of bearings; you can also estimate its distance from you. The thunder and lightning occur simultaneously at the point of lightning discharge, but we see the lightning discharge much sooner than we hear the thunder. Time this interval, in seconds. Multiply the number of seconds by 0.2; the result will be approximate distance off in statute miles (multiply by 0.34, or divide by 3, to get distance off in kilometers). Take care, however, that you are properly associating a particular flash with its thunder; this may be difficult if there is nearly continuous lightning.

Tornados

When numerous thunderstorms are associated with a cold front, the storms are apt to be organized in a long, narrow band. The forward edge of this band is usually marked by a squall line, along which the cold downdrafts from a series of thunderstorms meet the warm air. Here the wind direction changes suddenly in vicious gusts, and a sharp drop in temperature occurs. The importance of squall lines is that they often spawn a most destructive type of storm, the deadly TORNADO. These occur in boating areas adjacent to the Atlantic and Gulf of Mexico coasts of North America, and in all inland boating areas east of the Rocky Mountains.

Tornadoes formed at squall lines often occur in families and move with the wind that prevails in the warm sector ahead of the cold front. This wind is usually from the southwest. The warm air typically consists of two layers, a very moist one (source: Gulf of Mexico) near the ground and a relatively dry layer above. The temperature decreases with altitude rather rapidly in each layer. When this combination of air layers is lifted along a squall line or cold front, excessive instability develops and violent updrafts are created.

A tornado is essentially an air whirlpool of small horizontal extent which extends downward from a cumulonimbus cloud and has a funnel-like appearance (see Figure 1412). The average diameter of the visible funnel cloud is about 250 yards (230 m) but the destructive effects of this system of whirling winds may extend outward from the tornado center as much as ½ mile (0.4 km) on each side. The wind speed near the core can only be estimated, but it undoubtedly is as high as 200 knots.

Waterspouts

The marine counterpart of the tornado is the WATERSPOUT. The conditions favoring the formation of waterspouts at sea are similar to those conducive to the formation of tornadoes over land. Waterspouts are much more frequent in the tropics than in middle latitudes. Although they are less violent than tornadoes, they are nevertheless a real danger to small craft.

A waterspout, like a tornado, forms under a cumulonimbus cloud. A funnel-shaped protuberance first appears at the base of the cumulonimbus and grows downward toward the sea. Beneath it the water becomes agitated and a cloud of spray forms. The funnel-shaped cloud descends until it merges with the spray; it then assumes the shape of a tube that stretches from the sea surface to the base of the cloud (see Figure 1413).

The diameter of a waterspout may vary from 20 to 200 feet (6-60 m) or more. Its length from the sea to the base

Figure 1412 **A tornado funnel is a cloud of water droplets mixed with dust and debris, which are drawn up because of the greatly reduced atmospheric pressure within the funnel. This pressure drop causes wind to whirl inward and upward.**

Figure 1413 **A waterspout over St. Louis Bay, off Henderson Point, Mississippi. Note the cloud of spray just above the sea surface. This is the marine equivalent of a tornado.**

of the cloud is usually between 1,000 and 2,000 feet (300-600 m). It may last from ten minutes to half an hour. Its upper part often travels at a different speed and in a different direction from its base, so that it becomes bent and stretched-out. Finally, the tube breaks at a point about one-third of the way up to the cloud base, and the "spout" at the sea surface quickly subsides.

Considerably reduced air pressure at the center of a waterspout is indicated by the visible variations of the water level. A mound of water, a foot or so high, sometimes appears at the core, because the atmospheric pressure inside the funnel is perhaps 30 to 40 millibars less than that on the surface of the water surrounding the spout. This difference causes the rise of water at the center.

Like the tornado, the visible part of a waterspout is composed mostly of tiny water droplets formed by the condensation of water vapor in the air. Considerable quantities of salt spray, however, picked up by the strong winds at the base of the spout, are sometimes carried far aloft. This has been verified by observations of a fall of salty rain following the passage of a waterspout.

Hurricanes

HURRICANE is the popular term for a TROPICAL CYCLONE in North America. Unlike the extra-tropical cyclone, it is not related to warm and cold fronts. A hurricane is defined as a storm of tropical origin with a counterclockwise circulation reaching a strength of 64 knots (75 mph) or more at the center. In its earlier stages with winds less than 33 knots but with a closed circulatory pattern, the term is a TROPICAL DEPRESSION. When the winds increase beyond 33 knots it is a TROPICAL STORM until it reaches hurricane strength. In the Western Pacific Ocean, the term used for tropical cyclones is TYPHOON; because of the greater expanse of ocean there, these storms often become even larger and more intense than hurricanes.

Frequency of Tropical Cyclones

Over the years much information has been accumulated on the frequency of tropical cyclones in various regions of the world. In the Far East typhoons may occur in any month, although they are most common in late summer and early autumn. In North American waters, the period from early December through May is usually hurricane-free. August, September, and October are the months of greatest frequency; in these months hurricanes form over the tropical Atlantic, mostly between latitudes 8° N and 20° N. The infrequent hurricanes of June and November almost always originate in the southwestern part of the Caribbean Sea. (Interestingly, hurricanes do *not* occur in the South Atlantic.)

Development of a Hurricane

The birthplace of a hurricane typically lies within a diffuse and fairly large area of relatively low pressure situated somewhere in the 8° N–20° N latitude belt. The winds around the low-pressure area are not particularly strong, and, although cumulonimbus clouds and showers are more numerous than is usual in these latitudes, there is no clearly organized "weather system." This poorly defined condition may persist for several days before hurricane development commences. The development of highly sophisticated WEATHER SATELLITES has greatly improved the ability of forecasters to watch vast ocean areas where these storms originate for any signs of conditions favorable for hurricane development.

Figure 1414 **The wall cloud around the eye of a hurricane, as seen from an aircraft within the eye. Upper edge of the cloud is at an altitude of nearly 40,000 feet.**

When development starts, however, it takes place suddenly. Within an interval of 12 hours or less the barometric pressure drops 15 millibars or more over a small, almost circular area. Winds of hurricane force spring up and form a ring around the area; the width of this ring is at first only 20 to 40 miles (32–64 km). The clouds and showers become well organized and show a spiral structure. At this stage the growing tropical cyclone acquires an EYE. This is the inner area enclosed by the ring of hurricane-force winds which has expanded to a width of 100 miles (160 km) or more by the time the cyclone reaches maturity. Within the eye we find the lowest barometric reading.

Hurricane eyes average about 15 miles (24 km) in diameter but may be as large as 25 miles (40 km). The wind velocities in the eye of a hurricane are seldom greater than 15 knots and often are less. Cloud conditions vary over a wide range. At times there are only scattered clouds, but usually there is more than 50% cloud cover. Through the openings the sky is visible overhead and at a distance the dense towering clouds of the hurricane ring can be seen extending to great heights (Figure 1414). This feature is the WALL CLOUD. Seas within the eye are heavy and confused; the calm is only with respect to winds.

Hurricane Tracks

The usual track of an Atlantic hurricane is a parabola around the semipermanent Azores-Bermuda high-pressure area. Thus, after forming, a hurricane will move westward on the southern side of the Azores-Bermuda High, at the same time tending to work away from the equator. When the hurricane reaches the western side of this High, it begins to follow a more northerly track, and its direction of advance changes progressively toward the right. The position where the westward movement changes to an eastward movement is known as the POINT OF CURVATURE.

Occasionally when a hurricane is in a position near the southeast Atlantic coast, the Azores-Bermuda High happens to have an abnormal northward extension. In this situation the hurricane may fail to execute a complete recurvature in the vicinity of Cape Hatteras. It will skirt the western side of the High and follow a path almost due north along the Atlantic Coast, as occurred in the case of the destructive New England hurricane of September 1938 (see Figure 1415).

The rate of movement of tropical cyclones while they are still in low latitudes and heading westward is about 15 knots, which is considerably slower than the usual rate of travel of extra-tropical cyclones. After recurving they begin to move faster and usually attain a forward speed of at least 25 knots, sometimes 50 to 60 knots, as in the case of the September 1938 hurricane. A small proportion of hurricanes, however, do not recurve at all, and a few of these follow highly irregular tracks which may include a complete loop.

Hurricanes gradually decrease in intensity after they reach middle and high latitudes and move over colder water. Many lose their identity by absorption into the wind circulation around the larger extra-tropical cyclones of the North Atlantic.

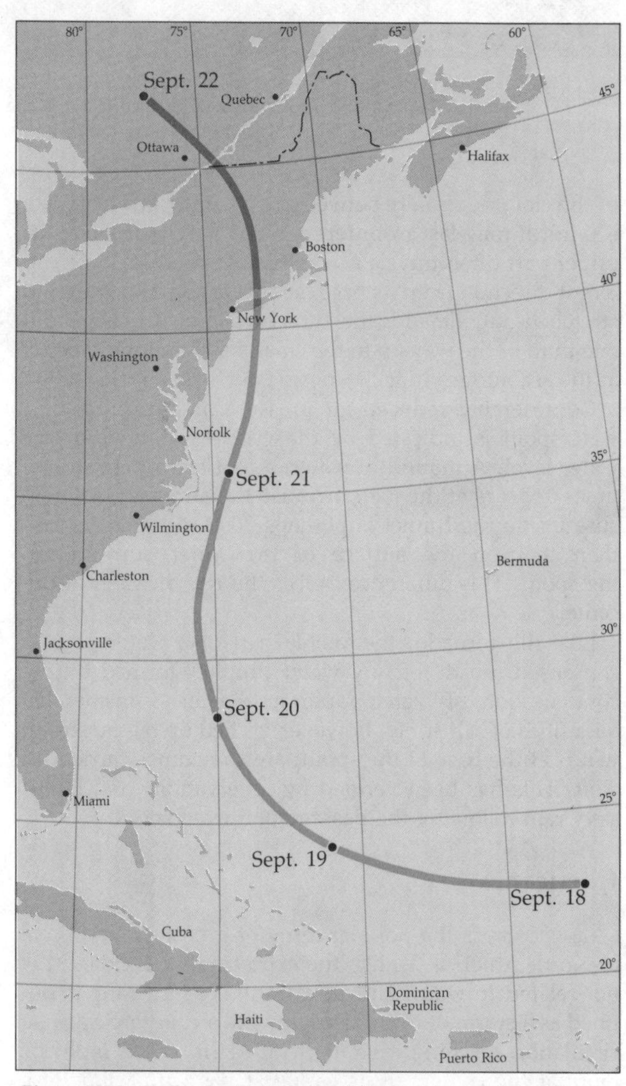

Figure 1415 **Path of a September 1938 hurricane shows the positions of its center at 0700 on succeeding days. A high-pressure system to the east kept it from following the more normal course to the northeast and back out to sea.**

Using Weather Forecasts

Some years ago, to be considered a fully qualified skipper you had to be able to take a series of complex daily weather maps and forecast your own weather. Now this is no longer necessary, although the ability to read and understand a simplified weather map will always be useful. The ready availability of weather forecasts by radio and TV does much of the work for you—but you still have the responsibility to get, and heed, the information.

Weather Maps

As with your navigation charts, information on weather maps is presented by use of symbols. Some of the more-frequently encountered symbols will be described below; we omit the many others now used only by professionals.

Station Models

A COMPLETE STATION MODEL (see Figure 1417) for a reporting location will tell you much more than you need to know. It includes information on wind direction and speed; temperature; visibility; cloud type, amount, and height; current pressure and past change plus current tendency; precipitation in past three hours; dew point temperatures; and current weather. Simplified models omit much of this and often show only current weather, temperature, and wind direction and speed.

Current weather is indicated by the appearance of the circle in the center of the station model—open for clear, partially or totally solid for degrees of cloudiness. Wind direction is indicated by a WEATHER VANE line and wind speed by the number, size, and shape of the FEATHERS on the end of the weather vane. Temperature will normally be given in degrees Farenheit (degrees Celsius in Canada and other areas using the metric system).

Air Masses and Fronts

Air masses are commonly shown as a large block letter H or L; occasionally the word High or Low will be spelled out, or a circle drawn around the letter. They are not commonly designated as continental or maritime, polar or tropical, as this can be inferred from their location and past movement.

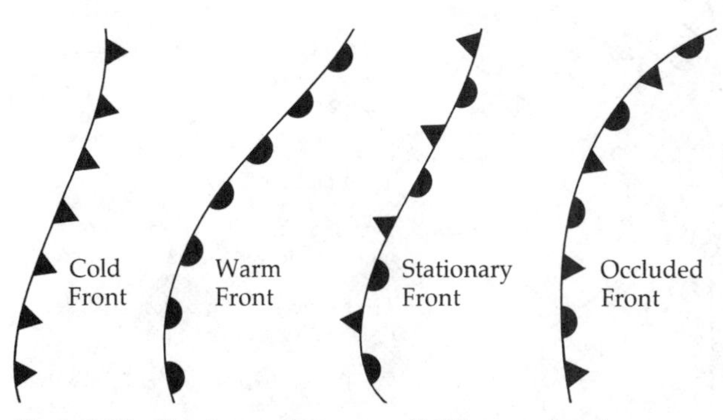

Figure 1416 **Fronts are shown as solid lines on weather maps, with triangles or half circles used singly or in combination to indicate the type of front. These symbols face in the direction that the pressure system and its front are moving.**

Fronts are shown as a heavy line. Cold fronts have a series of solid triangles on the line with one point in the direction of movement; see Figure 1416. Warm fronts have solid half-circles on the line, again the side that they are on indicating the direction of movement of the front. An occluded front has alternating triangles and half-circles on the *same side* of the line. A stationary front has the same alternating symbols but with the triangles on one side of the line and the half-circles on the other, indicating no movement.

In some sketches, a cold front is shown by a solid heavy line while a warm front is two parallel fine lines; an occluded front is thus a broken line with alternating solid and open segments. On less detailed maps, a front may be shown as merely a heavy line labeled "Cold" or "Warm," or "Stationary" Front. On TV weather maps in color, a warm front is normally shown in red, with a cold front in blue.

Isobars

A line connecting points of equal barometric pressure is termed an ISOBAR. Such lines are usually drawn on weather maps at intervals of four millibars of pressure—1020, 1024, 1028, etc, and are labeled along the line or at each end. On some weather maps, isobars will be labeled for pressure in terms of inches of mercury—29.97, 30.00, 30.03, etc.

Precipitation

Any precipitation, rain, snow, or hail is shown by one of a series of symbols; see Figure 1417. On less detailed maps, areas of precipitation are often shown by shading or crosshatching, with a descriptive word nearby or different forms of shading used to distinguish rain, snow, and other forms of precipitation.

Using Weather Maps

It is quite unlikely that you would have a weather map, yet not have a printed or broadcast forecast, but you might like to try your hand at making predictions from one or more weather maps.

Predictions from a single map are, of course, less accurate and reliable than those based on a series of weather maps over regular intervals of time. A forecast from a single map can only be made for 6 to 12 hours ahead—there are too many factors that can upset an orderly flow of events. (Make sure that you are using a map of actual conditions as of a specified recent time; many weather maps now printed in newspapers are for predicted conditions of the day of publication.) Assume that a front is advancing at roughly 20 miles an hour in the normal direction of movement. Sketch in the position of the front 12 hours after the date and time of the map—assuming that your map of actual conditions is recent enough that you still have time in which to make a usable 12-hour forecast. Then with the advanced front, and knowledge of the conditions accompanying this front, you can make your own forecast.

If you have a series of maps of existing conditions at

Weather Map Symbols

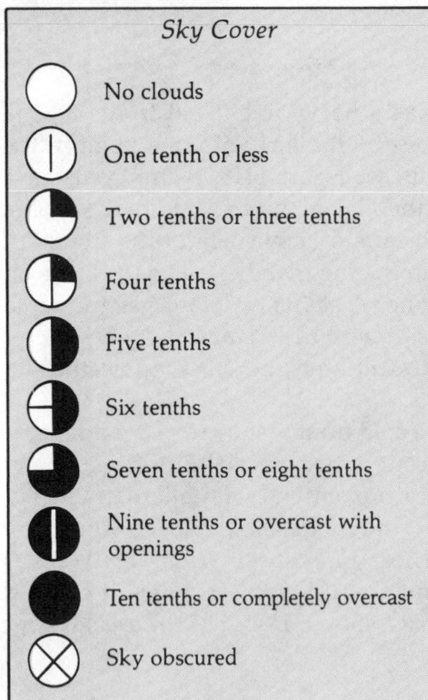

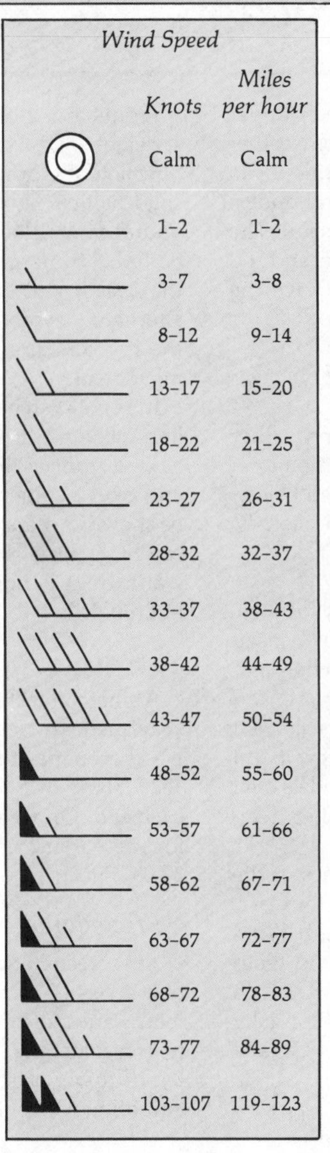

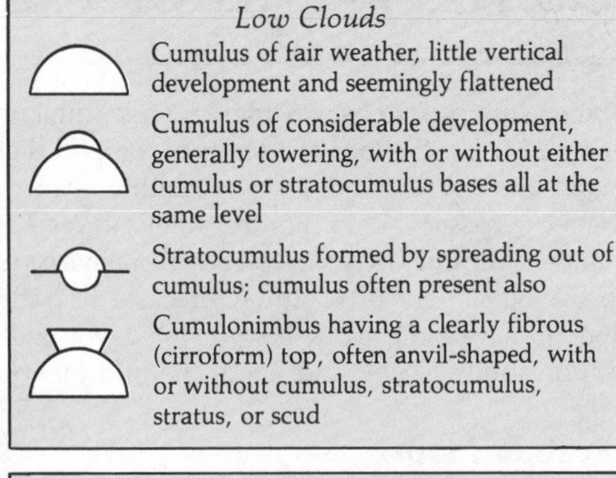

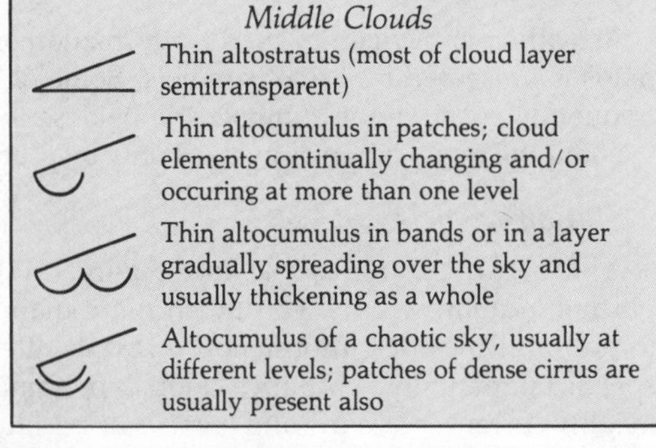

Present and Past Weather

Sandstorm or duststorm, or drifting snow

Fog, ice fog, thick haze or thick smoke

Drizzle

Rain

Snow, or rain and snow mixed, or ice pellets

Shower(s)

Thunderstorm, with or without precipitation

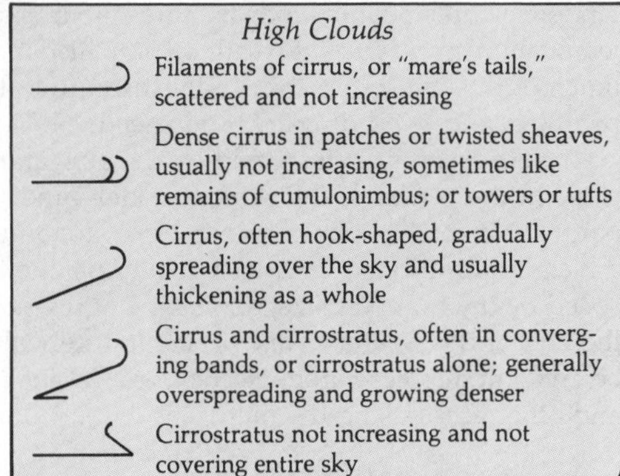

Specimen Station Model

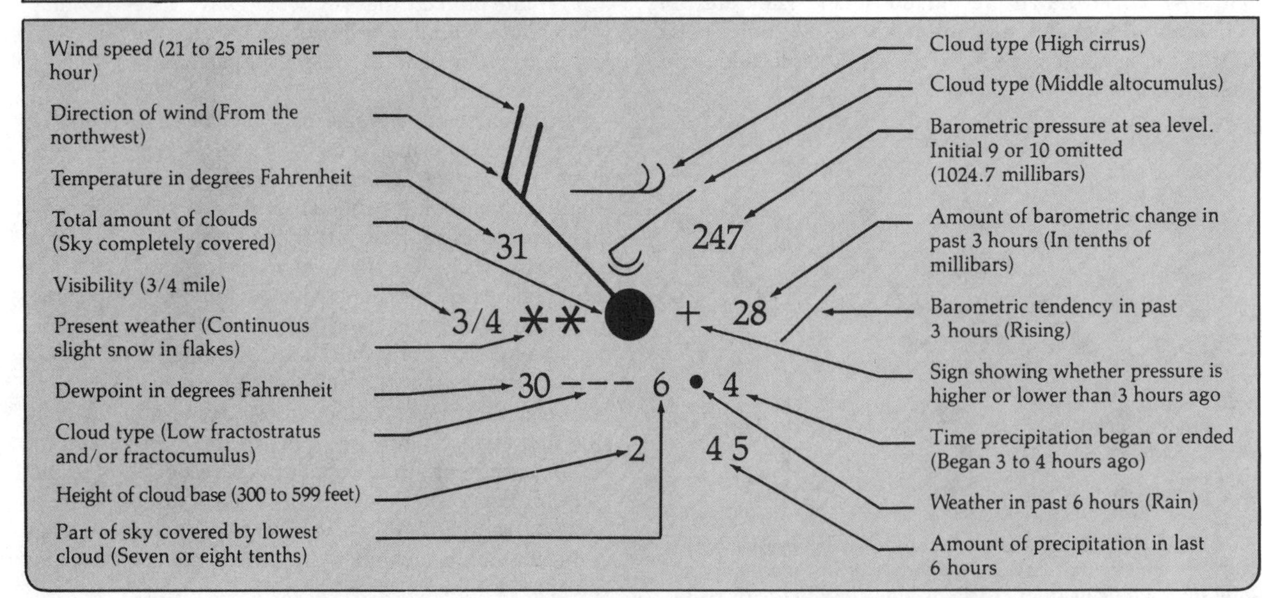

Figure 1417 **Here are the symbols most commonly found on weather maps issued by the National Weather Service. The station model shows how they indicate weather conditions.**

daily or half-daily intervals, you can make a much better forecast. Predict your frontal movement based on trends as well as actual events. If weather trends tend to repeat themselves in your areas, file sets of weather maps for typical frontal passages as additional guidance.

Newspaper Weather Maps

The weather maps printed in daily newspapers vary widely in how much information is presented and how details are shown; invariably these maps are less detailed than are official NWS weather maps. Basic information will be shown sufficiently to get a very generalized picture of nationwide weather, but little more than that. As noted above, maps are usually for predicted conditions rather than a report of actual events.

Television Weather Maps

Weather maps shown as part of local or network news programs also vary widely in level of detail. They do have advantages over newspaper maps in that they frequently show current conditions, and color is used for a more vivid presentation. Some TV weather reporters use animation to show predicted movement of fronts and weather patterns. Often satellite photographs, either still photos or time-lapse loops, are shown to explain conditions or trends. Current weather radar scans are also interesting and helpful in visualizing rain patterns out to about 125 miles (200 km).

Many local Cable TV systems carry "The Weather Channel"—a continuous live broadcast of weather conditions and forecasts. Information is highly detailed and several specialized services are included. Weather maps in a number of formats are shown, as well as satellite pictures and radar plots from areas of significant weather activity. Periodically, the national broadcast is interrupted for local reports and predictions.

Other Weather Maps

There are many other weather maps besides the ones usually seen by the public—the "surface analysis" map. These include upper-air maps for constant pressure levels, such as the 500-millibar map for an altitude of roughly 18,000 feet (5500 m). Winds at these levels do much to "steer" surface weather patterns, but use of upper-air maps requires specialized training and experience.

Weather Forecasts

The professional meteorologist takes great care in the preparation of his forecasts. Your responsibility is to read or listen carefully—don't see or hear something that isn't there, something that you were hoping for.

Most times on a boat you will get your forecast by listening to a radio transmission. If it is a scheduled broadcast, get a pencil and paper ready and take notes; if you have to use an abbreviated format to keep up with the flow of information, expand your fragments into complete form as soon as possible while the data are still fresh in your mind. Using a small tape recorder can help. If it is a continuous broadcast, still take notes but listen a second time to expand your initial notes; listen as many times as necessary to get the details fully and accurately. Special warnings on NWS stations are preceded by a ten-second high-pitched tone. If you hear this, grab paper and a pencil and stand by to record the information.

If you are watching a TV weather report and forecast, it is just as important to take notes, dividing your attention between the picture on the screen and your note pad.

Weather Warning Signals

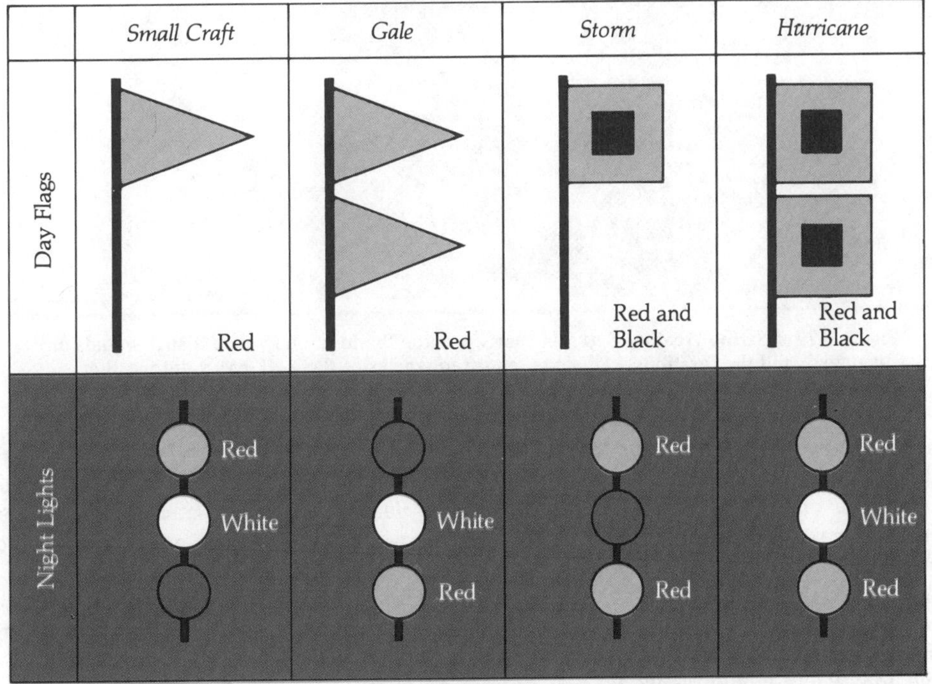

Figure 1418 **Day and night visual signals indicate adverse weather conditions. These are displayed at Coast Guard stations, municipal docks, and some marinas and yacht clubs. See text for details.**

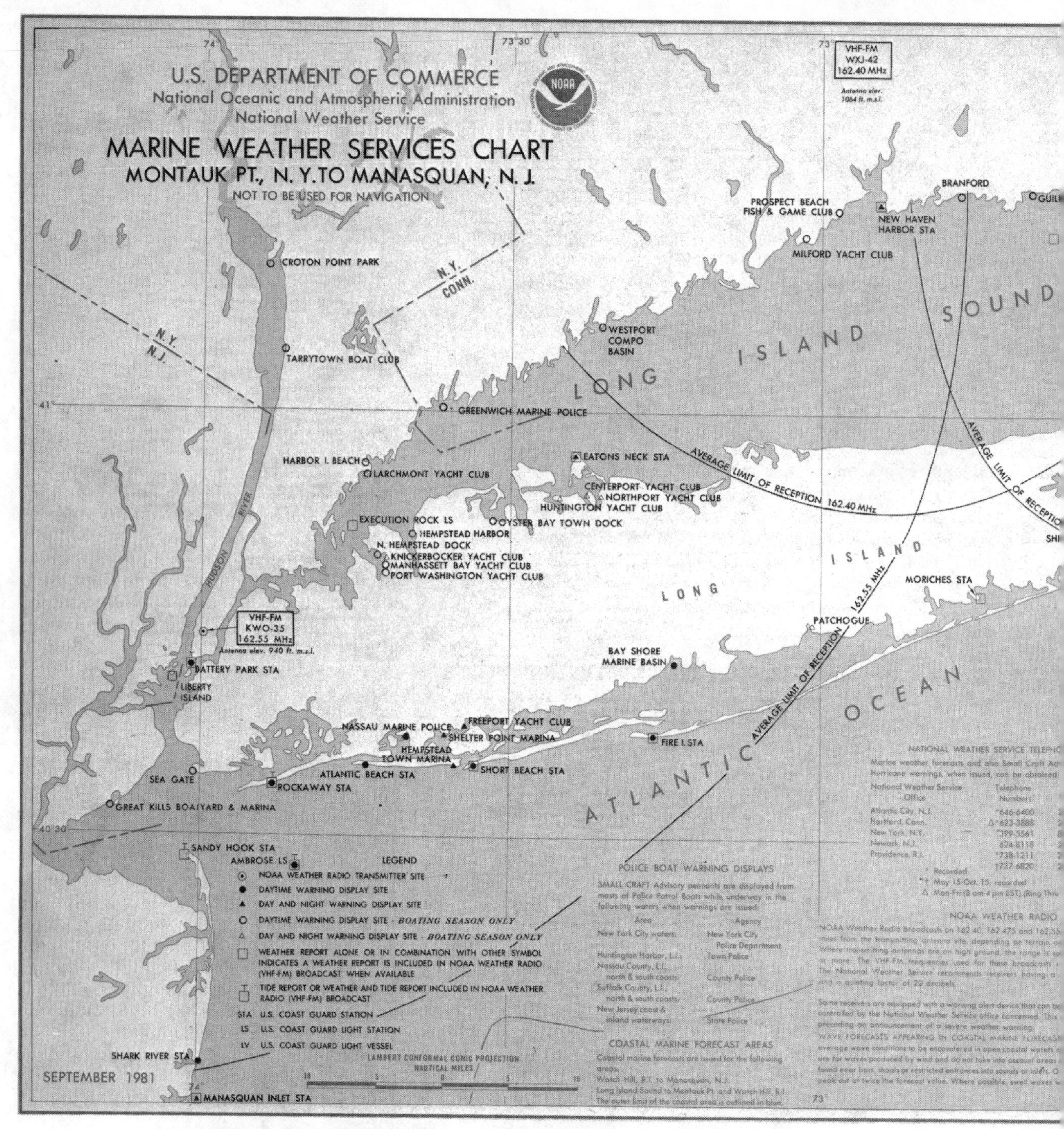

Figure 1419 **Marine Weather Services charts provide the locations where visual signals are displayed, and the locations, call signs, and frequencies of the VHF-FM radio weather broadcast stations in the charted area.**

In most instances, you will be able to get a forecast for your waters. If you can't, try to get a forecast for the region *westward* of you and apply your knowledge of weather movement to make a forecast for your own area.

Keep your weather-forecast notes and compare them with actual conditions later. You may be able to detect a pattern of error in the forecasts, such as weather frequently arriving 12 hours or so later than is forecast—apparently some forecasters would rather you be prepared a bit early for the arrival of a storm, than to be caught short.

Weather Forecast Terms

The National Weather Service has an ascending series of alerting messages—advisories, watches, and warnings—for mariners. These are keyed to increasingly hazardous weather and sea conditions.

Each advisory or warning condition has both a day and a night signal. These are displayed at prominent locations ashore, such as Coast Guard stations and lighthouses, marinas, and yacht clubs. The signals can be seen at some but not all such locations; check locally to know where to watch for them. Some locations display only the day sig-

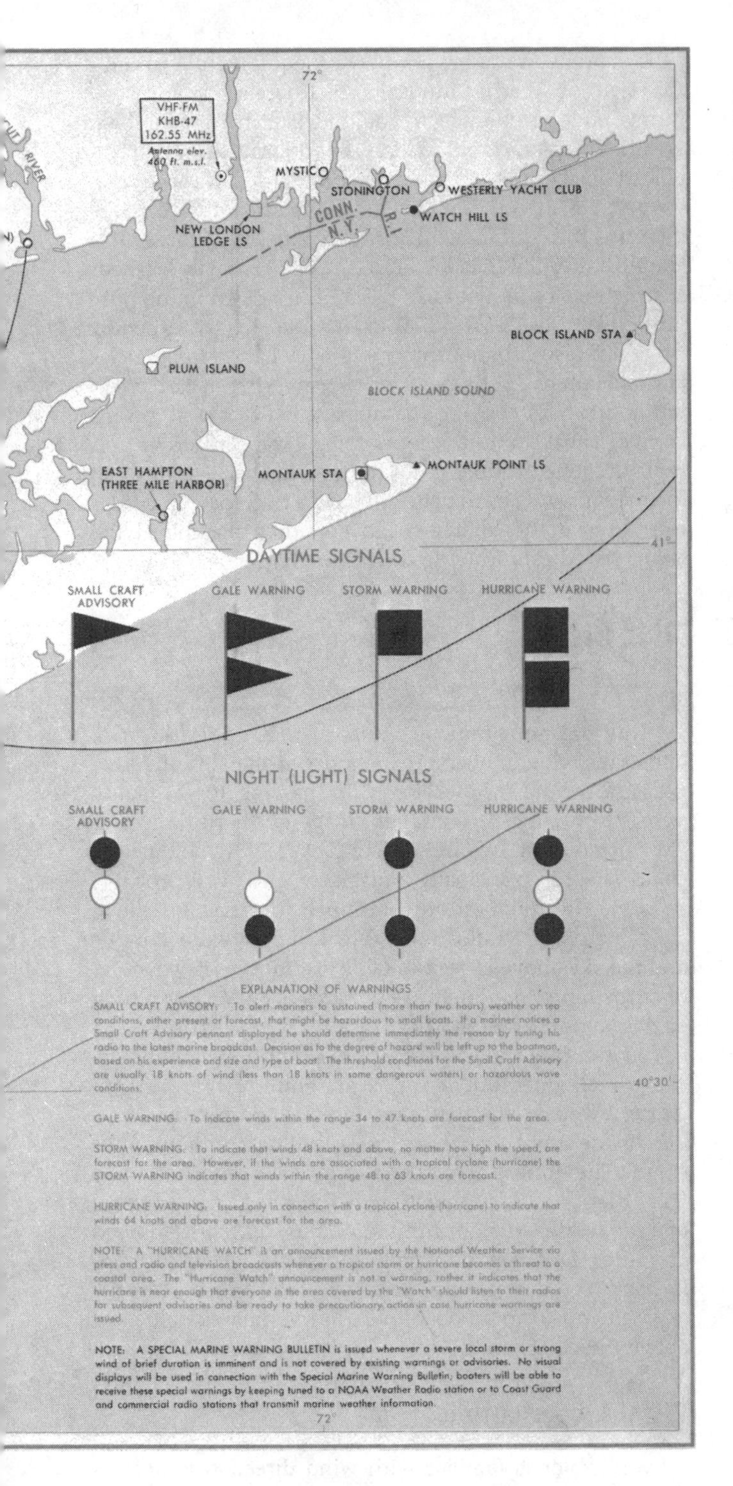

nal, others both the day and night signals. In some areas the small craft advisory red pennant is flown on local law enforcement patrol craft.

The Weather Service emphasizes that these visual storm warnings displayed along the coast are only *supplementary* to the written advisories and warnings given prompt and wide distribution by press, radio, and television. Important details of the forecasts and warnings in regard to the time, intensity, duration, and direction of storms cannot be given satisfactorily through visual signals alone.

Following is an explanation of the signals:

Small Craft Advisory One red pennant displayed by day and a red light above a white light at night, to indicate that fairly strong winds up to 33 knots (38 mph) and/or sea conditions dangerous to small craft operations are forecast for the area.

Gale Warning Two red pennants displayed by day and a white light above a red light at night, to indicate that winds ranging from 34 to 47 knots (39 to 54 mph) are forecast.

Storm Warning A single square red flag with black center displayed by day and two red lights at night, to indicate that winds 48 knots (55 mph) and above (*no matter how high the velocity*) are forecast. NOTE: If winds are associated with a tropical cyclone (hurricane) the *storm warning* display indicates forecast winds of 48 to 63 knots (55 to 73 mph). The *hurricane warning* is displayed only in connection with a tropical cyclone.

Hurricane Warning Two square red flags with black centers displayed by day and a white light between two red lights at night, to indicate that winds 64 knots (74 mph) and above are forecast. Hurricane warnings are not issued for the Great Lakes.

Use of Small Craft Advisories

The term SMALL CRAFT ADVISORY needs some explanation. Small craft as defined by the Weather Service are "small boats, yachts, tugs, barges with little freeboard, or any other low-powered craft."

A small-craft advisory does not distinguish between the expectation of a general, widespread, all-day blow of 25 knots or more and, for example, a forecast of isolated, late-afternoon thundersqualls in which winds dangerous to small craft will be localized and of short duration. It is up to you to deduce from your own observations, supplemented by any information you can obtain from radio broadcasts, which type of situation the advisory applies to and to plan your day's cruise accordingly.

Marine Weather Services Charts

The shore stations where storm warning signals are displayed are prominently marked on the Marine Weather Services Charts, a series which is published annually by the Weather Service. These charts, of which Figure 1419 is an example, also contain detailed information about the times of weather broadcasts from commercial stations, the radio frequencies of marine broadcast stations, the specific type of storm warnings issued and the visual display signals used in connection with the warnings. This series consists of 15 charts which cover the coastal waters of the United States, the Great Lakes, the Hawaiian Islands, and Puerto Rico and the Virgin Islands. They can be purchased from the Superintendent of Documents, Government Printing Office, Washington, DC 20402, or from some local sales agents for charts.

Boaters will find much interesting weather information on the various Pilot Charts of the North Atlantic and North Pacific Oceans issued by the DMA Hydrographic/Topographic Center. The Pilot Charts show average monthly wind and weather conditions over the oceans and in addition contain a vast amount of supplemental data on subjects closely allied to weather. See page 423.

Coast Pilot volumes also contain information on weather broadcasts.

Weather Maps at Sea

Just because you are offshore and not receiving your daily newspaper, you need not be without a reliable and up-to-date weather map. Facsimile radio transmissions can be received by a high-frequency receiver and processed in a unit that prints a weather map line by line. These transmissions, often called RADIOFAX, are broadcast on regular schedules by a number of shore stations. The receiver can be your regular communications set, or it can be built into the printer; the latter is more convenient to use, but more costly. A number of different charts are transmitted: surface analysis, prognosis (forecasts for 12, 24, 36, or 48 hours), upper-air winds, wave analysis, sea water temperature, satellite photographs, and others. These charts have significant advantages over voice announcements, as you have a specific "hard-copy" picture of current or forecast conditions to study in detail.

But you need not have expensive equipment to receive this type of weather information. It is also broadcast in Morse code, from which you can sketch in your own weather map on standard blank chart forms that are available at no charge from Port Meteorological Officers at major ports. And you don't even have to know the code! All data are the figures zero through nine, and these fit a simple pattern of five dots or dashes that is easily learned. Instructions on how to use the data are given in the publication *Worldwide Marine Weather Broadcasts* (see appendix B), which also contains information on station frequencies and schedules. The speed of transmission is not great, but accuracy in copying the figures, which are in groups of five, can be improved by using a tape recorder and replaying one or more times for double-checking or to fill in any gaps. It is even better to have a two-speed recorder recording at the higher speed and then playing at the lower speed as you write down the figures.

Reading Local Weather Signs

While weather forecasts are helpful, and you should always take care to get them, they are not completely dependable and they are normally made for a relatively large area. To supplement the official forecast, look about your boat frequently. Notice both the current weather and changes over the last hour or so. And finally, you should know how to interpret the weather signs that you see in clouds, winds, and pressure and temperature changes. If you are not in home waters, local advice may assist you with weather just as it does with navigation.

Cloud Observations

First, identify the cloud form (or forms) observed in the sky, and note whether they are *increasing* or *decreasing* in amount, and whether they are *lowering* or *lifting*. In general, thickening and lowering of a cloud layer, be it a layer or cirrostratus, altocumulus, or altostratus, is a sign of approaching *wet* weather. And when a layer of clouds shows signs of evaporation, that is, when holes or openings begin to appear in a layer of altostratus, or the elements of an altocumulus layer are frayed and indistinct at the edges, we have an indication of *improving* weather or, at least, delay of any development of foul weather.

Finally, note the *sequence of cloud forms* during the past few hours. Cirrus clouds are frequently the advance agents of an approaching extra-tropical cyclone, especially if they are followed by a layer of cirrostratus. In this case the problem is to forecast the track the low-pressure center will take to one side or the other of your location, the nearness of approach of the center, and the intensity of the low.

As the clouds thicken steadily from cirrus to cirrostratus and then to altostratus we naturally expect further development to nimbostratus with its rain or snow. There are usually contrary indications if such is not to be the case, or if the precipitation will arrive late, amount to little, and end soon, or will come in two brief periods separated by several hours of mild, more or less sunny weather.

If the northern horizon remains clear until a layer of altostratus clouds overspreads most of the sky, the low-pressure center will probably be passing to the south without bringing precipitation. If the northern horizon is slow in clouding up but becomes covered by the time the cloud sheet is principally altostratus there will probably be some precipitation but not much. If the cloud sheet, after increasing to altostratus, breaks up into altocumulus and the sky above is seen to have lost most of its covering of cirrostratus, the low-pressure area is either weakening or passing on to the north.

You can often detect the approach of a squall line accompanied by thunderstorms an hour or more in advance, by observing the thin white arch of the cirrus border to the anvil top of the approaching cumulonimbus cloud. The atmosphere ahead of a squall line is often so hazy that only this whitish arch will reveal the presence of a moderately distant cumulonimbus, for the shadowed air under the dense anvil will be invisible behind the sunlit hazy blue air near the observer. Thus this part of the sky will appear to be clear and will resemble the blue sky above the cirrus arch.

Wind Observations

Every sailor is familiar with wind direction indicators that can be mounted at the masthead. Because a boat anchored or underway can head in any direction all around the compass, a few mental calculations are necessary before we can use this indicator, or a yarn, or an owner's flag or club burgee, to determine the true direction of the wind. At anchor, the fly will give us the bearing of the wind relative to the boat's bow; this must be converted to the true bearing of the wind itself. We can use our compass to obtain these values, provided we know its deviation on our heading and know the variation for our anchorage. Also, we need to remember that wind direction is always stated as the true direction *from* which, not *toward* which, the wind is blowing.

For measuring wind strength we need something else.

This is an ANEMOMETER; the principal parts of the instrument are illustrated in Figure 1420 top. The anemometer is essentially a speedometer. It consists of a rotor with conical cups attached to the ends of spokes, and is designed for mounting at the masthead, where the wind is caught by the cups and causes them to turn at a speed proportional to wind speed. Indications of the rotor's speed are electrically transmitted to an indicator mounted in the cabin or elsewhere on your boat.

In the type of instrument illustrated in Figure 1420 bottom, the indicator is direct-reading, with its scale marked both in knots (0 to 50) and Beaufort force numbers (0 to 12). Battery drain (.09A) is insignificant. A 60-foot (18 m) cable is supplied to connect the rotor aloft with the illuminated indicator below. At anchor, the reading is the actual or true wind velocity at masthead height.

True and Apparent Wind

So far we have been considering the determination of the direction and speed of the wind while at anchor. Underway, the determination is more difficult, but the Oceanographic Office of the Navy has worked out the problem, so we may as well use their results. Table 14-1 is based on wind tables formerly published in *Bowditch.* Using our true course and speed, we can determine from this table the approximate TRUE WIND DIRECTION AND SPEED, provided we know the apparent direction and speed. Our wind indicator or owner's flag will give the apparent direction and our anemometer will give us the apparent strength. True wind direction and speed can also be determined using a Maneuvering Board form; see page 509.

The Beaufort Scale

Another method is to estimate the strength of the wind in terms of the BEAUFORT SCALE OF WIND FORCE. Visually we can observe the sea condition and note the Beaufort number to which it corresponds (see Table 14-2). The range of wind speed represented by each force number on the Beaufort scale is shown in knots, statute miles per hour, and kilometers per hour in the table.

Wind direction can also be judged by observing the direction from which the smallest ripples are coming, since these ripples always run with the wind, responding instantly to changes in wind direction.

You don't need to determine wind direction and strength with great precision, but reasonable estimates are helpful in preparing your own local forecasts. Estimating direction within 15 degrees is sufficient; within one Beaufort number is enough for strength. By observing wind direction and speed regularly and making a record of them you can obtain clues concerning potential weather developments as the wind shifts direction and changes its strength.

Wind Force—and Its Effect on the Sea

Boatmen frequently use the Beaufort Scale to log wind speed and the condition of the sea, but too often the description of sea conditions relies on verbal statements like "moderate waves," "large waves," "moderately high waves" (column 6, Table 14-2) which leaves the boatmen pretty much "at sea" in trying to visualize exactly what these relative terms imply.

Figure 1420 **Anemometer rotor is shown at the left on this sailboat's masthead fixture, *above.* It is connected by cable to the indicator, usually mounted in the cockpit. The anemometer display unit, *below,* shows wind speeds to 50 knots. Inner scale gives the Beaufort Scale force numbers.**

True Force and Direction of the Wind from Its Apparent Force and Direction on a Boat Under Way

	Apparent Wind Velocity (Knots)	*Speed of Boat*							
		5 Knots		*10 Knots*		*15 Knots*		*20 Knots*	
		True Wind		*True Wind*		*True Wind*		*True Wind*	
		Points off Bow	*Velocity Knots*	*Points off Bow*	*Velocity Knots*	*Points off Bow*	*Velocity Knots*	*Points off Bow*	*Velocity Knots*
I. Apparent Wind Direction Is Dead Ahead	Calm	D. As.	5 K.	D. As.	10 K.	D. As.	15 K.	D. As.	20 K.
	4 K.	D. As.	1 K.	D. As.	6 K.	D. As.	11 K.	D. As.	16 K.
	8 K.	D. Ah.	3 K.	D. As.	2 K.	D. As.	7 K.	D. As.	12 K.
	12 K.	D. Ah.	7 K.	D. Ah.	2 K.	D. As.	3 K.	D. As.	8 K.
	16 K.	D. Ah.	11 K.	D. Ah.	6 K.	D. Ah.	1 K.	D. As.	4 K.
	22 K.	D. Ah.	17 K.	D. Ah.	12 K.	D. Ah.	7 K.	D. Ah.	2 K.
	30 K.	D. Ah.	25 K.	D. Ah.	20 K.	D. Ah.	15 K.	D. Ah.	10 K.
	42 K.	D. Ah.	37 K.	D. Ah.	32 K.	D. Ah.	27 K.	D. Ah.	22 K.
	60 K.	D. Ah.	55 K.	D. Ah.	50 K.	D. Ah.	45 K.	D. Ah.	40 K.
II. Apparent Wind Direction Is 4 Points (Broad) Off the Bow	4 K.	11 pts.	4 K.	14 pts.	8 K.	15 pts.	12 K.	15 pts.	17 K.
	8 K.	7 pts.	6 K.	11 pts.	7 K.	13 pts.	11 K.	14 pts.	15 K.
	12 K.	6 pts.	9 K.	9 pts.	9 K.	11 pts.	11 K.	13 pts.	14 K.
	16 K.	5 pts.	13 K.	7 pts.	11 K.	10 pts.	12 K.	11 pts.	14 K.
	22 K.	5 pts.	19 K.	6 pts.	16 K.	8 pts.	15 K.	9 pts.	16 K.
	30 K.	5 pts.	27 K.	6 pts.	24 K.	7 pts.	22 K.	8 pts.	21 K.
	42 K.	4 pts.	39 K.	5 pts.	36 K.	6 pts.	33 K.	6 pts.	31 K.
	60 K.	4 pts.	57 K.	5 pts.	53 K.	5 pts.	51 K.	6 pts.	48 K.
III. Apparent Wind Direction Is 8 Points Off the Bow (Abeam)	4 K.	13 pts.	6 K.	14 pts.	11 K.	15 pts.	16 K.	15 pts.	20 K.
	8 K.	11 pts.	9 K.	13 pts.	13 K.	14 pts.	17 K.	14 pts.	22 K.
	12 K.	10 pts.	13 K.	12 pts.	16 K.	13 pts.	19 K.	13 pts.	23 K.
	16 K.	10 pts.	17 K.	11 pts.	19 K.	12 pts.	22 K.	13 pts.	26 K.
	22 K.	9 pts.	23 K.	10 pts.	24 K.	11 pts.	27 K.	12 pts.	30 K.
	30 K.	9 pts.	30 K.	10 pts.	32 K.	10 pts.	34 K.	11 pts.	36 K.
	42 K.	9 pts.	42 K.	9 pts.	43 K.	10 pts.	45 K.	10 pts.	47 K.
	60 K.	8 pts.	60 K.	9 pts.	61 K.	9 pts.	62 K.	10 pts.	63 K.
IV. Apparent Wind Direction Is 12 Poits Off the Bow (Broad on the Quarter)	4 K.	14 pts.	8 K.	15 pts.	13 K.	15 pts.	18 K.	15 pts.	23 K.
	8 K.	14 pts.	12 K.	14 pts.	17 K.	15 pts.	21 K.	15 pts.	26 K.
	12 K.	13 pts.	16 K.	14 pts.	20 K.	14 pts.	25 K.	15 pts.	30 K.
	16 K.	13 pts.	20 K.	14 pts.	24 K.	14 pts.	29 K.	14 pts.	33 K.
	22 K.	13 pts.	26 K.	13 pts.	30 K.	14 pts.	34 K.	14 pts.	39 K.
	30 K.	13 pts.	34 K.	13 pts.	38 K.	13 pts.	42 K.	14 pts.	46 K.
	42 K.	12 pts.	46 K.	13 pts.	50 K.	13 pts.	54 K.	13 pts.	58 K.
	60 K.	12 pts.	64 K.	13 pts.	67 K.	13 pts.	71 K.	13 pts.	75 K.

Conversion of Points Off Bow to True Direction of Wind

	Points Off Bow	*Boat's Heading—True*							
		000°	045°	090°	135°	180°	225°	270°	315°
I. When Wind Direction Obtained from Table Above Is Off Starboard Bow	Dead Ahead	N	NE	E	SE	S	SW	W	NW
	4 points	NE	E	SE	S	SW	W	NW	N
	8 points	E	SE	S	SW	W	NW	N	NE
	12 points	SE	S	SW	W	NW	N	NE	E
	Dead Astern	S	SW	W	NW	N	NE	E	SE
II. When Wind. Direction Obtained from Table Above Is Off Port Bow	Dead Ahead	N	NE	E	SE	S	SW	W	NW
	4 points	NW	N	NE	E	SE	S	SW	W
	8 points	W	NW	N	NE	E	SE	S	SW
	12 points	SW	W	NW	N	NE	E	SE	S
	Dead Astern	S	SW	W	NW	N	NE	E	SE

Abreviations: D. As. = Dead Astern. D. Ah. = Dead Ahead. K. = Knots. pts. = Points off bow.

To Use This Table

1. **With Wind Direction Indicator:** Determine Apparent Wind Direction off the Bow.
2. **With Anemometer:** Determine Apparent Wind Velocity, in Knots.
3. **Enter Upper Part of Table:** Use portion for nearest Apparent Wind Direction Opposite Apparent Wind Velocity and under nearest Speed of Boat, read Wind Direction in Points off Bow and True Wind Velocity in Knots. Note whether True Wind Direction is off Starboard or Port Bow.
4. **Enter Lower Part of Table:** Use portion for proper Bow: Starboard or Port. Opposite Points off Bow and under nearest Boat's True Heading, read True Wind Direction.
5. **Log:** Record True Wind Direction as obtained from Lower Part of Table and True Wind Velocity as obtained from Upper Part of Table in Boat's Weather Log.

Table 14-1

Beaufort Wind Scale

Beaufort Number or Force	*Wind Speed*			*World Meteorological Organization Description*	*Estimating Wind Speed*		
	Knots	*mph*	*km/hr*		*Effects Observed at Sea*	*Effects Observed Near Land*	*Effects Observed on Land*
0	under 1	under 1	under 1	Calm	Sea like a mirror	Calm	Calm; smoke rises vertically
1	1-3	1-3	1-5	Light Air	Ripples with appearance of scales; no foam crests	Small sailboat just has steerage way	Smoke drift indicates wind direction; vanes do not move
2	4-6	4-7	6-11	Light Breeze	Small wavelets; crests of glassy appearance, not breaking	Wind fills the sails of small boats which then travel at about 1—2 knots	Wind felt on face; leaves rustle; vanes begin to move
3	7-10	8-12	12-19	Gentle Breeze	Large wavelets; crests begin to break, scattered whitecaps	Sailboats begin to heel and travel at about 3—4 knots	Leaves, small twigs in constant motion; light flags extended
4	11-16	13-18	20-28	Moderate Breeze	Small waves 0.5—1.25 meters high, becoming longer; numerous whitecaps	Good working breeze, sailboats carry all sail with good heel	Dust, leaves, and loose paper raised up; small branches move
5	17-21	19-24	29-38	Fresh Breeze	Moderate waves of 1.25—2.5 meters taking longer form; many whitecaps; some spray	Sailboats shorten sail	Small trees in leaf begin to sway
6	22-27	25-31	39-49	Strong Breeze	Larger waves 2.5—4 meters forming; whitecaps everywhere; more spray	Sailboats have double reefed mainsails	Larger branches of trees in motion; whistling heard in wires
7	28-33	32-38	50-61	Near Gale	Sea heaps up, waves 4—6 meters; white foam from breaking waves begins to be blown in streaks	Boats remain in harbor; those at sea heave-to	Whole trees in motion; resistance felt in walking against wind
8	34-40	39-46	62-74	Gale	Moderately high (4—6 meter) waves of greater length; edges of crests begin to break into spindrift; foam is blown in well-marked streaks	All boats make for harbor, if near	Twigs and small branches broken off trees; progress generally impaired
9	41-47	47-54	75-88	Strong Gale	High waves (6 meters); sea begins to roll; dense streaks of foam; spray may reduce visibility		Slight structural damage occurs; slate blown from roofs
10	48-55	55-63	89-102	Storm	Very high waves (6—9 meters) with overhanging crests; sea takes a white appearance as foam is blown in very dense streaks; rolling is heavy and visibility is reduced		Seldom experienced on land; trees broken or uprooted; considerable structural damage occurs
11	56-63	64-72	103-117	Violent Storm	Exceptionally high (9—14 meters) waves; sea covered with white foam patches; visibility still more reduced		Very rarely experienced on land; usually accompanied by widespread damage
12	64 and over	73 and over	118 and over	Hurricane	Air filled with foam; waves over 14 meters; sea completely white with driving spray; visibility greatly reduced		

Table 14-2

Figure 1422 **This barometer is graduated only in inches of mercury; addition of a millibar scale makes such an instrument more useful when working with a weather map.**

The British Meteorological Office has solved this problem by issuing a State of Sea card (M.O. 688A) with photographs to accompany each of the descriptions of 13 wind forces of the Beaufort Scale. Thus the observer has a guide in estimating wind strength (in knots) when making weather reports or in logging sea conditions. Fetch, depth of water, swell, heavy rain, current, and the lag effect between the wind getting up and the sea increasing, may also affect the appearance of the sea. Range of wind speed and the mean wind speed are given for each force. By special permission, we reproduce (Figure 1421), six of these photographs (Forces 1, 3, 5, 8, 10, 12). Forces 0, 2, 4, 6, 7, 9, and 11, though not illustrated, may be estimated in relation to those above and below them in the scale.

Pressure Observations

Another weather instrument you should be familiar with is the ANEROID BAROMETER. The one illustrated (Figure 1422) has several interesting features.

First, there is the pressure scale. You probably are accustomed to thinking of barometric pressure in terms of inches of mercury, so the scale is graduated in these units. Weather maps are now printed with the pressures shown in millibars, and many radio weather reports specify this value, so an inner scale graduated in millibars would help; you thus would not have to worry about conversions between units. The "standard atmospheric pressure" of 29.92 inches of mercury is equal to 1013.2 millibars; 1 inch equals 33.86 millibars, or 1 millibar equals 0.03 inches of mercury.

Second, it is a rugged instrument and it has a high order of accuracy.

Third, it has a reference hand, for keeping track of changes in pressure.

The words "Fair–Change–Rain," in themselves, when they appear on the face of an aneroid barometer, are mainly decorative. It is not the actual barometric pressure that is important in forecasting; it is the direction and rate of change of pressure.

How a Barometer Is Used

A good barometer is a helpful instrument, provided you read it at regular intervals and keep a record of the readings, and provided you remember that there is much more to weather than just barometric pressure.

An individual reading of the barometer tells you only the pressure being exerted by the atmosphere on the earth's surface at a particular point of observation at that time. But, suppose you have logged the pressure readings at regular intervals, as follows:

Time	*Pressure (Inches)*	*Change*
0700	30.02	—
0800	30.00	−0.02
0900	29.97	−0.03
1000	29.93	−0.04
1100	29.88	−0.05
1200	29.82	−0.06

The pressure is falling, at an increasing rate. Trouble is brewing. A fall of 0.02 inch per hour is a low rate of fall; consequently, this figure would not be particularly disturbing. But a fall of 0.05 inch per hour is a rather high rate.

Next, there is a normal daily change in pressure. The pressure is usually at its maximum value about 1000 and 2200 each day, at its minimum value about 0400 and 1600. The variation between minimum and maximum may be as much as 0.05 inch change in these six-hour intervals (about 0.01 inch change per hour). Thus, when the pressure normally would increase about 0.03 inch (0700 to 1000) our pressure actually fell 0.09 inch.

Suppose, now, that at about 1200 you also observed that the wind was blowing from the NE with increasing force and that the barometer continued to fall at a high rate. A severe northeast gale is probably on its way. On the other hand, given the same barometer reading of about 29.80, rising rapidly with the wind going to west, you could expect improving weather. Quite a difference!

Barometer Rules

A few general rules are often helpful. First, foul weather is usually forecast by a falling barometer with winds from the east quadrants; clearing and fair weather is usually forecast by winds shifting to west quadrants with a rising barometer. Second, there are the rules formerly printed on NWS daily surface weather maps:

"When the wind sets in from points between south and southeast and the barometer falls steadily, a storm is approaching from the west of northwest, and its center will pass near or north of the observer within 12 to 24 hours, with the wind shifting to northwest by way of south and southwest.

"When the winds sets in from points between east and northeast and the barometer falls steadily, a storm is approaching from the south or southwest, and its center will pass near or to the south of the observer within 12 to 24 hours, with the wind shifting to northwest by way of north.

"The rapidity of the storm's approach and its intensity will be indicated by the rate and amount of fall in the barometer."

There are other generally useful barometer rules.

A falling barometer and a rising thermometer often forecast rain; barometer and thermometer rising together often forecast fine weather. A slowly rising barometer forecasts settled weather; a steady, slow fall of pressure indicates forthcoming unsettled or wet weather.

Barometric Changes and Wind Velocity

Let us now relate barometric changes to wind velocity. First, it is generally true that a rapidly falling barometer forecasts the development of strong winds. This is so because a falling barometer indicates the approach or development of a Low, and the pressure gradient is usually steep in the neighborhood of a low-pressure center. On the other hand, a rising barometer is associated with the prospect of lighter winds to come. This is true because a rising barometer indicates the approach or development of a High, and the pressure gradient is characteristically smaller in the neighborhood of a high-pressure center.

The barometer does not necessarily fall before or during a strong breeze. The wind often blows hard without any appreciable accompanying change in the barometer. This means that a steep pressure gradient exists (isobars close together, as seen on the weather map), but that the well-developed High or Low associated with the steep pressure gradient is practically stationary. In this case the wind may be expected to blow hard for some time; any slackening or change will take place gradually.

It not infrequently happens that the barometer falls quite rapidly, yet the wind remains comparatively light. If you remember the relation between wind velocity and pressure gradient, you can conclude that the gradient must be comparatively small (isobars relatively far apart). The rapid fall of the barometer must be accounted for, then, in either one or two ways. Either a Low with a weak pressure gradient on its forward side is approaching rapidly, or there is a rapid decrease of pressure taking place over the surrounding area, or both. In such a situation the pressure gradient at the rear of the Low is often steep, and in that case strong winds will set in as soon as the barometer commences to rise. (It will rise rapidly under these circumstances.) The fact that the barometer is now rising, however, indicates that decreasing winds may be expected soon.

The Barometer and Wind Shifts

Nearly all extra-tropical cyclones display an unsymmetrical distribution of pressure. The pressure gradients are seldom the same in the front and rear of an extra-tropical cyclone. During the approach of a Low the barometer alone gives no clue as to how much the wind will shift and what velocity it will have after the passage of the low-pressure system. This is particularly applicable to situations in which the wind blows from a southerly direction while the barometer is falling. The cessation of the fall of the barometer will coincide with a VEERING (gradual or sudden) shift of the wind to a more westerly direction. Unfortunately, if you have no information other than the variations in atmospheric pressure indicated by your barometer, you cannot foretell the exact features of the change.

In using barometric indications for local forecasting, remember that weather changes are influenced by the characteristics of the earth's surface in your locality. Check all rules against experience in your own cruising waters before you place full confidence in them.

Temperature Observations

Thermometer readings will not give as much information for weather predicting as data from other instruments, but they are not without some value.

Cold air carried down from a thunderstorm cloud with the rain may be felt as much as three miles in advance of the storm itself. Thus a warning is given and the approach of the storm is confirmed.

Judging the Likelihood of Fog

In order to judge the likelihood of fog formation, you can periodically measure the air temperature and dew-point temperature and see if the SPREAD (difference) between them is getting smaller.

Figure 1424 is a graph derived from a series of air and dew-point temperatures. By recording these temperatures and plotting their spread over a period of several hours, as is indicated by that part of the curve drawn as a solid line, you will have a basis for forecasting the time at which you are likely to be fog-bound. The dot-dash portion of the curve represents actual data, but it could just as easily have been drawn by extending the solid portion. If an error were made in this extrapolation, it would probably indicate that the fog would form at an earlier hour. This is on the safe side; you would be secure in your anchorage sometime before the 59th minute of the 11th hour!

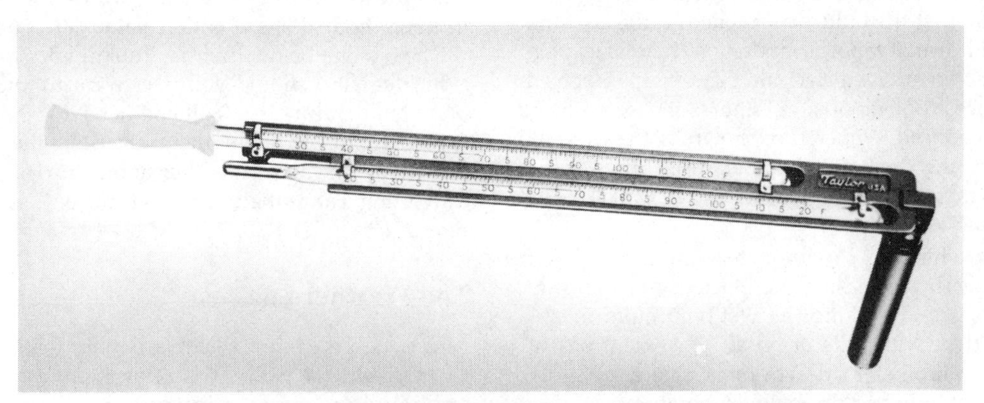

Figure 1423 **This pocket-type sling psychrometer has two five-inch tubes, etched with divisions reading from 20 degrees to 120 degrees Fahrenheit.**

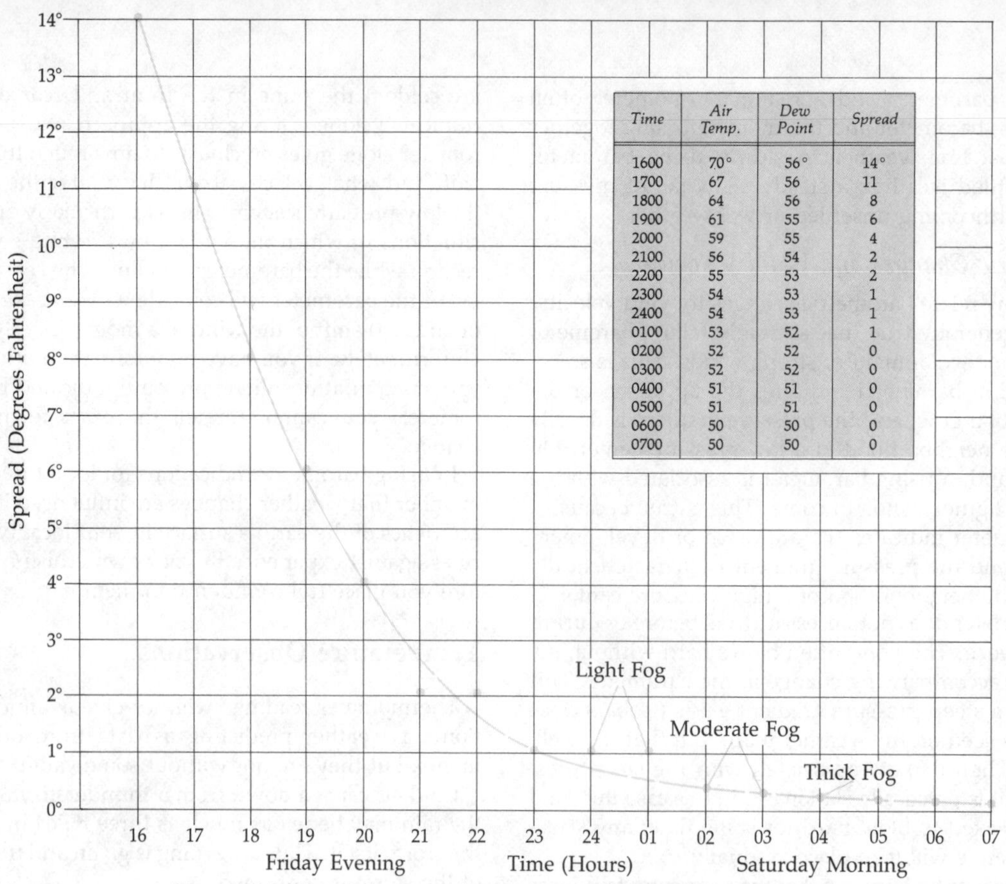

Time	Air Temp.	Dew Point	Spread
1600	70°	56°	14°
1700	67	56	11
1800	64	56	8
1900	61	55	6
2000	59	55	4
2100	56	54	2
2200	55	53	2
2300	54	53	1
2400	54	53	1
0100	53	52	1
0200	52	52	0
0300	52	52	0
0400	51	51	0
0500	51	51	0
0600	50	50	0
0700	50	50	0

Figure 1424 **Here is how a series of air and dewpoint temperatures is plotted (see text for details) based on the readings shown in the table.**

Note that while the average decrease in spread is about 1.5° F per hour, the decrease is at a much greater rate in the earlier hours of the day. So long as you do not make unreasonable allowances for a slowing up in the rate of change of the spread, this also will help to keep you on the safe side.

Determining the Dew-Point

You can determine the dew point by means of a simple-to-operate, inexpensive little gadget known as a SLING PSYCHROMETER (Figure 1423). A sling psychrometer is two thermometers mounted in a single holder with a handle that permits it to be whirled overhead. One thermometer, known as the DRY BULB, has its bulb of mercury exposed directly to the air and thus shows the actual temperature of the air. The other thermometer, the WET BULB, has its bulb covered with a piece of gauze: Soak this gauze in fresh water so that the bulb is moistened. If the air is not saturated with water vapor, evaporation then takes place from the wet-bulb thermometer, and the wet bulb is cooled, since the process of evaporation requires the expenditure of heat. The reduced temperature shown by the wet-bulb thermometer, the "wet-bulb temperature," thus represents the lowest temperature to which the air can be cooled by evaporating water into it.

When you whirl the psychrometer you create a draft around it, thereby increasing the efficiency of the evaporation process and making the wet-bulb more reliable than it would be with little or no air movement past it. Hence the psychrometer's design for whirling.

From the wet-bulb and dry-bulb temperatures, you can determine the dew point by referring to a suitable table. As you are far more interested in knowing the spread between air temperature and dew point, however, use Table 14-3, explained below.

If the air is already saturated with water vapor, no water can evaporate from the gauze and both thermometers will show the same value. The spread between air temperature and dew point is zero. But if the air is not already saturated with water vapor, subtract the wet-bulb temperature from the dry-bulb temperature. With this difference and the dry-bulb (the air) temperature, consult Table 14-3 and find the corresponding spread between the air temperature and the dew-point temperature. This is the figure you want.

If, in the late afternoon or early evening, the spread between the air temperature and dew point is *less* than approximately 6° F, and the air temperature is falling, you will probably encounter fog or greatly restricted visibility in a few hours. These critical values are emphasized by the heavy line below them in Table 14-3.

Incidentally, should you ever want to know the dew-point temperature itself, all you need do is to subtract the spread figure given in the table from the temperature shown by the dry-bulb thermometer. The formula for converting Fahrenheit temperature to Celsius temperature is: (°F − 32) × 0.56 = °C.

The Weather Log

Although the latest Weather Service forecasts are readily available via radio, it is often helpful to record your own cloud and weather observations. You can then check the reliability of the latest prediction. Occasionally the professional forecaster misjudges the future rate of travel of the weather pattern, which may move faster or slower than anticipated. Or a new, unforeseen development in the pattern may occur.

Figure 1425 left shows a form of Weather Log similar to that used in the Weather Course of the United States Power Squadrons. This is the face sheet; Figure 1425 right is the reverse side. The first weather items recorded are based on a reading of the latest weather map (if available) and a summary of radio reports received. Then use the reverse side to jot down your local observations. Sufficient columns are available to permit the entry of these data six times during one 24-hour day, at four-hour intervals. Entries may be made using the standard weather code symbols or any other way you choose, provided you use the same system consistently.

From these records you can estimate how and to what extent the actual weather during the next few hours might differ from those predicted in the official forecast.

Weather Proverbs

With all of modern weather forecasting's scientific instruments, high-speed computers, and highly trained professionals, don't overlook the guidance inherent in traditional WEATHER PROVERBS. Their origins have been lost in the passage of time, and their originators did not know why they were true, but their survival over centuries attests to their general validity. Use them, with caution, and adapt them to local waters.

Mackerel skies and mares' tails
Make tall ships carry low sails.

If high-flying cirrus clouds are few in the sky and resemble wisps in a mare's tail in the wind, this is a sign of fair weather. Only when the sky becomes heavy with cirrus, or mackerel, clouds—cirrocumulus that resemble wave-rippled sand on a beach—can you expect a storm. There is an exception to the mare's tail proverb, however. If cirrus clouds form as mares' tails with the hairs pointing

Air Temperature—Dewpoint Spread

(All Figures are in Degrees Fahrenheit at 30 Inches Barometric Pressure)

Difference Dry-bulb Minus Wet-bulb	*Air Temperature Shown by Dry-bulb Thermometer*												
	35	*40*	*45*	*50*	*55*	*60*	*65*	*70*	*75*	*80*	*85*	*90*	*95*
1	2	2	2	2	2	2	2	1	1	1	1	1	1
2	5	5	4	4	4	3	3	3	3	3	3	3	2
3	7	7	7	6	5	5	5	4	4	4	4	4	4
4	10	10	9	8	7	7	6	6	6	6	5	5	5
5	14	12	11	10	10	9	8	8	7	7	7	7	6
6	18	15	14	13	12	11	10	9	9	8	8	8	8
7	22	19	17	16	14	13	12	11	11	10	10	9	9
8	28	22	20	18	17	15	14	13	12	12	11	11	10
9	35	27	23	21	19	17	16	15	14	13	13	12	12
10	—	33	27	24	22	20	18	17	16	15	14	14	13
11	—	40	32	28	25	22	20	19	18	17	16	15	15
12	—	—	38	32	28	25	23	21	20	18	17	17	16
13	—	—	45	37	31	28	25	23	21	20	19	18	17
14	—	—	—	42	35	31	28	26	24	22	21	20	19
15	—	—	—	50	40	35	31	28	26	24	23	21	21

Opposite the figure for the difference between the dry-bulb and wet-bulb readings, and under the temperature shown by the dry-bulb thermometer, read the value of the dewpoint spread.

Based on U.S. Weather Service Psychometric Tables

Table 14-3

BOAT WEATHER LOG

Yacht_____ At/Passage_____ to _____
Day_____ Date_____ Time Zone_____ Skipper_____

1. Latest Weather Map: Date_____ Time_____ Summary of forecast and of principal regional weather features: _____
2. Radio Weather Reports Received (state source and time): _____
3. Local Weather Observations
4. Remarks and Local Forecast for Next_____ Hours (state time forecast effective): _____

LOCAL WEATHER OBSERVATIONS

Time
Latitude—degrees, minutes
Longitude—degrees, minutes
Course—degrees mag.
—degrees true
Speed—Knots
Barometer—in. or mb.
—tendency
Clouds—form
—moving from
—amount
—changing to
Sea—condition
—swells
—moving from
Temperatures—air, dry bulb
—dewpoint
—water
Visibility
Wind—direction, true
—shifting to
—velocity, true
—force (Beaufort)
Weather—present

Figure 1425 **Weather log, *left,* is used to record information developed from weather maps, or received by radio, and the forecast based on this data. Reverse side of page, *right,* is used for local weather observations. By recording this information and keeping the record, it is possible to develop considerable skill in making forecasts.**

upward or downward, the probability is for rain, even though the clouds may be scattered.

Red sky at morning
Sailors take warning;
Red sky at night
Sailors' delight.

There is a simple explanation for this old and quite reliable proverb. A red sunset results from viewing the sun through dusty particles in the air, the nuclei necessary for the formation of rain. This air probably would reach an observer the following day. Since weather tends to flow west to east in most places, if tomorrow's weather appears to the west as a line of wetness, the sun shining through the mass appears as a yellow or greyish orb. On the other hand if the weather lying to the west is dry, the sun will show at its reddest.

Red sky in the morning is caused by the rising, eastern sun lighting up the advance guard of high cirrus and cirrostratus, which will be followed later on by the lowering, frontal clouds. Red sky at night—a red-tinted sunset—often derives from the sky clearing at the western horizon, with the clouds overhead likely to pass before the night is done.

Rainbow in morning
Sailors take warning;
Rainbow toward night
Sailors' delight.

This is also quite an old weather jingle, and a little reasoning, particularly in view of the explanation above, would tell you it is true.

As noted before, storm centers usually move from the west. Thus, a morning rainbow would have to be viewed from its position, already in the west with the sun shining on it from the east. The storm would move in your direction and you could confidently expect rain. An evening rainbow, however, viewed to the east, would tell you that the storm has already passed. The following, another old seafarer's saying about rainbows, is certainly worth remembering because it is almost infallibly true:

Rainbow to windward, foul fall the day;
Rainbow to leeward, rain runs away.

If a rainbow is behind or with the direction of the prevailing wind, then you can expect its curtain of moisture to reach you. But if the rainbow appears to the lee of the wind, then you know rain has already passed and the grey line of showers is receding, moving away from you.

Winds that swing against the sun
And winds that bring the rain are one.
Winds that swing around the sun
Keep the rain storm on the run.

This old saying is based on the direction a weathervane is pointing, but observing a flag will do as well. It means that a wind that changes its direction so it moves from east to west, as the sun moves, almost always results in clear skies. But a wind that changes against the sun's movement, blowing first from the west, then the east, brings dirty weather with it.

A backing wind says storms are nigh;
But a veering wind will clear the sky.

This folklore observation generally refers to storms running in a southerly direction.

The shape and color of the moon as indicators of coming weather changes have long been a subject of controversy mostly among those meteorological experts who declare that the moon has no appreciable control over the weather beyond a very small tidal effect on the atmosphere. But for now, let's be content with the observation that as far as weather portents are concerned, the moon is one of the most visible and absolutely reliable signs of weather change. It is not the moon's influence that makes the following sayings ring with truth, it is other atmospheric modifications that influence the moon's appearance.

Sharp horns on the moon threaten high winds.

When you can clearly see the sharp horns or ends of a crescent moon with your naked eye, it means there are high-speed winds aloft which are sweeping away cloud forms. Inasmuch as these high winds always descend to earth, you can predict a windy day following.

When a halo rings the moon or sun
The rain will come upon the run.

Halos are excellent atmospheric signs of rain. Halos around the moon after a pale sun confirm the advent of rain, for you are viewing the moon through the ice crystals of high cirriform clouds. When the whole sky is covered with these cloud forms, a warm front is approaching, bringing a long, soft rain.

Let's look at two other halo sayings, which at first reading may seem to be contradictory:

A halo around the sun indicates the approach of a storm within three days, from the side which is the most brilliant.

Halos predict a storm at no great distance, and the open side of the halo tells the quarter from which it may be expected.

These two sayings are more explicit in their forecasting. As cirrus and cirrostratus fronts push across the sky in the region of the moon or sun, the halo first appears and subsequently becomes brightest in that part of the arc from which a low pressure system is approaching. Later, the halo becomes complete and the light is uniform throughout. As the storm advances, altostratus clouds arrive and obliterate the original, and for a time, the brightest part of the halo—that is, the side nearest the oncoming storm. Both sayings are useful but refer to different times in the life of the halo.

It is also true that when halos are double or triple, they signify that cirrostratus clouds are relatively thick, such as would be the case in a deep and well-developed storm. Broken halos indicate a much disturbed state in the upper atmosphere, with rain close at hand.

Now, to put any confusion at rest about the forecast persistence of rain by the appearance of sun and moon halos, the U.S. Weather Service has verified through repeated observations that sun halos will be followed by rain about 75% of the time. Halos around the moon have a rain forecasting accuracy of about 65%.

When boat horns sound hollow,
Rain will surely follow.

Anyone who has spent any time around boats knows the truth of this time-honored prophecy. Nor do you have to be sitting on a piling in a marina to notice the unusual sharpness of sounds on certain days—the more penetrating sound of a bell ringing or voices that carry longer distances are signs of the accoustical clarity when bad weather lowers the cloud ceiling toward earth. The tonal quality of sound is improved because the cloud layer bounces the sounds back, the way the walls of a canyon echo a cry. When the cloud barrier lifts, the same sounds dissipate in space. Another (English) version of this saying goes like this:

Sound traveling far and wide
A stormy day does like betide.

This suggests that you actually can hear bad weather approaching, say, when a faraway train whistle is audible when normally it would be faint. The reason the sound carries farther is that the whistle was blown under a lowering cloud ceiling whose extending barrier may not have reached your position yet.

Lightning from the west or northwest will reach you,
Lightning from the south or southeast will pass you by.

This is a true saying, if you live in the north temperate zone. Lightning comes hand in hand with storm clouds, and thunderheads always loom over the horizon from the west or northwest, and usually move east. So lightning anywhere from the south or southeast will pass you by.

210
240
W
300
330
N
S
CONSTELLATION
DANFORTH

15

Compass principles; how to choose, install, and adjust the instrument; variation and deviation calculations

THE MARINER'S COMPASS

Your boat's safety may well depend on her compass, and on your ability to use it properly. On long runs out of sight of land or aids to navigation, your compass enables you to steer accurately to your destination. In poor visibility you may have no other means of keeping on your course. Know the limitations of your magnetic compass, but within these, trust it to guide you safely.

Why It Works

MAGNETISM is a phenomenon that can only be known from its effects. It is a basic natural force and one which is essential to a vast number of devices, from incredibly complex electronic equipment to our boat's simple magnetic compass.

Magnetism appears as a physical force between two objects of metal, usually iron or an alloy of iron and other metals, at least one of which has been previously magnetized and has become a MAGNET. The area around a magnet in which its effect can be detected is called a MAGNETIC FIELD. This is commonly pictured as innumerable LINES OF FORCE, but these are purely a convention for illustration and actual lines as such do not exist; Figure 1501 is a typical illustration of a BAR MAGNET with some of the lines of force shown. (An electric current flowing in a wire also creates a magnetic field; more on this later.)

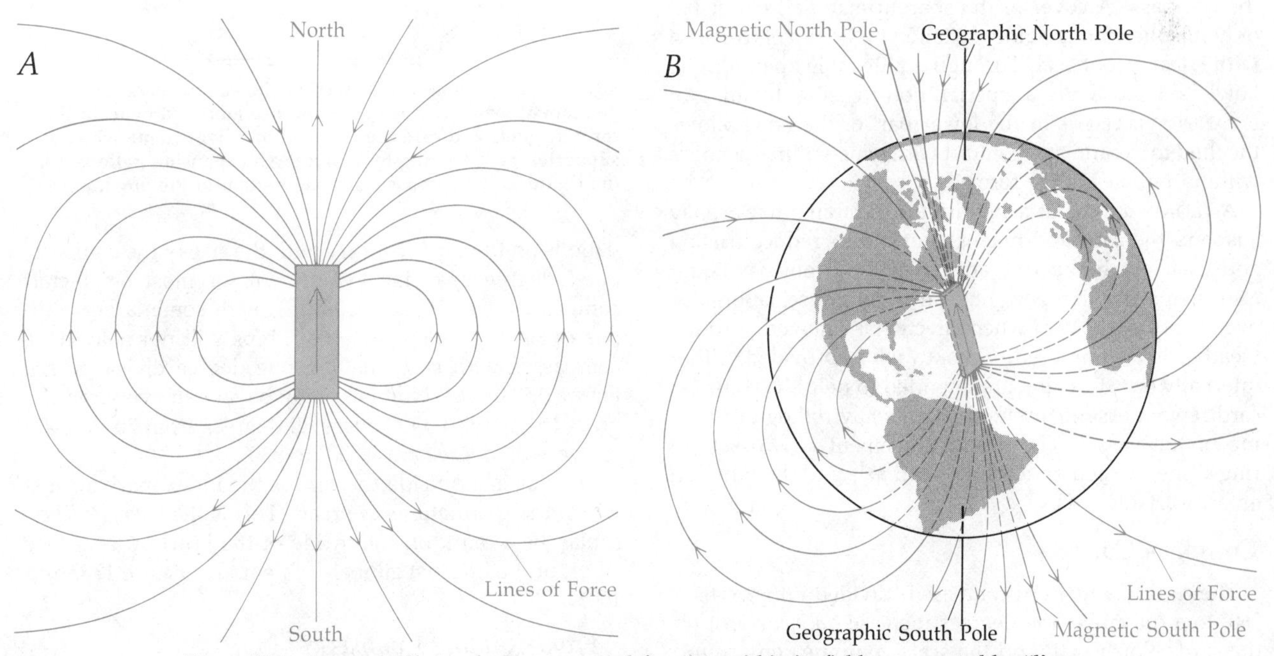

Figure 1501 **External effects of a bar magnet, *left,* exists within its field, represented by "lines of force." The magnetic properties of the earth can be visualized as a powerful magnet, *right,* at the earth's center, but not aligned with the earth's geographic axis.**

Basic Law of Magnetism

Each magnet, regardless of size, has two POLES where the magnetic action appears to be concentrated. These poles have opposite characteristics and are termed NORTH and SOUTH; see Figure 1501*A*. The basic law of magnetism is very simple: Opposites attract; likes repel. Thus an N pole is attracted to an S pole, but repels another N.

The Earth

The earth as a whole has magnetic properties and can be thought of as having a powerful bar magnet near its center, Figure 1501*B*. The magnetic properties of the earth are not uniformly distributed, and as a consequence its magnetic poles are *not* at the locations of geographic poles. The magnetic pole in the northern hemisphere is named the NORTH MAGNETIC POLE because of its location several hundred miles from the North Geographic pole. The SOUTH MAGNETIC POLE is the one located more than 800 miles from the South Geographic Pole.

The Basic Compass Principle

The basic principle of a magnetic compass is simple—a small bar magnet freely suspended in the earth's magnetic field will align itself parallel to the lines of force of that field, and thus establish a DIRECTION. The end of a magnet that points generally north is termed its NORTH POLE although it is actually the "north-seeking" pole; the other end is its SOUTH POLE. Normally, the compass will not point to geographic north, nor even provide the correct direction to the North magnetic pole. It is, however, a reliable and consistent direction, and for the purposes of navigation, it can be considered as being relatively constant over a period of several years.

The navigator usually has at hand information and rapid means for converting the reading of his compass to true geographic directions. He knows that if any local influences from iron or steel objects and electrical wires can be neutralized (or measured and recorded), his compass will be a dependable magnetic direction finder, needing neither mechanical nor electrical power.

Compass Construction

A typical compass has more than one bar magnet from which to derive its force to align with the earth's magnetic field. Usually, several long, thin permanent magnets are attached, parallel to each other, to a light non-magnetic wire frame; like poles of the magnets are all in the same direction. A light, round non-magnetic disc is mounted on the frame; this is the CARD of the compass and is marked around its circumference with graduations from which the direction can be read.

Centrally placed under the frame is a bearing which rides on a hard sharp point called the PIVOT; this in turn is supported from the outer case, which is the BOWL of the compass. A cover of transparent material, either flat or hemispherical, is rigidly fastened to the top of the bowl with a leak-proof seal. Through a pluggable opening, the bowl is filled with a special non-freezing liquid. An EXPANSION BELLOWS in the lower part of the bowl allows the fluid to expand and contract with temperatures changes without bubbles being formed.

A FLOAT may be attached under the frame to partially suspend the card and magnets, and thus reduce friction and wear on the bearing and pivot. The compass liquid also helps to dampen out any rapid oscillations or overswings of the card when direction is changed or rough weather is encountered. GIMBALS when provided, either internally or externally, are intended to help the compass card remain essentially level despite any rolling or pitching of the vessel. The best arrangement is two sets of rings, one pivoted on a fore-and-aft axis and the other on an athwartship axis.

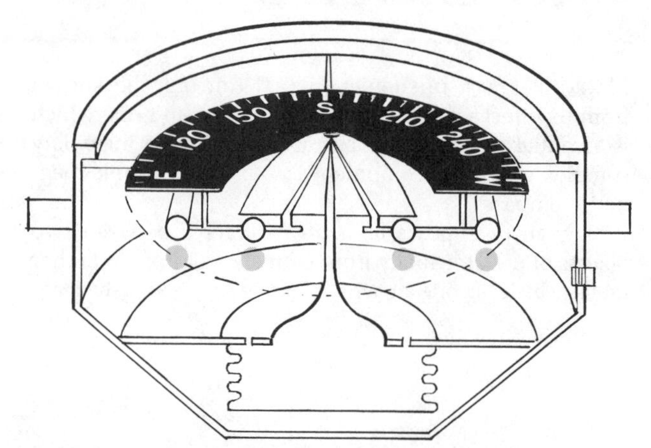

Figure 1502 **Compass cross-section shows four parallel permanent magnets (seen here end-on) mounted beneath the card; magnets and card are attached to a light frame which is supported on a bearing by a pivot. An expansion bellows for the liquid in the compass is at the bottom of the bowl.**

Compass Cards

The card of a modern compass is divided into DEGREES, 360° to a full circle; the degrees increase *clockwise* around the card. North is 0° and the scale continues on around through 90° (East), 180° (South), and 270° (West) back to North, which is 360° or 0°. The spacing between graduations on the card will vary with the size of the compass, but 5-degree spacings are normal for most small-craft compasses, (see Figure 1503 *left*), with some larger compasses having 2-degree cards. Ships will normally have compasses with cards having one degree divisions; see Figure 1503 *right*. Numbers may be shown every ten degrees from 10° to 350°; every 20 degrees from 20° to 340°; or every 30 degrees from 30° to 330°.

Over or just outside the card is the index mark against which the graduations are read. This is the LUBBER'S LINE; it may be a mark on the inside of the bowl or it may be a part of the internal gimbals system. See Figure 1502 and 1505.

Front-Reading Compasses

Some smaller and medium-size compasses are designed for mounting with direct reading from the front rather

Figure 1503 **Compass cards are graduated in various manners; those for most boats at five-degree intervals, *left,* with the ten-degree marks heavier for ease of reading. Numbers usually are at 30-degree intervals, and cardinal headings may be shown as letters rather than numbers. *Right:* Compasses for larger vessels have cards of larger diameter, sometimes with graduations down to one degree and with cardinal and intercardinal points shown. *Center:* When the compass is designed for front viewing, the lubber's line is on the side nearest the helmsman. The readings increase to the left, and a skipper accustomed to a conventional card must exercise care to avoid confusion when reading values between the numbered marks.**

than from above or nearly so. Here the lubber's line is nearer the helmsman rather than on the far side of the card. The card is graduated *counterclockwise,* and the greater readings are to the left of the index line rather than to the right. This fact can be confusing to a skipper accustomed to a more conventional flat-card compass, and must always be kept in mind when determining an exact value between the numbered marks on the card.

The Point System

In olden days, compass cards were subdivided into POINTS—32 points to a complete circle, one point equal to 11¼ degrees. These were named, not numbered, and a seaman early learned to "box the compass" by naming all the points. There were also half and quarter-points when a smaller subdivision was needed.

The CARDINAL points, North, East, South, and West—and the INTERCARDINAL points, Northeast, Southeast, Southwest, and Northwest—are still in common use as rough directions and as descriptions of wind direction. The COMBINATION points, such as NNE, ENE, and so on, are sometimes used for general directions, but the BY-POINTS, NxE (North by East), NNExN, NNExE, and their like, are rarely used.

The point system is now otherwise obsolete. For those who do not wish to forget the past, or want to learn points for ease in steering sailboats, a table of conversions between the point and degree systems is provided as Table 15-1.

Some modern compass cards label the intercardinal points as well as the cardinal headings, Figure 1503 *right,* and a few mark all points and quarter-points, Figure 1504 *top.* The examples and problems that follow, however, will all be in terms of degrees.

Conversion of Points and Degrees

Direction	Points	Degrees
North	N	0°00′
to	N by E	11°15′
East	NNE	22°30′
	NE by N	33°45′
	NE	45°00′
	NE by E	56°15′
	ENE	67°30′
	E by N	78°45′
East	E	90°00′
to	E by S	101°15′
South	ESE	112°30′
	SE by E	123°45′
	SE	135°00′
	SE by S	146°15′
	SSE	157°30′
	S by E	168°45′
South	S	180°00′
to	S by W	191°15′
West	SSW	202°30′
	SW by S	213°45′
	SW	225°00′
	SW by W	236°15′
	WSW	247°30′
	W by S	258°45′
West	W	270°00′
to	W by N	281°15′
North	WNW	292°30′
	NW by W	303°45′
	NW	315°00′
	NW by N	326°15′
	NNW	337°30′
	N by W	348°45′
	N	360°00′

Table 15-1

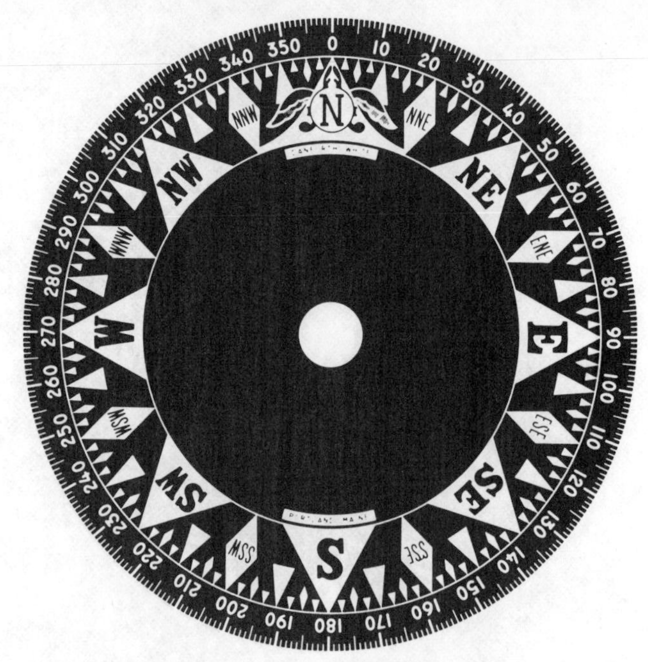

Figure 1504 **Some cards show all the points, and may be subdivided down to quarter-points, *above.* This is a large card that is also subdivided to one-degree intervals. *Below:* This bulkhead-mounted sailboat compass also indicates angle of heel on the scale below the bowl. Note the top-reading scale that is 180 degrees different so that it can be read against a forward lubber's line.**

Spherical Compasses

When the top over a compass is hemispherical, the dome serves to magnify and make the card appear much larger than it actually is. This aids in reading it closely, making it possible to steer a better course or take a better bearing. On some compasses with a hemispherical cover, the card itself is concave or dish-shaped rather than flat. This feature, combined with the shape of the dome, makes it possible to read the compass from a lower angle and at considerably greater distances; it is not necessary to stand more or less directly over the compass to read it. A compass with such a card of five-inch (12.7 cm) "apparent diameter" (actually somewhat smaller) can be read from ten feet (3m) or more away.

If the compass as a whole is spherical, bowl as well as dome, there is a considerable gain in stability of the card. The effect of a whole sphere is to permit the fluid inside to remain relatively undisturbed by roll, pitch, or yaw; the result is superior performance in rough seas.

Some compasses have special features for use on sailboats. These include more extensive gimbaling so as to permit free movement of the card at considerable angles of heel, and additional lubber's lines 45 degrees and 90 degrees to either side of the one aligned with the craft's keel as an aid in determining when to tack. These additional lubber's lines are also useful when the helmsman sits to one side, as when sailing with a tiller.

Binnacles

Compasses are sometimes mounted directly on a horizontal surface. More often, however, the compass is

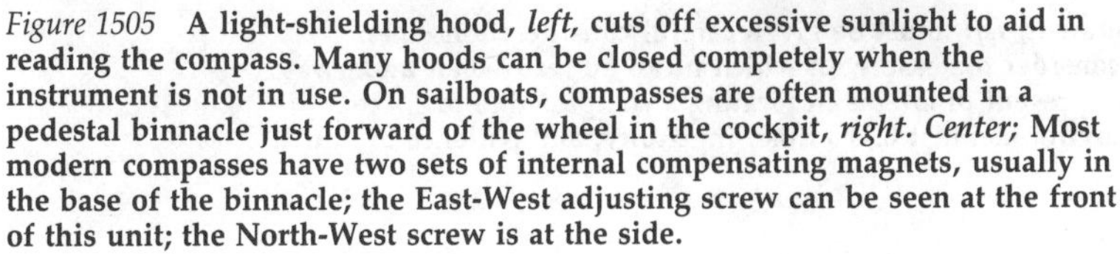

Figure 1505 **A light-shielding hood, *left,* cuts off excessive sunlight to aid in reading the compass. Many hoods can be closed completely when the instrument is not in use. On sailboats, compasses are often mounted in a pedestal binnacle just forward of the wheel in the cockpit, *right. Center;* Most modern compasses have two sets of internal compensating magnets, usually in the base of the binnacle; the East-West adjusting screw can be seen at the front of this unit; the North-West screw is at the side.**

mounted in an outer supporting case called a BINNACLE. This can be a simple case, sometimes with a light-shielding hood, which is then fastened down, Figure 1505 *left*. Or it can be a complete pedestal stand as might be used on larger sailboats, Figure 1505 *right*.

The outside of the binnacle or compass case, or any mounting bracket, should be black, preferably with a dull finish. There should be no shiny chromed parts that could reflect sunlight into the eyes of the helmsman.

Lighting

When compasses are lighted for night use, the lighting system is often a part of the compass case, cover, or binnacle. The light should be red in color and its intensity at a suitable level. If the light is not red, or is too bright, there is risk of loss of the helmsman's night vision; this will make it more difficult, or even impossible, for him to see other vessels, aids to navigation, or other objects. On the other hand, a too-dimly lit compass can be a source of eye-strain and a detriment to good steering.

Internal Compensators

As will be detailed later in this chapter, it usually is *not* possible to have a situation where there are no magnetic effects from iron and steel objects on the boat. The compass must then be COMPENSATED (ADJUSTED) with small additional bar magnets located and polarized so they counter-balance these unwanted magnetic influences. Such COMPENSATORS may be external to the compass and binnacle, but are more likely to be internal, usually a part of the binnacle unit; see Figure 1505 *center*.

Selection

Although a compass is a relatively simple mechanism in terms of its construction and operation, it is probably the most important navigational tool on any boat. Considerable thought should be given to its selection. This is not the place for a hasty decision or for saving money by buying a small, cheap unit. In fog or night, or any other form of reduced visibility, your compass is the basic instrument that shows direction—*it has to be right!*

Quality

Almost any new compass looks fine in the store, or aboard a boat in the quiet motion of a marina slip or at a mooring. But its behavior underway, when the sea makes up and the the craft pitches, rolls, and yaws, is of supreme importance. Will the card stick at some angle of heel? Will its apparent motion be jerky, or smooth and easy? Are the card graduations legible and are different headings easily distinguished? Is the compass going to be subjected to large temperature changes? What if a bubble appears under its glass which might distract the helmsman? The answers to these questions will depend upon the quality of the compass. Select a compass fully ade-

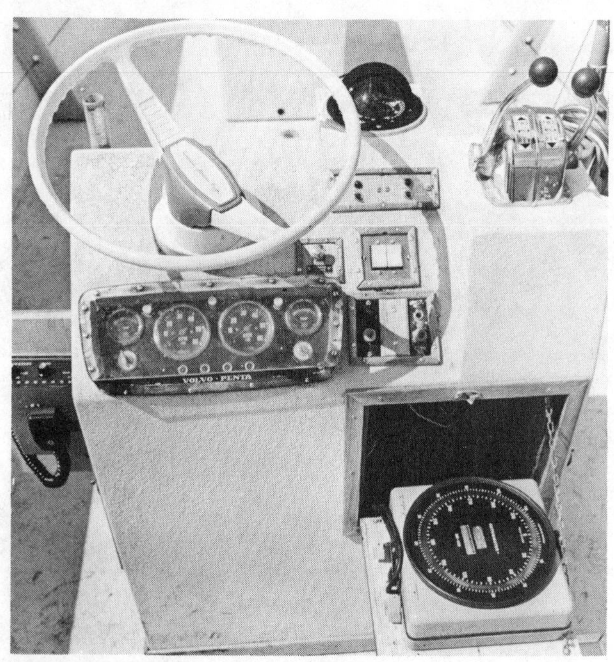

Figure 1506 **The magnetic environment, *left,* must be given careful consideration when selecting compass location. Also consider the ease with which it can be read when underway. Smaller boats, *right,* nearly always present problems in locating a compass due to space limitations and a poor magnetic environment. It takes time, ingenuity, and patience to find a suitable location for the instrument.**

quate for your expected needs, and if a choice must be made, choose the better quality. Examine a number of them before buying. Pick them up; tilt and turn them, simulating motions to which they would be subject in actual use afloat. The card should have a smooth stable reaction, coming to rest without oscillating about the lubber's line. Reasonable tilting, comparable to the rolling and pitching of your boat, should not appreciably affect the reading. In fairness to the compass, however, if it has internal correctors, they must be zeroed-in (to be covered later) before you make these tests.

Size

All too often, a small compass will be selected and installed on a small boat for the simple reason that the boat is just that, small. This is *not* the correct approach. Because the boat is small, it will be subjected to greater movement, both in normal operation and in rough waters. A small, cheap compass will have less stability and may even be completely unusable in severe conditions, perhaps just when it is most needed.

A compass for a boat should be as large as the physical limitations of mounting space sensibly allow—and the quality should be just as high as the owner's budget can possibly allow.

Card Selection

In selecting a compass for your boat, pay particular attention to the design of the card. Its graduations should be suited to the intended use. That a large craft may be held on course more easily and precisely than a small one is readily understandable. Thus the larger the vessel, the greater the number of divisions normally found around the outer edge of the card. Skippers of large ocean-going craft seem to prefer cards that can be read directly to two degrees, or even to single degrees. Studies have shown, however, that subdivisions smaller than five degrees may not be desirable. Cards with finer subdivisions may produce more eye-strain, require a higher intensity of illumination, and may result in greater steering action although no better course is made good. The experiences and preferences of the individual must govern here.

Choosing a Sailboat Compass Card

The helmsman of a sailboat is usually in an unprotected cockpit exposed to the weather. Spray and rain on the glass cover of the binnacle may obscure or distort the markings on the compass card. The smaller and more uniform the markings on the card, the easier it is for the helmsman to confuse one marking for another, lose his place on the card, and wander off course. This is especially true at night or offshore when he has no landmarks to warn him that he has been steering, for example, 055° instead of his intended course of 065°. And when he discovers his error, he may not know for how long he has been off course.

If the pattern on the card shows a large and prominent patch of contrasting color at the cardinal points and an even larger one at "North," even an inexperienced helmsman can steer with surprising accuracy by relying on the general position of this distinctive pattern as his guide, vague and blurry though it may be.

Testing

Although not always true, price is generally a fair indicator of quality. There are, however, simple tests that you can make yourself to aid in judging between different units. When making any of these tests, make sure that any other compasses are at least three or four feet (0.9-1.2 m) distant.

Test for minimum pivot friction (it should be zero) by turning the compass as a whole until a card graduation mark is exactly aligned with the lubber's line, then bring a small magnet or another compass or a steel object just close enough to the compass to cause the card to deflect two to five degrees to either side. Next, take it well away from the compass quickly, watching the card as you do so. The card should return to its former position *exactly*. This test is completed by drawing the card off to the other side—by moving the magnet, other compass, or steel object to the other side of the compass. Again the card should return to its initial position when the influence of the external magnet or steel object is removed. (Make sure that the object is moved far enough away each time for a fair test; turn a small magnet end-for-end—the card should not be affected.) Sticky pivots are rare, but they do occur and the test is too simple and easy to make to omit doing it. Do not purchase a compass that does not pass this test.

Repeat the above procedure, but this time draw the card off to each side about 20 to 30 degrees. Release the external influence abruptly and observe the amount of over-swing as the card returns back to, and past, its original position. A compass with proper dampening will have a minimum over-swing and will return to its proper position with a minimum of oscillation about that position.

This is relative; there are not absolute values to use as a standard. Thus these tests should be made on a comparative basis between several different compasses of the same model and between different models. Be fair, be careful—try to see that comparable tests are made on each compass.

Summary of Selection Criteria

1. The card should be easily read and graduated to your preference.
2. The card should remain level and not stick through reasonable angles of pitch and roll.
3. The card should move but slightly during any course change simulated by slowly and smoothly rotating the compass through 90 degrees or more. (Some small motion due to the inertia of the card is permissible, but the card should be dead-beat, swinging but once to a steady position, not oscillating about the lubber's line.)
4. The compass or its binnacle should have built-in correcting magnets.
5. The compass or binnacle should include provision for night lighting. (A means of adjusting the intensity of the light is desirable.)
6. The dome should be hemispherically shaped, rather than flat.
7. The compass should have a fully internally gimballed card and lubber's line.
8. The compass should have an expansion bellows in the assembly to prevent bubbles forming in the liquid due to temperature changes.
9. The compass or binnacle should be waterproof and have some form of light-controling hood.
10. Buy a quality instrument and pay the price. This is no place to cut corners; you get what you pay for.

Installation

The installation of a compass, from initial site selection through the actual mounting, is of critical importance to its performance. If any step is omitted or is not done carefully, it may never be possible to get the compass properly adjusted for precise navigation. Follow each step in detail and don't rush the work.

Locating the Compass

The first step is selecting the location in which to mount the compass. There are a number of factors to be considered, and some limitations that must be faced. Actually, you should consider where to place the unit *before* you buy the compass.

On Larger Craft

Ideally, the compass should be directly in front of the helmsman, placed so that he can read it without physical stress whether as he sits or stands. Give some thought to comfort in rough weather, and in conditions of poor visibility, day or night. His position is fairly well determined by the wheel or tiller. The compass has to be brought into what might be called his "zone of comfort." Too far away, he bends forward to watch it; too close, he bends backward for better vision. Much of the time he may be not only the helmsman but the forward and after lookout as well. So put the compass where he can bring his eyes to it for reading with a minimum of body movement. For the average person, a distance of 22 to 30 inches (56–76 cm) from his eyes with the head tilted forward not more than 20 degrees is about right. Additional details will be found in ABYC Safety Standard S-17.

On Smaller Boats

Locating a compass on a small boat may be more of a problem. It must, of course, be located where it can be easily seen by the helmsman, preferably directly before him. Such a location may not be physically possible, or it may be subject to undesirable magnetic influences that will cause DEVIATION in the compass readings (see next section). It may be necessary to locate the compass to one side of the helm position—on the centerline is common for smaller boats. If so, check carefully for PARALLAX ERROR—the change in *apparent reading* when the compass is viewed from one side as compared with directly behind. A slight difference in reading, perhaps up to five degrees, may be accepted, but more than that, which can easily occur, will result in defective and perhaps dangerous navigation. If you have this mounting problem, check several models of compasses, as some designs will have less parallax error than others.

Check for Magnetic Influences

With the best location selected from the viewpoint of visibility and use, now check for undesirable magnetic influences. The site should be at least two feet (0.6 m) from engine instruments, radios, bilge vapor indicators, electric gauges and instruments, and any iron or steel. (Stainless steel is nearly always non-magnetic, but there are many different alloys and each item of such metal should be checked.) Six or more feet (1.8 m) would be better than two, but almost always there must be compromises on boats. When one or more of these magnetic influences is too close, either it or the compass must be relocated. Vertical magnetic influence is to be especially avoided.

Test magnetic materials or influences that may be concealed from ready view. Make your tests with the compass itself. If the compass has built-in correctors, be sure before testing that they have been ZEROED-IN (see below). If this has not been done, motion of the case can make the card move, nullifying the test procedure.

Move the compass slowly and smoothly all around the proposed location without changing its orientation with the boat's centerline. Watch the card; one thing only will make it turn: a magnetic influence—find this with the compass. If the influence cannot be moved away or replaced with non-magnetic material, test to determine whether it is merely magnetic metal, a random piece of iron or steel, or a magnetized object. Successively bring the North and South poles of the compass near it. Both poles will be attracted if it is unmagnetized. If it attracts one pole and repels the other, the object is magnetized. Demagnetization may be attempted; see page 357.

Metal objects that move in normal operation should be moved. Turn the steering wheel; work the throttle, move the gear shift; open and close the windshield if it is near the compass. Try to duplicate all the changes that can occur in normal operation of the boat.

Check for Electrical Influences

Electrical currents flowing in wires near the proposed location can also exert undesired influences on the compass. Hold or tape down the compass at the proposed location, then test every circuit that might affect the compass. Switch on and off *all* the electrical loads, radio, bilge pump, depth sounder, lights, windshield wipers, etc.; don't overlook anything controlled by a switch on the panel or located near the proposed compass site. Start the engine so that its electrical instruments will be operating. If there is an auxiliary light plant, start it.

Make a Complete Check

In summary, check everything, *one item at a time,* that might influence the compass at the intended location, carefully watching the card as items are switched on and off, moved back and forth, started and stopped. When the card moves during any of these tests, the compass has been affected. Ideally, the compass should be relocated or the cause of the influence removed or demagnetized. In actual practice, however, you may have to settle for two separate states of magnetic environment; for example, with and without the windshield wipers working.

Vibration

Vibration can make a compass completely unusable at an otherwise desirable location. It can actually make a card slowly but continuously spin. Even if vibration did not cause the card to spin, it could result in excessive pivot wear. Mount the compass temporarily but firmly in the planned location and watch for vibration at all ranges of engine rpm and boat speed; vibration is unpredictable and may occur only at certain engine speeds.

Zeroing-In

A modern compass having internal compensators must be ZEROED-IN before being mounted on a boat. (If there is any doubt that a compass already mounted has been zeroed-in, it may be demounted and taken ashore for this purpose.) Zeroing-in is nothing more than adjusting the built-in compensators so that they have no effect on the compass. A compass that is subject only to the earth's field has no need for compensation; there are no deviations to remove. Thus before being installed any deviation caused by the improper position of the compensators themselves must be removed so that the unit can go aboard ready for whatever magnetic influences the boat may have. Setting the compensators by aligning screw slots with marks on the case or housing may not be sufficient for accurate navigation. Zeroing-in by trial-and-error is simple and effective; it does not require any knowledge of the direction of magnetic north.

Figure 1507 **A board with parallel edges and a large book are used to accomplish the zeroing-in process. A non-magnetic screwdriver or thin coin is used to turn adjusting screws.**

How to Do It

Zeroing-in must be done off the boat and in an area well away from any unknown magnetic influence such as iron or steel pipes and girders, radios, electric motors, refrigerators. Don't work wearing a steel belt buckle or watchband; keep all tools well away from the work area.

Using non-magnetic screws or tacks (check them with a small magnet to be sure) mount the compass temporarily on a small board with two exactly parallel edges. Line up the lubber's line with these parallel edges as closely as possible, but precise alignment is not mandatory. Place a large book and the compass on its small board on a level, flat surface—a plank will do fine, but not a table which may have screws or braces of magnetic materials. Move the board into close contact with the side of the book; see Figure 1507. Don't work on a concrete surface that may have internal reinforcing rods made of steel. Turn the book and board as a unit until the compass reads North exactly. Hold the book steady in this position; it serves as a fixed direction reference. Remove the compass and board, turn it end-for-end, and bring the opposite side of the board snugly against the book; the lubber's line has now been exactly reversed. If the compass now reads South exactly, the N-S internal compensator is already zeroed-in and requires no adjustment.

If, however, the reading is not exactly South, an adjustment is needed; read the difference from 180° and record it. With a *non-magnetic* screwdriver (a thin coin, such as a dime, will do for most compasses) slowly turn the N-S screw until *half* this difference is removed. If the difference increases when you first turn the screw, then turn it in the opposite direction until the desired result is obtained. Now slightly realign the book and compass board combination as a unit until the compass reads South exactly. Hold the book firmly and reverse the board.

The compass should now read North. If it does not, the difference will be much smaller than the initial difference at South. If there is a difference, again halve it by turning the N-S adjusting screw. Realign the book and compass, always keeping the board snugly against the book, until the reading is North exactly. Once again reverse the board and check to see that the compass reads South. If it does not read South exactly, again halve the difference. Continue this procedure until the reversal yields zero difference, or as nearly zero as possible. When this process has been completed, the North-South axis of the compass coincides with the magnetic meridian.

Next adjust the East-West compensators in the same way. Line up on either an East or West heading, reverse, and if the resultant heading is not exactly the opposite, then remove the *half* the difference by turning the E-W screw in the correct direction as determined by trial-and-error. Continue working on East and West headings just as on North and South until the desired results are obtained.

If exact reversals are not attainable, move the zeroing-in operations to a different location; some unseen magnetic influence may be present. If the compass manufacturer provided instructions for zeroing-in, they should be followed.

Do not discard the temporary mounting board. It may be useful in testing a possible permanent site on the boat.

Mounting

Once the compass has been zeroed-in and the prospective location checked for undesirable magnetic influences, you can proceed to the actual compass mounting. Certain basic requirements must be met.

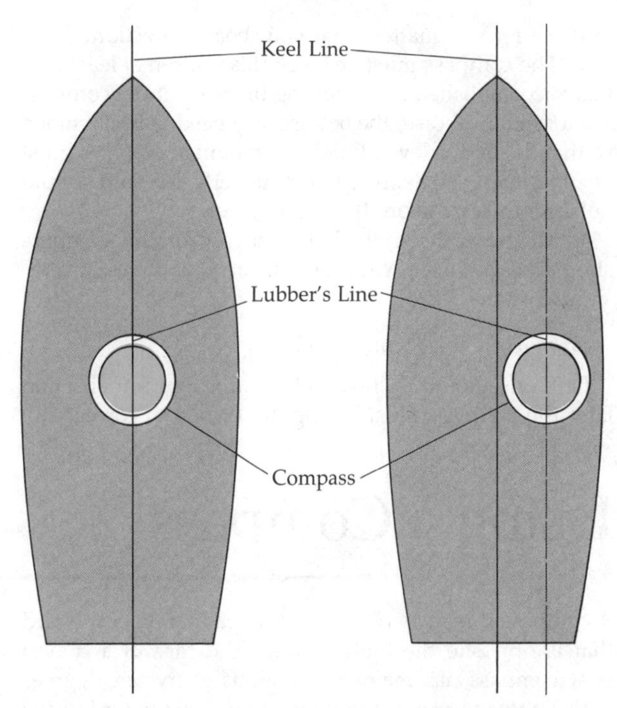

Figure 1508 **If the compass is not mounted directly over the keel line, take care to make sure that a line through the center of the card and the lubber's line is directly parallel to the centerline of the boat.**

1. The compass must be securely fastened to the boat.
2. A line from the center of the card through the lubber's line must be exactly parallel to the keel, the centerline of the boat, Figure 1508. (Ideally, the compass would be directly *over* the centerline. This usually is not practical, but *it must be kept at least parallel to* the centerline.)

Establishing this line parallel to your boat's centerline is quite simple, but it almost always takes a little more patience and time than you anticipate. Find the center of the transom using a metal measuring tape, and accurately mark that center on a piece of masking tape (so as not to permanently mark the boat). This is usually easy. Next you need a second center point at some convenient location forward of the compass position.

With help from some additional pairs of hands, stretch a string tightly between the two center points. Accurately measure out (athwartship) to the location you selected for your compass, and after marking off this distance at your transom and forward line, move the string to this new pair of points. If the string is higher than the area where the compass base will be, use a plumb bob to transfer the line. Of course the boat must be on an even keel if a plumb bob is used.

Once you have checked to make sure you have accurately established your compass mounting line parallel to the boat's centerline, you might just want to scratch or mark a short permanent line for future reference. You will install your compass so the lubber's line is exactly lined up with it. Many compasses have small alignment marks (forward and aft) where the housing touches the mounting surface, to aid in positioning on the scratch marks.

This alignment is extremely important, because an improperly aligned compass can never be properly adjusted for deviation, and will have a constant error (in addition to deviation) no matter what your boat's direction.

3. The compass must be level; this is often at least partially accomplished by mounting brackets; if the compass has a cylindrical case, the bottom may need to be shimmed so that the unit is level. Bulkhead-mount compasses must be mounted at 90 degrees to the lubber's line, and should not lean forward more than 15 degrees.

4. Magnetic material in the vicinity of the compass should be minimal. Wires near it, carrying direct current, should be twisted.

Fastening the Compass Down

Drill one mounting hole only. If during compensation later the compass must be slightly turned to correct any misalignment of the lubber's line, more than one hole could present problems. When compensation has been accomplished, the remaining hole(s) can be drilled and the fastening job completed; a check of the compensation should be repeated after this is done.

If Your Boat Already Has a Compass

If you have a boat and it already has a compass installed, or if you are buying a boat that comes equipped with a compass, you can still benefit from the above information because all too many factory- and owner-installed compasses have lubber's lines nowhere near parallel to the centerline.

Using a Compass

A compass is read by observing the graduation of the card that is opposite the lubber's line. The card of a typical boat compass has major graduations every ten degrees with lighter marks at intermediate five-degree intervals; see Figure 1503. On some models the width of the lubber's line is just about one degree. Thus it is possible to use this width to estimate a reading of one degree more or less than each card marking fairly accurately. A reading of slightly less or more than midway between the card markings gives the other individual degree values; see Figure 1509.

Always remember: when a boat turns, it is the boat, and the lubber's line, that turns—the compass card maintains the same orientation.

Figure 1509 **On some compasses, the width of the lubber's line equals one degree on the card. The *center* drawing shows how this can be used to read one degree to either side of a main graduation. As shown at *right*, a reading slightly more or less than half-way between marks would indicate two degrees more or less than the nearest mark.**

Using an Uncompensated Compass

A compass that has not been compensated—had deviation removed or else had a deviation table prepared—cannot be used to navigate with a chart, but there are simple and practical uses for it.

An uncalibrated compass may not measure direction correctly relative to magnetic north, but it does indicate directions in a stable, *repeatable* way if local conditions on the boat are unchanged. You can proceed from one point to another—for example, from one channel marker to the next—read the compass when you are right on course, and record that value. If you make the same run on another day, the reading will be the same, *provided* you haven't changed any magnetic influences near the compass. Thus if the day comes when it is foggy, or night catches you still out on the water, you can follow with confidence the compass heading that you recorded in good visibility.

It's a good idea to record compass headings on all the runs you regularly make, particularly a route you would follow during period of reduced visibility. Run the route in clear weather and record the compass headings. One word of caution, however: run the reverse course for *each* heading. You *cannot* add or subtract 180° from the forward run to get a reading for the return; residual deviation may give an erroneous reading.

Keep your compass course records handy and check the various headings frequently. If there is a change, look for the reason and eliminate it.

Using a compass for steering a boat and taking bearings will be covered in later chapters of this book.

Protection When Not in Use

The worst enemy of a compass is direct sunlight. When a boat sits day in and day out in a marina slip, something can and should be done about excess sunlight.

When a compass is not in use, it should be thoroughly shielded from the sun to prevent discoloration of the liquid, and perhaps of the card. Sometimes a binnacle will have a light-shielding hood; if so, keep it closed whenever the compass is not in use. If such a shield is not a part of your compass, improvise one of lightproof material; do not use cloth—even the tighest weave is not adequate.

Any bubbles that form, usually from cold temperatures, can be removed by adding liquid. This must be of the same type that is already in the compass to avoid further problems. This is a job that can only be done by the manufacturer's customer service department or by a professional compass repairman in his shop.

Compass Errors

Before a compass can be used for accurate navigation, you must understand COMPASS ERRORS—the differences between compass readings and true directions measured from geographic North. Normally, when working with your charts, you record and plot courses and bearings with respect to true North. Some experienced boatmen work in terms of magnetic directions.

Variation

The first, and most basic, compass error is VARIATION (V or VAR). For any given location on land or sea, this is the angle between the magnetic meridian and the geographic meridian; see Figure 1510. It is the angle between true north and north as indicated by a compass which is free from any nearby influences. Variation is designated as EAST or WEST in accordance with the way the compass needle is deflected. Any statement of variation, except zero, must have one of these labels.

Variation is a compass error about which a navigator can do nothing but recognize its existence and make allowances for it. Variation is the compass error that would exist on a boat entirely free from any on-board magnetic materials or other internal magnetic influences.

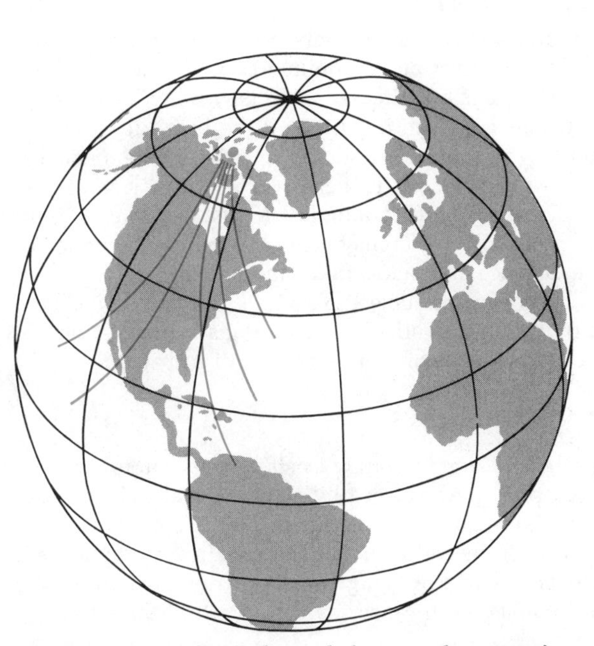

Figure 1510 **Variation is the angle between the magnetic meridian and the true meridian, and it can be either easterly or westerly. Variation depends upon geographic location, and in any part of the globe, above, it is the same for all vessels in any given vicinity. In the United States, *right,* it is generally easterly on the Pacific Coast and Gulf Coast, and westerly on the Atlantic Coast.**

General Variation

The earth's magnetic field is not uniform, and because the magnetic poles are not at the geographic poles (see Figure 1501), *variation changes with location;* see Figure 1510. Figure 1511 *top* illustrates the result of failure to apply different variation as a vessel moves from one location to another.

At any given place, the amount of variation is essentially constant. There is usually a small annual change, but if you are using a chart not more than two or three years old, this quite small amount of change can be neglected. The amount of variation, and its annual rate of change, is found within each compass rose on the chart; see Figure 1512.

On nearly all types of charts published by the National Ocean Service there will be found several COMPASS ROSES suitably placed for convenient plotting. These are circular figures several inches (perhaps 5 to 10 cm) in diameter with three concentric circular scales; see Figure 1512. The outer scale is graduated in degrees and it is oriented so that its zero point, indicated by a star symbol, points to true (geographic) north. The inner pair of scales is graduated in degrees and in the point system down to quarter-

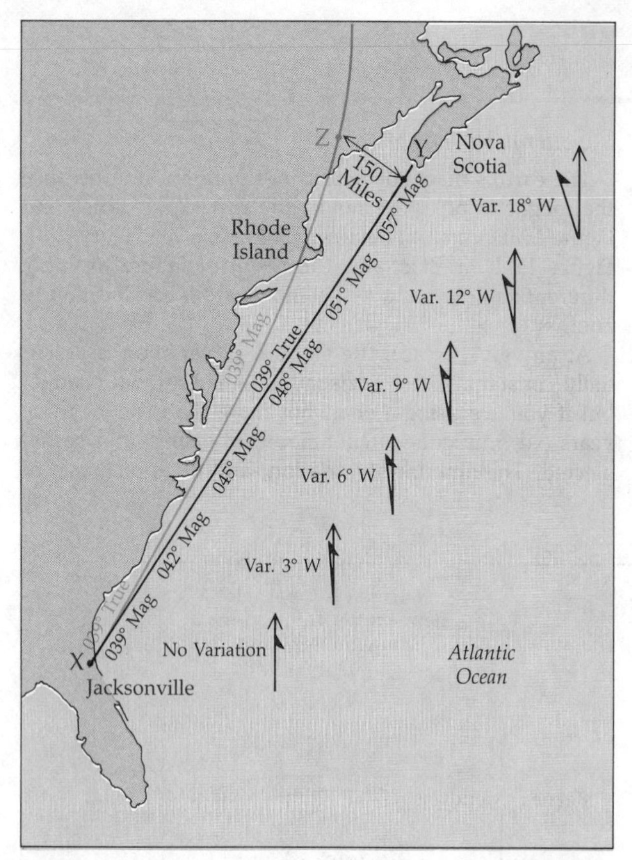

Figure 1511 **Do not overlook the change in variation with the change in location. Here the true and magnetic courses from X to Y are both 039° initially, but magnetic soon changes while the true course remains constant. By staying on the magnetic heading, an aviator would be 150 miles northwest of his goal, at point Z; the skipper of a boat, however, would run aground on the Rhode Island coast.**

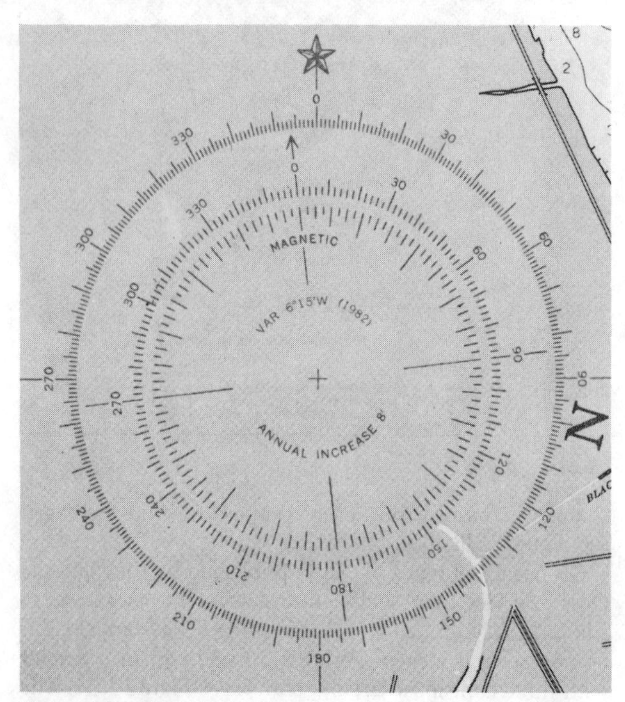

Figure 1512 **Most charts have several compass roses which show the variation graphically and descriptively. These are located so as to be convenient for plotting. On any chart, especially on small-scale charts of large areas, always use the compass rose nearest the vicinity concerned.**

points. The zero point of these scales, identified by an arrow symbol, shows the direction of the northern half of the magnetic meridian at that point. On some charts, the innermost scale in points is omitted.

The angle at the center of a compass rose between the star and arrow symbols is the variation *in that vicinity* for the year stated. This information is printed at the center of the rose to the nearest 15′; for example, "Var 6° 15′ W (1982)." The annual rate of change is noted to the nearest 1′, and whether the variation is increasing or decreasing; for example, "Annual Increase 2′ " or "Annual Decrease 6′ " or "No Annual Change" (see Figure 1512).

As noted above, magnetic variation changes with location. Thus, except for charts of a limited area such as a single harbor and its approaches, variation will most likely be different in various areas of a typical chart. For this reason, the variation graphically indicated by the several compass roses on any particular chart will be different. It is important *always* to refer to the compass rose *nearest your position* on the chart to read variation or to plot from its scales.

Local Attraction

In addition to the general overall magnetic variation, the navigator may also encounter LOCAL ATTRACTION. (Here "local" means a limited geographic vicinity, as opposed to magnetic influences on board the boat.) In a few localities, the compass is subject to irregular magnetic disturbances in the earth's field over relatively small areas. A striking example is found at Kingston, Ontario, Canada, where variation may change as much as 45 degrees in a distance of a mile and a half (2.4 km).

Charts of areas subject to such local attraction will bear warnings to this effect. Other means of navigation are needed when within range of such disturbances.

Deviation

A boat's compass rarely exists in an environment that is completely free of nearby magnetic materials or influences. Normally it is subject to magnetic forces in addition to those of the earth's field. Material already magnetized, or even capable of being magnetized by the magnets of the compass, will cause the compass to deviate from its

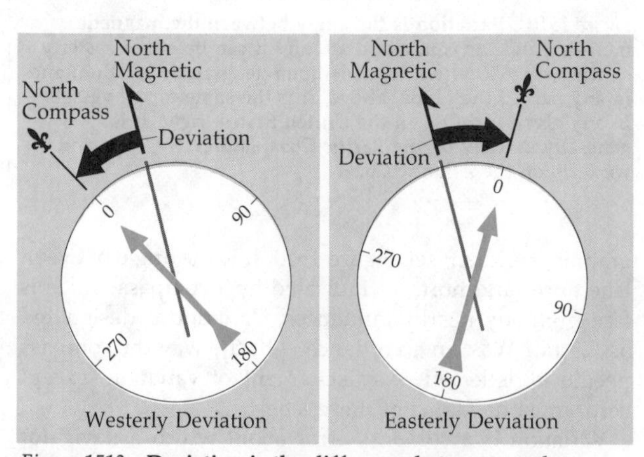

Figure 1513 **Deviation is the difference between north as indicated by the compass, and magnetic variation. It can be easterly or westerly, and depends on the magnetic conditions on a vessel. It changes with boat heading, and is not affected by the vessel's geographic location.**

ideal position along the magnetic meridian. Currents flowing in improperly installed electrical wiring can have the same effect.

The deflection of the compass from its magnetic orientation is called DEVIATION (D or DEV). It is the angle between the magnetic meridian and a line from the pivot through the north point of the compass card; the angle between the direction the compass would point if there were no deviating influences and the direction in which it actually does point. Theoretically, it can range from 0° to 180°, but in practice large values cannot be tolerated. Just as for variation, deviation can be EAST or WEST depending on whether compass north lies to the east or west of magnetic north. Deviation *must* carry one of these labels, unless it is zero. See Figure 1513.

While variation changes with geographic location, *deviation changes with the craft's heading* and does not change noticeably in any given geographic area. (Causes for this effect will be explained later, see page 357.) To cope with these changes a vessel needs a DEVIATION TABLE—a compilation of deviations, usually for each 15 degrees of heading by the compass; Table 15-2. For added convenience, a deviation table can also be prepared in terms of magnetic headings, Table 15-3. (Note: In these examples, the values of deviation are large for purposes of illustration. They are typical of an uncompensated compass.)

Deviation Determination by Compass Headings

Boat's Compass Heading	Range Bearing by Compass	Magnetic Range Bearing	Deviation
000°	082°	087°	5°E
015°	086°	087°	1°E
030°	091°	087°	4°W
045°	096°	087°	9°W
060°	100°	087°	13°W
075°	104°	087°	17°W
090°	106°	087°	19°W
105°	106°	087°	19°W
120°	104°	087°	17°W
135°	101°	087°	14°W
150°	097°	087°	10°W
165°	093°	087°	6°W
180°	089°	087°	2°W
195°	085°	087°	2°E
210°	082°	087°	5°E
225°	079°	087°	8°E
240°	076°	087°	11°E
255°	073°	087°	14°E
270°	070°	087°	17°E
285°	069°	087°	18°E
300°	070°	087°	17°E
315°	072°	087°	15°E
330°	075°	087°	12°E
345°	078°	087°	9°E

Table 15-2 **The navigator takes a bearing on a range for each 15 degrees of heading. By comparison with the known magnetic direction, he determines the deviation of his compass for each heading. (The actual values should be much smaller than those shown here for illustration.)**

Deviation Determination by Magnetic Headings

Boat's Magnetic Heading	Deviation
000°	7°E
015°	1°E
030°	6°W
045°	12°W
060°	17°W
075°	19°W
090°	18°W
105°	17°W
120°	14°W
135°	11°W
150°	8°W
165°	5°W
180°	2°W
195°	2°E
210°	4°E
225°	6°E
240°	9°E
255°	11°E
270°	14°E
285°	16°E
300°	17°E
315°	17°E
330°	15°E
345°	11°E

Table 15-3 **Deviation in terms of magnetic headings, rather than compass headings, is needed to determine the compass course to be steered. Here the values have been calculated and tabulated for each 15-degree magnetic heading.**

Using Deviation Tables

For ordinary navigation, it is normally sufficient to use the value of deviation from the line of the table that is nearest to the craft's actual heading, either compass or magnetic. This is particularly true for compensated compasses where the deviations will rarely exceed a few degress. (Typically, a small craft cannot be held on course closer than two or three degrees at best in moderate or rough waters.)

For more precise navigation, such as in contests and perhaps some races, it will be necessary to *interpolate* between tabular entries to get a more precise value of deviation; see page 359 for interpolation procedures.

Changes in Deviation

Note that any deviation table is valid *only* on the boat for which it was prepared, and *only* for the magnetic conditions prevailing at the time the table was prepared. If magnetic materials within the radius of effect—some three to five feet (0.9-1.2 m)—are added, taken away, or relocated, a new deviation table must be prepared, or as a minimum, sufficient checks must be made of the old table to ensure that its values are still correct. If any electrical or electronic equipment is installed, removed, or relocated, check to see if a new table is needed.

Compass Calculations

A heading or course is the angle which the centerline of the boat, or a line on a chart, makes with some other line of reference. Three lines of reference have been established: the direction of true north or the true meridian; the direction of the magnetic meridian; and the direction of the north point of the compass. Thus *there are three ways to name a course or heading:* TRUE, MAGNETIC, or COMPASS. Figure 1514 *left* illustrates the three reference lines and the three measurements of direction for one heading. (The amount of variation and deviation in this Figure is exaggerated for greater clarity.) The above remarks also apply to the measurement of directions for bearings.

The Application of Variation and Deviation

To be able to use your compass fully, you must develop an ability to CONVERT directions of one type to the other two quickly and accurately. These directions include headings, courses, and bearings. You use compass courses in steering your boat; you get compass bearings when you take a reading across your compass to a distant aid to navigation or landmark. You normally plot true directions on your chart and record them in your log. There are also magnetic directions—the intermediate steps in going from compass to true—which are also sometimes directly useful in themselves. Hence, there is need for continual interconversion between the three systems of naming direction.

Single-Step Calculations

Although most compass calculations involve the double conversion from true to magnetic to compass, and vice versa, it is best to consider single-step computations first.

The term CORRECT and CORRECTING are used in easily remembered phrases to help determine whether to add or subtract deviation and variation. This is done by considering true directions as being more "correct" than magnetic, and magnetic as being more correct than compass directions. This can be rationalized and easily remembered by observing that true has no errors, magnetic has one error (variation), and compass includes two errors (variation and deviation). Going from magnetic to true—or compass to magnetic—can thus be termed correcting. Conversely, the term UNCORRECTING is used when going from true to magnetic, or from magnetic to compass.

The basic rule is: *When correcting, add easterly errors,* which can be shortened to *Correcting add east* and memorized as C-A-E.

EXAMPLE 1

The magnetic course is 061°; the variation is 11°E.

Required The true course, TC.

The conversion is one of correcting and the variation is easterly, so the basic rule is used, *Correcting add east.*

$$061° + 11° = 072°$$

Answer TC 072°; see Figure 1514 *right.*

The basic rule is easily altered for application to the three other operations of conversion. Remember to *always change two, but only two, words* in the phrase for any application; see the following examples.

EXAMPLE 2

The magnetic course is 068°; the variation is 14° West.

Required The true course, TC.

The conversion is correcting and the variation is westerly. In the basic rule, EAST has been changed to WEST;

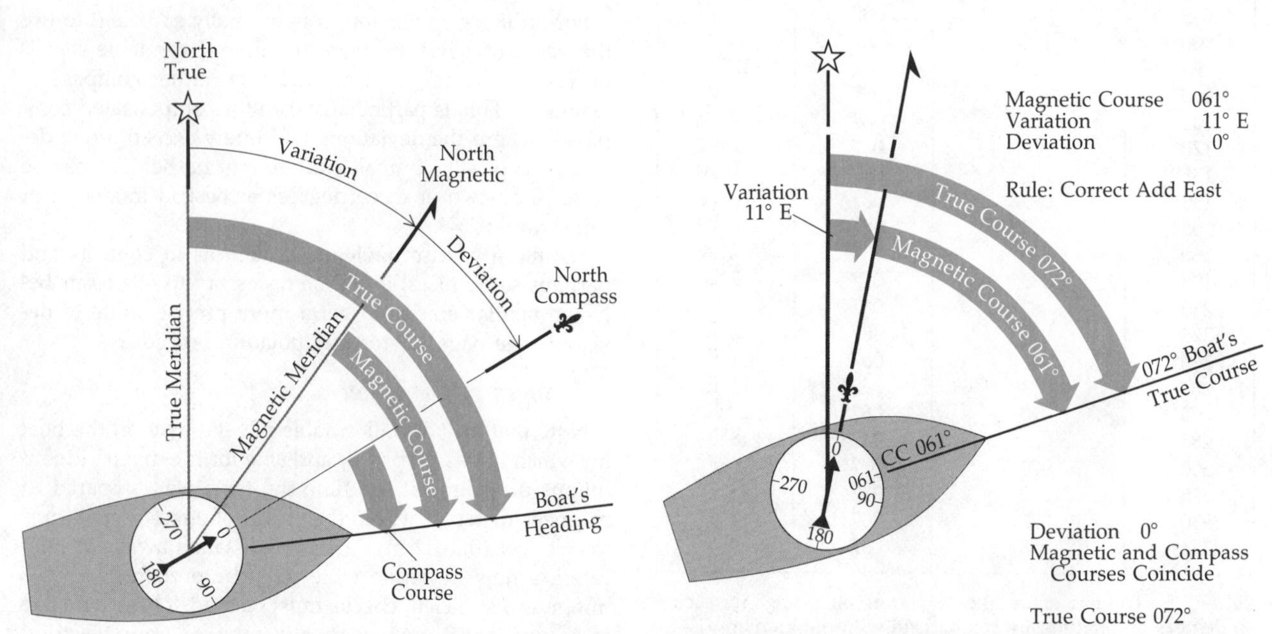

Figure 1514 ***Left:*** **A course, heading, or bearing can be named in any one of three sytems based on the reference direction used. It is, of course, the same line any way it is named. When "correcting" a magnetic course with easterly variation to a true course, add the variation,** ***right.***

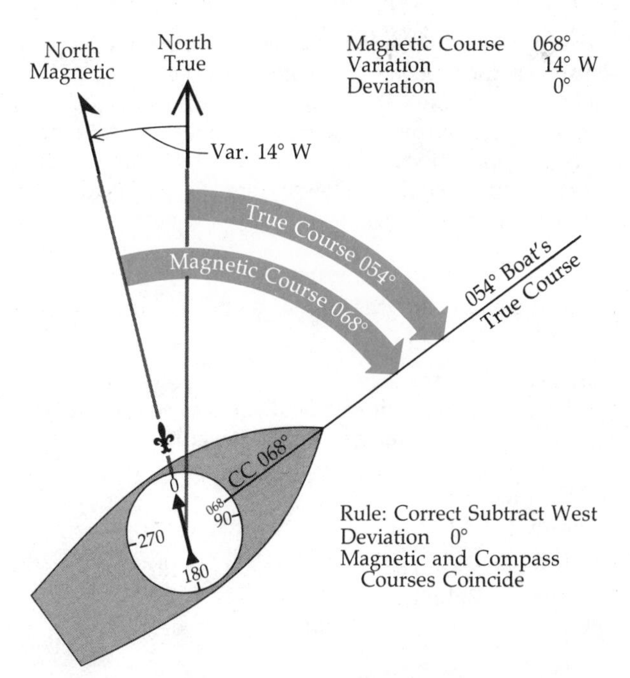

Figure 1515 **When "correcting" a magnetic course with westerly variation to a true course, subtract the variation.**

one word has been changed, another must be. Since it is correcting, "add" becomes "subtract." The rule to be applied here is thus: *Correcting subtract west.*

068° − 14° = 054°

Answer TC 054°; see Figure 1515.

EXAMPLE 3

The course of the boat by its compass is 212°; the deviation for that heading is 5°E.

Required The magnetic course, MC.

The conversion is correcting and the deviation is easterly; the applicable rule is *Correcting add east.*

212° + 5° = 217°

Answer MC 217°

Now consider the reverse process, uncorrecting. We again change two, but only two, words to the basic rule and we get: *Uncorrecting add west* and *Uncorrecting subtract east.* (Don't worry about remembering all four rules. It's easier to remember the basic rule of *C-A-E* and how to change it to fit the other situations.)

EXAMPLE 4

The true course is 351° and the variation is 12°W.

Required The magnetic course, MC.

The conversion is uncorrecting and the variation is westerly. The applicable rule is *Uncorrecting add west.*

351° + 12° = 363° = 003°

Answer MC 003°

EXAMPLE 5

The magnetic course is 172° and the deviation for that heading is 7° E.

Required The compass course, CC.

The conversion is uncorrecting and the deviation is easterly. The applicable rule is *Uncorrecting subtract* east.

172° − 7° = 165°

Answer CC 165°.

Two-Step Conversions

The same rules apply when two-step conversions are made from true to compass, or from compass to true. It is important to remember that the proper rule must be used for *each step separately.* It will always be correcting or uncorrecting for both steps, *but* the addition or subtraction may not be the same as this is determined individually by the east or west nature of the variation and deviation.

EXAMPLE 6

The true course is 088°; the variation is 18°W and the deviation is 12°W.

Required The compass course, CC.

Both conversions are uncorrecting and both errors are westerly. The rule for both calculations in *Uncorrecting add west.*

088° + 18° + 12° = 118°

Answer CC 118°; see Figure 1516.

EXAMPLE 7

The compass bearing is 107°; the variation is 6°E and the deviation for the heading that the boat is on is 2°W.

Required The true bearing, TB.

Both conversions are correcting; the variation is easterly but the deviation is westerly, so two rules are required: *Correcting add east, subtract west.*

107° + 6° − 2° = 111°

Answer TB 111°.

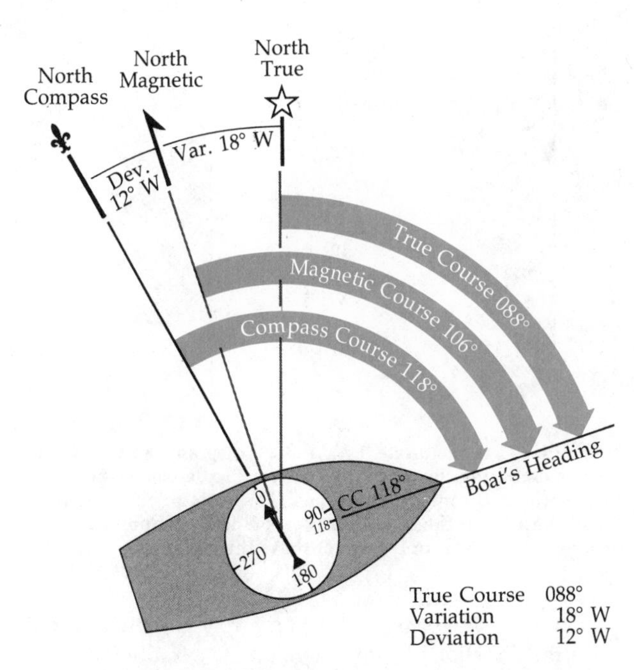

Figure 1516 **Here a true course is "uncorrected" to a compass course, with variation and deviation both westerly.**

Other Rules

The basic rule of *Correcting add east* and its variations is only one set of memory-aids for compass calculations: there are many more. You need to know only one; study several and pick the one that is easiest for you to remember and apply.

Another that is easily memorized and used is *Compass least, error east; Compass best, error west.* This can be used for conversions such as those shown in the examples above, but it is of particular value when the difference between compass and magnetic is known numerically and it must be decided whether the deviation is east or west. For example: if the magnetic course is known to be 192° but the compass reads 190°, what is the deviation? The numerical difference is 2° and the compass is "least"—190° is less than 192°—so the error is east; deviation is 2°E. If on the same course, the compass had read 195°, the compass would then have been "best"—195° being more (better) than 192°—and the deviation would be 3°W.

When considering variation as the difference between true and magnetic, apply the same phrase but with magnetic substituted for compass—the rhyme is not as good, but the principle is the same. For example: if a true bearing is 310° and the variation is 4° west, then the magnetic bearing will be "best" (more than the true value) or 314°. If the variation had been easterly, then magnetic would have been "least" or 4° less, for 306°.

Visual Aids

Many boatmen find a pictorial device a better memory-aid than a phrase or sentence. Letters are arranged vertically, representing the three ways of naming a direction—true, magnetic, and compass—with the respective differences—variation and deviation—properly placed between them. The arrangement can be memorized in several ways, such as *True Virtue Makes Dull Company* or in the reverse direction as *Can Dead Men Vote Twice.*

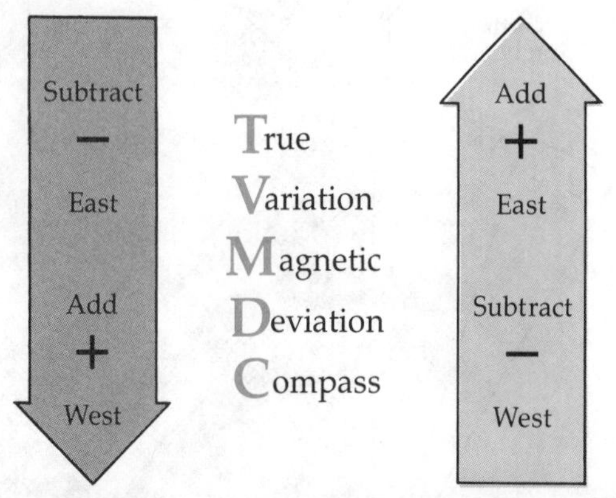

Figure 1517 **A pictorial "T V M D C" diagram can be an aid in remembering the rules from converting between true, magnetic, and compass directions. The basic rule is "Down Add West," but the other rules can be derived from this by changing two, but only two, of these three words.**

On each side of the letters, an arrow is drawn to indicate the direction of conversion; on the left, down for true to compass, and on the right, up for compass to true. Also on each side, the appropriate arithmetic process is indicated for use with easterly or westerly errors. See Figure 1517. The right side indicates correcting; the left side illustrates uncorrecting. If one memorizes *D-A-W* for *Down add west,* then the correct signs can always be added to the letters and arrows. (The three other actions follow the same pattern as above—always change two, but only two, words. Thus it would be Down subtract east, Up add east, and Up subtract west.)

EXAMPLE 8A

From the chart, the true course is 107° and the variation in the locality is 5°E.

Required The magnetic course, MC.

The conversion is down the diagram; use the left side. The variation is easterly so it is to be subtracted.

$$107° - 5° = 102°$$

Answer MC 102°.

Down	T	107°
Subtract	V (−)	5°E
East	M	102°

EXAMPLE 8B

For this magnetic heading, the deviation aboard this boat is 4°W.

Required The compass course, CC.

The conversion is still downward on the diagram; continue to use the left side. The deviation is westerly so it must be added.

$$102° + 4° = 106°$$

Answer CC 106°.

Down	M	102°
Add	D (+)	4°W
West	C	106°

Conversions such as this could be done in a single two-step operation.

EXAMPLE 9

A boat is on compass course (CC) 193°. Its deviation on this heading is 6°E; the variation for the vicinity is 9°W.

Required The true course, TC.

The conversion is upward on the diagram; use the right side. Easterly deviation is to be added; westerly variation is to be subtracted.

T	190°	UP
V (−)	9°W	ADD
M	199°	EAST
D (+)	6°E	SUBTRACT
C	193	WEST

Answer TC 190°.

Always Trust Your Compass

If you must depend on your compass for navigation, make a quick check for any objects near it that could cause additional, unmeasured deviations; typical objects that get placed too close to a compass include knives, beer and soft drink cans (if not aluminum), small radios, some cameras, your flashlight, and tools of all sorts. If your magnetic situation has been checked and is all clear, then trust your compass to direct you to your destination—it is better than your instincts!

The Importance of Accuracy

Keep in mind the results of not running an accurate course. If your compass or compass calculations should be in error by only 5 degrees, you will be a full mile off course for every 11½ miles run (1 km for every 11½ km). This could be hazardous when making a landfall or a coastal run at night or in poor visibility. Also, running a narrow channel in poor visibility with an inaccurate compass would be dangerous. Note the following table:

Error in Course	*Number of feet off course after sailing one nautical mile*	*Miles sailed to be one mile off course*
1°	106	57.3
2°	212	28.6
3°	318	19.1
4°	424	14.3
5°	530	11.5
6°	635	9.6
7°	740	8.2
8°	846	7.2
9°	950	6.4
10°	1055	5.7

Determining Deviation

Deviation is the angle between the compass-card axis and the magnetic meridian, Figure 1513.

Aboard any boat the direction of compass north is easily learned by a quick glance at the compass card. Not so the direction of the magnetic meridian—magnetic north. Thus it is not possible to compare visually the angle between these two reference directions, to determine the deviation on that heading. But the difference between an observed compass bearing and the known correct magnetic value for that bearing yields that same angle.

Deviation by Bow Bearings

The bearing is sighted over the compass with the boat headed directly toward the object or range. The magnetic direction is taken from the chart. The distant target can be a range of two objects, or it can be a single object if the location of the boat can be fixed precisely, such as close aboard an aid to navigation with the object far enough away that any slight offset of the boat is immaterial.

The range technique is the more accurate and is done as follows: Select two accurately-charted, visible objects, preferably ashore or fixed to the bottom. From the chart, measure and record the magnetic direction from the nearer to the farther mark. (The best ranges are those in the *Light List*; their true direction is given precisely, just apply variation to get magnetic.) See Figure 1518; A and B are two visible objects with B visible behind A. The magnetic direction of this range is 075°. The boat is run straight toward A on the range, keeping A and B visibly in line. While steady on this heading, the compass is read and recorded. In this example it reads 060°; the amount of the difference is the deviation, 15 degrees. Reference to any of the conversion rules makes it clear that the deviation is east. You can set up other ranges and find the deviation on such headings by repeating the bow bearing procedure as above. Use the side of a long object (pier or large building) as a range, if the direction can be determined from the chart. Use a buoy and a fixed object, or even two buoys, if a range of fixed objects is unavailable. Remember that the charted position of a buoy is that of its anchor; wind or current, or both, may swing a buoy around its charted position. Large ships and tugs sometimes move buoys off-station inadvertently by collision in thick weather. Buoys thus give less accurate deviation table results; replace the data as soon as you can with a table made under better conditions.

Deviation by Courses

Steering a visual course from one aid to navigation to another and noting the compass heading will, if the observation is made *at the start of each leg*, give reasonably good values of deviation when compared with the magnetic direction from the chart; see Figure 1519. Sighting down a long straight channel defined by aids to navigation or banks close by on either side is another excellent means of getting a compass heading to compare with the chart.

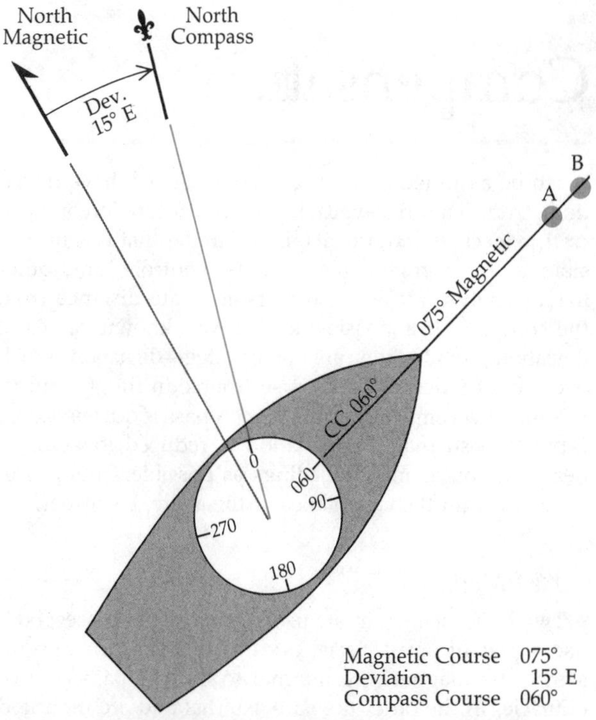

Figure 1518 **The simplest way to determine deviation on any heading is to locate two objects to use as a range whose magnetic direction can be established from the chart. Head the boat toward this range and read the compass; the difference, if any, is the deviation on that heading.**

Deviation Table by Swinging Ship

A precise technique for developing a deviation table is to SWING SHIP about a compass by running different headings across a range, recording the compass bearing of the range on each crossing. For example, two prominent charted objects are on a range with a magnetic bearing measuring 087°. On a calm day, the boat *Water Witch* sails across this range on compass headings successively 15 degrees apart. Using sight vanes and an azimuth circle mounted on the compass, her skipper takes bearing of the range for each crossing, and records and tabulates the results.

The skipper averages the various compass bearings and gets 086.9°. He notes that this is very close to the magnetic bearing of the range and decides he has a good set of data.

The first and last columns of this table now constitute a deviation table for *Water Witch* provided *no changes are made in the magnetic environment of the compass;* See Table 15-2. Note that the table's figures are much larger than would be normally tolerated in actual piloting. If deviations are this large, the compass needs compensating to reduce these deviations to zero, or nearly so. Note also that the table's deviations are in terms of *compass* headings. These are not, of course, the same as magnetic headings; when on a heading 045° *magnetic* the boat's bow points in a direction quite different from when she heads 045° by *compass*. This leads to a problem. The skipper determines from the chart his magnetic course; he then wants to know that deviation to apply to get the compass course he should steer. Use of Table 15-2 for that purpose requires time-consuming trial-and-error steps. To avoid this, he makes a second table listing the deviations in terms of *magnetic* headings; see Table 15-3. This can be done by trial-and-error method (but needs doing only once to make the table, rather than each time such deviation is needed) or the desired information can be taken from a Napier diagram (see page 359).

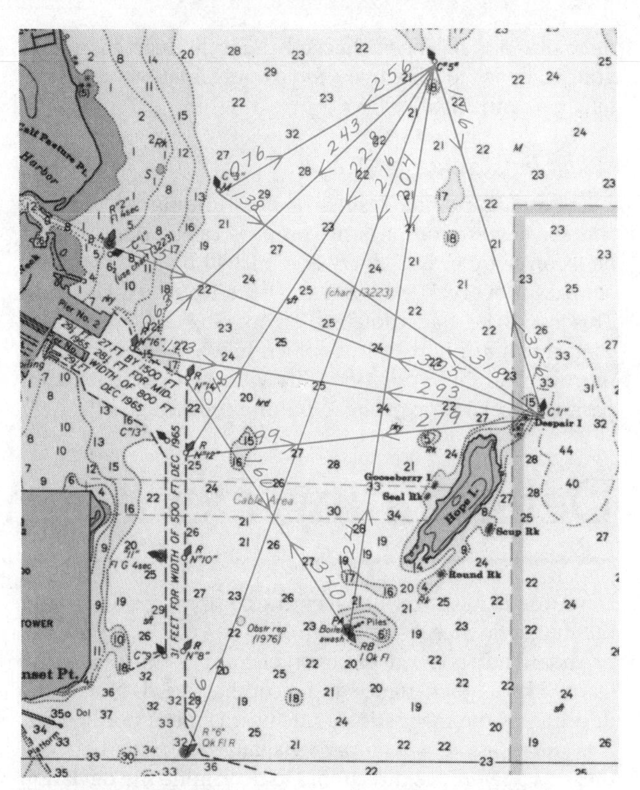

Figure 1519 **Carefully run courses between two charted positions can be used to determine deviation on a number of headings. Deviation will not be the same for reciprocal courses, and each must be run individually.**

Compensation

It can be assumed that a boat's compass will have deviation errors. The area available for a control station on most craft, especially motorboats, is so limited that it is impossible to have engine instruments, controls, and other magnetically undesirable objects at a safe distance from the compass. It is possible to live with known values of deviation provided they are not too large—deviations much over about 6 degrees can cause trouble in rough waters. It is, however, much better if the compass is COMPENSATED (ADJUSTED) so that the deviation is reduced to zero, or nearly so, on as many headings as possible. This procedure is within the capabilities of the average skipper.

Two Methods

Two basic techniques are used to adjust compasses; both use a pair of small COMPENSATING MAGNETS. In one instance the magnets are internal to the compass case or binnacle, in the other they are external and are mounted near the compass location. The objective of both methods is the same—to provide weak magnetic fields that cancel out the disturbing influences. Compensation consists of adjusting the effect of each of these magnets to achieve the counterbalances of their fields with the fields of the disturbing influences.

Using Internal Compensators

Most modern boat compasses now have internal compensating magnets. The two magnets, or pairs of magnets, are installed at right angles to each other. Usually there are two slotted screw heads at the edge of the compass; one marked *E-W*, and the other *N-S;* see Figure 1505.

Preliminary Steps

As discussed on page 342, the internal compensators should be zeroed-in before the compensation procedures are begun. Check also that the line from the center of the compass through the lubber's line is *exactly* parallel to your boat's centerline (see page 343). Select an object at least one-half mile distant. With your boat held absolutely motionless, sight on that object both over the compass and down the centerline.

Running Reciprocal Courses

The process of compensating a compass by running

reciprocal courses is much like that of zeroing-in. In that procedure, the book and board served to make certain that the lubber's line was turned exactly 180° at each reversal of the compass. It was as if the compass had been aboard a vessel which, on each reversal, sailed an exactly reciprocal magnetic course.

Zeroing-in ashore began by using an unknown magnetic direction or course, identified only as north (000°) by the compass. Zeroing-in was successful because this, or any unknown magnetic heading, could be accurately reversed. Afloat, compensation can be started in the same way. Afloat, the heart of the problem is achieving accurate reversal, putting and holding the boat on a course over the bottom which is exactly the reciprocal of the original unknown magnetic heading. This is not as difficult as might be expected.

Setting the Courses Between any two visible marks, or on a range, reversing the magnetic heading is obviously easy. In the procedures of compensation, however, there is need to run at least the four cardinal *magnetic* directions, East and West, North, and South. Seldom will you find in your local waters natural marks or aids to navigation on these exact magnetic headings. If you are fortunate, you may find special ranges on these headings which have been established for compass adjustment purposes. More likely, though, you must make your own temporary basic courses.

Departing from a fixed mark (a buoy will do, but a daybeacon or light is better), run on a steady compass heading until ready to reverse your course. Drop a disposable buoy (see below) creating the desired range. Next, execute a "buttonhook" turn (known technically as a Williamson turn), line up the disposable buoy with the object marking the original departure point, steady on this range visually, and *ignoring the compass,* head for the initial point. Run down the disposable buoy, and continue back toward the departure point; see Figure 1520. Make the turn as tightly as possible, keeping the buoy in sight, so as to complete the turn before the buoy has had time to drift from the spot which it was dropped. Choose a right or left turn, whichever can be made tighter if there is any difference; and swing enough to the *opposite* side before making the main turn so that when the main turn is completed, you can pick up the desired reciprocal course smoothly without overshooting and having to apply opposite rudder. This maneuver is not difficult to learn; all that is required is a few trials and then some practice. It has other uses, too; for example, it is the quickest and most accurate procedure for returning to the spot where a person or object has been lost overboard.

This maneuver has put the boat accurately on a reciprocal course, providing that the buoy remained where it was dropped and that no wind or current set the boat to either side of her heading on the *outward* run from the starting point. Compensation will be inaccurate if the boat has moved crabwise, headed one way but actually going another. The various courses of compass compensation procedures must therefore be run when wind and water conditions produce no leeway.

The skipper who hesitates to run directly over the disposable buoy should take it very close alongside, touching it if possible. On a range of ½ nautical mile, when the buoy is 10 feet (3 m) to one side of the boat's centerline, the course error will be only 11′, less than ⅕th of a degree.

Disposable Buoys Excellent disposable buoys can be made of cardboard milk cartons or plastic bleach bottles ballasted down with some sand or dirt, Figure 1521. Newspapers wadded into a ball about the size of a bas-

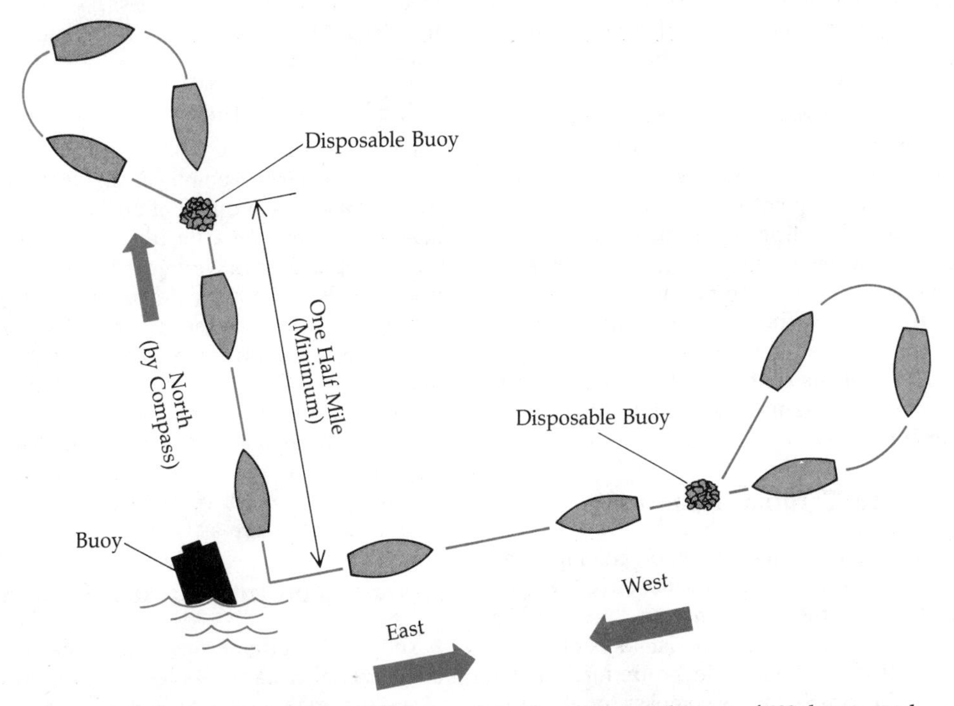

Figure 1520 **With a little practice, a skipper can learn to make smooth turns of 180 degrees and come back directly over the spot where the turn was started. Swing out far enough initially on the opposite side to the main turn so that the reciprocal course can be picked up without overshooting it.**

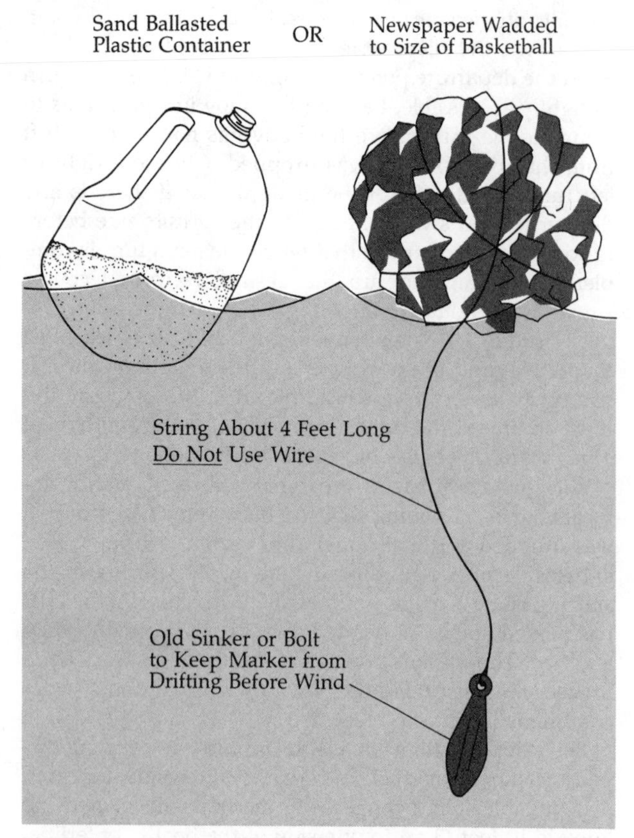

Figure 1521 **The spot where the turn was started becomes the near mark of a temporary range; identify this spot with a disposable buoy dropped off the stern just before starting the 180-degree turn toward the initial departure point.**

ketball and tied with string will work if they are weighted with a small heavy object such as a large bolt or old spark plug at the end of three or four feet (0.6-0.9 m) of string; wire or rope should not be used for such a buoy.

Other Preparations

Have a *non-magnetic screwdriver* available to adjust the screws of the compensating magnets. (Use a dime or a piece of heavy sheet copper or brass, or a brass machine screw filed down to a flat point.)

Store every magnetic article of the boat's gear in its regular place, put the windshield wiper blades in their normal "off" position, and keep no magnetic material near the compass. Steel eyeglass frames, steel partial dentures, or even the steel grommet in a yachting cap, if brought within a few inches of the compass, might cause deviation; someone may be working that close to the compass during adjustment.

Compensation on Cardinal Headings

As noted above, compensation must be accomplished when wind and current will not set the boat to either side of the intended course; the winds and seas must also be such that accurate steering is possible. First steer 090° (or 270°) by compass from the chosen departure mark, holding rigorously to this course by the compass for at least a half-mile; a mile would be better. It makes no difference whether the initial run is toward the east or west; this can be determined by the available water area and object used for a departure mark. Have someone drop a disposable buoy over the stern, execute the turn described above, and run the reciprocal course, Figure 1520. Run this course *not* by the compass, but by heading for the original departure point. Read the compass and record the figures.

If the compass reading is now 270° (or 090°), there is no deviation on east or west headings and compensation is not needed; the E-W internal compensator should not be touched.

If the compass reading for a return to the starting point is not 270° (or 090°), compensation is required. Exactly as in the process of zeroing-in, remove *half* the difference between 270° (or 090°) and the observed reading. Halving the difference is based on the assumption that the deviation of reciprocal courses is equal and opposite. This may not be exactly so, but it serves as a good initial approach.

Assume, for example, that the compass reading was 290°; then one-half of the 20-degree difference, or 10 degrees, is to be removed while the reciprocal course is carefully run. Turn the E-W adjusting screw, and watch the compass closely. If the compass reading increases rather than decreases, the screw has been turned in the wrong direction (there is no way to tell in advance which is the correct direction); turn the screw in the opposite direction. Move the screw slowly until the compass reading is 280° while the boat is exactly on the reciprocal course.

Return to the starting mark and make another E-W run. Head out again on a compass course of 090° (or 270°; the track through the water will *not* be the same as for the first run), drop a new disposable buoy, turn, and line up for the reciprocal course back to the point of departure. If the compass reading is not now 270° (or 090°)—it should be close—again remove half the difference by turning the screw of the compensating magnets.

Repeat these east-west runs until the deviation is eliminated or reduced to the smallest possible amount on these headings. Do not touch the E-W adjusting screw again after it is finally set.

North-South Adjustments

Follow a similar procedure on north-south headings. Run from a starting point—it can be the same as for E-W runs, or it can be a different one as required by the available water area—on a heading of 000° (or 180°) by compass, drop a disposable buoy, make the turn, and head back visually toward the departure point. Remove half of the error by adjusting the screw marked N-S. Repeat such runs as necessary just as was done for the east-west compensation.

When the north-south compensation has been completed, make a quick check for any east-west deviation; it should not have changed, but it doesn't hurt to be sure. The deviation on all cardinal headings should now be zero, or as near zero as is attainable in this boat.

Residuals on Intercardinal Headings

You can determine any deviations remaining on the intercardinal headings—NE, SE, SW, and NW—by any of the methods discussed earlier in this chapter; or, you can estimate them by making reciprocal runs on these headings. In this method, the deviation will be half the difference between the outward course steered by com-

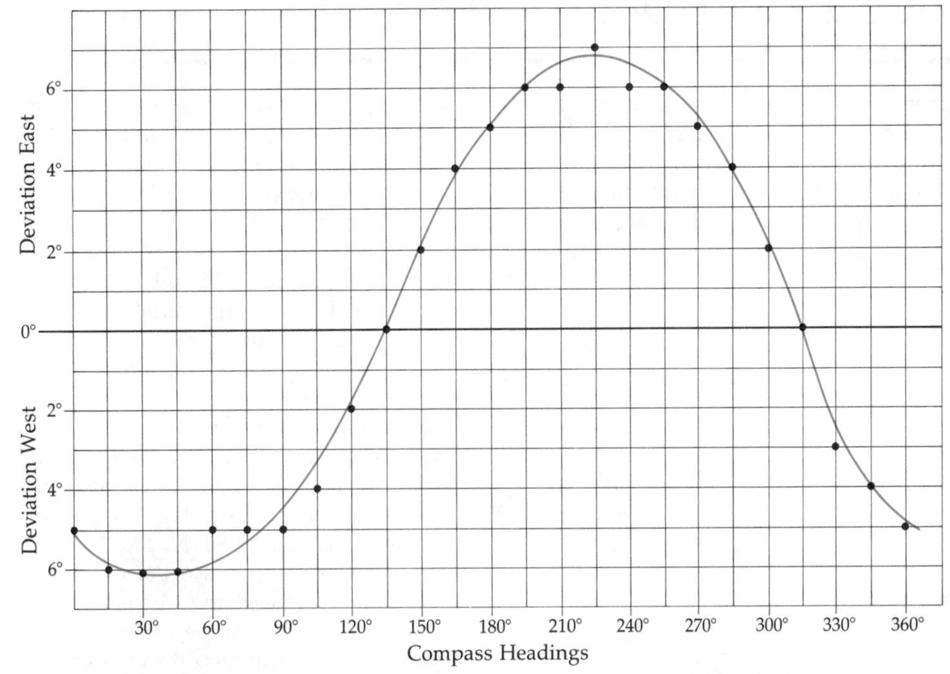

Figure 1522 **Before making a final deviation table, make a simple plot of the values to spot any that are not consistent with the others.**

pass and the reading of the compass when the turn has been made and the craft pointed back to the starting point. Be sure to note the direction of the deviation, east or west. *Do not touch the N-S or E-W adjustment screws.*

Failure to Achieve Compenstion

When compensation cannot be achieved satisfactorily, you undoubtedly have an undetected magnetic field on your boat. Prime suspects are the tachometer cable(s) and the steering mechanism. Test them with a thin piece of steel; a machinist's thickness gauge for a few thousands of an inch (0.05 to 0.10 mm) is ideal. Make sure that this thin piece of steel is not itself magnetic; bring it close to the compass; one end should exert a slight attraction on *both* N and S marks on the card; turning the gauge end-for-end at either N or S should not show a reversal of effect on the compass.

Touch one end of this thin piece of steel to the part being tested for magnetism and gently pull it away. If it tends to stick to the part, the latter is magnetized. Test all around thoroughly. Open the wheel housing or look under the instrument panel; test all metal parts found.

If a magnet is located, demagnetization will be required. You may need professional advice and assistance, but try your own hand at the job first, using the procedures on page 357.

The Final Deviation Table

When the compass has been adjusted to the greatest possible extent, make a complete deviation check and prepare a table for any deviations found; use any of the methods described earlier. Make this check under all conditions in which the compass might be used, check with the navigation lights on and off (they may be distant from the compass but wires running to their switch will be nearby), with the windshield wipers operating and not operating, and with the electronic depth sounder, radio direction finder, and radio each on and off. Look also for other electrical equipment whose operation could affect the compass. Remember to check on two cardinal headings 90 degrees apart. Most of this equipment will not disturb the compass, *but some could,* and you must be aware of it. If any checked items have to be on when the compass is in use, you may need to have more than one deviation table. You may also sometimes have to turn on certain equipment, whether needed or not, to recreate the magnetic environment that existed when the deviation table was prepared. A sailboat may need different deviation tables for port and starboard tacks because of different heeling errors. If two or more deviation tables are necessary, use different color paper for each.

Before preparing the final deviation table, plot a simple graph of the values to be used. Lay out a horizontal base line for compass headings from 0° to 360°; plot easterly deviation vertically above the base line and westerly deviations below that line at a suitable scale. When all values of deviation have been plotted, draw a *smooth* curve through the points, but don't expect all values to be exactly on that line since the deviations have been measured only to the nearest whole degree; see Figure 1522. You should be able to get a smooth curve through or near all of the points; if any value lies significantly off the curve, it is probably in error and should be rechecked. If you can draw a smooth curve, but its center axis does not coincide with the horizontal base line for 0° deviation, the lubber's line may be out of alignment; recheck carefully.

The final deviation table can be prepared as a direct-reading table of critical values; see page 359. It can be used with confidence if care is taken to see that the magnetic environment is not altered. Check it each year, even though you think that no changes have been made that would upset the deviation table.

Advanced Compass Topics

The following sections on various topics on magnetism and compasses are generally beyond the "need-to-know" of the average skipper. Some, however, will be of value to those with special problems, and many others will be of interest to those skippers who want to have a bit more than the minimum knowledge.

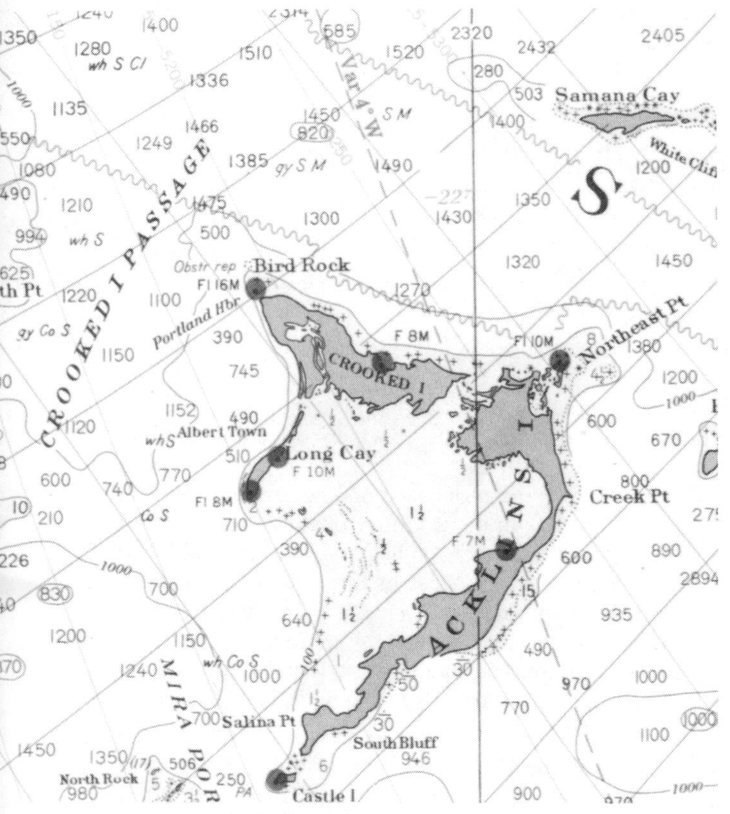

Figure 1523 **Charts of large areas show magnetic variation by a series of lines for each degree of variation. Each such *isogonic* line is labelled with its variation.**

Magnetism

Contrary to some popular ideas, the location of the earth's MAGNETIC POLES is of little interest to the navigator. (Should he ever get to their vicinity, he would then have to rely for directional information on some instrument other than his magnetic compass.)

Variation is *not* the angle between the direction to the true and magnetic poles. This concept, a fictitious visualization useful in helping students to realize that true and magnetic north are usually different directions, has been accepted by many as truth. This must be recognized as merely a learning aid, not a factual representation. The magnetic pole does not control the compass. The controling force is the earth's magnetic field. The compass magnet does not necessarily point in the precise direction of the magnetic pole.

Actually, two specific points on the earth's surface that are the magnetic poles do not exist. There are north and south MAGNETIC POLAR AREAS containing many apparent magnetic poles, places where a dip needle would stand vertically. If this seems strange, consider the problem of pinpointing the precise point in the end of a bar magnet that is the pole.

For scientific purposes, approximate positions of the theoretical North and South Magnetic Poles have been computed from a large number of continuing observations made over the world for many years. They are not diametrically opposite each other as are the geographic poles. The magnetic poles do not stay in the same place; they wander about with both short-term and long-term cycles. They are relatively distant from the geographic poles. Very roughly, the North Magnetic Pole is at 76° N, 100° W with the South Magnetic Pole at 66° S, 139° E. A magnetic compass at 80° N, 100° W would show north in a direction truly southward of the compass.

A magnetic meridian is generally not a segment of a great circle passing through the magnetic poles, since only one great circle can pass through any two surface points not at the ends of a diameter. Not so obvious, but just as true, the magnetic meridian is generally not part of a great circle from a given point to the adjacent magnetic pole.

Isogonic Lines

On smaller scale charts covering larger areas, variation may be shown by ISOGONIC LINES; every point on such a line has the same variation. Each line is labeled with the amount and direction of variation, the date, and the annual rate of change. The line joining all points having zero variation is called the AGONIC LINE. See Figure 1510, right.

Induced Magnetism

Masses of iron or related metals that do not show any magnetic properties under usual conditions will acquire INDUCED MAGNETISM when brought near a magnet and into its field. The polarity of an induced pole is *opposite* to that of the nearest pole of the magnet that caused it. When the inducing field is removed, any magnetism that remains is termed RESIDUAL. If the object retains magnetism for a long period of time without appreciable reduction in strength, this is PERMANENT MAGNETISM.

Magnetism From Electrical Wires

Although the copper of electrical wires is a non-magnetic material, currents flowing in them produce magnetic fields. It is this effect that makes possible electrical motors, generators, relays, electromagnets, etc. A direct current in a wire results in a field that can affect a vessel's compass. (Alternating-current electricity also produces magnetic fields, but not of the type to disturb a compass.) The polarity of a field around a wire carrying DC is related to the direction of the current, a fact that can be used to advantage. The two wires carrying power to any load have currents in the opposite directions, thus producing opposing fields. If these wires are twisted together, the field around one wire cancels out the field produced by the other, and the net effect on the compass will be *zero*, or nearly so. Always use two wires, don't use a common ground for return current, and always twist pairs of wires that pass near a compass. Don't fail, however, also to make an actual check by observing the compass closely as the electrical circuit is switched on and off; make this test on two cardinal headings 90 degrees apart.

Demagnetization

If tests with a thin strip of steel (see page 342) have shown that some metal object at the control station is magnetized, it *may* be possible to DEMAGNETIZE it with a device used by electronic repair shops such as a degaussing coil for TV sets or a magnetic-tape bulk eraser. If the magnetized object cannot be removed from the boat for this effort, it is *mandatory* that the compass be taken off lest the effectiveness of its magnets be destroyed.

The demagnetizing device is operated from 117-volt shore power; it uses alternating currents to produce magnetic fields that reverse at the 60 Hz frequency. Hold the device about a foot (0.3 m) from the target object and turn on the power; do not turn off the device until the full process has been carried out. Advance the device slowly toward the object until it touches; move it around slowly over all the accessible surfaces of the object. Then remove the device slowly away until it is at least the starting distance away and then turn it off. (If power to the demagnetizing device was interrupted at any time during the process, repeat all steps.) Take care not to demagnetize anything that is properly magnetic, such as electrical instruments or radio speakers.

Now test the disturbing object again with the thin strip of steel. If it has been possible to reach all of the object, it should be demagnetized. Often, however, complete access is not possible, and the object may have to be disassembled and/or removed from the boat for complete demagnetization.

After the demagnetization process is completed, remount the compass, and in *all* cases make an entirely new deviation table.

Why Deviation Depends on the Boat's Heading

Aboard a boat, the compass is subject to two magnetic forces, that of the earth and that of objects on the craft. If the boat were absolutely free of magnetism—no permanent magnets, no material subject to acquiring induced magnetism, and no fields from electrical currents—there would be no deviation on any heading, Figure 1524. The earth's force depends upon geographic location; the compasses of all vessels in the same harbor are subject to the same variation. The deviations aboard these ships and boats will be unalike because of the differences in the magnetic characteristics of each vessel. Thus if all of them put in identical magnetic headings, their compasses would differ, except in the remote circumstance that each compass was completely compensated.

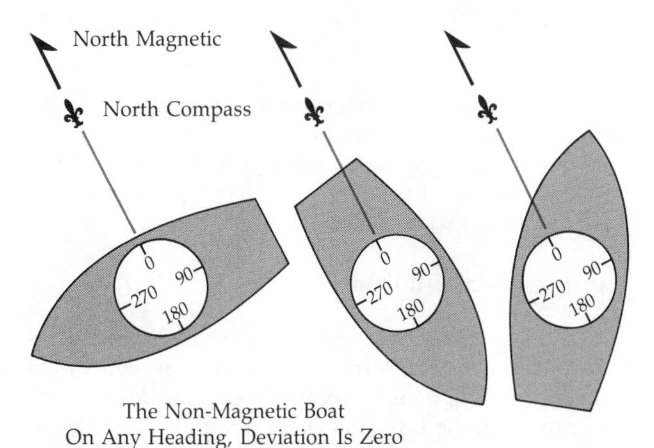

Figure 1524 **If there are no magnetic influences aboard a boat to disturb its compass, there is no deviation on any heading; however, this is rarely the case.**

DEVIATION, it will be remembered, *varies with the craft's heading* because of magnetic material on board. A few examples may make this clearer. Refer again to Figure 1524, the non-magnetic boat. No heading of this craft affects its compass card; its north point always lies in the magnetic meridian; the compass has no deviation. As the boat changes heading, the lubber's line, turning with the bow, indicates the magnetic heading on the card.

Now put aboard the craft, in the vicinity of the compass, items such as radios, electronic depth sounders, and electrical gauges and other instruments. Also on the boat, but usually farther from the compass, are metallic masses such as anchors, chain, fuel and water tanks, and other magnetic items, including the engine (the engine is by far the largest mass of magnetizable material but it is relatively distant from the compass). The boat has now acquired a magnetic character of its own; in addition to any permanent magnetism of some objects, there are now masses of unmagnetized material that can affect the compass by acquiring induced magnetism. The effects on the compass of permanent magnetism and induced magnetism are different and must be considered separately.

Effect of Permanent Magnetism

On typical boats of wood or fiberglass, the effect of permanent magnetism is by far the greater. To visualize this, let us assume that the same net effect would be exerted by a single permanent magnet located aft of the compass and slightly askew of the boat's centerline. This representation by an imaginary magnet is quite applicable; it simulates conditions found on most boats other than those with a steel hull. (The exact position of the imaginary magnet, its polarity, strength, and angle of skew are not important in this illustration of principles.)

Two forces, that of the field of the earth and that of the field of the imaginary magnet, now affect the compass; its poles are attracted or repelled in accordance with the basic law of magnetism. Without the influence of the imaginary magnet, the compass magnets and card will line up with the magnet meridian. The 0° mark of the card is an N (north-seeking) pole, and the 180° mark is at an S pole. Now, in addition to the effect of the earth's magnet poles on them, the N and S poles of the compass magnets will be attracted or repelled by the S pole of the imaginary magnet (the pole nearer the compass in these examples). The result is easterly deviation on some headings, westerly on others, and zero on those where the sign of the deviation changes from east to west, or from west to east.

Consider a typical boat, *Morning Star,* headed north (000°) MAGNETIC, Figure 1525 *left*. The S pole of the compass, 180° on the card, is repelled by the S pole of the imaginary

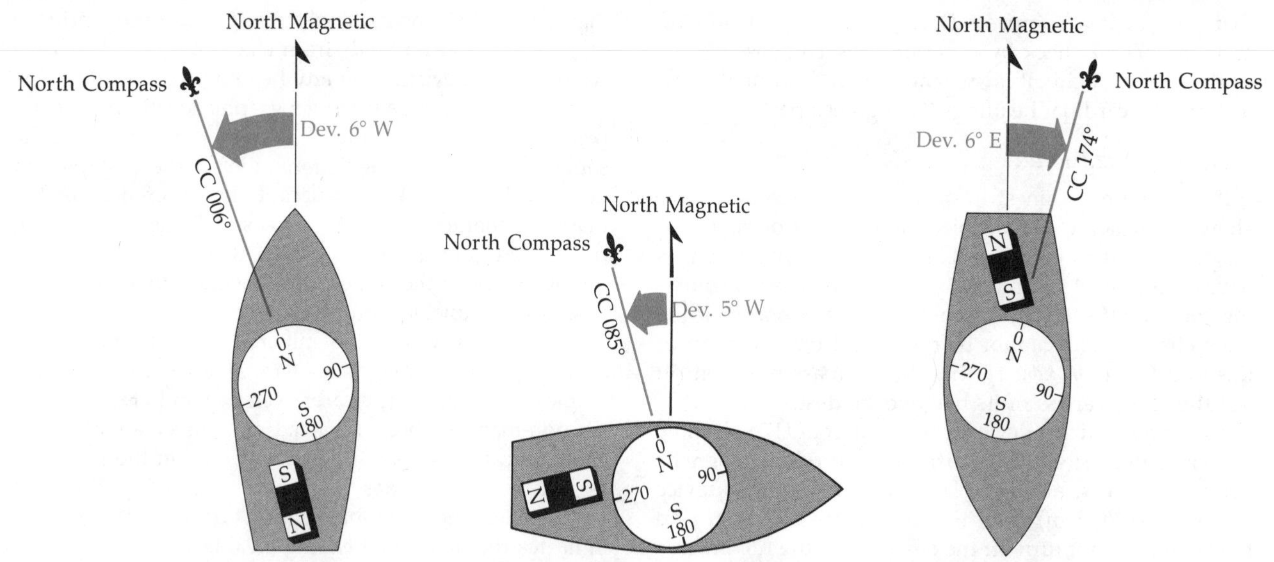

Figure 1525 **On a north magnetic heading, *left,* the S pole of the compass (180 on the card) is repelled by the S pole of the imaginary magnet; the card is deflected counterclockwise; there is westerly deviation. On an east magnetic heading, *center,* the N pole of the compass is again nearer the S pole of the imaginary magnet, and is attracted to it. The deflection is still counterclockwise, and the deviation is westerly. On a south magnetic heading, *right,* the N pole of the compass (000) is nearest the S pole of the imaginary magnet and is attracted to it. The card is deflected clockwise, and there is easterly deviation. (Deviation angles are exaggerated for clarity.)**

magnet. The result is that the card is rotated counterclockwise; the compass reading is increased, 006° in this instance, and there is westerly deviation.

On a south MAGNETIC heading, Figure 1525 *right,* the N pole of the compass magnet, 000° on the card, is nearer the S pole of the imaginary magnet and is attracted by it. The card is deflected clockwise and the compass reading decreases, to 174° and deviation is easterly.

If the magnetic heading of *Morning Star* is to the east, Figure 1525 *center,* the N pole of the compass magnet is the nearer pole to the S pole of the imaginary magnet and it is attracted. This results in a counterclockwise rotation of the compass card giving westerly deviation. On a west magnetic heading, the deflection of the card is clockwise for easterly deviation. (It is a good exercise to diagram this situation for yourself.)

In the above examples, the deviation is westerly on an east magnetic heading, but easterly on a south magnetic heading. Thus there must be some intermediate heading where the deviation passes through zero as the sign changes from west to east. Figure 1526 *left* approximates this situation. There is still attraction of the N pole of the compass by the S pole of the imaginary magnet, but on this heading the compass card has its N pole as close to the S pole of the magnet as it can get. Thus there is no deflection of the card in either direction, no deviation.

As has been illustrated above, deviation resulting from permanent magnetism will normally swing through one cycle of west to east and back to west in 360°, with two headings having zero deviation.

Do not assume that the location, polarity, and angle of skew of the imaginary magnet would be the same on any other boat as in the above examples, which are for the sole purpose of illustrating principles. A change of one or more of these factors would result in an entirely different set of deviations.

Effect of Induced Magnetism

The effect of induced magnetism on a compass can be visualized in a series of examples much like the above with an imaginary mass of magnetizable material. In this case, the disturbing pole would be the one induced in that mass and its sign would change as determined by the changing sign of the nearer pole of the inducing magnet. The sign of the deviation would change *four* times in 360 degrees, with *four* headings having zero deviation.

If the effects of permanent and induced magnetism are combined, a complex situation results; fortunately, this is unlikely for most small craft.

Effect of Geographic Position

Deviation can change with changes in MAGNETIC LATITUDE of a vessel. The strengths of most magnetic influences on a boat are unchanged at different geographic positions, but the horizontal component of the earth's field lessens in higher latitudes. Hence the interaction of these two forces, the cause of deviation, changes with varying magnetic latitudes.

Compass Calculations

COMPASS ERROR *(CE)* is sometimes used in compass calculations as a specific term. It is the algebraic sum of the variation and deviation. Because variation depends on geographic location, the deviation upon the craft's heading, there are various possible combinations of these two quantities, Figure 1526 *right.* Similarly to variation and deviation, it is expressed in degrees and direction, for

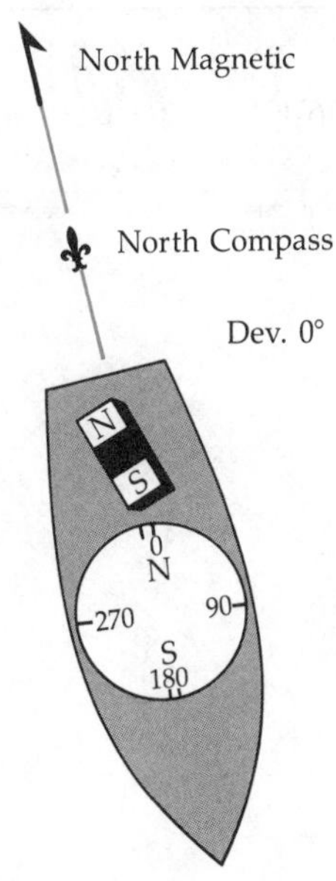

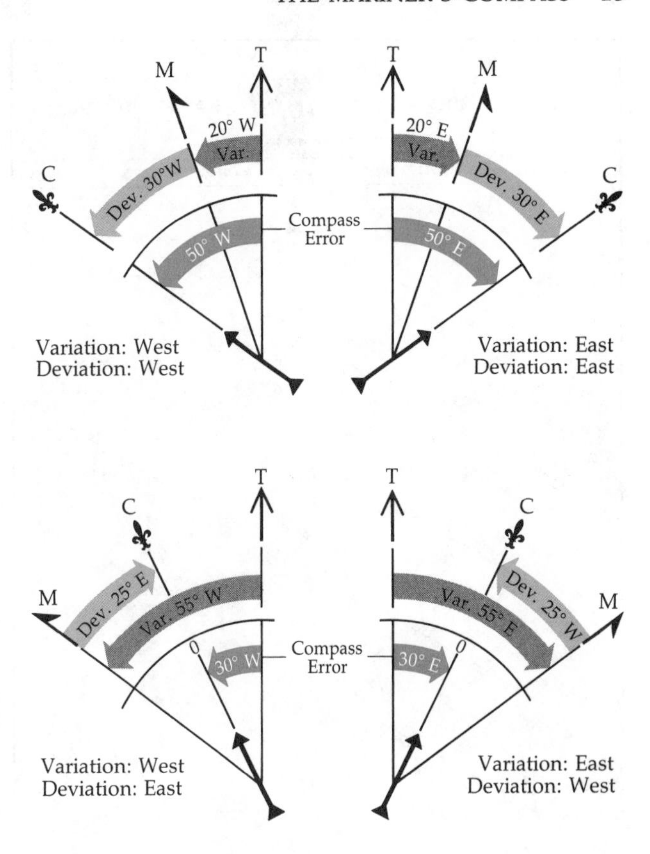

Figure 1526 ***Left:*** **On this heading, roughly 165°, the N pole of the compass is attracted to the S pole of the imaginary magnet, but it is as close as it can get; hence there is no deflection, and thus no deviation on this heading.** ***Right:*** **The two factors of variation and deviation are sometimes combined algebraically into a single value, termed "compass error (CE)." These drawings show the four possible combinations of easterly and westerly deviation and variation.**

example, 5°E or 7° west; in boating, it is given to the nearest whole degree.

This is a term used for convenience. The compass is not actually in error; it is operating according to the forces that control its behavior. Compass error is merely the angle between true north and compass north.

Interpolation

The process of determining intermediate values between tabular entries is called INTERPOLATION. It is most easily explained by several examples. Consider the following extract from a deviation table:

Compass Heading	*Deviation*
090°	7°E
105°	4°E

Assume that we need the deviation for a compass heading of 100°. The interval between tabular entries is 15 degrees; the difference between deviation values is 3 degrees. Since 100° is 10/15ths of the way between 090° and 105°, we multiply the value difference by that fraction: 3° × 10/15 = 2°. Since the deviation is decreasing, this is subtracted: 7°E − 2° = 5°E, the deviation for 100°.

If the values are not so even as above, an answer with a fraction will result. If we had desired the deviation for 098°, the fraction would have been 8/15, and the result 3° × 8/15 = 1.6°; then 7° − 1.6 = 5.4°E; but for boat use deviation is not used more precisely than the nearest whole degree and thus the deviation for 098° would be recorded and used as 5°E. If the fractional value is .5°, use the nearest even, not odd, degree.

Caution must be used when the two values of deviation are of *opposite* name. The value difference between 3°W and 1°E is *four* degrees, not two.

Critical Values Tables

A skipper will need deviation values in terms of both compass headings (for correcting calculations) and magnetic headings (for uncorrecting). A simple way to combine both of these, and have a more easily used table as well, is to prepare a direct-reading table of CRITICAL VALUES. Such a table is prepared in basic terms of deviations rather than headings. There is a line in the table for each whole degree of deviation; opposite this is listed on one side the range of compass headings for which that value of deviation is applicable; on the other side is the range of magnetic headings which have that value of deviation. See the below partial example.

Magnetic	*Deviation*	*Compass*
133°–144°	0°	133°–145°
145°–153°	1°E	146°–153°
154°–163°	2°E	154°–162°
164°–176°	3°E	163°–174°
177°–187°	4°E	175°–185°

Napier Diagrams

A NAPIER DIAGRAM (named after the man who developed it more than a century ago) is a graphic plot of deviation values which permits rapid and accurate determination of intermediate values as well as changes back and forth between compass and magnetic headings. It consists of a vertical base line marked from 000° to 360° with dots every degree, every fifth dot being slightly heavier for ease of reading. Usually the base line is printed in two sections of 180 degrees, sometimes with extensions of a few degrees at either end to facilitate use in these areas.

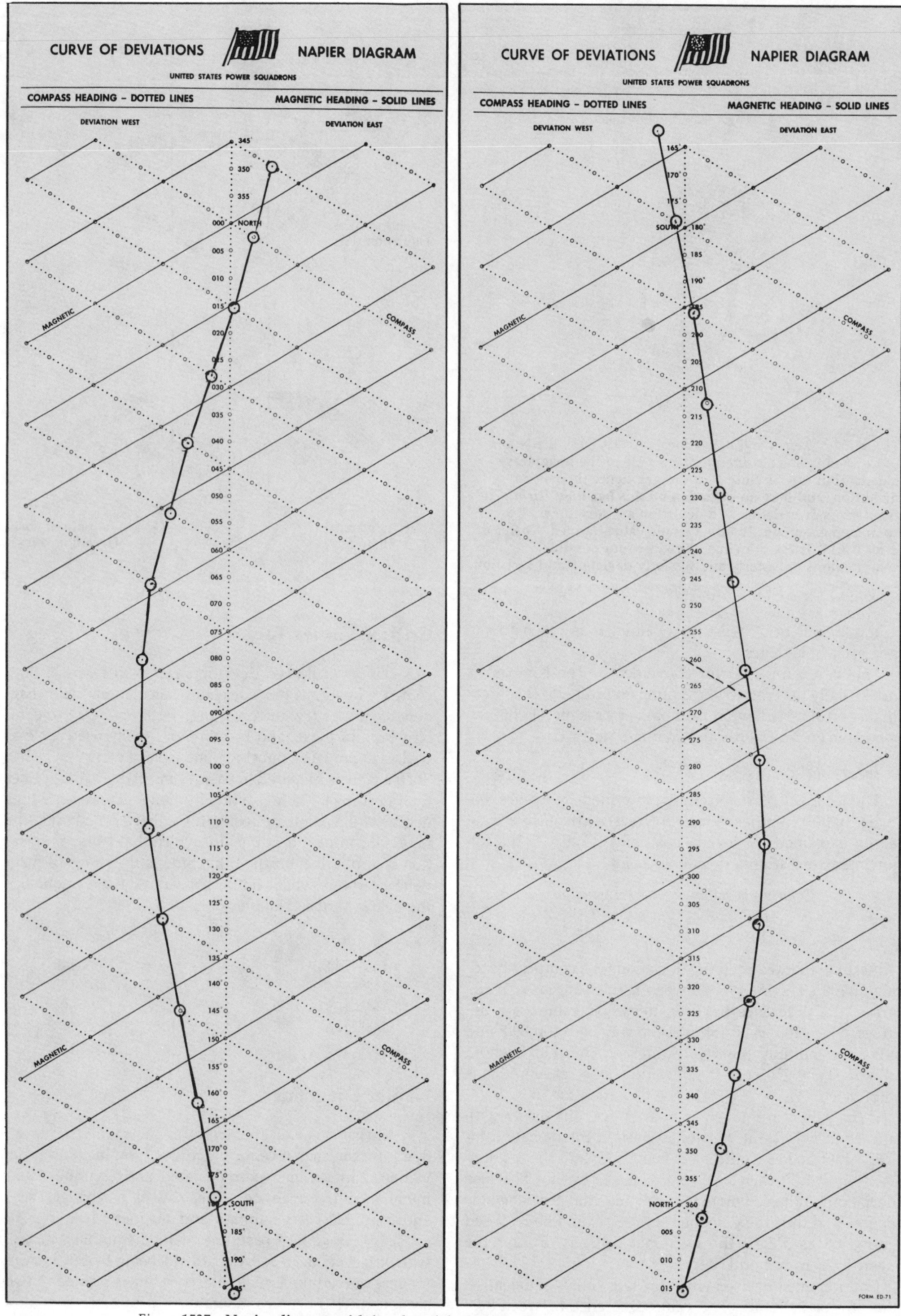

Figure 1527 **Napier diagram with its plot of deviation values provides a means of converting between magnetic and compass courses without separate tables for each conversion. See text.**

There are two series of cross lines making angles of 60 degrees with the base line and with each other; the lines appear at intervals of 15 degrees along the base line and those sloping downward to the left are solid. Deviations for compass headings are plotted along the dotted cross lines, easterly to the right of the base line and westerly to its left. A curve is drawn through the various plotted points; see Figure 1527.

To convert a magnetic heading to a compass heading find the value of the magnetic heading on the centerline, move along or parallel to a *solid* cross line until the curve is reached, then return to the center line by moving parallel to a *dotted* line. The compass heading is the value of the centerline at the point of return. The reverse process is used to convert a compass heading to a magnetic heading.

A Napier diagram is of particular value when the deviations are large and changing rapidly. If the deviations have been reduced to relatively small values by successful compensation, a Napier diagram is not needed, except perhaps briefly for the construction of a critical value table which will then become the practical source of deviation information for use in piloting.

Special Compensation Situations

The procedures of compensation described earlier in this chapter will fit most, but not all, recreational boats and their compasses. Special situations will be discussed below.

Using External Magnets

If the boat's compass (or binnacle) does not have internal compensating magnets, it still can be adjusted to near zero deviation if there is space to place external magnets where they are needed.

Two compensating magnets will be needed. Each is an encased permanent magnet having two holes for screw fastening. The ends of the magnets will either be marked N and S or be colored red on the north end and blue on the south. Masking tape is good for temporary fastening; be sure that the screws used for final fastening are non-magnetic. Carefully mark the longitudinal center of each magnet. In the areas where the external magnets will be placed, carefully mark out in chalk two lines through the compass center, one fore-and-aft and one athwartships; see Figure 1528 *top*.

General Procedures The compensation process consists of making outward runs on cardinal headings by compass, dropping a disposable buoy, turning, and heading back toward the point of departure. All this is done in the same manner as was done for internal compensators. Keep all external magnets far from the compass until each one is to be used.

Adjustment of E-W Headings If compensation is shown to be needed—for example, the compass reads 290° on the reciprocal of a 090° outward run—the correction is made by placing an external compensating magnet in a fore-and-aft position centered on an athwartship chalked line. This can be placed on either side of the compass in position A or A′ of Figure 1528 *bottom*. Should the compass reading now become more than 290°, the compensation is increasing the deviation; move the magnet to the other side of the compass or turn it end-for-end (but don't do both). The card will then move in the desired direction, toward 270°.

Should the card stop moving before 280° (the reading for removal of half of the error), the compensator, in a position such as at D, is too far from the compass. On the other hand, if the card moves past 280°, the compen-

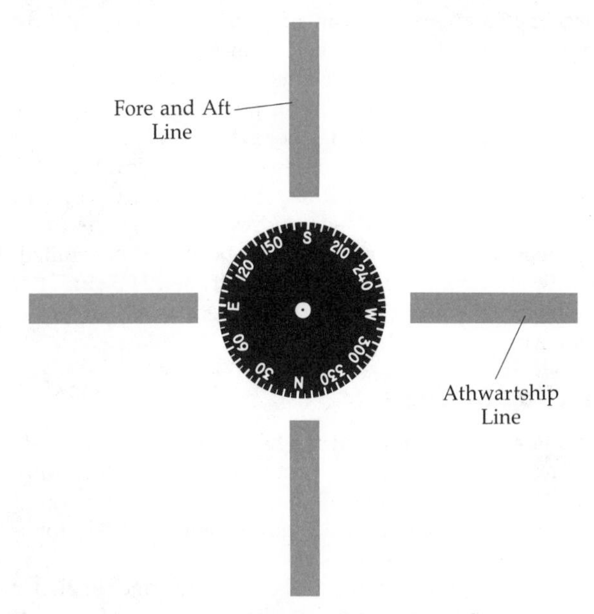

Figure 1528 **Before installing external compensating magnets, chalk in two lines, *above,* through the center of the compass mounting location. The magnets must always be placed so their centers are on one of these lines. *Below:* Magnets are placed forward or aft of the compass and on one side, as shown. When final placement is determined they are fastened with non-magnetic tacks or screws. See text for details on how they are used.**

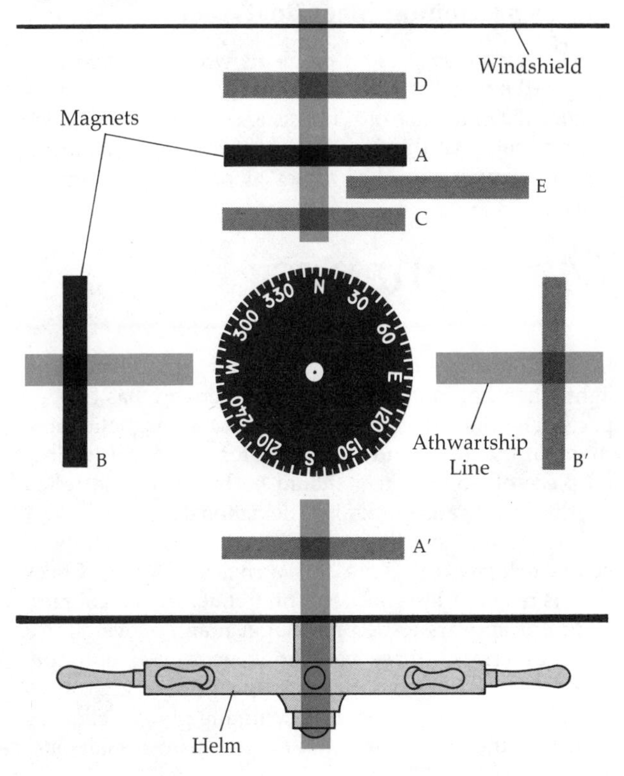

sator is too close to the compass, as at C, and must be moved farther away. Move the external magnet aft or forward as necessary until the desired reading is attained. Now tape the compensator down; no screws yet. *Do not let it get off the centerline as at E.* As with internal compensators, a second E-W run is then made, perhaps more, until zero deviation is achieved.

N-S Compensation After the turn back on a north or south run, one-half of any deviation is adjusted out by placing a compensator in an athwartship position *centered* on the fore-and-aft line, such as at B in Figure 1528. If there is no room for the magnet in front of the compass, place it aft of the instrument in the vicinity of B′. Move it closer to the compass or farther away—and reverse ends if needed—in the same manner as on east-west runs.

When, after sufficient runs, the deviation is zero, or reduced to a minimum, screw down the magnets (nonmagnetic screws) with care that their position is not shifted. It is wise to make one more E-W and then N-S run after the magnets have been secured in place as a final test of your work.

Heeling Magnets

Sailing craft, heeling out of a horizontal trim, frequently require HEELING MAGNETS below the compass to reduce their deviations. Small-craft skippers can make a rough compensation, but an exact adjustment is a job for a professional compass adjuster.

Place your sailboat on a N-S heading and heel her slightly, maintaining a constant heading. Place a correction magnet vertically under the compass so as to eliminate *only* the deviation induced by heeling; reverse the ends of the magnet if necessary. Even if you are unable to make this adjustment, try heeling your sailboat to see the effect on your compass.

Compensation on Steel Boats

Steel hulls may present problems whose solution may require the installation of soft iron QUADRANTIAL SPHERES (Figure 1529), the use of FLINDERS BARS, and possibly heeling magnets. All this is for the professional adjuster, or for the skipper who has achieved professional competence.

Iron and steel vessels are subject to changes in deviation upon large changes in latitude. This must be considered when cruising in such craft.

Compensation by Shadow Pin

A method of sailing reciprocal courses without setting up temporary ranges with disposable buoys uses a vertical shadow pin in the center of a horizontal disc graduated similar to a compass card; the disc should be gimbal mounted.

As the craft heads east (or west) by the compass, the disc is rotated by hand so that the shadow of the pin caused by the sun falls on the 090° (270°) mark. The turn is made and the boat is steadied on such a heading that the shadow falls on the opposite side of the disc, exactly on the 270° (090°) mark. If the compass now reads 270° (090°), there is no deviation on E-W headings. If not 270° (090°), the previously described procedures are followed for adjustment. A similar technique is used for north-south headings.

This procedure has advantages. The boat need not run far on any heading. No departure mark or disposable buoy is required. Neither the wind nor current will affect the result. But there is one complicating factor—the sun and its shadow do not stand still. Regardless of how steady the boat is being held on a heading, the shadow is moving. Time becomes a factor in the execution of the procedure; compensators must be adjusted before the shadow moves enough to upset the process, taking the boat off a true reciprocal heading.

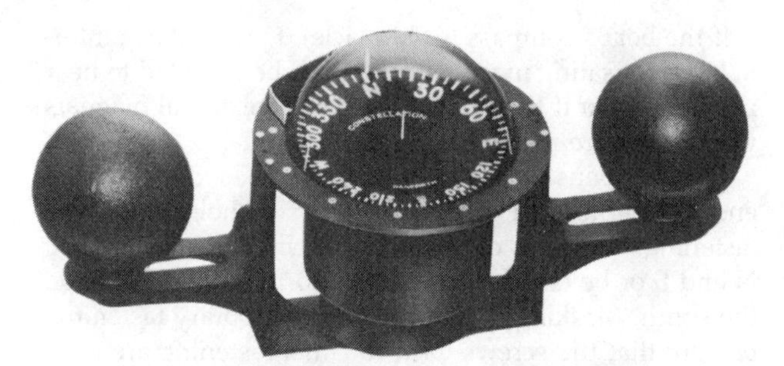

Figure 1529 **On vessels with steel hulls, it normally is necessary to mount two quadrantial spheres of soft iron on the binnacle, and adjust them to compensate the compass.**

Maintenance

In addition to protecting the instrument from direct sunlight when not in use, compass maintenance has two aspects. The first is the preservation of the magnetic environment that surrounds it. Except for occasional testing, no piece of iron or steel should be brought or installed near it, lest it cause unknown deviations.

The second aspect of maintenance consists mainly of getting to know your compass. Watch how it swings. Check that its readings are consistent on frequently run courses. Note if it appears to become sluggish and above all if it becomes erratic—these conditions warn you of undetected disturbances or a damaged pivot bearing.

Test for a damaged bearing or undue pivot friction by deflecting the card two or three degrees with a small magnet or piece of iron. If the card does not return to its former position, the compass should be removed from the boat and taken to a reputable shop for tests and probably repair.

As mentioned, a bubble can be removed by adding some liquid, but the liquid must be the same as that with which the compass is filled. This is best left to a compass repair shop.

Lightning and electric welding aboard may change the craft's own magnetic field. After exposure to either of these, check deviations.

Do read and follow the manufacturer's instructions for winter storage if the boat is taken out of regular service. They may add years to the compass's useful life.

Other Instruments

Several other direction instruments will be of general interest to boatmen although few skippers will have more than one or two of them.

Hand Bearing Compass

A small compass (some equipped with a handle) designed to be easily held in front of one's face is termed a HAND BEARING COMPASS; see page 437. This instrument is normally used for taking bearings, especially those that cannot be conveniently sighted across the steering compass. In an emergency a hand bearing compass can be used for steering the boat.

Remote-Reading Electronic Compass

There are magnetic compasses available with separate units for direction sensing and indicating, with wires connecting the two devices. The sensing unit can be located anywhere in the boat where deviating forces are at a minimum. The indicator can be located anywhere desired at the helm as it is unaffected by local magnetic fields.

Indicators are available with a dial-type face that has an extra pair of lines, which can be set so that the pointer is between them when on the desired heading.

Also available are models with a digital readout. These permit the dialing-in of variation so that indications are in true directions. Both of these types of remote-reading electronic compasses are capable of operating more than one indicator for use on boats with flying bridges or to meet the needs of a skipper who wants a direction-indicating device belowdecks (see following section).

Electronic compasses are also very desirable for use with satellite navigation receivers. Such sets require an input of the vessel's course (and speed) for computation of position using the radio signals from space.

Tell-Tale Compass

A TELL-TALE COMPASS is one located belowdecks, usually overhead above the skipper's bunk. These go back to the days when the captain used one to keep an eye on the course steered when he was not on deck. Today, they can serve the same purpose on long passages, but the more likely use is to watch out for wind and current shifts when riding at anchor for the night.

A tell-tale compass is an ordinary magnetic compass built upside-down so that the bottom of the card can be read. Some models have markings on both sides of their cards so that they can be used equally well in normal fashion or inverted as a tell-tale mounted overhead.

Pelorus

A PELORUS (see pages 437-438) is an instrument having sight vanes and a compass-like card, either of which can be clamped in a fixed position. Sometimes referred to as a DUMB COMPASS, it is used to take bearings when it is not possible to do so across the steering compass. The pelorus is placed in a location suitable for observation; its zero point is adjusted to either compass north or the boat's current heading, depending upon whether compass or relative bearings are desired.

Gyro-Compass

Details on GYRO-COMPASSES are beyond the scope of this book. Interested readers are referred to *Publication No. 9—Bowditch*—of the Defense Mapping Agency Hydrographic/Topographic Center, or to *Dutton's Navigation and Piloting* published by the U.S. Naval Institute. The gyro-compass is an expensive complex instrument designed to maintain a fixed direction in space. Contrary to the usual belief, it is not without errors. Aboard ship, allowances have to be made of changes in latitude and for rapid changes in the speed of the vessel. Its great advantage, however, is the absence of magnetic effects, variation and deviation; gyro-compasses are also suitable for integrating into complex navigation systems using computers. Gyro error is termed east or west and is applied exactly as is magnetic compass error.

Sun Compass

A SUN COMPASS is not a true compass; it is an instrument for taking bearings of the sun. The observer looks not at the sun, but at a shadow cast by a small pin in the center of the card. The card is graduated clockwise through 360° starting with zero at south. The shadow thus marks the sun's bearing. If the instrument is properly oriented with the vessel's heading, the shadow will show the ship's true course. The difference between its reading and the ship's compass is the compass error, CE. Applying variation yields the deviation of the ship's compass on the heading.

Accuracy depends upon knowledge of the sun's true bearing. Since latitude, date, and time affect this, and since the rate of change of the sun's true bearing is not constant, the procedure is complex and requires experience for satisfactory results. Some spherical compasses have a shadow pin mounted on the card directly above the pivot. A knowledgeable skipper can use this in checking deviation by azimuths of the sun. (This pin is also very useful when bearings are taken directly over the compass.)

16

How to make the best use of the buoys, lights, daymarks, and other aids established as the signposts of our waterways

AIDS TO NAVIGATION

Unlike the roads and highways that we drive on, the waterways we go boating on do not have road signs that tell us our location, the route or distance to a destination, or of hazards along the way. Instead the waterways have AIDS TO NAVIGATION, all those man-made objects used by mariners to determine position or a safe course. The aids also assist mariners in making landfalls, mark isolated dangers, enable pilots to follow channels, and provide a continuous chain of charted marks for precise piloting in coastal waters.

The term "aid to navigation" includes buoys, daybeacons, lights, lightships, radiobeacons, fog signals, and Loran and other electronic systems. It covers all the visible, audible, and electronic signals that are established by government and private authorities for piloting purposes. The Coast Guard uses the acronym ATON.

Note that while the marks, systems, and signals above may also be called NAVIGATIONAL AIDS, this second term also includes charts, instruments, and methods that assist in boat navigation.

The United States has been using a "lateral system" of buoyage (see page 371), meaning that wherever you travel within the country the basic system is the same; you needn't learn a new system for new waters. However, changes are being made so it will conform to the buoyage system of the International Association of Lighthouse Authorities (IALA), see page 371.

All aids to navigation are shown on the appropriate charts. Good piloting starts with the ability to correlate what appears on the charts with what you actually see on the water.

Aids established by the federal government are placed only where the amount of traffic justifies their cost and upkeep. Within bounds of necessity and cost, each aid is designed to be seen or heard over the greatest possible area.

Operating Agencies

The United States Coast Guard is the agency responsible for maintaining navigational aids on U.S. waters that are under federal jurisdiction or that serve the needs of the U.S. armed forces. Each Coast Guard District has supply and buoy depots, and special vessels for maintaining the aids.

State-Maintained Aids

On bodies of water wholly within the boundaries of a single state, and not navigable to the sea, the state is responsible for establishing and maintaining aids to navigation.

Although each state keeps authority over its waters, a uniform system of aids and regulatory markers has been agreed to by all; see pages 376-378.

"Private" Aids to Navigation

With prior approval, aids to navigation may be established in federal waters by individuals or agencies other than the Coast Guard; see any Coast Guard District office for information and procedures. These PRIVATE aids must be patterned after federal aids, and any fixed structure planned for navigable waters requires a permit from the Army Corps of Engineers.

All such aids to navigation—whether established by an individual, a corporation, a state or local government, or even a federal agency other than the Coast Guard, like the Navy—are private aids. They have the same appearance as Coast Guard-maintained aids, but are specially designated in the *Light Lists;* see page 393.

Protection by Law

Whether or not established by the Coast Guard, all aids navigation are protected by law. It is a criminal offense to cause any damage or hindrance to the proper operation of any aid. Do not deface, alter, move, or destroy any aid to navigation. Never tie your boat to a buoy, daybeacon, or light structure. Avoid anchoring so close to a buoy that you obscure the aid from the sight of passing vessels.

If you should unintentionally or unavoidably collide with or damage an aid to navigation, report the fact to the nearest Coast Guard unit without delay.

Suggestions for improvements to aids to navigation should be submitted by mail to the Commandant (G-NSR), U.S. Coast Guard Headquarters, Washington, D.C. 20593.

For your own safety and that of others, report any missing or malfunctioning aid immediately after returning to port. If the safety of navigation is threatened seriously, make the report by radio if you can.

Types of Aids to Navigation

The term "aid to navigation" encompasses a wide range of fixed and floating objects, from a single pile with a sign to lighthouses with an array of visible, audible, and electronic signals. Informal aids such as bush stakes marking natural channels or hazards in a creek are not a part of the organized system of aids to navigation.

Prominent buildings, cupolas, smokestacks, and other structures ashore as well as unique land features, also can be used as signposts for navigation. These are LANDMARKS, as distinguished from aids established solely for navigation.

Major Types

Buoys are floating objects—other than lightships—that are anchored to the bottom as aids to navigation. They have distinctive shapes and colors as determined by location and purpose, and may have visual, audible, and/or electronic signals; see color section, pages 577-580.

Daybeacons are unlighted fixed structures in the water. They may be a single-pile or a multiple-pile structure. Clusters of piles are called DOLPHINS. Daybeacons are equipped with one or more signboards, DAYMARKS, of a distinctive shape and color as determined by the information they convey; see color section, pages 577–580.

Lights are aids with active visual signals. They are fixed aids (lighted floating aids are designated as lighted buoys or lightships), and are classified by the Coast Guard and other authorities as PRIMARY SEACOAST LIGHTS, SECONDARY LIGHTS, or MINOR LIGHTS, as determined by their location, importance, and physical characteristics. A light's range and intensity vary with its classification. The shape and color of its supporting structure may be distinctive to identify it, but do not convey information as they do for buoys. The unofficial term "lighthouse" is often applied to primary seacoast lights and to some secondary lights.

Fog signals are audible signals sounded to assist mariners during periods of low visibility. They may occasionally be separate aids, as when located on the end of a jetty, but are generally part of a buoy, light, or larger aid to navigation.

Ranges are pairs of unlighted or lighted fixed aids that when observed in line show the pilot to be on the centerline of a channel. Individual structures may also serve to mark a turn in a channel; see page 579.

Lightships are specially equipped vessels anchored at specific locations to serve as aids to navigation. They are of distinctive shape and color, and have lights, sound signals, and radiobeacons.

Radiobeacons are transmitters broadcasting a characteristic signal specifically to aid navigation at night, in fog, or at distances exceeding normal visibility. These, too, are usually at another aid, but may be located separately.

Electronic navigation systems are radio transmitters, usually in groups, that emit special signals for use in navigation in fog or when beyond sight of land or offshore aids. The systems include *Loran-C* and *Omega* operated by the Coast Guard, aeronautical radiobeacons and *Omni* stations run by the Federal Aviation Administration, and other systems operated by government or private interests. In general, only USCG and FAA systems will be of use to boatmen.

Buoys

Buoys are anchored to the bottom at specific locations, and are shown on charts by special symbols and lettering that indicate their shape, color, and visual and/or sound signals. They vary widely in size. Buoys are secured to the bottom, using chain, to heavy concrete "sinkers" weighing up to six tons or more. The length of the chain will vary with the location, but may be as much as three times the depth of the water.

The buoyage system adopted for United States waters consists of several different types of buoys, each kind designed to serve under definite conditions. Broadly speaking, all buoys serve as daytime aids; many have lights and/or sound signals so they may be used at night and in periods of poor visibility.

A buoy's shape, color, and light characteristic, if any, give a pilot information about his location and about the safe guidance of his vessel. A buoy's size is usually determined by the importance of the waterway and size of vessels using it.

Buoy Characteristics

Buoys may be unlighted or lighted, sound buoys, or combination buoys (having both an audible and a visual signal). The Coast Guard maintains about 20,000 unlighted and 4,100 lighted and combination buoys in waters under its jurisdiction. Additional buoys are maintained by state and private agencies in non-federal waters.

Buoy Shapes

Unlighted buoys may be further classified by their shape.

Can buoys have a cylindrical above-water appearance, like a can or drum floating with its axis vertical and flat end upward; see Figure 1601. Two lifting lugs may project slightly above the flat top of a can buoy, but they do not significantly alter its appearance.

Nun buoys have an above-water appearance like that of a cylinder topped with a cone, pointed end up; Figure 1602. The cone may come to a point or be slightly rounded. Smaller nun buoys have a single lifting ring at the top; larger buoys have several lugs around the sides.

Unlighted buoys come in standardized sizes; a nun's above-water portion may vary from 30 inches (0.76 m) to 14 feet (4.27 m); can buoys range from roughly 18 inches (0.41 m) to nearly ten feet (2.97 m) above the waterline. Boaters should remember that a considerable portion of a buoy is under water, so that it is much larger and heavier than would appear from casual observation. Some smaller temporary buoys are now made of plastic materials.

Other buoys of special shapes, generally spherical, will be found in use as markers sometimes, but these are not regular aids to navigation. The Coast Guard has now eliminated the use of SPAR buoys, but they may be used in some other nations or in private systems. They are usually large logs, trimmed, shaped, and appropriately painted; they are anchored at one end by a chain.

Lighted, sound, and combination buoys are described by their visual and/or audible signals rather than by their shape, as discussed below.

Sound Buoys

A separate category of unlighted buoys includes those with a characteristic sound signal to aid in their location in fog or other reduced visibility. Different sound signals are used to distinguish between different aids within audible range of each other.

Bell buoys are steel floats surmounted by short skeleton towers in which a bell is mounted; Figure 1603. They are effective day and night, and especially in fog; they are much used because of their moderate maintenance costs. Bell buoys are operated by motion of the sea using four

Figure 1601 **Can buoy.**

tappers, loosely hung *externally* around the bell. When the buoy rolls in waves, wakes, or ground swells, a single note is heard at irregular intervals. Since bell buoys require some sea motion to operate, they are not normally used in sheltered waters; a horn buoy is used there instead if a fog signal is needed.

Gong buoys are similar in construction to bell buoys, except they have gongs instead of a bell; Figure 1604. Gong buoys give a distinctive characteristic when there are several sound buoys in one vicinity. They have four gongs of different tones with one tapper for each gong. As the sea rocks the buoy, the tappers strike against their gongs, sounding four different notes in an irregular sequence.

Whistle buoys have a whistle sounded by compressed air that is produced by sea motion. Whistle buoys are thus used principally in open and exposed locations where a ground swell normally exists, Figure 1605.

Horn buoys are rather infrequently used. They differ from whistle buoys in that they are electrically powered.

Lighted Buoys

Buoys may be equipped with lights of various colors, intensities, and flashing characteristics (now called RHYTHMS). Colors and characteristics of the light convey

Figure 1602 **Nun buoy.**

Figure 1603 **Bell buoy.**

Figure 1604 **Lighted gong buoy.**

Figure 1605 **Lighted whistle buoy.**

specific information to the mariner. Intensity depends upon the distance at which the aid must be detected, as influenced by such factors as background lighting, and normal atmospheric clarity. See Figure 1606.

Lighted buoys are metal floats with a battery-powered light atop a short skeleton tower; see Figure 1606. Lighted buoys can operate for many months without servicing, and have "daylight controls" that automatically turn the light on and off as darkness falls and lifts. The Coast Guard has begun a program of replacing the heavy, bulky primary batteries with storage batteries charged by panels of solar cells on top of the buoy.

Lights on buoys may be red, green, white, or yellow according to the specific function of the buoy; the use of these colors will be discussed later in this chapter.

Combination Buoys

Buoys with both a light and sound signal are designated COMBINATION BUOYS. Typical of these are "lighted bell buoys," and "lighted gong buoys." There is only one type of sound signal on a specific buoy. See Figure 1605.

Large Navigational Buoys

As a replacement for lightships, and for use at important offshore locations, the Coast Guard has developed LARGE NAVIGATIONAL BUOYS. These aids—referred to in some official publications as "super buoys"—combine a light and sound signal, and often also have a radiobeacon. This type of buoy can be roughly 40 feet (12.2 m) in diameter with a superstructure rising to 30 feet (9.1 m) or more; see Figure 1607.

Color of Buoys

Buoys may be of one solid color, or have a combination of two colors in horizontal bands or vertical stripes. The colors now used are red, green, white, yellow, black, orange, and blue; the latter three colors are being phased out as the IALA system is implemented in U.S. waters. The specific application of colors is discussed later in this chapter for each buoyage system.

Figure 1606 **Lighted buoy.**

Optical Reflectors

Almost all unlighted buoys are fitted with areas of reflective material, to help boatmen find them at night with searchlights. The material may be red, green, or yellow in banding or square patches and it has the same significance as lights of these colors. It can also be placed on the numbers or letters of lighted and unlighted buoys of all types.

Radar Reflectors

Many buoys have radar reflectors—vertical metal plates set at right angles to each other so as to greatly increase the echo returned to a radar receiver on a ship or boat.

Figure 1607 **Large navigational buoy. Called "super buoys" in some official publications, these aids have replaced lightships in some locations. Base is 40 feet in diameter; the light is more than 30 feet above sea level.**

Figure 1608 **Vertical metal plates arranged at right angles to each other provide a good radar target, *above*, on many buoys. Various light phase characteristics, *below*, help convey information on significance of buoys, as well as to help distinguish between buoys located near each other.**

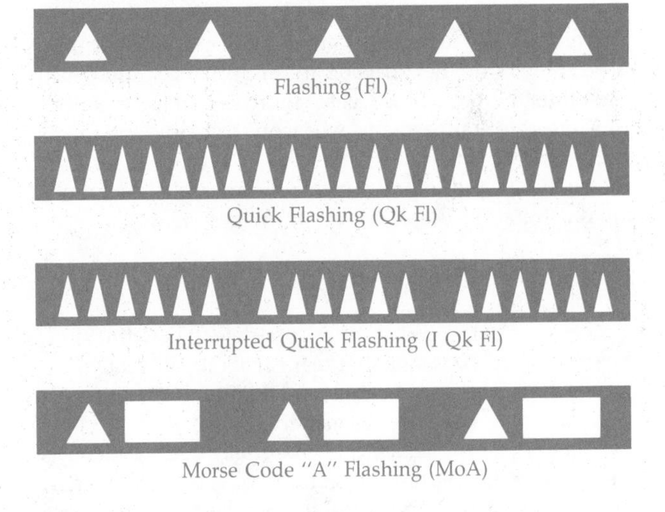

The plates are shaped and mounted to preserve the overall characteristic shape of an unlighted buoy or the general appearance of a lighted buoy. See Figure 1601.

Light Rhythms

The lights on lighted buoys will generally flash in one of several specific rhythms, for these reasons:

1. Flashing conserves the energy source within the buoy.
2. The light can be detected against a background of other lights.
3. The light can signal specific information during hours of darkness, like the need for special caution at a certain point in a channel.
4. Flashing patterns can differ from each other to distinguish clearly between buoys of similar functions that are within visible range of each other.

Flashing lights are those that come on for a single brief flash at regular intervals; the time of light is always *less* than the time of darkness. Coast Guard–maintained flashing buoys will flash their light not more than 30 times per minute; this is the more generally used characteristic. It is sometimes termed "slow flashing," but the official and correct description is simply "flashing."

Quick flashing lights will flash not fewer than 60 times each minute. These buoys are used for special situations where they can be more quickly spotted and where particular attention to piloting is required.

Interrupted quick flashing lights show a group of six quick flashes followed by a dark interval of five seconds, repeated every ten seconds. These are used in special situations as will be described later, but they are now being phased out.

Group Flashing (2 + 1) lights show two brief flashes, a brief interval, a single brief flash, and then an interval of darkness of several seconds' duration.

Morse code "A" flashing lights have a cycle of a short flash, and a brief dark interval, then a longer flash and a longer dark interval, repeated every eight seconds. This is the "dot-dash" of the letter "A" in Morse code; the color of the light is always white.

The above light rhythms are shown diagramatically in Figure 1608. The PERIOD of a light is the time it takes to complete one full cycle of flash and dark interval, or flashes and dark intervals. A light described as "Flashing 4 seconds" has a period of four seconds. One flash and one dark interval lasts just that long before the cycle is repeated. Three standard periods are used—flashing at intervals of 2.5, 4, and 6 seconds.

The term CHARACTERISTICS as applied to a lighted aid to navigation includes the color as well as the rhythm, and may also cover physical features such as nominal range.

Cautions in Using Buoys

Do not count on floating aids always maintaining their precise charted positions, or unerringly displaying their characteristics. The Coast Guard works constantly to keep aids functioning properly, but obstacles to perfect performance are so great that complete reliability is impossible.

Buoys are heavily anchored, but may shift, be carried away, or sunk by storms or ships. Heavy storms may also cause shoals to shift relative to their buoys.

Lighted buoys may malfunction and show no light, or show improper light characteristics. Audible signals on buoys are operated by action of the sea, and may be silent in calm water or may fail to sound because of a broken mechanism.

A buoy does not maintain its position directly over its sinker, as it must have some scope on its anchor chain. Under the influences of current and wind, it swings in small circles around the sinker, which is the charted location. (See page 412 for the charting of buoys.) This swinging is unpredictable, and a boat attempting to pass too close risks collision with a yawing buoy. In extremely strong current a buoy may be pulled beneath the water surface.

Buoys may also be temporarily removed for dredging operations, and/or in northern waters they may be discontinued for the winter, or changed to prevent damage or loss from ice flows. The *Light Lists* show dates for changes or for seasonal buoys, but these are only approximate and may be changed by weather or other conditions.

Temporary or permanent changes in buoys may be made between editions of charts. Keep informed of existing conditions through reading Notices to Mariners or Local Notices to Mariners—see pages 396-397.

All buoys (especially those in exposed positions) should, therefore, be regarded as warnings or guides, and not as infallible navigation marks. Whenever possible, navigate with bearings or angles on fixed objects on shore (see Chapter 21), and by soundings, rather than by total reliance on buoys.

Daybeacons

Daybeacons are unlighted aids that are fixed, rather than floating like buoys. They may be either on shore or in waters up to about 15 feet deep.

Daybeacons vary greatly in design and construction, depending upon their location and the distance from which they must be seen. The Coast Guard is making a continuing effort to standardize daybeacon structures and markings for easier identification as aids. Daybeacons for U.S. International and Inland waters, and their chart symbols, are illustrated in the color section, pages 577–579.

Daybeacon Characteristics

The simplest DAYBEACON is a single pile with signboards, called DAYMARKS, at or near its top, usually two facing in opposite directions; see Figure 1609. The pile may be wood, concrete, or metal.

A larger, more visible, and more sturdy daybeacon is the "three-pile dolphin" type: three piles a few feet apart at their lower ends, bound tightly together with wire cable at their tops. There are also some five-pile dolphins (four piles around one central pile.) The Coast Guard maintains approximately 10,000 daybeacons.

Figure 1609 **This minor light consists of a single pile, two daymarks, the light mechanism at the top, and a battery box on one side. Without the light, this aid to navigation would be a "daybeacon."**

Daymarks

To serve its purpose as an aid to navigation, a daymark usually bears a number, occasionally a letter or a number plus a letter.

Daymarks are normally either square or triangular, corresponding to can and nun buoys. Square daymarks are green with green reflective border. Triangular daymarks are red with red reflective border. The number or letters will also be of reflective material. In special applications, a daymark may also be octagonal (eight-sided) or diamond-shaped, carrying a brief warning or notice.

The Uses of Daybeacons

For obvious reasons the use of daybeacons is restricted to relatively shallow waters. Within this limitation, however, it is often more desirable than a buoy because it is firmly fixed in position, often easier to see and identify and easier to maintain.

Daybeacons are used primarily for channel marking, and they serve in the same manner as buoys in the buoyage systems to be described later in this chapter.

Minor Lights

Just as daybeacons are sometimes substituted for unlighted buoys, so may lighted buoys be replaced with MINOR LIGHTS. These are fixed structures of the same overall physical features as daybeacons, but equipped with a light generally similar in characteristics to those found on buoys. Most minor lights are part of a series marking a channel, river or harbor; also included, however, are some isolated single lights if they are of the same general size and characteristics. The term "minor light" does not include the more important lights marking harbors, peninsulas, major shoals, etc.; these have lights of greater intensity and/or special characteristics—these are designated as "secondary" or "primary seacoast" lights and are dscussed in detail later in this chapter.

Features of Minor Lights

Minor lights are placed on single piles (Figure 1609b), on multiple-pile dolphins, or on other structures in the water or on shore. MINOR LIGHTS carry daymarks for identification, and reflective material for nighttime safety should the light be extinguished.

Light Characteristics

A minor light normally has the same color and flashes with the same phase characteristics as a lighted buoy. Intensity will generally approximate that of a lighted buoy, but visibility may be increased by its greater height above water and its more stable platform. A combination of storage batteries and solar cells is gradually replacing the primary batteries used in past years.

Sound Signals

Minor lights may, in some locations, have an audible fog signal—an electrically operated horn or siren. In some cases the signal operates continuously for months when fog is expected.

Buoyage Systems

Most maritime nations use either the LATERAL SYSTEM OF BUOYAGE or the CARDINAL SYSTEM, or both. In the lateral system, the buoys indicate the direction to a danger relative to the course that should be followed. In the cardinal system, characteristics of buoys indicate location of the danger relative to the buoy itself. The term "cardinal" relates to cardinal points of the compass; see Chapter 15, pages 337-338.

The Basic U.S. System

In the United States, the lateral system of buoyage is uniformly used in all federal-jurisdiction areas and on many other bodies of water where it can be applied. In this system, the shape, coloring, numbering, and light characteristics of buoys are determined by their position with respect to the navigable channel, natural or dredged, as such channels are entered and followed *from seaward* toward the head of navigation.

As not all channels lead from seaward, certain arbitrary assumptions are used in order that the lateral system may be consistently applied. In coloring and numbering offshore buoys along the coasts, the following system has been adopted: proceeding in a southerly direction along the Atlantic Coast, in a northerly then westerly direction along the Gulf Coast, and in a northerly direction along the Pacific Coast will be considered the same as coming in from seaward. This can be remembered as proceeding around the coastline of the United States in a *clockwise* direction.

On the Great Lakes, offshore buoys are colored and numbered as proceeding from the outlet end of each lake toward its upper end. This will be generally westerly and northward on the Lakes, except on Lake Michigan where it will be southward. Buoys marking channels into harbors are colored and numbered just as for channels leading into coastal ports from seaward.

On the Mississippi and Ohio rivers and their tributaries, characteristics of aids to navigation are determined as proceeding from seaward toward the head of navigation, although local terminology describes "left bank" and "right bank" as proceeding with the flow of the river.

The IALA-B System

On 15 April 1982 the United States signed an agreement to change its buoyage system to that of the International Association of Lighthouse Authorities (IALA), designated as "System B—The Combined Cardinal and Lateral System (red to starboard)." System B will be used in North and South America, plus Japan, Korea, and the Phillipines. System A will be used in the rest of the world; see page 378.

System B uses five types of "marks"—those of lateral and cardinal systems, plus three different marks for isolated dangers and safe waters, and for special indications.

IALA-B Lateral System of Buoyage

*Returning from sea**	*Color*	*Number*	*Unlighted Buoy Shape*	*Lights or Lighted Buoys*		*Daymark Shape*
				Light Color	*Light Rhythm*	
Right side of channel	Red	Even	Nun	Red	Flashing or quick flashing	Triangular
Left side of channel	Green	Odd	Can	Green	Flashing or quick flashing	Square
Channel junction or obstruction	Red-and-green horizontally banded†	Not numbered; may be lettered	Nun or can†	Red or green†	Group flashing (2 + 1)	Triangular or square†
Midchannel or fairway	Red-and-white vertically striped	May be lettered	Spherical	White	Morse Code "A" Flashing	Octagonal

*or entering a harbor from a larger body of water, such as a lake.
†Preferred channel is indicated by color of uppermost band, shape of unlighted buoy, and color of light, if any.

Table 16-1 **The new IALA-B Buoyage System. Conversion to this system is underway now and will be completed in 1989.**

Not all types are used in all waters, but those that are used will conform to System B with respect to shapes, colors, topmarks, and light color and rhythm (characteristic). The IALA-B system makes no mention of sound signals such as gongs or whistles on buoys, and the present U.S. usage will not be changed.

System B generally resembles the old U.S. system, but there are some differences. Excellent progress is being made in the changeover, and completion will be not later than 1989. As the new system is gradually coming into predominance, it will be the basis for the following descriptions, with the former U.S. system's differences noted where appropriate.

Coloring

All buoys are painted distinctive colors to indicate which side you should pass them on or to show their special purpose. In the lateral system the significance of colors is as shown below; the traditional phrase "red, right, returning" helps in remembering the system.

Green buoys mark the port (left) side of a channel when entering from seaward, or a wreck or other obstruction that must be passed by keeping the buoy on the left hand. These buoys will be green in the IALA-B system.

Red buoys mark the starboard side of a channel when entering from seaward, or a hazard that you must pass by keeping the buoy to starboard.

Red-and-green horizontally banded buoys mark junctions in a channel, or hazards that you may pass on *either* side. If the topmost band is black (green in the new IALA-B system), the preferred channel is with the buoy to port of your boat. If the topmost band is red, the preferred channel is with the buoy to starboard. (Note: When proceeding *toward* the sea, it may *not* be possible to pass such buoys safely on either side. This is particularly true in situations where you are following one channel downstream and another channel joins in from the side, see Figure 1610, when such a buoy is spotted, be sure to consult the chart for the area.)

Red-and-white vertically striped buoys mark the fairway or midchannel. They are also used to divide the "in" and "out" channels of a Traffic Separation Scheme. These

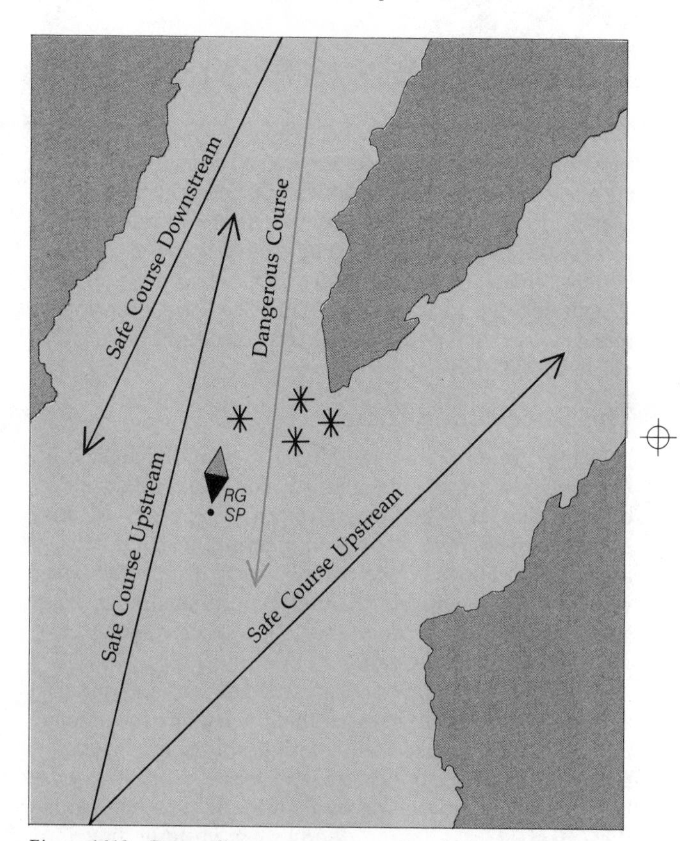

Figure 1610 **Proceeding upstream, a red-and-black horizontally banded junction buoy may be passed safely on either side. This is not true, however, when going downstream, as the illustration here shows.**

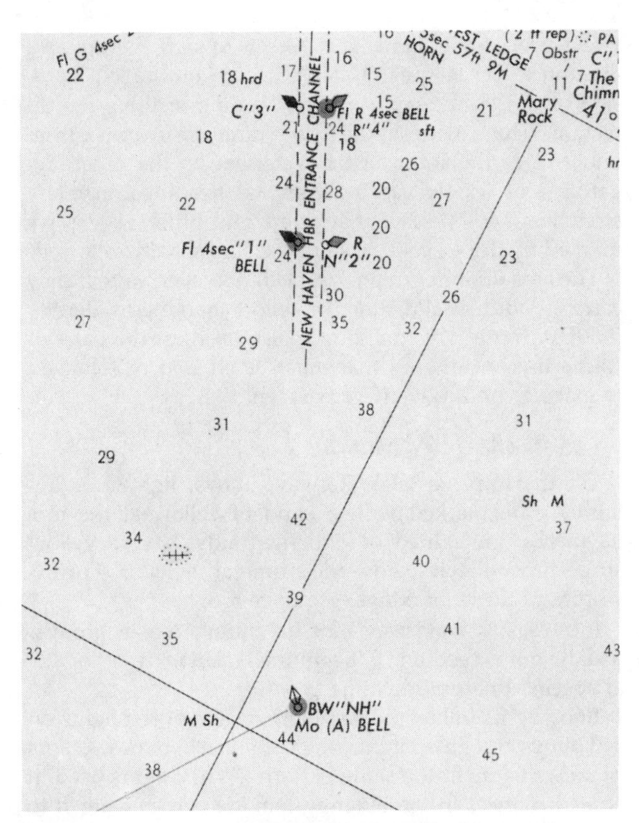

Figure 1611 **A red-and-white vertically striped buoy, often lighted and with a sound signal, is frequently used to assist in locating an entrance channel. Such a buoy is sometimes referred to as a "sea buoy."**

are spherical in shape if unlighted; if lighted there will normally be a spherical TOPMARK above the light that is roughly one-fifth the diameter of the buoy. In the older system, these buoys are black and white, either can or nun shape if unlighted, and no topmarks are used.

Note that areas of color running horizontally are "bands"; arranged vertically they are "stripes." These terms are used in other color combinations in the case of special-purpose buoys.

Shapes of Buoys

In the lateral system of buoyage, unlighted buoys have definite shape characteristics that indicate which side of the channel they mark. Distinguishing shapes are particularly valuable when you first sight a buoy in line with the sun and can see only its silhouette, rather than its color.

Can buoys, painted green (black until changed from the older system), mark the port (left) side of the channel when returning from seaward, or a hazard that you must pass by keeping the buoy to port.

Nun buoys, painted red, mark the starboard (right) side of channels, or hazards that must be passed by keeping the buoy to starboard.

Channel junction and obstruction buoys may be of *either* can or nun shape. If the uppermost band is *green* (*black*) the buoy is a *can;* if the uppermost band is *red* it is a *nun.*

Red-and-white vertically striped buoys are spherical in shape. (In the older system, they were black and white and could be either can or nun shape without differing significance.)

No special significance is to be attached to the shape of bell buoys, gong buoys, whistle buoys, lighted buoys, or combination buoys. The purpose of these is indicated by their coloring, number, or the rhythm of the light.

Numbering

Most buoys have "numbers" that actually may be numbers, letters, or a number-letter combination to help you find and identify them on charts. In the lateral system, numbers serve as yet another indication of which side the buoy should be passed. The system is as follows:

Odd-numbered buoys mark the port (left) side of a channel leading in from seaward. In accordance with the rules stated above, these will be green (black) buoys, cans if they are unlighted.

Even-numbered buoys mark the starboard (right) side of a channel; these will be red (nun) buoys.

Numbers increase from seaward and are kept in approximate sequence on the two sides of a channel by omitting numbers as appropriate if buoys are not uniformly placed in pairs. Occasionally, numbers will be omitted on longer stretches without buoys, to allow for possible later additions. Numbers followed by letters, such as 24A and 24B, are buoys added to a channel with the series not yet renumbered.

A buoy marking a wreck will often carry a number derived from the number of the buoy next downstream from it, preceded by the letters "WR". Thus, a buoy marking a wreck on a channel's left-hand side between buoys 17 and 19 would be "WR17A." A wreck buoy not related to a channel may be designated by one or two letters relating to the name of the wrecked vessel or a geographic location.

Letters without numbers are sometimes used for red-and-white (black-and-white) vertically striped buoys marking fairways, and for green-and-red (black-and-red) horizontally banded buoys with the green (black) band uppermost.

Numbers followed by letters may be used on buoys marking offshore dangers. For the buoy marked "2TL," for instance, the number has the usual sequential significance and the letters "TL" indicate a shoal known as "Turner's Lump."

Color of Lights

For all lighted buoys in the lateral system, the following system of colors is used:

Green lights on buoys mark the left-hand side of a channel returning from seaward; these are on green (black) odd-numbered buoys, or green-and-red (black-and-red) horizontally banded buoys with the green (black) band uppermost.

Red lights on buoys mark the right-hand side of a channel when entering from sea (red even-numbered buoys) or red-and-green (red-and-black) horizontally banded buoys with the red band uppermost.

White lights in the IALA-B system are used only on mid-channel and fairway buoys with the Morse letter "A" rhythm. (In the older U.S. system, they were used on a number of different types of aids where a light of greater visibility was desired.)

Light Rhythms

Flashing lights are placed only on green (black) or red buoys, or on special purpose buoys.

Quick flashing lights are placed only on channel-edge-marking green (black) and red buoys; these are used to indicate that *special caution* in piloting is required, as at sharp turns or changes in the width of the waterway, or to mark hazards that must be passed only on one side.

Group Flashing (2 + 1) lights are used only on buoys with red-and-green horizontal bands. These are the buoys at channel junctions and at wrecks that can be passed on either side. (In the older U.S. system, the lights displayed interrupted quick flashing and the buoys had red-and-black bands.)

Morse code "A" flashing lights are placed only on red-and-white (black-and-white) vertically striped buoys that mark a fairway or midchannel; these are passed close to on either side, and are always white.

Lateral lights on the Mississippi and other western rivers are being converted to the IALA-B system, but until that is completed some will flash with characteristics unique to those waters; see page 201.

Daybeacons and Minor Lights

The lateral system of buoyage has been described in terms of unlighted and lighted buoys, but the descriptions apply fully to comparable daybeacons and minor lights.

Daybeacons with red triangular daymarks may be substituted for nun buoys, or ones with green square daymarks may replace can buoys. Minor lights may be used in place of lighted or combination buoys. Structures subject to being repeatedly struck by vessels may be set back from the channel edge, as noted in the *Light List.*

If a daybeacon is used to indicate a channel junction or obstruction, the daymark will be red-and-green horizontally banded, with the color of the uppermost band indicating the main or preferred channel. The daymark shape will be either square or triangular as determined by the color or the top band, as with a can or nun buoy used for this purpose.

Daybeacons marking fairways or the midchannel have an octagonal-shaped daymark, painted red and white (black and white), divided in half vertically down the middle.

A diamond-shaped daymark has *no* significance in the lateral system. A typical application might be to increase the daytime detectability of a minor light which is not a part of a channel or waterway series. These are often used to mark a shoal, rock, submerged object, or other hazard; they are also used to mark prohibited areas where boats must not enter.

Intracoastal Waterway Aids

The Coast Guard maintains a system of aids to navigation along the Atlantic and Gulf Intracoastal Waterway (ICW). The coloring and numbering of buoys and daybeacons, and the color of lights on buoys and light structures, conforms with the lateral system of buoyage. The system is applied by considering passage from north to south along the Atlantic coast, and from south to north and east to west along the Gulf coast, as corresponding to returning from sea in an entrance channel. Thus, red buoys and daymarks are on the right side of the channel when proceeding from New Jersey toward Florida and then on to Texas; green (black) aids are on the left hand side of the Waterway when proceeding in the same direction. This rule is applied in a uniform manner from one end of the Intracoastal Waterway to the other, regardless of widely varying compass headings on many stretches, and the fact that rivers and other waterways marked by the seacoast system are occasionally followed.

The aids do differ in one respect, however, in that they carry an additional distinctive yellow marking to identify the ICW route. This marking is also used for the parts of connecting waterways that must be crossed or followed to make a continuous ICW passage.

Distinctive ICW Markings

On the Intracoastal Waterway, buoys, lighted or unlighted, are marked with a band of yellow at the top; daymarks, on lighted or unlighted aids, have a yellow stripe immediately below the numeral or letter. For examples of these markings, see page 578.

Intracoastal Waterway aids are numbered in groups, usually not exceeding 99, beginning again with "1" or "2" at specified natural dividing points.

Buoy lights follow the standard system of red lights on red buoys and green lights on green (black) buoys. Colors of minor lights fit the same pattern. (White lights used in some instances in the older system are being changed to the appropriate color.) Range lights, not being a part of the lateral system, may be any of the three standard colors.

Dual Marking

So that vessels may readily follow the Intracoastal Waterway where it coincides with another route (such as an important river marked on the seacoast system), the ICW uses *dual markings* for the aids that mark the river channel for other traffic. Special marks consist of a yellow square or triangle on a conspicuous part of the dual-purpose aid. The yellow square, in outline similar to a can buoy, indicates that the aid on which it is placed should be kept on the left-hand side when following the ICW in the direction from New Jersey to Texas. The yellow triangle, in outline similar to a nun buoy, indicates that the aid should be kept on the right hand side when traveling in the same direction.

The yellow square may appear on *either* a green (black) can or a red nun of the river channel marks (or on the daymarks of comparable daybeacons and light structures). Similarly, the yellow triangle dual markings may appear on any type of lateral aid. These similarly contradictory markings result from the fact that in some situations a southbound ICW route (red nuns on the right side) will be *up* a river channel from seaward (red nuns on the right side); and in other situations, the same ICW route will be *down* a river toward the sea where green (black) cans will be on the skipper's right side. In both of these situations, however, the yellow triangle will be kept to the right.

Where the yellow squares and triangles are added to regular river or harbor aids to navigation, the ICW yellow band or border is omitted.

Where dual markings are employed, the ICW skipper disregards the basic shape and coloring of the aid on which the yellow square or triangle is placed and pilots his craft *solely* by the shape of the yellow markings. The numbers on the aids will be those of the river's lateral system, and in some instances where the southward ICW proceeds down a river, the numbers will be temporarily decreasing rather than increasing. See page 578.

Wreck Buoys

Buoys that mark dangerous wrecks are generally lighted, and placed on the seaward or channel side of the obstruction and as near to it as possible. Wreck buoys are solid red or green (black) if they can be safely passed on only one side; horizontally banded otherwise, and numbered in regular sequence preceded by "WR." Be careful around wreck buoys, because sea action may have shifted the wreck since the last Coast Guard check.

Station Buoys

Important buoys, usually large combination buoys with both light and sound signals, are sometimes accompanied by smaller, unlighted STATION BUOYS. Station buoys are colored and numbered the same as the main buoy and placed on the side away from marine traffic. Station buoys are not charted, but their existence is noted in the appropriate *Light Lists.*

The station buoy is a back-up; it will mark the location should the main buoy be sunk or carried away, and aids in replacing the main buoy.

RACONs

Some major aids to navigation are equipped with RACONs. These are radar beacons which, when triggered by pulses from a vessel's radar, transmit a reply that results in a better defined display on that vessel's radarscope, thus increasing the accuracy of range and bearing measurements.

The reply may be coded to facilitate identification, in which case it will consist of a series of dots and dashes (short and/or long intensifications of the radar blips beginning at and extending beyond the RACON's position on the radar screen). The range is the measurement on the radarscope to the first dot or dash nearest its center. If the RACON is not coded, the beacon's signal will appear as a radial line extending from just beyond the reflected echo of the aid, or from just beyond where the echo would be seen if detected. Details of RACON coding will be found in the *Light Lists.* The coded response of a RACON may not be received if the radar set is adjusted to remove interference or sea return from the scope; interference controls should be turned off when reception of a RACON signal is desired.

Reporting Discrepancies

All boaters should realize that the Coast Guard cannot keep the many thousands of aids to navigation under constant observation, and for that reason, it is impossible to maintain every light, buoy, daybeacon, and fog signal operating properly and on its charted position at all times. Thus the safety of all who use the waters will be enhanced if every person who discovers an aid missing, off station, or operating improperly will notify the nearest Coast Guard unit of the situation that has been observed. Use radio, land telephone, or mail, as dictated by the urgency of the report.

Special Purpose Buoys

The Coast Guard also maintains several types of special purpose buoys, with no lateral significance, to mark anchorages, fish net areas, dredging limits, etc.

In the IALA-B system of buoyage, these are all yellow, regardless of the usage to which they are being put. These may be of can or nun shape, and, if lighted, the light will be yellow with any rhythm except those being used on aids in the lateral system. Daymarks on a daybeacon or minor light will be diamond shape and all yellow in color. (Under the former U.S. system a variety of colors were used on such buoys—they were white, white with green tops, white-and-black horizontally banded, white and international orange, and yellow-and-black vertically striped; these will be all consolidated into yellow aids.)

White buoys mark *anchorage* areas.

Note the difference between an *anchorage* buoy and a *mooring* buoy—an anchorage buoy is an aid marking the boundary of an area in which vessels may be anchored in accordance with prescribed regulations; a mooring buoy is a strong, heavily anchored buoy to which vessels can be made fast in lieu of anchoring.

White buoys with green tops are used in connection with *dredging* and *survey* operations.

White-and-black horizontally banded buoys mark *fish net* areas. Particularly, but not exclusively, in the Chesapeake Bay area, such buoys mark areas in which fish nets and traps may be placed at or near the surface; such areas and buoys are indicated on charts. These buoys may be cans or nuns (shape has no significance) and they carry identification numbers and letters.

White-and-international-orange buoys, with either horizontal bands or vertical stripes, are used for *special* purposes to which neither the lateral-system colors nor the other special-purpose colors apply. (See Figure 1612.)

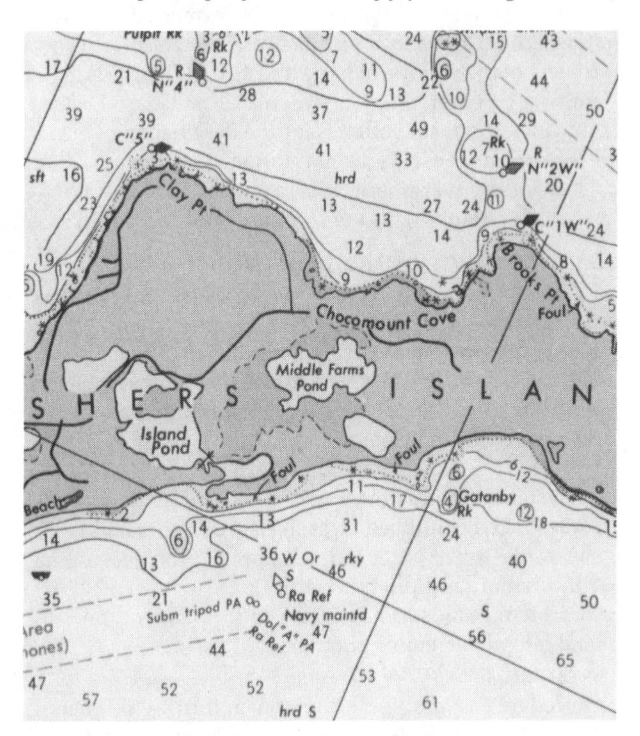

Figure 1612 **White-and-international-orange buoys are used for special purposes where lateral system colors, or other special-purpose colors, do not apply. These will be yellow.**

Typical applications are the marking of artificial "fish havens" and offshore oil drilling "subsea" installations.

Yellow-and-black vertically-striped buoys mark *seaplane* operating areas and have no marine significance other than to indicate the need for caution and a sharp lookout for aircraft.

Special-purpose buoys are illustrated in color in the color section. They may be lighted or unlighted, and some may have a fog signal. These buoys may be of any shape if unlighted—can, nun, or spherical.

IALA-B System: As the new buoyage system is implemented, *all* special-purpose buoys will be yellow; lights, if used, will also be yellow.

Special-Type Buoys

Special-type buoys may be encountered in all waters. Typical of these in inland waters are the buoys used in current surveys. The shape, size, color, lights (if any), and markings vary widely, and are listed in the public notices that are always published before such buoys are placed in navigable waters.

Offshore skippers may occasionally encounter large sea buoys used for gathering meteorological or oceanographic data, or those employed in naval defense operations.

Articulated Lights

A recent new type of aid to navigation is the ARTICULATED LIGHT; this is something of a cross between a minor light and a lighted buoy. A sealed hollow metal cylinder a foot or so in diameter and up to 50 or more feet in length is attached at one end by a swivel to a normal buoy "sinker." Because the cylinder is buoyant, it floats in an essentially vertical position (in some locations, an additional buoyancy collar may be attached to the cylinder just below the surface). Because no scope of chain is used, the aid is always very close to the established position with a negligible swinging circle. The length of the cylinder is selected to be the normal depth of the water, plus tidal range, plus 10 to 15 feet above the surface. At the top is mounted a typical light mechanism and daymarks in a manner similar to a minor light.

Articulated lights will be used where a position must be marked more precisely than is possible with a buoy, yet the depth of water, up to 40 feet, is too great for a normal pile or dolphin minor-light structure.

Uniform State Buoyage System

Each state has authority over control of navigation on waters that lie wholly within its boundaries and one not subject to federal jurisdiction (see page 39). This includes the responsibility for establishing and maintaining aids to navigation. Each state is free to mark its waters by any system it chooses, but the states have adopted a single consistent system of markers. With trailer-borne boats traveling freely over the highways from state to state, the gain for boatmen from the uniform system is obvious.

By Act of Congress, the UNIFORM STATE WATERWAY MARKING SYSTEM (USWMS) may be extended to cover waters subject to federal jurisdiction, but which have not yet been marked with aids by the Coast Guard. Agreement may be entered into between a Coast Guard District Commander and state officials for the designation of "state waters for private aids to navigation." In such waters, state or local governments may establish aids to navigation which comply with the standards of the USWMS.

Major Features of the USWMS

The UNIFORM STATE WATERWAY MARKING SYSTEM has been developed to indicate safe boating channels by marking the presence of either natural or artificial hazards. The system is designed to satisfy the needs of all types of small vessels, and it supplements and is generally compatible with the Coast Guard lateral system.

The USWMS consists of two categories of aids:

1. A system of regulatory markers to indicate dangerous or restricted controlled areas like speed zones and areas set aside for a particular use, or to provide general information and directions.
2. A system of aids to navigation to supplement the federal lateral system of buoyage.

Regulatory Markers

On federal waters, the boatman can turn to his charts, *Light Lists, Coast Pilots,* and other publications for information on natural hazards, zoned areas, directions, distances to supplement the knowledge that he gets from buoys, daybeacons, and other aids. On state waters, he has a uniform system of water signs or markers that, in themselves, convey their message without reference to any publication.

Just as Intracoastal Waterway markers are distinguished by their special yellow borders or other yellow marks, state REGULATORY MARKERS are identified by international-orange-and-white colors. On buoys, an orange band is near the top and bottom; on the white area between these bands there is a geometric shape, also in orange. An open diamond shape indicates danger. A diamond with a cross inside marks a prohibited area, and excludes vessels from it. A circle signifies zoning or control; vessels operating in such areas are subject to certain operating restrictions. A square or rectangular shape signals the conveying of information, the details of which are spelled out within the shape. (See page 580.)

Where the regulatory marker consists of a square or rectangular sign displayed from a structure, the sign is white with an international-orange border. If a diamond or circular shape is associated with the meaning of the marker, it will be centered on the signboard.

The geometric shape displayed on a regulatory marker conveys the basic idea of danger or control, so that the boatman can tell at a distance whether he should stay away or may safely approach for more information. To convey a specific meaning, spelled-out words or recognized abbreviations appear within the shape. The sole exception to this is the cross within the diamond shape, which absolutely prohibits boats from entering an area, because of danger to the boat, swimmers in a protected area, or for any other reason sufficient to warrant exclusion by law.

To minimize the risk of misinterpretation, initials, symbols, and silhouettes are not used. In some cases, words may be needed outside the geometric shape to give the reason, authority, or some clarification of meaning.

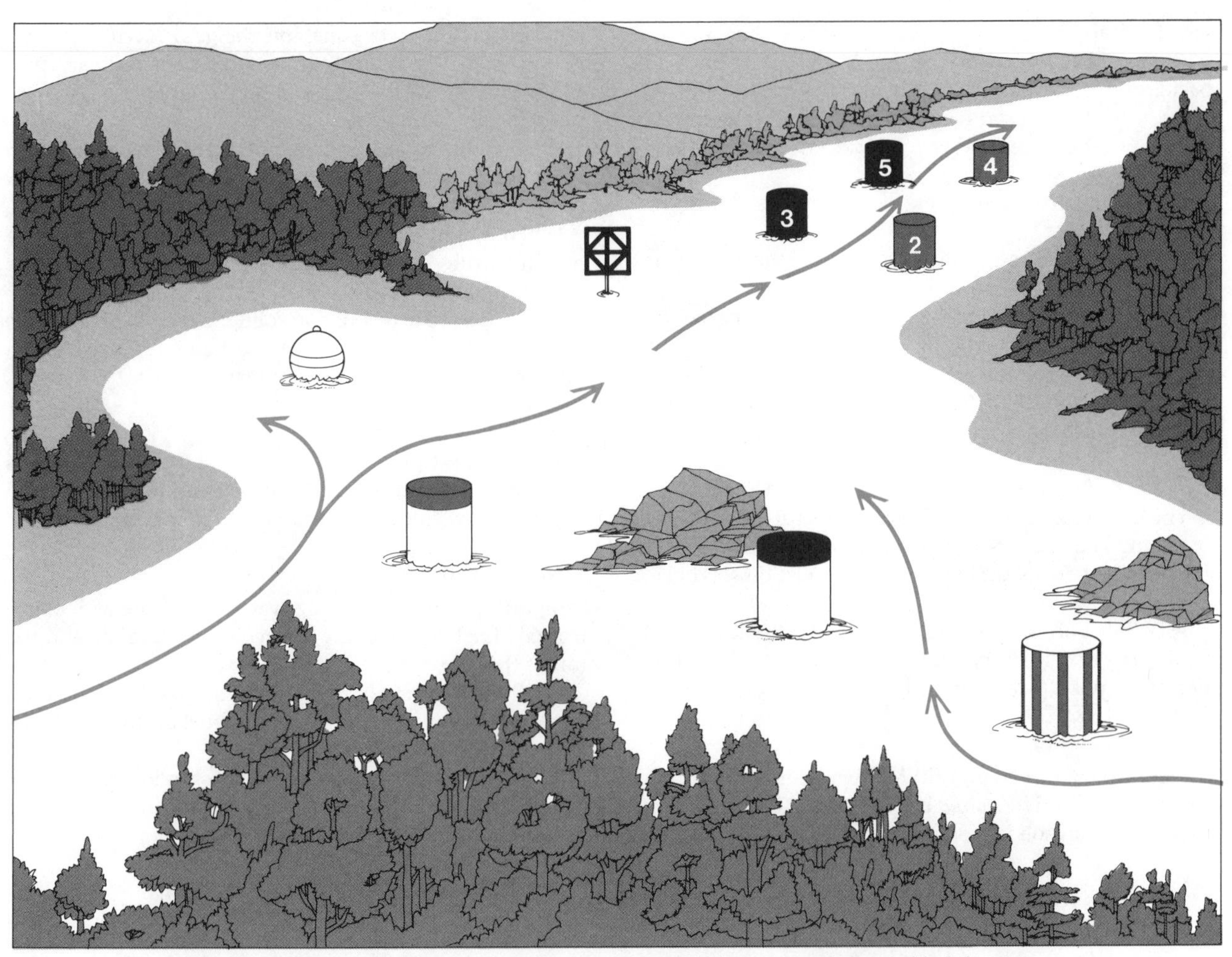

Figure 1613 **Running upstream in state-marked waters, a boat passes between black buoys to port and red buoys to starboard, outboard of a red-and-white obstruction buoy, south or west of a red-topped white buoy, and north or east of a white buoy with a black top.**

Aids to Navigation

The second category of marker in the USWMS is the AID TO NAVIGATION having lateral or cardinal meaning. In selecting types of buoys for use on waters not marked by the Coast Guard, the principal objective was to make the state system compatible with the federal.

On a well-defined channel, including a river or other relatively narrow natural or improved waterway, an aid to navigation of the USWMS is normally a solid-colored buoy. A buoy that marks the left side of a channel viewed *looking upstream* or toward the head of navigation is colored *solid black.* A buoy that marks the *right* side, looking in the same direction, is colored *solid red.* (No plans have yet been made to change to the green of the IALA-B system.) On a well-defined channel, solid-colored buoys will be found in pairs, one for each side of the channel that they mark, and opposite each other, leaving no doubt that the channel lies between the red and the black buoys.

On irregularly defined channels, solid-colored buoys may be used singly in staggered fashion on alternate sides provided that they are spaced at sufficiently close intervals to clearly inform the skipper that the channel lies between the buoys and that he should pass between them.

In the USWMS, shape is less important than color. Occasionally, solid red buoys will be of can shape rather than nun.

Where there is no well-defined channel, or when a body of water is obstructed by objects whose nature or location is such that the hazard might be approached from more than one direction, a *cardinal* system is used. The use of the cardinal system is strictly limited to waters wholly subject to state jurisdiction or waters covered by a Coast Guard–state agreement.

A white buoy with red top indicates that a boat must pass to the south or west of the buoy.

A white buoy with black top indicates that the safe water is to the *north* or *east* of the buoy.

A buoy with alternate red and white vertical stripes indicates that an obstruction extends between that buoy and the nearest shore. The number of red and white stripes may vary, but the width of the white stripes will always be twice that of the red ones. See Figure 1613.

Characteristics of Markers and Aids

The size, shape, material, and construction of all markers, fixed or floating, is at the discretion of the state authorities. They must, however, be visible under normal conditions from a safe distance.

Numbers

State aids and regulatory markers may carry numbers, letters, or words. These are placed so they are clearly visible to approaching and passing boats. Numbers and letters on red or black backgrounds are white; on white backgrounds they are black. The use of numbers on buoys is optional, but if used conform to the following system:

Odd numbers are used on *solid black* buoys and on *black-topped* buoys.

Even numbers are used on *solid red* buoys and on *red-topped* buoys.

All numbers *increase* in an *upstream* direction or toward the head of navigation.

Letters may be used only to identify regulatory markers and red-and-white vertically striped obstruction buoys. They must follow an alphabetical sequence in the up-stream direction; the letters "I" and "O" are not used, to prevent confusion with numbers.

Reflectors

The use of reflectors or reflective tape is at the discretion of the local authorities. Red reflectors or reflective material may be used on solid red buoys, and green on solid black buoys.

All other buoys have white reflectors or reflective material, as do regulatory markers, except that orange reflectors may be placed on their orange portions.

Lights

Lights may be used on USWMS regulatory markers and aids to navigation if desired by the state authorities. When used, lights on solid colored buoys are regularly flashing, regularly occulting, or equal interval. (An "occulting" light is one that is on more than it is off.) For ordinary purposes, the rate of flashing will not be more than 30 times per minute.

When lights have a distinct cautionary significance, such as at sharp bends in a channel, the light may be quick flashing—not fewer than 60 flashes each minute.

When a light is placed on a cardinal-system buoy or a red-and-white vertically striped buoy, it will always be quick flashing.

Red lights are used only on solid red buoys, green lights only on solid black buoys. White lights are used on all other buoys and on regulatory markers.

Ownership Markings

The use and placement of ownership identification is optional. Such markings are worded and positioned so as not to detract from the meaning of the marker or aid.

Mooring Buoys

Mooring buoys in USWMS waters are white with a horizontal blue band midway between the waterline and the top of the buoy.

A lighted mooring buoy normally shows a slow flashing white light, but if it is itself an obstruction to vessels, its light is quick flashing white.

A mooring buoy may carry ownership identification if the marking does not detract from the meaning conveyed by the color scheme and identification letter, if assigned.

Other Buoyage Systems

Buoyage on the "Western Rivers of the United States" (the Mississippi and its tributaries, and certain other designated rivers) conforms to the lateral system, but includes some additional shapes and daymarks not found in other areas. See the illustration in the color section.

On these rivers, unlighted buoys are not numbered, while the numbers on lighted buoys have no lateral significance; rather, they indicate the number of miles upstream from a designated reference point.

Additional details on the buoyage of the Western Rivers will be found in Chapter 9.

Other Nations

In Canadian waters, the lateral system of buoyage is essentially the same as in the United States. Minor differences in the physical appearance of buoys may be noted, but these are not great enough to cause confusion. Chart symbols, likewise, may be slightly different from those standardized for use on U.S. charts.

The nations of northwestern Europe—England and France, west to Ireland, north to Norway and Sweden, east to Poland and Russia—have made considerable progress in implementing a uniform buoyage system known as "System A—The combined Cardinal and Lateral System (red to port)."

Skipper's Responsibility

Familiarize yourself with the system of buoyage you will encounter before entering coastal waters of another nation. Consult the appropriate official publications or cruising guides for the necessary information.

Primary Seacoast and Secondary Lights

PRIMARY SEACOAST and SECONDARY LIGHTS in the U.S. are so designated because of their greater importance as aids to navigation. In general, they differ from the minor lights previously considered by their physical size, intensity of light, and complexity of light characteristics. These lights are more individual in nature than minor lights and buoys; only broad, general statements can be made about them as a group.

Primary seacoast lights warn the high-seas navigator of the proximity of land. They are the first aids seen when making a landfall (except where there may be a lightship, light tower, or large navigational buoy). A coastwise pilot can use these lights to keep farther offshore at night than by using other visual aids. These are the most powerful and distinctive lights in the U.S. system of aids to navigation.

Figure 1614 **Lighthouse structures show considerable variance in structure, based on location, terrain (if on shore), and strength and prevalence of storms in the area.**

Figure 1615 **Bands or stripes of color are often used to help distinguish lights from their backgrounds, and as an aid in their identification.**

Primary seacoast lights may be located on the mainland or offshore on islands and shoals. Offshore, they may mark a specific hazard or they may serve merely as a marker for ships approaching a major harbor.

Many primary seacoast lights are so classified from the importance of their location, the intensity of the light, and the prominence of the structure. Other aids are classed as secondary lights because of their lesser qualities in one or more of these characteristics. The dividing line is not clear however, and lights that seem to be more properly in one category may be classified in the other group in the *Light Lists* (see page 393). The difference in classification is of no real significance to boatmen and can be ignored in practical piloting situations.

Structures

The physical structure of a primary seacoast light and of many secondary lights is generally termed a LIGHTHOUSE although this is not an official designation used in the *Light Lists.* The structure's principal purpose is to support a light source and lens at a considerable height above water. The same structure may also house a fog signal, radiobeacon, other equipment, and quarters for the operating personnel. Auxiliary equipment and personnel are sometimes housed instead in a group of buildings nearby called a LIGHT STATION.

Lighthouses vary greatly in their outward appearances, depending on where they are, whether they are in the water or on shore, the light importance, the ground they stand on, and the prevalence of violent storms; see Figure 1614.

Lighthouse structures also vary with the range of visibility they need; a great range requires a tall tower or a high point of land, with a light of high candlepower. At points intermediate to principal lights, however, and where ship traffic is light, long range is not so necessary and a simpler structure can be used.

Coloring of Structures

Lighthouses and other light structures are marked with colors, bands, stripes, and other patterns to distinguish them against their backgrounds and to assist in their identification. See Figure 1615.

Light Characteristics

Primary seacoast and secondary lights have distinctive light characteristics—lights of different colors, and lights that show continuously while others show in patterns. Their three standard colors are white, red, and green.

Light Rhythm

Varying the intervals of light and darkness in both simple and complex ways yields many different rhythms for major lights.

Light Phase Characteristics

Light Pattern	*Abbreviations and Meanings*		*Phase Description*
	Lights That Do Not Change Color	*Lights That Show Color Variations*	
	F. = Fixed	**Alt.** = Alternating	A continuous steady light
	F. Fl. = Fixed and flashing	**Alt. F. Fl.** = Alternating fixed and flashing	A fixed light varied at regular intervals by a flash of greater brilliance
	F. Gp. Fl. = Fixed and group flashing	**Alt. F. Gp. Fl.** = Alternating fixed and group flashing	A fixed light varied at regular intervals by groups of 2 or more flashes of greater brilliance
	Fl. = Flashing	**Alt. Fl.** = Alternating flashing	Showing a single flash at regular intervals, the duration of light always being less than the duration of darkness
	Gp. Fl. = Group flashing	**Alt. Gp. Fl.** = Alternating group flashing	Showing at regular intervals groups of 2 or more flashes
	Gp. Fl. (1+2) = Composite group flashing	—	Light flashes are combined in alternate groups of different numbers
	E. Int. = Equal interval	—	Light with all durations of light and darkness equal
	Occ. = Occulting	**Alt. Occ.** = Alternating occulting	A light totally eclipsed at regular intervals, the duration of light always greater than the duration of darkness
	Gp. Occ. = Group Occulting	—	A light with a group of 2 or more eclipses at regular intervals
	Gp. Occ. (2+3) = Composite group occulting	—	A light in which the occultations are combined in alternate groups of different numbers

Table 16-2 **Light phase characteristics of primary and secondary lights permit rapid and positive identification at night. One full cycle of changes is the light's period.**

The term "flashing" has already been defined as a light that is on less than it is off in a regular sequence of single flashes occurring less than 30 times each minute. Some primary seacoast and secondary lights will "flash" in accordance with this definition although their characteristics will have no relation to the flashes of buoys and minor lights. In general, a flashing major light will have a longer period (time of one complete cycle of the characteristic) and may have a longer flash; for example, Cape Hatteras Light flashes once every 15 seconds with a 3-second flash. On the other hand, certain newer aids have xenon-discharge-tube lights that give a very bright, but very brief flash. The intensity of these lights can be varied for periods of good or bad visibility. A few major lights have a "fixed" characteristic (a continuous light without change of intensity or color) but light phase characteristics of primary seacoast or secondary lights are generally more complex, as described below and illustrated in Table 16-2.

Group Flashing (Fl [n]). The cycle of the light characteristic consists of two or more flashes separated by brief intervals and followed by a longer interval of darkness.

Alternating Flashing (Al Fl) Flashes of alternating color, usually white and red, or white and green. (Formerly "Alt. Fl.")

Occulting (Oc) The light is on more than it is off; the interval of time that the light is *lighted* is greater than the time that it is *eclipsed*. (Formerly "Occ.")

Isophase (Iso) Periods of light and darkness are equal; the light is described in the *Light Lists* or on charts in terms of the period, the lighted and eclipsed portions each being just half of that time interval. (Formerly "Equal Interval"—"E. Int.")

Group Occulting (Oc [n]) Intervals of light regularly broken by a series of two or more eclipses. This characteristic may have all eclipses of equal length, or one greater than the others. (Formerly "Gp. Occ.")

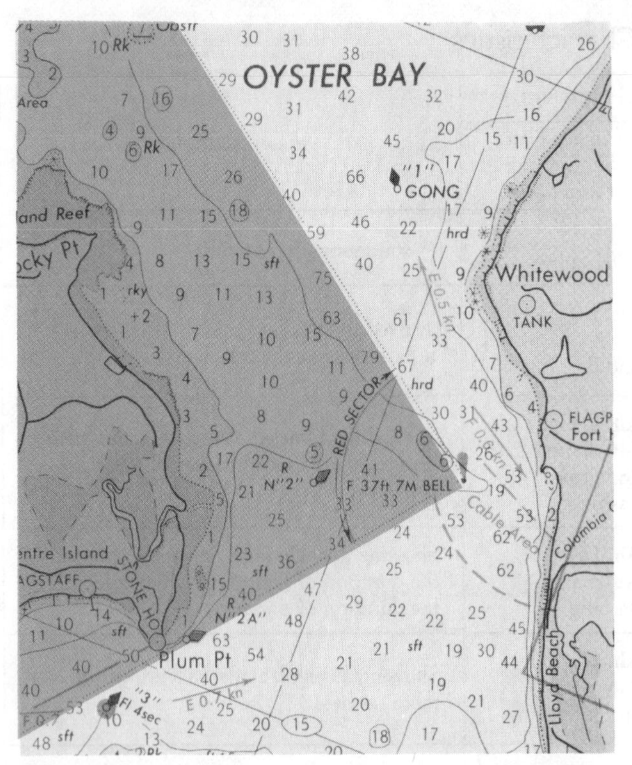

Figure 1616 **A red sector on a chart, *above*, identifies an area covered by a red light on the structure. This covers shoals or other hazards. The red light will have the same characteristics as those for the white light around the remainder of the horizon. *Below*: Some lights, particularly those on shore, have obscured sectors, in which they cannot be seen at all. These are shown on charts, and described in the *Light Lists*.**

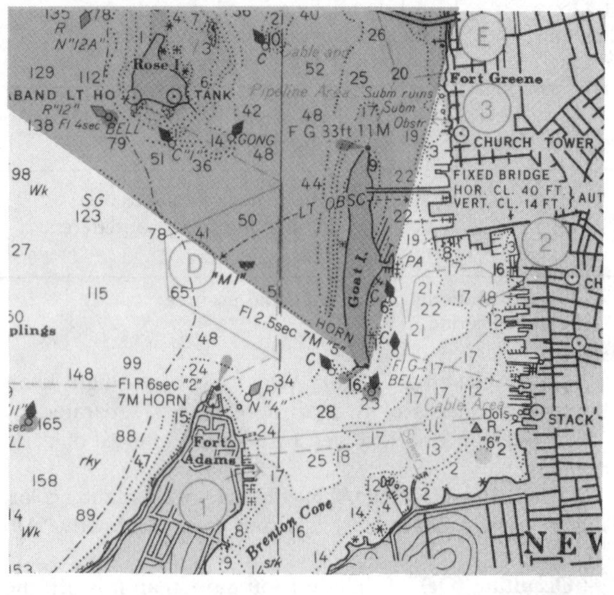

Fixed and Flashing (F and Fl) A fixed light varied at regular intervals by a flash of greater intensity. The flash may be of the same color as the fixed light (usually white) or of another color. This characteristic may also appear as *fixed and group flashing (F. and F1(n)).*

Complex Characteristics

The above light phase characteristics may be combined. Examples might include "Group flashing white, alternating flashing red" (Fl W [3], Fl R)—Gay Head Light—with three white and one red flash in each 40-second period; or "Fl W (1 + 4 + 3)"—Minots Ledge Light, where 1½-second flashes occur at 1½-second intervals in groups of one, four, and three separated by 5-second intervals and followed by a 15½-second longer interval to indicate the proper starting point of the 45-second period; or others of a generally similar nature.

Sectors

Many lights will have SECTORS—portions of their all-around arc of visibility in which the normally white light is seen as red. These sectors mark shoals or other hazards, or warn of land nearby.

Lights so equipped show one color from most directions, but a different color or colors over definite arcs of the horizon as indicated on charts and in the *Light Lists.* A sector changes the color of a light when viewed from certain directions, but *not* the flashing or occulting characteristic. For example, a fixed white light with a red sector, when viewed from within the sector, will appear as fixed red, Figure 1616 top.

Sectors may be a few degrees in width, as when marking a shoal or rock, or wide enough to extend from the direction of deep water to the shore. Bearings referring to sectors are expressed in degrees as observed *from a vessel toward the light.*

You should almost always avoid water areas covered by red sectors, but you should also check your chart to learn the extent of the hazard. Some lights are basically red (for danger) with one or more white sectors indicating the direction of safe passage, but a narrow white sector in others may also simply mark a turning point in a channel.

Lights may also have sectors in which the light is *obscured* and cannot be seen. These will be shown graphically on charts and described in the *Light Lists.*

The Visibility of Lights

A light's theoretical visibility in clear weather depends on two factors: its intensity and its height above water. It's intensity fixed its NOMINAL RANGE, which is defined in *Light Lists* as "the maximum distance at which the light may be seen in clear weather (meteorological visibility of 10 nautical miles)." Height is important because of the earth's curvature; height determines the GEOGRAPHIC RANGE at which the light can be seen. It is not affected by the intensity (provided that the light is bright enough to be seen out to the full distance of the geographic range).

The nominal range of major lights is generally greater than the geographic, and the distance from which such aids can be seen is limited only by the earth's curvature. Such lights are often termed "strong"; a light limited by its luminous range is a "weak light."

The glare, or LOOM, of strong lights is often seen far beyond the normal geographic range, and under rare atmospheric conditions the light itself may be visible at unusual distances. The range of visibility is obviously also *lessened* by rain, fog, snow, haze, or smoke.

Light Lists show the nominal range for all lighted aids except navigation ranges and directional lights (see page 393), and show how to convert nominal range to LUMINOUS RANGE—the maximum distance at which a light may be seen in *existing* visibility. Both nominal and luminous ranges take no account of elevation, observer's height of eye, or the curvature of the earth. For lights of complex characteristics, nominal ranges are given for each color and/or intensity.

The geographic range of lights is not given in the *Light Lists* but can be determined from the given height of the light source. The distance to the horizon may be taken from a table in the front pages of each *Light List* volume, or calculated using the equation $D = 1.17 \sqrt{H}$ (where H is the height in feet and the distance is in nautical miles) or $D = 2.12 \sqrt{H}$ (where H is in meters and distance is still in nautical miles). For any situation, the light's range as limited by the earth's curvature will be the distance to the horizon for the light *plus* the distance to the horizon for *your* height of eye—determine each distance separately, then add; *do not add heights and make a single calculation.* Boatmen should know their height of eye when at the controls of their boat (and the height of any other position on board to which they can climb to see farther).

Lights on inland waters, where their radius of usefulness is not great, are frequently "weak" lights whose intensity need not reach the full limit of their geographic range.

Identification of Lights

Charts can only briefly describe the characteristics of a primary seacoast or secondary light, by means of abbreviations and a notation of the total period of the light cycle. You will often need to consult the *Light List* for details of the characteristic to help you identify it positively.

When you first see a light, note its color and time, its full cycle of light changes. If color, period, and number of flashes per cycle match the *Light List* information, the light has been identified. As a precaution, however, check the charts and *Light List* to be sure that no other light in the vicinity has similar characteristics.

Cautions in Using Lights

Complex lights with several luminous ranges may appear differently at extreme distances where, for example, a white fixed (or flashing) light could be seen but a red flash of the same light was not yet within luminous range. Examination of a *Light List* will show that usually the nominal range of a red or green light is 15% to 30% *less* than that of the white light from the same aid. Be cautious when identifying lights like these.

The effect of fog, rain, snow, and haze on the visibility of lights is obvious. Colored lights are more quickly lost to sight in poor weather than are white lights, but refraction may also cause a light to be visible from a greater distance than normal.

Be cautious also when using light sectors in your navigation. The actual boundaries between the colors are not so distinct as the chart suggests; the lights shade gradually from one color into the other.

Note too that the increasing use of brilliant shore lights for advertising, illuminating bridges, and other purposes may cause marine navigational lights, particularly those in densely populated areas, to be outshone and difficult to distinguish from the background lighting.

A light can also be out for some reason. Unattended lights that are broken may not be immediately detected and corrected. If you do not see a light reasonably soon after your course and speed suggest, you should, check the situation carefully. Do not rely on any *one* light, except perhaps for making a landfall. For positive identification, use several lights *together* as a system, checking each against the others.

High Intensity Lights

At some locations, there is a "high intensity light" which has a greater intensity than the "main light" normally operated from that location. This is operated only under conditions of reduced visibility.

Emergency Lights

Emergency lights of reduced nominal range are displayed from many light stations when the main light is inoperative. These standby lights may or may not have the same characteristics as the main light. The existence of the standby light (if any) and its characteristics (if different) is noted in the *Light Lists*.

Fog Signals

Any sound-producing instrument operated in time of fog from a definite point shown on a chart serves as a useful fog signal. To use it effectively as an aid to navigation, you must be able to identify it and know its location.

As mentioned, the simpler fog signals used on buoys and at minor lights are operated by sea action and thus you may have difficulty identifying them. Fog signals at all lighthouses and lightships are electrically or mechanically operated on definite time schedules, however, so are easier to identify positively.

Signal Characteristics

Fog signal characteristics are described in terms of the length of a total cycle of one or more BLASTS of specific length and one or more SILENT INTERVALS, also of definite lengths. These times are shown in the *Light Lists* to aid in identification. (Normally, only the type of fog signal, without further details, is indicated on charts.) When you are counting the blasts and timing the cycle, refer to the *Light List*, for its details.

Fog Signal Equipment

Fog signals also differ from each other in tone, which helps in identification. The signal type for each station is shown in the *Light Lists* and on charts.

Diaphones produce sound by means of a slotted reciprocating piston activated by compressed air. Blasts may consist of two tones of different pitch, in which case the first part of the blast is higher pitch and the latter is lower. These alternate-pitch signals are termed "two-tone."

Diaphragm horns produce sound by means of a disc diaphragm vibrated by compressed air or electricity. Duplex or triplex horn units of different pitch are sometimes combined to produce a more musical signal.

Sirens produce sound by means of either a disc or cup-shaped rotor actuated by compressed air, steam, or electricity. These should not be confused with "police sirens"; these aids may produce a sound of constant pitch much like a diaphragm horn or a whistle.

Whistles produce sound by compressed air, emitted through a slot into a cylindrical bell chamber.

Operation of Signals

Fog signals at stations where a continuous watch is maintained are placed in operation whenever the visibility decreases below a limit set for that location; typically, this might be five miles. Fog signals at such stations are also sounded whenever the fog signal of a passing vessel is heard.

Fog signals at locations without a continuous watch may not always be sounded promptly when fog conditions occur, or may operate erratically due to mechanical difficulties.

Where fog signals are operated continuously on a seasonal basis or throughout the year, this information will be found in the *Light Lists.*

On buoys where fog signals are operated by sea motion, the signals may be heard in any condition of visibility.

Cautions in Using Fog Signals

Fog signals obviously depend upon the transmission of sound through the air. As aids to navigation, they thus have inherent limitations that you *must* consider. Because sound travels through air in a variable and unpredictable manner, you should note that:

1. The distance at which a fog signal can be heard may vary at any given instant with the bearing of the signal, and may be different on different occasions.
2. Under certain atmospheric conditions you may hear only part of a fog signal that is a siren or that has a combination of high and low tones.
3. There are sometimes areas close to a signal where you will not hear it, perhaps when the signal is screened by intervening land masses or other obstructions, or when it is on a high cliff.
4. The apparent loudness of a fog signal may be greater at a distance than in its immediate vicinity.
5. A patch of fog may exist at a short distance from a manned station but not be seen from it. Thus the signal may not be placed in operation.
6. Some fog signals require a start-up interval.
7. You may not hear a fog signal with your boat's engine on, but you may hear it suddenly when it is off or if you go forward away from its noise.

In summary, fog signals are valuable as warnings, but do not place implicit reliance upon them in navigating your vessel.

Based on the above established facts, you must *not* assume:

1. That you are out of the ordinary hearing distance of a fog signal because you do not hear it.

2. That because you hear a fog signal faintly, you are at a distance from it.

3. That you are near to the fog signal because you hear it clearly.

4. That the fog signal is not sounding because you do not hear it, even when you know that you are nearby.

5. That the detection distance and sound intensity under any one set of conditions is an infallible guide for any future occasion.

Standby Fog Signals

Standby fog signals are sounded at some of the more important stations when the main signal is inoperative. The standby signals may be of a different type and characteristic than the main signal. If different, details are given in the *Light Lists.*

Offshore Lights

LIGHTSHIPS are vessels of distinctive design and markings, equipped with lights, fog signals, and radiobeacons. They are anchored at specific, charted locations to serve as aids to navigation.

The Coast Guard has now retired the last U.S. lightship—at Nantucket Shoals—and all offshore stations are now marked by OFFSHORE LIGHT TOWERS or LARGE NAVIGATIONAL BUOYS.

Offshore Light Towers

A typical offshore light structure is shown in Figure 1617, bottom. It has a typical tower deckhouse 60 feet (18.3 m) above water, 80 feet (24.4 m) square, supported by steel lag pilings driven nearly 300 feet (91.4 m) into the ocean bottom. It accommodates living space, radiobeacon, communications, and oceanographic equipment. The deckhouse top is a landing platform that takes the largest helicopters flown by the Coast Guard. On one corner of the deckhouse is a 32-foot (9.8 m) radio tower supporting the radiobeacon antenna and a 3½-million candlepower light. At an elevation of 130 feet (39.6 m) above the water, it is visible for 18 nautical miles.

A light tower has several advantages over a lightship—lower operating costs, greater light range (from the greater height and stability), better fog signal projection, more accurate guidance for vessels (doesn't swing on a long-scope anchor chain and can't get blown of station), and longer life before replacement is required.

Unmanned Offshore Aids

In addition to the manned offshore light stations, there are several unmanned light towers, such as the one at Brenton Reef, and an increasing number of "large navigational buoys" (see page 368). The buoys have even lower operating costs, but lack the height of light and precise-

Figure 1617 **The last lightship in U.S. waters, *above,* was at Nantucket Shoals, with the primary vessel and the relief ship showing the same identification and characteristics. A typical light tower, *below,* has a helicopter landing pad, living quarters, and oceanographic equipment in addition to radiobeacon, fog signal, and the light itself.**

ness of location; they are used where the depth of water or other conditions preclude the construction of towers.

Cautions in Using Offshore Aids

To pass an offshore light tower or large navigational buoy safely, set your course to clear it comfortably, with no risk of collision. Bear in mind that buoys are anchored by a long scope of chain and that their swinging circle is large. The charted position is the location of the anchor.

A skipper steering toward a radiobeacon on an offshore aid to navigation, "homing" on it, should exercise particular care. *Never* rely solely on sighting the aid or hearing its fog signal in time to prevent hitting it. You can lessen the risk of collision by ensuring that the radio bearing does *not* remain constant, which would indicate that you were headed directly toward it.

Remember, too, that in extremely heavy weather a large navigational buoy may be carried off station, perhaps without that fact being known to the Coast Guard. Such a buoy may still have its radiobeacon and other aids operating. A navigator should not, therefore, rely on this type of aid during and immediately after a severe storm.

Ranges and Directional Lights

Ranges and directional lights serve to indicate the centerline of a channel and thus aid in the safe piloting of a vessel. Although they are used in connection with channels and other restricted waterways, and shown on all the appropriate charts, they are not a part of the lateral system of buoyage.

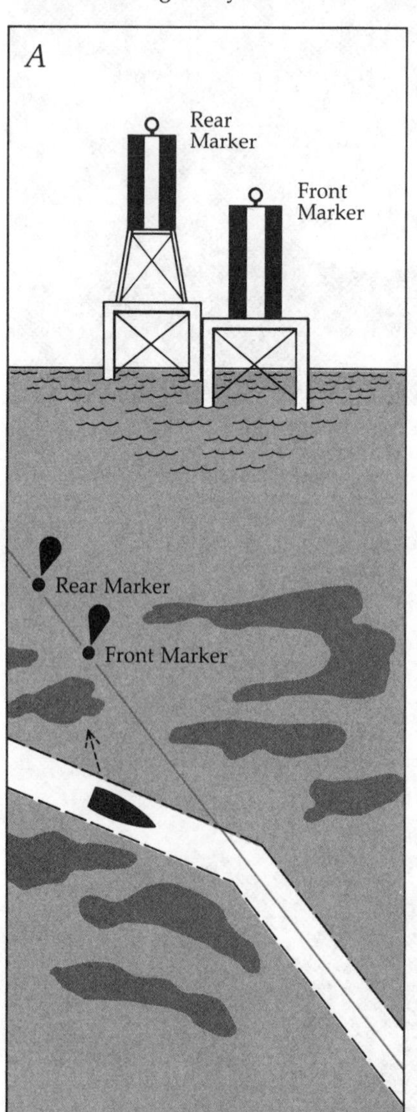

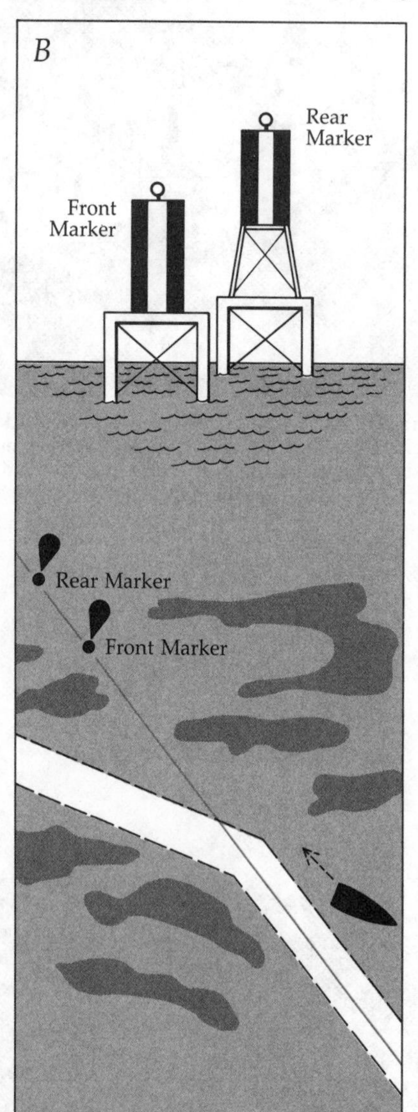

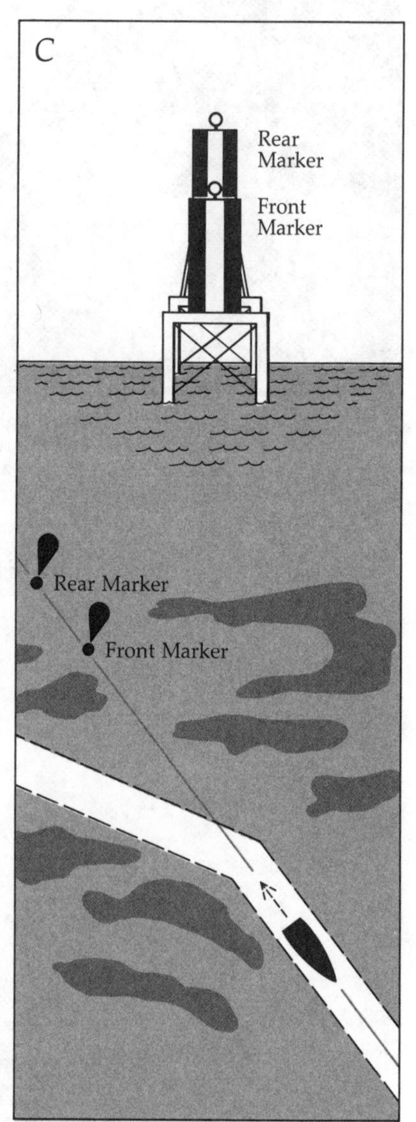

Figure 1618 **The skipper of a boat approaching a channel marked by a range, A, will first see the rear range marker to the left of the front marker. If he goes too far, B, he will see the rear marker to the right of the front one. By turning when the two are in line, C, he heads into the channel, and by keeping them in line maintains his course along the portion of the channel marked by the range.**

Ranges

A RANGE consists of two fixed aids to navigation so positioned with respect to each other that when seen in line they indicate that the observer's craft *may* be in safe waters, Figure 1618. The aids may be lighted or unlighted as determined by the importance of the range.

The conditional phrase "may be in safe waters" is used because observation of the two markers in line is *not* an absolute determination of safety. A range is "safe" only within specific limits of distance from the front marker; a vessel too close or too far away may be in a dangerous area. The aids that comprise the range do not in themselves indicate the usable portion of the range; check your chart and other aids.

Ranges are described in the *Light Lists* by first giving the position of the front marker, usually in terms of geographic coordinates—latitude and longitude—and then stating the location of the rear marker in terms of direction and distance from the front marker. This direction, given in degrees and minutes, true, need not be used in ordinary navigation, but is useful in making checks of compass deviation. The rear daymark (and light, if used) is always *higher* than the one on the front aid.

Because of their fixed nature, and the accuracy with which a vessel can be positioned by using them, ranges are among the best aids to navigation. Use a range whenever one is available; and use a buoy only to determine the beginning and end of the usable portion of the range.

Unlighted Ranges

Although any two objects may be used as a range, the term is properly applied only to those pairs of structures built specifically for that purpose. Special shapes and markings are used for the front and rear aids of a range for easier identification and more accurate alignment. Differing designs have been used in the past, but the Coast Guard has now standardized on the use of rectangular daymarks, longer dimension vertical, painted in vertical stripes of contrasting colors, see page 579. The design of specific range daymarks will normally be found in the *Light Lists.*

Lighted Ranges

Because of their importance and high accuracy in piloting, most range markers are equipped with lights, in addition to the usual daymarks, to extend their usefulness through the hours of darkness. Entrance channels are frequently marked with range lights; the Delaware River on the Atlantic Coast and the Columbia River on the Pacific Coast are examples of this.

Range lights may be of any color used with aids to navigation—white, red, or green—and may show any of several characteristics. The principal requirement is that they be easily distinguished from shore backgrounds and from other lights. Front and rear lights will, however, normally be of the same color (white is frequently used because of its greater visibility range), with different phase characteristics. Since both lights must be observed together for the proper steering of the craft, range lights often have a greater "on" interval than other lights do. Range rear lights are normally on more than their front counterparts; many ranges now show an isophase (equal interval) rear light and a quick-flashing front light.

Many range lights are fitted with special lenses that give a much greater intensity on the range centerline than off of it; the lights rapidly decrease in brilliance when observed from only a few degrees to either side. In some cases, the light will be visible only from on or very near to the range line; in other cases, a separate, lower light of lesser intensity may be seen all around the horizon—this can be either from the main light source or from a small auxiliary "passing" light. Light is shown around the horizon when the front aid also serves to mark the side of a channel at a turn of direction.

Some range lights will be of such high intensity that they can be seen and used for piloting in the daytime, being of more value than the painted daymarks.

Directional Lights

The establishment of a range requires suitable locations for two aids, separated adequately both horizontally and vertically. In some areas, this may not be possible and a single light of special characteristics will be employed.

A DIRECTIONAL LIGHT is a single light source fitted with a special lens so as to show a white light in a narrow beam along a desired direction, with red and green showing to either side. Width of the sectors will depend upon the local situation, but red will be seen if the pilot is to the right of the centerline as he approaches the aid from seaward, and green if he is to the left of the desired track. A typical directional light, at Deer Island, Massachusetts, shows white for a sector of 2.4° with red and green showing for 8.5° to either side. Another directional light in Delaware Bay has a white beam width of 1° 50′ with red and green sectors of 6°30′ to either side.

Directional lights will normally have an occulting or equal interval characteristic, so that they are easily followed.

Caution Regarding Directional Lights

A skipper should not place too great reliance on the various colors of a directional light for safe positional information. As noted for light sectors, the boundaries between colors are not sharp and clear; the light shades imperceptibly from one color to the other along the stated, and charted, dividing lines.

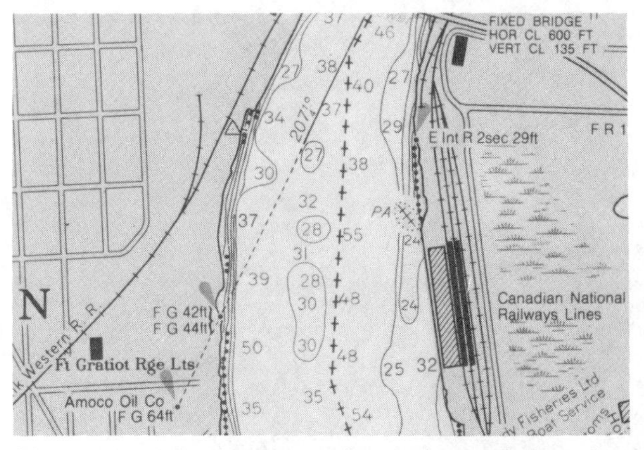

Figure 1619 **Lighted ranges are often used to mark channels in rivers. This is in the St. Clair river, and its true bearing (207°) is shown above the solid portion of the range line. Note the crossed dashed line marking the boundary between the U.S. and Canada.**

Coast Pilot
4
ATLANTIC AND GULF COAST
NOTICE TO MARINERS
TIDAL CURRENT CHARTS
East Coast of North and South America
ATLANTIC AND GULF COASTS
TRANSPORTATION
COAST GUARD

The useful books, tables, pamphlets, catalogs, and notices that can make your piloting more accurate

GOVERNMENT PUBLICATIONS

The typical boater undoubtedly thinks first of charts when considering government publications designed to help make his boating safe. It is certainly a valid thought—so much so that Chapter 18 is devoted entirely to charts—but he should not overlook the many other publications that can make his boating safer and more enjoyable. Generally, these are available from the same agencies that issue charts.

Publishing Agencies

Agencies of the federal government that issue publications valuable to the boatman include:

National Ocean Service (NOS—formerly National Ocean Survey), National Oceanic and Atmospheric Administration (NOAA), Department of Commerce.

United States Coast Guard, a part of the Department of Transportation.

The Defense Mapping Agency Hydrographic/Topographic Center (DMAHTC).

The U.S. Naval Observatory.

U.S. Army Corps of Engineers (District Offices).

National Weather Service, also a part of NOAA.

Government Printing Office

The Government Printing Office, an independent agency, does much of the actual printing. The Office of the Superintendent of Documents sells most of the publications, but there seems to be no rule about whether a publication is sold by GPO or by the government agency that prepared it.

Publications sold by the GPO can be bought by mail from: Superintendent of Documents, Government Printing Office, Washington, DC 20402. You can also buy them at a retail bookstore in the GPO building, North Capital and H Streets in Washington, and by mail and in person from GPO Regional Bookstores in many major cities. Make your checks payable to the "Superintendent of Documents," or put charges on your VISA or MasterCard account.

State Agencies

State agencies also produce publications of interest to boatmen, but there are too many of them to list in this book. Check with authorities in your own state, and write ahead to other states when you expect to cruise in new waters. See Appendix C for the name and address for inquiries in each state. Be as specific as possible in requests for information and literature. Information may be obtained that will add to safety and convenience, possibly avoiding legal embarrassment as well—remember, "Ignorance is no excuse" applies afloat as well as on shore.

Sales Agents

Government agencies have designated certain boating supply stores and marinas as official SALES AGENTS for their publications. Authorized sales agents may carry publications of the National Ocean Service, the Defense Mapping Agency Hydrographic/Topographic Center, or Coast Guard, or combinations of these. It does not hold true that because an establishment is an agent for one source of publications it will necessarily have the others.

Other Information

Additional information on where to obtain certain specific government publications is found in Appendix B.

National Ocean Service Publications

The National Ocean Service is charged with the survey of the coast, harbors, and tidal estuaries of the United States and its insular possessions. It issues the following publications relating to these waters as guides to navigation: Charts, *Coast Pilots, Tide Tables, Tidal Current Tables, Tidal Current Diagrams, Tidal Current Charts,* and *Chart Catalogs.*

Tide Tables

Tide Tables, Figure 1701, are of great value in determining the *predicted* height of the water at almost any place at any time. The tables are calculated in advance and are

Figure 1701 ***Tide Tables*** **published by the National Ocean Service provide necessary data for predicting height of tide for any desired day and hour at thousands of coastal points. Separate volumes cover Atlantic-Gulf and Pacific coasts.**

Figure 1702 ***Tidal Current Tables*** **are similar in format to the** ***Tide Tables.*** **A limited number of reference points together with thousands of subordinate stations permits calculation of current strengths and slacks at virtually any location.**

published annually in four volumes, one of which covers the East Coast of North and South America and another the West Coast of these continents. The other volumes are for Asian and European coasts.

The *Tide Tables* give the *predicted* times and heights of high and low waters for each day of the year at a number of important points known as REFERENCE STATIONS. The East Coast tables include 48 sets of detailed listings for such points as Portland, Boston, Sandy Hook, Baltimore, Miami, Pensacola, and Galveston. Comparable reference stations appear in the West Coast tables. Additional data show the differences in times and heights between these reference stations and thousands of other points, termed SUBORDINATE STATIONS. From these tables, a skipper can figure the tide at any time of any day at virtually any point along the coast.

THE NOS *Tide Tables* also include useful data including tables of sunrise and sunset, reduction of local mean time to standard zone time, moonrise and moonset, and astronomical events such as the phases of the moon. The tables and their use in piloting are discussed at length in Chapter 20.

Tidal Current Tables

The *Tidal Current Tables* have much the same format of reference and subordinate stations as the *Tide Tables;* see Figure 1702. Instead of times of high and low waters, however, these tables give the times and directions of minimum currents or slacks when the current reverses. These times do *not* correspond to times of high and low tides, and the *Tide Tables* cannot be used for current predictions. Velocity of the current at maximum strength is given in knots (nautical miles per hour).

Tidal Current Tables are published in two volumes: Atlantic Coast of North America, and Pacific Coast of North America and Asia. Each volume includes tables for calculating current velocity at any intermediate time, and the duration of slack water or weak currents.

Tidal Current Tables and their use in piloting are covered in detail in Chapter 20.

Tidal Current Charts

Tidal Current Charts are available for 12 bodies of water: Boston Harbor, Narragansett Bay to Nantucket Sound (two sets), Long Island and Block Island sounds, New York Harbor, Delaware Bay and River, Upper Chesapeake Bay, Charleston Harbor, Tampa Bay, San Francisco Bay, and Puget Sound (two sets). Each set consists of a series of 11 reproductions of the chart of the area, with the direction and velocity of tidal currents (shown graphically) for a specific hour with respect to the tide or current prediction for a major reference station. Currents in the various passages are indicated with red arrows and velocities are noted at numerous points; see Figure 1703. By following through the sequence of charts, hourly changes in velocity and direction are easily seen. These charts make it possible to visualize how tidal currents act at various points throughout every part of the entire 12-hour-plus cycle.

Where currents run at considerable velocity, and especially where tidal currents are complex, often flowing in opposite directions at the same stage of tide, you can gain a real advantage by consulting one of these charts. A few minutes' study may show you how to follow a favorable current through an entire passage, instead of needlessly bucking an opposing flow.

As for all current predictions, remember that these charts show normal conditions; a strong wind may influence both the strength of currents and the times that maximums are reached.

The Narragansett Bay *Tidal Current Chart* is used in conjunction with annual tide tables. All other *Tidal Current Charts* are used with the annual *Tidal Current Tables* for their respective areas. See page 469.

Tidal Current Diagrams

The *Tidal Current Diagrams* are a series of 12 monthly diagrams to be used with the *Tidal Current Charts* for Boston Harbor, Block Island Sound and Eastern Long Island Sound, New York Harbor, and Upper Chesapeake Bay to determine current flow on a particular day. These are republished for each year; make sure you have the proper edition.

Coast Pilots

Information on nautical charts is limited by space and by the system of symbols used. You will often need additional information for safe and convenient navigation. The National Ocean Service publishes such information in the *Coast Pilots*, covering the United States coastlines and the Great Lakes in nine separate volumes.

Each *Coast Pilot* (see Figure 1704) contains sailing directions between ports in its respective area, including recommended courses and distances. It describes channels with their controlling depths and all dangers and obstructions. It lists harbors and anchorages with information on boat supplies and marine repairs. It gives valuable information regarding canals, bridges, and the Intracoastal Waterways where applicable.

The *Coast Pilot* volumes cover the following areas:

Atlantic Coast:
- No. 1 Eastport to Cape Cod
- No. 2 Cape Cod to Sandy Hook
- No. 3 Sandy Hook to Cape Henry
- No. 4 Cape Henry to Key West
- No. 5 Gulf of Mexico, Puerto Rico, and Virgin Islands

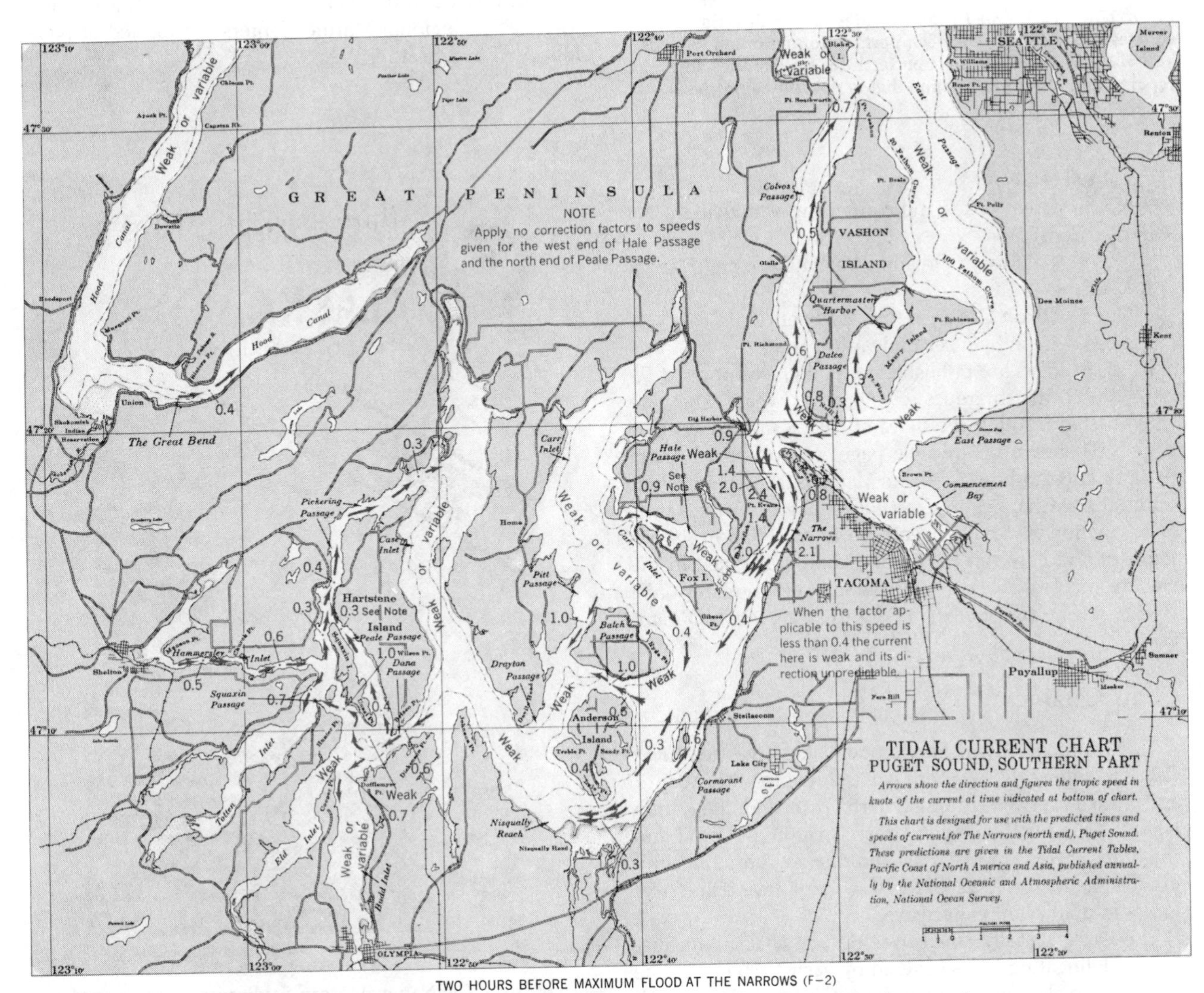

Figure 1703 ***Tidal Current Charts*** **are available for a number of major bodies of water where such flows are of significance in piloting. A series of 12 charts, for hourly intervals in the cycle of ebb and flow, make up each set. See text for the areas that are covered.**

Figure 1704 **Each *Coast Pilot* includes sailing directions, courses and distances between ports, information on hazards and aids to navigation, data on facilities for supplies and repairs, and other information that is not shown on charts.**

Great Lakes:
  No. 6 Great Lakes and connecting waterways
Pacific Coast:
  No. 7 California, Oregon, Washington, and Hawaii
Alaska:
  No. 8 Dixon Entrance to Cape Spencer
  No. 9 Cape Spencer to Beaufort Sea

All *Coast Pilots* are published annually, except *Coast Pilots* 8 and 9, which are issued every two years. They are all corrected through the dates of *Notices to Mariners* (see page 398) shown on the title page, so do not use them without checking the *Notices to Mariners* issued after their publication. Changes to *Coast Pilots* that affect the safety of navigation and have been reported to NOS in the interim period between new editions are published in the weekly and *Local Notices to Mariners,* except those for *Coast Pilot* 6, which are published only in *Local Notices to Mariners.*

Chart Catalogs

The National Ocean Service publishes five free *Chart Catalogs,* Figure 1705. There is a volume for the Atlantic and Gulf coasts, including Puerto Rico and the Virgin Islands; one for the Pacific Coast, including Hawaii and the other U.S. Pacific islands; and one for Alaska, including the Aleutian Islands. A fourth volume covers the Great Lakes and adjacent waterways.

These are actually small-scale outline charts with diagrams delineating areas covered by each NOS chart. The catalogs are also sources of additional information regarding charts and publications of other agencies; they also contain a listing, by states, of the names and addresses of local sales agents for charts and other publications.

A somewhat different catalog is NOS *Map and Chart Catalog 5:* Bathymetric Maps and Special Purpose Charts. This publication indexes such items as *Tidal Current Charts, Marine Weather Services Charts,* and *Storm Evacuation Maps.* The index charts show Fishery Conservation Zone boundaries 200 miles offshore, Continental Shelf limits (200-meter depth curve), and the outer edge of the Continental Slope (2500-meter depth curve).

NOS Offices

Nautical charts, *Tide* and *Tidal Current Tables, Tidal Current Diagrams, Coast Pilots,* and other publications of the National Ocean Service can be bought at NOS headquarters in Rockville, MD, or at the offices of the Distribution Division, 6501 Lafayette Avenue, Riverdale, MD 20737. (Both Rockville and Riverdale are suburbs of Washington, DC). Publications can be bought by mail from Riverdale if payment is sent with the order.

Other Offices

For information about NOS charts, publications, and activities, write to the National Ocean Service, Rockville, MD 20852. Regional Marine Centers are located at Norfolk, VA, and Seattle, WA.

Figure 1705 **The National Ocean Service issues catalogs of their nautical charts; these also include listings of other useful publications of the NOS and other agencies. These catalogs show the area covered by each chart, the scale, and the price.**

Coast Guard Publications

The United States Coast Guard prepares one major navigational publication and a number of minor ones, all worth consideration.

Light Lists

The Coast Guard has a series of publications for the coastal and inland waters known as the *Light Lists.* These books provide more complete information concerning aids to navigation than can be shown on charts (see Figure 1706) but they should *not* be used for navigation in place of charts and *Coast Pilots.* Volumes I-IV are published annually, Volume V every two years.

The *Light Lists* describe the lights (all classes), buoys, daybeacons, radiobeacons, and Loran-C chains maintained in all navigable waters of the United States by the Coast Guard and various private agencies. (In this usage, the Navy and all other non-USCG governmental organizations are considered to be "private agencies.") The data shown in the *Lists* include the official name of the aid, the characteristics of its light, sound, and radio signals, its structural appearance, position, and other significant factors.

Light Lists are published in five volumes as follows:

Volume I Atlantic Coast from St. Croix River, Maine, to Little River, South Carolina
Volume II Atlantic and Gulf Coasts from Little River, South Carolina, to Rio Grande, Texas
Volume III Pacific Coast and Pacific Islands
Volume IV Great Lakes
Volume V Mississippi River System

Within each volume, aids to navigation are listed by Coast Guard Districts in the following order: seacoast, major channels minor channels, and Intracoastal Waterway (if applicable). Lighted and unlighted aids appear together in their geographic order, with amplifying data on the same page.

Seacoast aids are listed in sequence from north to south along the Atlantic Coast, from south to north and east to west along the Gulf Coast, and from south to north along the Pacific Coast. Great Lakes aids are listed in a generally westerly direction. On the Atlantic and Gulf Coasts, aids along the Intracoastal Waterway are listed in the same sequence. For rivers and estuaries, the aids to navigation are shown from seaward to the head of navigation. Where an aid serves both a channel leading in from sea and the ICW, it will be listed separately in both sequences.

All volumes of the USCG *Light Lists* are for sale by the Superintendent of Documents, Government Printing Office; the volume for a particular area is sold by authorized chart and publication sales agents within that area. Often the *Light Lists* for other areas also will be available.

Navigation Rules

The Coast Guard has prepared an excellent booklet on the two sets of Rules of the Road. This is *Navigation Rules, International-Inland* officially designated as Commandant Instruction M16672.2A. It contains all changes except the

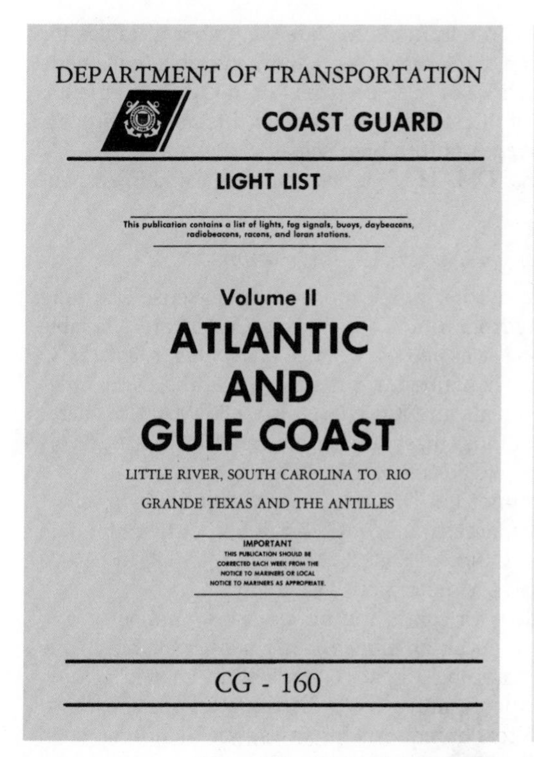

APALACHICOLA BAY — **FLORIDA** — EIGHTH DISTRICT

(1) No.	(2) Name / Characteristic	(3) Location Lat. N. Long. W.	(4) Nominal Range	(5) Ht. above water	(6) Structure Ht. above ground / Daymark	(7) Remarks	Year
	CHATTAHOOCHEE RIVER	River is marked with numerous unlisted daybeacons and unlighted buoys from mile 1.0 to Mile 154.7 which are periodically relocated to meet minor changing conditions.					
	CHEWALLA CREEK	Channel is marked with unlisted daybeacons and unlighted buoys from mile 0.0 to mile 1.0 which are periodically relocated to meet minor changing conditions.					
1562	GEORGIA POWER CO. TOWER LIGHTS. **(8) Occ. W.,** 1.5s (1s fl)	Mile 97.1	....	15	Skeleton tower	Private aid.	1964
	CAPE SAN BLAS (Chart 11401)						
	— Shoals Inner Buoy 1	In 33 feet 29 37.5 85 19.8	....	....	Black can	Green reflector.	
	— Shoals Inner Buoy 3	In 30 feet 29 38.5 85 22.6	....	....	Black can	Green reflector.	
	ST. JOSEPH BAY (Chart 11393)						
1563 140	— *Lighted Whistle Buoy SJ* ... **Mo. (A) W.**	In 41 feet 29 52.0 85 29.5	6	....	Black and white vertical stripes.	Ra ref.	
1564	— *Lighted Buoy 2* .... **Gp. Fl. R.,** 5s (2 Flashes)	In 39 feet 29 52.2 85 28.9	4	....	Red		
1565	— RANGE A FRONT LIGHT ... **Qk. Fl. W.**	In 27 feet 29 54.4 85 24.3	....	30	KRW on concrete pile structure.	Visible all around; higher intensity on rangeline.	1902—1960
1566 139	— RANGE A REAR LIGHT .... **E. Int. W.,** 6s	On shore, 2,930 yards 062°08′ from front light.	....	78	KRW on skeleton tower		1902—1955
1567	— *Lighted Buoy 3* .... **Fl. G.,** 4s	In 35 feet	4	....	Black		
	— Buoy 4	In 37 feet	....	....	Red nun	Red reflector.	
	— Buoy 5	In 33 feet	....	....	Black can	Green reflector.	
1568	— *Lighted Buoy 6* .... **Fl. R.,**2.5s	In 40 feet	4	....	Red		
	— Buoy 7	In 38 feet	....	....	Black can	Green reflector.	
	— Buoy 8	In 38 feet	....	....	Red nun	Red reflector.	
1569	— *Lighted Buoy 9* .... **Fl. G.,** 2.5s	In 32 feet	4	....	Black		

Figure 1706 ***Light Lists,*** **published by the U.S. Coast Guard, provide more detailed information on all types of aids to navigation than can be shown conveniently on charts.**

minor ones that came into effect on 30 October 1984.

This booklet presents the various rules and annexes side-by-side—International on the left, Inland on the right. Illustrations for specific rules appear near the applicable text. There are many illustrations of lights and shapes, each adjacent to the applicable rule.

The demarcation lines separating U.S. Inland Rules waters from International Rules waters are described in detail. The booklet also describes penalties for violations of rules or regulations; regulations relating to regattas and marine parades; and gives information on Vessel Traffic Services and Vessel Separation Schemes at many major ports and harbor entrances.

Although copies of earlier equivalent booklets were distributed by the Coast Guard without charge, budgetary limitations now require that *Navigation Rules, International-Inland* be sold. Order copies by mail from the Superintendent of Documents, Government Printing Office; or from local sales agents for charts and other nautical publications.

Other USCG Publications

The Coast Guard also issues publications about the safety of navigation and covering such topics as aids to navigation, applicable rules and regulations, and general safety matters. See Appendix B for a more complete listing. Some of these are free, others are available for a nominal charge.

CG-290 is a small pamphlet covering federal requirements for recreational boats. Topics include laws and regulations, numbering and documentation, reporting accidents, and approved equipment.

These publications and others can be obtained by writing USCG Headquarters (G-BA), Washington, DC 20593. Copies may also be available at Coast Guard District offices and some local USCG stations.

DMAHTC Publications

The Defense Mapping Agency Hydrographic/Topographic Center has several publications of interest to boatmen. Charts, publications, and other products of DMAHTC are identified with a prefix to the individual chart or publication number.

The nautical chart identification system uses a number of one to five digits, without prefix, as determined by the scale and assigned on a basis of regions and subregions. Publications and some other items will continue to have an "H.O." (Hydrographic Office) prefix until reprinted with a DMAHTC number.

Oceanographic products, primarily publications related to the dynamic nature of the oceans or the scientific aspects of oceanography, are printed by the Naval Oceanographic Office and carry numbers with prefixes that reflect the nature of the particular publication.

Bowditch (Publication No. 9)

The *American Practical Navigator,* originally written by Nathaniel Bowditch in 1799, and generally referred to simply as *Bowditch,* is an extensive treatise on piloting, celestial navigation, and other nautical matters. It is Publication No. 9 of DMAHTC.

Bowditch has long been accepted as an authority on questions of piloting and other forms of navigation. The current edition consists of two volumes. The 1981 Volume II contains tables, data, equations, and instructions needed to perform navigational calculations. The 1984 Volume I contains basic reference material. Volume II alone will meet the needs of most boatmen.

Other DMAHTC Publications

Other Hydrographic/Topographic Center publications that may be found useful to a boatman include Pubs. 117 A and B, *Radio Navigational Aids;* and Pub. 102, *International Code of Signals* (see page 534).

The Hydrographic/Topographic Center publishes a series of *Sailing Directions* which provide supplementary information for foreign coasts and ports in a manner generally similar to the *Coast Pilots* for U.S. waters.

The *Lists of Lights* published by this agency likewise cover foreign waters and so do not duplicate the Coast Guard *Light Lists.* These are DMAHTC Pubs. No. 110 through 116.

DMAHTC charts and publications are listed in the *DMA Catalog of Maps, Charts, and Related Products, Part 2.* There are nine regional volumes (I–IX), plus *Volume X, Miscellaneous and Special Purpose Navigational Charts, Sheets, and Tables.*

DMAHTC publications include a number of tables for the reduction of celestial observations, and tables and charts for plotting lines of position from Loran-C measurements. These are of interest only to those yachtsmen making extensive voyages on the high seas.

Additional DMAHTC publications are listed in Appendix B.

Obtaining DMAHTC Publications

Charts and other publications of the Defense Mapping Agency Hydrographic/Topographic Center are available through local sales agents. Listings of agents for DMAHTC products are contained in the volumes of the *DMA Catalog.* (Some agents for NOS charts also sell DMAHTC charts and publications; these are so indicated in the four volumes of the *NOS Nautical Chart Catalog.*)

If you cannot get DMAHTC charts and other publications from a local sales agent, you can order them by mail from DMA Office of Distribution Services (DDCP), 6500 Brookes Lane, Washington, DC 20315.

Mail orders for charts and publications must be accompanied by a check or money order made payable to the "Defense Mapping Agency" and will be mailed at government expense in regular printed-matter postal service. The added cost of requests for any special handling, such as air mail or special delivery, must be borne by the purchaser.

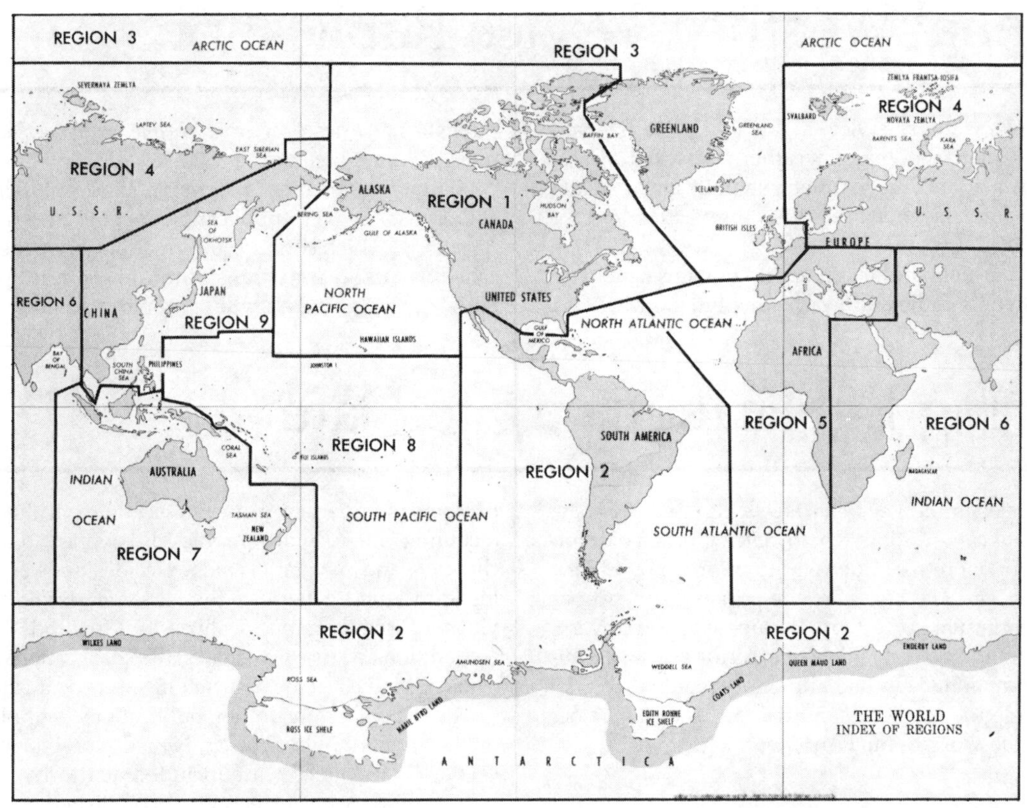

Figure 1707 **Charts published by the Defense Mapping Agency Hydrographic/Topographic Center are numbered according to a system of worldwide regions and subregions.**

Naval Observatory Publications

The Naval Observatory publishes the *Nautical Almanac* in annual editions and the *Air Almanac* for six-month periods; there is also the annual *Almanac for Computers*, useful with programmable calculators and microcomputers. These books contain astronomical data needed for celestial navigation.

The Naval Observatory also participates in and assists the publication of other navigational documents such as the *Tide Tables* and celestial sight reduction tables, but it is not the agency directly responsible for them.

Army Corps of Engineers Publications

The Corps of Engineers of the U.S. Army has the responsibility for navigational and informational publications on major inland (non-tidal) rivers such as the Tennessee, Ohio, and Mississippi, and many lakes and reservoirs behind the large dams.

Intracoastal Waterway Booklets

The Army Engineers has prepared two paperbound booklets on the Intracoastal Waterway, which comes under its jurisdiction. These booklets contain descriptive material, photographs, small-scale charts, and tabulated data.

Unfortunately, these publications are *not* periodically updated, and they are of far less value than the corresponding volumes of the NOS *Coast Pilots* with their annual supplements.

Bulletins on the Intracoastal Waterways are issued periodically by the Engineers District Offices. Addresses of these offices are given in Appendix B.

Rivers and Lakes Information

Regulations relating to the use of many rivers and lakes (reservoirs), *Navigational Bulletins*, and *Notices to Navigation Interests* are issued by various offices of the Corps of Engineers, as listed in Appendix B. Other government publications relating to inland river and lake boating are also listed on that page, together with information on their availability.

National Weather Service Publications

The National Weather Service, a part of NOAA, Department of Commerce, prepares weather maps that appear in many newspapers, but boatmen generally make greater use of radio and television broadcasts for weather and sea conditions.

To assist mariners and boatmen in knowing when and where to listen for radio and TV weather broadcasts, NWS publishes a series of *Marine Weather Services Charts* in annual editions. (See pages 322-323.)

Boatmen who cruise far offshore and in foreign waters should use the NWS publication *Worldwide Marine Weather Broadcasts*, which contains information on frequencies and schedules of stations transmitting in radiotelephone, radiotelegraph, radioteletype, and radiofacsimile modes.

Keeping Publications up to Date

You should be sure to keep charts and certain other navigational publications fully up to date. Outdated information can be more harmful than no information at all.

Coast Pilots and *Light Lists* are the primary publications that need continual correction. Fortunately, the government has provided a convenient means for executing this important function. The time and effort required are not great, provided a skipper keeps at it regularly and does not permit the work to build up a backlog.

Notice to Mariners

The Defense Mapping Agency Hydrographic/Topographic Center publishes a *Notice to Mariners* which is prepared jointly with the National Ocean Service and the Coast Guard; see Figure 1708. These pamphlets advise mariners of important matters affecting navigational safety, including new hydrographic discoveries, changes in channels and navigation aids, etc. Besides keeping mariners informed generally, the *Notice to Mariners* also provides information specifically useful for updating the latest editions of nautical charts and publications.

Each issue contains instructions on how it is to be used to correct charts and other publications. Supplementary information is published as Notice No. 1 of each year; lists of charts affected are included quarterly.

A "Summary of Corrections" is published semiannually covering charts, *Coast Pilots*, and *Sailing Directions*.

No. 7

NOTICE TO MARINERS

PUBLISHED WEEKLY BY THE
DEFENSE MAPPING AGENCY HYDROGRAPHIC/TOPOGRAPHIC CENTER

PREPARED JOINTLY WITH THE
NATIONAL OCEAN SURVEY AND U.S. COAST GUARD

CONTENTS

		PAGE
SEC. I	Chart Corrections	I-1.1
	Coast Pilots/Sailing Directions/Fleet Guides	I-2.1
	Catalog Corrections—New Charts and Pubs	I-3.1
	Chartlets/Depth Tabulations/Notes	I-4.1
SEC. II	Light List Corrections	II-1.1
	Radio Navigational Aids Corrections	II-2.1
	Other Pub. Corrections	II-3.1
SEC. III	Broadcast Warnings	III-1.1
	Marine Information—Miscellaneous	III-2.1

13 FEBRUARY

Figure 1708 ***Notice to Mariners,*** **published weekly by the DMAHTC, is used to correct charts and other navigational publications. A single volume covers the entire world.**

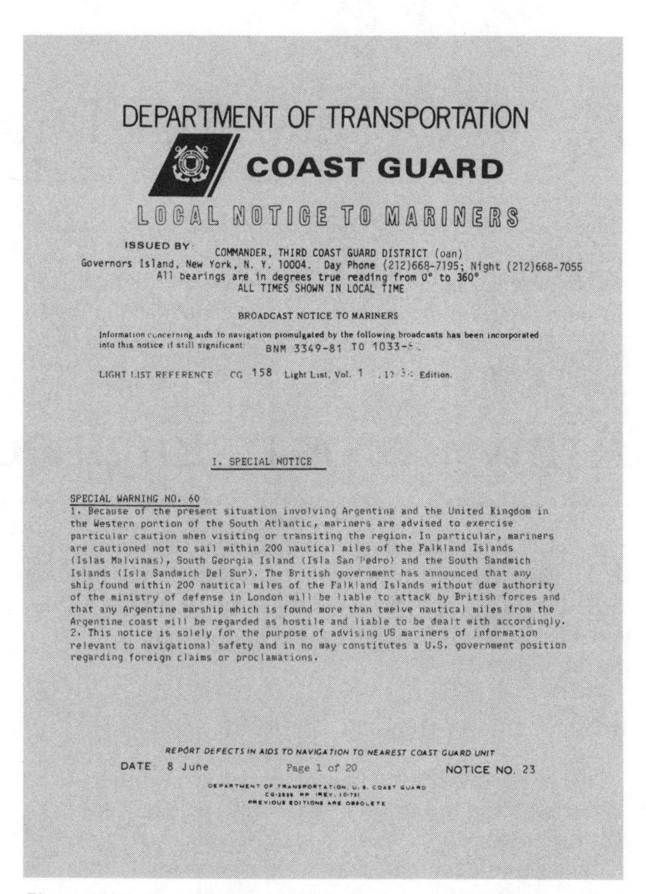

DEPARTMENT OF TRANSPORTATION
COAST GUARD

LOCAL NOTICE TO MARINERS

ISSUED BY: COMMANDER, THIRD COAST GUARD DISTRICT (oan)
Governors Island, New York, N. Y. 10004. Day Phone (212)668-7195; Night (212)668-7055
All bearings are in degrees true reading from 0° to 360°
ALL TIMES SHOWN IN LOCAL TIME

BROADCAST NOTICE TO MARINERS

Information concerning aids to navigation promulgated by the following broadcasts has been incorporated into this notice if still significant: BNM 3349-81 TO 1033-[illegible]

LIGHT LIST REFERENCE CG 158 Light List, Vol. 1 , 1[illegible] Edition.

I. SPECIAL NOTICE

SPECIAL WARNING NO. 60

1. Because of the present situation involving Argentina and the United Kingdom in the Western portion of the South Atlantic, mariners are advised to exercise particular caution when visiting or transiting the region. In particular, mariners are cautioned not to sail within 200 nautical miles of the Falkland Islands (Islas Malvinas), South Georgia Island (Isla San Pedro) and the South Sandwich Islands (Isla Sandwich Del Sur). The British government has announced that any ship found within 200 nautical miles of the Falkland Islands without due authority of the ministry of defense in London will be liable to attack by British forces and that any Argentine warship which is found more than twelve nautical miles from the Argentine coast will be regarded as hostile and liable to be dealt with accordingly.
2. This notice is solely for the purpose of advising US mariners of information relevant to navigational safety and in no way constitutes a U.S. government position regarding foreign claims or proclamations.

REPORT DEFECTS IN AIDS TO NAVIGATION TO NEAREST COAST GUARD UNIT

DATE: 8 June Page 1 of 20 NOTICE NO. 23

DEPARTMENT OF TRANSPORTATION, U.S. COAST GUARD
CG-2835 RR (REV. 10-73)
PREVIOUS EDITIONS ARE OBSOLETE

Figure 1709 ***Local Notices to Mariners*** **are published by each Coast Guard District on a "when required" basis. Local yacht clubs and marinas keep them on file.**

Local Notice to Mariners

The Commander of each Coast Guard District issues *Local Notices to Mariners;* Figure 1709. These are reproduced and mailed from the respective District offices, bringing information to users several weeks in advance of the weekly *Notices* that are printed in Washington, and mailed from there.

Local Notices are of particular interest to small-craft skippers, as the *Notices* from Washington do not carry information regarding inland and other waters not used by large ocean-going vessels.

How to Obtain Copies

Boaters can get on the mailing list for *Local Notices to Mariners* by sending a written request to the Commander (oan), of the applicable Coast Guard District. See Chapter 26 for district boundaries; you can obtain the District office mailing address from various Coast Guard publications, or your local Coast Guard or Coast Guard Auxiliary unit.

Getting on the distribution list for the weekly Notices to Mariners mailed from Washington, DC, is more difficult, and is rarely necessary. Application, with adequate justification, is made to Office of Distribution Services (IMA), Defense Mapping Agency, Washington, DC 20315.

Even if you don't get your own copies, information in *Notices* and *Local Notices* is readily available. Local sales agents receive copies and maintain files that you can read. Many yacht clubs and marinas also receive and post copies of *Local Notices*.

Report All Useful Information

All boatmen are urged to help governmental agencies maintain the buoyage system by reporting any damage to aids, malfunctioning of lights, shifting of shoals and channels, or new hazards. Use a radio if the matter is urgent; otherwise, suggestions for the improvement of aids to navigation should be sent to the Commandant, U.S. Coast Guard (G-N), Washington, DC 20593.

Data concerning dangers to navigation, changes in shoals and channels, and similar information affecting charts or publications of the NOS should be sent to the Director, National Ocean Service, Rockville, MD 20852.

DMAHTC also publishes a semi-annual *Summary of Corrections* in regional volumes. Each issue contains the full text of all accumulated corrections from *Notices to Mariners,* except for *Light Lists* and certain other navigational publications. It is easier to use than a file of *Notices* for bringing up to date a chart that has not been kept corrected. The *Summaries* do, however, require a paid subscription.

Quasi-Governmental Publications

In addition to the governmental agencies and their publications noted above, several activities best described as "quasi-governmental" produce publications of interest to boatmen. Two of these are discussed below.

Radio Technical Commission for Maritime Services

This organization is better known by its initials, RTCM. It includes individuals from governmental agencies such as the FCC, the Coast Guard, NOAA, Maritime Administration, and others—user organizations such as ocean steamship operating groups, Great Lakes shipping interests, and the United States Power Squadrons; equipment manufacturers; labor organizations; and communications companies such as the Bell System. The RTCM does not have authority to make binding decisions, but it wields considerable influence through its function as a meeting place for and resolution of conflicting views and interests.

The RTCM publishes the *Marine Radiotelephone Users Handbook,* an authoritative pamphlet on the FCC Rules and Regulations which also has useful information on the selection, installation, and proper use of radios on boats; see page 530. This useful booklet can be ordered from the Radio Technical Commission for Marine Services, c/o FCC, P.O. Box 19087, Washington, DC 20036.

Naval Institute Publications

The United States Naval Institute is not a governmental agency; rather it is a private association "for the advancement of professional, literary, and scientific knowledge in the Navy." One of its principal activities is the publication of books on naval and maritime matters.

The Naval Institute publishes a large number of books; most familiar to boatmen will be *Dutton's Navigation and Piloting.* Like *Bowditch,* this volume has come to be an accepted authority on matters of piloting and navigation. The current volume is an outgrowth of years of development from an early work; the book was initially known as *Navigation and Nautical Astronomy* by Benjamin Dutton. This useful book—now in its 14th edition—can be purchased in boating supply and book stores, or it can be ordered directly by mail from the U.S. Naval Institute, Annapolis, MD 21402.

MAR DEL NORT
Tropicus Cancri
CUBA
ESPANOLA
Iamaico
Pto Rico
La Bermuda
Golfo de Venecuela
L. de Maracaibo
Milliaria Germanica
Leucæ Hispanicæ
10 20 30 40 50 60 70
Amstelodami

18

Chart types and what they show; why they are important to the pilot, and how to use them to best advantage

THE NAUTICAL CHART

To travel anywhere safely in his boat, a skipper must have knowledge of water depths, shoals, and channels. He must also know the location of aids to navigation and landmarks, and where ports and harbors can be found. At any given position, he can generally measure the depth, and he may see some landmarks; but for true safety he has to know the depth ahead, the actual location of the aids to navigation he can see, and where more aids lie on the course he will follow. To plan the best route to his destination, he must know the dangers to navigation along the way. This information can best be determined from up-to-date nautical charts. The skipper must not only have the required charts on board, he must also know how to use them.

Charts vs. Maps

A NAUTICAL CHART, Figure 1801 top, is a representation in miniature, on a plane surface, of a portion of the earth's surface emphasizing the water and natural and man-made features of particular interest to a navigator. A MAP, Figure 1801 bottom, is a similar miniature representation for use on land in which the emphasis is on roads, cities, and political boundaries.

A chart includes information about depth of water, obstructions and other dangers to navigation, and the location and type of aids to navigation. Adjacent land areas are only portrayed in details that aid a navigator—the shoreline, harbor facilities, and prominent natural or man-made features. Charts are printed on heavy-weight, durable paper so that they may be used as worksheets on which courses may be plotted and positions determined. For skippers of small craft, there are even special charts with details on marinas and similar facilities.

A chart's basic purpose is to give the navigator information that enables him to make the *right decision in time to avoid danger.* Charts differ from road maps both in the kinds of information and in the precision of their details: for safety, charts must be extremely accurate. Even a small error in charting the position of a submerged obstruction can be a serious hazard to navigation.

At least one major oil company produces "cruising guides" for important boating areas. These are useful in planning a nautical trip, but they lack the detail precision and provision for revisions between printings (by means of *Notices to Mariners*) required for actual navigational use. Do *not* try to substitute them for official government charts.

Geographic Coordinates

Charts show a grid of intersecting lines to aid in describing a specific position on the water. These lines are charted representations of a system of GEOGRAPHIC COORDINATES that exist on the earth's surface.

The earth is nearly spherical in shape—it is slightly flattened along the polar axis, but the distortion is minimal and need concern only those who construct the charts, not boaters. A GREAT CIRCLE, Figure 1802 top, is the line traced out on the surface of a sphere by a plane cutting through the sphere at its center. It is the largest circle that can be drawn on the surface of a sphere. A SMALL CIRCLE, Figure 1802 bottom, is one marked on the surface of a sphere by a plane that does not pass through its center.

Meridians and Parallels

Geographic coordinates are defined by two sets of great and small circles. One is a set of great circles each of which passes through the north and south geographic poles—these are the MERIDIANS OF LONGITUDE, Figure 1803 top. The other set is a series of circles each established by a plane cutting through the earth perpendicular to the polar axis. The largest of these is midway between the poles and thus passes through the center of the earth, becoming a great circle; this is the EQUATOR, Figure 1803 bottom. Other parallel planes form small circles known as the PARALLELS OF LATITUDE, Figure 1803 bottom.

Geographic coordinates are measured in terms of DEGREES (one degree is 1/360th of a complete circle). The meridian that passes through Greenwich, England, is the reference for all measurements of longitude and is designated as the PRIME MERIDIAN, or 0 degrees. The longitude of any position on earth is described as ___ degrees East or West of Greenwich, to a maximum in either direction of 180°. The measurement can be thought of as either the angle at the North and South Poles between the meridian of the place being described and the prime meridian, or as the arc along the equator between these meridians, Figure 1804 left. The designation of "E" or "W" is an essential part of any statement of longitude, abbreviated as "Long." or "Lo" or as "λ" (the Greek letter *lambda*).

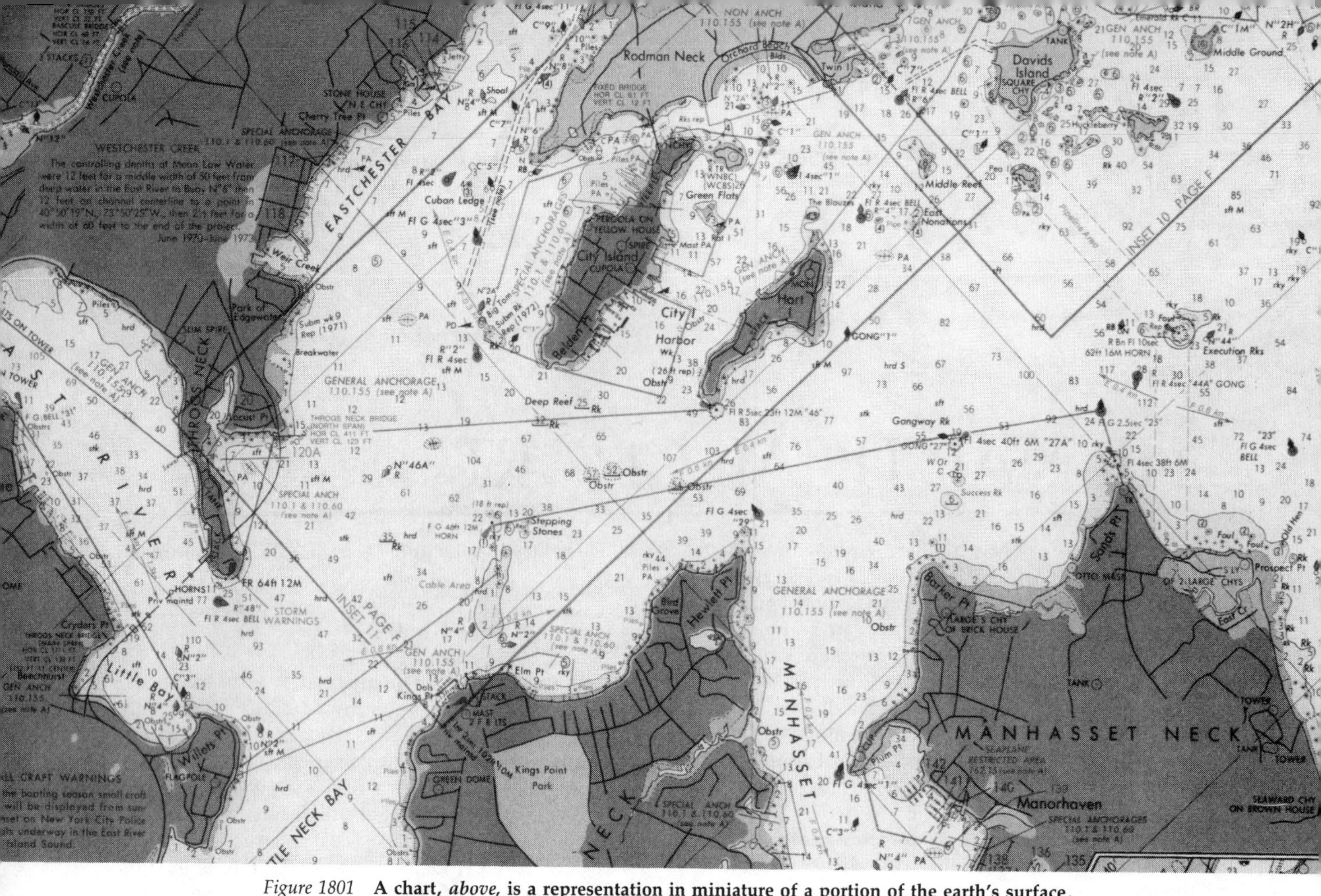

Figure 1801 **A chart, *above,* is a representation in miniature of a portion of the earth's surface, with emphasis on natural and man-made features that are of particular interest to boatmen. A map, *below,* is also a miniature representation of a portion of the earth's surface, but with emphasis on towns, roads, and features of particular interest to motorists.**

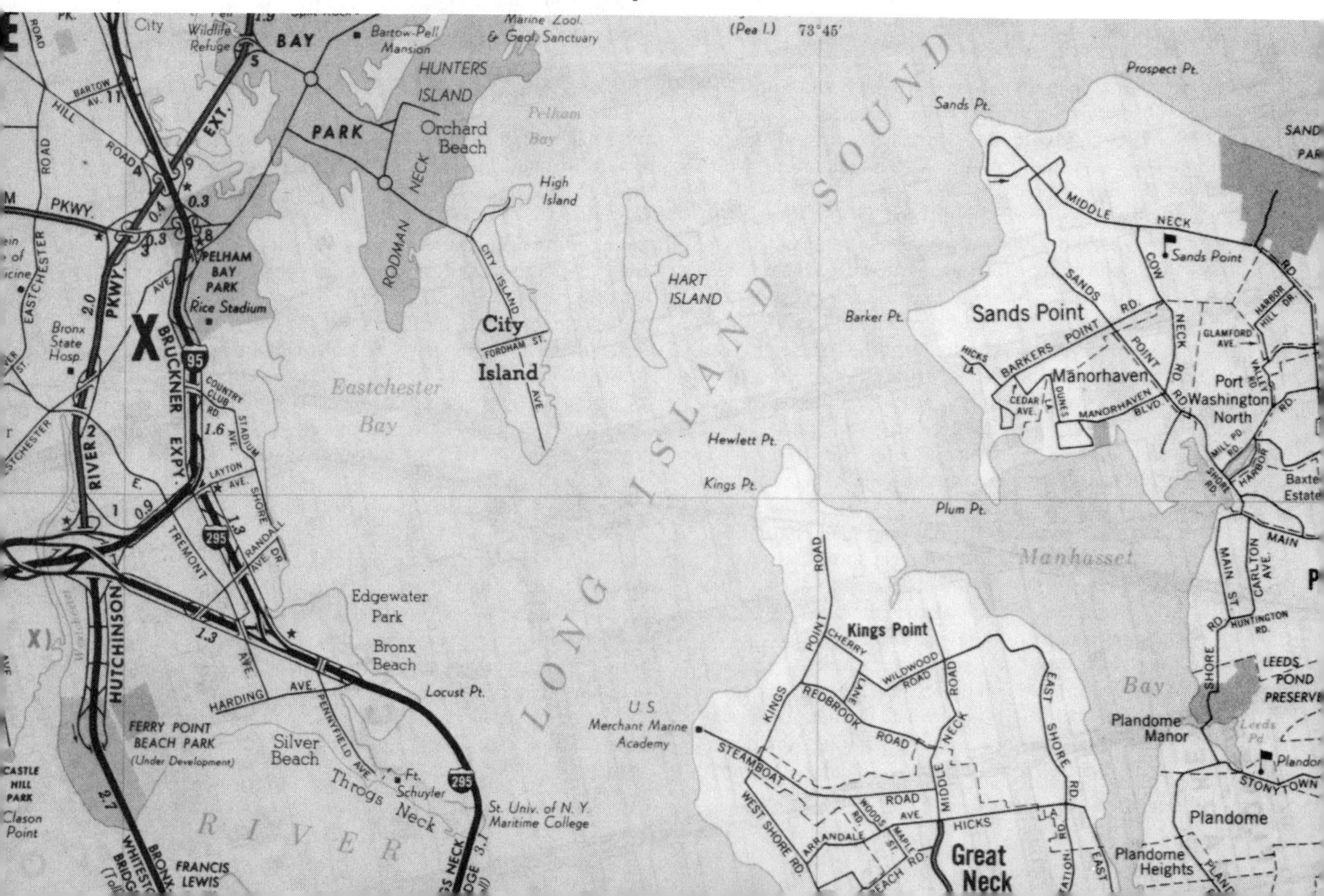

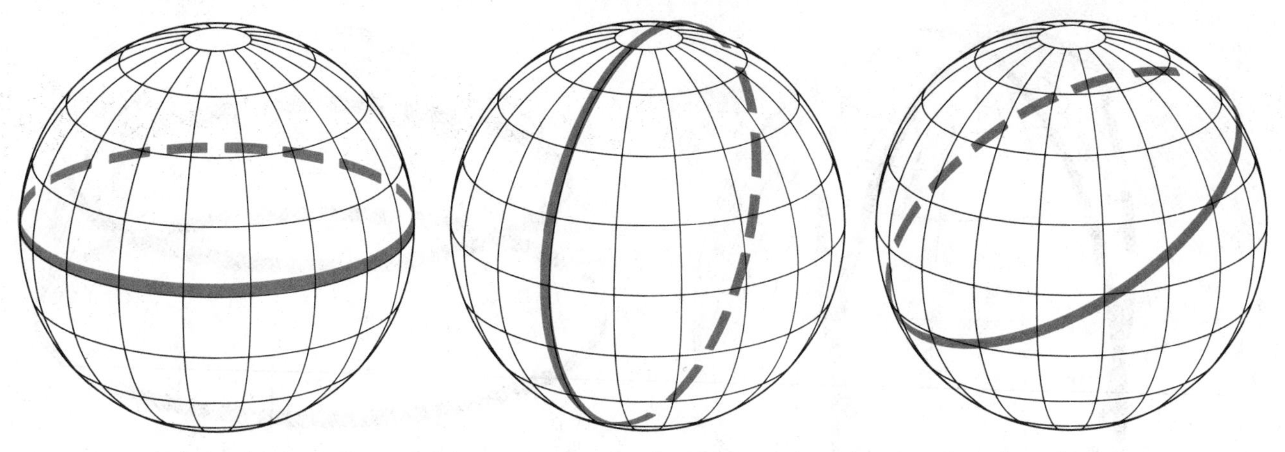

Figure 1802 **A great circle, *above,* is the line traced on the surface of a sphere by a plane that cuts through the center of the sphere. A small circle, *below,* is a line traced on the surface of a sphere by a plane that does not cut through the sphere's center.**

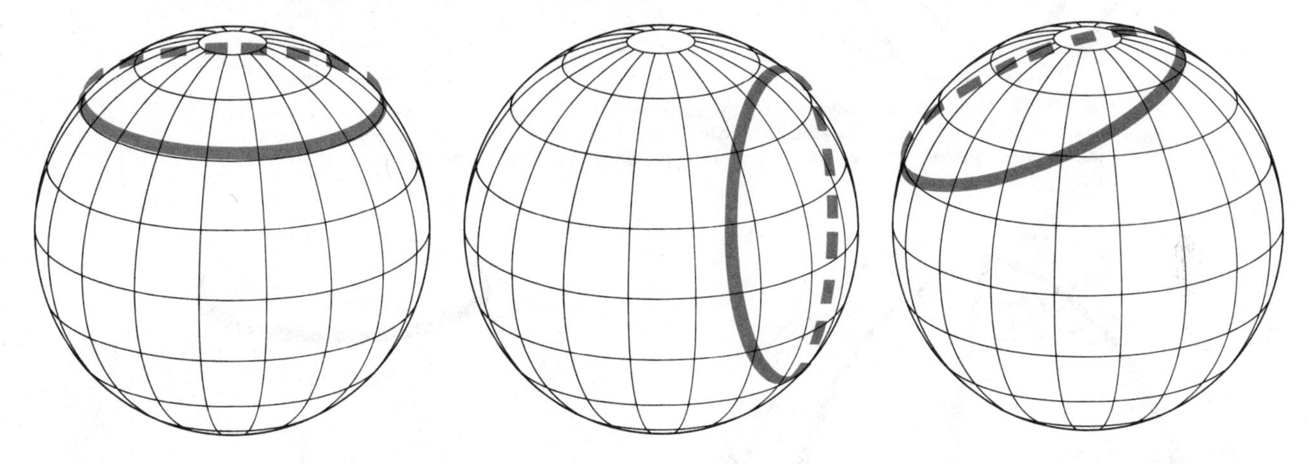

Parallels of latitude are measured in degrees north or south from the equator, from 0° at the equator to 90° at each pole. The designation of latitude (abbreviated as "Lat." or "L") as "N" or "S" is necessary for a complete position; see Figure 1804 right.

For greater precision in position definition, degrees are subdivided into MINUTES (60 minutes = 1 degree) and SECONDS (60 seconds = 1 minute). In some instances, minutes are divided decimally in tenths.

From Figure 1804, you can see that the meridians of longitude get closer together as one moves away from the equator in either direction, and converge at the poles. Thus the *distance* on the earth's surface between adjacent meridians is not a fixed quantity but varies with latitude. On the other hand, the parallels of latitude are equally spaced and the distance between successive parallels is the same. One degree of latitude is, for all practical purposes, 60 nautical miles; *1 minute of latitude is used as 1 nautical mile,* a relationship which we will later see is quite useful.

Chart Construction

The construction of a chart presents the problem of representing a spherical, three-dimensional surface on a plane, a two-dimensional sheet of paper. It is actually impossible to accomplish this exactly, and a certain amount of distortion is inevitable, but various methods, called PROJECTIONS, can provide practical and sufficiently accurate results.

The transfer of information from the sphere to the chart's flat surface should be accomplished with as little distortion as possible in the shape and size of land and water areas, the angular relation of positions, the distance between points, and other more technical properties. Each projection is superior to others in one or more of these qualities; none is superior in all characteristics. In all projections, as the area covered by the chart is decreased the distortion diminishes, and the difference between types of projections lessens.

Of the many projection techniques that are used, two are of primary interest to boatmen. The MERCATOR PROJECTION is an example of the most common; it is used for charts of ocean and coastal areas. The POLYCONIC PROJECTION is employed for charts of the Great Lakes and inland rivers. The average skipper can quite safely navigate his boat using either type of chart without a deep knowledge of the techniques of projection. (For those who would like to know more about the various projection methods, see pages 419-423.)

Direction

DIRECTION is defined as the angle between a line connecting one point with another point and a base, or reference, line extending from the original point toward the True or Magnetic North Pole; the angle is measured in

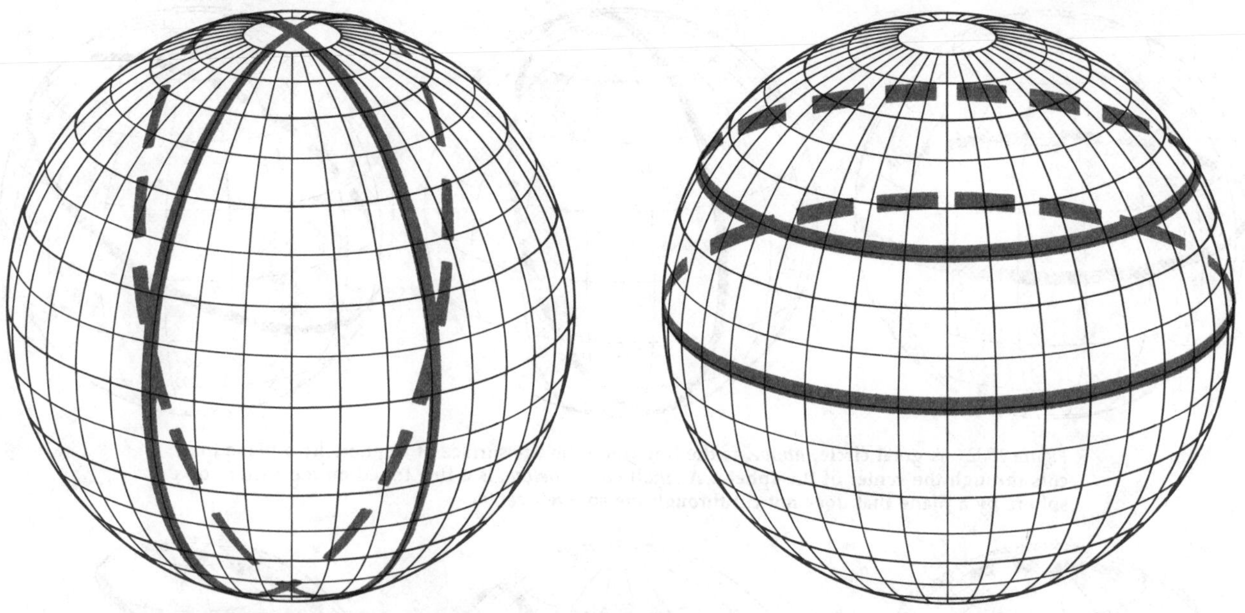

Figure 1803 **Meridians of longitude on the earth are great circles, *left,* which pass through both the north and south poles. Parallels of latitude, *right,* are small circles that are parallel to the plane of the equator. The equator is a great circle whose plane is perpendicular to the earth's axis.**

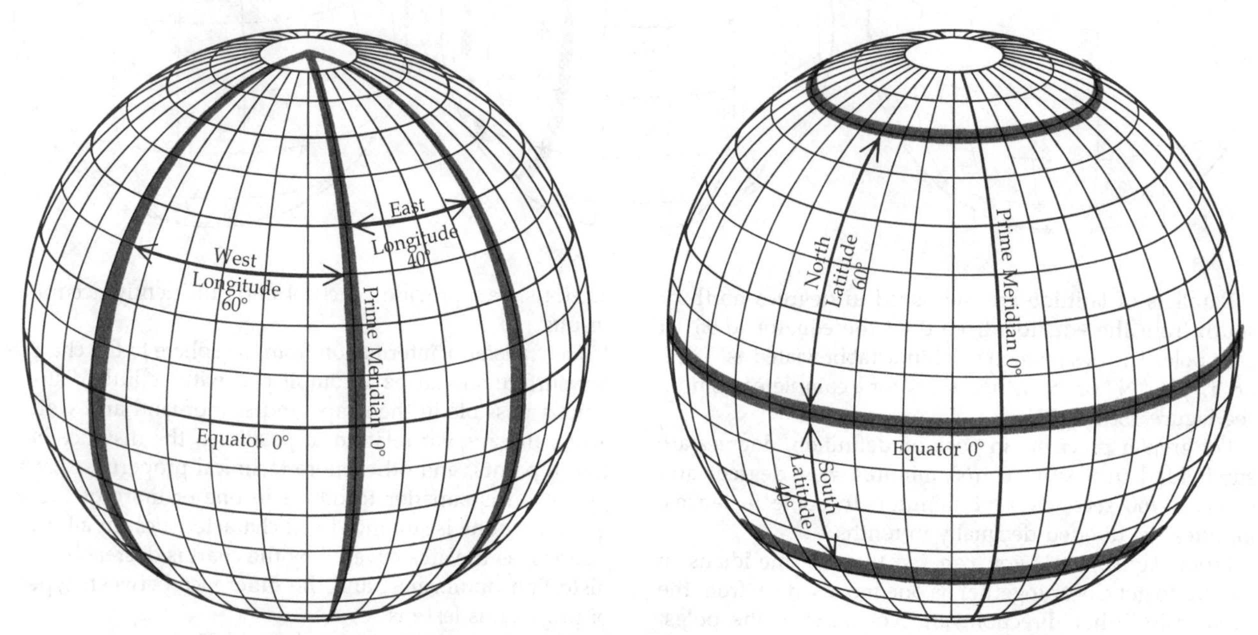

Figure 1804 **Longitude, *left,* is measured from the Prime Meridian (0 degrees), which passes through Greenwich, England, east or west to a maximum of 180 degrees. Latitude, *right,* is measured north or south from the equator (0 degrees) to the poles (90 degrees).**

degrees clockwise from the reference line. Thus direction on charts may be described as so many degrees TRUE (T) or so many degrees MAGNETIC (M). The difference between these directions is VARIATION and must be allowed for as described in Chapter 15. The principal difference in the use of charts on the Mercator projection from those on the polyconic projection lies in the techniques for measuring direction; this will be covered in Chapter 19.

Measurement of Direction

To facilitate the measurement of direction, as in plotting bearings and laying out courses, most charts have COMPASS ROSES printed on them. A compass rose, Figure 1805, consists of two or three concentric circles, several inches in diameter and accurately subdivided. The outer circle has its zero at *true* north; this is emphasized with a star. The next circle or circles are oriented to magnetic north. The middle circle, if there are three, is magnetic direction expressed in degrees, with an arrow printed over the zero point to indicate magnetic north. The innermost circle is also magnetic direction, but in terms of "points," and halves and quarters thereof; its use by modern boatmen will be limited. (One point = 11¼ degrees.) The use of points with the mariner's compass is covered in Chapter 15.

The difference between the orientation of the two sets

of circles is, of course, the magnetic variation at the location of the compass rose. The amount of the variation and its direction (Easterly or Westerly) is given in words and figures in the center of the rose, together with a statement of the year that such variation existed and the annual rate of change. When using a chart in a year much later than the compass rose date, it *may* be necessary to modify the variation shown by applying the annual rate of change. Such cases are relatively rare as rates are quite small and differences of a small fraction of a degree may generally be ignored.

Each chart has several compass roses printed on it in locations where they do not conflict with navigational information.

Until a skipper has thoroughly mastered the handling of compass "errors," he should use only true directions and the true (outer) compass rose. Later, the magnetic rose may be used *directly*, thus simplifying computations.

Several cautions are necessary when measuring directions on charts. For large-area charts, the magnetic variation can differ for various portions of the chart. Check each chart when you first start to use it, and to be sure, always use the compass rose *nearest* the area for which you are plotting. Depending upon a chart's type and scale, graduations on its compass rose circles may be for intervals of 1 degree, 2 degrees, or 5 degrees. On some charts, the outer (true) is subdivided into units of 1 degree while the inner (magnetic) circle, being smaller, is subdivided into steps of 2 degrees. Always check to determine the interval between adjacent marks on each compass rose scale.

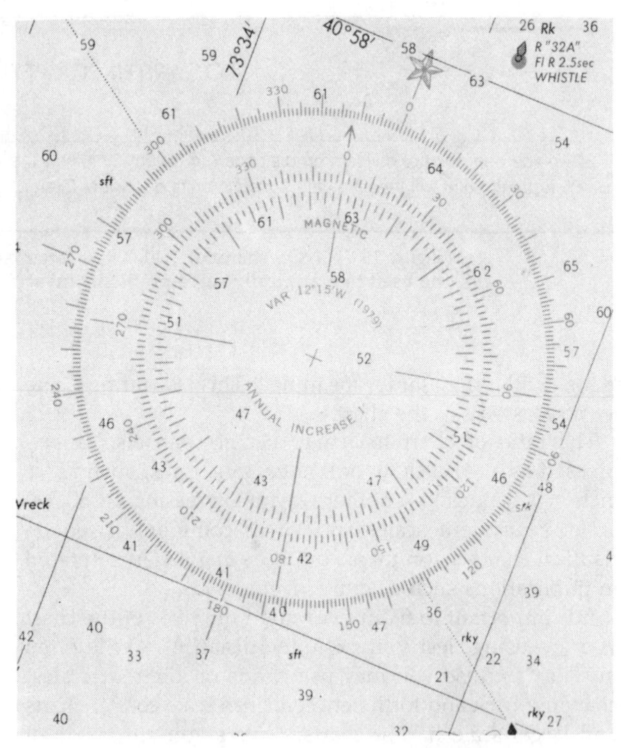

Figure 1805 **A compass rose illustrates true and magnetic north directions. The outer circle is in degrees with zero at true north; the inner circles are in degrees and "points" with their zero at magnetic north. Several compass roses are located conveniently on each chart for plotting courses and bearings. Some charts of the Small Craft series omit the innermost magnetic circle with its "point" system of subdivision.**

Distance

DISTANCES on charts are measured in statute or in nautical miles. Use of the STATUTE (LAND) MILE of 5,280 feet is limited to the Great Lakes, inland rivers, and the Atlantic and Gulf Intracoastal Waterways. The NAUTICAL MILE of 6,076.1 feet (1,852 m) is used on ocean and coastal waters.

You may sometimes need to convert from one unit to the other. This is not difficult—1 nautical mile = 1.15 statute miles, or roughly 7 nautical miles = 8 statute miles.

In navigation, distances of up to a mile or so usually are expressed in YARDS, a unit that is the same no matter which "mile" is used on the chart. METER may come into wider use for short distances; 1 meter = 1.094 yards. In some instances, to simplify calculations, 1 nautical mile is used as 2,000 yards; the error is not great, only about 1¼%.

Scale

Because a chart is a miniature representation of navigable water area, actual distances must be scaled down to much shorter dimensions on paper. This reduction is the SCALE of the chart, and it may be expressed as a ratio, 1:80,000 meaning that 1 unit on the chart represents 80,000 units on the actual land or water surface, or as a fraction

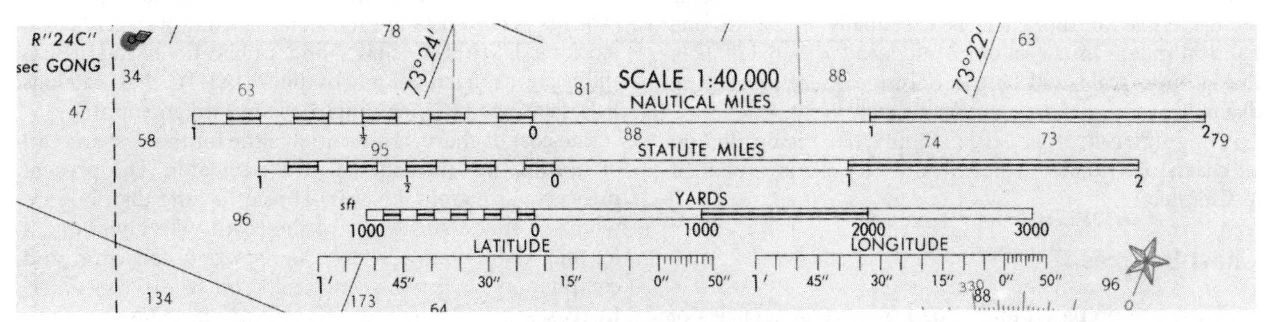

Figure 1806 **NOS charts of scales of 1:80,000 and larger will have a scale of nautical miles and one of yards; those of 1:40,000 and those of the Intracoastal Waterway will also carry a scale of statute miles. This Small Craft chart, at 1:40,000, also has latitude and longitude scales showing subdivisions of 1 degree.**

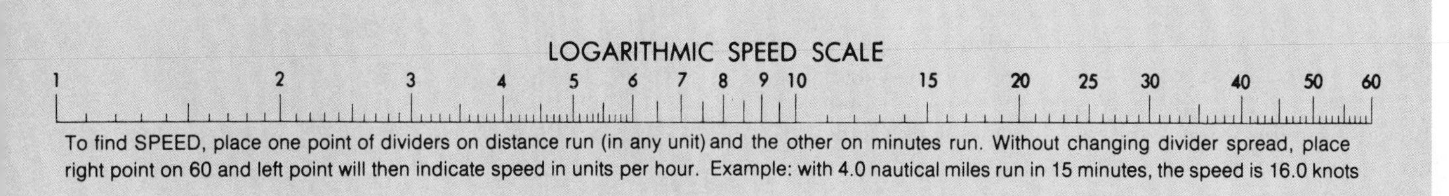

Figure 1807 **NOS charts of 1:40,000 or larger scale carry a Logarithmic Speed Scale which can be used to graphically solve problems involving time, speed, and distance.**

1/80,000 with the same meaning. This is termed the NATURAL SCALE of the chart.

The ratio of chart to actual distance can also be expressed as a NUMERICAL or EQUIVALENT SCALE, such as "1 inch = 1.1 miles,"—another way of expressing a 1:80,000 scale. Equivalent scales are not as commonly used on nautical charts as on maps, but they may be encountered in publications such as cruising guides.

It is important to fix in your mind the scale of the chart you are using, lest you misjudge distances. Quite often in a day's cruise you may use charts of different scales, changing back and forth between small-scale coastal charts and larger-scale harbor charts. Unless you are aware of the differing scales, you may find yourself in a dangerous position.

Large-Scale and Small-Scale

When chart scales are expressed fractionally, confusion sometimes results from the use of the terms "large-scale" and "small-scale." Since the number that is varied to change the scale is in the *denominator* of the fraction as it gets larger, the fraction, and hence the scale, gets smaller. For example, 1/80,000 is a smaller fraction than 1/40,000, so a chart to the former scale is termed a smaller-scale chart. The terms "large-scale" and "small-scale" are relative and have no limiting definitions. Scales may be as large as 1:5,000 for detailed harbor charts, or as small as 1 to several million for charts of large areas of the world.

Charts at a scale of 1:80,000 or larger will normally carry, in addition to a statement of scale, two sets of GRAPHIC SCALES, Figure 1806, each subdivided into conveniently and commonly used units. Note that one basic unit is placed to the *left* of the scale's zero point and is subdivided more finely than is the main part of the scale. The use of these graphic scales will be covered in Chapter 19.

When using Mercator charts, the navigator can take advantage of the fact that one minute of the *latitude* scale on each *side* of the chart is essentially equal to one nautical mile. On charts of a scale smaller than 1:80,000, the latitude scale will be the only means of measuring distance.

A LOGARITHMIC SPEED SCALE, Figure 1807, is printed on all charts of 1:40,000 or larger scale. Its use is explained in Chapter 19.

Chart Sources

Charts are prepared and issued by several agencies of the federal government. This is not duplication, however, because different agencies are responsible for different areas and types of charts. Most boatmen use charts prepared by the National Ocean Service of the National Oceanic and Atmospheric Administration, Department of Commerce. NOS charts cover the Great Lakes and the coastal waters of the United States, including harbors and rivers extending inland to the head of tidal action.

The Defense Mapping Agency Hydrographic/Topographic Center publishes charts of the high seas and foreign waters based on its own and other nations' surveys. Boatmen will use DMAHTC charts, for example, when cruising in the Bahamas. Canadian waters are charted by that country's Hydrographic Service.

Charts of major inland rivers such as the Mississippi and Ohio are issued by the U.S. Army Corps of Engineers. Also available are charts of many inland lakes and canal systems.

Chart Catalogs

The catalogs from issuing agencies indicate the area covered by each chart, the scale used, the price, and the type of electronic navigation system charted, if any. These catalogs are useful for planning a cruise into unfamiliar waters.

The National Ocean Service publishes five free chart catalogs, as listed in Chapter 17.

The DMAHTC chart catalog is discussed in detail in Chapter 17.

Where to Buy Charts

Charts may be purchased directly from the headquarters or field offices of the issuing agencies, or from retail sales agents. The addresses of NOS and DMAHTC distribution offices are given on pages 392 and 394. Sales agents are widely located in boating and shipping centers. The names and addresses of local sales agents for NOS charts and publications are listed in the various NOS chart catalogs. These lists are footnoted to indicate agents also stocking DMAHTC charts and publications and USCG publications. Each volume of the DMAHTC chart catalogs lists sales agents, including those in foreign countries.

The cost of charts is amazingly little for the vast amount of information furnished to the navigator. The present prices cover merely the costs of printing and distribution, which is only a small part of the cost to the government for their production. Costs of surveys, data collection, and compilation have not been passed on to the consumer. Increases in chart prices reflect only the general rise in costs.

NOS and DMAHTC charts are sold by government offices and sales agents located in most boating areas; a discount is allowed to each dealer for his quantity purchases. It is also possible for a boat or yacht club or other boating organization to purchase charts in quantity for a discount.

You should keep a full set of charts aboard for the waters you cruise, and regularly replace worn-out or outdated charts with new ones. Charts are still one of your best boating bargains.

What Charts Show

Charts include much information that you should study well before you actually use the chart for navigation.

Basic Information

Located on the chart where space is available is the GENERAL INFORMATION BLOCK, Figure 1808. Here is the chart title describing the waters covered (the chart number does not appear here, but rather in several places around the margins); a statement of the type of projection used and the scale; the unit of depth measurement (feet or fathoms—one fathom equals six feet—or meters), and the datum plane for such soundings. (Caution: if the chart has INSETS—"blow-ups" of areas of special interest—these will be at a larger scale than the chart as a whole.) There will also be a reference to the applicable volume of the *Coast Pilot* giving supplementary information.

Elsewhere on the chart where space is available (normally in land areas), you will find other information like the meaning of certain commonly used abbreviations, units and the datum for heights above water, notes of caution regarding dangers, tidal information, references to anchorage areas. Read *all* notes on charts; they may cover important information that cannot be graphically presented.

Editions and Revisions

The edition number and date appear in the margin at the lower left-hand corner; immediately following these figures are the date of the latest revised printing, if any; see Figure 1810. Most nautical charts are printed to supply a normal demand of one or two years for active areas, and from four to 12 years in areas where few changes occur.

Charts may be printed as-is when the stock runs low, but a REVISED PRINT is more likely if a new edition is not published. Revisions include all changes that have been printed in *Notices to Mariners* since the last revision. When major changes occur, such as significant differences between charted depths and actual conditions revealed by new surveys, an unscheduled NEW EDITION may be published. This will also include all other changes which have been made in aids to navigation, etc.

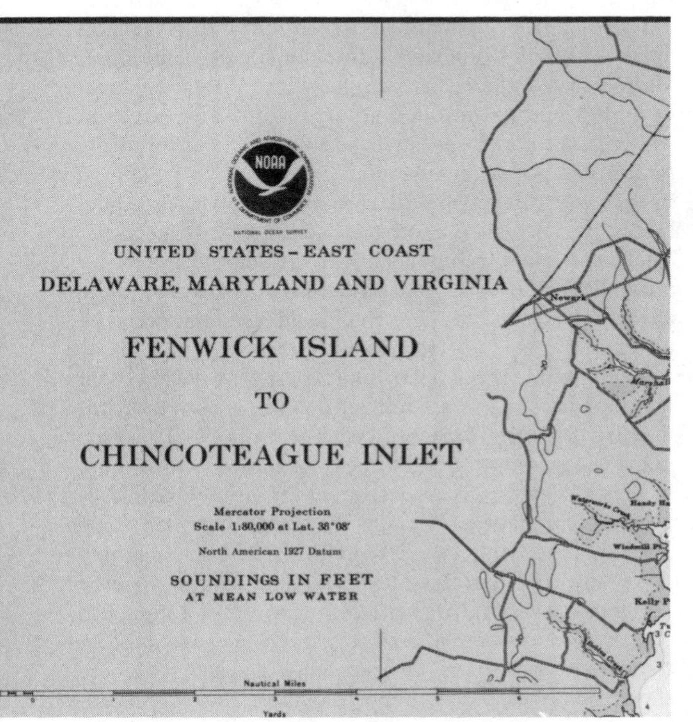

Figure 1808 **The title block shows the official name of the chart, the type of projection, scale, and datum and unit of measurement for depths. Printed nearby is much valuable information; be sure to read all notes before using any chart.**

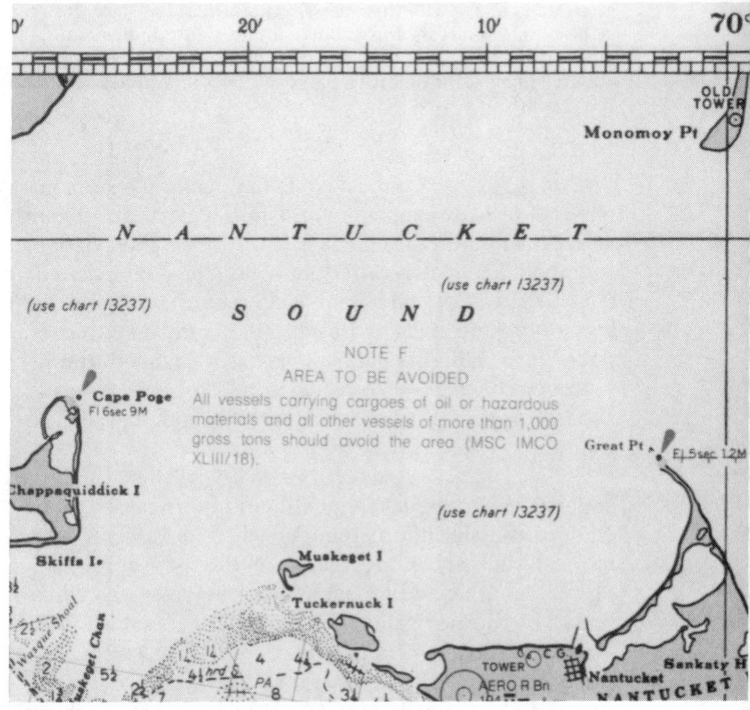

Figure 1809 **Notes printed on charts may concern navigation regulations, hazardous conditions at inlets, information on controlling depth that cannot be printed conveniently alongside a channel, or other matters that concern navigation.**

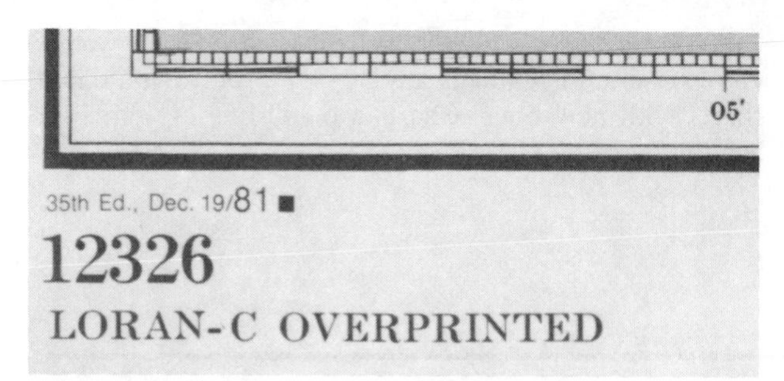

Figure 1810 **The edition number and date of each NOS chart are printed in the margin, at the lower left corner along with the chart number. Between editions charts may be revised; the date of revision also will be shown in the margin.**

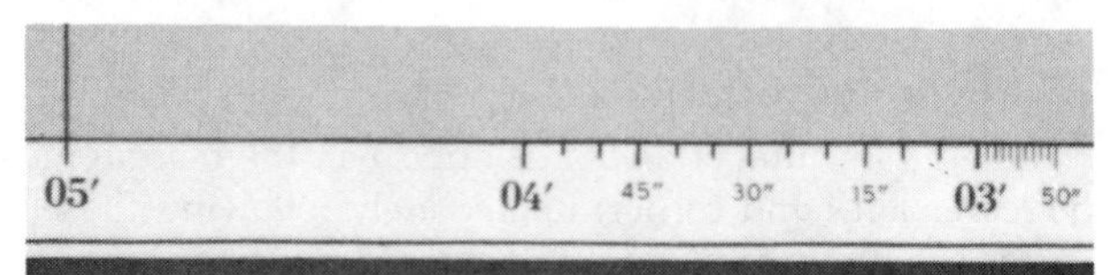

Figure 1811 **On charts of scales of 1:40,000 or larger, latitude and longitude scales in the borders are subdivided into minutes and seconds of arc. On this extract from a Harbor Chart, the meridians are drawn at 5-minute intervals, tick marks are placed at 1-minute intervals, and one 1-minute interval is subdivided into units of 5 seconds each. Note the ten units of 1-second of longitude placed to the right of the 03-minute figure.**

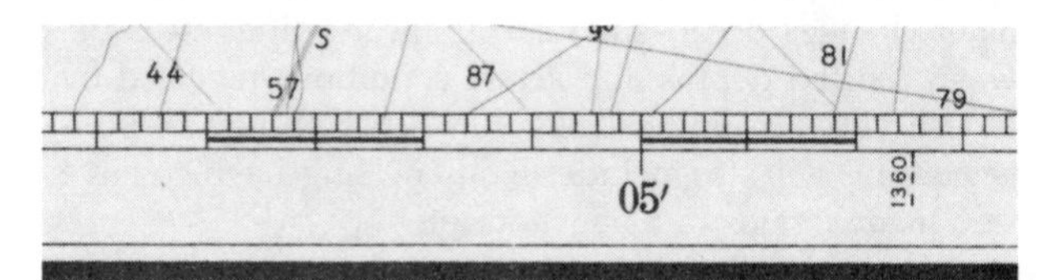

Figure 1812 **On charts of relatively small scale, the latitude and longitude border markings are in minutes and fractions of minutes of arc. ON this 1:80,000 chart, meridians are drawn at 10-minute intervals and subdivisions are in minutes and tenths of minutes. On smaller-scale charts, the smallest subdivisions might be fifths, halves, or whole minutes of arc.**

In late 1981 NOS published its first nautical chart compiled using an automated information system. Every item of information on the chart—every symbol, abbreviation, sounding, line, color, *everything*—was stored on magnetic discs as bits of digital data. Negatives for the printing of each color were made by laser beams controlled by a computer. Since a digital data base is easily updated, the revision of charts for new editions will become an easier and faster process as this technique is extended to all NOS charts.

Use only the latest edition of a chart. All new editions supersede older issues, which should be discarded. New editions contain information published in *Notices to Mariners,* plus all other corrections from extensive application of hydrographic and topographic surveys considered essential for safe navigation but not published in the *Notices.* To ensure that you know what are the latest editions, check the small NOS booklet *Dates of Latest Editions,* issued quarterly and found at local sales agents for charts and other NOS publications.

Between editions, correct your own charts from information published in *Notices to Mariners* and *Local Notices to Mariners* (see Chapter 17). Charts are not kept corrected by the National Ocean Survey while in stock awaiting sale. When you buy a new chart, check all *Notices* subsequent to the printed edition date and enter all applicable corrections.

Latitude and Longitude Scales

Conventional nautical charts with the geographical north direction toward the top of the sheet have latitude scales in each side border and longitude scales in the top and bottom borders. Meridians and parallels are drawn across the chart as fine black lines, usually at intervals of 2′, 5′, or 10′ as determined by the scale of the particular chart.

On NOS charts with a scale of 1:50,000 and larger, such as on harbor charts, the subdivisions in the border scales are in terms of minutes and seconds of latitude and longitude, Figure 1811.

On smaller-scale charts, the subdivisions are in minutes and fractions of minutes—charts at a scale of 1:80,000, such as 1210Tr, use minutes and tenths of minutes, Figure 1812. Smaller-scale charts use fifths or halves of minutes.

Where skewed projections are used, and north is not at the top of the sheet, as on Intracoastal Waterway and some small-craft charts, the subdivisions of latitude and longitude are indicated along parallels and meridians at several convenient places, or separately near the graphic scales.

Use of Color

Most charts use color to emphasize various features and so facilitate reading and interpretation. The colors vary with the agency publishing the chart and its intended use.

The NOS color system uses five multipurpose colors in either solid color or shades—black, magenta, gold, blue, and green. Land areas are a screened tint of gold (urban or built-up areas are often shown in a darker screened tint of that color); water areas are white (the color of the paper), except for the shallower regions, which are shown in a screened blue. Areas that are submerged at some tidal stages but uncovered at others, like sand bars, mud flats, coral reefs, and marshes, are green. On some charts, water areas that have been swept with wire drags to ensure the absence of isolated rocks or coral heads may be shown by a screened green with the depth of the sweep indicated.

Magenta ink is used for many purposes on charts; it has good visibility under red light, which is used for reading charts during darkness because it does not destroy night vision as white light does. Red buoys are printed in purple as are red daybeacon symbols. Lighted buoys of any color have a magenta disc over the small circle portion of the symbol to assist in identifying it as a lighted aid. A magenta flare symbol extending from a position dot is used with lights, lighted ranges, etc. Caution and danger symbols and notes are printed in magenta; also compass roses, usually, and recommended courses where shown. Black is used for most symbols and printed information.

The use of colors on DMAHTC charts is generally the same as described above for NOS charts, except that gray (screened black) is used for land areas.

Lettering Styles

To convey as much information as possible in the clearest form, certain classes of information are printed in one style of lettering and other classes in another style. By knowing what type of lettering is used for which class of information, you can more easily and quickly grasp the data being presented.

Vertical lettering is used for features that are dry at high water and not affected by movement of the water (except for the height of the feature above the water, which may be changed by tidal action). See the use of vertical lettering for the landmark stacks and spires, and the horn signal in Figure 1813.

Leaning, or slanting, letters such as those in Figure 1813 are used for water, underwater, and floating features, except depth figures. Note the use in Figure 1813 of slanted lettering for bottom features and buoy characteristics.

On smaller-scale charts, a small reef (covering and uncovering with tidal action) often cannot be distinguished by symbol from a small islet (always above water); the proper name for either might be "_____ Rock." The feature in doubt is an islet if the name is in vertical letters, but is a reef if lettered in slanting characters.

Similarly, a piling visible above water at all tidal stages is charted as "Pile," but one beneath the surface is noted as *"subm pile."*

Periods after abbreviations are omitted in water and in land areas, but the lower-case *i* and *j* are dotted. Periods are used only where needed for clarity, as, for example, in certain notes.

Water Features

The information shown on charts is a combination of the natural features of the water and land areas and various selected man-made objects and features. Each item shown is carefully chosen for its value to those who navigate vessels of all sizes.

Depths

The principal feature of concern to boaters is the *depth.* For any system of depth information there must be a reference plane, or DATUM. This is obvious in coastal areas where depths may change hourly as a result of tidal action; it is likewise true in inland areas where lake or river levels may also change, though more slowly on a seasonal basis. Each chart has on it a statement of the datum from which all depths, also called SOUNDINGS, are measured. The choice of the reference plane is based on many factors, most of them technical, but the primary consideration is that of selecting a datum near to normal low-water levels.

Planes of Reference On Pacific Coast charts the datum for the depths has long been MEAN LOWER LOW WATER, each tidal day having two low tides of different heights; see page 453. The National Ocean Service has changed the tidal datum on the Gulf of Mexico Coast to mean lower low water, and will eventually make the same change on the Atlantic Coast so that all U.S. charts will be consistent. These revisions are technical only and do not change any charted depths.

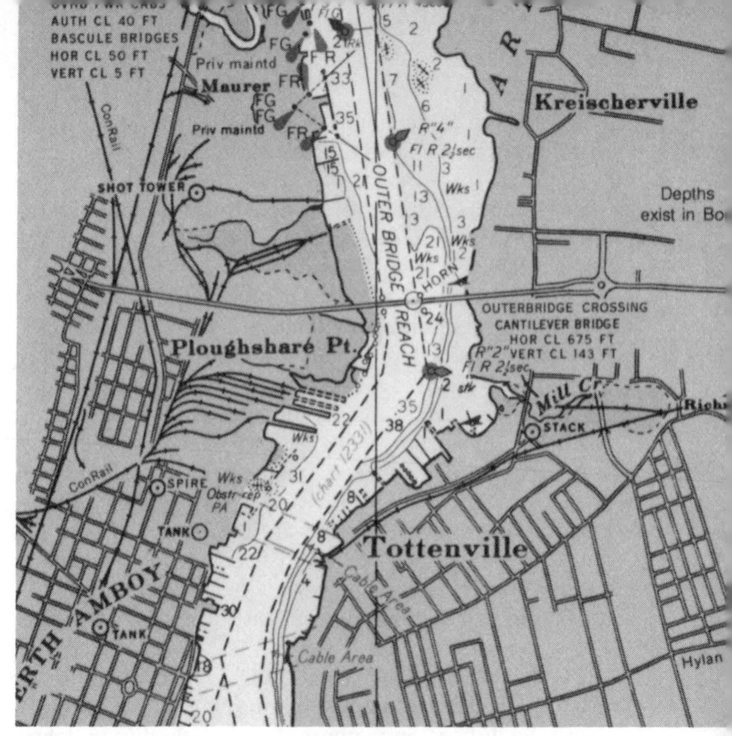

Figure 1813 **This chart excerpt illustrates use of vertical lettering for features above water—"TANK," "SPIRE"—and leaning letters (italics) for underwater features—"Wks" and "Rk" for wrecks and rocky bottom.**

By definition, "mean lower low water" is an average of all lowest water levels for tidal days over a period of time (usually 19 years). Thus on some days the lower low tide will be *below* the datum. This will result in *actual* depths being *shallower* than the charted figures. Many charts will have a small box with a tabulation of the extreme variations from charted depths that may be expected for various points, Figure 1814. Prolonged winds from certain directions, or persistent extremes of barometric pressure, may cause temporary local differences from charted depths. *Remember that there are exceptional conditions at which times the water may be much shallower than indicated on the chart.*

The datum for water depths on the Great lakes and other inland bodies of water is some arbitrarily established plane, usually at or near long-term low averages.

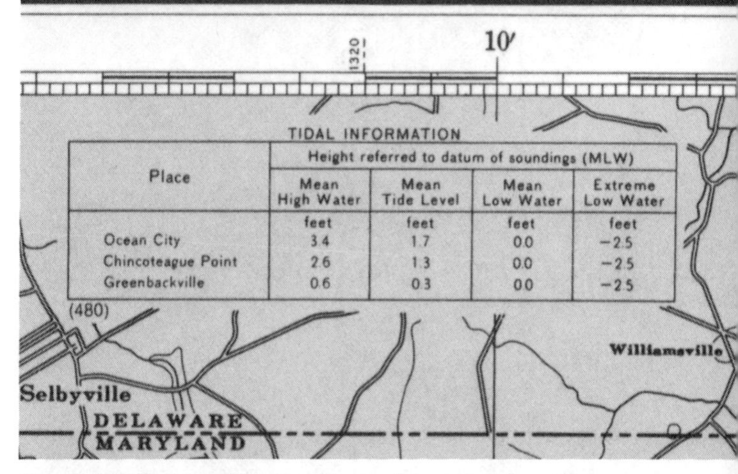

Figure 1814 **Coast and Harbor Charts often carry information on the normal range of tides, and the extreme variations from charted depths that may be expected. Check all newly purchased charts for this important information.**

This datum is usually also used for heights on such charts and will be clearly indicated on the chart.

How Depths Are Shown Depth information is shown on a chart by many small printed figures. These indicate the depth at that point, usually measured in feet or fathoms. A few newer charts have depths measured in meters and decimeters (tenths of a meter); 1 fathom is 1.83 meters. The printed depth figures are only a small fraction of the many soundings taken by the survey team. Only the more significant and representative are selected for use on the final chart.

A skipper can form some opinion of the characteristics of the bottom by noting the density of the depth information. Where depth figures are rather widely spaced, he can be assured of a reasonably flat or uniformly sloping bottom. Wherever the depths vary irregularly or abruptly, the figures will be more frequent and more closely spaced.

Depth Curves Most charts will have contour lines, usually called DEPTH CURVES, connecting points of equal depth. Such lines will appear at certain depths as determined by the scale of the chart, the relative range of depths, and the type of vessel expected to use the chart. Typically, depth curves are shown for 6, 12, 18, 30, and 60 feet, and multiples of 60 feet. Depth curves are shown as continuous solid lines with depth labels or various combinations of dots and dashes to code the depth along each line, but it is often easier to learn a line's significance by inspection of the depth figures on either side of it.

On many charts, a blue tint is shown in water areas out to the curve that is considered to be the danger curve for the majority of important marine traffic that is expected to use that particular chart. In general, the 6-foot curve is considered the danger curve for small-craft and Intracoastal Waterway charts, the 12- or 18-foot curve for harbor charts, and the 30-foot curve for coast and general charts. (These types of charts are discussed in more detail on pages 413-417.) In some instances the area between the 18-foot curves may be tinted in a lighter shade of blue than the shallower areas. Thus it can be seen that while blue tint means shallow water, this coloring does not have exactly the same meaning on all charts. Check each chart you plan to use to determine at just what depth the coloring changes.

Charts without depth curves must be used with caution, as soundings may be too scarce to allow the lines to be drawn accurately.

Avoid an isolated sounding that is shallower than surrounding depths, particularly with a solid or dotted line ring (depth curve) around it, as it may be doubtful how closely the spot has been examined and whether the least depth has been found.

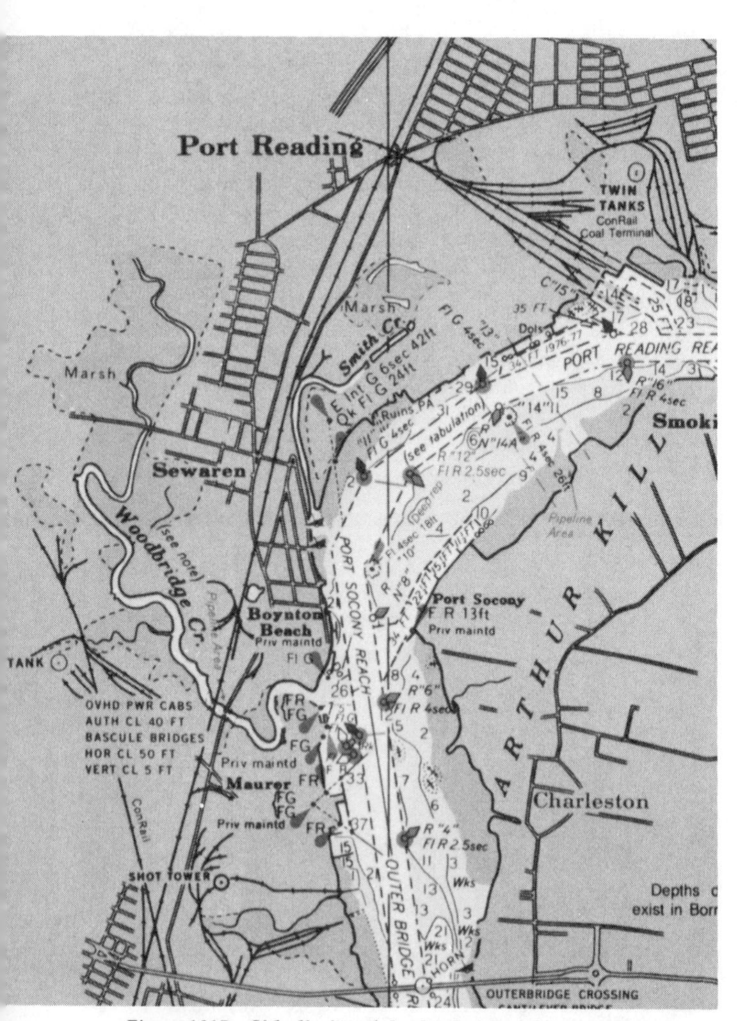

Figure 1815 **Side limits of dredged channels are marked by dashed lines. Information on the depth, and its date of measurement, is often printed between or alongside.**

Dredged Channels

Dredged channels are shown on a chart by two dashed lines to represent the side limits of the improvement. The channel's depth and the date on which such data were obtained are often shown within the lines or close alongside, see Figure 1815. A dredged basin will be similarly outlined with printed information on depths and date. The depth shown, such as "6 feet, Oct. 1982," is the controlling depth through the channel on the date shown, but does not mean this depth exists over the full width of the channel; some charts may indicate that the stated depth is only "on centerline."

Channels are sometimes described in terms of specific width as well as depth; for example, "8 feet for width of 100 feet." Depths may have subsequently changed from either shoaling or further dredging, so if your boat's draft is close to the depth shown for the channel, get local information if you can before entering. Detailed information for many dredged channels is shown in tabular form on the applicable charts, with revisions of the data published in *Notices to Mariners* or *Local Notices to Mariners* as changes occur.

Nature of the Bottom

The nature of the bottom, such as sand, rocky, mud, grass, or "hard" or "soft," is indicated for many areas by abbreviations. This information is especially valuable when you are anchoring, so take advantage of it wherever it appears. The meanings of these and other abbreviations are usually given on the face of the chart near the basic identification block; many are self-evident.

The Shoreline

The shoreline shown on charts is the MEAN HIGH-WATER LINE (HIGH-WATER LINE in lakes and non-tidal areas) except in marsh or mangrove areas where the outer edge of vegetation is used. Natural shoreline is represented by a slightly heavier line than man-made shoreline. Unsurveyed shoreline, or shoreline connecting two surveys that do not join satisfactorily, is shown by a dashed line. The low-water line is marked by a single row of dots. The outer limits of marsh are indicated by a fine solid line. The region between the high- and low-water lines is tinted green, and may be labeled "Grass," "Mud," "Sand," etc.

Features of Land Areas

Features and characteristics of land areas are shown on nautical charts in only such detail as will assist a navigator on the water. Details are usually confined to those near the shoreline or of such a prominent nature as to be clearly visible for some distance offshore.

How Topography Is Shown

The general topography of land areas is indicated by contours, form lines, or hachures. CONTOURS are lines connecting points of equal elevation. Their specific height, usually measured in feet, may be shown by figures placed at suitable points along the lines. The interval of height between adjacent contours is uniform over any one chart.

FORM LINES, or SKETCH CONTOURS, are shown by broken lines and are contour approximations meant to indicate terrain formations without giving exact information on height. They are used in areas where accurate data are not available to do otherwise. The interval between form lines is not necessarily uniform and no height figures are given.

HACHURES are short lines or groups of lines that indicate the approximate location of steep slopes. The lines follow the general direction of the slope, with the length of the lines indicating the height of the slope.

Cliffs, Vegetation, and the Shore

Cliffs are represented by bands of irregular hachures. The symbol is not an exact "plan view," but rather somewhat of a "side elevation"; its extent is roughly proportional to the height of the cliff. For example, a perpendicular cliff of 100 feet height will be shown by a hachured band wider than one representing a cliff of 15 feet with slope.

Spot elevations are normally given on nautical charts only for summits or the tops of conspicuous landmarks, Figure 1816. Heights are measured from mean high water (or other datum established for inland bodies of water).

The type of vegetation on land will sometimes be indicated by symbols or wording where this information may be useful to mariners.

The nature of the shore is sometimes indicated by various symbols—rows of fine dots denote a sandy beach, small circles indicate gravel, or irregular shapes mean boulders.

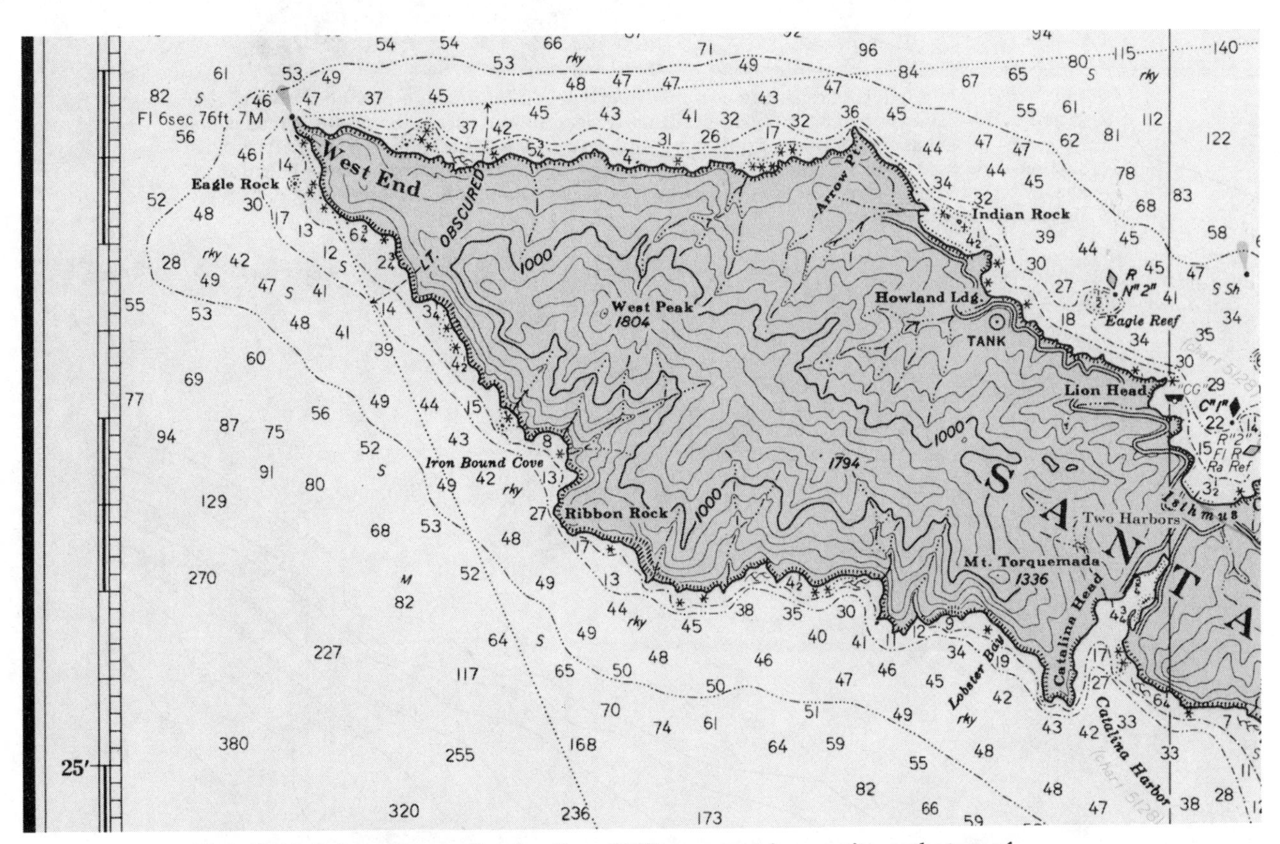

Figure 1816 **Information on the elevation of hills or mountain summits, or the tops of conspicuous landmarks, is often printed on nautical charts. Heights usually are measured from mean high water in feet, but meters and decimeters (tenths of meters) are now on some charts.**

Figure 1817 **On relatively large-scale charts, *above,* detailed information may be shown of the streets and buildings of a city or town, particularly near the waterfront. Some street names may be given, as well as the location of public buildings such as a customhouse or post office. On smaller-scale charts, *below,* cities, towns, and built-up areas are indicated by cross-hatching as shown, or by heavier screening that shows up as a darker, but not black, area. Single, heavy lines indicate principal roads.**

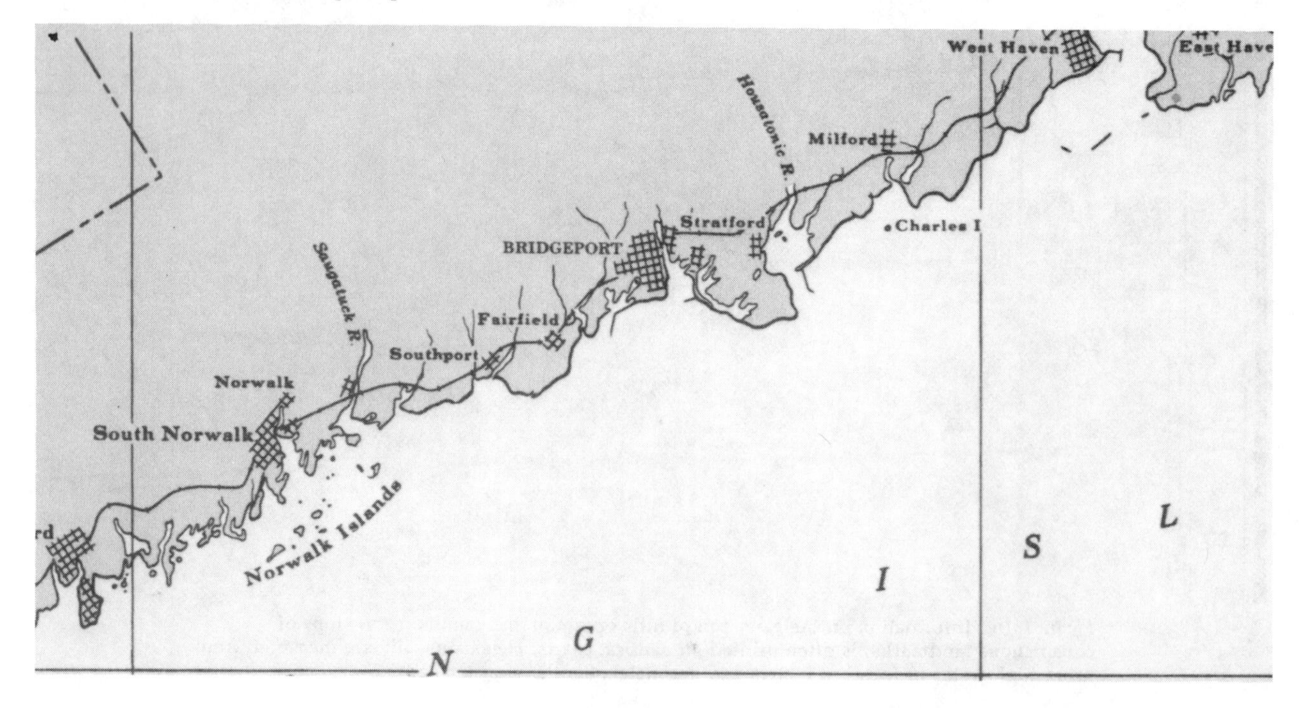

Man-Made Features

Man-made features on land are shown in detail where they relate directly to water-borne traffic. Examples are piers, bridges, overhead power cables, and breakwaters. Other man-made features on land, such as built-up areas, roads, and streets, may be shown in some detail or generalized as determined by their usefulness to navigation and the scale of the chart. On large-scale charts the actual network of streets may be shown with public buildings such as the post office and customs house individually identified. Figure 1817 top. On less detailed charts, the town or city may be represented by a cross-hatched area for the approximate limits of the built-up area, with major streets and roads shown by single heavy lines; see Figure 1817 bottom. Locations of prominent isolated objects, tanks, stacks, spires, etc., are shown accurately so they may be used for taking bearings.

Specific descriptive names have been given to certain types of landmark objects to standardize terminology. Among the more often used are the following:

BUILDING or HOUSE—the appropriate term is used when the entire structure is a landmark, rather than any individual feature of it.

SPIRE—a slender, pointed structure extending above a building. It is seldom less than two-thirds of the entire height of the structure, and its lines are rarely broken by intermediate structures. Spires are typically found on churches.

DOME—a large, rounded, hemispherical structure rising above a building; for example, the dome of the United States Capitol in Washington.

CUPOLA—a dome-shaped tower or turret rising from a building, generally small in comparison with the building.

CHIMNEY—a relatively small projection for conveying smoke from a building to the atmosphere. This term is used when the building is more prominent than the chimney, but a better bearing can be taken on the smaller feature.

STACK—a tall smokestack or chimney. This term is used when the stack is more prominent as a landmark than the accompanying buildings.

FLAGPOLE—a single staff from which flags are displayed. This term is used when the pole is not attached to a building.

FLAGSTAFF—a flagpole arising from a building.

RADIO TOWER—a tall pole or structure for elevating radio antennas.

RADIO MAST—a relatively short pole or slender structure for elevating radio antennas; usually found in groups.

TOWER—any structure with its base on the ground and high in proportion to its base, or that part of a structure higher than the rest, but having essentially vertical sides for the greater part of its height.

LOOKOUT STATION or WATCH TOWER—a tower surmounted by a small house from which a watch is regularly kept.

TANK—a water tank elevated high above ground by a tall skeleton framework. GAS TANK and OIL TANK are terms used for distinctive structures of specialized design, and are usually lower than a water tank and not supported by a skeleton framework.

STANDPIPE—a tall cylindrical structure whose height is several times its diameter.

TREE—an isolated, conspicuous tree useful as a navigational landmark (seldom used).

Bridges over navigable waterways are shown with the type of bridge—bascule, swing, suspension, etc.—and both horizontal and vertical clearances in feet, the latter measured from mean *high* water (or other shoreline plane of reference used for heights of objects).

When two similar objects are so located that separate landmark symbols cannot be used, the word "TWIN" is added to the identifying name or abbreviation. When only one of a group of similar objects is charted, a descriptive legend is added in parentheses; for example, "(TALLEST OF FOUR)" or "(NORTHEAST OF THREE)".

Radio broadcasting station (AM) antennas are shown on charts where they may be used for taking visual or radio bearings. The call letters and frequency are often shown adjacent to the symbol marking the location of the towers.

Stacks, radio towers, and other tall towers are required to have lights to indicate their present to aircraft. When these lights are useful for marine navigation, they appear on charts as "FR," "Occ R," or "Fl R." Also seen are multiple very high-intensity, very-short-flash lights that are charted with the common designation of "STROBE." These lights flash in daylight as well as at night and eliminate the need for the tower or stack to be painted with alternating red and white bands.

Symbols and Abbreviations

The vast amount of information shown on a chart, and the closeness of many items, necessitate the extensive use of symbols and abbreviations. You should be familiar with all symbols and abbreviations on the charts you use. You must be able to read and interpret your charts quickly and accurately; the safety of your boat may depend on this ability.

International Standards

SYMBOLS are conventional shapes and designs indicating the presence of a certain feature or object at the location shown on the chart. No attempt is made at an accurate or detailed representation of the object, but the correct location is shown. Symbols and abbreviations used on NOS and DMAHTC charts are standardized and appear in a small pamphlet as Chart No. 1. Generally similar information is printed on the reverse side of Training Chart 1210T. The symbols and abbreviations are in general conformance with world-wide usage as adopted by the International Hydrographic Organization.

The standardized symbols and abbreviations are shown at the end of this chapter.

Basic Symbols & Abbreviations

Simple inspection of many symbols will reveal a pattern in the way that they are formed. If you know the general principles of chart symbols, you'll have an easier time learning the details.

Buoys

BUOYS, except mooring buoys, are shown by a diamond-shaped symbol and a small open circle indicating the position. (The circle replaced a dot used on older charts; the dot implied a too-precise location for a buoy that can swing about its anchor on a length of chain.) To avoid interference with other features on the chart, it is often necessary to show the diamond shape at various angles to the dot or circle.

On charts using the normal number of colors, RED BUOYS are in mangenta; the letter "R" may also be shown adjacent to the symbol GREEN BUOYS are shown in that color with the letter "G" nearby. (BLACK BUOYS are shown as a solid black symbol without further identification of color.) Other buoys are shown by open outline symbols with the color indicated by an appropriate abbreviation.

A buoy symbol with a line across its *shorter* axis indicates a HORIZONTALLY BANDED BUOY. For a junction buoy both colors are used; magenta (for red) over green, or green over magenta, as the buoy itself is painted. The letters "RG" or "GR," respectively, appear near the symbol. (Until IALA-B conversion is completed, "green" will be "black" and "G" will be "B.") Special purpose buoys are shown by an uncolored symbol and the letter "Y" (until conversion, other appropriate abbreviations will be used, such as "W Or").

A buoy symbol with a line across its *longer* axis represents a VERTICALLY STRIPED BUOY. No colors are used on this symbol; the colors are indicated by the abbreviation "RW" for red-and-white ("BW" if not yet converted to the IALA-B system).

The type and shape of UNLIGHTED BUOYS is normally indicated by an abbreviation such as "C" for can or "N" for nun.

Because they are a potential hazard to navigation, "superbuoys" are charted with a special symbol, see Figure 1818. This category includes large navigational buoys, offshore data collection buoys, and buoys for mooring tankers offshore while loading or unloading.

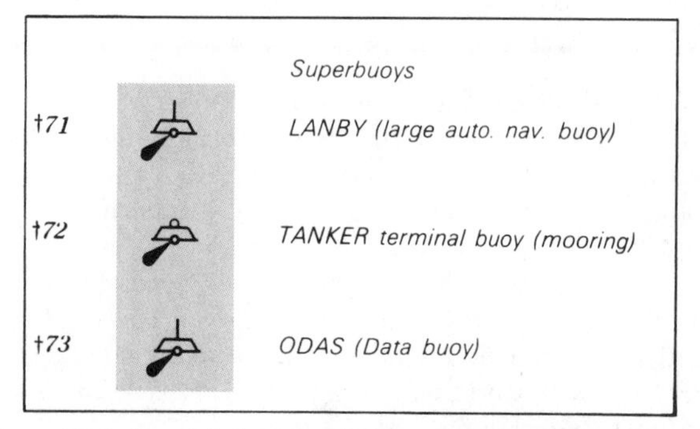

Figure 1818 **Special chart symbols are used to indicate the position of large automatic navigation buoys, tanker terminal buoys, and data buoys—all types known as "superbuoys."**

LIGHTED BUOYS are indicated by a small magenta disc over the small circle (or dot) that marks the buoy's position. The color and rhythm of the light, and the "hull" colors are indicated by abbreviations near the symbol.

Daybeacons

The symbol for daybeacons—unlighted fixed aids to navigation—may be either a small triangle or square.

The square symbol, colored green and with the letter "G" nearby, is used for daybeacons that have a solid green daymark of this shape.

The triangular symbol, colored magenta and with the letter "R" nearby, is used for daybeacons with solid red daymarks of this shape.

The symbol for a daybeacon with red-over-green triangular daymarks is an open triangle with the letters "RG"; for a daybeacon with green-over-red square daymarks it is an open square symbol with the letters "GR."

All octagonal, diamond-shaped, round, or rectangular daymarks will be represented by open square symbols and letter abbreviations as appropriate for the colors concerned.

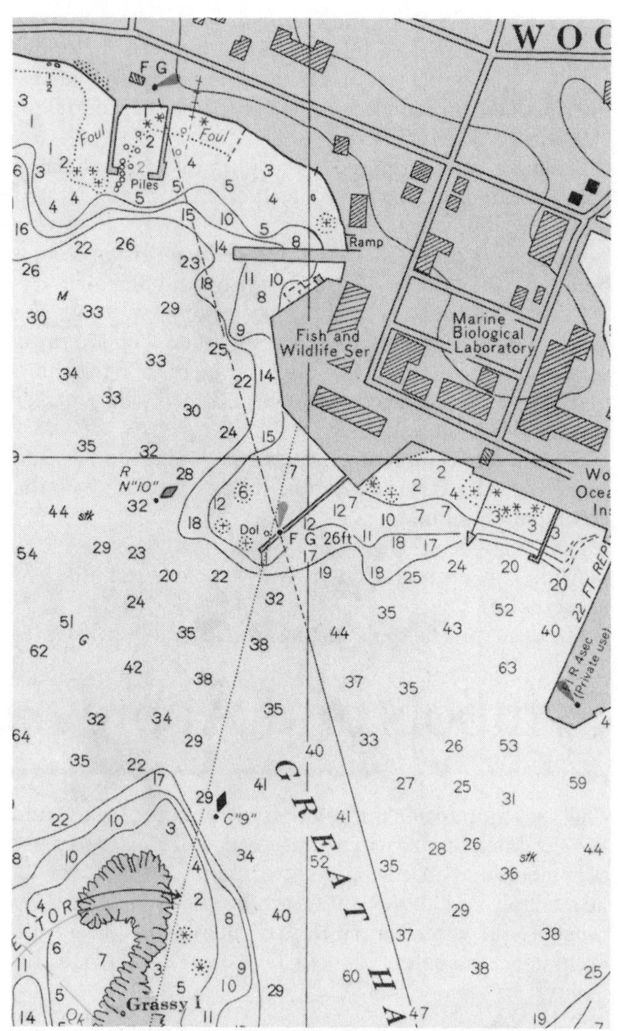

Figure 1819 **Ranges are excellent aids to navigation; they are charted by showing the front and rear markers with a line that denotes the range. The line is solid over the portion that is to be navigated, and dashed where it should not be followed.**

Lights, All Types

The chart symbol for lights of all sizes—from the simple light on a single pile in inland waters to the largest of primary seacoast lights—is the same. This is a black position dot with a magenta "flare" giving much the appearance of a large exclamation mark. In addition to color and characteristics, there may be information on the height of the light and its nominal range (no allowance made for curvature of the earth or observer's height of eye).

Changes for New Buoyage System

Implementation of the International Association of Lighthouse Authorities (IALA) Maritime Buoyage System "B" was begun in the spring of 1982; it is proceeding at a steady pace with completion scheduled for not later than 1989. See Chapter 16 for details.

Changes in aids to navigation are announced by *Notices to Mariners* and *Local Notices to Mariners* as soon as they occur. Such changes, however, will not show up on your charts until the next edition *unless you make them yourself—keep your charts up to date!*

Fog Signals

The type of FOG SIGNAL on buoys and lights so equipped is indicated by a descriptive word or abbreviation adjacent to the chart symbol.

Identification by Number

Buoys and lights are usually NUMBERED (or less frequently, designated with letters or combinations of letters and numbers). This identification is placed on the chart near the symbol and is enclosed in quotation marks to distinguish the figures from depth data or other numbers.

Primary and some secondary lights are named; the words, abbreviated as necessary, are printed near the symbol where space permits.

Ranges

RANGES are indicated by the two symbols of the front and rear markers (lights or daybeacons), plus a line joining them and extending beyond. This line is solid over the distance for which the range is to be used for navigation; it continues on as a dashed line to the front marker and on to the rear marker, see Figure 1819.

Radiobeacons

RADIOBEACONS are indicated by a magenta circle around the basic symbol for the aid to navigation at which the beacon is located, plus the abbreviation "R Bn." Aeronautical radiobeacons are shown only if they are useful for marine navigation; the same symbol is used and the identification "AERO" is added. The frequency, in kilohertz, is given, as is the identifying signal in dots and dashes of the Morse code.

Dangers to Navigation

Symbols are also used for many types of DANGERS to navigation. Differentiation is made between rocks that are awash at times and those which remain below the surface at all tides, between visible wrecks and submerged ones, and between hazards that have been definitely located and those whose position is doubtful. There are a number of symbols and abbreviations for objects and areas dangerous to navigation. Spend adequate time studying them, with emphasis on the types commonly found in your home waters.

Chart Numbering System

All NOS and DMAHTC charts are numbered in a common system. This is based on REGIONS and SUBREGIONS. Boaters will generally be concerned only with charts having *five-digit* numbers; such charts have a scale of 1:2,000,000 or larger. The first digit refers to a region of the world, and the second, together with the first, to a subregion; the final three digits, assigned systematically within the subregion, denote the specific chart area.

Region 1 includes the waters in and around the United States and Canada. Region 2 covers Central and South America, Mexico, the Bahamas, and the West Indies.

Region 1 has nine subregions designated counterclockwise around North America from Subregion 11 for the Gulf of Mexico and the Atlantic Coast up to Cape Hatteras. Subregion 12 extends from there to the eastern tip of Long Island, and 13 goes on to the Canadian border. Subregion 14 covers the Great Lakes; 18 is the U.S. Pacific Coast.

The final three digits of a five-digit number are assigned counterclockwise around the subregion or along the coast. Many numbers are left unassigned so that future charts can be fitted into the system.

Five NOS Series

As previously mentioned, charts are published in a wide range of scales. For general convenience of reference, the

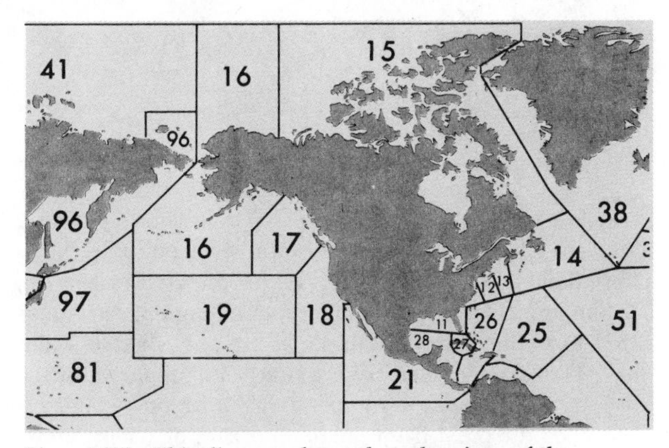

Figure 1820 **This diagram shows the subregions of the DMAHTC and NOS chart regions 1 and 2, which cover Canada and the United States. The number of the subregion forms the first two digits of a five-digit chart number.**

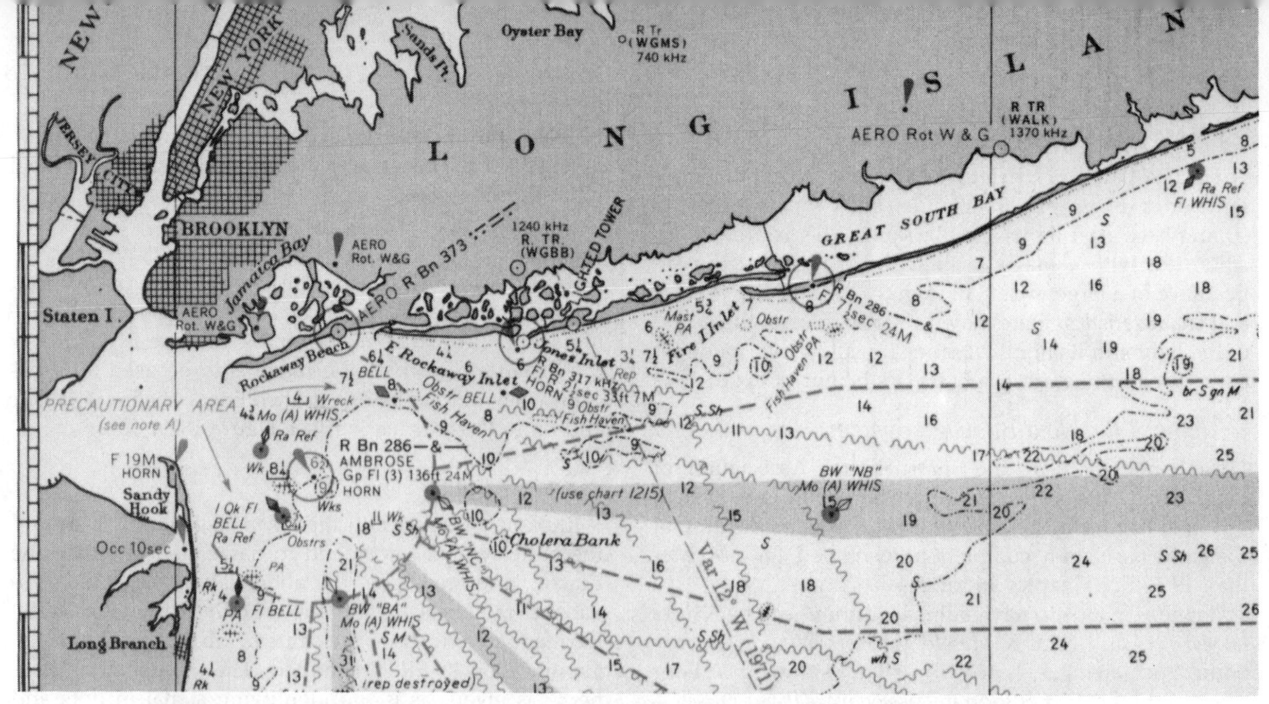

Figure 1821 **This is a section of a "Sailing Chart," *above.* With a scale of 1:600,000 or less, this is the smallest scale series; such charts cover long stretches of coastline. They are used by vessels approaching from the high seas, or between distant coast ports. Figure 1822 shows portions of the same area in increasing scales. *Below:* A small section of "General Chart" 12300: This is at a larger scale than the Sailing Chart, but like it should be used only for offshore navigation as it shows only major aids.**

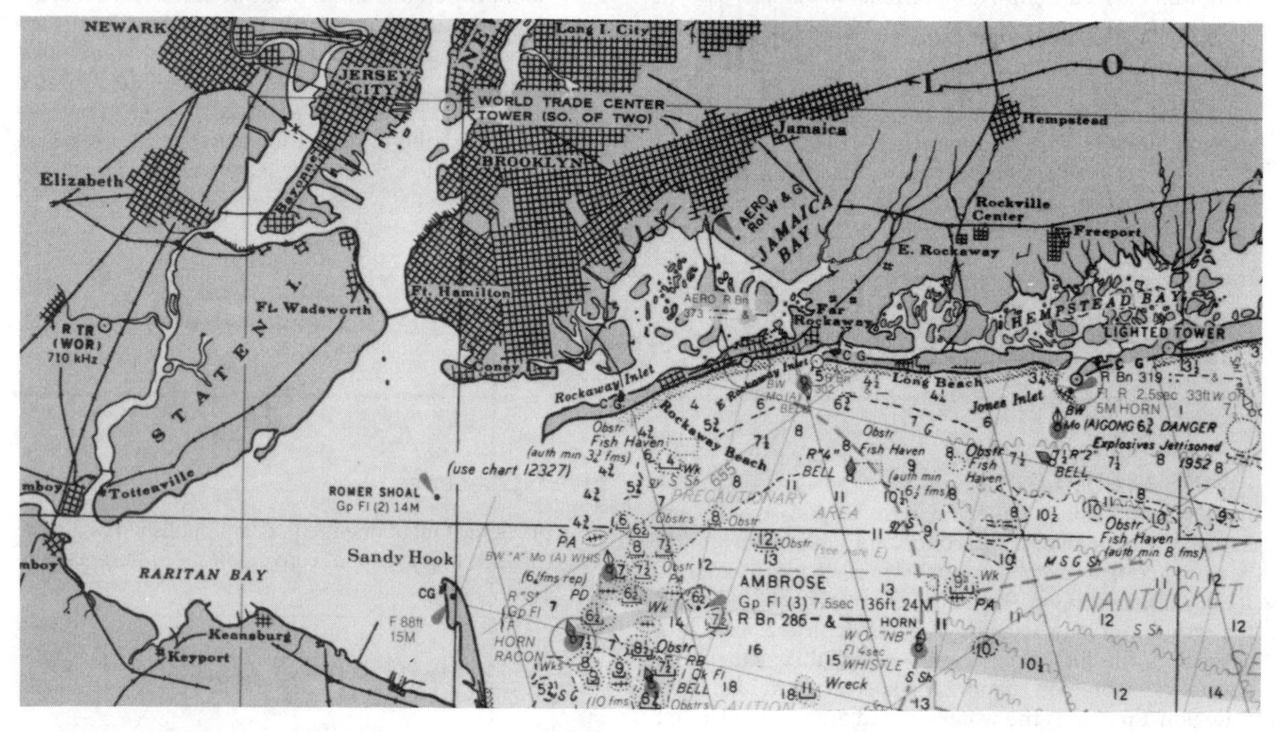

NOS has classified charts into "series" as follows:

1. Sailing Charts—the smallest scale charts covering long stretches of coastline; for example, Cape Sable, Newfoundland to Cape Hatteras, NC; or the Gulf of Mexico; or San Francisco to Cape Flattery, Washington; see Figure 1821 top. The charts of this series are published at scales of 1:600,000 or smaller. Sailing charts are prepared for the use of the navigator in fixing his position as he approaches the coast from the open ocean, or when sailing between distant coast ports. They show the offshore soundings, the principal lights and outer buoys, and landmarks visible at great distances. Other than for ocean cruising races, the average boater will have little use for charts in this series, except perhaps to plot the path of hurricanes and other tropical disturbances.

2. General Charts—The second series comprises charts with scales in the range of 1:150,000 to 1:600,000. These cover more limited areas, such as Cape May, NJ to Cape Hatteras; or Mississippi River to Galveston Bay; or San Francisco to Point Arena, CA. General charts are intended for coastwise navigation outside of offshore reefs and shoals when the vessel's course is mostly within sight of land and her position can be fixed by landmarks, lights, buoys, and soundings. See Figure 1821 bottom.

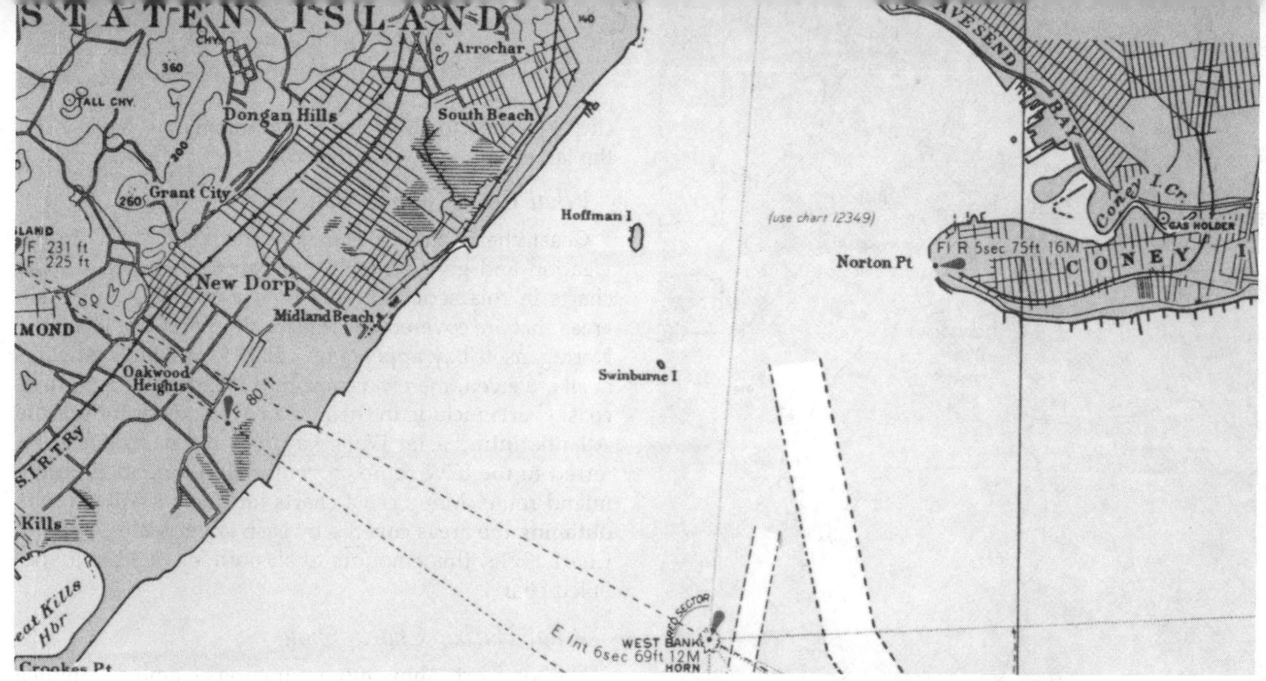

Figure 1822 **Above: a portion of "Coast Chart" 12326: These charts, usually at a scale of 1:80,000, are used for close-in coastwise piloting, entering and leaving harbors, and for cruising on some large inland bodies of water. More buoys and other aids to navigation are shown than on smaller scale charts, and more soundings are shown; these are in feet rather than fathoms or meters. "Harbor Chart" 12327, *below,* is at a scale of 1:40,000. Charts of this type show all hazards and aids to navigation in the area covered. More land features are also shown.**

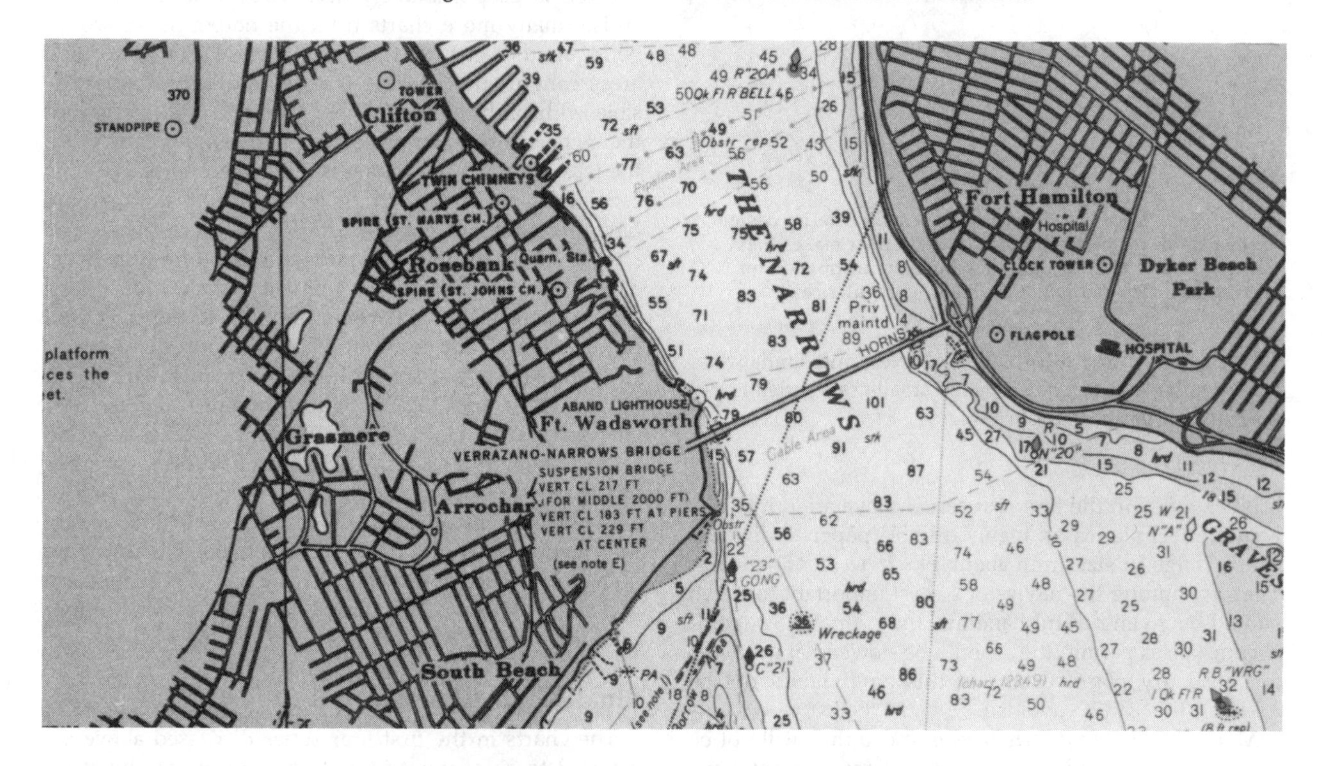

3. Coast Charts—This next larger scale series consists of charts for close-in coastwise navigation, for entering and leaving harbors, and for navigating large inland bodies of water. The scales used range from 1:50,000 to 1:150,000 with most at 1:80,000. See Figure 1822 top. Typical examples of coast charts are the widely used training chart No. 1210Tr, and such navigational charts as the series of five which cover Chesapeake Bay or 18746 which takes the California skipper from Long Beach or Newport to Santa Catalina Island and back. The average boatman may use several charts from this series.

4. Harbor Charts—This is the largest-scale and most-detailed series; see Figure 1822 bottom. Scales range from 1:50,000 to 1:5,000 with an occasional inset of even larger scale. The scale used for any specific chart is determined by the need for showing detail and by the area to be covered by a single sheet.

5. Small-Craft Charts—These compact charts provide the small-craft skipper with a convenient, folded-format, small-size chart designed primarily for use in confined spaces. One of the formats is designed to cover long, narrow waterways; another is a folded multi-page chart covering larger areas. All small-craft charts include a tabulation of public marine facilities available to the boatman,

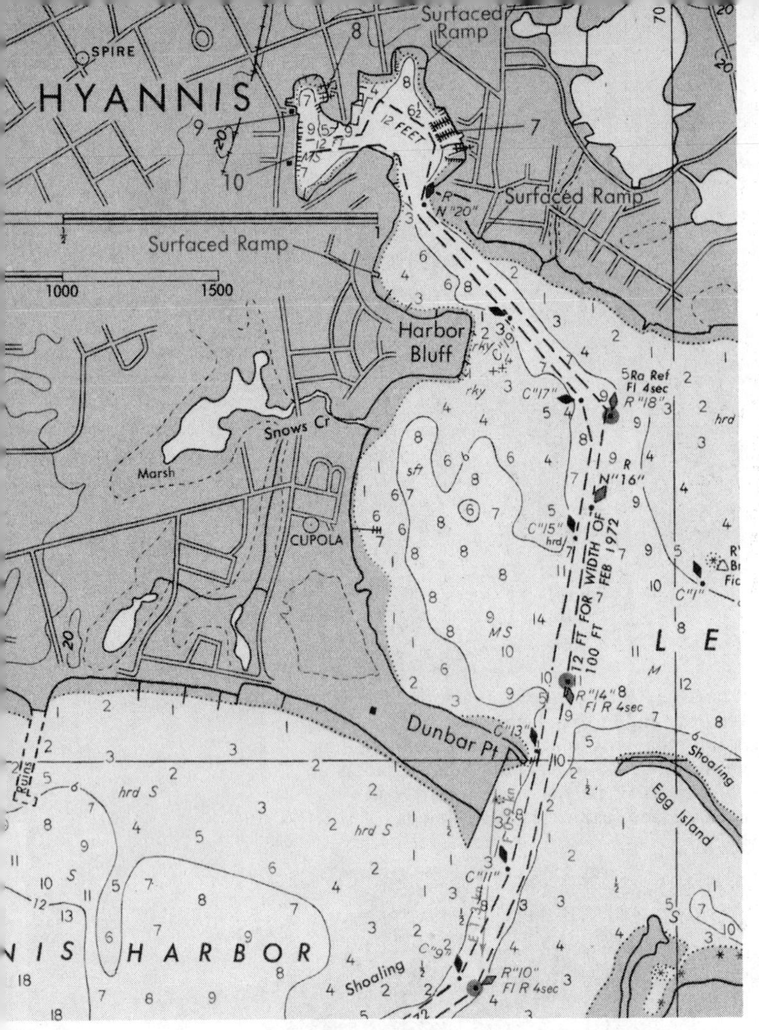

Figure 1823 **"Small Craft" charts are a special series designed for owners of recreational boats. Their format makes them easy to use and to stow, and they include information on boating facilities that is not included on regular charts.**

tide tables, weather information sources, and similar data of particular use to boaters. This small-craft series is described in greater detail below.

Stowage and Use

NOS charts in the first four series above are printed by accurate techniques on highly durable paper. Individual charts range in size from about 19×26 to 36×54 inches. They are among the navigator's most important tools, so should be given careful handling and proper stowage. If circumstances permit, they should be stowed flat or rolled, and in a dry place. Charts of this type should not be folded if this can be avoided.

Make any permanent corrections in ink so they will not be inadvertently erased; make all other lines and notations lightly in pencil so they may be erased without damaging the chart.

Selecting the Proper Chart

From a consideration of the five categories of charts discussed above, you can see that most boating areas will appear on two charts of different series, and that some areas will be covered by three or four charts of different scales. Such charts will vary widely in the extent of the area covered and the amount of detail shown. Choosing the proper chart for your use is important. In general, the closer you are to shoal water and dangers to navigation, the larger you will want the scale of your chart.

What Coast Charts Show

Coast charts show the major hazards and aids to navigation and give general information on depths. Some charts in this series entirely omit any details in certain areas that are covered by larger-scale charts. For example, Narragansett Bay appears on Chart 13218, but no details at all are given, merely a small note "(Chart 13221)." Other coast charts include in their area coverage portions of the Atlantic Intracoastal Waterway, but the navigator is referred to the ICW route charts for all information on the inland route. Many coast charts include a small diagram outlining the areas covered by each larger scale chart. On Chart 13218, this amounts to all portions of 13 more-detailed charts.

What Harbor Charts Show

Harbor charts show more numerous soundings and *all* aids to navigation, and permit the most accurate fixing of position from plotted bearings. The question may be asked, why ever select any but the largest-scale chart? The answer lies in the fact that as the scale is increased the area covered is proportionately decreased. Thus for a given cruise, many more charts from the harbor series would be required than from the coast series. Further, in some areas continuous coverage from port to port is not possible with harbor charts alone. Another problem is that the increased number of harbor charts would complicate the task of laying out a long run between ports.

Selecting the proper charts will usually mean that you have a mixture of coast charts for the longer runs and harbor charts for entering ports and exploring up rivers and creeks. For some areas, you will find it useful to have one or more general charts in addition to the coast and harbor charts. For example, the best overall route up Chesapeake Bay is more easily plotted on just two general charts, 12220 and 12260, than on a series of five coast charts, 12221 to 12273. The coast charts will be desirable for the actual trip, however, when used with some harbor charts.

In the margin of many charts, you will find helpful information regarding the next chart to use when you are going in a particular direction. This note will take the form of a statement such as "(Joins Chart 13233)" or "(Continued on Chart 13236)."

Small-Craft Charts

The charts in the first four series discussed above are referred to as conventional charts and are intended for flat or rolled storage. The fifth series, small-craft charts, is quite different, designed for more convenient use in the limited space available on boats, and for folded storage; see Figure 1823. There are approximately 90 small-craft charts, each numbered in the normal five-digit style.

Types of Small-Craft Charts

Small-craft charts are printed in three general formats termed FOLIO, ROUTE, and AREA as follows:

Small-craft folio charts, consisting of three or four sheets printed front and back, accordion folded, and bound in a suitable cover.

		Tides		Depth		Services					Boat Rental					Supplies				
No	Location	Mean Range-Ft.	Diff (Hrs) Hampton Roads	Approach-Feet (Reported)	Alongside-Feet (Reported)	Electricity-Moorings-Berths (Transients)	Ramp Surfaced-Natural	Repairs Hull-Motor-Radio	Marine Railway-Feet	Lift Capacity-Tons	Canoe-Row-Motor	Charter-House-Sail	Food-Lodging-Camping	Toilets-Showers-Laundry	Pump-Out Station / Winter Storage Wet-Dry	Nautical Chart Sales	Water-Ice	Groceries-Hardware	Bait-Tackle	Diesel Oil-Gasoline
1	ORIENTAL			8	8	BME	S	HM					F	TSL	W	C	WI	GH	BT	DG
1A	WINDMILL POINT			5	8	B E	S						F		WD		WI			DG
1B	DEWEY POINT					B	S										WI	GH	BT	G
2	ORIENTAL			9	10	BME	S	HM	60				FL	TSL			WI	GH	BT	DG
3C	WHITTAKER CREEK			10	8	BME		HMR			M	CHS		TSL	WD	C	WI	GH		DG
3D	WHITTAKER CREEK			10	8			HMR		15				TS		C	WI	H		
3E	WHITTAKER CREEK			8	8			HM		10		C S						H		
3F	BURTON CREEK			5	5	BME	S						FLC	TS	WD		WI	GH	BT	G
4	CORE CREEK BRIDGE	2.5	-1	8	3		S						FLC	TS			WI	G	BT	G
4A	ADAMS CREEK CANAL			5	5	BME	S							TS	W		WI			G
6	BEAUFORT	2.5	-1	17	20									T			W	H		DG
10	RADIO ISLAND	2.8	-1	10	6			HM												
10A	RADIO ISLAND	2.8	-1	4	4	BME		HM	40		M	HS			WD					
11	RADIO ISLAND	2.8	-1	5	5		S	HM		6				T	WD		WI	H	T	G
11B	RADIO ISLAND			4		BME	S							T	W		WI		BT	G
12	MOREHEAD CITY	2.8	-1	2	7	B					M	HS		TS	W		WI	G		DG

Figure 1824 **Table on a Small Craft chart lists tide ranges, depths, and the services and supplies available at marine facilities in the area covered. The numbers at the left are keyed to the chart.**

Small-craft route charts, consisting of a single sheet printed front and back and accordion folded; some are slipped into a suitable jacket.

Small-craft area charts, usually consisting of a conventional chart printed on lighter-weight paper with additional data for the small boat skipper. Half the chart is printed on each side of the paper with a slight area of overlap. The chart is accordion folded and may be issued in a protective jacket.

Route and area charts are being redesigned into a five-inch by 10-inch "pocket fold" format without a cover jacket. All information once shown on the separate jackets is now included on the margins of the chart itself. The elimination of the jacket allows a lower price.

Facilities Data

A unique feature of these small-craft charts is the variety of data printed on the chart or the protective jacket, Figure 1824. Repair yard and marina locations are clearly marked on the chart, and the available services and supplies are tabulated. A tide table for the year, marine weather information, Rules of the Road, whistle signals, and warning notes are included for ready reference.

Small-craft charts make frequent use of insets to show such features as small creeks and harbors in greater detail at a larger scale. Figure 1826 shows an inset from Chart 12285 of the Potomac River.

Courses Indicated

Many of the folio and route types of small-craft charts indicate a recommended track to be followed. The longer stretches of these tracks are marked as to *true* course and distance in miles and tenths. Route charts of the Intracoastal Waterway also have numbered marks every five *statute* miles indicating the accumulated distance southward from Norfolk, VA to Florida, and eastward and westward from Harvey Lock, LA (also westward across the Okeechobee Waterway and northward along the Florida Gulf Coast); see Figure 1825. Facilities along the ICW are designated in accordance with a numbering system that starts fresh with "1" on each chart of the series.

Periodic Revision

Small-craft charts are revised and reissued annually or biannually, usually to coincide with the start of the boating season in each locality. These charts are *not* hand corrected by the NOS after they are printed and placed in stock. Keep your chart up-to-date between yearly editions by applying all critical changes as published in *Notices to Mariners* and *Local Notices to Mariners.* This is not a great chore, *if* you keep up with the changes and don't get behind.

CANOE CHARTS covering the Minnesota–Ontario Border Lakes are designed to meet the needs of operators of small, shallow-draft craft.

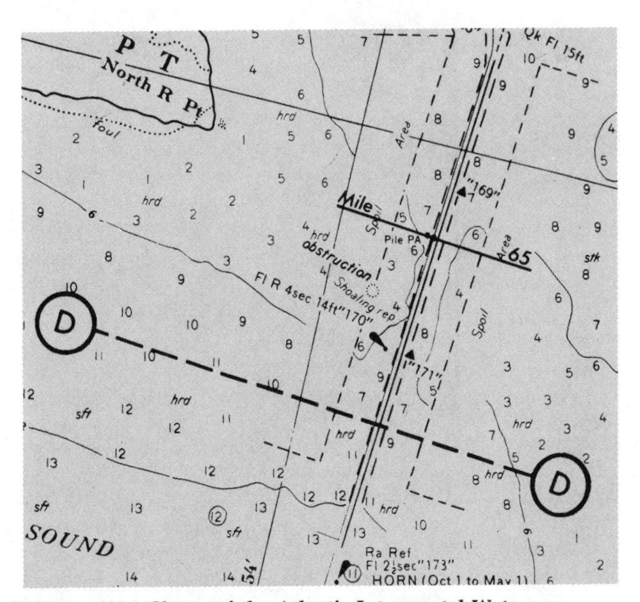

Figure 1825 **Charts of the Atlantic Intracoastal Waterway show a fine magenta line that indicates the route to be followed. Tick marks are placed at five-mile intervals along this course line and are labelled with the accumulated mileage south from Norfolk, Virginia. The "D-D" line on this excerpt is a matching indicator to facilitate shifting to the adjoining chart.**

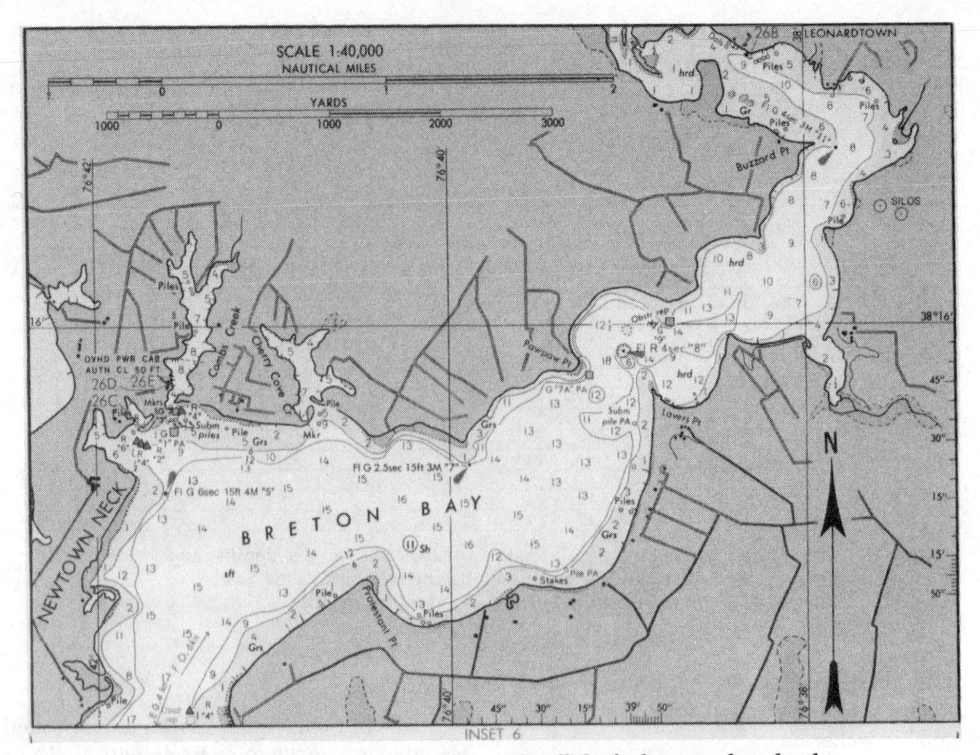

Figure 1826 **Insets are used liberally on Small Craft charts to show local cruising areas in greater detail. This inset, *above,* No. 6 from Chart 12285 (see *below*) for Breton Bay, is typical. On these charts, depths less than six feet are tinted blue, channel marks (triangles and squares in this illustration) are in green and magenta.**

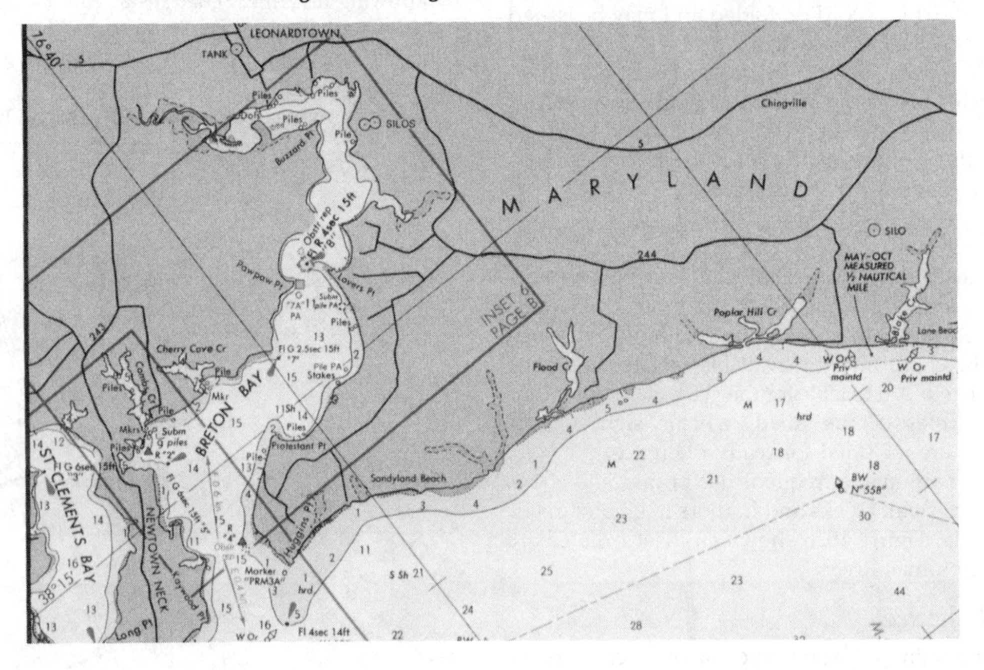

DMAHTC Charts

Charts from the DMAHTC are used by skippers making long ocean voyages or visiting waters of other nations (except Canada). The way of showing information does not differ much from the more familiar NOS charts; symbols and abbreviations will be familiar to the coastal boatman, but land areas are shaded grey rather than gold. Symbols are different for lighted buoys and those with radar reflectors. Some charts will show depths and heights in meters and fractions, rather than in feet or fathoms.

The DMAHTC often issues special editions for some of the major ocean sailing races. These are regular editions of the applicable charts overprinted with additional information for the yachtsman, including the direct rhumb line, typical sailing tracks for seasonal winds, additional current data, and other useful items. The publication of these special charts is well publicized in *Notices to Mariners* and some boating periodicals.

Remember that many DMAHTC charts are based on

surveys done by other nations, Figure 1827. The authority for the charted information is always given, as is the date of the surveys.

Great Lakes Charts

Polyconic projection (see below) is used for most of the charts of the Great Lakes; a few smaller-scale charts are also published in Mercator projection editions, as are all those published in metric editions. Other small variations between these and coastal charts may be noted. Often courses and distances (in statue miles) will be shown for runs between important points.

On the Great Lakes and connecting waters, special editions of charts for small craft are available in a number of boating areas. Some of these are individual pocket-fold charts, but most are in sets bound into "Small Craft Book Charts."

Inland River Charts

Boaters on inland rivers use charts that differ in many respects from those used in coastal waters. Often the inland river charts are issued in book form with several pages covering successive stretches of a river.

Probably the most obvious difference is the usual lack of depth figures. In lieu of these, there is generally a broken line designating the route to be followed. To make the best use of each paper sheet, pages may be oriented differently; north is seldom toward the top and its actual direction is shown by an arrow. Some symbols may vary slightly in appearance from those on "salt water" charts, and additional ones may be used as required by local conditions. Distances are often designated in terms of statute miles *upriver* from a specified origin point.

More detailed information on river charts will be found in Chapter 9; see pages 201-202. An extensive listing of where river charts and related publications can be obtained is presented in Appendix B.

Cautions Regarding Use

Producing charts for the vast coastline and contiguous waterways of the United States is a major undertaking. The U.S. Atlantic coastline exceeds 24,500 nautical miles, the Gulf Coast 15,000 miles, the Pacific Coast 7,000 miles, and the Alaskan and Hawaiian shorelines total more than 30,000 nautical miles. NOS publishes almost 1,000 charts

Figure 1827 **Many charts issued by the DMAHTC are based on surveys done by other nations, and this is noted in the title block. Such charts should be used with caution, particularly if the surveys date back many years, which is often the case.**

covering over 3.6 million square miles, and both of these figures increase each year. In meeting its global responsibilities, the DMA Hydrographic/Topographic Center puts out charts numbered in the thousands, and there are, in addition, the many Army Engineers charts.

Keeping so many charts up-to-date is obviously a staggering task. Surveys are constantly being made in new areas and must be rechecked in old areas. NOS has an extensive program of cooperative reporting by boaters to supplement its own information-gathering capability. Formal programs are established in the United States Power Squadrons and the U.S. Coast Guard Auxiliary, but *all* individual skippers are encouraged to report any corrections, additions, or comments to the chart's issuing agency. Comments are also desired on other publications such as the *Coast Pilots*. Send comments to the Director, National Ocean Survey, Rockville, MD 20852.

Charting agencies make every effort to keep their products accurate and up-to-date with changing editions. Major disturbances of nature such as hurricanes along the Atlantic Coast and earthquakes in the Pacific Northwest cause sudden and extensive changes in hydrography, and destroy aids to navigation. The everyday forces of wind and wave cause slower and less obvious changes in channels and shoals.

Be alert to the possibility of changes. Most charts will cite the authorities for the information presented and frequently the date of the information. Use additional caution when the surveys date back many years.

Chart Projections

You can safely navigate your boat without knowledge of the various types of projection used in the preparation of charts. As in almost any field, however, greater knowledge will assist in understanding and using nautical charts. Hence the following paragraphs offer additional information on chart projections.

The Mercator projection used in ocean and coastal waters, and the polyconic projection used in inland lakes and rivers, have both been mentioned earlier in this chapter. These, plus the GNOMONIC PROJECTION used in polar regions, will now be presented in more detail. There are other systems of projection, but each has limited application and need not be considered here.

Mercator Projection

The Mercator projection is often illustrated as a projection onto a cylinder. Actually, the chart is developed mathematically to allow for the known shape of the earth, which is not quite a true sphere. The meridians appear as straight, vertical lines, Figure 1828 top. Here is our first example of distortion—the meridians no longer converge, but are now shown as being parallel to each other. This changes the representation of the shape of objects by stretching out their dimensions in an east-west direction.

To minimize the distortion of shape, one of the qualities that must be preserved as much as possible, there must be a stretching-out of dimensions in a north-south direction. The parallels of latitude appear as straight lines intersecting the meridians at right angles. Their spacing increases northward from the Equator, Figure 1828, in accordance with a mathematical formula that recognizes the slightly oblate shape of the earth. This increase in spacing is not obvious in the case of charts of relatively small areas, such as in the harbor and coastal series, but it is quite apparent in Mercator projections of the world.

The Mercator projection is said to be CONFORMAL, which means that directions can be determined correctly, and distances measured to the same scale in all directions.

By the Mercator technique of distortion, then counter-distortion, the shape of areas in high latitudes is correctly shown, but their size appears greater than that of similar

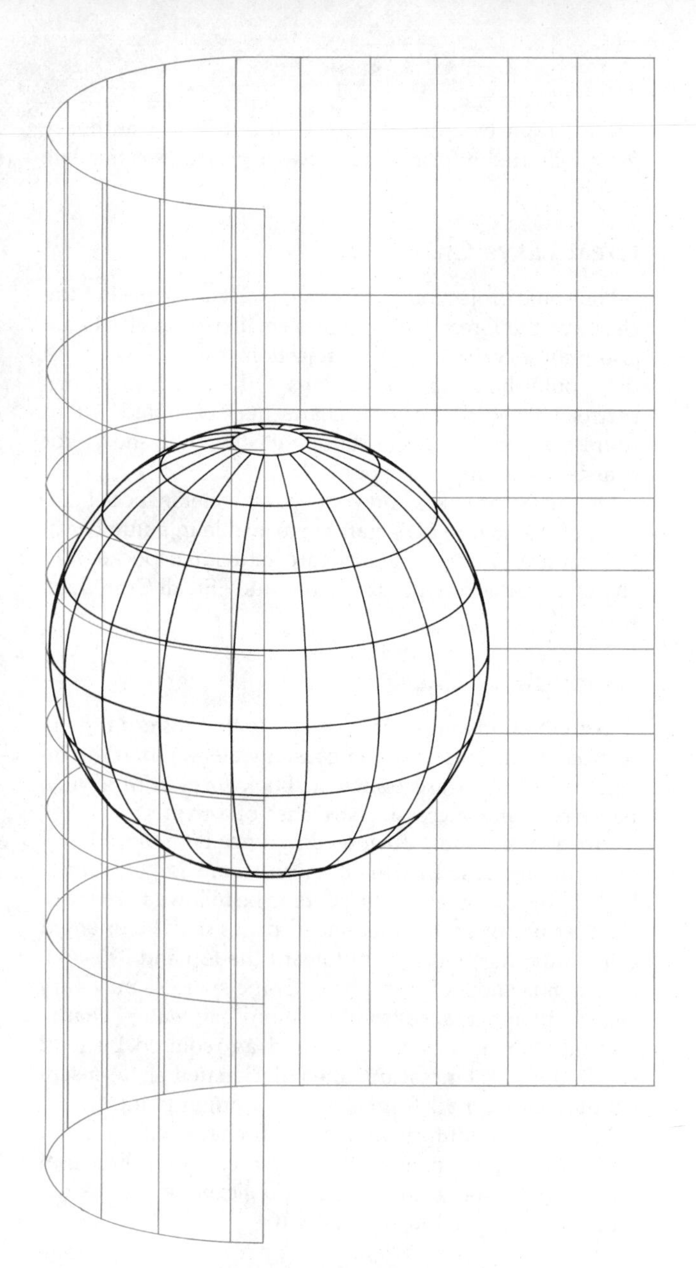

Figure 1828 **A Mercator projection, *right* and *opposite above*, can be visualized as the placement of a cylinder around the earth, parallel to the polar axis, and touching the earth at the equator. The meridians, projected out onto the cylinder, appear as straight lines; the parallels of latitude, projected to the cylinder, intersect the meridian lines at right angles. Note how a mercator projection of the world shows considerable distortion in near-polar latitudes.**

Figure 1829 **Polyconic projections, *below left*, are developed onto a series of cones, each cone tangent to a different parallel of latitude. This drawing shows three cones tangent at 40, 50, and 60 degrees north latitude. For clarity, the projection of the earth's surface onto the cones is not shown. The polyconic projection of a large area, *below right*, emphasizes the curved characteristics of the parallels and meridians on this type of chart. On charts of small areas, this curvature exists, but it is so slight that it is not so noticeable.**

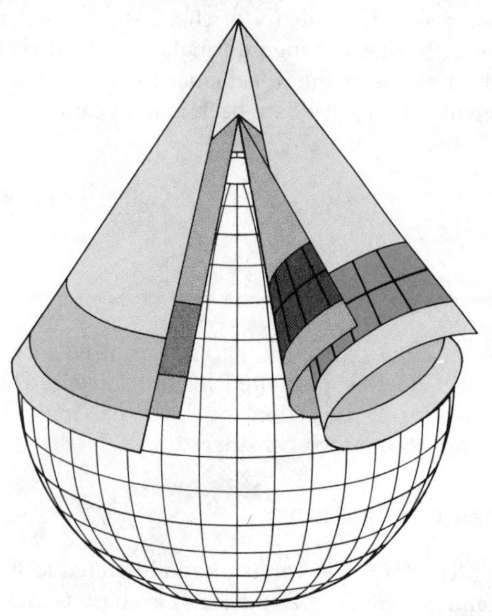

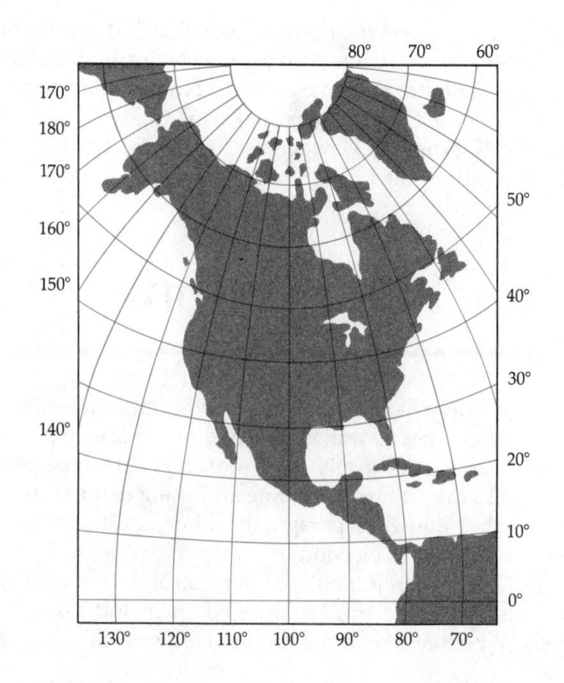

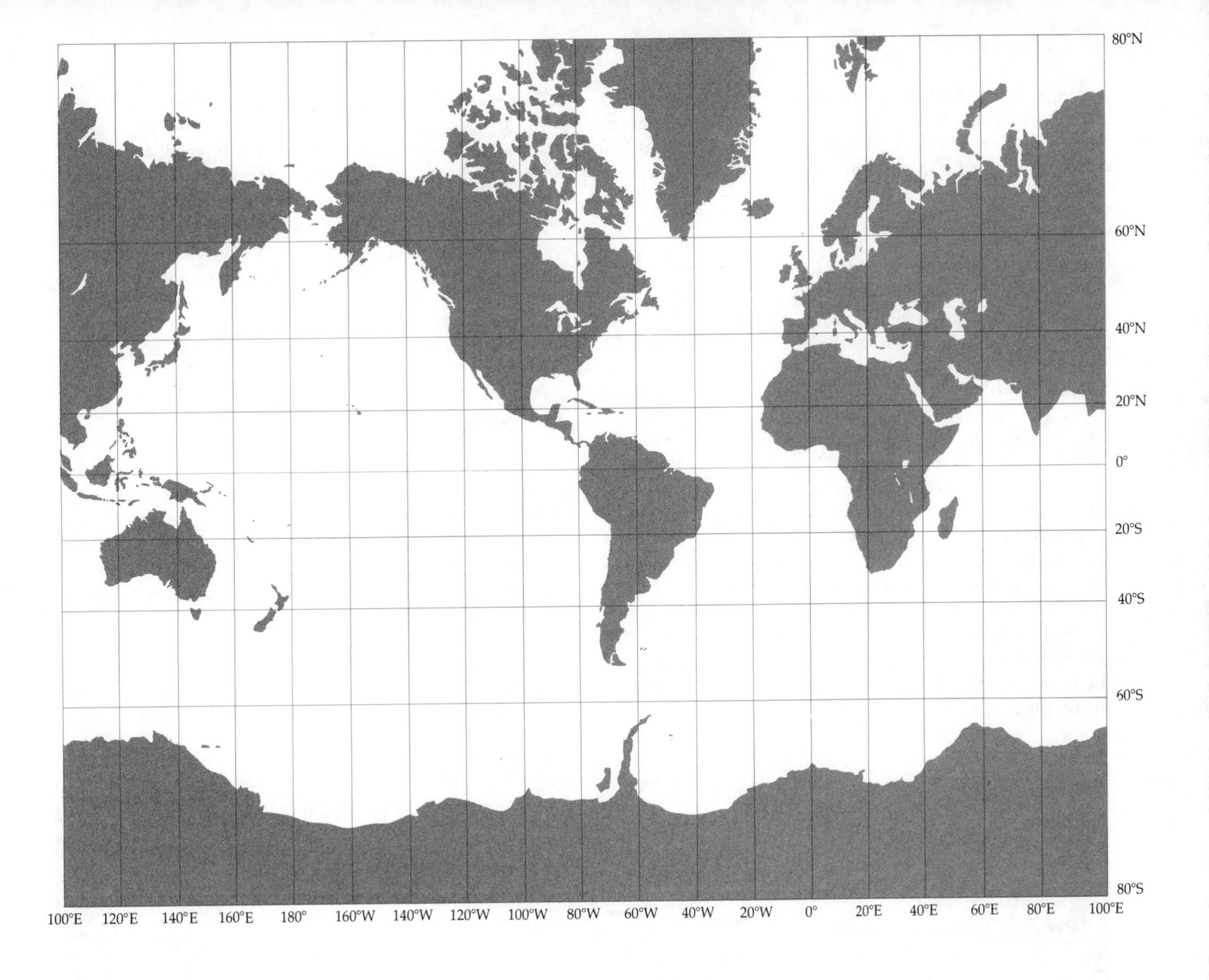

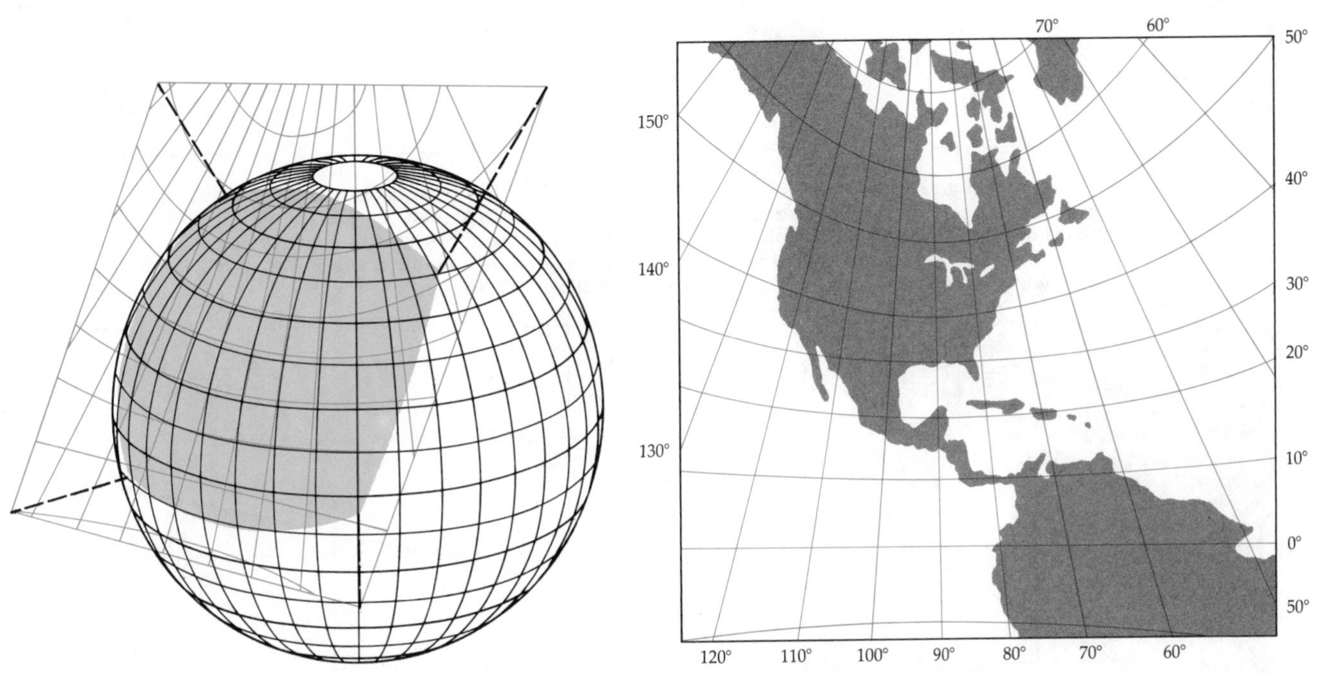

Figure 1830 **A Gnomonic projection, *left,* is made by placing a plane surface tangent to the earth at any location. Points on the earth's surface are then projected onto the plane. A chart made on the Gnomonic projection, *right,* shows meridians as straight lines converging toward the nearer pole. Parallels, other than the equator, appear as curved lines.**

areas in lower latitudes. An island, for example, in 60 degrees latitude (Alaska) would appear considerably larger than an island of the same size located at 25 degrees latitude (Florida). Its shape, however, would still be true to the actual proportions.

Advantages and Disadvantages

The great value of the Mercator chart is that straight-line meridians of longitude intersect straight-line parallels of latitude at right angles to form an easily used rectangular grid. Directions can be measured with reference to any meridian or parallel, or any compass rose. The geographic coordinates of a position can easily be measured from the scales along the four borders of the chart. *We can draw upon a Mercator chart a straight line between two points and actually run that course* by determining the compass direction between them; the heading is the same all along the line. Such a line is called a RHUMB LINE. However, a great circle, the shortest distance between two points on the earth's surface, is a curved line on a Mercator chart; this is more difficult to calculate and plot. For moderate runs, the added distance of a rhumb line is insignificant, so the rhumb is the track that is used. Radio waves follow great circle paths and radio bearings on stations more than about 50 miles distant will require correction before being plotted on a Mercator chart.

The scale of a Mercator chart varies with the distance away from the equator as a result of the N-S expansion. The change is unimportant on charts of small areas such as harbor charts, and the graphic scale may be used. The change in scale with latitude does become significant, however, in charts covering greater areas, as on general, coastal, and sailing charts. On such charts, when you measure distances using the latitude scale on either side margin, *take care that distance is measured at a point on the latitude scale directly opposite the region of the chart being used. Never use the longitude scale at the top and bottom of the chart for measuring distance.*

Polyconic Projection

Another form of chart construction is the *polyconic* projection. This method is based on the development of the earth's surface upon a series of cones, a different one being used for each parallel of latitude, Figure 1828 bottom left. The vertex of the cone is at the point where a tangent to the earth at the specified latitude intersects the earth's axis extended.

The polyconic projection yields little distortion in shape, and relative sizes are more correctly preserved than in the Mercator projection. The scale is correct along any parallel and along the central meridian of the projection. Along other me-

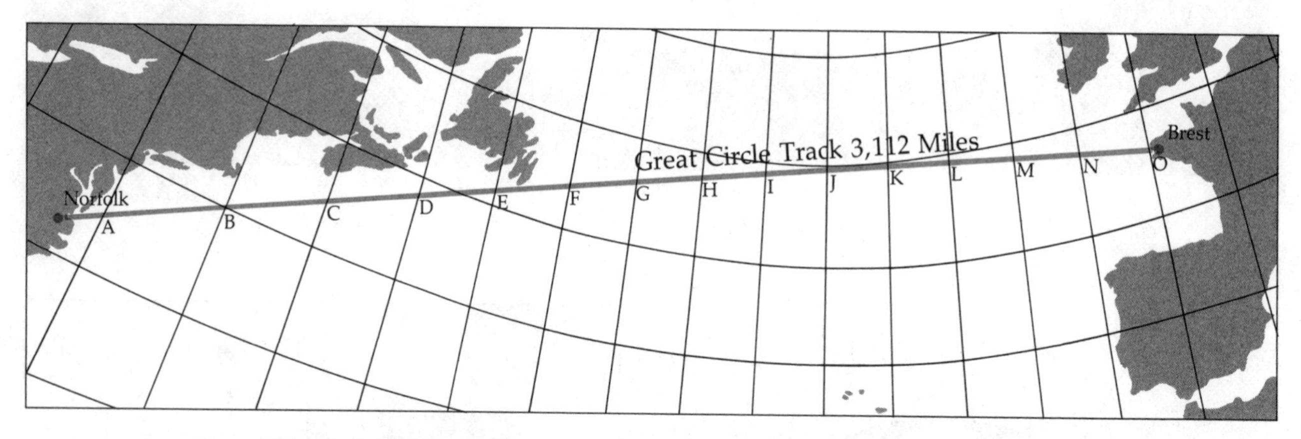

Figure 1831 **A great circle provides the shortest distance between any two points on the earth's surface, and can be plotted as a straight line, *above,* on a Gnomonic chart. To plot it on a Mercator chart, *below,* the geographic coordinates are noted at each meridian on the Gnomonic chart, and these points, transferred to the Mercator chart, are connected by a series of straight lines which appear to form a curve and which can be steered without difficulty.**

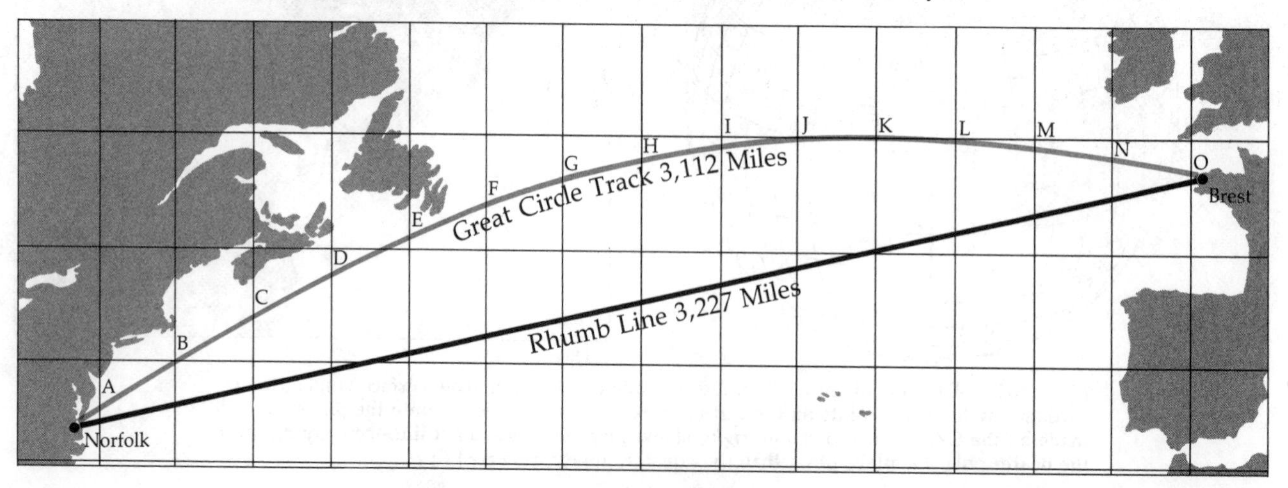

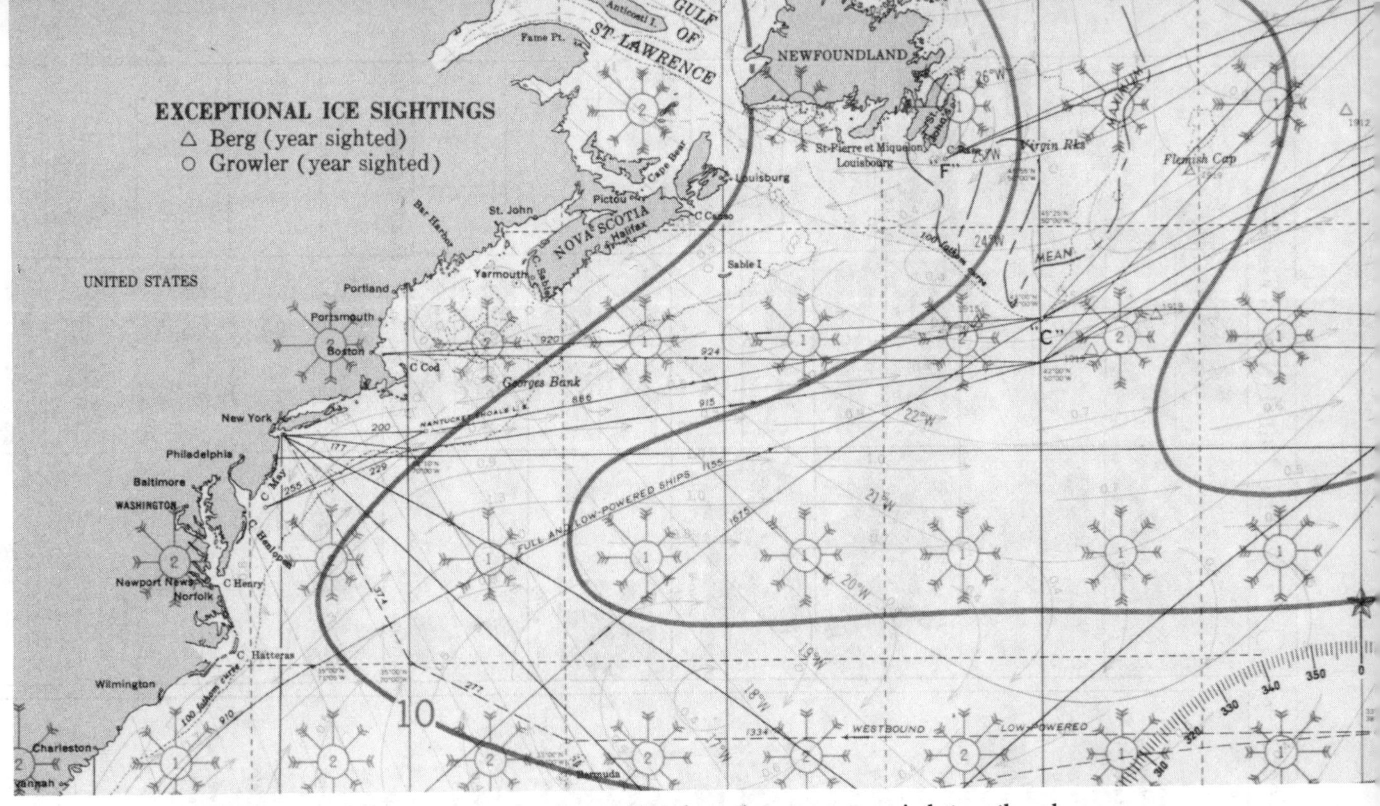

Figure 1832 **Pilot Charts, one per month, provide information on average wind strength and direction, current set and drift, icebergs, normal shipping lanes, fog, weather, and much other useful information that cannot be provided on normal navigation charts.**

ridians, the scale increases with increased difference in longitude from the central meridian.

Parallels appear as nonconcentric arcs of circles and meridians as curved lines converging toward the pole, concave toward the central meridian; see Figure 1828 bottom right. These characteristics contrast with the straight-line parallels and meridians of Mercator charts, and are the reasons why this projection is not so widely used in marine navigation. Directions from any point should be measured relative to the meridian passing through that point; in actual practice, the nearest compass rose is used. Great Lakes charts have a graphic PLOTTING INTERPOLATOR for close measurements of latitude and longitude.

A variation of this type is the LAMBERT CONFORMAL PROJECTION based on one cone that *intersects* the earth's surface at two parallels. It is used mostly by aviators because a straight line nearly approximates a great circle, and radio bearings can be plotted without the corrections needed when using a Mercator chart.

Gnomonic Projection

A gnomonic chart results when the meridians and parallels of latitude are projected onto a plane surface tangent to the earth at one point, Figure 1829. Meridians appear as straight lines converging toward the nearer pole; the parallels of latitude, except for the equator, appear as curves.

Distortion is great, but this projection is used in special cases because of its unique advantage—*great circles appear as straight lines.* Probably the easiest way to obtain a great circle track on a Mercator or polyconic chart is to draw it as a straight line on a gnomonic chart and then transfer points along the line to the other chart using the geographic coordinates for each point. The points so transferred are then connected with short rhumb lines and the result will approximate, closely enough, the great circle path, Figure 1830 top and bottom.

A special case of gnomonic chart projection occurs when a geographic pole is selected as the point of tangency. Now all meridians will appear as straight lines, and the parallels as concentric circles. The result is a chart easily used for polar regions where ordinary Mercator charts cannot be used.

Pilot Charts

No discussion of nautical charts would be complete without a mention of a unique but valuable chart issued quarterly by the DMAHTC. These PILOT CHARTS present in graphic form information on ocean currents and weather probabilities for each month, plus other data of interest to a navigator. Articles of general navigational interest are printed on the reverse side of each chart. They are issued in two editions: *Chart No. 16, North Atlantic Ocean,* and *Chart No. 55, North Pacific Ocean.*

Chart No. 1, reproduced in part on the following eight pages, is a booklet that shows the symbols and abbreviations used on charts issued by this agency. It illustrates U.S. aids to navigation as shown in the color section of this book, as well as those of the International Association of Lighthouse Authorities (IALA) System A, which are used in most European waters, and other parts of the world.

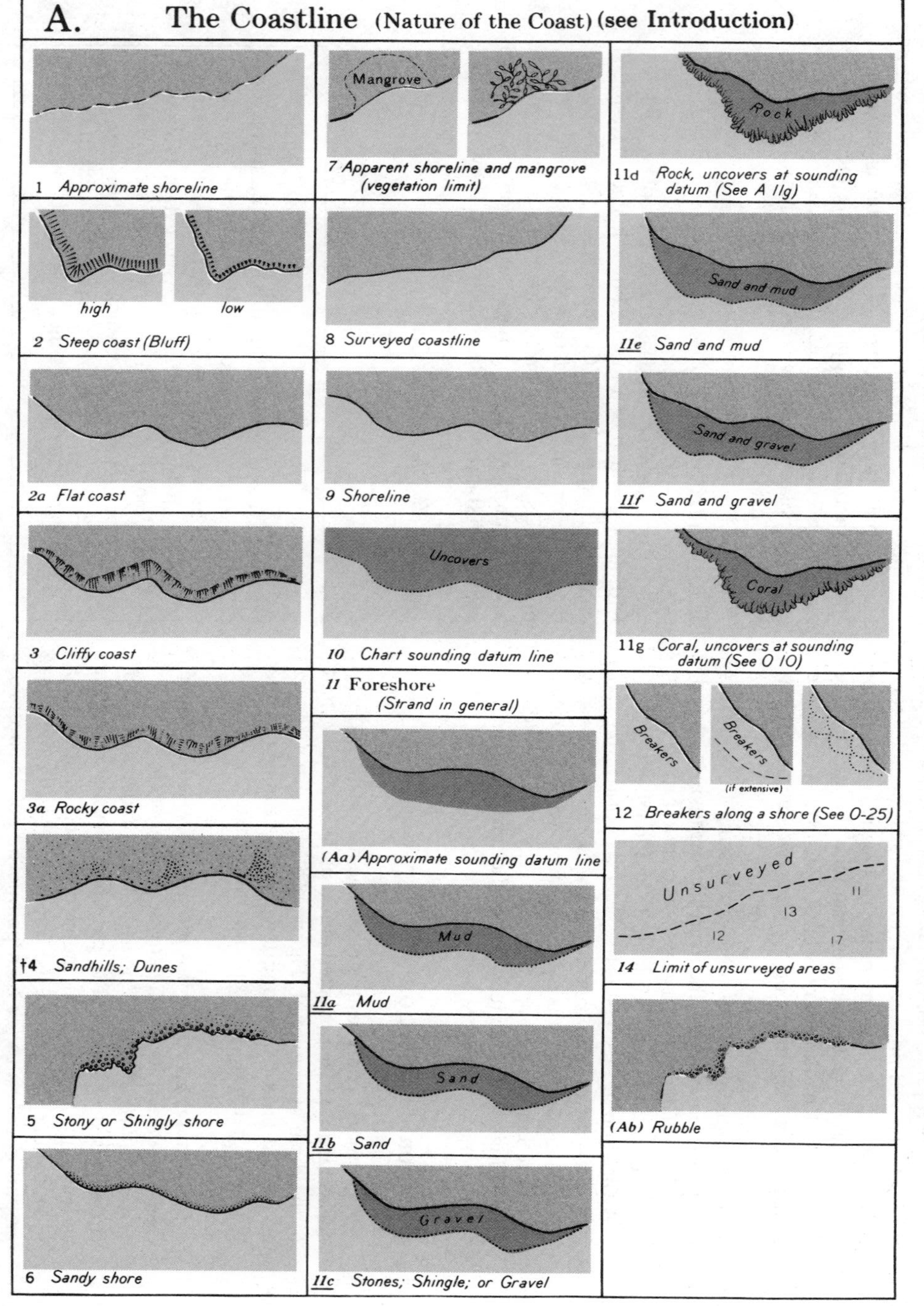

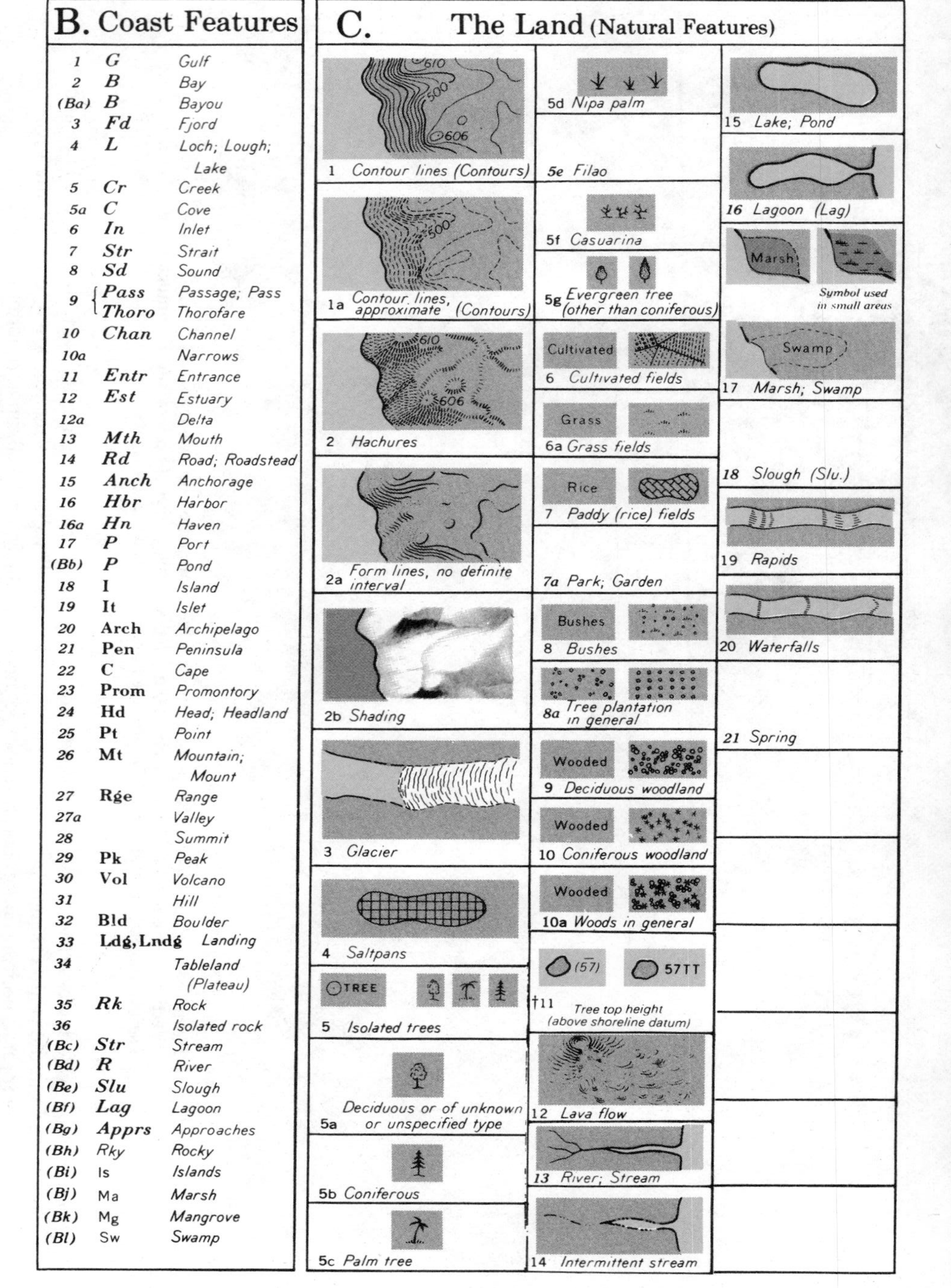

B. Coast Features

No.	Abbr.	Feature
1	G	Gulf
2	B	Bay
(Ba)	B	Bayou
3	Fd	Fjord
4	L	Loch; Lough; Lake
5	Cr	Creek
5a	C	Cove
6	In	Inlet
7	Str	Strait
8	Sd	Sound
9	Pass	Passage; Pass
	Thoro	Thorofare
10	Chan	Channel
10a		Narrows
11	Entr	Entrance
12	Est	Estuary
12a		Delta
13	Mth	Mouth
14	Rd	Road; Roadstead
15	Anch	Anchorage
16	Hbr	Harbor
16a	Hn	Haven
17	P	Port
(Bb)	P	Pond
18	I	Island
19	It	Islet
20	Arch	Archipelago
21	Pen	Peninsula
22	C	Cape
23	Prom	Promontory
24	Hd	Head; Headland
25	Pt	Point
26	Mt	Mountain; Mount
27	Rge	Range
27a		Valley
28		Summit
29	Pk	Peak
30	Vol	Volcano
31		Hill
32	Bld	Boulder
33	Ldg, Lndg	Landing
34		Tableland (Plateau)
35	Rk	Rock
36		Isolated rock
(Bc)	Str	Stream
(Bd)	R	River
(Be)	Slu	Slough
(Bf)	Lag	Lagoon
(Bg)	Apprs	Approaches
(Bh)	Rky	Rocky
(Bi)	Is	Islands
(Bj)	Ma	Marsh
(Bk)	Mg	Mangrove
(Bl)	Sw	Swamp

D. Control Points

No.	Symbol / Abbreviation		Meaning
1			Triangulation point (station)
1a			Astronomic station
2		(See In)	Fixed point (landmark, position accurate)
(Da)		(See Io)	Fixed point (landmark, position approx.)
3	· 256		Summit of height (Peak) (when not a landmark)
(Db)	256		Peak, accentuated by contours
(Dc)	256		Peak, accentuated by hachures
(Dd)			Peak, elevation not determined
(De)	256		Peak, when a landmark
4		Obs Spot	Observation spot
*5		BM	Bench mark
6	View X		View point
7			Datum point for grid of a plan
8			Graphical triangulation point
9		Astro	Astronomical
10		Tri	Triangulation
(Df)		C of E	Corps of Engineers
12			Great trigonometrical survey station
13			Traverse station
14		Bdy Mon	Boundary monument
(Dg)			International boundary monument

E. Units

No.	Abbreviation	Meaning
1	hr, h	Hour
2	m, min	Minute (of time)
3	sec, s	Second (of time)
4	m	Meter
4a	dm	Decimeter
4b	cm	Centimeter
4c	mm	Millimeter
4d	m^2	Square meter
4e	m^3	Cubic meter
5	km	Kilometer(s)
6	in, ins	Inch(es)
7	ft	Foot, feet
8	yd, yds	Yard(s)
9	fm, fms	Fathom(s)
10	cbl	Cable length
11	M, Mi, NMi, NM	Nautical mile(s)
12	kn	Knot(s)
12a	t	Tonne (metric ton = 2,204.6 lbs.)
12b	cd	Candela (new candle)
13	lat	Latitude
14	long	Longitude
14a		Greenwich
15	pub	Publication
16	Ed	Edition
17	corr	Correction
18	alt	Altitude
19	ht; elev	Height; Elevation
20	°	Degree
21	′	Minute (of arc)
22	″	Second (of arc)
23	No	Number
(Ea)	St M, St Mi	Statute mile
(Eb)	μsec, μs	Microsecond
(Ec)	Hz	Hertz (cps)
(Ed)	kHz	Kilohertz (kc)
(Ee)	MHz	Megahertz (Mc)
(Ef)	cps, c/s	Cycles/second (Hz)
(Eg)	kc	Kilocycle (kHz)
(Eh)	Mc	Megacycle (MHz)
(Ei)	T	Ton (U.S. short ton = 2,000 lbs.)

F. Adjectives, Adverbs, Nouns, and Other Words

No.	Abbreviation	Meaning
1	gt	Great
2	lit	Little
3	Lrg	Large
4	sml	Small
5		Outer
6		Inner
7	mid	Middle
8		Old
9	anc	Ancient
10		New
11	St	Saint
12	CONSPIC	Conspicuous
13		Remarkable
14	D, Destr	Destroyed
15		Projected
16	dist	Distant
17	abt	About
18		See chart
18a		See plan
19		Lighted; Luminous
20	sub	Submarine
21		Eventual
22	AERO	Aeronautical
23		Higher
23a		Lower
24	exper	Experimental
25	discontd	Discontinued
26	prohib	Prohibited
27	explos	Explosive
28	estab	Established
29	elec	Electric
30	priv	Private, Privately
31	prom	Prominent
32	std	Standard
33	subm	Submerged
34	approx	Approximate
35		Maritime
36	maintd	Maintained
37	aband	Abandoned
38	temp	Temporary
39	occas	Occasional
40	extr	Extreme
41		Navigable
42	N M	Notice to Mariners
(Fa)	L N M	Local Notice to Mariners
43		Sailing Directions
44		List of Lights
(Fb)	unverd	Unverified
(Fc)	AUTH	Authorized
(Fd)	CL	Clearance
(Fe)	cor	Corner
(Ff)	concr	Concrete
(Fg)	fl	Flood
(Fh)	mod	Moderate
(Fi)	bet	Between
(Fj)	1st	First
(Fk)	2nd, 2d	Second
(Fl)	3rd, 3d	Third
(Fm)	4th	Fourth
(Fn)	DW	Deep Water
(Fo)	min	Minimum
(Fp)	max	Maximum
(Fq)	N'ly	Northerly
(Fr)	S'ly	Southerly
(Fs)	E'ly	Easterly
(Ft)	W'ly	Westerly
(Fu)	Sk	Stroke
(Fv)	Restr	Restricted
†(Fw)	Bl	Blast
†(Fx)	CFR	Code of Federal Regulations
†(Fy)	COLREGS	International Regulations for Preventing Collisions at Sea, 1972
†(Fz)	IWW	Intracoastal Waterway

G. Ports and Harbors

No.	Symbol	Abbreviation	Meaning
1		Anch	Anchorage (large vessels)
2		Anch	Anchorage (small vessels)
3		Hbr	Harbor
4		Hn	Haven
5		P	Port
6		Bkw	Breakwater
6a			Dike
7			Mole
8			Jetty (partly below MHW)
8a			Submerged jetty
(Ga)			Jetty (small scale)
9		Pier	Pier
10			Spit
11			Groin (partly below MHW)
†12	ANCH PROHIBITED	ANCH PROHIB	Anchorage prohibited (screen optional) (See P 25)
12a			Anchorage reserved
12b	QUARANTINE ANCHORAGE	QUAR ANCH	Quarantine anchorage
13	Spoil Area		Spoil ground (See P 11) (Dump Site)
(Gb)	Dumping Ground		Dumping ground (depths may be less than indicated) (Dump Site)
(Gc)	80 83 85 Disposal Area depths from survey of JUNE 1972 90 98		Disposal area (Dump Site)
(Gd)	Ⓟ		Pump-out facilities
14		Fsh stks	Fisheries, Fishing stakes
14a			Fish trap; Fish weirs (actual shape charted)
14b			Duck blind
15			Tuna nets (See G 14a)
15a	Oys	Oys	Oyster bed
16		Ldg, Lndg	Landing place
17			Watering place
18		Whf	Wharf
19			Quay
20			Berth
20a	14		Anchoring berth
20b	3		Berth number
21		Dol	Dolphin
†22			Bollard
23			Mooring ring
24			Crane
25			Landing stage
25a			Landing stairs
26		Quar	Quarantine
27			Lazaret
†28		Hbr Mr / Harbor Master	Harbormaster's office
29		Cus Ho	Customhouse
30			Fishing harbor
31			Winter harbor
32			Refuge harbor
33		B Hbr	Boat harbor
34			Stranding harbor (uncovers at LW)
35			Dock
36			Drydock (actual shape on large-scale charts)
37			Floating dock (actual shape on large-scale charts)
38			Gridiron; Careening grid
39			Patent slip; Slipway; Marine railway
39a		Ramp	Ramp
40	Lock		Lock (point upstream) (See H 13)
41			Wetdock
42			Shipyard
43			Lumber yard
44		Health Office	Health officer's office
45		Hk	Hulk (actual shape on large-scale charts) (See O 11)
46	PROHIBITED AREA	PROHIB AREA	Prohibited area (screen optional)
46a	10		Calling-in point for vessel traffic control
47			Anchorage for seaplanes
48			Seaplane landing area
49	Under construction		Work in progress
50	Under construction		Under construction
51			Work projected
(Ge)	Subm ruins		Submerged ruins
†(Gf)	Dump Site		Dump Site

H. Topography (Artificial Features)

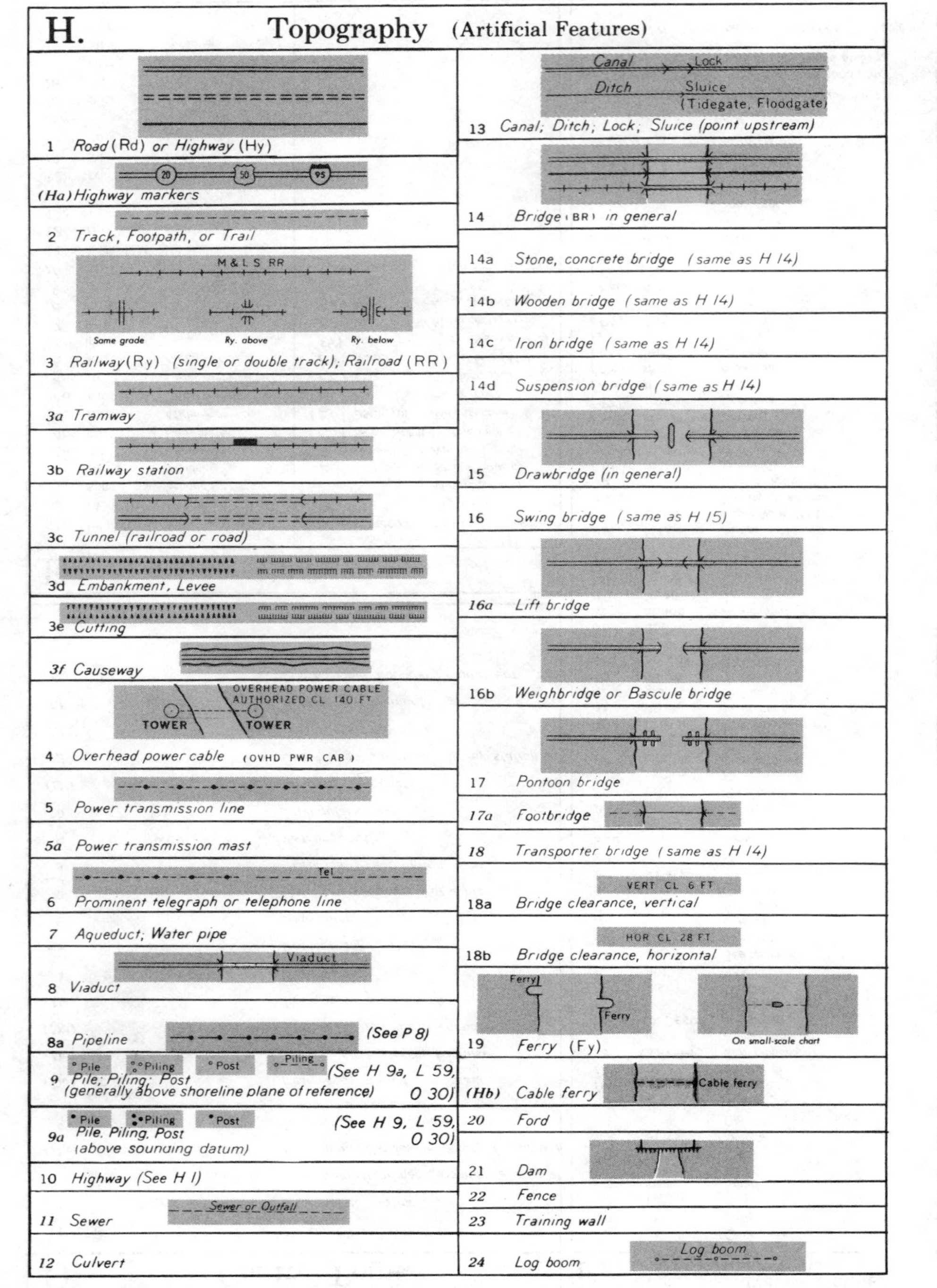

No.	Feature	Text in symbol
1	Road (Rd) or Highway (Hy)	
(Ha)	Highway markers	20, 50, 95
2	Track, Footpath, or Trail	
3	Railway (Ry) (single or double track); Railroad (RR)	M & L S RR; Same grade; Ry. above; Ry. below
3a	Tramway	
3b	Railway station	
3c	Tunnel (railroad or road)	
3d	Embankment, Levee	
3e	Cutting	
3f	Causeway	
4	Overhead power cable (OVHD PWR CAB)	OVERHEAD POWER CABLE AUTHORIZED CL 140 FT; TOWER; TOWER
5	Power transmission line	
5a	Power transmission mast	
6	Prominent telegraph or telephone line	Tel
7	Aqueduct; Water pipe	
8	Viaduct	Viaduct
8a	Pipeline (See P 8)	
9	Pile; Piling; Post (generally above shoreline plane of reference) (See H 9a, L 59, O 30)	Pile; Piling; Post; Piling
9a	Pile, Piling, Post (above sounding datum) (See H 9, L 59, O 30)	Pile; Piling; Post
10	Highway (See H 1)	
11	Sewer	Sewer or Outfall
12	Culvert	
13	Canal; Ditch; Lock; Sluice (point upstream)	Canal; Lock; Ditch; Sluice (Tidegate, Floodgate)
14	Bridge (BR) in general	
14a	Stone, concrete bridge (same as H 14)	
14b	Wooden bridge (same as H 14)	
14c	Iron bridge (same as H 14)	
14d	Suspension bridge (same as H 14)	
15	Drawbridge (in general)	
16	Swing bridge (same as H 15)	
16a	Lift bridge	
16b	Weighbridge or Bascule bridge	
17	Pontoon bridge	
17a	Footbridge	
18	Transporter bridge (same as H 14)	
18a	Bridge clearance, vertical	VERT CL 6 FT
18b	Bridge clearance, horizontal	HOR CL 28 FT
19	Ferry (Fy)	Ferry; Ferry; On small-scale chart
(Hb)	Cable ferry	Cable ferry
20	Ford	
21	Dam	
22	Fence	
23	Training wall	
24	Log boom	Log boom

I. Buildings and Structures (see Introduction)

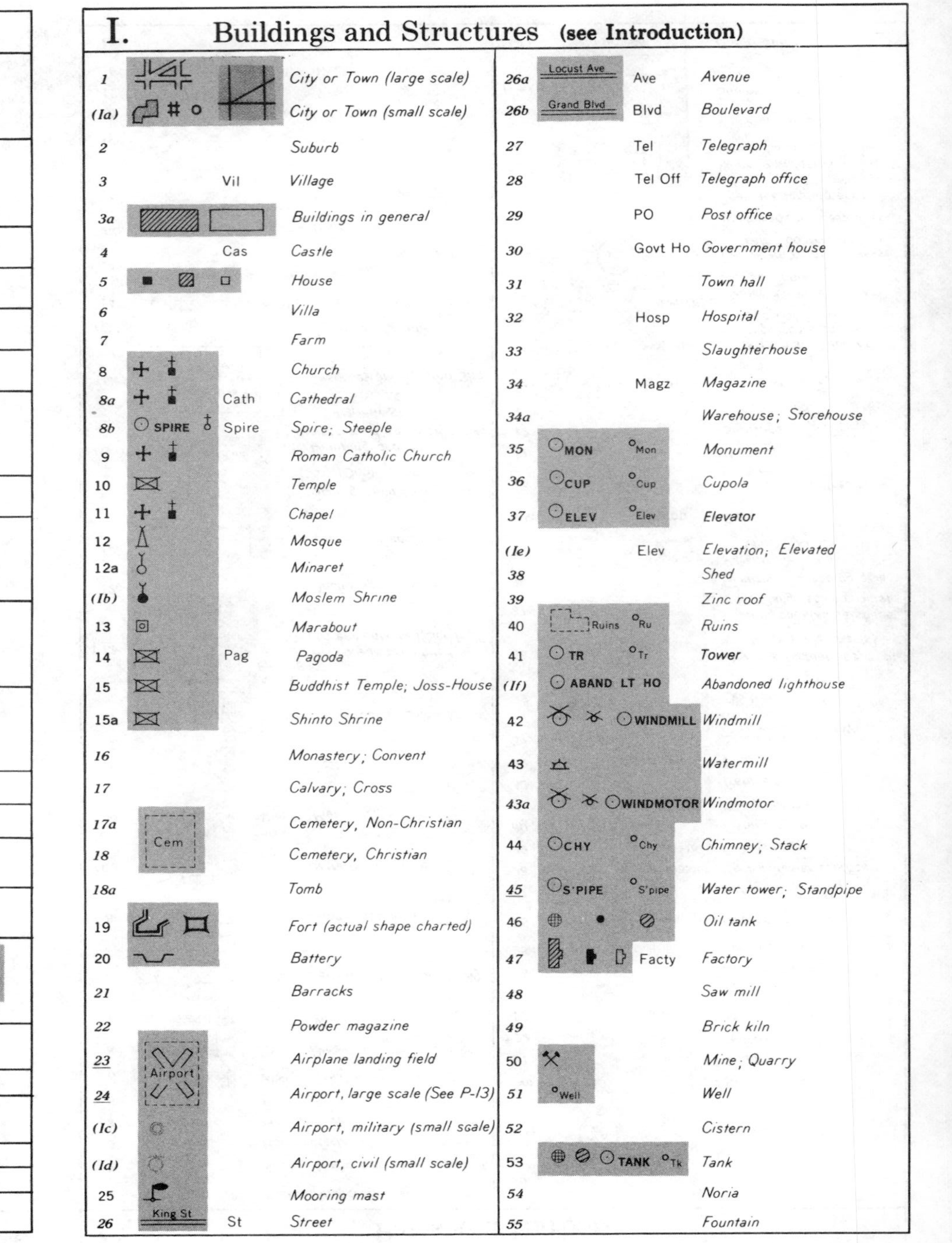

No.	Symbol / Abbreviation	Feature
1		City or Town (large scale)
(Ia)		City or Town (small scale)
2		Suburb
3	Vil	Village
3a		Buildings in general
4	Cas	Castle
5		House
6		Villa
7		Farm
8		Church
8a	Cath	Cathedral
8b	SPIRE; Spire	Spire; Steeple
9		Roman Catholic Church
10		Temple
11		Chapel
12		Mosque
12a		Minaret
(Ib)		Moslem Shrine
13		Marabout
14	Pag	Pagoda
15		Buddhist Temple; Joss-House
15a		Shinto Shrine
16		Monastery; Convent
17		Calvary; Cross
17a	Cem	Cemetery, Non-Christian
18		Cemetery, Christian
18a		Tomb
19		Fort (actual shape charted)
20		Battery
21		Barracks
22		Powder magazine
23	Airport	Airplane landing field
24		Airport, large scale (See P-13)
(Ic)		Airport, military (small scale)
(Id)		Airport, civil (small scale)
25		Mooring mast
26	King St; St	Street
26a	Locust Ave; Ave	Avenue
26b	Grand Blvd; Blvd	Boulevard
27	Tel	Telegraph
28	Tel Off	Telegraph office
29	PO	Post office
30	Govt Ho	Government house
31		Town hall
32	Hosp	Hospital
33		Slaughterhouse
34	Magz	Magazine
34a		Warehouse; Storehouse
35	MON; Mon	Monument
36	CUP; Cup	Cupola
37	ELEV; Elev	Elevator
(Ie)	Elev	Elevation; Elevated
38		Shed
39		Zinc roof
40	Ruins; Ru	Ruins
41	TR; Tr	Tower
(If)	ABAND LT HO	Abandoned lighthouse
42	WINDMILL	Windmill
43		Watermill
43a	WINDMOTOR	Windmotor
44	CHY; Chy	Chimney; Stack
45	S'PIPE; S'pipe	Water tower; Standpipe
46		Oil tank
47	Facty	Factory
48		Saw mill
49		Brick kiln
50		Mine; Quarry
51	Well	Well
52		Cistern
53	TANK; Tk	Tank
54		Noria
55		Fountain

I. Buildings and Structures (continued)

No.	Symbol/Abbreviation	Description
61	Inst	Institute
62		Establishment
63		Bathing establishment
64	Ct Ho	Courthouse
65	Sch	School
(Ig)	HS	High school
(Ih)	Univ	University
66	Bldg	Building
67	Pav	Pavilion
68		Hut
69		Stadium
70	T	Telephone
71		Gas tank; Gasometer
72	GAB Gab	Gable
73		Wall
74		Pyramid
75		Pillar
†76		Oil derrick
(Ii)	Ltd	Limited
(Ij)	Apt	Apartment
(Ik)	Cap	Capitol
(Il)	Co	Company
(Im)	Corp	Corporation
(In)		Landmark (position accurate)(See D 2)
(Io)		Landmark (position approximate)(See Da)

J. Miscellaneous Stations

No.	Symbol/Abbreviation	Description
1	Sta	Any kind of station
2	Sta	Station
3	C G	Coast Guard station (similar to Lifesaving Station, J 6)
(Ja)	R TR C G WALLIS SANDS	Coast Guard station (when landmark)
4	LOOK TR	Lookout station; Watch tower
5		Lifeboat station
6	LS S	Lifesaving station (See J 3)
7	Rkt Sta	Rocket station
† 8	Pilots PIL STA	Pilot station/Pilots
9	Sig Sta	Signal station
10	Sem	Semaphore
11	S Sig Sta	Storm signal station
12		Weather signal station
(Jb)	NWS SIG STA	Nat'l Weather Service signal sta
13		Tide signal station
14		Stream signal station
15		Ice signal station
16		Time signal station
16a		Manned oceanographic station
16b		Unmanned oceanographic station
17		Time ball
18		Signal mast
18a	Mast	Mast
19	FS FS FP FP	Flagstaff; Flagpole
19a	F TR FTr	Flag tower
20		Signal
21	Obsy	Observatory
22	Off	Office
(Jc)	BELL	Bell (on land)
(Jd)	HECP	Harbor entrance control post
(Je)	MARINE POLICE	Marine police station
(Jf)	FIREBOAT STATION	Fireboat station

K. Lights

No.	Symbol/Abbreviation	Description
1		Position of light
2	Lt	Light
(Ka)		Riprap surrounding light
3	Lt Ho	Lighthouse
4	AERO AERO	Aeronautical light (See F-22)
4a		Marine and air navigation light
5	Bn Bn	Light beacon
6		Light vessel; Lightship
8		Lantern
9		Street lamp
10	REF	Reflector
11		Leading light
12	RED RED	Sector light
13	RED GREEN RED GREEN	Directional light
14		Harbor light
15		Fishing light
16		Tidal light
17	Priv maintd	Private light (maintained by private interests; to be used with caution)
21	F	Fixed (steady light)
†22	Occ; Oc	Occulting (total duration of light more than dark)
23	Fl	Single-Flashing (total duration of light less than dark)
† (Kb)	L Fl	Long-Flashing (2s or longer)
	Fl (2+1)	Composite group-flashing
23a	Iso E Int	Isophase (light and dark equal)
†24	Qk Fl; Q	Continuous Quick Flashing (50 to 79 per minute; 60 in U.S.)
	Q (3)	Group Quick
25	Int Qk Fl; I Qk Fl; IQ	Interrupted Quick Flashing
† (Kc)	V Qk Fl; VQ	Continuous Very Quick Flashing (80 to 159-usually either 100 or 120 per minute)
	VQ (3)	Group Very Quick
	IVQ	Interrupted Very Quick
	UQ	Continuous Ultra Quick (160 or more-usually 240 to 300 flashes per minute)
	IUQ	Interrupted Ultra Quick
25a	S Fl	Short Flashing
†26	Alt; Al;	Alternating
†27	Gp Occ; Oc (2) Oc (2+3)	Group-Occulting Composite group occulting
†28	Gp Fl; Fl (3)	Group Flashing
28a	S-L Fl	Short-Long Flashing
28b		Group-Short Flashing
29	F Fl	Fixed and Flashing
30	F Gp Fl	Fixed and Group Flashing
30a	Mo (A)	Morse Code light (with flashes grouped as in letter A)
31	Rot	Revolving or Rotating light
41		Period
42		Every
43		With
44		Visible (range)
(Kd)	M; Mi; N Mi	Nautical mile (See E-11)
(Ke)	m min	Minutes (See E-2)
(Kf)	s; sec	Seconds (See E-3)
45	Fl	Flash
46	Occ	Occultation
46a		Eclipse
47	Gp	Group
48	Occ	Intermittent light
49	SEC	Sector
50		Color of sector
51	Aux	Auxiliary light
52		Varied
61	Vi	Violet
62		Purple
63	Bu	Blue
64	G	Green
65	Or; Y	Orange
66	R	Red
67	W	White
67a	Am	Amber
(Ko)	Y	Yellow
68	OBSC	Obscured light
68a	Fog Det Lt	Fog detector light (See Nb)

Figures in parentheses are examples

K. Lights (continued)

No.	Abbr.	Description
69		Unwatched light
70	Occas	Occasional light
71	Irreg	Irregular light
72	Prov	Provisional light
73	Temp	Temporary light
(Kg)	D; Destr	Destroyed
74	Exting	Extinguished light
75		Faint light
76		Upper light
77		Lower light
78		Rear light
79		Front light
80	Vert	Vertical lights
81	Hor	Horizontal lights
(Kh)	VB	Vertical beam
(Ki)	RGE	Range
(Kj)	Exper	Experimental light
(Kk)	TRLB	Temporarily replaced by lighted buoy showing the same characteristics
(Kl)	TRUB	Temporarily replaced by unlighted buoy
(Km)	TLB	Temporary lighted buoy
(Kn)	TUB	Temporary unlighted buoy

L. Buoys and Beacons (see Introduction)

No.	Symbol	Description
1		Approximate position of buoy
2		**Light buoy**
3	BELL BELL BELL	**Bell buoy**
3a	GONG GONG GONG	**Gong buoy**
4	WHIS WHIS	Whistle buoy
5	C C	Can or Cylindrical buoy
6	N N	Nun or Conical buoy
7	SP SP	Spherical buoy
8	S S	Spar buoy
†8a	P P	Pillar or Spindle buoy
9		Buoy with topmark (ball) (see L-57)
10		Barrel or Ton buoy
(La)		Color unknown
(Lb)	FLOAT FLOAT	Float
12	FLOAT FLOAT FLOAT	Lightfloat
13		Outer or Landfall buoy
14	BW BW	Fairway buoy (BWVS)
14a	BW BW	Midchannel buoy (BWVS)
15	R "2" R "2" R "2"	Starboard-hand buoy (entering from seaward)
16	"1" "1"	Port-hand buoy (entering from seaward)
†17	RB BR RB RB	Bifurcation buoy
†18	RB BR RB RB	Junction buoy
†19	RB BR RB RB	Isolated danger buoy
†20	RB BR RB G G G G G	Wreck buoy
†20a	RB BR RB G G	Obstruction buoy
21	Tel Tel	Telegraph-cable buoy
22		Mooring buoy (colors of mooring buoys never carried)
22a		Mooring
22b	Tel Tel	Mooring buoy with telegraphic communications
22c	T T	Mooring buoy with telephonic communications
23		Warping buoy
24	Y Y	Quarantine buoy
24a		Practice area buoy
25	Explos Anch Explos Anch	Explosive anchorage buoy
25a	AERO AERO	Aeronautical anchorage buoy
26	Deviation Deviation	Compass adjustment buoy
27	BW BW	Fish trap (area) buoy (BWHB)
27a		Spoil ground buoy
28	W W	Anchorage buoy (marks limits)
29	Priv maintd Priv maintd	Private aid to navigation (buoy) (maintained by private interests, use with caution)

L. Buoys and Beacons (continued)

No.	Symbol	Abbr.	Description
30			Temporary buoy (See Ki, j, k, l)
30a			Winter buoy
31		HB	Horizontal stripes or bands
32		VS	Vertical stripes
33		Chec	Checkered
33a		Diag	Diagonal bands
41		W	White
42		B	Black
43		R	Red
44		Y	Yellow
45		G	Green
46		Br	Brown
47		Gy	Gray
48		Bu	Blue
48a		Am	Amber
48b		Or	Orange
51			Floating beacon
†52	RW Bn, W Bn, R Bn, G Bn, G Bn		Fixed beacon (unlighted or daybeacon)
	Bn		Black beacon
	Bn		Color unknown
(Lc)	MARKER, Marker		Private aid to navigation
53		Bn	Beacon, in general (See L 52)
54			Tower beacon
55			Cardinal marking system
56	Deviation Bn		Compass adjustment beacon
†57	etc		Topmarks (See L 9)
58			Telegraph-cable (landing) beacon
59	Piles Piles		Piles (See O 30; H 9, 9a)
			Stakes
	Stumps		Stumps (See O 30)
			Perches
61	CAIRN Cairn		Cairn
62			Painted patches
63	TR		Landmark (position accurate) (See D 2)
(Ld)	Tr		Landmark (position approximate)
64		REF	Reflector
65	MARKER		Range targets, markers
(Le)	W Or W Or W Or W Or		Special-purpose buoys
66			Oil installation buoy
67			Drilling platform (See Of, Og)
70			NOTE: Refer to new L-70, on page 11, for aid system used in certain foreign waters.
(Lf)	Ra Ref		Radar reflector (See M 13) (not charted on IALA Sys. "A" marks see L-70)
			Superbuoys
†71			LANBY (large auto. nav. buoy)
†72			TANKER terminal buoy (mooring)
†73			ODAS (Data buoy)

L.70 Buoys and Beacons IALA* Buoyage System "A"

The combined Cardinal and Lateral System (Red to Port)

Where in force, System 'A' applies to all fixed and floating marks other than lighthouses, sector lights and leading-marks, lightships and 'lighthouse buoys.' There are no special characteristics reserved for marking wrecks.

UNLIT MARKS	LIGHTED MARKS

Lateral, generally marking the limits of well-defined channels.

Port Hand

All red
Topmark (if any): can — R R R R — Fl.R, Occ.R etc — Red light (any characteristic)

Symbol used to indicate buoyage direction where not obvious; size and orientation varied to suit its situation.

Starboard Hand

All green or black
Topmark (if any): cone — G B G B — Fl.G, Occ.G etc — Green Light (any characteristic)

Cardinal, indicating navigable water to the named side of the mark.

Topmarks: 2 black cones

NW — North Mark, Black above yellow (BY) — NE

West Mark (YBY) — Point of interest — East Mark (BYB)

Yellow with black band — Black with yellow band

SW — South Mark, Yellow above black (YB) — SE

White light — Time (seconds) 0 5 10 15

North Mark — V Qk Fl or Qk Fl

East Mark — V Qk Fl(3)5s or Qk Fl(3)10s — Period

South Mark — V Qk Fl(6)+L Fl.10s or Qk Fl(6)+L Fl.15s — Period

West Mark — V Qk Fl(9)10s or Qk Fl(9)15s — Period

The same abbreviations are used for lights on spar buoys.
The periods, 5s, 10s and 15s, may not always be charted.

Isolated danger, stationed over a danger with navigable water around it.

Body: black with red horizontal band(s)
Topmarks: 2 black spheres — BRB BRB — Gp Fl(2), Gp Fl(2) — White light

Safe water, such as mid-channel and landfall marks.

Body: red and white vertical stripes
Topmark (if any): red sphere — RW RW RW — Iso,Occ or L Fl; Iso,Occ or L Fl; Iso,Occ or L Fl — White light

Special, not primarily to assist navigation but to indicate special features.

Body (shape optional): yellow
Topmark (if any): yellow X — Y Y Y etc — Gp Fl(4) Y or Fl Y; Gp Fl(4) Y or Fl Y; Gp Fl(4) Y or Fl Y — Yellow light

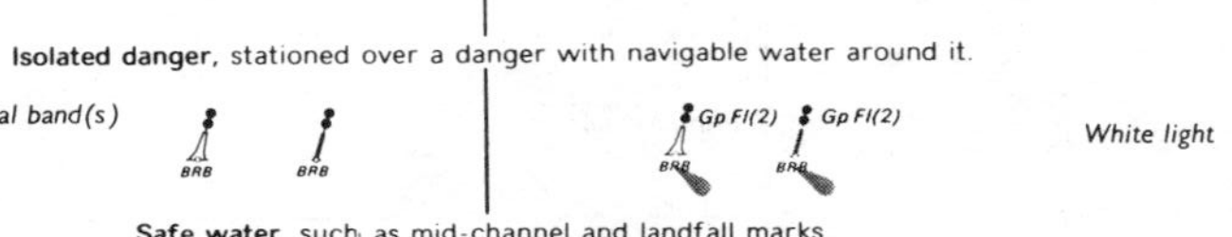

NOTES

STANDARD BUOY SHAPES are can, conical, spherical, pillar (including high focal plane), and spar, but variations may occur.
COLOR ABBREVIATIONS under buoy symbols, especially spar buoys, may sometimes be omitted.
PERIODS of lights, where charted, are shown thus: 10s (for 10 seconds).
RADAR REFECTORS are not charted.
LIGHT FLARES AND ARROWS are purple.
*International Association of Lighthouse Authorities

M. Radio and Radar Stations

No.	Symbol	Description
1	R Sta	Radio telegraph station
2	RT	Radio telephone station
3	R Bn	Radiobeacon
4	R Bn	Circular radiobeacon
† 5	RD 072°30' RD	Directional radiobeacon; Radio range
6		Rotating loop radiobeacon
7	RDF	Radio direction finding station
(Ma)	ANTENNA (TELEM) / TELEM ANT	Telemetry antenna
(Mb)	R RELAY MAST	Radio relay mast
(Mc)	MICRO TR	Microwave tower
9	R MAST / R TR	Radio mast / Radio tower
9a	TV TR	Television mast; Television tower
10	R TR (WBAL) 1090 kHz	Radio broadcasting station (commercial)
10a	R Sta	QTG radio station
11	Ra	Radar station
12	Racon	Radar responder beacon
13	Ra Ref	Radar reflector (See L-Lf) (not charted on IALA Sys. "A" marks; see L-70)
14	Ra (conspic)	Radar conspicuous object
14a		Ramark
15	D F S	Distance finding station (synchronized signals)
16	AERO R Bn 302	Aeronautical radiobeacon
17	Decca Sta	Decca station
18	Loran Sta Venice	Loran station (name)
19	CONSOL Bn 190 kHz MMF	Consol (Consolan) station
(Md)	AERO R Rge 342	Aeronautical radio range
(Me)	Ra Ref Calibration Bn	Radar calibration beacon
(Mf)	LORAN TR SPRING ISLAND	Loran tower (name)
(Mg)	R TR F R Lt	Obstruction light
(Mh)	RA DOME, Ra Dome / DOME (RADAR), Dome (Radar)	Radar dome
(Mi)	uhf	Ultrahigh frequency
(Mj)	vhf	Very high frequency

N. Fog Signals

No.	Abbreviation	Description
† 1	Fog Sig	Fog-signal station
2		Radio fog-signal station
3	GUN	Explosive fog signal
4		Submarine fog signal
5	SUB-BELL	Submarine fog bell (action of waves)
6	SUB-BELL	Submarine fog bell (mechanical)
7	SUB-OSC	Submarine oscillator
8	NAUTO	Nautophone
9	DIA	Diaphone
10	GUN	Fog gun
11	SIREN	Fog siren
12	HORN	Fog trumpet
13	HORN	Air (foghorn)
13a	HORN	Electric (foghorn)
14	BELL	Fog bell
15	WHIS	Fog whistle
16	HORN	Reed horn
17	GONG	Fog gong
18		Submarine sound signal not connected to the shore (See N 5, 6, 7)
18a		Submarine sound signal connected to the shore (See N 5, 6, 7)
(Na)	HORN	Typhon
(Nb)	Fog Det Lt	Fog detector light (See K 68a)
† (Nc)	Mo	Morse Code fog signal

O. Dangers

No.	Symbol	Description
†1	(25) (25)	Rock which does not cover (height above MHW)
2	*Uncov 2 ft, Uncov 2 ft, *(2), (2)	Rock which covers and uncovers with height above chart sounding datum (see Introduction)
3		Rock awash at (near) level of chart sounding datum
		Dotted line emphasizes danger to navigation
(Oa)	*	Rock awash (height unknown)
		Dotted line emphasizes danger to navigation
4	+	Submerged rock (depth unknown)
		Dotted line emphasizes danger to navigation
5	(5) Rk	Shoal sounding on isolated rock
6		Submerged rock not dangerous to surface navigation (See O 4)
6a	21 Rk, 21 Obstr	Sunken danger with depth cleared by wire drag
7	Reef	Reef of unknown extent
8	Sub Vol	Submarine volcano
9	Discol Water	Discolored water
10	Coral, Co, Co, Co	Coral reef, detached (uncovers at sounding datum)
	Co, 3, Reef line	Coral or Rocky reef, covered at sounding datum (See A-IId, IIg)
11		Wreck showing any portion of hull or superstructure (above sounding datum)
12	Masts	Wreck with only masts visible (above sounding datum)
13		Old symbols for wrecks
13a	PA	Wreck always partially submerged
14		Sunken wreck dangerous to surface navigation (less than 11 fathoms over wreck) (See O 6a)
15	(5) Wk	Wreck over which depth is known
15a	21 Wk	Wreck with depth cleared by wire drag
15b	(8) Wk	Unsurveyed wreck over which the exact depth is unknown, but is considered to have a safe clearance to the depth shown
16		Sunken wreck, not dangerous to surface navigation
17	Foul	Foul ground, Foul bottom (fb)
18	Tide Rips (Symbol used only in small areas)	Overfalls or Tide rips
19	Eddies (Symbol used only in small areas)	Eddies
20	Kelp (Symbol used only in small areas)	Kelp, Seaweed
21	Bk	Bank
22	Shl	Shoal
23	Rf	Reef (See A IId, IIg, O 10)
23a		Ridge
24	Le	Ledge
25		Breakers (See A 12)
26		Submerged rock (See O 4),
27	(5) Obstr	Obstruction
(Oh)	Obstr, Well, Obstr, Well	Subm well; Subm well (buoyed)
	Obstruction (fish haven)	
(Oc)		Fish haven (artificial fishing reef)
28		Wreck (See O 11 to 16)
29	Wreckage, Wks	Wreckage
29a		Wreck remains (dangerous only for anchoring)
	Subm piles, Subm piling	
30		Submerged piling (See H-9, 9a; L 59)
30a	Snags, Stumps	Snags; Submerged stumps (See L 59)
31		Lesser depth possible
32	Uncov	Dries (See A 10; O 2, 10)
33	Cov	Covers (See O 2, 10)
34	Uncov	Uncovers (See A 10; O 2, 10)
	(3) Rep (1958)	Reported (with date)
35	Eagle Rk (rep 1958)	Reported (with name and date)
36	Discol	Discolored (See O 9)
37		Isolated danger
38		Limiting danger line
39	rky	Limit of rocky area
41	PA	Position approximate
42	PD	Position doubtful
43	ED	Existence doubtful
44	P Pos	Position
45	D	Doubtful
46		Unexamined
(Od)	LD	Least Depth
(Oe)	Subm Crib; Crib (above water)	Crib
(Of)	Platform (lighted) HORN	Offshore platform (unnamed)
(Og)	Hazel (lighted) HORN	Offshore platform (named)

† P. Various Limits, etc.

No.	Symbol	Description
1		Leading line; Range line
2		Transit
3		In line with
4		Limit of sector
†5		Channel, Course, Track recommended (marked by buoys or beacons) (see P 21)
†5a	DW	Recommended track for deep draft vessels (defined by fixed mark(s))
†5b	DW83 ft, DW76 ft	Depth is shown where it has been obtained by the cognizant authority
		Alternate course
	Ra—Ra	Radar-guided track
†6a		Established traffic separation scheme. One-way traffic lanes (separated by line or zone)
†6b		Established traffic separation scheme: Roundabout
		If no separation zone exists, the center of the roundabout is shown by a circle
†6c		Recommended direction of traffic flow
7		Submarine cable (power telegraph, telephone, etc.)
7a	Cable Area	Submarine cable area
7b		Abandoned submarine cable (includes disused cable)
8		Submarine pipeline
8a	Pipeline Area	Submarine pipeline area
8b		Abandoned submarine pipeline
9		Maritime limit in general
(Pb)	RESTRICTED AREA	Limit of restricted area
†10		Limits of national fishing zones
(Pc)		U.S. Harbor Line
11		Limit of dumping ground, spoil ground (See P 9, G 13)
12		Anchorage limit
13		Limit of airport (See I 23, 24)
13a		Limit of military practice areas
14		Limit of sovereignty (Territorial waters)
15		Customs boundary
16		International boundary (also State boundary)
17		Stream limit
18		Ice limit
19		Limit of tide
20		Limit of Navigation
21		Recommended track (not marked by buoys or beacons)
†21a	DW, DW, DW, DW	Recommended track for deep draft vessels (track not defined by fixed mark(s))
†21b	DW83 ft, DW76 ft, DW83 ft, DW76 ft	Depth is shown where it has been obtained by the cognizant authority
22		District or province limit
23		Reservation line (Options)
24	COURSE 053°00' TRUE; MARKERS; MARKERS	Measured distance
25	PROHIBITED AREA	Prohibited area (see G 12, 46) (Screen optional)
(Pd)	SAFETY FAIRWAY	Shipping safety fairway (two-way traffic)
†(Pe)		Limits of former mine danger area
†(Pf)	26133	Reference larger scale chart.
(Pg)		Limit of fishing areas (fish trap areas)
†(Ph)		3-mile Territorial Sea Boundary; 12-mile Contiguous Zone Boundary; headland to headland line
†(Pi)		COLREGS demarcation line

Q. Soundings

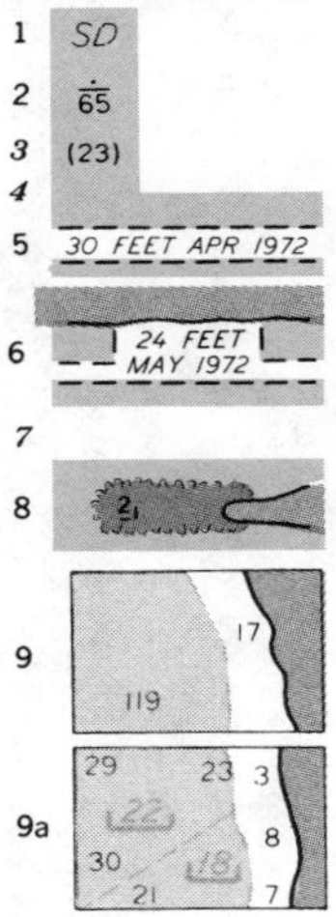

No.	Symbol	Description
1	SD	Doubtful sounding
2	65	No bottom found
3	(23)	Out of position
4		Least depth in narrow channels
5	30 FEET APR 1972	Dredged channel (with controlling depth indicated)
6	24 FEET MAY 1972	Dredged area
7		Swept channel (See Q 9)
8	2_1	**Drying (or uncovering) heights above chart sounding datum**
9	17 119	Swept area, not adequately sounded (shown by purple or green tint
9a	29 23 3 22 30 18 8 21 7	Swept area adequately sounded (swept by wire drag to depth indicated)
10		Hairline depth figures
10a	8_2 19	Figures for ordinary soundings
11		Soundings taken from foreign charts
12		Soundings taken from older surveys (or smaller scale chts)
13	8_2 19	Echo soundings
14	8_2 19	Sloping figures
15	8_2 19	Upright figures (See Q 10a)
16	(25) (2)	Bracketed figures (See O 1, 2)
17	6	Underlined sounding figures (See Q 8)
18	3_2 6_1	Soundings expressed in fathoms and feet
22		Unsounded area
(Qa)	6 5 2ft	Stream

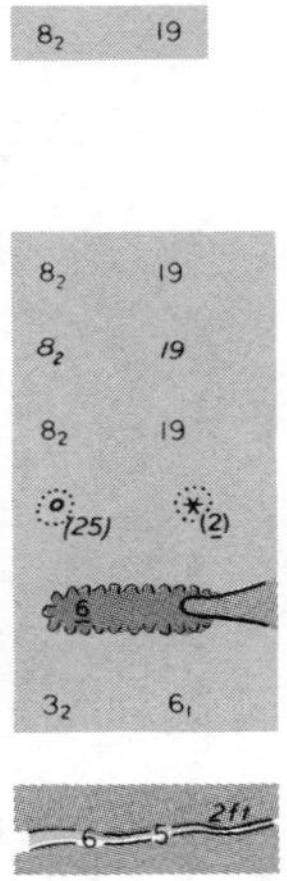

R. Depth Contours and Tints (see General Remarks)

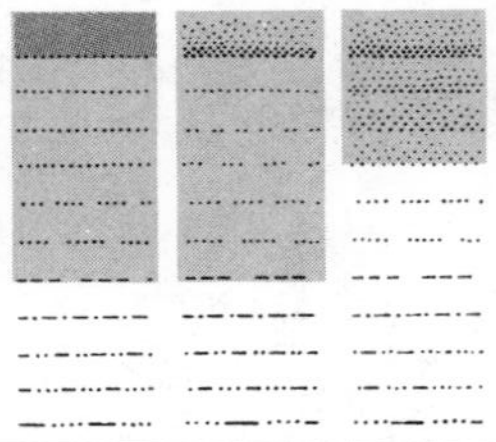

Feet	Fm / Meters
0	0
6	1
12	2
18	3
24	4
30	5
36	6
60	10
120	20
180	30
240	40

Feet	Fm / Meters
300	50
600	100
1,200	200
1,800	300
2,400	400
3,000	500
6,000	1,000
12,000	2,000
18,000	3,000

Or continuous lines, with values — 5 — (blue or black) — 100 —

S. Quality of the Bottom

1	Grd	Ground
2	S	Sand
3	M	Mud; Muddy
4	Oz	Ooze
5	Ml	Marl
6	Cl	Clay
7	G	Gravel
8	Sn	Shingle
9	P	Pebbles
10	St	Stones
11	Rk; rky	Rock; Rocky
11a	Blds	Boulders
12	Ck	Chalk
12a	Ca	Calcareous
13	Qz	Quartz
13a	Sch	Schist
14	Co	Coral
(Sa)	Co Hd	Coral head
15	Mds	Madrepores
16	Vol	Volcanic
(Sb)	Vol Ash	Volcanic ash
17	La	Lava
18	Pm	Pumice
19	T	Tufa
20	Sc	Scoriae
21	Cn	Cinders
21a		Ash
22	Mn	Manganese
23	Sh	Shells
24	Oys	Oysters
25	Ms	Mussels
26	Spg	Sponge
27	K	Kelp
28	Wd	Seaweed
28	Grs	Grass
29	Stg	Sea-tangle

S. Quality of the Bottom (continued)

31	Spi	Spicules
32	Fr	Foraminifera
33	Gl	Globigerina
34	Di	Diatoms
35	Rd	Radiolaria
36	Pt	Pteropods
37	Po	Polyzoa
38	Cir	Cirripedia
38a	Fu	Fucus
38b	Ma	Mattes
39	fne	Fine
40	crs	Coarse
41	sft	Soft
42	hrd	Hard
43	stf	Stiff
44	sml	Small
45	lrg	Large
46	stk	Sticky
47	brk	Broken
47a	grd	Ground (Shells)
48	rt	Rotten
49	str	Streaky
50	spk	Speckled
51	gty	Gritty
52	dec	Decayed
53	fly	Flinty
54	glac	Glacial
55	ten	Tenacious
56	wh	White
57	bk	Black
58	vi	Violet
59	bu	Blue
60	gn	Green
61	yl	Yellow
62	or	Orange
63	rd	Red
64	br	Brown
65	ch	Chocolate
66	gy	Gray
67	lt	Light
68	dk	Dark
70	vard	Varied
71	unev	Uneven
(Sc)	S/M	Surface layer and Under layer
76		Freshwater springs in seabed
†(Sd)	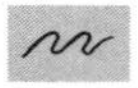	Mobile bottom (sand waves)

T. Tides and Currents

1	HW	High water
1a	HHW	Higher high water
2	LW	Low water
(Ta)	LWD	Low-water datum
2a	LLW	Lower low water
3	MTL	Mean tide level
4	MSL	Mean sea level
4a		Elevation of mean sea level above chart (sounding) datum
5		Chart datum (datum for sounding reduction)
6	Sp	Spring tide
7	Np	Neap tide
7a	MHW	Mean high water
8	MHWS	Mean high-water springs
8a	MHWN	Mean high-water neaps
8b	MHHW	Mean higher high water
8c	MLW	Mean low water
9	MLWS	Mean low-water springs
9a	MLWN	Mean low-water neaps
9b	MLLW	Mean lower low water
10	ISLW	Indian spring low water
11	HWF & C	High-water full and change (vulgar establishment of the port)
12	LWF & C	Low-water full and change
13		Mean establishment of the port
13a		Establishment of the port
14		Unit of height
15		Equinoctial
16		Quarter; Quadrature
17	Str	Stream
18	2 kn	Current, general, with rate
19	2 kn	Flood stream (current) with rate
20	2 kn	Ebb stream (current) with rate
21	Tide gauge	Tide gauge; Tidepole; Automatic tide gauge
23	vel	Velocity; Rate
24	kn	Knots
25	ht	Height
26		Tide
27		New moon
28		Full moon
29		Ordinary
30		Syzygy
31	fl	Flood
32		Ebb
33		Tidal stream diagram
34	 A B	Place for which tabulated tidal stream data are given
35		Range (of tide)
36		Phase lag
(Tb)	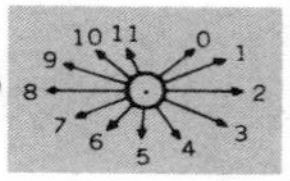 0 1 2 3 4 5 6 7 8 9 10 11	Current diagram, with explanatory note
†(Tc)	CRD	Columbia River Datum
†(Td)	GCLWD	Gulf Coast Low Water Datum

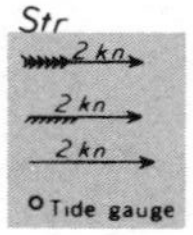

19

The dimensions and tools of piloting and their use in dead reckoning; how to establish a speed curve for your boat

BASIC PILOTING PROCEDURES

PILOTING is the use of landmarks, aids to navigation, and soundings to conduct a vessel safely through channels and harbors, and along coasts where depths of water and dangers to navigation require constant attention to the boat's position and course.

Piloting is one of the principal subdivisions of NAVIGATION—the science and art of directing the movements of a vessel from one position to another in a safe and efficient manner. It is a science because it uses principles and procedures based on centuries of observation, analysis, and study; it is an art because interpretations of observations and other information require individual judgment and skill.

An adjunct to piloting is DEAD RECKONING, a procedure by which a boat's approximate location is determined at any time by its movements since the last accurate determination of position. The term is said to have evolved from *deduced* reckoning, often abbreviated "de'ed" reckoning in old ships' logs.

Other subdivisions of navigation are ELECTRONIC and CELESTIAL. Electronic systems applicable to small craft will be covered in the following chapters; celestial navigation is covered in more advanced texts.

The Importance of Piloting

Piloting is used by boatmen in rivers, bays, lakes, sounds and close alongshore when on the open ocean or large lakes—all areas known as PILOT WATERS. The navigator of a boat of any size in pilot waters must have adequate training and knowledge; he must be constantly alert, and give his task close attention. Frequent determinations of position are usually essential, and course or speed changes may be necessary at relatively short intervals.

The High Seas vs. Pilot Waters

On the high seas, or well offshore in such large bodies of water as the Great Lakes, your navigation can be more leisurely and relaxed. An error or uncertainty of position of a few miles presents no immediate hazard to the safety of boat and crew. But when you approach PILOT WATERS, you need greater accuracy. An error of only a few yards can result in your running aground with possibly serious consequences. The presence of other vessels nearby underscores your constant need to know where the dangers lie and where your own boat can be steered safely.

The Enjoyment of Piloting

You will enjoy piloting most when you can do it without anxiety; which means that you need a background of study and practice. Try "overnavigating" in times of fair weather so you acquire the skill needed to direct your boat safely through fog, rain, or night without fear or strain.

The Dimensions of Piloting

The basic dimensions of piloting are direction, distance, and time. Other quantities which must be measured, calculated, or used include speed, position, and depths and heights. You must have a ready understanding of how each of these dimensions is measured, expressed in units, used in calculations, and plotted on charts.

Direction

DIRECTION is the position of one point relative to another point without reference to the distance between them. As discussed earlier in the chapters on compasses and charts, modern navigation uses the system of angular measurement in which a complete circle is divided into 360 units called "degrees." In some piloting procedures degrees may be subdivided into minutes, or into common or decimal fractions.

True, Magnetic, or Compass

Also as discussed, directions are normally referenced to a base line running from the origin point toward the geographic North Pole. Such directions can be measured on a chart with reference to the meridians of longitude and are called TRUE DIRECTIONS. Measurements made with respect to the direction of the earth's magnetic field at that point are termed MAGNETIC DIRECTIONS, and those referred to local magnetic conditions as measured by the boat's compass are designated COMPASS DIRECTIONS, see

Figure 1901. *It is essential that you always designate the directorial reference used: true (T), magnetic (M), or compass (C).*

Directions and Angles

The basic system of measurement uses the reference direction as 0° (north) and measures clockwise through 90° (east), 180° (south), and 270° (west) around to 360° which is north again. Directions are expressed in three-digit form such as 005°, 030°, 150°. Note that zeros are added before the direction figures to make a three-digit number; for example, 005° or 055°, (Figure 1902 top left). *Angles* that are *not* directions are expressed in one, two, or three digits as appropriate; 5 degrees, 30 degrees, 150 degrees, (see Figure 1902 top right).

Reciprocals

For any given direction, there is its RECIPROCAL direction: its direct opposite, differing by 180 degrees. Thus the reciprocal of 030° is 210°, and the reciprocal of 300° is 120°. To find the reciprocal of any direction, simply add 180 degrees if the given direction is less than that amount; or subtract 180 degrees if it is more, as shown in Figure 1902 bottom.

Distance

DISTANCE is defined as the spacial separation between two points without regard to the direction of one from the other. It is the length of the shortest line that can be drawn between the two points.

The basic unit of distance in piloting in the U.S. is the MILE, but as noted in Chapter 18, a boatman may encounter two types of miles. The STATUTE MILE is used on inland bodies of water such as the Mississippi River and its tributaries, the Great Lakes, and the Atlantic and Gulf Intracoastal Waterways. It is 5,280 feet in length; the same mile as commonly used on land. On the high seas and connecting tidal waters, the unit of measurement is the NAUTICAL MILE of 6,076.1 feet. This "salt-water" mile is equal to one minute of *latitude,* and this relationship is often used in navigation. The conversion between nautical and statute miles is a factor of 1.15; roughly, seven nautical miles equals eight statute miles.

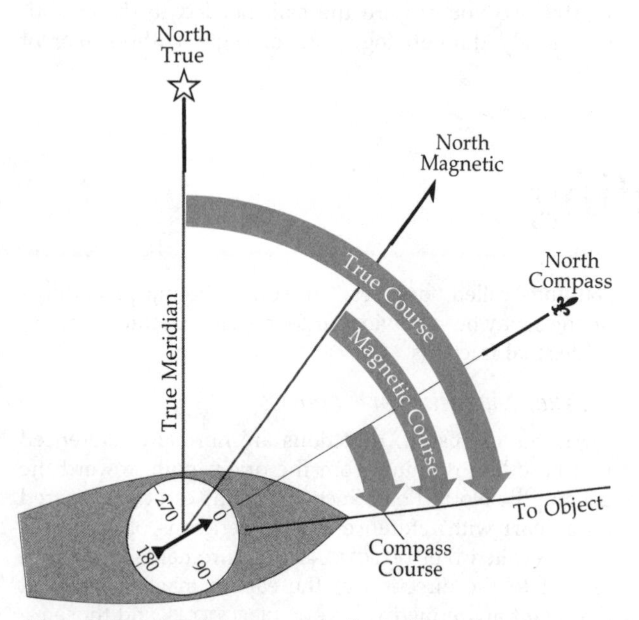

Figure 1901 **Direction, one of the basic dimensions of piloting, can be measured with respect to True, Magnetic, or Compass North; always designate the reference used.**

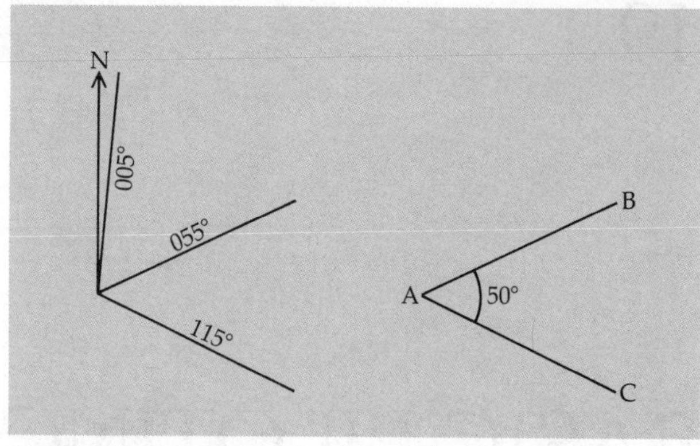

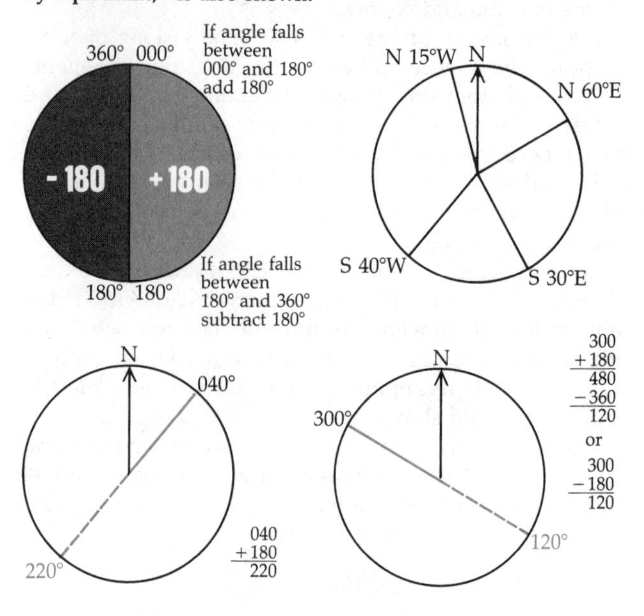

Figure 1902 **Directions are always designated by a three-digit number; add zeros before a single- or double-digit number, *left.* An angle between two directions, *right,* as distinguished from the directions themselves, is not expressed as a three-digit number. The reciprocal of a direction is found by adding 180 degrees to the given direction. If the total exceeds 360 degrees, subtract that amount—or simply subtract 180 degrees from the given direction. An old form of direction notation, by "quadrant," is also shown.**

You may cruise from an area using one kind of mile into waters where the other kind is used. Be sure to determine which type of mile is to be used in your calculations. For shorter distances, a few miles or less, the unit of measurement may be the YARD; feet are seldom used.

With the coming of the metric system, boatmen must be prepared for greater use of METERS (m) and KILOMETERS (km); they must be familiar with these units and conversion to and from conventional units. Factors are: 1 foot = 0.3048 meter; 1 yard = 0.9144 meter; 1 statute mile = 1.609 kilometers; 1 nautical mile = 1.852 kilometers.

Time

Although the pilot does not need so accurate a knowledge of the exact time of day as does a celestial navigator, ability to determine the passage of time and to perform calculations with such ELAPSED TIME is essential.

Units of Time

The units of time used in boating are the everyday ones of hours and minutes. In piloting, measurements are seldom carried to the precision of seconds of time although decimal fractions of minutes may occur occasionally in calculations. Seconds and fractions of minutes may be used in competitive events such as races and predicted log contests, but seldom otherwise.

The 24-Hour Clock System

In navigation, including piloting, the time of day is expressed in a 24-hour system that eliminates the designations of "am" and "pm." Time is written in four-digit figures; the first two are the hour and the last two are the minutes. The day starts at 0000 or midnight; the next minute is 0001, 12:01 am in the "shore" system; 0100 would be 1 am, and so on to 1200 at noon. The second half of the day continues in the same pattern with 1300 being 1 pm, 1832 being 6:32 pm, etc., to 2400 for midnight. (The time 2400 for one day is the same as 0000 for the next day.) Time is spoken as "zero seven hundred" or "fifteen forty." *The word "hours" is not used.* The times of 1000 and 2000 are correctly spoken as "ten hundred" and "twenty hundred," *not* as "one thousand" or "two thousand."

When performing arithmetic computations with time, remember that when "borrowing" or "carrying" there are *60* minutes in an hour, and not 100. As obvious and simple a matter as this may be, the mistake occurs all too often.

Time Zones

A navigator in pilot waters must also be alert to TIME ZONES. Even in coastal or inland waters, you can cruise from one zone to another and need to reset clocks.

A further complication in time is the prevalence of daylight time during the summer months. Government publications are in standard time; local sources of information such as newspapers and radio broadcasts use daylight time if it is in effect. Daylight time is one hour *later* than standard time. When converting from standard time to daylight, *add* one hour; from daylight to standard, *subtract* one hour.

Speed

No matter what your boat is, SPEED is an essential dimension of piloting. Speed is defined as the number of units of distance traveled in a specified unit of time.

The basic unit of speed is MILES PER HOUR, whether these be nautical or statute miles, as determined by location. A KNOT is a nautical mile per hour. (It is a mark of ignorance to say "knots per hour.")

Conversion of Knots and MPH

The conversion factors between statute miles per hour (MPH) and knots are the same as for the corresponding units of distance—1 knot = 1.15 MPH, or 7 knots roughly equals 8 MPH. (1 knot = 1.852 km/h).

The unit "knot" may be abbreviated either as "kn." or "kt."; the former has a somewhat greater usage.

Position

The ability to describe the position of his vessel accurately is an essential requirement for a pilot, and one that marks him as well qualified. To realize the importance and the difficulties of this seemingly simple task, listen to your radio on VHF Channel 16 or 22 on a weekend afternoon during the boating season. The hesitant, and inaccurate, attempts by skippers to simply say where they are surely must irritate the Coast Guard and embarrass competent boatmen.

Relative and Geographic Coordinates

POSITION may be described in relative terms or by geographic coordinates. To state your boat's RELATIVE POSITION, describe it as being a certain distance and direction from an identifiable point such as a landmark or aid to navigation. Precision depends on the accuracy of the data on which you base the position; you might say that you were "about two miles southwest of Brenton Reef Light"; or if you had the capability of being more precise, you could say "I'm 2.2 miles, 230° true from Brenton Reef Light." Be careful not to confuse the direction you measure *toward* the identifiable point with your report of position *from* the object. In relative positioning, the distance can, of course, be essentially zero, as would be the case with a position report like "I am at Lighted Whistle Buoy 2."

These examples of position description have all used visible identifiable objects. You can also state your GEOGRAPHIC POSITION in terms of LATITUDE and LONGITUDE using those straight, uniformly spaced lines on charts. Using this procedure, measure the position from the markings on either side and the top or bottom of your chart.

The units of measurement for geographic coordinates are DEGREES and MINUTES, but this degree is not the same unit as used in the measurement of direction, nor is this minute the same as the unit of time. Be careful not to

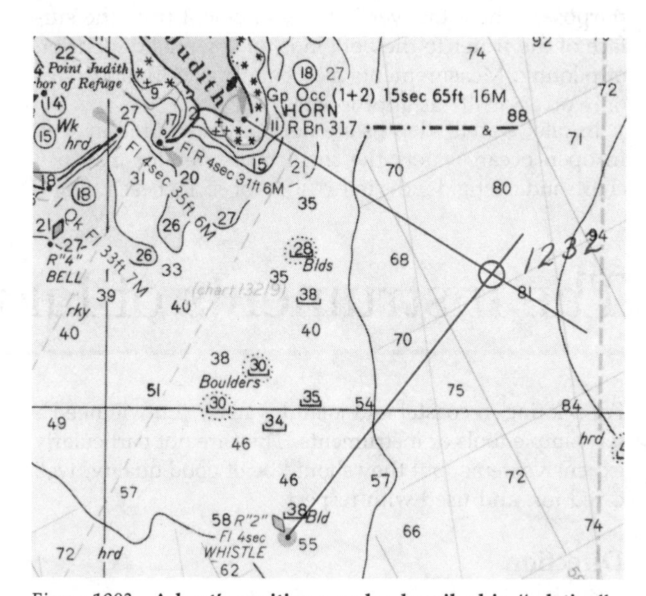

Figure 1903 **A boat's position may be described in "relative" terms, such as at a certain distance and direction from an identifiable object such as an aid to navigation. In this example it would be "1.8 miles, 120 degrees true, from Point Judith Light."**

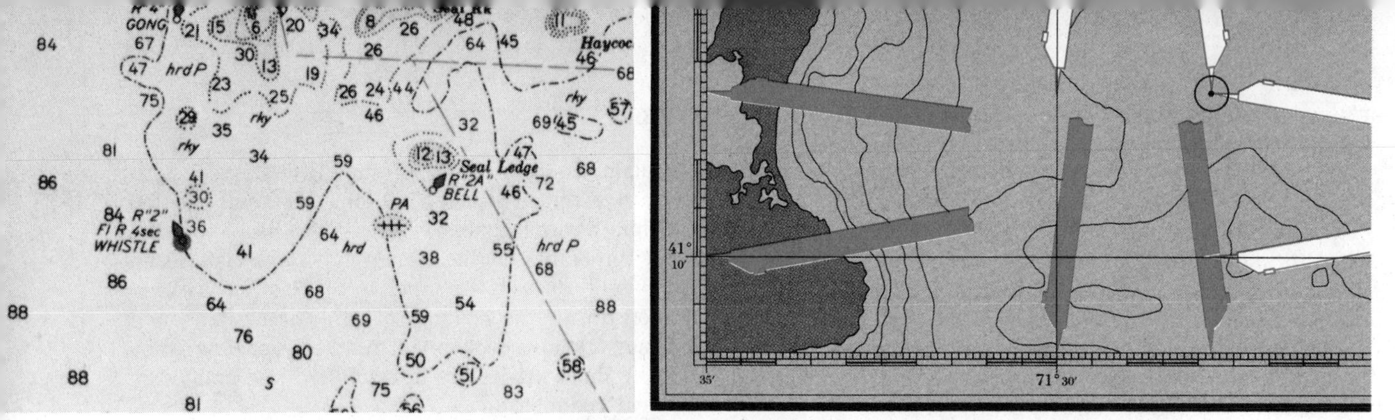

Figure 1904 **Depth is another dimension of piloting; be sure you know whether depths on the chart you are using, *left,* are in feet, fathoms, or on some charts, in meters. Geographic coordinates are often used to describe a boat's position. Using the subdivisions of the chart's borders, *right,* latitude and longitude are measured as shown here, and recorded with the time.**

confuse these units of position with other units with similar names. For precise position definition you can use either SECONDS or TENTHS OF MINUTES, as determined by your chart. The smallest unit on the marginal scales of NOS coast charts is tenths of minutes, but for harbor charts it is seconds; see page 406 and Figures 1811 and 1812.

To state your boat's geographic position, give the latitude first, before longitude, and follow the figures by "north" or "south" as appropriate. In U.S. waters the latitude is, of course, always north. Similarly, longitude must be designated "east" or "west" to be complete; all U.S. waters are in west longitude; Figure 1904 right.

Skippers of boats with Loran-C often describe their positions in terms of Loran readings; this is useful in rendezvousing with other craft, and is acceptable to the Coast Guard in search and rescue operations.

Depths

The DEPTH of the water is important both for the safety of a boat in preventing grounding, and for navigational purposes. Thus, this vertical measurement from the surface of the water to the bottom is an essential dimension of piloting. Measurements may be made continuously or only occasionally as appropriate.

In pilot waters, depths are normally measured in FEET. In open ocean waters the small boat operator may use charts indicating depths in FATHOMS of six feet each. Some charts of other nations may use METERS and DECIMETERS. Check each chart when you buy it, and again when you use it, to note its unit of depth measurement.

Depths often fluctuate from the chart's printed figures because of tidal changes. Charts indicate the DATUM used—the reference plane, such as mean lower low water, from which measurements were made.

Heights

The height or elevation of objects is also of concern to the pilot. The height of some landmarks and lighted aids to navigation may determine their range of visibility. Of more critical importance, however, are such vertical measurements as the clearance under a bridge; see Figure 2002. HEIGHTS, or vertical dimensions upward from the surface of the water, are measured in feet (or meters). You will find that in tidal areas the plane of reference for heights is *not* the same as for depths.

The usual datum for height measurement is MEAN HIGH WATER. This will be an imaginary plane surface above the mean low-water datum for depths by an amount equal to the MEAN TIDE RANGE; charts usually show the height of mean high water (see Figure 1814).

Note that mean high water is the *average of all highs,* and there will be many instances of normal highs being *above* the datum, with correspondingly *lesser* vertical clearances.

The Instruments of Piloting

For piloting in coastal and inland waters, you will need a few simple tools or instruments. They are not particularly expensive items, but they should be of good quality, well cared for, and used with respect.

Direction

Many instruments are used for measuring direction—some directly and others on a chart. Still other instruments measure direction both from a chart and from observation.

Determining Direction

The basic instrument for determining direction is a COMPASS. With rare exceptions it will be a MAGNETIC compass as described in Chapter 15. Its directions are "compass" directions and thus require correction for deviation and variation (see pages 348-351).

The compass is used primarily to determine the direction in which the boat is headed. Depending upon its mounting and its location in the boat, you can also use the steering compass to determine the direction of other objects from your boat. Readings may be taken by sighting across the compass itself and estimating the reading of the compass card. Steering compass sights are generally accurate enough, but physical obstructions usually limit them to objects forward of a beam.

Hand Bearing Compass You can take compass sights more flexibly with a HAND BEARING COMPASS, which is held in the hand and not in a stationary mounting. These come in many styles but each is basically a small liquid-filled compass with a suitable grip for holding in front of your eye, plus a set of sights and/or a prismatic optical system for simultaneously observing a distant object through the sights and reading the compass card. A typical hand bearing compass is shown in Figure 1905; this model uses a prism in lieu of sights. Figure 1906 shows how a small steering compass can often be removed from its mounting and taken to where there is an unobstructed line of sight to a distant object. It is also possible to fit a small compass in front of one binocular lens to have magnification and direct compass reading simultaneously.

A hand bearing compass is normally used on deck away from the usual causes of magnetic deviation as discussed in Chapter 15, page 357. Take care, however, not to get too close to objects of iron or similar material, such as anchors or stays and shrouds. While your boat is stationary in a known position, take bearings on distant objects from several places on board and compare them with directions from a chart. Watch out, also, for magnetic objects on your person—metal-frame eyeglasses and things in your shirt pockets.

Pelorus Another instrument for measuring directions from the boat is a PELORUS, or "dumb compass." This simple device consists of a set of sighting vanes mounted over a circular scale calibrated in degrees like a compass card. You will need some way to mount or temporarily position the pelorus so that it can be oriented to the fore-and-aft centerline of your boat.

The circular scale of a pelorus is usually made so that it may be rotated to a desired position and there clamped in place. With the scale so set that 000° is dead ahead, over or parallel to the centerline of the boat, the directions that are then measured are termed RELATIVE BEARINGS. Convert these to compass directions or compass bearings by adding the heading of the boat as read from the steering compass at the instant of observation (subtract 360 degrees from this sum if it exceeds that amount). You need a steady hand at the wheel, and the observer should call "Mark" at the time that he reads the pelorus so that the helmsman simultaneously may read the boat's compass. Alternatively, the helmsman may call a series of "marks" as he is on the prescribed heading, remaining silent if the boat falls off to either side. When using this technique, the observer takes his reading only when the helmsman indicates that he is on the correct heading.

A second method of using a pelorus is to set the scale so that it matches the compass for the heading being steered. Readings then taken with the pelorus are direct compass bearings without conversion. Two cautions must be observed—the boat must be directly on course at the time of observation; and, when correcting for *deviation*, the value used is that for the *heading* of the boat, *not* that for the observed direction.

Figure 1905 **A hand-bearing compass, *above*, can be used from almost any point on a boat, but be careful to keep away from large masses of magnetic material that could introduce deviation errors. A pelorus may be purchased, or easily homemade, *below*. A compass card like the one shown in Figure 1503 is cut out and mounted on a wood or metal base. A set of rotatable sighting vanes completes the instrument.**

Figure 1906 **Some steering compasses are so mounted that they can be unshipped and held at eye level for bearings.**

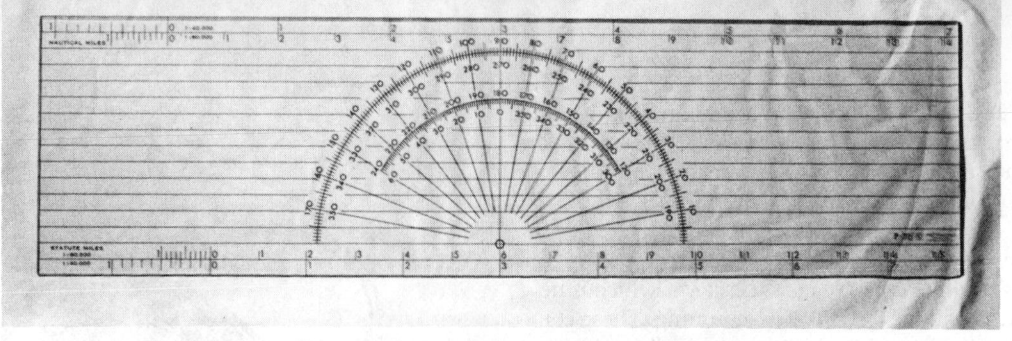

Figure 1907 **A course plotter is a single rectangular piece of transparent plastic ruled with a set of lines parallel to the long edges, and semi-circular scales. This is similar to the model used by the U.S. Power Squadrons.**

Plotting Directions

Once you have determined a direction by compass, pelorus, or other device, you must plot it on the chart. There are several instruments for doing this, and they are the same tools used for determining direction from the chart itself.

Course Plotters These are pieces of clear plastic, usually rectangular, which have one or more semi-circular angular scales marked on them; see Figure 1907. The center of the scales is at or near the center of one of the longer sides of the plotter and is usually emphasized with a small circle or bull's eye. Plotters normally have two main scales, one from 000° to 180° and the other from 180° to 360°; each calibrated in degrees. There may also be smaller AUXILIARY SCALES which are offset 90 degrees from the main scales. Lines are marked on the plotter parallel to the longer sides.

Course plotters are used in the following manner:

To determine the direction of a course or bearing from a given point. Place the plotter on the chart so that one of its longer sides is along your course or bearing line, and slide the plotter until the bull's eye is over a MERIDIAN (longitude lines running north-south). Read the true direction on the scale where it is intersected by the meridian. Easterly courses are read on the scale that reads from 000° to 180° and westerly courses on the other main scale; see Figure 1908 left.

If it is more convenient, find your plotted course or bearing with one of the plotter's marked parallel lines rather than its edge. It is not essential that you actually draw in the line connecting the two points (the plotter can be aligned using only the two points concerned) but you will usually find it easier and safer to draw in the connecting line.

When the direction to be measured is within 20 degrees or so of due north or south, it may be difficult to reach a meridian by sliding the course plotter across the chart. The small inner auxiliary scales have been included on the plotter for just such cases. Slide the plotter until the bull's eye intersects a PARALLEL OF LATITUDE (east-west) line. The intersection of this line with the appropriate *auxiliary* scale indicates the direction of the course or bearings; see Figure 1908 right.

To plot a specified direction (course or bearing) from a given point. Put a pencil on the origin point, keep one of the longer edges of the course plotter snug against the pencil, and slide the plotter around until the center bull's eye and the desired mark on the appropriate main scale

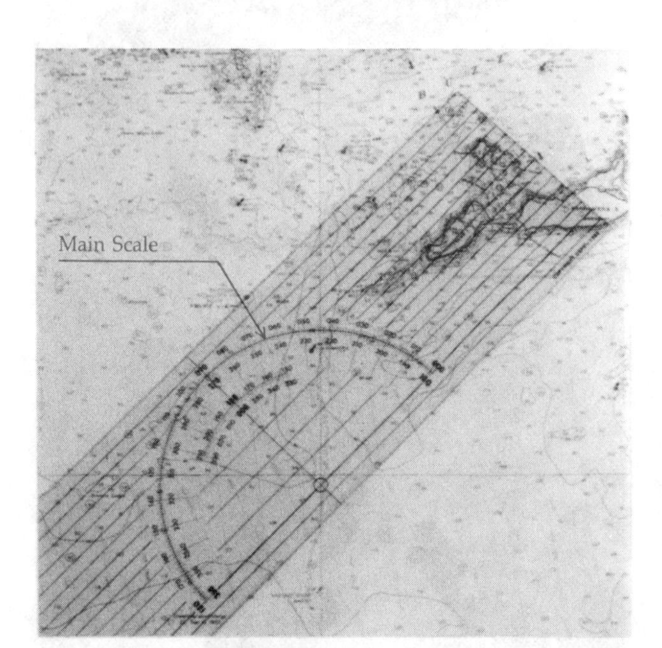

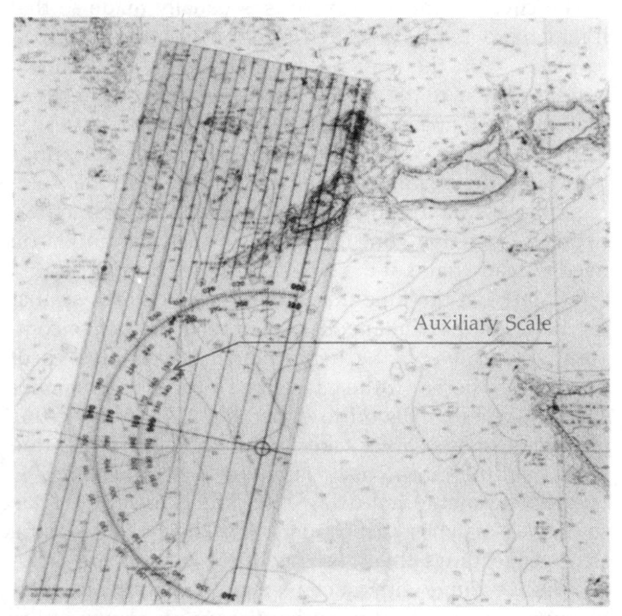

Figure 1908 **The course plotter is lined up with the course line along one of the longer sides, left, and then is moved until a meridian cuts through the "bull's eye." Direction can then be read from the appropriate main scale. In the case of courses within a few degrees of North or South, use the auxiliary inner scales and a parallel of latitude, *right*.**

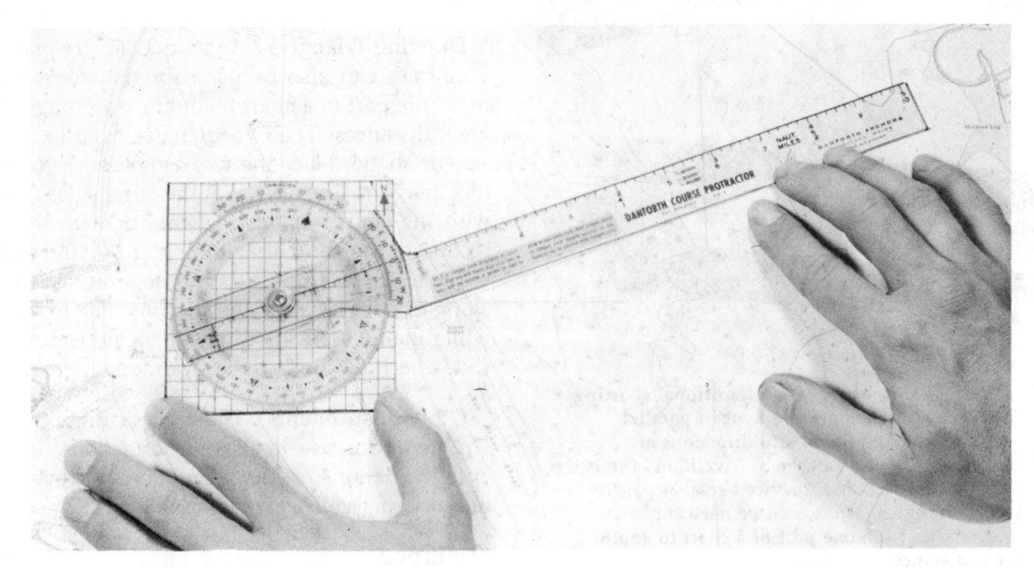

Figure 1909 **Some navigators prefer to use a course protractor for their plotting tool. The long arm can be moved with respect to the square grid.**

both lie along the same meridian. With the plotter thus positioned, draw in the specified direction from the given point.

Alternatively, you can first position the plotter using the bull's-eye and scale markings without regard to the specified origin point, then slide the plotter up or down the meridian until one of the longer edges is over the origin point and then draw in the direction line.

For directions nearly north or south, use one of the small auxiliary scales on a parallel of latitude.

To extend a line that must be longer than the length of the course plotter. Place a pair of dividers, opened to three or four inches, tightly against the edge of the plotter and then slide the plotter along using the divider points as guides. Draw in the extension of the course or bearing line after the plotter has been advanced.

To draw a new line parallel to an existing course or bearing line. Use the parallel lines marked on the course plotter as guides.

Some course plotter models do more than just measure directions. One, called a "Nautical Ruler," has distance scales along its longer sides for both nautical and statute miles at common chart scales. Loran-C interpolator scales are along the shorter edges and considerable other information is also printed on this device.

Course Protractors Many pilots use a COURSE PROTRACTOR, Figure 1909, as their primary plotting tool. This instrument, with its moving parts, is not as easy to use as the course plotter, but used with care it gives as satisfactory results.

To measure the direction of a course or bearing. Place the center of the course protractor on the chart exactly over the specified origin point, such as your boat's position or an aid to navigation. Then swing the protractor's arm around to the nearest compass rose on the chart, making the upper edge of the arm (which is in line with the center of the compass part of the course protractor) pass directly over the center of the compass rose.

Holding the course protractor arm firmly in this position, turn the compass part of the protractor around until the arm's upper edge cuts across the same degree marking of the protractor compass as it does at the compass rose. The compass and the rose are now parallel. Holding the protractor compass firmly against the chart, move the protractor arm around until its edge cuts across the second point involved in the course or bearing. You can now read the direction in degrees directly from the protractor compass scale.

To lay off a line in a specified direction from the given point. Line up the protractor rose with the chart's compass rose as in the preceding instructions. Then rotate the arm until the desired direction is indicated on the compass scale, and draw in the line; extend the line back to the origin point after you have listed the course and the protractor has been lifted off the chart.

A variation of the course protractor—the Pocket Instant Navigator™ (PIN)—consists of two concentric discs with degree scales around the outer edge; one disc has a grid of lines at right angles to each other, and there is a separate pointer that can rotate with respect to both discs. Rotate the disc without the grid lines so as to graphically allow for magnetic variation. Then place the PIN on the chart centered on your position with the grid lines parallel to a convenient latitude or longitude line on the chart. The moveable pointer can now be aligned to read directly a true or magnetic direction.

Parallel Rulers A traditional instrument for measuring and plotting directions on charts is a set of PARALLEL RULERS. Parallel rulers may be made of black (as shown in Figure 1910 top) or clear transparent plastic. The two rulers are connected by linkages that keep their edges parallel. To measure the direction of a line, line up one ruler with the desired objects on the chart and then "walk" the pair across the chart to the nearest compass rose by alternately holding one ruler and moving the other. To plot a line of stated direction, reverse the process; start at the compass rose and walk to the desired origin point. Take care that the rulers do not slip.

Figure 1910 **Parallel rules,** ***above,*** **are a traditional charting instrument. The two straightedges are kept in parallel alignment by the connecting links, and directions are transferred from one place to another by "walking" the rulers from course line to compass rose, or vice versa. A pair of ordinary drawing triangles,** ***below,*** **can be used for transferring a direction from one part of a chart to another, but not for great distances.**

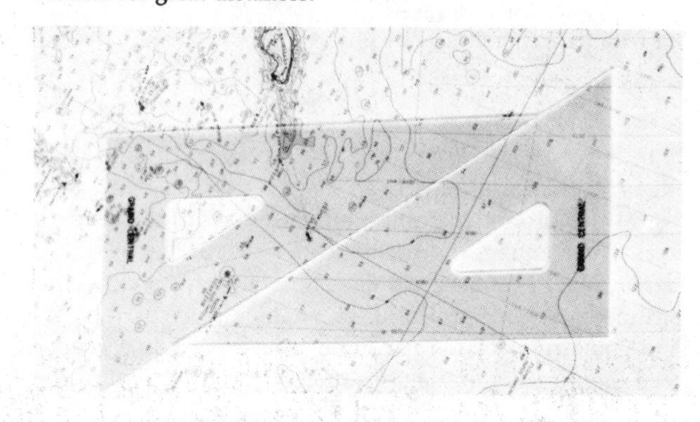

Figure 1911 **A modern patent log,** ***above,*** **has a small impeller mounted on the bottom of the boat. This is connected to a dialed or digital instrument that may show both speed and distance. A "Courser,"** ***below,*** **is a sheet of soft, flexible transparent plastic on which parallel lines have been ruled. It is simple and convenient to use in smaller boats, and its accuracy is adequate for small boat applications.**

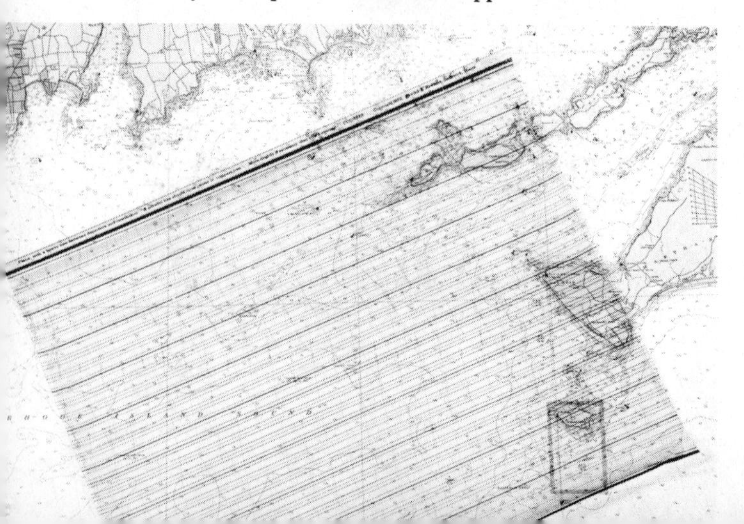

Drawing Triangles A pair of ordinary plastic DRAWING TRIANGLES can also be used for transferring a direction from one part of a chart to another, although not for very great distances. The two triangles need not be similar in size or shape. Place the two hypotenuses (longest sides) together, and line up one of the other sides of one triangle with the course or bearing line, or with the desired direction at the compass rose; Figure 1910. Hold the other triangle firmly in place as a base, and slide the first one along its edge carrying the specified line to a new position while maintaining its direction. If necessary, alternately slide and hold the triangles for moving greater distances.

Other Instruments There are a number of other instruments for use in plotting directions, including patented variations on those mentioned above. You should use any that you feel comfortable with.

Distance

Distance to an object can be measured directly by RADAR, and distance traveled can be measured directly by a SUMMING LOG, but most boats have neither of these items of equipment.

A hand-held OPTICAL RANGEFINDER can be used to measure distance to an object up to a thousand yards or so. Most often in piloting, however, distances are measured by taking them from a chart.

Chart Measurements for Distance

Dividers Distance is measured on a chart with a pair of DIVIDERS; Figure 1912. Open the two arms and the friction at the pivot is sufficient to hold the separation between the points. Most dividers have some means for adjusting this friction; it should be enough to hold the arms in place, but not so much as to make opening or closing difficult. A special type of dividers has a center cross-piece (like the horizontal part of the capital letter "A") which can be rotated by a knurled knob to set and maintain the opening between the arms; Figure 1912 center; thus the distance between the points cannot accidentally change. This type of dividers is particularly useful if kept set to some standard distance, such as one mile to the scale of the chart being used.

To measure distance with dividers, first open them to the distance between two points on the chart, then transfer them without change to the chart's graphic scale. Note that the zero point on this scale is not at the left-hand end, but rather is one basic unit up the scale. This unit to the left of zero is more finely divided than are the remaining basic units. To measure any distance, set the right-hand point of the dividers on the basic unit mark so that the left-hand point falls somewhere on the more-finely divided unit to the left of the zero. The distance measured is the sum of the basic units counted off to the right from the scale's zero point plus the fractional unit to the left from zero. Be sure to note the type of fraction used, such as seconds of arc.

If the distance on the chart cannot be spanned with the dividers opened widely (about 60 degrees is the maximum practical opening), set them at a convenient opening for a whole number of units on the graphic scale or latitude subdivisions, step this off the necessary number of times,

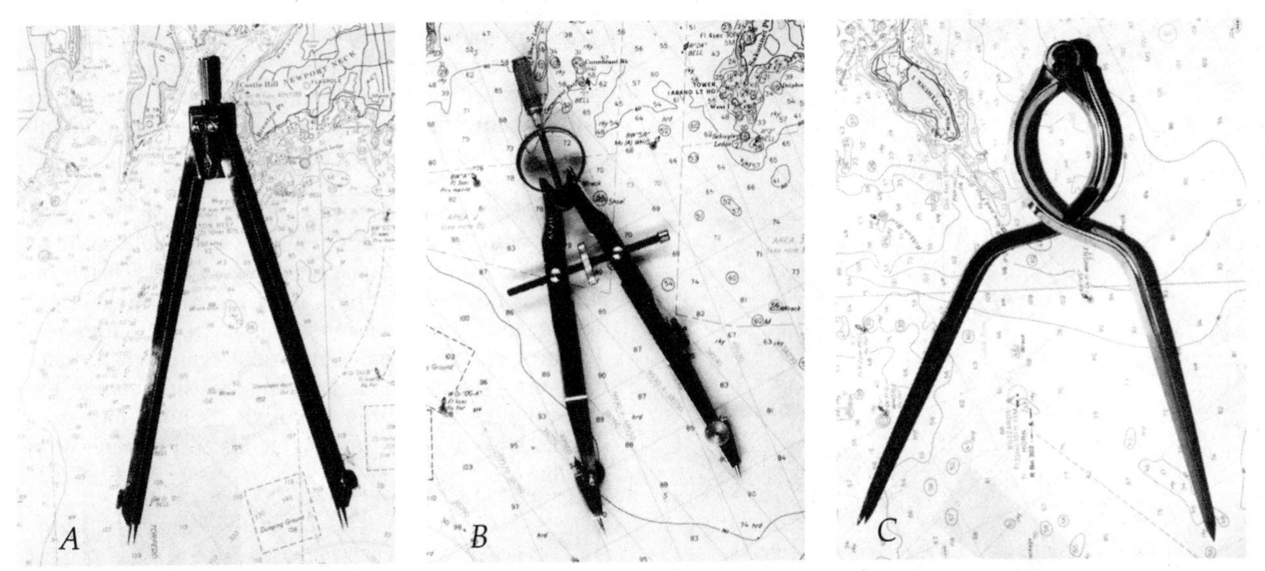

Figure 1912 **Divider types: *A*—Friction at the pivot holds the arms to the desired opening. *B*—An adjustable center cross-arm maintains the separation and avoids accidental changes. *C*—Long-legged dividers can be used for measuring long runs.**

then measure the odd remainder. The total distance is then the simple sum of the parts stepped off and measured separately, as in Figure 1913.

To mark off a desired distance on the chart, set the right point of the dividers on the nearest lower whole number of units, and the left point on the remaining fractional part of a unit measured leftward from zero on the scale. The dividers are now properly set for the specified distance at the scale of the chart being used and can be applied to the chart. If the distance is too great for one setting of the dividers, step it off in increments.

Charts at a scale smaller than 1:50,000 do not have a graphic scale; distances are measured on the *latitude* scales at either side of the chart. Take care to measure on these scales near the same latitude as the portion of the chart being used; in other words, move directly horizontally across the chart to either of its sides to use the scales.

With a little experience you can set and use dividers with one hand. Making sure that the friction at the pivot is properly adjusted, practice this technique until it is easy for you; it will add convenience and speed to your piloting work.

Figure 1913 **When a chart distance is too great to be measured with a single setting of the dividers, open the points to a convenient whole number of units. Step these along the chart for the required number of times and then measure any small left-over distance in the usual manner.**

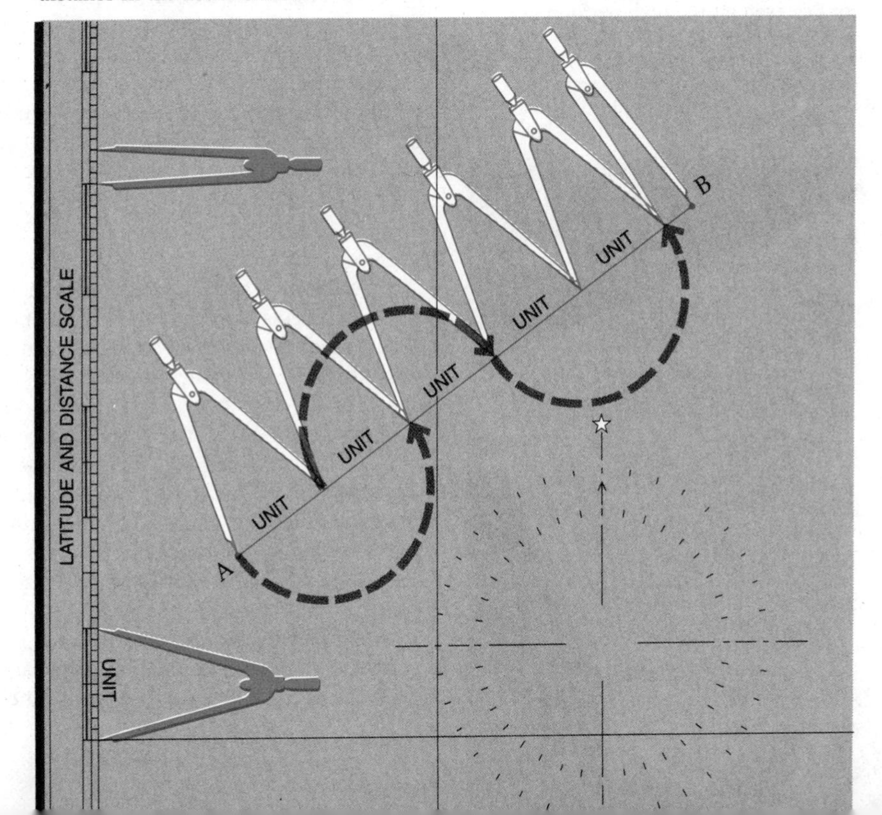

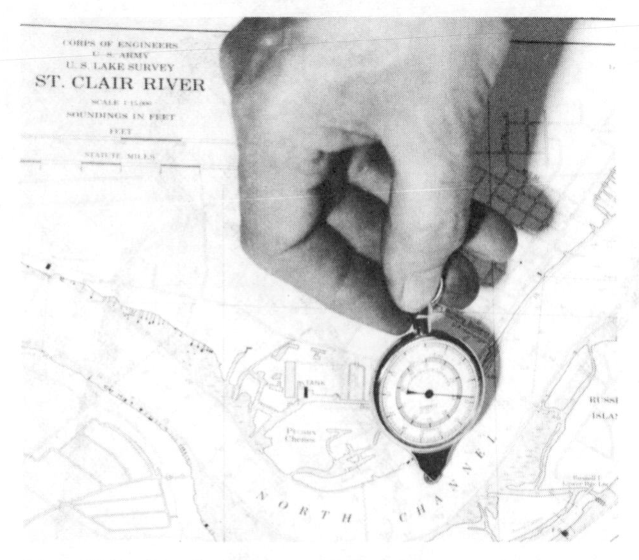

Figure 1914 **If a chart measurer is rolled across a chart between two points, an indicating dial shows the distance directly for typical chart scales. It is particularly useful in measuring distances on rivers on other waterways where there are many bends and curves.**

Figure 1915 **Modern speedometers are available in models suitable for sailboat installations, *above,* scaled so that even slight changes in speed show clearly. For powerboats, there are now instruments for use even on the high-performance craft which attain speeds of 50 knots or more. This calculator, *below,* enables the pilot to solve time, speed, and distance problems, as well as other conversions or measurements. There are special aids on it for sailors, and a plotting board on the reverse side.**

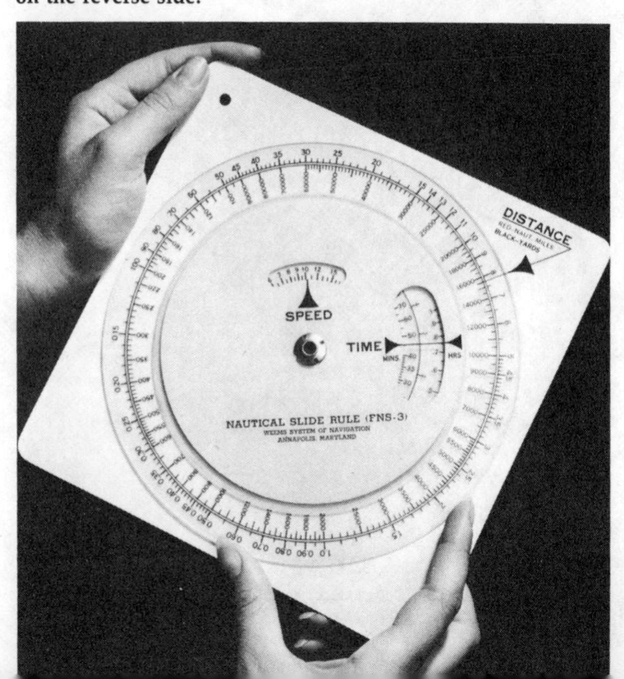

An instrument that looks much like a pair of dividers, except that a pencil lead or pen is substituted for one point, is called a COMPASS (or DRAWING COMPASS to distinguish it from a magnetic compass). This is used primarily for drawing circles or arcs of circles.

Chart Measurer Distance can also be measured across a chart with reduced, but generally acceptable, accuracy by using a CHART MEASURER, as in Figure 1914. This device has a small wheel that rolls along the chart and is internally geared to an indicating dial. Chart measurers read distances directly in miles at various typical chart scales; there are separate models for charts using statute and nautical miles. They are particularly useful in measuring distances up rivers with many bends and changes of direction.

Time

Every skipper, no matter how small his boat, should have a dependable and reasonably accurate timepiece, whether a clock, wrist watch, or pocket watch. Long-term accuracy is of secondary importance in piloting, but short-term errors, those accumulated over the length of a day or half-day, should be small. A knowledge of the time of day within a few minutes is usually all you will need. If you use a clock, mount it where it is clearly visible from the helm-seat or plotting table.

ELAPSED TIME is often of greater interest than ABSOLUTE TIME; a stop watch is handy for this. Most stop watches have a second hand that makes one revolution per minute, and this is the simplest kind to use, but there are some that sweep completely around in 30 seconds, and even a few that have a 10-second period. Know for sure the type of stop watch that you are using.

Also quite useful is a COUNTDOWN TIMER with alarm. This can be either a simple kitchen timer from the galley, or a wrist watch with the countdown feature. Countdown timers save you the trouble of keeping an eye on the clock to be ready for a course change or an expected sighting of an aid to navigation.

Digital wrist watches are excellent for use in piloting. Some newer models have many features in addition to ordinary timekeeping, such as settable alarm, stopwatch, and countdown timer with alarm.

Speed

Speed is a dimension that can either be measured directly or calculated from knowledge of distance and time. Direct reading instruments are convenient, but they give only the relative speed through the water, and not the speed made good over the bottom.

Speed through the water used to be measured by use of a CHIP LOG. More modern marine speedometers use the pressure built up in a small tube by motion of the boat or the rotation of an impeller. There are many different models with varying speed ranges, from those for sailboats (Figure 1915 top) to models reading high enough for fast speedboats.

Speed can be manually calculated from distance covered and elapsed time, as discussed later in this chapter, or with special calculating devices. These calculators are adaptations of general mathematical slide rules; they may be either linear or circular in form; Figure 1915 bottom. Small electronic calculators (see page 496) are also very useful for speed-time-distance computations.

Depth

Depth can be measured manually or electronically. A hand LEAD LINE is simple, accurate, and not subject to breakdowns, but it is awkward to use, inconvenient in bad weather, and can give only one or two readings per minute; a lead line can be used only at quite slow speeds, or to check depths around an anchored boat.

Many small boats today have an ELECTRONIC DEPTH SOUNDER—a convenient device that gives clear and accurate measurements of the depth of water beneath the boat. A depth sounder gives readings many times each second, so frequently that they appear to be a smooth, continuous depth measurement. For more information on these relatively inexpensive but most useful piloting aids, see pages 537-539.

Miscellaneous Piloting Tools

Among the most important of all piloting tools are ordinary PENCILS and ERASERS. Pencils should be neither too hard nor too soft. If too hard they tend to score into the chart paper and if too soft they smudge. A medium (No. 2) pencil works well; experiment to see what hardness is best for you. Keep several pencils on hand, well sharpened and handy to the plotting table. Fine lead (0.5 mm) mechanical pencils are excellent.

A soft eraser of the Pink Pearl type is good for most erasures; an art gum eraser works well for general chart cleaning.

Binoculars Good binoculars are essential for most piloting situations. In choosing binoculars remember that higher powers give greater magnification, bringing distant objects closer, but only at the cost of a more limited field of view. An adequate field of view is essential on small boats with their rapid and sometimes violent motion. Binoculars are designated by two figures, such as "6×30" or "10×50." The first figure indicates the power of magnification, the second is the diameter of the front lens in millimeters. This latter characteristic is important in night use. Most authorities recommend 7×50 binoculars as best for marine use.

Binoculars may be individually focused (IF) for each eye, or centrally focused (CF) for both eyes, with a minor adjustment on one eyepiece to balance any difference between a person's two eyes. Both types are fine for marine use; choice is based on personal preference.

Keep binoculars in their case when you are not using them, and keep their strap around your neck when you are. Be careful when you put them down that they cannot slide off and be damaged.

Flashlights Keep several FLASHLIGHTS or ELECTRIC LANTERNS on your boat for emergency use. One should have a red lens or filter so you can use it for reading charts at night (white lights cause a temporary loss of night vision). Hand-held magnifiers with built-in red illumination are excellent for reading chart details underway at night.

Measurements

In considering the measurement of various quantities in piloting, and the calculations in which they are used, you must first consider the standards of accuracy and precision.

"Accuracy" and "precision" are *not* synonymous. PRECISION relates to the degree of fineness of measurement of the value under consideration. ACCURACY relates to how closely the stated value is to the true exact value.

Statements of distance as 32 miles or 32.0 miles are not quite the same thing. The first merely says that to the best of observation and measurement the distance is not 31 nor 33 miles; the second says that it is not 31.9 nor 32.1 miles. Note the difference in the degree of *preciseness* of these two statements of the same distance. Never write 32.0 for 32 unless your measurements are sufficiently precise to warrant it. The accuracy of the value recorded, regardless of how precisely it may or may not be stated, is determined by the tools and/or techniques used to measure it.

Standard Limits of Precision

Navigating various sizes of vessels naturally involves different standards of precision and accuracy, as befitting the conditions encountered. The piloting of small boats does not permit so high a degree of accuracy as on large ships that offer a more stable platform.

Direction

Direction is measured in small-craft navigation to the nearest whole degree. It is not reasonable to measure or calculate directions to a finer degree of precision when a boat is seldom steered closer than 2 degrees or 3 degrees to the desired course.

Distance

Distances are normally expressed to the nearest tenth of a mile. This degree of precision, roughly 200 yards for a nautical mile, is reasonable in consideration of the size of the vessel and other measurement standards.

Time

Time is measured and calculated to the nearest minute. Fractions of a minute are rarely of any significance in routine piloting. In contests, however, time is calculated to decimal fractions and used in terms of seconds.

Speed

Speed is calculated to the nearest tenth of a knot or mile per hour. It is seldom measurable to such fine units, but a calculation to the nearest tenth is not inconsistent with the expressed standards of precision of distance and time. This same degree of precision is used in calculations of current velocity.

Position

Geographic coordinates are expressed to the nearest tenth of a minute of latitude and longitude, or to the nearest second, as determined by the scale of the chart. As explained in Chapter 18, latitude and longitude markings are subdivided into minutes and seconds on the larger-scale charts (1:49,000 and larger) and in fractions of minutes on smaller-scale charts (1:50,000 and smaller).

Depths and Heights of Tide

Tidal variations in the depth of water are normally tabulated to the nearest tenth of a foot (or meter); calculations are carried out to the same degree of precision. Remember, however, that the effects of tidal action by winds and atmospheric pressure make this degree of precision hardly warranted.

Rounding of Numbers

Any mathematical expression of a quantity has a certain number of "significant figures." The quantity 4 has one significant figure, for instance; 4.2 or 14 each have two significant figures; 5.12, 43.8, and 609 each have three significant figures.

The process of reducing the number of significant figures is called ROUNDING. To have uniform results, rules have been established for the rounding of numbers.

1. If the digit to be rounded off is 4 or less, it is dropped or changed to a zero.

8.23 is rounded to 8.2
432 is rounded to 430

2. If the digit to be rounded off is 6 or larger, the preceding digit is raised to the next higher value and the rounded digit is dropped or changed to a zero.

8.27 is rounded to 8.3
439 is rounded to 440

3. If the digit to be rounded off is a 5, it is desirable to round to the nearest *even* value, up or down.

8.25 is rounded to 8.2
435 is rounded to 440

This rule may seem arbitrary,but it is followed for consistency in results; it has the advantage that when two such rounded figures are added together and divided by two for an average, the result will not present a new need for rounding.

Note, however, that slightly different figures will result from using a calculator. When a calculator is set to display a fixed number of decimal places, a 5 to be rounded off always results in the preceding digit being changed to the next higher—even or odd—number. (Internal calculations use the full range of numbers without rounding.) Computers, with their greater capabilities, can be programmed to round up only half of the time, when the preceding digit is odd; they round down for other numbers—this ensures less distortion of the final results.

4. Rounding can be applied to more than one final digit; but all such rounding must be done in *one* step. For example: 6148 is rounded to 6100 in a single action; do not round 6148 to 6150, and then round 6150 to 6200.

In many of the statements of precision requirements in the preceding section, the phrase "to the nearest_____" has been used. Rounding is used to reduce various quantities to such limitations. For example: if you calculated the distance traveled in 5 minutes at 13 knots, you would get 1.08 miles. However, distance is normally stated to the nearest tenth of a mile, and so the proper expression for distance run would be 1.1 miles.

Dead Reckoning

When operating his boat in large bodies of water, a pilot should have at all times at least a rough knowledge of his position on the chart. Basic to such knowledge is a technique of navigation known as DEAD RECKONING, usually abbreviated to *DR*. This is the advancement of the boat's position on the chart from its last accurately determined location, using the courses steered and the speeds through the water. No allowance is made for the effects of wind, waves, current, or steering errors.

Terms Used in Dead Reckoning

The DR TRACK (or DR TRACK LINE) is the path that a boat is expected to follow as represented on the chart by a line drawn from the last known position using courses and distances through the water. The path the boat actually travels may be different due to one or more offsetting influences, to be considered in Chapter 21.

COURSE, abbreviated "C," is the direction in which a boat is to be steered, or is being steered; the direction of travel through the water. Courses are normally plotted as true directions labeled with three-digit figures with zeros added as necessary—8° becomes 008; 42° becomes 042. Some skippers may plot and label courses as magnetic or compass directions.

HEADING, the direction in which a boat is pointed at any given moment, is often given in terms of magnetic or compass directions; these values are not part of a plot.

SPEED, abbreviated "S," is the rate of travel through the water. This is the DR track speed; it is used, with elapsed time, to determine DR POSITIONS along the track line.

DISTANCE, abbreviated "D," may be used with a DR plot of a future intended track. (Speed is not normally shown on a plot except for a vessel underway.)

The Basic Principles of Dead Reckoning

You should follow these basic principles of dead reckoning:

1. *A DR track is always started from a known position.*

2. *Only true courses steered are used for determining a DR track.*

3. *Only the speed through the water is used for determining distance traveled and a DR position along the track.*

Although it may appear unusual to ignore the effects of a current that is known to exist, this is done for reasons that will be established later.

The Importance of Dead Reckoning

You should always plot a DR track when navigating in large, open bodies of water. It is the primary representation of your boat's path, the base to which other factors, such as the effect of current, are applied. Dead reckoning is the basic method of navigation to which you will apply corrections and adjustments from other sources of information.

At the same time, remember that this DR track rarely represents your boat's actual progress. If there were no steering errors, speed errors, or external influences, the DR track could be used as a means of determining your boat's position at any time, as well as the ETA (Estimated Time of Arrival) at a destination. But even with errors and external influences a DR plot is always an important safety measure in the event of unexpected variations in current, and if you encounter fog or other loss of visibility.

Plotting

Fundamental to the use of dead reckoning is the use of charts and plots of a boat's intended and actual positions. You should use standard SYMBOLS and LABELS so your chart work will be understandable to another person, and to yourself at a later date.

Basic Requirements

The basic requirements of plotting are *accuracy, neatness,* and *completeness.* All measurements taken from the chart must be made carefully, all direct observations must be made as accurately as conditions on a small boat permit, and all calculations should be made in full and in writing. If time permits, each of these actions should be repeated as a check; errors can be costly!

Neatness in plotting is essential to avoid confusion of information on the chart. Drawing extra or overly long lines on charts or scribbling extraneous notes on them may obscure vital information.

Information on a chart must also be complete. You will often need to refer back to information that you placed on the chart hours or days ago, and it can be dangerous to rely on memory to supply details.

Labeling

Draw the lines on your charts lightly and no longer than necessary. Keep your straightedge slightly off the desired position of the line you are drawing, to allow for the thickness of the pencil point, no matter how fine it may be.

The requirements of neatness and completeness combine to establish a need for LABELING.

Immediately after drawing any line on a chart, or plotting any point, label it. The basic rules for labeling are:

1. *The label for any line is placed along that line.*

2. *The label for any point should not be along any line;* it should make an angle with any line so that its nature as the label of a point will be unmistakably clear.

The above basic rules are applied in the labeling of DR plots in the following manner:

1. The direction label is placed *above* the track line as a three digit number, preceded by "C" for Course and followed by "T," "M," or "C" to indicate True, Magnetic, or Compass. Note that the degree symbol "°" and periods are omitted.

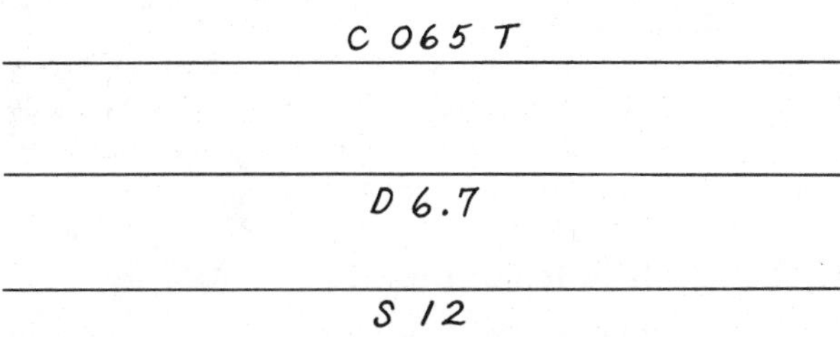

Figure 1916 **The course is labeled above the line as shown here, with direction as a three-digit number (add zeros as necessary), followed by "T," "M," or "C" for True, Magnetic, or Compass as appropriate. Speed is labeled below the line, with the letter "S" in front of it. Distance may also be shown below the course line, with the designator "D." Note the space between the numbers and the letters.**

2. The speed along the track is indicated by numerals placed *under* the track line, usually directly beneath the direction and preceded by the letter "S"; see Figure 1916. Units (knots or MPH) are omitted.

3. A known position at the start of a DR track, a FIX (see Chapter 21), is shown as a circle across the line; a small dot may be placed on the line for emphasis. It is labeled with time *placed horizontally;* the word "Fix" is understood and is not shown; see Figure 1917.

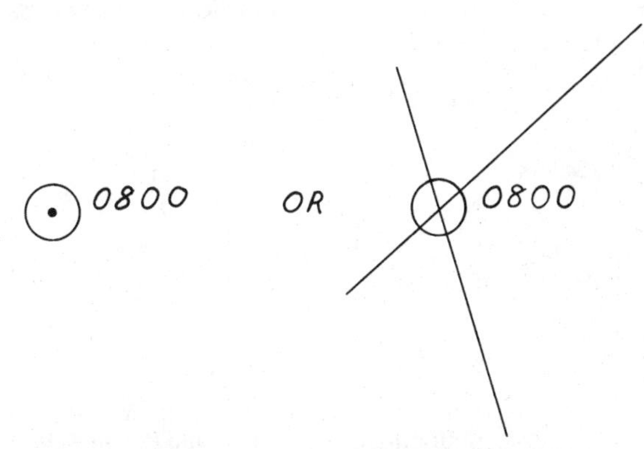

Figure 1917 **A known position of the boat is plotted as a dot with a small circle around it. If it is along a line, the dot will not show separately. Label the position with the time as a four-digit number in the 24-hour system.**

4. A DR position, calculated as a distance along the track at the set speed through the water, is shown as a half-circle (with a dot) along the track line; it is labeled with the time *placed at an angle to the horizontal.* "DR" is understood and is not written in; see Figure 1918.

5. When planning and pre-plotting a run, speed, which is often affected by sea conditions, may not be known in advance. In this case, distance, D, may be labeled *below*

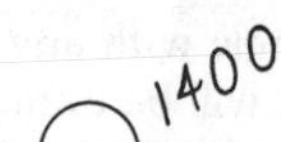

Figure 1918 **A dead reckoning position along a track is plotted as a half-circle around the point on the line. Add the time in the same manner as for a known position.**

the course line in lieu of speed; units (nautical or statute miles) are not shown.

Further applications of the basic rules of labeling will be given as additional piloting procedures and situations are introduced in subsequent chapters.

All lines on a chart should be erased when no longer needed, to keep the chart clear. Erasures should be made as lightly as possible to avoid damaging the chart and obscuring its printed information.

D, T, and S Calculations

As mentioned earlier, calculations involving distance (D), time (T), and speed (S) are often made with a small calculator. Use of a calculator is acceptable, but you should also be able to make your calculations accurately and quickly without one, using only a simple set of equations and ordinary arithmetic. The three basic equations are:

$$D=ST \qquad S=\frac{D}{T} \qquad T=\frac{D}{S}$$

Where D is distance in miles, T is time in hours, and S is speed in knots or miles per hour as determined by the type of mile being used. Note carefully that T is in *hours* in these basic equations. To use time in *minutes,* as is more normally the case, the equations are modified to read:

$$D=\frac{ST}{60} \qquad S=\frac{60D}{T} \qquad T=\frac{60D}{S}$$

Examples of the use of these practical equations may serve to make them clearer:

1. You are cruising at 14 knots; how far will you travel in 40 minutes?

$$D=\frac{ST}{60} \qquad D=\frac{14\times 40}{60}=9.3 \text{ miles}$$

Note that the calculated answer of 9.33 is rounded to the nearest tenth according to the rule for the degree of precision to be used in stating distance.

2. On one of the Great Lakes, it took you 40 minutes to travel 11 miles; what is your speed?

$$S=\frac{60D}{T} \qquad S=\frac{60\times 11}{40} = 16.5 \text{ MPH}$$

3. You have 9½ miles to go to reach your destination, on a broad reach you are sailing at 6.5 knots; how long will it take you to get there?

$$T=\frac{60D}{S} \qquad T=\frac{60\times 9.5}{6.5} = 88 \text{ minutes}$$

Note again the rounding of results; the calculated answer of 87.6 minutes is used as 88 minutes. In powerboat piloting contests it would probably be used as 87 minutes 36 seconds.

The equations are also usable with kilometers and km/h. Memorize the three equations for distance, speed, and time. Practice using them until you are thoroughly familiar with them, and can get correct answers quickly.

Use of Logarithmic Scale on Charts

Charts of the National Ocean Service at scales of 1:40,000 and larger have printed on them a logarithmic speed scale. To find speed, place one point of your dividers on the mark on the scale indicating the distance in nautical miles, and the other point on the number corresponding to the time in minutes; Figure 1919. Without changing the spread between the divider arms, place the right point on the "60" at the right end of the scale; the left point will then indicate on the scale the speed in knots.

The same logarithmic scale can also be used to *determine the time* required to cover a given distance at a specified speed (for situations not exceeding one hour). Set the two divider points on the scale marks representing speed in knots and distance in miles. Move the dividers, without changing the spread, until the right point is at "60" on the scale; the other point will indicate the time in minutes.

Likewise, *distance can be determined* from this logarithmic scale using knowledge of time and speed. Set the right point of the dividers on "60" and the left point at the mark on the scale corresponding to the speed in knots. Then, without changing the spread, move the right point to the mark on the scale representing the time in minutes; the left point indicates the distance in nautical miles.

The logarithmic scale is used in the same manner with distances in statute miles and speeds in MPH.

Figure 1919 **To use a logarithmic speed scale, set one point of the dividers on the scale division indicating miles traveled and the other to the number corresponding to the time, in minutes. With the dividers maintaining the same spread, transfer them so the right point is on the "60" of the scale; the left point then indicates boat speed in knots or mph.**

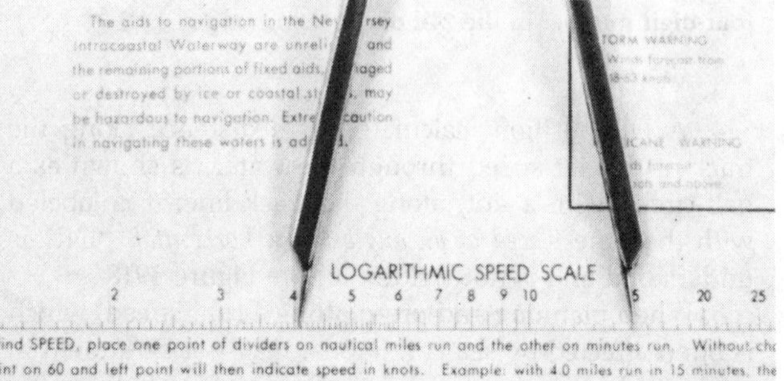

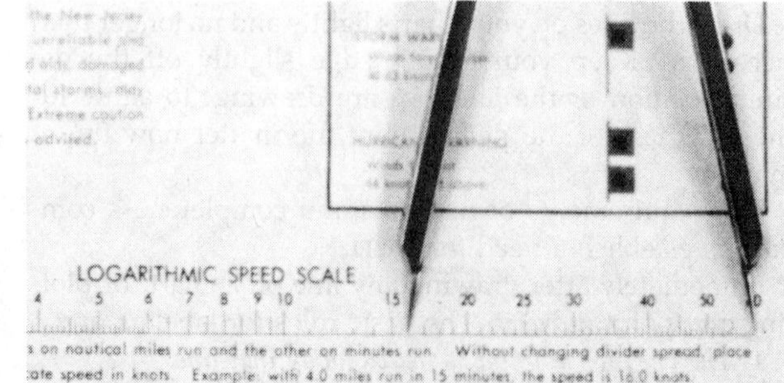

NOS charts have instructions for determining speed printed beneath the logarithmic scale, but not the procedures for determining distance or time. In all cases, you must know two of the three quantities in order to determine the other.

Use of S-D-T Calculators

It is not possible here to give detailed instructions for operating all models of speed-distance-time calculators. In general, they will have two or more scales, each logarithmically subdivided. The calculator will be set using two of the factors and the answer, the third factor, will be read off at an index mark. See Figure 1915.

If you have a calculator for S-D-T problems, read the instructions carefully and practice with it sufficiently, using simple, self-evident problems, to be sure you can get reliable results later even in emergency.

Speed Curves

Although some boats may have marine speedometers, the more-often used method for determining speed of powerboats is through the use of engine speed as measured by the TACHOMETER in revolutions per minute (RPM). A SPEED CURVE is prepared as a plot on cross-section (graph) paper of the boat's speed in knots or MPH for various engine speeds in RPM.

Factors Affecting Speed Curves

The boat's speed at a specified engine setting may be affected by several factors. The extent of each effect will vary with the size of the boat, type of hull, and other characteristics.

LOAD is a primary factor influencing a boat's speed. The number of people aboard, the amount of fuel and water in the tanks, and the amount and location of other weights on board will affect the depth to which the hull sinks in the water and the angular trim. Both displacement and trim may be expected to have an effect on speed.

Another major factor affecting speed is the UNDERWATER HULL CONDITION. Fouling growth like barnacles or moss increases the drag (the resistance to movement through the water), and slows the boat at any speed. Fouling on the propellor itself will drastically affect performance.

Whenever preparing speed data on a boat, note the loading and underwater hull conditions as well as the figures for RPM and speed. If you make a speed curve at the start of the season, when the bottom is clean, check it later in the season if your boat is used in waters where fouling is a problem. You may need a new speed curve, or you may be able to determine a small correction that can give you a more accurate measure of speed. You should also know what speed differences to expect from full tanks to half to nearly empty; the differences can be surprising.

Obtaining Speed Curves

Speed curves are obtained by making repeated runs over a known distance using different throttle settings and timing each run accurately. You can use any reasonable distance, but it should not be less than a half-mile so small timing errors will not excessively influence the results; it need not be more than a mile, to avoid excessive time and fuel requirements for the trials.

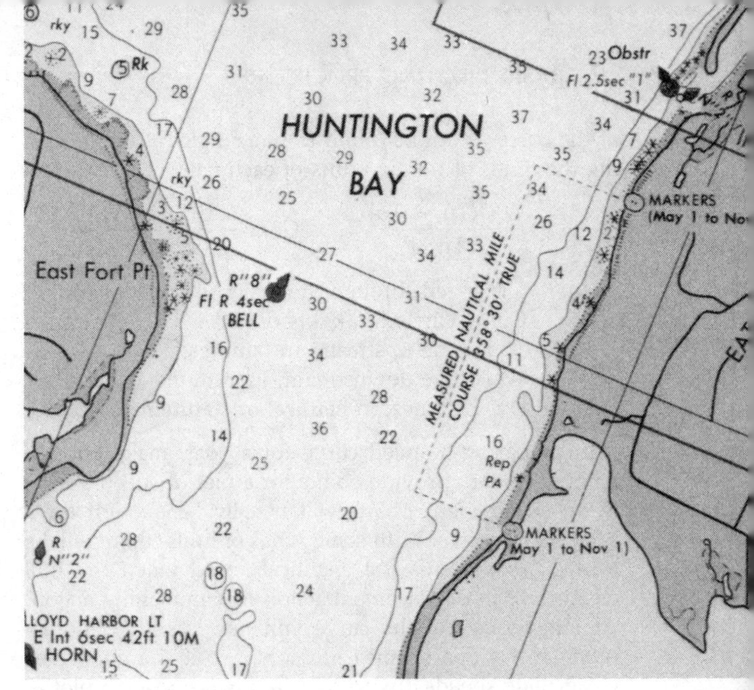

Figure 1920 **Measured miles have been established in many areas so that accurate speed trials can be made. These will be shown on the chart of the area.**

The run need not be an even half-mile or mile if the distance is accurately known. Do not depend upon floating aids to navigation—they may be slightly off station, and, in any event, they have some scope on their anchor chains and will swing about under the effects of wind and current. Many areas will have MEASURED MILES (or half-miles), Figure 1920. These are accurately surveyed distances with each end marked by ranges. Use these courses whenever possible; they are accurate and calculations are easier with the even-mile distance. But do not let the absence of a measured mile keep you from making a speed curve. Wharves, fixed aids to navigation, or points of land will also give you the accurate distance you need.

In most speed trials you will need to run the known distance twice, once in each direction, in order to allow for the effects of current. Even in waters not affected by currents, you should make round-trip runs for each throttle setting to allow for wind effects.

For each *one-way* run, measure the time and steer your boat carefully to make the most direct run. Compute the speed for each run by the equations on page 446. Tables are available that give speeds for various elapsed time over a measured mile. If the measured distance is an exact half-mile, just divide the tabulated speeds by two. Then average the *speeds* of each pair of runs at a given RPM, for the true speed of the boat through the water. The strength of the current is one-half the difference between the speed in the two directions of any pair of runs. *Caution:* do *not* average the *times* of a pair of runs to get a single time for use in the calculations; this will *not* give the correct value for speed through the water.

If time is measured with a regular clock or watch, be careful in making the subtractions to get elapsed time. Remember that there are 60 seconds in each minute, not 100, and likewise 60 minutes in one hour. Most people are so used to decimal calculations that they make errors when "borrowing" in the subtraction of clock times.

If one is willing to use a slightly more complex equation, the boat's speed through the water (or the strength of the current) can be found from a single calculation using the times of the two runs of each pair.

$$S = \frac{60D(Tu + Td)}{2TuTd} \qquad Cur = \frac{60D(Tu - Td)}{2TuTd}$$

Where S is speed through the water in knots or MPH
Cur is current in knots or MPH
Tu is time upstream, in minutes
Td is time downstream, in minutes
D is distance, in nautical or statute miles

In preparing a speed curve for a boat, make enough pairs of runs to provide points for a plot of speed versus RPM; six or eight points will usually be enough for a satisfactory curve. With some types of hulls, there will be a break in the curve at a critical speed when the hull changes from displacement action to semiplaning action. At this portion of the curve you may need additional, more closely spaced measurements, so it is a good idea to calculate speeds during runs and make a rough plot as you go along.

You may also want to calculate the current's strength for each pair of runs. The current values will probably vary during the speed trials, but the variations should be small and in a consistent direction, either steadily increasing or decreasing, or going through a slack period. You will get the best results by running your trials at a time of minimum current.

Example of a Speed Curve

A set of speed trials was run for the motor yacht *Trident* over the measured mile off Kent Island in Chesapeake Bay. This is an excellent course as it is marked both by buoys offshore as well as by ranges on land. The presence of the buoys aids in steering a straight run from one end of the course to the other; the ranges are used for accuracy in timing. (See Figure 1920.)

On this particular day, it was not convenient to wait for slack water, but a time was selected that would result in something less than maximum ebbing current. A table was set up in the log, and runs were made in each direction at speeds of normal interest from 900 RPM to 2150 RPM, which was maximum for the 6-71 diesels.

The results of the runs are shown in Figure 1921 top. An entry was also made in the log that these trials were

	NORTH-SOUTH		SOUTH-NORTH		AVERAGE	
RPM	TIME	SPEED	TIME	SPEED	SPEED	CURRENT
900	7M 54S	7.60	11M 25S	5.26	6.43	1.17
1100	6M 53S	8.72	9M 14.2S	6.50	7.61	1.11
1300	6M 08.8S	9.76	7M 36.8S	7.88	8.82	.94
1500	5M 35.8S	10.72	6M 40.4S	9.00	9.86	.86
1700	5M 10.8S	11.59	5M 54.2S	10.17	10.88	.71
1900	4M 38.0S	12.96	5M 06.4S	11.76	12.36	.60
2150	3M 48.6S	15.75	4M 05.6S	14.73	15.24	.51

Figure 1921 **Tabulated results, *above,* of speed trials made for *Trident.* Runs were made in each direction to account for the effect of current. Entries in the boat's log showed its bottom condition, and fuel and water loads aboard. The speed curve, *below,* plotted from this data is accurate only for load and hull conditions similar to those during the speed trials.**

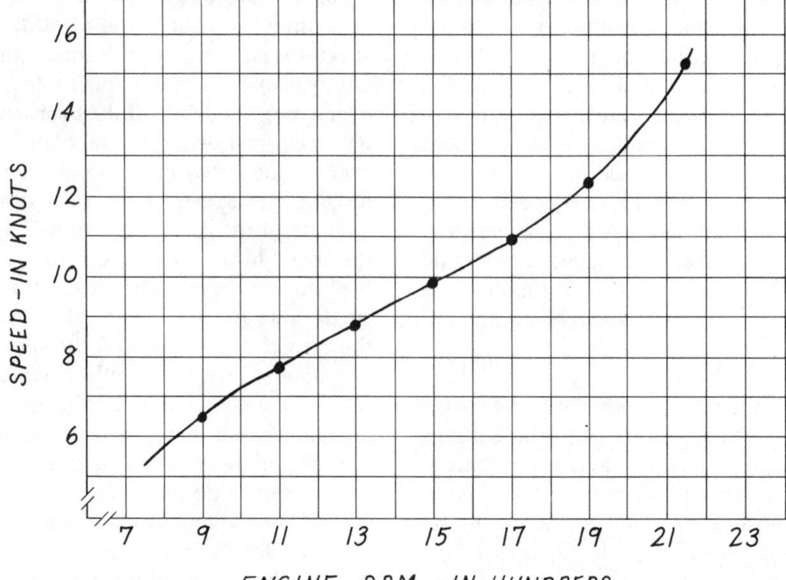

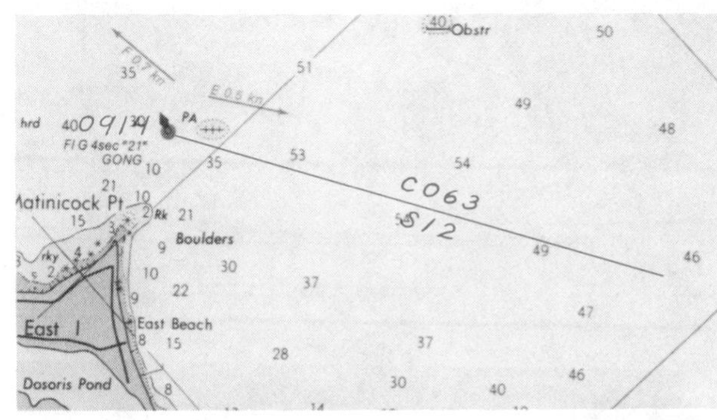

Figure 1922 **A dead reckoning plot is started when leaving a known position. This is plotted as a fix, with the time; course and speed are labeled along the DR track.**

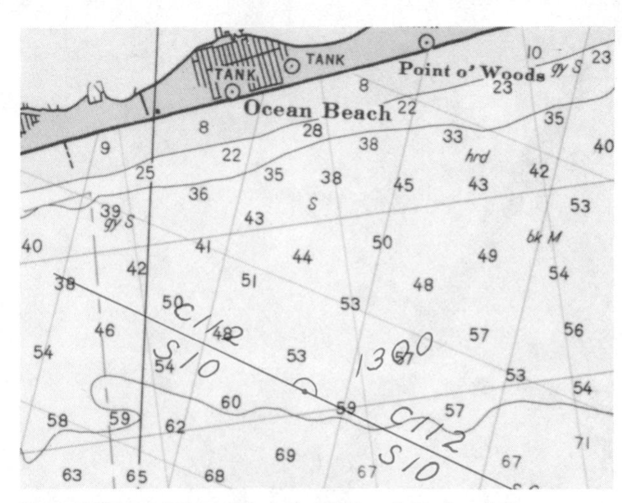

Figure 1924 **A DR position should be plotted each hour on the hour, *above*, even if there is no change in course or speed. Whenever the boat's position is fixed, such as by passage close to a buoy, *below*, a DR position is plotted for that time, calculated from speed and distance information. The fix is also plotted and a new DR track is started.**

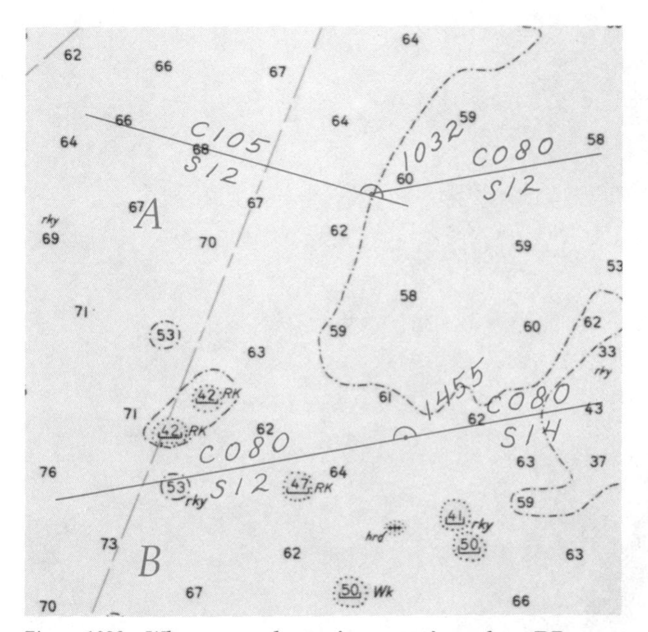

Figure 1923 **Whenever a change in course is made, a DR position is plotted for that time; the new course and speed are labeled along the new DR track. If a variation is made in speed without a change in direction, a DR position is plotted for the time of the change, and new course and speed labels are entered following it.**

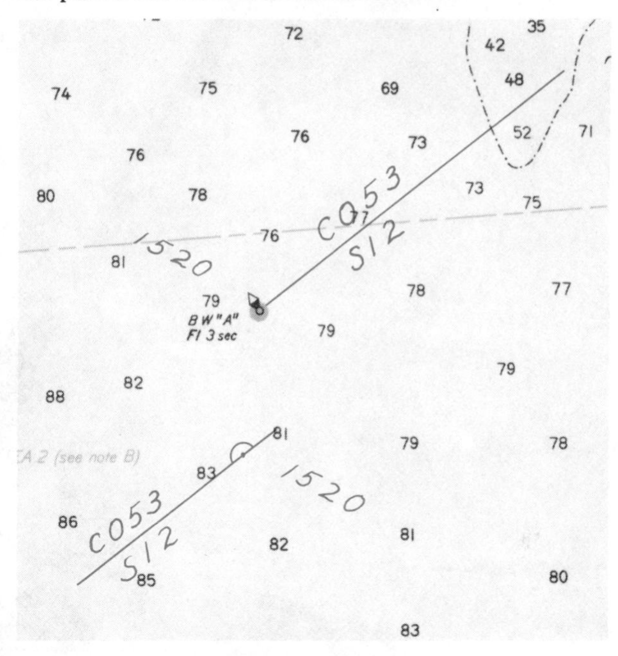

made with fuel tanks 0.4 full, the water tanks approximately ⅓ full, and with a clean bottom. The column of the table marked "Current" is not necessary, but it serves as a flag to quickly expose any inconsistent data. Note that on these trials the current is decreasing at a reasonably consistent rate.

After the runs had been completed, a plot was made on cross-section paper of the boat's speed as a function of engine RPM. This resulted in the speed curve shown as Figure 1921 bottom.

Dead Reckoning Plots

With knowledge of dead reckoning terms and principles, the rules for labeling points and lines, and the procedures for making calculations involving distance, time, and speed, you can now consider the use of DR plots.

There are several specific rules for making and using DR plots.

1. *A DR plot should be started when leaving a known position; Figure 1922.*
2. *A DR position should be shown whenever a change is made in course, Figure 1923A, or in speed, Figure 1923B.*
3. *A DR position should be plotted each hour on the hour (more frequently under conditions of reduced visibility); Figure 1924 top.*
4. *A new DR track should be started each time the boat's position is fixed. The old DR position for the same time as the fix should also be shown at the end of the old DR track, see Figure 1924 bottom.*

CT 2430 C

20

How to determine the state of tide or tidal current flow in coastal waters; how to allow for current and determine its strength

TIDES AND CURRENTS

An understanding of tides and currents is important to the skipper in coastal waters, as these can to some extent affect where he can travel or anchor safely, how long it will take him to get there—or the speed he will need to arrive at a given time, and the heading he must maintain to make good a given course over the bottom.

Here we should clarify the proper meaning and use of some terms that often are used incorrectly. TIDE is the rise and fall of the ocean level as a result of changes in the gravitational attraction between the earth, moon, and sun. It is a *vertical* motion only. CURRENT is the *horizontal* motion of water from any cause. TIDAL CURRENT is the flow of water from one point to another that results from a difference in tidal heights at those points. To say "The tide certainly is running strongly today!" is not correct, for tides may be high or low, but they do not "run." The correct expression would be "The tidal current certainly is strong today." Remember—tide is vertical change; current is horizontal flow.

Tides

Tides originate in the open oceans and seas, but are only noticeable and significant close to shore. The effect of tides will be observed along coastal beaches, in bays and sounds, and up rivers generally as far as the first rapids, waterfall, or dam. Curiously, the effect of tides may be more noticeable a hundred miles up a river than it is at the river's mouth, because water piles up higher in its narrower stretches. Coastal regions in which the water levels are subject to tidal action are often referred to as "tidewater" areas.

Definition of Terms

In addition to the basic definition of tide as given above, certain other terms used in connection with tidal action must be defined. The HEIGHT OF TIDE at any specified time is the vertical measurement between the surface of the water and the TIDAL DATUM, or reference plane. Do not confuse "height of tide" with "depth of water." The latter is the total distance from the surface to the bottom. The tidal datum for an area is selected so that the heights of tide are normally positive values, but the height can at times be a small negative number when the water level falls below the datum.

HIGH WATER, or HIGH TIDE, is the highest level reached by an ascending tide. Correspondingly, LOW WATER, or LOW TIDE, is the lowest level reached by a descending tide; Figure 2002. The difference between high and low waters is the RANGE of the tide.

The change in tidal level does not occur at a uniform rate; starting from low water, the level builds up slowly, then at an increasing rate which in turn tapers off as high water is reached. The decrease in tidal stage from high water to low follows a corresponding pattern of a slow buildup to a maximum rate roughly midway between stages, followed by a decreasing rate. At both high and low tides, there will be periods of relatively no change in level; these are termed STAND. MEAN SEA LEVEL is the average level of the open ocean, and corresponds closely to mid-tide levels offshore.

Tidal Theory

Tidal theory involves the interaction of gravitational and centrifugal forces—the inward attractions of the earth on one hand and the sun and the moon on the other, balanced by the outward forces resulting from the revolution of the earth in its orbit. The gravitational and centrifugal forces are in balance as a whole—otherwise the bodies would fly apart from each other or else crash together—but they are not quite in balance at most points on the earth's surface, and this is what causes the tides. The effects of the sun and moon will be described separately, even though, of course, they act simultaneously.

Earth-Sun Effects

Although not precisely the case, the earth can be thought of as revolving around the sun, and just as a stone tied to the end of a string tends to sail off when a young boy whirls it about his head, so the earth tends to fly off into space. This effect is known as centrifugal force; it is shown

Figure 2001 **The rise and fall of ocean tidal levels cause a flow first into and then out of inland water bodies such as bays, sounds, and the lower reaches of rivers. Because they can set a boat off its course, these tidal currents may have important effects on piloting.**

at the left in Figure 2003. (Remember that we are talking about the centrifugal force related to the sun-earth system, and not that of the spinning of the earth on its own axis.)

The earth is kept from flying off into space by the gravitational attraction of the sun, shown at the right in Figure 2003. These forces are in overall balance.

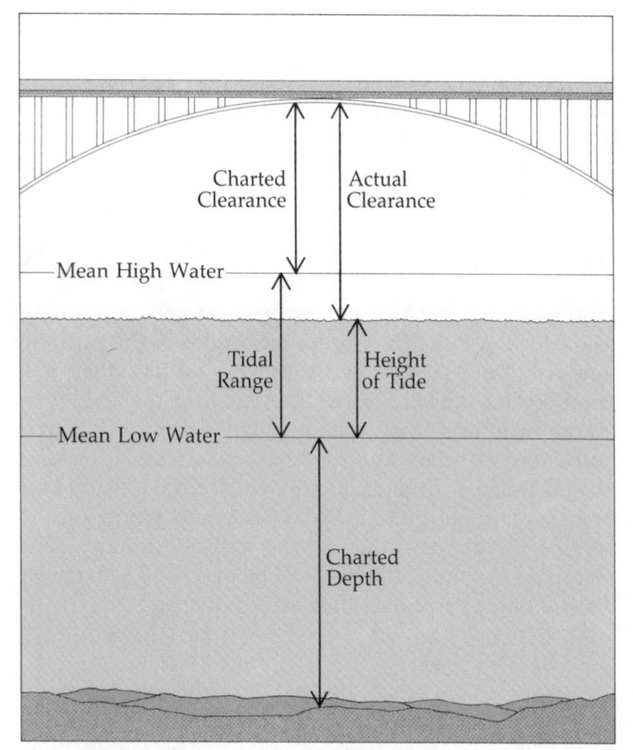

Figure 2002 **Diagram illustrates the relationship of some of the terms used to describe tidal conditions. Mean lower low water is the reference datum for many charts.**

This balance is not exact at all points. The centrifugal force is everywhere the same, and parallel to a line from the center of the earth to the center of the sun. Gravitational forces are *not* everywhere equal and parallel; from each point on the earth's surface, they extend toward the center of the sun, greater on the nearer side as a result of the lesser distance. Figure 2003 shows these forces at representative points, and the resultant force from their combination.

In theory, we can think of the earth as being a smooth sphere uniformly covered with water (no land areas). Figure 2004 shows how the resultant forces cause the water to flow toward the areas of the earth's surface that are *both nearest* and *farthest* from the sun; here there will be "high tides." As this water flows, the areas from which it comes will have less, and hence "low tides." (The tides on the side of the earth nearer the sun are slightly greater than those on the far side, but the difference is not great, only about 5%.)

As the earth rotates on its axis, once every 24 hours, the line of direction to the sun constantly changes. Thus each point of the earth's surface will have two high and two low tides each day. As a result of the tilt of the earth's axis, the pairs of highs, and those of the lows, will not normally be of exactly the same level.

Earth-Moon Effects

The moon is commonly thought of as revolving about the earth; actually, the two bodies revolve around a common point on a monthly cycle. This point is located about 2,900 miles (4,667 km) from the center of the earth toward the moon (about 1,100 miles or 1,770 km deep *inside* the earth). Both tend to fly away from this common point (centrifugal force), but the mutual gravitational attraction acts as a counterbalance, and they remain the same distance apart. However, the gravitation pull of the moon effects the waters of the earth in the same manner as the pull of the sun.

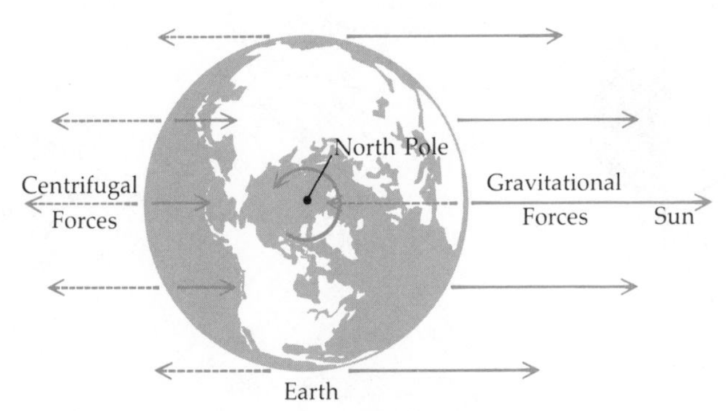

Figure 2003 **Tides result from the differences between centrifugal forces and gravitational forces. The forces illustrated here represent the interaction of the earth and sun; corresponding forces result from the relationship between the earth and the moon.**

Spring and Neap Tides

Now combine the earth-sun and earth-moon systems. Since the moon is much closer to the earth (about 238,860 miles, 384,400 km, away), its tidal forces are roughly 2¼ times greater than those of the sun (about 92,900,000 miles or 149,500,000 km away). The result is that the observed tide usually "follows the moon," but the action is somewhat modified by the sun's relative position. The two high and two low waters each day occur about 50 minutes later than the corresponding tides of the previous day.

In the course of any one month, the three bodies line up in sun-moon-earth (at the left in Figure 2005) and sun-earth-moon (at the right). These are the times of the new and full moon, respectively. In both cases, the sun's effect lines up with and reinforces the moon's effect, tending to result in greater-than-average tidal ranges (about 20%) called SPRING TIDES; note that this name has nothing to do with the season of the year.

At top and bottom positions in Figure 2005, when the moon is at its first and third quarters, the tidal "bulge" caused by the sun is at right angles to that caused by the moon (they are said to be "in QUADRATURE"). The two tidal effects are in conflict and partially cancel each other, resulting in smaller-than-average ranges (again about 20%); these are NEAP TIDES.

Note that the tidal range of any given point varies from month to month, and from year to year. The monthly variation results from the fact that the earth is not at the center of the moon's orbit. When the moon is closest to the earth (at perigee), the lunar influence is maximum and tides will have the greatest ranges. Conversely, when the moon is farthest from the earth (at apogee), its effect and tidal ranges are least.

In a similar manner, the yearly variations in the daily ranges of the tides are caused by the changing gravitational effects of the sun as that body's distance from the earth becomes greater or less.

Actual Tides

The actual tide that we observe often seems to be at odds with the theoretical forces that govern it. Here are the main reasons for this:

1. Great masses of land, the continents, irregularly shaped and irregularly placed, act to interrupt, restrict, and reflect tidal movements.
2. Water, although generally appearing to flow freely, is actually a somewhat viscous substance that lags in its response to tidal forces.
3. Friction is present as the ocean waters "rub" against the ocean bottom.
4. The depth to the bottom of the sea, varying widely, influences the speed of the horizontal tidal motion.
5. The depths of the ocean areas and the restrictions of the continents often result in "basins" that have their own way of responding to tidal forces.

Although these reasons account for great differences between theoretical tidal forces and actual observed tides, there nevertheless remain definite, constant relationships between the two at any particular location. By observing the tide, and relating these observations with the movements of the sun, moon, and earth, these constant relationships can be determined. With this information, tides can be predicted for any future date at a given place.

Types of Tides

A tide which each day has two high waters approximately equal in height, and two low waters also about equal, is known as a SEMIDIURNAL tide. This is the most common tide, and, in the United States, occurs along the East Coast (Figure 2006, New York).

In a monthly cycle, the moon moves north and south of the equator. Figure 2007 illustrates the importance of this action to tides. Point A is under a bulge in the envelope. One half-day later, at point B, it is again under the bulge but the height is not as large as at A. This situation, combined with coastal characteristics, tends to give rise to a "twice daily" tide with unequal high and/or low waters in some areas. This is known as the MIXED type of tide; see Figure 2006, San Francisco. The term "low water" may be modified to indicate the more pronounced of the two lows. This tidal stage is termed LOWER LOW WATER and is averaged to determine the MEAN LOWER LOW WATER (MLLW); this is used as the tidal datum in many areas subject to mixed tides. Likewise, the more

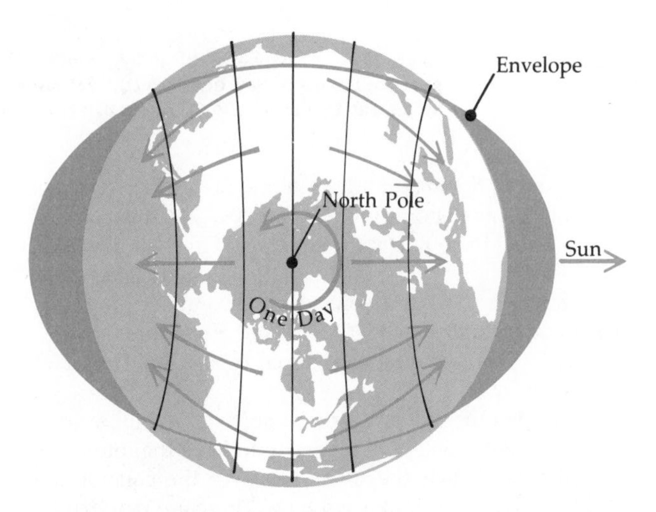

Figure 2004 **Inequalities of gravitational and centrifugal forces result in high tides on *opposite* sides of the earth at the same time, on the side away from the sun as well as the side toward it. Low tides are midway between these.**

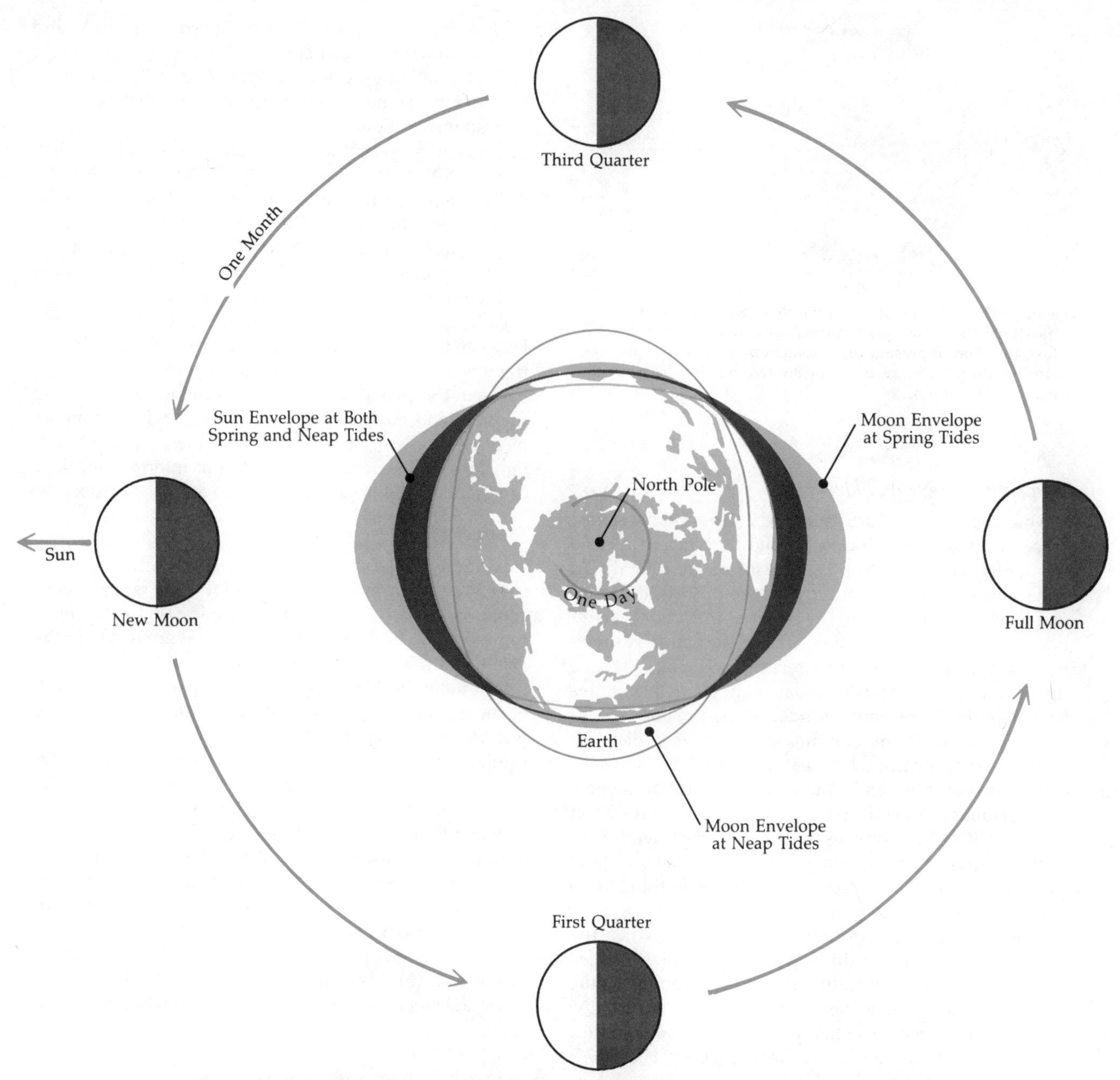

Figure 2005 **At new and full moon, combined gravitational pull of the sun and moon acts to produce the maximum tidal effect; tidal ranges are greatest. At first and third quarters of the moon, the two gravitational forces offset each other partially and the net effect is minimum.**

significant of the higher tides is termed HIGHER HIGH WATER.

Now consider point C in Figure 2007. At this place, it is still under the bulge of the envelope. One half-day later, at point D, however, it is above the low part. Hence the tidal forces tend to cause only one high and one low water each day (actually each 24 hours and 50 minutes approximately). This is the DIURNAL type, typified by Pensacola in Figure 2006.

Whenever the moon is farthest north (as in Figure 2007) or south, there will be a tendency to have the diurnal or mixed type. When the moon lies over the equator, we tend to have the semidiurnal type. These are EQUATORIAL TIDES and the tendency toward producing inequality is then at a minimum.

These theoretical considerations are modified, however, by many practical factors such as the general configuration of the coastline.

Special Tidal Situations

Peculiarities in the tide can be found almost everywhere, but none compare with those in the Bay of Fundy. Twice each day, the waters surge in and out of the Bay producing, at Burntcoat Head, the highest tidal range in the world; a typical rise and fall of nearly 44 feet (13.4 meters). At normal spring tides, it rises 51½ feet (15.7 m) and on perigee springs, as much as 53 feet or 16.2 meters.

The great range is often attributed to the funnel-like shape of the Bay, but this is not the main cause. Just as the water in a wash basin will slosh when you move your hand back and forth in just the right period of time, depending upon the depth of the water and the shape of the basin, so the tide will attempt to oscillate water in bays in cycles of 12 hours and 25 minutes. It would be a coincidence indeed if a bay were of such a shape and depth as to have a complete oscillation with a period of exactly 12 hours and 25 minutes. A bay can easily have a *part* of such an oscillation, however; such is the case in the Bay of Fundy.

A further factor in the large tidal ranges in the Bay of Fundy is that they are controlled by the Gulf of Maine tides which, in turn, are controlled by the open ocean tides. The relationships between these tides are such as to exaggerate their ranges.

Reference Planes

As mentioned, the heights of tides are reckoned from the specific reference plane, or datum. The datum long used on the Pacific Coast, where mixed tides widely prevail, is mean lower low water (MLLW), as discussed on page 453, and the datum on the Atlantic Coast has long been mean low water (MLW). The Gulf Coast low water datum is now MLLW; this was a technical change only, and no depths on charts were affected. In future years, the National Ocean Service (NOS) will change the tidal datum on the Atlantic Coast to MLLW so that all areas will be the same.

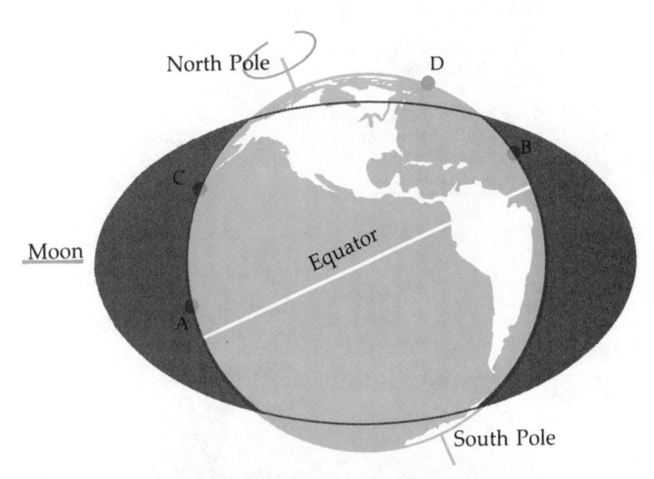

Figure 2007 **As the moon travels north and south of the earth's equatorial plane, it results in variations of the daily tidal cycle at any given location. Similar, but lesser, effects result from the changes in the sun's position.**

The Importance of Tides

A good knowledge of tidal action is essential for safe navigation. The skipper of a boat of any size will be faced many times with the need to know the time of high water or low water, and their probable heights. He may want to cross some shoal area, passable at certain tidal stages but not at others. He may be about to anchor, and the scope to pay out will be affected by the tide's range. He may be going to tie up to a pier or wharf in a strange harbor, and need to know about the tides to adjust his lines properly for overnight.

Sources of Tidal Information

The basic source of information on the time of high and low water, and their heights above (or below) the datum, are the TIDE TABLES published by the National Ocean Service, Figure 2008. Any predictions appearing in newspapers or broadcast over radio and TV stations will have been extracted from these tables (and converted to daylight time if in effect). Small-craft charts include tidal information in their margins or on the cover jacket if one is used; see Figure 2009.

Not to be overlooked is the possibility of "local knowledge." Never hesitate to ask experienced local watermen when you are in unfamiliar waters and need information of any kind. The best of tables prepared by electronic computers sometimes cannot compare with a knowledge of what to expect that is based on years of local experience.

Tide Tables

The NOS *Tide Tables* are of great value in determining the height of water at any place at a given time. They are calculated in advance and published annually in four volumes, one of which covers the East Coast of North and South America, and another the West Coast of these continents.

The *Tables* can usually be bought at any authorized sales agent for NOS charts, or by mail from the Superintendent of Documents, Government Printing Office, Washington, DC 20402, or from GPO Regional Bookstores in many cities.

The *Tide Tables* give the predicted times and heights of high and low water for each day of the year at important points known as REFERENCE STATIONS. Portland, Boston, Sandy Hook, and Key West are examples of points for which detailed information is given in the East Coast *Tables*. Reference stations in the West Coast *Tables* include

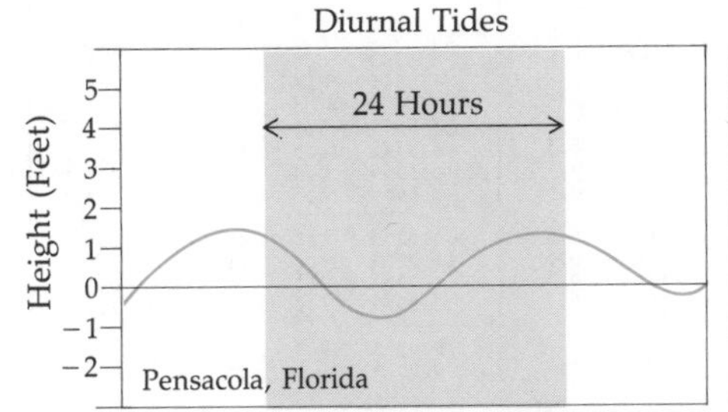

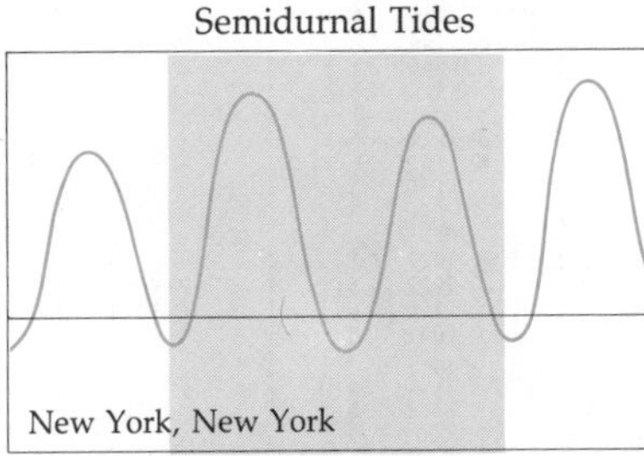

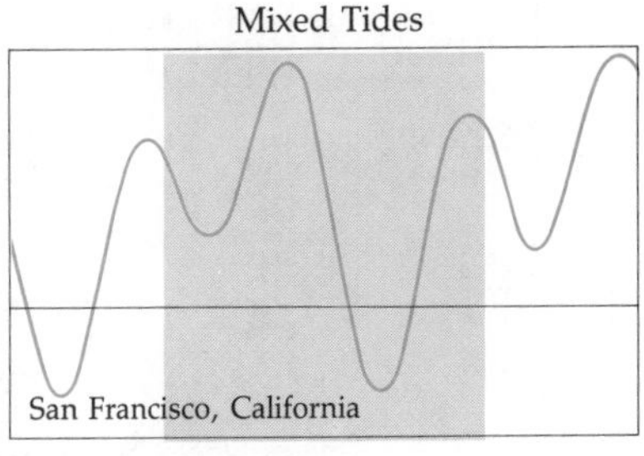

Figure 2006 **Characteristics of the daily cycle of tides vary widely at different places. Shown here are the three basic types: diurnal, *left*; semi-diurnal, *center*; mixed, *right*.**

Figure 2008 **The National Ocean Service uses computerized equipment to prepare advanced predictions of tidal levels at many points; these are published in the annual Tide Tables.**

San Diego, the Golden Gate at San Francisco, and Aberdeen, Washington.

Additional data show the difference in times and heights between these reference stations and thousands of other points termed SUBORDINATE STATIONS. From these tables, the tide at virtually any point of significance along the coasts can easily be computed. The West Coast *Tide Tables,* for example, contain predictions for 40 reference stations, and differences for about 1,100 subordinate stations in North and South America.

Other nations have similar sets of tables for their waters.

All factors that can be determined in advance are taken into account in tide predictions, but remember when using them that other factors can also influence the height of tides greatly. Such influences include barometric pressure and wind. In many areas the effect of a prolonged gale from a certain quarter can offset all other factors. Tidal rivers may be affected by the volume of water flowing down from the watershed. Normal seasonal variations of flow are allowed for in the predictions, but unexpected prolonged wet or dry spells may cause significant changes to tidal height predictions. Intense rainfall upriver may change both heights and times of tides and the effects may not appear downstream for several days.

You should therefore be careful whenever using the *Tide Tables,* especially since low waters can go considerably lower than the level predicted.

Explanation of the Tables

Table 1 of the *Tide Tables,* Figure 2010, gives the predicted times and heights of high and low water at the main reference stations and is practically self-explanatory; note that heights are given in both feet and meters. Where no sign is given before the predicted height, the quantity is positive and is to be added to the depths as given on the chart. When the value is preceded by a minus (−) sign, the "heights" are to be *subtracted* from charted depths.

Time is given in the four-digit system from 0001 to 2400 (see Chapter 19). In many areas you must be careful regarding *daylight time;* the *Tide Tables* are published in local *standard* time, and you must correct for it if you are in a DT locality.

Where there are normally two high and two low tides each date, they are roughly a little less than an hour later each succeeding day. Consequently, a high or low tide may skip a calendar day, as indicated by a blank space in the *Tide Tables,* Figure 2011. If it is a low tide that is skipped, for example, you will note that the previous corresponding low occurred late in the foregoing day, and the next one early in the following day; see the sequence for late evening low tides on 2, 3, and 4 November in Figure 2010.

Note also from a review of the *Tide Tables* that at some places there will be only one high and one low tide on some days, with the usual four tides on other days. This is not a diurnal tide situation, where a single high and low would occur every day. The condition considered here arises when the configuration of the land and the periods

BRIDGEPORT, CONN.

Predicted times and heights of high and low water-Eastern Standard Time. For Daylight Saving Time, add 1 hour.
To predict local tide, apply the time difference listed in the facility tabulations to these tide predictions.

FEBRUARY 1982						MARCH 1982						APRIL 1982						MAY 1982					
Day	Time h.m.	Ht. ft.	Day	Time h.m.	Ht. ft.	Day	Time h.m.	Ht. ft.	Day	Time h.m.	Ht. ft.	Day	Time h.m.	Ht. ft.	Day	Time h.m.	Ht. ft.	Day	Time h.m.	Ht. ft.	Day	Time h.m.	Ht. ft.
1	0404	6.6	16	0513	6.0	1	0245	6.9	16	0333	6.2	1	0434	6.6	16	0434	6.2	1	0533	6.6	16	0447	6.4
M	1031	-0.2	Tu	1141	0.5	M	0912	-0.5	Tu	0959	0.4	Th	1109	0.0	F	1103	0.9	Sa	1200	0.1	Su	1112	0.8
	1635	5.9		1744	5.3		1517	6.1		1604	5.6		1720	5.9		1710	5.9		1815	6.5		1722	6.5
	2247	0.0		2355	0.7		2128	-0.1		2215	0.9		2333	0.5		2324	1.3					2340	1.0
2	0506	6.6	17	0608	6.0	2	0343	6.7	17	0426	6.1	2	0546	6.5	17	0530	6.2	2	0031	0.4	17	0543	6.4
Tu	1136	-0.2	W	1236	0.5	Tu	1012	-0.3	W	1054	0.7	F	1218	0.0	Sa	1157	0.8	Su	0642	6.6	M	1203	0.6
	1740	5.7		1841	5.4		1620	5.8		1658	5.5		1829	6.0		1805	6.1		1303	0.1		1815	6.7
	2352	0.0					2231	0.2		2310	1.0								1916	6.7			
3	0610	6.7	18	0050	0.8	3	0447	6.6	18	0520	6.0	3	0044	0.4	18	0021	1.1	3	0134	0.1	18	0036	0.7
W	1244	-0.3	Th	0703	6.1	W	1121	-0.1	Th	1150	0.8	Sa	0656	6.6	Su	0627	6.3	M	0743	6.6	Tu	0639	6.5
	1848	5.7		1330	0.4		1728	5.7		1755	5.5		1322	-0.1		1250	0.7		1359	0.0		1255	0.4
				1935	5.5		2340	0.3					1933	6.3		1900	6.4		2010	7.0		1908	7.0
4	0058	0.0	19	0143	0.7	4	0556	6.5	19	0007	1.1	4	0148	0.1	19	0115	0.8	4	0230	-0.2	19	0129	0.3
Th	0716	6.8	F	0753	6.3	Th	1231	-0.1	F	0617	6.1	Su	0800	6.7	M	0721	6.5	Tu	0839	6.7	W	0733	6.6
	1348	-0.5		1420	0.3		1839	5.7		1245	0.7		1422	-0.2		1340	0.4		1451	-0.1		1345	0.1
	1954	5.9		2025	5.8					1851	5.7		2031	6.6		1948	6.7		2059	7.2		1959	7.4
5	0204	-0.2	20	0233	0.5	5	0050	0.2	20	0103	1.0	5	0247	-0.3	20	0206	0.3	5	0321	-0.5	20	0222	-0.2
F	0818	7.0	Sa	0841	6.5	F	0706	6.6	Sa	0711	6.3	M	0857	6.9	Tu	0812	6.7	W	0929	6.7	Th	0827	6.8

Figure 2009 **Small-craft charts give annual tidal predictions for reference stations covered by the chart. Information tabulated for marinas and other boating facilities include tidal ranges at such stations, and time differences from the nearest reference station.**

SEATTLE, WASHINGTON 95

Times and Heights of High and Low Waters

OCTOBER

Day	Time h m	Height ft	Height m	Day	Time h m	Height ft	Height m
1 F	0315	9.5	2.9	16 Sa	0416	10.5	3.2
	0934	1.4	0.4		1008	2.5	0.8
	1610	11.1	3.4		1617	11.5	3.5
	2213	3.2	1.0		2242	0.9	0.3
2 Sa	0400	9.9	3.0	17 Su	0507	10.8	3.3
	1011	1.7	0.5		1050	3.3	1.0
	1630	11.2	3.4		1645	11.2	3.4
	2245	2.2	0.7		2316	0.3	0.1
3 Su	0444	10.4	3.2	18 M	0553	11.0	3.4
	1050	2.2	0.7		1129	4.1	1.2
	1655	11.3	3.4		1712	10.9	3.3
	2317	1.2	0.4		2351	0.0	0.0
4 M	0531	10.7	3.3	19 Tu	0642	11.1	3.4
	1125	2.9	0.9		1212	4.9	1.5
	1721	11.3	3.4		1739	10.4	3.2
	2356	0.3	0.1				

NOVEMBER

Day	Time h m	Height ft	Height m	Day	Time h m	Height ft	Height m
1 M	0445	11.0	3.4	16 Tu	0558	11.5	3.5
	1024	4.3	1.3		1117	6.1	1.9
	1600	11.5	3.5		1622	10.6	3.2
	2248	-0.8	-0.2		2323	-0.9	-0.3
2 Tu	0531	11.5	3.5	17 W	0640	11.7	3.6
	1107	5.0	1.5		1159	6.5	2.0
	1629	11.5	3.5		1651	10.3	3.1
	2326	-1.7	-0.5		2356	-0.9	-0.3
3 W	0623	11.9	3.6	18 Th	0721	11.8	3.6
	1153	5.7	1.7		1241	6.9	2.1
	1704	11.4	3.5		1722	9.9	3.0
4 Th	0008	-2.1	-0.6	19 F	0032	-0.7	-0.2
	0715	12.1	3.7		0804	11.8	3.6
	1241	6.3	1.9		1328	7.1	2.2
	1742	11.1	3.4		1754	9.4	2.9

DECEMBER

Day	Time h m	Height ft	Height m	Day	Time h m	Height ft	Height m
1 W	0535	12.3	3.7	16 Th	0636	12.2	3.7
	1050	6.7	2.0		1145	7.4	2.3
	1550	11.8	3.6		1617	10.2	3.1
	2304	-3.0	-0.9		2333	-1.2	-0.4
2 Th	0624	12.7	3.9	17 F	0709	12.3	3.7
	1140	7.1	2.2		1227	7.4	2.3
	1631	11.7	3.6		1653	9.9	3.0
	2349	-3.2	-1.0				
3 F	0713	13.0	4.0	18 Sa	0009	-1.0	-0.3
	1233	7.2	2.2		0744	12.3	3.7
	1716	11.3	3.4		1310	7.3	2.2
					1731	9.6	2.9
4 Sa	0036	-2.9	-0.9	19 Su	0044	-0.6	-0.2
	0805	13.0	4.0		0822	12.3	3.7
	1330	7.2	2.2		1358	7.2	2.2
	1809	10.6	3.2		1817	9.1	2.8

Figure 2010 **Table 1 of the *Tide Tables* lists the daily predictions of time and height for high and low waters at all the selected reference stations, for the entire year.**

between successive tides are such that one tide is reflected back from the shore and alters the effect of the succeeding tide.

Sometimes the difference in the height of the two high or two low waters of a day is so increased as to cause only one low water each day. These tides are not unusual in the tropics and consequently are called TROPIC TIDES.

Table 2, Tidal Differences and Constants, Figure 2012, gives the information necessary to find the time and height of tide for thousands of subordinate stations by applying simple corrections to the data given for the main reference stations. The name of the applicable reference station appears at the head of the particular section in which the subordinate station is listed. After the subordinate station number and name, the following information is given in the columns of Table 2:

1. Latitude and longitude of the subordinate station.
2. Differences in time and height of high (and low) waters at the subordinate station and its designated reference station.
3. Mean and spring (or diurnal) tidal ranges.
4. Mean tide level.

As mentioned, note that the existence of a "minus tide" means that the actual depths of water will be *less* than the figures on the chart.

To determine the time of high or low water at any station in Table 2, use the column marked Differences, Time. This gives the hours and minutes to be added to (+) or subtracted from (−) the time of the respective high or low water at the reference station shown in bold-face type next *above* the listing of the subordinate station. Be careful in making calculations near midnight. Applying the time difference may cause you to cross the line from one day to another. Simply add or subtract 24 hours as necessary.

The height of the tide at a station in Table 2 is determined by applying the height difference, or in some cases the RATIO. A plus sign (+) indicates that the difference increases the height given for the designated reference station; a minus sign (−) indicates that it decreases the Table 1 value. Differences are not given in metric values; if needed, conversions from feet can be made using Table 7.

Where height differences would give unsatisfactory predictions, ratios may be substituted for heights. Ratios are identified by an asterisk, and are given as a decimal fraction by which the height at the reference station is to be multiplied to determine the height at the subordinate station.

In the columns headed "Ranges," the MEAN RANGE is the difference in height between mean high water and mean low water. This figure is useful in many areas where it may be added to mean low water to get mean high water, the datum commonly used for measuring vertical heights above water for bridge and other vertical clearances. The SPRING RANGE is the average semidiurnal range occurring twice monthly when the moon is new or full. It is larger than the mean range where the type of tide is either semidiurnal or mixed, and is of no practical significance where the tide is of the diurnal type. Where this is the situation, the tables give the DIURNAL RANGE, which is the difference in height between mean higher high water and mean lower low water.

Day	Time h m	Height ft	Height m	Day	Time h m	Height ft	Height m	Day	Time h m	Height ft	Height m
16 Tu	0435	6.7	2.0	1 M	0217	4.2	1.3	16 Tu	0306	6.1	1.9
	1023	10.1	3.1		0819	11.7	3.6		0835	9.7	3.0
	1749	1.9	0.6		1514	0.9	0.3		1543	1.6	0.5
					2153	9.6	2.9		2313	9.4	2.9
17 W	0128	9.3	2.8	2 Tu	0309	5.6	1.7	17 W	0412	6.9	2.1
	0559	7.4	2.3		0901	11.3	3.4		0919	9.2	2.8
	1111	9.7	3.0		1613	0.4	0.1		1643	1.6	0.5
	1849	1.4	0.4		2323	9.7	3.0				
18 Th	0237	10.0	3.0	3 W	0418	6.8	2.1	18 Th	0039	9.7	3.0
	0727	7.7	2.3		0951	10.9	3.3		0539	7.4	2.3
	1203	9.5	2.9		1719	0.0	0.0		1012	8.8	2.7
	1940	0.9	0.3						1747	1.5	0.5

Figure 2011 **Table 1 extract shows how a high or low tide may be omitted near midnight (see text); this might occur about once each two weeks.**

TABLE 2. - TIDAL DIFFERENCES AND OTHER CONSTANTS 175

NO.	PLACE	POSITION Lat.	Long.	DIFFERENCES Time High water	Time Low water	Height High water	Height Low water	RANGES Mean	Diurnal	Mean Tide Level
		° ' N	° ' W	h. m.	h. m.	ft	ft	ft	ft	ft
	Puget Sound Time meridian, 120°W			on SEATTLE, p.92						
912	Hansville	47 55	122 33	-0 07	+0 04	-1.0	-0.1	6.7	10.3	6.1
913	Point No Point	47 55	122 32	-0 16	-0 16	*0.92	*0.92	6.7	10.4	6.1
914	Edmonds	47 49	122 23	-0 05	+0 02	-0.4	0.0	7.2	10.9	6.4
915	Port Madison	47 42	122 32	-0 08	-0 08	+0.1	0.0	7.7	11.4	6.6
917	Poulsbo, Liberty Bay	47 44	122 39	+0 02	+0 08	+0.6	+0.1	8.1	11.9	6.9
919	Brownsville, Port Orchard	47 39	122 37	+0 02	+0 08	+0.4	0.0	8.0	11.7	6.8
921	SEATTLE (Madison St.), Elliott Bay	47 36	122 20	Daily predictions				7.6	11.3	6.6
923	Eighth Ave. S., Duwamish Waterway	47 32	122 19	+0 05	+0 07	-0.1	0.0	7.5	11.1	6.5
925	Port Blakely	47 36	122 30	+0 02	+0 03	+0.2	0.0	7.8	11.5	6.7
927	Pleasant Beach, Rich Passage	47 36	122 32	+0 01	+0 07	+0.2	0.0	7.8	11.5	6.7
929	Bremerton, Port Orchard	47 33	122 38	+0 07	+0 12	+0.4	0.0	8.0	11.7	6.8
931	Tracyton, Dyes Inlet	47 37	122 40	+0 30	+0 56	+1.0	0.0	8.6	12.3	7.1
933	South Colby, Yukon Harbor	47 31	122 32	+0 01	+0 07	+0.3	0.0	7.9	11.6	6.7
935	Des Moines	47 24	122 20	+0 03	+0 09	+0.4	0.0	8.0	11.7	6.8
937	Burton, Quartermaster Harbor	47 23	122 28	+0 07	+0 13	+0.6	0.0	8.2	11.9	6.9
939	Gig Harbor	47 20	122 35	+0 06	+0 14	+0.6	0.0	8.2	11.8	6.9
941	Tacoma, Commencement Bay	47 17	122 25	+0 07	+0 06	+0.5	0.0	8.1	11.8	6.8
943	Arletta, Hale Passage	47 17	122 39	+0 23	+0 36	+1.7	0.0	9.3	13.0	7.4
945	Home, Von Geldern Cove, Carr Inlet	47 16	122 45	+0 27	+0 39	+2.3	+0.2	9.7	13.6	7.8
947	Wauna, Carr Inlet	47 23	122 38	+0 20	+0 36	+1.8	0.0	9.4	13.1	7.5
949	Stellacoom	47 10	122 36	+0 22	+0 35	+1.8	0.0	9.4	13.1	7.5
951	Hyde Point, McNeil Island	47 12	122 39	+0 23	+0 41	+2.1	+0.2	9.5	13.4	7.7
953	Sequalitchew Creek, Nisqually Reach	47 07	122 40	+0 24	+0 42	+2.1	+0.1	9.6	13.4	7.7
955	Longbranch, Filucy Bay	47 13	122 45	+0 26	+0 39	+2.2	+0.1	9.7	13.5	7.7
957	Henderson Inlet	47 09	122 50	+0 27	+0 45	+2.6	+0.2	10.0	14.0	8.0
959	Vaughn, Case Inlet	47 20	122 46	+0 35	+0 47	+2.8	+0.2	10.2	14.1	8.1
961	Allyn, Case Inlet	47 23	122 49	+0 27	+0 46	+2.8	+0.2	10.2	14.1	8.1
963	Walkers Landing, Pickering Passage	47 17	122 56	+0 35	+0 47	+2.9	+0.2	10.3	14.3	8.1
965	Arcadia, Pickering Passage	47 12	122 56	+0 35	+0 54	+3.0	+0.2	10.4	14.4	8.2
967	Shelton, Oakland Bay	47 13	123 05	+1 12	+1 54	+2.8	-0.2	10.6	14.2	7.9
969	Burns Point, Totten Inlet	47 07	123 03	+0 36	+0 54	+3.6	+0.2	11.0	15.0	8.5
971	Rocky Point, Eld Inlet	47 04	123 01	+0 34	+0 52	+3.3	+0.3	10.6	14.7	8.4
973	Dofflemyer Point, Budd Inlet	47 08	122 54	+0 29	+0 47	+3.1	+0.2	10.5	14.4	8.2
975	Olympia, Budd Inlet	47 03	122 54	+0 31	+0 46	+3.1	+0.2	10.5	14.4	8.2
	Possession Sound and Port Susan									
977	Mukilteo	47 57	122 18	-0 08	-0 12	-0.3	-0.1	7.4	11.0	6.4
979	Everett	47 59	122 13	-0 09	-0 11	-0.2	0.0	7.4	11.1	6.5
981	Tulalip	48 04	122 17	+0 02	-0 02	0.0	0.0	7.6	11.2	6.6
982	Kayak Point	48 08	122 22	-0 04	-0 01	-0.4	-0.1	7.3	10.9	6.3
983	Stanwood, Stillaguamish River <7>	48 14	122 22	+0 18	+2 10	*0.63	*0.29	5.7	7.4	3.6
	Saratoga Passage and Skagit Bay									
984	Greenbank, Whidbey Island	48 06	122 34	-0 01	-0 08	0.0	0.0	7.6	11.3	6.6
985	Coupeville, Penn Cove	48 13	122 41	+0 05	0 00	+0.2	0.0	7.8	11.5	6.7
987	La Conner, Swinomish Channel <8>	48 23	122 30	+0 22	+0 34	*0.88	*0.93	6.5	10.0	5.9
989	Ala Spit	48 24	122 35	+0 12	+0 27	*0.92	*0.92	6.9	10.5	6.1
991	Yokeko Point, Deception Pass	48 25	122 37	+0 21	+0 34	-0.9	-0.2	6.9	10.5	6.1
993	Cornet Bay, Deception Pass	48 24	122 37	+0 15	+0 26	*0.89	*0.89	6.6	10.2	6.0

Figure 2012 **Table 2 gives data for hundreds of subordinate stations so that predictions can be worked out for almost any point of significance to navigation.**

Table 3, Figure 2013, is provided in order that detailed calculations can be made for the height of the tide at any desired moment between the times of high and low waters. It is equally usable for either the reference stations in Table 1 or the subordinate stations of Table 2. Note that Table 3 is not a complete set of variations from one low to one high. Since the rise and fall are assumed to be symmetrical, only a half-table need be printed. Calculations are made from a high or low water, whichever is nearer to the specified time. In using Table 3 the nearest tabular values are used; interpolation is not necessary.

If the degree of precision of Table 3 is not required (it seldom is in practical piloting situations), a much simpler and quicker estimation can be made by using the following one-two-three rule of thumb. The tide may be assumed to rise or fall 1/12 of the full range during the first and sixth hours after high and low water stands, 2/12 during the second and fifth hours, and 3/12 during the third and fourth hours. The results obtained by this rule will suffice for essentially all situations and locations, but should be compared with Table 3 calculations as a check when entering new areas.

The *Tide Tables* also include four other minor tables which, although not directly related to tidal calculations, are often useful. Table 4 provides sunrise and sunset data at five-day intervals for various latitudes. Table 5 lists corrections to convert the local mean times of Table 4 to standard zone time. Table 6 tabulates times of moonrise and moonset for certain selected locations. Table 7 allows direct conversion of feet to meters. The inside back cover of the publication lists other useful data such as the phases of the moon, solar equinoxes, and solstices.

Tables 1 through 6 are each preceded by informative material which should be read carefully prior to use of the table concerned.

Tidal Effects on Vertical Clearances

The tide's rise and fall changes the vertical clearance under fixed structures such as bridges or overhead power cables. These clearances are stated on charts and in *Coast*

Pilots as heights measured from a datum which is *not* the same plane as used for depths and tidal predictions. The datum for heights is normally mean high water, the average of all high water levels. The use of this datum ensures that clearances and heights are normally greater than charted or *Coast Pilot* values.

It will thus be necessary to determine the height of MHW above the tidal datum. In any area where the tidal datum is mean lower low water, the plane of mean high water is above MLLW by an amount equal to the sum of the "mean tide level" plus one-half of the "mean range"; both of these values are listed in Table 2 of the *Tide Tables* for all stations.

If the tidal datum for any given locations is mean low water, mean high water is above MLW by the "mean tide range."

If the tide level at any given moment is below MHW, the vertical clearance under a bridge or other fixed structure is then greater than the figures shown on the chart; but if the tide height is *above* the level of MHW, then the clearance is *less*. Calculate the vertical clearance in advance if you anticipate a tight situation, but also observe the clearance gauges usually found at bridges. Clearances will normally be greater than the charted MHW values, but will occasionally be less.

Examples of Tidal Calculations

The instructions in the *Tide Tables* should be fully adequate for solving any problem. However, examples will be worked out here for various situations as guides to the use of the various individual tables. Comments and cautions relating to the solution of practical problems involving the *Tide Tables* will also be given.

EXAMPLE 1

Determining the time and height of a high or low tide at a reference station.

Problem What is the predicted time and height of the evening low tide at Seattle on Sunday, 3 October?

Solution Using Table 1, Figure 2010, it will be seen that the evening low water occurs at 2317 Pacific Standard Time on this date and that the height is 1.2 feet above the tidal datum of mean lower low water.

Notes 1. Observe that if the date had been 2 November, the tide level at low water stand would have been a *minus* figure, −1.7 feet. At this time, the water level is predicted to be *below* the datum and actual depths will be *less* than those printed on the charts of this area.

2. Observe that if the date had been 18 November, there

TABLE 3.—HEIGHT OF TIDE AT ANY TIME 191

Duration of rise or fall, see footnote	Time from the nearest high water or low water														
h. m.	*h. m.*	*h. m.*	*h. m.*	*h. m.*	*h. m.*	*h. m.*	*h. m.*	*h. m.*	*h. m.*	*h. m.*	*h. m.*	*h. m.*	*h. m.*	*h. m.*	*h. m.*
4 00	0 08	0 16	0 24	0 32	0 40	0 48	0 56	1 04	1 12	1 20	1 28	1 36	1 44	1 52	2 00
4 20	0 09	0 17	0 26	0 35	0 43	0 52	1 01	1 09	1 18	1 27	1 35	1 44	1 53	2 01	2 10
4 40	0 09	0 19	0 28	0 37	0 47	0 56	1 05	1 15	1 24	1 33	1 43	1 52	2 01	2 11	2 20
5 00	0 10	0 20	0 30	0 40	0 50	1 00	1 10	1 20	1 30	1 40	1 50	2 00	2 10	2 20	2 30
5 20	0 11	0 21	0 32	0 43	0 53	1 04	1 15	1 25	1 36	1 47	1 57	2 08	2 19	2 29	2 40
5 40	0 11	0 23	0 34	0 45	0 57	1 08	1 19	1 31	1 42	1 53	2 05	2 16	2 27	2 39	2 50
6 00	0 12	0 24	0 36	0 48	1 00	1 12	1 24	1 36	1 48	2 00	2 12	2 24	2 36	2 48	3 00
6 20	0 13	0 25	0 38	0 51	1 03	1 16	1 29	1 41	1 54	2 07	2 19	2 32	2 45	2 57	3 10
6 40	0 13	0 27	0 40	0 53	1 07	1 20	1 33	1 47	2 00	2 13	2 27	2 40	2 53	3 07	3 20
7 00	0 14	0 28	0 42	0 56	1 10	1 24	1 38	1 52	2 06	2 20	2 34	2 48	3 02	3 16	3 30
7 20	0 15	0 29	0 44	0 59	1 13	1 28	1 43	1 57	2 12	2 27	2 41	2 56	3 11	3 25	3 40
7 40	0 15	0 31	0 46	1 01	1 17	1 32	1 47	2 03	2 18	2 33	2 49	3 04	3 19	3 35	3 50
8 00	0 16	0 32	0 48	1 04	1 20	1 36	1 52	2 08	2 24	2 40	2 56	3 12	3 28	3 44	4 00
8 20	0 17	0 33	0 50	1 07	1 23	1 40	1 57	2 13	2 30	2 47	3 03	3 20	3 37	3 53	4 10
8 40	0 17	0 35	0 52	1 09	1 27	1 44	2 01	2 19	2 36	2 53	3 11	3 28	3 45	4 03	4 20
9 00	0 18	0 36	0 54	1 12	1 30	1 48	2 06	2 24	2 42	3 00	3 18	3 36	3 54	4 12	4 30
9 20	0 19	0 37	0 56	1 15	1 33	1 52	2 11	2 29	2 48	3 07	3 25	3 44	4 03	4 21	4 40
9 40	0 19	0 39	0 58	1 17	1 37	1 56	2 15	2 35	2 54	3 13	3 33	3 52	4 11	4 31	4 50
10 00	0 20	0 40	1 00	1 20	1 40	2 00	2 20	2 40	3 00	3 20	3 40	4 00	4 20	4 40	5 00
10 20	0 21	0 41	1 02	1 23	1 43	2 04	2 25	2 45	3 06	3 27	3 47	4 08	4 29	4 49	5 10
10 40	0 21	0 43	1 04	1 25	1 47	2 08	2 29	2 51	3 12	3 33	3 55	4 16	4 37	4 59	5 20

e, see footnote	Correction to height														
Ft.	*Ft.*	*Ft.*	*Ft.*	*Ft.*	*Ft.*	*Ft.*	*Ft.*	*Ft.*	*Ft.*	*Ft.*	*Ft.*	*Ft.*	*Ft.*	*Ft.*	*Ft.*
0.5	0.0	0.0	0.0	0.0	0.0	0.0	0.1	0.1	0.1	0.1	0.1	0.2	0.2	0.2	0.2
1.0	0.0	0.0	0.0	0.0	0.1	0.1	0.1	0.2	0.2	0.2	0.3	0.3	0.4	0.4	0.5
1.5	0.0	0.0	0.0	0.1	0.1	0.1	0.2	0.2	0.3	0.4	0.4	0.5	0.6	0.7	0.8
2.0	0.0	0.0	0.0	0.1	0.1	0.2	0.3	0.3	0.4	0.5	0.6	0.7	0.8	0.9	1.0
2.5	0.0	0.0	0.1	0.1	0.2	0.2	0.3	0.4	0.5	0.6	0.7	0.9	1.0	1.1	1.2
3.0	0.0	0.0	0.1	0.1	0.2	0.3	0.4	0.5	0.6	0.8	0.9	1.0	1.2	1.3	1.5
3.5	0.0	0.0	0.1	0.2	0.2	0.3	0.4	0.6	0.7	0.9	1.0	1.2	1.4	1.6	1.8
4.0	0.0	0.0	0.1	0.2	0.3	0.4	0.5	0.7	0.8	1.0	1.2	1.4	1.6	1.8	2.0
4.5	0.0	0.0	0.1	0.2	0.3	0.4	0.6	0.7	0.9	1.1	1.3	1.6	1.8	2.0	2.2
5.0	0.0	0.1	0.1	0.2	0.3	0.5	0.6	0.8	1.0	1.2	1.5	1.7	2.0	2.2	2.5
5.5	0.0	0.1	0.1	0.2	0.4	0.5	0.7	0.9	1.1	1.4	1.6	1.9	2.2	2.5	2.8
6.0	0.0	0.1	0.1	0.3	0.4	0.6	0.8	1.0	1.2	1.5	1.8	2.1	2.4	2.7	3.0
6.5	0.0	0.1	0.2	0.3	0.4	0.6	0.8	1.1	1.3	1.6	1.9	2.2	2.6	2.9	3.2
7.0	0.0	0.1	0.2	0.3	0.5	0.7	0.9	1.2	1.4	1.8	2.1	2.4	2.8	3.1	3.5
7.5	0.0	0.1	0.2	0.3	0.5	0.7	1.0	1.2	1.5	1.9	2.2	2.6	3.0	3.4	3.8
8.0	0.0	0.1	0.2	0.3	0.5	0.8	1.0	1.3	1.6	2.0	2.4	2.8	3.2	3.6	4.0
8.5	0.0	0.1	0.2	0.4	0.6	0.8	1.1	1.4	1.8	2.1	2.5	2.9	3.4	3.8	4.2
9.0	0.0	0.1	0.2	0.4	0.6	0.9	1.2	1.5	1.9	2.2	2.7	3.1	3.6	4.0	4.5
9.5	0.0	0.1	0.2	0.4	0.6	0.9	1.2	1.6	2.0	2.4	2.8	3.3	3.8	4.3	4.8
10.0	0.0	0.1	0.2	0.4	0.7	1.0	1.3	1.7	2.1	2.5	3.0	3.5	4.0	4.5	5.0

Figure 2013 **Table 3, Height of Tide at Any Time, is used to determine tide level at intermediate times between high and low tides.**

would have been no solution as the normal progression of the tides on a cycle of roughly 24 hours and 50 minutes has moved the normal evening low tide past midnight and into the next day. Thus Thursday, 18 November, has only three tides rather than the usual four.

3. Also note that the reference stations are listed in Table 2 as well and additional information is shown there. This table (Figure 2012) indicates, for reference stations as well as subordinate stations, the specific location for which predictions are given, the mean and spring (or diurnal) tidal ranges, and the mean tide level.

4. Remember to add one hour if you are in an area using daylight saving time.

EXAMPLE 2

Determining the time and height of a high or low tide at a subordinate station.

Problem What is the predicted time and height of the morning low water at Shelton on Oakland Bay, Puget Sound, on Sunday, 19 December?

Solution From the Index to Table 2 in the back of the *Tide Tables,* Shelton is found to be Subordinate Station No. 967. It is then located in Table 2 (Figure 2012). First, note that the reference station shown next *above* this place is "Seattle." Then note the time and height differences for Shelton for *low* waters; be sure to use the correct columns.

The differences are then applied as follows:

Time		*Height*
00 44	At Seattle	−0.6
+1:54	Differences	−0.2
02 38	At Shelton	−0.8

Thus, at Shelton on 19 December, the predicted morning low will occur at 0238 PST with a height of −0.8 feet (0.8 feet *below* the tidal datum).

Notes 1. If the date had been 3 October, the morning low water at Seattle for that date is at 1050. Adding the time difference of $1^h\ 54^m$ would have resulted in a time prediction at Shelton of 1244, which is *not a morning tide* at the subordinate station. In this case it would be necessary to use the low water at the reference station that occurs before midnight on the *preceding day.* At Seattle on 2 October, there is a low water at 2245; adding the time difference gives a morning low tide at Shelton of 0039 on 3 October.

2. Observe that if the subordinate station had been Point No Point, the height would be determined by multiplying the height at the reference station by the *ratio* of 0.92 rather than by adding or subtracting a difference in feet. (Ratios may be either greater or less than 1.) Note also that here the time differences are *negative;* the high and low waters at the subordinate station occur *earlier* than at the reference station of Table 1.

3. The use of minus time differences with predictions of tides to occur shortly after midnight at the reference station may result in a change of date *backward* into the preceding day.

EXAMPLE 3

Determining the level of the tide at a reference station at a given time between high and low waters.

Problem What is the predicted height of the tide at Seattle at 1840 on Thursday, 16 December?

Solution From Table 1, note that the given time of 1840 falls between a high tide at 1617 and a low tide at 2333. We compute the time difference and range as follows:

Time	*Height*
16 17	10.2 Feet
23 33	−1.2
7:16 Time difference	11.4 Feet range

Thus, our desired tide level is on a falling tide whose range is 11.4 feet (since the low water height is a negative value, we are subtracting a minus number, which is arithmetically equivalent to adding the numerical values).

The desired time is nearer to the time of the high water, so calculations will be made in Table 3 (Figure 2013), using this starting point.

18 40	Desired time
16 17	Time of nearest high or low water
2:23	Difference

The given time is $2^h\ 23^m$ after the nearest high water. Table 3 is used to the nearest tabulated value; do not interpolate. Entering the *upper* part of the table on the line for Duration of rise or fall of $7^h\ 20^m$ (nearest value to $7^h\ 12^m$) and read across to the entry nearest $2^h\ 23^m$; this is $2^h\ 27^m$ in the tenth column from the left. Follow *down* this column into the *lower* part of Table 3 to the line for Range of Tide of 11.5 feet (nearest value to actual range 11.4 feet). At the intersection of this line and column is found the correction to the height of the tide; here, 2.9 feet.

Since we have noted that in this example the tide is falling, and we are calculating from high water, the correction is subtracted from the height of high water: 10.2-2.9 = 7.3.

The predicted height of the tide at Seattle at 1840 PST on 16 December is 7.3 feet above the tidal datum.

Notes 1. Be sure that calculations are made for the right pair of high and low tides; be sure that the calculations are made for the *nearest* high *or* low water.

2. Be careful to apply the final correction to the nearest high or low water as used in its computation; do not apply it to the range; apply it in the right direction, down from a high or up from a low.

EXAMPLE 4

Determining the height of tide at a subordinate station at a given time.

Problem What is the height of the tide at Point No Point at 0900 on Tuesday, 2 November?

Solution First, the times of the high and low waters on either side of the stated time must be calculated for the subordinate station using Tables 1 and 2. This is done as follows:

High water	05 31	At Seattle	11.5 Feet
	−:16	Difference/Ratio	0.92
	05 15	At Point No Point	10.6 Feet
Low water	11 07	At Seattle	5.0 Feet
	−:16	Difference/Ratio	0.92
	10 51	At Point No Point	4.6 Feet

Next, we calculate the time difference and range:

10 51		10.6
05 15		4.6
5:36	Time difference	6.0 Feet range

The time from the nearest high or low is calculated

10 51	
09 00	
1:51	Time from nearest low

With the data from the above calculations, enter Table 3 for a duration of rise or fall of $5^h\ 36^m$ (use $5^h\ 40^m$), a time from nearest high or low of $1^h\ 51^m$ (use $1^h\ 53^m$), and a range of 6.0 feet (use 6.0 feet). From these data, we find a correction to height of tide of 1.5 feet. We know that the tide is falling and that the nearest time of stand was the low water at 1051. From these facts, we can see that the correction is to be added to the low water height; 4.6 + 1.5 = 6.1.

The height of the tide at Point No Point at 0900 PST on 2 November is predicted to be 6.1 feet above the datum.

Notes 1. Be sure to use the high and low tides occurring *at the subordinate station* on either side of the given time. It may be that in some instances with large time differences, a correction of times from the reference station to the subordinate station will show that you have selected the incorrect pair of tides; in this case select another high or low tide so that the pair used at the subordinate station will bracket the given time.

2. In this case the ratios for high and low waters were the same; this will not always be the case.

EXAMPLE 5

Determining the time of the tide reaching a given height at a reference station.

Problem At what time on the afternoon of 17 October will the height of the rising tide reach 10 feet at Seattle?

Solution This is essentially Example 3 in reverse. First, determine the range and duration of rise (or fall).

16 45	High water	11.2 Feet
10 50	Low water	3.3
5:55	Duration/Range	7.9 Feet

It is noted that the desired difference in height of tide is 11.2-10.0 = 1.2 feet. Enter the *lower* part of Table 3 on the line for a range of 8.0 feet (nearest value to 7.9) and find the column in which the correction nearest 1.2 feet is tabulated; in this case, the nearest value (1.3) is found in the ninth column from the left. Proceed *up* this column to the line in the upper part of the table for a Duration of $6^h\ 00^m$ (nearest value to $5^h\ 55^m$). The time from nearest high or low found on this line is $1^h\ 36^m$. Since our desired level is nearer to high water than low, this time difference is subtracted from the time of high water; 1645-1:36 = 1509.

Thus the desired tidal height of 10 feet above datum is predicted to occur at 1509 on 17 October.

Note A similar calculation can be made for a subordinate station by first determining the applicable high and low water times and heights at that station.

EXAMPLE 6

Determination of vertical clearance.

Problem What will be the vertical clearance under the fixed west span of a bridge across a wide canal near Ala Split, Washington, at the time of morning high tide on 2 December?

Solution Chart 18445 states the clearance to be 35 feet. The datum for heights is mean high water; the tidal datum is mean lower low water.

From Tables 1 and 2, we determine the predicted height of the tide at the specified time.

At Seattle	12.7 Feet
Ratio	0.92
At Ala Split	11.7 Feet

From Table 2, we calculate the height of mean high water for Ala Split:

Mean tide level	5.9 Feet
½ Mean range	3.4
Mean high water	9.3 Feet above tidal datum

The difference between predicted highwater at the specified time and MHW is 11.7-9.3=2.4 feet. The tide is above MHW and the bridge clearance is reduced, 35 − 2.4 = 33 feet, to nearest whole foot.

On the morning of 2 December, the clearance at high tide under the fixed west span of the bridge across the Canal near Ala Split is predicted to be 33 feet; this is *less* than the clearance printed on the chart.

Currents

CURRENT is the horizontal motion of water. It may result from any one of several factors, or from a combination of two or three. Although certain of these causes are more important to a boatman than others are, he should have a general understanding of all.

Tidal Currents

Boatmen in coastal areas will be affected most by TIDAL CURRENTS. The rise and fall of tidal levels is a result of the flow of water to and from a given locality. This flow results in tidal current effects.

The normal type of tidal current, in bays and rivers, is the REVERSING current that flows alternately in one direction and then the opposite. Off shore, tidal currents may be of the ROTARY type, flowing with little change in strength, but slowly and steadily changing direction.

A special form of tidal current is the HYDRAULIC type such as flows in a waterway connecting two bodies of water. Differences in the time and height of the high and

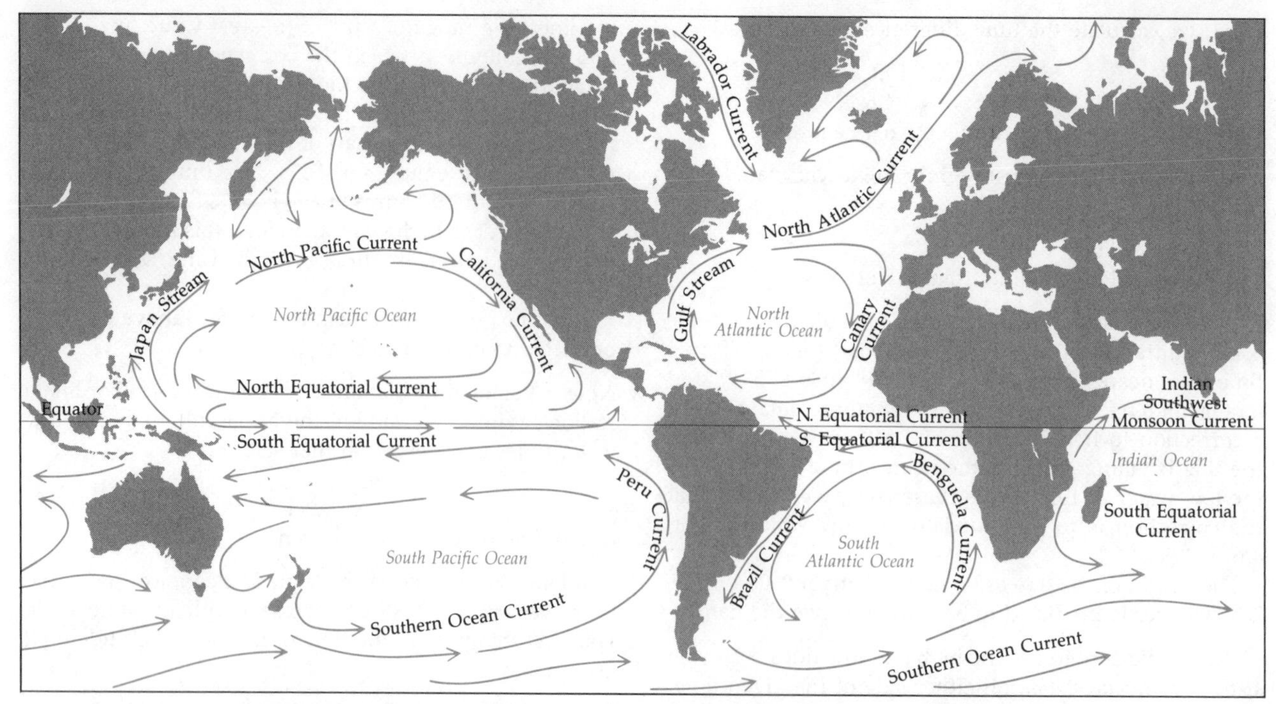

Figure 2014 **These are the major ocean currents of the world; the Gulf Stream off the East Coast and the California Current along the Pacific Coast are those that are of most interest to boaters in the United States, Canada, and Mexico.**

low waters of the two bays, sounds, or other tidal waters cause a flow from one to the other and back again. A typical example of hydraulic current is the flow through the Cape Cod Canal between Massachusetts Bay and Buzzards Bay.

River Currents

Boatmen on rivers above the head of tidal action must take into account RIVER CURRENTS. (Where tidal influences are felt, river currents are merged into tidal currents and are not considered separately.) River currents vary considerably with the width and depth of the stream, the season, and recent rainfall in the river basin; see Chapter 9 for details.

Ocean Currents

Offshore piloting will frequently require knowledge and consideration of OCEAN CURRENTS. These currents result from relatively constant winds such as the "trade winds" and "prevailing westerlies." The rotation of the earth and variations in water density are also factors in the patterns of ocean currents.

The ocean currents of greatest interest to boatmen are the Gulf Stream and the California Current; Figure 2014. The Gulf Stream is a northerly and easterly flow of warm water along the Atlantic Coast of the United States. It is quite close to shore along the southern part of Florida, but moves progressively further to sea as it flows northward, where it both broadens and slows.

The California Current flows generally southward and a bit eastward along the Pacific Coast of Canada and the United States, turning sharply westward off Baja California (Mexico). It is a flow of colder water and, in general, is slower and less sharply defined than the Gulf Stream.

Wind-Driven Currents

In addition to the consistent ocean currents caused by sustained wind patterns, temporary conditions may create local WIND-DRIVEN CURRENTS. Wind blowing across the sea causes the surface water to move. The extent of this effect varies with many factors, but generally a steady wind for 12 hours or longer will result in a discernible current.

For a rough rule of thumb, the strength of a wind-driven current can be taken as 2% of the wind's velocity. The direction of the current will *not* be the same as that of the wind, a result of the earth's rotation. In the Northern Hemisphere, the current will be deflected to the right to a degree determined by the latitude and the depth of the water. The deflection may be as small as 15 degrees in shallow coastal areas, or as great as 45 degrees on the high seas; it is greater in the higher latitudes.

Definitions of Current Terms

Currents have both strength and direction. The proper terms should be used in describing each of these characteristics.

The SET of a current is the direction *toward* which it is flowing. A current that flows from north to south is termed a southerly current and has a set of 180 degrees. (Note the difference here from the manner in which wind direction is described—it is exactly the opposite: a wind from north to south is called a northerly wind with a direction of 000 degrees.)

The DRIFT of a current is its speed, normally in knots, except for river currents, which are in MPH (1 knot = 1.15 MPH = 1.85 km/hr). Current drift is stated to the nearest tenth of a knot.

A tidal current is said to FLOOD when it flows in from the sea and results in higher tidal stages. Conversely, a tidal current EBBS when the flow is seaward and water levels fall.

Slack vs. Stand

As these currents reverse, there are brief periods of no discernible flow, called SLACK, or SLACK WATER. The time of occurrence of slack is *not* the same as the time of STAND, when vertical rise or fall of the tide has stopped. Tidal currents do *not* automatically slack and reverse direction when tide levels stand at high or low water. High water at a given point simply means that the level there will not get any higher. Further up the bay or river, the tide will not have reached its maximum height and water must therefore continue to flow in so that it can continue to rise. The current can still be flooding after stand has been passed at a given point and the level has started to fall.

For example, let us consider the tides and currents on Chesapeake Bay. High tide occurs at Baltimore some seven hours after it does at Smith Point, roughly half way up the 140 miles from Cape Henry at the entrance to Baltimore. On a certain day, high water occurs at 1126 at Smith Point, but slack water does not occur until 1304. The flooding current has thus continued for 1^h 38^m after high water was reached.

Corresponding time intervals occur in the case of low water stand and the slack between ebb and flood currents.

In many places, the time lag between a low or high water stand and slack water is not a matter of minutes but of hours. At the Narrows in New York Harbor, flood current continues for about two hours after high water is reached and the tide begins to fall, ebb for roughly 2½ hours after low water stand. After slack, the current increases until midflood or mid-ebb, then gradually decreases. Where ebb and flood last for about six hours—as along the Atlantic seaboard—current will be strongest about three hours after slack. Thus, the skipper who figures his passage out through the Narrows from the time of high water, rather than slack, will start about two and a half hours too soon and will run into a current at nearly its maximum strength.

Current Effects

Effect on Course and Speed Made Good

A current directly in line with a boat's motion through the water will have a maximum effect on the speed made good, but with no off-course influence. The effect can be of real significance in figuring your ETA at your destination. It can even affect the safety of your boat and its crew if you have figured your fuel too closely and run into a bow-on current.

A current that is nearly at a right angle to your course through the water will have a maximum effect on the course made good and a minor effect on the distance you must travel to reach your destination.

Knowledge of current set and drift can assist your cruising. Select departure times to take advantage of favorable currents, or at least to minimize adverse effects. A 12-knot boat speed and a 2-knot current, reasonably typical situations, can combine to result in either a 10-knot or a 14-knot speed made good—the 40% gain of a favorable current over an opposing one is significant in terms of both time and fuel. For slower craft the gains are even greater—50% for a 10-knot boat, 67% for an 8-knot craft, and 100% for one making only 6 knots.

Even lesser currents have some significance. A half-knot current would hinder a swimmer and make rowing a boat noticeably more difficult. A one-knot current can seriously affect a sailboat in light breezes.

Difficult Locations

In many boating areas there are locations where current conditions can be critical.

Numerous ocean inlets are difficult or dangerous in certain combinations of current and onshore surf. In general, difficult surf conditions will be made more hazardous by an outward-flowing (ebbing) current. The topic of inlet seamanship is covered in more detail on page 218.

In many narrow bodies of water, the maximum current velocity is so high that passage is impossible for boats of limited power, and substantially slowed for boats of greater engine power; Figure 2015. Such narrow passages are particularly characteristic of Pacific Northwest boating areas, but do occur elsewhere. Currents in New York City's East River reach 4.6 knots, and they are more than 5 knots at the Golden Gate of San Francisco. Velocities of 3½ to 4 knots are common in much-traveled passages like Woods Hole, Massachusetts, and Plum Gut, at the eastern end of Long Island, New York

Tidal Current Predictions

Without experience or official information, local current prediction is always risky. East of Badgers Island in Portsmouth, Maine, for example, the average ebb current flows at a maximum of less than a half knot. Yet southwest of the same island it averages 3.7 knots.

One rule is fairly safe for most locations—the ebb is stronger and lasts longer than the flood. Eighty percent of all reference stations on the Atlantic, Gulf, and Pacific coasts of the U.S. report currents stronger at the ebb. This is normal because river flow adds to the ebb, but hinders the flood.

On the Atlantic Coast, expect to find two approximately equal flood currents and two similar ebb currents in a cycle of roughly 25 hours. On the Pacific Coast, however, two floods and ebbs differ markedly. On the Gulf Coast, there may be just one flood and one ebb in 25 hours. In each case these patterns are, of course, generally similar to tidal action in their respective areas.

Don't try to predict current velocity from the time that it takes a high tide to reach a given point from the sea's entance. Dividing the distance from Cape Henry to Baltimore by the time that it takes high water to work its way up Chesapeake Bay gives a speed of 13 knots. True maximum flood current strength is only about one knot.

Another useful fact about tidal currents is that tidal currents at different places *cannot* be forecast from their tidal ranges. You would expect strong currents at Eastport, Maine, where the difference between successive high and low waters reaches as much as 20 feet. And you would be right; there are three-knot currents there. But Galveston, Texas, with only a two-foot range of tides, has cur-

Figure 2015 **In some narrow passages, tidal currents are so strong that a lightly powered boat could not make headway against them, and must wait for slack water. Here, an auxiliary races through Cobscook Falls in Maine, propelled by a 10-knot "downhill" current.**

rents up to more than two knots. So has Miami with a three-foot range, and Charleston, South Carolina, with a six-foot range—these are stronger currents than Boston, where the range is often more than 10 feet and as strong as at Anchorage, Alaska, where it's as much as 35 feet from some highs to the next low.

A good forecasting rule for all oceans: expect strong tidal currents where two bays meet. The reason: tidal ranges and high water times in the two bodies of water are likely to be different.

For the coasting skipper, here is another tidal current fact that may be useful near the beach: flood and ebb don't usually set to and from shore, but rather parallel with it. This is as true off New Jersey and Florida as it is off California and Oregon. A few miles offshore, however, and in some very large bays, the current behaves quite differently—the rotary current mentioned previously in this chapter.

Tidal Current Tables

At any given place, current strength varies with the phases of the moon and its distance from the earth. It will be strongest when tidal ranges are greatest—near new and full moon—and weakest when tidal ranges are least—near first and last quarters. Current speed may vary as much as 40% above and below its average value.

The relationship between currents and tides makes possible the prediction of tidal currents. NOS publishes two volumes of predictions annually; one covers the Atlantic Coast of North America and the other the Pacific Coast of North America and Asia.

Each volume includes predictions of tidal currents in bays, sounds, and rivers, plus ocean currents such as the Gulf Stream. General information on wind-driven currents is also included, although these of course result from temporary, local conditions and so cannot be predicted a year or more ahead. Your own past experience and local knowledge are the best source of information about how storm winds affect local waters.

TIDAL CURRENT TABLES are available at authorized NOS sales agents, and from the Superintendent of Documents, Government Printing Office, Washington, DC 20402.

Description of Tables

The format and layout of the *Tidal Current Tables* is much the same as for the *Tide Tables* discussed earlier in this chapter. A system of reference stations, plus constants and differences for subordinate stations, is used to calculate the predictions for many points.

Table 1 There are 22 reference stations in the Atlantic Coast volume and 36 for the Pacific Coast; the Gulf of Mexico is included in the Atlantic Coast volume. For each station, there are tabulated the predicted times and strengths of maximum flood and ebb currents, plus the times of slack water. The direction of the flood and ebb currents is also listed, see Figure 2016.

Table 2 Time differences and velocity ratios are listed for hundreds of subordinate stations. Figure 2017. Following the station number and descriptive location, there is the latitude and longitude to the nearest minute. Time differences are given for maximum flood and ebb, and for minimum current (usually slack) before flood and before ebb. Speed ratios are tabulated for maximum current in both directions (given in degrees, true, for direction *toward* which current flows). Also listed are average speeds and directions, including speed at "slack" as currents do not always decrease fully to zero velocity. A few stations will have only the entry "Current weak and variable"; this information, even though negative in nature, is useful in planning a cruise. A number of "endnotes" are used to explain special conditions at various stations; these are found at the end of Table 2.

Table 3 This table, Figure 2018, provides a convenient means for determining the current's strength at times intermediate between slack and maximum velocity. Use nearest tabulated values without interpolation.

68 CHESAPEAKE BAY ENTRANCE, VIRGINIA

F-Flood, Dir. 305° True E-Ebb, Dir. 125° True

SEPTEMBER								OCTOBER							
Day	Slack Water Time h.m.	Maximum Current Time h.m.	Maximum Current Vel. knots	Day	Slack Water Time h.m.	Maximum Current Time h.m.	Maximum Current Vel. knots	Day	Slack Water Time h.m.	Maximum Current Time h.m.	Maximum Current Vel. knots	Day	Slack Water Time h.m.	Maximum Current Time h.m.	Maximum Current Vel. knots
1		0142	1.2E	16		0153	1.6E	1		0141	1.4E	16		0211	1.7E
W	0502	0715	0.6F	Th	0508	0741	1.0F	F	0456	0726	0.9F	Sa	0528	0814	1.2F
	0939	1340	1.5E		1023	1405	1.9E		1003	1350	1.6E		1107	1437	1.7E
	1702	1947	1.0F		1725	2013	1.4F		1712	1951	1.1F		1758	2034	1.1F
	2253				2318				2244				2321		
2		0218	1.3E	17		0238	1.7E	2		0216	1.5E	17		0252	1.7E
Th	0534	0754	0.7F	F	0551	0829	1.2F	Sa	0529	0802	1.0F	Su	0606	0851	1.2F
	1021	1417	1.5E		1115	1453	1.9E		1046	1429	1.6E		1152	1520	1.7E
	1740	2024	1.1F		1813	2058	1.3F		1751	2027	1.1F		1841	2111	1.0F
	2326				2357				2316				2352		
3		0252	1.3E	18		0319	1.7E	3		0249	1.5E	18		0328	1.6E
F	0606	0828	0.8F	Sa	0633	0912	1.2F	Su	0602	0843	1.2F	M	0644	0932	1.2F
	1102	1455	1.6E		1204	1539	1.8E		1130	1507	1.7E		1233	1601	1.6E
	1817	2059	1.1F		1859	2138	1.2F		1831	2103	1.1F		1923	2145	0.9F
	2356								2348						
4		0323	1.4E	19	0032	0359	1.7E	4		0325	1.6E	19	0021	0406	1.6E
Sa	0638	0907	0.9F	Su	0713	0956	1.2F	M	0639	0922	1.3F	Tu	0723	1008	1.2F
	1142	1531	1.6E		1249	1622	1.7E		1215	1549	1.7E		1313	1642	1.4E
	1855	2132	1.1F		1944	2216	1.1F		1912	2141	1.1F		2005	2220	0.7F
5	0027	0356	1.4E	20	0105	0439	1.6E	5	0022	0358	1.7E	20	0049	0442	1.5E
Su	0712	0944	1.0F	M	0755	1037	1.1F	Tu	0718	1005	1.3F	W	0802	1046	1.1F
	1224	1609	1.6E		1333	1706	1.6E		1302	1634	1.6E		1352	1723	1.3E
	1934	2210	1.1F		2030	2254	0.9F		1957	2224	1.0F		2049	2257	0.6F
6	0057	0431	1.5E	21	0136	0518	1.5E	6	0057	0439	1.7E	21	0117	0520	1.3E
M	0748	1025	1.1F	Tu	0837	1118	1.1F	W	0803	1050	1.4F	Th	0845	1125	1.0F
	1309	1650	1.6E		1417	1754	1.4E		1353	1722	1.5E		1432	1809	1.1E
	2017	2249	1.0F		2117	2334	0.7F		2047	2308	0.9F		2137	2338	0.5F

Figure 2016 **In Table 1 of the *Tidal Current Tables,* the daily predictions are listed for times and strengths of maximum flood and ebb currents, and time of slack water, at major reference stations for the calendar year.**

TABLE 2. - CURRENT DIFFERENCES AND OTHER CONSTANTS

NO.	PLACE	METER DEPTH	POSITION Lat.	POSITION Long.	TIME DIFFERENCES Min. before Flood	TIME DIFFERENCES Flood	TIME DIFFERENCES Min. before Ebb	TIME DIFFERENCES Ebb	SPEED RATIOS Flood	SPEED RATIOS Ebb	Minimum before Flood knots	Minimum before Flood deg.	Maximum Flood knots	Maximum Flood deg.	Minimum before Ebb knots	Minimum before Ebb deg.	Maximum Ebb knots	Maximum Ebb deg.
		ft	° ' N	° ' W	h. m.	h. m.	h. m.	h. m.			knots	deg.	knots	deg.	knots	deg.	knots	deg.
	DEL., MD. and VA. COAST Time meridian, 75°W				on CHESAPEAKE BAY ENTRANCE, p.64													
3100	Cape Charles, 70 miles east of..........		37 05	74 51	See table 5.													
3105	Smith Island Shoal, southeast of........	7	37 05.3	75 43.5	-2 14	-2 12	-2 04	-2 05	0.3	0.3	0.0	- -	0.3	298	0.0	- -	0.4	068
3110	Chesapeake Light, 4.4 miles northeast of		36 59	75 42	See table 5.													
3115	Cape Henry Light, 2.2 miles southeast of		36 53.9	75 58.7	-1 54	-1 18	-0 39	-1 41	1.0	0.6	0.0	- -	1.0	346	0.0	- -	0.9	165
	CHESAPEAKE BAY																	
3120	Cape Henry Light, 1 mile north of.......		36 56.4	76 00.5	+0 04	-0 25	-0 08	-0 25	1.1	1.3	0.0	- -	1.1	280	0.0	- -	2.0	090
3125	Cape Henry Light, 1.8 miles north of....		36 57.4	76 00.1	-0 23	-0 11	+0 10	-0 17	1.2	1.0	0.0	- -	1.2	292	0.0	- -	1.5	099
3130	CHESAPEAKE BAY ENTRANCE................	7	36 58.8	76 00.4	Daily predictions						0.0	- -	1.0	306	0.0	- -	1.5	126
3135	Cape Henry Light, 4.6 miles north of....		37 00.1	75 59.3	-1 05	-0 46	-0 10	-0 54	1.3	0.9	0.0	- -	1.3	294	0.0	- -	1.3	104
3140	Cape Charles Light, 9.5 mi. WSW of......	6	37 03.7	76 05.4	-0 12	+0 08	+0 32	-0 05	1.5	0.9	0.0	- -	1.5	319	0.0	- -	1.4	126
3145	Cape Henry Light, 8.3 mi. northwest of..		37 02.2	76 06.6	-0 22	-0 12	+0 16	-0 05	1.0	0.7	0.0	- -	1.0	329	0.0	- -	1.1	133
3150	Lynnhaven Roads.........................		36 55.1	76 04.9	-0 58	-0 37	-0 14	-0 41	0.8	0.6	0.0	- -	0.8	280	0.0	- -	0.9	070
3155	Lynnhaven Inlet bridge..................		36 54.4	76 05.6	-1 56	-2 05	-2 12	-3 01	0.6	0.9	0.0	- -	0.6	180	0.0	- -	1.4	000
	Chesapeake Bay Bridge Tunnel																	
3160	Chesapeake Beach, 1.5 miles north of.		36 56.69	76 07.33	-0 09	-0 07	-0 23	-0 31	0.8	0.6	0.0	- -	0.8	305	0.0	- -	0.9	100
3165	Thimble Shoal Channel...............		36 58.33	76 06.67	-0 53	-0 46	-0 24	-0 39	1.4	0.9	0.0	- -	1.4	310	0.0	- -	1.3	095
3170	Tail of the Horseshoe...............		36 59.57	76 06.20	-0 33	-0 25	-0 13	-0 59	0.9	0.7	0.0	- -	0.9	300	0.0	- -	1.0	110
3175	Middle Ground, channel west of.......		37 03.00	76 05.00	-0 10	-0 20	-0 36	+0 04	1.6	0.9	0.0	- -	1.6	335	0.0	- -	1.3	150
3180	Chesapeake Channel..................		37 02.50	76 04.33	-0 33	-0 17	+0 03	-0 12	1.8	1.0	0.0	- -	1.8	335	0.0	- -	1.5	145
3185	Fisherman Island, 3.2 miles WSW of...		37 04.00	76 02.25	-1 00	-1 07	-0 46	-1 07	1.2	1.1	0.0	- -	1.2	330	0.0	- -	1.6	135
3190	Fisherman Island, 1.4 miles WSW of...		37 04.78	76 00.25	-1 47	-0 57	-0 41	-1 33	1.8	0.7	0.0	- -	1.8	330	0.0	- -	1.1	140
3195	Fisherman I., 1.8 miles south of.....		37 03.58	75 58.77	-1 04	-1 00	-0 27	-1 24	1.6	0.9	0.0	- -	1.6	320	0.0	- -	1.4	120
3200	Fisherman I., 0.4 mile west of.......		37 05.57	75 59.33	-0 59	-1 03	-0 35	-1 13	2.0	1.3	0.0	- -	2.0	005	0.0	- -	2.0	175
3205	Fisherman I., 1.1 miles northwest of.		37 06.50	76 00.00	-1 17	-0 35	-0 06	-0 50	1.8	1.1	0.0	- -	1.8	355	0.0	- -	1.6	165
3210	Cape Charles, off Wise Point.........	5	37 06.88	75 58.30	-0 29	-0 18	+0 27	+0 49	0.7	0.1	0.0	- -	0.7	305	0.0	- -	0.2	075
	Little Creek																	
3215	North of east jetty..................	10	36 56.05	76 10.60	-2 00	-2 02	-1 42	-1 59	0.9	0.7	0.0	- -	0.9	280	0.0	- -	1.0	076
3220	0.5 mile north of west jetty.........	10	36 56.32	76 10.81	-1 37	-1 03	-0 42	-1 31	0.9	0.6	0.0	- -	0.9	274	0.0	- -	0.9	108
3225	Old Plantation Flats Light, west of.....		37 14.0	76 04.1	+0 53	+1 06	+1 26	+0 35	1.2	0.9	0.0	- -	1.2	005	0.0	- -	1.3	175
3230	York Spit Channel.......................	7	37 12.9	76 08.5	+0 55	+0 55	+0 55	+0 55	0.8	0.7	0.0	- -	0.8	010	0.0	- -	1.1	195
3235	Wolf Trap Light, 0.5 mile west of.......		37 23.4	76 11.9	+1 05	+1 05	+1 05	+1 05	1.0	0.8	0.0	- -	1.0	015	0.0	- -	1.2	190
3240	Wolf Trap Light, 5.8 miles east of......		37 23.1	76 04.3	+1 45	+1 45	+1 45	+1 45	0.9	0.9	0.0	- -	0.9	015	0.0	- -	1.3	175
3245	Stingray Point, 5.5 miles east of.......		37 35.0	76 10.4	+1 50	+2 41	+2 52	+2 01	1.0	0.6	0.0	- -	1.0	343	0.0	- -	0.9	179
3250	Stingray Point, 12.5 miles east of......		37 33.8	76 02.3	+1 40	+2 05	+1 40	+2 05	1.0	0.5	0.0	- -	1.0	030	0.0	- -	0.8	175
3255	Smith Point, 4.5 miles east of..........		37 52.9	76 08.6	+3 11	+3 14	+3 14	+3 15	0.7	0.5	0.0	- -	0.7	352	0.0	- -	0.8	163

Figure 2017 **Table 2 shows current differences and other constants for hundreds of subordinate stations so that current predictions can be made at all these locations.**

Table 4 Although slack water is only a momentary event, there is a period of time on either side of slack during which the current is so weak as to be negligible for practical piloting purposes. This period, naturally, varies with the maximum strength of the current, being longer for weak currents. Two sub-tables, Figure 2019, predict the duration of currents from 0.1 to 0.5 knots by tenths for normal reversing currents and for the hydraulic currents found at certain specified locations.

Table 5 For the Atlantic Coast only, information is given on rotary tidal currents at various offshore points

TABLE A

Interval between slack and desired time	Interval between slack and maximum current													
	h. m. 1 20	*h. m.* 1 40	*h. m.* 2 00	*h. m.* 2 20	*h. m.* 2 40	*h. m.* 3 00	*h. m.* 3 20	*h. m.* 3 40	*h. m.* 4 00	*h. m.* 4 20	*h. m.* 4 40	*h. m.* 5 00	*h. m.* 5 20	*h. m.* 5 40
h. m.	*f.*	*f.*	*f.*	*f.*	*f.*	*f.*	*f.*	*f.*	*f.*	*f.*	*f.*	*f.*	*f.*	*f.*
0 20	0.4	0.3	0.3	0.2	0.2	0.2	0.2	0.1	0.1	0.1	0.1	0.1	0.1	0.1
0 40	0.7	0.6	0.5	0.4	0.4	0.3	0.3	0.3	0.3	0.2	0.2	0.2	0.2	0.2
1 00	0.9	0.8	0.7	0.6	0.6	0.5	0.5	0.4	0.4	0.4	0.3	0.3	0.3	0.3
1 20	1.0	1.0	0.9	0.8	0.7	0.6	0.6	0.5	0.5	0.5	0.4	0.4	0.4	0.4
1 40	------	1.0	1.0	0.9	0.8	0.8	0.7	0.7	0.6	0.6	0.5	0.5	0.5	0.4
2 00	------	------	1.0	1.0	0.9	0.9	0.8	0.8	0.7	0.7	0.6	0.6	0.6	0.5
2 20	------	------	------	1.0	1.0	0.9	0.9	0.8	0.8	0.7	0.7	0.7	0.6	0.6
2 40	------	------	------	------	1.0	1.0	1.0	0.9	0.9	0.8	0.8	0.7	0.7	0.7
3 00	------	------	------	------	------	1.0	1.0	1.0	0.9	0.9	0.8	0.8	0.8	0.7
3 20	------	------	------	------	------	------	1.0	1.0	1.0	0.9	0.9	0.9	0.8	0.8
3 40	------	------	------	------	------	------	------	1.0	1.0	1.0	0.9	0.9	0.9	0.9
4 00	------	------	------	------	------	------	------	------	1.0	1.0	1.0	1.0	0.9	0.9
4 20	------	------	------	------	------	------	------	------	------	1.0	1.0	1.0	1.0	0.9
4 40	------	------	------	------	------	------	------	------	------	------	1.0	1.0	1.0	1.0
5 00	------	------	------	------	------	------	------	------	------	------	------	1.0	1.0	1.0
5 20	------	------	------	------	------	------	------	------	------	------	------	------	1.0	1.0
5 40	------	------	------	------	------	------	------	------	------	------	------	------	------	1.0

Figure 2018 **By using the appropriate part of Table 3, it is possible to determine current velocity at any time between slack and maximum strength at ebb or flood. Table A, shown, is used for all stations except Cape Cod Canal, Hell Gate, and the Chesapeake and Delaware Canal.**

of navigational interest. These points are described in terms of general location and specific geographic coordinates. Predictions of velocity and direction are referred to times after maximum flood at designated reference stations.

The inside back cover of the *Tidal Current Tables* has the same astronomical data as is found in the *Tide Tables;* see page 458.

Current Diagrams

For a number of major tidal waterways of the United States, *Tidal Current Tables* give CURRENT DIAGRAMS (such as Figure 2020)—graphic means for selecting a favorable time for traveling in either direction along these routes.

Time

The *Tidal Current Tables* list all predictions in *local standard time.* Be sure to make an appropriate conversion to daylight time if this is in effect. Subtract an hour from your watch time before using the tables, and add an hour to the results of your calculations.

Cautions

As with tidal predictions, the data in the *Tidal Current Tables* may often be upset by sustained abnormal local conditions such as winds or rainfall. Use the current predictions with caution during and immediately after such weather abnormalities.

Note also that tidal current predictions are generally for a *spot location only;* the set and drift may be quite different only a mile or less away. This is at variance from predictions of high and low tides, which can usually be used over fairly wide areas around the reference station.

TABLE A

Maximum current	Period with a velocity not more than—				
	0.1 knot	0.2 knot	0.3 knot	0.4 knot	0.5 knot
Knots	*Minutes*	*Minutes*	*Minutes*	*Minutes*	*Minutes*
1.0	23	46	70	94	120
1.5	15	31	46	62	78
2.0	11	23	35	46	58
3.0	8	15	23	31	38
4.0	6	11	17	23	29
5.0	5	9	14	18	23
6.0	4	8	11	15	19
7.0	3	7	10	13	16
8.0	3	6	9	11	14
9.0	3	5	8	10	13
10.0	2	5	7	9	11

Figure 2019 **Table 4 is used to determine duration of slack at a station. Although actual slack water is only momentary, current velocity is quite small for a significant period while current direction is reversing. Part A or Part B of the table is used in the same situations as for Table 3.**

Examples of Tidal Current Calculations

The *Tidal Current Tables* contain all information needed for determining such predicted conditions as the time of maximum current and its strength, the time of slack water, and the duration of slack (actually, the duration of the very weak current conditions). Examples and solutions for typical problems follow with comments and cautions.

EXAMPLE 1

Determining the time and strength of maximum current, and the time of slack, at a reference station.

Problem What is the time and strength of the maximum ebb current at Chesapeake Bay Entrance during the afternoon of 16 October?

Solution As with the *Tide Tables,* the answer is available for a reference station by direct inspection. Figure 2016 is a typical page from the *Tidal Current Tables;* we can see that the maximum ebb current on the specified afternoon is 1.7 knots setting 125° true; it is predicted to occur at 1437 EST (1537 EDT).

Problem What is the time of the first slack before flood at this station on 21 October?

Solution Table 1 does not directly identify the slacks as being "slack before ebb" or "slack before flood"; this must be determined by comparing the slack time with the nature of the next maximum current that occurs.

From Figure 2016, we can see that the earliest slack that will be followed by a flooding current at Chesapeake Bay Entrance on 21 October is predicted for 0845 EST (0945 EDT).

Notes 1. Times obtained from the *Tidal Current Tables* are *standard;* add one hour for daylight time if in effect.

2. The set of the current for a reference station is found at the *top* of the page in Table 1; it is also given in Table 2, which lists further information like the geographic coordinates of the station, and the average velocity of maximum flood and ebb currents.

3. The normal day at Chesapeake Bay Entrance, where the tide is of the semi-diurnal type, will have four slacks and four maximums. The tidal cycle of 24^h and 50^m will result in the occasional omission of a slack or maximum.

EXAMPLE 2

Determining the time and strength of maximum current, and the time of slack, at a subordinate station.

Problem What is the time and strength of the morning flood current in Lynnhaven Inlet at the bridge on 21 September?

Solution Table 2, Figure 2017, gives time differences and velocity ratios to be applied to the predictions at the appropriate reference station. There is an Index to Table 2 in the rear of the *Tidal Current Tables* if it is needed to locate the given subordinate station.

In this problem, the time difference and velocity ratio are applied as follows:

Time		*Velocity*
11 18	At Chesapeake Bay Entrance	1.1 Knot
−2:05	Difference/Ratio	0.6
09 13	At Lynnhaven Inlet	0.7 Knot

The set (direction) of the current is also noted from the appropriate column of Table 2; in this case, it is 180° true.

Thus the predictions are for a maximum current of 0.7 knot setting 180° true at Lynnhaven Inlet bridge at 0913 EST (actually used as 1013 EDT) on 21 September.

Problem What is the time of the first afternoon slack water at Lynnhaven Inlet bridge on 5 September?

Solution From Table 1, it is noted that the first afternoon slack for this date is a slack before flood begins. From Table 2, the time difference is found to be −1:56. This is applied to the time of slack at the reference station:

19 34	At Chespeake Bay Entrance
−1:56	Difference
17 38	At Lynnhaven Inlet bridge

The time of the first afternoon slack water at Lynnhaven Inlet bridge on 5 September is 1738 EST, or 1838 EDT. (The first afternoon slack at the reference station, at 1224, was *not* used as the application of the time difference made this a *morning* slack at the subordinate station.)

Notes Observe that Table 2 shows separate time differences for the four events of a tidal current cycle; occasionally two or more of these may be the same, but always carefully check the column headings and select the proper time difference. Note also that the speed ratios for maximum flood and ebb currents are normally different.

2. Note that the direction of the current at a subordinate station must be taken from Table 2. It will nearly always differ from that at the reference station. No statement of current is complete without giving direction as well as strength.

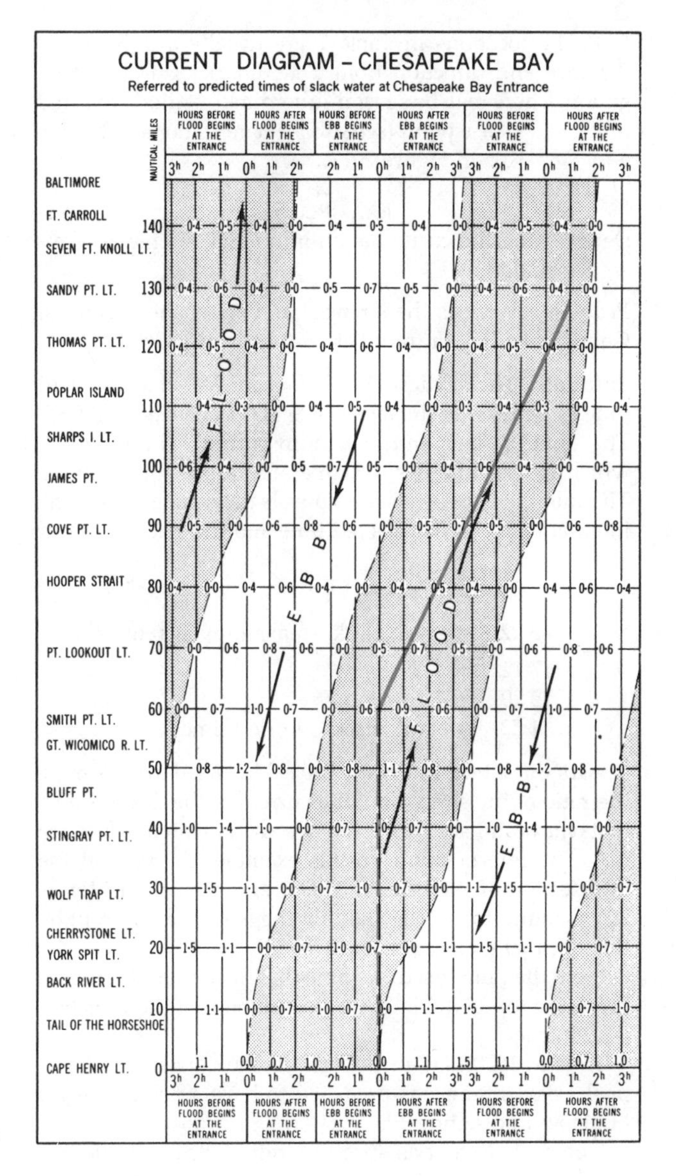

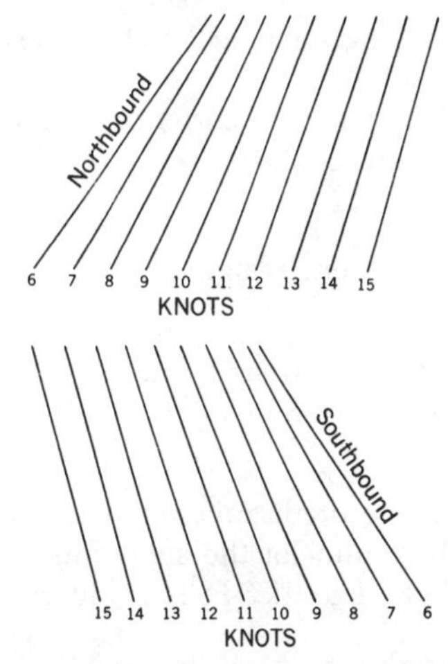

Figure 2020 **Current Diagrams from the *Tidal Current Tables* provide a quick and simple means of determining the best speed and time of departure for a run up or down the given waterway. The heavy dashed and solid lines are for the problem given in Example 6, page 468.**

3. The locations in Table 2 are usually a point some distance and direction from a landmark or aid to navigation. There may be several subordinate stations referred to the same base point (see subordinate stations 3120-3145 and 3185-3205 in Figure 2017).

EXAMPLE 3

Determining the current at an intermediate time at a reference station.

Problem What is the strength and set of the current at Chesapeake Bay Entrance at 1620 EST on 1 October?

Solution The daylight time is first converted to standard time for entry into the table; 1620 EDT becomes 1520 EST. The time of slack and maximum current (ebb or flood) which bracket the desired time are found from Table 1. The interval between these times is determined, as is the interval between the desired time and the slack.

17 12
13 50
3:22 Interval, slack—maximum current

17 12
15 20
1:52 Interval, slack—desired time

With these time intervals, Table 3A is used to determine the ratio of the strength of the current at the desired time to its maximum strength. The nearest tabulated values are used, no interpolation. In this example, the ratio at the intersection of the line for $2^h\ 00^m$ and the column for $3^h\ 20^m$ is found to be 0.8. Multiply the maximum current by this decimal factor, $1.6 \times 0.8 = 1.3$.

From the times used, we note that the current is ebbing. From the top of Table 2, we determine that the direction is 125° true.

Thus at 1620 EDT on 1 October, the current at Chesapeake Bay Entrance is predicted to have a velocity of 1.3 knots and be setting 125° true.

Notes 1. Except as specially indicated, use Table 3A, the upper portion of Table 3. The lower (B) portion is for use in designated waterways *only*.

2. Be sure that the interval is calculated between the desired time and the time of *slack*, whether or not this time is nearer to the given time than the time of maximum current.

3. Note that calculations of current strength are rounded to the nearest tenth of a knot.

EXAMPLE 4

Determining the current at an intermediate time at a subordinate station.

Problem What is the strength and set of the current at a point 5½ miles east of Stingray Point at 0935 EDT on 16 October?

Solution First, the predictions for slack and maximum current must be found for the subordinate station after converting 0935 EDT to 0835 EST for entering the tables.

Slack		*Maximum*	
05 28	At Chesapeake Bay Entrance	08 14	1.2 Knots, Flood
+1:50	Difference/Ratio	+2:41	1.0
07 18	At subordinate station	10 55	1.2

With the information developed above, and the desired time, further calculations are made as follows:

10 55
07 18
3:37 Interval, slack—maximum current

08 35 (EST)
07 18
1:17 Interval, slack—desired time

Using Table 3A, Figure 2018, the speed ratio is found to be 0.5; then $1.2 \times 0.5 = 0.6$ from Table 2, the direction is seen to be 343° true.

The current at 0935 EDT at a point 5½ miles east of Stingray Point on 16 October is predicted to be 0.6 knot setting 343° true.

Note Calculations for the strength at an intermediate time of a hydraulic current, such as in the Cape Cod Canal or at Hell Gate in East River, New York City, are handled exactly as above, except that the lower (B) part of Table 3 is used.

EXAMPLE 5

Determining the duration of slack (weak current) at a designated point.

Problem For how long will the current be less than 0.4 knots around the time of first slack before ebb 2 October at Chesapeake Bay Entrance?

Solution From Table 1 (Figure 2016), for this date, the maximum currents on either side of this slack (at 0529) are 1.5 knots ebb at 0216, and 1.0 knots flood at 0802.

Using Table 4A (Figure 2019), the duration of current less than 0.4 knot is determined for each maximum. One-half of each such duration is used for the period from 0.4 to 0 knots and then 0 to 0.4 knots.

The value for the ending ebb current is one half of 62, or 31 minutes; for the beginning flood current, it is one-half of 94, or 47 minutes.

On 2 October at Chesapeake Bay Entrance, the predicted duration of a current less than 0.4 knots is 31 + 47 = 78 minutes, from 0529-31 = 0458 until 0529 + 47 = 0616.

Notes 1. If maximum strengths of the ebb and flood currents are essentially the same, only one figure need be taken from Table 4. Interpolate as necessary.

2. Use the A or B portion of Table 4 in the same manner as for Table 3.

EXAMPLE 6

Use of a Current Diagram.

Problem For an afternoon run up Chesapeake Bay from Smith Point to Sandy Point Light at ten knots on 21 October, what time should you depart from Smith Point for the most favorable current conditions?

Solution Use the current diagram for Chesapeake Bay and a graphic solution, Figure 2020. Draw a line on the diagram parallel to the ten-knot northbound speed line so that it fits generally in the center of the shaded area marked "Flood." Project downward from the intersection of this line with the horizontal line marked Smith Point Light to the scale at the bottom of the diagram. The mark here is "0^h after ebb begins at the entrance."

Referring to Table 1, it will be seen that on the given date, the afternoon slack before ebb occurs at 1432.

For a run up Chesapeake Bay to Sandy Point Light at ten knots on the afternoon of 21 October, it is predicted that the most favorable current conditions will be obtained if you leave Smith Point Light at about 1432 EST or 1532 EDT.

Notes 1. Similar solutions can be worked out for southbound trips, but on longer runs you will probably be faced with both favorable and unfavorable current conditions. Choose a starting time to minimize adverse conditions.

2. Conditions shown on Current Diagrams are averages and for typical conditions; small variations should be expected in specific situations.

Tidal Current Charts

NOS also publishes a series of TIDAL CURRENT CHARTS; see Figure 2021. These are available for 12 bodies of water, from Boston Harbor around to Puget Sound (see complete listing on page 390).

Tidal Current Charts are made up in the form of a series of 12 reproductions of a small-scale chart of the area. Each chart depicts the direction and strength of the current for a specific time in terms of hours after the predicted time of the beginning of flood or the beginning of ebb at the appropriate reference station. (*Tidal Current Charts* for Narragansett Bay are used with *Tide Tables* rather than *Tidal Current Tables*.) The currents in various passages and portions of the body of water are indicated by arrows and numbers. By following through the sequence of the charts, the hourly changes in strength and direction are easily seen.

These charts must be used with caution as tidal current strengths may vary widely between points separated by only a short distance.

Tidal Current Charts are updated when required by new surveys; use only the latest edition.

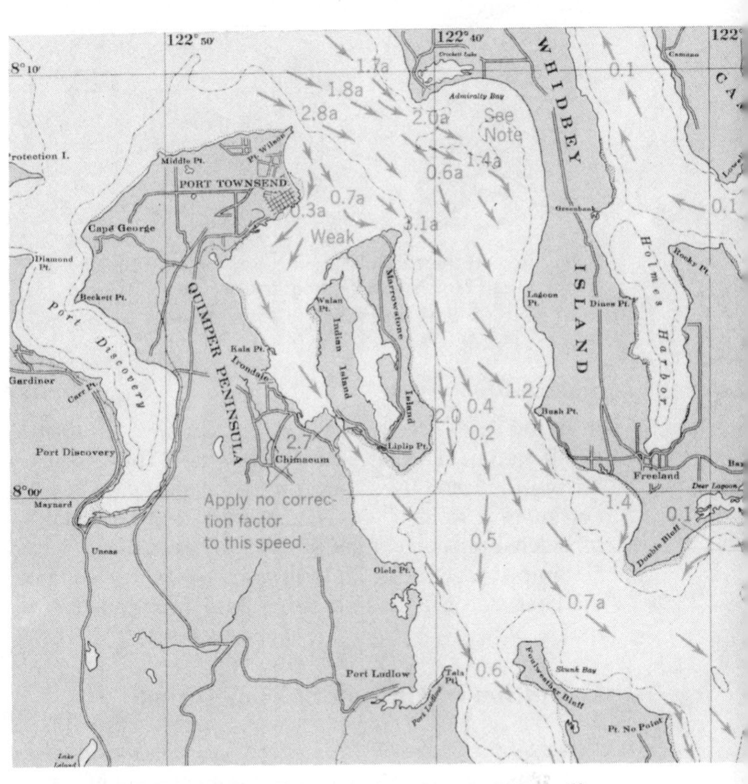

Figure 2021 ***Tidal Current Charts* consist of a set of 12 small-scale chartlets showing currents existing at hourly intervals throughout a complete cycle of flood and ebb.**

Tidal Current Diagrams

A development of the computer age, TIDAL CURRENT DIAGRAMS are a series of 12 monthly graphs to be used with *Tidal Current Charts* instead of the *Tidal Current Tables*. The diagram method is more convenient as the graphs indicate directly, from the date and time, the chart to be used and the speed correction factor to be applied.

Tidal Current Diagrams are available only for use with the *Tidal Current Charts* of Boston Harbor, New York Harbor, Upper Chesapeake Bay, and Block and Eastern Long Island sounds.

Supplementary Sources of Information

The ultimate source of current information is your own eyesight and past experience. These are most helpful even where there are predictions from the NOS tables. As noted before, the tabular data are to be expected under "normal" conditions, and may be easily upset by unusual circumstances. Strong winds will, for example, drive water into or out of bays and modify tidal levels and currents.

Currents and Piloting

One of the most interesting problemss in small-boat piloting is the matter of currents, their effect upon boat speed, the determination of courses whicvh must be steered to make good a desired track, and the time required to reach a destination. This is known as CURRENT SAILING.

As a boat is propelled and steered through the water, it moves with respect to it. At the same time the water may be moving with respect to the bottom and the shore because of current. The resultant motion of the boat is the net effect of these two motions combined, with regard to both speed and direction. The actual course made good over the bottom will not be the same as the DR track, in terms of either course or speed.

The importance of tidal currents should not be underestimated. Unexpected current is always a threat to the skipper because it can carry his boat off course, possibly into dangerous water. The risk is greater with slower boat speeds and under conditions of reduced visibility. A prediction of current effect can be added to a plot of a DR track to obtain an ESTIMATED POSITION (EP), plotted as a small square with a dot in the center, Figure 2022.

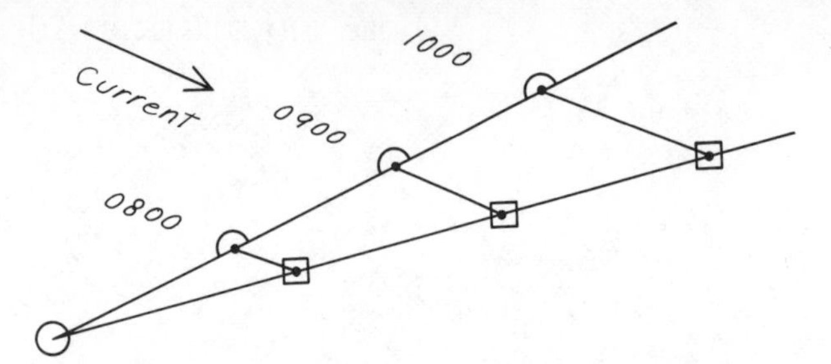

Figure 2022 **If the current is known, or can be estimated, a DR plot (half circles) can be modified to show a series of Estimated Positions (EPs—plotted as squares).**

Leeway

By definition, leeway is the leeward (away from the wind) motion of a vessel due to the wind. While sailboats are most affected by it, larger motorboats and yachts are not immune to its action. The wind's effect need not be considered separately from current, however; the two may be lumped together, with such factors as wave action on the boat, and the total offsetting influence termed "current."

Definition of Current Sailing Terms

The terms "Course" and "Speed" are used in DR plots for the motion of the boat *through the water* without regard to current. Now this most important influence will be studied and additional terms must be introduced.

The INTENDED TRACK is the expected path of the boat, as plotted on a chart, after consideration has been given to the effect of current.

TRACK, abbreviated as TR, is the direction (true) of the intended track line.

SPEED OF ADVANCE, SOA, is the intended rate of travel along the intended track line.

Note that the intended track will not always be the actual track, and so two more terms are needed.

COURSE OVER THE GROUND, COG, is the direction of the *actual* path of the boat, the track made good; sometimes termed "Course made good."

SPEED OVER THE GROUND, SOG, is the actual rate of travel along this track; sometimes termed "Speed made good."

Current Situations

A study of the effects of current resolves itself into two basic situations, as follows:

1. When the set of the current is in the same direction as the boat's motion, or is in exactly the opposite direction.
2. When the direction of the current is at an angle to the boat's course, either a right or an oblique angle.

The first situation is, of course, the simplest and most easily solved. The speed of the current, the DRIFT, is added to or subtracted from the speed through the water to obtain the speed over the ground. The course over the ground (or intended track) is the same as the DR course—COG equals C, as does TR.

Current Diagrams

When the boat's motion and the set of the current form an angle with each other, the solution for the resultant course and speed is more complex, but still not difficult. Several methods may be used, but a graphic solution using a CURRENT DIAGRAM is usually the easiest to understand.

Basically, a current diagram represents the two component motions separately, as if they occurred independently and sequentially, which, of course, they do not. These diagrams can be drawn in terms of velocities or distances. The former is easier and is usually used; current diagrams in this book will be drawn in terms of speed and labeled as shown in Figure 2023 top. If distances are plotted, be sure to use the same period of time for each component motion—one hour is commonly used since the units of distance will then be the same numerically as the units of speed.

Accuracy of Current Diagrams

The accuracy with which the resultant course and speed can be determined depends largely on the accuracy with which the current has been determined. Values of the current usually must be taken from tidal current tables or charts, or estimated by the skipper from visual observations.

Current diagrams may also be called "vector triangles of velocity." The term "vector" in mathematics means a quantity that has both magnitude and direction—directed quantities. In current sailing, the directed quantities are the motions of the boat and the water (the current).

Vectors

A vector may be represented graphically by an arrow, a segment of a straight line with an arrowhead indicating the direction, and the length of the line scaled to the speed, Figure 2023 bottom. If we specify that a certain unit of length is equal to a certain unit of speed—e.g., that 1 inch equals 1 knot—then two such vectors can represent

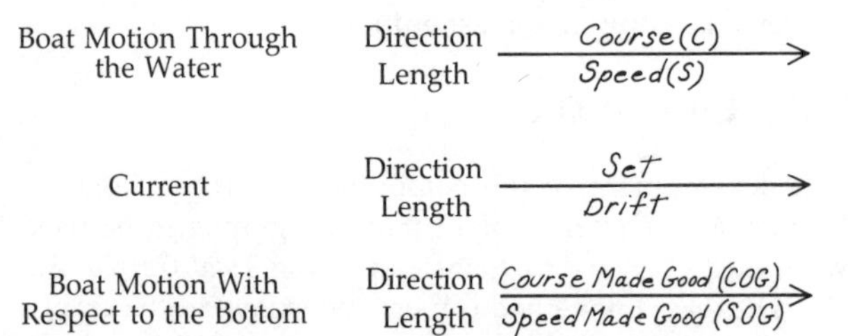

Figure 2023 **The component lines of a current diagram must always be correctly and completely labeled. Direction is shown above the line and velocity is shown below the line. Typical labels are shown *above*. A vector, *below*, is a line representing a directed quantity. It has direction as shown by the arrowhead, and magnitude, as indicated by its length. Any convenient scale may be selected to show this magnitude.**

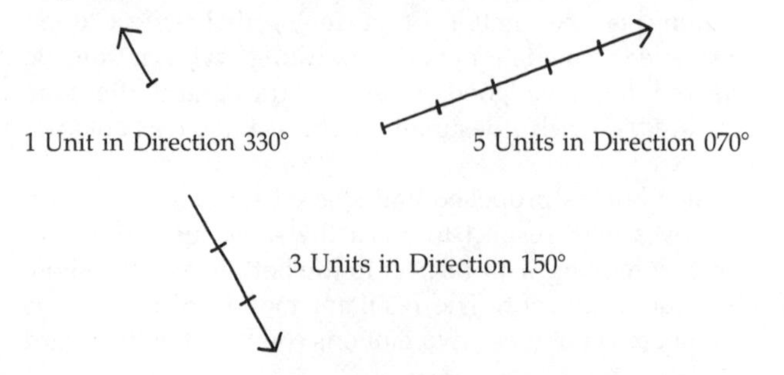

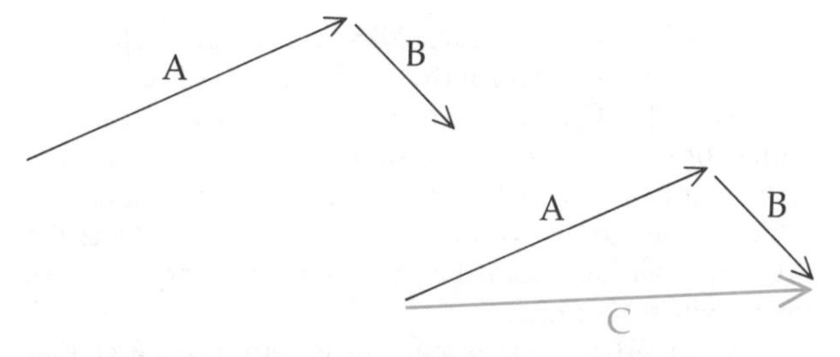

Figure 2024 **Vectors may be combined graphically to determine the vector sum. Typically, vectors A and B above, which represent component motions, are combined as vector C, the resultant motion of the two components.**

graphically two different velocities. Any speed scale may be used, the larger the better for accuracy. The size of the available paper and working space will normally control the scale.

Current diagrams may be drawn on a chart either as part of the plot or separately. They may also be drawn on plain paper; in this case, it is wise to draw in a north line as the reference for measuring directions.

Since boats are subject to two distinct motions—the boat through the water, and the water with respect to the bottom—we will now consider how the resultant motion, or vector sum, is obtained by current diagrams.

Vector Triangles

When the two motions are not in line with each other, they form two sides of a triangle, Figure 2024. Completing the triangle gives the third side which will be a vector representing the resultant motion or velocity. Thus, when any two velocity vectors are drawn to the same scale and form the sides of a velocity triangle, the third side will be the resultant velocity vector (the vector sum of the other two), and its direction and magnitude may be measured from the diagram.

It may be an aid to the visualization of the component motions if a time period of one hour is used and the points that are the corners of the triangle are considered as positions of the boat before and after certain motions, as follows:

- O The origin.
- DR The DR position of the boat as a result solely of its motion through the water.
- W The position of the boat solely as a result of the motion of the water.
- P The position (intended or actual) of the boat as a result of the combined action of the component motions.

It should be noted that, in some cases, two of the above letters are applicable to a position; it is customary to use only one.

"Tail-to-Head" Relationships Note carefully how the vectors for boat motion through the water and the current are drawn. These vectors are always drawn "tail-to-head," *not* so that both are directed out from the same source. (This rule applies only when one of the vectors is current; not when they both represent boat's motion.)

If both boat motion through the water and current are known, either may be drawn first from the origin; Figure 2025*B* will give the same resultant motion as Figure 2025*A*.

The Four "Cases" of Current Problems There are four typical current problems, different combinations of known and unknown factors. For convenience, we will call them Case 1, 2, 3, and 4.

Case 1

Known: Boat's course (C) and speed (S) through the water; current set and drift.

To be determined: The course and speed over the ground (COG and SOG).

This is the determination of what the effect of a known current will be if no allowance is made for it while running the DR course.

Case 2

Known: Boat's course (C) and speed (S) through the water; the course (COG) and speed over the ground (SOG).

To be determined: The set and drift of the current.

This is the determination of the nature of an unknown current from observation of its effect.

Case 3

Known: Boat's speed through the water (S), the set and drift of the current, and the intended track (TR).

To be determined: The course to be steered (C) and the speed of advance (SOA) along the intended track.

This is the determination of corrected course to be steered, but without regard for the effect on speed or ETA.

Case 4

Known: The set and drift of the current, the intended track (TR), and the speed of advance desired (SOA).

To be determined: The course to be steered (C) and speed to be run through the water (S).

This is the "rendezvous" or "contest" case where the destination and the time of arrival are specified.

Illustrative Examples of Current Problems

Case 1—the effect of current on a boat's course and speed. If the set of the current is as shown in Figure 2026, then the boat will be set off to the right of the direction in which

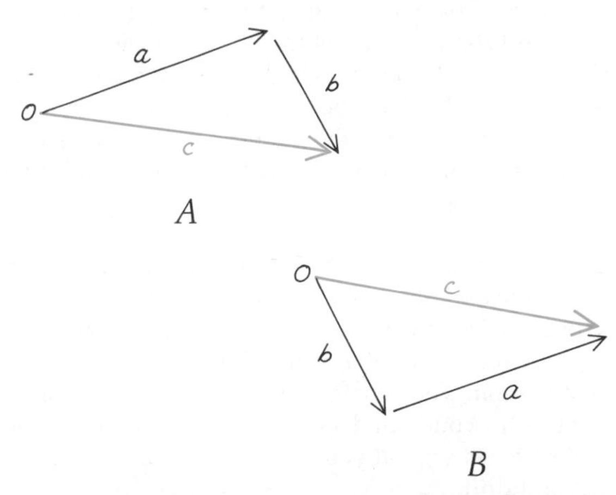

Figure 2025 **The *A* and *B* vector triangles here appear to be different because in *A* vector *a* was drawn before vector *b*; and in the *B* triangle, vector *b* was drawn before vector *a*. The resultant, vector, *c*, is the same for either procedure.**

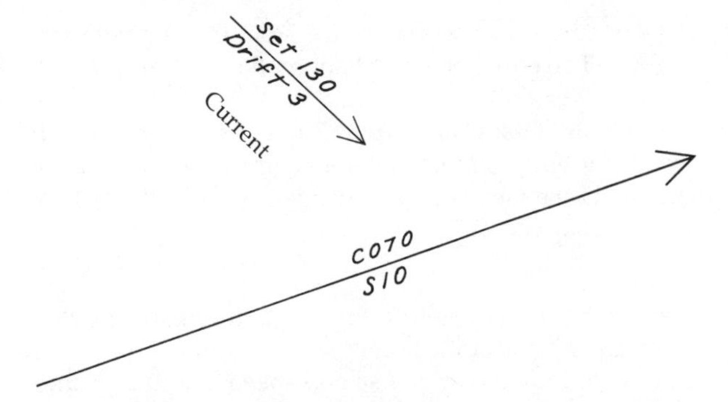

Figure 2026 **If a boat is traveling through the water on a course of 070°, and the current is setting in the direction 130°, a current diagram will show how far, and in what direction, the boat will be set off course to starboard.**

it is being steered. We will use a current diagram to determine the exact extent of this effect, the path the boat can be expected to follow, and its speed along that path (the intended track and speed of advance).

We first draw in a north line as a reference for measuring directions, Figure 2027*A*, and then the vector for the boat's speed through the water, O-DR, a line drawn from the origin in the direction 070° for a length of 10 units, Figure 2027*B*. Next we add the vector for current, DR-W, three units in the direction 130°, Figure 2027*C*. Note that we have observed the "tail-to-head" relationship rule. (Either of these two vectors could have been drawn first; the triangle would appear differently, but the result will be the same.) Because these vectors form two sides of a triangle of velocities, the third side O-W, Figure 2027*D*, is the resultant velocity at which the boat moves with respect to the bottom under the combined influences of its propulsion and the current. The point W can now be relabeled "P." The intended (or expected) track (TR) and the speed of advance (SOA) can be measured from the O-P line. In this example, it turns out that the boat can be expected to sail a course over the ground of 083° and to have a speed of advance of 11.8 knots. Note that the directions of these vectors are plotted as *true* directions.

Summarizing briefly, we have drawn vectors to indicate independently the motion of the boat from two different influences, its own propulsion and the current. Actually, of course, the boat will *not* go first from O to DR and then on to P. All the time, it will travel directly along the intended track O-P. The boat is steered on course C, the direction of O-DR, but due to the effect of current it is expected to travel along the intended track O-P. This is the route that must be considered for shoals and other hazards to navigation.

Case 2—A situation in which you know the course you have steered and the speed through the water from either the speed curve of your boat or a marine speedometer. It is also obvious to you that you did not arrive at your DR position. From your chart plot, you have been able to determine the course and speed over the ground. Current has acted to set you off your course—you desire to know its set and drift.

After again drawing a north reference line, plot vectors for your motion through the water—C 255°, S 12 knots; and your motion with respect to the bottom—COG 245°, SOG 13.4 knots. See Figure 2028 top. These vectors are both drawn outward from the origin, O, to points DR and P respectively. The "tail-to-head" rule is not applicable as neither of these vectors represents current. The action of the current has been to offset your boat from DR to P (which is also point W in this case), thus the set is the direction *from* DR *toward* P, and the drift is the length of this line in scale units.

In Figure 2028 top, the current is found to be setting 192° with a drift of 2.6 knots. This is the *average* current for the time period and location of the run from O to P for which the calculations were made; it is *not* the current *at* P. For the next leg of your cruise, these current values can be used as is, or modified as required by the passage of time and/or the continuing change in position of the boat.

Case 3—A typical cruising situation. Here you know the track you desire to make good (TR 080°), you have decided to run at your normal cruising speed through the water (S 10 knots), and you have calculated (or estimated) the current's set (140°) and drift (4 knots). What you desire to know is the course to be steered (C) and the speed of

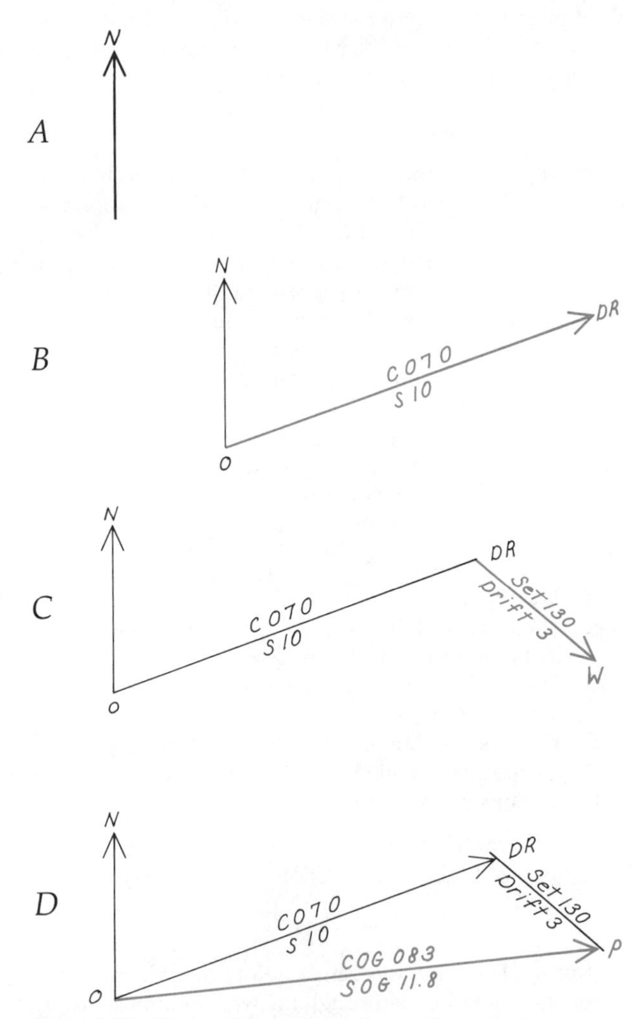

Figure 2027 **Current diagram for Case 1: *A* shows the reference north linee (not needed if the diagram is plotted directly on a chart); *B* shows the vector line for the course and speed. At *C* the vector for current set and drift is added. At *D* connecting vector from O to P shows the resultant motion and speed of the boat over the bottom.**

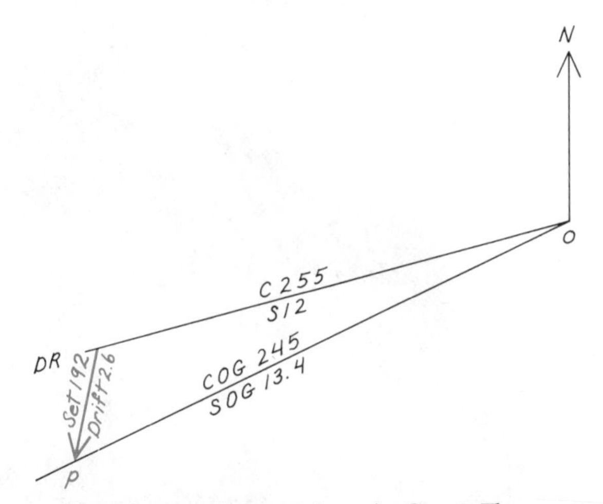

Figure 2028 **Current diagram, *above,* for Case 2: The prevailing current (actually the net effect of all offsetting influences) can be found graphically by plotting vectors for the DR track of the boat and its actual course and speed for a given time, then adding the vector from DR to P. *Below:* Case 3 Current Diagram: If you know your desired track, and your speed through the water, a graphic solution can be used to determine the course to be steered and the speed of advance. For safety, a DR plot is added.**

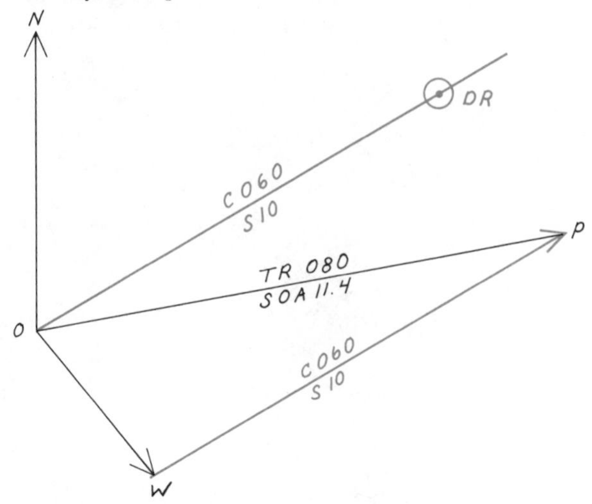

advance (SOA) which can be used to figure your estimated time of arrival (ETA).

Draw in a north line and measure directions from it, Figure 2028 bottom. Plot the current vector, O-W, at the specified direction and length, and a line (not a vector yet) in the direction of the intended track—this line should be of indefinite length at this time. From point W, swing an arc, with dividers or drawing compass, equal in length to the speed through the water in scale units. The point at which this arc intersects the intended track line is point P and the vector triangle has been completed. The direction of W-P is the course to be steered (C 060°); the length of the vector O-P is the speed over the ground (SOA 11.4) and from this the ETA can be calculated.

Let us look at the reasoning behind this graphic solution of Case 3. Again considering the component motions separately for the sake of simplicity, the boat is moved by current from O to W. It is to move from W at the specified speed through the water, but must get back on the intended track line; the problem is to find the point P on the track which is "S" units from point W. The solution is found by swinging an arc as described above.

Remember that all vectors are plotted as *true* directions, including C, the course to be steered in the preceding solution. This must be changed to a compass direction (course) for actual use at the helm.

In accordance with the principles of dead reckoning, it is desirable to plot the DR track even though a current is known to exist. This line, drawn from point O in the direction C and with a length of S scale units, forms a basis for consideration of possible hazards if the current is not as calculated or estimated.

Case 4—A variation of Case 3, in which you desire to arrive at a specified point at a given time. It may be that you are competing in a predicted log contest, or merely that you have agreed to meet friends at that time and place. In addition to the data on current that you have, you have decided upon your track and your speed of advance. For Figure 2029, let us assume that the current sets 205° at 3 knots. You need to make good a track of 320°, and a quick distance-time-speed calculation sets your required speed of advance at 11.5 knots.

From the north reference line, draw the current vector O-W and the intended track vector O-P. Complete the triangle with the vector W-P which will give you the course to be steered (C 332 True) and the speed to run through the water (S 13.1) to arrive at the destination at the desired time. Line O-DR should be drawn in from the origin as a dead reckoning track for the sake of safety.

Solutions by Calculator

Solutions of Cases 1,2,3, and 4 by use of a calculator, rather than graphic plots, are shown in Chapter 21. These should *not* be used without a clear understanding of the fundamentals of current sailing, the component motions, and their relationships to each other. These fundamentals, and the terms and symbols used, should be learned from simple graphic solutions as presented above.

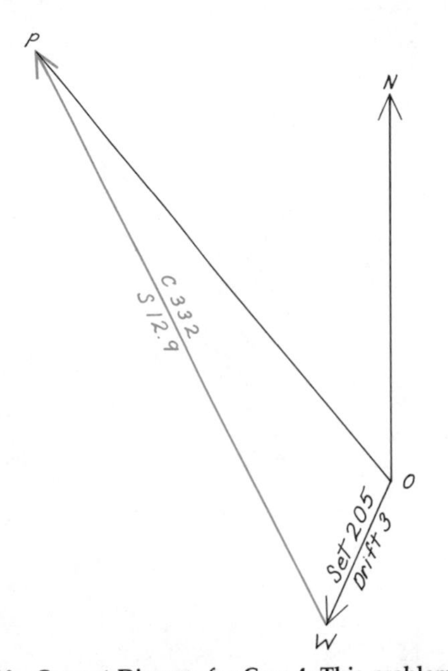

Figure 2029 **Current Diagram for Case 4: This problem concerns the course to be steered and the speed at which to run in order to arrive at a specified destination at a predetermined time. This could be used in establishing a rendezvous with another boat, or in predicted log contests.**

21

How to establish lines or circles of position, and use these to determine the location of your boat; advanced piloting techniques

PILOTING: POSITION DETERMINATION

The art of piloting reaches its climax in POSITION DETERMINATION. Underway on a body of water of any size, where the safety of your boat and its crew is at stake, it's not "where you ought to be," or "where you think you are," but your knowledge of "where you are for sure" that counts. Developing an ability to determine your position quickly and accurately under a wide range of conditions should be one of your primary goals.

The Need for Position Determination

In everyday cruising, the strength and direction of the actual current will be somewhat different from that calculated or estimated. Local wind conditions or abnormal rainfall frequently upset tabular predictions; estimates based on visual observations of current are inaccurate unless the boatman has had considerable experience. Because of these uncertainties, current sailing solutions discussed in the previous chapter can be considered only as *estimates* of present positions, and of future motions and positions. While buoy positions are usually reliable, they are not infallible.

For these reasons, the boat's position should be fixed as often as possible so that its safety can be verified, or corrective action taken promptly. The true effects of the current may also become known when a position is newly fixed. And a new DR plot can be begun, fresher and more accurate than the old one.

Knowledge of where you are, extended from a *recent* position determination, is essential if you must call for help. Or, if another boat has trouble, you can set a direct course to render assistance if you are sure of where you are.

The Skipper's Responsibilities

It is your duty as skipper to fix the position of your craft as precisely and as frequently as required by the proximity to hazards. This is true whether you are doing the piloting, or someone else is. You can delegate the function but not the responsibility.

The actual procedures in position determination vary widely in practice. When you are proceeding down a narrow channel, positioning is informal and a chart plot may not be maintained. But position determination is not being omitted; rather it is being done continuously by visual reference to the aids to navigation. On the other hand, during an open ocean passage, a plot will be maintained, but positions will be determined, other than by DR calculations, perhaps only three or four times each day, or even only by a daily noon sight.

Between these extremes are the normal cruising situations in pilot waters. Cruising just offshore, or in the larger inland bodies of water, a skipper will usually maintain a plot of his track with periodic checks on its accuracy, perhaps every 15 or 20 minutes, perhaps at hourly intervals.

The ability to determine the position of your boat with acceptable accuracy under any visibility conditions is essential. The absence of this ability, or any limitation on it, should restrict the extent of your boating activities, setting the boundaries of the water areas and weather conditions you will enter.

Definition of Terms

A LINE OF POSITION (LOP) is a line, in actuality or drawn on a chart, at some point along which an observer is presumed to be located, Figure 2101. An LOP may result

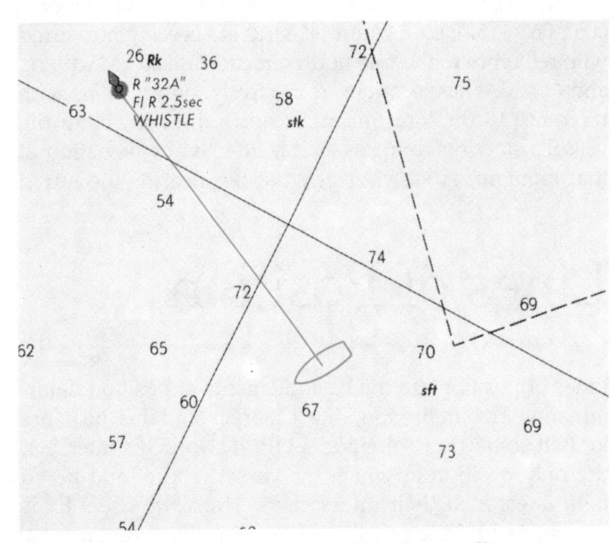

Figure 2101 **A line of position is a line, in actuality or on a chart, along which the observer is presumed to be located. A single line will not determine position, but it does tell the observer where he is *not* located, and such information can be useful in itself in many situations.**

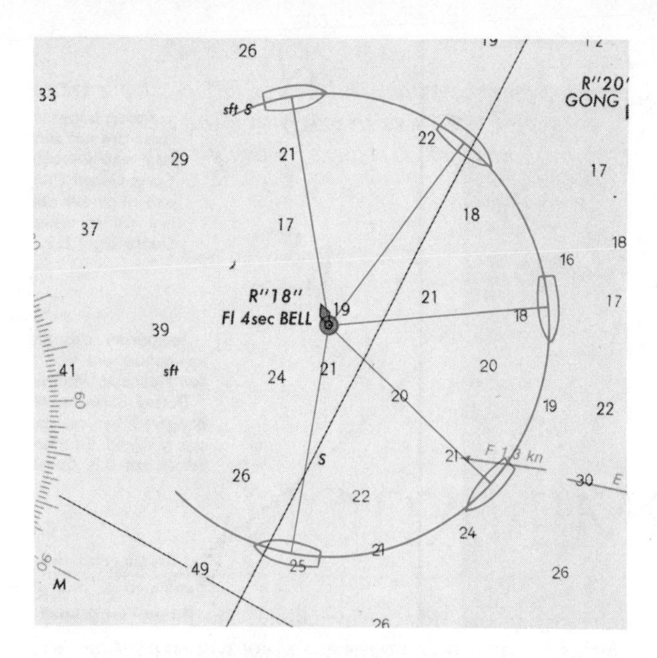

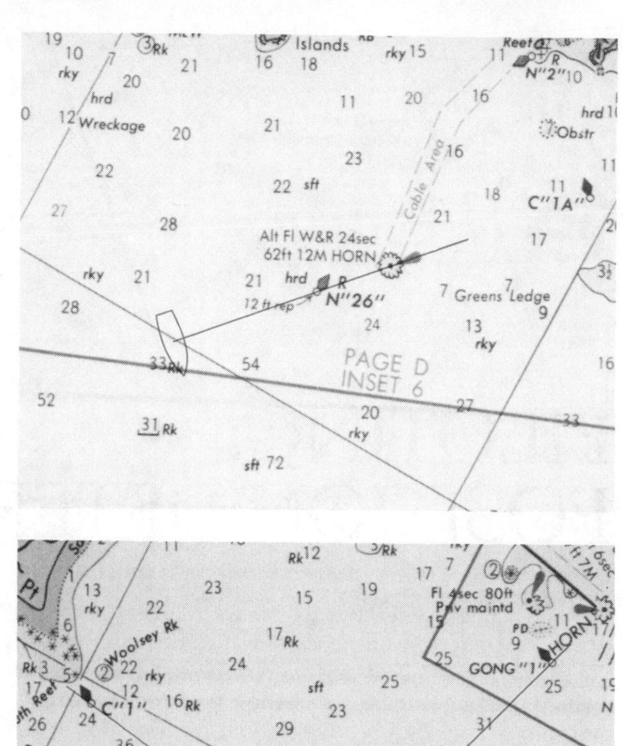

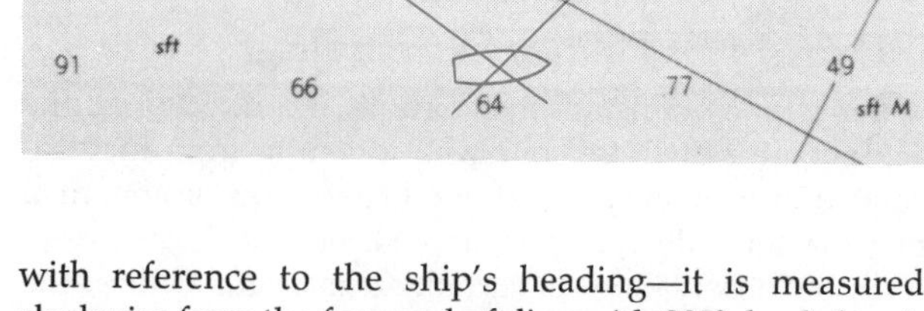

Figure 2102 **Lines of position may be curved, *above,* as well as straight. A measurement of distance from an identified object yields a circular LOP. It may be plotted as a complete circle, or as an arc that covers the pertinent area. When two objects can be observed in line, *above right,* an excellent line of position is established. Such "ranges" may have been set up specifically as aids to navigation, or they may be any two identifiable objects, including aids to navigation, landmarks, or other recognizable features. A "fix" is an accurately determined position for the observer and his boat, *right.* It is determined from currently observed lines of position, or other data obtained by the pilot.**

from observation or measurement; from visual, electronic, or celestial sources. It may be straight or curved; a circular LOP is sometimes referred to as a CIRCLE OF POSITION, Figure 2102 left. A line of position may be ADVANCED (moved forward) in time according to the movement of the vessel during the time interval involved. In rare cases, an LOP may be RETIRED (moved backward) in time.

A BEARING is the direction of an object from the observer, expressed in degrees as a three-digit number—005°, 062°, 157°, etc. A TRUE BEARING is a bearing measured with reference to the true north direction as 000°. MAGNETIC and COMPASS BEARINGS are, respectively, observations with reference to the local magnetic north direction or to the vessel's steering compass as it is affected by deviation at that moment. A RELATIVE BEARING is a bearing measured with reference to the ship's heading—it is measured clockwise from the fore-and-aft line with 000° dead ahead, 090° broad on the starboard beam, 180° dead astern etc.

A RANGE consists of two objects that can be observed in line with each other and the observer, Figure 2102 top right.

A FIX is an accurately located position determined without reference to any prior position, Figure 2102 bottom right. A RUNNING FIX is a position that has been determined from LOPs at least one of which has been taken at a different time and advanced (or, rarely, retarded) to the time of the other observation.

An ESTIMATED POSITION (EP) is the best position obtainable short of a fix or good-quality running fix. It is the most probable position, determined from incomplete data, or from data of questionable accuracy.

Lines of Position

Lines of position are the basic elements of position determination. By definition, the observer and his boat are located somewhere along an LOP. If two LOPs intersect, the only position at which the vessel can be, and be on both lines, is at their intersection. Thus, the usual fix is determined by the crossing of two lines of position.

Labeling

Since LOPs are also lines that are drawn on a chart, they should be labeled immediately, and preferably to a standardized system. A label should contain all information necessary for identification, but nothing further that might cause confusion or clutter up the chart. The information to be recorded for each line of position is the time that it was observed or measured and its basic dimension, such as direction toward, or distance from, the object used.

A bearing is an LOP that has time and direction. Figure 2103 left shows several examples of correctly labeled bearings. Time is always shown *above* the line of the bearing,

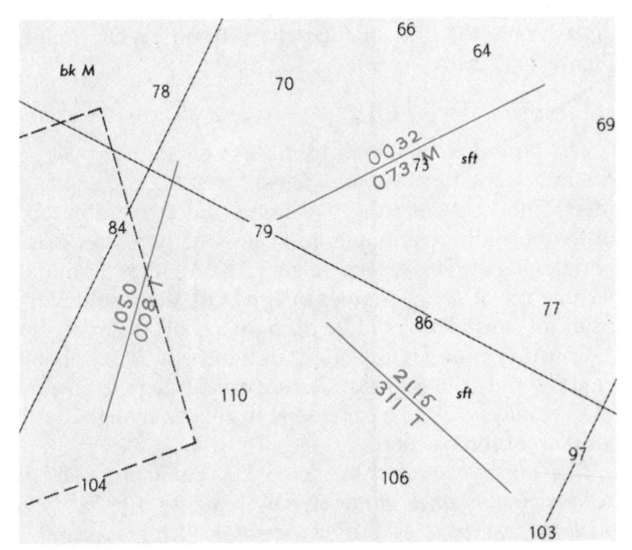

Figure 2103 **The correct labeling of lines of position is important; unlabeled or mislabeled lines cause confusion. Time is shown above the line in the 24-hour clock system, and direction as a three-digit number below the line, *above.* Note that the "°" symbol for degrees is not used. Circles of position, *right,* are labeled in the same manner as lines of position, with time above the line, and distance below as it would normally be viewed and read. Thus the time may be "inside" or "outside" the circle, as determined by the curvature of the arc.**

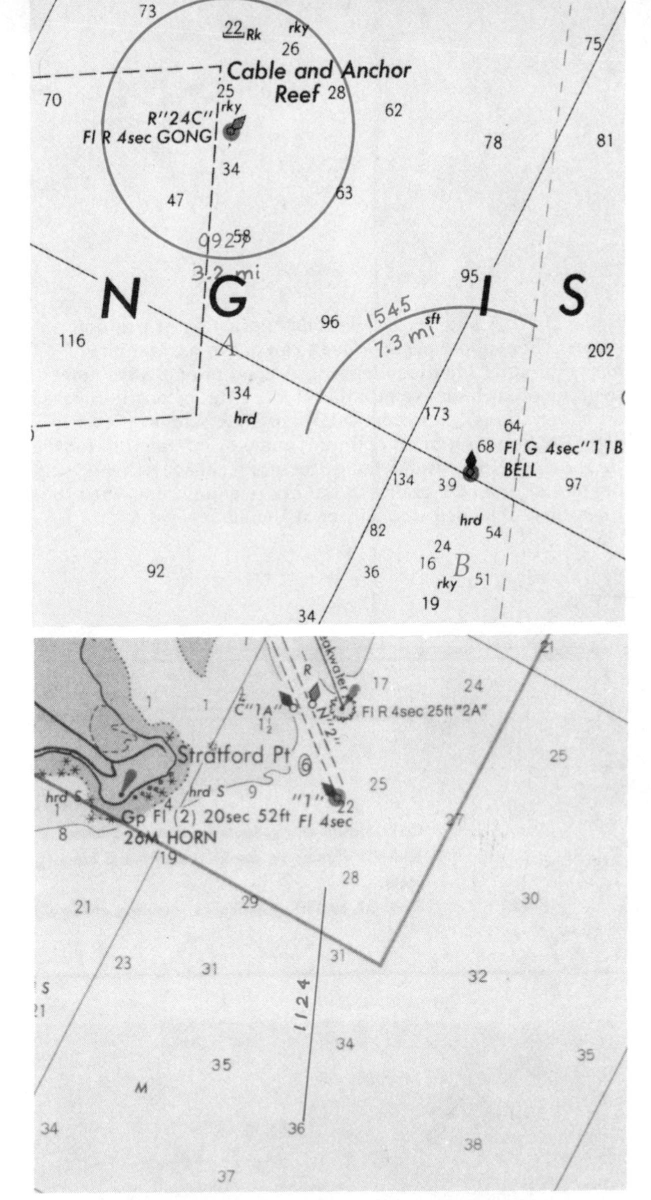

Figure 2104 **Since a range is based on the relationship between two sighted objects, it is often used without a measurement of its actual direction. Ranges are labeled only with the time of the observation, *above. Below:* A line of position may be "advanced" to a later time, as detailed in the text. An advanced LOP is labeled with the original time, and the time for which it is replotted. The direction, of course, remains the same.**

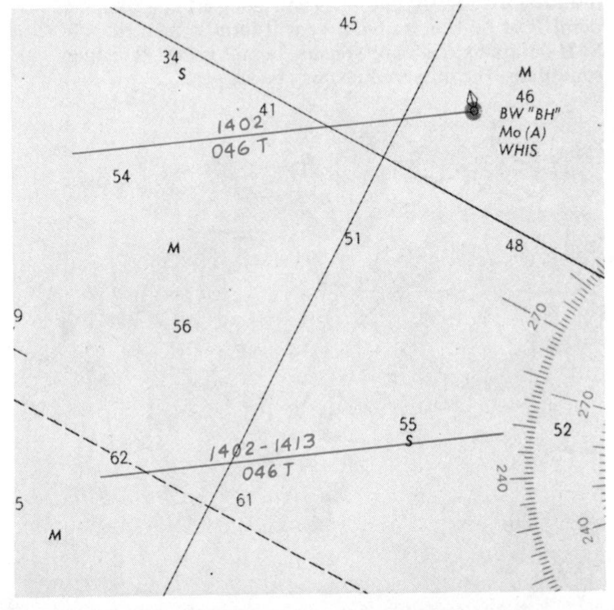

and direction *below* the line. Time is given as a four-digit figure in the 24-hour system. Directions may be true ("T") or magnetic ("M") and are written as three-digit group with zeros prefixed as necessary. Customarily the degree symbol ° is not used in this labelling, since all three-digit figures will be for directions.

A circle of position has dimensions of time and distance; it may be plotted as a complete circle, or as an arc, Figure 2103 right. Time is labeled above the curved line and distance, with units, is shown below the line.

A range is a line of position whose direction is self-evident from the two points which define it. In this case, only time need be shown; it is placed above the line as for other LOPs; see Figure 2104 top.

Do not draw the line completely through the chart symbols for the two objects used as the range; this will avoid erasing across the symbols later. Draw the line neatly and no longer than necessary; it need not extend all the way to the front range marker or other object if the two features used to form the range are clearly evident.

A line of position that has been advanced in time is labeled with *both* times above the line, the time of original observation or measurement and the time to which it has been advanced. The time of the original LOP is written first, followed by a dash and then the second time, Figure 2104 bottom.

Fixes

As defined, a fix is an accurately located position for a vessel. On many occasions in small-boat piloting, position will be established by passing close by an identifiable object, usually an aid to navigation. This is, of course, a fix of the highest possible accuracy, and a skipper should

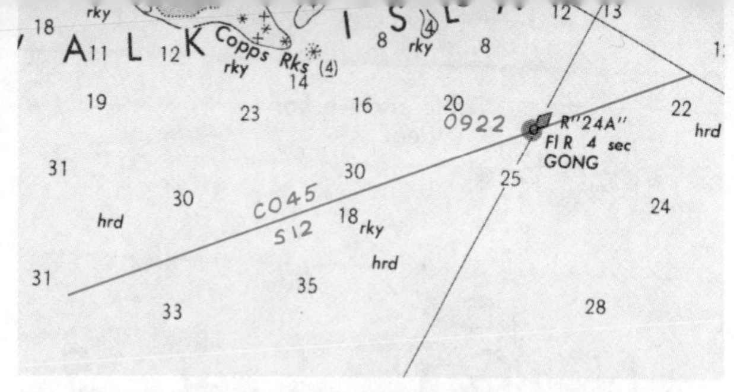

Figure 2105 **A simple, excellent determination of position occurs when a boat passes close to an aid to navigation, *above,* or other identifiable point. A good pilot always notes the time of such an event on his chart. Lines of position based on observations taken from small boats are seldom precisely accurate. Selection of an optimum angle of intersection for the lines helps. Note, *below,* the difference a change of two degrees in one LOP makes in the intersection point when the lines cross at 90 degrees, left, or at 30 degrees, *right.***

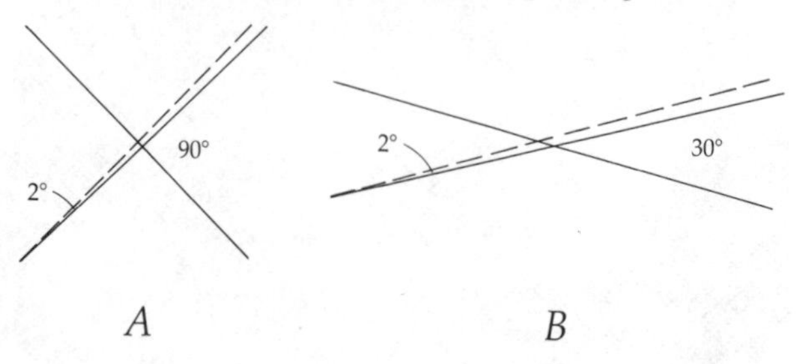

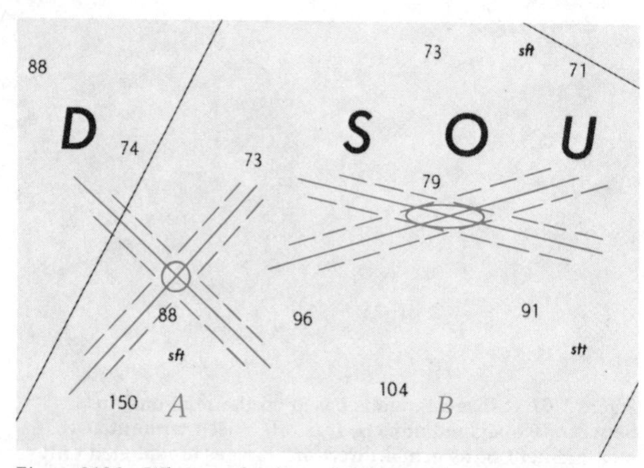

Figure 2106 **When each of two LOPs may have an inaccuracy of several degrees, there is an area of uncertainty around the intersection point, *above.* This area is smallest at angles near 90 degrees. Use caution with angles less than 60 degrees. *Below:* Use of more than two LOPs increases accuracy of the resulting fix. Ideally, the lines will intersect at a common point as at A. More often, they will form a small triangle as at B. If the triangle is unreasonably large for the prevailing conditions, the observations may be suspect.**

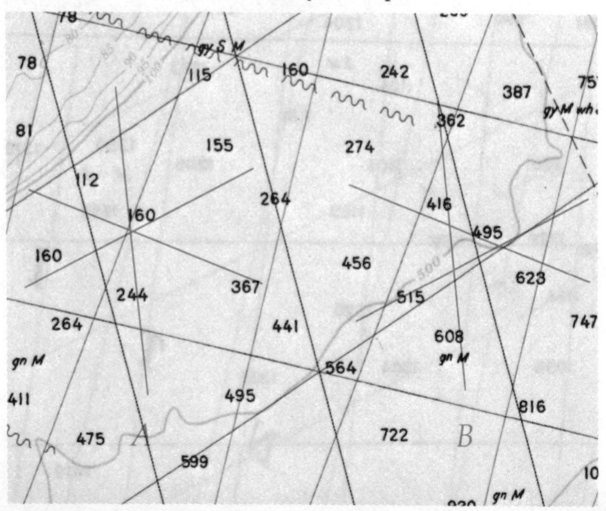

always note the time of such an event on his chart plot, Figure 2105 top.

Fix From Two LOPs

The typical fix obtained from lines of position will be the intersection of two such lines. Note that the angle of intersection between the two LOPs affects the accuracy of the position determination. Where the two lines cross at right angles, 90 degrees, (Figure 2105*A*), if there should be an error of say, 2 degrees in one LOP, the resulting fix from this mistake would be off only a short distance.

Consider now Figure 2105*B* in which the intersection angle is only 30 degrees. An error of 2 degrees in one LOP would result in a considerably greater change in the location of intersection.

Two lines of position should intersect as nearly as possible at right angles; the angle of intersection should never be less than 60 degrees nor more than 120 degrees if this can be avoided.

Where there is no alternative, a fix from two lines of position intersecting at angles of 30 degrees to 60 degrees (120 degrees to 150 degrees) can be used, but only with caution. A position from two LOPs intersecting at less than 30 degrees (more than 150 degrees) is of dubious accuracy.

Areas of Uncertainty Lines of position will normally be stated as having a specific direction, such as 072°, 147°, etc. In actuality, there may be an uncertainty in each line of two or three or even more degrees. This condition can be shown graphically on the chart by additional lines drawn lightly on either side of the basic LOP; this should always be done if such uncertainty could mean the difference between a safe position and a hazardous one.

Lines of position from various sources and techniques will have different degrees of uncertainty. Only experience gives the pilot the ability to assign relative values of probable accuracy to the various methods.

If each of two LOPs is each drawn with its graphic representation of uncertainty, the result is an AREA OF UNCERTAINTY at their intersection. Figures 2106*A* and *B*, top show how the angle of intersection alters this area of uncertainty and why it is desirable to have the LOPs cross as nearly as possible at an angle of 90 degrees. The area of uncertainty actually is elliptical (circular if the LOPs cross at 90 degrees) rather than the quadrilateral formed by the outside limiting lines for each bearing. A detailed discussion of this topic, including the additional complexities that arise if one line of position can be considered to be more accurate than the other, is in *Bowditch.*

More Than Two LOPs

Although most lines of position obtained in small-craft piloting are reasonably accurate, it is desirable to use more than two LOPs to reduce the uncertainty of position. There is no theoretical upper limit to the number of LOPs that can be used, but practical considerations normally limit you to three.

Ideally, since the observer is presumed to be on all three lines of position, they would intersect in a common point, Figure 2106*A* bottom. The much more likely result is that the plotted lines will form a triangle, as at *B*.

For optimum results, the three lines should each differ in direction as closely as possible by 60 degrees (or 120

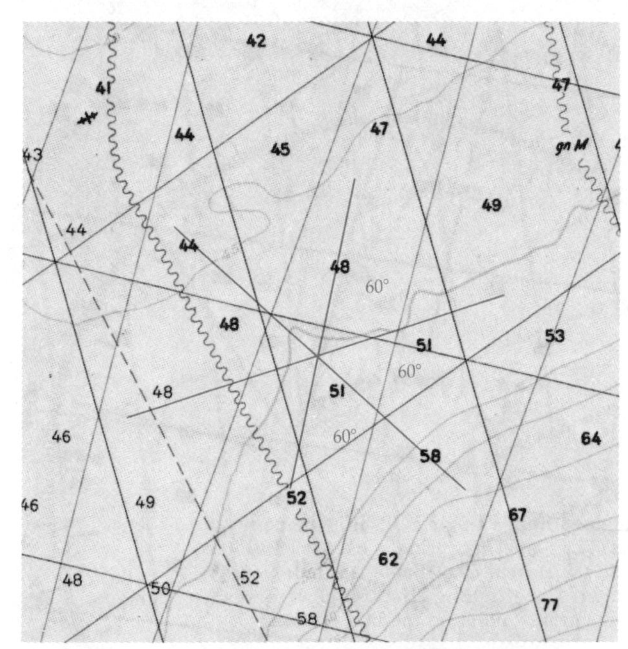

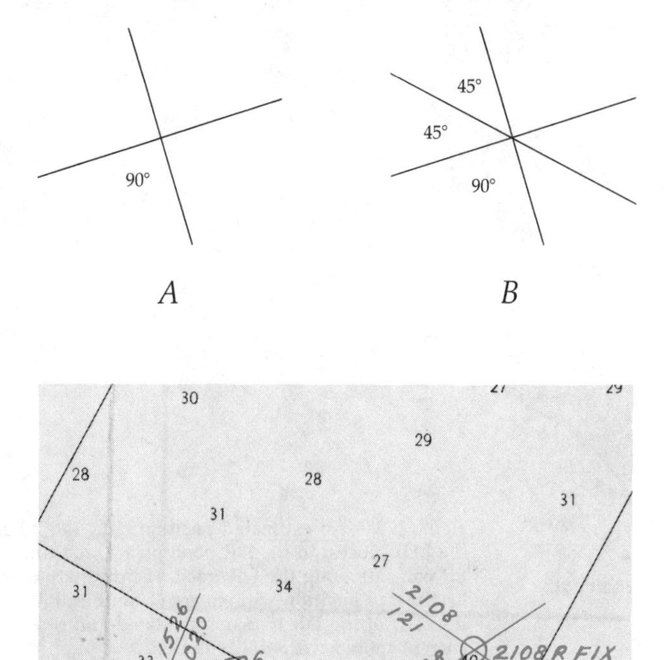

Figure 2107 **For an optimum set of three angles, the lines should cross at 60 degrees to each other. That is seldom possible, but it is the most desirable arrangement. An alternative procedure, *above right,* is to select two objects that will provide LOPs at an angle close to 90 degrees, and add a third whose LOP will be at 45 degrees to the others, as at B. If there is not time to take the third sight, the first two provide a near-optimum two bearing fix. *Right:* A fix is labeled with the time, but the word "fix" is not used. The circle itself signifies this is a fix, as at A. For a running fix, however, the label "R FIX" is added along with time of the second observation.**

degrees), Figure 2107 left. In this configuration, an inaccuracy in any of the LOPs will result in the minimum net error of position.

An alternative technique to the above is to obtain two lines of position at nearly 90 degrees to each other, and then a third LOP making an angle of approximately 45 degrees with those, Figures 2107*A* and *B* top right. This procedure has the advantage of providing an optimum two-LOP plot in the event that a change of visibility or other circumstance should suddenly prevent the third observation.

Triangles of Position

As noted above, three lines of position will usually form a TRIANGLE OF POSITION. If this triangle is large, there is probably a significant error in one or more of the LOPs, and a check should be made of the observations, calculations, and plotting.

If, however, the triangle is relatively small (and "relatively" will have to be defined in each situation as determined by the techniques used, the experience of the observer, the size of the boat, sea state, and other conditions), it is customary to use the center of the triangle of position, as estimated by eye, for the fix.

There are, however, some situations in which the true position may lie *outside* the triangle; see *Dutton's Navigation and Piloting* for a discussion of this more complex case.

Labeling Fixes

A fix is plotted on a chart as a small circle, with a dot in its center if required; it is labeled with time placed horizontally; the word "Fix" is not necessary and should be omitted to avoid clutter; see Figure 2107*A* bottom right. A running fix uses the same symbol but adds "R Fix"; see Figure 2107*B* bottom right.

If there are three lines that do not meet in one point a dot is placed in the triangle and then circled. If the fix is obtained by passing close to a buoy or other aid to navigation, the dot or center of the symbol may be used for the plot and the small circle omitted. The usual distance off, 50 to 100 yards (45-90 m) or so, is not significant at typical chart scales; greater distances are estimated and the fix is plotted in the correct direction at a scaled distance.

The Value of a Single LOP

While a fix based on two or three lines of position is the most desirable situation, the value of a single LOP should not be overlooked.

Basically, a single line of position cannot tell a pilot where he is at the moment, but it can, within the limitations of its accuracy, tell him where he is *not.* If he is presumed to be somewhere along the LOP, then he cannot be at a position appreciably distant from the line. This information, although lacking in detail and negative in nature, can often be of value to a navigator concerned about his craft possibly being in dangerous waters.

A single line of position can frequently be combined with a DR position to obtain an ESTIMATED POSITION. This

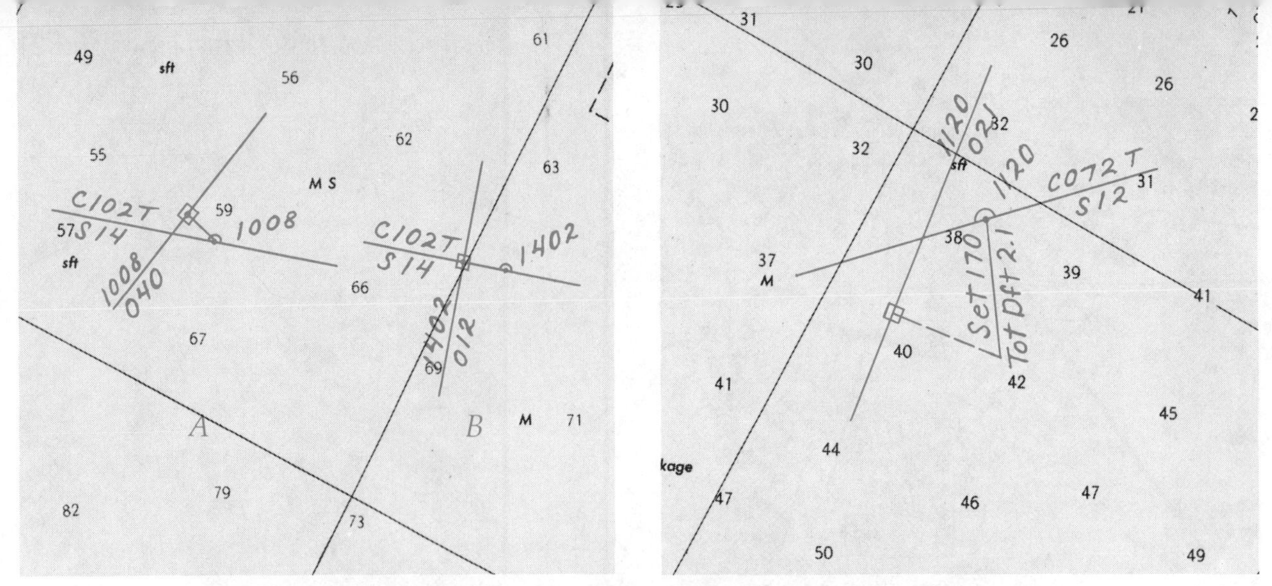

Figure 2108 **An estimated position, *left,* can be obtained from a single LOP. It is the point on the LOP closest to the DR position for that time; see A. If a beam bearing is used, as at B, the EP will fall along the DR track. If information about the current is available, an improved "estimated position with current" can be plotted, *right.* The offsetting effect of the current since the start of the DR track is calculated and plotted; the nearest point on the LOP from this current-influenced point is the EP.**

EP is the point along the LOP that is closest to the DR position for the same time. A line is drawn from the DR position perpendicular to the LOP until it intersects that line. An estimated position is marked by a dot within a square; the line can go through the square. As a square is used only for estimated positions, it is not necessary to add "EP"; time may be omitted as it appears on the LOP.

It is also possible to obtain an "estimated position with current" from a single LOP if the set and drift of the current have been determined from predictions or from its effect on the boat's course and speed made good. From the DR position for the specified time, a line is drawn to scale to represent the effect of the current since the DR track was started from the last fix. This line is drawn in the direction of the set of the current and for a length equal to the total drift, the distance found by multiplying the drift by the elapsed time since the last fix. This total drift line is labeled and from its end another line is drawn perpendicular to the LOP until it intersects that line; see Figure 2108 right. The intersection is the EP WITH CURRENT; it is labeled in the same manner as before.

In using an EP, make generous allowances for the uncertainty of the position; always err on the side of safety.

In some piloting situations, a single bearing on an object can be advanced and crossed with a later LOP on the same object. This will result in a running fix, not as desirable as a fix from two independent lines of position, but far better than a DR position or your "best guess."

Running fixes, and the use of a single LOP with depth measurements, will be discussed later in this chapter.

Visual Observations

On a typical motorboat or sailboat, your primary source of lines of position will be visual observations. These will include direct and relative bearings, ranges, and horizontal and vertical angle measurements. Correct identification of the sighted object is essential. Other methods and equipment may lead to LOPs such as those from radio direction finding, radar, and depth measurements, as discussed later, but these will normally be secondary to visual observations.

Bearings Dead Ahead

On a powerboat, bearings taken directly ahead require no auxiliary equipment; the boat's steering compass will suffice. Only insignificant effects will be made on the DR track if the bow is momentarily swung off course and pointed toward an object of the bearing. The boat should be held on this new heading just long enough for the helmsman to line up on the object and for the compass to settle down and be read. This is normally a matter of less than a minute, perhaps only 20 or 30 seconds.

To use this technique, the pilot must be sure enough of his general location that the off-course swing can be made safely. But if such is the situation, he should not hesitate to use this quick and simple way of getting a line of position. The accuracy of bearings obtained by this method is normally considerably greater than those of the more complex techniques to be discussed later.

This procedure has an additional advantage in that it can be accomplished by the helmsman alone. Other methods generally require the services of a second individual as the bearing-taker while the helmsman continues his normal duties, plus reading the steering compass at the time of the observation.

Beam Bearings

If the observed aid to navigation or landmark is well around on either beam, it may not be feasible to make so great a deviation from the heading being steered in order to take a dead-ahead bearing. Or perhaps the limited width of the available deep water does not permit so large an excursion from the DR track. In this case the other simple bearing technique may be used, and it works on sailboats as well.

Determine some portion of the boat that is at a right angle to the fore-and-aft axis of the craft. Among the possibilities are bulkheads, seat backs, deck seams, etc. Sights taken along these features will determine when the observed object is directly abeam—a relative bearing of 090° or 270°.

The boat may be held on course and the passage of time awaited until sighting along the pre-selected portion of the boat indicates that the aid to navigation or other object is exactly broad on the beam. Its direction then is, of course, 90 degrees greater or less than the direction read from the boat's compass at that moment.

If waiting until the object comes abeam is not feasible or desirable, then the heading of the craft can sometimes be temporarily altered slightly to bring the sighted object on the beam more quickly. Should such a temporary change of heading be made, be sure that the 90° is added to or subtracted from the compass reading at the moment the sighted object is abeam; do not use the normal base course.

Correction of Compass Bearings

Remember that in either of these types of observations the bearings taken are COMPASS BEARINGS. It will be necessary to correct them for deviation and variation (see Chapter 15) and then plot them as TRUE BEARINGS. Any of the plotting instruments and techniques described in Chapter 19 may be used.

Beam bearings are obtained by adding or subtracting 90 degrees from the boat's true heading. Be sure to correct the boat's compass heading for deviation and variation *first,* and *then* add the 90 degrees if the observation was to starboard or subtract the 90 degrees if the sighting was on the port beam. Do not use the direction of the bearing for entering the boat's deviation table; use the *heading.*

Plotting Bearings

The direction measured by the visual observation is *from the boat toward the object.* When plotting is being done, the position of the boat is not known, but it is still possible to draw the line so that it will have the correct direction and lead toward, to, or through the chart symbol for the sighted object. The particular plotting instrument used will determine the exact procedures used.

It is also possible to plot a line of position outward from the sighted object by using the RECIPROCAL of the corrected observed bearing. Here, be sure *first* to correct the observed compass (or relative) bearing to a true direction, and then to add or subtract 180 degrees to obtain the reciprocal; do not reverse this sequence.

More Sophisticated Bearings

The procedures just described for bearings dead ahead and broad on the beam suffice in many situations, but it may be neither safe nor convenient to alter the boat's heading to take a bearing. In that case bearings can also be taken in any direction *without* a change in the boat's heading. Such bearings involve the use of the boat's compass, a pelorus, or a hand-bearing compass, or some combination of these instruments.

Using the Boat's Compass

On many boats, it is possible to take bearings directly over the steering compass, at least through a limited angle on either side of the bow. As determined by the construction of the compass, it may be feasible to sight directly over the card and get bearings of acceptable accuracy. On other models, a set of sighting vanes may be placed on the compass for greater precision in direction measurement. The directions in which bearings may be taken using the boat's steering compass may be limited by the nearby superstructure of the boat. Be sure to correct for deviation of the boat's heading, *not* that for the bearing direction, and for variation in the locality.

Using a Hand-Held Compass

Direct visual bearings can also be taken using a hand bearing compass, or even with the boat's compass if it is a model that can be quickly dismounted with accurate remounting. Normally it is not feasible to prepare a deviation table for such hand compass use, but careful tests should be made to determine the locations about the boat, if any, at which a hand-held device can be used without deviation errors. These tests are best made from known positions by taking bearings on objects whose actual direction can be established from the chart. If deviation-free

Figure 2109 **A simple, accurate way to take a bearing is to "aim" the boat directly at the sighted object and read the compass. Brief off-course swings for this purpose will not affect the DR plot materially, but be sure there are no hazards alongside your course. Correct compass bearings to true bearings before plotting them on your chart.**

Figure 2110 **Sometimes it is possible to sight across the steering compass to take a bearing, but in a small boat such as this, only a rough reading may be obtained.**

locations can be found on board, then correction will be required for magnetic variation only, to obtain true bearings for plotting.

Using a Pelorus

Alternatively, a pelorus may be used to measure directions, usually as relative bearings. This instrument may be used anywhere from which the desired object can be seen. The caution in this case is that the pelorus must be accurately aligned with the fore-and-aft axis of the vessel. Work out the correct positioning of the pelorus for several locations on the boat so that sights may be taken on an object regardless of its position relative to the boat.

The scale of the pelorus can be aligned to the boat's heading and observations made directly in terms of compass bearings, but this procedure is less desirable than the technique of relative bearings to be described below. In all cases, the compass bearings must be converted to true bearings before plotting.

It is possible to convert relative bearings to magnetic bearings and plot them as such, using the magnetic directions circle of the chart compass roses. Because the chart latitude and longitude lines cannot be used as direction references, and magnetic directions must be plotted for all purposes—courses, ranges, current set—to avoid confusion, this procedure is not recommended.

Relative Bearings

Relative bearings are those taken with respect to the vessel alone, without reference to geographic directions. Relative bearings offer some advantages, but also have some built-in problems.

The instrument for taking relative bearings is the pelorus. As noted it has the advantage of being usable anywhere on the boat if properly oriented. Its disadvantage is that close coordination is required between two persons—the bearing-taker and the helmsman. The heading of the boat at the instant of observation must be known, and direct communication between these two individuals is required.

To take a relative bearing, the scale of the pelorus is set with 000° dead ahead. The helmsman is alerted to the fact that a bearing is about to be taken; he concentrates on steady steering and continuously reads the compass. When the observation is made, the bearing-taker calls out "Mark," and reads the scale of the pelorus. (It is very helpful if the bearing-taker calls out "Stand by" a few seconds before his "Mark.") At the word "Mark," the helmsman notes the reading of the steering comapss and calls it off to the observer or other person who will compute the

Figure 2111 **A hand bearing compass, *left,* can be used to take direct readings on objects that cannot be sighted over the boat's steering compass. Make checks to find locations where this compass is least subject to local magnetic influences. A pelorus, *right,* eliminates the magnetic influences that may be experienced with a hand-held compass. It can be used at any point from which the target object can be seen. The instrument, however, must be properly oriented to the fore-and-aft axis of the boat.**

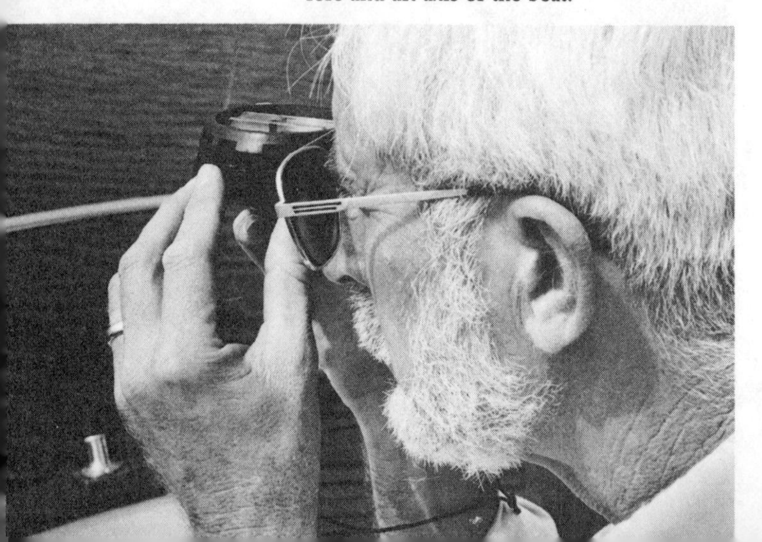

bearing. If the compass should be swinging at the moment, so that an accurate reading cannot be taken, the helmsman calls out that information and the pelorus reading is discarded; another attempt is then made to take the bearing.

An alternative technique is to have the helmsman call a series of "Marks" when, but only when, he has the boat directly on course. The bearing-taker makes his observation only when he is told that the boat is on the specified heading.

When the relative bearing has been taken by either of the above techniques, it must be converted to a true bearing for plotting. This is done by adding the numerical values of the relative bearing and the *true* heading of the boat at the instant of observation. If the sum so obtained is more than 360 degrees, that amount is subtracted. The true bearing can then be plotted and labeled.

Always be sure to use the deviation value for the heading of the boat, and not for the bearing angle. This is a common error, and one to be continually guarded against.

In Figure 2112*A*, an observer on a boat heading 047° true takes a relative bearing on a buoy of 062°. To determine the true bearing of this aid to navigation, these two numbers are added, 047° + 062° = 109°.

In Figure 2112*B*, the boat is heading 306° by its compass when a relative bearing of 321° is measured. The variation is 6° W and the deviation (for the compass heading of 306°) is 2° E. The true heading of the boat is thus 306° − 6° + 2° = 302°. The sum of the relative bearing, 321° and the true heading 302° is 623°; therefore 360° is subtracted and a value of 263° is found to be the true bearing.

If the pilot has multiple objects on which to take bearings, he should select the nearer ones (provided, of course, that desirable angles of intersection between the LOPs will result). An angular error of 1 degree will result in a lateral error of about 100 feet at a distance of one mile—at two miles, the *same* 1 degree angular error will cause a lateral displacement of 200 feet; at three miles, 300 feet, etc. (17 m at 1 km, 35 m at 2 km, etc.).

Ranges

Lines of position from ranges are of exceptional value in position determination. They are free from all of the magnetic effects that might cause errors in bearings taken with a compass or a pelorus. Such LOPs are also easily obtained, much more so than those from bearings. Ranges can be absolutely accurate, as accurate as the charted positions of the two objects lined up in the observation. No matter how small the boat, how rough the water, or how poor the visibility, if you can see the objects come into line, you can plot a good line of position on your chart and so have half of an accurate fix.

Ranges can be classified into two groups. First, there are those that consist of two aids to navigation constructed specifically to serve as a range and are charted with special symbols, see Figure 1819, page 413. The true direction of such ranges can be determined from the information in the *Light List* for the rear marker. These ranges are particularly useful in determining compass deviation because of their high accuracy.

However, any two objects that can be identified by sight and on the chart form a natural range and can be used in the same manner as one consisting of two formally established navigational aids. Among such objects are ordinary aids to navigation of all types; spires, radio towers, flagpoles, stacks, etc.; identifiable portions of bridges, such as center spans; and other prominent, isolated features; see Figure 2113. Be careful in using points of land as either a front or rear range mark; errors may be made in attempting to sight on the exact end of the point, particularly if it is of low elevation.

Taking an LOP from a range requires no more than the observation of the time that the two objects came into line.

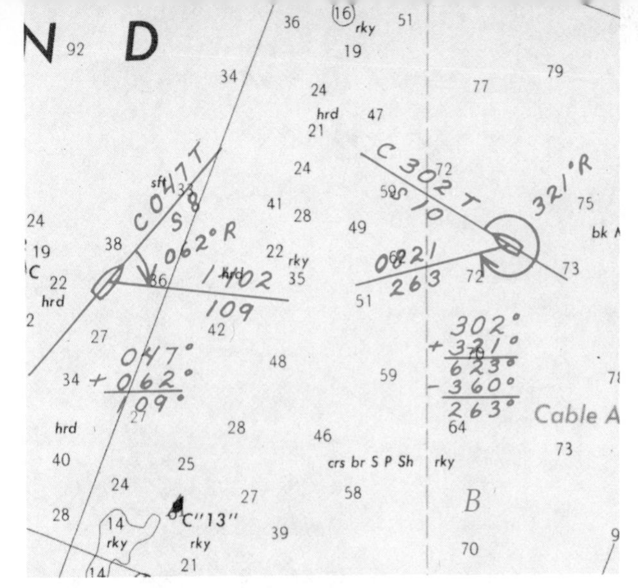

Figure 2112 **Relative bearings must be converted to true bearings before they are plotted. Add the relative bearing to the boat's true heading at the instant of the observation. If the sum exceeds 360 degrees, subtract that amount.**

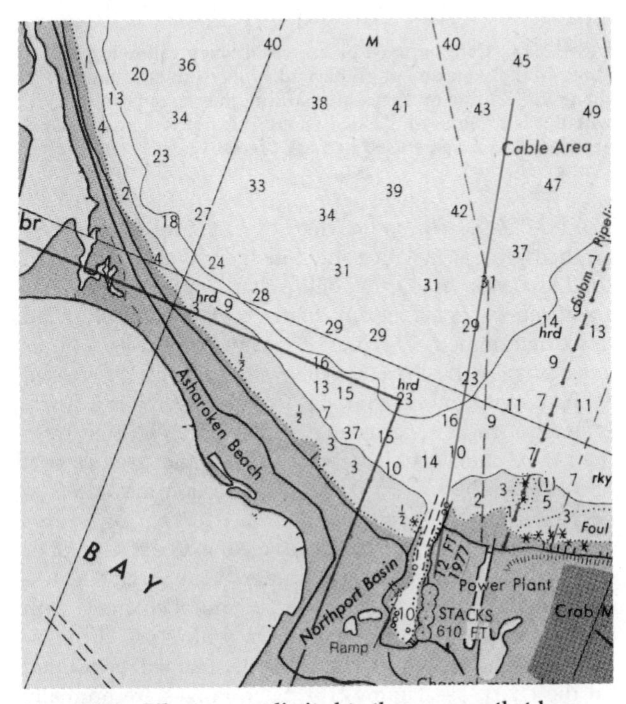

Figure 2113 **Pilots are not limited to those ranges that have been established as such by the Coast Guard. Here four tall smokestacks, shown on the chart, make an excellent range when aligned. These stacks, at Northport, Long Island, provide a range that is within half a degree of magnetic south.**

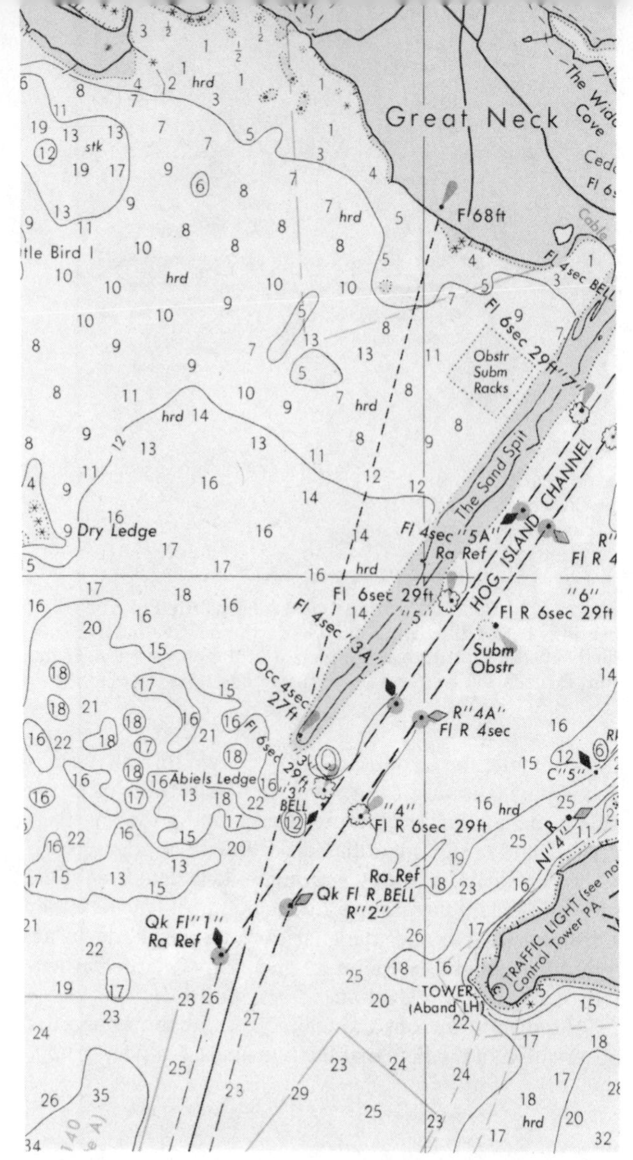

Figure 2114 **Caution must be exercised when following a range so that you do not go beyond proper limits of the channel. The line on the chart marking the range is solid where it is to be used, dashed where it is not to be used. One or two buoys may be used to mark the start or end of a range-marked channel.**

The LOP is plotted on the chart by lining up the symbols for the two objects with a straightedge and drawing a light solid line over the portion of the chart where the LOP has significance. Its actual direction will not ordinarily need to be determined. The line should be labeled as soon as it is drawn on the chart. This line of position can be crossed for a fix with another range, or with any other form of LOP. If such is not available at the time, the LOP from the range often may be advanced later and used as part of a running fix; in this case, the direction may need to be labeled on the original LOP.

Ranges are often established by the Coast Guard or other authority to mark the center of important waterways, usually dredged channels or natural channels with hazards close on either side, often with cross currents. When so established, the range center line will be printed on the chart, see Figure 2114. Such ranges are normally used for direct steering rather than as a line of position, but either use is valuable.

Observe two cautions steering up or down a channel marked by a range. A vessel traveling in the opposite direction on the range will be on a collision course, or will pass too close for comfort. Secondly, at the beginning and end of the range, note the ends of the solid portion of the line on the chart—a range can be followed for too great a distance on either end. Buoys will often be used to mark the spot at which to turn onto or off of the range line, Figure 2114. Study your chart carefully when using such a range.

Horizontal Angles

A fix may also be obtained by measuring the two horizontal angles between the lines of sight to three identifiable objects, without a measurement of the relative or geographic direction to any of them. This procedure avoids the inaccuracies involved in relative or compass bearings. Such horizontal angles can be measured to a high degree of accuracy (as close as half a degree) with a sextant.

For plotting these horizontal angles, a THREE-ARM PROTRACTOR, Figure 2115, is the preferred instrument. Each side arm is set for the angle measured to its side of the line of sight to the center object. Then the instrument is moved about on the chart until the arms line up with the symbols for the objects sighted upon in measuring the angles. There is a small hole in the center of the protractor through which a pencil point can be placed to mark the fix on the chart.

The same general technique can be used without a special instrument by drawing lines with the proper angles between them on a piece of transparent paper, Figure 2116 top. This is then moved about as before until the lines and points are properly related.

Be careful in selecting the objects between which the angles will be measured. Three points, not in a straight line, will all lie on the circumference of a specific circle, such as points *X*, *Y*, and *Z* in Figure 2116 bottom. An angle measured between *X* and *Y*, or *Y* and *Z*, will be the same measured at *A* or *B* or at any point on the circular arc from *X* around through *A* and *B* to *Z*. Thus if the observer should be on the circle, his position will be indeterminate. This situation, known as a "revolver," can occur if the center object sighted upon is farther away from the observer than the other two. By selecting three objects so that they are essentially in line, or that the center one is closer to the observer, you can avoid an indeterminate situation.

A "revolver" should be avoided, but if one does develop it can be made determinate by the addition of another LOP, such as a single bearing on one of the three objects, or on any other point.

The geographic coordinates of a position can be calculated from the known latitude and longitude of the three objects and the measured angles between them, but the complexity of the mathematics makes the use of a calculator or microcomputer program necessary.

LOPs From Horizontal Angles

A fix can also be determined from lines of position derived from two horizontal angles that may or may not share a common point. Each angle can be used to establish a circle of position. The circle contains both of the objects sighted upon and the position of the observer. The radius of the circle is the distance between the two objects divided by twice the sine of the measured angle; after the distance is measured from the chart, this is an easy computation with an electronic calculator. Set a pair of dividers—or better yet, a drafting compass—for the computed radius, scribe an arc from each observed point, and where the two arcs intersect is the center of the circle. Using this center, draw in as much of the circular LOP as you need. Do this for two angles and the intersection of the circles fixes your position. (There will be two intersections, but common sense, or a single rough bearing, will suffice to eliminate the ambiguity.)

This procedure is useful on any boat, but it can be especially valuable on sailboats in rough weather or when racing. Under such conditions, taking accurate direct bearings across the steering compass, or with a hand bearing compass, may be very difficult. A simple plastic sextant is not expensive and is all that is required. This procedure is also a good introduction to the use of a sextant in more advanced navigation techniques.

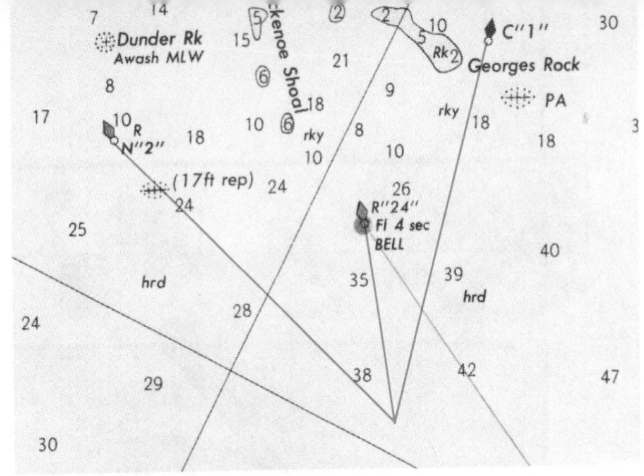

Figure 2115 **If a three-arm protractor is not available, the angles can be drawn on a sheet of transparent paper, *above*, which is then moved about on the chart until correctly positioned. In using the two-horizontal-angle method, *below*, the center of the three objects, Y, should not be farther away from the observer than the other two, or the results may be a "revolver," as shown here, where the boat may be at any point on a circle containing the three objects.**

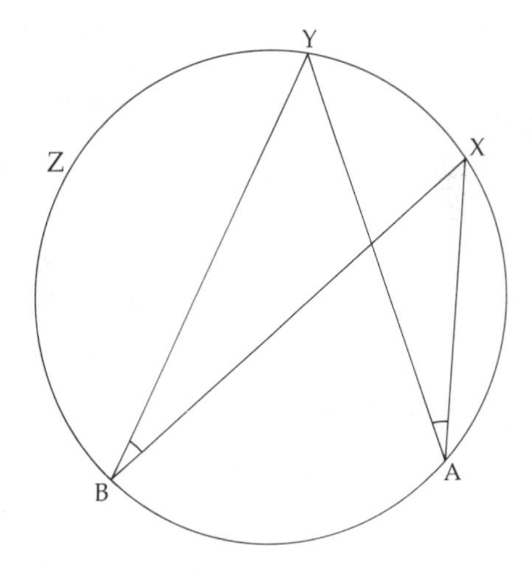

Vertical Angles

A circular line of position represents points at a constant distance from a specific object. Such distances are often found by the measurement of vertical angles with a marine sextant, or with a specialized device for such purposes alone.

The height of many natural and man-made features like lighthouses, bridges, and radio towers is shown on charts. Be sure to determine if the height shown is that above the structure's base on land or the vertical distance above the chart's datum for heights, usually mean high water. Frequently, a correction may be required for the tidal level at the time of observation, its difference from MHW, see Chapter 20. On such features as lighthouses, the height usually given is that of the *light*, not that of the top of the structure.

Figure 2116 **A three-arm protractor is a specialized plotting instrument used with the two-horizontal-angles technique of position determination. When the angles have been set, the arms are lined up with the appropriate chart symbols, and position is marked with a pencil point through the hole in the center of the scaled protractor.**

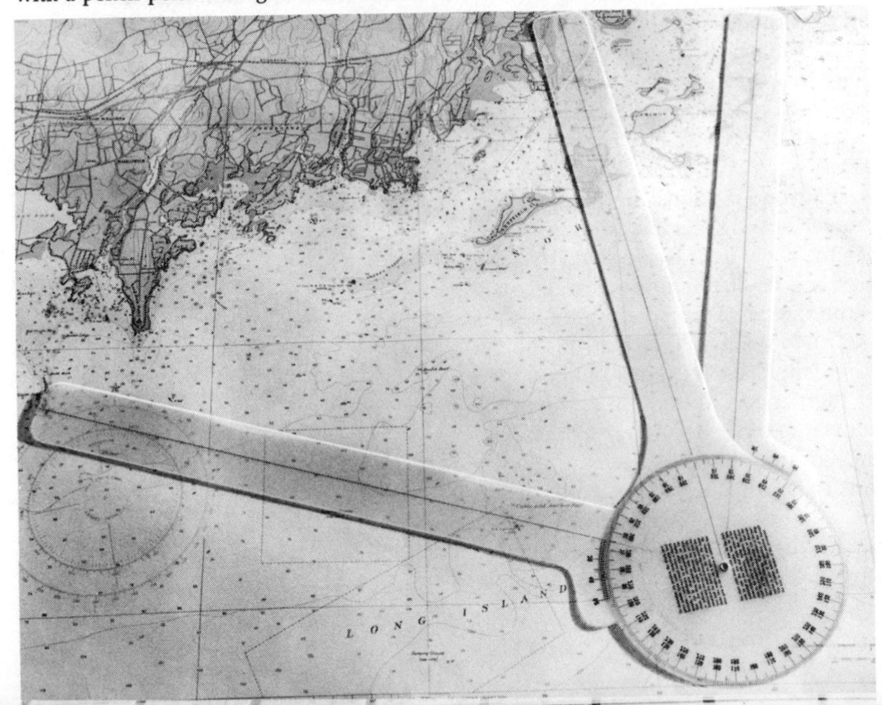

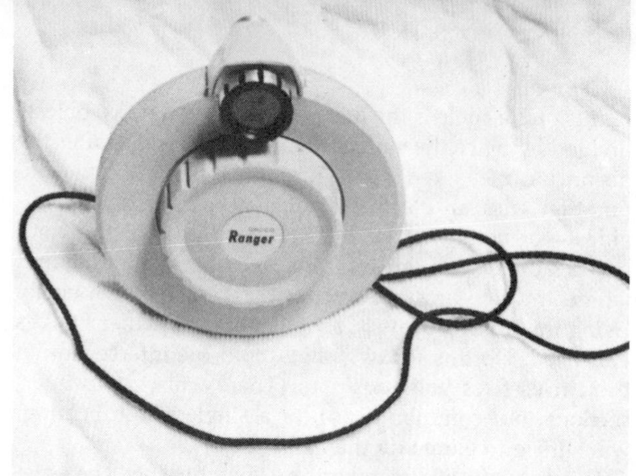

Figure 2117 **A marine sextant, *left,* most commonly used for taking celestial navigation sights, may also be used in piloting for the measurement of horizontal or vertical angles. If the vertical or horizontal dimension of an object is known, this optical rangefinger, *above,* can make an accurate measurement of the distance up to a mile.**

Using a sextant, the vertical angle is measured between the base and the top or other part of the structure. This angle (A) and the known height (h) can be used to determine the distance (d) from the observer to the object by a simple trigonometric equation: $d = h \div \tan A$. Use trig tables from *Bowditch,* Volume II, or an electronic calculator.

Special range-finder instruments are marketed which use the principle of matching images much like the focusing mechanism of many cameras.

Danger Bearings and Angles

The safety of navigation often can be ensured without a complete fix. As noted earlier, a single line of position has value—if it is of reasonable accuracy, it can assure you that you are somewhere along it and not elsewhere. A line of position can be chosen, in many piloting situations, that will keep a boat in safe water without precisely defining its position.

Danger Bearings

A bearing line can be established so that positions on one side will assure the boat's being in safe waters, while a position on the other side may signify a hazardous situation; such a *danger bearing* is shown in Figure 2118*A*. Here, a shoal, unmarked by any aid to navigation, lies offshore; the problem is to pass it safely. A lighthouse is observed on shore just beyond the shoal, and is identified on the chart. The danger bearing is a line from the lighthouse just tangent to the shoal on the safe side. This line is drawn on the chart and its direction is measured. The line is labeled with the direction preceded by the letters "NMT" (not more than). No time is shown as it is not an actual observation. For emphasis, hachures can be added on the side toward the danger, and using a red pencil will make it stand out from other lines on the chart.

As the boat approaches the hazardous area to port, a series of observations are made on the selected object. In the example shown, any bearing on the lighthouse numerically *less* than the danger bearing indicates a *safe* position; any bearing *greater* than the danger bearing indicates that the boat is in or approaching the shoal area if it continues on the same course. Should the danger area lie to starboard, rather than to port as in Figure 2118*B*, the "greater" and "less" factors above would be reversed and the label on the line would be "NMT" (not more than).

Danger bearings cannot always be established, but their use should be considered whenever conditions permit. It is necessary to have a prominent object, although it need not be an aid to navigation, that can be seen from the boat and positively identified on the chart. This object should lie beyond the danger area in the same general direction as the course of the boat as it approaches and passes the area to be avoided.

Figure 2118 **A danger bearing can be established to avoid an unmarked hazardous area. In A, any bearing on object X that is more than 337° indicates potential danger for the boat. In B, the reverse is true; a bearing less than 008° indicates possible danger if the boat continues on course.**

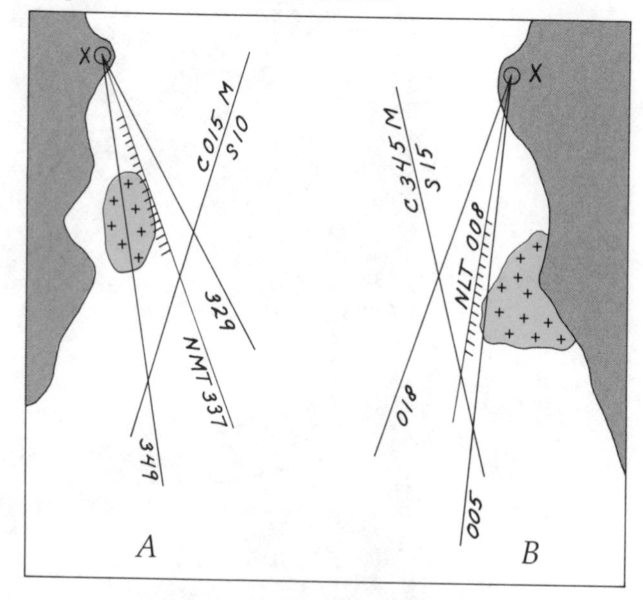

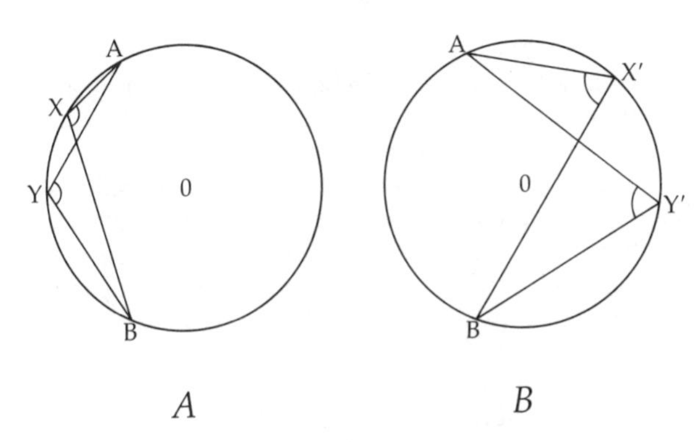

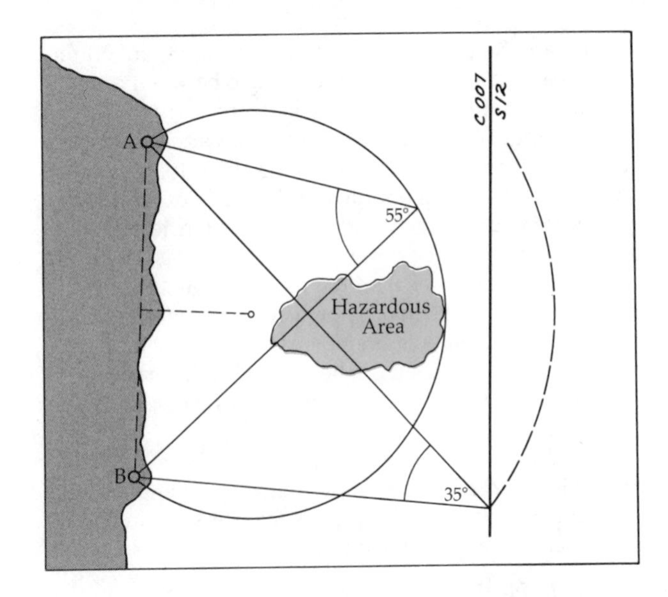

Figure 2119 ***Left:*** **The various points at which there is a constant angle between the lines of sight to two objects form a circle. The angle between the lines to A and B is the same at X or Y, as well as at points X′ and Y′. Horizontal angles between A and B, *right,* are measured as the boat approaches the hazardous area. Any angle *less* than 55 degrees indicates that the boat is in safe waters.**

Horizontal Danger Angles

A horizontal angle measured between two identifiable fixed objects defines a circle of position. In Figure 2119 left, an observer at X or Y would measure the same angle between A and B; he would also measure the same angle at any point along that semi-circular arc. If the observer were on the other side of the center of the circle, at X′ or Y′, or other point on that arc, a constant angle also would be observed. This angle, however, is not the same as that measured across the circle at X and Y. Note that the angle will be greater than 90 degrees if the sighted objects are on the same side of the circle as the observer, as in Figure 2119*A* left, and less than 90 degrees if they are on the opposite side as in Figure 2119*B* left.

Such a circle may be established to indicate the boundary between positions of safety and those of possible danger. When such an LOP is set up, the angle defined by the circle is termed the HORIZONTAL DANGER ANGLE.

Figure 2119 right illustrates the use of a single horizontal danger angle to avoid an unmarked shoal area. The problem is to stay sufficiently offshore to miss the hazard. The horizontal danger angle is found by drawing a circle that includes two prominent identifiable objects ashore, A and B in Figure 2119 right, and the shoal area. The circle is established by drawing a line between the sighted objects, and then drawing a second line at right angles to the first line at its midpoint. The center of the circle is found by trial-and-error along this second line so that the arc includes the desired points.

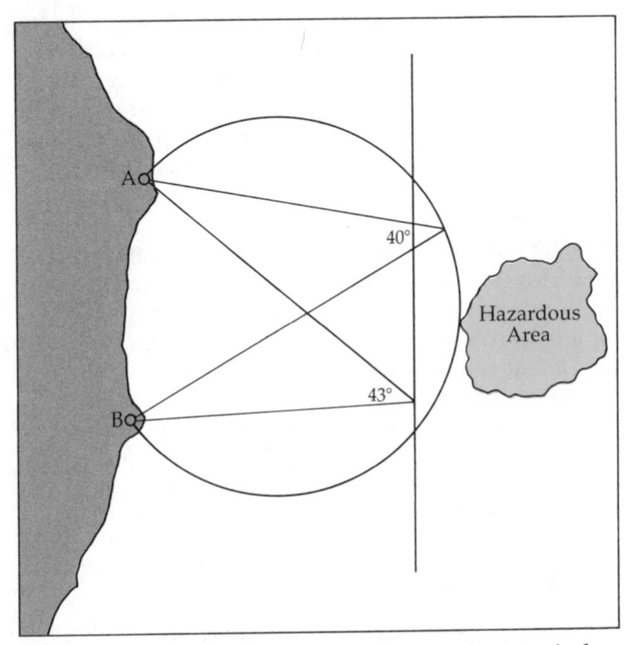

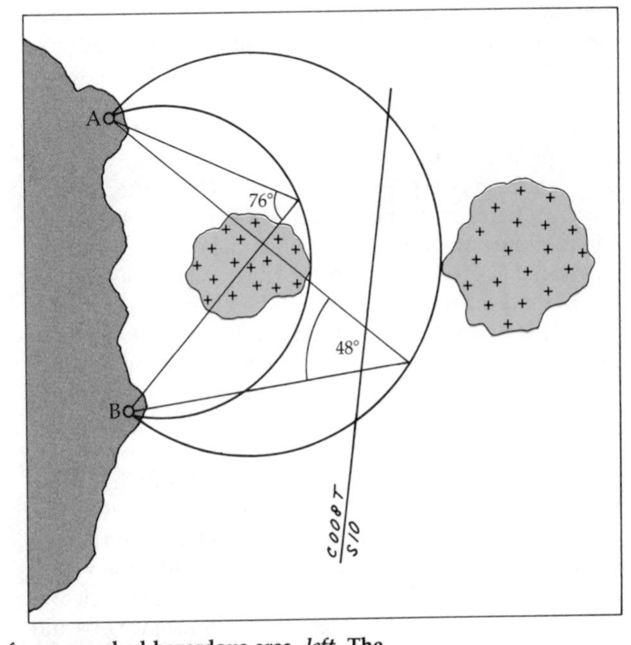

Figure 2120 **Here it is desired to pass inshore of an unmarked hazardous area, *left.* The horizontal danger angle is established, but now the safe measurement are those that are *greater* than the horizontal danger angle. For passage between two unmarked hazardous areas, *right,* double horizontal danger angles may be used. This technique is the combination of the two preceding situations; a safe passage is indicated by horizontal angles *between* specified upper and lower limits.**

Lines are then drawn to A and B from any point on the circle near the shoal area and the angle between them is measured; in this example, it is 55 degrees.

If the angle between the objects, measured from the boat as it approaches the area, is *less* than the horizontal danger angle, then the radius of the circle on which the boat is located is larger, and the boat is farther offshore in *safe* waters. On the other hand, if the measured horizontal angle is greater than the danger angle, the boat is closer in and may be in or approaching a hazardous area. The horizontal angle is preferably measured with a sextant, but in most cases it can be determined by calculating the difference between two compass or relative bearings on the objects.

Should the situation require passing inshore of a shoal area, Figure 2120 left, a single horizontal danger angle can again be used. In this case, *safe* waters are indicated by angles *greater* than the danger angle.

Double Horizontal Danger Angles

The two situations described above can be combined into DOUBLE HORIZONTAL DANGER ANGLES where the requirement is for a safe passage between two offshore hazardous areas. Figure 2120 right is essentially a combination of the two preceding illustrations. The principles are the same; here the safe horizontal angles are those *between* upper and lower danger limits.

Vertical Danger Angles

A circle of position, obtained from a vertical angle measurement, can also be used to mark the boundary between safe and hazardous waters. The advantage here is that only a single prominent object is required. With that object identified visually and on the chart, a circle is drawn using that object as the center and having a radius that will just include all of the hazardous area. The radius is measured from the chart and converted to a vertical angle by geometric formula or Table 9 in *Bowditch*, Volume II. The danger circle is then labeled with the VERTICAL DANGER ANGLE in the same manner that a horizontal danger angle is labeled.

As the boat approaches the dangerous area, a series of vertical angle measurements is taken on the selected object. Measurement of a vertical angle *less* than that specified as the vertical danger angle indicates that the observer is farther offshore, and consequently he is in *safe* waters.

Corresponding situations would prevail for safe passage inshore of a hazardous area, or between two danger areas, as was shown for horizontal danger angles.

Measurements of distance directly from an optical range finder can be substituted for vertical angles if such a device is on board.

Positioning Procedures

As mentioned earlier, the "basic" fix is obtained by crossing two lines of position, and a third LOP is desirable. It is assumed for the fix that the observations or measurements for these LOPs are made simultaneously, but the normal situation on a small boat will be that of a single bearing-taker and one piece of equipment. Observations are therefore taken sequentially rather than concurrently. If the observations are taken quickly, the distance traveled between them is negligible, too small to be plotted on the chart. Adherence to the procedures to be described here will minimize the error from sequential observations.

As a result of the movement of the boat along its course, the observed bearing angles will be changing. Consider the relative *rate of change* of the bearing to each object to be sighted upon; those on the beam will be changing more rapidly than those more forward or aft of the beam. Bearings on close objects will change more rapidly than those at greater distances.

For the most accurate determination of a fix from two lines of position, first take a sight on the object with the *least* rate of change, then on the other object, and finally a repeat on the first object. Use the second observation and the average of the first and third bearings (which are, of course, on the same object). If circumstances permit only one observation on each object, it is usually preferable to take the more rapidly changing bearing last and then make an immediate chart plot of both bearings.

If you are able to take observations on three objects, take them in descending order of rate of change of bearing and consider taking a repeat of the first bearing, if possible, for averaging as before.

The time of the fix should be that of the approximate middle of the series of observations. At typical boat cruising speeds and chart scales, a difference of a minute or two will not be significant. At 12 knots, a boat will travel ⅕ mile in one minute; on a 1:80,000 scale chart, this distance of approximately 400 yards is less than 3⁄16 inch.

Positioning Standards of Precision

As set forth in Chapter 19, there are generally accepted standards of precision in describing the position of a vessel. If geographic coordinates are used, latitude and longitude, in that sequence, are stated to the nearest tenth of a minute on charts with scales of 1:50,000 or smaller, and to the nearest second on charts of larger scale.

If the position is stated with respect to some aid to navigation or landmark, direction from that point is given to the nearest degree (true) and distance is given to the nearest tenth of a mile.

The Running Fix

The lack of a second object on which to make an observation may at times prevent the immediate determination of the boat's position by a normal fix, even though one good line of position has been established. In such situations, a somewhat less accurate determination of position may be made by a RUNNING FIX (R FIX). In this tech-

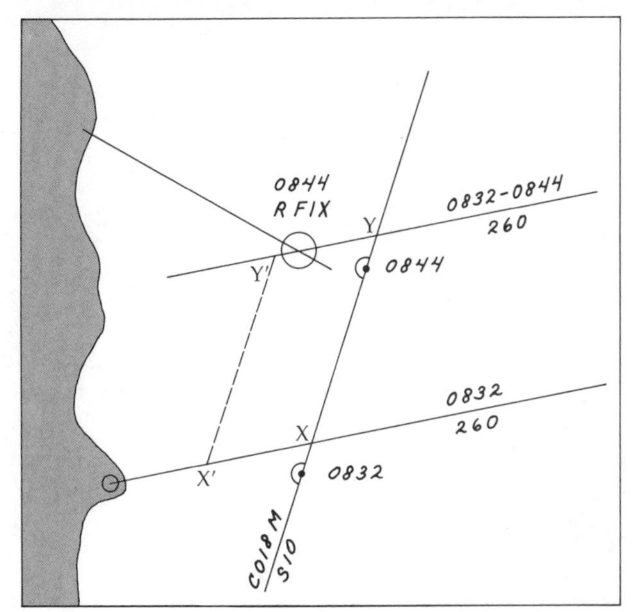

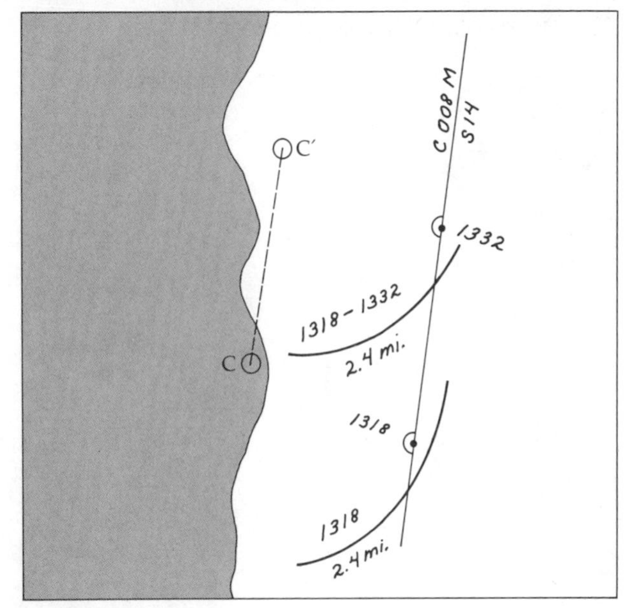

Figure 2121 **A line of position is advanced by moving forward any point on it an amount equal to the boat's motion during the time interval, *above*, and re-drawing the line through the advanced point. The intersection of this LOP and another LOP for this time is a running fix. A circular LOP, *right*, is advanced by merely moving the center point an amount equal in direction and distance to the motion of the boat during the time interval, and re-drawing the circle on the new center. C—C′ is equal to the spacing between the two DR positions.**

nique there is one line of position at the time of the running fix and another from a different time, usually earlier. This latter LOP must be replotted from its position at the time of its measurement to account for the movement of the boat during the intervening time. The act of advancing the first LOP brings it to a common time with the second observation so that their intersection may be considered as a determination of position. Note that this line so advanced may have resulted from an observation on the *same* object as that used for the second sighting at the time of the running fix, or on a different object.

Advancing a Line of Position

A line of position is advanced as follows:

1. A point is selected on the original LOP. This can be the intersection of the line with the DR track, X in Figure 2121 left, or any point on the LOP, such as X′.
2. The selected point is moved in an amount equal in distance and direction to the boat's motion during the time interval involved. This is most easily measured from the DR position at the time of observation to the DR position for the advanced time.
3. The new LOP is drawn in through the advanced point parallel to the original LOP.
4. The advanced LOP is labeled as soon as it is drawn: both times are shown above the line and the direction is shown beneath it.

Advancing a Circular LOP

If the LOP to be advanced is circular, as would be the case with a distance measurement by vertical angle or range finder, the center of the circle is the point advanced by the procedure described above, Figure 2121 right. The circle or arc is redrawn using the new center, and is then labeled as for any other advanced LOP.

Advancing an LOP With Course and/or Speed Changes

The advancement of the selected point on the initial LOP must take into account all changes in course and/or speed of the boat during the time interval. It is frequently convenient to calculate and plot the DR positions for the time of the initial observation and for the advanced time. The movement of the selected point on the initial LOP should parallel and equal a line drawn between these two DR positions. This technique is particularly useful if there has been more than one change in course and/or speed, Figure 2122 left.

Advancing an LOP With Current

The advancement of a line of position may take into account current effects if they are adequately known or can be estimated. The selected point on the initial LOP is first advanced for DR speeds and courses, and then for the set and drift of the current, both calculations using the elapsed time interval between the two observations; see Figure 2122 right.

The Accuracy of Advanced LOPs

The accuracy of an advanced line of position can be only as good as the accuracy of the initial observation *decreased* by any errors or uncertainties in the navigator's data for the boat's course and speed (and current data, if used). Obviously, the longer the time interval over which the sight is advanced, the greater the uncertainties become. The interval should, therefore, be the minimum possible, and, in piloting, should rarely exceed 30 minutes. (In navigation on the high seas, celestial lines of position are often advaced several hours, but the accuracy of positioning is less critical in these circumstances.)

The same criteria for the angle of intersection of lines of position apply to running fixes as were discussed previously for normal fixes.

Lines of position may be retired (moved back to an earlier time) in a similar manner, but such movement is quite rare in piloting.

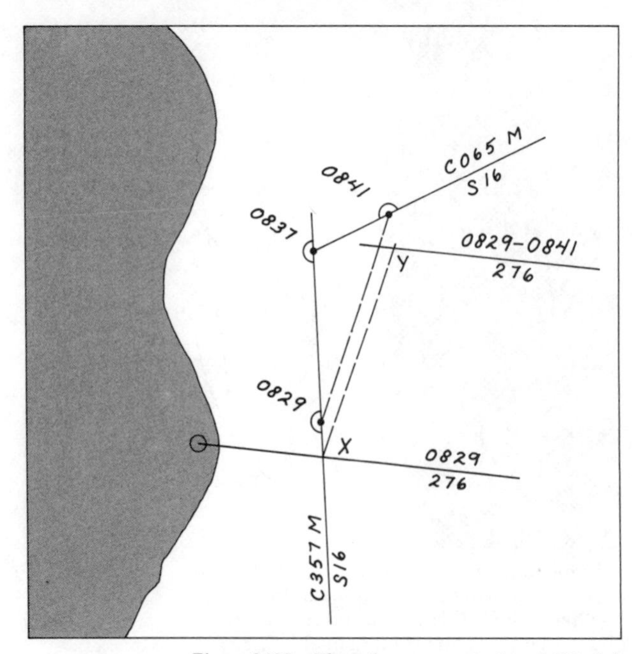

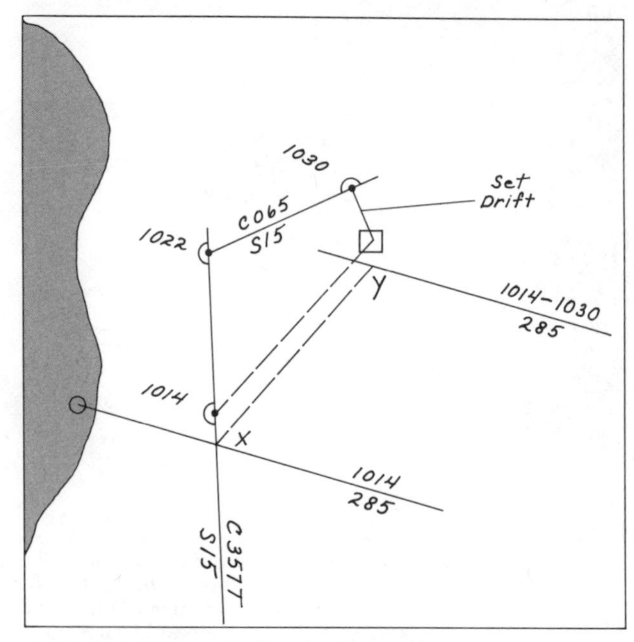

Figure 2122 **The advancement of an LOP, *left,* must take into account all changes of course and/or speed during the time interval. The net effect can be determined by drawing a light line between the two DR positions for the times concerned. The selected point on the original LOP is advanced in the same directon and distance as this net effect line. If the LOP is to be advanced with allowance for current, *right,* the same procedure as before is followed, except that first a further movement is made from the second DR position for the current during the time interval, and this is the estimated position at the time the LOP is advanced.**

Running Fixes and DR Plots

The DR plot may be interrupted and restarted from a running fix if it is considered of reasonable accuracy. In all cases, however, a fix obtained from two or more essentially simultaneous observations or measurements is to be preferred over a running fix.

Observations on a Single Object

Position information can be obtained from successive observations on a single object by means of several specialized procedures. These include bow-and-beam bearings, doubling the angle on the bow, two bearings and run between, and two relative bearings.

In each case, it is important to note that the angles and distance run be measured with respect to the course being made good over the bottom; corrections to the DR track and all relative bearings must be made for current or other offsetting influences if such information is known or can be estimated.

Bow-and-Beam Bearings

Position determination is easily accomplished by the technique of taking two particular successive bearings on a single object to one side or the other of the boat's course. The first bearing is taken when the object sighted upon bears either 45 degrees to starboard (RB 045°) or to port (RB 315°). The second bearing is taken when the same object is broad on the respective beam (RB 090° or 270°), Figure 2123 left. The time interval between bearings is found, and then the distance traveled between the two sightings is calculated using the boat's speed. Because of the nature of the triangle established by the two bearings and the course line, the boat is as far from the sighted object at the time of the second (beam) bearing as the distance traveled during the time interval. Since there is an LOP (the beam bearing) and a distance, the boat's position is fixed.

The principal advantage of this procedure is the ease with which it can be accomplished. Beam bearings are easily taken, and 45-degree relative bearings can be quickly set up using crude sights oriented with an ordinary plastic drafting triangle. On some boats, it may be possible to make the 45-degree sight over the steering compass. The bow-and-beam technique often can be done by the helmsman without assistance.

Doubling the Angle on the Bow

This is a more generalized application of the same principles as those involved in bow-and-beam bearings. Geometric principles establish that whenever an angle on the bow is doubled, the distance from the position at which the second bearing is taken to the object sighted upon is equal to the distance traveled (over the bottom) during the time interval between the two observations, Figure 2123 right. The angle is the relative bearing to starboard or to port. A relative bearing of, say, 340 degrees, measured in the conventional manner, must be converted to an angle of 20 degrees to port for the purpose of "doubling"; in this case, the doubled angle would be RB 320° (40 degrees to port).

This doubling-of-the-bow-angle technique has the advantage of determining the boat's position *before* the sighted object is abeam. With information about the boat's position at such an earlier moment, the course can be projected ahead on the chart. Should the boat be traveling too close to shore for passing a headland, a more adequate warning

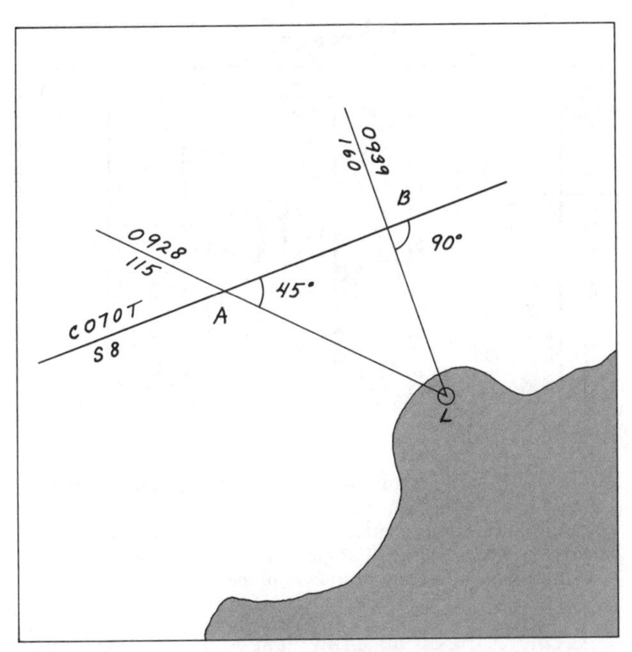

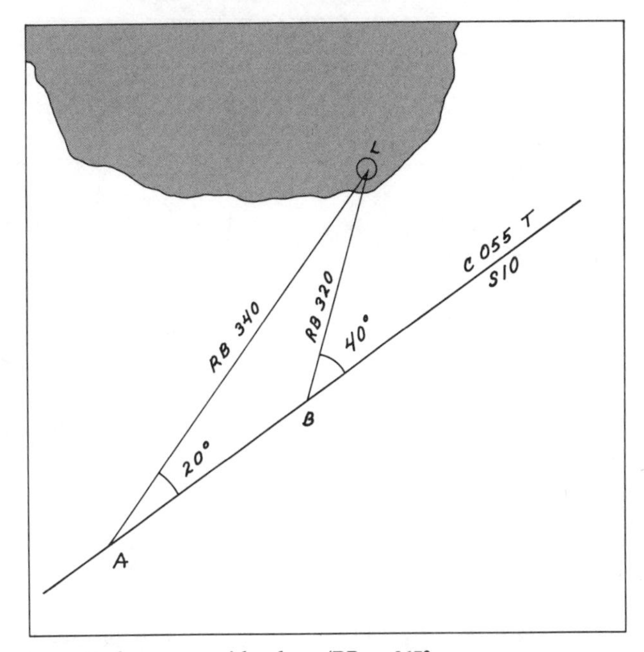

Figure 2123 **If a bearing is taken when an object bears 45 degrees on either bow (RB = 045° or 315°), *left,* and again when the same object is broad on the beam (RB = 090° or 270°), the distance to the object at the time of the second bearing is equal to the distance run over the bottom between the bearings; distance B-L = A-B. *Right:* If two observations are made so that the second relative bearing is twice that of the first, as measured from the bow to either starboard or port, the distance to the sighted object at the time of the second bearing is the same as the distance made good between the sightings: L-B = A-B.**

of the dangerous course will thus be in hand, and corrective action can be taken sooner.

This technique does have a disadvantage, however, in that it requires the possession and use of a pelorus or other instrument for measuring relative bearings with reasonable accuracy. Another disadvantage is that a second person is needed as a bearing-taker.

Two Relative Bearings

A more generalized solution from two relative bearings can be used if a copy of *Bowditch,* Volume II, is on board. Table 7 of that book uses two items of information—the angle between the course and the first bearing (the first relative bearing) and the difference between the course and the second bearing (the second relative bearing). Columns of the Table are in terms of the first item and lines are in terms of the second item above; the interval between tabular entries is two degrees in both cases.

For any combination of the two relative bearings within the limits of Table 7, two factors will be found. The first number is a factor by which the distance run between the bearings is multiplied to obtain the distance away from the sighted object at the time of the second bearing. The second factor of the same entry in Table 7 is the multiplier to be used to determine the distance off when the object is abeam, assuming, of course, that the same course and speed are maintained. See Figure 2125.

Two-Bearings-and-Run-Between

With only a single object upon which sights can be taken, another method, known as "two-bearings-and-run-between," may be used. In this procedure a bearing is taken on the object, the boat proceeds along her course, and a second bearing is taken after the angle has changed

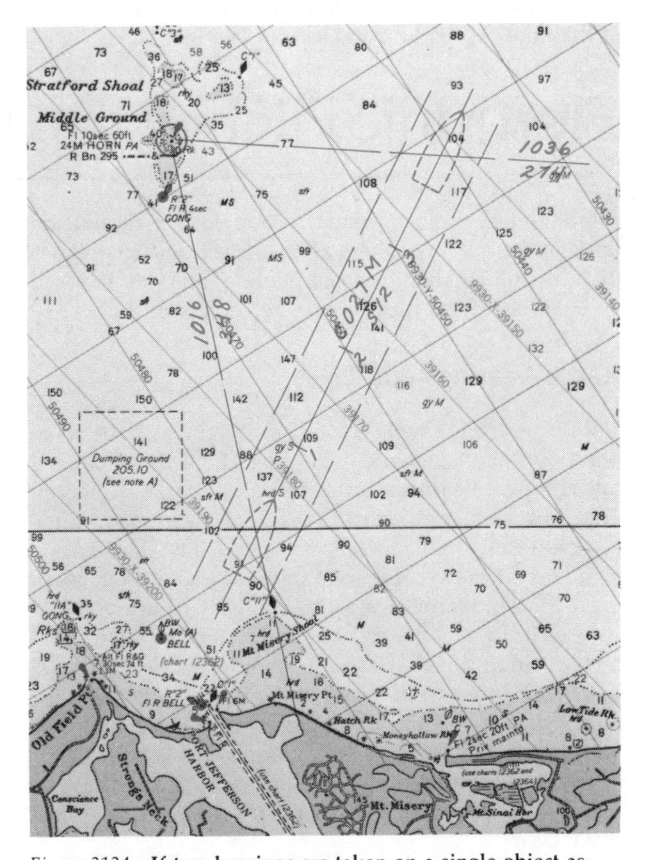

Figure 2124 **If two bearings are taken on a single object as the boat passes it, and are plotted on the chart, only one point can be found on each LOP where the boat's course line and distance made good will fit. The dashed lines to either side of the course line show how improperly placed course lines would be too short or too long.**

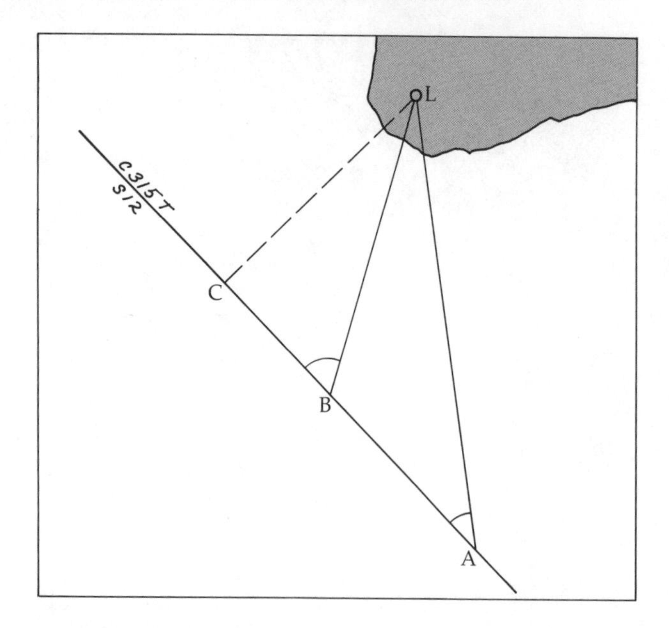

TABLE 7
Distance of an Object by Two Bearings

Difference between the course and second bearing	Difference between the course and first bearing													
	34°		36°		38°		40°		42°		44°		46°	
°														
44	3. 22	2. 24												
46	2. 69	1. 93	3. 39	2. 43										
48	2. 31	1. 72	2. 83	2. 10	3. 55	2. 63								
50	2. 03	1. 55	2. 43	1. 86	2. 96	2. 27	3. 70	2. 84						
52	1. 81	1. 43	2. 13	1. 68	2. 54	2. 01	3. 09	2. 44	3. 85	3. 04				
54	1. 63	1. 32	1. 90	1. 54	2. 23	1. 81	2. 66	2. 15	3. 22	2. 60	4. 00	3. 24		
56	1. 49	1. 24	1. 72	1. 42	1. 99	1. 65	2. 33	1. 93	2. 77	2. 29	3. 34	2. 77	4. 14	3. 43
58	1. 37	1. 17	1. 57	1. 33	1. 80	1. 53	2. 08	1. 76	2. 43	2. 06	2. 87	2. 44	3. 46	2. 93
60	1. 28	1. 10	1. 45	1. 25	1. 64	1. 42	1. 88	1. 63	2. 17	1. 88	2. 52	2. 18	2. 97	2. 57
62	1. 19	1. 05	1. 34	1. 18	1. 51	1. 34	1. 72	1. 52	1. 96	1. 73	2. 25	1. 98	2. 61	2. 30
64	1. 12	1. 01	1. 25	1. 13	1. 40	1. 26	1. 58	1. 42	1. 79	1. 61	2. 03	1. 83	2. 33	2. 09
66	1. 06	0. 96	1. 18	1. 07	1. 31	1. 20	1. 47	1. 34	1. 65	1. 51	1. 85	1. 69	2. 10	1. 92
68	1. 00	0. 93	1. 11	1. 03	1. 23	1. 14	1. 37	1. 27	1. 53	1. 42	1. 71	1. 58	1. 92	1. 78
70	0. 95	0. 89	1. 05	0. 99	1. 16	1. 09	1. 29	1. 21	1. 43	1. 34	1. 58	1. 49	1. 77	1. 66
72	0. 91	0. 86	1. 00	0. 95	1. 10	1. 05	1. 21	1. 15	1. 34	1. 27	1. 48	1. 41	1. 64	1. 56
74	0. 87	0. 84	0. 95	0. 92	1. 05	1. 01	1. 15	1. 10	1. 26	1. 21	1. 39	1. 34	1. 53	1. 47
76	0. 84	0. 81	0. 91	0. 89	1. 00	0. 97	1. 09	1. 06	1. 20	1. 16	1. 31	1. 27	1. 44	1. 40
78	0. 80	0. 79	0. 88	0. 86	0. 96	0. 94	1. 04	1. 02	1. 14	1. 11	1. 24	1. 22	1. 36	1. 33
80	0. 78	0. 77	0. 85	0. 83	0. 92	0. 91	1. 00	0. 98	1. 09	1. 07	1. 18	1. 16	1. 28	1. 27

Figure 2125 ***Left:*** **Distance off at the time of second bearing L-B and distance off when abeam, L-C, can be found by applying multiplying factors to the distance run, A-B. These factors for various pairs of angles are given in Bowditch Table 7. A sample of this table is shown *above*.**

by at least 30 degrees. This second bearing may be taken before or after the sighted object is passed abeam. From the times of each bearing, the time interval is calculated, and from this, the distance run is determined.

Both bearings are plotted, as is the course of the boat, Figure 2124. A pair of dividers is opened to the distance run between the two bearings and is then moved parallel to the course line until the points fall on the bearing lines. The divider points now indicate the positions of the boat at the times of the first and second bearings.

As before, the distance run must be that *over the bottom;* suitable corrections must be made for any current. The accuracy of this technique depends upon the accuracy of each of the two bearings, the accuracy of the calculation of the distance run over the bottom, and the accuracy with which the boat was steered during the interval between the two sightings. The net effect is to make the positions so determined somewhat less certain than the other techniques described above.

Electronic Piloting

Although electronic navigation is a topic in itself, outside the scope of piloting, there are two electronic techniques that are comparable to visual piloting, and can be used on boats that have the appropriate equipment.

Radio Bearings and Fixes

With relatively simple equipment, bearings can be taken on radio stations, even though the stations lie far beyond visual range. Bearings can be taken on radiobeacons operated by the Coast Guard as aids to navigation; on some aeronautical ranges and beacons operated by the FAA; on standard AM radio broadcast stations; and on stations in the 2-3 MHz marine band (but SSB signals are poor for RDF use). In general, the desirability for use in radio direction finding is in descending order as listed above. Radio bearings can be taken under any condition of visibility, and on stations at ranges as great as 100 miles or more; accuracies are, of course, more favorable with nearer stations. Direction finding can also be done on VHF signals, but with separate equipment and at much lesser distances.

Radio bearings are taken with a RADIO DIRECTION FINDER, usually referred to as an RDF if it is manually operated, or as an ADF if the bearings are taken and displayed automatically. These items of electronic equipment are discussed more fully in Chapter 23.

Using Radio Bearings

Radio bearings are plotted in essentially the same manner as visual bearings. They are not as sharp as visual observations, however; an accuracy of 2 to 3 degrees is the best that can be expected, and this level will be achieved only with experience. Accordingly, it is advisable to obtain three radio LOPs, and even so, the position so determined should not be considered as precise as one from visual bearings.

Accurate identification of the station being received is essential, and the exact location of the transmitting antenna must be plotted on the nautical chart if it is not already shown there. Radiobeacons, both marine and aeronautical, are normally so plotted, although sometimes additional aeronautical beacons can be transcribed from aviation to nautical charts. The antenna location of a few standard AM broadcasting radio stations will be found on charts identified by call letters and frequency, but often others can be added.

Radio bearings taken on a station more than approximately 50 miles distant must be corrected before being plotted on a Mercator chart. (Distance will vary with latitude and the relative position of the vessel and the radio station.) This correction is required because radio waves travel via a great circle path, and this does not plot as a straight line on a Mercator chart; see page 422. Table 1 in *Bowditch,* Volume II, provides the correction factor and instructions on how it should be applied. No correction is needed for nearer stations, or on stations at any distance if the plotting is done on a gnomonic chart.

A sequence or "round" of several radio bearings on different radiobeacons or other stations can normally be taken quickly enough so that any movement of the boat during this time can be ignored and the lines of position can be plotted as of the mid-time of the sequence. If any appreciable time delays are encountered, however, an RDF line of position can be advanced or retarded in the same manner as a visual LOP.

To assist in plotting RDF bearings, special charts are available for some areas. These are commercial adaptations of government charts with compass roses entered on each radiobeacon or other usable station. Alternatively, a skipper may apply adhesive, transparent plastic compass roses to regular nautical charts at the location of suitable radiobeacons and stations.

In considering the accuracy of radio bearings, it must be kept in mind that these are referenced to the boat's heading as determined by its magnetic compass. Compass deviation must be known and taken into consideration. Any sloppy steering or yawing off-course as a result of rough water will directly affect the RDF bearings.

Labeling RDF Bearings and Fixes

In view of the lesser accuracy of RDF bearings as compared to visual sighting, and particularly those radio bearings taken on distant stations, it is desirable that the fix obtained from such lines of position be so identified. A circle symbol, with a dot in the center if necessary, is used—just as for a visual fix—but it is clearly identified as a radio direction finder position by "RDF" after the time; see Figure 2126 top.

If RDF bearings must be used to establish a position in possibly hazardous waters, it is advisable to plot additional lines two or three degrees (or more if conditions make for a reduction of normal accuracy) on either side of the basic LOP. These will clearly show the possible *area*, rather than specific position, in which the RDF-equipped boat is located; see Figure 2309 bottom.

Although it is of much lesser accuracy, an RDF running fix is better than no position information beyond dead reckoning if only one radiobeacon or other station can be used. Just as with visual techniques, an initial radio bearing is taken and plotted. Then after sufficient movement of the boat to ensure an adequate change in direction, a second radio bearing is taken on the same signal source and is plotted. The first LOP can be advanced to a common time and thus an RDF running fix is established.

Combined Lines of Position

It is often possible, or necessary, to combine an RDF line of position with a visual bearing or range; this might occur only when one visual LOP could be obtained and it was necessary to cross it with another LOP from any source to get a fix. If radio lines of position are mixed with visual bearings or ranges, the former should be identified by adding the letters "RDF" above the line following the time figures; see Figure 2126 bottom.

Radio direction finding may also help the skipper whose boat is *not* equipped with an RDF. After establishing communications, an RDF-equipped boat or Coast Guard unit

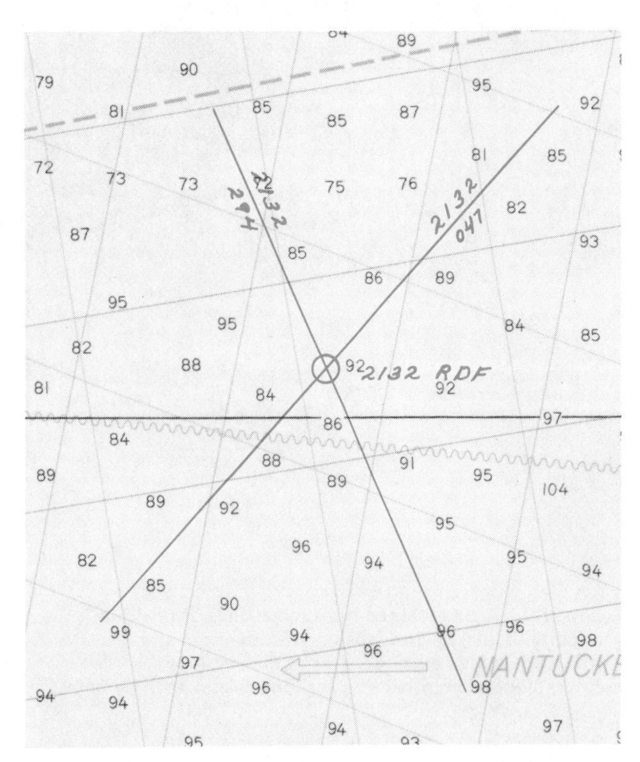

Figure 2126 **Because of the accuracy of a fix, *above,* derived from radio bearings is usually less than that of one obtained from visual observatons, it should be labled "RDF." If a combination of LOPs includes one from an RDF measurement, *below,* it may be well to add the letters "RDF" above the line to this bearing measurement.**

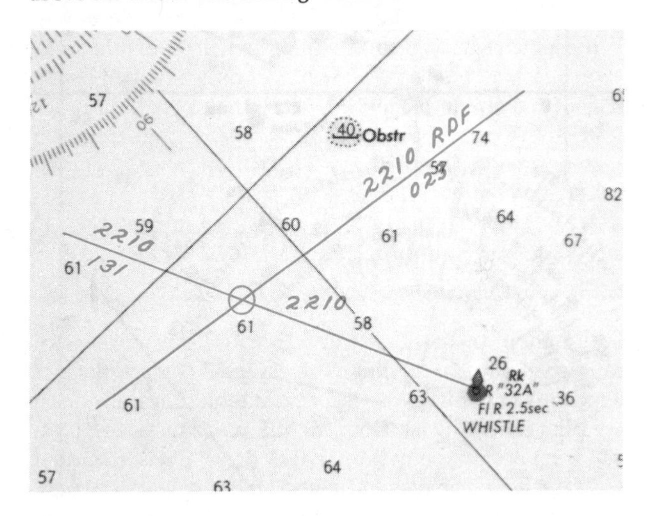

can take bearings on his signals as he gives a "long count" and then advise him of his position or home in on him to render assistance. There are, however, some limitations on this procedure. The newer single-sideband signals are not as good for direction finding purposes as the older double-sideband signals were, and VHF has a far lesser range.

Radar Bearings and Distances

Radar is another item of electronic equipment that is covered in greater detail in Chapter 23, but which will be considered here in connection with piloting. Radar has unique advantages in that a single instrument can measure

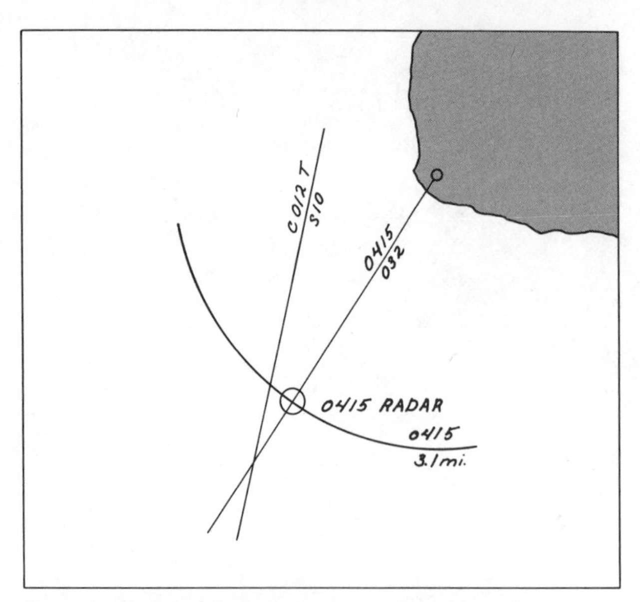

Figure 2127 **A fix obtained from radar lines of position is plotted with a circle and labeled with the time and "RADAR"; in this example, a fix is the result of a measurement of both bearing and distance to a single object.**

both direction and distance (range), to moving or still targets, and that these measurements can be accomplished under any visibility conditions.

Radar Observations

Measurements of direction by radar are *not* as accurate as those by visual sightings, but radar will penetrate darkness; light and moderate rain without difficulty, and heavy rain with some reduction in effectiveness; fog; and other restrictions that would prohibit visual bearings or range measurements.

On the other hand, measurements of distance by radar can be quite accurate, much more so than such observations taken by vertical sextant angles or with simple optical range finders.

Use of Radar Information

Radar data are used in the same manner as information from visual observations. Lines of position are combined to obtain fixes—typical combinations include two bearings, a bearing with a distance measurement to the same or another object, or best of all, two distances measurements. A fix obtained by radar bearings and/or distance measurements is labeled with time and "RADAR"; see Figure 2127.

When visibility conditions permit, a frequent combination, and an excellent one, is that of a visual bearing on an object and a radar measurement of distance to that same point.

Radar distance measurements also may be used in the same manner as distance determinations by vertical sextant angle to avoid a shoal area delineated by a circular LOP from prominent objects identifiable on the radarscope.

Radar observations on isolated objects such as buoys, offshore lighthouses, etc., usually are made without difficulty in identification of the object on the radarscope. When the radar target is on shore, however, some problems may be encountered in picking out the exact object on which the bearing and range information is to be obtained. Navigate with caution when using radar information from such objects.

Even more so than other types of piloting, radar requires practice and the building up of experience. If you have a radar, work out problems with it often in noncritical times so that you can employ it confidently when its use is essential.

Depth Information in Piloting

Information on the depth of water under a vessel is of course valuable in piloting. The essential characteristic of depth information is the fact that although it cannot tell the pilot where he is, it can tell him *positively* where he is *not*. If an accurate measurement of depth gives a reading of 26 feet, you may be at any number of places where that is the depth, but you certainly are not at a position where the depth is significantly greater or less.

How Depth Information Is Obtained

Data on the depth of the water beneath a vessel can be obtained manually or electronically. A hand lead line, see page 443, is a time-honored instrument of piloting, and every boat should have one of a length suitable for the waters usually cruised. It is accurate and dependable in use, although far from efficient or convenient. Electronic depth sounders provide information at a much faster rate and with the convenience of reading a simple dial or screen. They can also measure great depths that would be impracticable with a lead line. Electronic sounders are discussed in detail in Chapter 23.

Use of Depth Information

In using depth information, regardless of how obtained, a correction for the height of tide may be required. The use of *Tide Tables* and the necessary calculations with their data are covered in Chapter 20.

Depth data can be combined with a line or lines of position obtained from other means to yield position information. Under some circumstances, depth information alone may be of value in position determination.

Depth Information and a Single LOP

If the bottom of the body of water being navigated has some slope, and this slope is reasonably uniform, it may be possible to get position information from a single LOP, typically a beam bearing, and a depth reading. The bearing line is plotted and then examined for a spot where the depth figure on the chart agrees with the measured depth as corrected for the tidal stage; be sure that there aren't several such depths along the LOP, Figure 2128. Such a location should be considered as an estimated position (EP) rather than a Fix.

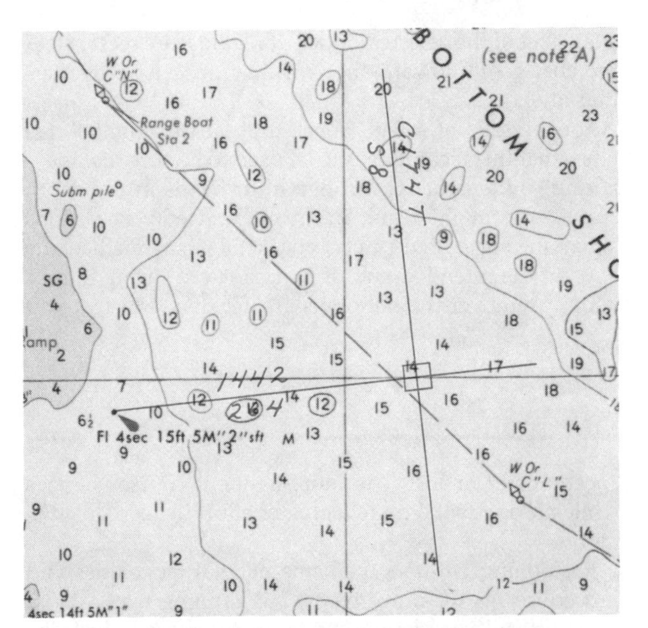

Figure 2128 **Depth information can often be used with a single LOP, such as a beam bearing, to determine an EP.**

Matching Measured Depths to the Chart

In some instances, a rough estimate of a boat's position can be obtained by matching a series of depth readings, corrected for the prevailing height of the tide, with depths printed on the chart. Depth readings are taken and recorded at regular time intervals, such as those corresponding to intervals of distance of one-tenth to one-half mile, as determined by the scale of the chart concerned, the density of printed depth figures, etc. (Electronic depth sounders can provide far more depth measurements than can be used.) These depths are marked on a piece of transparent paper at intervals determined by the scale of the chart, Figure 2129. This piece of paper is then moved about on the chart, keeping the line of soundings parallel to a line representing the course steered while they were being taken. A maximum degree of match is sought between the observed depths and the charted depths; exact concurrence should not be expected.

In the example shown here, the 60-foot depth as measured is aligned with the 60-foot curve on the chart, and a search for a match is started. If the line of soundings is moved north of the position shown in Figure 2129, the depths at the outer end would still be in general agreement, but the mismatch at the inshore end would generally show that the line was incorrectly placed. If the line were moved down the chart, keeping it parallel to the course sailed, then the 37-foot measurement near the inshore end would mismatch with the actual 48-foot depth. The matching found by the location shown in Figure 2129 is as close as may be expected, and this position of the line can be considered to fairly well indicate the boat's track as the soundings were made.

Position determination by the use of depth data is not well suited to shorelines that are foul with offshore rocks, or areas that have varying, irregular depths. Nor can this technique be used where the depth is quite uniform with few or no variations. There are, however, many situations where it can, and should, be used.

Position Checking by Depth Data

If an EP has been determined by some other method or combination of methods, you can check it by a depth measurement. Remember that confirmation cannot be positive, but denial can be quite certain if there is a significant difference between the depth at the boat's actual location and that charted for the estimated position. If depth readings indicate that this be the case, a further

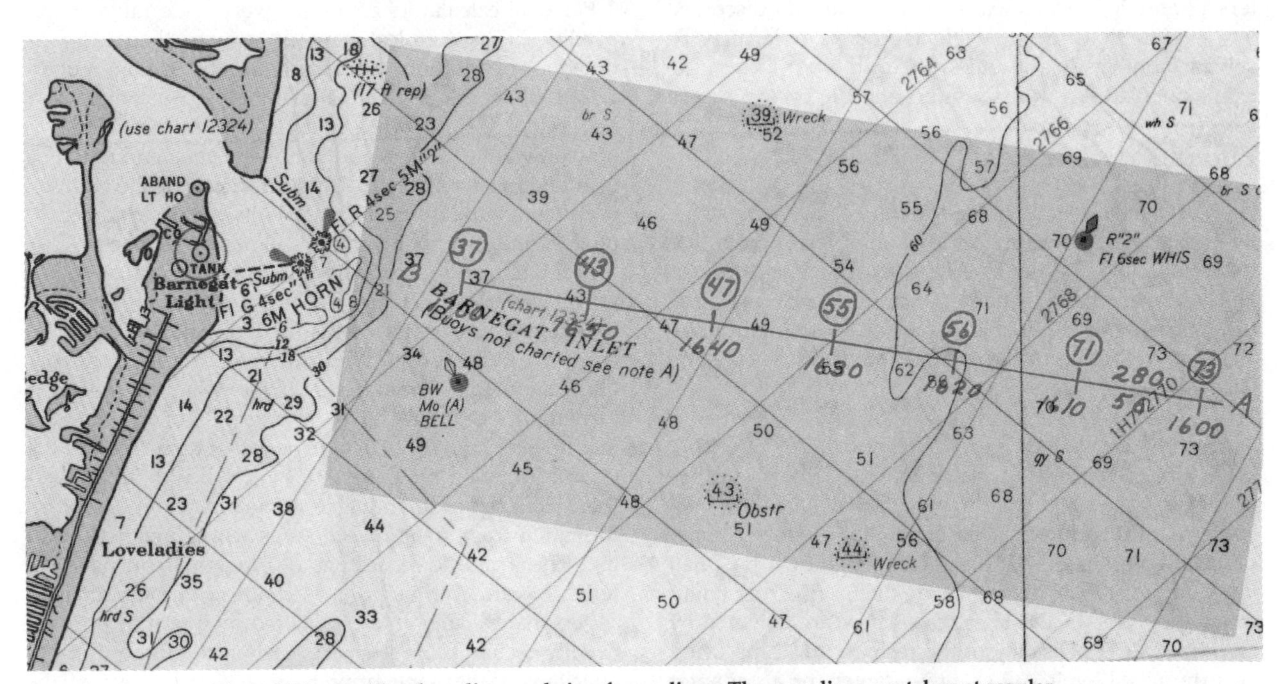

Figure 2129 **Position by a line or chain of soundings: The soundings are taken at regular intervals and plotted on a sheet of transparent paper. This sheet is then moved about on the chart until measured depth correspond with charted depths. See text for details.**

verification of the estimated position is a necessity, perhaps even an urgent necessity!

Fathom Curve Sailing

When cruising along a coast that has a fairly uniformly sloping bottom offshore, piloting can often be simplified by an examination of the chart for a fathom curve that will keep the vessel in safe waters generally and avoid any specific hazards. The boat is then steered so as to keep the reading of the electronic depth sounder within a few feet of the selected depth. If readings increase, steer the boat gently toward shore; if they decrease, edge farther offshore.

For the sake of safety, check the chart for the *full* distance that this technique is to be used to be sure that there are no sudden changes in depth, or bends in the depth curve, that might result in an unexpectedly hazardous situation. For added safety, check the chart for hazards beyond the intended end of fathom-curve sailing in case your speed over the bottom is greater than anticipated.

Piloting Under Sail

The task of piloting a sailboat is not greatly different from the same activity in a motorboat, but it can be more "work." Quite often under sail you cannot proceed directly toward your destination—if it lies generally to windward you will have to make a series of tacks. This adds to the task of dead reckoning, and makes that form of piloting even more important as a part of knowing where you are at all times.

Allow for the greater leeway made by a sailboat, and take fixes more often. Under sail, speed through the water must be estimated by eye if the boat is not fitted with a speedometer or log. The more-frequent course changes while tacking will require a greater effort to keep up a DR plot.

Positioning will use the same general techniques and procedures as for powerboats, but running fixes will be more difficult and less accurate for the reasons just discussed. Radio direction finding and the use of radar in piloting will be no different from the same activity aboard a powerboat. The electronic depth sounder, however, may require special precautions because of heeling.

Specialized Techniques

The preceding pages have provided basic information required for the safe navigation of a boat in pilot waters. Beyond these are techniques that are useful in special situations, or that may provide a quicker or easier solution to a piloting problem. These techniques should not be approached without a thorough understanding of Chapters 19 and 20, and the basic methods already discussed in Chapter 21. Current problems, for example, may be solved more easily and quickly with the use of an electronic calculator—but first understand the principles.

Piloting by Calculator

The availability of integrated circuit "chips" with steadily expanding capabilities, at lower and lower costs, has made possible the small ELECTRONIC CALCULATOR variously termed "hand," "pocket," or "personal." This small device has made it possible for persons with little understanding of math to solve complex problems quickly and easily.

The Versatility of Calculators

Personal electronic calculators are available in a range of physical sizes and computational capabilities. All contain several integrated circuit chips plus a small number of other components. The answers are shown by small figures produced by light-emitting diodes (LED) or liquid crystal displays (LCD). These calculators are powered by internal batteries, usually of the rechargeable type using an external 117-volt AC charger-adapter, or an adapter cord from a 12-volt DC cigarette lighter socket. Some calculators are smaller than a pack of cigarettes, but more typical models measure roughly 4 to 5½ inches (100-140 mm) by 2½ to 3½ inches (64-89 mm) and are less than 1 inch (25 mm) thick; units with a capability to print out answers on a paper tape are, of course, somewhat larger.

Types of Calculators

Personal calculators come in a very wide range of capabilities. The simplest have little more than the four functions of addition, subtraction, multiplication, and division, plus perhaps a single memory register; there may be a "change sign" key and a percent key. Their price is now only a few dollars and many are small enough to be included in a wristwatch. The next step up is the addition of trigonometric functions, the ability to raise numbers to powers and take roots, and additional memories.

Programmable calculators add considerable computational power—the steps of a mathematical process may be entered as a series of instructions, and then run as many times as desired with different data for each run. On some models the program is lost when the power is turned off, but others have the capability of retaining the instructions for a considerable period of time. To save the effort of entering programs—which can be quite lengthy—there are card-programmable models that can accept information from small plastic strips with a magnetic coating. These can be recordings of programs you have entered, or cards can be purchased already programmed.

Personal calculators are powered by internal batteries, usually rechargeable. Some models include a built-in or attachable paper-tape printer for a "hard copy" of the calculations, but these are larger and require more battery power, lessening their desirability for boating use.

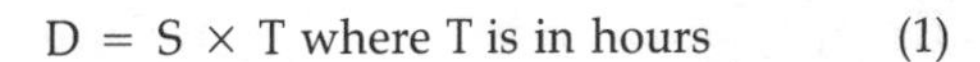

Microcomputers Afloat

With the steadily decreasing physical size and power requirements of microcomputers, more and more of them are now "going boating." There are many small enough to be described as "briefcase" or "lap-portable" models. Power is obtained from internal batteries or connection to a 12-volt DC source. These use LCD displays rather than larger and more fragile cathode-ray tubes.

A personal computer on a boat can solve navigational problems, keep "ship's records," and even let the skipper work on his office problems if he has to take his work with him! The capabilities of these portable units extend far beyond what would be expected from their small size.

Applications for Piloting

The first and most important point to note about the use of calculators and personal computers in piloting is that the problem to be solved is *not* changed. The calculator only makes your efforts easier, quicker, and more accurate—*you* must still solve the problem. This means that you must have the same knowledge of input data and procedures to be used as if you were getting the answer by paper and pencil.

Distance-Time-Speed

Even the simplest calculator will be sufficient for problems of distance, time, and speed. The usual equations are:

$$D = S \times T \text{ where T is in hours} \quad (1)$$

$$D = \frac{S \times T}{60} \text{ where T is in minutes} \quad (2)$$

These equations can, of course, be algebraically transposed in calculating S or T when the other factors are known; see page 446.

Remember that when subtracting two times to get elapsed time for a run there are 60 minutes in an hour and not 100 as in ordinary mathematics—hours and minutes of 24-hour time cannot be subtracted decimally. Time may be calculated in minutes for use in equation 2 by multiplying the hours by 60 and then adding the minutes. Alternatively, equation 1 can be used by dividing minutes by 60 to obtain a decimal fraction which is appended to the whole number of hours, if any, of the elapsed time.

Calculations Involving Tides and Currents

A personal calculator that has the trigonometric functions—sine, cosine, and tangent—and their inverse functions, arc sin or $\sin^{-1}$, etc.—can greatly ease the solution of problems involving tides and currents.

Height of Tide Calculation of the height of tide at an intermediate time between high water and low water can be done by use of Table 3 of the *Tide Tables*—*if* you happen to have a copy handy. If you don't, you can use your calculator and the times of high and lower water from the local newspaper or the data now included on small-craft editions of charts.

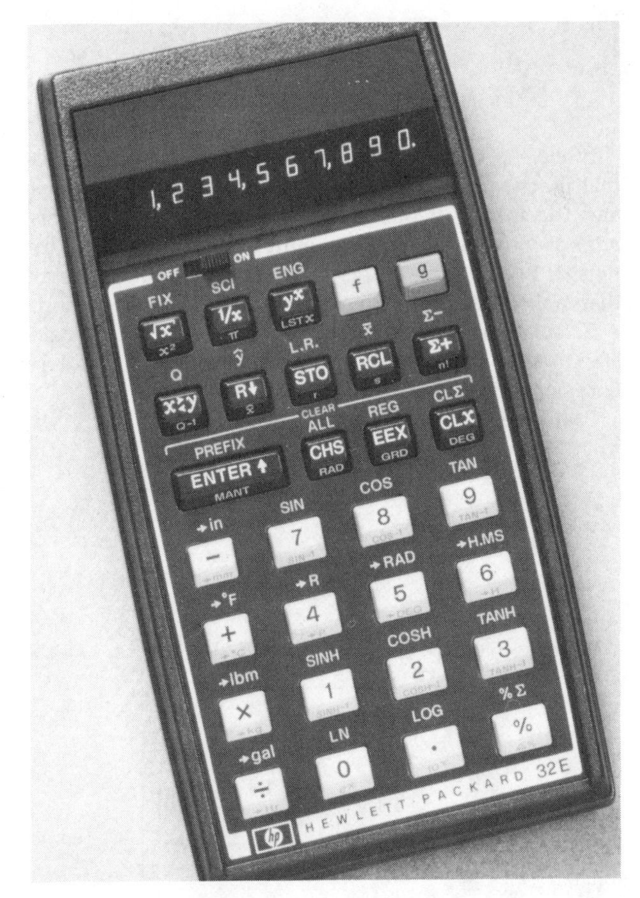

Figure 2130 **These moderately priced calculators are typical of "slide rule" or "scientific" models; they can solve many problems for a boatman. The TI-35, *above,* is capable of all arithmetic operations, plus powers, roots, and reciprocals. It can also perform trigonometric and inverse trig functions, and store intermediate results in a memory register. The HP-24C, *right,* has full "scientific" capability, and can also store a "program," a series of steps used to solve a problem such as current sailing. This feature facilitates the successive running of similar problems. This calculator has eight "addressable" memories for the storage of constants or intermediate results in a long problem.**

The equation used is based on the variation of the height of tide from low to high or high to low following a cosine curve. This is the procedure used by the National Ocean Service in the preparation of Table 3, but it must be remembered that actual conditions of time and height will not always match predictions, due to short-term local conditions.

The predicted height of tide at any desired time is the low water predicted height plus a correction.

$$Ht_D = Ht_{LW} + Corr$$

The correction is equal to the product of a factor (F), which varies with time, and the range of the tide.

$$Corr = F(Ht_{HW} - Ht_{LW})$$

The correction factor is determined by use of a HAVERSINE relationship:

$$F = \frac{1 - \cos\left(\frac{T_{LW} \sim T_D}{T_{HW} \sim T_{LW}} \times 180°\right)}{2}$$

where T_{LW} is the time of low water,
T_{HW} is the time of high water,
T_D is desired time.

The symbol ~ means absolute difference, the lesser quantity subtracted from the larger; it is used to avoid negative numbers.

Caution: Times may be in terms of hours and minutes for subtraction, but must be expressed decimally for division.

Thus the equation for the height of tide at a desired time is

$$Ht_D = Ht_{LW} + (Ht_{HW} - Ht_{LW})\left[\frac{1 - \cos\left(\frac{T_{LW} \sim T_D}{T_{HW} \sim T_{LW}} \times 180°\right)}{2}\right]$$

While the mathematics may appear formidable, in actual practice it is merely the punching of a few buttons and the instantaneous appearance of the answer! This answer may vary a small amount from that obtained by normal use of Table 3, but it is actually more precise as that table is used to nearest values without interpolation.

Calculations may be similarly made for a subordinate station after applying time differences and height differences or ratios.

Example:

$$Ht_D = -1.2 + [10.2 - (-1.2)] \left[\frac{1 - \cos\left(\frac{2{,}333 - 1{,}840}{2{,}333 - 1{,}617} \times 180°\right)}{2}\right]$$

$$= -1.2 + 11.4\left[\frac{1 - \cos\left(\frac{4.83}{7.27} \times 180°\right)}{2}\right]$$

$$= -1.2 + 11.4\left[\frac{1 - \cos(-119.6°)}{2}\right]$$

$$= -1.2 + 11.4\left[\frac{1 - (-.494)}{2}\right]$$

$$= -1.2 + (11.4 \times .747) = -1.2$$

$$+ 8.52 = 7.32 = 7.3 \text{ feet}$$

Strength of Current A generally similar method is used for determining the strength of current at a desired time. In this instance, however, the time interval is between a maximum current and slack water, rather than between one maximum and the opposite maximum. The equation used is:

$$S_D = S_M \times \cos\left[90° - \left(\frac{T_D \sim T_S}{T_M \sim T_S} \times 90°\right)\right]$$

where
T_D is the desired time
T_S is the time of slack water
T_M is time of maximum current, ebb or flood
S_M is strength of maximum current, ebb or flood
S_D is strength of current at the desired time

The direction of the current is the same as the maximum used, ebb or flood; ignore any negative signs appearing in the answer. This equation is to be used only with the normal reversing type of tidal current found in bays, sounds, and rivers. It is not for use with hydraulic currents that exist in connecting waterways, such as Hell's Gate in New York, or the Cape Cod Canal.

Example:

$$S_D = 1.6 \cos\left[90° - \left(\frac{1{,}712 - 1{,}520}{1{,}712 - 1{,}350} \times 90°\right)\right]$$

$$= 1.6 \cos\left[90° - \left(\frac{1.87}{3.37} \times 90°\right)\right]$$

$$= 1.6 \cos(90° - 49.94°)$$

$$= 1.6 \cos 40.06° = 1.6 \times .765 = 1.224$$

$$= 1.2 \text{ knots}$$

Current Sailing A personal calculator is very useful in solving problems of current sailing. Rather than using the graphic techniques of pages 472-473, addition and subtraction of the vectors can be accomplished by pushing a few buttons, and a more accurate result obtained. The general procedure is the resolution of two vectors into their north-south and east-west components, the addition (or subtraction) of these components, and the establishment of the third vector from these net values.

For comparison purposes, the illustrative examples shown below will be the same as those solved graphically on pages 472-473.

Case 1 (Figure 2131) This is a situation of adding two vectors:

$$\overrightarrow{\text{Path over ground}} = \overrightarrow{\text{Path thru water}} + \overrightarrow{\text{Current}}$$

$$\begin{aligned}
\text{COG} &= \tan^{-1}\left(\frac{\text{S sin C} + \text{Drift sin Set}}{\text{S cos C} + \text{Drift cos Set}}\right)\\
&= \tan^{-1}\left(\frac{10 \sin 70° + 3 \sin 130°}{10 \cos 70° + 3 \cos 130°}\right)\\
&= \tan^{-1}\left(\frac{9.40 + 2.30}{3.42\ (-1.93)}\right)\\
&= \tan^{-1}\left(\frac{11.70}{1.49}\right) = 82.74°\\
&= 083°
\end{aligned}$$

$$\begin{aligned}
\text{SOG} &= \frac{\text{S sin C} + \text{Dft sin Set}}{\text{sin COG}}\\
&= \frac{9.40 + 2.30}{0.99} = 11.82\\
&= 11.8 \text{ knots}
\end{aligned}$$

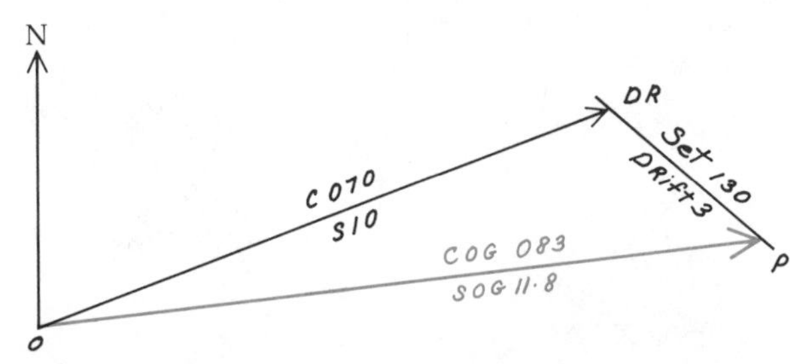

Figure 2131 **Current Sailing—Case 1: The black lines represent the known data: boat speed and course through the water, plus current set and drift. The blue line shows the course and speed over the ground as determined by use of the hand calculator. Although the formula looks quite complex, its application is very quick and simple.**

In the above and all other computations by electronic calculators, do not be misled by the long detailed readings of the machine. The value for sin 70° may be shown as 0.9396926208, and this will be used in internal operations of the machine. The result may be computed to be 11.78982612 knots, but this rounds to 11.8 knots, or, more simply, 12 knots.

These equations will work when the times involved straddle a midnight and fall into two days, but be careful. The simplest method is to add 24 hours to the times of the second day so that they are numerically larger than those of the first day. For example, a low water time of 2254 and a desired time of 0115 the following day is best handled as 2515 − 2254 = 2:21. (It is desirable to actually write down the later time with 24 hours added rather than trying to accomplish the addition then subtraction mentally.)

Some calculators have a key for changing angles in degrees, minutes, and seconds (D.MS) to degrees and decimal fractions (DD.d). This will function equally well for converting hours, minutes, and seconds (H.MS) to hours and decimal fractions. (HH.h). The conversion process is usable in either direction and is very helpful in equations such as those used here where times must be both subtracted and divided.

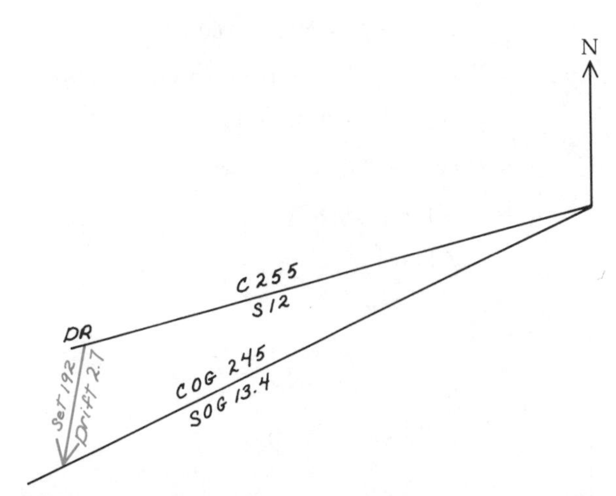

Figure 2132 **Current Sailing—Case 2: The known values are boat course and speed through the water, plus the course and speed made good over the bottom. The blue line indicates the calculator solution to the set and drift of the current, which is actually the sum of all offsetting influences.**

Case 2 (Figure 2132) Again, two vectors are known, subtraction will yield the other.

$$\overrightarrow{\text{Current}} = \overrightarrow{\text{Path over ground}} - \overrightarrow{\text{Path thru water}}$$

$$\begin{aligned}
\text{Set} &= \tan^{-1}\left(\frac{\text{SOG sin COG} - \text{S sin C}}{\text{SOG cos COG} - \text{S cos C}}\right)\\
&= \tan^{-1}\left(\frac{13.4 \sin 245° - 12.0 \sin 255°}{13.4 \cos 245° - 12.0 \cos 255°}\right)\\
&= \tan^{-1}\left(\frac{-0.55}{-2.56}\right)\\
&= \tan^{-1} 0.216 = 12°*\\
&= 192°
\end{aligned}$$

$$\begin{aligned}
\text{Dft} &= \left(\frac{\text{SOG sin COG} - \text{S sin C}}{\text{sin Set}}\right)\\
&= \left(\frac{13.4 \sin 245° - 12 \sin 255°}{\sin 192°}\right)\\
&= 2.617 = 2.6 \text{ knots}
\end{aligned}$$

*As the calculator result will be an angle between 0 degrees and 90 degrees, this answer is "angle in quadrant." From inspection of a rough sketch or the vectors, it will be seen that the set is 180° + 12°, or 192°.

Case 3 (Figure 2133) In this instance, one vector is known; this is the set and drift of the current. Boat speed through the water is also known, as well as the intended track. To find the course to be steered and speed of advance, a different type of solution is required.

$$C = TR - \sin^{-1}\left[\left(\frac{\text{Drift}}{S}\right)\sin(\text{Set} - TR)\right]$$

$$= 080° - \sin^{-1}\left[\left(\frac{4}{10}\right)\sin(140 - 80°)\right]$$

$$= 080° - \sin^{-1}(.346)$$

$$= 080° - 20.3° = 59.7 = 060°$$

$$SOA = [S\cos(C - TR)] + [\text{Drift}\cos(\text{Set} - TR)]$$

$$= [10\cos(060 - 080)] + [4\cos(140 - 080)]$$

$$= (10\cos - 20°) + (4\cos 60°)$$

$$= (9.40) + (2.00)$$

$$= 11.4 \text{ knots}$$

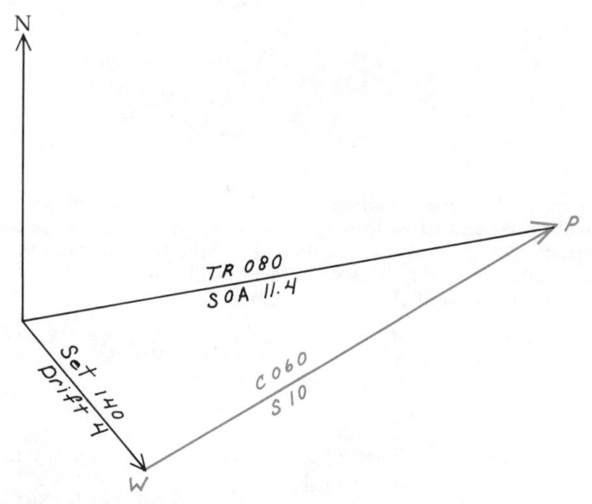

Figure 2133 **Current Sailing—Case 3: Here the known values are set and drift of the current, the boat's speed through the water, and the direction of the track along which it is desired to run. The calculator easily and quickly solves the vector triangle for the course to be steered, and the speed of advance along the intended track.**

Case 4 (Figure 2134) This is again the subtraction of one known vector from another to obtain the third, unknown vector.

$$\overrightarrow{\text{Path thru water}} = \overrightarrow{\text{Path over ground}} - \overrightarrow{\text{Current}}$$

$$C = \tan^{-1}\left[\frac{SOA\sin TR - \text{Drift}\sin\text{Set}}{SOA\cos TR - \text{Drift}\cos\text{Set}}\right]$$

$$= \tan^{-1}\left[\frac{11.5\sin 320° - 3\sin 205°}{11.5\cos 320° - 3\cos 205°}\right]$$

$$= \tan^{-1}\left[\frac{-7.39 - (-1.27)}{8.81 - (-2.72)}\right]$$

$$= \tan^{-1}\left[\frac{-6.12}{11.53}\right] = 28.0°$$

Again this is an angle in quadrant, and C = 360° − 28.0° = 332°.

$$S = \frac{SOA\sin TR - \text{Drift}\sin\text{Set}}{\sin C}$$

$$= \frac{-6.12}{\sin 332°} = 13.05$$

$$= 13.0 \text{ knots}$$

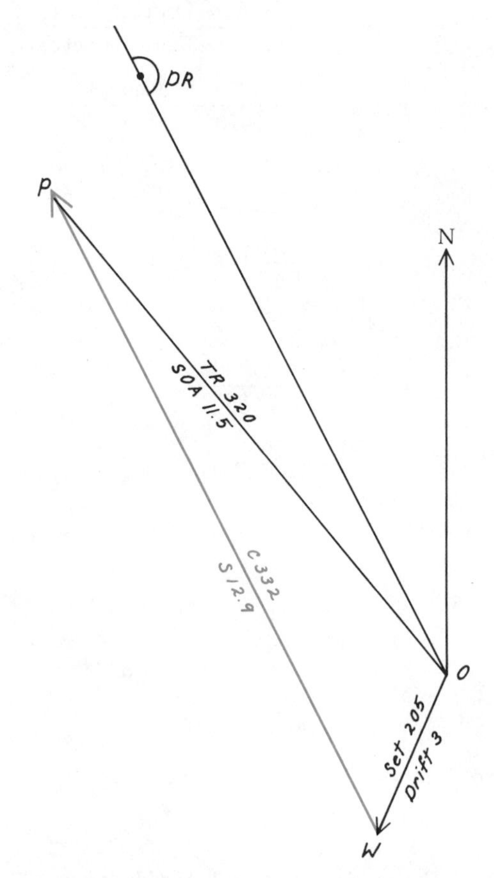

Figure 2134 **Current Sailing—Case 4: The vector triangle is solved here for the course to be steered and speed to be run through the water in order to reach a specified destination at a predetermined time. This problem is typical of predicted log contests, and it is useful when a rendezvous with another boat has been set.**

Some more advanced models of calculators have a key for conversion from rectangular coordinates to polar coordinates. This use of an internally stored program makes short and simple the task of resolving a vector into its N-S and E-W components, which are then added to or subtracted from the components of another vector. After the net components are determined, the inverse procedure is used to obtain the resultant vector.

Bearings and Distances

Information on a vessel's position can be obtained from two successive bearings on a single object. In some instances, these can be handled quite simply as cases of "doubling the angle on the bow," or bow and beam bearings as discussed on pages 491-492, or as special pairs of angles as considered on page 502. More often, however, it is impracticable or inconvenient to wait until two particular angles can be measured—a more general solution is desirable or necessary. If a copy of Table 7 from *Bowditch,* Volume II, is on board, a solution can be made for bearings at 2-degree intervals. The most generalized solution accepts any value for either bearing and gives an accurate solution quite quickly—it uses a personal calculator.

In the following calculations, an input element that critically affects the accuracy of the solution is the distance traveled during the interval between bearings. In computing this, the boat's course and speed must be combined with estimated current to obtain Course over Ground (COG) and Speed over Ground (SOG); this latter figure is combined with time to obtain distance traveled—all of these computations can be made by calculators.

In Figure 2125, the first bearing is the angle at A, the second bearing is the angle at B, the distance traveled is AB, and the distance off when abeam is LC.

Although the distance LA could be calculated, this cannot be done until the boat has reached B, where the second bearing is taken—hence, it is of little interest and the position at B is the one normally determined. Quite often, the point of nearest approach to L is of interest, the distance off when abeam L. It, too, can be easily calculated, and *in advance* of reaching point C.

$$LB = \frac{D \sin B_1}{\sin(B_2 - B_1)}$$

$$LC = LB \sin B_2$$

where D is distance run between the two bearings

B_1 is the first bearing (at A)

B_2 is second bearing (at B)

Note carefully that B_1 and B_2 are angles to the Course over Ground (COG); relative bearings must be corrected for any difference between COG and C, the course being steered.

Example: (Figure 2125): First bearing 37°, second bearing 61°. Distance run between bearings 2.3 miles.

$$LB = \frac{2.3 \sin 37^\circ}{\sin(61 - 37^\circ)}$$

$$= \frac{2.3 \times 0.602}{0.407} = 3.40$$

$$= 3.4 \text{ miles}$$

$$LC = 3.40 \times \sin 61^\circ = 3.40 \times 0.874$$

$$= 2.9 \text{ miles}$$

Distance to Horizon The distance to the horizon for a given height of eye can be determined easily on any calculator having a square root capability.

$$D = 1.17\sqrt{h}$$

where D is in nautical miles, and h is in feet, or

$$D = 2.12\sqrt{h}$$

where D is in nautical miles and h is in meters.

If the distance is required in statute miles, multiply D by 1.15; if required in kilometers, multiply D by 1.852.

Traverse Sailing

When a boat is running a course with several legs of different directions and lengths, the net change of position can be calculated by means of a traverse table. This situation arises in some navigational contests, when sailing, or when a boat is trolling for fish out of sight of land or aids to navigation. Again, solution is possible from *Bowditch,* Volume II, Table 3 in this instance. With or without a copy of *Bowditch,* the problem is easily solved by calculator and the simplest of trigonometric equations.

The various legs are broken down into north-south and east-west components, the components are summed up taking care that southerly and westerly components are given a negative sign, and the resultant single course and direction determined from the net N-S and E-W components. This is simply vector addition of as many legs as have been run.

Other Navigational Applications

A skipper can develop many specialized applications to meet his own piloting needs once he has gained some experience and developed a moderate level of skill with his personal calculator or microcomputer. Typical of these uses are: position determination by two nonsimultaneous bearings with a run between them (on separate objects rather than on a single object as described above), determining course over ground from three bearings on a single object (without knowledge of distance run between times of each bearing), compass deviation table preparation, true and apparent wind, and many others limited only by the skipper's needs, imagination, and abilities with his calculator

General Arithmetic Use

A calculator can also be put to good use in boating for general arithmetic problems. These include speed over a measured distance, fuel consumption and cost, and the height of tide at intermediate times using all the factors given in the *Tide Tables.* A calculator is also useful for many problems of simple subtraction or addition, such as checking on a yard repair bill.

Calculators can be helpful in converting between various units such as feet, yards, and miles; they are particularly convenient in making conversions in both directions between customary and metric units.

Calculators Aboard Boats

The small size and light weight of personal calculators make them suitable for any type of boat and boating; the many applications discussed above make one a desirable "tool" for navigation. They are, however, complex electronic devices and should be protected from the harsher aspects of a marine environment.

A possible problem is the matter of electric power, although some models have batteries that will provide about a thousand hours of operation. Some of the simplest calculators operate from replaceable dry-cell batteries; many of these can be, and usually are, converted to rechargeable batteries. Most calculators are used with an adapter-charger for direct operation from a 117-volt, AC source plus concurrent recharging of an internal battery.

As "shore-type" AC power is not normally available underway, the question is usually the operating life of the

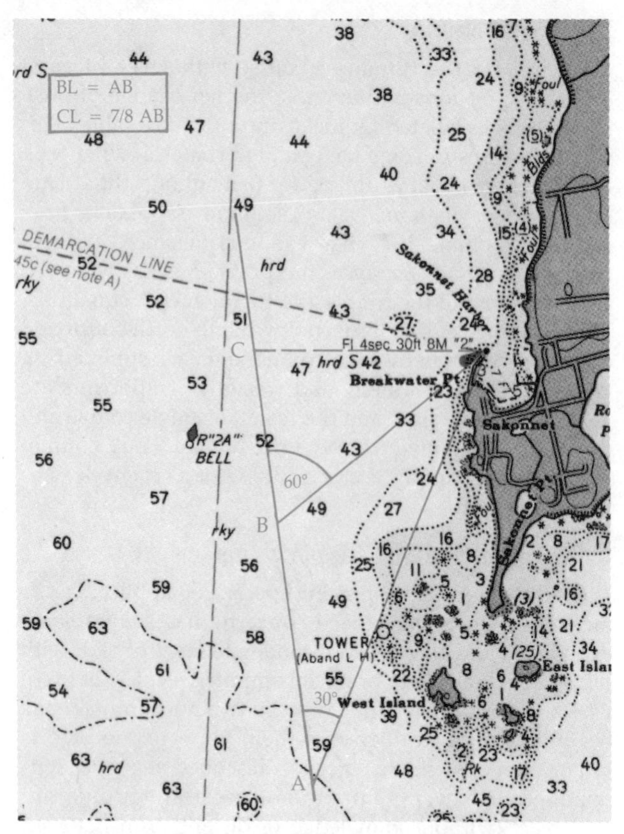

Figure 2135 **If observations are made when the relative bearings of a fixed object are 030° and 060°, and the distance made good is calculated, the pilot can determine the distance from the object both at the time of the second bearing and when it is abeam by using this "7/8th Rule."**

internal rechargeable battery. The life of a fully charged battery will differ with the calculator model and with the on-off time ratio, but can be expected to be from roughly six hours for simpler units to perhaps three hours for larger models. Much of the power is used for display digits; battery drain can be much reduced by immediately "clearing" the display to a single "0" as soon as the results of a calculation have been noted. The greatest saving of battery power is achieved by turning the calculator off between periods of use. Although generally more expensive, a model that retains its program and data in its memory after being turned off is usually desirable.

For short-term use on boats, a spare battery can be carried and installed when needed. For longer-term cruising with most calculators, it is possible to purchase, or make, a charger-adapter that permits operation from the boat's 12-volt DC electrical system. The drain is negligible—on the order of ¼ ampere or less.

Specialized Positioning Procedures

In addition to the generally used procedures for position determination, other, more specialized techniques can be employed.

Combinations of Relative Bearings

The bow-and-beam-bearings technique has already been described, Figure 2123 left, using a set of two relative bearings, 045° and 090° (or 315° and 270°). It was shown because of its simplicity and the ease with which such a pair of bearings can be obtained. There are other combinations of relative bearings, not so easily obtained, but quite easily used.

Special Pairs of Bearings The following sets of bearings have such a relationship to each other that the run between the first bearing and the second will nearly equal the distance away from the sighted object when it is passed abeam:

20°–30°	21°–32°	22°–34°
23°–36°	24°–39°	25°–41°
27°–46°	29°–51°	30°–54°
31°–56°	32°–59°	34°–64°
35°–67°	36°–69°	37°–71°
38°–74°	39°–77°	40°–79°
41°–81°	43°–86°	44°–88°

Note that these are pairs of relative bearings to port as well as to starboard. In the table above, "20° – 30°" can be either relative bearings 020° and 030°, or 340° and 330°; "31° – 56°" can be either RB 031° and 056°, or RB 329° and 304°; etc.

The Seven-Eighths Rule If observations are made when the relative bearings are 30° and 60° on either bow, simple calculations will give two useful items of information. The distance run between the two bearings is equal to the distance to the object at the time of the second bearing (doubling the angle on the bow). Also, this same distance, Figure 2135, multiplied by 7/8 is the distance that the boat will be off from the sighted object when it is broad on the beam, provided that the vessel's course has not changed.

The Seven-Tenths Rule A situation comparable to the preceding is that which exists when the two relative bearings are 22½° and 45° to port or starboard (assuming that bearings can be taken to a half-degree). In this case, as before, the distance away from the sighted object at the time of the second bearing is equal to the distance run between bearings, but the multiplier is 7/10 to determine the distance off the sighted object when it is abeam.

Three Bearings and Run Between

The technique of using two bearings and the run between them was illustrated in Figure 2124. Position information derived from this procedure depends on accurate determination of the course and speed made good over the bottom during the run. Along strange coasts, without current predictions or means of estimating current effects, course and speed on the ground cannot be accurately judged.

By taking a *third* bearing and timing a second run (between the second and third bearings), enough additional known factors are entered into the problem so that no assumption need be made about the course made good nor the actual speed over the bottom. All that is required for this technique is a prominent object or mark on shore, plus the means for taking bearings and measuring time intervals.

As soon as you spot the object, "X" in Figure 2136 top, take a bearing on it and note the exact time; record both items of information. Plot the bearing line on your chart, line XA in Figure 2136*A*. Next, when the object is exactly broad on the beam, note the time. Plot this line of position on the chart. Later, when the bearing to X has changed enough to give a good angle of intersection, take a third bearing and again note the time carefully. Plot this line, XC. Be sure that throughout this procedure you are maintaining a steady course and speed through the water.

Since you have recorded the time of each bearing, you can determine the two elapsed times, and compute the *ratio* between these intervals.

For example, if you ran for 32½ minutes between the first and second (beam) bearings, and then for 22½ minutes more to the third bearing, the ratio would be 32½/22½, which can be reduced to 6.5/4.5.

Through a convenient point, B, on the beam bearing line, draw a light "construction" line in the direction of your course (C) as steered. Using point "B" as the reference or zero point, measure off to each side, along the construction line, distances proportional to the ratio of the time intervals using any convenient scale, Figure 2136*B*.

Erect perpendiculars to the construction line at the spots determined by these proportionate distances. These lines, drawn at 90-degree angles to the construction line, intersect the bearing lines XA and XC at points A′ and C′ respectively, Figure 2136*C*.

Connect points A′ and C′ with a line and you have your answer. The direction of line A′–C′ is the course you are making good and includes the effects of currents, wind, and errors in steering. Its direction may be taken from the chart and compared with the intended course. It should be noted that this is *not the actual track*—the selection of the scale to represent the two time intervals would determine the location of points A′ and C′, and hence the line's location, although not its direction.

It is because ratios are used, and not actual speeds, that an accurate speed over the bottom is not needed here as it was in the case of the technique of two bearings and run between. Although it does not allow the navigator to determine his position or his speed over the bottom, this technique does show plainly whether he is making good his intended course, or is being set onto or off the shore; it helps determine whether the actual track parallels the shore or not.

Relative Motion

The easiest way to acquire what is often called a "sailor's eye" is to learn the simple but basic principles of RELATIVE MOTION. It is one of the most useful skills that a boatman can acquire, because it gives quick and accurate answers to problems involving moving vessels.

For example, you find that your boat is converging on another boat that has the right of way. If you hold your present course and speed, will you clear her? Or your sailboat is beating to windward on a port tack, and you find a boat coming in on the starboard tack. Will you cross safely ahead, or must you pass astern? Or you have laid a course for a buoy, making allowance for the current. Have you made the right allowance? Relative motion will give you the answers to all such questions.

When a boat is underway, its movement across the water is termed ACTUAL MOTION. If you are anchored, the movements of another boat appear in its actual relationship to

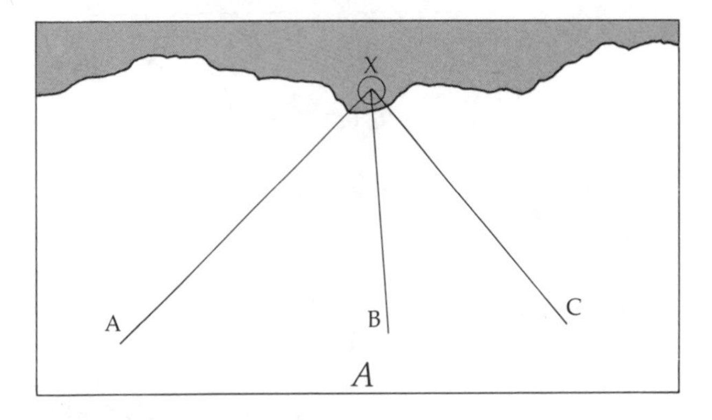

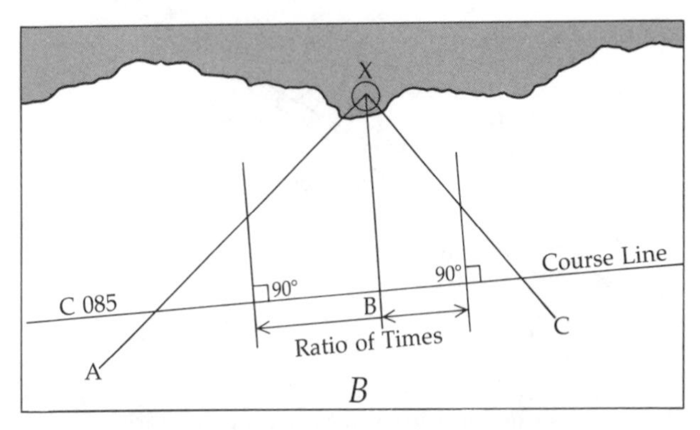

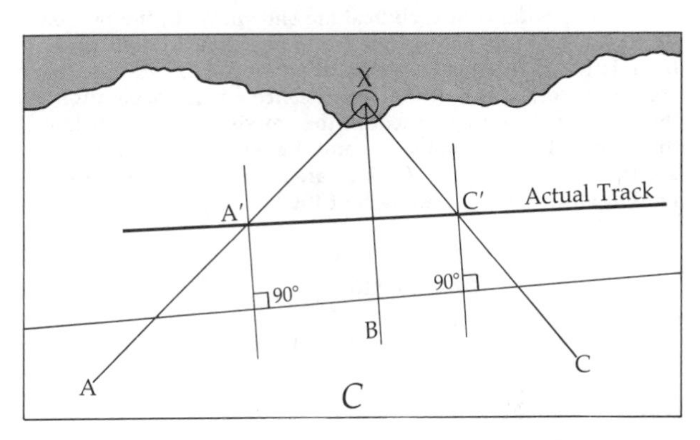

Figure 2136 **The course of a boat under unknown conditions of current can be determined by taking three bearings on a single object, *above,* See text for the details of this technique. If a boat's heading changes, the relative bearing of an object is changed by the same amount, *below.* From the situation at A to that at B, the true heading of the boat has changed from 050° to 020°. The relative bearing has changed by the same amount, from 090° to 120°. Note that the sum of the heading and relative bearing remains constant.**

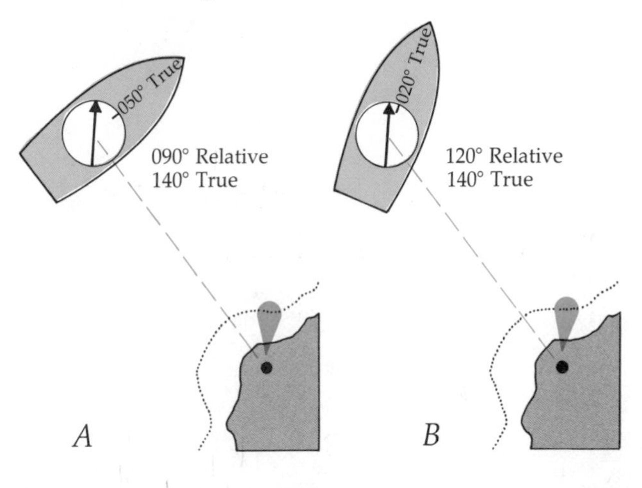

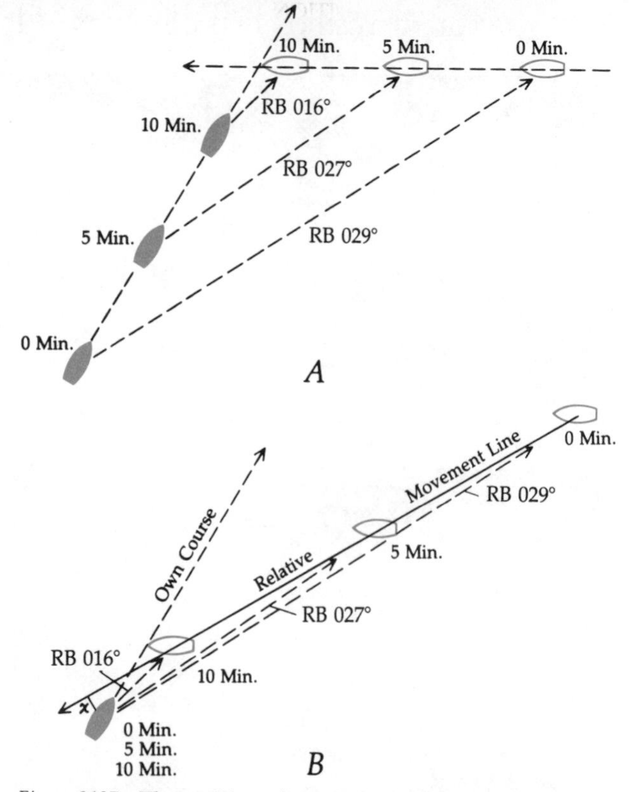

Figure 2137 **The position of two boats is shown, *above,* for three intervals of time separated by five minutes each. At A, the actual positions of each boat are shown. At B, the motion of the other boat is shown relative to our boat. In both cases the relative bearings move forward on our boat, and the other craft will cross ahead of us. The distance *x* is the separation at the point of closest approach. In the crossing situation *below,* the relative bearings move aft, and the other boat will pass astern of us. The actual positions are shown in A, and relative motion is shown in B, for each of the bearings.**

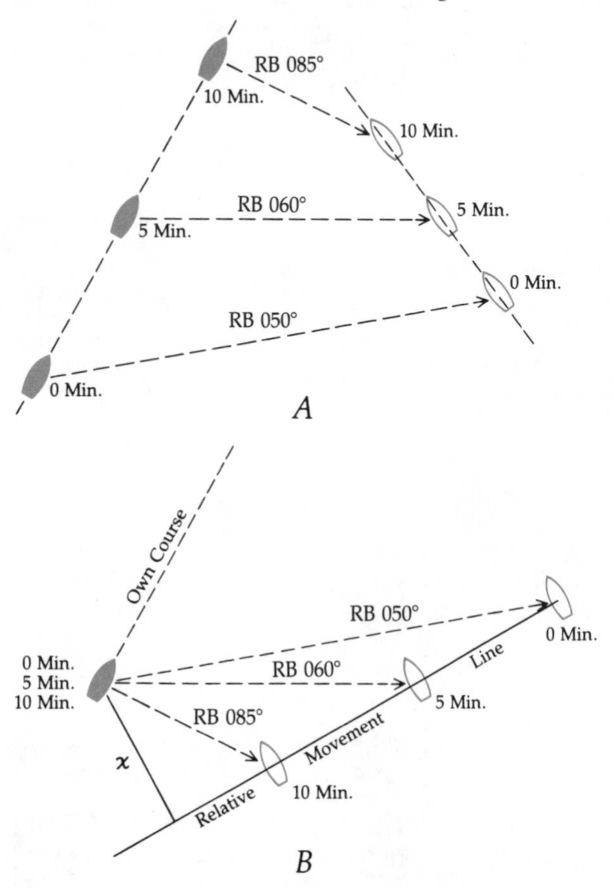

the earth's surface; you are observing that boat's actual motion. But if you get underway, the movement of the other boat appears to be different because you are now observing relative motion. Some objects, especially moving ones, seem to be doing things that are not actually happening.

Suppose that you are underway and proceeding north at 5 knots up a marked channel. You sight another boat coming directly toward you; it is also making 5 knots. The two boats are converging at a RELATIVE SPEED of 10 knots, which is simply the sum of your speed and his. Both are passing the stationary channel markers at the same actual speed, but the distance between the craft is lessening at a rate that is the sum of the individual speeds.

As the other boat approaches, you recognize the skipper as a friend and turn around to join him on his southerly course. As soon as the boats are alongside each other on the same course and at the same speed, relative motion ceases to exist because the other boat stays in the same place relative to your boat. Only actual motion remains—that of your boat past the buoys and over the bottom. If the other boat develops engine trouble and slows down there is again relative motion between the two vessels. It will appear to be moving aft because of its reduced speed. Thus, the rule may be stated as follows: *relative motion is present only when the actual movements of two or more objects are not the same.*

Relative Bearings

The bearing of an object is usually taken as a relative bearing. With practice, the relative bearing of an object can be estimated within five or ten degrees. It must be remembered that a change in course of your own boat will change the relative bearings of all objects around it. Thus, if you are on a course of 050° true and sight an object bearing 090° relative, the true bearing of that object will be 140°, Figure 2136*A* bottom. Come left to a course of 020°, and although the true bearing will remain unchanged at 140°, the relative bearing will now become 120°, Figure 2136*B* bottom.

In small boat piloting, approximations will usually suffice in relative motion problems and special instruments are not needed. A steady hand at the wheel or tiller, a clear eye, and a rough mental diagram are all that are needed.

To Cross or Not to Cross

When determining the relative motion of another boat, it is necessary either to convert relative bearings to compass (or true) bearings, or to maintain a steady course. In practice, it is easier to maintain a steady course with the result that any change in relative bearing results from a change in the positions of the boats relative to each other, rather than from your boat's changing course. Since the desired information is the changing positions of the boats relative to each other, the actual value of the relative bearing is of far less importance than the direction and rate of change, if any.

Collision Bearings What we want to know in any crossing or converging situation is: will we cross ahead of, or astern of, the other boat, or will we hit her? Because the two boats are considered relative to each other, it is convenient to think of one's own boat as stationary—relative to the other boat. Thus, we have three basic situations:

1. The relative bearing of the other boat moves ahead (toward the bow), or
2. The relative bearing of the other boat moves aft, or
3. The relative bearing of the other boat remains constant.

Figure 2137*A* top illustrates the first situation, that of a boat passing ahead. This illustration indicates the actual successive positions of the two boats. Figure 2137*B* top represents the same situation shown in terms of motion relative to our own boat as stationary. The bearing change is then more clearly seen.

The line connecting the positions of the other boat in Figure 2137*B* is the LINE OF RELATIVE MOVEMENT, the course and distance traveled by the other boat in relation to our boat. Provided both boats maintain course and speed, it is a straight line. In this situation, when the line of relative motion is extended, it passes the bow of our "stationary" boat; therefore, the other boat following that line relative to us will cross ahead. The distance "x," Figure 2137*B*, is the distance apart that the two boats will be at the CLOSEST POINT OF APPROACH (CPA).

The second case, in which the relative bearing moves aft, is the opposite of the first situation. The relative motion line passes astern, and so will the other boat. Figures 2137*A* and *B* bottom illustrate this situation.

If the relative bearing does not change, the line of relative motion will pass through our boat's position, the distance "x" is zero, and the two boats will collide unless one (or both) changes course or speed. See Figure 2138.

It is important to note that Figures 2137-2138 are intended to illustrate situations rather than actual chart plots on a boat. The position of the other boat could not be plotted unless distance as well as bearing information was available, and this is unlikely unless your boat has radar. The significance here is that *bearing information only*, which is available to any pilot, will by its change, if any, tell him whether he will pass ahead, astern, or "through" the other vessel. The closest point of approach can be determined only if an actual plot based on distances is made, but lack of this information is not significant.

Rules for Crossing Situations The foregoing discussion, formulated into rules, results in the following procedures for a crossing situation:

1. Maintain a reasonably steady course and speed.
2. Observe the relative bearing of the other boat only when you are on your specified compass course. You must be on the same course each time you take a relative bearing.
3. Watch the other boat for changes in course or speed which would obviously upset the relative motion conditions.
4. If the relative bearing moves ahead, the other boat will pass ahead.
   If the relative bearing moves aft, the other boat will pass astern.
   If the relative bearing is steady, there is danger of collision.

Offsetting Effects of Current or Wind There is another useful application of relative motion in the case of allowances for offsetting the effects of current and/or wind.

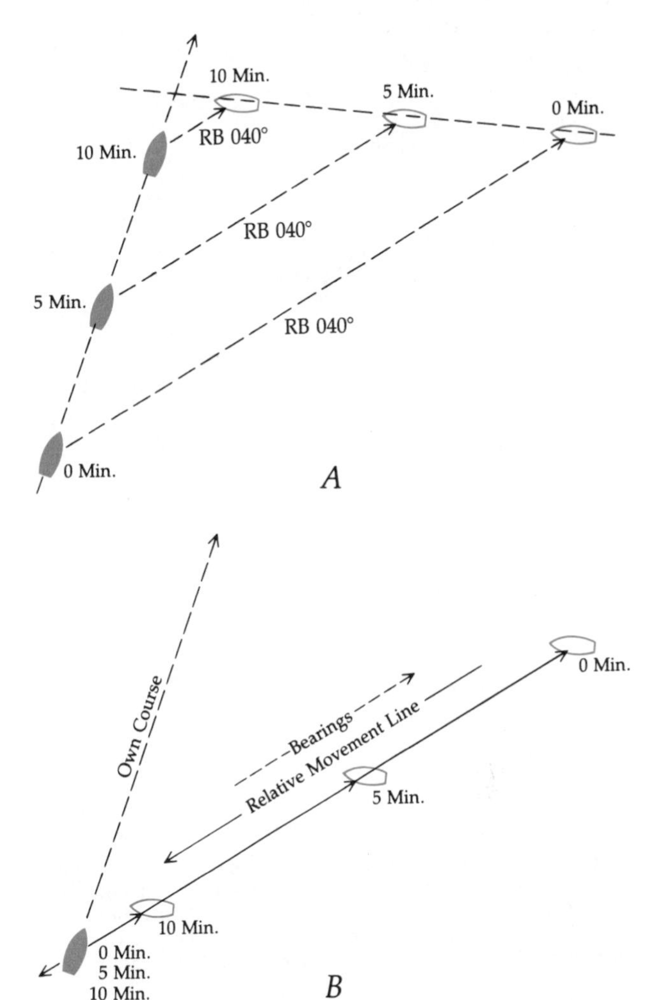

Figure 2138 **In this crossing situation, the relative bearing does not change. A collision will result if both boats maintain course and speed. Note that in B the line of relative movement pass through our boat, and x is zero.**

In Figure 2139, the boat has been put on a course that her skipper believes will put her close aboard the buoy marking the shoal. At the time of the first bearing, "0" minutes, the buoys bears 340° relative. Five minutes later, the relative bearing has changed to 350°. It is clear from this change that the boat is being set down more than expected and that the relative movement line of the buoy will pass ahead of the boat, i.e., the boat will pass on the shoal side of the buoy. If the bearings on the buoy remained steady, the boat would pass close to it; and if the bearings shifted gradually away from the bow, the skipper would know that he would clear the shoal safely.

In Figure 2139*A*, the problem is shown in terms of the boat moving relative to the buoy. In Figure 2139*B*, the boat is considered to be stationary and the relative movement line is that of the buoy. Note that the relative movement lines in the two representations are parallel, equal, and opposite.

The Maneuvering Board

The Defense Mapping Agency Hydrographic/Topographic Center publishes a plotting sheet known as the

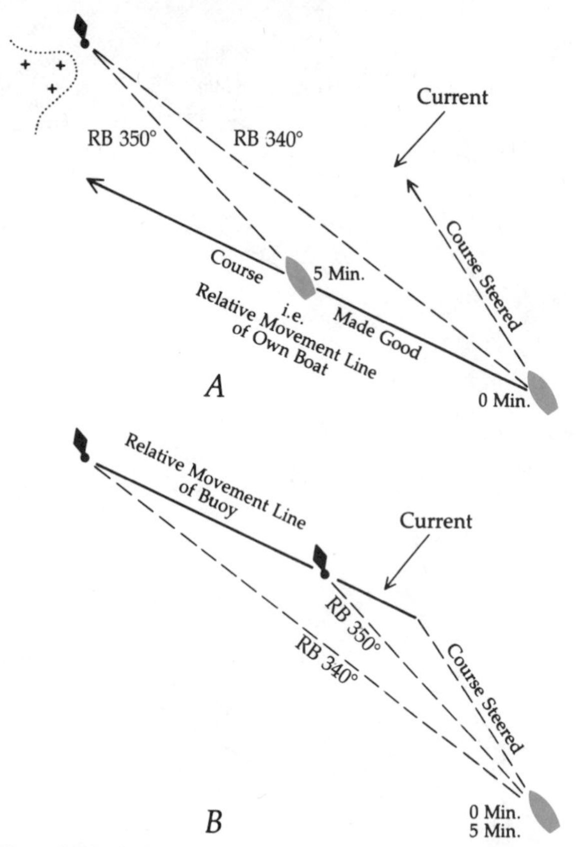

Figure 2139 **Relative motion considerations also apply when the other object is stationary, such as a buoy. The change, if any, of the relative bearings can be used to determine on which side of it you will pass. In A and B, the motion is relative to the buoy and to your boat, respectively.**

Maneuvering Board (in pads of 50 sheets; No. 5090, or 5091 in larger size). These forms are used by ships at sea for the solution of relative motion problems. Complicated movements of large task forces, as well as the tracking of a single ship, are plotted on maneuvering boards to facilitate the solution of problems of interception and the determination of the course and speed of radar contacts.

Use on Boats

Here we will consider a few of the simpler, though useful, applications of the maneuvering board, using only the plotting sheet, parallel rulers or equivalent, and a pair of dividers. For the skipper who has radar aboard, however, the applications are increased ten-fold.

Description of the Sheet

To observe the physical appearance of a maneuvering board, examine Figure 2140 top, which shows one reduced about 40% from its original size of approximately 12 inches square. On it is printed, in green, a large circle with bearing lines radiating outward from the center every ten degrees, and ten concentric circles a half-inch apart to indicate speed or distance. At the right and left sides are lines of scale with numbered marks spaced equal to the distance between the concentric circles. Thus, a speed line drawn out to the fifth circle could indicate 5, 10, 15, 20, or 25 knots as determined by your choice of scale.

At the bottom of the sheet are three lines divided into logarithmic scales for use in solving time, distance, and speed problems. If two of these quantities are known the third may be found by drawing a line between the known values on the appropriate scales, and finding the answer at the intersection of this line with the third scale.

In Figure 2140, observe the solution to three problems; (1) if you run 6 miles in 30 minutes, your speed is 12 knots; (2) to run 6 miles at 12 knots will require 30 minutes; or (3) running at 12 knots for 30 minutes, you will cover 6 miles. Distance, time, and speeds—knowing two, you can quickly, easily, and accurately solve for the third. This alone makes the maneuvering board a handy device, but it is only a small sample of its usefulness.

The bearing lines and circles of distance facilitate the plotting of positions of other boats or objects relative to yours. These radial lines and concentric circles can also be used for vector diagrams with lines drawn to represent actual and relative motion. Each vector, according to its length and direction, depicts a statement of fact in the problem. For example: a vector line to indicate a course of 040° and a speed of 5 knots would be drawn outward along the 40° bearing line to a length of 5 units, i.e., to the fifth circle. If the speed is greater than 10 units (knots or MPH), use is made of the 2:1, 3:1, 4:1, or 5:1 scales at the sides of the chart.

A vector line indicates graphically the direction and velocity of an element of the problem, such as the course and speed of your boat, another boat, the wind, or the current. In some instances, the vector will not be drawn from the center of the sheet, but its direction is always measured with reference to the center point and the outer circular scale.

Points on diagrams of positions are labeled with uppercase (capital) letters; points on vector diagrams are labeled with lower-case letters.

Maneuvering board sheets are printed on both sides, and the paper is sufficiently heavy that both sides may be used. The cover sheets of both the 5090 and 5091 pads include instructions for their use in various problems.

Use of the Maneuvering Board

The 5090 or 5091 sheets may be used for plots of the relative motion of two vessels in a crossing situation as shown in Figures 2137-2138. These sheets are particularly useful on board radar-equipped vessels where distance as well as bearing can be measured. Figure 2141 top, for example, is a re-plot of Figure 2137*B* bottom.

From a plot such as Figure 2141 top, with successive relative positions of the other boat plotted at recorded times, the direction of relative movement (DRM) and speed of relative movement (SRM) can be found. SRM is determined from the logarithmic scales at the bottom of the sheet using distance and time as the known factors. Knowing these relative motion quantities, and the course and speed of your own boat, a VECTOR TRIANGLE can be used to determine the *actual* course and speed of the other boat. The relationships between these quantities are shown in Figure 2141 bottom. Point *e*, always at the center of the maneuvering board, represents earth, the reference for actual motion. Point *r* is the outer end of the speed vector, *e-r*, for the reference vessel (your boat). The vector *r-m* represents the relative motion of the maneuvering (other) boat. By completing the triangle with the vector *e-m*, always outward from the point *e* at the center, the actual

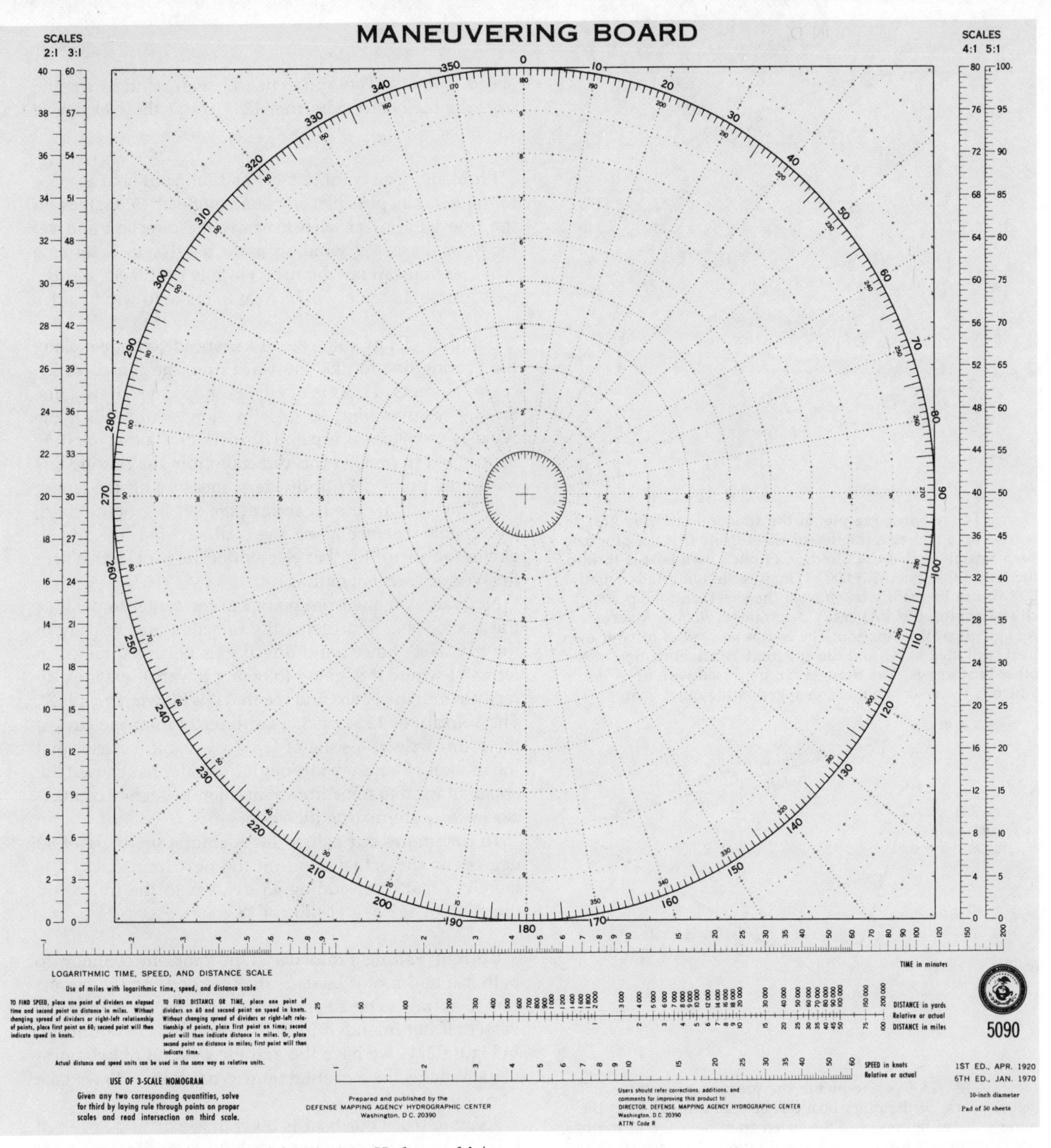

Figure 2140 **The Defense Mapping Agency Hydrographic/ Topographic Center publishes a useful plotting sheet known as a maneuvering board. These sheets, *above,* which come in pads of 50 sheets of either of two sizes, are particularly handy when plotting situations of relative motion. Logarithmic time-distance-speed scales are printed at the bottom of the sheet. The problem shown by the line drawn across the scales is discussed in the text.**

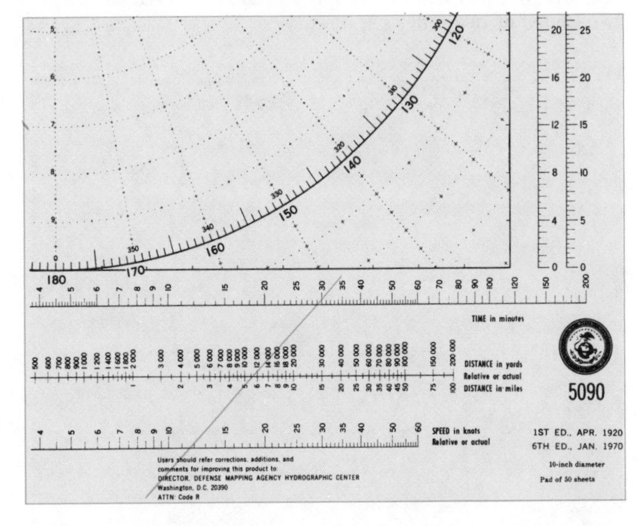

course and speed of the other boat are found. When such a vector triangle is drawn on a maneuvering board sheet, as in Figure 2142, the actual course can be read directly from the outer circular scale, and the actual speed can be read from the convenient concentric circles of the sheet.

True Wind Problem Figure 2143 illustrates a TRUE WIND PROBLEM.

Situation Your sailboat is reaching on true course 080°,

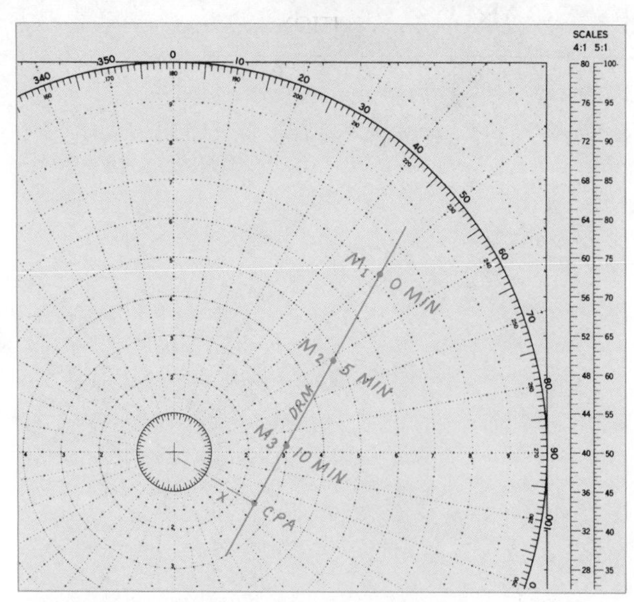

Figure 2141 **Above:** **a re-plot of the situation in** ***Figure 2137B bottom*** **on a maneuvering board sheet. Note that the plot has been rotated so that the direction of one's own boat is toward the top of the sheet (RB 000°). The direction of relative motion (DRM) can be readily taken from the sheet, and the point of closest approach (CPA) easily determined.** ***Below:*** **A vector triangle of relative motion: The course and speed of your own boat is plotted as** ***e-r*** **and the apparent (relative) motion of the other boat as** ***r-m.*** **The triangle can be closed with line** ***e-m,*** **which is the actual course and speed of the other boat.**

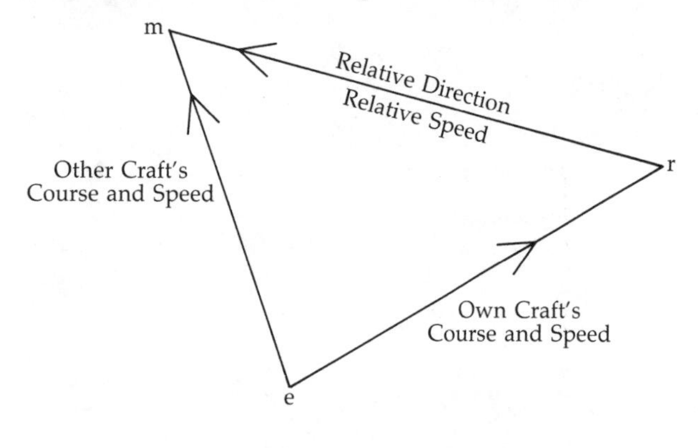

Figure 2142 **This is a re-plot of the preceding illustration. The actual course of the other boat can be read directly from the outer scale of figures, and the speed from the concentric rings; note that in this example the speed ratio is 2:1; each circle is two knots, or two miles per hour.**

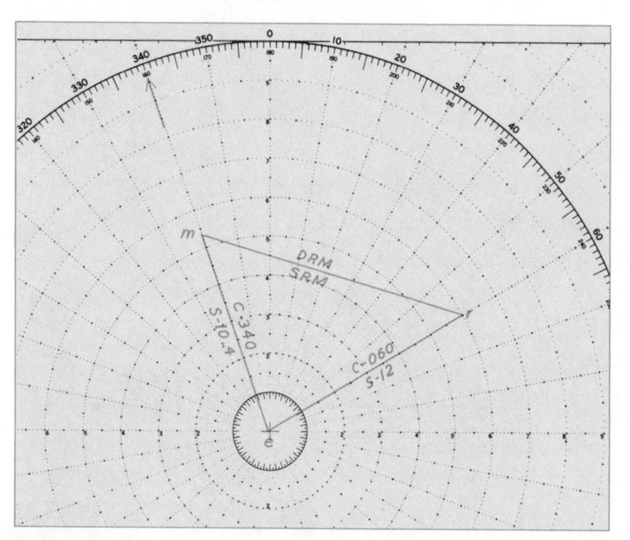

speed 5 knots. The apparent (relative) wind is coming over the starboard side from 125° true with a velocity of 15 knots.

Problem You want to tack, and sail your boat as close to the wind as possible, say at an angle of 45 degrees off the true wind. You wish to know the direction and velocity of the *actual* wind so as to be able to determine whether you can lay the mark on this next tack, and the best course that you will be able to make to windward.

Solution Draw your course and speed vector, *e-r*, along the bearing line for 080° outward from the center for 2½ circles (5 knots at a scale of 2:1). From point *r*, draw a line for the relative wind in the direction *toward* which the wind is traveling; use parallel rulers or another plotting instrument to transfer this direction from the bearing line out to the point *r*. With dividers, measure off the length for 15 knots along the 2:1 scale at the left side of the chart; lay off this distance along the relative wind line and so obtain the vector for that element of the problem. Where this vector ends is point *w*.

Next, draw a line from point *e* at the center to point *w*; this vector, *e-w*, represents the direction and velocity of the *true wind*. Since wind direction is never expressed in terms of where it is going (blowing toward), we look directly across the board and see that it is blowing *from* 142°. The velocity of 12 knots is read directly from the circles, the sixth circle at a scale of 2:1. Now it can be seen that the motion of your boat through the water has caused the apparent wind to come from a direction 17 degrees counterclockwise from its true direction.

To determine our course for sailing at an angle of 45 degrees to the actual wind on the next (port) tack, it is merely a matter of adding 45 degrees to the true wind direction of 142 degrees to get 187 degrees.

Current Sailing Problems One problem common to both sail and motor boats is that of choosing the correct course in traversing an area where the current will set the vessel off her intended track if allowances are not made. In Figure 2144 we have the graphic solution of two questions: What is the current doing? What course do we take to correct for it?

Assume that your boat is on course 320° at a speed of 6 knots to pass close aboard a light tower that now, at 0800, is dead ahead, 9 miles distant. As you progress through the water on your specified course, you notice that the light tower appears to be moving to your left. This indicates that a current is at work on your port side.

To learn the set and drift of the current, and the course needed to correct for it: Plot the 0800 position of the light tower from you—320°, 9 miles. Fix your present position and determine from the chart the new relative position of the tower. In Figure 2144 this is 310° and 7 miles; label this position with the time, 0829.

Now, draw in your boat's course and speed vector, *e-r*, of 320° and 6 knots, using a 2:1 scale ratio and the third circle. Note that the same scale ratio does not have to be used for both distance and speeds; it must, however, be constant within either category.

With parallel rulers or other plotting instrument, transfer the direction of the line of relative movement between

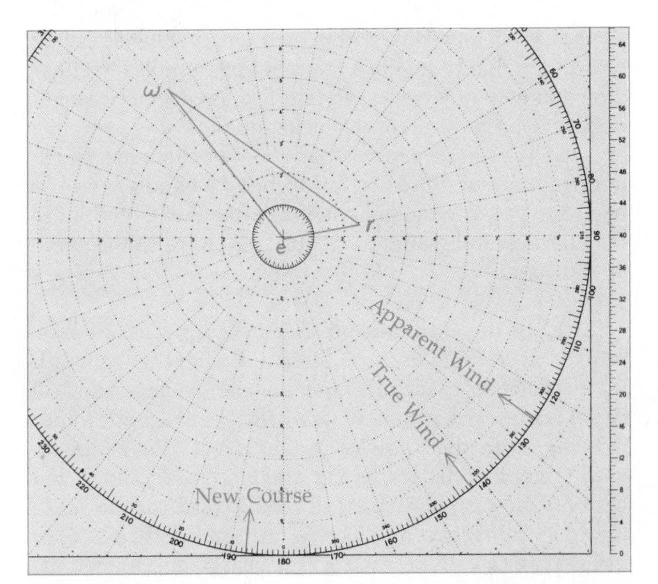

Figure 2143 **The maneuvering board can be used to determine the velocity and direction of the true wind to aid in setting a new course on the next tack. See the text for details.**

the two plots of the light tower's position to point *r* and draw in a line to some convenient length. After measuring the distance "traveled" by the tower solve for its relative speed using the distance-time-speed logarithmic scales. In this case, 2.4 miles in 29 minutes gives 5.2 knots.

Using the speed scale factor of 2:1, the speed of 5.2 knots is laid off along the line drawn from *r*. The point so found is labeled *m* and a line drawn *from m* to *e* at the center represents the set and drift of the current. This is found to be 080° and 3 knots.

With this information, we can now determine a new course to correct for the effect of the current. Remembering that our vessel *never* leaves the center of the maneuvering board, we must make the light tower "come to us." A line drawn from the 0829 position of the tower, the broken line in Figure 2144, is the line that we wish it to follow on its way to us.

Draw a line parallel to this, starting at a convenient distance out, beyond the third circle (6 knots at our 2:1 speed scale), and draw toward point *m* in the same direction as the intended track of the tower; this is line *a-m* in the illustration. Where this relative movement line crosses the third circle is the course to steer for a speed of 6 knots—287°. It should be noted that there are many possible combinations of courses and speeds available along this line, and that any increase in speed would bring a lesser course change, and vice versa. For example: looking out along the line *a-m*, we would find that for 8 knots, the course would be 293°; for 10 knots, 297°; for 12 knots, 299°.

Other Uses for the Maneuvering Board The examples above have barely touched on the many uses of the maneuvering board sheets. There are many more applications—all interesting, but of lesser direct use in small boats. These are detailed in the pages of *Bowditch* and *Dutton's*.

A pad of maneuvering board sheets, either 5090 or 5091, may be obtained almost anywhere charts and publications of the DMAHTC are sold.

Longshore Piloting

There are a number of specialized techniques that can be a part of every skipper's piloting skills. Their usefulness has been tested thoroughly.

Deliberate Offset of Course

One of these specialized piloting techniques is used in making a landfall. The essence of it is this: Lay your course, *not* for your objective but decidedly to one side to allow for possible inaccuracies in the offsetting effect of the current, or to account for possible uncertainties of position. The advantage of this procedure lies in the fact that should you not arrive at your destination at the scheduled time, you have a near-certainty, rather than considerable doubt, as to which way you should turn to reach your objective.

The technique is best explained by use of an example.

Situation Let us assume that you have been fishing somewhere in the area between Block Island and Martha's Vineyard. After several hours of trolling, drifting, and just circling around, your position is quite uncertain. Just as you decide that your fishing luck has run out for the day and it is time to head for Block Island Harbor, fog sets in.

Your best estimate in the absence of any positive information, is that you are somewhere to the east of bell buoy "1", marked as point *B* in Figure 2145; you believe that you are somewhere in the vicinity of point *A* on the chart.

Solution With Deliberate Offset of Course It's a good plan to lay your course for the off-lying bell buoy—that's what such buoys are there for—but even if there were no buoy, there is merit in laying the course decidedly off to one side of the ultimate objective, Block Island Harbor.

Suppose you lay your course AB to the bell buoy and miss it by ⅛ mile to the north. When you pick up the 3-fathom (18-foot) curve, at C, you know you are north of the harbor—because you would have picked it up much sooner if you were south of the harbor entrance.

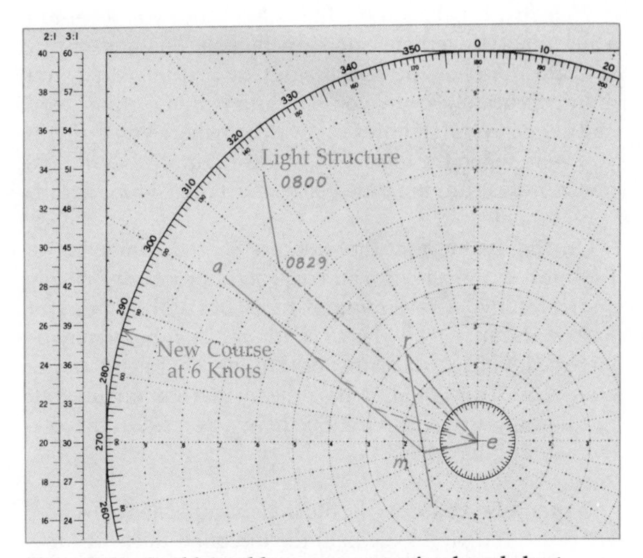

Figure 2144 **In this problem, a maneuvering board sheet enables the pilot to determine the current that is setting the boat off the desired track. It is also possible to determine graphically the necessary new course to be steered for any given speed.**

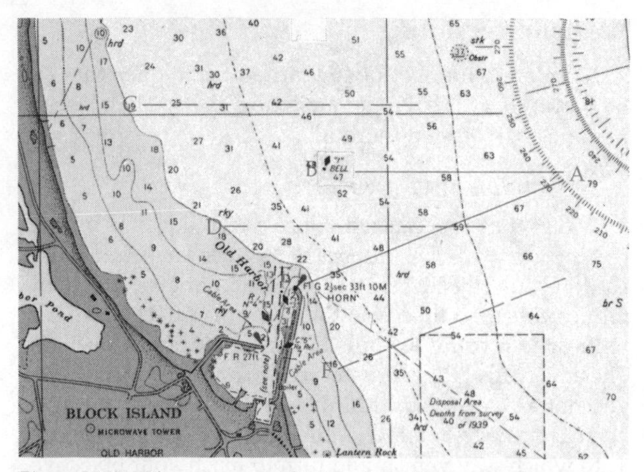

Figure 2145 **If you are caught in a fog offshore in the area of point A, the inclination might be to head directly for the harbor entrace at E. It is much safer to deliberately head for a point more to the north on a course that will take you past the bell buoy at B, as explained in the text.**

Had you missed it by ⅛ mile to the south, and picked up the 3-fathom curve at *D*, you could still follow the curve southeastward to buoy "3." Even if your position were ¼ mile south of *A* when you laid your course westward, you would still pick up the 3-fathom curve at the harbor entrance.

You find added assurance of your general position above the harbor entrance when you find depths holding generally at 3 fathoms on your course southeastward. If your calculations were completely wrong and you ran southeastward from any point below the harbor entrance, depths would increase.

Solution Without Deliberate Offset of Course Consider, however, what could happen if you missed your objective *E* by ¼ mile to the south if the course had been laid direct to buoy "3." Picking up the 3-fathom curve at *F*, you could not be sure whether you were north or south of the entrance.

Your first conclusion might be that, since you have run a few minutes overtime before reaching the 3-fathom curve, you are north of the harbor. But don't forget that you didn't accurately know your starting point. Suppose you had been several minutes eastward of where you thought you were when the fog set in. In this event, the additional few minutes running time before reaching 18-foot depths is reasonable.

On the assumption that you are indeed somewhat to the north of the harbor, you turn south, proceeding slowly and carefully. A few minutes pass, but still no buoy or harbor entrance. How far to continue? That buoy may be just ahead, obscured by the fog. On the other hand, doubt creeps in. You couldn't have been that far off in your reckoning—or could you? So there you are, in a fog—literally and figuratively.

General Procedures In the first example above (where we laid a course for the bell) the intended track was laid about ⅜ mile to the north of the objective (the harbor entrance). How much deliberate offset of course is made in any given case varies within limits determined by what you feel would be a maximum error under the circumstances. On a long run in from offshore, making a landfall on a beach that trends in a straight line for miles in either direction with few marks of identification even if visibility is good, you may prefer to make an allowance of a mile or more. It adds little to the total run, but eliminates much uncertainty. But don't use this technique blindly, without regard to possible dangers that may line the beach. Study the chart carefully and adapt the procedure to the situation at hand.

Crossing the Gulf Stream This same procedure may be used in crossing the Gulf Stream. Rather than attempting to make an exact allowance for the distance you will be set northward, a somewhat overly generous allowance is made, with the knowledge that if you don't pick up your target landmark after the allotted passage time, you have only to turn northward. It can be comforting to know the proper direction in which to turn, rather than have to make a choice with perhaps a 50-50 chance of being wrong.

The same technique can be used in recrossing the Gulf Stream on your westward passage back to Florida. Again, over-allow for the northward offsetting influence of the Stream and plan to turn northward as you approach the coast if you see no identifiable landmarks.

Night Piloting by Timetable

Another specialized technique that will increase the pleasure of nighttime piloting is to establish a "timetable" for your particular cruise.

Preparing the Timetable When you anticipate a night run, don't wait until you are on your way before studying out courses, distances, running times, etc. Go over the entire trip in advance and acquaint yourself thoroughly with the separate legs of the cruise, the aids to navigation that you will pass, the characteristics of the lights, etc.

Then set up a timetable, assuming that you will leave your point of departure at exactly 0000, and note in orderly fashion the predicted time of arrival abeam of every light and buoy on both sides of your intended track. Alongside each entry, show the characteristics of the navigational aid and the compass course at the time. You can also enter into your timetable the approximate times that major lights can be expected to become visible.

Using the Timetable Invest in an alarm clock or watch with a luminous dial and set it for 12 o'clock when you start your run from the point of departure on which your predetermined timetable is based; a digital dial, if lighted, can be set at 0000 and then read directly as elapsed time. Then as the flashing, occulting, or other lights successively appear over the horizon, you won't have to dash below to puzzle out characteristics and identify each in turn. The elapsed time from your start will clue you as to their identities.

The timetable will aid you in knowing where you are at any time, what light you just passed, what the characteristics will be of the next light to appear, and when that should happen. The use of elapsed time, rather than actual clock time, allows flexibility in the starting time; it is not necessary to revise your entire timetable if you take your departure earlier or later than originally planned. If any discrepancy creeps in consistently between when

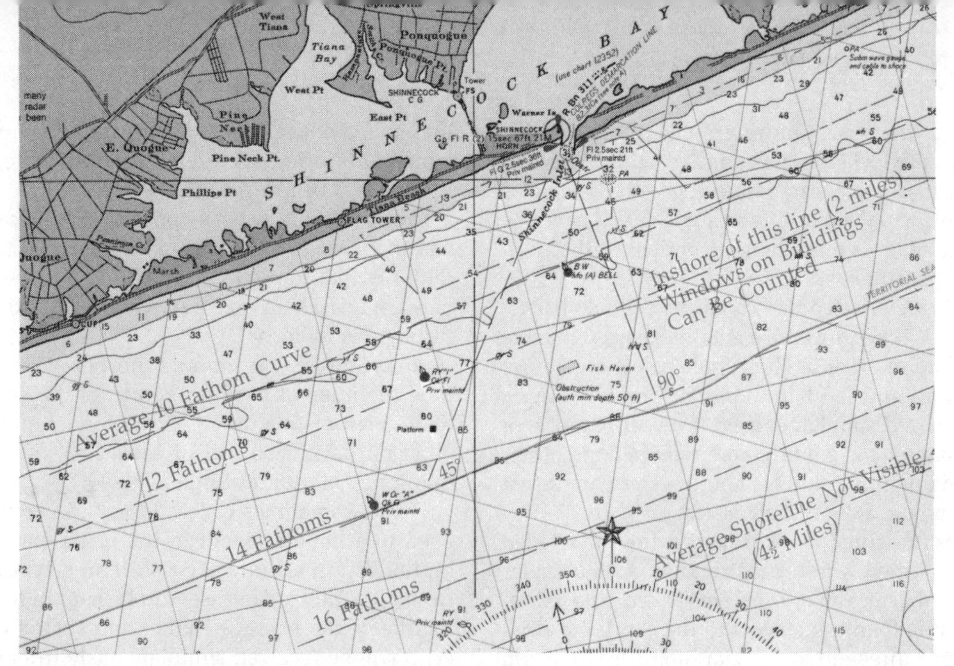

Figure 2146 **It is possible, for your height of eye on your boat, to determine the distance at which you can see the beach, and certain features of buildings. These distances should not vary much from those shown here for a typical boat.**

events should happen and when they actually do, the reading of the clock or watch can be shifted as needed.

Night Coastwise Piloting by Shore Lights

For night piloting alongshore, the lights of towns and settlements along the beach can be used to estimate distances offshore. An experienced boatman reports that he has been able to maintain a course approximately 5½ miles offshore going down the New Jersey Coast simply by keeping in sight of the reflected glow in the sky above lights in towns along the shore.

The appearance of direct rays of light would be a signal indicating that the distance offshore was decreasing to 5 miles or less, whereupon it would be advisable to haul off until just the reflected glow was visible once more. The objective of keeping so far offshore here is to avoid the extensive areas of fish net stakes, and yet not lose contact with land.

The distance that the direct rays will be seen will be determined by the height of the observer's eye, and should be determined by the individual skipper for his own boat. The principles involved remain the same for all vessels.

Following the Beach

Elsewhere, we have considered how a boat may be piloted along a coast by following a constant depth, or fathom curve. You may not have a sounder, however, or it might be broken or not available for some other reason. It is well to have another means of following the shoreline approximately when visibility conditions permit the observation of objects on the beach.

Let's say that you wish to parallel the New Jersey or Long Island beach (or any average coast where there are no hills or mountains ashore to stand out prominently at long range). Figure 2146 shows such a coastline. From the deck of a typical small boat at sea, you will be unable to see the beach if you are more than about 4½ miles offshore, due to the earth's curvature. If you can just see the beach, you are roughly 4 miles out. This, of course, will not hold true if there is fog or haze.

Distance Off by Visibility of Details As you follow the coast, you may trend in closer without realizing that you are off your intended track. Buildings appear, and you find that you can distinguish detail enough to make out individual windows of houses. This means that you are within roughly 2 miles of the beach.

Thus you have established two limits—by keeping within sight of the beach, but far enough off that you cannot count windows, you are averaging 2-4 miles off shore. Meticulous pilots may object that this is too loose, but there are many cruises that do not require any higher degree of precision.

Check for Your Height-of-Eye The 2-mile distance at which windows can be counted will not vary with the observer's height of eye. This is a matter of distance and detail. The 4- to 4½-mile limits of visibility for the beach, however, are subject to variation from one boat to another. These are average figures; for reasonable accuracy, establish more exact distances for your own boat by using the bow-and-beam technique or any other positioning procedure.

Using a Position Finder

Earlier in this chapter, we considered a positioning technique using two horizontal angles between the lines of sight to three identifiable objects. See Figures 2115-2116. Although this is a standard method, it has not been widely used on small boats because a sextant is normally required to measure the angles and a three-arm protractor to plot them accurately. As a sextant would be a relatively large investment for this use only, those who do not venture onto the high seas and use celestial navigation do not normally have one aboard.

The Weems Position-Finder is a combination instrument that can be used both to measure the angles and do the chart plotting. It eliminates the need for a sextant for making observations, and the attendant steps of reading angles from the sextant scale and setting these values on the protractor. This results in a saving of time, effort, and risk of error. As the observations are taken, the arms of this instrument are locked in position, and the device becomes, in effect, a three-arm protractor with the angles already set. The actual values of the angles need never be known. The solution is entirely independent of the compass or other instruments.

Three objects are selected that can be identified visually and on the chart. The procedure is to measure the angle between the line of sight to the center object and those to the right-hand and left-hand objects. The two angles cannot be measured simultaneously. If it is apparent that one angle is changing at a more rapid rate than the other, the angle changing more slowly should be measured first and angle changing more rapidly last.

Each angle is measured in accordance with the detailed instructions furnished with the instrument. As the angles are measured, the movable arms are locked into place. The instrument is then transferred to the chart and used as a conventional three-arm protractor, lining up each arm with the respective chart symbol and marking the position with a pencil through the hole in the center of the device. Figure 2147 shows the Position-Finder in place on a chart.

Echo Piloting

An approximate method of determining distance, in passages on bodies of water where there are sheer cliffs that will produce echoes, is to sound a short blast of the whistle or horn, and, preferably with a stop watch, time the interval before the echo is received.

Divide this interval in seconds by two (because the sound has to travel to the shore and return as the echo), and multiply that figure by 1100, a rough value for the speed of sound in air in feet per second. (Multiply by 340 for distance in meters.)

For example, you sound a short blast on your horn and time the interval until the return of the echo as five seconds. Half of this time is 2½ seconds; multiply by 1100, and you have 2650 feet or a distance off of slightly less than one-half nautical mile.

This is sometimes called "dog-bark navigation" and is used to a limited extent in the inside passage to Alaska on the British Columbia coast. If an echo can be received from both shores, a vessel can be kept in the approximate middle of the passage by holding such a course that an echo will be received simultaneously from both sides.

Figure 2147 **A position finder in place on a chart: The arms are set directly as the observations are made, and no angles need to be read in order to determine the fix. The three arms are aligned with the chart symbols for the objects sighted; your position is at the center of the protractor.**

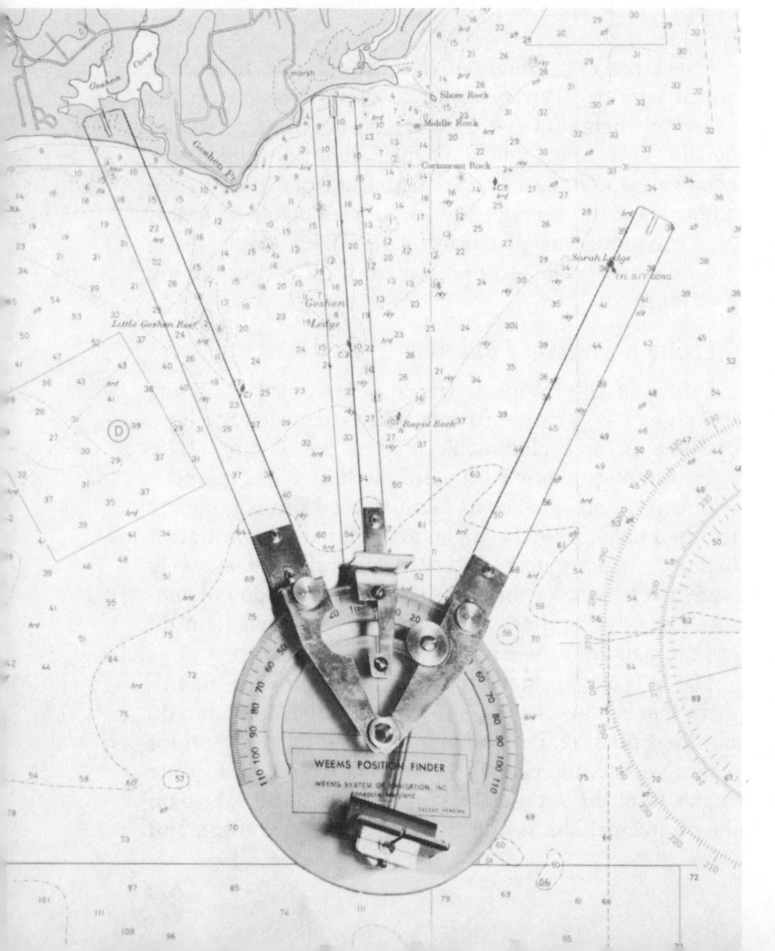

The Rule-of-Sixty

Another useful specialized piloting technique is the RULE-OF-SIXTY. It provides a simple, practical way of changing course to clear an off-lying danger area without a lot of chart work. Again, the technique is best explained with an example.

Let's assume in this case that you have come out of Portsmouth Harbor and are running a southerly course down the coast to Cape Ann. See Figure 2148. Somewhere out off Newburyport you pick up dead ahead the light on Straitsmouth Island, marked *C* in the figure. Your course made good along the line *AC* has been 164° true. A bell buoy has been placed a mile and a half eastward of the light to mark a number of rocks and ledges that must be cleared. It is obvious that you must make a change of course for safety. The problem is how much?

Instead of getting out the chart and plotting a position and a new course to clear the bell buoy, you can apply the Rule-of-Sixty and get a new course with a simple mental calculation. You know from the chart or *Light List* that the light on Straitsmouth Island is visible for 8 miles, assuming good visibility. (With less than perfect visibility, you will have to establish your distance from the light by other means. But on this night, the visibility is good and you can accept the distance as 8 miles when the light appears on the horizon.)

The procedure is to divide 60 (the rule) by 8 (the distance); the result is 7.5. Since you want to clear the light by a mile and a half, you multiply 7.5 by 1.5 and get 11.25. This is rounded to 11, and is the number of degrees that you must change your course. Thus, your new course is 164°-11° = 153° true. Long before you need to be concerned about the rocks and ledges, you will pick up the bell buoy north of Flat Ground and then the light on the bell buoy at *D*.

Although this technique is quite valid and acceptable as a specialized procedure, it should not be considered a full substitute for the basic and conventional procedure of accurately determining your position and plotting a revised course. There may be circumstances where the Rule-of-Sixty will be useful, such as on a short-handed passage or under conditions too rough for accurate plot-

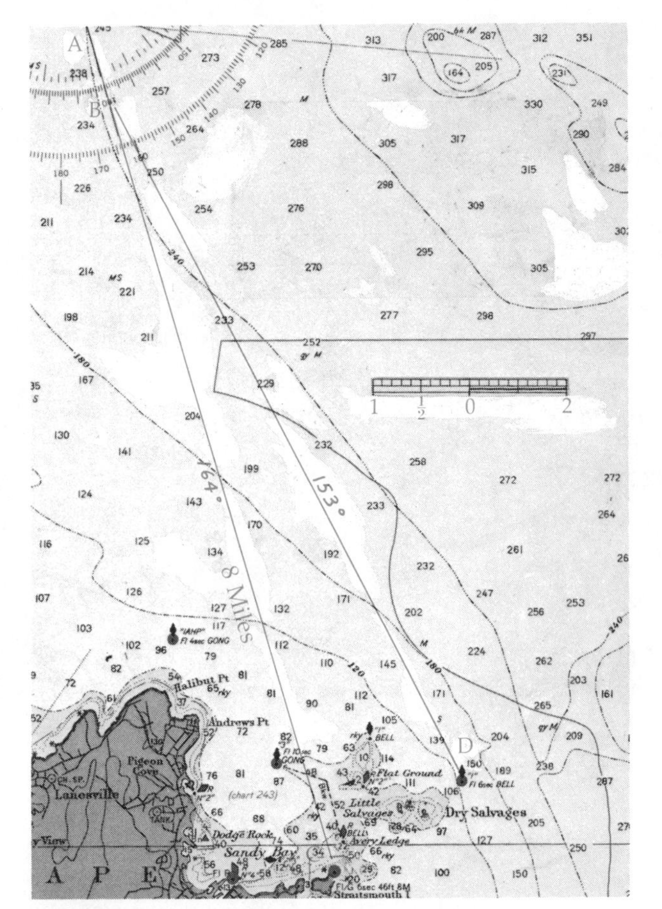

Figure 2148 **The "Rule of 60" can guide you in determining how much to change course to clear an obstacle seen ahead at a known distance. The rule can be applied without making a chart plot of the situation.**

ting, but consider it a "secondary" method rather than a first-choice method.

Determining the Distance Around a Headland

A pilot may be confronted with the problem of determining the total distance to be covered when rounding a headland or point of land in a series or short, straight courses. There is a graphic method that will give a quick and accurate solution.

In Figure 2149, the boat is at *A*. You want to know the total run to *F* on the other side of the point if you proceed to *B*, then *C,D*, and *E* in turn.

In the figure, lines are drawn from point to point, but after you become familiar with the technique, you can merely swing the legs of your dividers from point to point so as to ensure that the invisible line determined by the points will be far enough offshore to be safe.

Step-by-Step Procedures

The starting point is *A*; use a pair of dividers and measure the distance to *B*. Keep the dividers' leg point that is at *B* on that point, and swing the other leg around from *A* to the right until it touches the broken line extended backwards from *C* to *B* at A^1. Hold this dividers' point on A^1, and extend the other leg from *B* to *C*.

Now hold the point at *C* fast and swing the other point around from A^1 until it touches the broken line extended backward from *D* to *C* at A^2. Hold this point fast at A^2 and open the dividers further, as before, until the point at *C* reaches to *D*.

This same step-by-step procedure is repeated to find A^3 and A^4. When the dividers have been finally extended from A^4 to *F*, they are set to measure the total distance around the headland using the chart's graphic scale or latitude markings.

This technique saves much time and is more accurate than the usual procedure of measuring each leg separately and adding the distance together to get the total run.

Practice—and More Practice

Position determination is the one part of piloting that truly cannot be totally learned "from the book." Study is important—it is indeed essential—but you must put into actual practice the various procedures and techniques. You must know the rules and principles involved, but you must also be able to *apply* them. Make a habit of "over-navigating" during daylight and in good weather so that you will be experienced, capable, and confident at night or in foul weather.

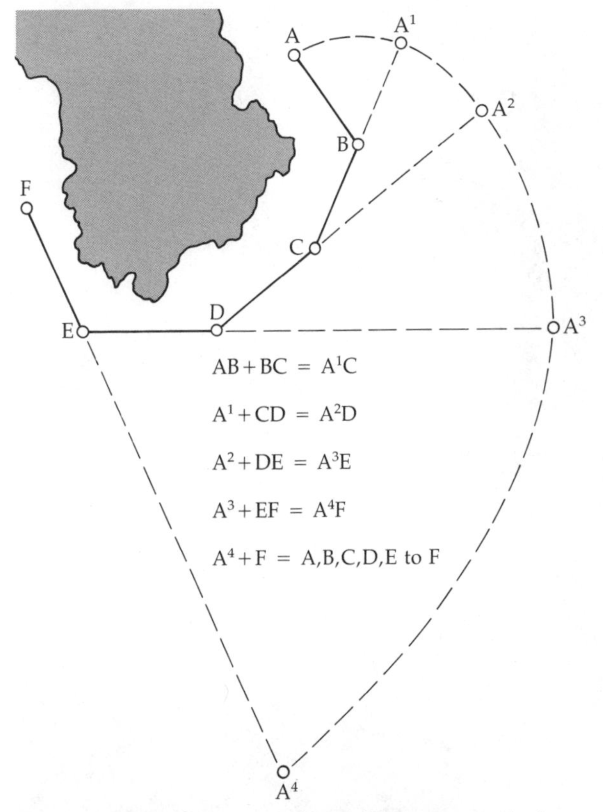

Figure 2149 **Length of several straight courses required to pass around a point of land can be added graphically and measured all at one time, rather than measuring them separately and adding the distances numerically.**

22

Selection and operation of radiotelephones for local and long-distance communications; emergency procedures; loud hailers and intercoms for boats

ELECTRONIC EQUIPMENT: COMMUNICATIONS

The ever-increasing variety of electronic equipment available for small craft makes it necessary for all boatmen to know more about this subject. "Electricity" and "electronics" are generally unfamiliar topics to the typical boatman, yet if he fails to put them to work for him on his boat, he is neglecting valuable assistance easily available to him.

The owner of a boat of any size should know four things about electronic equipment - (1) what is available, (2) how, in very general terms, it operates, (3) how it should be used for greatest effectiveness, and (4) what he should and should *not* do regarding its maintenance. Technical language will be kept at a minimum; only enough details will be included to make the necessary explanations.

ELECTRICAL and ELECTRONIC are terms often used in an overlapping manner. In this chapter, *electrical equipment* includes such things as motors, generators, and lights; *electronic equipment* is that employing tubes, transistors, and integrated circuits. The items of electronic equipment to be covered are radiotelephones (in this chapter), depth sounders, radio direction finders, radar, Loran-C, Omega, autopilots, and several minor devices (Chapter 23). Communication by visual signals is also covered in this chapter.

Electrical Systems

Electrical power is the life-blood of all electronic equipment. Although a few transistorized items may operate from self-contained batteries, most electronic equipment is powered from the boat's electrical system. With many of today's outboard motors having alternators, there is almost no lower limit to the size of boat that can have one or more electronic devices, but the total load must not exceed the electrical system's capacity.

Voltages

Electronic equipment is available for input voltages of 12, 32, or 117, the first two being for direct current (DC) systems, and the last for alternating current (AC) operation. By far the most popular voltage is 12; its primary advantages are the widest selection of items and generally the lowest cost.

Six-volt systems are now obsolete; the lower voltage requires higher currents for the same power output as a 12-volt system has, and this in turn requires heavier wiring to avoid excessive voltage drops between the battery and the load. The use of 32-volt equipment has the advantage of requiring even smaller currents for a given load than do 12-volt systems, but many items such as low and medium power radios, depth sounders, etc., are largely unavailable in this voltage rating. A 32-volt electrical system also has the disadvantage of requiring a larger and more expensive battery.

The current popularity of 12-volt systems in automobiles makes economical the use of 12-volt alternators, batteries, and regulators on boats. Equipment powered by 117 volts AC can be used, but the primary power must be continuously generated because AC cannot be stored in batteries as can DC. Thus, except where specifically mentioned as being different, it will be assumed throughout this chapter that items of electronic equipment being discussed are powered from a 12-volt DC electrical system.

Generators and Alternators

On essentially all boats, a basic source of electrical power will be a generator or an alternator driven by a main engine; see Figure 2201. The alternator is the more modern of the two and has the advantage of a greater output at low engine RPM. The external result is the same with both devices: a DC voltage somewhat greater than that of the battery (about 14 volts for a 12-volt system), so that energy can be put into the battery against its natural tendency to discharge into a completed circuit.

The flow of electricity from the generator or alternator into the battery is through a VOLTAGE REGULATOR. This device prevents an alternator from charging at a rate in excess of its capacity, prevents the battery from being overcharged, and cuts out a generator from the circuit when the engine is not running, to prevent the generator from running down the battery. For more on the operation of such generators and alternators, see books and pamphlets on engine electrical systems; an excellent source of such information is *Your Boat's Electrical System, 1981-82* from Hearst Marine Books.

Boats that are in port with people aboard much of the time may need a BATTERY CHARGER, sometimes called a POWER CONVERTER. This device takes 117-volt AC shore

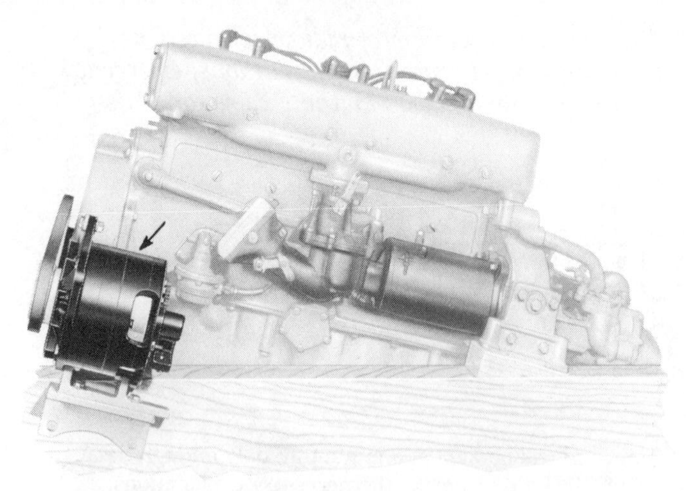

Figure 2201 **Basic source of electrical power is the engine's alternator or generator. Most boats have 12 volt DC systems, although some large boats may have a 32-volt system.**

power, *transforms* it to a lower voltage, and then *rectifies* it to DC. Thus as energy is taken from the boat's battery, it is replaced from shore power and the battery is completely or partially protected from becoming discharged.

Storage Batteries

Fortunately, DC electrical energy can be stored in lead-acid batteries, and electronic equipment can be used without the operation of the main engine.

There are two types of electrical loads on a boat. There are the relatively long-term, moderate loads of lights, pumps, electronic equipment, etc., and there is the very brief, but very heavy, load of engine starting. Normal automotive-type batteries can be used for both, but the first category is better served by a design called a "deep-cycle" battery. In either case, a true marine storage battery is preferable with its thicker and stronger case and special internal construction features.

For electronic equipment, the basic question may be stated as "Is there enough storage capacity in the boat's battery or batteries for all present equipment, plus any new items that might be added?" Capacity is measured in AMPERE-HOURS—the product of current drain in amperes multiplied by time in hours. You must compute this for all electrical loads (lights, motors, pumps, etc.), as well as for electronic equipment. And there must be enough left over to start the main engine if you have only one battery installed. If demand exceeds capacity you will need a larger battery, or a second battery. Installing a larger battery may require a larger alternator and new regulator; your ratio of running time to engine-off time will determine your needs. Batteries must be properly installed in any boat; see page 270.

With a two-battery installation, there should be a vapor-proof master battery switch with settings that allow you to connect the load to either battery separately, or to both. There is also an "off" position, which is a good way to insure against unauthorized use or accidental battery drain when you leave the boat. Instructions supplied with the switch show how to hook it up. Two-battery systems charged from a single alternator often include a BATTERY ISOLATOR consisting of two heavy-duty diodes connected so as to prevent one battery from discharging into the other, as might happen if a fault developed in one or the other.

Generally, one battery is designated for starting only; switch to the other (perhaps a deep-discharge type) when at anchor or dockside so you are not discharging the starting battery. Use the switch in the "both" position when extra power is needed, and for charging.

Wiring

If the generator/alternator and battery are the heart of a boat's electrical system, then the wires supplying power to the electronic equipment and other loads are the veins and arteries. The wires must be heavy enough—of sufficient cross-sectional area—to carry the current of the loads connected to that circuit. Adequate size is determined by two factors: heating effect and voltage drop. Current passing through a wire increases its temperature; obviously it must not become hot enough to be a fire hazard. As a general rule, a wire should not become warm to the touch when carrying its full load.

The VOLTAGE DROP problem is the more common one. The heating losses mentioned above result in a lower voltage being delivered to the load than is put in at the battery end. The voltage drop increases with a greater load, being directly proportional to the current in the wire. If the conductors are not of adequate size, the voltage delivered may be too low to operate the electronic device or other load efficiently.

Most boats are delivered with wiring adequate for the *then-installed* equipment. A problem may arise as further accessories are installed. The solution is either to replace existing wires with larger conductors, or to install additional main circuits back to the battery (assuming, always, that the various load combinations have been calculated and the battery capacity is adequate).

A wire's gauge number runs opposite to its size; the larger the wire, the smaller the gauge number. Wires of smaller diameter than 18 gauge should not be used for any purpose in a boat's electrical system; most circuits should have heavier wire. The voltage drop is proportional to the length of the circuit between source and load, as well as to the amount of the load. The longer the circuit, the larger the wire must be to prevent an excessive drop. Table 22-1 provides information about the minimum size of wire appropriate for various loads and lengths in a 12-volt system. These wire sizes will prevent a voltage drop greater than 3%; this is required by critical electronic equipment. Lights and pumps can tolerate a lower voltage, and wires can be selected in accordance with Table 12-1 on page 271. For a 32-volt system, you may add 4 to the gauge numbers in either table, but never use wire smaller than #18. Remember, the distance used in these tables is not direct, but rather as the wires must be routed—it will be longer than you think!

The insulation of wiring on boats should be suitable for the damp conditions belowdecks; thermoplastic insulation is best. Stranded wire should be used rather than solid; all wiring should be well secured and protected from abrasion and chafing.

Switches, Fuses, Circuit Breakers

Of the utmost importance in the electrical system are the various protective devices, including SWITCHES, FUSES, and CIRCUIT BREAKERS. Every marine electrical system should have a MAIN DISTRIBUTION CENTER where there are switches and overload protective devices. In addition to the switches on the individual pieces of electronic gear that are used to turn them on and off in regular operation, the various loads should be connected together into several BRANCH CIRCUITS, each with a protective element and a means of shutting off power. A short-circuit or other failure on one branch circuit will thus not require turning off all electrical power. Note, too, that each circuit should be identified on the master panel. A wiring diagram kept with the boat's papers is also advisable. See page 559.

Any combination of loads into branch circuits should be carefully considered, as should selecting the existing circuit to which a new accessory is to be connected. Loads likely to be used simultaneously should be connected to different branch circuits; likewise, loads that are most vital to the boat's safety should be connected to different branch circuits.

Further, there should be a MASTER SWITCH that cuts off all electrical power in case of fire, for working on the electrical system, or when the boat is to be left unattended. An excellent type of master switch is an enclosed, explosion-proof, heavy-duty switch that can be *mounted directly at the battery* and operated remotely from the instrument panel (or other location outside the engine compartment) by means of a cable such as is used on engine chokes and throttles.

Overload protection may be by means of either fuses or circuit breakers. Fuses are less expensive initially, but are a one-time-use device. The cartridge type that fits into clips is preferred over the screw-in plug type often found on shore; the cartridge type is less subject to loosening under vibration, with resulting poor contact. On circuits with motors, be sure to use a "slow-blow" type of fuse to withstand the initial starting current surge without going to an overly large capacity fuse. Circuit breakers are a greater initial expense, but there are no spares to be bought and carried aboard as there are with fuses. Many types of circuit breakers can be tripped manually and thus can additionally serve as switches.

Recommended Wire Gauge Sizes
12-Volt Systems
Length of wire in feet—Source to Load

Current in Amps	*10 or less*	*15*	*20*	*30*	*40*	*50*
5	14	12	12	10	8	8
10	12	10	8	6	6	4
15	10	8	6	4	4	3
20	8	6	6	4	2	2
25	8	6	4	2	2	1

Table 22-1 **Table, *above*, based on ABYC Standard E-9, specifies the wire gauge to be used versus conductor length for a voltage drop of not more than 3% in a 12-volt system.**

Figure 2202 **Each item of electrical equipment should be on its own separate fuse or circuit breaker, below, so that trouble in one unit will not put other devices out of operation.**

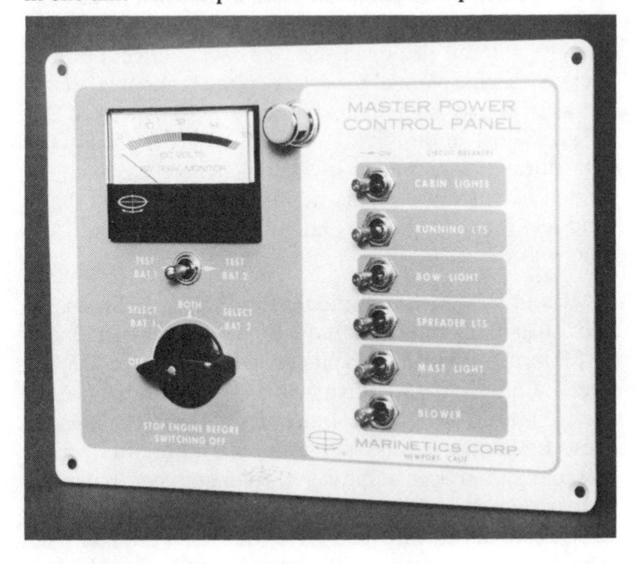

Coast Guard Safety Standards

The Coast Guard has established safety standards for manufacturers to follow for electrical systems of boats powered with gasoline engines (except outboard engines). The standards are designed to prevent fires and explosions; they are not primarily focused on the efficient operation of a boat's electrical and electronic devices. The standards are not retroactive; they apply only to new boats manufactured after the several effective dates of the various sections of the regulations.

With respect to electronic equipment, and associated wiring and overload protection equipment, the Coast Guard standards provide a most desirable level of safety, but wire sizes may not be as large as recommended in Table 22-1. Even on new boats built to the Coast Guard standard, you should measure the voltage *at each item of electronic equipment* with an accurate voltmeter when the unit is operating at maximum current drain (a radio while transmitting, for example).

Radiotelephones

The principal purpose of a radio on any recreational boat is *safety*. Certain other uses are authorized, but by law these are secondary to safety communications.

The rules and regulations of the Federal Communications Commission (FCC) uniformly refer to "ship stations"—this applies to vessels of all sizes, even the smallest of boats.

Safety Communications

A boat's radio can be used to summon assistance in many kinds of situations. You may have a leak with risk of sinking; your motor may have failed in the face of worsening weather; or someone aboard may be ill or injured. The possibilities of radio's adding to the safety of your boat and crew are virtually unlimited.

Many stations listen on the distress frequency—Coast Guard shore facilities, vessels of all sizes, and many commercial stations on shore. The FCC rules require that a listening watch be maintained on Channel 16 (see page 521) if the VHF radio of a boat is turned on but is not actively being used to communicate with another station. Thus the chances are excellent that someone will hear your distress call and either come to your assistance or get help for you.

Operational Communications

Beyond its safety value, a boat's radio may be used for contacts with other boats for OPERATIONAL communications. The very large number of marine band stations, and the limited number of radio channels available, have made it necessary for the government to impose severe restrictions on the use of radios. As defined by the FCC, operational communications are limited to matters relating to "navigation, movement, and management." NAVIGATION includes the actual piloting of a vessel, while MOVEMENT relates to future moves of a boat such as might occur during a club cruise. Radio messages of the MANAGEMENT category pertain to getting fuel, dockage, repairs, etc., and are limited to matters too urgent for handling by slower means of communication.

Business Communications

A third type of operational radio traffic is called BUSINESS COMMUNICATIONS, but this is limited to commercial and government craft, and so need not be considered here. Note that for talking between recreational boats the *only* permissible kinds of radio transmission are SAFETY and OPERATIONAL communications as narrowly defined by the FCC. Social and personal conversations between boats—any "superflous" communications—are strictly prohibited on marine frequencies.

Ship-to-Shore Communications

In most areas of the United States, a boat's radio can also provide contact with commercial shore stations on channels designated for that purpose. Through many of these stations, the boat becomes part of the nationwide telephone system; calls may be placed to, or received from, any home or business telephone. The restrictions placed on boat-to-boat contacts do not apply on these channels; calls placed through a MARINE OPERATOR may be of a personal or social nature. Charges for ship-to-shore calls vary among areas. In many cases, two charges will be billed—a "linkage" charge for connecting the boat's radio with the land lines, the other for the call itself on land.

It is possible for two boats, out of range of each other, to communicate via a shore-based marine operator; this is not often done, but don't overlook the possibility. This can even involve two marine operators in different ports with a land-line connection between them, each using radio to a boat in her service area.

Equipment

Without going too deeply into technicalities, you will need to consider certain characteristics of radio equipment to understand the proper selection, licensing, and use of radiotelephones. These characteristics include frequency bands, types of modulation, power, antennas, and operating range. Most sets for small boats combine the transmitter and receiver into a single unit called a "transceiver."

Frequency Bands

Radio frequencies are measured in kilohertz (kHz) and megahertz (MHz); one MHz = 1,000 kHz; these have replaced the older terms of "kilocycle" and "megacycle."

The total radiofrequency spectrum is divided into various bands—from very low frequency (VLF) to extremely high frequency (EHF); see Figure 2203. Various communications and navigation systems work in one or more of these bands.

Radio communications for boats are in the medium frequency (MF), high frequency (HF), and very high frequency (VHF) bands. The characteristics of each band are rather different and the use of each is determined by its characteristics—principally its operating range.

VHF The very high frequency portion of the spectrum extends from 30 to 300 MHz, but the portion of interest to boatmen lies between 156 and 163 MHz. VHF radio signals are often described as "line of sight," but actually the range is somewhat greater due to a slight degree of bending over the horizon. While visual distance to the horizon is $1.17 \sqrt{H}$ (where H is height above the surface in feet and distance is in nautical miles), a formula often used for VHF radio range calculations is $1.4 \sqrt{H}$. (The formula would be $2.5 \sqrt{H}$ where height is in meters with distance still in nautical miles.)

These formulas do not yield an exact or precise determination and variations are to be expected. However, it can be seen that height of antenna is a primary factor in determining communications range at very high frequencies; the amount of power used has some effect but it is relatively slight. Normal maximum range is about 20 miles, but signals may sometimes be heard at distances of 200 miles or more under "freak" conditions. This usually occurs during temperature inversions where a "duct" is formed between the earth and air layers at several thousand feet that are warmer, rather than colder, than conditions at the surface.

MF and HF Medium frequencies (MF) lie between 300 and 3,000 kHz, and high frequences (HF) between 3 and 30 MHz. Of interest to boatmen is a section of MF between 2 and 3 MHz and small slices of HF near 4, 6, 8, 12, 16, and 22 MHz. The transmission characteristics of these frequency bands vary from one to another, from day to night, and, in some cases, from summer to winter. Propagation is both by GROUND WAVES, whereby the signal follows around the surface of the earth gradually diminishing in strength, and by SKY WAVES, where the signal is radiated straight outward at an angle to the horizon, encounters an IONIZED layer in the upper atmosphere, and is reflected back to earth. These ionized layers (there are more than one) are at heights varying from 30 to 215 miles as determined by daylight/darkness conditions, season of the year, and the status of the 11-year cycle of sunspots (radiation from the sun causes the ionization). Communications may be established over distances up to a few hundred miles by use of the ground waves, or out to thousands of miles using sky waves. There will normally be a "skip zone" between where the ground wave fades

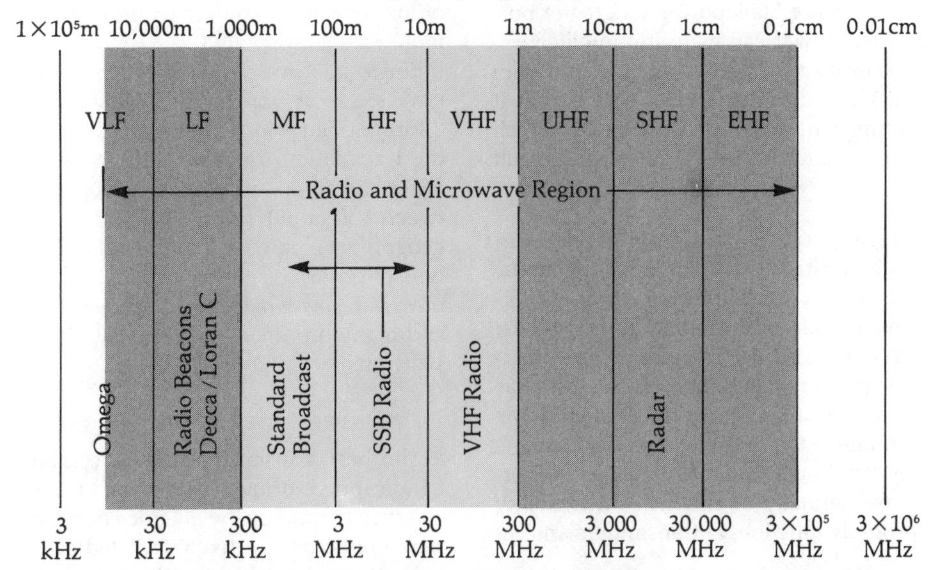

Figure 2203 **The radio frequency spectrum is divided into broad "bands" from very low frequency (VLF) to extremely high frequency (EHF). In boating, all bands from VLF (Omega navigation system) to SHF (radar) may be used. Radio communications are in the MF, HF, and VHF bands, shown in blue.**

out and the nearest range of reception of sky waves; see Figure 2204.

Selecting a frequency band for communications with a distant ship or shore station is a complex matter affected by the distance, time of day, and season of the year. Specific instructions will be found in the operation manuals of radios used for high-seas HF communications, and experience gained by use of such equipment is of great assistance in selecting the optimum frequencies. In general terms, the following guidance can be offered:

1. The 2 MHz band is usable for shorter distances of 20 to 100 miles day or night, with somewhat greater distances being reached at night. VHF, however, must be used for short distances.
2. The 4 MHz band is a short-to-medium-distance band 20 to 250 miles—during the day, but can open out to longer distances at night—150 to 1,500 miles.
3. The 8 MHz band is a medium-distance, 250 to 1,500 miles band during the day and a long-distance 400 to 3,000 miles band at night.
4. The 12, 16, and 22 MHz bands are for use over thousands of miles.
5. The highest frequency band that can be used is normally the best choice, because signals will be stronger and atmospheric noise least.

An excellent way to determine which frequency band to use is to listen; if you can hear the distant station you want to communicate with, he'll probably be able to hear you after the channel has been cleared. If you don't hear it on a specific frequency, try channels on higher and lower frequency bands.

Types of Modulation

The radio signal transmitted must be MODULATED to carry the information to the other end (except for Morse code). The voice or other information can be transmitted by either amplitude modulation (AM) or frequency modulation (FM).

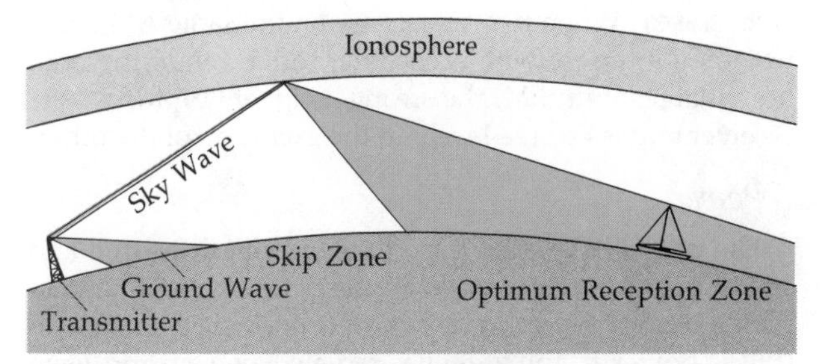

Figure 2204 **Signals are transmitted both by groundwaves that extend out along the earth for a few hundred miles, and by sky waves that are reflected by the ionosphere and return to earth farther out. Between the reception zones for ground and sky waves is the "skip zone," an area of no signals.**

AM AMPLITUDE MODULATION is the technique used by conventional standard broadcast stations between 550 and 1600 kHz. In this procedure, the signal is varied in strength (amplitude) at voice or music frequency rates. The result of amplitude modulations is a complex signal consisting of a CARRIER wave, and two SIDEBANDS, one on either side. The simplest form of transmitter sends out the entire signal, carrier plus both upper and lower sidebands; this is DOUBLE SIDEBAND (DSB) transmission, designated as A3E (formerly A3) in the FCC rules and regulations. This full signal is not needed for communications and can be reduced to just one sideband, with little or no carrier; this is SINGLE SIDEBAND transmission (SSB), designated as J3E (formerly A3J) for fully suppressed carrier, as R3E (formerly A3A) for reduced carrier (a "pilot" carrier is used for some specialized applications), H3E (formerly A3H) for full carrier (for use by receivers designed for DSB signals). H3E is sometimes referred to as "AME" for AM equivalent, or as "compatible AM." In marine communications, the upper sideband is the only one used.

The advantages of SSB transmission are that its "narrower" signals take up less space in a crowded spectrum and so allow more channels to be used within a given band, and that it gives more "talk power"; a greater portion of the transmitted power carries useful intelligence. (The carrier wave bears no intelligence, and a second sideband is redundant.) The disadvantage of SSB is that it requires a more complex transmitter and receiver, which are more expensive to purchase and a bit more difficult to operate.

FM With FREQUENCY MODULATED signals the strength of the radio waves is constant (no carrier, no sidebands) but its *frequency* is varied at an audio rate by the information (voice or music) being transmitted; within limits, the extent to which it is varied is a measure of the "loudness" of the audio signal. Frequency modulation, too, has its advantages and disadvantages. On the plus side, the signal is relatively immune to atmospheric noise ("static") and man-made electrical noise such as ignition interference. FM signals are normally of excellent intelligibility and clarity with properly functioning transmitters and receivers.

On the negative side, FM signals are much broader than AM signals and take up more spectrum space. FM is thus used only at VHF and higher frequencies where adequate space is available.

An unusual feature of frequency modulation is the CAPTURE EFFECT. When two signals are on the same AM channel, they interfere with each other, often rendering both unreadable. With FM, the stronger signal "captures" the receiver and is received well, to the exclusion of the other.

Power

The power of a radio transmitter can be measured and stated in a number of ways; input power to the final amplifier, output power, carrier power, peak envelope power (PEP), power to the antenna, radiated power, and even "effective radiated power."

VHF/FM transmitters are rated in OUTPUT POWER from the set at the point to which the antenna feedline is connected; it is normally one-half of the input power to the final amplifier. VHF sets on boats are limited to 25 watts and must have a front-panel control for readily reducing output to 1 watt or less; see Figure 2205. (Operating rules of the FCC require that the lower power be used for short-range communications where such is sufficient; the use of low power for contact with nearby boats may provide more readable signals.)

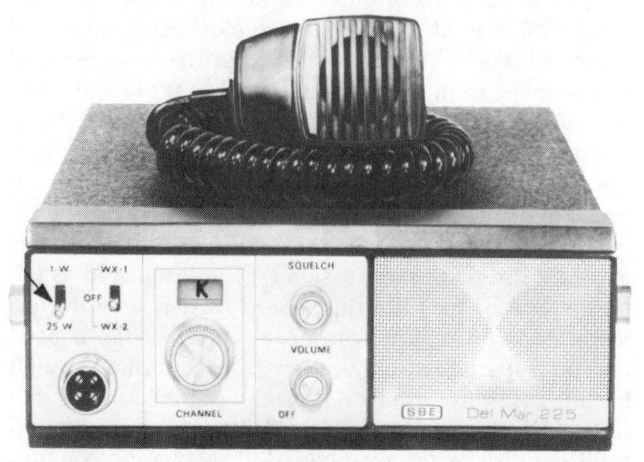

Figure 2205 **All VHF transmitters must have a switch (arrow) to reduce power to one watt or less. Low power should be used wherever it is sufficient, especially for short distances. It must be used on Channel 13.**

Single-sideband radiotelephones are measured as to their PEAK ENVELOPE POWER (PEP), the average power output "during one radio frequency cycle at the highest crest of the modulation under conditions of normal operation." SSB transmitters on boats are subject to power limits between 150 and 1,150 watts, PEP, as determined by the geographic area concerned, the frequency used, and the type of communications involved. For practical purposes, however, few small boats will need more than 150 watts and many can operate effectively with sets rated at 50 to 100 watts.

Antennas

The best transmitter-receiver combination is severely handicapped, or even worthless, if it is not connected to an efficient antenna by an efficient feedline. The outgoing signal must be effectively radiated and the incoming signal adequately picked up; the transmitting requirement is normally much more severe.

VHF A short antenna fed by coaxial cable is used on boats; no ground connection is needed. At 156-163 MHz, this is only about 36 inches (0.9 m) long. Antennas for VHF sets are simple from a physical viewpoint and can be made longer for greater GAIN—greater effective radiated power—while still remaining small enough for practical use on boats. The gain of an antenna, its ability to increase effective radiated power, is measured in decibels (dB). This is a logarithmic ratio in which 3 dB is equivalent to doubling power; another 3, making 6 dB, doubles again, quadrupling the original power; and 9 dB is power multiplied by a factor of eight.

In general, antennas are described in advertisments and promotional literature as having 3 dB more gain than is strictly true. To avoid confusion, these generally accepted terms will be used in the discussion below.

The basic VHF antenna, often called a "sailboat" antenna because of its primary use at a masthead, is a short fiberglass "whip" (with a wire inside) roughly 3½ to 4½ feet (1.1-1.4 m) in length. This antenna is usually described as a "3 dB" model although it is the smallest marine antenna used, the basic design to which others are compared, and thus actually should be termed 0 dB. This same basic antenna is sometimes used on smaller power boats, but more often the antenna is 8 to 9 feet (2.4-2.7 m) in length, a "6 dB" model. The largest antenna normally used on boats is 16 to 20 feet (4.9-6.1 m) in length and termed an "8" or "9 dB" design.

The use of a gain antenna is desirable, provided it is not overdone. The gain of a VHF communications antenna is achieved by "squeezing down" the radiation pattern so that power is, in effect, "pushed out" in a useful direction rather than being radiated in all directions, and thus wasted; see Figure 2206 top. The disadvantage of higher gain antennas is that the sharpened beam becomes somewhat like a searchlight beam, and in rough water (which often exists in emergency situations when the radio is needed most) it may move off the other station from time to time causing fading and intermittent reception; see Figure 2206 bottom.

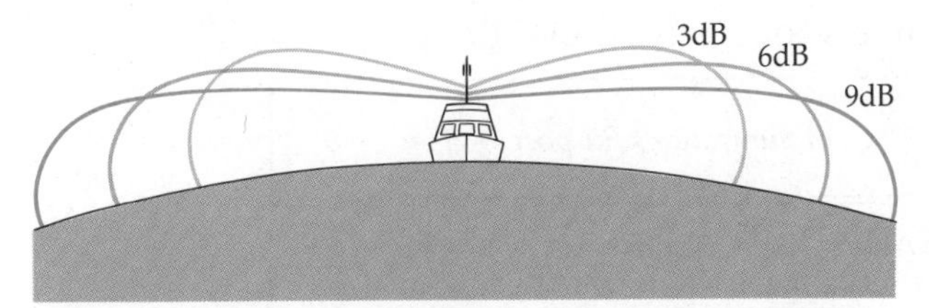

Figure 2206 **Higher gain antennas achieve greater range by "squeezing down" the radiation pattern, *above,* increasing signals at lower, desirable angles and decreasing the power that is wasted in upward, unproductive angles. An antenna with too much gain—too flat a radiation pattern—may be disadvantageous in rough water. As the boat rolls and pitches, *below,* the radiated signal may go above the receiving boat or down into the water. Either will cause fading and intermittent reception.**

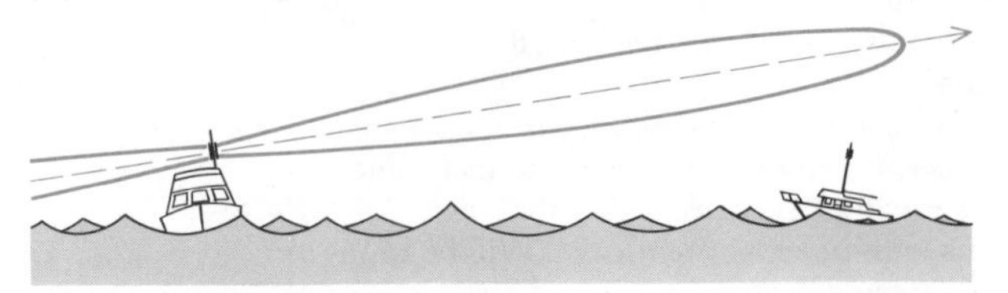

A VHF antenna should be as high as physically practicable and provide as much gain as can be obtained without possible detrimental effects. For sailboats, the amount of heel and masthead motion limits antennas to the small, simple types, but the mast top height provides the desirable range. For powerboats up to about 35 feet (11 m) in length, the best antenna is the 8-foot "6 dB" model. Larger boats could physically mount a "9 dB" 18-foot antenna but probably would do better with an 8-foot antenna mounted at the upper end of a 10-foot "extender" tube—thus height is gained while keeping to a reasonable amount of gain and directivity. (Sailboats using a masthead antenna should carry a simple secondary antenna for mounting on deck if dismasted; this may be when communication is most needed, and the masthead antenna is down with the mast.)

The antenna feedline is important in VHF radio. There is a loss of power that is proportional to length; co-ax cables are rated as so many "dB per foot." For short runs, up to 20 feet (6 m), typical of power craft, the smaller RG-58 cable is satisfactory; for the greater distances up to sailboat mastheads, the use of larger RG-8 will conserve an appreciable amount of power for radiation by the antenna.

MF and HF The antenna requirement for SSB transmitters is a difficult one to meet on small craft, especially motorboats. Except for sailing craft, the antenna is normally a self-supporting vertical pole of fiberglass with a wire conductor inside. This is electrically one-quarter of a wavelength long; "electrically" because an actual physical quarter wavelength would be a structural impossibility, at 2 MHz about 117 feet (35.6m) high! In actual practice, a LOADING COIL is built into the antenna near the midpoint so that it appears to the radio waves to be a quarter wavelength while physically being much smaller. On sailboats, it is common to install an insulator near the upper and lower ends of a backstay and use the intervening length of metal wire as an antenna. On larger sailing craft, this makes a quite efficient antenna after loading and tuning adjustments have been made.

Most SSB radio installations on boats present difficult problems of antennas and grounding (a proper ground connection is a vital part of a quarter-wavelength antenna). The problem is not in the type of modulation but in the fact that SSB sets normally cover a very wide range of frequencies, 2 MHz to 12 or 16 MHz, all with the same antenna. (A few installations may use two different antennas for more efficient operations.) This requires the use of an ANTENNA COUPLER and often considerable effort by a highly skilled technician. The results obtained are worth every bit of effort. A new type of coupler is available that changes antenna loading automatically with frequency changes.

Modes of Operation

Two different modes of operation are used in marine radio communications. In SIMPLEX operation, both stations transmit and receive on the same frequency. For SEMI-DUPLEX (or HALF-DUPLEX) operation, two frequencies are used, each station trnsmitting on one and receiving on the other alternately. (FULL-DUPLEX allows a station to transmit and receive simultaneously, as with a conventional telephone on land; this is not possible with boat radio equipment.)

The Marine Radio System

For recreational boats, the maritime communications system is divided into two parts—short range, and medium and long range. Each of these portions involves different equipment—VHF/FM for short range, and MF-HF/SSB for greater distances. By regulations of the FCC, a boat must first be fitted with VHF before SSB equipment can be installed and licensed; for the vast majority of privately owned small craft, VHF is all that is needed.

The VHF/FM Radio Service

Frequencies between 156 and 163 MHz have been allocated to maritime use by international agreements. Within this band, CHANNELS are established at 25 kHz spacing. Some channels are assigned a single frequency for simplex operation, others have two frequencies for semi-duplex operation. A standardized sysem of channel numbering has been adopted, and a boatman will rarely need to concern himself with specific frequencies in megahertz; see Table 22-2.

VHF/FM radio operations are much facilitated by having enough channels to assign separate ones for the different types of communications, and even for various types of vessels. A *primary advantage of VHF/FM radio communication is lost to a skipper if his set is not equipped to operate on an adequate number of channels.*

The primary VHF channel is 16; this is the distress, safety, and calling frequency required by law to be on every vessel. The other mandatory channel is 6 (sometimes called 06), limited to intership *safety* communications; this frequency is *not* to be used for ordinary operational or navigational exchanges, and certainly not for personal communications.

Features of the VHF service are the assignment of *different channels* to commercial and non-commercial vessels for operational communications, and the limited range of VHF which excludes interference from stations 30 to 50

Priority List of VHF-FM Channels for Recreational Boats

Channel Number	*Frequency (MHz) Transmit*	*Receive*	*Communications Purpose*
16	156.800	156.800	DISTRESS SAFETY and CALLING (mandatory)
06	156.300	156.300	Intership safety communications (mandatory)
22	157.100	157.100	Primary liaison with USCG vessels and USCG shore stations, and for Coast Guard marine information broadcasts
68	156.425	156.425	Non-commercial intership and ship to coast (marinas, yacht clubs, etc.)
09	156.450	156.450	Commercial and non-commercial intership and ship to coast (commercial docks, marinas, & some clubs)
26	157.300	161.900	Public telephone, first priority
28	157.400	162.000	Public telephone, first priority
25	157.250	161.850	Public telephone (Also 24, 84, 85, 86, 87, 88)
27	157.350	161.950	Public telephone
13	156.650	156.650	Navigational—Bridge to Bridge (1 watt only). Mandatory for ocean vessels, dredges in channels, and large tugs while towing. Army installing for communications with boats in their locks. Will be found, also, on Army operated bridges
14	156.700	156.700	Port Operations channel for communications with bridge and lock tenders. Some Coast Guard shore stations have this channel for working
70	156.525	156.525	Non-commercial only, intership
12	156.600	156.600	Port Operations—traffic advisory—still being used as channel to work USCG shore stations
72	156.625	156.625	Non-commercial intership (2nd priority)
WX-1		162.550	Weather broadcasts
WX-2		162.400	Weather broadcasts
WX-3		162.475	Weather broadcasts
69	156.475	156.475	Non-commercial intership and ship to coast
71	156.575	156.575	Non-commercial intership and ship to coast
78	156.925	156.925	Non-commercial intership and ship to coast

Table 22-2 **This table lists all the required and recommended VHF channels for recreational (non-commercial) boats. Not all are required if your number of frequencies is limited, particularly public telephone channels. Select the one or ones needed for your cruising area. If your radio has only one weather channel, consider use of a separate weather receiver for the second and third channels. Coast Guard Auxiliary boats may use Channel 83 (157.175 MHz) for some official activities, and under limited conditions, aircraft may use Channels 6, 8, 9, 16, 18, 67, 68, and 70 for communications with ships and boats.**

miles or more away. Non-commercial craft are assigned six exclusive channels for intership traffic; commercial vessels have 12 other channels of their own. The most used intership non-commercial channels are 68 and 70. Channel 9 (or 09) can be used by both non-commercial and commercial vessels; it is the only channel properly available for any necessary intercommunication between these two types of vessels.

Ship-shore communications to marine operators are possible on a number of channels. Certain ones, such as 26 and 28, are assigned in preference to others so as to limit the number of different frequencies on which a boat must be capable of operating. Ship-shore radio contacts are also possible on VHF with yacht clubs, marinas, and similar LIMITED SHORE STATIONS as they are designated in the FCC Rules; these use other channels such as 9 or 68.

Channel 22 is basically a U.S. government frequency, but non-government vessels, including recreational boats, are allowed to use it for non-distress communications with the Coast Guard. *Every boat should be fitted with this channel.*

Channel 13 is restricted to "bridge-to-bridge" communications. The term "bridge" originally meant only the control station of a vessel and the channel was used for directing the safe passage of vessels past one another. More recently, this channel has been authorized for use for communications with the operating personnel of drawbridges to request and acknowledge openings (see page 102). Large vessels, tugs, and dredges in channels must *guard* (maintain a listening watch on) this in addition to Channel 16. It is used with abbreviated operating procedures to ensure navigational safety while maneuvering in close quarters. It must not be used for any other purpose, and power is limited to 1 watt. It is not required for recreational craft but may often be of use, especially where drawbridges are radio equipped.

Other channels are used in specified areas solely for such purposes as vessel traffic control, and bridge and lock operations.

Under limited conditions, aircraft are authorized to use Channels 6 and 16; 9, 68, 70, and 72; and 8, 18 and 67 for communications with ships and boats.

A word of caution here—certain channels used internationally for two-frequency semi-duplex operation have been converted by the United States to one-frequency simplex operation (using the lower of the two frequencies). In this case, the letter "A" is sometimes, but not always, appended to the channel number. Typical of this is Channel 22 which properly should be called 22A. In this book, however, the simpler designation without the "A" will be used.

Channel assignments in Canada are generally the same as in the United States as both nations follow the international agreements. There are, however, slight differences and a U.S. boater planning to cruise into Canadian waters should get specific information for the area to be visited; also, weather channels are on different frequencies than in the U.S.

MF-HF Single-Sideband Service

If a boat is fitted with a VHF/FM radio, and the owner can show that his cruising will take him farther offshore than this short-range set can accommodate, he may install single-sideband equipment for the medium and high frequencies. Communication services are generally of the same nature as on VHF. Specific frequencies are allocated in various "bands" near 2, 4, 6, 8, 12, 16, and 22 MHz. The type of SSB modulations—J3E, R3E, or H3E—will vary with the frequency and the use being made of it.

The MF distress, safety, and calling frequency is 2182 kHz; this must be installed in every set. Modulation is J3E for U.S. vessels in U.S. waters; H3E modulation is allowed for contacts with ships and shore stations of other nations, if needed.

The frequency 2670 kHz, using J3E or R3E modulation, is used for communications with the U.S. Coast Guard. For such contacts at greater ranges, there are Coast Guard frequencies in the 4, 6, 8,12, and 16 MHz bands. (When calling on the HF channels, response may be briefly delayed if the Coast Guard station is busy on another frequency.)

Various frequencies in the 2 MHz band are authorized for intership contacts, using SSB in its most efficient mode, J3E. Frequencies may be limited to certain geographic areas, types of vessels, or non-interference to other services. The frequencies 2082.5 and 2638 kHz have no limitations and are widely used. Other ship-to-ship channels will be found in the 4, 6, 8, 12, and 16 MHz bands, two or three in each band.

Operating frequencies are often referred to by a "channel number" rather than a value in megahertz. These numbers consist of a digit or digits from the band frequency followed by a letter; thus 4143.6 kHz becomes 4B, 8291.1 kHz is called 8A, etc.

Ship-shore communications to marine operators are divided into two "services." The COASTAL SERVICE operates in the 2 and 4 MHz bands with stations in 20 to 30 major U.S. port cities, plus, of course, many foreign ports. Ranges are up to roughly 200 miles on 2 MHz and several times that distance on 4 MHz at night. The HIGH SEAS SERVICE uses frequencies in the 4, 8, 12, and 16 MHz bands. Because of the long distances covered, there are U.S. shore stations only at limited points—New York, Miami (Ft. Lauderdale), Mobile, San Francisco, and Honolulu. On both services, semi-duplex mode is used with J3E modulation (older sets with R3E may use that mode), and care must be taken regarding the frequency used for calling a marine operator—considering time of day, distance, and, to a lesser degree, the direction of the shore station from the boat.

Equipment Selection

Selecting a specific VHF/FM radiotelephone for a recreational boat basically involves power, channel capacity, and cost.

Power The maximum power for VHF is 25 watts output (which equates roughtly to 50 watts input, for comparison with other methods of measuring transmitter power). Power is not a major factor in operating range, but can make the difference in the "capture effect" of VHF/FM communications that essentially prevents interference between the signals being received at the same time. All sets on the market are rated at 25 watts except for portable and hand-held units.

All sets have a "low power" switch to reduce output to 1 watt or less as required by FCC regulations.

Channel Capacity As previously mentioned, you need an adequate number of channels to make the most effective use of a VHF radio. Although the FCC requires only three channels (16, 6, and one "working" frequency), you should have at least 12-channel capacity. Fortunately, this presents no difficulties with new sets. Crystal-controlled sets with a limited number of channels, such as 12, are now rarely manufactured and sold—the bulk of all VHF radios, even hand-held models, now use synthesized frequency control.

The actual number of channels available will vary from set to set, and will probably be different for transmit and receive, but in all cases the quantity will be more than adequate for any recreational or commercial boater. Typical numbers of channels vary from 54 to 90 or more; even the smallest of these numbers is more than adequate for use in U.S. waters, covering all channels that are legally available. Channel selection may be by rotating a dial or pressing buttons on a key pad similar to a telephone.

The more-expensive models of synthesized VHF radios may have additional desirable features. These include, but are not limited to, scanning of all or a selected number of channels; dual receivers to permit simultaneous monitoring of Channel 16 (required by law) plus a working channel (desirable when cruising with a group, or when expecting a call from onshore); individual push-buttons for quicker selection of the more often used channels; and radio direction-finding capability.

Accessory Equipment Some boats will need a remote-control unit so the radio can be operated from either of

Figure 2207 **An "all-channel" VHF radio offers the maximum operational flexibility. Care must be taken to not use an unauthorized channel. The set shown here has a dual receiving capability so that both Channel 16 and any other desired frequency can be monitored simultaneously.**

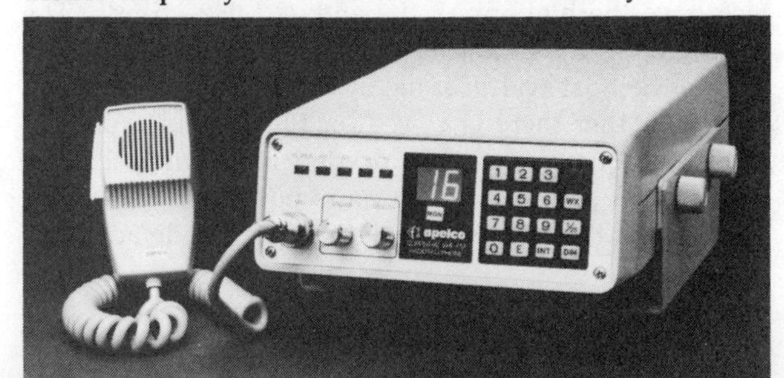

two steering positions. FCC regulations require that a remote-control unit have the capability of channel selection, to avoid the temptation not to switch to a working channel after Channel 16 has been used to establish contact. It must also be capable of switching between high and low power.

As many VHF radios have neither a dual receiver nor a scanning capability, it is often useful to install a SCANNER as a second receiver. This compact unit can scan across additional channels, from 6 to as many as 12 in some models, stopping to receive on any frequency if there are communications in progress, or going rapidly to the next if that channel is quiet (or when transmissions cease on a channel at which it has stopped). The scan rate is adjustable on some models; switches can be set to bypass any channel, or to stop scanning action and operate on a selected frequency. Channels can be controlled by crystal or synthesizer.

The FCC Rules provide for an "associated portable ship unit" with a regular boat radio station. This is a small, hand-held, two-way VHF set that can be taken into a dinghy. Strict limitations have been placed on this unit: it must have Channel 16 and at least one intership channel; it can be used *only* to communicate with the "mother" vessel; it can *not* be used on or from the shore. The call sign is that of the mother craft plus an appropriate unit designator. Power is limited to *one watt,* which *excludes* many hand-held units marketed with two to five watts power; such units, however, can still be separately licensed for use as non-installed, self-contained radios on daysailers, small runabouts, etc.

A multi-test meter can be installed in the feedline between a VHF set and its antenna. This will give a positive indication of power from the transmitter (the front-panel lights that glow when the microphone switch is pressed are generally meaningless as to actual transmission). This meter can also indicate antenna condition and power output in watts; a short piece of wire from the unit will pick up enough signal so that it can function as a field strength meter. There is negligible insertion loss and the tester can be left in the line as a constant monitor; this will give continuing assurance of proper operation and eliminate the need for on-the-air "radio checks."

SSB Equipment

SSB sets can easily cost three to five, or more, times the price of VHF transceivers, due to their more complex circuitry. Many SSB sets have the 150 watts of power allowed for most use. Often, however, a set of 50 to 100 watts PEP will provide fully adequate service for a recreational boatman.

Major considerations in selecting SSB equipment are frequency range and number of channels. In general, all sets cover the MF band and the lower HF bands, typically up through the 8 MHz frequencies. More expensive sets also have the higher—12, 16, and 22 MHz—bands. The requirements for various bands depend primarily on what waters the boat will use, the distances to stations it will need, whether there is a need for full 24-hour capability (higher frequencies will be needed for daytime distant contacts). A set of 75 to 100 watts power, limited to the 2, 4, 6, and 8 Mhz bands, will usually provide adequate

Figure 2208 **A single sideband (SSB) radio transmitter and receiver are required for communications beyond VHF range. Frequencies are available for various distances and day or night operation in various bands from 2 to 22 MHz.**

communications for the skipper who is going not too far beyond VHF range, and at a considerable savings in cost. Sets limited to 2 MHz frequencies are available, but the slight additional savings hardly compensate for the lack of capability on higher frequencies.

The number of channels is related to the number of bands used, with fewer required on a four-band SSB radio than with a set covering six or seven bands. It is important, however, to have *enough* channels to meet one's needs for communications with the Coast Guard, other boats and ships, and with marine operators ashore, keeping in mind that different bands must be used at different times of the day and at different distances. With a four-band set, ten channels would probably be a minimum.

The latest development in single-sideband transceivers is the fully synthesized set which can transmit and receive on every frequency from 2 to 29 MHz in 0.1 kHz steps. Not every frequency can be used, of course; only a few hundred are authorized for marine stations.

These sets have a keyboard-entry frequency selection system, but in most instances frequencies are not punched in directly. To avoid getting on an unauthorized frequency, the radio is internally programmed for some 40 to 99 specific frequencies or frequency pairs; these are then keyed up in terms of channel numbers. With automatic antenna tuners, these sets offer the ultimate in SSB radio flexibility, but at a price of several thousand dollars.

Installation and Maintenance

A VHF radio with synthesized frequency control need not be installed by a licensed technician, and the same is true for a crystal-controlled set provided that its crystals were put in the set by the manufacturer and the frequencies checked at that time. The later addition of crystals for other channels may only be done by a technician holding an FCC *General Class* commercial radio operator license (formerly designated as "First Class" or "Second Class" licenses).

Installating a SSB set requires certain on-board tests made by a licensed technician. Radio installations on gasoline-powered boats may require some form of engine noise suppression; normally a job requiring the skills of a technician. With regard to maintenance, an unlicensed person is limited to matters which will not affect the quality of the signal on the air. For example, he can replace bad fuses or tubes, but cannot change crystals or adjust antenna loading.

The crystals used to control the frequency of a radio—VHF (including synthesizer sets which contain one or two crystals) or MF-HF—are subject to "aging," with resultant gradual shift in frequency. The licensee of a station is responsible for any off-frequency operation that could result in a citation from the FCC. There is no exact guideline for how often a frequency check should be made—this requires a licensed technician and sophisticated equipment—but every two years would be a maximum interval, plus after receiving reports on the air of weak or "fuzzy" signals.

Radio Licenses

With over 400,000 marine radio stations on the air, the need for licensing and regulations can easily be seen. The FCC issues its Rules and Regulations in various parts, of which "Part 83—Stations on Shipboard in the Maritime Service"—applies to recreational boats.

The FCC uses a system of licenses controlling the use of radio stations, to hold down interference and permit emergency and essential communications. Recognizing that harmful interference could result from either malfunctioning equipment or from misuse of a properly operating set, licenses are required for both the station and the person operating it. Although it is termed a station license, FCC authorization relates to the transmitting component only. The set owner need not concern himself with the many technical requirements for equipment provided that he has a set that is "type accepted."

The Station License

A station license may be issued to a U.S. citizen, corporation, or an alien individual, but not to the government of another nation, or its representative. Application is made on Form 506 which should be mailed to the FCC, P.O. Box 1040, Gettysburg, PA 17325*. Issuing the license may take as long as 60 days; if you need immediate use of the radio, complete Form 506-A for "temporary operating authority" and keep it aboard. This form allows legal use of your set and provides a temporary call sign based on your boat's documentation or state registration number.

Radio station licenses in the U.S. are issued in the name of the *owner* and the *vessel.* A station license is not automatically transferred to another person upon sale of the boat, nor may a license be moved with the radio set to a new boat owned by the same person. A simple change in the name of the boat or licensee (but not a change in ownership) or his address does *not* require license modification. Just send a letter to the FCC advising them of the change; a copy of this letter must be posted on board with the license.

*At the present time no fee is charged for an original station license or renewal (nor for an operator permit), but this may change in the future.

FCC regulations require that a station license be conspicuously posted aboard the vessel. You must apply for renewal before expiration of its five-year term, using Form 504-B. If you did make timely aplication for renewal, operation may continue even if you have not received the renewed license before the expiration date. If the use of the radio station is ever permanently discontinued, you must return the license to the FCC in Washington for cancellation.

The Operator's Permit

A personal license, a radio OPERATOR PERMIT, is not required for the operation of a VHF set on a "voluntarily equipped" vessel—a recreational boat, or one carrying six or less passengers for hire, "on a domestic voyage." (An exact interpretation of "domestic voyage" for boaters who go well offshore has not yet been established.) For other operation of such craft, the lowest grade of radio operator license—a RESTRICTED RADIOTELEPHONE OPERATOR PERMIT—is adequate. For vessels carrying more than six passengers in commercial service, the operator must have a MARINE (formerly "Third Class") RADIOTELEPHONE OPERATOR PERMIT. Other "compulsorily equipped" vessels include those required to have a set under the Vessel Bridge-to-Bridge Radiotelephone Act—large vessels, tugs, and dredges in channels. A higher class license is available for people with technical training and experience, but they are needed only for making tuning adjustments and repairs. An unlicensed person may talk into the microphone of a radio, but a licensed operator must be present and responsible for the proper use of the station.

An applicant for any grade of U.S. operator's license may be a citizen of any nation. A Restricted Permit is obtained by submitting an application on FCC Form 753, which is mailed to the FCC, P.O. Box 1050, Gettysburg, PA 17325. You do not need to appear in person at any FCC office. The permit is issued, without test or examination, by declaration. The applicant must be at least 14 years old and "certify" that he (1) can receive and transmit spoken messages in the English language; (2) can keep a rough log in English, or in a foreign language translatable into English; (3) is familiar with the applicable laws, treaty provisions, rules, and regulations; and (4) understands his responsibility to keep currently informed of the regulations, etc. The Restricted Permit is valid for the lifetime of the person to whom issued, unless it is suspended or revoked. A temporary operator permit is also obtained when Form 506-A is filed for a temporary station license.

For the Marine Radio Operator Permit, there is no age limit, but an examination is required and so you must visit an FCC office. This test is non-technical, covering only operating rules and procedures; questions are of the multiple-choice type. You will find the examination not at all difficult if you prepare for it properly. A free Study Guide is available from FCC offices. For skippers of recreational boats the privileges of this higher class license are no greater than those of a Restricted Permit, but for many it is a matter of pride to qualify and post it on their craft.

If your radio operator permit is lost or becomes illegible, you should immediately apply for a duplicate. Use the same form as for an original. State the circumstances fully, and, if the license has been lost, you must certify that a reasonable search has been made. Continued operation is authorized if a signed copy of the application for a duplicate is posted. Should a lost license be found later, send either it or the duplicate at once to the FCC for cancellation.

Radio Operating Rules

Radio operating rules have been established to reduce the interference that would result from the overcrowded conditions on the few frequencies available. Actually, the legally required procedures are only a bare minimum and must be supplemented with voluntary procedures.

As stated before, radio communications (other than ship-shore telephone calls) must be necessary and of a "safety" or "operational" nature. Even within these limitations, a system of priorities has been established to make sure that the more important messages get through. First, of course, are DISTRESS calls and related follow-up messages; these are identified by the signal "Mayday" and receive "absolute priority" over other communcations. Second are URGENT messages—those relating to the safety of a ship, aircraft, or other vessel, or of some person on board or in sight. Urgent communications are identified by "Pan-Pan" (pronounced "Pahn-Pahn"). Third are SAFETY messages—those concerning the safety of navigation or giving important weather warnings; and identified by the signal "Security" (pronounced "Saycuritay").

The system of priorities has been established to ensure that the message that *has* to get through does so without delay. No station or operator has any exclusive rights to any frequency; the nature of his traffic determines whether he should transmit or keep silent. One of the most important FCC regulations requires that an operator *listen before transmitting* to ensure that he will not interfere with the communications of others, and particularly not with distress or other priority traffic.

Frequencies for Calling and Distress

The most important channel on the VHF band is Channel 16—156.80 MHz—the distress, safety, and calling frequency. In order to keep one channel relatively clear of traffic so that even a weak distress call can be heard, the use of this frequency is limited to distress traffic and the *initial* contact between vessels. This latter use as a "calling frequency" is permitted to ensure that when a station is not working on another channel, it is listening on Channel 16 for calls. Thus many stations are constantly monitoring the distress frequency and a call for help is much more likely to be heard.

Listening on Channel 16 is not only a logical procedure, it is legally required. A "voluntarily equipped" boat need not have its radio turned on, but if it is on, it must be tuned to the distress and calling frequency when not being actively used on another channel.

On the MF band, 2182 kHz is used in a similar manner for distress, safety, and calling; plans are being developed to establish one distress and safety frequency on each of the 4,6,8,12, and 16 MHz bands. It should be carefully noted that if the distance is relatively short, contact on VHF *must* be attempted first before a MF or HF channel is used.

Ship-to-Ship Communications

After making their initial contact on Channel 16, boats using the VHF band *must* shift from the calling channel to a "working" channel to communicate with each other. *No conversation, no matter how brief, is authorized on Channel 16.*

For recreational craft, the working channels are 68, 70, 72, 69, 71, and 78. Any of these may be used; they are listed here in sequence of probable availability on boats. (Here the advantage of an "all-channel" VHF set becomes apparent; two boats so equipped can talk on a channel that is not usually installed on most 12-channel radios and thus has little traffic on it.)

Channel 9 must be used if communications are with a commercial vessel. Channel 22 will be used if with a Coast Guard boat or cutter; Channel 12 is an alternate.

Several ship-to-ship simplex channels are provided on each MF or HF band for SSB communications.

Ship-to-Shore Communications

VHF communications with marine operators are normally begun directly on the working channel without an initial contact on Channel 16. Even if the call is originating from a shore station and the operator makes a call on "16," the boat should respond only on the working channel, such as 26, 28, 27, 25, etc., as the marine operator may not be listening on 16. Consult local boatmen, cruising guides, or other publications to find the channels (there may be more than one) used by the marine operator in any specific area.

VHF stations are authorized for yacht clubs, marinas, and the like. These facilities have sets much like those on boats, and initial calls on Channel 16 should be followed by a shift to a working channel. The channels most often used are 9 and 68 (despite their heavy use for boat-to-boat communications); other available channels are 69, 71, and 78, but not 70 or 72.

The VHF working frequency to Coast Guard shore stations is Channel 22; an alternate is Channel 12.

Several ship-to-shore semi-duplex channels are assigned on each MF and HF band for marine operator contacts.

SSB working frequencies with the Coast Guard include 2670 kHz simplex, plus semi-duplex HF channels; not all Coast Guard radio stations will guard all assigned frequencies; consult a listing of stations and frequencies.

Special Communications

Recreational boatmen seldom need to use Channel 13, the bridge-to-bridge frequency. This channel may be monitored but is best left for use by ships and tugs. It is sometimes of value, however, as when passing a dredge or a tug and tow in restricted waters. In this case use only the lower power (1 watt or less) output; call directly on 13, not 16; use the vessel's names only, without call signs. (In some areas commercial vessels guarding bridge-to-bridge and vessel movement control channels do not have to maintain a watch on Channel 16 and may not do so.)

Communications with drawbridges and canal locks may involve special procedures that are best learned locally.

Legal Requirements

The licensee of a radio station on a voluntarily equipped

boat is not required to have a copy of the FCC Rules and Regulations, but this does *not* relieve him of the responsibility of knowing and obeying them. Simplified and condensed copies of the rules can be obtained from a number of sources.

The licensee is likewise not required to maintain a radio station log, but a prudent skipper will take notes on any significant radio traffic that he hears, especially any relating to a distress situation. He likewise will keep a record of any maintenance work performed on his radio, noting the nature of the work and who did it.

The operators of compulsorily equipped vessels are required to have a copy of the rules and keep a radio log. Logs should be retained for one year in normal operations, longer if involved in special circumstances as described in the rules.

A Restricted Radiotelephone Operator Permit, if required, must be carried on one's person. A higher grade of license must be posted, and a photocopy carried if it is to be used elsewhere.

Rules Governing Transmissions

Transmissions on the calling channel must be limited to determining the working frequency to be used, and this must not take longer than two minutes. Any one calling transmission must not take longer than 30 seconds and, if no reply is heard, wait two minutes before calling again. Another two minutes must separate a second and a third attempt to make contact and, if these fail, delay to 15 minutes before starting a new calling cycle (the 15 minutes may be reduced to three if other stations will not receive interference from the calls.) These restrictions do not apply to an emergency involving safety.

Once contact is made on a working frequency, the exchange of transmissions between boats must (1) be of legally permissible nature, and (2) be of the *minimum length possible.* (On MF working frequencies, the exchange must not exceed three minutes and must not be resumed until another ten minutes have elapsed; this does not apply to communications relating to safety of life or property.) Use no more than the minimum power that will get through—1 watt on VHF if possible.

The Communications Act of 1934, together with the FCC Rules, strictly prohibit the use on the air of any language which is "obscene, indecent, or profane." This is watched closely by various government monitoring stations and violators have often been taken to court. Special, more severe, penalties are specifically provided for this offense.

Procedure for Station Identification

The FCC has established procedures for identifying your station on the air. Such identification must be given in the English language by stating the official call sign. Phonetic words for the letters may be used, but are not legally required and should not be used unless they are necessary as they will lengthen the transmission; any understandable word may be used, but the international aviation and military alphabet (page 582) is probably the best.

Radio stations must be identified, as a minimum, at the beginning and end of an exchange of transmissions with another station, but identification of each transmission is *not* required. If both Channel 16 (or 2182 kHz) and a working frequency are used, identification by call sign need be given only at the beginning on the calling frequency and at the end on the working frequency. Identification must be made of each transmission for any other purpose, such as a test. If transmissions continue for more than 15 minutes, the station must be identified at intervals of 15 minutes or less, except that on a ship-shore telephone call this may be deferred until the end of the call.

Secrecy Requirements

The Communication Act of 1934 and the Rules of the FCC protect the secrecy of communications. No person may divulge to another person, except the addressee or his authorized agent, any knowledge gained from receiving or intercepting radio transmissions not addressed to himself, *nor shall he use such knowledge for his own benefit.* This basic point of law should be carefully noted by all who operate radios on boats. It does not apply to distress communications or to broadcasts for the general use of the public, but it does apply to *all* other conversations heard on the air.

Operating Procedures

The FCC Rules contain only a few instances of specific operating procedures, chiefly in connection with emergency communications. The various examples of operating procedures given on the following pages not only conform with the regulations, they are offered as examples of good practices on the air.

How to Make a Call

Listen carefully to make sure that the channel you want to use is not busy. If it is busy, you will hear voices, or from some public shore stations, an intermittent busy tone. Except in a safety emergency, don't interrupt.

When the conversation is to take place on a ship-to-ship frequency—unless you have reached an agreement in advance as to the time and frequency—establish contact on Channel 16 (or 2182 kHz) and then shift to an agreed-upon intership channel.

When the conversation is to take place through a commercial shore station, make your initial contact on a working frequency of that station.

Both of these practices are designed to relieve the load on Channel 16 and 2182 kHz so that their availability for safety purposes will not be jeopardized.

Steps to Follow in Making a Call
(Other Than a Distress, Urgency, or Safety Call)

Boat-to-Boat Calls Listen to working channels to find one not being used. Listen to Channel 16 (or 2182 kHz) to make sure that it is not in use. If it is free, put your transmitter on the air and say (for this example your boat's name is *Fayaway* and its call sign is WAB 1234, and you are calling *Moonfleet*): "*Moonfleet* this is *Fayaway,* WAB 1234, Over." (You will be asked for an account number or billing instructions.) To avoid confusion, always observe the proper sequence of call signs—state the name or call sign of the other station *first,* then give your own identification after saying "This is"—don't reverse the sequence. End your call with the word "Over." (If necessary, give the identification of the station called two or three times, but not more; you will rarely need to give your own boat name and call sign more than once.) The entire calling transmission must not take longer than 30 seconds.

Listen for a reply. If no contact is made, repeat the above *after an interval of at least two minutes.*

Ship-to-ship contacts on HF working channels will have to be by prior arrangement as there are no specified calling frequencies on the 4 MHz and higher bands.

After establishing contact on Channel 16 (or 2182 kHz), switch to the agreed intership working channel. After the conversation is completed, say: "This is *Fayaway,* WAB 1234, Out."

If the two stations have been in recent communications, the distance is short, and the other skipper is expecting the call, or similar circumstances, the operating procedures can be considerably abbreviated. For example, the calling station can propose the working channel in his first transmission, and the called station merely says "Roger" or repeats the working channel number to agree to the shift, and gives his call sign. The procedure word "Over" indicating that you are ending your transmission and are awaiting a response can usually be omitted when using VHF with "solid" contact between stations; its omission does no harm and speeds up communications. Never combine "Over" and "Out" as these have different meanings.

Ship-to-Shore Service Listen to make sure the working channel you wish to use is not busy. (Again, your boat is *Fayaway,* and you are calling a Miami number.) If the channel is clear, put your transmitter on the air and say:

"Miami marine operator this is *Fayaway,* WAB 1234, Over."

Listen for a reply. If no contact is made, repeat after an interval of at least two minutes.

When the marine operator answers, say:

"This is *Fayaway,* WAB 1234, calling 234-5678 in Miami, Over."

After the telephone conversation is completed, say: "This is *Fayaway* WAB 1234, Out."

When using SSB to Marine Operators you should use J3E mode, but R3E may be used if that is what is installed on an older set.

How to Receive a Call

Your boat can be reached only when your receiver is turned on and tuned to the frequency over which you expect to receive calls. Some, but not all, marine operators will make an announcement on Channel 16 that they have traffic for a specific vessel; such calls should be answered on that operator's working frequency, *not* on 16.

The receiver you use to maintain watch on Channel 16 will ensure that you get calls addressed to you by other boats; it *may* give you information that a marine operator has a call for you. When you are using your set on a working channel, however, you will miss any marine operator calls. It is thus advisable, if you are expecting a telephone call from shore, to call the operator periodically on her working channel to "check for traffic"—some shore stations will have regular schedules to announce the names of vessels for which calls are being held.

Boat-to-Boat Calls When you hear your boat (using the boat names in the first example) called, put your transmitter on the air and say—"*Moonfleet,* This is *Fayaway,* WAB 1234, Over." Again note that the proper sequence is the name of the called boat, "this is," the name of the calling boat, and its call sign.

Switch to the agreed-upon intership channel. After the conversation is completed, say—"This is *Fayaway,* WAB 1234, Out."

Shore-to-Ship Calls When you hear the name of your boat called, put your transmitter on the air and say—"Miami marine operator, this is *Fayaway,* WAB 1234, Over."

After the conversation is completed, say—"This is *Fayaway,* WAB 1234, Out."

Again, good contact between stations can result in abbreviated procedures including the elimination of "Over." (Except perhaps in ship-shore calls patched to a shore station where the other person is not familiar with marine semi-duplex operation and will need to be told to wait for each "over" as a cue to when to start talking.)

Distress, Urgency, Safety

In an emergency, as part of the marine safety and communications system, you have help on Channel 16 (or 2182 kHz) at your fingertips wherever you may be. (On the high seas far offshore, a call on a Coast Guard 4 or 6 MHz or other HF channel using pure SSB, J3E), may be more effective than a call on 2182 kHz using DSB or compatible SSB, H3E.)

Only when grave and imminent danger threatens your boat and immediate help is required, should you use the distress procedure—radiotelephone alarm signal (if available)—and MAYDAY. Transmitted on Channel 16, it should be heard by many boats, as well as by the Coast Guard and other shore stations within range.

The RADIOTELEPHONE ALARM SIGNAL consists of two audio tones of different pitch, 1300 and 2200 Hz, transmitted alternately ¼ second of each tone. The purpose of this signal is to attract the attention of people on watch, and, at some stations, to activate automatic devices giving an alarm. It must be used *only* to announce that a distress call or message is about to follow. (Exception: it may be used with the urgency signal in two specified instances; see FCC Rules.) Some Coast Guard shore stations are now equipped with devices to generate this signal, and you should know what it means if you hear it.

The Distress Procedure—"Mayday"

Distress communications include the following actions:

1. The *Radiotelephone Alarm Signal* (whenever possible for at least 30 but not more than 60 seconds) followed by—
2. The *Distress Call.*
3. The *Distress Message.*
4. Acknowledgment of Receipt of Distress Message.
5. Further Distress Messages and other communications.
6. Transmission of the Distress Procedure by a boat or shore station not itself in distress.
7. Termination of Distress Situation.

The DISTRESS CALL consists of:

- the Distress Signal *Mayday,* spoken three times;
- the words *this is*

• the identification (name and call sign) of the vessel in distress, spoken three times.

The DISTRESS MESSAGE follows immediately and consists of:

• the Distress Signal *Mayday*, spoken once;

• the identification of the craft;

• particulars of its position (latitude and longitude, or true bearing and distance in miles from a known geographical position);

• the nature of the distress and the kind of assistance desired;

• any other information that might facilitate the rescue; especially a description of the boat; length, color, type, etc.; and the number of people aboard; *over.*

Example of Distress Procedure

"(Alarm signal, if available for 30 to 60 seconds). Mayday, Mayday, Mayday. This is Yacht *Stardust,* yacht *Stardust,* yacht *Stardust,* WZY 1234.

"Mayday, Yacht *Stardust,* WYZ 1234. 133 degrees true, 12 miles from Montauk Point. Struck submerged object, taking on water fast, engine disabled. Estimate cannot stay afloat more than one hour. Four persons on board. *Stardust* is a 32-foot sport fisherman, white hull, green trim. Maintaining watch on Channel 16. This is Yacht *Stardust,* WYZ 1234. Over."

General Rules for Distress Signaling

With your life at stake, you have a far better chance by following the correct procedure, but the provisions of the International Radio Regulations authorize a vessel in distress to use *any means* at its disposal to attract attention, make known its position, and obtain help.

Stay on Channel 16 (or 2182 kHz), repeating the call several times if no answer is immediately received. If no response is received even then, however, repeat the distress call on any other available frequency on which you might attract attention.

Speak slowly and distinctly. Use phonetic words for letters when necessary, especially when giving the letters of your call sign.

You may be requested to transmit a "long count" or other suitable signals to permit direction finding stations to determine your position; always end your transmission with the name and call sign of your boat.

If you have to abandon ship, it may be desirable to lock your VHF transmitter on the air (by taping down the mike button) to provide a signal for homing by rescue units. This is best done on a channel other than 16, and after coordination with searching vessels or aircraft. If you have already been visually located, do not lock your transmitter on the air as the signal would interfere with rescue operations.

All vessels having knowledge of distress traffic, and which cannot themselves assist, are *forbidden* to transmit on the frequency of the distress traffic; but they should listen and follow the situation until it is evident that assistance is being provided. Always listen before transmitting. It is *unlawful* for any radio operator to *willfully* or *maliciously interfere with* or cause interference to *any radio communication* or *signal.* No person shall knowingly transmit, or cause to be transmitted, any *false* or *fraudulent signal of distress* or *communication* relating thereto.

Radio Silence

The signal SEELONCE (silence) MAYDAY has been adopted internationally to control transmissions on the distress frequency, telling all other stations to leave the air and maintain radio silence. This signal is to be used *only* by the unit in distress or the station controlling the distress traffic. Any other station that considers it necessary to advise one or more other stations of the need to keep off the air should use the signal "Seelonce Distress" followed by its identification.

The signal to indicate the end of radio silence and permission to resume normal operation is SEELONCE FEENEE (finis.)

The oddly spelled words above are phonetic equivalents from the French as adapted for international use.

If You Hear a Mayday Call

If you are not in distress but hear a Mayday call, this is what you should do:

1. *Listen—Do Not Transmit.*

2. Try to determine if your boat is in the best position to take action, or if some other vessel is better located or better equipped.

3. *If* yours is the logical boat to render assistance (using the distress example above), reply with a call to the distressed vessel as follows—

"*Stardust, Stardust, Stardust,* this is *Fayaway, Fayaway, Fayaway,* WAB 1234. Received Mayday."

When the other vessel has acknowledged your call, continue with your offer of assistance by giving your position, your speed toward the scene of distress, and the estimated time to get there. But be sure before you transmit that you will not be interfering with the signal of another vessel better situated to render immediate assistance.

4. If yours is not the logical boat to take action, maintain radio silence but monitor the frequency closely for any further development. Start making notes so that you can write up the events in your boat's log.

When another station retransmits a distress message, the words MAYDAY RELAY must be spoken three times before station identification.

Use Mayday Sparingly

The distress signal MAYDAY should be used *only* when the vessel is "threatened with *grave and immediate danger* and requests *immediate assistance.*" It should *not* be used for situations such as being out of fuel, running aground, or engine failure under conditions of no immediate danger. Reserve MAYDAY for true emergencies of real hazard to life and property; don't "cry wolf." Make an ordinary call for assistance, or if appropriate, use the urgency signal Pan-Pan three times before the call.

The safety signal Security is spoken three times before transmission such as those relating to defects in aids to navigation and weather warnings; it is used mainly by shore stations.

Test Transmissions

Always remember, when making tests, to take every precaution not to cause interference to others. Listen before testing to make sure that the frequency is not busy.

You must obtain the consent of any station affected before proceeding. If the air appears clear, put your transmitter on the air and say—

"This is *Fayaway*, WAB 1234, test."

If you hear no station tell you to "Wait," you may proceed, saying "Testing" followed by a number count or other phraseology that will not confuse listeners.

At the end of the test, announce the name and call sign of your boat.

The FCC Rules prohibit "general calls"—calls not to a specific vessel, but rather to "any boat." Calls for a routine radio check are rarely necessary; modern solid-state radios are very reliable. If your set worked last weekend, and you are now hearing other boats, it is highly probable that your set is still functioning satisfactorily. But if you feel that a test is required, follow these procedures. Listen on a working channel, and when you hear two boats finish a conversation, call one of them and ask for your radio check. Don't call on Channel 16 unless you hear no conversations on a working frequency; if you do call on "16," don't try to get your report there—shift to a working channel to have the other station tell you how he heard you. A report of reception should be short and specific—there's nothing better than "loud and clear" if such is the case.

Calls *on Channel 16* (and 2182 kHz) to Coast Guard stations and units for radio checks are prohibited by FCC regulations.

RTCM Booklet

Radio operations aboard recreational boats is very well explained in the small booklet *Marine Radiotelephone Users Handbook*, published by the Radio Technical Commission for Marine Services. This publication is now accompanied by a copy of Subpart CC of Part 83, FCC Rules, which can fill the legal requirements for a copy of such rules. See page 397 for additional information.

Information Broadcasts

In the U.S., marine weather is broadcast by NOAA stations continuously from many locations; see page 301. Broadcast stations and marine operators transmit weather information of varying degrees of accuracy.

Coast Guard stations on VHF and MF/HF bands broadcast marine information or regular schedules. VHF transmissions are on Channel 22 following an alerting announcement on Channel 16; special broadcasts may be made at any time for safety purposes. Coast Guard broadcasts may include weather summaries and forecasts.

Waters of Other Nations

Marine radios on vessels may be operated in waters of other nations—Canadian boats visiting the United States, U.S. craft cruising in Canadian or Bahamian waters, etc. Additional licensing is not required, but skippers should be aware that there may be different restrictions or limitations on frequencies, permissible communication, or other regulated matters. Check with local authorities at your first port in each foreign country.

Violations and Penalties

You should never be in violation of the FCC Rules and you should have no need for knowledge of the procedures and penalties involved. Yet things do not always work out that way, so it is just as well to be informed.

If you receive a CITATION OF VIOLATION from the FCC, you must reply in duplicate within ten days, to the office that issued the citation. If a complete reply cannot be given in that time, send an interim reply and supplement it as soon as possible. If for reasons beyond your control you cannot reply at all within ten days, do so at the earliest date practicable, and fully support your reasons for the delay. Each letter to the FCC must be complete and contain all the facts without cross-reference to other correspondence. The answer must contain a full explanation of the incident and describe the actions taken to prevent a recurrence of it. If personnel errors are involved, your reply must state the name and license number of the operator concerned.

You may be lucky, however, and receive a WARNING NOTICE rather than a citation. Generally in this instance, no reply is required. The FCC form that you receive will indicate whether or not an answer is necessary. If one is required, don't get yourself into further trouble by failing to answer *within ten days.*

Revocation and Suspension of License

A station license may be revoked for any one of a number of specified violations of the Communications Act or the FCC Rules. Operator licenses and permits normally are not revoked but are suspended for varying periods of time, up to the balance of the license term. Notice of suspension must be given in writing and is not effective until 15 days after receipt. Within this period, you can apply for a hearing, and this automatically defers the suspension until after the hearing has been held and the FCC has ruled.

Fines Imposed

In addition to the revocation or suspension of licenses, the FCC can prosecute violators in the Federal District Courts. Any violation of the Communications Act may be punished by a fine of not more than $10,000, or imprisonment for not more than one year, or both. A second offense, not necessarily a repeat of the first, increases the maximum limit on the prison term to two years. For a violation of any FCC rule, regulation, restriction, or condition, or of any treaty provision, a court may additionally impose a fine of not more than $500 for *each* and *every day* during which the violation occurred.

Administrative Forfeitures

To avoid the delays, costs, and cumbersome procedures of formal court prosecutions, the FCC has the authority to levy its own ADMINISTRATIVE FORFEITURES—actually small fines—for 12 specified violations. These are in addition to any other penalties that may be imposed by law.

Among these twelve, the violations of principal concern to the operators of radio stations on boats are:

- Transmission of any unauthorized communications on a distress or calling frequency.
- Failure to identify the station at the times and in the

manner prescribed by the Rules.

- Interference with a distress call or distress communications.
- Operation of a station without a valid permit or license of the proper grade.
- Transmission of any false call contrary to the Rules.
- Failure to respond to official communications from the FCC.

The maximum forfeiture for a single violation, or a series of violations all falling within a single category as listed above, is $100. If more than one category is involved, however, the maximum liability is raised to $500 for a station licensee, or $400 for an individual operator. Note that the term "operator" may be applied to any person using the equipment, whether or not licensed by the FCC. In some cases, if two different persons are involved in an incident involving one station, forfeitures may be assessed against both the station licensee and the person who was operating the station.

The procedures for the imposition of these administrative fines have been kept simple, yet ample protection is afforded to the rights of individuals. Upon receipt of a NOTICE OF APPARENT LIABILITY FOR FORFEITURE, the addressed person has three possible courses of action. He can pay the fine and so close the incident; or he may, within 30 days, submit a written statement to the FCC giving reasons why he should be allowed to pay a lesser fine, or none at all; or he may request an interview with an FCC official. These last two courses of action can be combined, submitting both an explanation and a request for an interview. If either or both of these actions are taken, the FCC will review all information relating to the case, and make a final determination.

There is no judicial appeal from the FCC's ruling, and you had best pay up as there are established procedures for turning over cases of non-payment to a District Attorney for prosecution.

Citizens Band Radio

Citizens Band (CB) radio operations must be considered separately from those on the VHF and MF-HF bands as they come under an entirely different set of FCC Rules. CB sets are relatively inexpensive, easy to install, and simple to operate. See Figure 2209.

The Citizens Band Radio Service is unique; it is not intended to duplicate any other radio service. It is *not* a substitute for the safety features of regular marine radio service in the VHF band, nor for the ship-to-shore connections into the public telephone system. Likewise, it is *not* a new type of hobby or amateur service for casual contacts at great distances. The CB Service has its own particular functions and these do not overlap or conflict with other established services.

FCC licenses are not required for CB operations. All previously issued licenses have been cancelled.

Authorized Operations

The Citizens Band consists of 40 channels between the limits of 26.965 and 27.405 MHz. Although specific frequencies are listed in the FCC Rules, these are commonly referred to by channel number, 1 to 40. Any channel may be used for either double or single sideband amplitude modulation. (Older 23-channel sets are fully compatible with Channels 1–23 on the newer 40-channel transceivers. Any channel may be used for communications between units of the same station or different stations, subject to the limitations placed on Channel 9 as described below. No channel is assigned to any particular user or group of users; all must be shared.

Figure 2209 **A Citizens Band radio can be used for personal communications of a nature not proper for the VHF band, and to maintain contact with various shore-based stations.**

Channel 9 is reserved for "emergency communications involving the immediate safety of life or the immediate protection of property." It is also available for "traveler assistance"; this term is not defined in FCC Rules, but presumably it could be applied to a boatman needing help but not in an emergency situation. (Other channels can be used for emergency communications, but will probably be subjected to much more interference.) Some Coast Guard stations now guard CB Channel 9 on a *secondary* basis, but this is *not* a substitute for the regular marine distress frequencies of VHF Channel 16 and 2182 kHz.

Citizen Band stations are limited to 4 watts carrier (output) power on AM and 12 watts peak power on single sideband; external power amplifiers are specifically and strictly prohibited. On boats, a "marine-type" antenna must be used; an automobile antenna will not work.

The usual reliable range is on the order of 5 miles (8 km), but may extend to 15 miles (24 km) or more. These are ground wave ranges; sky wave transmission will often occur and CB signals will be heard at distances of several thousand miles. The FCC Rules prohibit exchanges between stations more than 250 km (155.3 miles) apart.

A CB set must be "type accepted" by the FCC. Repairs, and any modifications authorized by the manufacturer, must be done by or under the immediate supervision of a person "certified as technically qualified to perform transmitter installation, operation, maintenance, and repair duties in the private land mobile services and fixed services by an organization or committee representative of users in those services." An owner-operator may adjust an antenna to his set and make test transmissions, but that is all.

Operating Procedures

Operation of a Citizens Band station is quite simple. No specific operating procedures need to be learned and used; no log is required. *However, the elimination of licenses does not mean the total deregulation of CB operations.* There are a few Rules that must be known and obeyed.

The FCC Rules require that *all* communications be restricted to the *minimum practicable transmission time,* but no time limits are placed on intercommunications between units of the same station. Between different stations, however, an exchange must be limited to a maximum of 5 minutes; after its completion, both stations must remain off the air for an additional 1 minute. These time limitations do not apply in the case of emergency communications; all other rules, however, remain applicable.

Station identification is not required, but is "encouraged" by the FCC. Suggested forms are the previously assigned CB call sign, the letter "K" followed by the operator's initials and residence ZIP code, or an organization name and unit number. Operators are encouraged to use a CB "handle" only in conjunction with one of the above forms of identification.

Communications can be in any language, but the use of codes (other than the "ten-code" operating signals commonly employed) is prohibited. A CB station is for personal and business communications, and emergencies, including traveler assistance and civil defense drills and tests; it may also be used for voice paging. You may not receive direct or indirect compensation for operating a CB station, but this does not prohibit your use of CB communications in connection with your rendering other commercial services as long as you are paid for the other service and not for the actual use of a CB station.

A CB station may *not* be used for any illegal activity, to transmit music or sound effects, to advertise or solicit the sale of goods or services, to intentionally interfere with another CB station, or any of several other actions set forth in the FCC Rules. Profane, obscene, or indecent language is expressly prohibited. Also forbidden are communications with stations in foreign countries, except Canada.

Amateur Radio

Amateur ("ham") radio is another form of communications available to the boatman. It is most useful to those who cruise to places distant from their home port; it can be used for personal, but not business, communications under many circumstances. Ham radio does *not* provide a specific distress channel, but there are several frequencies—7,268 and 14,313 kHz, for example—where long usage has unofficially established their value in emergencies; there are many recorded instances where the amateur bands were used in emergencies on the high seas. No service is provided to marine operators, but friendly hams ashore are usually glad to pass along personal messages. There are no restrictions on boat-to-boat ham communications (except in the waters of some other countries).

The amateur radio service has many bands of frequencies from 1.8 to 24,000 MHz. In this extremely wide spread, there are suitable frequencies for communications at any distance desired. Radiotelegraph (Morse code), radiotelephone (AM or FM), radioteletype, and television signals may be sent in specified sub-bands; some hams even communicate by "moon-bounce" signals. Equipment may be purchased or home-made, and installation and maintenance can be by the owner-operator as the required licenses cover appropriate technical knowledge.

Licenses

Amateur radio operation requires both a station license and an operator license; these are separate from any commercial license or permit. Ham radio operator licenses are issued in various grades—Novice, Technician, General, Advanced, and Extra. A knowledge of Morse code is required even though all planned operation will be by voice. Five words per minute sending and receiving speed will get you started as a Novice with limited radiotelegraph privileges, but you need not go through this stage. A General Class license will be needed for voice communications on HF bands; this requires a 13 words per minute code speed, plus a written exam on regulations and radio fundamentals.

A ham license is not easily obtained, but neither is it impossible for anyone who seriously desires it and will devote some time to study and code practice.

Additional information can be obtained from local stores that sell ham radio equipment, or by mail from the American Radio Relay League, Newington, CT 06111.

Other Communications

There is often need for communications at very short distances, a few yards to perhaps a quarter- or half-mile from the boat, or on board the craft itself. Two systems are available to meet this requirement.

Hailers

For voice communications from boat-to-boat or boat-to-shore over distances up to several hundred yards or more, a LOUD HAILER is most useful. This may be either an installed item of equipment or a portable POWER MEGAPHONE; the installed unit will have the most power and greatest range, but a hand-held model will be more convenient on smaller motorboats and on sailboats.

Various models of installed hailers offer additional useful features such as auxiliary use as a fog horn, perhaps sounding automatically at selected intervals; or a listening capability whereby the loudspeaker can also act as a microphone to pick up distant sounds, aiding in *two-way* voice communications. In selecting equipment, be sure to get a unit with adequate power; 25 to as much as 80 watts is desirable. (The "P.A." output from many CB sets is only 2 to 4 watts and is not enough for boats except for on-board use.)

Proper installation is essential. This requires adequate primary power wiring (a powerful hailer will draw 10 amps or more, check voltage at the unit), adequate wiring from the amplifier to the speaker (remember the many watts being carried), and proper location and mounting of the speaker (to disperse the sound in the desired direction and prevent feedback into the microphone). Note that despite what is seen on many boats, proper mounting is with the long axis of a rectangular speaker opening vertical, not horizontal; see Figure 2210. An excellent feature of some installations is mounting the speaker either on the searchlight or on a separate mount that can be both elevated and trained in direction in the same manner as a searchlight. This permits the most efficient use of the hailer to vessels off to one side or a bridge tender high above the boat. The combination of a hailer that can pick up and amplify distant sounds with a directional speaker on a trainable mount will permit "audio direction finding" on fog signals to a surprising degree of precision—a real help in zero-visibility piloting situations.

Figure 2210 **An installed loud hailer is excellent for direct voice communications over distances from a few yards to a half-mile or more. Some models will pick up incoming sounds to make two-way communication possible. This hailer is properly installed for maximum horizontal sound coverage.**

Intercom Systems

While most people think of a boat as being too small and compact to have internal communications problems, there are numerous possible applications for simple intercom systems. The interior of many larger cruisers, particularly those of the double-cabin type, are so divided that voices do not carry well from one compartment to another, and engine noise adds to the problem. A simple three-station intercom system connecting the bridge, forward cabin, and after cabin can facilitate conversation between these locations and save many steps. On a boat with a small crew, such as a couple living aboard by themselves, a quick and easy means of internal communications is truly a safety item, as the helmsman need not leave the controls or divert his attention to carry on a conversation or summon assistance.

On a larger boat, the intercom system can be extended to include a unit in a weather-proof box on the foredeck. Then you can have constant contact between the person at the boat's controls and the person raising or lowering the anchor, or handling the docking lines. Just try to get up a badly fouled anchor without communicating with the helmsman, and you will quickly appreciate this small electronic device.

Signaling

Although radio has been for years the dominant means of ship-to-ship and ship-to-shore communications, boatmen often need to supplement radio communications with other methods. Reasons for this may be incompatibility of language, lack of a common radio frequency or other technical factor, or an emergency involving failure of the radio equipment. Various supplementary means of communication will be considered under the overall term of SIGNALING.

The Value of Visual Signaling

The time taken to acquire a rudimentary knowledge of visual signaling can help a boatman in at least two ways. For one thing, he'll have the satisfaction of knowing what is going on and being able to explain it to others.

The greater value in a knowledge of signaling, however, lies in the margin of safety it provides. Boatmen are so dependent on radio-telephones today that they tend to overlook the possibility of its not being available when needed. It need not be an electronic failure that precludes your communicating with another craft; it could be merely a lack of a common frequency or type of modulation.

The International Code of Signals

The INTERNATIONAL CODE OF SIGNALS is printed as DMAHTC *Publication No. 102,* which can be bought from local sales agents or ordered by mail (see Appendix B).

Pub. 102 includes general procedures applicable to all forms of communication, plus specific rules for flag signaling, flashing light signaling, sound signaling, radiotelegraphy, radiotelephony, and signaling by hand flags, using either semaphore or the Morse code.

The signals consist of: (1) single-letter signals allocated to meanings which are urgent, important, or of common use; (2) two-letter signals for general messages; and (3) three-letter signals (all beginning with "M") for medical messages.

In certain cases, COMPLEMENTS are used—an added numeral following the basic two-letter signal. Complements express: (1) variations in the meaning of a basic signal; (2) questions and answers related to the meaning of the basic signal; and (3) supplementary, specific, or more detailed meanings.

Typical Code-Signals From Pub. 102

The simple single-letter signal codes are shown beneath the flags on page 582. Typical two-letter groups are listed in Figure 2211.

Special Procedures

When names must be sent, or other words which have no signal groups in the International Code, they must be spelled out. The signal group YZ *may* be used to indicate that the groups that follow are plain language words rather than code signals; it can, however, be omitted if the spelling is obvious.

The International Code provides special procedures for many purposes such as signaling times, courses and bearings, and geographical coordinates.

Expressions in any one of nine languages are taken from the International Code, and certain principles must be followed in its use as outlined in *Pub. No. 102*. Read them carefully before attempting to use the code.

Flag Hoist Signaling

The set of International Code flags prescribed for flaghoist signaling consists of 26 alphabetical flags, ten numeral pennants, three substitutes, and the answering pennant. SUBSTITUTE and REPEATER pennants (interchangeable terms) are used as replacements for flags (or pennants) already used once in a hoist; each set has only one of each letter or number.

Five standardized colors are used for signal flags—red, white, blue, yellow, and black. Most of the flags are of two colors, arranged for maximum contrast. Two flags are of a single color, several use three colors, and one uses four. Each has a line sewn into its hoist edge, extending slightly above the cloth to a metal ring, and for several inches below it to a snap hook. This "tail line" provides the spacing between the flags of a hoist; the snaps and rings permit them to be joined together rapidly and connected to a signal halyard.

Signal flags are flown singly and in combinations of two or more; flags and pennants may be mixed as required. In the procedures of the International Code, substitutes repeat the same *class* of flag that precedes them; following alphabet flags, they repeat them, but if used with numeral pennants, they repeat such pennants. The answering pennant is used as a decimal point.

AE	I must abandon my vessel.
CJ	Do you require assistance?
CN	I am unable to give assistance.
JI	Are you aground?
JL	You are running the risk of going aground.
JW	I have sprung a leak.
KN	I cannot take you in tow.
LN	Light (name follows) has been extinguished.
LO	I am not in my correct position. (To be used by a lightship.)
LR	Bar is not dangerous.
LS	Bar is dangerous.
MF	Course to reach me is . . .
MG	You should steer course . . .
NF	You are running into danger.
NG	You are in a dangerous position.

Figure 2211 **Typical two-letter signals from the International Code of Signals, Pub. No. 102. These may be made by flag hoist, sound, flashing light, or other means.**

General Yachting Uses of Signal Flags

Most boats will not carry a full set of signal flags and pennants, but may carry one or more for special use under certain conditions.

The flying of the single flag "Q" when returning from a foreign port is correct in accordance with its International Code meaning "My vessel is healthy and I request free pratique." PRATIQUE is the formal term for permission to use a port.

The code flag "T" is often seen flying on boats at yacht club moorings. Its meaning at such times is a request for "transportation"—a call for the club launch to come alongside and pick up passengers for shore. This is not in accordance with the International Code of Signals, but has general acceptance and understanding through wide usage.

Another single code-flag signal is the letter "M" flown to signify "Doctor on board"; this differs from the revised International Code meaning. Not yet in general use, its value is gradually being recognized and the practice is spreading. While most doctors look to their boating hours as time "to get away from it all," knowing that a neighboring boat has a doctor on board can be helpful if an accident happens.

Flashing Light Signaling

Signaling by short and long flashes of light is widely used by naval forces. It is much faster than hoists of code flags and has a greater range than semaphore signaling; from a military viewpoint, it is more secure than radio transmissions.

Most boatmen have little regular use for flashing light signaling, but it could help in emergencies. Perhaps a boat is aground at night, and it is unsafe for another boat to approach close enough for shouting back and forth; flashing light signaling would permit communication from a safe distance. It has the advantage of needing no special equipment (an ordinary flashlight suffices), but the disadvantage that it requires a knowledge of the Morse code.

The International Morse Code

As noted, the International Morse code of dots and dashes is used for flashing light signaling. The flashes and the spaces between them are defined in terms of UNITS—a dot is one unit long; a dash is three units; the space between the dots and dashes of a single character (letter

Figure 2212 **At many anchorages, a "T" (Transportation) flag is displayed to signal for a launch so that persons on board may be taken ashore. A sound signal may also be given.**

The International Morse Code

A	•—	M	——	Y	—•——
B	—•••	N	—•	Z	——••
C	—•—•	O	———	1	•————
D	—••	P	•——•	2	••———
E	•	Q	——•—	3	•••——
F	••—•	R	•—•	4	••••—
G	——•	S	•••	5	•••••
H	••••	T	—	6	—••••
I	••	U	••—	7	——•••
J	•———	V	•••—	8	———••
K	—•—	W	•——	9	————•
L	•—••	X	—••—	0	—————

Period •—•—•—
Comma ——••——
Interrogative •—• ——•— (RQ)
Distress Call •••———••• ($\overline{\text{SOS}}$)
From —•• • (DE)
Invitation to transmit (go ahead) —•— (K)
Wait •—••• ($\overline{\text{AS}}$)
Error •••••••• ($\overline{\text{EEEE}}$ etc.)
Received •—•(R)
End of each message •—•—• ($\overline{\text{AR}}$)

A dash is equal to three dots.
The space between parts of the same letter is equal to one dot.
The space between two letters is equal to three dots.
The space between two words is equal to five dots.

Figure 2213 **The International Morse Code, with certain simple procedural signals.**

or number) is one unit long, between characters it is three, and between words or groups it is five. The length of a "unit" is roughly set by the skill of the operator and the speed at which he can transmit accurately; it is best to transmit more slowly than your maximum capability, to minimize errors, and to exaggerate the lengths of dashes and spaces slightly rather than shorten them.

Morse code letters, numbers, and punctuation marks are signaled by combinations of dots and dashes. Letters have from one to four components, numbers have five, and punctuation symbols have six. The basic Morse code is shown in Figure 2213.

Memorizing the Morse code characters takes time, but it *can* be done by anyone, and it is well worth the effort. Once learned, and occasionally used, it will not be forgotten.

MK 2 INTEGRATED NAVIGATION SYSTEM
NCS

23

Selection and operation of radio direction finders, depth sounders, radar, Loran, Omega, satellite navigation systems, and other specialized gear

ELECTRONIC EQUIPMENT: NAVIGATION AND OTHER USES

While most of the attention to marine electronic equipment is normally focused on radiotelephones, there are many other devices ranging from those that are highly desirable, even necessary, to units that may be classed as merely useful for convenience or comfort. This chapter will examine a number of these items, of marine electronic equipment as used for navigation, safety, and other purposes.

Depth Sounders

DEPTH SOUNDERS (Figure 2301) closely rival radiotelephones as the most popular item of electronic equipment for boats; they have a wide range of use—lakes, rivers, bays, and offshore. Your sounder can be one of the most interesting and useful devices on your boat. The term "Fathometer" is often used for these devices, but it is a registered trademark of one manufacturer, Raytheon, and is properly applied only to sounders from that source.

Depth sounders are a modern replacement for the hand-held lead line used for centuries to determine the depth of water beneath a ship. This electronic device furnishes a vastly greater amount of information, and does it with far greater ease, especially in nasty weather. It provides safety and convenience, so is doubly advantageous to have on board.

How Depth Is Measured

Depth is determined by measuring the round-trip time for a pulse of ultrasonic energy to travel from the boat to the bottom of the water and be reflected back to the boat. See Figure 2302. The frequency of the audio pulses generally lies between 50,000 and 200,000 Hertz (cycles per second), too high to be heard by human ears. Use of the lower frequency permits greater depths to be measured; use of higher frequencies results in a sharper beam and better resolution of the bottom and fish.

The average velocity of sound through the water is approximately 4,800 feet per second; slight variations in speed occur between salt and fresh water, and with different temperatures. The resulting small errors, however, can be safely ignored for the relatively shallow depths of interest to the operators of recreational boats.

Probably the greatest advantage of the electronic device over the hand-held lead line is the continuous nature of the information gained. Depth sounders vary widely in the rate at which readings are taken, but in all cases many more soundings are taken than could be accomplished by hand. Typical units take readings at rates between 1 and 30 per second, with an accuracy of about ±5%; fully sufficient for normal navigation.

Components of a Depth Sounder

The major components of a depth sounder are a source of energy (transmitter), a means of sending out the pulses and picking up the echoes (transducer), a receiver to amplify the weak echoes, and a visual presentation of the information. The transducer unit usually takes the form of a round block of hard ceramic material several inches in diameter and an inch or so thick. In many cases, it is streamlined to reduce drag. It is normally mounted through the hull roughly amidship, but sometimes it is mounted internally or on a transom bracket. The disadvantage of an internal mounting is a loss of maximum depths that can be read, up to 50% or more. The disadvantage of a transom mount is an inability to take measurements while underway at speed because of turbulence in the water.

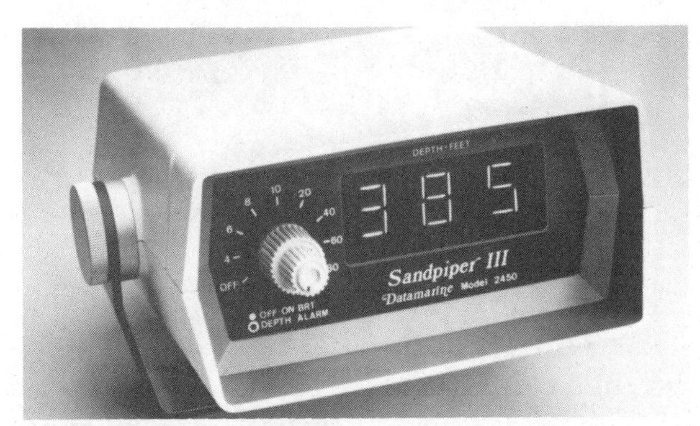

Figure 2301 **Electronic depth sounders are useful piloting tools on all waters, and on boats of most sizes and types. This digital model gives a clear indication of depth.**

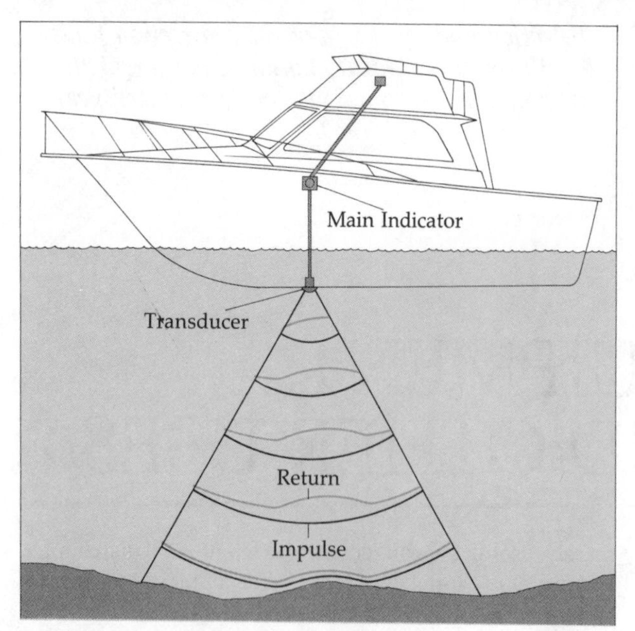

Figure 2302 **An electronic depth sounder measures distance to the bottom by the time it takes for pulses of high-frequency signals to travel to the bottom and back.**

The visual presentation of information on the water's depth is accomplished by an INDICATOR, a RECORDER, or a VIDEO DISPLAY SCREEN. The indicator provides a non-permanent indication of the depth by one of three methods: flashing light, digital display, or electrical meter. The flashing light is mounted on the end of an arm that rotates around a scale much like the second hand of a clock, only much faster. (A recent solid-state "flasher" has a mark that appears to rotate, but actually the unit has no moving parts.) The zero of the scale is usually at the top of the dial and a flash of light occurs there when the outgoing pulse leaves the transducer on the boat's bottom. A second flash occurs when the pulse is received back at the transducer. The deeper the water, the longer it will take for the echo to return to the boat; the longer this takes, the farther the arm will have rotated around the dial. (Figure 2303 left.) Thus the scale of depths increases clockwise around the face of the indicator.

Because of its ease in use, the digital display is the most popular type. Most such sounders use a liquid-crystal display (LCD), which is easily read in daylight and does not require a light shield in bright-light conditions like an open boat or a flying bridge; flashing-light models usually have a shield or hood. LCD displays must be illuminated by an internal light for reading at night.

Depth sounders with an electrical-meter type of read-out are still seen on some boats. These are quite useful, but their manufacture has ceased in favor of the more easily read digital displays.

Many depth sounders can read on two or more depth scales regardless of the type of read-out. Some models have a 1:2 ratio such as 0–60 and 0–120 feet; others have a 1:6 ratio such as 0–60 feet and 0–60 fathoms (360 feet). Some models now available have digital displays that can be set to read in feet, fathoms, or meters, reflecting the growing use of the metric system on charts.

Principles of Electronic Depth Sounding

An example of how a flashing-light sounder operates may clarify the general principles. If the unit has a depth range of 240 feet, at full scale the sound pulses will have traveled a round-trip distance of 480 feet. Taking the speed of sound in water at 4,800 feet per second, the round-trip will have taken 1/10 second. Therefore the arm carrying the flashing light must make one full rotation in 1/30 second; this is 10 RPS or 600 RPM.

If the depth is only, say, 80 feet, the echo will cause the light to flash when the arm has made only 1/3 of a full revolution, and the flash will occur at a position corresponding to four o'clock on a clock dial. In many models, it is possible to detect greater than full-scale depths by turning the sensitivity control until a flash is seen on the "second trip around" of the rotating arm. Using the same sounder as in the preceding example, if the water was known to be quite deep, and a weak reading of 60 feet was seen with the sensitivity control well advanced, it *could* mean an actual depth of 300 feet (full scale plus 60

Figure 2303 **Flasher type sounder, *left,* has scale to 100 feet. Greater depths can be measured by adding 100 to the reading. Other models may be calibrated for two ranges. With a recording type sounder, *right,* depth is measured many times each second, and presented as a series of markings on a slowly-moving scaled tape.**

feet). Some models are now calibrated with two sets of numbers around the dial for first and second revolutions of the flashing light. *Be careful to avoid reading the depth as much greater than it actually is.*

Digital-display and meter types of sounders substitute complex electronic circuitry—often in the form of integrated-current "chips"—for the mechanical action of a rotating arm with flashing light. Microprocessors are often used to average depth echoes to avoid excessive fluctuation of the read-out. Some digital-display depth sounders also include read-outs of one or more additional data such as speed, distance traveled, and water temperature. One model will even "talk" to you, giving depths in synthesized speech.

With the recording type of sounder, a permanent record is made of the depths, and notations as to the boat's position can be made directly on the paper tape. See Figure 2303 right. The paper moves horizontally from a supply roll to a take-up spool at the rate of one inch in several minutes. A dry method of chemically marking the paper is normally used to avoid the messiness of ink. A recording type of depth sounder has the advantages of presenting a permanent record of the bottom, but is more expensive. Some models now have with both flashing light or digital-display plus a permanent paper-type record.

A recent development is the video display sounder, which presents a visual "elevation" picture of the bottom, plus underwater objects such as large fish, or schools of fish. The display changes constantly as the boat moves along its course; information can be recorded on a video tape system for "replay," if desired. Some models provide a color picture which aids in fish identification; some interface with Loran equipment and show position information on the screen. Many models make provision for expanding or reducing scales, or showing just a portion of the depth being measured. While many of these features are geared to the fisherman, such sounders are, of course, excellent for piloting purposes.

Correction to "Zero" Depth

If you have a sounder on your boat, you must know its "zero" depth. The transducer is generally mounted on the hull several feet below the water surface but some distance up from the lowest point of the keel. As depths are measured from the transducer, the readings will be *neither* the full depth of the water from the surface nor the clearance between the keel and the bottom. On some models, the zero reading can be offset plus or minus a few feet so that the depth indications can be either the true water depth or the clearance under the boat without adding or subtracting a fixed amount from each reading. Water-depth readings are advantageous for navigation, but some skippers feel better with a direct reading of clearance under the keel—the choice is yours if your sounder has the zero-offset feature.

Flashing-light and video display sounders permit a rough determination of the nature of the bottom from the appearance of the flash or display which indicates the depth. Learn how to use this valuable additional capability of these types; you will gain a surprising amount of information. Sharp, clear indications mean a hard bottom. Conversely, broad and fuzzy flashes or displays indicate a soft, muddy bottom. Multiple flashes or displays, each

Figure 2304 **A video type sounder presents a display on a tube face similar to a television set. Some models have a "replay" or hold feature, and may interface with Loran equipment to show position information as well as depth.**

fairly sharp, can result from rocky bottoms; the ultrasonic pulses are reflected more or less horizontally to adjacent rocks before being returned upward. The added time delays for these sideward bounces make them show up as if they were at greater depths.

Additional flashes or displays at multiples of the least depth indicated may result, in shallow water with a hard bottom, from pulses being reflected downward again by the craft's hull following the first round-trip. Two or even three round-trips may occur; the cure for this is to turn back the sensitivity control until only the first indication is seen. (Some units have automatic gain controls to eliminate this.) In *all* cases of multiple flashes, play it safe and assume that the *minimum* reading is the true depth! As the depth indication on flashing-light sounders will always have some width, be sure to read the "trailing" edge, or least indicated depth. This is the correct value.

Application in Piloting

The primary application of an electronic depth sounder is to assist in the safe navigation of the boat in unfamiliar waters. It can provide much useful information, but must be used with care as *it makes no predictions ahead.* (There are larger, much more expensive *scanning sonar* devices that look ahead and to each side, presenting a radar-like view of underwater conditions.)

Information on water depth is reassuring in itself; furthermore, it can be used to assist in determining the boat's position. Details are given on pages 494-496.

Depth Alarms

Many electronic depth sounder models have audible alarm signals that are activated whenever the depth decreases to a predetermined value. On some units this is one or more fixed depths; on others the alarm can be set for any depth. There are also alarm-only units, which can be added to most installed sounders that lack this feature.

A more recent development is the multiple depth alarm. This feature is especially useful when the boat is anchored. With one alarm set just below the present depth and the other just above it, with allowance made for tidal changes, there will be an audible alarm given if the boat drags anchor toward *either* shallower or deeper water. Solid state circuitry is used so that battery drain is held to a minimum during the period at anchor.

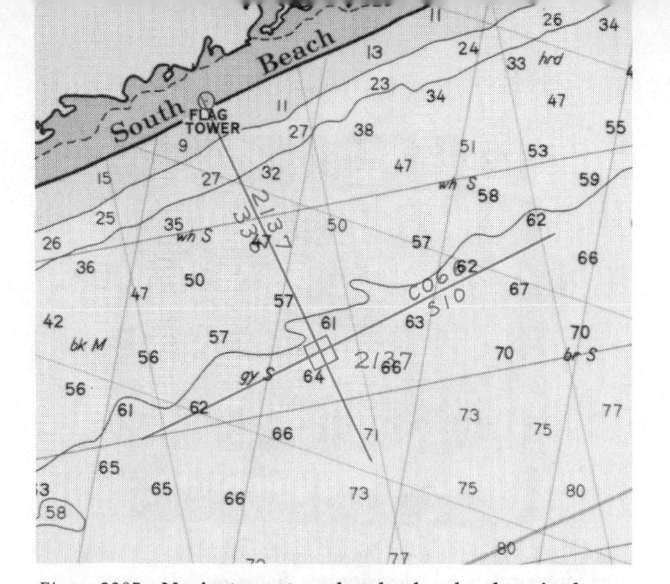

Figure 2305 **Navigator notes when landmark ashore is abeam, but can get no other line of position. The depth sounder reading, corrected for state of tide, can be plotted on the beam LOP for an estimated position at 2137.**

Another recent development is the so-called "forward-looking" depth sounder. Actually, depths continue to be measured directly downward only, but a microprocessor evaluates the *trend* of recent soundings and predicts the depth ahead for a brief interval—typically 10 or 20 seconds—and sounds the alarm if appropriate. Obviously, such models are of most value over a gently sloping bottom and would give *no warning* of an isolated rock, coral head, or similar underwater obstruction.

Fish Finding

Locating fish with a depth sounder is possible, but you need experience to get reliable results. Detection with most equipment will generally be limited to schools of fish or quite large single fish. This requires a flashing-light or recording type of sounder; it is not possible with digital displays.

Radio Direction Finders

On the seacoasts of the United States, and on the Great Lakes and other large inland bodies of water, a RADIO DIRECTION FINDER (RDF) is an important piece of electronic equipment. See Figure 2306. Installed primarily as a safety item, it can also be a great convenience to the boat operator. It is the primary radio aid to navigation for small craft.

A complete radio direction finding system has four essential components:

1. One or more radio transmitters at known locations.
2. An RDF set on the boat.
3. Charts covering both the location of the transmitters and the area of operation of the boat.
4. A person who knows the operation of the system.

To be fully effective the RDF system must be used with competence and confidence—an incorrect radio bearing can lead to disaster; a correct bearing that is ignored because of mistrust can be equally disastrous.

Special RDF Features

Basically, an RDF is a radio receiver with two additional features. First is the DIRECTIONAL ANTENNA. Usually, this

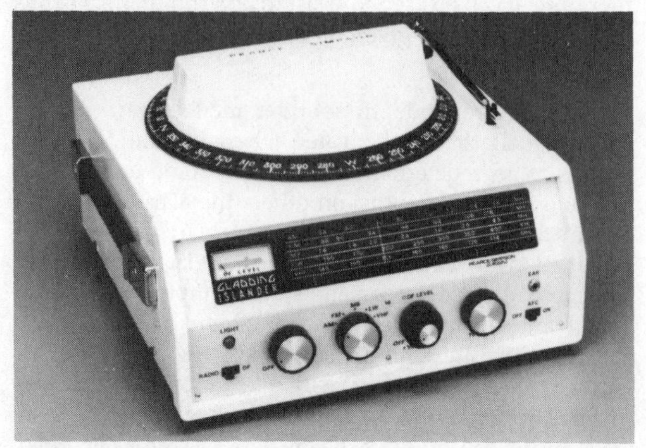

Figure 2306 **When visibility is poor, a radio direction finder can help fix position. A navigator also can use it to "home in" on a transmitter near his destination.**

antenna can rotate so the set can be secured firmly in a convenient place. An RDF set's directional antenna is an improved version of the simple loop used on portable receivers, the directional characteristics of which are familiar to most boaters. The antenna may take the form of an open loop a foot or so in diameter, or it may appear as a plastic bar measuring about an inch square by some six inches in length. Both types are normally mounted on top of the set; either will do the job.

As the antenna is rotated through 360 degrees, the directional antenna shows two positions of maximum signal strength, and two positions of minimum received signal called "nulls." With properly balanced construction and no local interfering objects, the two maximum signal positions will be separated by 180 degrees, as will be the nulls, which are 90 degrees in either direction from the maximums. It is characteristic of such antennas that the maximum signal points are broad and poorly defined, while the nulls are marked and relatively precise. Thus it is the nulls that are used for direction finding.

On some models the antenna is rotated by turning a front-panel knob; this is a desirable feature because the null can be affected if the user's hand is placed on the antenna. If the loop must be handled, it should be touched only at the outer edge of its base.

The second special feature of RDFs is a VISUAL NULL INDICATOR. While the operator can judge by ear the position of the antenna at minimum signal with fair accuracy, he can get a more precise bearing by observing a visual indicator. This is normally a small electric meter, read for either a maximum or a minimum deflection of its needle in accordance with the set's instructions.

Some modern RDFs now have a digital frequency display that is easily read for precise tuning.

RDF Frequency Bands

RDFs normally cover two frequency bands—a low frequency (LF) beacon band and the standard AM radio broadcast band. The LF beacon band is of primary interest

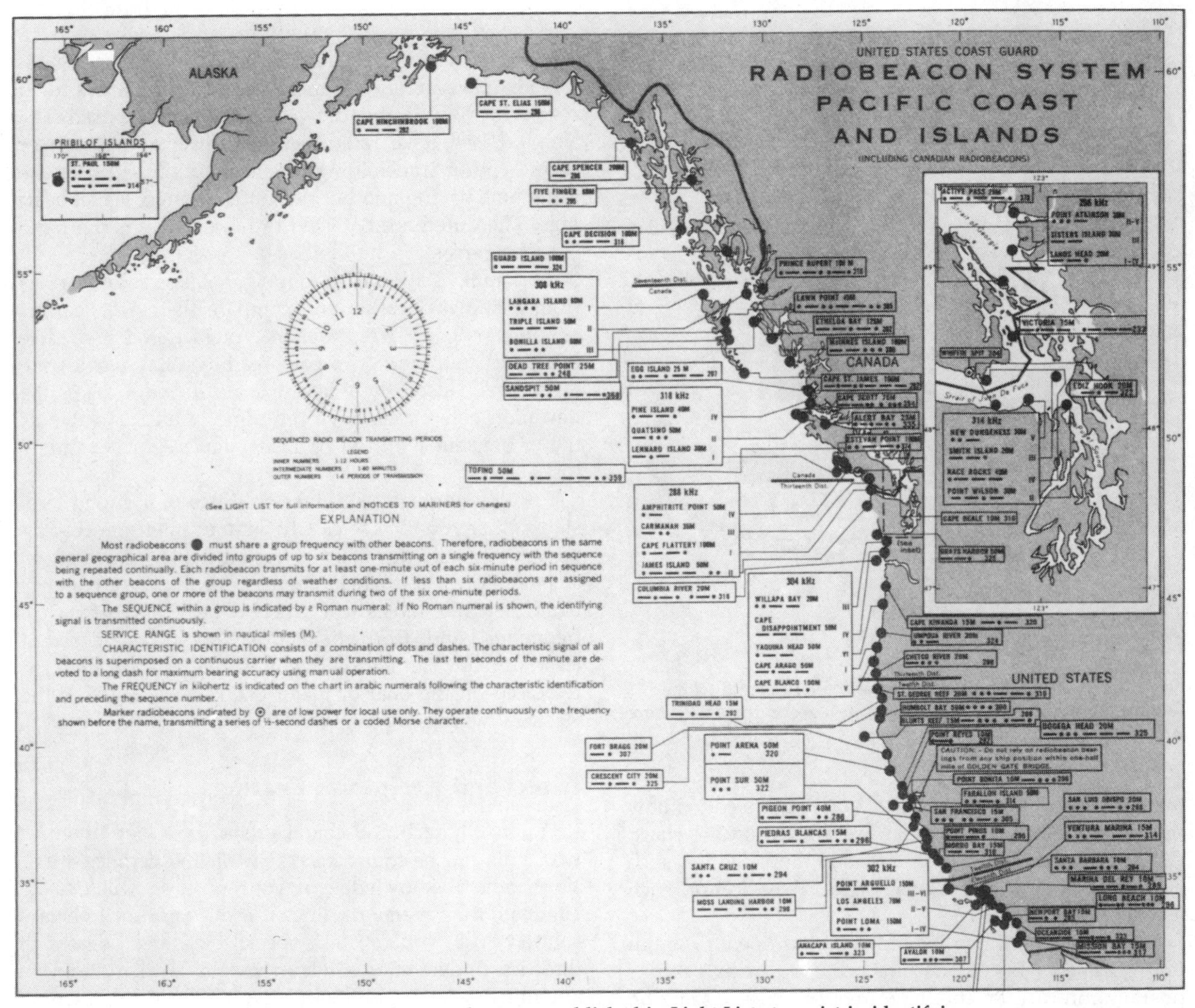

Figure 2307 **Radiobeacon charts are published in Light Lists to assist in identifying radiobeacons. Frequencies, schedules of operation, and identifying signals are given.**

for the marine radiobeacons operated by the Coast Guard on frequencies between 285 and 325 kHz at locations along or just off U.S. coastlines and the Great Lakes. See Figure 2307. Aeronautical beacons, somewhat lower and higher in frequency, are also within the tuning range of these sets and can be used for direction finding. (These aerobeacons, or RANGES as they are called, are an excellent source of weather information at 15 and 45 minutes past each hour.)

Coverage of the standard AM radio broadcast band is desirable because of the large number of stations on which bearings can be taken. Note, however, that the accuracy of bearings on this band will not be quite as good as on the LF beacon band.

Direction finders are available, although more expensive, for the VHF band (see page 544), and some transceivers have an additional capability of indicating the relative direction from which signals are being received. (A few LF-MF direction finders can receive VHF signals; some can pick up only weather broadcasts, others can be tuned to any marine band frequency. These RDFs can be used for monitoring a second channel, but they do not have any direction-finding capability in this mode. Other models may also be used to listen to CB channels or FM broadcast stations.)

Another type of RDF is the hand-held unit. These are popular because they can be moved about the boat to any point where radio deviation (see page 542) is absent or at a minimum. Hand-held RDFs are powered by internal dry-cells, usually cover the beacon band only, and have an integral small magnetic compass. Some have a null meter; others depend on an audible indication of the null.

In direction-finding the location of the transmitting antenna must be known. The location of beacons operated by the Coast Guard is shown on standard charts and in the *Light List*. Note that land-based Coast Guard VHF transmitters may not be at the site of the listed Coast Guard station. Charts show the location of radiobeacons and major broadcast station antennas, together with call letters and frequencies.

Marine Radiobeacons

Radiobeacons are rated in terms of "service range" from 200 miles to as little as 10 miles for beacons of limited local interest.

Radiobeacons are identified by one, two, or sometimes

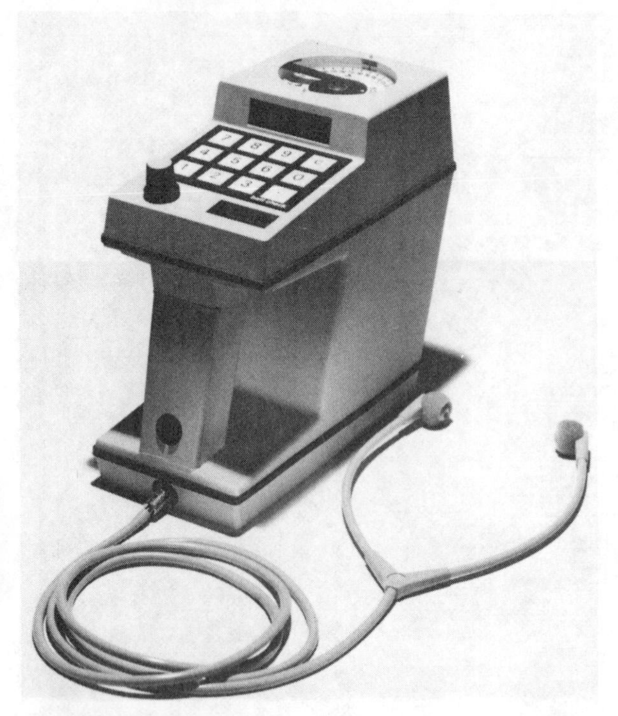

Figure 2308 **A hand-held RDF is particularly useful when homing in on a signal. Try to locate an area on the boat where it is least subject to deviation error.**

three letters sent in Morse code. You don't need to know the code; the dots and dashes are printed on the chart. Often the letters relate to the location they identify, such AC for Atlantic City—but sometimes they do not, such as U for Miami.

Most, but not all, marine radiobeacons operate continuously. The others are "sequenced," up to six beacons operating in turn on the same frequency, each being on the air for one minute and then off while the others take their turn. The sequence of operation can be determined from the charts or by reference to the *Light List*. It is indicated by a Roman numeral immediately following the frequency. If, for example, the figure is "I," the radiobeacon transmits during the 1st, 7th, 13th, 19th, 25th, etc., minute of each hour; if the figure is "II," the transmitting minutes are the 2nd, 8th, 14th, etc.; and so on for the other sequences indicated by "III," "IV," etc. If there is no Roman numeral shown on the chart or in the *Light List*, the radiobeacon operates continuously. Each radiobeacon sends its characteristic letter(s) in Morse code for 50 seconds then a long dash for 10 seconds.

Radio Deviation

All RDFs are subject to local DEVIATION errors caused by the absorption and reradiation of the radio signals by nearby wires and other metallic objects on the boat. This is generally similar to the effect of masses of iron on a magnetic compass. One important difference, however, is that the radio bearing errors vary with respect to the *relative* bearing angle; the boat's heading is *not* the deciding factor as it is with compass deviation. A table, or curve, must be prepared to show the CORRECTIONS for RDF error on various relative bearings. Separate tables will generally be required for each of the frequency bands.

Determining RDF Deviation Error

RDF deviation errors are determined by comparing observed radio bearings on various visible objects with known correct bearings. Either of two methods may be used. The simplest is to take radio bearings on a radiobeacon or other type of transmitting antenna within sight, at the same time having another person take direct visual bearings. The difference between these readings is the radio deviation error.

If no radio station is within sight, take radio bearings from a known location and compare them with correct bearings as taken from a chart. Avoid doing this while made fast to a pier; erroneous readings may result from reflections caused by adjacent boats. In both cases, the deviation corrections have the same numerical value as the errors, but with the opposite sign, + for − and − for +.

To calibrate your RDF, take readings of a strong local station that you receive clearly without interference. Prepare a complete deviation correction table, preferably in steps of not more than 15 degrees, for the low frequency band. Separate tables will probably be required for each frequency band. Correction tables should be checked at least annually, and additionally if the direction finder is relocated on the boat or major changes are made in the electrical wiring, rigging, etc.

Direct and Reciprocal Bearings

Due to the technical characteristics of a loop antenna, two nulls can be found *approximately* 180 degrees apart. Your general knowledge of your position will usually eliminate the reverse reading. If it does not, or to be absolutely sure, you can take special measures. Nearly all direction finders have either an integral auxiliary "sense" antenna or a connection terminal for an external antenna. By operating a switch, this antenna is connected into the circuit to change the directional characteristics to a single null and single maximum pattern. Simple procedures will then indicate which of the previous nulls is the direct bearing and which is the reciprocal reading. (The sense antenna is not left in the circuit all the time, because more precise readings are obtained from the loop alone.)

Always use a portable RDF set from the same position on your boat; changing its location may require a completely new deviation table. As an RDF contains strong magnets in the speaker and meter, keep it at least several feet away from your compass, whether turned on or off.

How to Take a Radio Bearing

To take a radio bearing, follow these steps:

1. Set the scale built into the set, usually around the base of the antenna, so that 000° is dead ahead.
2. Rotate the directional antenna (Figure 2306) until a null point is precisely located, read the angle from the scale; this is the uncorrected relative radio bearing. Caution: the boat must be directly on course at the moment that the bearing is taken, as any error in heading will be reflected in the resultant radio bearing.
3. If there is any doubt about whether the reading just taken is the direct or reciprocal bearing, use the sense antenna to identify it. If the reading is the reciprocal,

do *not* add or subtract 180 degrees—take a new bearing.
4. Having determined the direct bearing angle, apply the proper deviation correction; the sum is the corrected relative radio bearing.
5. Add the boat's *true* heading, subtracting 360 degrees if the sum exceeds that amount. This is now the true radio bearing from the boat; plot in the same manner as with a visual bearing.

The steps above outline the basic procedures to be followed in taking a radio bearing. For any particular RDF set, however, study the manufacturer's manual and follow its instructions closely.

Hand-held RFDs combine the radio direction finder with an integral compass. The entire unit is made with a "pistol grip" handle and is held to the eye and "aimed" at the source of radio signals. Direct radio bearings are read in terms of compass direction, but care must be taken for both radio and compass deviation.

Plotting the Radio Bearing

If the transmitting station is more than 50 miles away, and the plotting is to be done on a Mercator chart, a further correction is required because the great circle path along which radio waves travel does not plot as a straight line on this type of chart. The amount of correction, and the procedure for applying it, can be found in Table 1 of *Bowditch*, or in any *Coast Pilot*.

Radio bearings are plotted in the same manner as visual bearings, and the use of an RDF requires the same degree of piloting knowledge and plotting skill. Radio bearings do not have as high a degree of precision as visual bearings, and for this reason a three-bearing fix is desirable. See Figure 2309 top. The best probable accuracy for radio bearings is on the order of 2 degrees, and you need experience to reach this. In situations involving proximity to dangerous waters, draw additional lines 2 degrees to either side of the corrected bearing; these will indicate graphically the probable limits of your position. See Figure 2309 bottom.

Position-Finding by Radio Bearings

Radio bearings can also be used in any of the other forms of position plotting—bow-and-beam bearings, doubling the angle, danger angles, etc. If you can get only a single radio bearing, you can combine this with a visual bearing, an astronomical line of position, or a depth reading to determine an estimated position. Remember that this EP will be no better than the accuracy of the poorest method used, and navigate accordingly.

To ensure the most accurate position from radio bearings:
1. Take bearings from as strong signals as possible.
2. Use the low frequency beacon band in preference to higher frequencies.
3. Avoid taking bearings on radio stations located inland, particularly those behind coastal hills or mountains.
4. Take three or more bearings on different stations if possible; be sure to identify the station correctly and know the loction of the transmitting antenna.
5. Take bearings so that the intersection angle is at least 30 degrees, preferably 90 degrees, for two-station fixes, or 60 degrees or 120 degrees for three-station fixes.

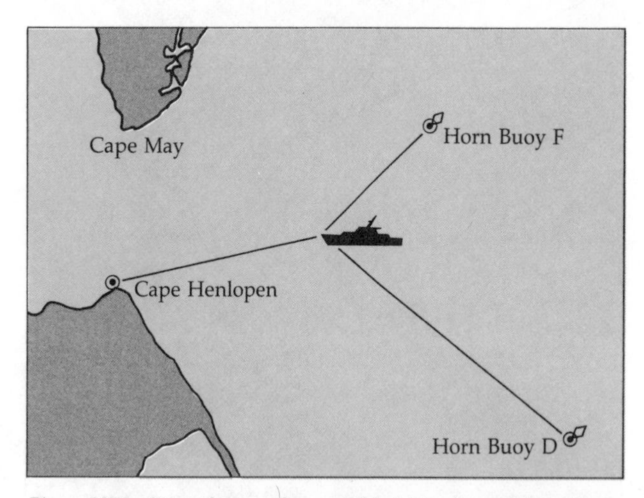

Figure 2309 **Boat shown *above* could determine position by bearings on any two radiobeacons at 1, 2, and 3. A better fix is obtained by using all three beacons. Radio bearings are not as precise as visual bearings; the best accuracy that can be expected is ±2 degrees in optimum conditions, *below*. Here the boat could be at any point within the tinted area where the bearings cross.**

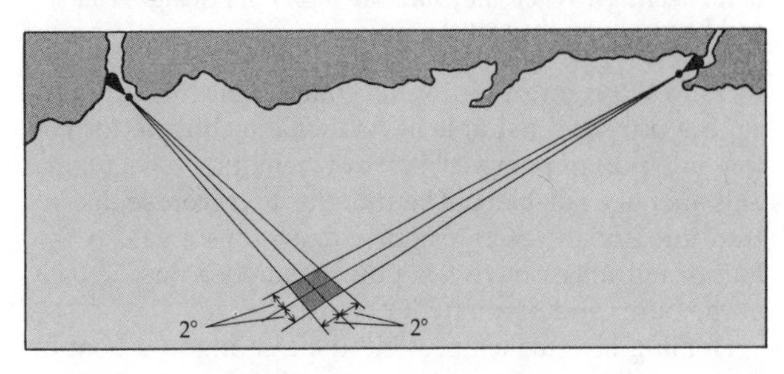

More on position-finding with RDF bearings will be found on pages 492-493.

Factors Affecting Accuracy

Two circumstances tend to degrade the accuracy of radio bearings. "Night effect" is the term applied to the fact that all radio bearings taken at night, but particularly those made near sunset and sunrise, show a broader and sometimes shifting null. This lessens the accuracy of any bearing taken at these times, so you should try to take bearings under these conditions only on stations less than 30 to 40 miles distant. If you cannot, take a series of readings on each station and average them for a mean reading—but still consider the accuracy to be less than that of a reading taken in the daytime.

The second special circumstance of lessened accuracy comes into effect when the radio waves travel approximately parallel to a coastline for any appreciable part of the distance from the transmitting station to the RDF. There is a bending effect on the signals that can be as much as 10 degrees, pulling the waves in toward the land. This makes the boat appear closer to land than it is.

Homing With an RDF

Aside from determining a boat's position, an RDF has several other interesting uses. When the directional antenna is set on 000° relative, a craft may be directed toward a radio station by steering to keep the signal strength at the deepest point of the null. One RDF model has a direct-reading meter which shows the helmsman whether

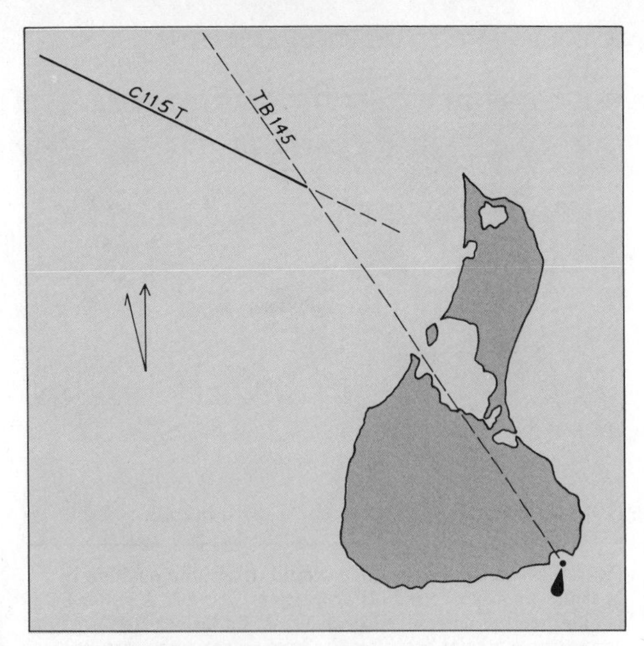

Figure 2310 **From his chart, a skipper notes that a bearing of 145° on Southeast Point, Block Island, will bring him safely to Great Salt Pond. He holds his 115° heading, taking frequent radio bearings. When they increase to 145° he changes course and homes in on the signal.**

he is to the right or left of his course. This homing procedure provides a simple navigational technique for getting into port in poor visibility or when other navigational aids are not available; but be sure the direct course doesn't lead into shoal waters or other dangerous areas. A few harbor entrances have low-power radiobeacons, usually at the outer end of a jetty if one exists.

Homing also makes possible the steering of a boat on the most direct course to the scene of a distress situation. Such action is normally only possible on VHF channels with RDFs for such frequencies or a transceiver having a DF capability.

Be sure not to run directly into the source of the signals (this may seem like rather obvious advice, but it has happened!).

Use and Care of an RDF

When you are considering the purchase, installation, and use of an RDF, remember that it is first of all a safety device. Any other planned use, such as an auxiliary receiver, is a bonus feature only, and must not detract from readiness for its primary use when needed. Most portable RDFs are powered by small dry cell (flashlight) batteries; many models have a switch position for the null meter for checking battery strength at any time. Carry a set of spare batteries on cruises to guard against losing RDF capability if the unit is accidentally left on for an extended period of time. Replace batteries annually, whether they seem to need it or not.

Use your RDF often in periods of good visibility, to gain familiarity with its operation under circumstances permitting visual checks on the radio bearings you get.

Automatic Radio Direction Finders

An automatic radio direction finder (ADF) indicates on a dial the direction to a transmitter once it has been tuned in—no swinging of a loop, no 180 degree ambiguity.

ADFs cover the same frequency bands as manual RDFs and use the same types of transmitting stations. They are more complex in circuitry and thus more expensive. The antenna is continuously rotated, either mechanically or electronically, whenever the set is turned on. Often such equipment is a fixed installation with a remote antenna, but portable models are available.

The advantages of ADFs lie in their ease and speed of operation. They are, however, subject to the same radio deviation as manual RDFs and a correction table must be prepared.

Some ADFs (and some manual models) display the frequency to which they are tuned in a digital readout. ADFs may also show the direction toward the source of the signals digitally.

VHF Radio Direction Finding

Direction finders for VHF are of the automatic (ADF) type. Most models are self-contained, complete units, but a growing number are designed to function with a VHF radio used for communications. All VHF direction finders require a separate special antenna, and all are several times more expensive than LF-MF models.

The indication of direction may be by a pointer rotating around a dial (as used by many ADFs for the LF and MF bands) or by a ring of LEDs around a circular scale, one even 5° or 10°, one simplified model can only be used for homing and gives a left-right indication by the needle on a meter. The accuracy of VHF radio bearings is approximately the same as on the lower frequencies.

The Coast Guard does not operate any radiobeacons on the VHF band, but the continuous transmissions of the NOAA weather stations provide a good signal for determination of a line of position; the range would be the same as in regular reception of weather information. This will normally provide only one LOP, but in limited areas it may be possible to receive two weather stations and

Figure 2311 **Some radio direction finders are of the automatic type. They require a special antenna, and yield about the same accuracy as those used for lower frequencies.**

thus get a fix. Coast Guard stations and Marine Operators normally have multiple antenna sites, and it is *not* possible to use their signals for taking bearings. The usual applications, however, of a VHF direction finder will be for homing or for taking bearings on other boats and ships, or on suitable shore stations.

Radar

Radar is an excellent means of marine navigation and it is used on vessels of all sizes down to boats of about 30 feet (9.1 m) length. Although space and cost limit its use on recreational boats, boaters should know its capabilities and limitations, for their own safety when cruising on waters navigated by radar-equipped vessels.

Radar Principles

A radar set sends out brief pulses of super-high frequency radio waves that are reflected by objects at a distance. The time that it takes for the pulse to go out and the echo to return is a measure of the distance to the reflecting object. See Figure 2312. In broad principles, this is the same technique as described for depth sounders, except that transmission is through air rather than water, and radio waves have been substituted for ultrasonic pulses. A refinement has been made in that the radar pulses are sent out in a very narrow beam, which can be pointed in any direction around the horizon and used to determine direction as well as distance.

Components of a Radar Set

The major components of a radar set are:

1. The TRANSMITTER, which generates the radio waves; it includes the MODULATOR which causes the energy to be sent out in brief pulses.

2. The ANTENNA, which radiates the pulses and collects the returning echoes. The antenna is highly directional in its horizontal characteristics, but eight to ten times wider vertically. The beam pattern can be thought of as being like a fan turned up on edge. The beam's narrow horizontal directivity gives it a fairly good angle-measuring capability, while its broadness in the vertical plane helps keep the beam on an object despite any rolling or pitching of the vessel. The antenna may be exposed or covered by a plastic RADOME which is transparent to radar waves; never paint a radome except in accordance with the manufacturer's instructions. An exposed antenna is generally used on powerboats, and is somewhat larger than one in a radome with a resultant narrower beam and slightly better resolution in bearings. A dome-type enclosure is often used on sailboats to eliminate possible entanglement of lines and sails with the rotating antenna. A radar antenna on a boat should be located so that no one is in line with the radiated energy and close to the antenna, so as to avoid any possible health hazard.

3. The RECEIVER, which detects the returned reflections and amplifies them to a usable strength.

4. The INDICATOR, which provides a visual display of objects sending back reflections.

Radars operate at frequencies far above the usual radio communications bands. At such super-high frequencies radar pulses act much like light waves in that they travel in essentially straight lines. They travel at the speed of light, 186,000 statute miles per second. For each nautical mile of distance to the target, only a fraction more than 12 microseconds (millionths of a second) are required for the round-trip of the outgoing pulse and the returned echo. Pulses, each of which lasts for only a fraction of one microsecond, are sent out at a rate of from 600 to 4,000 each second depending upon the design of the equipment. The directional antenna rotates at a rate of one revolution in about 2 to 4 seconds. The round-trip time for a pulse is so short that the antenna has not appreciably moved before the reflection is returned.

The Plan Position Indicator (PPI)

Marine navigational radars use a PLAN POSITION INDICATOR (PPI) type of display. A circular cathode ray tube of a special type from 5 to 20 inches (12.77 to 50.8 cm) in diameter is used, see Figure 2314; smaller scopes may have a magnifying lens added. The center of the face represents the position of the radar-equipped vessel and the presentation is roughly like a navigational chart.

A bright radial line on the face of the tube represents the radar beam; it rotates in synchronism with the antenna. Reflections show up as points or patches of light depending upon the size of the echo-producing object. The persistence of the screen is such that the points and patches of light do not completely fade out before the antenna has made another rotation and they are restored to brilliance. Thus the picture on the radarscope is "repainted" every few seconds. Some units now available provide a color display that helps to distinguish various types of targets.

A recent development in radar indicators is "raster scan," in which the radar echoes are digitally processed by integrated circuits and the information is presented on a square screen much like that of a television set. The presentation is "refreshed" more often and is easier to use. The display can be temporarily "frozen" for more accurate measurement of distance and bearing.

Radar Range Scales

The relative bearing of an object is indicated directly on the screen; a position corresponding to the "12" on a clock face is directly ahead. The distance to the object is proportional to the distance from the center of the screen to the point of light which is the echo. Most radars use concentric circles of light as range markers, to make estimat-

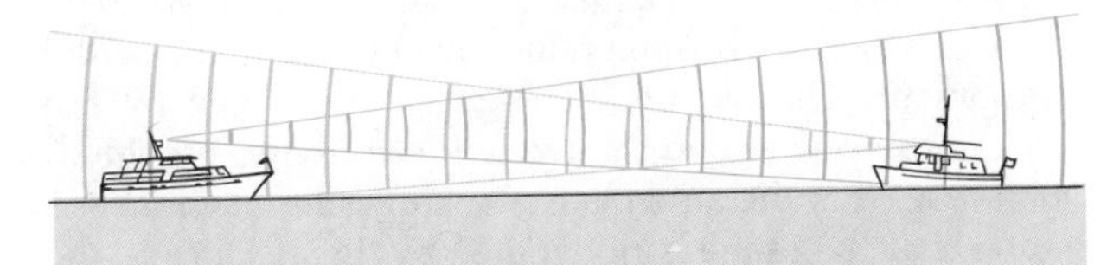

Figure 2312 **A marine radar set operates by sending brief pulses of super high frequency radio waves, which are reflected back by other ships, navigational aids, land masses, or other targets. The speed is so great that the echo is received before the next pulse is transmitted.**

Figure 2313 **On a sailboat, the antenna, shown here mounted on the mizzen mast, must be housed in a radome to prevent any possibility of fouling the rigging or sails.**

ing distances both easier and more accurate. All radar sets have multiple range scales that are selected to suit the purpose for which the radar is being used. Longer range scales provide coverage of greater areas, but at a cost of less detail and poorer definition. Typical ranges for a small-craft radar might be ½, 1, 2, 4, 8, and 16 miles.

Radar sets have both a maximum and a minimum range, each of which is of importance in the operation of the equipment. The maximum range is determined by the transmitter power and the receiver sensitivity, provided, of course, that the antenna is at a sufficient height above water that the range is not limited by the distance to the horizon. See Figure 2315. (The radar pulses normally travel with just a slight amount of bending; thus the radar horizon is about 7% farther away than the visual horizon.)

Because a radar pulse has a definite duration, and therefore occupies a definite length in space as it moves outward from the antenna, there is a minimum range within which objects cannot be detected. This minimum range, usually between 20 and 50 yards (18 to 46 m), is important when maneuvering in close quarters, as when passing buoys at the side of a narrow channel.

Radar sets are rated in terms of peak pulse power; for boats, sets will generally be 3-10 kw. The average power consumption from the craft's electrical system is, of course, much less—50-250 watts.

Units for Small Craft

Radar sets for small boats usually consist of two units. Modern design makes it possible to combine the antenna, transmitter, and a portion of the receiver into a single unit installed on a mast or on the pilothouse. This unit, usually 35 to 100 pounds (15.9 to 45.4 kg) should be located as high as practicable. Not much is gained by increasing the height of a boat's radar antenna by just a few feet, and attention must be given to any stability problems that may arise. Radar targets that will be of interest at distances approaching the horizon will be large ships or land masses, both of which will be considerably above the line-of-sight horizon. The antenna should have as unobstructed a "look" as possible in all directions. The remainder of the receiver and the indicator are located near the helmsman's position. Some indicator units are so small that they can be fitted into a pilot-house in any number of positions.

Because radar sets radiate radio frequency energy, they must be licensed by the FCC. No license is required to operate the equipment, but for its installation and maintenance the technician must have a commercial radio operator's license with a special "ship radar" endorsement. The owner and station licensee of a marine radar installation is responsible that a properly licensed individual does all the technical work on the equipment.

Principal Applications

Radars have two principal applications aboard vessels. They are often thought of primarily as anti-collision devices, but are even more often used to assist in piloting the vessel.

Figure 2314 **Information on ships, buoys, shorelines, and other targets are presented as patches of light on the PPI scope. The position of the boat on which the set is mounted is at the center of the scope.**

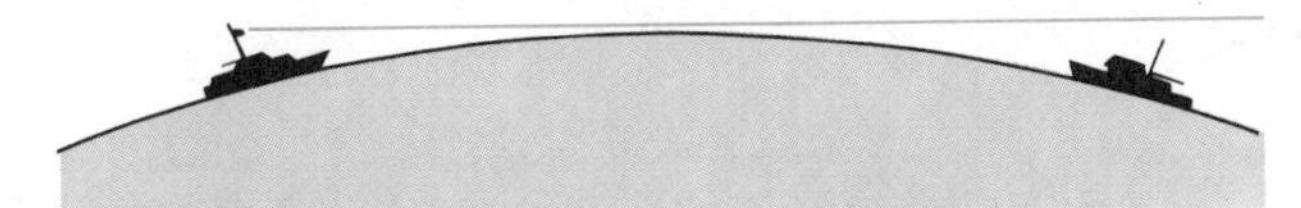

Figure 2315 **Maximum range of radar set on a small boat is likely to be limited by antenna height, rather than by its power. Radar waves bend slightly, but distance to the radar horizon is only 7% greater than the optical horizon.**

Radar offers an excellent means of extending the coverage of a visual lookout, especially at night and under conditions of reduced visibility. This greater range of detection affords more time for a vessel to maneuver to avoid another craft or an obstacle. With a radar plot, an early determination can be made of another vessel's course and speed. More information on the anti-collision use of radar can be found in *Dutton's Navigation and Piloting* and DMAHTC Pub. No. 1310, *Radar Navigation Manual.*

Radar also helps when making a landfall from offshore; running down a coast and picking up landmarks for fixes; or traveling in confined waters, entering an inlet, and the like. Radar has real advantages even in the daytime, and, of course, becomes particularly helpful at night or in fog, or other circumstances of reduced visibility.

Fixes by Radar

The daytime advantages of radar stem from its ability to measure distance. You will generally be able to see a landmark or navigation aid and take a visual bearing easily. If you can see only one such object, however, you can get only a single line of position and a fix is not possible. With a measurement of distance by radar, a fix becomes possible. See Figure 2316 left. The bearing could have been made by radar simultaneously with the range measurement, but a visual bearing is more precise and should be used if possible. Being able to get a fix from a single object can often be important.

With radar you can also get fixes in low visibility situations—from a single radar bearing and range measurement, from crossed bearings (but this is not too accurate because of the finite width of the radar beam), or preferably from distance measurements to two or more points that can be identified on both the radar screen and the chart; Figure 2316 right. Plot more than two ranges or bearings against the chance of a false fix resulting from one of the echoes being incorrectly identified. In all cases, the use of radar lines of position and fixes is the same as for visual plotting, giving due consideration to the lessened accuracy of bearing information.

Coastline Pictures on the PPI

It is not safe to use radar to determine your boat's position by observing the presentation of a coastline on a radarscope. The image seen on the PPI often varies appreciably from the appearance of a chart. Because the radar beam, as narrow (1 to 4 degrees) as it is, does have a finite width, and because the pulse, as brief (0.1 to 0.5 microsecond) as it is, does have a definite length in space, the picture on a radar PPI scope is not an exact replica of the land area or other object returning the echoes. Also, radar echoes will often be returned from buildings, hills, etc., some distance *inland* from the shorelines shown on charts.

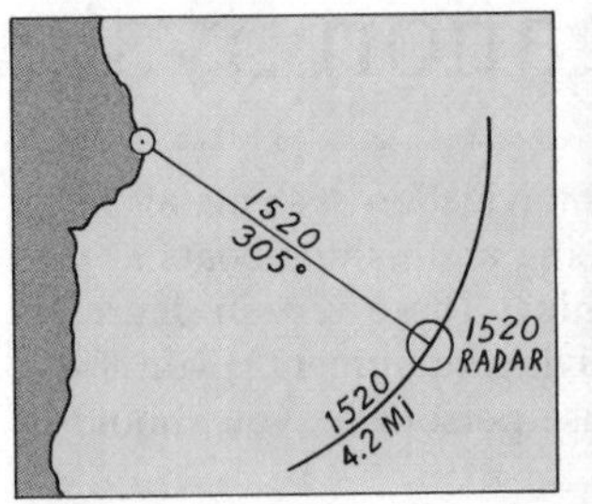

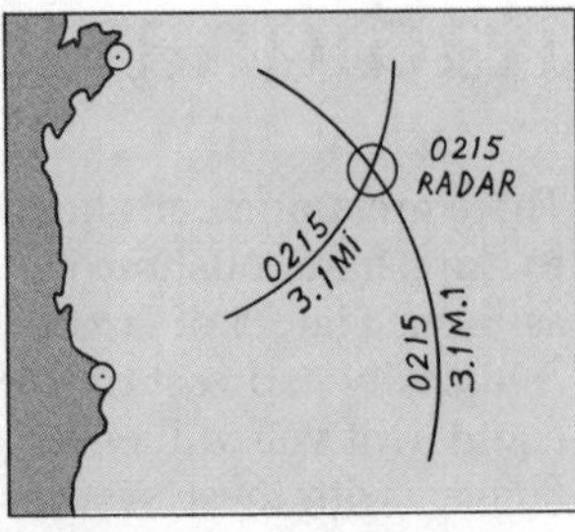

Figure 2316 **Radar fix may be obtained from a single object, using both distance and bearing measurements, *left,* or by plotting the distance to two identifiable objects, *right.* Radar distance measurements are more accurate than radar bearings. Ranges to more than two objects are desirable.**

Passive Radar Reflectors

A boat owner who is without radar can still increase his safety—he can equip his craft with a PASSIVE RADAR REFLECTOR; see Figure 2317. This simple, inexpensive item consists of thin metal sheets or fine-mesh screening arranged in mutually perpendicular planes. These may fold for storage, but must be flat and rigid with respect to each other when open for use. A reflector with each surface only about two feet (60 cm) square can provide a strong return signal if properly used. For maximum effectiveness, hoist it up from three corners so that one of the eight "pockets" is straight up; this is sometimes termed the "rain catching" position. A properly used reflector will enhance the chance of your boat being detected on the radars of passing ships, Coast Guard vessels, and other radar-equipped rescue craft.

Radar Direction Finders

Small units are now available that will detect an operating radar on a nearby ship or boat. These operate much like radar detectors used in cars on highways to avoid police speed-measuring units, but the marine models have the added ability to take accurate relative bearings. Upon detection of a radar signal, both audible and visual warnings are given; signals from multiple radars received at the same time can be differentiated.

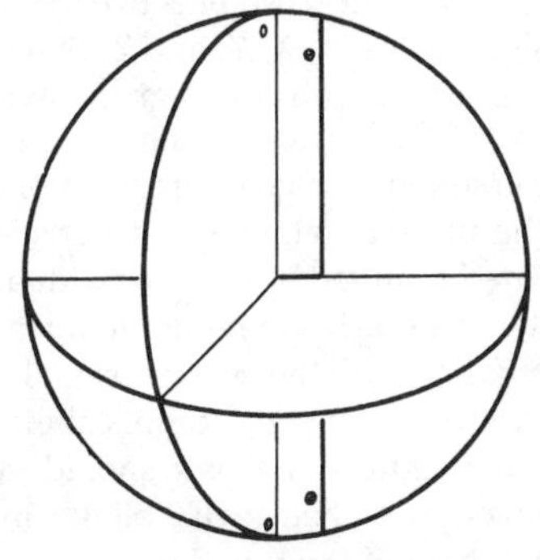

Figure 2317 **In waters frequented by radar-equipped ships, it is wise to rig a radar reflector. It makes the boat's echo show up much brighter on a radar scope.**

Radionavigation Systems

There are a number of radionavigation systems available to skippers of offshore cruising and fishing boats as well as to navigators of larger ships. These vary in degree of complexity and cost of receiving equipment. Even if it is equipment you will never use personally, you should be familiar with these systems.

The radionavigation system most widely used by boatmen in U.S. waters is LORAN-C. Another system is OMEGA, now operational with nearly global coverage.

DECCA is a short-range, high-frequency radionavigation system available only in limited areas; it is more widely used in Europe than North America. Decca is unique in that in the United States it is commercially operated, rather than by a government agency. Oil companies use it to locate offshore drilling platforms.

VHF OMNIRANGE (VOR for short), an aeronautical navigation system that is useful to high-altitude aircraft at a range of hundreds of miles from the transmitting antenna, is a line-of-sight system. It is sometimes used by boats, but at sea level its useful range is limited to a few tens of miles from the transmitting antenna.

SATELLITE navigation is becoming more widely used by ocean-cruising boats as the cost of receivers steadily declines. The NavSat, or Transit, system has been operational for many years. The more advanced NavStar, or GPS, system is still under development and test, with a scheduled availability date of the late 1980s.

Loran

LORAN—the name is derived from LOng RAnge Navigation—is an electronic system using shore-based radio transmitters and shipboard receivers to allow mariners to determine their position at sea. Loran works in all kinds of weather, 24 hours a day.

The original form of this electronic navigation system, LORAN-A, was developed during World War II for military applications. At the end of that war the Coast Guard was operating 70 transmitting stations; by 1971 the number had grown to 83. By the early 1970s, however, advances in electronic technology had made possible an improved system. The Loran-A system has now been phased out in U.S. waters, although operation continues in the northwest Pacific Ocean in the general area of Japan.

Loran-C Principles

The transmitters of Loran-C all operate on 100 kHz in CHAINS of a MASTER STATION (M) plus two to four SECONDARY STATIONS designated as W, X, Y, and Z. Stations of a chain are located so as to provide signal coverage over a wide coastal area. Each station transmits groups of pulses on the same frequency; signals from secondaries follow those from the master station at very precise time intervals. Chains are identified by their individual pulse group repetition intervals (GRI)—the time, in microseconds, between transmissions of the master signal. All Loran-C transmitters are frequency and time stabilized by atomic standards; if the signals of a pair should get out of tolerance, however, the chain will "blink" in a code that indicates the nature of the trouble.

The *difference* in the time of arrival at the receiver of pulse groups from the master and each secondary is measured precisely by electronic circuitry, and this information is used to determine a line of position. The use of two or more master-secondary pairs of signals yields the same number of LOPs and thus a Loran-C fix.

Loran-C has a ground-wave (most reliable) range of up to 1,200 miles, with sky-wave reception out to as far as 3,000 miles. The accuracy of Loran-C positions from ground-wave signals varies from 0.1 to 0.25 miles (0.2 to 0.5 km) depending upon where the receiver is in the coverage area; positional errors on sky-wave signals may be as much as eight times greater. A significant feature of Loran-C is the "repeatability" of positions obtained with this system. Ground-wave signals are very stable and a boat should be able to return to a prior position within 50 to 200 feet (15-60 m).

Loran-C readings are highly accurate on or near the "baseline" between the master and secondary stations, but are subject to significant errors on or near the extensions of the baseline beyond each station. Using sky-waves within 250 miles of a station being used is not recommended.

As of mid-1985, there were 44 Loran-C stations operating in 14 chains. Basic ground-wave coverage exists for the U.S. and Canadian Atlantic and Pacific Coasts, the upper North Atlantic Ocean, the Mediterranean Sea, small areas in the North Pacific west of Hawaii and south of Japan, and at the Suez Canal. Reception of sky-wave signals extends these coverage areas, but at lessened positioning accuracy.

Loran-C Receivers

Loran-C receivers automatically acquire and track the signals from the chain to which the set is tuned. The system is complex, but such complexity is handled internally by advanced solid-state circuits—all the operator has to do is turn the set on and tune to the proper GRI. (The coverage of a single chain is quite great and many boats will never need to change the GRI setting.) The Loran-C receiver will take several minutes to "settle down" and give steady readings; then the operator merely reads out the measured time differences and uses this information to plot his position on a chart that has been overprinted with Loran-C lines of position.

Proper installation is critical to good performance of a Loran-C receiver. The antenna should be away from all stays, metal masts, and other antennas; try several possible locations and settle on the one that provides the best

Figure 2318 **Loran-C receivers feature automatic signal acquisition and tracking with cycle matching. The model shown here can provide up to four lines of position, displaying the readings sequentially on its readout.**

signal-to-noise ratio and quickest "settle-down" time. Use only an antenna recommended by the manufacturer.

Equally important in installation is good grounding for the antenna coupler and the receiver. On wooden and fiberglass craft, make a connection with a copper strap 1 inch (2.5 cm) wide.

Interference to Loran-C signal reception must be eliminated or reduced as much as possible. Sources of the reference include engine ignition systems, alternators, and some other items of electronic equipment, especially a TV set or fluorescent light.

Loran-C receivers are subject to interference from other signals on nearby frequencies. Such unwanted noise can seriously degrade the receiver's performance, or even make time-difference measurements impossible. This interference is eliminated by "notch filters" sharply tuned to the unwanted signals. Loran-C receivers have from two to four such filters; these may be factory-tuned or operator-tunable, or a combination of both. The number and type of filters needed will depend upon the interference conditions in the area(s) in which the receiver is used. In general, a combination of two preset (internally tuned) and two front-panel, operator-adjustable filters will provide excellent interference protection and adequate flexibility even if cruising from one area to another.

Using Loran-C Information

Loran-C charts are at a scale of 1:80,000 or smaller—charts in the coastal and sailing series (see Chapter 18). Loran-C lines are not placed on harbor charts—scales of 1:40,000 and larger—as the accuracy of this system is not sufficient for navigation down a channel or close to hazards. You can and should, however, record your own Loran-C readings at specific points on harbor charts and take full advantage of the excellent capability of repeatability. A series of specific points could describe the desired track from harbor entrance to home berth, and would be very valuable under conditions of reduced visibility.

Loran-C charts published by the National Ocean Service have their grid of time-difference lines corrected for the mixed over-land/over-water path of the signals. Initially the position of these lines is based on theoretical radio-propagation characteristics; as actual field data are collected, the lattices are shifted for improved accuracy—a chart note will tell you which is the case for any specific chart. DMAHTC charts are gridded only for all-over-water signal paths. For full accuracy, observed time differences must be adjusted with small values taken from Loran-C Correction Tables; this is also required to get comparability between DMAHTC and NOS charts for the same area. DMAHTC charts for areas where sky waves are normally received and used will often have small blocks of numbers printed at intervals. These are correction values to be applied to the signals from indicated stations; separate day and night corrections may be provided. The operating manual for your Loran-C receiver should be consulted as to the procedures to be used for inserting these corrections.

Loran-C receivers generally have a dual display that permits simultaneous readouts for the two values of time difference required for a fix. Often a receiver will automatically make measurements on three or four station-pairs with these additional readings being displayed sequentially.

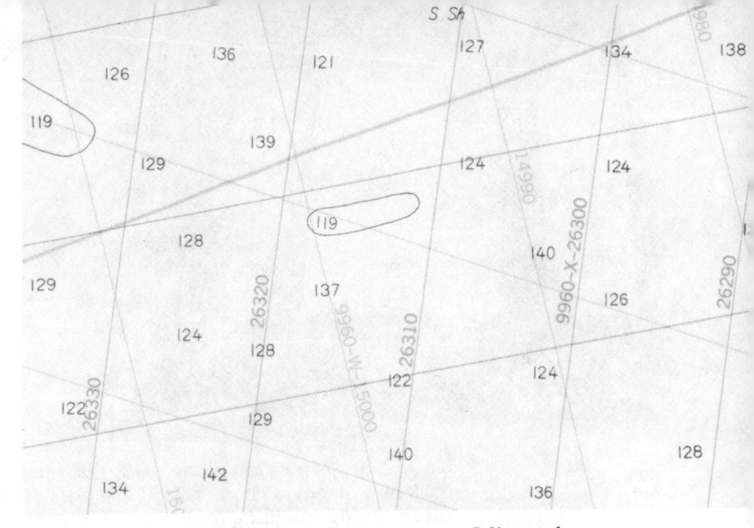

Figure 2319 **Portion of a chart showing Loran-C lines of position. Different colors are used for each set of lines. Although position can be determined from Loran tables, direct plotting on a chart is the simplest and quickest way of getting a fix. Similar charts show Omega navigation system lines of position.**

Most Loran-C receivers include a microprocessor that gives a direct readout in latitude and longitude, plus direction and distance to a pre-set destination; also speed along the track, cross-track error, time to go, and even a destination arrival signal when almost there. From 8 to as many as 100 intermediate destinations—called "way-points"—are typically available. Some caution must be used with Loran-C sets that give position in latitude and longitude, because their computations may be based only on theoretical propagation factors. Some models allow a manual input of values from the Loran-C Correction Tables or chart; a few sets do this automatically from internally stored data. Position information in terms of time-difference readings can be exchanged between receivers on different vessels with excellent results. If, however, latitude-longitude data are used, lesser accuracy may result as different receivers may use different coordinate conversion programs.

Locating a Loran-C receiver so that cross-track error and other steering information is readily available to the helmsman is often a problem for sailboats and smaller powered craft. Models are now available with a display panel that can be mounted separately at the helm, with the bulk of the set located at any convenient place, connected to the display with a multi-connector cable.

The latest development in Loran navigation is an "interface unit" that permits positional information from a Loran-C receiver, converted in most cases to cross-track error, to be fed to the vessel's autopilot, with the result that she is brought back onto the intended track. The nearly continuous flow of corrective signals causes the boat to automatically "crab into" any offsetting current. Cross-track errors of only a few hundred feet or less can be sensed and corrective instructions generated. One model of Loran/autopilot interface operates in a slightly different manner—here the "electronic brain" senses any difference in the direction of the preset destination from the craft's actual position and from where she would be if on course at that time. Interface units include delay times to "smooth" the flow of data and make course changes more gradual.

Some Loran-C receivers can stand an "anchor watch" for your boat. After putting the hook down, the set is

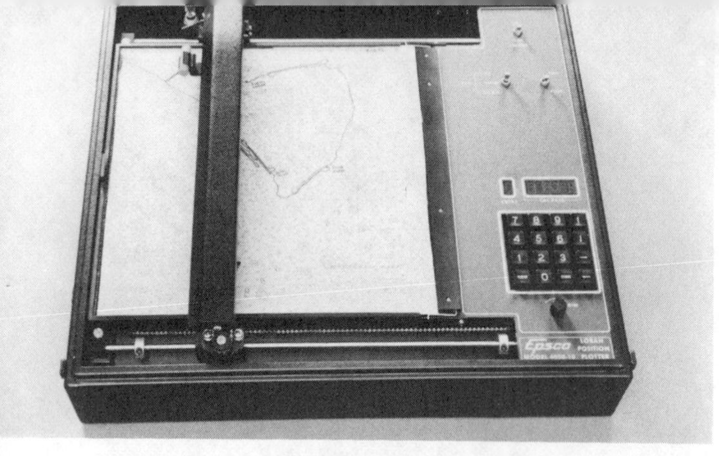

Figure 2320 **A position plotter, *left,* that interfaces with Loran equipment will trace the boat's track on a standard chart. The VLF signals of the Omega system, *right,* now cover almost the entire world, providing fixes of good accuracy for vessels on the high seas. Receiver outputs show either lines of position, or latitude and longitude directly.**

switched to that mode so that a small change in Loran-C readings of a specified amount in any direction will sound an audible signal.

Also available are Loran-C PLOTTERS. These draw a trace on a chart using data from the receiver—you can see your course and speed made good as you go along.

Omega

The OMEGA electronic navigation system uses very-low-frequency (VLF) radio waves; four different frequencies from 10.2 to 13.6 kHz are used. Such VLF signals have considerable range and stability over day and night paths. An advantage of the system is that complete global coverage can be obtained by the use of only six transmitters properly situated. Ideally, stations would be located at the North and South Poles and 90 degrees apart on the equator; such a requirement must, of course, be modified to meet practical considerations. In actual practice, the Omega network has eight transmitting sites to allow for possible equipment failures and off-air time for routine maintenance. The transmitters are located approximately 6,000 miles apart and at any point signals from at least four stations should be usable. Stations in the U.S. are operated by USCG personnel.

The Omega System

The Omega system was originally developed by the U.S. Navy for its submarines, surfaced or submerged (VLF signals can be received while under water), as well as for surface vessels and for aircraft. Receiving equipment is now available for civilian ships and aircraft, including recreational boats. Omega equipment is relatively expensive, but advancing technology and increased production continue to bring price reductions.

Omega Fixes

Omega stations transmit continuous-wave signals, rather than pulses, for approximately one second out of every ten seconds on each frequency used. Signals from a single pair of stations on a single frequency can furnish a hyperbolic line of position, but rough position knowledge is required to within about four miles to identify the set of lines, called a "lane," within which the receiver is located. Use of a second frequency reduces the need for position knowledge to 12 miles, and use of a third frequency extends this to 36 miles, and the fourth to 144 miles. Note, however, that Omega receivers used on ships and boats are almost all single-frequency (10.2 kHz) models; multi-frequency sets are for use on high-speed aircraft where a reset position cannot be so precisely established.

Two or more lines of position are combined in the normal manner to obtain an Omega fix. Station pairs should be selected so as to get lines crossing at large angles, as near 90 degrees as possible for two lines, or 60 degrees for three lines. As with Loran, special charts are used with over-printed Omega lines of position. Receivers have "lane counters" to keep a record of the number of lanes crossed since the counter was reset after a fix was established. Omega receivers must be operated continuously to keep the lane count running; if operation is interrupted, a fix must be obtained by some other means to reset the unit and resume the Omega count. More expensive models of receivers can internally compute and display position directly in latitude and longitude.

Omega signals are affected by sky-wave propagation conditions and it is necessary to refer to published correction tables in the use of this system. The nominal all-weather accuracy is one mile in the daytime and two miles at night. Special techniques are available within local areas for increased precision in position fixing; this is known as DIFFERENTIAL OMEGA, and is useful for high-accuracy work.

The signals from Omega transmitters are controlled by atomic frequency standards and can be used for adjusting or calibrating less accurate frequency standards.

All eight stations of the Omega Navigation System are now in operation and coverage is essentially worldwide. It is still necessary, however, to have "validation surveys" to reconcile differences between theoretical propagation factors and those actually encountered in the field. The system is being declared "fully operational" area by area as these surveys are completed. The publication *Omega Global Radionavigation—A Guide for Users* is available for purchase from the Government Printing Office.

Figure 2321 **Satellite navigation receiver detects frequency shifts as the satellite approaches and then passes overhead. For a good fix, the satellite must be between ten degrees and 20 degrees above the horizon.**

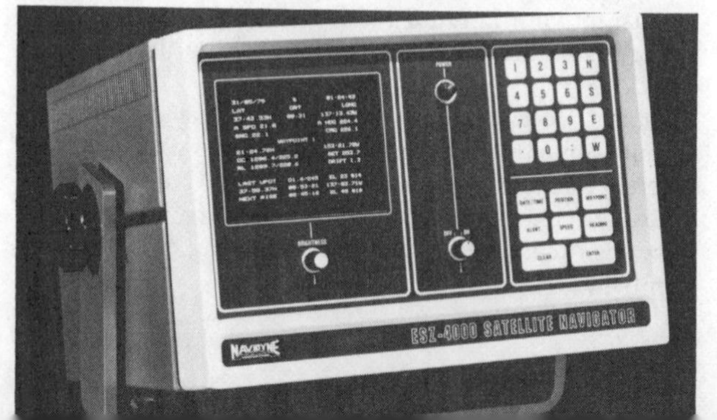

Satellite Navigation

The Navy Navigational Satellite System—often called TRANSIT—became operational in 1964 and was released for civilian use in 1967. Signals are transmitted from five or six "birds" in polar orbits on two frequencies, 150 and 400 MHz; only the 400 MHz signal is used on non-military sets. The satellites are roughly 580 miles above the earth, circling it every 107 minutes.

The set of satellite orbits forms a "birdcage" within which the Earth rotates. Wherever a satellite rises above the horizon the opportunity exists to get a fix. The spacings of the orbits, however, is not uniform. The mean time between fixes varies from 30 to 95 minutes as a function of latitude, being greater nearer the equator, and there is the "worst-case" possibility of 12 hours between Transit fixes.

Originally SatNav receivers were very expensive, but more recently technological advances, and competition among an increasing number of manufacturers, have brought this equipment within the price range of more and more boaters. Additional features have been added, including the ability to program multiple waypoints with readouts of directions and courses to them, either great circle or rhumb line; displays of the times of upcoming satellite passes; set and drift since last fix; and date and time.

SatNav Fixes

A Transit fix requires reception of a satellite's signals for at least two minutes and preferably longer. During this time, the vessel's direction and speed must be entered into the receiver's internal microprocessor; the lower the accuracy of this information, the poorer the fix. With good input data, however, SatNav fixes can be accurate to about 100 yards (91 m); receiver output is in the form of digital displays of latitude and longitude. A SatNav computer works on the Doppler principle, detecting frequency shifts as the satellite approaches and then passes away. For a good fix, a satellite must be between 10 and 20 degrees above the horizon.

Combined Systems

A recent development is a receiver that combines the positional information from Loran-C signals with similar information from a navigational satellite for what is called "LorSat" navigation. The unit can display positions from either system independently, or their combined output. The Loran information is useful for providing the needed course and speed data to the satellite receiver, and for navigating between satellite fixes. The satellite output in turn is useful to the Loran-C receiver in that it greatly extends its range of operation by permitting the more accurate use of sky-waves.

Other receiver combinations include Omega and SatNav, where Omega provides positional data between satellite fixes, and SatNav ensures the accuracy of Omega lane counts, plus a precise fix if the count is lost and has to be restarted.

Automatic Steering Devices

AUTOMATIC STEERING DEVICES are electrical or electro-hydraulic equipment used to hold a boat on a predetermined heading. They are often called AUTOMATIC PILOTS or AUTOPILOTS, but this is not technically correct as they do no part of the "piloting" of a boat.

These devices use electric or hydraulic motors to move the rudder to the desired angle and to hold it there. The control of these motors is made automatic by the addition of a direction-sensing element and suitable electronic amplifiers. Either magnetic or gyroscopic compasses can be used to sense heading changes, but the lower cost magnetic is nearly always used. For sailboats, there are models that work with either wheel or tiller steering.

Most automatic steering devices have manual controls that allow temporary over-riding of the automatic controls; the boat returns to the present heading when the over-ride switch is released. This manual over-ride feature permits brief course changes to avoid obstructions or other boats without having to reset the heading. Control boxes on long cords permit a helmsman to steer the boat while away from the helm. In almost all instances, throttle and gear changes must be made in the normal manner, but units are now available that permit these functions, as well as manual steering, to be controlled from remote locations.

The latest designs of automatic pilots are described as "nonhunting proportional rate." Corrections are applied immediately whenever the boat strays off the set heading, but small errors are corrected gently and large errors are corrected vigorously. Correcting errors in proportion to the speed and amount of error results in smooth steering action, eliminating overcorrection.

Using an automatic steering device is simple. It does much to relieve boredom and fatigue on long stretches, and can, under most circumstances, steer a straighter course than a human helmsman. The disadvantage in this equipment lies in its misuse: excessive reliance on the device and failure to keep a person at the helm as a lookout, ready to take over immediately if a dangerous situation develops. Remember, too, that the sensing element is usually no more "intelligent" than a magnetic compass; it is subject to all the same outside corrupting influences such as nearby tools, portable radios, and beer cans.

Miscellaneous Electronic Equipment

There are many small items of electronic equipment that can be installed on a boat for the greater safety and convenience of the owner and his guests. Just remember that each item will add to the load on the battery and electrical wiring.

Emergency Position Indicating Radiobeacons (EPIRB)

Every boat that goes offshore beyond reliable VHF radio range, roughly 20 miles, should carry a Class A or Class

B EMERGENCY POSITION INDICATING RADIOBEACON (EPIRB). This small device transmits a distinctive tone signal on two *aircraft* frequencies—121.5 MHz, the emergency channel of civil aviation, and 243.0 MHz, the "guard" channel for military aircraft. This signal serves first to alert passing aircraft that an emergency exists and later to guide searchers to the scene of distress. The transmitter is not powerful—a fraction of one watt—but its signals can often be heard out to as far as 200 miles by high-flying aircraft. If the signals are heard by more than one passing plane, it is often possible to get a very rough fix that will at least indicate the general area to be searched. There are now in orbit satellites of several nations that monitor EPIRB frequencies and relay signals to earth stations.

EPIRBs are battery-powered and are required by the FCC to operate continuously with rated power for at least 48 hours (many will operate for as long as a week); the battery must be marked with dates of manufacture and proper replacement (50% of life remaining). The transmitter is completely transistorized and suitable for storage in a marine environment over a long period of time.

Class A EPIRBs are for certain types of vessels on which they must, by law, be carried. The less-expensive Class B units are for voluntary carriage by other vessels such as recreational boats. This simpler design need not be capable of floating by itself, but a flotation collar is usually added if the unit is not buoyant; this type of EPIRB can be activated either manually or automatically upon floating in the water. A Class A EPIRB must be capable of floating free from a sinking vessel and activating automatically, including extending its antenna from the stored position.

EPIRBs have a test switch for checking readiness for use. This must be done for *one second only* during the first five minutes of every hour. Every effort must be made to avoid "false alarms."

Once an EPIRB is turned on in an emergency situation, *it must be left on.* Turning it off for various periods "to save the battery" destroys its effectiveness for homing on the signals by rescue air and surface craft.

EPIRBs for recreational boats are available for less than $300; they are an essential safety item for any boat going offshore. There is no assurance that the signal will be picked up and the authorities notified; many aircraft monitor 121.5 or 243.0 MHz, but this is not required by law or regulation. Many actual cases, however, have occurred where an EPIRB was the direct and only means of getting assistance to the survivors of a sunken boat. As these devices send out radio waves, they must be licensed; this is easily done by modification of the radio station license (if no other radio is licensed on the boat, a separate EPIRB license must be obtained).

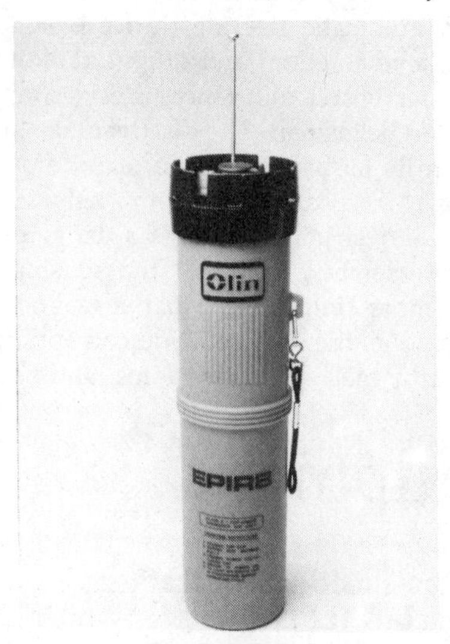

Figure 2322 **A Class C EPIRB is designed for use on boats that operate close to shore; its signals are broadcast on Channels 15 and 16, alternately, for a 24-hour period.**

Class C EPIRBs

A new and different type of Emergency Position Indicating Radiobeacon was authorized in 1979 and became available in 1981. These are the Class "C" units, operating on different frequencies and intended for use in *coastal* waters out to roughly 20 miles from shore—this is the area covered by Coast Guard and VHF shore stations.

A Class C EPIRB operates on VHF Channels 16 and 15, transmitting the radiotelephone alarm signal (see page 528). A very brief alerting transmission is made on Channel 16—on this channel because it is the one monitored by the Coast Guard and by ship and shore stations, brief so as to not interfere with essential communications on this frequency. The alarm signal is then transmitted at greater length on Channel 15 for verification and possible use for direction finding. A complex cycle of transmissions on these two channels, plus periods of no signal, continues automatically for 24 hours after which it goes silent unless the activation switch is manually operated.

Class C EPIRBs are generally similar in appearance and design to Class A and B units with respect to size, weight, solid-state circuitry, power source, flotability, and antenna. Some models also include a strobe light whose flash may help searchers find the craft in distress. The lower cost of these units—under $200—is intended to make them attractive for boats not equipped with two-way VHF radios. (If carried on a radio-equipped boat that set must first be used in an attempt to get assistance *before* the EPIRB is activated.)

The Class C units do not duplicate the operational use of Class A and B EPIRBs, and it would not be redundant to carry both. If the boat has no radio station license, one must be obtained as the Class C EPIRB is an active radio transmitter.*

Fuel Vapor Detectors

A FUEL VAPOR DETECTOR is considered by many boatmen to be an essential piece of safety equipment aboard any craft using gasoline as fuel. See Figure 2323. These devices provide a visual, and in some units an audible, warning of the build-up of any dangerous concentration of fuel vapors in the bilges. Remember, however, that these units are *not an absolutely positive* means of warning of a hazardous condition; they are not as infallible as the human nose. In doubt, open the hatches and use your nose—it is the best "bilge sniffer" of them all.

Fuel vapor detectors come in several designs, each with good points and disadvantages. These include the hot wire, cold sensors, and semiconductor types of detector

*Class D and E EPIRBs are specialized models for the same frequencies as Class C units; these are required on certain classes of commercial vessels on the Great Lakes and other specifically designated waters.

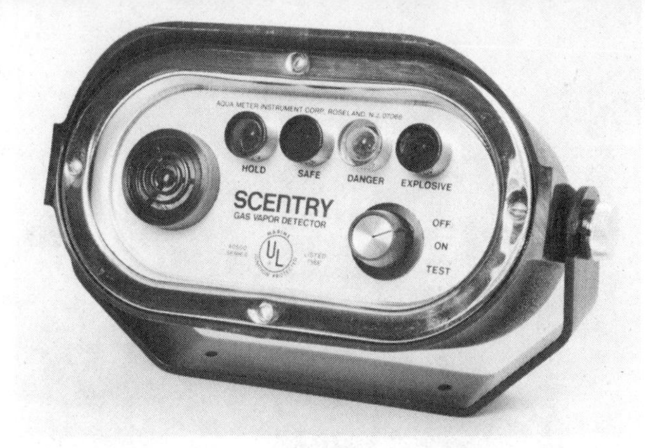

Figure 2323 **A properly installed and used fuel vapor detector will warn of explosive fume buildup in the bilge. Models are available with audible and visual signals.**

units. Price is not a factor for these types, and a choice should be made on the general features and suitability for your particular installation. Some sensors will also detect hazardous levels of other gases such as LPG (propane), alcohol, or hydrogen (from battery charging).

Installing a fuel vapor detector is not difficult electrically or mechanically, but it *must* be done correctly. The location of the detector unit is important. Have your installation checked by a surveyor or safety inspector from your insurance company. Even the best fuel vapor detector is no substitute for electric bilge blowers—forced ventilation fans that pull air and fumes out of the bilges.

A recent addition to the family of "bilge sniffers" is a small pocket-sized portable unit its detector element on a flexible extension that can be poked into any nook or cranny. It is non-specific for vapors and has both audible and visual alarms.

Marine Converters

The growing use of electrical equipment on boats has resulted in a steadily increasing demand for power from their batteries. The alternator on a boat's engine will keep the batteries charged if the engine is operated a large-enough percentage of time that the electrical equipment is in use. There is no way to state a general figure for what proportion of the time is required, because individual conditions vary too widely.

If you find that all too often your battery is down so far it won't start the engine, then you are a candidate for a MARINE CONVERTER. This relatively simple electrical device converts 117-volt AC dockside electricity into the 6, 12, or 32 volts DC required for the boat's electrical system. It is essentially a continuous-duty battery-charger with control circuits to permit it to operate safely if left unattended. Do not use an automotive charger.

Inverters

There is no way to store 117-volt AC in a battery, and if AC is needed away from the shore the only means of getting it (without installing an auxiliary generating plant) is to "invert" the DC power from the batteries into AC.

Alternating current aboard a boat of any size is useful for powering television or hi-fi sets, tape recorders, electric razors, electric mixers, or other small appliances. Transistorized INVERTERS are available that can supply small or moderate amounts of "shore-type" electrical power; other models use rotating machinery, a DC motor turning an AC generator. These DC-to-AC inverters are *not* suitable for use with electrical devices whose function is to produce heat, such as toasters and irons.

Auxiliary Generating Plants

Solid-state or rotary inverters are usually limited to continuous output power ratings of 500 to 1,000 watts, with briefly higher loads possible for starting electrical motors; at these outputs a considerable drain is placed on a boat's 12- or 32-volt batteries. If you need greater amounts of 117-volt AC power for tasks like cooking, heating, or air conditioning, you can consider installing a small AUXILIARY GENERATING PLANT. These units are available in sizes from 500 watts to 15 kilowatts or more, with either gasoline or diesel engines.

Calculating the size of generator you need is not hard, because nearly all electrical equipment is plainly rated in watts of power consumed. Be careful, however, about loads that are not constant. For example, an electric refrigerator unit may consume only about 100 watts *average* power but while it is running it may require 400 watts, and the starting load will be higher still.

Rudder Position Indicators

A simple and inexpensive, but very useful, item is a RUDDER POSITION INDICATOR. It is standard equipment on large vessels and tugboats, but seldom found on recreational boats, probably because their owners know little about the device and how it works. A small unit is installed at the rudder post and wires are run forward to a meter located within easy view of the helmsman. This meter is calibrated in degrees of right or left rudder. A quick glance will tell the exact position of the rudder at that moment; such information is of considerable value when maneuvering with little steerageway.

Weather Receivers

When on board, you can get the continuous NWS broadcasts (see page 522) on one of the weather channels of your boat's VHF radio. To get the same weather information *at home,* you can use one of the small, inexpenisve sets available from sources such as Radio Shack. These will tune to the weather channel used in your area; Most are powered from an internal battery that will last a long time. Others operate from house AC power with an internal battery for backup. Some sets are turned on automatically by a "tone alert" signal which is transmitted ahead of all special warning messages.

Facsimile Receivers

Government radio stations of the United States and other nations transmit weather maps by radiofacsimile on regular schedules (see page 323). These are on various HF channels suitable to the time of day of the transmission. Maps may show recent surface conditions, or forecast conditions for 12 to 48 hours ahead; special maps forecast wave/swell conditions, surface temperatures, ice, etc.

These broadcast maps are printed out on your boat, by a special device that either connects to a suitable receiver or is self-contained with its own receiver. Units such as these can supply excellent weather information, but a level of knowledge and experience is required to interpret the charts received.

Owner's Manual
Down East
CHAPMAN'S LOG &
OWNER'S MANUAL

Log-keeping and other paper work that can add to your efficiency as a skipper

BOAT MANAGEMENT

A boat of any size is a complex item that must be properly used and maintained for safety, efficiency, maximum enjoyment, and the protection of one's investment. This is "boat management," and documents and records make up a vital part of a good management system. Some persons may not enjoy doing paperwork, but it can be made less burdensome and more effective by an orderly approach.

It is necessary, however, for each owner-skipper to work out the specific actions and records for his boat, because every boat and its use are highly individualized matters. What is best for a cabin cruiser just won't be right for an outboard runabout or a sailboat; what is adequate for the weekend user of a Class 1 boat will fall short of meeting the needs of a cruising yachtsman. But in *all* cases, there is a requirement for a certain amount of management.

Certificates and Documents

The term SHIP'S PAPERS has a long and honored history. Nearly all vessels, including the smallest of boats, have some official or legal papers relating to them. Even a rowing dinghy or a sailboat of any size without mechanical propulsion (and thus exempt from any requirement for registration and numbering except in certain states) will have a bill of sale or purchase receipt.

Many items can be included in a listing of ship's papers for a recreational vessel. Among these are the Certificate of Number or Certificate of Documentation, radio licenses, survey reports, insurance policies, and personal certificates.

Required Papers

As discussed fully in Chapter 2, the owner of a boat registered by the Coast Guard, or by a state under a numbering plan approved in accordance with the Federal Boat Safety Act of 1971, is issued a Certificate of Number. This must be on board whenever the boat is in use (with a few specified exceptions), and so the law states that it shall be of "pocket size." If it is lost or destroyed, immediate action must be taken to obtain a duplicate, but use of the boat can continue meanwhile.

In some states a sticker or decal for the current year or registration period must be attached to the boat in a specified location. Broadly speaking, this small item can be considered a part of the ship's papers. It is visual evidence that the boat's registration is valid.

Documents

Boats over a certain size *may* be "documented" by the Coast Guard in lieu of registration and numbering. Information on this procedure is given in Chapter 2.

Documented boats must have their Certificate of Documentation on board at all times, and a Certificate of Number or Documentation must be produced by the boat's owner or operator for inspection by a law-enforcement official upon his request.

Radio Licenses

A radio station aboard a boat must be licensed by the Federal Communications Commission. This includes VHF radiotelephones, single sideband (SSB) radios, and radars. No license is required for small "handietalkies" on CB frequencies provided that their power is not over 0.1 watt. No license is required for electronic equipment such as radio direction finders, Loran receivers, electronic depth sounders, and other types of equipment that do not *transmit* radio signals.

For the operation of SSB radios in all areas, and the operation of VHF sets under some circumstances, an individual must have a FCC operator's permit. None is required for CB sets or radar.

Posting of Licenses The radio station license must be on board the boat, posted in a prominent location as close as possible to the radio. A Restricted Radiotelephone Operator Permit, if the person has one, must be carried on that individual's person; any higher class of permit or license must be posted or a photocopy carried.

Copy of FCC Rules

Licensees of radios on voluntarily equipped boats—recreational craft and those carrying six or less passengers for hire—are not required to have a copy of the FCC Rules and Regulations, *but such a person is still required to know and comply with the applicable provisions.* The full text of the official rules is very lengthy, complex, and technical, but there are available unofficial simplified versions in the form of slim pamphlets. The prudent skipper will obtain a copy of one of these, study it, and then keep it handy for later reference.

Licensees of compulsorily equipped vessels—those carrying more than six passengers for hire or engaging in other commercial operations—must have a copy of Part 83 of the FCC Rules and Regulations, and, except under unusual circumstances, it should be on board the vessel.

Additional information on the paperwork related to marine radio stations will be found in Chapter 22.

Optional Boat Papers

Although not required by any law or regulation, an annual Courtesy Marine Examination by the Coast Guard Auxiliary is an excellent check on a boat's safety and "readiness for sea." If this examination is passed a sticker is awarded and the inspection form signed by the Auxiliarist and given to the skipper. This signed form should be stowed safely away with other ship's papers.

Personal Certificates

If a motorboat is not carrying "passengers for hire" (see page 50) there is no requirement for the operator to have a Coast Guard license—state requrements may differ and should be checked. (The term "noncommercial operation" is often used in this connection, but it is not strictly correct because a person commercially fishing, crabbing, lobstering, etc., does not need a USCG operator's license provided he has no passengers for hire on board. He may need other permits, however.)

It has been held that a salesman, broker, or owner demonstrating a boat to a prospective buyer must have an operator's license or else place the boat under the command of a licensed individual.

Many skippers get a USCG license in one or more categories, just for the sake of having one, rather than from any regular need for it. If so, it should be posted where it can readily be seen; if nothing more, it may serve to reassure any nervous guests aboard for a cruise.

Likewise, the skipper's membership certificate in the United States Power Squadrons, with its endorsements for qualifications in Advanced Grades and Elective Courses, is often posted on his boat rather than ashore at home. The same, too, is frequently done with one's certificate of course graduation or membership in the United States Coast Guard Auxiliary. As with a USCG operator's license, these documents will attest to the owner's qualifications and abilities.

Insurance Papers

Most boats are covered by insurance policies; see page 75. The owner may keep his policy either on board or ashore in a secure place such as a safe deposit box.

If the policy is not carried on board, there should be a memorandum in the ship's papers listing the insurance company, policy number, coverages, expiration date, and other pertinent information such as the agent's name, address, and telephone numbers, both office and residence.

Logs and Log-Keeping

An important part of the paperwork of boat management is maintaining the legally required or otherwise desirable logs, such as those for navigational, engine, and radio operations. You can, of course, keep these all in the same covers or binder if you like.

Developing Your Log Format

The first step in log-keeping is to have logs that fit your needs. It is pointless to follow a format that provides space for information that does not apply to your style of boating—for example, the nature of the bottom as brought up by an "armed" lead, or true headings if you regularly use magnetic readings from the chart. Trying to use an unsuitable log format is a bad practice, because it may discourage one from making any entries at all. Many printed logs simply do not meet an individual's requirements—if they do, fine; but if they do not, then develop your own style for the data that you find necessary and useful at a later date.

A log should be kept in columnar form as much as possible, rather than in paragraph style, but, as discussed below, consider your needs; columns should not be set up for data that are only occasionally entered.

Printed log books are available from Hearst Marine Books (*Chapman's Log and Owner's Manual*) and at most marine supply stores. These may be used as they are, or column headings may be modified. In general, using a loose-leaf notebook for a log should be avoided because pages can be ripped out and lost. Much more desirable is a bound, hard-cover book—often a Record book, available from office supply stores, will do nicely for the boatman who creates his own columns. It is quite possible that your deck log needs will be met by such a book ruled in just three columns and headed: Date, Time, and Remarks.

Determining Needs

It may take some time to determine what is actually needed and what is not. Your real requirements for log data may not even be clear until your second cruise over some of the same areas months later when you will notice any omissions.

In considering which format to use, or in designing one of your own, be systematic. List the items (such as Date, Time, and Course) that you wish to enter; consider their relative value. Sort possible entries into two categories: frequent and occasional. The first group will guide you in establishing specific columns; the second category can be put in a general column headed Remarks. Once you

CRUISING LOG

Time	Place Abeam: Buoy, Landmark or Other	Distance off	COURSE TO NEXT OBJECTIVE			Motor R.P.M.	Estimated Speed	Estimated Time to Next Objective	Actual Time Between Objectives	Remarks
			Course Steered	Magnetic	Distance					
0810	Underway									
0822	Sea Buoy R"2"		077		18.5	3500	8.5	2:10	2:12	
1036	Twin Towers	2 mi								Anchored for fishing
1445	Underway		253		18.5	3500	8.5	2:10	2:14	
1659	Sea Buoy R"2"									
1710	Back in regular slip									
	at marina - boat									
	secured									
	Tom Hooper									

Figure 2401 **It is important that the log entries for each day's run, or for each period of time the boat spends in port, be validated by the skipper's signature. This may have important legal aspects if the log is ever needed to establish facts in a court of law.**

have worked out your needs, reduce your log-keeping to those items only, and buy or prepare a form most convenient for recording them. Be careful, however, that you don't overlook data (such as weather conditions, state of current or tide, etc.) for which you may have only infrequent use but which would be invaluable in case of accident, breakdown, tax or insurance inquiry, etc.—be sensible but don't be *too* brief!

Deck Logs

A DECK LOG is a record of navigational data—the time you started a day's run and the time you reached your destination, plus time of passing important navigational aids or landmarks and towns. These latter items are particularly useful when you are cruising in inland waters, where a constant speed cannot be maintained and you might later want to know the running time between such points. If your cruising takes you offshore, you should include your various headings and speeds, with the time of any change in either.

Other important information for the deck log is the names of any persons aboard other than the usual family crew. Entries might also be made of the weather—wind, visibility, sea conditions, etc.—at the start of the day's run and again when there are any significant changes. Detailed entries should also be made of any unusual events on board or seen from the boat, like trouble with your engine, or assistance rendered to another vessel. These entries can greatly assist your memory at a later date.

It is useful as well to record your operating expenses, such as fuel and dockage charges. Such records must be kept of course, if you plan to take advantage of any business-related tax deductions. It is often of interest to record fuel prices at stops where no fuel is taken on, or overnight dockage fees at brief daytime stops; this information may be useful when planning future cruises.

To help keep entries accurate, a boat's deck log should account for *every* day, in port or underway. A simple entry like this one will do it: "8-12 June. In regular slip; no one aboard."

Validating the Log

Regardless of a deck log's format, certain essential elements should be entered. Number each page and add the boat's name on each one, usually at the top. Sign each day's entries, or a combined entry for several days in port, as you complete them. This authenticated record may be needed in connection with an insurance claim, a law suit, or other investigation. If a boat owner can state under oath that it is his practice to keep a daily log and then present a signed entry for the day in question, he has gone a long way toward legally establishing the situation as seen by him. Be sure that you *never* make erasures in a log—if you need to correct an item, rule out the old material without making it illegible, and then write in the correct entry if there is space, or make reference to where it will be found elsewhere in the log. Initial the correction and add the date if it is made on a later day.

Why Not an Illustrated Log?

With a bit of proper planning, log-keeping can be interesting and fun—and the deck log itself will be of interest and use long after a cruise is over. Why not consider an illustrated log? Use marginal sketches if anyone has the artistic ability; or put in photographs of scenes, events, and people.

Engine Logs

An engine log is kept in terms of hours of operation, and is valuable in timing oil changes and other preventive maintenance. Record all routine maintenance and other significant events such as any troubles or repairs. If an engine log records the addition of crankcase oil, for instance, consumption can be accurately computed. Any increase in the use of oil may be a valuable warning of engine troubles. Likewise, fuel consumption can be periodically figured in terms of engine hours, and any change noted.

Hours of engine use can be estimated, or roughly figured from the deck log, but the easiest and by far the best way is by using an electric engine-hour meter. This instrument is not expensive, and the average skipper can install one in an hour or two.

Radio Log

Although a radio log is not required for voluntarily equipped vessels, it is wise to make notes on unusual

Figure 2402 **An excellent way to keep notes for logs while underway is to record information while it is fresh, using a small portable tape recorder. With this procedure, the actual log can be written up later, at the end of the day's run, with no sacrifice of accuracy. It will be much more legible than a log written at the helm while the boat is underway, particularly if the boat is pounding in rough seas.**

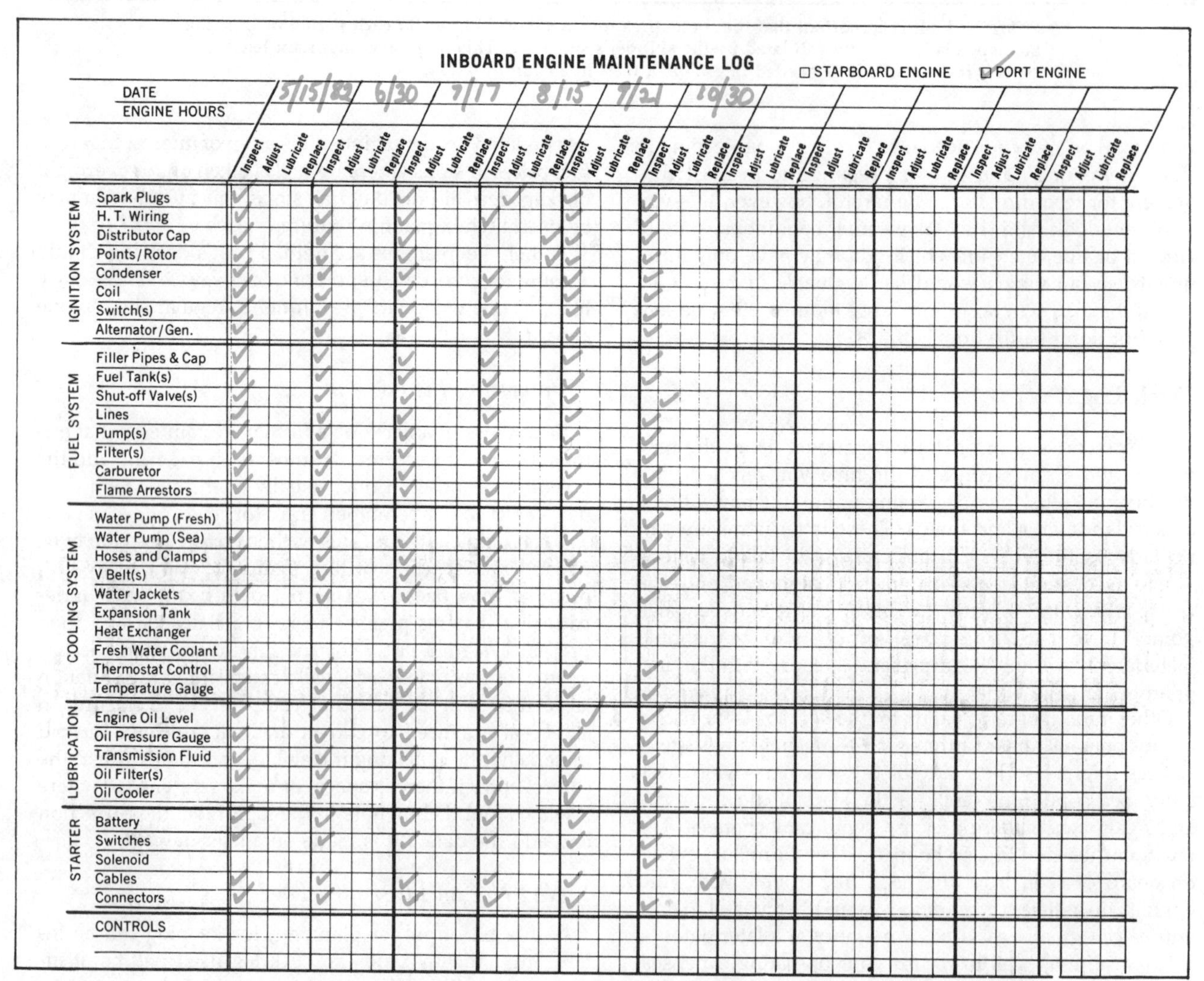

INBOARD ENGINE MAINTENANCE LOG ☐ STARBOARD ENGINE ☑ PORT ENGINE

DATE	5/15/82	6/30	7/17	8/15	9/21	10/30					
ENGINE HOURS											
	Inspect / Adjust / Lubricate / Replace	Inspect / Adjust / Lubricate / Replace	Inspect / Adjust / Lubricate / Replace	Inspect / Adjust / Lubricate / Replace	Inspect / Adjust / Lubricate / Replace	Inspect / Adjust / Lubricate / Replace	Inspect / Adjust / Lubricate / Replace	Inspect / Adjust / Lubricate / Replace	Inspect / Adjust / Lubricate / Replace	Inspect / Adjust / Lubricate / Replace	Inspect / Adjust / Lubricate / Replace

IGNITION SYSTEM: Spark Plugs; H. T. Wiring; Distributor Cap; Points/Rotor; Condenser; Coil; Switch(s); Alternator/Gen.

FUEL SYSTEM: Filler Pipes & Cap; Fuel Tank(s); Shut-off Valve(s); Lines; Pump(s); Filter(s); Carburetor; Flame Arrestors

COOLING SYSTEM: Water Pump (Fresh); Water Pump (Sea); Hoses and Clamps; V Belt(s); Water Jackets; Expansion Tank; Heat Exchanger; Fresh Water Coolant; Thermostat Control; Temperature Gauge

LUBRICATION: Engine Oil Level; Oil Pressure Gauge; Transmission Fluid; Oil Filter(s); Oil Cooler

STARTER: Battery; Switches; Solenoid; Cables; Connectors

CONTROLS

Figure 2403 **An engine maintenance log that covers all routine adjustments and inspections will ensure that all necessary tasks are carried out on a regular basis.**

events, especially those that might later be a part of a legal action. These can be recorded in the deck log, perhaps with some special indication to make them easier to find.

A compulsorily equipped vessel must keep a radio log; the required entries will be found in Part 83 of the FCC Rules and Regulations. It may be desirable to record additional items, such as contacts with the Coast Guard, Marine Operators, and other craft; there are no restrictions on entries beyond the required minimum.

Log-Keeping Techniques

A most important feature of good log-keeping is to make entries while the facts are fresh in your mind—the very best time is *immediately* after an event. On a small boat this may present problems—you may not be able to write neatly or even legibly with the boat's motion, or the cockpit of a sailboat may be far too wet for paperwork. One solution is to record information on a small portable tape recorder that can be kept in a plastic bag if there is spray or rain at the helm. Make permanent log entries later that day at the end of the run, or when there is time to spare. However, do not delay too long in making the entries in the permanent log.

One caution—watch your tape supply; don't get caught at the end of a day with no data because you ran to the end of a cassette and didn't notice! It is a good idea to every so often rewind for a few seconds and then listen to what is on the tape so as to be sure you are getting the information properly recorded.

Retention of Logs

Since maintaining a deck or engine log is voluntary on a recreational boat, saving it after the season is also at the owner's discretion. It should in any case be kept in a safe place until any possibility of litigation has passed; many skippers keep their logs permanently. A radio log, however, *must be retained* for the period specified by the current regulations.

Instruction Books

A boat owner should have instruction manuals or data sheets for all equipment installed on his boat. Such papers are valuable for repairs and routine maintenance, and normally include a "parts list" which facilitates getting replacement components.

The prudent skipper has a *complete* set of instruction books. He starts with the main engines and goes straight through the inventory to the smallest motor pump or horn. There should be no item categorized as "too small," "too insignificant," or "too dependable" for inclusion in this check-up.

Rare, indeed, is the boat that has a complete set of manuals when purchased, either new or used. Those that are not on board must be obtained—sometimes from a local dealer, or directly from a manufacturer. The boat owner can usually determine the manufacturer's address from the name plate on the equipment or from advertisements in *Motor Boating & Sailing* or other publications.

Model and Serial Numbers

When making this equipment check, make a note of both model *and* serial numbers for all items. These can be recorded in a special notebook or in the back of a log book. The latter is convenient, but you'll need to transcribe the information when one volume of a log is filled and another is started.

When writing away for an instruction book or replacement part for any piece of equipment, be sure to include *both* model and serial numbers, because there often are variations in design after a certain serial number, without there being a change in model designation.

Parts List

Parts lists must include, as a minimum, identification of the items that normally wear out within a season or so, and whose replacement is therefore a matter of routine maintenance.

It is in fact desirable to have an instruction manual or parts book so complete that it identifies *all* parts and subassemblies, to help in ordering replacement parts in the event of a breakdown.

Names and Addresses

Instruction manuals, parts lists, and the like should be annotated with the name, address, and telephone number of the supply sources and service technicians.

Company addresses, and sometimes names, do change. Be alert for such changes as you read advertisements in boating magazines, receive catalogs, attend boat shows, etc., and keep your listings up to date. Post offices retain forwarding instructions for only one year.

Sails

Sailboats should have complete information on each sail carried—manufacturer, date made, cloth type and weight, and all dimensions, including those relating to battens. It is much easier to take measurements under ideal condi-

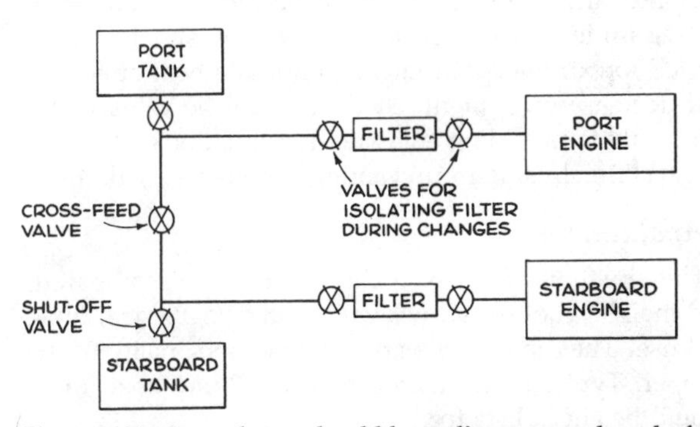

Figure 2404 **Large boats should have diagrams on board of each electrical system, and the network of tanks, valves, and lines of the fuel system. These will be of great value when troubles in these systems must be found and fixed.**

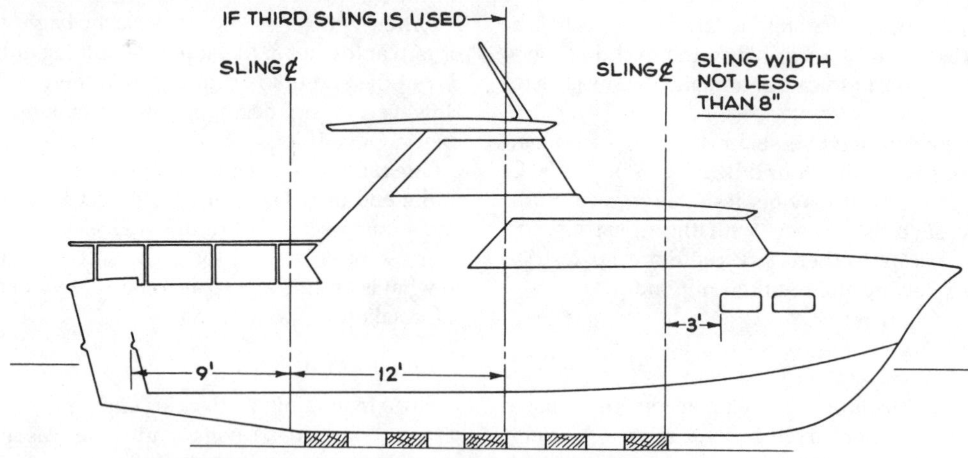

Figure 2405 **A "docking plan" is a sketch indicating where a boat's keel ends, and where shore blocks are to be placed for a marine railway or elevator. Plan also shows where lifting slings must be located.**

tions at one's convenience, and then have the information in a notebook or log, than it is to have to measure under adverse conditions, perhaps in an emergency.

Information about the Boat

Very few boats are delivered with what might be termed an instruction book on the craft itself. Such a document would include electrical wiring diagrams for each system of different voltage; a drawing of the fresh water system's pumps, lines, and valves; a diagram of the fuel lines and valves; information on the location of all seacocks and other through-hull fittings.

If not found on board the boat when it is purchased, some or all of the above documents can possibly be obtained from the manufacturer of the craft, being sure to include all available identification data including hull number. On a used boat, all such information, either coming with the boat or later obtained from the builder, should be *verified* because previous owners may have modified or added to one or more of the systems. In all probability the new owner will have to search out and record much of the information discussed above.

Docking Plan

For boats longer than roughly 30 feet (9.1 m) every skipper should have a docking plan. This drawing shows where slings should be placed if the boat is lifted, or where shoring blocks must be placed if she is to be hauled on a marine railway. See Figure 2405. Close attention to this plan will prevent excessive stress on the hull when the vessel is taken out of the water, and prevent possible damage to the rudder.

Safe Stowage of Books

It is important that all instruction books, parts lists, and similar publications be on board, but kept safe from damage or loss. It is generally most convenient to put them in a waterproof plastic envelope and then put that in a secure place. On larger craft, where water damage is less likely, a large heavy-paper case with lettered or numbered subdividers makes a good file for these documents.

Check Lists

As the size of a boat increases, its operation becomes more complex, and more details must be remembered when getting underway or docking. While they should not be carried to extremes that might complicate boat operation and decrease enjoyment, check lists can be valuable for safety and for avoiding embarrassing situations. They can also have real value in preventing expensive accidents.

Applications

The situations for which check lists can be prepared, and their level of detail, will vary with a boat's size, type, and use. They may well vary with the personality of the skipper. Typically, for a cruising Class 2 or 3 boat, there should be check lists for:

1. Getting underway
2. Approaching a pier or mooring
3. Securing after a day's operations
4. Departure from the boat for a day or more
5. Periodic routine preventive maintenance

Importance for Short-Handed Operation

Check lists are of particular importance for the person who is aboard by himself. Short-handed operation of a boat requires maximum thought and planning, because once a boat is underway many tasks become substantially more difficult—if not impossible—than they would have been before lines were cast off. A check list will minimize reliance on memory during the final minutes before casting off, and reduce the penalties for forgetfulness.

Development of Check Lists

The contents of any check list will vary so much with the individual boat and her skipper that no more than simple check-list items can be given here. Perhaps the most significant characteristic of any check list is that it is a living document—it is never completed and final; it must be under review each time it is used. A check list should be specifically reviewed and dated roughly once a year, even if no changes are made.

Getting Underway

Before getting underway from a marina berth, many disconnections must be made—electric cords, water hose, possibly a telephone connection, and, of course, the lines that have secured the boat. Forget just one of these and you're in trouble!

Even the most elementary "housekeeping" chores are candidates for a pre-departure check list. Trash and garbage should be put in proper containers ashore rather than taken along for a day's cruise. Water tanks should be topped off, if appropriate, and certainly fuel tanks should be checked.

Not too silly an item for family groups is a final head-count of children and pets. Even if someone is missed before the boat gets far away, it can mean another docking and undocking, a needless waste of time.

Such tasks as setting switches and valves properly, raising or lowering a radio antenna, or even setting the boat's colors are all minor, but they are best done before getting underway. A check list helps prevent inadvertently omitting a task because each person thought someone else had done it.

Coming into Port

A check list also helps when approaching a pier or wharf. You will make a much better landing, with less chance for embarrassment, if the proper docking lines are laid out and fenders are tied in place ready to use. Once the boat is made fast to shore, there will be switches to be turned and valves to be opened or closed. Log entries must be made while the information is still fresh. Supplies of fuel and water should be checked so that refills can be arranged if needed.

Departure From the Boat

The check list for leaving a boat unattended is very important because items overlooked often will not be remembered until you are far from the boat and corrective actions are impractical or impossible. Primary choices for this list are items relating to the safety and security of the unattended craft—turning off fuel valves, the proper settings for electrical switches, pumping out the bilge and leaving the switch on automatic (or arranging for periodic pumping out). Other departure check list items are securing ports, windows, hatches, and doors.

Routine Maintenance

Routine maintenance check lists should include items based on how much the boat is used (usually in terms of engine hours) and on calendar dates (weekly, monthly, or seasonal checks). Typical of the former are oil level checks and changes, and oil and fuel filter changes.

On a calendar basis the lists should note such matters as electrolyte levels in storage-batteries, pressure gauges on dry-chemical fire extinguishers, and all navigation lights. Check the operation of automatic bilge alarms or pump switches by running water into the boat. Periodically close and open seacocks several times to ensure their free and easy operation in case they are needed in an emergency. Equipment and supplies carried on board for emergencies should be inspected for any signs of deterioration.

Check Lists Must Be Used

The most important aspect of check lists is that they must be *used*, and used *regularly*. A list prepared and filed away is of little value, and it may in fact create a false sense of security. For safety and to avoid embarrassment, the boatman should train himself to reach for the list whenever he is ready to start an operation for which one has been established. He should be like the aircraft pilot who goes through such a check list every time, no matter how many years of experience he has or how many times he has taken off and landed that day.

The value of a check list lies in its use, not in its mere existence!

RF
G
ARAB

25

Courtesy afloat; cruising with guests aboard; proper display of ensigns and other flags on boats and ashore

BOATING CUSTOMS

Boating is a relaxed, informal kind of recreation, no longer bound up in the ceremonial routines that once were considered essential elements of proper "yachting etiquette." However, there are many occasions when proper observances and behavior are expected, even if it is only a matter of common courtesy to your boating neighbors.

You don't need to belong to a yacht club or similar organization, but it helps to have an understanding of club practices which you might encounter, as well as normal good manners afloat. You should be able to recognize a race or regatta in progress, and take whatever steps are necessary to avoid interfering with it. You should be able to drop your hook in an anchorage, tie up at a marina, raft up with other boats, or cruise in company with others without causing friction. You should know and follow the proper procedures for displaying flags on your boat.

General Boating Customs

The customs and traditions that remain a part of pleasure boating are practices that evolved from the "safest" or "most appropriate" means of doing a thing in given circumstances. In some cases they had their origin in naval practices, particularly in elements relating to discipline and authority, but in all cases they represent good sense and good manners afloat.

Cruising

Except when cruising the high seas, or some remote inland waters, you will generally be within sight of other boats; and often you will be close enough that any maneuvers you make, even activities on board your boat, will have an impact on others.

When Underway

A faster boat overtaking and passing a slower one in a narrow channel should slow down *sufficiently* to cause no damage or discomfort. Often overlooked is the fact that it may be necessary for the *slower* boat itself to reduce speed. For example, if the slower boat is making 8 knots, the faster boat can slow only to about 10 knots in order to have enough speed differential left to get past. But at that speed the passing boat may make a wake that is uncomfortable to the other craft. In such cases, the overtaken boat should slow to about 4 knots to allow herself to be passed at 6 or 7 knots with little wake.

If adequate depth of water extends outward on one or both sides of the course, it is the courteous thing for the passing boat to swing well out to the safe side to minimize the discomfort to the overtaken boat. A power boat, of course, should pass a sailboat well to leeward, or astern of it in a crossing situation. These actions are beyond those required by the Rules of the Road, which apply only when a danger of collision exists. These are a matter of good manners, of causing the least interference to the vessel with the most limited maneuverability.

Regattas and Races

A fleet of small sailboats all headed in the same general direction is a sure indication of a race in progress. While it might be possible to pass through such a fleet "legally," observing the proper Rules of the Road, this action might interfere with the courses and tactics of the racing skippers. Stay well clear so that neither your wake nor your "wind shadow" will have any effect on the race boats.

Cruising-type sailboats engaged in a long-distance race may not be easy to identify as competitors in an event, but whether such a boat is racing or just cruising, it is best to take no action which would interfere with her progress.

Marine parades, powerboat races, and other marine events may be held, with a regatta permit from the Coast Guard, in waters otherwise subject to normal marine traffic. The Coast Guard, Coast Guard Auxiliary, or the organization sponsoring the event may provide patrol boats to keep non-participating craft clear of the event. It makes good sense to obey instructions given from any such patrol boat; it could be dangerous to blunder out into the path of speeding race boats, as well as discourteous to interfere with the performance of any type of regatta.

Figure 2501 **Races and regattas may be patroled by private craft, *above,* as well as by Coast Guard, Coast Guard Auxiliary, or local law enforcement vessels. Follow instructions given from any such patrol boat, and stay clear of the action. *Below:* At an anchorage or marina, turn down radios or similar equipment and keep other noise levels low. Many boating crews turn in early after a day's cruise.**

Cruising in Company

When in a group cruise, maintain speed and keep a constant distance from the other boats, according to a prearranged plan. Agree in advance about places to stop for fuel, supplies, sightseeing, etc. Do not use VHF-FM marine radio for social communication between boats; it is for safety and operational traffic only. CB radio is suitable for social use.

Guests Aboard

If you plan to invite guests along on a cruise for a day, a weekend, or a more extended period, let them know in advance what is expected of them, particularly if they are without previous cruising experience.

Tell them to bring a minimum of clothes, with due regard to the season, in collapsible containers. Duffle bags are available in so many sizes and price ranges there's no need for anyone to bring a standard suitcase onto a boat, where there's no place to stow it. Let your guests know what clothing is best suited for the cruising, including swim suits, and even shore wear if you will be stopping at yacht clubs or sightseeing.

Assign each guest a locker, which you have cleared, for the clothing brought aboard; let guests know they are not to leave clothes or gear scattered about as it may interfere with the operation of your boat.

When you give a sailing time, explain that this is based on tide, currents, normal weather patterns, the length of the planned trip, or whatever might be the case, and that

guests should be on board and ready to leave *before* this sailing time.

Explain also that rising and bedtimes are a matter of convenience to everyone aboard, due to the limited washing and toilet facilities. As skipper, you should be the first to rise in the morning, so the others, hearing you up and about, will know it's time to get up. In the evening, let it be known that when you announce it's time to retire, everyone should do so.

Let guests know why they must stop smoking when fuel is taken on, and why care must be taken even when smoking is permissible. A carelessly flipped cigarette ash or butt can start a fire in a berth, awning, or compartment. Let cigar- or pipe-smoking guests know they can enjoy their smokes in the open air (when smoking is permissible), but that in confined spaces of cabins these may be offensive to others; cigars, especially, tend to leave a particularly unpleasant after-odor.

Point out that small particles like pipe tobacco and ashes, peanut shells, bits of potato chips, crumbs have a way of getting into cracks, and corners, and there defy the ordinary cleaning facilities found on a boat.

Note that you cannot arrange to have your guests pay for part of the fuel or provisions for a cruise; this is viewed by the Coast Guard as a partial charter and commercial operation (see page 50).

Anchorages, Moorings, Marinas

Be a good boating neighbor when you drop the hook in an anchorage, lie at a mooring, or tie up at a yacht club or marina pier. Don't anchor so close to other boats that rodes foul as boats change position in a wind shift. Consider the state of the tide and the effect of its range on your swinging radius. Use a guest mooring only with permission; it may be reserved for another boat, or it may be unsuitable for your vessel. Tie up only briefly to fuel piers.

In the evening hours, take care not to disturb people on other boats in the area. Sound travels exceptionally well across water, and many cruising boat crews turn in early for dawn departures. Keep voices down and play radios or stereo equipment only at low levels. If you should be one of the early departees, leave with an absolute minimum of noise.

Rafting

When rafting, the boat that will be last to leave is the one to drop the hook (provided ground tackle is adequate), take the mooring, or lie alongside the pier or wharf. The others should lash up so that the first to leave will be farthest outboard, and then the others in order of departure.

Avoid crossing from one boat to another unnecessarily, and unless you are on very informal terms with your neighbors, ask permission before you do so. Be sure to supply your share of fenders, fender boards, and lines—as well as ingredients for parties and meals that will be part of the rafting scene.

Don't throw trash and garbage overboard, even if it is biodegradable; it is unsightly in the meantime. Secure flapping halyards; they can be a most annoying source of noise for some distance (and can chafe the surface of the mast). When coming into or leaving an anchorage or boat basin of any type, do so at dead slow speed to keep your wake at an absolute minimum. Again, this is done to disturb other boats as little as possible. Also, you are responsible for any damage caused by your wake.

Exercise pets ashore in authorized areas, or well away from normal traffic areas in any case. Don't go off and leave clothes in an untended laundry machine.

Clubs and Organizations

If you are a member of a recognized yacht club, it is customary for other yacht clubs to extend courtesies such as use of guest slips and moorings and clubhouse facilities when you visit. Again, it is proper procedure to ask permission to use such facilities, even if you just wish to tie up briefly at the club dock to pick up a member who is joining you.

When visiting a yacht club of which you are not a member, observe the actions and routines of the local owner-members, and particularly the club officers. This is especially important in respect to evening "colors" (flags; see below); not all clubs strictly calculate the daily time of sunset, and some sunset signals for colors may be earlier than you would normally expect.

Ceremonies and Routines

Yacht clubs and organizations such as the United States Power Squadrons generally have by-laws or handbooks that spell out requirements for ceremonial procedures, such as the exchange of salutes, daily color ceremonies, salutes between vessels, precedence in boarding or leaving launches, and flag courtesies.

In many cases there's an 0800 gun sounded as a signal to raise colors and a sunset gun for lowering colors (the procedures are detailed later in this chapter). If you expect to be away from your boat at the time of sunset—in the club for dinner, for example—take in your flags before you leave the boat.

If hired personnel are not on hand when a boat is coming in to tie up near your position, it is good manners to offer to help with the docking lines. Always ask permission before boarding another boat. A holdover from naval practice is to bring a dinghy or tender alongside the starboard side of a boat at anchor or on a mooring.

Flag Display Afloat and Ashore

There is no legislation governing the flying of any flag on numbered, undocumented, or unlicensed vessels. Documented yachts are expected to fly the yacht ensign (see below), but this regulation is not enforced.

However, through the years, customs have been established for the types of flags that may be flown, and when and where they are to be displayed. In recent years new procedures have evolved, such as flying the National En-

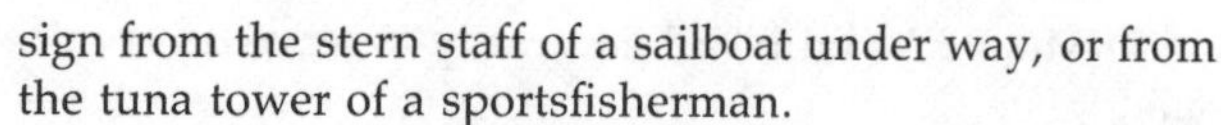
sign from the stern staff of a sailboat under way, or from the tuna tower of a sportsfisherman.

Where the sections below specify "at the starboard spreader" it means the spreader that is on the most forward mast if there is more than one, and by the most outboard hoist on that spreader.

Where the sections below specify a vessel "at anchor" this also applies to any other non-underway status—at a mooring or made fast to the shore.

The term *colors* actually applies only to the flag at the stern of a vessel to denote her nationality. In practice, however, it has come to be used for all flags flown, and it will be used in that context in this chapter.

It is normally the proper practice *not* to fly more than one flag on a single hoist. There are limited exceptions, however, for officer's flags on certain types of craft.

Here are the flags, pennants, and burgees that may be flown from U.S.-owned recreational boats, and the proper methods for their display. See Appendix F, page 610.

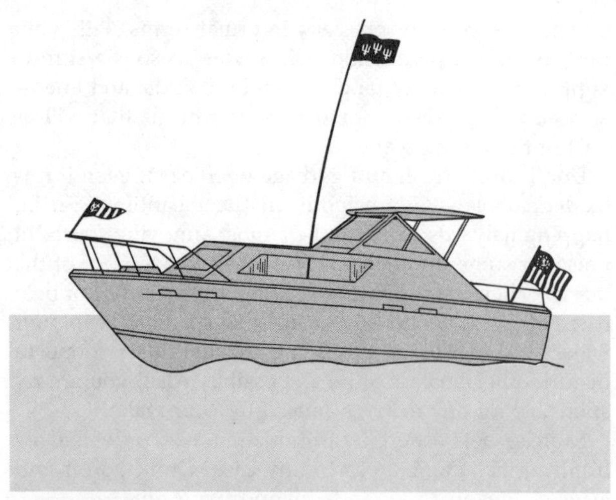

Figure 2502 **On a typical small cruiser with radio antenna, a USPS officer's flag or past officer's signal, or a USCG Auxiliary ensign may be flown from the antenna at the same height as if on a signal mast. Flags at bow and stern are the same as above.**

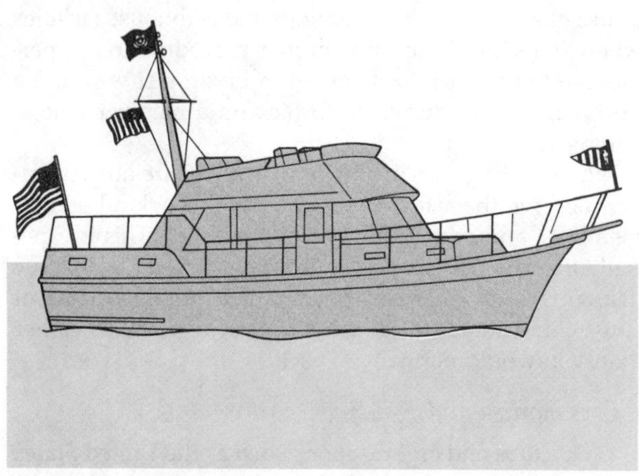

Figure 2503 **On a cruiser with signal mast with spreaders, the U.S., USPS, or Yacht Ensign is flown from the stern staff, and the Squadron pennant or club burgee is at the bow staff. The officer's flag, private signal, or USCG Auxiliary Ensign is at the masthead. An owner-absent or guest flag, or the USPS Ensign, is displayed at the starboard spreader.**

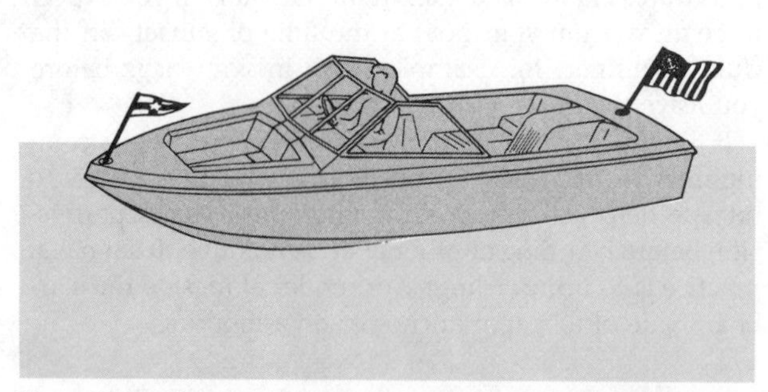

Figure 2504 **On a mastless motorboat the U.S. Ensign (or if not in foreign waters, the USPS or Yacht Ensign) is flown from the bow staff.**

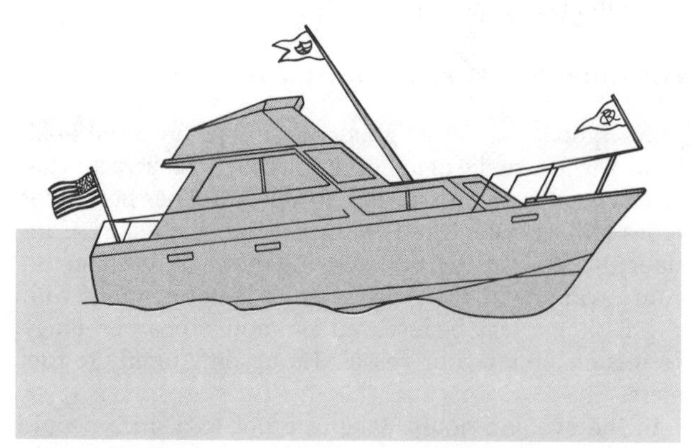

Figure 2505 **On a small cruiser with a signal mast, flags are displayed in the same manner as for Figure 2502.**

Figure 2506 **When a gaff is added to the signal mast, it is the place for the U.S., or Yacht Ensign while underway. At anchor, these are flown at the stern.**

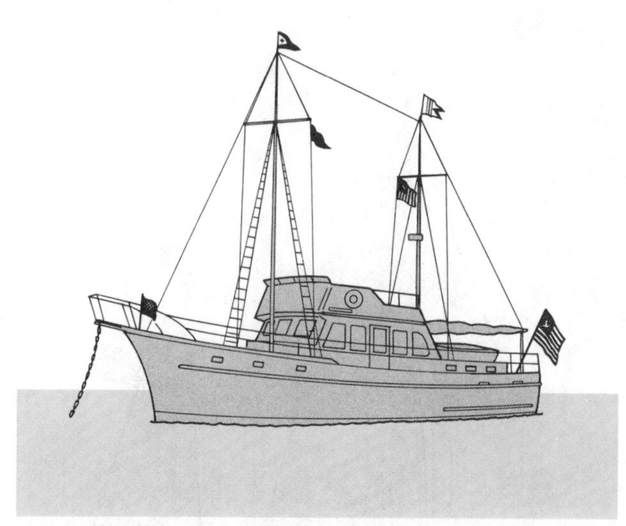

Color Illustrations

Flags and pennants commonly flown on boats, the International Code Flags, and the flags, pennants, and uniform insignia of the USPS and USCG Auxiliary are shown on pages 581-583, 586-588.

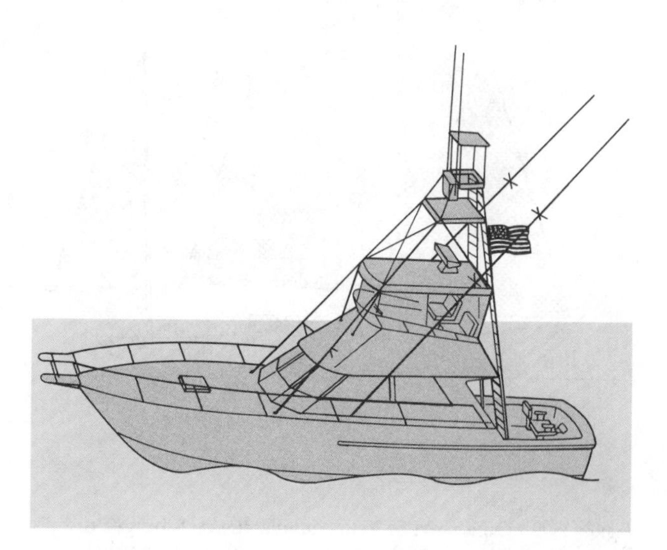

Figure 2508 **A documented two-masted motorboat with spreaders on each mast, *above,* at anchor: The U.S., or Yacht Ensign is at the stern staff; club burgee, Squadron pennant, or USCG Auxiliary pennant at the forward masthead, and private signal or officer's flag at the forward starboard. The USPS Ensign, if not at the stern, is at the after starboard spreader, and crew's meal pennant (seldom seen) is at the port forward spreader. The Union Jack is at the bow (jack) staff; this is displayed only when the vessel is anchored. On sportfishermen, *right,* the U.S., USPS, or Yacht Ensign is flown abaft the tuna tower to keep the cockpit coaming clear at the stern for fishing lines.**

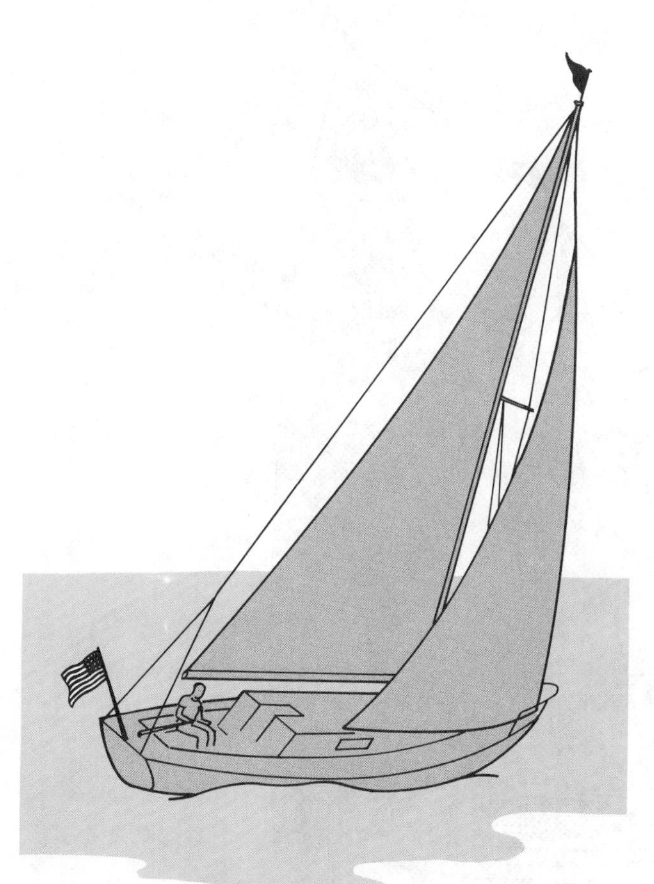

Figure 2507 **A single-masted sailboat underway may fly the U.S., USPS, or Yacht Ensign at a stern staff, *left,* or at the leech of the mainsail approximately ⅔ the length of the leech above the clew, *right.* The club burgee, Squadron pennant, private signal, or officer's flag is at the masthead.**

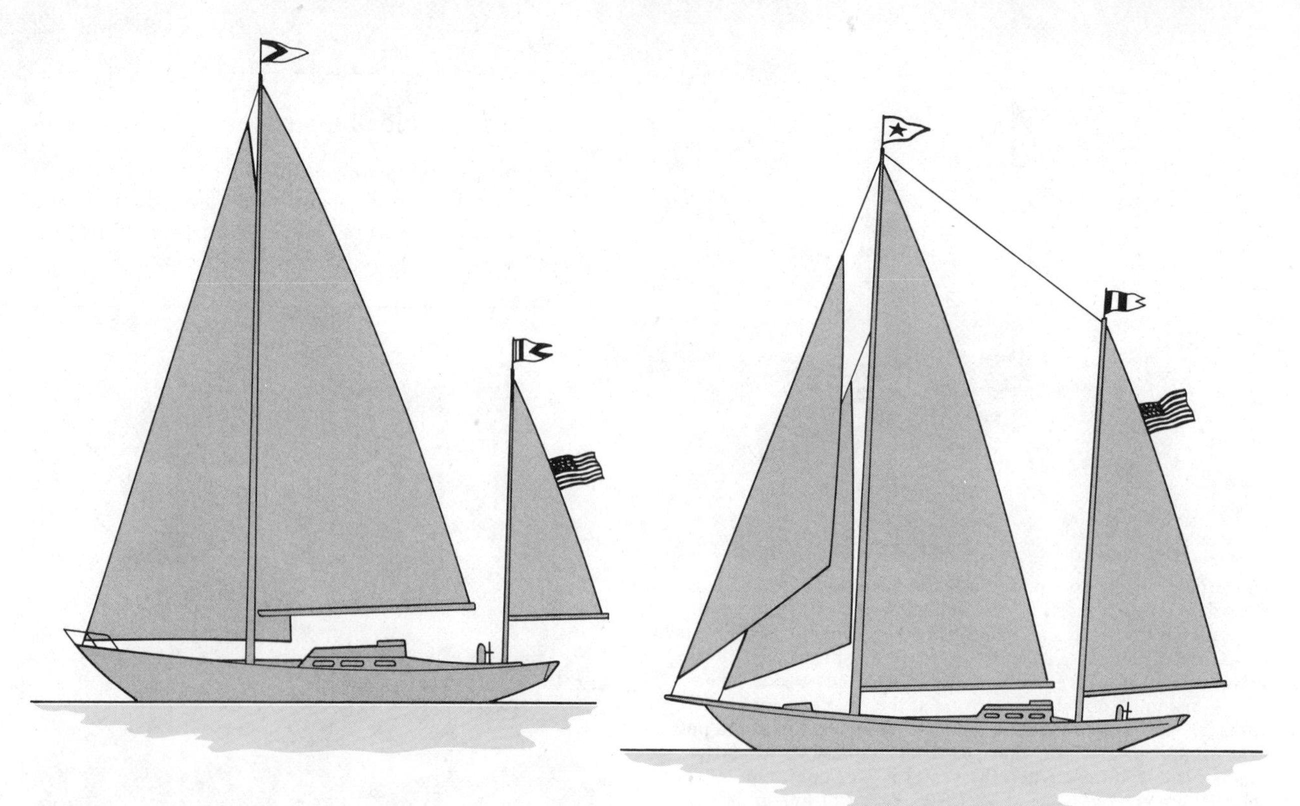

Figure 2509 **On a yawl or ketch, *above,* the club burgee or Squadron pennant is at the main masthead, and the private signal or officer's flag is at the mizzen masthead. At anchor or in port, the U.S. or Yacht Ensign is flown from the stern staff; and the USPS Ensign at the starboard main spreader, if the boat is operated by a member. On a schooner underway, *below,* flag arrangement is the same as for a yawl or ketch, even though the relative height of the masts is reversed. The U.S. or Yacht Ensign may be flown from a stern staff while underway (not racing) if the length of the aftermost boom permits.**

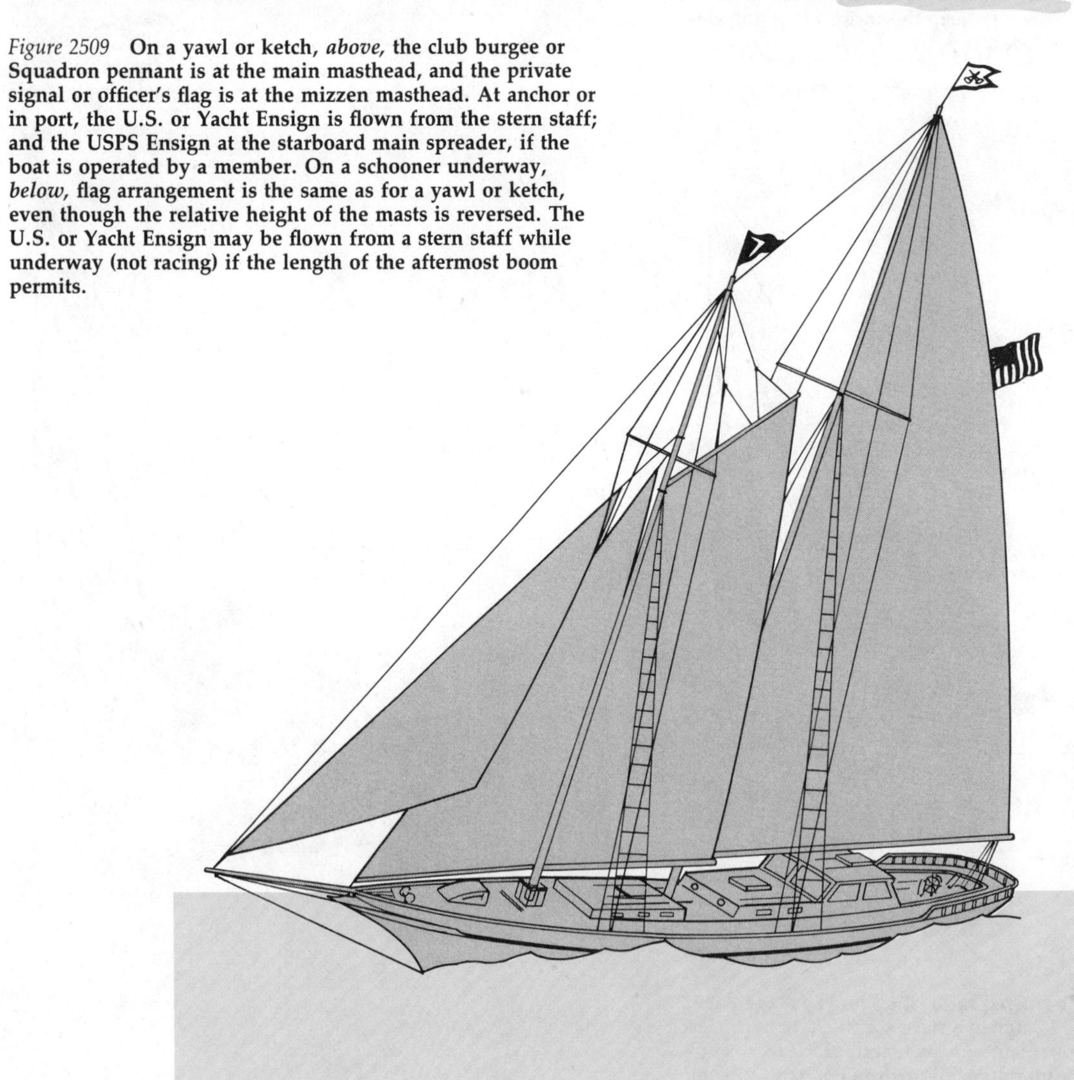

United States Ensign

The U.S. Ensign is proper for all U.S. yachts, without reservation. This is "Old Glory," with 50 stars and 13 stripes. All boats, when at anchor, fly it from the stern staff. It is flown from the stern staff of powerboats underway on inland waters. On the high seas it need be flown only when meeting or passing other vessels. If the powerboat has a mast and gaff, the proper display is at the gaff. On a sportsfisherman, where a stern staff would be in the way of the action, the practice is to fly the Ensign from a halyard rigged just behind the tuna tower.

On Marconi-rigged sailboats under sail alone, the practice for many years had been to fly the ensign from the leech of the aftermost sail, approximately ⅔ the length of the leech above the clew. This puts it in about the same position it would occupy if the boat were gaff-rigged, and on gaff-rigged sailboats it is still proper to fly the Ensign from the peak of the aftermost gaff.

The advent of the modern high-aspect-ratio rig, with the boom end well inboard of the stern, has made it possible to fly the Ensign from the stern staff of a sailboat underway, and this is now accepted practice. However, the Ensign should never be displayed while the boat is racing. Under power alone, or at anchor or made fast, the Ensign should be flown from the stern staff of all sailboats. If an overhanging boom requires that the staff be off center, it should be on the starboard side.

United States Yacht Ensign

This is the 13-star "Betsy Ross" flag, with a fouled anchor in the union. Originally restricted to documented vessels of a specific classification, it is now flown on recreational boats of all types and sizes instead of the National Flag. Many yacht clubs now provide in their bylaws that the Yacht Ensign be flown regardless of boat size or documentation status. Whenever a boat is taken into international or foreign waters, however, the 50-star U.S. Ensign is the proper flag to display.

United States Power Squadron Ensign

The USPS Ensign is flown as a signal to others that the boat is commanded by a member of the USPS in good standing. The USPS is a national fraternity of boatmen dedicated to better and safer boating through education and civic service (see Chapter 27).

The preferred location for flying the USPS Ensign is the starboard yardarm or spreader, underway or at anchor, or made fast to shore, on motor and sailing craft. It may be flown from the stern staff in place of the U.S. or Yacht Ensign, but this is usually done only on smaller boats that lack a mast. On sailboats underway, it may be flown from the aftermost peak or leech in place of other ensigns.

The USPS Ensign may be flown at its proper location on boats displaying the USCG Auxiliary Ensign to indicate the owner is a member of both organizations; however, it cannot be flown if the craft is under Coast Guard orders.

United States Coast Guard Auxiliary Ensign

Known as the "Blue Ensign," the USCG Aux flag is flown on a boat that has been approved as a "Facility" by the organization for the current year. It is flown both day and night.

On a vessel without a mast, the Blue Ensign is flown at the bow staff; if there is one mast, it is flown at the masthead. On a vessel with two or more masts, the USCG Aux Ensign is displayed at the main masthead. It is *never* flown in place of the National Ensign.

When the Auxiliary Ensign is displayed, it is improper to hoist a guest, owner absent, meal, cocktail, or novelty flag.

Coast Guard Auxiliary Operational Ensign

The boat of a member of the USCG Auxiliary that meets a particularly high standard of equipment and availability is called an "Operational Facility," and can be called on for use under Coast Guard orders in assistance and patrol missions. When operating under USCG orders, these boats fly the Coast Guard Auxiliary Operational Ensign—white with the Coast Guard's "racing stripes" of red and blue—in place of the "Blue Ensign."

Yacht Club Burgee

Generally triangular in shape although sometimes swallow-tailed, the yacht club burgee may be flown by day only, or day and night, as set by the rules of the yacht club concerned. It is flown from the bow staff of mastless and single-masted motorboats, at the foremost masthead of vessels with two or more masts, and the main masthead of ketches and yawls. The burgee may be flown while underway (but not racing) and at anchor. It is permissible to substitute the owner's private signal for the burgee on single-masted yachts without bow staff, when underway.

Squadron Pennant

A distinguishing Squadron pennant which has been authorized by the USPS may be flown in lieu of a club burgee and from the same positions. This pennant may be flown by day only, or both day and night.

Owner's Private Signal

This is generally swallow-tailed in shape, but it may be rectangular or pennant-shaped. It is flown from the masthead of a single-masted motorboat or sailboat, or from the aftermost mast of motor or sailing vessels with two or more masts. It may be flown by day only, or day and night.

Mastless motor boats may fly this signal from the bow staff in place of a club burgee.

Officer Flags

Flags designating yacht club or USPS officers are rectangular in shape, blue (with white design) for senior officers; red for next lower in rank; and white (with blue design) for lower ranks. Other officers' flags (except fleet captain and fleet surgeon) may be swallow-tailed or triangular in shape, as provided in the regulations of those organizations making provisions for such flags.

An officer flag is flown in place of the owner's private signal on all rigs of motor and sailing vessels except single-masted sailboats, when it is flown in place of the club burgee at the masthead. On smaller motorboats without a signal mast, a USPS officer flag may be flown from a radio antenna either singly or beneath the USPS Ensign.

USCG Auxiliary Officer Flags

The flag of a USCG Aux officer flies day and night when the officer is on board. On a vessel without a mast, it is flown at the bow staff in place of the Auxiliary Ensign; on a vessel with a mast, it is flown at the starboard spreader. Past officers' burgees are displayed in the same manner.

Only one officer's pennant or burgee may be flown at one time. An incumbent officer's pennant takes precedence. When the Auxiliary Ensign is displayed, it is improper to hoist a guest, owner absent, meal, cocktail, or novelty flag.

Miscellaneous Flags

There are a number of flags that once were used on large yachts with professional crews; while it is still permissible to display such flags under the appropriate conditions, such flags are seldom seen. These include the Union Jack, owner absent flag, guest flag, and owner's and crew meal flags.

Other Flags

Diver Down There are two flags flown in connection with diving operations, a red flag with a single diagonal stripe of white, and a rigid replica of the International Code flag "A." The use of these on boats is discussed on page 86; it is not proper for them to be flown on shore.

Race Committee or Regatta Committee Flag This is a blue rectangular flag with a single vertical fouled anchor in white with the letters "R" and "C" alongside the anchor in white or red. This flag is frequently larger than normal and may be flown at any position for greatest visibility. Often other flags, which might cause confusion, are taken down.

Quarantine Flag International Code letter "Q"—is flown when entering a foreign port (except Canada and a few others), or returning to a U.S. port from a foreign cruise. It signals that the vessel is "healthy" and requests clearance into the port; it is taken down after customs and immigration formalities have been completed.

Transportation or Tender Flag Used in many harbors where boats lie at moorings, and yacht clubs or commercial operators provide launch (water taxi) service to and from the shore. This is the International Code flag "T"; it is normally used with a sound signal.

Protest Flag International Code "B" will often be seen at sailing or other contests. It signals that the vessel flying it will file a protest on another vessel (or vessels) at the conclusion of the event.

Man Overboard The generally recognized signal is the International Code flag for the letter "O"; this flag is often fixed to a staff which in turn is attached to a life ring.

Distress Although it is not official, the flying of an inverted U.S. Ensign will normally be recognized as a signal of distress. (Note that the flags of many countries have no readily recognizable "top" and "bottom" and thus cannot be inverted as a distress signal.)

Size of Flags

Although flags come in a fixed, standardized series of sizes, there are guidelines which will help in selecting the proper size for your boat.

Flags are more often too small than too large; use the rules below and round upward to the nearest *larger* standard size.

The flag at the stern of your boat—U.S. Ensign, Yacht Ensign, or USPS Ensign—should be one inch on the fly for each foot of overall length. The hoist will normally be two-thirds of the fly, but some flags such as the Canadian Ensign and the USCG Aux Ensign have different proportions.

Other flags such as club burgees, officers' flags, and private signals for use on sailboats should be approximately ½ inch on the fly for each foot of the highest mast above the water. For flying on powerboats, these flags should be roughly ⅝ inch on the fly for each foot of overall length. The shape and proportions of pennants and burgees will be as prescribed by the organization to which they relate.

Raising and Lowering Flags

"Colors are made" each morning at 0800; as mentioned, at yacht club and similar organization docks or anchorages, this may be signalled by a morning gun. The National Ensign or Yacht Ensign is hoisted at the stern (or set in place on its staff). This is followed by the USPS Ensign at the starboard spreader (if not already flying on a day-and-night basis) provided the skipper is a member of USPS. Then comes the club burgee or Squadron pennant at the bow, and the private signal at the masthead. (An officer's flag, if flown in place of a private signal, would be flown continuously.)

If the boat bears a valid USCG Aux Facility decal, it would be flying the Auxiliary Ensign at the masthead, day and night. A USCG Aux officer's pennant or burgee may be flown, day and night, at the starboard spreader.

At sunset, colors not properly flown on a day-and-night basis should be lowered in reverse sequence, the Ensign at the stern always being the last to be secured. A cannon report also may be used as a sunset signal.

Dressing Ship

On national holidays, at regattas, and on other special occasions, yachts often "dress ship" with International Code signal flags. Officer's flags, club burgees, and national flags are not used. The ship is dressed at 0800, and remains so dressed until evening colors (while at anchor only, except for a vessel's maiden and final voyages, and participation in a marine parade or other unique situation). See Figure 2510.

In dressing ship, the Yacht Ensign is hoisted at the stern staff, and the Union Jack may be displayed at the jack (bow) staff. A rainbow of flags of the International Code is arranged, reaching from the waterline forward to the waterline aft, by way of the bowsprit end (or stem if there's no bowsprit) and the masthead(s). Flags and pennants are bent on alternately, rather than in any indiscriminate manner. Since there are twice as many letter flags as nu-

Figure 2510 **To dress ship, flags of the International Code are arrangement in the sequence given in the text. The Ensign is at the stern staff, Union Jack at the bow (jack) staff, club burgee or Squadron pennant at the forward masthead, and officer's flag or private signal at the after masthead. On this schooner, a crew's meal pennant is shown at the forward port spreader, and a guest flag at the starboard main spreader.**

meral pennants, it is good practice, as in the Navy, to follow a sequence of two flags, one pennant, two flags, one pennant, throughout. The sequence recommended here provides a harmonious color pattern throughout:

Starting from forward: AB2, UJ1, KE3, GH6, IV5, FL4, DM7, PO Third Repeater, RN First Repeater, ST Zero, CX9, WQ8, ZY Second Repeater.

Honoring Other National Flags

As a matter of courtesy, it is proper to fly the flag of a foreign nation on your boat when you enter and operate on its waters. There are only a limited number of positions from which flags may be displayed, and consequently when a flag of another nation is flown, it usually must displace one of the flags commonly displayed in home waters.

Customs observed in various foreign waters differ from each other; in case of doubt, inquire locally or observe other craft from your country.

As noted previously, it is preferable for U.S. vessels while in international or foreign waters to fly the U.S. Ensign (50-star flag) at the stern or gaff or leech, rather than the USPS Ensign or the Yacht Ensign. When the starboard spreader is used for the "courtesy ensign" of the foreign country, the USPS Ensign or similar flag may be flown from the port spreader; if the vessel has multiple flag halyards on the starboard spreader, the USPS Ensign is flown there, inboard from the courtesy ensign. Figures 2511 and 2512 show proper placement of the foreign flag on vessels of common types.

The U.S. Ensign, club burgee, officer's flag, and private signal are flown as in home waters.

Don't fly a foreign courtesy ensign after you have returned to U.S. waters. It may show that you've "been there," but it is *not* proper flag etiquette.

Display of State Flags

Any citizen of any state may fly the flag of that state unless doing so is specifically prohibited by law. On a vessel with one or more masts, the state flag is flown at the main masthead in place of the private signal, officer's flag, or USCG Aux Ensign (and in this case an Auxiliary officer cannot fly his officer's pennant from the starboard spreader). When the state flag displaces a yacht club or USPS officer's flag, the officer's flag cannot be flown from any other hoist. On a mastless boat, the state flag can be flown at the bow staff in lieu of a club burgee, or it may replace any flag flown from a suitable radio antenna.

The flying of a state flag at the stern of a boat is *not* proper; nor is it proper to fly from this place of honor any Confederate, "pirate," or other "gag" flag and certainly not the flag of a foreign country of one's heritage.

Flag Displays Ashore

The flagpole or mast of a yacht club, or a similar organization, is considered to represent the mast of a vessel,

Figure 2511 **On a mastless powerboat, *left,* the courtesy flag of another nation replaces any flag normally flown at the bow. When a motorboat has a mast with spreaders, *right,* the courtesy flag is flown at the starboard spreader.**

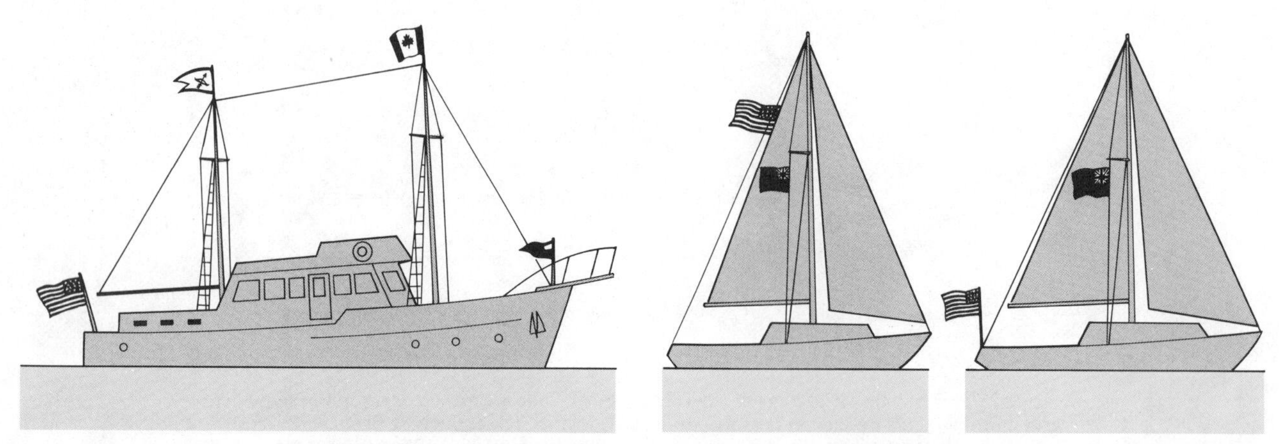

Figure 2512 **On a two-masted motorboat, *left,* the courtesy flag displaces any flag normally flown at the forward masthead. On a sailboat, *right,* the courtesy flag is flown at the starboard spreader, whether the U.S. Ensign is at the stern staff, or flown from the leech. If there is more than one mast, the courtesy flag is flown from the starboard spreader of the forward mast.**

and the peak of the gaff, if one is used, is the place of honor from which the U.S. Ensign is flown, just as it would be on a gaff-rigged boat.

There has been some confusion because proper flag etiquette requires no other flag to be flown *above* the U.S. Ensign, and obviously another flag, such as a yacht club burgee at the masthead, will be *higher than* the U.S. flag when the latter is at the gaff. This is entirely proper because "above," in flag etiquette, means "directly on top of," as would be the case if the burgee were above the U.S. Ensign on the same hoist.

Figure 2513 shows the standard mast and pole displays as if you were ashore, facing seaward.

Note that signal flags, such as storm advisory flags, should be flown from a conspicuous hoist; that on national holidays and on days of special yachting significance it is permissible to fly the flags of the International Code; and that flags flown ashore at private homes may follow the code used for yacht clubs and similar organizations.

Half-Masting Flags

A flag is flown at half-mast (or half-staff) in respect for a deceased person or persons. Although there are no laws governing the half-masting of flags on private vessels, or at private homes and clubs, most citizens follow the flag display customs that are used on U.S. Government buildings and ships.

The U.S. Ensign is flown at half-mast to reflect *national* mourning as declared by a Presidential proclamation. The duration varies from a day or so up to 30 days, determined by the deceased person's position. On Memorial Day, the U.S. flag is flown at half-mast until 1220, the time of the final gun of the traditional 21-gun salute commencing at noon.

On a simple flagstaff—as at the stern of a vessel or a flagpole ashore—the "half-mast" position is approximately three-fourths the way up to the top. If the flagpole has a yardarm, or yardarm and gaff, the half-mast position is that which is level with the yardarm.

When the U.S. flag is displayed at half-mast on a vessel, other flags remain at their normal position. When it is half-masted ashore, fly only a private signal or club burgee at the masthead of a gaff-rigged mast with it.

When the U.S. Ensign is to be flown at half-mast, it should be hoisted fully and smartly, then lowered ceremoniously to the half-mast position. Before lowering, it is again raised to full height and lowered from there.

Some yacht clubs follow the practice of flying their burgee at half-mast for a period of mourning on the death of a club member. A private signal may be flown at half-mast on the death of the owner of that vessel.

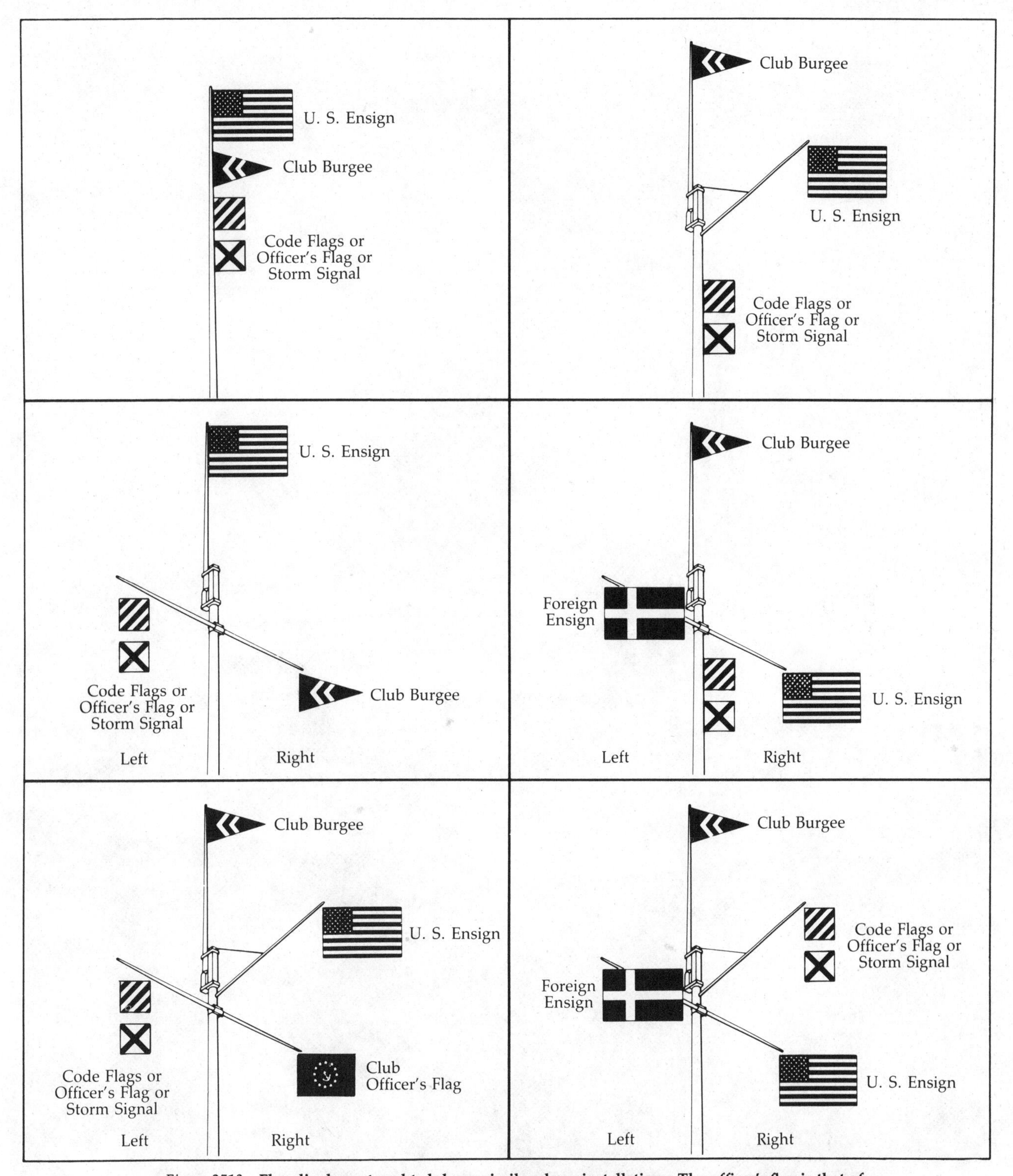

Figure 2513 **Flag displays at yacht clubs or similar shore installations: The officer's flag is that of the senior officer present on the grounds, or aboard his boat in the club anchorage. Storm signals may be substituted for the International Code Flags, which are displayed on national holidays or other days of special yachting significance.**

COAST GUARD

26

A government agency and its volunteer civilian auxiliary that work together to promote safety afloat

THE UNITED STATES COAST GUARD AND COAST GUARD AUXILIARY

The never-ending work of the United States Coast Guard (USCG) touches upon the activities of the vast majority of U.S. boatmen. Even far from salt water, far up the major rivers of this nation, the work of Coast Guardsmen assists and protects the boating public. In many of its functions, the Coast Guard is assisted by its civilian supporting organization, the Coast Guard Auxiliary (USCG Aux). (Note the correct abbreviation; it is *not* "USCGA," which stands for the Coast Guard Academy.)

The principal activities of the Coast Guard relating to boating are in the areas of law enforcement and safety. These take many forms as will be discussed in this chapter. Coast Guard Auxiliarists do not have law-enforcement powers, but they do engage in a wide variety of activities, afloat and ashore, that further the cause of safety on the water.

The United States Coast Guard

The U.S. Coast Guard is a military service and a branch of the Armed Forces. It is *not,* however, normally a part of the Department of Defense. In peacetime, the USCG functions as an agency of the Department of Transportation. In time of war or national emergency, however, the Coast Guard, or units thereof, may be transferred to the Navy.

The Coast Guard Ensign

The flag of the Coast Guard, its distinctive Ensign, is closely connected with its history. The red and white *vertical* stripes number *16,* the number of states in the Union when the flag was first authorized in 1799.

The Coast Guard Ensign is illustrated in the color section. It has the unusual feature that it is flown *day and night* on active USCG units afloat and ashore—a constant reminder that the Coast Guard is always on duty to render assistance.

Coast Guard cutters, ships, boats, aircraft, and vehicles are easily identified by the distinctive red and blue slanted stripes, with the USCG emblem on the red.

Functions Relating to Boating

Boatmen come into contact with the Coast Guard in both of its major areas of law enforcement and safety. Many of the federal laws relating to boating are implemented by regulations issued by the Commandant of the Coast Guard.

Boating Regulations

The Federal Boat Safety Act of 1971 is the basis for the registration and numbering of motorboats. Although the Coast Guard performs this task itself in only two states and one other jurisdiction, the boat registration systems of those states that carry out this function for themselves must be approved by the Commandant. Coast Guard patrol vessels may check the registration papers of boats on all U.S. waters.

Boats of more than a specified minimum size may be "documented" rather than registered and numbered. In *all* states and territories, this function is performed by officials of the Coast Guard.

Coast Guard regulations also spell out the details of the safety equipment required by the Act of 1971. USCG personnel are authorized to make inspections of boats to determine the adequacy of such equipment.

Boardings may be made without a search warrant to check for illegal activities.

Safety Activities

Coast Guard vessels make frequent patrols to keep a watchful eye for any reckless operation of boats and all hazards to navigation. Many races, regattas, and other marine events are patroled to ensure the safety of both participants and spectators.

The Coast Guard often has booths and exhibits at boat shows and other marine events. At these, boatmen can get information, advice, and assistance on problems relating to safety on the water—the Coast Guard would much rather prevent an accident than rescue victims!

Aids to Navigation

The Coast Guard is responsible for operating thousands of aids to navigation, from unlighted buoys and daybeacons to large lighthouses. It also operates hundreds of radiobeacons and many stations of the Loran-C and Omega electronic navigation systems.

This is a quiet and unspectacular, but most necessary, service to all who travel on the water. The installation and maintenance of lighted and unlighted aids to navigation is a major function of the Coast Guard, and one that affects boatmen every time they leave their moorings.

Search and Rescue

The rescue of mariners in distress is probably the most dramatic activity of the Coast Guard, the one that makes the headlines when a ship goes down in a storm. Less publicized, but equally important, are the many instances when the USCG comes to the aid of the boatman who has lost his way at sea, gone aground, suffered dismasting or engine failure, or merely run out of fuel. Increasingly, however, requests for assitance in non-life-threatening situations, such as mechanical or fuel problems in stable weather conditions, are being referred to commercial operations. Many boatmen venture out on the waters, offshore or inland, with a greater sense of security knowing that the Coast Guard is standing by to help, living up to its motto "Semper Paratus"—"Always ready."

Figure 2601 **Coast Guard mobile boarding teams can conduct limited rescue operations in emergencies, but their primary role is to extend law enforcement and general boating safety activities to inland areas of federal navigable waterways.**

Figure 2602 **Small Coast Guard cutters, *right,* are equipped to handle many types of emergencies as well as search and rescue operations. Large cutters, various types of specialty craft, and helicopters, *left,* handle other Coast Guard operations.**

Search and rescue surface units include ships known as "cutters" and small craft of many sizes. Fixed-wing aircraft and helicopters both extend the search capabilities of surface vessels and in many cases perform rescue missions themselves when wind and sea conditions permit.

Organization

The Commandant of the Coast Guard, an Admiral, and his staff are located at USCG HEADQUARTERS in Washington, D.C. Operational activities are grouped geographically into DISTRICTS each under its Commander, a Rear Admiral. Figure 2603 shows the boundaries of the Districts of the contiguous 48 states; the 14th District is in Hawaii and the 17th District covers Alaskan waters. For operational matters, the Districts of the Atlantic/Gulf of Mexico and Pacific coasts have been placed under Atlantic and Pacific Area commands, respectively.

The operating units of the Coast Guard—the individual bases, stations, cutters, boats, aircraft units, etc.—may be under the direct control of a District, or these may be under an intermediate level, the GROUP.

Officers and Enlisted Personnel

Personnel of the Coast Guard include commissioned officers, warrant officers, and enlisted men; ranks are the same as in the U.S. Navy. Uniforms are generally similar also except for being of a distinctive lighter shade of "Coast Guard Blue" cloth. The distinguishing USCG device is a small shield worn above an officer's stripes and on the lower right sleeve of an enlisted man's uniform shirt or jacket.

There is a Coast Guard Reserve to support the regular establishment. Women serve in both commissioned and enlisted ranks in the regular Coast Guard as well as the Reserves.

(Text continued on page 595)

AIDS TO NAVIGATION ON NAVIGABLE WATERS
except Western Rivers and Intracoastal Waterway

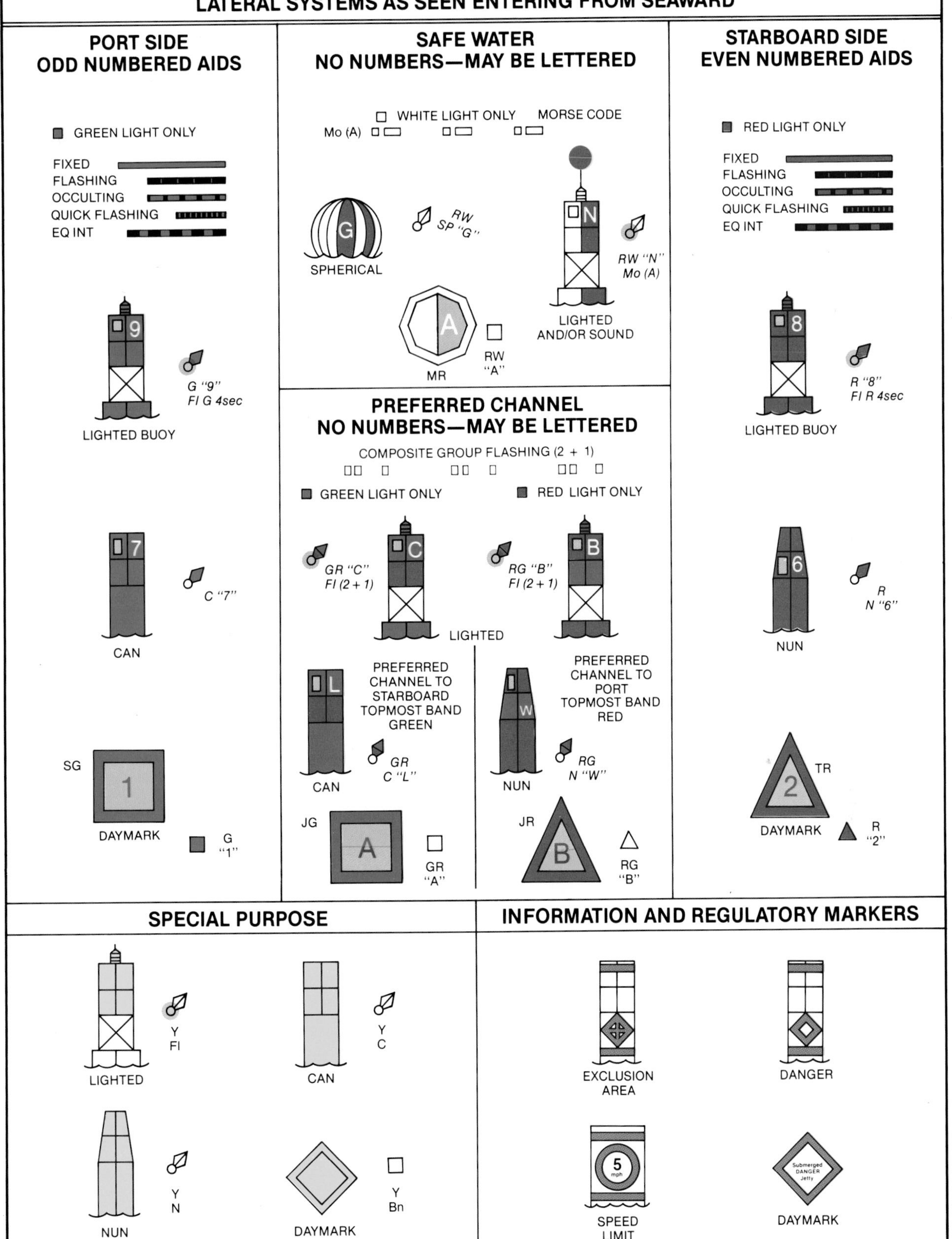

Aids are shown as they will appear in the new IALA-B system which is being implemented at this time, with completion scheduled by 1989. Under the old system, buoys shown here as green are black, and lights on such buoys are white. Buoys shown here as green and red are black and red in the old system, with white lights where green lights are indicated here.

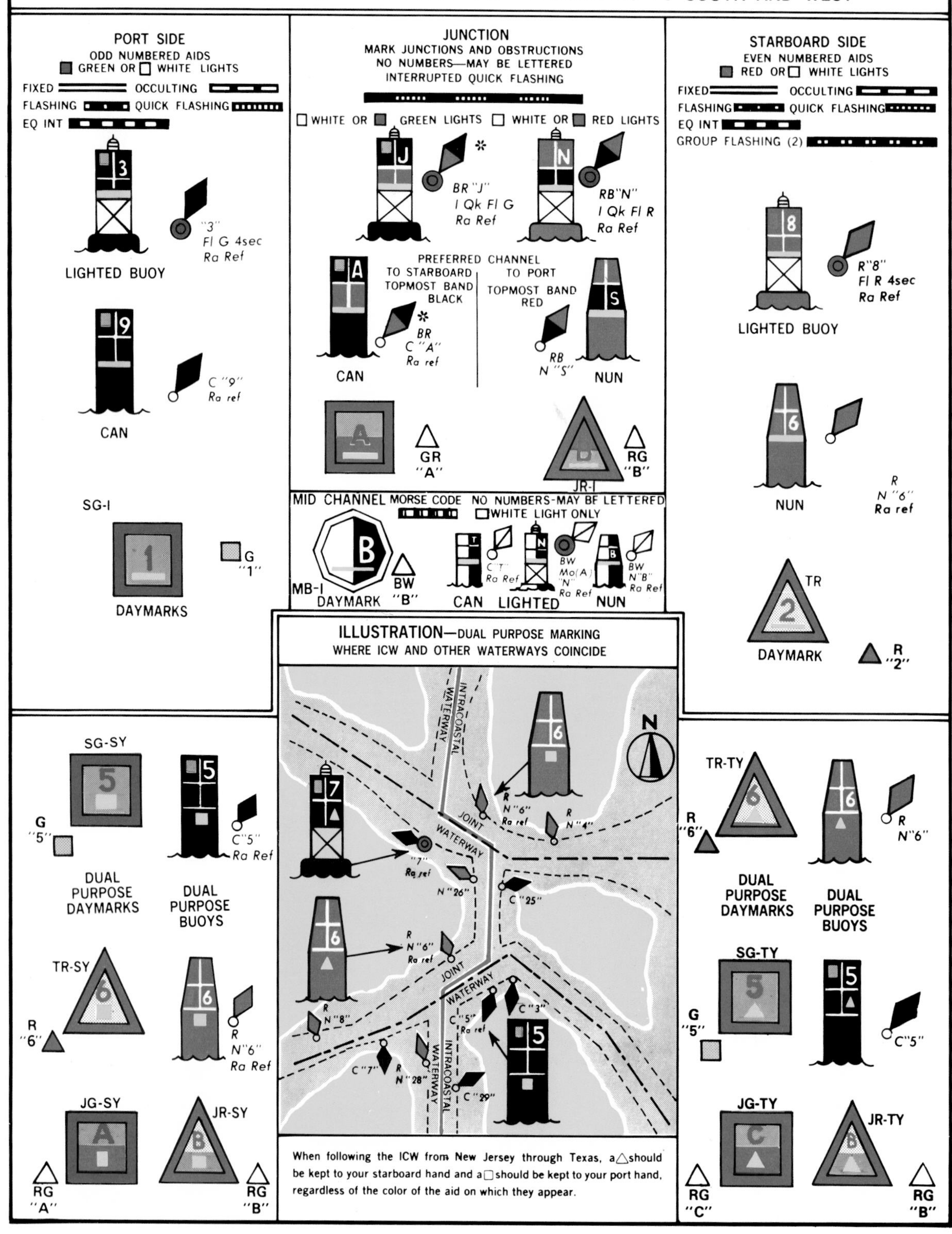

Black aids to navigation will be changed to green to conform to the IALA system. White lights will be changed to green or red to match the color of the aid. The distinctive yellow ICW marks will be unchanged.

AIDS TO NAVIGATION ON WESTERN RIVERS

AS SEEN ENTERING FROM SEAWARD

PORT SIDE

GREEN OR WHITE LIGHTS
FLASHING

LIGHTED BUOY

CAN

SG

PASSING DAYMARK

CG

CROSSING DAYMARK

176.9

MILE BOARD

JUNCTION

MARK JUNCTIONS AND OBSTRUCTIONS
INTERRUPTED QUICK FLASHING

PREFERRED CHANNEL TO STARBOARD
TOPMOST BAND BLACK

PREFERRED CHANNEL TO PORT
TOPMOST BAND RED

WHITE OR GREEN LIGHTS

WHITE OR RED LIGHTS

LIGHTED

CAN

NUN

JG

JR

STARBOARD SIDE

RED OR WHITE LIGHTS
GROUP FLASHING (2)

LIGHTED BUOY

NUN

TR

PASSING DAYMARK

CR

CROSSING DAYMARK

123.5

MILE BOARD

RANGE DAYMARKS AS FOUND ON

NAVIGABLE WATERS EXCEPT – ICW – MAY BE LETTERED

KWB KWR KRW KRB KBW KBR KGB KBG KGR KRG

INTRACOASTAL WATERWAY – MAY BE LETTERED

KWB-I KWR-I KRW-I KRB-I KBW-I KBR-I KGB-I KBG-I KGR-I KRG-I

Black aids to navigation will be changed to green to conform to the IALA system. White lights will be changed to green or red to match the color of the aid.

UNIFORM STATE WATERWAY MARKING SYSTEM

AIDS TO NAVIGATION

PORT (left) SIDE

COLOR — Black
NUMBERS — Odd
LIGHTS — Flashing green
REFLECTORS — Green

THE LATERAL SYSTEM—In well-defined channels and narrow waterways, USWMS aids to navigation normally are solid-colored buoys. Though a can and a nun are illustrated here, SHAPES may vary. COLOR is the significant feature. When proceeding UPSTREAM or toward the head of navigation, BLACK BUOYS ← mark the *left* side of the channel and must be kept on the left (port) hand. RED BUOYS → mark the *right* side of the channel and must be kept on the right (starboard) side. This conforms with practice on other federal waterways. On waters having no well-defined inlet or outlet, arbitrary assumptions may be made. Inquire in the locality for further information and charts when available.

STARBOARD (right) SIDE

COLOR — Red
NUMBERS — Even
LIGHTS — Flashing red
REFLECTORS — Red

Note: In some areas red buoys may be of "can" shape.

THE CARDINAL SYSTEM—Used where there is no well-defined channel or where an obstruction may be approached from more than one direction.

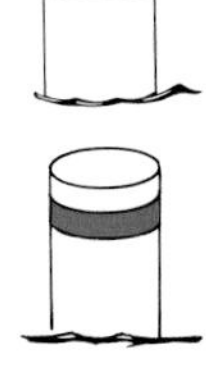

BLACK-TOPPED WHITE BUOY indicates boat should pass to NORTH or EAST of it. Reflector or light, if used, is white, the light quick-flashing.

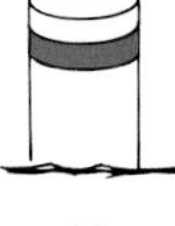

RED-TOPPED WHITE BUOY indicates boat should pass to SOUTH or WEST of it. Reflector or light, if used, is white, the light quick-flashing.

RED-AND-WHITE VERTICALLY STRIPED BUOY indicates boat should not pass between buoy and nearest shore. Used when reef or obstruction requires boat to go *outside* buoy (away from shore). White stripes are twice the width of red stripes. Reflector or light, if used, is white, the light quick-flashing.

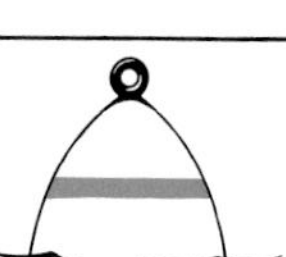

MOORING BUOY—White with horizontal blue band. If lighted, shows slow-flashing light unless it constitutes an obstruction at night, when light would be quick-flashing.

NOTE—The use of lights, reflectors, numbers and letters on USWMS aids is discretionary.

LIGHTS—On solid-colored (red or black) buoys, lights when used are flashing, occulting, or equal interval. For ordinary purposes, *slow-flashing* (not more than 30 per minute). *Quick-flashing* (not less than 60 per minute) used at turns, constrictions, or obstructions to indicate *caution.*

REFLECTORS—On lateral-type buoys, *red* reflectors or retro-reflective materials are used on solid-red buoys, *green* reflectors on solid-black buoys, *white* on all others including regulatory markers (except that *orange* may be used on orange portions of markers).

NUMBERS—*White* on red or black backgrounds. *Black* on white backgrounds. Numbers increase in an upstream direction.

LETTERS—When used on regulatory and white-and-red striped obstruction markers, letters are in alphabetical sequence in an upstream direction. (Letters I and O omitted.)

UNIFORM STATE REGULATORY MARKERS

Diamond shape warns of DANGER! Suggested wording for specific dangers: ROCK (illustrated), DAM, SNAG, DREDGE, WING-DAM, FERRY CABLE, MARINE CONSTRUCTION, etc.

Circle marks CONTROLLED AREA "as illustrated." Suggested wording to control or prohibit boating activities: 5 MPH (illustrated), NO FISHING, NO SKI, NO SWIM, NO SCUBA, NO PROP BOATS, SKI ONLY, FISHING ONLY, SKIN DIVERS ONLY, etc.

Diamond shape with cross means BOATS KEEP OUT! Explanatory reasons may be indicated outside the crossed diamond shape, for example SWIM AREA (illustrated), DAM, WATERFALL, RAPIDS, DOMESTIC WATER, etc.

Square or rectangle gives INFORMATION, names, activities. May give place names, distances, arrows indicating directions, availability of gas, oil, groceries, marine repairs, etc.

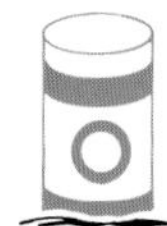

REGULATORY MARKERS are *white* with *international orange* geometric shapes. Buoys may be used as regulatory markers. Such buoys are *white* with two horizontal bands of *international orange*—one at the top, another just above the waterline. Geometric shapes, colored *international orange,* are placed on the white body of the buoy between the orange bands. When square or rectangular *signs* are displayed on structures as regulatory markers, they are *white* with *international orange borders.* Diamond and circular shapes, when used, are centered on the signboard.

NATIONAL AND YACHTING FLAGS

U.S. NATIONAL ENSIGN AND MERCHANT FLAG

U.S. YACHT ENSIGN

U.S. CUSTOMS

U.S. COAST GUARD

MEXICAN NATIONAL ENSIGN

CANADIAN NATIONAL ENSIGN

BAHAMAS NATIONAL ENSIGN

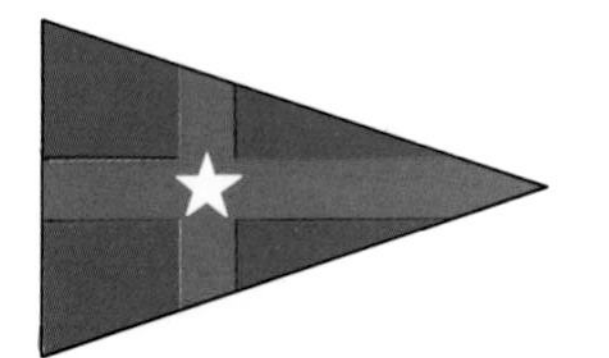

YACHT CLUB BURGEE

YACHT CLUB COMMODORE

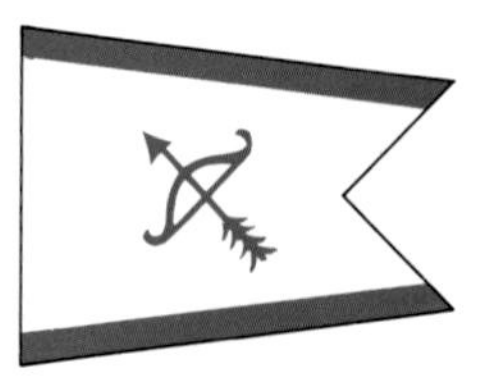

YACHT PRIVATE SIGNAL

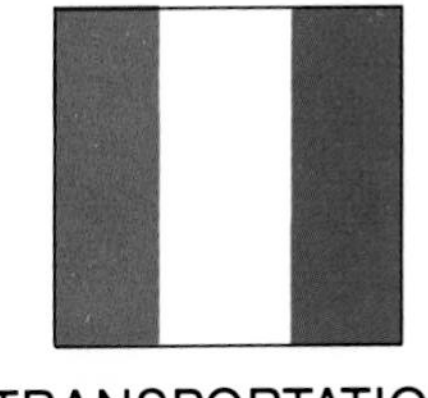

TRANSPORTATION

YACHT PROTEST

DIVER

INTERNATIONAL A UNDERWATER OPERATIONS

INTERNATIONAL FLAGS AND PENNANTS

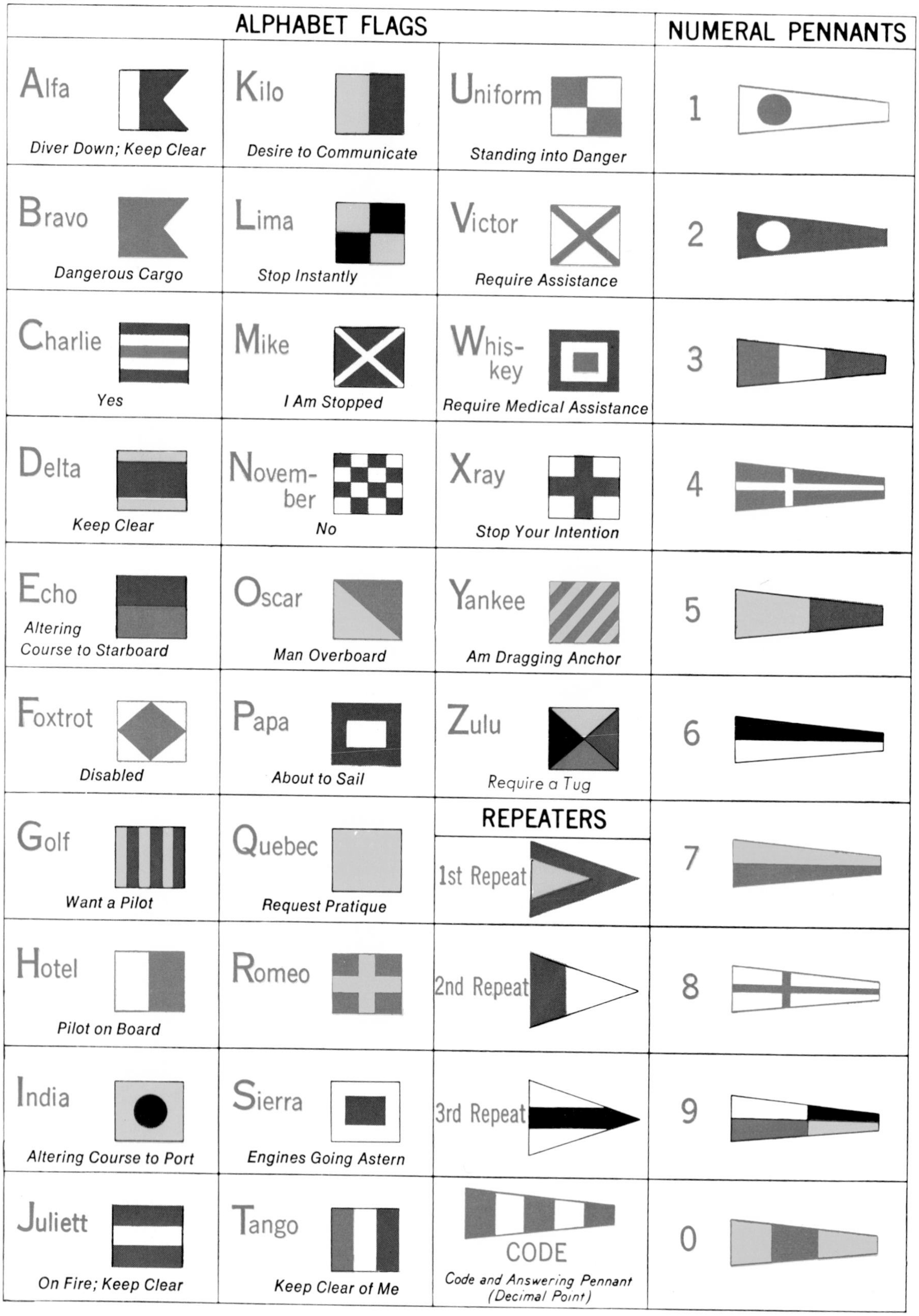

ALPHABET FLAGS			NUMERAL PENNANTS
Alfa — *Diver Down; Keep Clear*	Kilo — *Desire to Communicate*	Uniform — *Standing into Danger*	1
Bravo — *Dangerous Cargo*	Lima — *Stop Instantly*	Victor — *Require Assistance*	2
Charlie — *Yes*	Mike — *I Am Stopped*	Whis-key — *Require Medical Assistance*	3
Delta — *Keep Clear*	Novem-ber — *No*	Xray — *Stop Your Intention*	4
Echo — *Altering Course to Starboard*	Oscar — *Man Overboard*	Yankee — *Am Dragging Anchor*	5
Foxtrot — *Disabled*	Papa — *About to Sail*	Zulu — *Require a Tug*	6
Golf — *Want a Pilot*	Quebec — *Request Pratique*	**REPEATERS** 1st Repeat	7
Hotel — *Pilot on Board*	Romeo	2nd Repeat	8
India — *Altering Course to Port*	Sierra — *Engines Going Astern*	3rd Repeat	9
Juliett — *On Fire; Keep Clear*	Tango — *Keep Clear of Me*	CODE — *Code and Answering Pennant (Decimal Point)*	0

How International Code Flags are Used in Signaling

1, 2, and 3 CHARACTER SIGNALS

SINGLE LETTER

B

B – I am taking on, or discharging, or carrying dangerous goods

Very urgent, important, or commonly-used signals

TWO LETTERS

K

N

KN – I cannot take you in tow

General messages

TWO LETTERS AND NUMERAL

K

N

1

KN 1 – I cannot take you in tow but I will report you and ask for immediate assistance

Numeral added to general message to provide variation in meaning, to ask or answer a question, or to supplement the basic message

REPEATERS (Substitutes)

B

FIRST REPEATER Repeats uppermost flag

C

SECOND REPEATER Repeats second flag from top.

BBCB

T

1

3

SECOND REPEATER

0

T1330

A repeater (substitute) repeats the class of flags which it immediately follows – in this case a numeral pennant

ANSWERING PENNANT

AT THE DIP Hoisted by receiving vessel as each hoist of transmitting ship is seen

CLOSE UP Receiving vessel indicates she understands the hoist. At end of signal, indicates message is complete

PLAIN LANGUAGE MESSAGE, SPELLED OUT

Y

Z

The words which follow are in plain language. Use is optional; may be omitted if spelling in plain language is obvious.

R

E

A

D

M

O

T

O

SECOND REPEATER

R

B

O

A

T

I

N

G

Special signals are no longer used to indicate separation between spelled words or the end of spelling. Words should be selected and/or divided so that the meaning of the hoist is clear.

Race Signals of the United States Yacht Racing Union

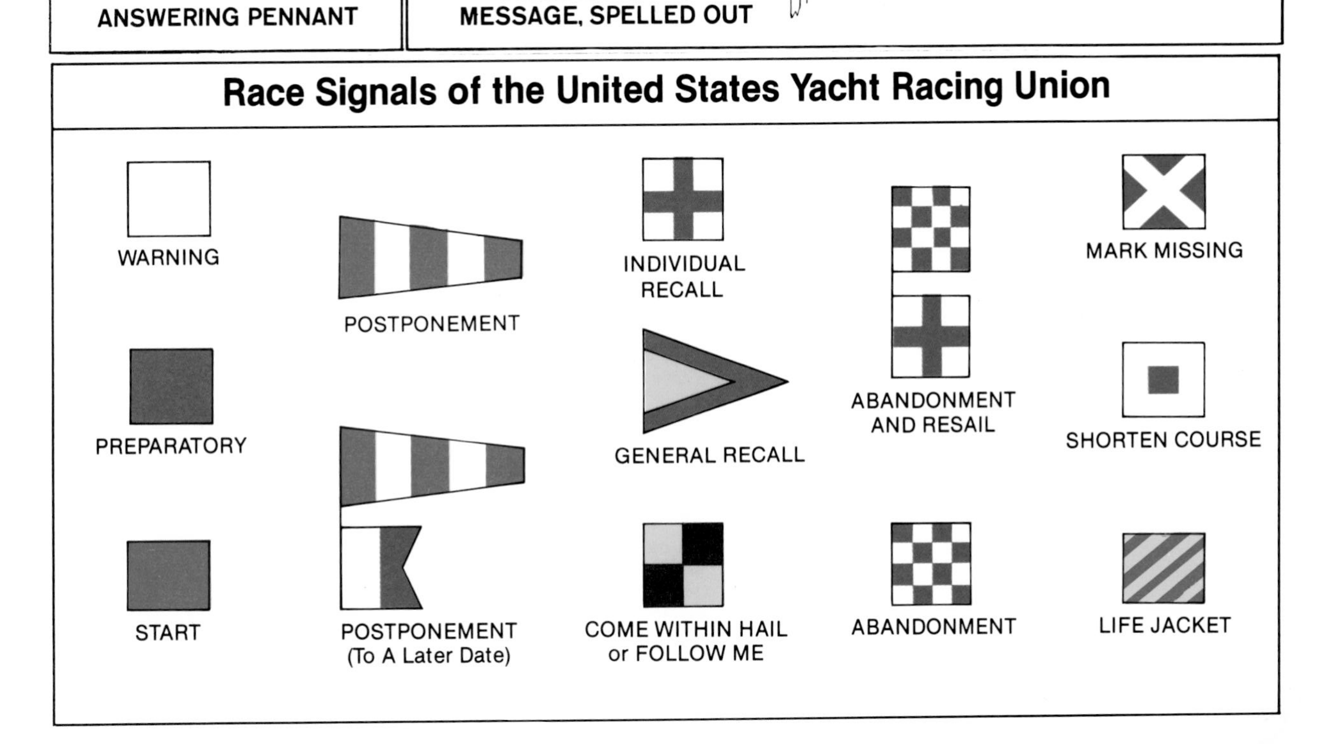

Reading the Clouds

Cirrostratus is a high veil of ice crystal cloud through which the sun can be seen, along with a well-defined halo; it may herald an approaching storm system.

Cirrocumulus cloud puffs are smaller in apparent size than the sun or moon, which they would not obscure. This is sometimes called a "mackerel sky."

Cirrostratus covers the entire sky, with some thicker patches of altostratus below it. As cirrus clouds thicken, the lower portions become water droplets, not ice.

More cirrus: Flat layer indicates a stable air condition with warm air over cooler at a point where water vapor, a gas, condenses into ice crystals.

Another cirrus sky. If sun or moon were in the picture, a halo would be seen. This is a sure indication of the structure of these clouds.

Ground haze, mixed with industrial smog, dissipates in morning sunlight which is not blocked by a thin layer of cirrostratus cloud.

Altostratus represents warm air riding over cold air ahead of a warm front. Layer will grow thicker and lower, with continuous rain or snow.

Altostratus patches illuminated by the setting sun are the remnants of cumulus clouds that rise, flatten, and finally dissipate as the air cools.

Nimbostratus, just prior to the onset of a steady rainfall. If the sun were in the picture, its image would be blurred, and there would be no halo.

Stratocumulus associated with a low pressure system may remain in warm sector after the warm front has passed through an area. Cirrus is also present.

Stratocumulus is white with dark patches in thick portions of the clouds. This type may appear as rounded masses or rolls which may or may not be merged.

Altostratus and altocumulus clouds cover the entire sky. The sun would appear as if viewed through ground glass if it were in the picture.

Altostratus here is a thick, almost uniform cloud layer about 10,000 feet high. If this thickens and lowers, precipitation is on the way.

Stratocumulus is a thick, solid layer with a lumpy base. It is often formed by the spreading out of cumulus, and may be followed by clearing at night.

More altostratus that covers the entire sky, with widely scattered cumulus beneath it. This followed a day of warm, hazy sunshine in late May.

Stratus has a thick, almost uniform look. This indicates stable air, and as there is little turbulence in this, any precipitation will be fine drizzle.

Cumulus clouds normally are associated with fair weather, but local hot spots and extreme development can change them into storm cumulonimbus.

Cumulus of little development indicate continued fair weather; following the passage of a cold front, they may be accompanied by brisk winds.

Cumulus late in the afternoon tends to thin and flatten as the earth's surface cools and the supply of rising warm, moist air diminishes.

More fair-weather cumulus. If they form in warm air just ahead of a cold front, however, they can be accompanied by squalls and thunderstorms.

Cumulus development culminates in an anvil-topped thunderhead that may tower 30,000 to 50,000 feet above the surface of the earth. Air in the cloud is turbulent.

Cumulonimbus base as a line of heavy thunderstorms approach. Squall winds reached 50 knots in gusts, and the accompanying rain was torrential.

Cumulonimbus associated with summer showers results from local development of cumulus clouds. Rainfall is likely to be brief, but moderately heavy.

Fog is a cloud with its base resting on the earth's surface. Here it is formed by warm, moist air that settled over the cooler surface of Lake Erie.

UNITED STATES POWER SQUADRONS FLAGS and INSIGNIA

USPS ENSIGN

The Ensign of United States Power Squadrons may be displayed only by enrolled members of USPS. It is an outward and visible sign that the vessel displaying it is under the charge of a person who has made a study of piloting and small boat handling, and will recognize the rights of others and the traditions of the sea. The Squadrons' Ensign also marks a craft as being under the command of a man who has met certain minimum requirements and is so honored for meeting them. This honor may not be bought, sold, rented, loaned or given away. The USPS Ensign is displayed, during the hours from 0800 to sundown from the stern, or day-and-night from the starboard spreader of a power vessel. On craft under sail, it is flown from the leach or peak of the mainsail or aftermost sail.

USPS OBJECTIVES

Shall be to establish a high standard of skill in handling and navigation of yachts, to encourage the study of the science of navigation, and small boat handling, to cooperate with the agencies of the United States Government charged with the enforcement of the laws and regulations relating to navigation and to stimulate interest in activities which will tend to the upbuilding of our Army, Navy, Coast Guard and Merchant Marine.

NATIONAL OFFICERS

OFFICE	INSIGNIA	FLAG	SLEEVE
CHIEF COMMANDER			
VICE COMMANDERS			
REAR COMMANDERS			
STAFF COMMANDERS			
NATIONAL FLAG LIEUTENANT			
AIDES TO C/C CHAPLAIN			

DISTRICT OFFICERS

OFFICE	INSIGNIA	FLAG	SLEEVE
DISTRICT COMMANDER			
DISTRICT LIEUTENANT COMMANDERS			
DISTRICT 1st LIEUTENANTS			
DISTRICT LIEUTENANTS			
DISTRICT FLAG LIEUTENANT			
DISTRICT AIDES CHAPLAIN			

SQUADRON OFFICERS

OFFICE	INSIGNIA	FLAG	SLEEVE
SQUADRON COMMANDER			
SQUADRON LIEUTENANT COMMANDERS			
SQUADRON 1st LIEUTENANTS			
SQUADRON LIEUTENANTS, CHAPLAIN			
SQUADRON FLAG LIEUTENANT			

PAST OFFICERS' SIGNALS

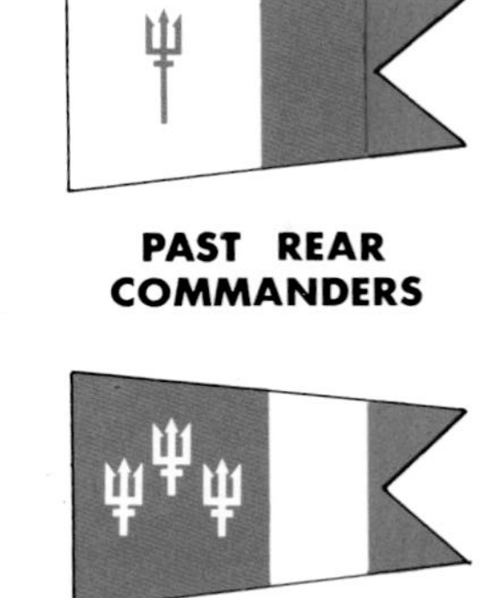

PAST CHIEF COMMANDERS

PAST VICE COMMANDERS

PAST REAR COMMANDERS

PAST STAFF COMMANDERS

PAST DISTRICT COMMANDERS

PAST SQUADRON COMMANDERS

EDUCATIONAL INSIGNIA

NATIONAL COMMITTEES

RULES

MEMBERS

ELECTIVE COURSES

ADVANCED GRADES

GOVERNING BOARD GENERAL MEMBERS

LOCAL BOARDS

CHAIRMAN

MEMBERS

ELECTIVE COURSES

ADVANCED GRADES

OIC PENNANT

A blue pennant, six inches on the hoist and 36 inches on the fly, tapering to a point from the hoist, may be flown above the officer's flag by the designated officer in charge of a USPS, District or Squadron activity afloat.

SEAMAN

PILOT

ADVANCED PILOT

JUNIOR NAVIGATOR

NAVIGATOR

Note: A Senior Member who has completed all courses may wear a gold border around the Navigator and Senior Member insignia.

EMERITUS MEMBER GOVERNING BOARD

CAP DEVICE

SENIOR MEMBER

MERIT MARKS

WORN ON LEFT SLEEVE ONLY

U.S. COAST GUARD AUXILIARY FLAGS AND INSIGNIA

COAST GUARD AUXILIARY ENSIGN

NATIONAL COMMODORE

OPERATIONAL FACILITY PENNANT

NATIONAL VICE COMMODORE

NATIONAL REAR COMMODORE

DISTRICT COMMODORE

DIVISION CAPTAIN

FLOTILLA COMMANDER

DISTRICT VICE COMMODORE
(White field, blue markings, Dist. Rear Commodore)

DIVISION VICE CAPTAIN

FLOTILLA VICE COMMANDER

CHIEF OF DEPARTMENT
3 Bars—Division Chief
2 Bars—Branch Chief and Aide to National Commodore
1 Bar—Branch Assistant

DISTRICT STAFF OFFICER
1½ Bars—Assistant DSO

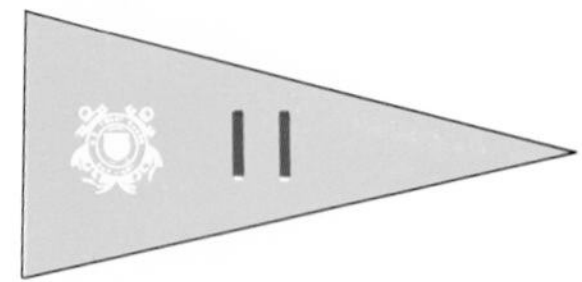
DIVISION STAFF OFFICER

FLOTILLA STAFF OFFICER
(Also Aide to District Commodore)

CAP DEVICE

SHIELD
Elected Officer and Member

SHIELD
Staff Officer

SHIELD
Past Officer

AUXOP DEVICE

AUXILIARY AVIATOR

	METAL COLLAR INSIGNIA[1]	SHOULDER BOARDS[2]		METAL COLLAR INSIGNIA[1]	SHOULDER BOARDS[2]
National Commodore			Division Vice Captain With Red "A": Branch Chief; Dist. Staff Officer; Aide to NACO DCO Admin. Officer		
National Vice Commodore		 	Aide to DCO Admin. Asst. to DCO Flotilla Commander With Red "A": Branch Asst.; Asst. Dist. Staff Officer		
National Rear Commodore		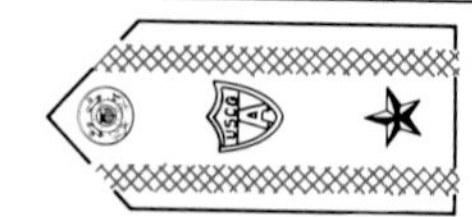	Flotilla Vice Commander With Red "A": Division Staff Officer		
District Vice Commodore With Red "A": District Rear Commodore Department Chief	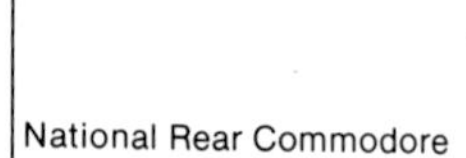 		Flotilla Staff Officer; Aide to DCO		
Division Captain With Red "A": Division Chief NACO Admin. Officer			Member		

[1]Large metal shoulder insignia is worn on blue raincoats; the same insignia in smaller size is worn on khaki, blue flannel, and khaki tropical shirts.
[2]Shoulder Marks are worn on khaki coats, blue overcoats, and white tropical shirts.

Running Lights

Illustrations Are Keyed to Table A (Numbers) and Table B (Letters)

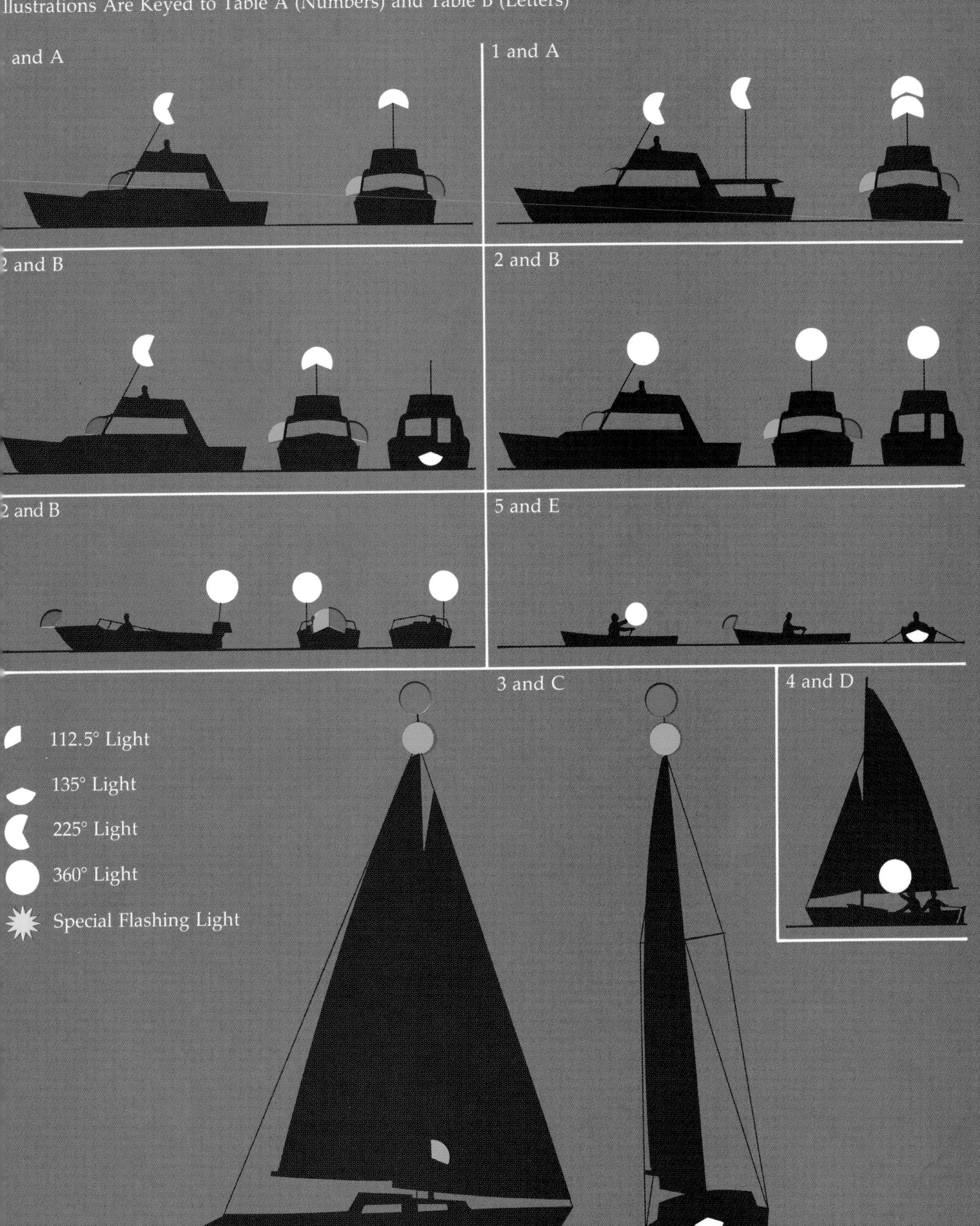

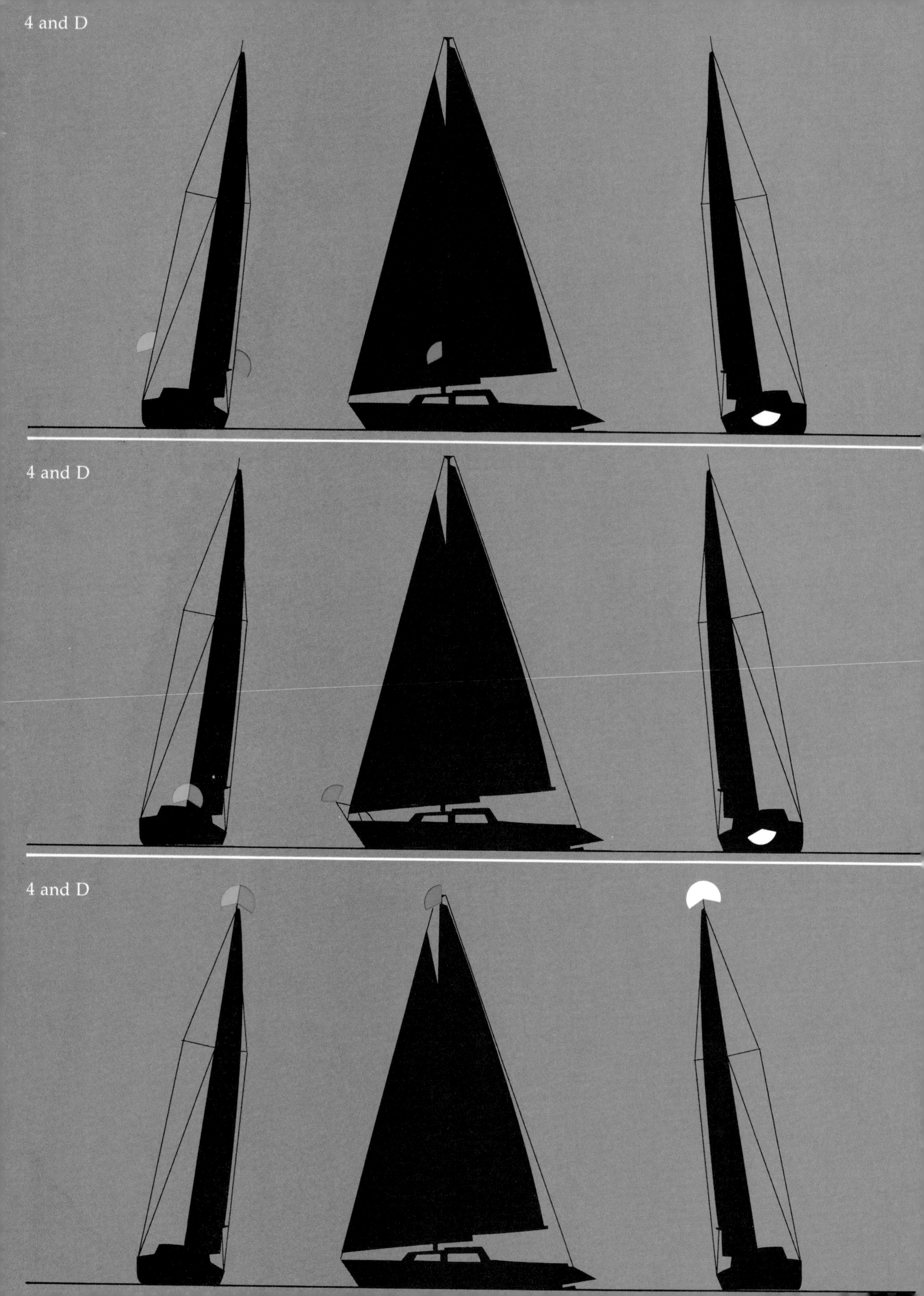
4 and D
4 and D
4 and D

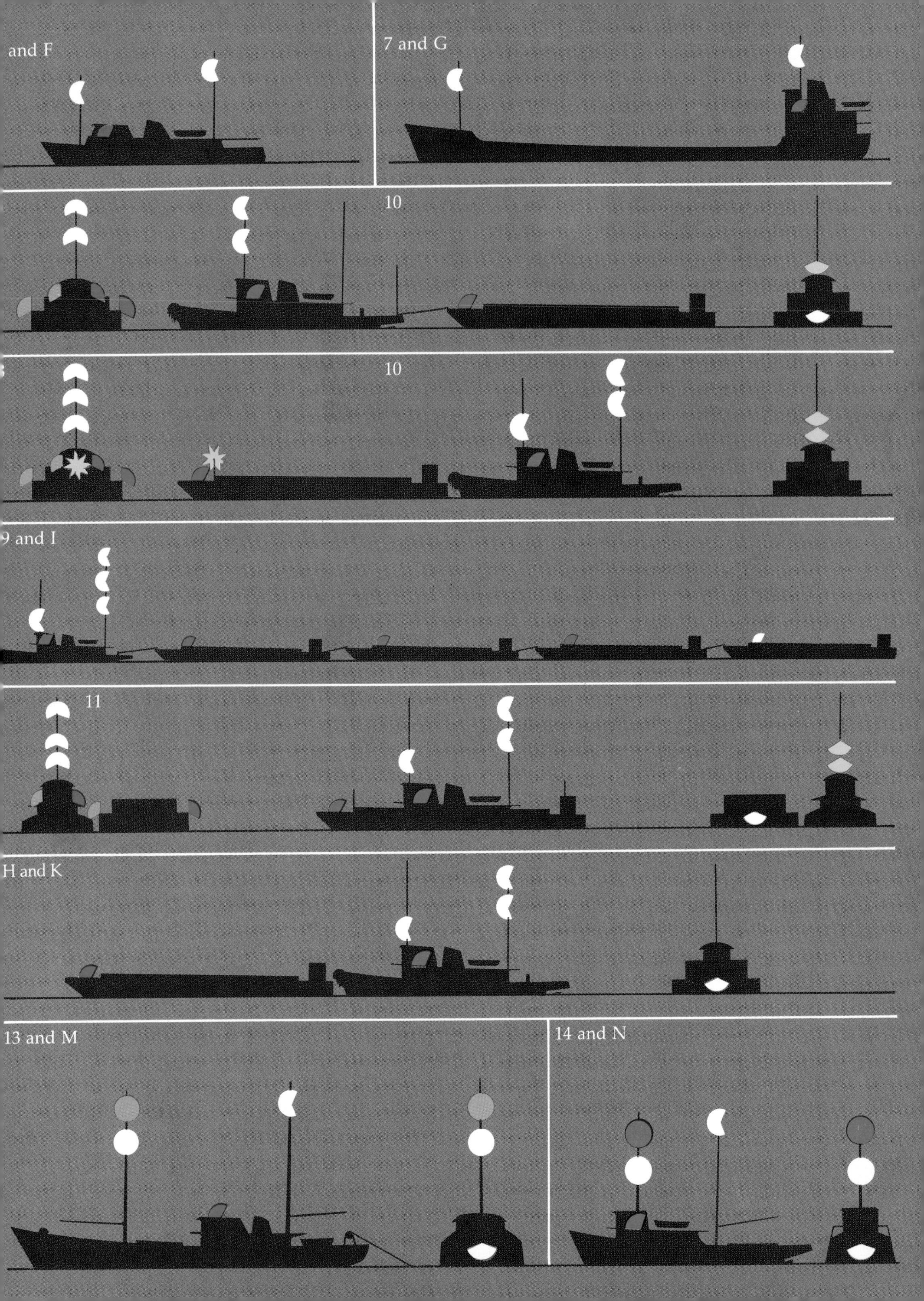

and F
7 and G
10
10
9 and I
11
H and K
13 and M
14 and N

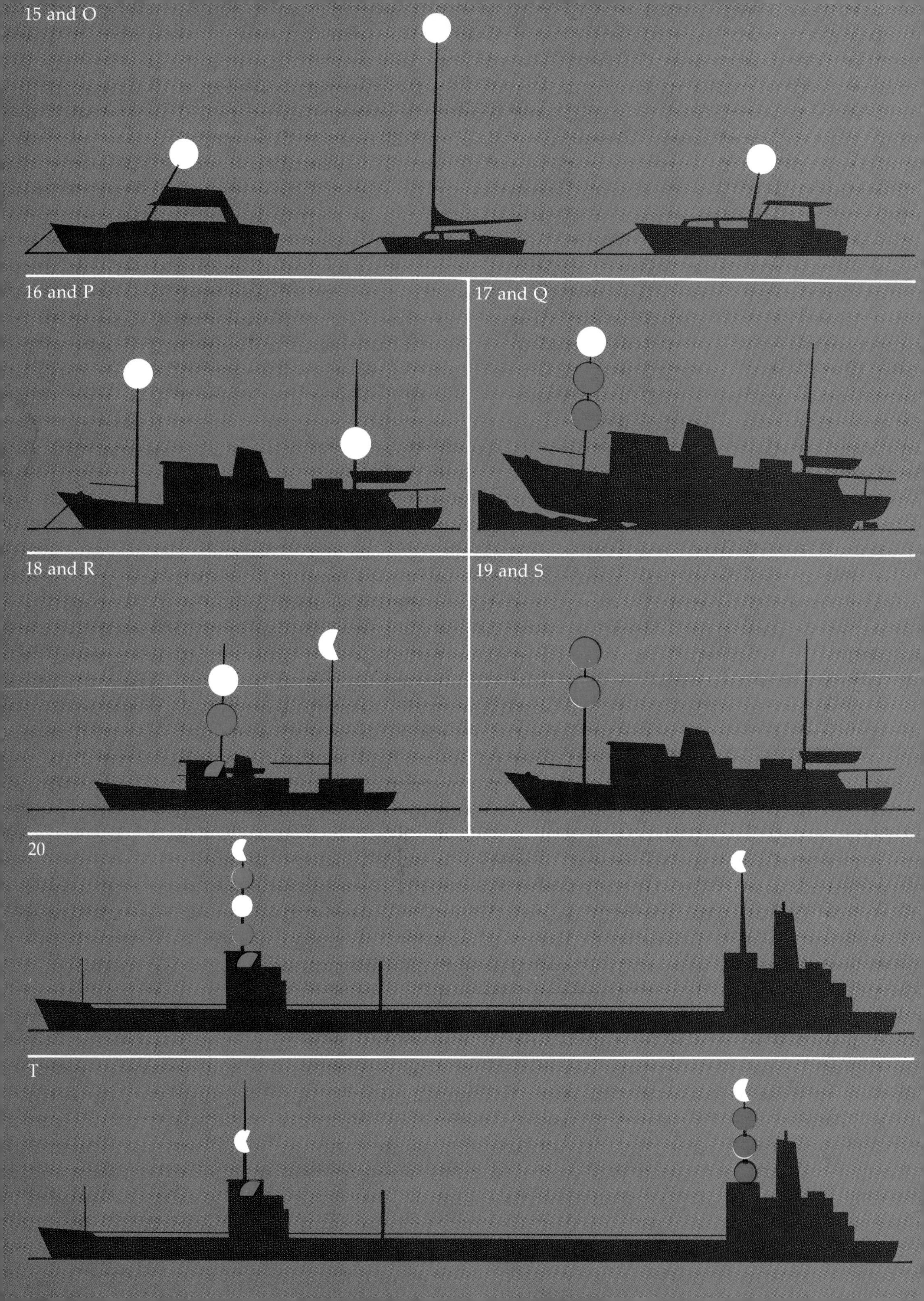
15 and O
16 and P
17 and Q
18 and R
19 and S
20
T

Table A Lights for Various Types of Vessels—1980 Inland Rules

Also see color illustrations, pages 589–592

(Effective 24 December 1981)

Color Diagram	*Vessel*	*Masthead (forward)*	*Side*	*Stern*	*Additional lights or Remarks*
1	Power-driven vessel 12 m but less than 20 m in length	White, 225°, Vis. 3 mi. At least 2.5 m above gunwale[3]	Separate red and green, 112½°, or combination, vis. 2 mi. Above hull at least 1 m below masthead light[2]	White 135°, vis. 2 mi.	[1]After masthead light may be shown but not required. (Exception allowed on Great Lakes) [2]Fitted with inboard screens if necessary to prevent being seen across bow
2	Power-driven vessel less than 12 m in length	Can be less than 2.5 m above gunwale, but at least 1 m above side lights[1,3]	Separate red and green, 112½°, or combination, vis. 1 mi. Above hull at least 1 m below masthead light[2,3]	White 135°, vis. 2 mi.	[3]Less than 12 m in length, need only have all-round white light, vis. 2 mi. but should have side lights
3	Sailing vessel under 20 m in length	None	Separate red and green, 112½°, or combination, vis. 2 mi.[2,4]	White 135°, vis. 2 mi.	Optional—two all-round lights at or near top of mast, red over green, separated at least 1 m, vis. 2 mi.
4	Sailing vessel under 12 m in length	None	Separate red and green, 112½°, or combination, vis. 1 mi.[2,4,5]	White, 135°, vis. 2 mi.[4,5]	[4]May be combined into triple combination light at masthead [5]Less than 7 m, need only have flashlight or lantern to show
5	Vessel propelled by oars	None	May show separate red and green, 112½°, or combination, vis. 1 mi.[6]	May show white, 135°, vis. 2 mi.[6]	[6]Need only have flashlight or lantern to show white light
6	Power-driven vessel 20 but less than 50 m in length	White, 225°, vis. 5 mi. Not more than ½ of length aft from stem; 6 m, or beam (up to 10 m) above hull	Red and green, 112½°, vis. 2 mi. At or near sides of vessel; above hull at least 1 m below masthead light	White, 135°, vis. 2 mi.	After masthead light may be shown; at least 4.5 m higher than forward masthead light
7	Power-driven vessel 50 m or more in length	Not more than ½ of length aft from stem; 6 m or beam (up to 10 m) above hull	Red and green, 112½°, vis. 3 mi. At or near sides of vessel; above hull at least 1 m below masthead light	White, 135°, vis. 3 mi.	After masthead light required; at least 4.5 m higher and ¼ of vessel length (up to 50 m) aft of forward masthead light
8	Vessel towing; tow less than 200 m overall from stern of towing vessel. (Also towing alongside or pushing ahead)	Two white, arranged vertically, 225°, vis. determined by length of vessel (not required pushing ahead or towing alongside on Western Rivers)	Normal for size of vessel		Towing astern: Towing light[7] over stern light. Pushing ahead or towing alongside: Two towing lights[7] vertically [7]Vis. 3 mi. for vessels 50 m or more in length; 2 mi. for shorter vessels
9	Vessel towing; tow 200 m or more overall length	Three white, arranged vertically, 225°, vis. determined by length of vessel	Normal for size of vessel		Towing light: yellow, 135°, above sternlight[7]
10	Vessel being towed astern, if manned	None	Normal for size of vessel		
11	Vessel being towed alongside or pushed ahead	None	Normal for size of vessel; at forward end	Normal for size of vessel (not used for pushed ahead)	Also "special flashing light" at center or forward end. A group of vessels is lighted as a single vessel
12	Vessel engaged in trawling or drift fishing		[8]Show only normal lights of power-driven or sailing vessel		
13	Vessel engaged in trawling	None[12]	When making way through the water, normal for size of vessel		Underway or at anchor, two all-round lights, green over white[7,9,10,11] [9]Vertical spacing 1 m [10]Lower light not less than 4 m (2 m if under 20 m in length) above hull [11]Lower light above sidelights at least twice vertical spacing
14	Vessel engaged in fishing, other than trawling (or trolling)	None[12]	When making way through the water, normal for size of vessel		Underway or at anchor, two all-round lights, red over white[7,9,10,11] [12]When not actually fishing, show normal masthead lights for vessel its size
15	Vessel at anchor, less than 50 m in length	None	None	None	White, all-round light where can best be seen. Vis. 2 mi. (not required if less than 7 m in length and not anchored in a narrow channel or where vessels normally navigate)
16	Vessel at anchor; 50 m or more in length	None	None	None	White, all-round light in fore part of vessel not less than 6 m above hull. A second white, all-round light in after part, not less than 4.5 m lower than forward anchor light. Vis. 3 mi.
17	Vessel aground	None	None	None	Anchor light(s) as line 15 or 16, plus two red all-round lights of same visibility range[7,9,10] (not required if less than 12 m in length)
18	Pilot vessel	None if on pilot duty; normal if underway and not on pilot duty	When underway, normal for size of vessel	When underway, normal for size of vessel	Two all-round lights, white over red, at masthead[7,9,10] If at anchor, normal anchor light(s); line 15 or 16
19	Vessel not under command	None	If making way through the water, normal for size of vessel	If making way through the water, normal for size of vessel	Two red all-round lights, vertically where best can be seen[7,9,10]
20	Vessel restricted in ability to maneuver	None	When making way through the water, normal for size of vessel	When making way through the water, normal for size of vessel	Three all-round lights vertically, red-white-red.[7,9] If at anchor, normal anchor light(s). (Not required if less than 12 m in length)

Equivalent measurements in customary units: 50 m = 164 ft; 6 m = 19.7 ft; 7 m = 23.0 ft; 1 m = 3.3 ft; 2 m = 6.6 ft; 100 m = 328 ft; 200 m = 656 ft; 12 m = 39.4 ft; 20 m = 65.6 ft; 2.5 m = 8.2 ft; 4.5 m = 14.8 ft

Table B Lights for Various Types of Vessels—1972 International Rules

Also see color illustrations, pages 589–592

Color Diagram	*Vessel*	*Masthead (forward)*	*Side*	*Stern*	*Additional lights or Remarks*
A	Power-driven vessel 12 m but less than 20 m in length	White, 225°, vis. 3 mi. At least 2.5 m above gunwale[3]	Separate red and green, 112½°, or combination, vis. 2 mi. Above hull at least 1 m below masthead light[2]	White, 135°, vis. 2 mi.	[1]After masthead light may be shown but not required. [2]Fitted with inboard screens if necessary to prevent being seen across bow
B	Power-driven vessel less than 12 m in length	White, 225°, vis. 2 mi. Can be less than 2.5 m above gunwale, but at least 1 m above side lights[3]	Separate red and green, 112½°, or combination, vis. 1 mi. Above hull at least 1 m below masthead light[2]	White, 135°, vis. 2 mi.	[3]Less than 7 m and less than 7 kt max speed need only have all-round white light, vis. 2 mi. but should have sidelights
C	Sailing vessel under 20 m in length	None	Separate red and green, 112½°, or combination, vis. 2 mi.	White, 135°, vis. 2 mi.	Optional—two all-round lights at or near top of mast, red over green, separated at least 1 m, vis. 2 mi.
D	Sailing vessel under 12 m in length	None	Separate red and green, 112½°, or combination, vis. 1 mi.[2,4]	White, 135°, vis. 2 mi.[4,5]	[4]May be combined into triple combination light at masthead [5]Less than 7 m, need only have flashlight or lantern to show
E	Vessel propelled by oars	None	May show separate red and green, 112½°, or combination, vis. 1 mi.[6]	May show white, 135°, vis. 2 mi.[6]	[6]Need only have flashlight or lantern to show white light
F	Power-driven vessel 20 but less than 50 m in length	White, 225°, vis. 5 mi. Not more than ¼ of length aft from stem; 6 m or beam (up to 12 m) above hull	Red and green, 112½°, vis. 2 mi. At or near sides of vessel; not more than ¾ height of masthead light	White, 135°, vis. 2 mi.	After masthead light may be shown; at least 4.5 m higher than forward masthead light
G	Power-driven vessel 50 m or more in length	White, 225°, vis. 6 mi. Not more than ¼ of length aft from stem; 6 m or beam (up to 12 m) above hull	Red and green, 112½°, vis. 3 mi. At or near sides of vessel; not more than ¾ height of forward masthead light	White, 135°, vis. 3 mi.	After masthead light required; at least 4.5 m higher and half of vessel length (up to 100 m) aft of forward masthead light
H	Vessel towing; tow less than 200 m overall vessel. (Also towing alongside of pushing ahead)	Two white, arranged vertically, 225°, vis. determined by length of vessel	Normal for size of vessel		Towing light[7] over sternlight (not shown when towing alongside or pushing ahead) [7]Vis. 3 mi. for vessels 50 m or more in length; 2 mi. for shorter vessels
I	Vessel towing; tow 200 m or more overall length	Three white, arranged vertically, 225°, vis. determined by length of vessel	Normal for size of vessel		Towing light[7] over sternlight
J	Vessel being towed astern, if manned	None	Normal for size of vessel		
K	Vessel being towed alongside or pushed ahead	None	Normal for size of vessel; at forward end	Normal for size of vessel (not used for pushed ahead)	A group of vessels is lighted as a single vessel
L	Vessel engaged in trawling or drift fishing		[8]Show only normal lights of power-driven or sailing vessel		
M	Vessel engaged in trawling	None[12]	When making way through the water, normal for size of vessel		Underway or at anchor, two all-round lights, green over white[7,9,10,11] [9]Vertical spacing 2 m for vessels 20 m or more in length, 1 m for shorter vessels [10]Lower light not less than 4 m (2 m if under 20 m in length) above hull [11]Lower light above sidelights at least twice vertical spacing
N	Vessel engaged in fishing, other than trawling (or trolling)	None[12]	When making way through the water, normal for size of vessel		Underway or at anchor, two all-round lights, red over white[7,9,10,11] [12]When not actually fishing, show normal masthead lights for vessel its size
O	Vessel at anchor, less than 50 m in length	None	None	None	White, all-round light where can best be seen. Vis. 2 mi. (not required if less than 7 m in length and not anchored in a narrow channel or where vessels normally navigate)
P	Vessel at anchor; 50 m or more in length	None	None	None	White, all-round light in fore part of vessel not less than 6 m above hull. A second white, all-round light in after part, not less than 4.5 m lower than forward anchor light. Vis. 3 mi.
Q	Vessel aground	None	None	None	Normal anchor light(s) plus two red all-round lights of same visibility range
R	Pilot vessel	None if on pilot duty; normal if underway and not on pilot duty	When underway, normal for size of vessel		Two all-round lights, white over red, at masthead[7,9,10] If at anchor, normal anchor light(s); line 15 or 16
S	Vessel not under command	None	If making way through the water, normal for size of vessel		Two red all-round lights, vertically where best can be seen[7,9,10]
T	Vessel constrained by her draft	Normal for size of vessel	Normal for size of vessel		Three red all-round lights, arranged vertically and equally spaced[7,9,10]

Equivalent measurements in customary units: 50 m = 164 ft; 6 m = 19.7 ft; 7 m = 23.0 ft; 1 m = 3.3 ft; 2 m = 6.6 ft; 100 m = 328 ft; 200 m = 656 ft; 12 m = 39.4 ft; 20 m = 65.6 ft; 2.5 m = 8.2 ft; 4.5 m = 14.8 ft

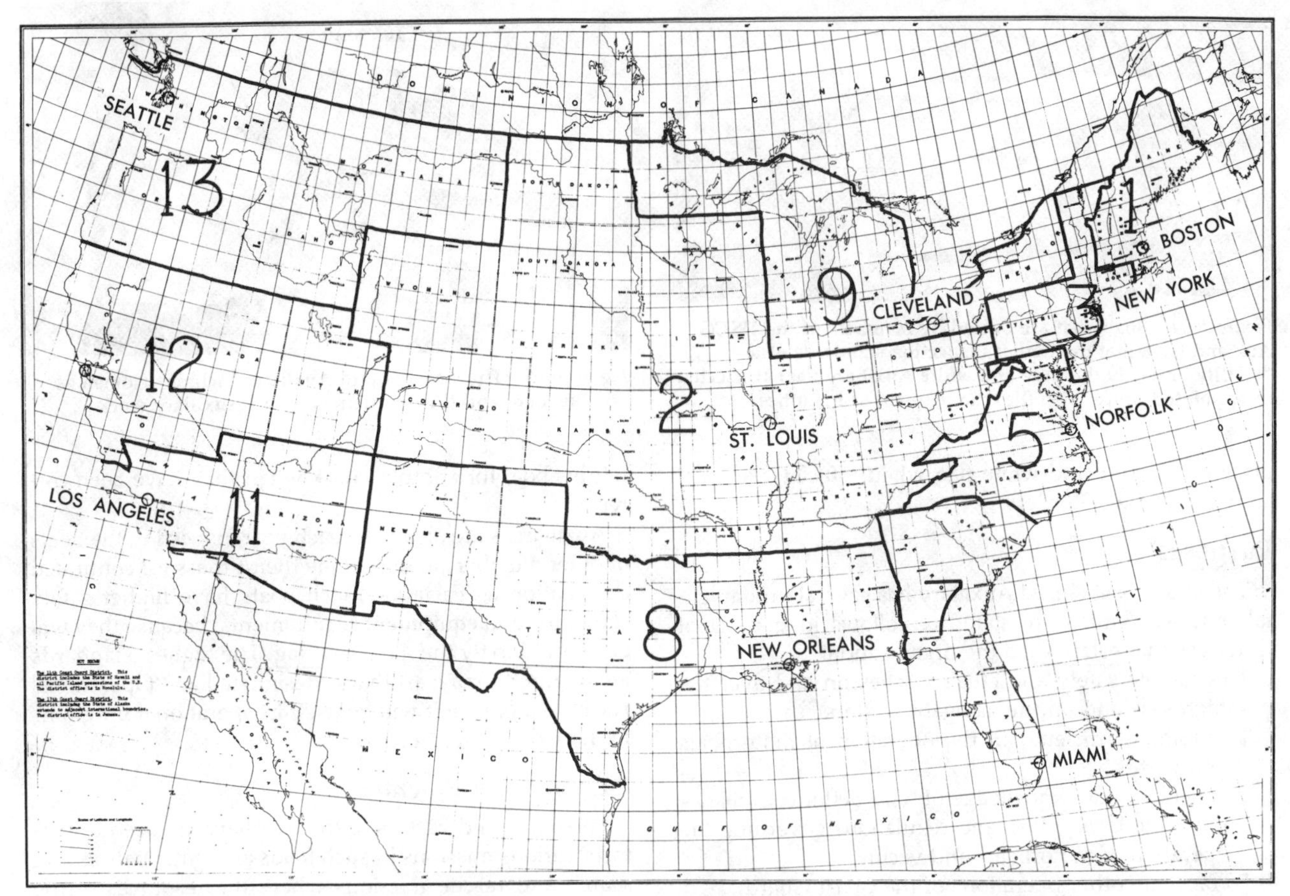

Figure 2603 **There are ten Coast Guard Districts in the continental United States, with Hawaii the 14th District and Alaska the 17th. Note that there are no districts 4, 6, 10, or 15.**

(Text continued from page 576)

Other Coast Guard Functions

Functions of the Coast Guard not directly related to boating include the Merchant Marine Inspection Program that sets and enforces safety standards for vessels and crews and pollution control for spills of oil and toxic chemicals. Icebreakers and cutters of reinforced construction keep harbors open for commercial navigation in winter, and each spring cutters and aircraft work with the International Ice Patrol tracking icebergs in the North Atlantic shipping lanes. In 1977 the Coast Guard became responsible for enforcement of offshore fishery laws out to 200 miles (370 km). In recent years, the interdiction of drug smugglers and illegal aliens has become a primary task in certain coastal areas.

The final major function of the Coast Guard is to maintain at all times a high state of readiness to function as a specialized service in the Navy in time of war, at which time it operates under the Department of the Navy.

The Coast Guard Auxiliary

The Coast Guard Auxiliary is a *civilian* organization functioning under the direction of the Commandant of the Coast Guard. It is composed of persons interested in the Coast Guard and its principles, dedicated to the interests of their country, and concerned about the safety and welfare of their fellow men. Despite its uniforms and insignia, the Auxiliary is a non-military body.

Eligibility for the Auxiliary extends to persons, male or female, U.S. citizens, who are at least 17 years old, and who own at least 25% of a boat, aircraft, or land-fixed or mobile radio station (or who have some other needed special qualification). To become a Basically Qualified Member, one must meet the eligibility requirements and pass the BQ examination and practical demonstrations. Members are encouraged to take advanced training toward the higher status of AUXOP and/or "Auxiliary Coxswain." The boat, aircraft, or radio station is termed a "Facility."

The Auxiliary Flag

A boat that has qualified as a Facility may fly the Coast Guard Auxiliary flag. This is a rectangular blue flag with a white diagonal slash on which the USCG Aux emblem appears; see color section.

The Auxiliary flag shares with the regular USCG Ensign the distinction of being authorized to fly from boats *day and night.* See Chapter 25.

The Coast Guard Auxiliary Operational Ensign is white with the Coast Guard red-and-blue slanted stripes; the Auxiliary emblem is on the red stripe. This is flown in

Figure 2604 **Boats that are owned by members of the USCG Aux and have met the higher inspection standards of a "facility" may fly the blue and white Auxiliary flag; an elected or appointed officer also flies the pennant of his office.**

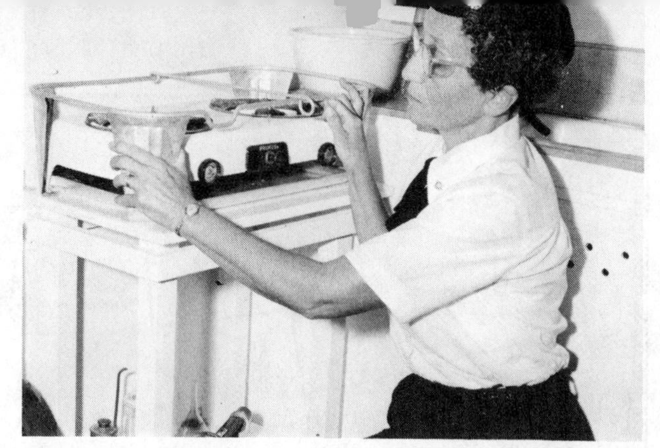

Figure 2605 **The Coast Guard Auxiliary includes women as well as men, and they participate in all Auxiliary work.**

place of the blue-and-white flag when a facility is operating under USCG orders.

Purposes

The U.S. Coast Guard Auxiliary has several fundamental purposes which are stated as follows in the Act of Congress that established the organization:

To promote safety and effect rescues on and over the high seas and on the navigable waters.

To promote efficiency in the operation of motorboats and yachts.

To foster a wider knowledge of, and better compliance with, the laws, rules, and regulations governing the operation of motorboats and yachts.

To facilitate other operations of the Coast Guard.

Activities

In carrying out its stated purposes, the Auxiliary has three basic program areas—Public Education, Vessel Examinations, and Operations. All of these are focused on the objective of greater safety for those who go on the water in small craft.

Education

The Auxiliary offers several courses to the boating public, tailored to meet the needs of various types of boating—power or sail, large or small. Courses consist of from one to 13 lessons; classes are held in the evenings or on weekends. Constant attention is given to the improvement of these courses and the development of new lessons or courses as required.

Members of the Auxiliary are actively encouraged to take "specialty" courses to improve their knowledge and increase their value to the organization. Typical subjects include communications, search and rescue, patrol procedures, weather, and others.

Vessel Examinations

Many members of the Coast Guard Auxiliary are active in the Vessel Examination program. As discussed more fully in Chapter 3, boat owners are urged each year to request a "Courtesy Marine Examination." This purely voluntary action ensures that a thorough check is made of all safety-related equipment. Boats that pass are awarded the CME sticker for the current calendar year. Boats that do not pass are not reported to any authority—the owner is advised of the deficiencies and is encouraged to resubmit his boat for another check when they have been corrected.

Auxiliarists are also busy each year inspecting the boats of other members to determine their fitness for continued designation as Facilities. Such boats have higher safety standards and equipment requirements, because they will be allowed to fly the Auxiliary flag. The highest standards are required for a craft that is designated an "Operational Facility," because it will be used in operational programs under Coast Guard orders.

Operational Programs

Personnel and vessels of the Auxiliary are used to perform various missions in such fields as regatta patrols and search and rescue missions where the resources of the regular Coast Guard are not sufficient to meet the demands placed upon them. This often occurs on weekends and during holiday periods. For such duties, orders are issued by a designated Coast Guard officer. To improve operational efficiency, the Auxiliary conducts an extensive Boat Crew Qualification Program for those of its members engaged in such work. This leads progressively to designation as "Crewman," "Vessel Operator," and "Auxiliary Coxswain."

Auxiliary members have *no law enforcement powers.* In those cases where such is needed, a Coast Guard officer or petty officer will be embarked in the Auxiliary vessel and the boat will fly the regular Coast Guard Ensign.

Figure 2606 **Thousands of voluntary courtesy checks are made each year by members of the Auxiliary on boats of all types and sizes. Deficiencies are brought to the attention of the owner, but are not reported to any law-enforcement officials.**

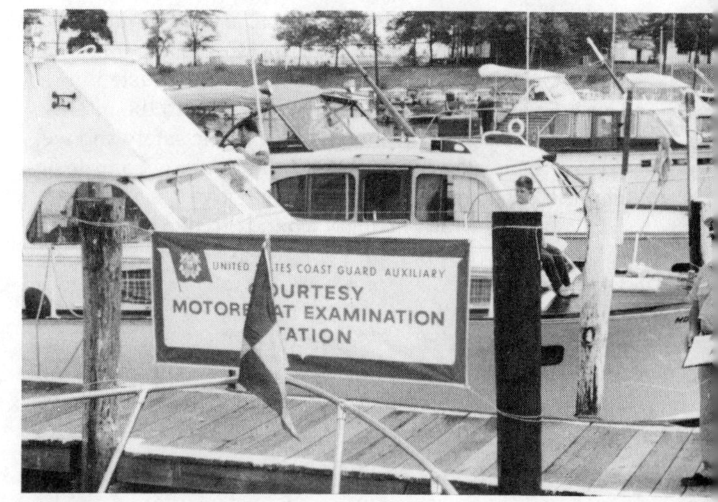

Figure 2607 **Coast Guard Auxiliarists donate many hours of their time to take part in safety patrols, and assist in search and rescue operations along with the regular Coast Guard.**

Auxiliary vessels also conduct informal "safety patrols" without official orders. Other types of patrols include those to check on aids to navigation, or those for reporting errors and needed changes on charts, as a part of the Chart Updating Program of the National Ocean Service.

Auxiliary aircraft may be used to supplement regular Coast Guard units in search and rescue operations and administrative support missions. In some areas Auxiliary aircraft make "twilight patrols" at the end of days of heavy boating activity to spot any boat that may need aid in reaching port safely.

Auxiliarists donate their time and abilities to the cause of safety on the water; there is no compensation for services rendered, except personal satisfaction, and only minor reimbursement of expenses, limited to activities under official orders.

Fellowship Activities

Units of the Auxiliary work hard at their basic responsibilities, but not to the complete exclusion of social activities. Rendezvous, parties, dances, and other events which tend to promote goodwill and fellowship are held on appropriate occasions.

Organization

The basic unit of the Coast Guard Auxiliary is the FLOTILLA. This is a local group of members and their facilities who work together in the various programs. Membership varies from the tens to the hundreds of individuals.

Flotillas are grouped geographically into DIVISIONS; this is the normal intermediate level between the local flotilla and the DISTRICT. In some cases, however, a District is divided into two or more REGIONS for better control.

Officers

Auxiliary units are headed by their own elected officers—FLOTILLA COMMANDER, DIVISION CAPTAIN, or DISTRICT COMMODORE. These are assisted by other elected and appointed officers. The USCG Aux as a whole is headed by a NATIONAL COMMODORE, with a NATIONAL VICE COMMODORE and three NATIONAL REAR COMMODORES.

At all organizational levels, there are appointed staff officers for the various operational and administrative functions.

Uniforms and Insignia

The uniforms have evolved through several distinct phases over recent years, and are now more "military" than "yachting" although the organization remains basically civilian in nature.

Various Uniforms

The USCG Aux has Dinner Dress, Service Dress, and Working uniforms; plus Tropical Blue and Tropical White outfits. Members do not need all of these, only those appropriate to their area and normal activities. There are also uniform jackets and raincoats. Uniforms are required for certain activities, such as official patrols and when doing Courtesy Marine examinations; instructors normally wear uniforms. The purchase and maintenance of uniforms is at the individual Auxiliarist's expense.

Officers' Insignia

The office held by members is indicated by sleeve braid, shoulder boards, and pin-on insignia, all patterned after those worn by regular Coast Guard officers, but with differences that distinguish the Auxiliarist.

Braid on sleeves and shoulder boards is of silver rather than gold, and is worn in stripes, half-stripes, and broad stripes in a series comparable to those worn by Ensigns to Vice Admirals in the Coast Guard and Navy. The Auxiliary, however, does not use rank titles like those of the regular and reserve services; insignia is prescribed for the office held rather than for rank titles.

Shoulder boards, worn on most uniforms, carry a varying number of stripes. Shoulder boards of the most senior Auxiliary officers are of solid silver braid with one, two, or three stars as appropriate to the office that they hold. The Auxiliary shield is placed above the stripes or stars of shoulder boards.

Collar insignia consists of the same series of designs used for officers of the U.S. Armed Forces, but with a letter "A" in blue or red superimposed. Similar insignia, in larger size, is worn on uniform jackets (windbreakers) and raincoats.

Rank insignia of the Coast Guard Auxiliary are illustrated on page 588.

Uniform Caps

Officers and members of the Auxiliary wear frame caps similar to those used by regular officers. Chin straps of members' frame caps are black; those of officers are silver. Visor ornamentation in silver is worn on the caps of senior officers.

A "working cap" of boating style is authorized in Coast Guard Blue. The fore-and-aft "garrison cap" may be worn with service dress or working uniforms when more convenient than the frame cap.

Why Not Join?

If you're an eligible boater, why not join the Auxiliary? You are sure to learn much for your boating safety and enjoyment, and there is nothing finer than the sense of satisfaction that comes from helping others. Ask any Coast Guard unit or Auxiliarist where *you* can "sign up."

27

How this organization provides public basic boating courses, as well as advanced and specialized classes for its members

THE UNITED STATES POWER SQUADRONS

The United States Power Squadrons is a non-governmental, private membership organization, self-supporting in its efforts to enhance boating safety through education. Somewhat contrary to its name, which has more historical significance than current-day literal meaning, the United States Power Squadrons now has many sailboat skippers as members, as well as many members who do not own a boat at all.

Purposes

The purposes of USPS may be described by quoting the "Objects" of the organization as stated in its constitution.

"To selectively associate congenial citizens of the United States of America of good character having a common love and appreciation of yachting as a nation-wide fraternity of boatmen.

"To encourage and promote yachting, power and sail, and to provide through local Squadrons and otherwise a practical means to foster fraternal and social relationships among citizens of the United States of America interested in yachting.

"To encourage and promote a high amateur standard of skill in the handling and navigation of yachts, power and sail; to encourage and promote the study of the science and art of navigation, seamanship, and small boat handling; to develop and promote instructional programs for the benefit of members; and to stimulate members to increase their knowledge of and skill in yachting, through instruction, self-education, and participation in marine sports events and competitions.

"To encourage its members to abide by recognized yachting traditions, customs, and etiquette.

"To encourage its members to render such altruistic, patriotic, or other civil service as it may from time to time determine or elect."

The USPS Ensign

The USPS Ensign consists of seven *blue* and six *white vertical* stripes with a union of *red* on which is the same white fouled anchor and circle of stars as on the yacht ensign; see color illustration on page 586.

The Ensign of the United States Power Squadrons may be flown on boats only when they are skippered by members of the organization. Rules for where and when it can be flown are given in Chapter 25 and on page 610.

In the form of decals and stickers, automobile emblems, boating cap insignia, etc., the USPS design can be worn or displayed only by members in good standing.

The Ensign is the name of the USPS monthly publication.

Historical Development

In 1912, the motorboat was beginning to challenge the sailboat for a place in the field of recreational boating. A group in the Boston Yacht Club felt that there was a serious lack of knowledge on the part of some who were taking up this new form of boating, and decided to improve the situation by conducting classes. This program led to formation of the "Power Squadron of the Boston Yacht Club," and set the educational basis for the USPS.

On 2 February 1914, with Charles F. Chapman in attendance, a meeting was held at the New York Yacht Club that resulted in the formation of the "United States Power Squadrons." Following World War I, the programs of the USPS were reorganized into a new format that emphasized instruction as a service to boatmen and boating in general. Growth was rapid and steady with Squadrons being organized in numerous coastal and inland boating areas.

With the end of World War II, recreational boating boomed, and with it the USPS Squadrons were formed in many new areas, including Japan and Okinawa, as well as Hawaii, Alaska, Puerto Rico, and the Canal Zone.

Membership

The USPS is a private organization and membership is by invitation. Members must be male or female U.S. citizens over 18 years of age who have met entrance qualifications set by the national organization and who have been elected by the local unit. Because the USPS is a volunteer organization dependent upon its membership for support, an important aspect of prospective members' qualifications is their willingness to "give" of their time and talents as well as to "take" the knowledge and improved skills that USPS offers.

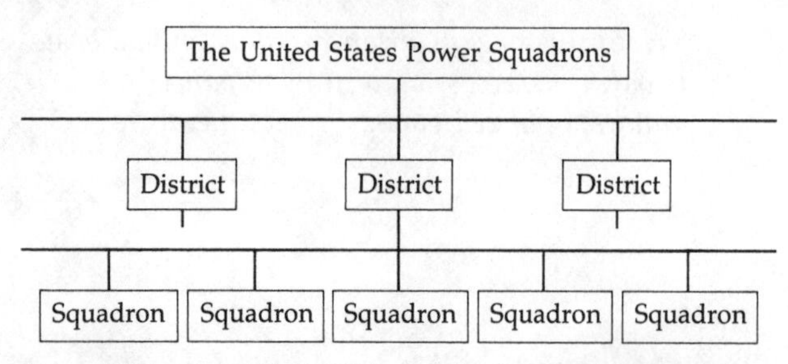

Figure 2701 **Simplified USPS organizational diagram: There are 30 Districts and more than 380 Squadrons. The number of Squadrons in a District varies from five to more than 30, and each echelon is led by elected officers.**

Family group membership is available. There must be at least one "Active Member" (either husband or wife). Others between the ages of 12 and 25, male or female, living at home or full-time students living at school with residence listed as the family address may join as "Family Members." These persons may take courses, teach, and/or serve on educational and social committees, but may not vote, hold office, or wear uniforms, including the USPS blazer. There may be more than one Active Member in a family group; for example, both husband and wife may have that status.

There are provisions for invitation to Apprentice status for youths of U.S. citizenship who are at least 12 years of age. This status may be retained until their 21st birthday. Apprentices may take USPS Advanced Grades and Elective Courses, but have no membership rights.

Organization

The basic unit of the USPS is the local SQUADRON. These have names of geographic significance and may have as few as 20 members or 500 or more.

Squadrons are grouped into DISTRICTS which are numbered; this is the intermediate level between the local unit and the NATIONAL organization.

Officers

All levels of organization are led by elected officers. Squadrons and Districts have a COMMANDER, plus several LIEUTENANT COMMANDERS and LIEUTENANTS. The USPS as a whole is directed by a CHIEF COMMANDER and five department heads with the rank of VICE COMMANDER. National Committee Chairmen and other national staff officers hold the rank of REAR COMMANDER or STAFF COMMANDER. All officers are unpaid volunteers; there is a small paid office staff at the National Headquarters at Raleigh, N.C.

Uniforms

Although the United States Power Squadrons is not a military organization, it has winter and summer, formal and informal uniforms for men and women. These are not required to be purchased by all members, and many do not. Most officers will wear uniforms for meetings, rendezvous ceremonies, and other special occasions. Uniforms may or may not be worn to various USPS educational classes.

Figure 2702 **As a public service function, each Squadron may annually present the USPS Boating Course for the general public. This covers all aspects of boating safely. USPS members may wear any one of a series of blue and white uniforms as determined by the nature of the Squadron activity and the climate. A blazer is available for informal occasions. While officers are expected to have them, all members are not required to purchase uniforms.**

Insignia of Rank

Officer's rank in the United States Power Squadrons is indicated on the uniform coat by sleeve braid and embroidered insignia. A varying number of dark blue stripes is used, with three different widths of braid used in combination. White shirts carry the sleeve insignia but not the braid.

Uniform Cap

The uniform cap can be worn both with the various uniforms and by itself with informal boating clothes. The rank of various Squadron, District, and National officers is also indicated on the cap. A USPS cap ornament may be worn on non-uniform boating caps, but in such cases no insignia of rank is shown.

Further Information

Details of the USPS uniforms and insignia of rank are spelled out in the organization's Operations Manual. An illustrated table of information on officers' sleeve braid and insignia, cap insignia, and flags will be found on pages 586 and 587.

The USPS Today

In 1985 the United States Power Squadrons comprised more than 52,000 members. There were more than 450 Squadrons assigned to 33 Districts.

Educational Programs

The educational programs of the USPS are a major activity. There are three major areas of educational effort. One of these is for the general public. The other programs are for members only, and include courses for further personal study and the internal training of the membership to improve their teaching abilities.

The USPS Boating Course

Each local Squadron is encouraged to give each year at least one series of classes of the USPS BOATING COURSE for the general public. Many Squadrons give the classes two or more times each year, varying the place and/or night of the week so as to reach the greatest number of boatmen and their families.

The USPS Boating Course is focused on the vast, and steadily growing, number of men and their families who go out on the water each year in small craft. The course covers such basic topics as:

- Know Your Boats
- Equipment, Regulations, and Safe Operation
- Boat Handling
- Elementary Seamanship
- Charts and Aids to Navigation
- Basic Navigation
- Boat Trailering
- Weather and Regional Boating
- Engine Troubleshooting
- Sailing
- Piloting

The course material is prepared and distributed on a national basis, as is the end-of-course examination. Instruction is by local USPS members, supplemented in some instances by outside experts. The instruction is given free of charge by the local Squadron, although students will usually find it desirable to purchase course material, a recommended textbook (*Chapman's*), notebook binder, plotting instruments, etc., if they do not already have them. In some areas, the course is presented as a part of the adult education program of a local school system and a small registration fee may be charged by that organization.

Family groups are encouraged to take the USPS Boating Course together; the minimum age is 12 years. Boating is widely enjoyed as family recreation, and both safety and pleasure are enhanced if all are knowledgeable.

The Advanced Grades Program

In keeping with its stated purpose of encouraging the study of the science of navigation, the USPS offers its members a progressive series of five courses.

The Piloting Course is a broad introductory course which prepares USPS members for their further studies in the Advanced Grades program. Topics covered include Safety Afloat, Charting, Piloting, Rules of the Road, Manners and Customs, and Radiotelephone. Experienced boatmen who have been taken into membership may elect to qualify in Piloting by taking the examination without attending classes.

The Seamanship Course has as its primary purpose to provide basic information in this subject to people who have had little or no experience on the water. Knowledgeable boatmen, however, will find it quite valuable as a refresher course. The material covered in "S" applies to both motorboats and sailboats. The course includes instruction in "marlinespike seamanship" to develop a working knowledge of knots, bends, hitches, and splices.

The Advanced Piloting Course covers the basic principles and more important practices of pilotage, including modern electronic methods. The lessons are of great interest and practical value to the small boat skipper because much of the knowledge can be applied at once to the use of his own boat. "AP" includes a thorough study of the mariner's compass and its errors. The student is instructed in the use of government publications concerned with piloting. He learns how to determine the height of the tide or the strength of a tidal current at any specified time. Further instruction is given in the use of charts, in the laying of courses, and in the determination of position by bearings, angles, and soundings.

The Junior Navigator Course teaches the "sailings"—the mathematical counterpart of plotting. The practical use of the marine sextant is covered, including the taking and reduction of sights on the sun, moon, planets, and stars. A main objective of "JN" is to provide instruction sufficiently complete in itself that the graduate could navigate out of sight of land and bring his boat safely back to port.

The Navigator Course deals with alternative methods, special situations, and more advanced techniques. It aims to develop greater skill in taking sights and higher precision in position finding. Throughout, the course emphasizes orderly methods of carrying out the day's work of a navigator at sea. The grade of "N" is the highest awarded by the USPS—its "PhD" in the field of boating study. The only higher status is the "Educational Achievement Award" given to those who have qualified in *all* Advanced Grades and Elective Courses.

Figure 2703 **The Advanced Grades program of USPS education culminates in the Junior Navigator and Navigator courses. These cover all aspects of celestial navigation, and the final grade of "N" is highly esteemed by USPS members.**

Figure 2704 **Engine maintenance is one of several elective courses available to members; others cover weather, sailing, marine electronics, and instructor qualification.**

The Elective Courses Program

The USPS also offers to its Active and Family members and Apprentices, five elective courses in specialty subjects; these do not constitute a series and may be taken in any sequence.

Engine Maintenance familiarizes students with the general construction, operating principles, and simple maintenance of marine gasoline and diesel engines, including outboard motors. Because one of the major objectives of the course is to make the skipper a self-reliant and effective trouble-shooter, the diagnosis of all types of engine troubles is emphasized. "EM" is *not* intended to produce trained *mechanics,* only well-informed and resourceful *users* of marine engines.

Marine Electronics focuses on the legally proper and most effective *use* of marine radios and other electronic equipment that can be installed on boats for greater safety and convenience. This is *not* a technical course, and is well within the capabilities of the average USPS member.

The Sail Course is proof that the USPS is not just for the motorboat skipper! It covers sail terminology, types of rigs and hulls, the theory of sailing with emphasis on the balance of hull and sails, stability, and other related topics.

The Weather Course is concerned with what the weather is all about and how its many variations can be predicted. Throughout the course, the student is encouraged to make observations and predictions on his own, so that he may apply the principles learned in class to his activities on the water.

The Instructor Qualification Course The USPS depends almost entirely upon its own membership as a source of instructors for all courses. This course can be of great value for members who would like to teach, yet hesitate to do so for lack of experience—and for those already teaching who would like to increase their classroom effectiveness. It covers such topic areas as the principles of teaching and the preparation of effective lesson plans. It also guides the novice teacher in the use of a wide range of teaching aids including films, slides, overhead projector transparencies, chalkboards, flip charts, and others.

Supplemental Programs

To provide further educational "packages" for members who have completed their basic boating education, the USPS has developed a series of SUPPLEMENTAL PROGRAMS. These generally are shorter and below the level of USPS courses.

Topics covered include such diverse subjects as the use and care of hand tools, predicted log racing, the magnetic compass, boat insurance, and the use of calculators for navigation.

Other Activities

Cooperative Charting Program

The USPS participates in a COOPERATIVE CHARTING PROGRAM with the National Ocean Service (formerly National Ocean Survey). In this effort, the boating activities of thousands of USPS members are coordinated toward the reporting of errors, omissions, and changes for nautical charts and related publications. The limited field facilities of NOS are thus much strengthened and expanded, resulting in more accurate charts.

Many Squadrons also participate in events of a local nature such as boat shows and the annual observance of National Safe Boating Week.

Social Activities

It's not "all work and no play" in the USPS. Squadrons hold a wide variety of social activities both afloat and ashore. Each District has a Fall and a Spring Conference. These are working sessions, but they are usually preceded or followed by evening social gatherings. The national-level Annual Meeting each January, and Spring and Fall Governing Board Meetings, combine both intense official events with those of a much lighter nature.

The Canadian Power Squadrons

In Canada, the Canadian Power Squadrons (CPS) is organized along lines generally similar to the USPS and conducts roughly equivalent training programs. There are close lines of communication and a well-developed sense of brotherhood between these two organizations.

Appendix A Canadian Boating Laws and Regulations

Boaters in Canada are governed by different laws and regulations than those discussed in Chapter 2 for U.S. owners and skippers. In most cases the Canadian regulations do not apply to U.S. boaters temporarily visiting in Canadian waters.

Boat Licenses

Licenses are required for all vessels not exceeding 15 registered tons (and each passenger craft not exceeding 20 registered tons) equipped with a motor or motors of 10 horsepower or more. Any Canada Customs office can provide a license free on request. Before the boat is operated, the license number must be marked in block characters, not less than 75 mm high and in a color that contrasts with the background, on each side of the bow or on a board permanently attached as close to the bow as possible. The number must be clearly visible from each side.

Construction Standards

Canadian Small Vessel Regulations require that certain classes of boats be built to specific current safety standards covering minimum requirements for hull construction, flotation, ventilation of explosive fumes, and fuel and electrical systems.

Basically, these standards apply to power-driven pleasure boats not longer than 6 meters and without an enclosed cabin for sleeping, and to *all* pleasure craft, regardless of length or type of accommodation, if fitted with one or more gasoline engines for propulsion or generation of electrical power.

Boats that meet the prescribed standards will have an appropriate decal prominently displayed.

Safe Loading and Engine Power

Canadian law requires every pleasure boat 5 meters or less in length, powered with an outboard motor of 10 horsepower or more, to carry a plate issued by Transport Canada stating the recommended maximum load and engine power. The recommended load capacity includes weight of passengers, engine, fuel and fuel tanks, and all other equipment.

Applications for these plates can be obtained from any Canada Customs or Coast Guard Ship Safety office. The completed form and appropriate fee should be sent to Canadian Coast Guard, Ship Safety, Place de Ville, Ottawa, Ontario K1A 0N7.

Ventilation

Canadian Small Vessel Regulations require any enclosed space in which an inboard gasoline engine is installed to be ventilated efficiently by suitable ventilators and an exhaust fan. Although this specific regulation applies only to inboard engines, all enclosed spaces in both inboard and outboard powerboats should be well ventilated if they contain fuel tanks or other containers of gasoline.

The number, size, and other details of ventilation ducts and fans are generally the same as the U.S. requirements discussed on pages 66-68.

Reckless Operation

All Rules of the Road must be obeyed, and the operation of any craft must not be "reckless."

Anyone operating a boat, air-cushion vehicle, water skis, surfboard, or any towed object in a manner dangerous to navigation, life, or limb is guilty of an indictable offense and liable to imprisonment or punishment on summary conviction.

Under the Criminal Code of Canada, this offense includes:

- operating a vessel when impaired,
- towing a person on water skis after dark or without another person keeping watch, and
- failing to stop at the scene of an accident.

Charges can be laid against a reckless operator by "laying an information," a procedure that requires making a sworn statement before a Magistrate or Justice of the Peace.

Equipment Requirements for Canadian Boats

The list of required equipment for boats licensed in Canada is more extensive than it is for their counterparts in the United States. Canadian craft are divided into five classes based on overall length in metric units.

Vessels Not Longer Than 5.5 Meters

1. One approved small vessel life jacket, or approved PFD, or approved lifesaving cushion for each person on board.
2. Two oars and rowlocks or two paddles.
3. One bailer or one manual pump.
4. One Class B-I fire extinguisher if the vessel has an inboard motor, permanently fixed or built-in fuel tanks, or a cooking or heating appliance that burns liquid or gaseous fuel.*
5. Navigation lights, if permanently fitted, that comply with the Collision Regulations.
6. Some means of making an efficient sound signal.

Vessels Longer Than 5.5 Meters But Not Longer Than 8 Meters

1. One approved small vessel life jacket or approved PFD for each person on board.
2. Two oars and rowlocks or two paddles, or one anchor with not less than 15 m of cable, rope, or chain.
3. One bailer or one manual pump.
4. One Class B-I fire extinguisher if the vessel is power-driven, or has a cooking or heating appliance that burns liquid or gaseous fuel.*
5. One throwing device that may be either an approved lifesaving cushion, a buoyant heaving line (recommended minimum length 15 m), or an approved lifebuoy 508, 610, or 762 mm in diameter.

*Details on various acceptable fire extinguishers and pyrotechnic distress signals will be found in the *Canadian Coast Guard Boating Handbook*.

6. Six approved pyrotechnic distress signals (flares) of which at least three must be Types A, B, or C, and the remaining three Types A, B, C, or D. (Note: These are not required if the vessel is engaged in or preparing for racing competition and has no sleeping accommodation, or is operating in a river, canal, or lake in which the boat can never be more than one mile from shore, or is propelled solely by oars or paddles.)
7. Navigation lights, if permanently fitted, that comply with the requirements of the Collision Regulations.
8. Some means of making an efficient sound signal.

Vessels Longer Than 8 Meters But Not Longer Than 12 Meters

1. One approved lifejacket or PFD for each person on board. (Note: Sailing vessels without an enclosed sleeping cabin may instead carry one approved PFD for each person on board.)
2. One approved lifebuoy 610 or 762 mm in diameter.
3. One buoyant heaving line at least 15 m long.
4. One bilge pump and one manual bailer.
5. Twelve approved pyrotechnic distress signals of which at least six must be Types A, B, or C, and the remaining six Types A, B, C, or D.
6. One anchor with at least 15 m of cable, rope, or chain.
7. One Class B-II fire extinguisher if the vessel is power-driven, or has a cooking or heating appliance that burns liquid or gaseous fuel.
8. Navigation lights and sound signalling apparatus that permit the vessel to comply with the Collision Regulations.

Vessels Longer Than 12 Meters But Not Longer Than 20 Meters

1. One approved standard life jacket or one approved small vessel life jacket for each person on board.
2. One approved lifebuoy 762 mm in diameter, or two approved lifebuoys 610 mm in diameter.
3. One buoyant heaving line at least 15 m long.
4. Twelve approved pyrotechnic distress signals of which at least six must by Types A, B, or C and the remaining six can be Type A, B, C, or D.
5. One anchor with at least 15 m of cable, rope, or chain.
6. Two fire buckets or other effective means of carrying water to any part of the vessel to extinguish a fire.
7. (a) A manual or power-driven pump outside the machinery space with one fire hose and nozzle that can direct a jet of water into any part of the vessel.
   (b) Two Class B-II fire extinguishers, one of which is next to the sleeping cabin entrance and the other next to the machinery space entrance.
8. Efficient bilge-pumping system.
9. One additional Class B-II fire extinguisher if the vessel is power driven, or has a cooking or heating appliance that burns liquid or gaseous fuel.
10. One fire axe.
11. Navigation lights and sound signaling apparatus that permit the vessel to comply with the Collision Regulations.

Other Vessels

Sailboards and water scooters must have one approved PFD for each person.

Vessels longer than 20 m require larger amounts of equipment as specified in the *Canadian Coast Guard Boating Handbook*.

Radar Reflectors

Every pleasure boat shorter than 20 m *or* constructed primarily of materials other than metal must have a passive radar reflector that provides a response in the 3-cm marine radar band equivalent to an effective reflecting area not less than 10 m^2 through 360 degrees; this must be located, if possible, at least 4 m above the water.

Appendix B Chart and Cruising Information Sources

Charts of Various Waterways

U.S. Coastal Waters and Great Lakes Atlantic, Pacific, and Gulf Coasts; the Atlantic and Gulf Intracoastal Waterways; the Hudson River north to Troy, New York; the Great Lakes and connecting rivers; Lake Champlain; New York State Canals; and the Minnesota-Ontario Border Lakes: National Ocean Service, Distribution Division (NCG33), Riverdale, MD 20737, or local sales agents listed in chart catalogs.

New York State Canals Bound booklet of charts of the Champlain, Erie, Oswego, and Cayuga-Seneca canals: National Ocean Service Distribution Division, (N/CG33) Riverdale, MD 20737 or local sales agents.

Mississippi River and Tributaries Middle & Upper Mississippi, Cairo, Illinois, to Minneapolis, Minnesota; Middle Missisipi River from Cairo, Illinois, to Grafton, Illinois; Mississippi River from Cairo, Illinois, to Gulf of Mexico; Small Boat Chart, Alton, Illinois, to Clarksville on the Mississippi River and Grafton, Illinois, to LaGrange, Illinois, on the Illinois River; Illinois Waterways from Grafton, Illinois, to Lake Michigan at Chicago and Calumet Harbor: U.S. Army Engineer District, 210 Tucker Boulevard North, St. Louis, MO 63101.

Mississippi River and Tributaries, below Ohio River Mississippi River Commission, P.O. Box 80, Vicksburg, MS 39180. This office also has a free booklet, *Mississippi River Navigation,* which discusses its history, development, and navigation.

Mississippi River and connecting waterways, north of the Ohio River U.S. Army Engineer Division, North Central, 219 S. Dearborn St., Chicago, IL 60605.

Ohio River and Tributaries; Pittsburgh, Pennsylvania, to the Mississippi River U.S. Army Engineer Division, P.O. Box 1159, Cincinnati, OH 45201.

Tennessee and Cumberland Rivers U.S. Army Engineer District, P.O. Box 1070, Nashville, TN 37202; also Tennessee Valley Authority, Maps and Engineering Records Section, 102A Union Building, Knoxville, TN.

Missouri River and Tributaries U.S. Army Engineers, Missouri River Division, P.O. Box 103 DTS, Omaha, NB 68101.

Canadian Waters Chart Distribution Office, Department of the Environment, P.O. Box 8080, 1675 Russel Road, Ottawa, Ontario K1G 3H6. Charts include coastal waters; Canadian sections of the Great Lakes including Georgian Bay; the St. Lawrence River; Richelieu River; Ottawa River; the Rideau Waterway; and other Canadian lakes and waterways. Indexes of charts for any area are free, and chart prices and details are given in a Coastal and Inland Waters catalog.

Waters of Other Nations Defense Mapping Agency Hydrographic/Topographic Center, Office of Distribution Services (DDCP), 6500 Brooks Lane, Washington, DC 20315, or authorized sales agents. A general catalog (free) and nine regional catalogs are available.

Notices to Mariners

Notice to Mariners Weekly publication: Defense Mapping Agency Hydrographic/Topographic Center, Washington, DC 20390.

Local Notices to Mariners Issued as necessary by the Commanders of Coast Guard Districts, and available at District offices.

Coast Pilots

U.S. Waters Atlantic, Gulf, and Pacific Coasts; Atlantic and Gulf Intracoastal Waterways; the Great Lakes: National Ocean Service, Rockville, MD 20852, its distribution offices, or sales agents as listed semi-annually in *Notice to Mariners*.

Canadian Waters Chart Distribution Office, Department of the Environment, P.O. Box 8080, 1675 Russel Road, Ottawa, Ontario K1G 3H6. Coastal and Inland Waters Catalog gives details. Descriptive list of *Pilots* and *Sailing Directions* is also available.

Light Lists

Light Lists published by the U.S. Coast Guard; Superintendent of Documents, Washington, DC 20402, or many of the sales agents listed in NOS Chart Catalogs.

Tide and Current Tables

Tide Tables, Current Tables, Tidal Current Charts, Tidal Current Diagrams National Ocean Service, Rockville, MD 20852, its distribution offices, or many of the sales agents listed in the NOS chart catalogs.

Rules of the Road

Rules of the Road COMDTINST 16672.2A. Booklet contains full text of the 1972 International Rules of the Road and the 1980 U.S. Inland Navigational Rules, and all Annexes; illustrated. Also contains information on Demarcation Lines separating the two sets of Rules, and the text of the Bridge-to-Bridge Radiotelephone Act. Sold by local chart sales agents and Regional Bookstores of the Government Printing Office.

Handbook of Boating Laws Separate editions for Southern, Northeastern, North Central, and Western states list state requirements for registering, numbering, equipment, and small boat operation; fuel tax laws, and applicable federal regulations; Outboard Boating Club, 401 N. Michigan Ave., Chicago, IL 60611.

Cruising Guides

Cruising the Canals Free booklet with map of the New York State Canal system; Waterways Maintenance Division, State Department of Transportation, 1220 Washington St., Albany, NY 12232.

Waterway Guide Publication provides detailed information on inland waterways in four editions: Northern, Maine to New York; Middle Atlantic, New York to Sea Island, Georgia; Southern, Sea Island, Georgia, to Florida and Gulf Coast to New Orleans; Great Lakes, New York to Great Lakes, with connecting Canadian canals and rivers; Waterway Guide, Inc., 93 Main St., Annapolis, MD 21401.

Intracoastal Waterway Booklets Comprehensive descriptions of the Intracoastal Waterway, with data on navigation, charts, distances. Prepared by U.S. Army Corps of Engineers in two sections; (1), Atlantic, Boston to Key West; (2), Gulf, Key West to Brownsville, Texas. Superintendent of Documents, Washington, DC 20402.

Intracoastal Waterway Bulletins Frequent bulletins giving latest information on the condition of the Intracoastal Waterway are published by the U.S. Army Corps of Engineers; available from District Offices at: 803 Front St., Norfolk, VA 23510; P.O. Box 1890, Wilmington, NC 28401; P.O. Box 919, Charleston, SC 29402; P.O. Box 889, Savannah, GA 31402; P.O. Box 4970, Jacksonville, FL 32201; P.O. Box 2288, Mobile, AL 36601; and P.O. Box 1229, Galveston, TX 77551.

Cruising the Pacific Coast, Acapulco to Skagway Carolyn and John West, W.W. Norton & Co., 500 Fifth Ave., New York, NY 10110.

Cruising the San Juan Islands Bruce Calhoun, W. W. Norton & Co., 500 Fifth Ave., New York, NY 10110.

Yachtsman's Guide to the Greater Antilles By Harry Kline. Virgin Islands, Puerto Rico, Dominican Republic, and Haiti. Tropic Isle Publishers, Inc., P.O. Box 611141, North Miami, FL 33161.

Cruising Charts, Guides, Booklets Texas Travel Service, P.O. Box 1459, Houston, TX 77001.

Quimby's Harbor Guide The navigable Mississippi from Minneapolis to New Orleans, plus harbors on the Illinois and Arkansas waterways, and St. Croix and Black rivers. Covers harbors, services, cities, towns, transportation, tourist interests, some history, locks and dams, sources of navigation charts and books, hazards. Mildred Quimby, P.O. Box 85, Prairie du Chien, WS 53821. (319) 873-2369.

Cruising Guide to Lake Ontario Harbor information, facilities, Lake Ontario and the Thousand Islands area of the St. Lawrence River; Welland Canal lock information. Cruising Guide to Lake Ontario, Box 338, Youngstown, NY 14174.

California Coastal Passages By Brian M. Fagan. Facilities and sailing directions, San Francisco, California, to Ensenada, Mexico. Capra Press and ChartGuide, Ltd., P.O. Box 2068, Santa Barbara, CA 93120.

Guide for Cruising Maryland Waters Prepared by Maryland Dept. of Natural Resources, Tidewater Administration. Twenty full-color charts, with more than 200 courses and distances plotted; marina and facility information reference listing. Department of Natural Resources, Tawes State Office Bldg., Annapolis, MD 21401. A Maryland Basic Boating Course is available from the same address.

U.S. Coast Guard Publications

This Is the Seal of Safety. . .Get a Free Motorboat Examination Aux-204. Flyer explains the Auxiliary Courtesy Marine Examination and the standards which must be met to be awarded a CME decal. Commandant (G-BAU) U.S. Coast Guard, Washington, DC 20593.

Federal Requirements for Recreational Boats Digest of boating laws and regulations covering numbering, accidents, sales to aliens, law enforcement, documentation, and equipment requirements, plus safety suggestions. Available at all Coast Guard offices.

Visual Distress Signals CG-152. Illustrated booklet provides description and guidance for use of distress signals suitable for boats. Free from Coast Guard District offices, or Headquarters, U.S. Coast Guard, Washington, DC 20593.

The Skipper's Course Booklet is a "do-it-yourself" program in basic boating safety, with informal text and an exam at the end of the course. A certificate is awarded to all who pass. Bookstore #15, P.O. Box 713, Pueblo, CO 81002.

Modifications (for a new look in U.S. Aids to Navigation) ANSC SN 3022. Illustrated information on the changes to U.S. aids to navigation that are being made to bring them into conformity with the IALA-B system. Copies free from Coast Guard offices or Commandant (G-NSR-1), Coast Guard Headquarters, Washington, DC 20593.

Publications of the Defense Mapping Agency Hydrographic/Topographic Center

DMAHTC publications are available from its Office of Distribution Services (DDCP), 6500 Brooks Lane, Washington, DC 20315, or its authorized sales agents.

DMA Catalog of Maps, Charts and Related Products, Part 2, Vol. X Catalog lists some of the more popular index charts, world charts, general nautical charts, magnetic charts, oceanographic and bottom sediment charts, aeronautical charts, Loran charts published by the agency.

Sailing Directions Books supplementing DMAHTC charts contain descriptions of coastlines, harbors, dangers, aids, port facilities, and other data that cannot be shown conveniently on charts.

Daily Memorandum A synopsis of the latest information relating to dangers and aids to navigation, together with advance items that will appear in *Notices to Mariners*. Issued locally by Branch Hydrographic Offices. Most urgent items are also broadcast by radio.

Pilot Charts

No. 16 Pilot Chart of the North Atlantic Ocean (monthly).

No. 55 Pilot Chart of the North Pacific Ocean (monthly).

No. 106 Atlas of Pilot Charts, South Atlantic Ocean and Central America waters.

No. 107 Atlas of Pilot Charts, South Pacific and Indian Oceans.

Lists of Lights and Fog Signals

Pub. 110 and 111 Coast of North and South America (only the seacoast lights of the United States), the West Indies, and Hawaiian Islands.

Pub. 112 Islands of the Pacific and Indian Oceans, Australia, Asia, and the East Coast of Africa.

Pub. 113 West Coasts of Europe and Africa, the Mediterranean Sea, Black Sea, and the Sea of Azov.

Pub. 114 Britain and Ireland, English Channel, and North Sea.

Pub. 115 Norway, Iceland, and Arctic Ocean.

Pub. 116 Baltic Sea with Kattegat, Belts and Sound, and Gulf of Bosnia.

National Oceanographic Office

Pub. 234 Breakers and Surf; Principles in Forecasting.

Pub. 601 Wind, Sea, and Swells; Theory of Relations in Forecasting.

Pub. 602 Wind, Waves at Sea, Breakers, and Surf.

Miscellaneous DMAHTC Publications

Pub. 9 American Practical Navigator, originally by Nathaniel Bowditch. Volume I (1984); Volume II (1981).

Pub. 102 International Code of Signals.

Pub. 117A, 117B Radio Navigation Aids, Marine Direction-Finding Stations, Radio Beacons, Time Signals, Navigational Warnings, Distress Signals, Medical Advice and Quarantine Stations, Loran, and Regulations Covering the Use of Radio in Territorial Waters.

Pub. 150 World Port Index.

Pub. 217 Maneuvering Board Manual.

Pub. 226 Handbook of Magnetic Compass Adjustment and Compensation.

Pub. 1310 Radar Navigation Manual.

Weather Publications

Marine Weather Service Charts A series of 15 charts contains broadcast schedules of radio stations, National Weather Service telephone numbers, and locations of warning display stations. Charts cover the Atlantic, Gulf, and Pacific coasts, and waters adjacent to Hawaii, Puerto Rico, the Virgin Islands, and Alaska. Distribution Division (N/CG33), National Ocean Service, 6501 Lafayette Ave., Riverdale, MD 20737.

Marine Weather Services and High Seas Storm Information Service Brochures contain marine weather information in capsule form. Distribution Division (N/CG33), National Ocean Service, 6501 Lafayette Ave., Riverdale, MD 20737.

Worldwide Marine Weather Broadcasts Broadcast schedules of marine weather information from all areas of the world where such service is available. Superintendent of Documents, U.S. Government Printing Office, Washington, DC 20402.

Hurricane, the Greatest Storm on Earth Updated hurricane information, 34 pages; Stock Number 003-018-00018-1. Superintendent of Documents, Government Printing Office, Washington, DC 20402.

The Daily Weather Map, Weekly Series Available by annual subscription. A complete explanation of the maps, including all symbols and tables, appears on the reverse side of the Sunday map only. Published by National Weather Service. Superintendent of Documents, Washington, DC 20402.

Miscellaneous Publications

The American Nautical Almanac Compact publication from the United States Naval Observatory contains all ephemeris material essential to the solution of problems of navigational position; star chart is included. Superintendent of Documents, Washington, DC 20402, or sales agents.

Dutton's Navigation and Piloting By Elbert S. Maloney; 14th Edition, 1985. U.S. Naval Institute, Annapolis, MD 21402.

A Mariner's Guide to Rules of the Road By William H. Tate. A concise and comprehensive presentation of the 1972 International Rules of the Road and current U.S. Inland Rules; extensively illustrated. U.S. Naval Institute, Annapolis, MD 21402.

First Aid Illustrated 160-page manual provides only first-aid instruction. Superintendent of Documents, Government Printing Office, Washington, DC 20402.

Miscellaneous Publication No. 9 The Ship's Medicine Chest and First Aid at Sea. Book from the United States Health Service includes special instructions for emergency treatment, and first aid by radio; 496 pages, illustrated. Superintendent of Documents, Washington, DC 20402.

Appendix C State Numbering Certificates

The Federal Boat Safety Act of 1971 continued the pattern previously established by the Federal Boating Act of 1958 whereby powered craft could be registered and numbered by state authorities if the state had set up a program for such actions in accordance with federal standards. Approval of the state system by the U.S. Coast Guard is required in each case for the sake of state-to-state uniformity.

The Coast Guard has officially approved the numbering systems of all states except Alaska and New Hampshire. In these jurisdictions application forms for Coast Guard registration numbers may be obtained through local post offices or any Coast Guard facility. In Guam and U.S. Samoa application should be made to the Officer in Charge, Marine Inspection, U.S. Coast Guard.

Boats principally used in states having an approved numbering law are numbered exclusively by the state; Coast Guard numbers are obsolete. In compliance with the Federal Act all of these state laws grant 60-day reciprocity to out-of-state boats awarded numbers pursuant to federal law or a federally approved state system.

The Outboard Boating Club of America (401 No. Michigan Ave., Chicago, IL 60611) has compiled a list of state agencies responsible for boat numbering laws.

Alabama Department of Conservation, State Administrative Building, Montgomery, AL 36130.

Alaska Department of Public Safety, P.O. Box 6188 Annex, Anchorage, AK 99502.

Arizona Game and Fish Department, 2222 W. Greenway Rd., Phoenix, AZ 85023.

Arkansas Licensing Division, Department of Finance & Administration, P.O. Box 1272, Little Rock, AR 72201.

California Vessel Registration Section, Department of Motor Vehicles, P.O. Box 11319, Sacramento, CA 95818.

Colorado Division of Parks & Outdoor Recreation, 13787 S. Highway 85, Littleton, CO 80125.

Connecticut Department of Environmental Protection, Rm 217, State Office Building, Hartford, CT 06106.

Delaware Department of Natural Resources, P.O. Box 1401, Dover, DE 19903.

Florida Department of Natural Resources, 3900 Commonwealth Blvd., Tallahassee, FL 32303.

Georgia Department of Natural Resources, 270 Washington St. S.W., Atlanta, GA 30334.

Hawaii Harbors Division, Department of Transportation, 79 S. Nimitz Highway, Honolulu, HI 96813.

Idaho Department of Parks & Recreation, 2177 Warm Springs Avenue, Statehouse Mail, Boise, ID 83720.

Indiana Department of Natural Resources, 606 State Office Building, Indianapolis, IN 46204.

Iowa State Conservation Commission, Wallace Building, Des Moines, IO 50319.

Kansas Forestry, Fish & Game Commission, R.R. #2, Box 54A, Pratt, KS 67124.

Kentucky Department of Natural Resources, Div. of Water Patrol, Capitol Plaza Tower, Frankfort, KY 40601.

Louisiana Department of Wildlife & Fisheries, P.O. Box 15570, Baton Rouge, LA 70895.

Maine Watercraft Section, Division of Recreational Safety & Registration, Department of Inland Fisheries & Wildlife, 284 State Street, Augusta, ME 04333.

Maryland Licensing & Consumer Services, Department of Natural Resources, P.O. Box 1869, Tawes State Office Building, Annapolis, MD 21404-1869.

Massachusetts Division of Marine & Recreational Vehicles, 100 Nashua St., Boston, MA 02114.

Michigan Bureau of Driver & Vehicle Services, Department of State, 7064 Crowner Drive, Lansing, MI 48918.

Minnesota Department of Natural Resources, 500 Lafayette Rd., Centennial Office Building, St. Paul, MN 55146.

Mississippi Department of Wildlife Cons., P.O. Box 451, Jackson, MS 39205.

Missouri State Water Patrol, Department of Public Safety, P.O. Box 603, Jefferson City, MO 65102.

Montana Registrar's Bureau, Motor Vehicle Division, Department of Justice, Deer Lodge, MT 59722.

Nebraska State Game & Parks Commission, P.O. Box 30370, Lincoln, NB 68503.

Nevada Division of Law Enforcement, Department of Wildlife, Box 10678, Reno, NV 89520.

New Hampshire Division of Safety Services, Department of Safety, Hazen Drive, Concord, NH 03305. The State registers powerboats operated on inland or nontidal waters. As this registration system does not comply with the Federal Boating Act of 1958, the Coast Guard has retained the responsibility for registering and numbering undocumented vessels with more than 10 horsepower chiefly used on navigable waters of the United States within the territorial limits of New Hampshire.

New Jersey Bureau of Marine Law Enforcement, Box 7068, West Trenton, NJ 08625.

New Mexico Bureau of Boating Safety, State Park & Recreation Division, Natural Resources Department, P.O. Box 1147, Santa Fe, NM 87504-1147.

New York Marine & Recreational Vehicles Agency, Building No. 1, Empire State Plaza, Albany, NY 12238.

North Carolina Division of Boating, Wildlife Resources Commission, Archdale Building, Raleigh, NC 27611.

North Dakota State Game & Fish Department, 2121 Lovett Avenue, Bismarck, ND 58505.

Ohio Watercraft Division, Department of Natural Resources, Fountain Square, Columbus, OH 43224.

Oklahoma State Tax Commission, 2501 N. Lincoln Blvd., Oklahoma City, OK 73194.

Oregon State Marine Board, No. 505, 3000 Market St. N.E., Salem, OR 97310.

Pennsylvania Pennsylvania Fish Commission, P.O. Box 1673, Harrisburg, PA 17105-1673.

Rhode Island Division of Boat Registration, Department of Environmental Management, 22 Hayes St., Providence, RI 02903.

South Carolina Division of Boating, Wildlife & Marine Resources Department, P.O. Box 12559, Charleston, SC 29412.

South Dakota Department of Game, Fish & Parks, 445 E. Capitol, Pierre, SD 57501.

Tennessee Wildlife Resources Agency, P.O. Box 40747, Nashville, TN 37204.

Texas Parks & Wildlife Department, 4200 Smith School Road, Austin, TX 78744.

Utah Motor Vehicle Division, 1095 Motor Ave., Salt Lake City, UT 84116.

Vermont Marine Division, Department of Public Safety, 103 Main St., P.O., Montpelier, VT 05602.

Virginia Game & Inland Fisheries Commission, P.O. Box 11104, Richmond, VA 23230-1104.

Washington Department of Licensing, P.O. Box 9909, Olympia, WA 98504.

West Virginia Department of Natural Resources, 1800 Washington St. E., Charleston, WV 25305.

Wisconsin Bureau of Law Enforcement, Department of Natural Resources, P.O. Box 7921, Madison, WI 53707.

Wyoming Game & Fish Department, 5400 Bishop Blvd., Cheyenne, WY 82002.

Dist. of Columbia Harbor Patrol, Metropolitan Police Department, 550 Water St., S.W., Washington, DC 20024.

Puerto Rico Maritime Department, Puerto Rico Ports Authority, GPO Box 2829, San Juan, PR 00936.

Virgin Islands Department of Conservation & Cultural Affairs, P.O. Box 4340, St. Thomas, VI 00803.

Appendix D

USPS-USCG Auxiliary Course References

USPS References for Instruction Courses

Boating Course

Section 1 Know Your Boats. Pages 15–20, 29–30, 43–43, 107–108, 283–285, 290
Section 2 Equipment, Regulations, and Safe Operation. Pages 54–55, 57–75, 119–120, 290
Section 3 Boat Handling. Pages 77–89, 91–105, 290, 589–594
Section 4 Elementary Seamanship. Pages 121–122, 140–169, 215–227, 233–253, 291
Section 5 Charts and Aids to Navigation. Pages 365–387, 399–431, 577–580
Section 6 Basic Navigation. Pages 335–346, 438–446
Section 7 Boat Trailering. Pages 124–137
Section 8 Weather; Regional Boating. Pages 271–273, 300–333, 584–585
Section 8A Coastal Boating. Pages 218, 451–469
Section 8B Great Lakes Boating. Pages 53–54, 213, 371
Section 8C Inland Boating. Pages 199–213
Section 9 Engine Troubleshooting. Pages 266–269, 276–277
Section 10 Sailing. Pages 171–195
Section 11 Piloting. Pages 346–351, 440–449, 475–480

Seamanship Course

Section 1 Marlinespike Seamanship. Pages 283–299
Section 2 Rules of the Road; Man Overboard. Pages 77–107, 226–228, 589–594
Section 3 Stability, Trim, and Hull Forms. Pages 17–18, 107–108, 116–120
Section 4 Boat Handling I. Pages 152–155, 162–163, 228–229, 269–273, 321–327
Section 5 Docking. Pages 140–152, 156–166, 190–193
Section 6 Anchoring, Rafting, and Mooring. Pages 233–255
Section 7 Boat Handling II. Pages 215–221, 278–281, 301–318
Section 8 Stranding, Towing, Damage Control. Pages 221–225, 229–230
Section 9 Radiotelephone. Pages 515–532, 582
Section 10 Boating Customs. Pages 563–573, 582–583, 586–588
Section 11 Maintenance. Pages 276–277
Section 12 Plotting Problems. Pages 444–449, 475–496

Piloting Course

Section 1 Equipment and Government Regulations. Pages 39–75
Section 2 Charts. Pages 399–431
Section 3 Rules of the Road. Pages 77–105, 589–594
Section 4 Aids to Navigation. Pages 365–387, 577–580
Section 5 Plotting and Labeling. Pages 444–449
Section 6 Compass. Pages 335–363
Section 7 Bearings. Pages 475–494
Section 8 Marine Radio for Small Craft. Pages 517–533
Section 9 A Day's Cruise. Pages 475–503

Advanced Piloting Course

Section 1 Charts. Pages 201–203, 213, 400–425
Section 2 Aids to Navigation. Pages 364–389
Section 3 The Compass. Pages 334–363
Section 4 Bearings. Pages 25–26, 33, 436–438, 474–489
Section 6 Time-Speed-Distance. Pages 434–435, 440–443, 496–497
Section 7 Tides and Tide Tables. Pages 451–461
Section 8 Tidal Currents and Tidal Current Tables. Pages 461–467
Section 9 Current Effects. Pages 469–473
Section 10 Positioning. Pages 475–496
Section 11 The Log. Pages 556–559

USCG Auxiliary References for Study Courses

Boating Skills and Seamanship Course

The Safe Way to Boating Enjoyment. Pages 106–124, 256–281, 321–323
Boater's Language and Trailering. Pages 14–37, 124–137, 195
Boating Handling. Pages 106–124, 138–169, 232–255
Legal Requirements. Pages 38–75
Rules of the Road. 76–105
Aids to Navigation. Pages 364–399, 577–580
Piloting. Pages 334–363, 400–451, 475–513
Marine Engines. Pages 109–112, 266–272
Marlinespike Seamanship. Pages 224–225, 282–299
Sailing. Pages 170–197
Weather. Pages 300–333
Radiotelephones. 514–535
Locks and Dams. Pages 209–212

Sailing and Seamanship Course

What Makes a Sailboat. Pages 171–172, 175–179
How a Boat Sails. Pages 187–190
Basic Sailboat Maneuvering. Pages 173–175, 190–193
Rigging and Boat Handling. Pages 176–179
Weather Forecasting and Heavy Weather Sailing. Pages 301–333
Rules of the Road and Legal Requirements. Pages 38–55, 77–80, 82, 91–92, 99
Sailing Seamanship. Pages 171–195
Engines for Sailboats. Page 190
Tuning and Variant Rigs. Pages 179–184
Equipment for You and Your Boat. Pages 56–75
Sailboat Piloting. Pages 194–195
Radiotelephones. Pages 517–531

Appendix E The Metric System

The Metric Conversion Act of 1975 declared the policy of the United States to increase the use of the metric system of measurement on a voluntary basis with the goal of "a nation predominantly, although not exclusively, metric." Accordingly, this book includes metric equivalents along with the customary (English) units wherever applicable.

The proper name is the International System of Units, abbreviated (from the French *System Internationale*) as SI. SI provides a logical and interrelated framework for measurements in science, industry, commerce, and other forms of human endeavor.

The modern metric system is based upon a foundation of base units, with multiples and submultiples expressed in a decimal system using various prefixes.

The Units

Those base units of most interest to boatmen are for length, weight, and temperature. These units and their symbols are:

Length	meter (m)
Mass (weight)	kilogram (kg)
Temperature	Celsius (°C)

Note that Celsius (formerly Centigrade) degrees are the same as degrees Kelvin, the scientific temperature scale of the metric system; the zero point of the Celsius scale is not the same as that for the Kelvin scale, however.

The symbols, shown in parentheses, are *not* abbreviations. They are written in lower case letters except for those units (such as Celsius) named after a person. Periods are not used with any symbols.

Derived Units

Other units in the metric system are derived from the base units. Typical of those expressed in combinations of base units are:

Area	square meter (m^2)
Volume	cubic meter (m^3)
Speed, velocity	meter per second (m/s)

Multiples and Submultiples

Units larger and smaller than the base units are formed by adding prefixes decimally to make multiples and submultiples. Symbols for the prefixes are added to the symbols for the base units.

1,000	10^3	kilo	k
100	10^2	hecto	h
10	10^1	deka	da
0.1	10^{-1}	deci	d
0.01	10^{-2}	centi	c
0.001	10^{-3}	milli	m

There are others, both larger multiples and smaller submultiples, but they are of little interest to boatmen.

Retained Customary Units

Certain units that are not part of the metric system are so widely used that there is no plan to abandon them. These include days, hours, minutes, and seconds as measurements of time; and degrees, minutes, and seconds as measurements of arc. (The radian is a metric measurement of arc, but its use has been limited to scientific applications.)

Other customary units, to be continued "for a limited time, subject to future review" include the nautical mile and knot as measurements of distance and speed; and bar and millibar as measures of atmospheric pressure.

Conversion

The conversion factors here are rounded-off for practical use, and do not, in most cases, yield exact results.

Customary Units to Metric

	Known Value	*Multiply by*	*To Find*	
in	inches	25.4	millimeters	mm
ft	feet	0.3048	meter	m
yd	yards	0.9144	meters	m
s mi	statute miles	1.609	kilometers	km
n mi	nautical miles	1.852	kilometers	km
oz	ounces (weight)	28.35	grams	g
lb	pounds	0.4536	kilograms	kg
oz	ounces (liquid)	30.28	liters	l
qt	quarts	0.9464	milliliters	ml
gal	gallons	3.785	liters	1
°F	Fahrenheit temperature	5/9 after subtracting 32	Celsius temperature	°C

Metric Units to Customary

	Known Value	*Multiply by*	*To Find*	
cm	centimeters	0.3937	inches	in
m	meters	3.281	feet	ft
m	meters	1.094	yards	yd
km	kilometers	0.6214	stat. miles	s mi
km	kilometers	0.5400	naut. miles	n mi
g	grams	0.03527	ounce (weight)	oz
kg	kilograms	2.205	pounds	lb
ml	milliliters	0.03302	ounce (liquid)	oz
l	liters	1.057	quarts	qt
1	liters	0.2642	gallon	gal
°C	Celsius temperature	9/5 then add 32	Fahrenheit temperature	°F

Appendix F USPS Guide to On-board Flag Display

FLAG	WHEN FLOWN	POWER YACHT WITHOUT MAST	POWER YACHT WITH SIGNAL MAST	SAILING YACHT WITH ONE MAST	POWER OR SAIL YACHT WITH TWO MASTS
U.S. ENSIGN, U.S. YACHT ENSIGN	0800 to sunset	Flag staff	Flag staff	Flag staff. Optional when underway: peak of gaff if so rigged or 2/3 up leech of mainsail	Flag staff. Optional when underway: peak of aftermost gaff if so rigged or 2/3 up leech of aftermost sail
FOREIGN ENSIGN	According to local custom when in foreign waters	Bow staff	Starboard spreader (outboard halyard)	Starboard spreader (outboard halyard)	Starboard spreader (outboard halyard) of foremost mast
	When foreign dignitary on board	Bow staff	Bow staff	Bow staff or forestay	Bow staff or forestay
USPS ENSIGN	Day and night *except* 0800 to sunset when flown in lieu of U.S. ensign* *Only when in commission and under command of USPS member*	Antenna amidships or, if no suitable antenna, from bow staff*	Starboard spreader. If foreign ensign flown: inboard starboard spreader halyard, if equipped, or port spreader*	As for power yacht with signal mast*	Foremost starboard spreader*
OFFICER (Incumbent or past)	Day and night when in commission	Radio antenna (beneath USPS ensign); bow staff	Masthead	Masthead	Aftermost masthead
PRIVATE SIGNAL (HOUSE FLAG)	Day and night when in commission	Bow staff	Masthead	Masthead	Aftermost masthead
SQUADRON BURGEE, YACHT CLUB BURGEE	Day and night when in commission	Bow staff	Bow staff	Bow staff if so equipped, or masthead	Foremost masthead
UNION JACK	0800 to sunset when not underway on Sundays or holidays or when dressing ship	(Not flown)	(Not flown)	(Not flown)	Jack staff
OFFICER-IN-CHARGE	Day and night during activity of which in charge	Above officer flag	Above officer flag	Above officer flag	Above officer flag
OWNER ABSENT	Day and night when owner not on board	(Not flown)	Starboard spreader but inboard of foreign or USPS ensign if flown (or at port spreader if necessary)	As for power yacht with signal mast	Foremost starboard spreader as for power yacht with signal mast
GUEST	Day and night when owner absent and guests in charge	(Not flown)	As for Owner Absent	As for Owner Absent	As for Owner Absent

* In U.S. waters the USPS ensign may be flown in lieu of (and at the same times and locations as) the U.S. ensign.

Reprinted from the March, 1983 issue of *The Ensign,* official publication of the United States Power Squadrons.

Appendix G **Photographs from the British Meteorological Office show wind forces and their effects on the sea: Beaufort Force 1 (calm air), *upper left;* Force 3 (gentle breeze), *center left;* Force 5 (fresh breeze), *bottom left;* Force 8 (gale), *upper right,* Force 10 (storm), *center right;* and Force 12 (hurricane), *bottom right.* Note here how the surface of the sea is completely obscured by driving foam.**

Acknowledgments and Credits

The following individuals and organizations were of particular assistance to the editors in checking facts, giving technical information and advice, offering facilities and materials for photographing, and in many other ways:

Albin Marine, Cos Cob, Connecticut
Lt. Bill Batson, Aids to Navigation Branch, 7th District, U.S. Coast Guard
Peggy Baxter, Waldman Graphics, Inc.
David Beach, National Marine Manufacturers Association
John Bottomley, New York
Ed Caroe, Caroe Marketing, Inc.
Linda Cullen, Hearst Marine Books
R. James Curran, Massapequa, New York
Gale Foster, Samson Ocean Systems
Dan Fales, *Motor Boating & Sailing* Magazine
James E. Gearhart, National Ocean Service
Sohei Hohrik, New York Yacht Club Library
Bill Husted, United States Power Squadrons
Henry Iken, Walken Graphics, Inc.
Peter A. Janssen, *Motor Boating & Sailing* Magazine
Lt. Cdr. Craig E. Jud, Atlantic Area Public Affairs, U.S. Coast Guard
Frank C. Leyden, Coast Guard Headquarters
James Lippman and William Rosenfeld, American Boat and Yacht Council
Martin Luray, *Sail* Magazine
James McGrew, Fid-O/McGrew Splicing Tool Co.
Frank MacLear, MacLear & Harris
Lt. C. Douglas Mason, National Ocean Service
CWO3 Richard L. Marini, U.S. Coast Guard Station, Eaton's Neck
Lt. Cdr. Dierck Meyer, Captree Power Squadron
Robert Ogg, Danforth Company
Len Pretty, Coast Guard Auxiliary
Edward S. Quest, Florida Wire & Rigging Works
Linda Readerman, Hearst Books
Richard L. Rath, *Yachting* Magazine
Cdr. Terry Ross, New York Power Squadron
Douglas Sethoms, United States Power Squadrons
Carl Spiegel, Marine Development & Research Corp.
Joseph Swift, Mercury Marine
Yale Cordage

Photo Credits

Attaway, Roy: 1013
Baker, Bob: 1506 left
Barlow, Peter: Frontispiece, 114, Chapter 5 Opener, 701, 703, 704, 803, 807, 822, 823, 826, 833, 1126, Chapter 12 Opener, 1219, Chapter 14 Opener, 1416, Chapter 16 Opener, 1604, 1605, Chapter 20 Opener
Beeston, Diane: 835
Bottomley, John A.: Color p. 584, second row left
Bottomley, Peg: Color p. 584, bottom row left, center, right; p. 585, second row center; third row left, center.
Bottomley, Tom: 111, 117, 130, Chapter 2 Opener, 209, 403-405, 412, 506, 507, 603, 607, 612, 613, 617, 621, 625, 630-632, 634, 637, 638, 640, 714, 808, 809, 829, 915, 1111, 1114 right, 1209 right, bottom, 1213, 1307, 1310, 1420, 1422, 1504, 1515, Chapter 17 Opener, 1609, 1918, Chapter 21 Opener, Chapter 22 Opener, Chapter 27 Opener, 2702, 2704, Color p. 584, top row right; second row center, right; third row all; p. 585, top row center, right; second row left, right; third row, all. Endpapers.
British Meteorological Office: 1421
Emmons, Gardner: 1404
Hooper, Queene: 815 bottom right, 1309 right, Chapter 25 Opener
Goldblatt, Stan: 817, 819
Koelbel, William: 2001
MacLear, Frank: 611
Maloney, Elbert: 128, 214, 215, 304, 316, 317, 633, 1116, 1204, 1506 right
Maloney, Mary: 2402
Mercury Marine: 129
Miller, Gary: 408
Moreland, Bob, St. Petersburg Times: 816
Orling, Alan S., 1615 bottom
Perkell, Jeff: 207, 213, Chapter 3 Opener, 321, 824 bottom right, bottom left, 1001 bottom, 1105, 1203, 1209 top left, top right, 1209 bottom left, Chapter 19 Opener, 2109-2111 left, 2118 left, 2210, 2212, Chapter 24 Opener
Rosenfeld, Morris & Stanley: 101, 103, 707, 830, 832, 1001 top, 1002, 1011, 1202, 1216, 1910 top, Chapter 26 Opener
Samsom Cordage: 1109 right
Stevenson, John: 802, 818, 1201, 2313, 2501 bottom
United States Coast Guard: 124, 127, 1607, 2501 top, 2601, 2604
United States Power Squadrons: Color p. 584, top row left, center; p. 585, third row right
Warren, M. E.: Chapter 4 Opener, 1614 right center
Whiting, John: 112, 610, 624, 627, 635, 639, 641, Chapter 7 Opener, 735 upper right, bottom; 744, 831, 836, 1005, 1129, Chapter 15 Opener, 1505 right, 1602, 1614 top left, bottom center; Chapter 23 Opener

Illustration Credits

Beechel, Allen: 104, 105, 108, 109, 115, 120, 125, 308, 309, 312, 402, 406, 407, 409-411, 502-515, 517, 609, 615, 622, 702, 705, 706, 708-710, 712-720, 722, 734, 736-740, 742, 743, 1007-1010, 1124, 1125, 1134, 1501, color pages 589-592
Borst, H. Shaw, Inc.: 901, 1802-1804, 1828-1830, 2014
Brotman, Adolph: 303, 619, 2003-2005, 2007
Burns, Dana: 116, 119, 508, 602, 616, 620, 628, 636, 1101, 1402, 1406-1410, 1424, 2002, 2302
United States Coast Guard: 2603
Walken Graphics, Inc.: 518, 801, 803-805, 810, 811, 814, 815, 820, 824, 825, 827, 836, 903, 913, 1004, 1104, 1115, 1118, 1120, 1122, 1123, 1127, 1132, 1213, 1214, 1418, 1508-1510, 1511 top, 1513-1522, 1524-1526, 1528, 1606, 1618, 1901-1904, 1913, 1916, 1917, 1920-1923, 2006, 2022-2029, 2101-2108, 2113-2116, 2119-2130, 2132-2147, 2149, 2150, 2203, 2204, 2206, 2213, 2305, 2309, 2310, 2312, 2315, 2316, 2502-2505, 2507
Wright, Phil: 828, 1301-1303, 1305, 1306, 1308, 1326

Index

C

G

H

M

N

O

P

Q

R

S

W

Y

Z

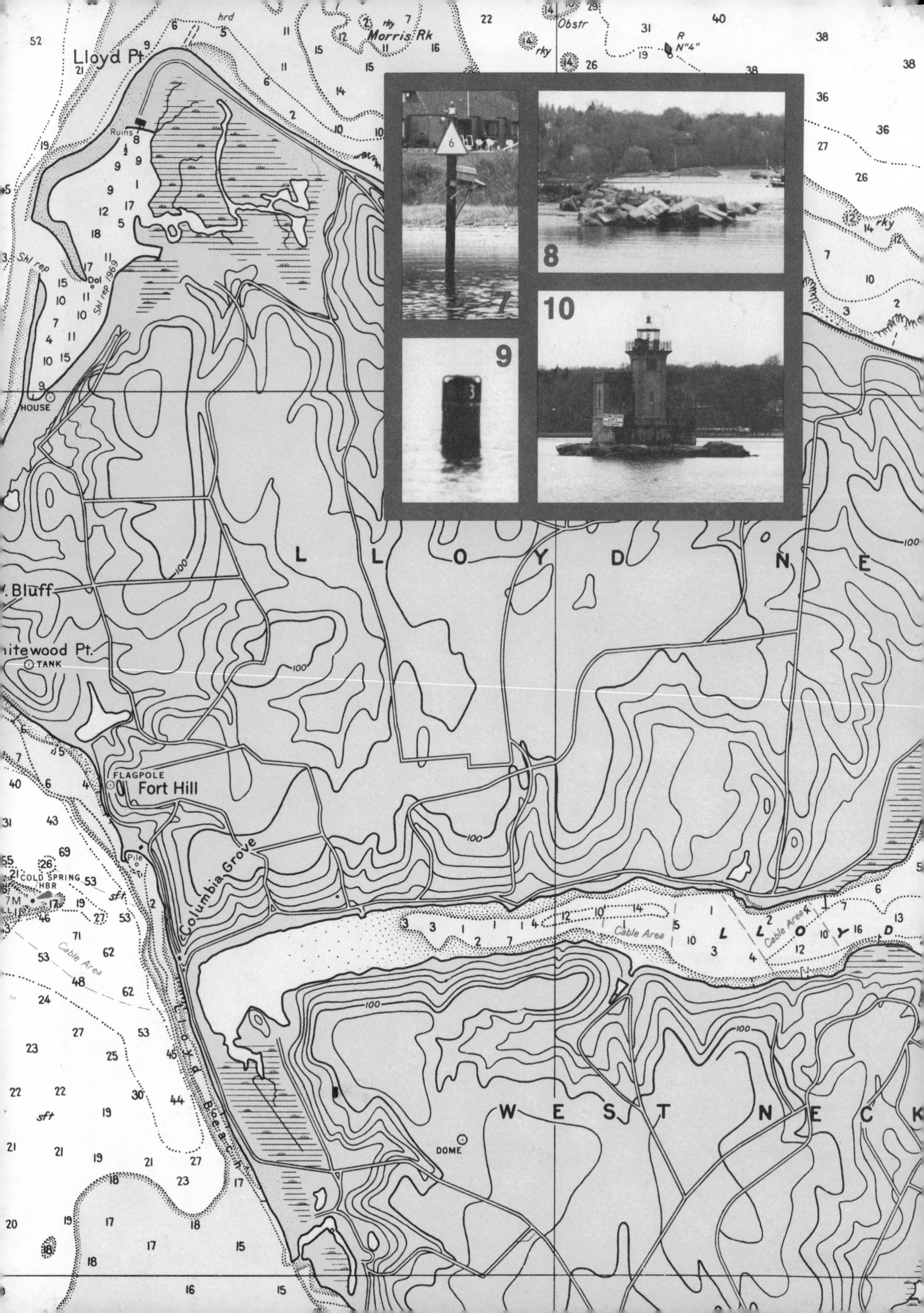

Lloyd Pt.
Morris Rk
Obstr
R N"4"
Ruins
Shl rep
Shl rep 1969
HOUSE
Bluff
hitewood Pt.
TANK
FLAGPOLE
Fort Hill
Pile
COLD SPRING HBR
Columbia Grove
Cable Area
Lloyd Beach
L L O Y D N E
W E S T N E C K
DOME
6
3
7
8
9
10

This book belongs to:

KRISTI YAMAGUCHI
Illustrated by Tim Bowers

Cover design by Jane Archer (www.psbella.com)

Published by Sourcebooks Jabberwocky, an imprint of Sourcebooks, Inc.
P.O. Box 4410, Naperville, Illinois 60567-4410
(630) 961-3900
Fax: (630) 961-2168
www.jabberwockykids.com

Library of Congress Cataloging-in-Publication data is on file with the publisher.

Source of Production: Bang Printing, Brainerd, MN, USA
Date of Production: April 2011
Run number: 14983

Printed and bound in the United States of America.
BG 10 9 8 7 6 5 4 3

Poppy was a pig.
A pot-bellied, waddling, toddling pig.
She was a pig with dreams. And she
was a pig who dreamed big!
She wanted to be a star.

Poppy had always dreamed of
being a posh prima ballerina.

She tried out for Swan Lake, a famous ballet.
But Poppy was not graceful. In fact,
she was quite clumsy.

"Follow your dreams!" said Poppy's mother, who loved her no matter what. "You go, girl!" said Poppy's grandparents, who were her biggest fans. "Dream big, pig!" said Poppy's best friend, Emma, who was always there for her.

"Dancing is just not for you," said
the people in charge of the ballet.
"Try something else!"

So Poppy tried out for Singing Stars,
a popular chorus competition.
She had always dreamed
of being a soulful singer.

But Poppy sang off-key. And to be honest,
she couldn't really carry a tune.

“You go, girl!” said
Poppy’s grandparents.

“Follow your
dreams!” said
Poppy’s mother.

“Dream big, pig!”
said Emma.

"Singing is just not for you," said the people in charge of the competition. "Try something else!"

So Poppy tried out for Supermodel Search.
She had always dreamed of being a big-time
splashy supermodel.

But Poppy
was not very glitzy
or glittery, and she even
tripped on her fancy gown.

"Follow your dreams!" said Poppy's mother.

"You go, girl!" said Poppy's grandparents.

"Dream big, pig!" said Emma.

"Modeling is just not for you," said the
people in charge of the search.
"Try something else!"

But Poppy didn't know what else to try.

And as she wandered through New Pork City,
she began to wonder if her dreams
would really come true.

Poppy was about to give up when she
heard her mother say, "Just follow your heart.
Remember, we love you no matter what."
And her grandparents cheer, "We're your biggest fans!"
And her best friend, Emma, squeal,
"We're here for you!"

Poppy smiled. She knew
just what to do!

When Poppy thought about all the things she truly loved—her friends and family were at the top of the list! So, the next day, Poppy invited Emma for a "pig's day out" in the park.

While giggling and strolling along, they spotted an ice rink.
Poppy and Emma watched the skaters skimming and spinning,
swooping and swizzling on the ice. Poppy realized it was the
most beautiful sight she had ever seen.
Her heart danced with joy!

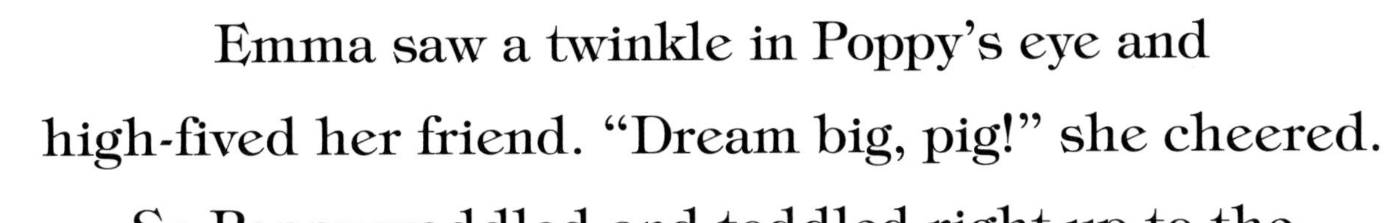

Emma saw a twinkle in Poppy's eye and high-fived her friend. "Dream big, pig!" she cheered. So Poppy waddled and toddled right up to the teacher and said, "I'd like to be a spectacular ice-skating star."

"A pig on ice?" the teacher pondered. "Honey, I don't know if that's possible."

"Anything's possible," responded Poppy. "I believe in dreaming big!"

The teacher shrugged. "As you wish," she said. "We'll see if the pig's got pizzazz."

Poppy laced up her skates. She slipped and slid all over the ice. She fell.

But, this time...

...Poppy got up.

Over and over and over, she shuffled and stumbled and fumbled and fell.

But by the time the rink closed for the night,
Poppy was skating more than she was falling.
And it felt like...magic!

Poppy returned to the rink the very next day.
Her cheeks were pink with winter
wind and excitement.

She was so happy gliding and sliding and tumbling and bumbling on the ice, she didn't even notice that she wasn't perfect.

And nobody else did either.

Now, a most persistent pig, Poppy learned to twirl and swirl and to do dips and lunges and splits. Poppy learned to do jumps and spirals and lifts.

Before she knew it, more and more skaters stopped to watch Poppy practice—she was quite a sight! She even had her picture on the front page of the newspaper. Poppy felt like a star!

Some of her fans made T-shirts
that read "FOLLOW YOUR DREAMS!" Others
wore hats that said "DREAM BIG, PIG!" And
tote bags declared "YOU GO, GIRL!"
Poppy's dreams had come true!

Time went by, but Poppy didn't stop dreaming.
One day, she decided to be a pilot. She wanted
to parachute and be the first sky-diving pig.
"When pigs fly!" said the other pilots.
But they did not know Poppy.
She was a pig who dreamed big.

To Mommy's Angels—
Keara and Emma, you give me more joy
than I can ever express. This is for you both
in hopes that you will always dream big!
I love you to infinity…Mom

To Rubin,
an encouraging
voice when I'm
dreaming big.
—T.B.

Acknowledgments

To Linda Oatman High, it was so amazing
collaborating ideas with you. Poppy is
certainly someone who is inspiring and
positive, and you helped bring her
to life. A heartfelt thank-you to
you for writing this book with
me. It's been such an honor.
—Kristi